Lecture Notes in Mechanical Engineering

Lecture Notes in Mechanical Engineering (LNME) publishes the latest developments in Mechanical Engineering—quickly, informally and with high quality. Original research reported in proceedings and post-proceedings represents the core of LNME. Volumes published in LNME embrace all aspects, subfields and new challenges of mechanical engineering. Topics in the series include:

- Engineering Design
- Machinery and Machine Elements
- Mechanical Structures and Stress Analysis
- Automotive Engineering
- Engine Technology
- Aerospace Technology and Astronautics
- Nanotechnology and Microengineering
- Control, Robotics, Mechatronics
- MEMS
- Theoretical and Applied Mechanics
- Dynamical Systems, Control
- Fluid Mechanics
- Engineering Thermodynamics, Heat and Mass Transfer
- Manufacturing
- Precision Engineering, Instrumentation, Measurement
- Materials Engineering
- Tribology and Surface Technology

To submit a proposal or request further information, please contact the Springer Editor of your location:

China: Dr. Mengchu Huang at mengchu.huang@springer.com
India: Priya Vyas at priya.vyas@springer.com
Rest of Asia, Australia, New Zealand: Swati Meherishi at swati.meherishi@springer.com
All other countries: Dr. Leontina Di Cecco at Leontina.dicecco@springer.com

To submit a proposal for a monograph, please check our Springer Tracts in Mechanical Engineering at http://www.springer.com/series/11693 or contact Leontina.dicecco@springer.com

Indexed by SCOPUS. The books of the series are submitted for indexing to Web of Science.

More information about this series at http://www.springer.com/series/11236

Numan M. Durakbasa · M. Güneş Gençyılmaz
Editors

Digital Conversion on the Way to Industry 4.0

Selected Papers from ISPR2020, September 24–26, 2020 Online - Turkey

Springer

Editors
Numan M. Durakbasa
Research Group for Production Metrology and Adaptronic Systems
TU Wien (Vienna University of Technology)
Vienna, Wien, Austria

M. Güneş Gençyılmaz
Faculty of Engineering, Industrial Engineering Department
İstanbul Aydın University
İstanbul, Turkey

ISSN 2195-4356 ISSN 2195-4364 (electronic)
Lecture Notes in Mechanical Engineering
ISBN 978-3-030-62783-6 ISBN 978-3-030-62784-3 (eBook)
https://doi.org/10.1007/978-3-030-62784-3

This Springer imprint is published by the registered company Springer Nature Switzerland AG
The registered company address is: Gewerbestrasse 11, 6330 Cham, Switzerland

Foreword

Dear Colleagues and Friends,

It is a great pleasure to welcome you to the "**International Symposium for Production Research - ISPR 2020**", Antalya/Turkey, from 24–26 September 2020, with the overall theme of "Digital Conversion on the way to Industry 4.0". As in all previous years we are particularly pleased with our continued successful collaboration with Society for Production Research, İstanbul, Turkey, in the organization of this event.

The purpose of the symposium is to bring together researchers, scientists and leading world experts at universities, companies, institutions, communities, associations and societies in the domain of interest from around the world to share ideas, experiences, research results and vision on production and operations management and technology. This organization shall also provide a forum for experts and professionals to discuss the challenges, opportunities and advance innovation of the theme of this year's symposium.

We hope that this "on-line" gathering will provide the participants an opportunity for scholarly exchange of ideas and questions despite the absence of physical presence and face-to-face discussions. We wish you all a productive symposium and we sincerely hope that this pandemic will be behind us soon and that we can all go back to our regular format of conferences and symposia in the near future.

September 2020

Kurt Matyas
Ayhan Toraman

Preface

This book consists of the selected papers presented at the **20th International Symposium for Production Research (ISPR2020)** that would have been held from September 24–26, 2020, at a beautiful resort in Antalya, Turkey. Unfortunately, this event, like many others in all corners of the world, had to be converted to a virtual symposium due to the COVID-19 pandemic. Thanks to the magnificent improvements in Internet technologies we could easily switch our face-to-face meetings to virtual events with the aid of a user-friendly software program. This technical support was provided generously by the Vice-Rectorate for Digitalization and Infrastructure of TU Wien and our heartfelt thanks and gratitude go to Univ. Prof. Dipl. -Ing. Dr.techn. Dr.h.c.mult. Josef Eberhardsteiner, Vice-Rector.

This symposium was organized by Society for Production Research İstanbul, Turkey, and TU Wien, Austria, for the fourth year in a row.

The generic theme of "**Industry 4.0**" was first adopted in the symposium held in 2016 and maintained in the following four symposiums in 2017, 2018, 2019, and 2020 but each time with an updated emphasis on the relevant developments and progress on the various aspects of this "Fourth Industrial Revolution." In other words, the symposiums maintained the same main theme from 2016 but studied a different aspect of production systems and production management, in the subsequent years with the purpose of drawing the attention of the researchers to the cases of Industry 4.0 applications of this new industrial era.

The world of science and technology is under increasing influence of the requirements of Industry 4.0. Transition to a new era seems inevitable for every sector of industry. Given the importance of this theme, ISPR2020 hosted numerous distinguished speakers from both the academia and the industry to hear their views on the applications of the Industry 4.0 on the various components of production systems.

A total of over 350 participants attended this year's symposium—academics, practitioners, and scientists from 17 countries, 10 invited speakers made invaluable contributions to the symposium. The symposium programme included keynote addresses (opening/closing session), breakout sessions and workshop discussions, a welcome concert, and a closing session on the final day.

This book contains 83 refereed selected papers in 22 categories shown in the contents of the book.

We would like to express our gratitude to Vice-Rector Prof. Kurt Matyas, also the Honorary Chairman of the Scientific Committee of this symposium for his leadership and generous support. We also thank to Prof. Dr. Ayhan Toraman, Past President of Society of Production Research and Honorary Chairman of the Scientific Committee; Prof. Christian Bauer, Dean of the Faculty of Mechanical and Industrial Engineering; Prof. Detlef Gerhard, former Dean of the Faculty of Mechanical and Industrial Engineering; and Prof. Friedrich Bleicher, Head of the Institute for Production Engineering and Photonic Technologies for their interest and support for this symposium.

We would like to thank all the keynote and invited speakers whose contributions enhanced the success of the symposium.

In organizing this event, our colleagues in Vienna and İstanbul contributed endless hours of hard work, energy and wisdom to make this event the success it was. On the Vienna side, our sincere thanks go to the staff of the Research Group Production Metrology and Adaptronic Systems of the Institute for Production Engineering and Photonic Technologies, in particular, to Dipl.-Ing. Erol Güçlü.

On the İstanbul side, we are very grateful to Assis. Prof. Dr. Nükhet Tunçbilek who were involved in every aspect of the organization from the very beginning. Our special thanks go to Ms. Tuğçe Beldek and Mr. Aziz Kemal Konyalıoğlu, our young colleagues, research assistants, and Ph.D. candidates at İstanbul Technical University and Galatasaray University and Comp. Eng. Tufan Tunçbilek who prepared and managed virtuall event together with Dipl. Ing. Erol Güçlü perfectly and, also research assistants at TU Wien, for their hard and dedicated work. We would like to thank to Prof. Dr. Hatice Camgöz Akdağ and Assist. Prof. Dr. Zeynep Girgin who moderated the symposium successfully.

We would like to express our gratitude to the board members of the Society for Production Research in İstanbul for their strong support and dedication to make the symposium a success.

Our very special thanks go to our colleagues, the participants of this symposium. Undoubtedly, they were the core component of this organization.

We would like to recognize and thank our dear colleagues who graciously accepted to join the honorary and scientific committees or who served as peers in this event.

Finally, no such event is possible without the generous support of patrons and sponsors. In this regard, we would like to thank Dr. Michael Ludwig, the Mayor of Vienna, for his continued generous support of this symposium series and to all the corporations and individuals who provided invaluable financial and intellectual contributions.

And last but not least, we are grateful to Ms. Petra Jantzen Editor from Springer Nature, for her competent guidance, professionalism, and patience.

M. Güneş Gençyılmaz
Numan M. Durakbasa

Key Technology Roadmap In Production of Future – Industry 4.0

Today, Industry 4.0 is spreading more and more worldwide to industrial companies and especially to SMEs. With a strong infrastructure and a rapid connection between the components of the entire factory, Industry 4.0 will introduce innovation and more robust, reliable, and sustainable production. Smart economic growth, which is based on high-quality training and on the promotion of research and innovation as well as the building of digital factories, is most crucial and central to the future developments. This will ensure a sustainable development which respects the ecology by using green technologies and which will improve the working environment thus creating new and better jobs for a wider population.

Comprehensive knowledge in the areas of market requirements, product and process development and design, intelligent metrology, and end-of-life management are important conditions to achieve rapid, agile, waste-free, and cost-effective production of innovative, customized complex products using next-generation materials. This will also protect the environment by making zero emissions and improve environmental sustainability and reduce the use of energy within the factory of futures.

Cyber-physical systems (CPS) improve resource productivity and efficiency and enable more flexible models of work organization. Companies that use CPS will have a clear advantage when it comes to recruiting the best employees since they can offer a better work–life balance. Augmented reality (AR) and multimodal interaction will be used to train the factory workers so that they will be able to deal with the complexity of cyber-physical production and enable new forms of collaboration by digital systems. The smart factory concept which includes the need for closer interaction with tele-operation and tele-presence as part of the process in networked production centers will spread to industrial companies worldwide.

The application of artificial intelligence (AI) in the production field has also imposed new technological demands. To meet the market demands in present and future global industrial world manufacturing enterprises of any kind and any size must be flexible and agile enough to respond quickly to product demand changes. With support of artificial intelligence and modern information technology, it is possible to realize modern, cost-effective, customer-driven production considering the importance and basic role of integrated systems and advanced technology.

This new concept, according to Industry 4.0, can be developed on the basis of advanced production systems and multi-functional integrated factory model as well as extensive use of the IT, AI, CPS, simulation, virtual reality (VR), intelligent quality systems, automation, robotics, sophisticated metrology, and advanced engineering data exchange techniques.

September 2020

Numan M. Durakbasa
M. Günes Gencyilmaz

Organization

ISPR2020 was organized by TU Wien, Austria, and the Society for Production Research, Turkey. The symposium took place in Turkey.

Editors

Numan M. Durakbasa
M. Güneş Gençyılmaz

Co-editors

Peter Kopacek
Ayhan Toraman
Selim Zaim
Jorge Martin Bauer
Serpil Erol
Semra Birgün
Kemal Güven Gülen
Jan Torgersen
Alptekin Erkollar
Mahmut Tekin

Honorary Chairs

Kurt Matyas	TU Wien, Austria
Ayhan Toraman	Society for Production Research, Turkey

Symposium Chairs

Numan M. Durakbasa	TU Wien, Austria
M. Güneş Gençyılmaz	İstanbul Aydın University, Turkey

International Honorary Committee

Albert Weckenmann	Friedrich Alexander Universität Erlangen, Germany
Alice E. Smith	Auburn University, USA
Andrew Kusiak	The University of Iowa, USA
Christian Bauer	TU Wien, Austria
Daniela Popescu	Technical University of Cluj Napoca, Romania
Detlef Gerhard	Ruhr-University Bochum, Germany
Duc Truong Pham	The University of Birmingham, UK
Friedrich Bleicher	TU Wien, Austria
Lubomir Šooš	Slovak University of Technology Slovakia
Sorin Popescu	Technical University of Cluj-Napoca, Romania
Stanislaw Adamczak	Kielce University of Technology, Poland
Vidosav Majstorovic	University of Belgrade, Serbia

Organizing Committee

Selim Zaim, Turkey
Güneş Gençyılmaz, Turkey
Erol Güçlü, Austria
Semra Birgün, Turkey
Serpil Erol, Turkey
Berna Dengiz, Turkey
Kemal Güven Gülen, Turkey
Alptekin Erkollar, Turkey
Mahmut Tekin, Turkey
Eva Maria Walcher, Austria
Ferhan Çebi, Turkey
Dursun Delen, USA
Günther Poszvek, Austria
Şakir Esnaf, Turkey
Alp Baray, Turkey
Lukas Kräuter, Austria
Nükhet Tunçbilek, Turkey
David Riepl, Austria
Alp Baray, Turkey
Hatice Camgöz Akdağ, Turkey
Haluk Soyuer, Turkey
Hür Bersam Bolat, Turkey
Ece Soyuer, Austria
Jorge Bauer, Austria
Tuğçe Beldek, Turkey
Aziz Kemal Konyalıoğlu, Turkey
Liane Höller, Austria
Gökçen Baş, Austria
Gamze Uğur Tuncer, Austria

This book is prepared for publishing by

Tuğçe Beldek, Turkey
Aziz Kemal Konyalıoğlu, Turkey

Scientific Committee

S. Adamczak	Kielce University of Technology, Poland
H. Camgöz Akdağ	Technical University, Turkey
A. Akdoğan	Yıldız Technical University, Turkey
K. Altinel	Bosphorus University, Turkey
İ. Ar	Yıldırım Beyazıt University, Turkey
B. Baki	Karadeniz Technical University, Turkey

Ş. Baray	İstanbul University, Turkey
G. Baş	TU Wien, Austria
N. Başoğlu	İzmir Institute of Technology, Turkey
J. Bauer	National Technological University Buenos Aires, Argentina
A. Baykasoğlu	Dokuz Eylül University, Turkey
E. Bayraktar	The American University of the Middle East, Kuwait
F. Berto	Norwegian University of Science and Technology, Norway
S. Birgün	İstanbul Gedik University, Turkey
P. Blecha	Brno University of Technology, Czechia
F. Bleicher	TU Wien, Austria
H. B. Bolat	İstanbul Technical University, Turkey
F. Bozbura	Bahçeşehir University, Turkey
İ. Böğrekci	Adnan Menderes University, Turkey
I. Budak	Faculty of Technical Sciences in Novi Sad, Serbia
A. Bulgak	Concordia University, Canada
İ. Çavuşoğlu	Marmara University, Turkey
L. A. Crisan	Technical University of Cluj Napoca, Romania
F. Cus	University of Maribor, Slovenia
F. Çebi	İstanbul Technical University, Turkey
D. Delen	Oklahoma State University, USA
M. Demirbağ	University of Essex, UK
P. Demircioğlu	Adnan Menderes University, Turkey
B. Dengiz	Başkent University, Turkey
T. Dereli	İskenderun University, Turkey
M. Dinçmen	İstanbul Technical University, Turkey
M. Dragomir	Technical University of Cluj Napoca, Romania
Á. Drégelyi Kiss	Óbuda University, Hungary
N. Durakbaşa	TU Wien, Austria
B. Durmuşoğlu	İstanbul Technical University, Turkey
A. Erkollar	Sakarya University, Turkey
S. Erol	Gazi University, Turkey
Ş. Esnaf	İstanbul University, Turkey
S. Firat	Marmara University, Turkey
W. Fisher	University of California, Berkeley, USA
G. Gençyılmaz	İstanbul Aydın University, Turkey
D. Gerhard	Ruhr-University Bochum, Germany
S. Gözlü	İstanbul Technical University, Turkey
S. Grozav	Technical University of Cluj Napoca, Romania
K. Gülen	Namık Kemal University, Turkey
A. Güngör	Pamukkale University, Turkey
E. Hekelová	City University Bratislava, Slovakia

F. Holesovsky	J. E. Purkyně University in Ústí nad Labem, Czech Republic
Z. Irani	University of Bradford, UK
V. Işler	Hasan Kalyoncu University, Turkey
C. Kahraman	İstanbul Technical University, Turkey
İ. Kara	Başkent University, Turkey
T. Katoka	Kindai University, Japan
M. Klumpp	Duisburg University, Germany
D. Kocaoğlu	Portland State University, USA
T. Koç	İstanbul Technical University, Turkey
P. Kopacek	TU Wien, Austria
P. Kovac	University of Novi Sad, Serbia
M. Králik	Slovak University of Technology in Bratislava, Slovakia
L. Kräuter	TU Wien, Austria
C. Kubat	Sakarya University, Turkey
U. Kula	American University of the Middle East, Kuwait
A. Kulakli	American University of the Middle East, Kuwait
J. Kundrak	University of Miskolc, Hungary
A. Kusiak	University of Iowa, USA
V. Majstorovic	University of Belgrade, Serbia
I. Mankova	Technical University of Kosice, Slovakia
A. Markopoulos	National Technical University of Athens, Greece
K. Matyas	TU Wien, Austria
M. Novák	J. E. Purkyně University in Ústí nad Labem, Czech Republic
Y. Omurtag	Robert Morris University, USA
N. Özçakar	İstanbul University, Turkey
D. Özdemir	Bilgi University, Turkey
A. Özok	Okan University, Turkey
E. Öztemel	Marmara University, Turkey
J. Peterka	Slovak University of Technology, Slovakia
D. Pham	University of Birmingham, UK
M. Pokusová	Slovak University of Technology in Bratislava, Slovakia
S. Popescu	Technical University of Cluj Napoca, Romania
D. Popescu	Technical University of Cluj Napoca, Romania
E. Pucher	TU Wien, Austria
A. Sağbaş	Namık Kemal University, Turkey
B. Sağbaş	Yıldız Technical University, Turkey
J. Sanz Juan	Polytechnic University of Valencia, Spain
A. Sharif	University of Bradford, UK
A. Smith	University of Auburn, USA
Ľ. Šooš	Slovakia University of Technology in Bratislava, Slovakia

A. Sorguç	Middle East Technical University, Turkey
H. Soyuer	Ege University, Turkey
K. Stepien	Kielce University of Technology, Poland
P. Tamas	Budapest University of Technology and Economics
B. Tan	Koç University, Turkey
M. Tanyaş	Maltepe University, Turkey
F. Taşgetiren	Yaşar University, Turkey
H. Taşkin	Sakarya University, Turkey
M. Tekin	Selçuk University, Turkey
A. Toraman	İstanbul Technical University, Turkey
J. Torgersen	Norwegian University of Science and Technology, Norway
O. Torkul	Sakarya University, Turkey
G. Ulusoy	Sabancı University, Turkey
G. Varga	University of Miskolc, Hungary
K. Velíšek	Slovak University of Technology, Slovakia
J. Wang	National University of Singapore
A. Weckenmann	University of Erlangen Nürnberg, Germany
M. Yalçintaş	İstanbul Commerce University, Turkey
M. Yenisey	İstanbul University, Turkey
M. E. Yurci	Yıldız Technical University, Turkey
S. Zaim	İstanbul Zaim University, Turkey
W. Zebala	Cracow University of Technology, Poland
M. Zerenler	Selçuk University, Turkey

Reviewers

Akdogan, A.
Aktin, T.
Baray, Ş.
Bauer, J.
Berto, F.
Beyca, Ö.
Börühan, G.
Blecha, P.
Budak, I.
Bulut, Ö.
Büyüksaatçi, S.
Camgöz-Akdağ, H.
Crisan, L.
Çebi, F.
Demircioğlu, P.
Dragomir, M.
Dregelyi-Kiss, A.
Durakbasa, N.
Ekinci, E.
Erkollar, A.
Ersoy, P.
Esnaf, Ş.
Gençyılmaz, G.
Gergin, Z.
Gökçe, M.
Grozav, S.
Gülen, K.
Kabadurmuş, Ö.
Kandiller, L.
Kazançoğlu, Y.
Kopacek, P.
Králik, M.
Kräuter, L.
Majstrovic, V.

Maňková, I.
Mullaoğlu, G.
Novák, M.
Öner, A.
Öner, E.
Özgür, A.
Özkan, S.
Öztop, H.
Öztürkoğlu, Y.
Öztürkoğlu, Ö.
Paldrak, M.
Pokusová, M.
Soyuer, H.
Staiou, E.
Torgersen, J.
Tunçbilek, N.
Üney-Yüksektepe, F.
Varga, G.
Yetkin, B.
Zaim, S.
Zebala, W.

Contents

Healthcare Systems and Management

Industrial Applications

Industrial Robotics in Production Systems and Industrial Engineering

Industry 4.0 Applications

Logistics Management

Optimization

Process Management

Productivity and Performance Management

Quality Management

R&D and Technology Management

Big Data and Analytics

A Hybrid Recommender System: Uniqueness of Choices by Using Machine Learning Technique

Sanjeev Kumar[1(✉)], Ashirbad Sarangi[1], and R. P. Mohanty[2]

[1] CSE, the ICFAI University, Dehradun, India
tanwar.sanjeev@gmail.com, rubi.ashirbad@gmail.com
[2] SOA University, Bhubaneswar, Odisha, India
rpmohanty@gmail.com

Abstract. The problem of recommending similar sets of items in the online business community is called item recommendation. An item recommendation aims to recommend a new item that matches the user's interests. Universally, recommendation amenities have become significant due to their support in e-commerce applications like online shopping, digital promotions, and various research domains. The collaborative approach of the recommender engine filters out the k-nearest neighbours and then the similarity is compared between the neighbourhoods. In this paper, an algorithm is proposed, to recommend the items to the users with respect to the uniqueness of users' choice. The result achieved is a mixture of both types of items, which are commonly and rarely bought by others. The proposed technique uses machine learning modules to learn actively and recommend.

Keywords: Recommender systems · k-Nearest neighbour · Uniqueness · Machine learning · Recommendation algorithms

1 Introduction

Machine learning is gaining popular acceptance in the present time as an important branch of artificial intelligence that allows computer systems to learn directly from examples, data, machines and also experiences etc. Many algorithms have been developed by computer science professionals and academia over the past few years and unfortunately, we are perplexed to make the right choice of the algorithm for the right problem. A recommender system is a type of information filtering system used in machine learning, which recommends some products to the users according to their choices and the availability of the items in the online platform. By drawing from huge data sets, the system's algorithm can pinpoint accurate user preferences. Once you know what your users like, you can recommend them new and relevant content. Such a system is being widely used and is changing significantly the interface between the end users and websites. It has a number of applications in many useful domains such as online shopping, education, economy, and scientific research etc. [1, 2]. The rapid change in technology and digital data sets [3] brings the concepts of customized search,

N. M. Durakbasa and M. G. Gençyılmaz (Eds.): ISPR 2020, LNME, pp. 3–13, 2021.
https://doi.org/10.1007/978-3-030-62784-3_1

filters the information and Google, Yahoo etc. are using massive data analysis methods [4–6]. Such approaches are categorized in four major domains: The Content-Based Filtering (CBF), The Graph-Based Method (GB), The Collaborative Filtering (CF) and Hybrid Recommend techniques. The comparison of these algorithms of recommendation systems are compiled in Table-1 [7, 8]. CBF works on user past records and based on user information, it develops a model [9]; and mapping to search keywords to the user profile for recommending the items. CF mainly work on the concept of item rating by the user and are recommended based on similarity neighborhood users [10]. Graph based techniques work on present choice and user past purchase choice, based on these two data sets future choice is predicted. Some of the research carried out in GB techniques [11], user and items are considered nodes whereas relation between users and items are considered edges. The popular random walk algorithm on the graph is used to get the relationship between user and item. For better forecasting to recommend the items to the users, the researchers have come up with idea where CF and CBF both techniques are used for recommending the items is called the hybrid techniques [12].In scientific research, recommender systems help a lot for recommending similar domain papers to the researchers [13–15].

Table 1. Comparison of algorithms used in recommender systems

Algorithm	Advantage	Disadvantage
Content based filtering	1. Recommendation results are relevant to users self-preferences 2. Similar neighbours for users.	1. Keyword relevance quality is ambiguous 2. New Item, new user problem 3. Fail to handle dynamic data set
Collaborative filtering	1. Recommendation results may be unforeseen 2.Strongly considers user's perspective	1. Suffer from new items and new users in order to build user profile 2. Due to Sparsity, it is difficult to find similar neighbours for users. 3. Privacy of user information's
Graph based method	1. Used different source to recommend 2. Can add social relation, faith in relationships between users and recommendation systems	1. Neither reacts on user interest nor item features.
Hybrid method	1. Works in combined fashion to overcome the CF, CBF, GB disadvantage	1. Privacy of user information's still a problem

2 Statement of the Problem

In the retail shops, only those items ("Item" is the general term used to denote what the system recommends to users) are displayed on the showcase, which has a high demand from the consumers, eventually hiding the greater lot of products that is present in the warehouse, ultimately resulting in a huge loss. Moreover, the consumer has to select from only a few of the variety of products that are displayed, although a large variety of products actually exist with the retailer in the warehouse. This problem is overcome by the online shopping. The online sites display the whole range of products that are present with them. But now, the user has a very large range of products to choose from, which again is a very tedious task.

To resolve this problem, recommender engines are introduced. These systems will recommend the user some selected items from the super set of items according to the user buying preference and interest. Thus, for an efficient recommendation, it is essential to have an engine that recommends the user about a finite number of items that will be focussing on selling the rarely bought items as much as possible with respect to the user's choice and interest. We propose an algorithm in the Sect. 3, which gives the result to the user as a combination of both, the frequently and rarely bought items according to the previous pattern of buying behaviour of the articles.

2.1 Definitions

2.1.1 Vocabulary

The items are denoted by *i*. The set of items that have been rated by an active user *a* is written as *I*. There are some more terms that are used in this context.

1) Buying Frequency (b_i): The frequency of buying of the items by the all the users in aggregate is the buying frequency of the item. The term b_{max} is the maximum buying frequency of the items.

$$b_i = \text{No. of times item i is bought}$$

2) Common Items and Unique Items: The set of items that are frequently bought by the users i.e. have a higher buying frequency are called Common Items. Similarly, the set of items which have a lower buying frequency i.e. are rarely bought are called Unique Items.
3) Threshold Value (b_t): The Common and Unique items are segregated through the measure. This is the threshold value for the items. Common Items have buying frequency greater than the threshold value and the Unique Items have buying frequency less than the threshold value.
4) Degree of Uniqueness of Items (u_i): This is the measure of the Uniqueness of the items. The items which are listed as unique or common are then computed in a quantitative manner to check how much unique they are. The u_i describes the degree of uniqueness of the item *i*.
5) Average rating of the User ($\overline{r_a}$): The users are allowed to rate the items bought by them on certain rating platform which is adopted by the online retailing site. That platform may be in form of stars or in the form of numbers. Every rating platform has a mean value around which the whole system can be centred around and which can be used to normalise all the ratings. That mean value is called the Universal Average. The user provides ratings as per the scale of the system that is adopted by the retailing site. That rating is then normalised and taken into consideration to find the user's average.

6) Accepted Set (A): This is the set of only those items which have a normalised weight greater than 0. It is found by the following formula.

$$i \in A \, if \, r_{i,a} - \bar{r}_a > 0$$

7) Degree of Uniqueness of Users (q_i): Similarly as of the items, this is the measure of uniqueness of the users. Every user has a certain degree of uniqueness. An assumption is made in this definition that the each and every user has a certain degree of uniqueness and that no user exists who is a completely unique or completely common.
8) Frequency of features: The items that are present in the online retail shops are broken down in form of features. Each item has number of features. If the feature k (f_k) is present in the any item that is present in the accepted set A of user a, then that feature is in the consideration for user a and hence is written as $f_{k,a}$.

2.1.2 Concepts

The results are received in a two-phase relation. Phase-I is the All Purpose Detail. In this phase, the concrete values are computed which are calculated only once. These are the common information that remains same for all of the users. These values change and need to be recalculated if any change occurs in any of the independent variables such as the buying frequency of items and ratings by the user. Unless and until the user alters the number of rated items these details need not be calculated. Thus for a passive user, the details remain the same for a long period of time and for an active user it changes frequently. This phase has 3 steps to find out the details. Phase-II is the User Specific Details. These details are calculated every time the user a logs into his/her account in the online retail sites. These details depend more upon the Phase-I values. Hence, a change in those variables will ultimately affect the recommendation. This phase has 6 steps for finding out the 'Final' set for recommendation.

3 Proposed Recommender System Algorithm

The proposed algorithm consists of two phases, in Phase; I data, preparation and analysis is carried out with respect to items and users. Whereas in Phase II, final recommended set of items are achieved (See Fig. 1).

Phase-I*: *All Purpose Details

*Step 1: **Computation of Threshold value*** (b_t)

The b_i of each item i is recorded. According to the buying frequency of each product, the graph is plotted. The graph is plotted between b_i vs i, i.e. the graph is plotted as buying frequency vs items. Threshold value is being computed by using below formula (see Fig. 2).

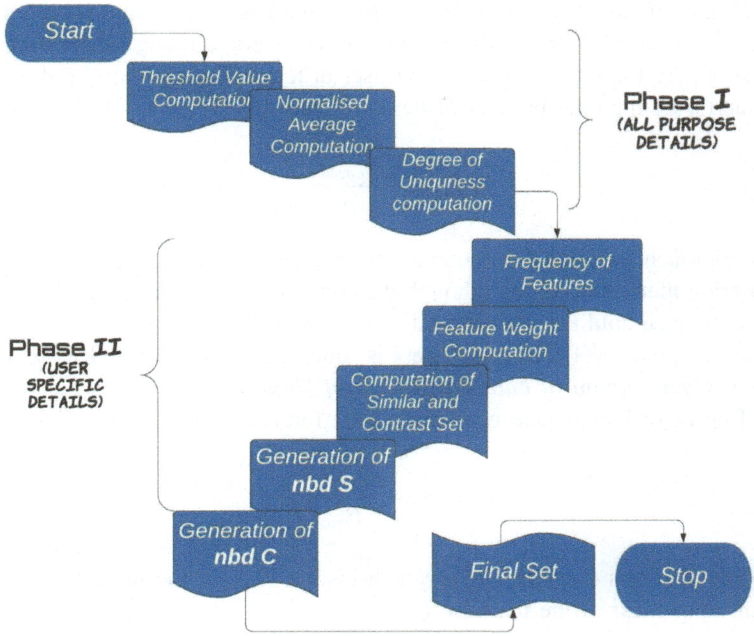

Fig. 1. Flow graph of proposed recommender systems

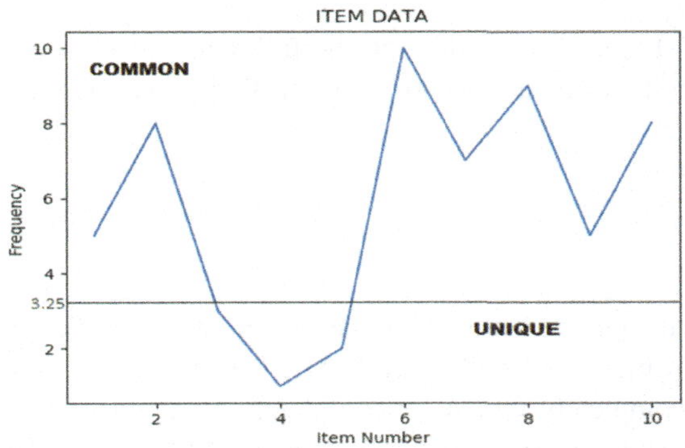

Fig. 2. Threshold value

The value b_t is calculated by the formula:
Area bounded by the curve $b_{j;\ j>t}$ = Area bounded by the curve by $b_{j;\ j<=t}$
After finding the b_t,
if $b_i > b_t$, $i \in$ Common Items else $b_i < b_t$, $i \in$ Unique Items for all items $i \in I$

Step 2: ***Computation of the Average rating of all the users and Accepted Set (A)***

As referred in the Sect. 2.A.5, the normalised average rating of the user is computed. Let $r_{i,a}$ be the rating given by the user in the online scale, v_a be the universal average and n be the total number of items in the set I.

$$\overline{r_a} = \frac{\sum r_{i,a} - v_a}{n}$$

This approach has less time complexity than the conventional user-user collaborative filtering method because, although the computation is done on the large user set, its done only once until the user doesn't change until the user a rates or reviews a new item. The accepted set A for all the users is found according to the Sect. 2.A.6.

Step 3: ***Computation of both the Degree of Uniqueness***

The Degree of Uniqueness of the items are found out by the following formula,

$$u_i = -\left(\frac{b_i - b_t}{b_{max}}\right)$$

According to this formula, the lower the value of b_i of the item, the more is the degree of uniqueness of the item. So for,

$$i \in \text{Common Items} \ : u_i < 0$$
$$i \in \text{Unique Items} \ : u_i > 0$$

A database of the items is made with the following attributes.

Similarly, with the Degree of Uniqueness of the Users. These are found out by the following formula,

$$\overline{q_a} = \overline{r_a} \frac{\sum r_i u_i}{n_A}$$

Whereas n_A is the number of items in the Accepted set A. The formula is devised as per the assumption made earlier in Sect. 2.A.7, that there exists no person among all the users who is completely unique or completely common. Hence, according to the assumption the measure of Degree of Uniqueness of the User ($\overline{q_a}$) (Sect. 2.A.7) is never an integer i.e. can never be −1 or +1.

Phase-II: User Specific Details

Step 1: ***Computation of frequency of features***

The frequency of features is calculated for all the items in accepted set A for an user a. As mentioned in Sect. 2.A.8, $f_{k,a}$ is calculated for the user according to the items he/she has previously bought and rated. Buying frequency of items are taken an for instance to test the approach.

$$f_{k,a} = \frac{\Sigma_{i \in A} No.\ of\ times\ f_k\ appears\ in\ item\ i}{\Sigma\ all\ features\ in\ A\ for\ user\ a}$$

Step 2: ***Computation of Weight of each feature***

The weight of each feature indicates how important the feature is for the user a. The more the importance, greater is the weight. It is computed in the following pattern.

$$w_{k,a} = \sum f_{k,a} \cdot u_i$$

Step 3: From Phase-I, Step 3, the degree of uniqueness of the user a ($\overline{q_a}$) is already calculated. All the weights of features which are similar to the sign of $\overline{q_a}$ or 0 are taken into the 'Similar' set and the features which are in the opposite sign of the $\overline{q_a}$ are considered to be in the 'Contrast' set. This can be otherwise written as,

$$\text{if } \bar{q}_a \cdot w_{k,a} \geqslant 0 : f_k \in \text{Similar}$$
$$\text{else } f_k \in \text{Contrast}$$

Step 4: All the items that are not present in the set I (Sect. 2.A) but have the f_k such that $f_k \in$ Similar set are selected as the neighbourhood of Similar Items (nbd S).

$$\text{if} \quad f_k \in \text{Similar} \quad \&\& \quad f_k \in i \quad \&\& \quad i \notin I$$
$$\text{Then } i \in nbd\ S$$

if an item i,

$$\frac{\Sigma f_{j-i}}{n_{Similar}} \geqslant 0.75 \text{ Then } i \in \text{Final}$$

where f_{j-i} indicates the feature j in item i, $n_{Similar}$ is the number of features in Similar set, 'Final' set is the set of items to be recommended.

Step 5: Similar to the previous, all the items that are not present in the set I (Sect. 2. A) but have f_k such that $f_k \in$ Contrast are selected as the neighbourhood of Contrast Items (nbd C).

$$\text{if} \quad f_k \in \text{Contrast} \quad \&\& \quad f_k \in i \quad \&\& \quad i \notin I$$
$$\text{Then } i \in nbd\ C$$

if an item i,

$$\frac{\Sigma f_{j-i}}{n_{Contrast}} \leqslant 0.25 \text{ Then } i \in \text{Final}$$

where f_{j-i} indicates the feature j in item i, $n_{Contrast}$ is the number of features in Contrast set, 'Final' set is the set of items to be recommended.

Step 6: The items in the 'Final' are recommended to the user a.

4 Results and Discussions

The proposed algorithm is implemented and tested on a database consisting of 10 items and 16 users. Items are assumed to be wearable t-shirts. Features of the t-shirt are deciding factors for the user. In our problem, we have taken 6 major features defined as

Full sleeve- f1, Half sleeve – f2, Collared- f3, V-Neck- f4, Printed– f5, and Non-Printed– f6. User rating of all the items and Normalized accepted set (*A*) are recorded in Table 2. Item database is recorded in Table 3. Users weighted features matrix data are computed and maintained in Table 4.

Table 2. Simple and normalized user rating

USER RATING OF ALL THE ITEMS												NORAMALIZED ACCEPTED SET (A)										
User\ Items	1	2	3	4	5	6	7	8	9	10	User Average	User\ Items	1	2	3	4	5	6	7	8	9	10
1	4				4.5		2			3	3.375	1	0.625				1.125					
2		2				4			3		3	2						1			0	
3		1				2.5			1	2	1.625	3						0.9				0.4
4	2		1				3.5			3	2.375	4							1.1			0.6
5	1	3	3.5					1.5	2		2.2	5		0.8	1.3							
6		3.5				4.5		3		1.5	3.125	6		0.4				1.4				
7						3	4	4			3.667	7							0.3	0.3		
8						4.5		3.5	2		3.333	8						1.2		0.2		
9		2	2.5			4	1.5			2.5	2.5	9			0			1.5				0
10								4.5			4.5	10								0		
11		2.5			2	2	1.5	2			2	11		0.5			0	0		0		
12	1.5			3		3				2.5	2.5	12				0.5		0.5				0
13							1	2		4	2.333	13										1.7
14		4.5				3.5	4.5	3		2	3.5	14		1				0	1			
15	3.5					3			4		3.5	15	0								0.5	
16		1						2			1.5	16								0.5		

Table 3. Items database

Items (i)	Degree of uniqueness (u_i)	Features						Frequency
		f_1	f_2	f_3	f_4	f_5	f_6	
1	−0.175	1	0	1	0	0	1	5
2	−0.475	0	1	1	0	0	1	8
3	0.025	1	0	0	1	1	0	3
4	0.225	1	0	0	1	0	1	1
5	0.125	0	1	0	1	0	1	2
6	−0.6	1	0	1	0	0	1	10
7	−0.375	0	1	1	0	1	0	7
8	0.575	0	1	0	1	1	0	9
9	−0.175	1	0	1	0	1	0	5
10	−0.475	0	1	0	1	0	1	8

Table 4. Users features weights matrix

Items features weight matrix	Uniqueness of user (q_i)	f_1	f_2	f_3	f_4	f_5	f_6
1	−0.006	−0.029	0.021	−0.029	0.021	0.000	−0.217
2	0.000	−0.258	0.000	−0.258	0.000	−0.029	−0.100
3	−0.403	−0.100	−0.079	−0.100	−0.079	0.000	−0.358
4	0.225	0.000	−0.283	−0.063	−0.079	−0.063	−0.079
5	−0.013	0.004	−0.079	−0.079	0.004	0.004	−0.079
6	−0.053	0.021	−0.079	−0.117	0.000	0.000	−0.117
7	0.106	0.000	−0.317	0.063	−0.096	−0.317	0.000
8	−0.133	−0.100	−0.096	−0.100	−0.096	−0.096	−0.100
9	0.150	−0.158	−0.079	−0.100	−0.150	0.004	−0.100
10	0.000	0.000	−0.096	0.000	−0.096	−0.096	0.000
11	0.000	−0.100	−0.463	−0.358	−0.150	−0.096	−0.475
12	0.048	−0.125	0.000	−0.100	0.038	0.000	−0.125
13	0.528	0.000	−0.079	0.000	−0.079	0.000	−0.079
14	−0.142	−0.258	−0.063	−0.575	0.000	−0.063	−0.258
15	−0.022	−0.117	0.000	−0.117	0.000	−0.063	−0.029
16	0.432	0.000	−0.096	0.000	−0.096	−0.096	0.000

The users item features weight matrix suggests the weight of each feature for the user. The weight can be compared to the Uniqueness of User (q_i) to find the similar and contrast set according to the Sect. 3. Finally items are recommended by the proposed technique to respective user is listed in Table 5. Sign (✓) means that particular item is recommended to particular user.

Table 5. Final set of recommended items to a particular user

	← Set Items →										
		1	2	3	4	5	6	7	8	9	10
Users	1			✓	✓		✓			✓	
	2					✓			✓		✓
	3	✓		✓	✓	✓		✓	✓		
	4				✓	✓				✓	
	5					✓	✓	✓			✓
	6	✓		✓	✓	✓		✓		✓	
	7	✓			✓						
	8	✓	✓	✓	✓	✓		✓			✓
	9					✓			✓	✓	
	10	✓	✓	✓	✓		✓	✓		✓	
	11			✓	✓					✓	✓
	12			✓		✓		✓	✓		
	13	✓		✓			✓				
	14	✓		✓	✓	✓				✓	
	15		✓	✓	✓	✓		✓	✓		✓
	16	✓			✓	✓	✓			✓	

5 Conclusions

In this study, we have focused on developing an intelligent recommender system to recommend items to the prospective users. Our system is tested on a real data set and working effectively. The proposed recommender predicts the products depending upon their taste of choices - whether unique or not. It actively learns from the users' previous reviewed and rated products. It provides a hybrid dataset to the user, which has the items from both the unique (Items those are not purchased frequently) and common sets (Items those are frequently bought and rated by the users). So, in this way, every user is recommended at least one item that belongs to the rarely bought category, thus, fulfilling the primary aim of recommender to suggest items from the long tail whose features are similar to 75% or greater to user interested dataset.

This proposed algorithm overcomes some of the disadvantages of other algorithms presented in Table 1, and it has the following advantages: (1) Recommends all the users from the rarely bought items set, (2) It can handle dynamic dataset, (3) Keyword relevance quality is unambiguous, (4) It can add to social relationship, faith in relationships between users and recommendation systems and (5) Time complexity O(n).

The proposed algorithm can be implemented at an enterprise software application level, which may be utilized in various situations such as research paper selection and recommendations by the editors, on the basis of dynamic data set. It can be further extended to solve the cold start problem and the interference in privacy of user can be minimised. A more detailed study will allow making a common neighbourhood of similar types of users and hence reducing the dataset for computation, thus ultimately reducing the time taken to find the solution set. This algorithm can be hybridised with meta-heuristic algorithms for optimizing the search of similar neighbourhoods for users. Moreover, further scope exists to make relative comparison between other algorithms in terms of testing their efficacy and effectiveness. This algorithm, we hope will be of considerable importance in facilitating the process of rational decision making across variety of sectors such as e-government, e-learning, e-tourism, e-group tasks etc. The proposed algorithm is an innovative attempt to establish some degree of flexibility in variety of situations. However this algorithm may be made robust. Therefore, in terms of finality in refinement, we make no claim.

References

1. Yu, S., Liu, J., Yang, Z., Chen, Z., Jiang, H., Tolba, A., Xia, F.: Pave: personalized academic venue recommendation exploiting co-publication networks. J. Netw. Comput. Appl. **104**, 38–47 (2018)
2. Trappey, A.J.C., Trappey, C.V., Wu, C.Y., Fan, C.Y., Lin, Y.L.: Intelligent patent recommendation system for innovative design collaboration. J. Netw. Comput. Appl. **36**(6), 1441–1450 (2013)
3. Liu, Q., Zhou, M., Zhao, X.: Understanding news 2.0: a framework for explaining the number of comments from readers on online news. Inf. Manag. **52**(7), 764–776 (2015)
4. Liu, J., Kong, X., Xia, F., Bai, X., Wang, L., Qing, Q., Lee, I.: Artificial intelligence in the 21st century. IEEE Access **6**(99), 34403–34421 (2018)

5. Liu, H., Yang, Z., Lee, I., Xu, Z., Yu, S., Xia, F.: Car: INCORPORATING fiLTERED CITATION relations for scientific article recommendation. In: Proceedings of 2015 IEEE International Conference on Smart City/Social Com/Sustain Com (Smart City), pp. 513–518. IEEE (2015)
6. Bollen, J., Nelson, M.L., Geisler, G., Araujo, R.: Usage derived recommendations for a video digital library. J. Netw. Comput. Appl. **30**(3), 1059–1083 (2007)
7. Song, T., Yi, C., Huang, J.: Whose recommendations do you follow? an investigation of tie strength, shopping stage, and deal scarcity. Inf. Manag. **54**(8), 1072–1083 (2017)
8. Pazzani, M.J., Billsus, D.: Content-based recommendation systems. In: The Adaptive Web, pp. 325–341. Springer, Heidelberg (2007)
9. Miah, S.J., Vu, H.Q., Gammack, J., Mcgrath, M.: A big data analytics method for tourist behaviour analysis. Inf. Manag. **54**(6), 771–785 (2016)
10. Zhang, W., Wang, J., Feng, W.: Combining latent factor model with location features for event-based group recommendation. In: ACM SIGKDD International Conference on Knowledge Discovery and Data Mining, pp. 910–918 (2013)
11. Zhao, W., Wu, R., Liu, H.: Paper recommendation based on the knowledge gap between a researcher's background knowledge and research target. Inf. Process. Manag. **52**(5), 976–988 (2016)
12. Balabanovi´c, M., Shoham, Y.: Fab: content-based, collaborative recommendation. Commun. ACM **40**(3), 66–72 (1997)
13. Sugiyama, K., Kan, M.-Y.: Scholarly paper recommendation via user's recent research interests. In: Proceedings of the 10th Annual Joint Conference on Digital Libraries, pp. 29–38. ACM (2010)
14. Feng, H., Tian, J., Wang, H.J., Li, M.: Personalized recommendation based on time-weighted overlapping community detection. Inf. Manag. **52**(7), 789–800 (2015)
15. Bai, X., Wang, M., Lee, I., Yang, Z., Kong, X., Xia, F.: Scientific paper recommendation: a survey. IEEE Access **7**, 9324–9339 (2019)

IoT Based Real-World Emission Analysis of Motorcycles

Ernst Pucher(✉), Andreas Gruber, and Christian Spitzwieser

Vienna University of Technology, Vienna, Austria
{ernst.pucher,andreas.gruber}@tuwien.ac.at

Abstract. The presented study shows results of real driving emission (RDE) measurements of up-to-date Euro 4 four-stroke petrol engines of motorcycles. For the real-world research, a scooter rated at 17,5 kW at 8250 rpm, a small sport motorcycle rated at 32 kW at 9000 rpm and a touring motorcycle at 100 kW rated at 7750 rpm were selected to verify the range of exhaust emissions on the road. All motorcycles have a mixture preparation system by injection system and intake manifold injection. The vehicles were operated at city driving with typical stop-and-go phases areas as well as on highways at speeds up to 130 km/h.

The ultra-compact exhaust measurement system OBM 5.0 was modified for the motorbike application. This mobile exhaust lab offers now close-to-real-time analysis of exhaust components using IoT, cloud computing and V2X connectivity. The exhaust gas mass flow is calculated in real time from the parameters provided by the vehicle field bus and the gaseous components with a mass balance and a reaction kinetics model. The link with the vehicle speed provides the emission rates in grams per distance. Fuel consumption is calculated using the carbon balance. An integrated GPS sensor provides signals to link the geographic information of the measurement sections with the exhaust components and to calculate the resulting real driving resistances. The determined mass emissions of the pollutants, the average particle size and their number are output as a comprehensive data set every second. As a close to real-time check, the data can be assessed locally by the operator on a mobile device.

The installation of the whole system can be done in less than one hour and the equipment worked very satisfying. The investigations carried out showed that the methodology of RDE measurements also leads to good and reliable results in the two-wheeler sector. The three vehicles were dynamically driven in real traffic. It could, as it is already known from other vehicle classes, a wide spread of results within the same actual Euro 4 emission class be observed. With regard to the used three-way catalytic converters, noticeably high carbon monoxide and nitrogen oxide emissions were detected in one vehicle.

Keywords: Sustainability · Emission control · IoT · Big data analysis · NOx · Optimization

N. M. Durakbasa and M. G. Gençyılmaz (Eds.): ISPR 2020, LNME, pp. 14–23, 2021.
https://doi.org/10.1007/978-3-030-62784-3_2

1 Introduction

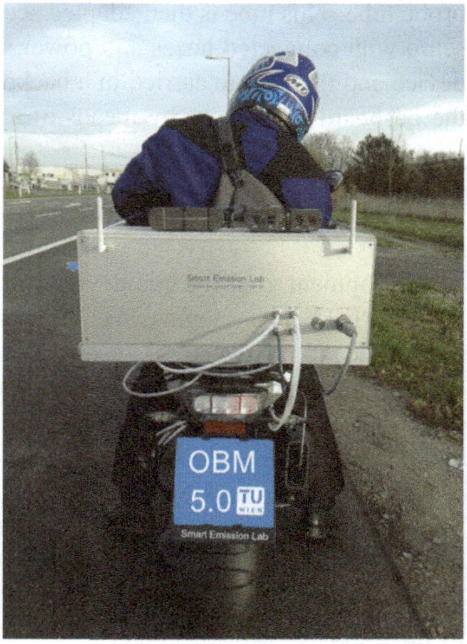

Fig. 1. Measuring setup 1

The OBM 5.0 is a portable measuring system which has already been used and tested in investigations on several different vehicle types [1–3]. The field of application is broad and will be extended in the context of this paper by application on motorcycles of class L3e with subsequent test drives. The OBM 5.0 has already been used for tests on a snowmobile [4]. The shock resistance of the measuring device is therefore sufficient for the application on single-track vehicles.

With the help of the measuring device, the mass emissions of the exhaust components can be determined in real time. The CO2, CO and HC concentrations are detected by NDIR sensors and the NO, NO2, NOx and O2 concentrations by electrochemical sensors. The measurement of the emitted particle number is carried out by means of a Diffusion charge sensor. The measuring device, in the version with integrated particle measuring system, weighs approximately 16 kg and can be operated with a 12 V vehicle power supply, whereby the energy consumption is less than 60 W. This advantageous energy consumption means that rechargeable batteries are sufficient for the power supply and the vehicle electrical system does not have to be used. The OBM 5.0 was originally developed for real-world exhaust gas measurements on passenger cars. The installation time for the application in passenger cars is about 30 min and can also be carried out in this time for motorcycles.

Fig. 2. Measuring setup 2

A partial flow of the exhaust gas is fed to the measuring device by means of a hose. The partial flow is taken directly at the outlet of the muffler. The OBM 5.0 is secured against any movement by means of a tension belt on an adapter console screwed onto the passenger grab handle or the chassis there. The power supply of the measuring device is provided by two battery packs, which are secured on the upper side by the tension belt. The hose, which conducts the partial exhaust gas flow into the measuring device, is connected to a probe tube. This tube is then adapted according to the position of the muffler. The entire setup with connected hoses and power supply is shown in Fig. 1 and Fig. 2. The GPS device and the PC are carried in a backpack. The OBD interface reader is connected to the on-board electronics by means of an adapter.

2 IOT Connection of the Emission Measuring Device

A development goal for this device was the application of state-of-the-art Internet-of-Things (IoT) technologies [5]. The first step was the implementation of wireless communication between the measurement device itself and its control-computer. They can communicate via WiFi 5-Protocol (former: IEEE 802.11ac). After this was established, the next step was to create a network-application to upload the gained data and make it accessible for the user through a web-browser based GUI. The final step is linking all those measurement systems together using a MQTT (Message Queuing Telemetry Transport) -based data platform. In using this widely acclaimed IoT standard protocol, it is possible to control, collect the data from, and service a huge amount of measurement systems. In Austria alone, if said measurement method is implemented as a standard, there is the demand for several thousand individual measurement devices.

The aim of the developed variant was to use and implement an architecture that meets all requirements for a scalable IoT system. Criteria for the selection continued to be an integrated solution to prevent data loss during a connection interruption, as well as an IoT system, which corresponds to a standard. The use of Open Standards allows for easy system extensibility, future-proofing, and compatible third-party system support.

The final version was developed based on a professional IoT architecture. It was the system IoT Watson I.B.M. selected, which was developed according to the IoT architecture MQTT. MQTT is the most used IoT standard. MQTT defines a publish- and subscribe system with a central broker and the associated protocols for connecting IoT nodes. IoT nodes are on the one hand devices which send the signals of the IoT sensors to the central broker. Data transfer to the broker is mentioned in the MQTT Architecture Event. The counterpart to this is a command, which is sent by the broker to all devices, for example to send a central start command. The second type of defined IoT nodes are applications. On the one hand, these receive the measurement data of the devices and on the other hand can send command to devices.

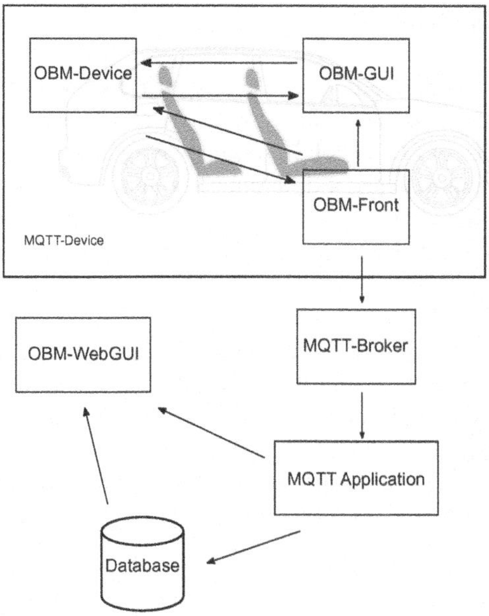

Fig. 3. IOT integration of the short test device

The logon of the IoT nodes at the central broker is defined in the MQTT protocol. IoT application nodes have the option of defining filters when sending the commands or receiving the events to specifically communicate with only a subset of the IoT nodes. The graphics in Fig. 3 shows the integration of the OBM emission meter into the IBM IoT Watson system.

In the software of the decentralized emission measuring device an MQTT device node has been implemented. This is responsible for sending the OBM measurement data to the central broker. As an MQTT device adapter an implementation for the programming language Go was chosen. This adapter offers the possibility to temporarily persist the measurement data in case of interruption of the data communication. Possible is an in-memory persistence or a persistence in the file system. If in-memory persistence is selected, the data will only be saved during runtime of the emissions meter software. Should the emissions meter software be terminated in the absence of a data connection, the metering data would not be sent to the central IoT broker and would be lost. If persisted in the file system, the measurement data would also persist after stopping and starting the emissions meter software. As soon as there is a connection between the emission measuring device and the IoT Broker, the data is sent automatically.

By registering the emission measuring device MQTT Device with the IoT Broker, the respective nodes can be uniquely identified. The MQTT transmission protocol fulfills all requirements for data transmission according to the current safety criteria. Through appropriate authorization, authentication and encryption of the measurement data during transport, these are protected against improper access.

3 Experimental Setup

When selecting the test objects, attention was first paid to which vehicle models of the EG class L3e were most frequently newly registered in Austria in the year 2018. In total, 26,041 motorcycles of EG class L3e were newly registered in Austria in 2018 [6]. Care is taken to select motorcycles of different sizes in order to show that the test setup can be installed just as easily on small single-track motor vehicles as on large motorcycles. The choice is a BMW R 1250 GS, a KTM 390 Duke and a Vespa GTS 300 ABS.

The test track chosen for the test drives consists of a city drive and a motorway drive. The exact course of the routes can be seen in Fig. 4. On the left side there is shown the city driving course and on the right side the course for motorway driving. The speeds travelled are marked in colour in the legend. On average, the target speed for the urban driving cycle is 20 km/h. The actual speed profile for driving the BMW R 1250 GS is shown in Fig. 5 as an example for all test vehicles. The maximum value is below 50 km/h and the driven average value is only 1.5% higher than the predicted value.

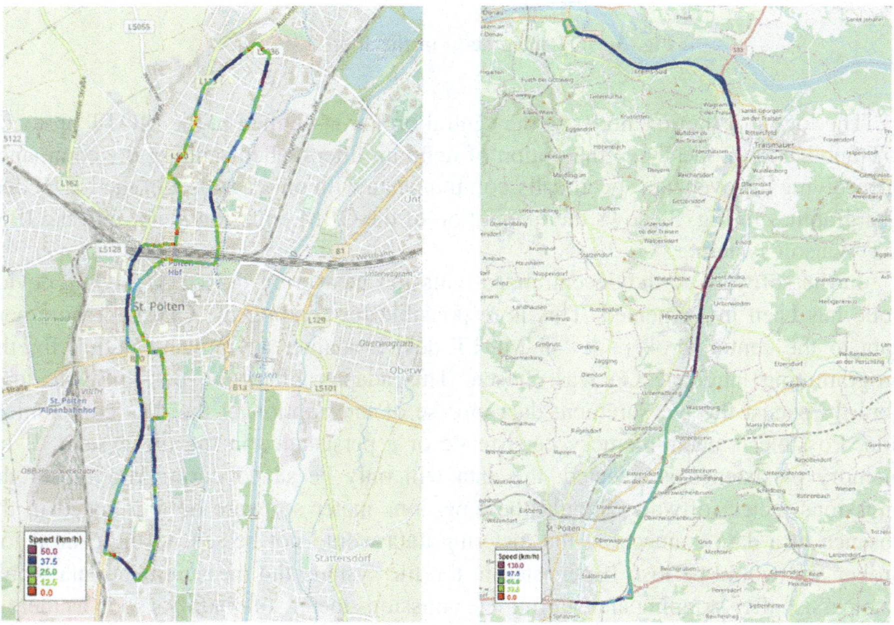

Fig. 4. Measuring route (V_{BMW})

The speed profile of the other vehicles does not differ significant from the one shown here. Figure 5 also shows how the gradient and therefore also the altitude conditions of the city trip look like. It is important that the starting point and end point of the measuring section are at the same sea level in order to compensate for the influences of the height difference.

The speed profile of the motorway journey is also shown in Fig. 4. Here, the data of the BMW are shown again as an example. Except for the 130 km/h section all test phases can be completed at an almost constant driving speed. The reason for the disturbance at about 500 s is due to traffic conditions. To ensure safe driving, occasional braking was necessary. This is also the case to a similar extent for the other motorcycles tested. The elevation profile of the motorway can again be seen in Fig. 5. In the first half of the motorway ride, a falling gradient of the height profile occurs for the most part, the second part is characterized by an ascending change in height. Since the individual sections were driven at the same speed in both directions, the influence of the height difference is minimized, as already mentioned. The measuring runs of the BMW and the KTM are carried out in relatively short time intervals at the same day, which also reduces the influence of the wind.

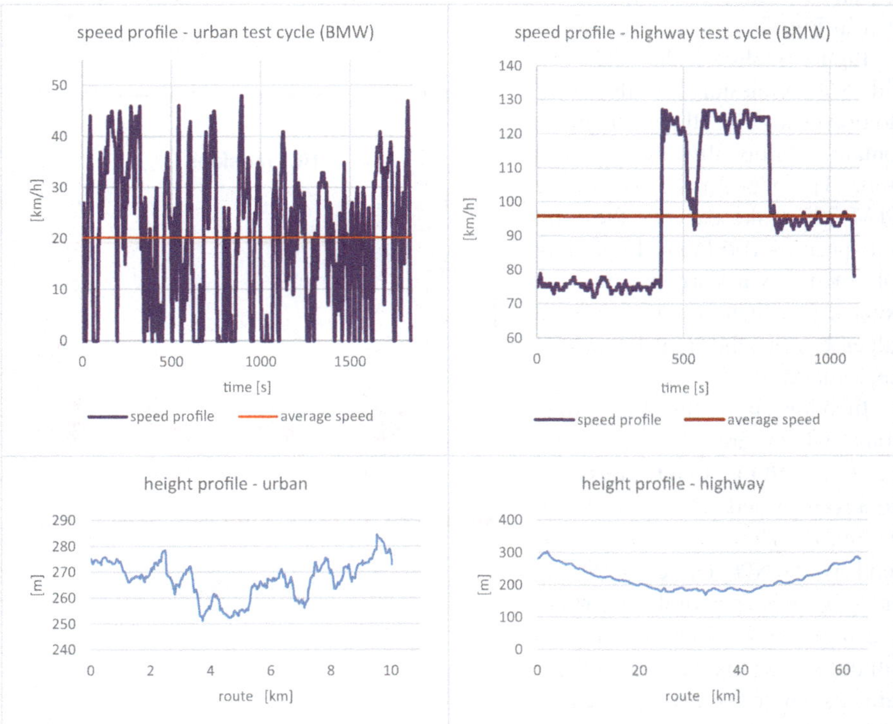

Fig. 5. Speed and height profile (BMW)

4 Result of the Emission Monitoring

Before going into the details of the measurement results, it must be mentioned that the limit values of the Euro standards are not limit values for real driving. In general, there are still no mandatory RDE emission limits for motorcycles. The WMTC consists of three phases, with the average speed of the third phase being 94.4 km/h. Four different test phases were investigated in this paper. In one of them, the 100 km/h phase, an average speed of around 94.2 km/h was achieved. As the two average speeds are close together, the Euro 4 limit value is shown in the three diagrams opposite only as a reference value and exclusively for this test phase.

A mean value was calculated from the measured real-time data. These average values are shown in all diagrams of Fig. 5, 6 and 7. On average, the constant speed test with the three speeds 80 km/h, 100 km/h and 130 km/h is driven for approximately 20 km. Fuel consumption and CO2 emissions go hand in hand, as can be clearly seen in Fig. 7.

Figure 6 shows the CO, HC and NOx emissions of the three motorcycles. The three diagrams contain additionally the Euro 4 limit. The CO-Emissions of the KTM 390 Duke exceeds the limit at a speed of 100 km/h. Regarding unburned hydrocarbons in the exhaust gas, there are no measured values beyond the limit for any of the vehicles.

In addition to the NO_x emissions, which are also shown in Fig. 6, the NO and NO_2 emissions are also recorded. These are shown in Fig. 8. They show a similar trend as the NO_x curve. Both, CO and NO_x emissions of the KTM are at a high level. However, it goes without saying that several of these vehicles would have to be tested in order to substantiate the exhaust gas values with sufficient certainty. The emission level of the Vespa only exceeds in the range of 130 km/h, but in this case to a large extent.

It is noticeable that the exhaust gas values of the Vespa are almost exclusively too high for the test phase 130 km/h. It is therefore

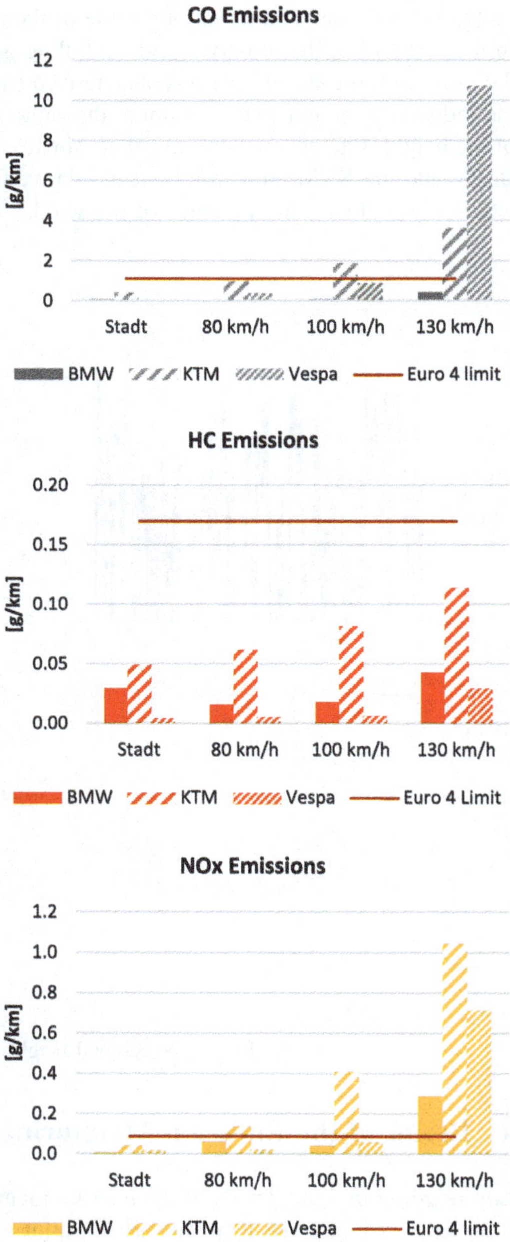

Fig. 6 CO, HC and NOX emissions

worth mentioning that the Vespa was constantly in the range of its maximum speed during this test phase and that the speed was electronically limited by the vehicle. The values of the particle number are also shown in Fig. 8 on the right side. The Vespa GTS 300 has a higher particle number than the other two vehicles for all of the test sections. The maxima of all three vehicles are in the 130 km/h section. This is especially true for the KTM and the Vespa. For the particle number, with increasing speed a falling trend was determined for the Vespa. For the KTM and the BMW, the number of particles emitted increases with increasing speed.

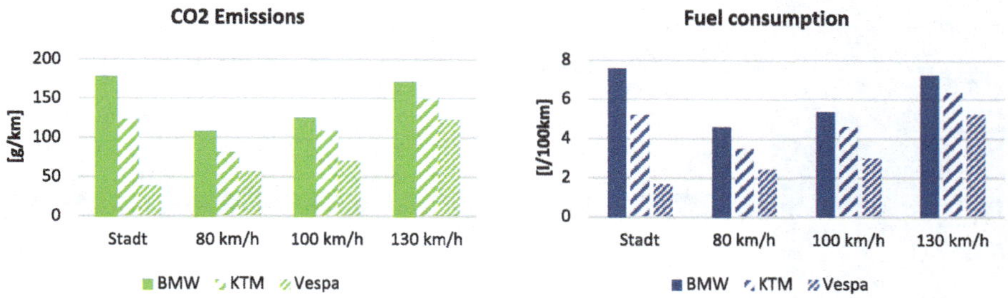

Fig. 7. CO2 emissions and fuel consumption

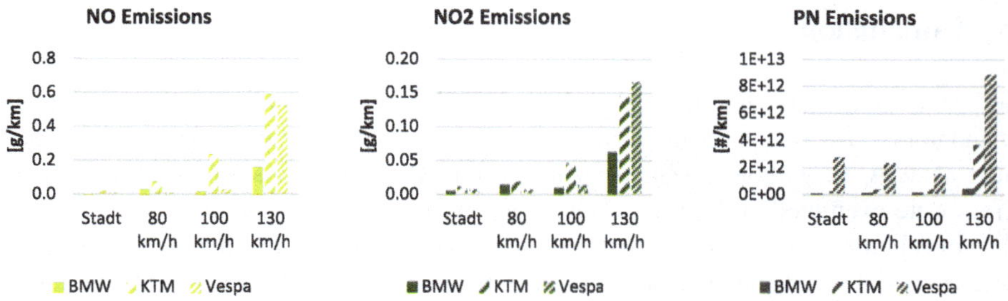

Fig. 8. NO, NO2 and PN emissions

The technical data of the tested motorcycles and the weather conditions during the investigations can be taken from the following Table 1.

Table 1. Vehicle data, average speed and weather conditions

vehicle data			
manufacturer	BMW	KTM	Piaggio
vehicle type	R 1250 GS	KTM 390 DUKE	Vespa GTS 300 GTS
dead load	249 kg	163 kg	160 kg
total weight	~349 kg	~263 kg	~260 kg
fuel	gasoline	gasoline	gasoline
gross weight	1254 cm^3	373 cm^3	278 cm^3
power	100 kW at 7750/min.	32 kW at 9000/min.	17,5 kW at 8250/min.
emission norm	Euro 4	Euro 4	Euro 4
initial registration	04-04-2019	02-07-2019	09-08-2019
$V_{average}$ [km/h]			
urban	20.3	20.5	19.5
80 km/h	76.1	77.8	76.1
100 km/h	95.5	96.2	91
130 km/h	121.8	124.9	117.2
weather conditions			
T_u [°C]	10	9	6
ϕ [%]	75	76	100
V_{wind} [km/h]	6	8	11
wind direction [-]	North	Northeast	Northeast

5 Conclusion

In the area of cars and trucks, real-world exhaust gas measurements are already mandatory. This is not yet the case for motorcycles. In order to show that such measurements work without problems with the OBM 5.0, three motorcycles of different sizes were examined in this paper. These are exclusively vehicles of the EG class L3e. The installation time of the measuring device is about 30 min if an adapter for the OBD interface is available, as it is the case for the application in passenger cars, but not yet for motorcycles.

The under real world conditions measured exhaust emissions of the three Euro 4 vehicles show a maximum in the 130 km/h phase for each exhaust component. As a reference for the 100 km/h phase, the Euro 4 limit is shown in the diagrams for CO, HC and NOx. In general, exhaust gas values rise with increasing speed. The CO2 emissions, which go hand in hand with fuel consumption, are higher for urban traffic than for the 80 km/h drive. The NO and NO2 curve show a similar trend as the NOx curve. For the particle number, with increasing speed a falling trend was determined for the Vespa. For the KTM and the BMW, the number of particles emitted increases with increasing speed.

The Vespa GTS 300 ABS, the smallest measured vehicle, is below the valid limits up to a speed of 100 km/h. However, the test phase 130 km/h is in the range of its vehicle-specific maximum speed, which is why a sudden increase of all exhaust components can be seen here. Only the HC emissions are unproblematic for all tested vehicles. In conclusion, it can be said that, as expected, the emission values increase at high speed.

Acknowledgement. We would like to thank in particular our former colleague Emanuel Garieri, as well as our partner from the Austrian Automobile Club with Roadside Assistance Service (ÖAMTC) and the Austrian Federal Ministry of Climate Action, Environment, Energy, Mobility, Innovation and Technology for the cooperation in real-world measurements.

References

1. Pucher, E., Schilk, A., Gruber, A., Hacker, A.: Real-world in-car emission measurements of euro 5 and euro 6 diesel passenger cars. In: 25th CRC Real World Emissions Workshop, Long Beach, California (2015)
2. Gruber, A., Pucher, E., Spitzwieser, C.: NOx and PN emissions from vehicles with SI-engines with stop-start systems in urban traffic. In: 29th CRC Real World Emissions Workshop, Long Beach, California (2019)
3. Pucher, E., Gruber, A., Eidmann, A.: Real-time V2I communication based CO2 and NOx emission monitoring. In: 22nd ITS World Congress, Strasbourg, France (2017)
4. Cachón, L., Pucher, E., et al.: Real-world emission performance of a snowmobile with dual-fuel gasoline and natural gas powertrain. In: "SETC2010CD", herausgegeben von: SAE International, (eingeladen), Paper-Nr. 2010-32-0125, 11 S (2010). ISBN: 978-0-7680-2086-1
5. Eidmann, A.: Vehicle to Infrastructure (V2I) real-time communication for mobile measurement systems. Doctoral thesis at the TU Wien, Institute for Powertrains and Automotive Technology (2018)
6. Bundesanstalt Statistik Österreich. Neuzulassungen in Österreich. Accessed 11 Sep 2019, https://www.statistik.at/wcm/idc/idcplg?IdcService=GET_NATIVE_FILE&RevisionSelectionMethod=LatestReleased&dDocName=116096

Decision Making

Using Best Worst Method for Location Selection of Piezoelectric Tiles

Ali Sarper Özer[1], Heliya Hasani[1], Elif Bahar Genç[2], Nilsu Kutlu[2], Gül Tekin Temur[1], and Çağlar Sivri[2(✉)]

[1] Industrial Engineering, Bahçeşehir University, Istanbul, Turkey
{alisarper.ozer,heliyahasani}@bahcesehir.edu.tr, gul.temur@eng.bau.edu.tr
[2] Engineering Management, Bahçeşehir University, Istanbul, Turkey
{elifbahar.genc,nilsu.kutlu}@bahcesehir.edu.tr, caglar.sivri@eng.bau.edu.tr

Abstract. This study aims to select the best location for implementation of piezoelectric tiles. This decision handles objective and subjective criteria simultaneously therefore it is beneficial to use a multi-criteria decision making tool for the problem. As a novel approach, Best-Worst Method is chosen to evaluate alternatives and affecting criteria. Surveys are sent to the experts who have valuable experience on this subject and asked to compare criteria and sub-criteria pair wisely. After gathering the answers, the surveys are checked for their consistencies and the method is applied to find the importance of criteria. A case study is conducted for a campus of a university. The proposed methodology can be a good supportive tool to choose the satisfactory locations for construction of piezoelectric tiles.

Keywords: Best Worst Method · Location selection · Multi-criteria decision making · Piezoelectricity

1 Introduction

According to Woodford (2009), piezoelectricity (also referred to the piezoelectric effect) is the emergence of a voltage cross the parts of a crystal once you subject it to mechanical push (by pressing it) [1], In the near future, the increasing consumption and demand for energy resources [2, 3] and environmental pollution have forced people to consider cleaner energy sources. It is possible to see the long-term benefits of using these energy sources. Explicitly, in a world where the use of electricity is increasing constantly, the significance of piezoelectricity increases as well, since it enables mechanical energy to be transformed into electrical energy.

The efficient utilization of piezoelectric tiles depends on many strategic decisions including location selection. Researchers have been attached high importance on location selection in the industry. It is not dependent on a few static factors but rather dynamically changing factors. Once a location is selected for a specific purpose and the actions are taken accordingly it becomes too expensive or inefficient to change. This is the reason why every aspect of the selection decision must be thought of. In the

N. M. Durakbasa and M. G. Gençyılmaz (Eds.): ISPR 2020, LNME, pp. 27–34, 2021.
https://doi.org/10.1007/978-3-030-62784-3_3

relevant literature, these aspects are handled by using multi criteria decision making (MCDM) approaches. For instance, Ozdemir and Sahin (2018) used one of the most common MCDM methods named Analytic Hierarchy Process (AHP) for the location selection of a solar PV power plant [4]. Senvar et al. (2016) has found the optimum location for a hospital site location in Istanbul using Hesitant Fuzzy TOPSIS [5]. Wu et al. (2007) used AHP to find the optimal selection of location for Taiwanese hospitals [6]. To our knowledge, there is no study focusing on location selection for construction of piezoelectric tiles. Therefore in this study we aim to fill this gap by using a novel MCDM approach named Best Worst Method (BWM). A case study is conducted for the campus of a famous university in Turkey.

In the following section the details of BWM are given. In Sect. 3, the proposed methodology is explained and the related criteria are determined. In Sect. 4, the application is done using preferences of decision-makers.

2 Best Worst Method as a Novel Multi-Criteriia Decision Making

In AHP, which is one of popular MCDM methods, due to the lack of required quantitative data users who apply the methodology are giving scores depending on their preferences. It is required to compare all criteria pair wisely; therefore it increases the followed steps in the procedure. On the other hand, in Best Worst Method there is still not enough quantitative data but while applying the method, the comparisons are made only between best criterion to the others and the worst criterion and the others. Therefore, it is less complicated compared to AHP.

Let's assume there is a MCDM problem that has n criteria and for evaluation, a scale from 1/9 to 9 is used. After the criteria are set, the decision-makers identify their preferences over criteria. Firstly, the Best-to-Others vector is created as XB = (xB1, xB2,…,xBn) and then, Others-to-Worst vector is created as Xw = (x1W,x2W,…, xnW)T. xBj refers to the best criteria to j, while xjW refers to the preference of criterion j to the worst criteria. The best weight is achieved by minimizing the values of | (WB/Wj) – xBj | and | (Wj /WW) – xjW |. Then the following optimization model is applied.

Min ξ
Subject to.
$| (W_B/W_{j)} - x_{Bj} | \leq \xi$ for all *j*.
$| (W_j /W_{W)} - x_{jw} | \leq \xi$ for all *j*.
$\Sigma W_j = 1.$
$W_j \geq 0$, for all *j*.

Solving the above equation yields the criteria weights and ξ value will be used to calculate consistency of decision-makers choices [7].

3 The Proposed Methodology

The explanation of the proposed methodology is given below:

1. Prepare surveys for evaluation of criteria, sub-criteria and alternatives.
2. Send the surveys to the authorized people to gather evaluations.
3. Check each consistency ratio and the survey is iterated if the consistency is not satisfactory.
4. Apply mathematical model given at previous section to obtain criteria.
5. Take the average of decision-makers' sub-criteria weights to find the local weights.
6. Take the average of decision-makers' criteria weights to find the local weights.
7. Take the average of decision-makers' weights for alternatives to find the local weights.
8. Multiply each sub criterion's local weight with parent main criteria weight to find the global weight that addresses the importance of sub criterion.
9. Then, multiply the global weight of each sub-criterion with each alternative location's weight and sum all these values for finding the weight of each alternative.
10. The weights of each alternative location are ranked, and the highest value is the best alternative while lowest value is the worst alternative.

3.1 Determination of the Criteria Set

Table 1 illustrates the set of criteria and sub-criteria evaluated for the piezoelectricity location selection. Because there is no study on location selection of piezoelectricity tiles in the literature, the criteria are figured out by interviewing with the experts and considering similar studies as [8, 9, 10, 11].

Table 1. Criteria and sub-criteria of the proposed methodology

Criteria	Sub-Criteria
Risk	- Technical risk - External risk - Organizational risk
Economic	- Investment cost - Maintenance cost - Payback period
Social	- Social benefits - Social attractiveness - Social awareness - Social connectivity
Technical	- Safety - Infrastructure - Accessibility - Expansion

(continued)

Table 1. *(continued)*

Criteria	Sub-Criteria
Demographics	- Current population - Population density - Population expansion
Strategic	- Top Management support - Reputation

1) Risk Criteria

   a) Technical Risk: It refers to strength of the tiles belongs to material used.
   b) External Risk: It refers to unforeseen events such as the displacement or harm of the pavements via floods and earthquakes etc.
   c) Organizational Risk: It refers to regulations affecting strategic decisions applied in an enterprise.

2) Economic Criteria

   a) Investment Cost: It refers to the installment cost of the pavements that changes belongs to quality.
   b) Maintenance Cost: It refers to the maintenance cost that changes belongs to quality of the used material.
   c) Payback-Period: It is the period passed until earning money. It is dependent on a lot of factors namely; energy produced per step, total number of steps and from which price you will be selling the energy produced.

3) Social Criteria

   a) Social Benefits: It refers to the future development parameters depending on efficient energy usage and sustainable growth.
   b) Social Awareness: It refers to environmental awareness and energy conversion awareness of the society where the system is applied. It affects the usability and continuity of the system.
   c) Social Attractiveness: It refers to location based popularity.
   d) Social Connectivity: It refers to the capability to help people interact with each other.

4) Technical Criteria

   a) Safety: It refers how the utilization of the location is safe for human life.
   b) Infrastructure: It refers to the field of the location that has impact on the installation capacity and efficiency.
   c) Accessibility: It addresses how well the area is placed so that it maximizes the step.
   d) Expansion: It refers to the capacity of the location for being expanded in the future.

5) Demographic Criteria

   a) Current population: It refers to the population that uses the alternative location. It is important to know which location is close to public transportation areas and crowded pedestrian zones.
   b) Population density: It refers to the density that has potential to effect energy producing capacity per a defined period.
   c) Population expansion: It refers to futural impact and improvement potential of the location.

6) Strategic Criteria

   a) Reputation: It refers to the popularity of the alternatives that affects the trust of the society.
   b) Top Management Support: It refers to the managerial support for executing the project and avoiding any future restrictions.

3.2 Computation of Weights and Application Results

The procedure given at previous part is followed to compute the weights. The surveys conducted by three experts were analyzed with BWM. According to the results given in Table 2, "Technical Criteria" is determined as the best main criterion, and "Demographic Criteria" is chosen as the worst main criterion. The averages of "Strategic Criteria", "Economical Criteria" and "Risk Criteria" have values close to each other.

Table 2. BWM results for main criteria

	Main Criteria					
	Risk	Economic	Social	Technical	Demographic	Strategic
Expert 1	0.0328	0.1532	0.0919	0.2298	0.1149	0.3775
Expert 2	0.2410	0.1607	0.0689	0.3982	0.0349	0.0964
Expert 3	0.0360	0.2518	0.1259	0.4137	0.1007	0.0719
Average	0.1033	0.1885	0.0956	0.3472	0.0835	0.1819

To identify the overall ranking among the sub-criteria, multiplication is done between sub and their own parent main criterion's weights which denominated as global weights in Table 2. It is significant to perceive the importance of all sub-criteria by comparing each other. When the first three stages of ranking have examined, it can be deduced that the "Infrastructure" (0.2085) of "Technical Criteria" is the most influential criteria. "Top Management Support" of "Strategic Criteria" follows subsequently. Third and also the fourth stage belong to "Economic Criteria" with a close difference than each other. On the other hand, "Population Expansion" from "Demographic Criteria" has the least importance. The global weights are given at Table 3.

Table 3. Global weights of sub criteria

Overall ranking	Sub-criteria	Global weights
1	Infrastructure	0.2085
2	Top management support	0.1637
3	Payback period	0.0888
4	Investment cost	0.0790
5	Technical risk	0.0626
6	Population density	0.0584
7	Safety	0.0562
8	Accessibility	0.0453
9	Social benefits	0.0399
10	Expansion	0.0372
11	Organizational risk	0.0320
12	Social attractiveness	0.0217
13	Social awareness	0.0211
14	Maintenance cost	0.0207
15	Current population	0.0199
16	Reputation	0.0182
17	Social connectivity	0.0128
18	External risk	0.0087
19	Population expansion	0.0051

The location selection procedure is conducted for one of popular campuses among Turkish universities. The procedure given in Sect. 3 was applied. The evaluations of the alternative locations were multiplied by global weights and the best alternative locations were found.

Six alternative locations (A1, A2, A3, A4, A5, A6) are chosen for the case study. The results are given at Table 4. According to the results, A6 is found as the best location for the selected campus.

Table 4. Weights of alternative locations

Criteria	Sub-criteria	Global weights	A1	A2	A3	A4	A5	A6
Risk	Technical risk	0.0626	0.0870	0.1045	0.0746	0.4343	0.2611	0.0385
	External risk	0.0087	0.0870	0.1045	0.0746	0.4343	0.2611	0.0385
	Organizational risk	0.0320	0.0337	0.1180	0.1573	0.2360	0.3876	0.0674
Economic	Investment cost	0.0790	0.0762	0.2667	0.0667	0.4588	0.0889	0.0427
	Maintenance cost	0.0207	0.1586	0.2380	0.3909	0.0595	0.1190	0.0340
	Payback period	0.0888	0.1586	0.0595	0.2380	0.0340	0.1190	0.3909

(*continued*)

Table 4. (*continued*)

Criteria	Sub-criteria	Global weights	A1	A2	A3	A4	A5	A6
Social	Social benefits	0.0399	0.1832	0.0687	0.2748	0.0611	0.1374	0.2748
	Social attractiveness	0.0217	0.1586	0.0595	0.2380	0.0340	0.1190	0.3909
	Social awareness	0.0211	0.1667	0.1667	0.1667	0.1667	0.1667	0.1667
	Social connectivity	0.0128	0.1586	0.0595	0.2380	0.0340	0.1190	0.3909
Technical	Safety	0.0562	0.0766	0.2680	0.0670	0.4571	0.0893	0.0420
	Infrastructure	0.2085	0.2360	0.0674	0.3876	0.1180	0.1573	0.0337
	Accessibility	0.0453	0.1532	0.0328	0.1149	0.0919	0.2298	0.3775
	Expansion	0.0372	0.1127	0.0376	0.1408	0.1878	0.0704	0.4507
Demographics	Current population	0.0199	0.1573	0.0674	0.2360	0.0337	0.1180	0.3876
	Population density	0.0584	0.1292	0.2585	0.0862	0.0369	0.0646	0.4246
	Population expansion	0.0051	0.1270	0.0363	0.1016	0.0635	0.2541	0.4174
Strategic	Top management support	0.1637	0.0358	0.1647	0.2471	0.0824	0.0618	0.4082
	Reputation	0.0182	0.1244	0.0355	0.0829	0.2488	0.0995	0.4088
Total points		0.0626	0.1311	0.1271	0.2148	0.1688	0.1344	0.2239

4 Conclusions

The continuous increase in human population brings the inevitable rise in resource consumption in its wake. Hence, the world has led to search for alternative energy sources. Piezoelectric application, which has started to take its place in the world, helps a large part of electrical energy production in the long term. Its ability to create a remarkable amount of electrical energy makes piezoelectric applications a worldwide investment in the future. The aim of this study is to propose a methodology to select the best location for piezoelectric tile construction. In order to show how it can be implemented in real life, a case study is conducted at a popular university in Turkey.

The selection of the location while setting up the system is crucial for piezoelectric energy storage and conservation. Six main criteria and their nineteen sub criteria are considers in the model, and the prepared surveys have been answered by experts to be evaluated with Best Worst Method which is a novel MCDM method. As a result, it is observed that the most important main criterion was chosen as "Technical Criteria", while the most important of the sub-criteria was determined as "Infrastructure" which is the sub-criteria of "Technical Criteria" as well. The infrastructure of the university should be controlled by the authorities and the panels should be made in places where they can stand firm against external factors will be beneficial in terms of extending the life of the product and will also contribute economically. For this reason, the product should be placed in places where the infrastructure is suitable for maximum efficiency. Subsequently, as a sub-criterion of "Technical Criteria"; "Safety", "Accessibility" and "Expansion" follow "Infrastructure" respectively. University should consider that it is safe to place the tile due to its deflection level. "Accessibility" concerns the scale of the tiles that placed in the selected area. It is always possible to expand, remove or relocate the whole tile through its easy setup and maintainability.

In the near future the researchers who are interested in this topic can extend this study by integrating fuzzy sets to BWM. Furthermore a sensitivity analysis can be applied in order to see how the changes on criteria affect the solution.

References

1. Erturk, A., Inman, D.J.: Piezoelectric Energy Harvesting. Wile, Hoboken (2011)
2. Gray, R.: Getting power to people where and when it's needed could rewrite the geopolitical rulebook. https://www.bbc.com/future/article/20170313-the-biggest-energy-challenges-facing-humanity/. Accessed 30 Dec 2019
3. Birol, F.: World Energy Outlook (2018). https://www.iea.org/reports/world-energy-outlook-2018/, Accessed 30 Dec 2019
4. Carbone, P.: Measurement: Journal of the International Measurement Confederation. Elsevier (2011)
5. Senvar, O., Otay, I., Bolturk, E.: Hospital site selection via hesitant fuzzy Topsis. IFAC **49**, 1140–1145 (2016)
6. Wu, C.R., Lin, C.T., Chen, H.C.: Build. Environ. **42**, 1431–1444 (2007)
7. Rezaei, J.: Best-worst multi-criteria decision-making method. Omega **53**, 49–57 (2015)
8. Mahadevan, R., Asafu-Adjaye, J.: Energy consumption, economic growth and prices: a reassessment using panel VECM for developed and developing countries. Energy Policy **35**, 2481–2490 (2007)
9. Kruger, P.: Alternative Energy Resources—A Quest for Sustainable Energy. Wiley, New Jersey (2006)
10. Arnau, A.: Piezoelectric Transducers and Applications. Springer, Heidelberg (2004)
11. Lu, C., Czanderna, A.: Applications of Piezoelectric Quartz Crystal Microbalances. Elsevier, Amsterdam (2012)

Facility and Capacity Management

Site Selection for a Training Centre Focused on Industry 4.0 by Using DEMATEL and COPRAS

Semra Birgün[1(✉)] and Mesut Ulu[2]

[1] Industrial Engineering Department, İstanbul Gedik University, Istanbul, Turkey
semrabirgun@gmail.com

[2] Occupational Health and Safety Department, Bandırma Onyedi Eylül University, Bandırma, Turkey
mulu@bandirma.edu.tr

Abstract. One of the most important elements of Industry 4.0 and digitalization is human. The change of technological infrastructures and processes will also change the qualities sought in human resources concomitantly. Requirements of the digital world will replace the existing knowledge and experience. For this reason, whether white- or blue-collar employees must acquire the competence and skills that will meet these requirements. While the existing processes show the characteristic of change; the white-collar personnel to manage these processes should be compatible with this transformation and well equipped with the capability of managing the relevant processes. On the other hand, although the number of blue-collar employees is expected to decrease; a great need for more qualified practitioners compatible with the digital world will arise. It is obvious that short term on-the-job trainings will not be sufficient. This is the duty of universities as well as the private sector. Beyond the programs of the existing vocational and technical education high schools and universities, the need is obvious for raising specialists that will implement and manage the digital technology and design it, if necessary. It was planned to establish a "digital transformation training/application center" totally within the scope of digital technology and Industry 4.0, in order to ensure that the students receive their trainings within the real system both from the trainers and practitioners in a practical way. In this study which is about the problem of selecting a site for this center, the selection criteria were determined by considering these needs and requirements. A novel hybrid methodology is applied in this study for the problem of multiple criteria decisions making. DEMATEL method is used to determine the weight of the criteria and the COPRAS method is used to evaluate the alternatives based on these weighted criteria and select the site of establishment.

Keywords: COPRAS · DEMATEL · Digitalization · Education · Industry 4.0 · Multi-criteria decision making · Site selection

N. M. Durakbasa and M. G. Gençyılmaz (Eds.): ISPR 2020, LNME, pp. 37–50, 2021.
https://doi.org/10.1007/978-3-030-62784-3_4

1 Introduction

Industry 4.0 and digitalization have been the focal point of the industry world and literature since 2011, as a result of the development of technology and engineering applications as of the industrial revolution. Today's industrial development has arisen as using the technologies such as the machines communicating with each other, smart systems, cloud informatics, cyber-physical systems in the field of production and service as well as using the internet in the field of production; in addition to the steps of ensuring mechanization with the invention of the steam machine, making the serial production possible with the involvement of the electrical energy, ensuring automation by using the information technologies and computer languages in production and allowing the establishment of dark factories.

When it comes to Industry 4.0, what we first encounter in the literature is its technological infrastructure. According to most authors [1–8] this infrastructure consists of the systems listed below:

Internet of Things (IoT)
System Integration
Cloud Computing
Virtual Reality (VR)
Autonomous Robots
Additive Manufacturing
Big Data and Analytics
Simulation
Cyber Physical Systems (CPS)
Augmented Reality (AR)
Cybersecurity

In order to use this and similar technologies in production, some other support systems such as MES, ERP, PLM, PDM, PPS, CAD/CAM, SCM, etc. are also needed. Frank, Dalenogare and Ayala [9] have proposed a structure related to Industry 4.0 technology layers as a result of the research they conducted in 2019. According to this structure, they performed leveling as front-end and base technologies. While the IoT, Cloud, Big Data and Analytics constitute the base technologies in the model; front-end technologies are presented as the smart supply chain, smart working, smart manufacturing and smart product. As shown in Fig. 1; in order to design, select, manage, realize and implement the different technologies needed, ensure their maintenance and sustainability; flexible, educated and qualified workforce is required.

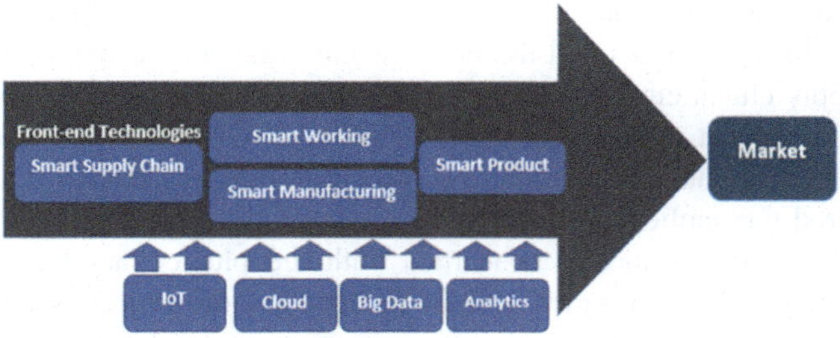

Fig. 1. Theoretical Framework of Industry 4.0 Technologies [9]

Competencies sought in human resources for Industry 4.0 [1, 10–14]:

- Analytical thinking and innovation
- Active learning and learning strategies,
- Creativity, originality and entrepreneurship,
- Technology design and programming,
- Critical thinking and analysis, complex problem-solving, leadership and social activity,
- Emotional intelligence,
- Reasoning, problem-solving and thinking
- System analysis and evaluation.

Very few of the existing white- and blue-collar employees have these competencies. For Industry 4.0 transformation and sustainability; education is an essential factor in order to ensure that both the existing employees and potential candidates of secondary and high education acquire these competencies. For this reason, an assessment was made in this study for the "Digital Transformation Training and Practice Center-DTTPC" that is planned to be established and the location.

2 Brief Description About the Digital Transformation Training and Practice Center

DTTPC focuses on both the current and future workforce. This center, in which training programs at different levels for the existing blue and white-collar personnel and the young generation will be included, is planned to be established in an organized industrial zone or near an industry cluster to allow the students to practice personally.

In the center, where training is planned at three different levels as the business management training, technical training and technology training; trainings at the business management level aim to enable the white-collar employees to learn how to design, implement and manage the new business models. In addition to the fundamental business management courses that are compatible with Industry 4.0, such as optimization, scheduling, production/process/inventory/energy management, ergonomics, supply chain management, strategy development, marketing, human resources management, organization and corporate culture, etc.; trainings will also be provided on the selection and management of technology and information systems, creating the smart manufacturing systems and the business environment, designing and managing the smart supply chain, etc. Other than these, students will be ensured to be informed about decision support systems such as ERP, MES, SCADA, etc.; Cloud, Big Data, AR/VR, robotic, lean and agile production, flexible working and the whole technology and be acquired the application skill.

Technical training is mostly concerned with the blue-collar employees. In this program, training will be provided at the operator level on subjects such as operating the machines, their maintenance and repair, ensuring sustainability of the information systems.

Technology level will be a training program open to both the white- and blue-collar employees. At this level; competence will be gained on subjects such as traceability of products, additive manufacturing, smart systems, cybersecurity, AR, VR, IoT, big data, robotics, communication.

The competencies to be gained by this Center in relation to Industry 4.0 in the areas of strategic, leadership, customer relations, product/service, operations, corporate culture, human resources, technology are listed below:

- Data usage
- Process, strategy, marketing channels design, management, monitoring
- Supply chain management
- Usage of Industry 4.0 infrastructure and tools (artificial intelligence, big data, AR, VR, drone, robotics, cloud, etc.)
- Strategy development, road map
- Use of information systems (decision support systems, tracking systems, big data, data storage, paperless data, etc.)
- Technology and innovation management
- Investment analysis
- Workforce planning
- Smart logistics and storage
- Collaborative working, etc.

In the center where the training durations will be different according to the past educational status and positions of students, the purpose is not to provide a superficial training, but to provide practical training both in the laboratory environments in the center and within the industry, which will bring competences fully compatible with the requirements of Industry 4.0. Trainers will be selected among the academicians and practitioners. On this ground, it is important to establish the center in environment/environments where the industry is intensive and R&D centers are present with the presence of digitalization in whole or in part. Proximity is also an important factor to ensure that those working in that cluster easily access to the training location. Within this context, the criteria determined as a result of the joint opinions of the two experts were taken into consideration for site selection as seen in Table 1.

Table 1. Site selection criteria

Criteria	Symbol	Criteria	Symbol
R&D Center	A	Smart supply chain	I
Presence of smart factories	B	Industry 4.0 applications**	J
Innovative applications	C	Businesses having the Industry 4.0 strategy	K
Data security*	D	Lean production	L
Robot applications	E	Internationalized companies	M
Embedded systems	F	Financial aids and incentives	N
Inter-machine integration and communication	G	University and industry cooperation	O
Digital production systems	H		

* Including cloud solutions
** IoT/big data/augmented reality/simulation/3D printer/etc.

The factors given above can be increased. In this study, the decision was first made with the help of DEMATEL method regarding which of the above-specified criteria will be effective in order to determine the site of the center in accordance with the purpose; and then site selection was carried out with the help of COPRAS according to the weighted criteria. This two-stage selection is described in the next chapter.

3 Site Selection for the Digital Transformation Training and Practice Center

In this section, DEMATEL and COPRAS, which are the multiple criteria decision-making techniques, are briefly summarized and stages of DTTPC site selection are explained. With DEMATEL method, weights of the criteria can be determined, and importance levels can be listed. It is an effective method for determining the relation between the criteria and determining the weight of criteria. COPRAS, on the other hand, is a simple method with shorter calculations compared to methods such as AHP, VIKOR [15]. Calculations can be made with COPRAS for the criteria desired to be maximized and minimized and for qualitative and quantitative criteria [16]. COPRAS shows the usefulness levels of the alternatives, makes comparisons and ranks the alternatives as good and bad in a percentage manner.

When multi-criteria decision-making techniques remain incapable in some cases; researchers that apply to a different multi-criteria decision making can obtain more serious and successful results [17]. For this reason, DEMATEL method was applied to give weight to the criteria and COPRAS method was applied to assess the options based on these weighted criteria in the study.

3.1 DEMATEL as the Decision-Making Technique and Determining the Criteria to Be Used in Site Selection

The DEMATEL method, developed by the Genova Battele Institute in order to analyze the complex "world problems", uses the knowledge of experts to organize a structural model [18–20]. Weights of the criteria can be determined and levels of importance of the criteria can be ranked with the DEMATEL method [21]. This method measures the degree of impact between the criteria beyond the relation of the system. The most important advantage of the DEMATEL method is that it covers indirect relations that contain a compromising cause-effect model [22]. In DEMATEL method, which is based on the graph theory, it is accepted that all criteria are related to each other. DEMATEL is a successful method in determining the weight of the criteria, but it should be used in conjunction with other multi-criteria decision-making method in order to grade the alternatives [23]. This method can be applied in many areas from marketing to human resources; where more than one criterion affects the preference between alternatives [20, 21, 23].

In the DEMATEL method to be applied to determine the weight of the criteria, the stages follow the steps below [21, 23–25] respectively:

Step 1: The relations between the criteria are determined by the expert group by using the binary comparison scale in Table 2. Scoring can be performed between 0–3 or 0–4. Numerical values show the extent to which one criterion affects the other.

Table 2. Binary comparison scale

Numerical value	Definition
0	Neutral
1	Low impact
2	Moderate impact
3	High impact
4	Very high impact

Step 2: In case there is more than one expert evaluating the criteria, the arithmetic average of the given points is taken. Afterwards, these values are placed in the matrix and an asymmetric matrix with "0" diagonals is obtained. This matrix we have obtained is called the direct relation matrix (X).

$$\mathbf{X} = \begin{vmatrix} 0 & \cdots & X_{1n} \\ \vdots & \ddots & \vdots \\ X_{n1} & \cdots & 0 \end{vmatrix}$$

Step 3: After obtaining the direct relation matrix, the largest of the sum of each row and column is found as equality 1 is shown.

$$s = \max(max\sum\nolimits_{j=1}^{n} X_{ij}, \sum\nolimits_{i=1}^{n} X_{ij}) \tag{1}$$

Thereafter, each element of the matrix is divided by the "s" value and the normalized direct relation matrix (C) is formed.

$$C = \frac{X}{s} \tag{2}$$

Step 4: As seen in Equality 3, the total relation matrix (F) is obtained.

$$\lim_{H \to \infty} C + C^2 + C^3 + \ldots + C^H$$

$$F = C + C^2 + C^3 + \ldots + C^H = C(I - C)^{-1} \tag{3}$$

Step 5: In order to determine the affecting and affected factor groups and calculate the net impact degrees in this step, row and column sums are found after determining the total relation matrix (F). These obtained values for each criterion:

The sum of each row (D_i) indicate that the criterion affects other criteria directly or indirectly,

The sum of each column (R_i) indicates the sum of direct or indirect affection of the criterion by other criteria.

$D_i + R_i$ for each criterion shows the sent and received total impact value,

$D_i - R_i$ for each criterion shows the total impact made by the criterion to the system.

$D_i + R_i$ indicates the importance of the criterion within the system.

If $D_i - R_i$ value is positive, it is qualified as the affecting and if negative, it is qualified as the affected.

Step 6: In this stage, the threshold value of the matrix is determined, and the impact-oriented scatter graph is drawn. Criteria above the threshold value are determined as the affecting and the impact direction on the diagram is indicated with an arrow. The case of being affected by any criteria is also shown in the diagram. Arrows are formed from the affecting to the affected. The threshold value can be determined by the experts. When this is not possible, it can also be determined by taking the average of the total relation Matrix (F).

Step 7: It is found with the help of the weight of the criteria (W_i); (4) and (5) formulas.

$$W_{ij} = \sqrt{(D_i - R_i)^2 + (D_i + R_i)^2} \tag{4}$$

$$W_i = \frac{w_{ia}}{\sum_{i=1}^{n} w_{ia}} \tag{5}$$

In this way, the weight of the criteria is found.

For DTTPC site selection, analysis of 15 factors listed in Table 1 with the DEMATEL method was given respectively.

The criteria were evaluated by agreement between the experts of the subject and using the binary comparison scale in Table 2. Evaluation results of the experts are given in Table 3.

Table 3. Direct relation matrix (X)

Criteria	A	B	C	D	E	F	G	H	I	J	K	L	M	N	O	
A	0	4	4	2	4	4	2	4	4	4	2	2	4	4	4	48
B	4	0	4	3	4	4	4	4	4	4	0	3	4	4	4	50
C	4	4	0	2	3	3	2	4	3	4	3	2	4	4	3	45
D	2	4	3	0	4	3	4	4	4	4	2	2	4	2	1	43
E	3	4	3	3	0	3	4	4	2	4	3	3	2	4	3	45
F	2	4	4	4	4	0	4	4	3	4	3	2	3	4	4	49
G	2	4	2	3	3	2	0	4	4	4	1	4	4	4	3	44
H	3	4	3	3	4	4	4	0	3	4	4	4	4	4	3	51
I	1	3	1	2	1	1	3	2	0	4	4	4	4	3	2	35
J	4	4	4	4	4	4	4	4	3	0	4	4	4	4	4	**55**
K	2	4	2	3	1	2	3	4	4	4	0	4	4	4	3	44
L	1	3	1	2	1	1	4	3	3	3	2	0	3	1	3	31
M	3	2	2	2	2	1	3	3	4	3	3	2	0	2	3	35
N	4	2	3	1	2	3	4	4	3	4	4	1	4	0	3	42
O	2	3	3	1	2	3	3	4	2	4	2	2	3	3	0	37
	37	49	39	35	39	38	48	52	46	54	37	39	51	47	43	

The value 55, which is the largest of the sums of the rows and columns of the Direct Relation Matrix was found as the "s value", the Normalized Direct Relation Matrix (C) was obtained and the transaction in Equality (2) was made, and the Total Relation Matrix (F) seen in Table 4 was obtained.

Table 4. Total relation matrix (F)

Cri.	A	B	C	D	E	F	G	H	I	J	K	L	M	N	O
A	0,201	0,321	0,276	0,221	0,276	0,271	0,287	0,337	0,307	0,347	0,235	0,243	0,334	0,316	0,296
B	0,275	0,262	0,282	0,244	0,284	0,278	0,328	0,346	0,316	0,357	0,208	0,266	0,343	0,323	0,304
C	0,258	0,307	0,196	0,211	0,248	0,244	0,273	0,322	0,279	0,332	0,240	0,232	0,320	0,302	0,268
D	0,216	0,298	0,238	0,171	0,257	0,234	0,297	0,311	0,286	0,320	0,214	0,226	0,309	0,259	0,224
E	0,242	0,309	0,248	0,229	0,197	0,245	0,307	0,324	0,264	0,333	0,238	0,250	0,289	0,303	0,268
F	0,240	0,327	0,280	0,259	0,281	0,208	0,326	0,344	0,297	0,354	0,254	0,248	0,324	0,321	0,300
G	0,216	0,296	0,221	0,220	0,239	0,218	0,230	0,311	0,286	0,321	0,199	0,258	0,310	0,290	0,258
H	0,261	0,334	0,268	0,248	0,285	0,280	0,334	0,283	0,305	0,362	0,276	0,288	0,348	0,327	0,292
I	0,162	0,233	0,165	0,170	0,165	0,162	0,235	0,229	0,175	0,269	0,211	0,221	0,261	0,228	0,199
J	0,293	0,354	0,301	0,278	0,302	0,296	0,352	0,372	0,324	0,316	0,291	0,303	0,369	0,346	0,325
K	0,215	0,296	0,220	0,220	0,205	0,217	0,281	0,310	0,287	0,320	0,179	0,258	0,310	0,290	0,257
L	0,146	0,216	0,151	0,156	0,152	0,148	0,232	0,225	0,209	0,232	0,162	0,138	0,225	0,178	0,199
M	0,196	0,220	0,184	0,171	0,185	0,165	0,235	0,247	0,244	0,256	0,197	0,189	0,195	0,215	0,217
N	0,245	0,260	0,235	0,184	0,218	0,231	0,290	0,306	0,265	0,315	0,245	0,205	0,305	0,220	0,254
O	0,194	0,251	0,215	0,166	0,199	0,212	0,250	0,280	0,225	0,288	0,192	0,200	0,262	0,247	0,181

The affecting and affected factor groups were determined by using the Total Relation Matrix data and net impact degrees were calculated and shown in Table 5. 0,258; which is the mean of the Total Relation Matrix, was determined as the threshold value.

Table 5. The affecting and affected factor groups/weights of the criteria and criteria priorities

D	R	D + R	D − R	Criteria	$\sqrt{(D+R)^2+(D-R)^2}$	Weights of criteria (W)	Rank of criteria
4,269	3,361	7,629	0,908	A	7,683	0,065	10
4,415	4,285	8,700	0,130	B	8,701	0,073	3
4,032	3,481	7,513	0,550	C	7,533	0,063	11
3,860	3,148	7,008	3,860	D	8,001	0,067	6
4,045	3,492	7,537	4,045	E	8,554	0,072	4
4,363	3,406	7,770	0,957	F	7,828	0,066	8
3,871	4,257	8,128	−0,386	G	8,137	0,069	5
4,491	4,549	9,041	−0,058	H	9,041	0,076	2
3,084	4,069	7,153	−0,985	I	7,221	0,061	13
4,822	4,721	9,543	0,101	J	9,544	0,080	1
3,865	3,340	7,205	0,525	K	7,224	0,061	12
2,769	3,522	6,291	−0,753	L	6,336	0,053	15
3,119	4,506	7,624	−1,387	M	7,749	0,065	9
3,779	4,166	7,945	−0,388	N	7,954	0,067	7
3,362	3,842	7,204	−0,480	O	7,220	0,061	14
				Total	118,726	1,000	

In the last stage of DEMATEL method, weights of the criteria were found and the criteria were sorted according to the degree of priority (J > H > B > E > G > D > N > F > M > A > C > K > I > O > L). Accordingly, it is seen that the criterion having the highest level of importance is *Industry 4.0 applications,* and *digital production systems* are the criterion having the second-largest importance. *The presence of smart factories* is also of great importance in terms of education. Although lean systems provide the right environments and economic applications in the digitalization process; lean production was determined as the least important criterion in DTTPC site selection.

These criteria weights obtained were used at the stage of choosing alternatives with the COPRAS method.

3.2 COPRAS as the Decision-Making Technique and Evaluation of Site Selection Alternatives

COPRAS- Complex Proportional Assessment method was presented in 1996 by Zavadskas and Kaklauskas [17, 26, 27] who were the researches of Vilnius Gediminas Technical University and it is one of the best methods used in the field of multiple criteria decisions making [16, 17, 26]. COPRAS is able to evaluate the qualitative and quantitative criteria and is a method preferred for performing optimal evaluation [17].

The steps to be followed in COPRAS method can be summarized as follows [17, 28–30]:

Step 1: A decision matrix as shown in Equality (6) is created first. The decision matrix shown with D consists of x_{ij} values.

$$D = \begin{bmatrix} x_{11} & \cdots & x_{1n} \\ \vdots & \ddots & \vdots \\ x_{m1} & \cdots & x_{mn} \end{bmatrix} \tag{6}$$

Step 2: The decision matrix normalized with the help of Equality (7) is obtained.

$$x_{ij}^{*} = \frac{x_{ij}}{\sum_{i=1}^{m} x_{ij}} \quad \forall j = 1, 2, ..., n \tag{7}$$

Step 3: By using the decision matrix normalized with the weight value (W_j) of each evaluation criterion; the decision d_{ij} matrix is created by calculating as the weighted normalized equality (8), symbolized as D′ including the elements.

$$D' = d_{ij} = x_{ij}^{*}.w_{ij} \tag{8}$$

Step 4: While the useful criteria express the criteria where higher values show a better status in achieving the desired result, the useless criteria express the criteria where lower values show a better status in achieving the desired result [29]. The sum of the values in the separately weighted normalized decision matrix is calculated for useful and useless criteria. Sum of the values in the weighted normalized decision matrix is shown as S_{i+} for the useful criteria and S_{i-} for the useless criteria.

$$S_{i+} = \sum_{j=1}^{f} d_{ij} \quad j = 1, 2, ..., /S_i = \sum_{j=f+1}^{n} d_{ij} \quad j = f+1, ..., n \tag{9}$$

Step 5: Relative importance value is calculated for each alternative Q_i by using Equality (10).

$$Q_i = S_{i+} + \frac{\sum_{i=1}^{m} S_{i-}}{S_{i-}.\sum_{i=1}^{m} 1/S_{i-}} \tag{10}$$

Step 6: Relative priority value with the highest value is found among the relative priority values with the help of Equality (11).

$$Q_{max} = maximum Q_i \;\; \forall i = 1, 2, \ldots, m \tag{11}$$

***Step** 7:* P_i performance index is calculated for each alternative by using Equality (12). The alternative with a performance index of 100 P_i indicates the best alternative.

$$P_i = \frac{Q_i}{Q_{max}}.100\% \quad (12)$$

In DTTPC site selection stage; questionnaires were sent to 13 clusters which were thought to comply with Industry 4.0 and determined to be highly efficient among others in the previously conducted study [31] by taking account of the clusters in the development and maturity phase and operate in Turkey. Four regions involved the clusters were taken into assessment among the returning six questionnaires. Two of those regions are in Ankara (I, IV) and the others are in Bursa (II) and İzmir (III) provinces. The reason for preferring the DTTPC site in the clustering region is that the clusters are capable of competing in the international environment, they are prone to apply the requirements of the era, have the potential of generating an expert workforce and are open to innovative approaches. In this way, it will be ensured that the compliance needs of those working within the scope of the cluster/clusters are fulfilled and support is received from the companies in this structure for training. By being near or in the clustering region; it will also be possible to support more companies and more employees.

In order to evaluate these four location site alternatives according to the weight of the criteria determined with DEMATEL method in the previous section, COPRAS method was applied. In the first step, the decision matrix generated from the numerical values obtained for four regions as a result of the questionnaire was given in Table 6.

Table 6. Decision matrix

Location	Criteria														
	A	B	C	D	E	F	G	H	I	J	K	L	M	N	O
I	3	3	4	4	3	3	3	3	3	3	4	4	4	5	4
II	2	3	3	2	3	2	3	4	4	3	3	3	4	5	3
III	3	1	3	3	3	3	2	3	1	2	2	4	3	4	4
IV	4	3	4	3	2	3	3	2	3	3	3	3	4	4	5

The decision matrix was normalized, and the weighted normalized decision matrix with the weight of criteria found in DEMATEL method was obtained as in Table 7.

J, H, B, E, G, D, N, F, M and A were determined as the useful criteria and C, K, I, O and L were determined as the useless criteria; S_{i+} and S_{i-} values calculated by using Equality (9), Q_i - relative importance values calculated for each alternative in Table 8 and by making use of Equality (11) and P_i - performance indexes were given in Table 9.

Table 7. Weighted normalized decision matrix

Location	Criteria														
	A	B	C	D	E	F	G	H	I	J	K	L	M	N	O
I	0,016	0,022	0,018	0,022	0,020	0,018	0,019	0,019	0,017	0,022	0,020	0,015	0,017	0,019	0,015
II	0,011	0,022	0,014	0,011	0,020	0,012	0,019	0,025	0,022	0,022	0,015	0,011	0,017	0,019	0,011
III	0,016	0,007	0,014	0,017	0,020	0,018	0,013	0,019	0,006	0,015	0,010	0,015	0,013	0,015	0,015
IV	0,022	0,022	0,018	0,017	0,013	0,018	0,019	0,013	0,017	0,022	0,015	0,011	0,017	0,015	0,019

Table 8. S_{i+} and S_{i-} values for each alternative

Location	S_{i+}	S_{i-}
I	0.194	0.085
II	0.177	0.074
III	0.152	0.060
IV	0.177	0.080

Table 9. Q_i and P_i values for each alternative

Location	Q_i	P_i	Preference order
I	**0.258**	100.000	1
II	0.252	97.632	2
III	0.244	94.567	4
IV	0.245	95.068	3

The alternative with the performance index of 100, which is symbolized as P_i, represents the best alternative according to the COPRAS method. In this way, the best region for the site of establishment was determined as the region I. The region II following this region appears to be the second-best alternative. Region III was in the last place.

When the result is evaluated; it is seen that there are three clusters within region I and when the criteria are taken into consideration; they have characteristics that may be complementary to each other. According to the importance levels of the criteria and the features provided by them; this region can be considered as ideal compared to other regions for the "Digital Transformation Training and Practice Center" to be established. Thus, it will be appropriate to locate the center near this clustering sited in Ankara-I.

4 Conclusion

In this study, the problem of selecting location for a Digital Transformation Training and Practice Center, which was designed to allow the human resources to gain the necessary competences and capabilities on the subjects of understanding, utilizing and implementing, designing, managing the different business models that arose with fourth industrial revolution, smart systems and their infrastructure; shortly the innovations arising for the changing and transforming manufacturing/service world was addressed. To weight the criteria and determine the levels of importance; DEMATEL was utilized among the multi-criteria decision-making techniques, and COPRAS was used to make a selection among the alternative regions.

The study has two original contributions. The first one is to design an educational institution that will provide training both at the associate, undergraduate and graduate levels as well as vocational and practical training within the scope of harmonization with Industry 4.0; and to determine the site selection criteria in line with the specified purpose to determine the site where this institution will be established. The selection criteria here have originality since they have different features compared to the classical criteria and these are important in terms of shedding light to those who will conduct researches on the subject. These criteria can be enriched with a more comprehensive study. The second contribution is to use DEMATEL and COPRAS methods together. Since the methods used in multi-criteria decision making remain weak in some aspects, it requires completing the research with different methods.

The difficulties experienced in data collection within the scope of the conducted study caused an inability to use more alternatives. While mentioning Industry 4.0 and big data on one hand; the inability to collect data constitutes a contradiction and generates the greatest constraint in such studies. This study is the preliminary step of a study with a broader scope and it will continue by planning the training center in detail.

Acknowledgement. We would like to thank to Mr. B. Tuğbay Gümüş who supported us to benefit from her research and to the valuable cluster officials who gave serious answers to the surveys we sent.

References

1. Büyükuslu, A.R.: Toplum 5.0 Süper Akıllı Toplum. Der Yayınları, İstanbul (2020)
2. Schumacher, A., Erol, S., Sihna, W.: Changeable, agile, reconfigurable & virtual production, a maturity model for assessing Industry 4.0 readiness and maturity of manufacturing enterprises. Procedia CIRP **52**, 161–166 (2016)
3. Ghobakhloo, M.: The future of manufacturing industry: a strategic roadmap toward Industry 4.0. J. Manuf. Technol. Manag. **29**, 910–936 (2018)
4. Gergin, Z., Üney-Yüksektepe, F., Gençyılmaz, M.G., Aktin, A.T., Gülen, K.G., İlhan, D.A., Dündar, U., Cebeci, Ö., Çavdarlı, A.I.: Industry 4.0 scorecard of Turkish SMEs. In: Proceedings of the International Symposium for Production Research 2018 (ISPR 2018), Switzerland, pp. 426–437 (2019)
5. Numanoğlu, N., Ince, F.H. (eds): Türkiye'nin Sanayide Dijital Dönüşüm Yetkinliği: Türkiye'nin 4. Sanayi Devrimi, TÜSİAD-T/2017, pp. 12–589. CB Basımevi, Istanbul (2017)
6. Kiraz, A., Canpolat, O., Erkan, E.F., Uygun, Ö.: IMPULS kriterleri ile Endüstri 4.0 eğiliminin değerlendirilmesi: Bir bulanık bilişsel harita uygulaması. Acad. Platform J. Eng. Sci. **7–1**, 14–23 (2019)
7. Büyükuslu, A.R.: Dijital Dönüşüm. Der Yayınları, Istanbul (2018)
8. Banger, G.: Endüstri 4.0: Uygulama ve Dönüşüm Rehberi. Dorlion Yayınları, Eskişehir (2018)
9. Frank, A.G., Dalenogare, L.S., Ayala, N.F.: Industry 4.0 technologies: ımplementation patterns in manufacturing companies. Int. J. Prod. Econ. **210**, 15–26 (2019)
10. Hecklau, F., Galeitzke, M., Flachs, S., Kohl, H.: Holistic approach for human resource management in Industry 4.0. Procedia CIRP **54**, 1–6 (2016)
11. Shamim, S., Cang, S., Yu, H., Li, Y.: Management approaches for Industry 4.0: a human resource management perspective. In: IEEE, pp. 5309–5316 (2016)
12. Filizöz, B., Orhan, U.: İnsan kaynakları yönetimi bağlamında Endüstri 4.0: Bir yazın çalışması. C.Ü. İktisadi ve İdari Bilimler Dergisi **19**, 110–117 (2018)
13. Cem, E.: Hastane yönetimi için yeni bir Endüstri 4.0 olgunluk modeli geliştirilmesi ve örnek bir uygulama. MSc. Thesis, Çukurova University, Adana, Türkiye (2019)

14. GE Türkiye Blog (2018). https://geturkiyeblog.com/2018/17.12.2018/
15. Sarıçalı, G.: Çok kriterli karar verme yöntemlerinden KEMIRA-M ve COPRAS yöntemlerinin mermer işletmesinde makine seçim sürecine uygulanması. MS Thesis, Pamukkale University, Denizli, Türkiye (2018)
16. Podvezko, V.: The comparative analysis of MCDA methods SAW and COPRAS. Inzinerine Ekonomika-Eng. Econ. **22**(2), 134–146 (2011)
17. Çakır, E., Kutlu Karabıyık, B.: Bütünleşik SWARA - COPRAS yöntemi kullanarak bulut depolama hizmet sağlayıcılarının değerlendirilmesi. Bilişim Teknolojileri Dergisi **10**, 417–434 (2017)
18. Liou, J.J.H., Yen, L., Tzeng, G.H.: Building an effective safety management system for airlines. J. Air Transp. Manag. **14**, 20–26 (2008)
19. Wu, W.-W., Lee, Y.-T.: Developing global managers' competencies using the fuzzy DEMATEL method. Exp. Syst. Appl. **32**, 499–507 (2007)
20. Suo, W.L., Feng, B., Fan, Z.P.: Extension of the DEMATEL method in an uncertain linguistic environment. Soft Comput. **16**, 471–483 (2012)
21. Uyanık, C., Tuzkaya, G., Kalender, Z.T., Oguztimur, S.: An integrated DEMATEL–IF-TOPSIS methodology for logistics centers' location selection problem: an application for Istanbul metropolitan area. Transport, p. 9 (2020). Article in press
22. Aksakal, E., Dağdeviren, M.: ANP ve DEMATEL yöntemleri ile personel seçimi problemine bütünleşik bir yaklaşım. J. Fac. Eng. Arch. Gazi Univ. **25**, 905–913 (2010)
23. Hsu, C.W., Kuo, T.C., Chen, S.H., Hud, A.H.: Using DEMATEL to develop a carbon management model of supplier selection in green supply chain management. J. Cleaner Prod. **56**, 164–172 (2013)
24. Çınar, Y.: Kariyer tercihi probleminin yapısal bir modeli ve riske karşı tutumlar: Olasılıklı DEMATEL yöntemi temelli bütünleşik bir yaklaşım. Sosyoekonomi **19**, 157–186 (2013)
25. Shieh, S.J.I., Wub, H.H., Huang, K.K.: A DEMATEL method in identifying key success factors of hospital service quality. Knowl. Based Syst. **23**, 277–282 (2019)
26. Hashemkhani Zolfani, S., Bahrami, M.: Investment prioritizing in high tech industries based on SWARA-COPRAS approach. Technol. Econ. Dev. Econ. **20**(3), 534–553 (2014)
27. Kaklauskas, A., Zavadskas, E.K., Trinkunas, V.: A multiple criteria decision support on-line system for construction. Eng. Appl. Artif. Intell. **20**, 163–175 (2007)
28. Kaklauskas, A., Zavadskas, E.K., Raslanas, S., Ginevicius, R., Komka, A., Malinauskas, P.: Selection of low-e windows in retrofit of public buildings by applying multiple criteria method COPRAS: a Lithuanian case. Energy Build. **38**, 454–462 (2006)
29. Özdağoğlu, A.: İmalat işletmeleri için eksantrik pres alternatiflerinin COPRAS yöntemi ile karşılaştırılması. Gümüşhane Üniversitesi Sosyal Bilimler Enstitüsü Electron. J. **4**, 1–22 (2013)
30. Özdağoğlu, A.: Çok ölçütlü karar verme modellerinde normalizasyon tekniklerinin sonuçlara etkisi: COPRAS örneği. Eskişehir Osmangazi Univ. İİBF J. **8**, 229–252 (2013)
31. Gümüş, T.B.: Uluslararasılaşan kümelerin Endüstri 4.0 ile ilişkisi. Otomotiv sektöründe bir çalişma, PhD Thesis, Beykent University, Istanbul, Turkey (2020)

Fuzzy Logic

Facility Location for Unmanned Aerial Vehicle Base Stations to Provide Uninterrupted Mobile Communication After Earthquakes

Gamze Sevdik[1(✉)], Sakir Esnaf[1], and Engin Baytürk[2]

[1] Industrial Engineering, Istanbul University-Cerrahpasa, Istanbul, Turkey
gamze.sevdik@ogr.iu.edu.tr, sesnaf@istanbul.edu.tr
[2] Engineering and Technology, American College of the Middle East, Egaila, Kuwait
Engin.Bayturk@acm.edu.kw

Abstract. Earthquakes are unforeseeable aberrations which could lead to significant loss of life and property and require external aid in management. According to the United States Geological Institute (USGS), more than a thousand earthquakes occur every year on earth, and about 15 of them are greater than 7 intensities. More than 800 thousand people died in earthquakes from 2000 to 2015 [5]. In order to diminish the loss of life, reaching accurate information promptly from disaster areas in the first hours of earthquakes bears high importance. Thus, post-disaster management operations have become increasingly worthy. In this study, facility location for the base station of unmanned aerial vehicles will be made by using Fuzzy C-Means and Fuzzy C-Means – Center of Gravity algorithm according to probabilistic demands in the regions where the communication network is interrupted.

Keywords: Unmanned aerial vehicle · Base station · Natural disaster · Weber problem · Probabilistic fuzzy C-Means · Probabilistic fuzzy C-Means – center of gravity

1 Introduction

Turkey, as North Anatolian Fault Line, which is extending from the Gulf of Saros to Van Lake, the East Anatolian Fault Line, which is extending from Hatay to Bingol, and the Western Anatolian Fault Line, which exists on the Aegean Region, is on the major fault lines. In case of a possible earthquake, uninterrupted communication with the disaster area is vital.

In the current era which is called information age, Unmanned Aerial Vehicles have frequently used thanks to their advantages such as low purchasing and manufacturing costs, long flight times, dynamic assignment, fastness, and flexibility. Unmanned Aerial Vehicles are utilized for exploration, data collection, surveillance, and armed conflicts in regards to military purposes; for monitoring, logistic, research, and development regarding civil purposes. However, the shortage of manpower, time constraints, and an urgent need for information in the event of a disaster have made the use of Unmanned Aerial Vehicles an advantageous option in disaster operations

N. M. Durakbasa and M. G. Gençyılmaz (Eds.): ISPR 2020, LNME, pp. 53–61, 2021.
https://doi.org/10.1007/978-3-030-62784-3_5

management. UAVs are used in functions such as establishing a temporary communication network, creating current maps of the disaster area, analysing the images collected, finding regions that are more suitable for rescue teams to reach disaster victims, and providing fast and quality information transmission in the management of disaster operations [3].

The problem of facility location as a remarkable branch of operations research extends to the Egyptians' specialization in the selection of the pyramids in 4000 BC or the Romans developing methods for the construction of the arena and temple in 100 BC. The facility location problem is often used by governments or private companies to make long-term and strategic decisions [5]. Choosing the right facility location is crucial for providing the best service to the max number of people, while also reducing costs.

The first use of UAVs in the world is accepted in 1849 as Austria's sending 200 pilots-free balloons over Venice [13]. In 2016, Mercedes Benz talked about the idea of a "drone-equipped truck" that uses "ceiling-mounted unmanned aerial vehicles (UAV)". In 2017, Google offered its service that distributed aid to the disaster area after the natural disaster named Wing [3]. Amazon, one of the e-commerce giants, announced that it was planned to deliver fast cargo using UAVs [7]. The UAVs have been used by world giants for daily work. Since 2011, Matternet company located in the USA has been testing UAV in the transportation of health products and blood samples. According to the CEO of the company, using UAVs in the health sector is both a cheaper and more time-saving method [3].

Golabi et al. [5] conducted a UAV base station location study using heuristic algorithms to communicate with the regions where the mobile network was interrupted after the earthquake in Tehran.

In 2018, device-to-device communication technology (D2D) was discussed for post-disaster management [10]. A sample scenario was also presented in the study aiming for the best communication between individuals and institutions after the disaster.

In 2018, Uysal et al. [14] tested the accuracy and usability of images obtained from UAV at Afyon Kocatepe University to facilitate search and rescue activities after disasters.

Kim et al. [7] developed a stochastic model to determine the facility location and carrying capacity of UAVs using the Benders decomposition from heuristic algorithms to make faster decisions in 2018.

The studies in which Unmanned Aerial Vehicles are used in humanitarian aid logistics by Yılmaz [15] have been examined in detail and it has been shown that UAVs are a promising technology in emergency and disaster logistics.

Gergin et al. [4] proposed an artificial bee colony-based clustering algorithm for the multi-facility location problem for health waste facilities in Istanbul.

A hybrid solution method was proposed by Guzelgoz et al. [6], consisting of Revised Weighted Fuzzy C-Means, Weiszfeld, Reinforced Weiszfeld, and Nelder-Mead algorithm to the uncapacitated multi-resource facility problem.

A stochastic model has been presented by Nasrabadi et al. [11] to determine the location and capacity of drone facilities to effectively deal with a disaster. In the study, an intuitive algorithm has been developed using Benders decomposition. The Linear programming rounding method is used to make the algorithm efficient.

Bayturk, in his study in 2020, has developed a solution to the problem of multi-facility location, which includes stochastic customer and demand points, with clustering algorithms. Based on the Fuzzy C-Means method, Fuzzy C-Means, Fuzzy C-Means -Gravity Center, Revised Weighted Fuzzy C-Means-Gravity Center, Revised Weighted Fuzzy C-Means - Nelder Mead methods were provided [1].

In this study, it is aimed to minimize loss of life in the event of a disaster by the UAV base station facility location in areas where mobile communication is interrupted after the earthquake. 226 Mobile device coordinate data in Manisa province are used. Probabilistic Fuzzy C-Means and Probabilistic Fuzzy C-Means- Center of Gravity algorithms that are used for the first time for UAV base station facility location, and the results of proposed algorithms are compared according to the total transportation cost.

The rest of paper is organized as follows, in the second section, Weber Problem is discussed. In the third section, proposed algorithms are provided. In the fourth section, details related to the case study are discussed. In the fifth section, the results of the proposed algorithms and comparison of algorithms are provided. The last section of the paper contains a discussion about conclusions and suggestions for future studies.

2 Weber Problem

In facility location problems, if the location where the facility will be established will be selected from the nominated locations, the Discrete Location Problem; If any point in Euclidean space is to be chosen, it is called the Continuous Location Problem [12].

Facility location problems are examined under 4 main headings according to solution space and capacity constraints [12].

- Uncapacitated Discrete Location Allocation Problem (UDLAP)
- Capacitated Discrete Location Allocation Problem (CDLAP)
- Uncapacitated Continuous Location Allocation Problem (UCLAP) (Weber problem)
- Capacitated Continuous Location Allocation Problem (CCLAP)

The Weber Problem (WP) is an operations research problem created by Alfred Weber in 1909 and used for location optimization purposes. This problem is mainly solved by minimizing the weighted sums of Euclidean distances. Continuous Location Allocation Problems are called the Weber Problem [2]. The solution to the Weber problem depends primarily on the number of facilities, such as Single Facility Location Problem (SFLP) and Multi-Facility Location Problem (MFLP) [8]. Another factor affecting the solution in the multiple-facility Location Problem is the fact that the facility is multi-source or single-source. The classification of the Weber problem under four main titles is given below [12]:

- Weber Problem (WP)
- Multi-Facility Weber Problem (MFWP)
- Single-source Capacitated Multi- Facility Weber Problem (SSCMFWP)
- Multi-source Capacitated Multi- Facility Weber Problem (MSCMFWP)

Weber Problem, seen in real-life problems, aims to determine the location of the facilities to be established to meet all demands and to minimize the cost. It is generally formulated as follows [12]:

$$\min_X \sum_{i=1}^{n} w_i d(x, a_i) \tag{1}$$

Subject to;

$$w_i \geq 0, \quad w_i \in \mathbb{R}, \quad \forall i$$
$$x \in \mathbb{R}^2$$

a_i: (a_{i1}, a_{i2}) current location for demands i = 1,…,n
x: facility location [x: (x_1, x_2)]
w_i: total demand of ith demand
$d(x, a_i)$: Euclidean distance between the ith demand and established facility

$$d(x, a_i) = \sqrt{(x_1 - a_{i1})^2 + (x_2 - a_{i2})^2} \tag{2}$$

3 Methodology

In this study, Fuzzy C-Means, Fuzzy C-Means - Gravity Center methods were used for the first time in the location of UAV base stations after the earthquake and Considering the probable demands in the Continuous Capacity Unconstrained Multi-Facility Weber problem, in the case that communication network is interrupted after the earthquake in Manisa, the microcell Unmanned Aerial Vehicle base station facility location will be selected.

3.1 Determination of Distance in Probabilistic Methods

Fuzzy C-Means and Revised Weighted Fuzzy C-Means methods are calculated by Euclidean distance. The formula used in calculating the distance to the facilities by accepting the customer location (P (x, y)) as constant is as follows [1]:

$$\mathrm{d(P, Q)} = \sqrt{(x_1 - y_1)^2 + (x_2 - y_2)^2} \tag{3}$$

When there are N parameters, the following formula is used:

$$\mathrm{d(P, Q)} = \sqrt{(x_1 - y_1)^2 + (x_2 - y_2)^2 + \ldots + (x_n - y_n)^2} \tag{4}$$

The distance of the non-fixed positions to the origin is found by the following formula:

$$d(P,Q) = \sqrt{\left(\frac{\bar{X}_1}{\sqrt{s_{11}}}\right)^2 + \left(\frac{\bar{X}_2}{\sqrt{s_{22}}}\right)^2} \tag{5}$$

or

$$d(P,Q) = \sqrt{\frac{\bar{X}_1^2}{s_{11}} + \frac{\bar{X}_2^2}{s_{22}}} \tag{6}$$

$\bar{X}_1$: Average of the client's position in x coordinate
$\bar{X}_2$: Average of the client's position in y coordinate
s_{11}: Varience of the client's position in x coordinate
s_{22}: Varience of the client's position in y coordinate

3.2 Determination of Demand in Probabilistic Methods

In the study, it will be discussed that the demand meets a specified level of service by taking the example of the work carried out by Baytürk in 2020. By the normal distribution, the probabilistic demand is calculated as follows [1]:

$$w_i = \overline{wi} + z\,\sigma_{wi} \tag{7}$$

w_i = probabilistic demand of ith customer
$\bar{w}_i$ = expected demand of ith customer
σ_{wi} = standard deviation of ith customer
z = number of standard deviation
$P(z <= \frac{wi - \overline{wi}}{\sigma\, wi})$ = T Percentage of meeting the ith customer's demand

3.3 Fuzzy C-Means

Some clustering algorithms assign each demand point to only one point, assuming well-defined boundaries between sets. However, in real cases, boundaries between clusters cannot be clearly defined. For this reason, fuzzy decision-making methods are frequently used. Fuzzy C-Means are seen as a fuzzy version of the k-means algorithm and is based on minimizing an objective function called the c-means function [9].

The steps of the Fuzzy C-Means algorithm are as follows [1]:

$$Jp(U,c) = \sum_{i=1}^{n} \sum_{j=1}^{k} w_i (u_{ij})^p ||a_i - c_j||^2 \tag{8}$$

Step 1: Initially, the membership matrix $U^{(0)}$ = [uij] is randomly assigned
Step 2: Cluster centers are calculated.

$$c_j = \frac{\sum_{i=1}^{n} (u_{ij})^p a_i}{\sum_{i=1}^{n} (u_{ij})^p} \tag{9}$$

Step 3: Membership matrix is updated.

$$u_{ij} = \frac{1}{\sum_{j=1}^{k} \left(\frac{a_i - c_j}{a_i - c_j}\right)^{\frac{2}{p-1}}} \tag{10}$$

Step 4: Continue until the difference between consecutive membership matrices and the total cost achieved is less than a certain error value (Jp).

c: number of cluster
n: number of customer
p: coefficient of membership
c_j: coordinate of cluster center j
[u_{ij}]: membership matrix
a_i: coordinate of ith customer
$J_p(U,c)$: cost function
$\|a_i - c_j\| = d(a_i - c_j)$: Euclidean distance between the ith demand and j^{th} facility

$$\|a_i - c_j\| = (a_i - c_j)^T A (a_i - c_j) \tag{11}$$

$\|a_i - c_j\|$ can be written in the objective function as $d(a_i - c_j)$

3.4 Fuzzy C-Means – Center of Gravity (FCM– COG)

Fuzzy C-Means is applied first and then the Center of Gravity method. In the first stage, the fuzzy centers are determined with the Fuzzy C-Means method and customer-facility assignments are made, and in the second stage, the Center of Gravity method is applied.

It is implemented as follows [1]:

Step 1: Fuzzy C-Means method determines the number of cluster centers (k), customers are assigned to suitable facilities. Subclusters are then created with the assigned customers.
Step 2: Cluster centers are recalculated using the Center of Gravity for each subcluster.

Step 3: The objective function is recalculated.
Step 4: If the difference (Jp) between two consecutive membership matrices (U(k), U(k + 1)) is greater than a specified error value, return to step 2, otherwise the algorithm will stop working.

When the algorithm is stopped, the cluster centers are determined. It is subdivided according to the determined cluster centers.

4 Case Study

Due to frequent earthquakes and on the fault line, Manisa was chosen as a pilot area. Coordinate information of 226 mobile device users in Manisa province was used. The data is collected from one of the telecommunication companies in Turkey. The outputs of PFCM, PFCM-COG methods for the cluster center numbers (N = 12, N = 15, N = 20, N = 25) determined in the study were obtained as total transportation cost and processing time. The customer service level is determined by 90%. The hybrid method is expected to give better results.

5 Results

The results of the case study are given in Table 1 and Table 2, respectively, as total transportation cost and processing time. It is desired to obtain results where the total transportation cost is less and the processing time is shorter.

Table 1. Total transportation cost for PCFM, PCFM-COG

Total transportation cost	PFCM	PFCM-COG
N = 12	7412.430	7450.046
N = 15	6546.856	6502.349
N = 20	5703.463	5625.220
N = 25	5130.046	5118.792

Table 2. CPU time for PFCM, PFCM-COG

CPU time (second)	PFCM	PFCM-COG
N = 12	0.076	0.247
N = 15	0.081	0.241
N = 20	0.089	0.303
N = 25	0.276	0.316

According to the total transportation cost, the PFCM-COG algorithm gives better results. Since the PFCM algorithm uses the weight of each customer equally (which is equal to 1 for all customers), it provides higher transportation costs.

As the number of cluster centers increases, the processing time increases. It is observed that the fastest response was PFCM in terms of processing time. Note that, processing times of each algorithm in Table 2 are given in seconds.

6 Conclusion

In this study, 226 Mobile device coordinate data in Manisa province are used. Probabilistic Fuzzy C-Means and Probabilistic Fuzzy C-Means- Center of Gravity algorithms are applied and the results of proposed algorithms are compared according to the total transportation cost.

It has been observed that the PFCM-COG algorithm gives better results in terms of the total transportation cost, while the PFCM algorithm gives better results in terms of the processing time.

In the future studies, Probabilistic Revised Weighted Fuzzy C-Means and Revised Weighted Fuzzy C-Means – Center of Gravity algorithms can be used as benchmarking methods and can be compared with the proposed algorithms in terms of the total transportation costs and CPU time. In addition to that weather and altitude might be taken into consideration, since these factors can also affect the results in real-life applications.

References

1. Bayturk, E.: A fuzzy clustering based model for continuous multiple facility location problem with stochastic demand. PhD Thesis, İstanbul University-Cerrahpaşa, Istanbul, Turkey (2020). (In Turkish)
2. Demir, E.: Determining the most suitable location for receiving support in emergency cases prior to takeoff at airports by applying the weber problem. Turk. Geogr. Rev. **70**, 81–85 (2018). (In Turkish)
3. Elmas, D.A.: On vehicle routing problem of unmanned air vehicles. Master's Thesis, Hacettepe University, Ankara, Turkey (2019). (In Turkish)
4. Gergin, Z., Tuncbilek, N., Esnaf, S.: Clustering approach using artificial bee colony algorithm for healthcare waste disposal facility location problem. Int. J. Oper. Res. Inf. Syst. (IJORIS) **10**(1), 56–75 (2019)
5. Golabi, M., Shavarani, S.M., Izbirak, G.: An edge-based stochastic facility location problem in UAV-supported humanitarian relief logistics: a case study of Tehran earthquake. Nat. Hazards **87**(3), 1545–1565 (2017)
6. Guzelgoz, I., Esnaf, S., Kucukdeniz, T.: Revised weighted fuzzy c-means and fortified weiszfeld hybrid method for uncapacitated multi-facility location problems. Int. J. Oper. Res. Inf. Syst. (IJORIS) **10**(4), 53–71 (2019)
7. Kim, D., Lee, K., Moon, I.: Stochastic facility location model for drones considering uncertain flight distance. Ann. Oper. Res. **283**, 1–20 (2018)
8. Kucukdeniz, T., Esnaf, S., Bayturk, E.: Extended single-iteration fuzzy c-means, and gustafson-kessel algorithms for medium-sized (106) multisource weber problem. Int. J. Oper. Res. Inf. Syst. (IJORIS) **10**(3), 1–15 (2019)

9. Kucukdeniz, T., Baray, A., Ecerkale, K., Esnaf, S.: Integrated use of fuzzy c-means and convex programming for capacitated multi-facility location problem. Exp. Syst. Appl. **39**(4), 4306–4314 (2012)
10. Marttin, V., Yüzgeç, U., Bayılmıs, C.: Device to device (d2d) communication technologies for post disaster communication. In: 2nd International Symposium on Natural Hazards and Disaster Management, Sakarya, Turkey, pp. 687–696 (2018)
11. Nasrabadi, A.M., Najafi, M., Zolfagharinia, H.: Considering short-term and long-term uncertainties in location and capacity planning of public healthcare facilities. Eur. J. Oper. Res. **281**(1), 152–173 (2020)
12. Tuncbilek, N.: Modified artificial bee colony algorithm for the capacitated muli-facility weber problem. Phd thesis, İstanbul University-Cerrahpaşa, Istanbul, Turkey (2018). (In Turkish)
13. Ucar, U.Ü., İşleyen, S.K.: A new solution approach for UAV routing problem with moving target – heterogeneous fleet. J. Polytech. **22**(4), 999–1016 (2019). (In Turkish)
14. Uysal, M., Yılmaz, M., Tiryakioglu, I., Polat, N.: The use of unmanned aerial vehicles in disaster management. Eskişehir Tech. Univ. J. Sci. Technol. B Theor. Sci. **6**, 219–224 (2018). (In Turkish)
15. Yılmaz, U.: Usage areas of unmanned aerial vehicles in humanitarian logistics activities. J. Vocat. Soc. Sci. Turk **2**, 43–54. (In Turkish)

Healthcare Systems and Management

A Guide Application in Case of Emergency Health: A Case of Turkey

Hatice Camgöz Akdağ, Selin Şişmanoğlu, Süleyman Çoban, Tuğçe Beldek(✉), and Aziz Kemal Konyalıoğlu

Department of Management Engineering, Istanbul Technical University, Istanbul, Turkey
{camgozakdag, sismanoglu, cobans, beldek, konyalioglua}@itu.edu.tr

Abstract. Health constitutes power and the societies who track developments in health sector hold power. The developed countries such as Canada lead the field and developing countries such Turkey follows them. The problems in Turkey healthcare system is getting be solved year by year but there are still shortcomings in established system like insufficient precautions for decreasing unreliable emergency callings and unnecessary intensity in emergency departments. This study took inspiration from Telehealth which is a component of Canadian healthcare system and focuses on reducing the rate of unreliable callings to 112 and intensity in emergency departments by designing a mobile health application for heart attack situation. In this study, it is aimed to design an application named as Docto'Ring' with giving information of how to use the application and how the emergency situation was determined which enables to call ambulances with appeared button. Besides, the user interface was prepared in detail. The testing process was done only by government due to the laws about health constraints.

Keywords: Healthcare · Application systems · Healthcare management · Emergency system

1 Introduction

Health is easily affected by the world's economic, environmental, technological, and demographic changes. World Health Organization (1948) explains health is "a state of complete physical, mental and social well-being and not merely the absence of disease or infirmity". The definition of health is changing day by day. The reasons could be aligned such as the changes in needs and importance of patients, diseases, health technologies. Nowadays, countries are making developments in health sector. Some countries develop new ideas, implement and sell to other countries. The power belongs to developed countries. At the same time, new models are proceeding to establish like efficient usage of ecosystem, convenient cures, enlightenment and awareness raising techniques for patients. Healthcare technologies of countries are differentiating through investments in health industry. However, the similarity will be occurring in the future due to usage of the similar innovations. The line of emergency which is 112 in Turkey

N. M. Durakbasa and M. G. Gençyılmaz (Eds.): ISPR 2020, LNME, pp. 65–79, 2021.
https://doi.org/10.1007/978-3-030-62784-3_6

is always ready for everybody who needs help for their situation of emergency health. However, the rate of unreliable emergency calls is very high. This problem causes wasting of time, failing to reach the patient in case of emergency and death.

The study is beginning by discovering problems in Turkey health and emergency system. Afterwards, world healthcare system is analysed. Canadian Healthcare System is based on being able to have an efficient health system. Then, the two countries' emergency health system are compared and Telehealth which is a Canadian health application is decided to adapt in Turkey. Also, the surveys and interviews with doctors, paramedics and private ambulance companies are done in order to make sure the existing problems and get their opinions about usage Telehealth in Turkey. With this project, it is hoped that the problems in Turkey emergency health system could be reduced by designing an app. The first step is Literature Research which is followed by surveys and interviews and at the end the proposed solution model which will be designed is presented.

The aim of the study is to determine and reduce the proportion of unreliable callings to 112 and the intensity of hospitals in the emergency room with finding the difficulties of supplying necessary materials, medicines and treatment obstructer regulations which will be provided by designing an app.

2 Literature Review

2.1 World Healthcare System

Innovation which is the infrastructure of health system is the key element. With developing technology, countries adopt and perform technological innovations in health industry. Health systems have an important role to provide better health to the population. Health indicators which take place in health industry such as the mortality, morbidity, environmental indicators guide countries to establish new health systems. Thus, countries gave importance on health expenditures. The data from OECD Health Spending 2018, which is Fig. 1, demonstrates that, The United States has the highest health spending rate. It could be understood that it is the best in health sector. However, while comparing other developed countries, with its quality, it ranks last.

Healthcare systems in Europe are separated into two categories which are National Health System (Beveridge model) and Social Security System (modelo Bismarck). The differentiation is occurred among the bureaucratic procedures like the wage system of doctors, tax systems. In National Health System, due to the existence of private sector, healthcare professionals are paid by allowances. However, financial resources pay the salaries of healthcare professionals in Social Security System (Ministerio de Sanidad, Servicios Sociales e Igualdad 2019, p. 7).

The contagious illnesses due to the increase of natality, change in living conditions and climate change complicate the health system efficiency in Asia. Thus, the new forms of hospitals, the mixture health system of public and private services, several kinds of insurances are established to reach all segments of society (Chongsuvivatwong et al. 2011, p. 429). Donev et al. state that USA apply Free-market private insurance model with letting to focus on private sector and exclude poor people (2013, p. 7). The

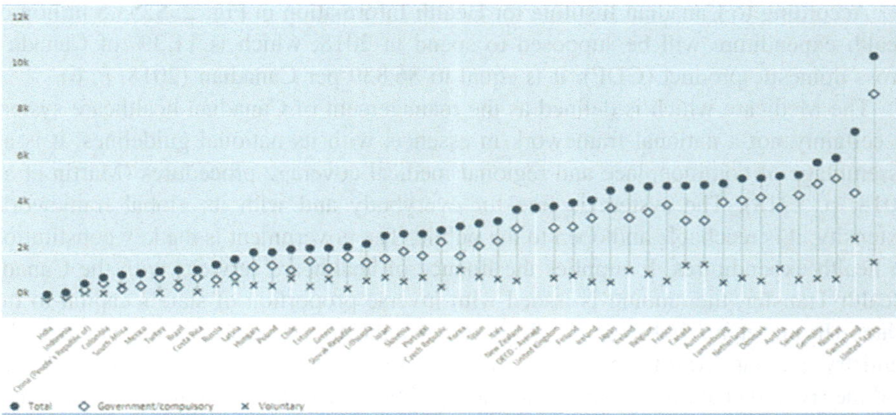

Fig. 1. Health Spending 2018 (Source: OECD)

three different types of world healthcare systems would like to reach the same goal with delivering different services. On the contrary of United States, Canada applies Beveridge Model and gives importance on technological developments in health industry. They use Telehealth which enables to remote registration, remote diagnosis and telecare. The features of Docto 'Ring' are similar to the Canadian telehealth that's why the Canadian healthcare system is executed.

2.2 Canadian Healthcare System

Canada is a country where people of many ethnic backgrounds live. It gives importance to provide all kinds of services equally, without distinguishing its citizens. Canada devotes a large part of the state budget to health expenditures as demonstrated in Fig. 2.

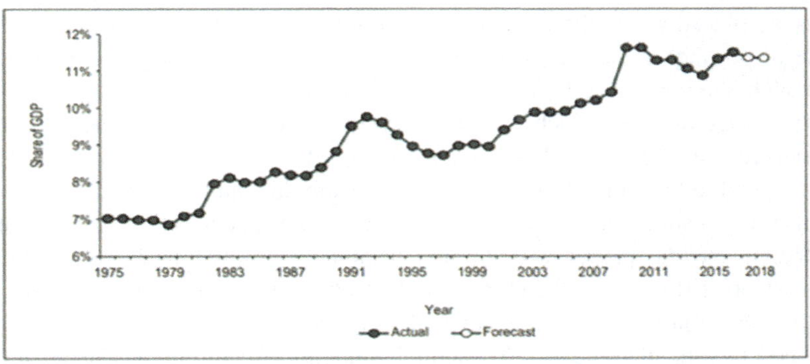

Fig. 2. Total health expenditure as a percentage of GDP, Canada, 1975 To 2018 (National health expenditure database, Canadian institute for health information)

According to Canadian Institute for Health Information in Fig. 2, $253.5 billion of health expenditure will be supposed to spend in 2018, which is 11.3% of Canada's gross domestic product (GDP). It is equal to $6,839 per Canadian (2018, p. 6).

The Medicare which is defined as the management of Canadian healthcare system is certainly not a national framework in essence, with its national guidelines, it is an assemblage of commonplace and regional medical coverage procedures (Martin et al. 2018, p. 1718). The system is free for everybody and with its global framework, extensity, it is reachable and open to the public. The government is the key constitution in health expenditures. It supplies the finance of healthcare services with the Canada Health Transfer, the amount is issued with inverse proportion of state's capital to the states (Irvine et al. 2013, p. 1). However, healthcare professionals are not paid by ministry, they are paid from public. Thus, the controlling way of costs is getting easy and the risks of catching disease is increasing. This situation is let to usage of insurance from everybody in the country.

The practice care settings are differentiating abide by patients' situation. The care of primary, emergency, ambulatory, urgent, chronic, palliative, community and hospitals, public health, telehealth, nursing stations, family health teams are various health services offered by Canada. Primary care is the first organization that people consult for their health. Emergency departments are always ready to help patients' in case of emergency health. When there is a need for complex medical treatments, patients are sent to ambulatory care organizations. Urgent care organizations are special places which are separated from emergency departments but they have similar properties. Chronic care is the extension of ambulatory care but the fractionation is occurred in the duration of disease. The end state of diseases is treated in palliative care organizations. Community care organizations are established for personal needs. Allin and Rudoler explain that hospitals are 2 types which are private in Ontario and public in some regions of Canada. The private hospitals are financing with their performances but public ones are financing from government (2017, p. 23). Telehealth is a new technology in health industry which Canada uses. It is a phone service and it enables to deliver health system afar with a healthcare professional from call center. Nursing stations are placed where there are no medical services or could not be accessed. It's a mixture of primary and emergency services. The last one is family health teams. It is similar with Turkey; every family has a family doctor.

Four levels of care which are primary, secondary, quaternary, tertiary care is implemented. Secondary level of care is coming after primary care which the health problem could not be solved by primary care doctors like ambulatory care. Profoundly particular administrations are contained by tertiary care specialists' in which treatments are done in specific places with limited healthcare professionals like cardiologists (Donev et al. 2013, p. 8). Academic centers offer the highest level of care which constitute quaternary care such as hospitals for sick kids.

Several factors affect health like alcohol, smoking. According to Canadian Cancer Society 2016 data in Fig. 3, cancer has the largest share of causes of mortality. Canada is still seeking new treatment techniques to reduce the mortality rate.

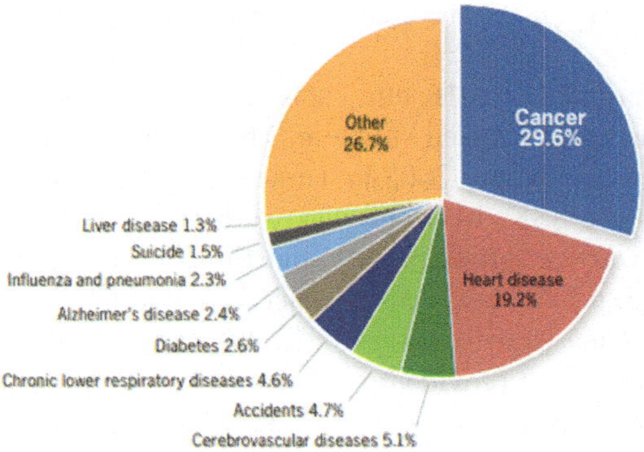

Fig. 3. Proportion of deaths due to cancer and other causes, Canada (Canadian cancer society 2019)

2.3 Turkish Healthcare System

The health system is differentiating from developed countries to developing countries. Developed countries seek new ways for developing their healthcare systems. Some of them are exporter of new ideas. Turkey is not in the category of developed countries. However, the remarkable growth has been seen in Turkish healthcare system. The Health Transformation Program is released in 2003 with the leadership of Ministry of Health which is the main responsible of Turkey health system such as establishing health policies, delivering better service quality. The purpose of the program is to progress the fundamental health results which are fallen behind as far as developed countries, to enhance the quality and productivity of Turkish health system, to give the right to everybody for reaching healthcare institutions and to get universal scope (Gürsoy 2015, p. 84).

The healthcare services are separated into three categories which are preventive, for therapeutic purpose and for rehabilitation purpose. Protecting healthy lives and increasing in life standards could be provided by reducing risks of several diseases in short and long term with protective healthcare services (Akdur 2006, p. 13). The system is applied towards people and environment. The examples could be aligned as taking precautions such as usage of filters in factories are took for the purpose of decrease environmental pollution, draining swamp, overpowering solid and liquid wastes, raising public awareness (Sargutan 2005, p. 401). The aim of preventive healthcare systems is preventing diseases which is the main feature of the system (Çelikay and Gümüş 2000, p. 184). The primary care health system in Turkey applies the preventive system. The healthcare services for therapeutic purpose needs complex equipment system and high cost. The system consists of investigation, medical examination and treatment process with becoming ill of people. (Yildirim 1994, p. 13). The ambulatory care, inpatient treatment and home care services are examples of the system. Primary, secondary and tertiary care settings are occurred in therapeutic purpose Turkish healthcare services. Family medicines are constituted primary care which ambulatory care is done, the private or public hospitals are formed secondary care, the oncologic, university hospitals are created tertiary care (Çelikay and Gümüş 2000, pp. 185–186).

Nowadays, the principles of Health Transformation Program are performed. The health system in Turkey is a mixed structure. Three different health insurance system were used by government before 2003 in Turkey with presenting distinct advantages, arrangements which are SSK, its scope was workers in public and private industry, Bag-Kur, its scope was for self-employed, and Emekli Sandigi, its scope was retired people (Gürsoy 2015, p.87). However, poor people could not be passed over the three healthcare systems. Then, some changes in health system are done which one is the establishment of Green Card provided poor people to access the health facilities. The other one is the scope of Bag-Kur is changed. Tatar states that the scope of Bag-Kur is enhanced into headmen, anybody who had not belonged to any Turkish health insurance system, housewives and people who works in agriculture (2011, p.126). The rate of private and public spending is arranged by the collaboration of Ministry of Health, Ministry of Finance and Treasury and Stable Assistance Fund for Children.

Primary, secondary and tertiary healthcare system are supplied by MOH and also preventive health services. The funding system for MOH and university hospital was same. Ministry of Finance supplied funds them with including wages of workers, costs and some fees which were taken from private people.

After implementation of Health Transformation Program, the funding system has changed. Reimbursement system is established and performed. The pharmacies which are supplied by SSK are closed. To the purpose of extinguish the inequalities among prior health insurance systems which were SSK, Bag-Kur, Emekli Sandigi, Health Implementation Practice (SUT) and Medula which is the benefiting system were launched in 2007 (Gürsoy 2015, p. 90).

The financing structure of health system in Turkey is differentiating. The first category is taking healthcare services with market economy which includes pay out of pocket and private health insurance. The second category is taking healthcare services with public economy which includes compulsory social insurance and tax incomes (Çelikay and Gümüş 2000, pp. 187–196). In market economy financing system, the cost of taking health system is determined according to the balance of supply and demand. Healthcare systems could be utilized by directly the own payments of people or private health insurance with the lack of social interference-intervention (McGregor 2001, p. 2). Çelikay and Gümüş explain that health care is purchased through out-of-pocket payment (2000, p. 188). There is not any intermediary firm, the relationship is constituted among patients and hospitals, dentists. The other component of market economy is private health insurance. Groupama, Allianz, Axa are examples of private insurance companies in Turkey. The premiums from individuals are collected by private health insurance companies and the payments are done to the health service provider in case of a possible health problem (Çelikay and Gümüş 2000, p. 190). In public economy finance system, the government has a key role due to the existence of poor people. Otherwise, they could not get health system. According to Mutlu and Işik, in cases where health services are provided free of charge by the government or under compulsory health insurance, the price of health services for the individual will decrease to zero and this will result in the price being ineffective on demand (2012, p. 82). The costs which are occurred from health problem are met by government. The last component is financing healthcare services with tax incomes. The finance of general tax revenues and healthcare services is a vital collective tool to decrease in social risk and provide health services to all individuals (Çelikay and Gümüş 2000, p. 195).

Gürsoy states that the pleasure of patients to the quality of healthcare services is increased with establishment of Health Transformation Program which enables to the decrease in the duration of waiting times and increase in the duration of consultation times. (2015, p. 101).

3 Methodology

3.1 Turkey's Mobile Health Application

The number of Turkey mobile health applications is increasing day by day. "MHRS", "E-nabiz", "İlk Yardim", Medline and Heart Rate Monitor are some examples of them.

MHRS is an appointment system developed for the hospitals affiliated to the ministry of health so that patients can make appointments from the doctor of their choice on the most appropriate day and time. The health information about patients such as blood pressure, the rate of sugar could be kept with "E-nabiz". Everybody could reach to the datas 24/7. They can access the system via the web as well as mobile devices with IOS and Android operating systems. Also, citizens can make an appointment via e-nabiz for bone marrow-blood and organ donation. Another example is İlk Yardim. It is prepared by the experts of Anadolu Medical Center. Mobile first aid application is free for iPhone users. It includes fainting, insect stings, respiratory blockages, fractures, foreign body ingestion into eyes, ears and nose, bleeding, injury, remittance, poisoning, artificial respiration, burns and freezes, heat strokes and first aid. The menu of the app is designed to provide quick access to information in case of emergency requiring first aid information. It is supported by explanatory visuals.

Heart Rate Monitor is a health monitoring application that provides information about the patient's' rhythm values and risks while entering the application by touching the tempo of heart on the screen. In addition, Medline is the application which enables remotely monitor the disease. It is established with the collaboration of Vodafone & Medline.

3.2 Needs Analysis

The statistical formula in (1) is used to calculate the determination of sample size which let to gets more reliable results statistically. Thus, the survey will be significant for research

$$n = \frac{z^2[\pi(1-\pi)]}{s.e^2} \quad (1)$$

Where, the symbols represent the values as follows;

n = *the sample size*
$z = 1.645$ *(at 90% confidence level)*
s.e (sampling error) $= 0.05$
$\pi = 0.5$

The value of π $(1-\pi)$ is the biggest number for $\pi = 0.5$. So π was selected 0.5. This provides more reliable results for survey. After these calculations, n value will be 270.6. Rounding up for 270.6 is 271. So, survey was solved by 271 people.

3.3 Survey Questions and Analysis

In the survey study, the ten questions were prepared. The fifth, sixth and ninth questions have sub-questions. The Objective of this survey is to get more information for emergency healthcare system in Turkey and Docto'Ring' with point of view of public.

Firstly, a filter question is used to eliminate people who calls private ambulance companies. The people who use private ambulances, the survey continues with sixth question for them. Contrary, the people who use 112, it is asked to the arrival time and reasons of calling ambulances for them. 5- point Likert Scale is used for the evaluation questions. Then, the situation of emergency departments is measured. At the end, after asking demographic questions like age and gender, Docto'Ring' is introduced. In appendix, the survey could be seen.

To be able to understand the efficiency of ambulances, the questions are asked below:

- Codifying easily for calling ambulances
- Short arrival time
- Knowledge of paramedics about the patients
- Unreliable emergency callings

Respondents are asked to answer questions about ambulances and Docto'Ring' in a 5-point Likert Scale.

- Definitely agree
- Agree
- Neither agree or disagree
- Disagree
- Definitely disagree

To be able to understand the efficiency of emergency departments, the questions are asked below:

- Knowledge and interest to patients of healthcare personnel
- Number of hospital beds
- Necessary medical equipment's such as an intensive care
- The duration of interference
- The rate of waiting patients in emergency departments

Respondents are asked to answer questions about emergency departments in a Rating Scale.

According to the survey, it could be seen that approximately 58% of the 271 people called ambulances. The high rate of calling ambulance indicates that there may be situations where everyone needs an ambulance. The ambulances should be in touch with public because no one knows when their health will deteriorate. Therefore, the ambulance system should be able to serve equally to everyone's right to life. The number of people could not be equal to the number of ambulances. The system will crash where everyone calls an ambulance at the same time. This situation can be in big disasters such as earthquakes and floods. The number of ambulances and the number of stations is very important to prevent the system from collapsing under normal conditions. In any case, with intensive calling, the system may be inadequate. Therefore, unnecessary callings must be avoided.

The emergency line is 112. It is free and also works 24 h in a day. When a person calls the ambulance, the ambulance comes from the nearest station to the scene. First intervention is done. Depending on the situation, the patient is taken to the hospital. The ambulances which are called from 112 are not private ambulance. This is the main reason why private ambulance companies appear low in the survey. Private ambulance companies are mainly used for patient transport. Approximately 6% people prefer several private ambulance companies for their health situation. However, the perception of public about 112 is an only way for health emergency. That's why approximately 94% of people prefer 112.

People call ambulances for a lot of reasons. The most common reason with 26,8% is heart attack. The serious injuries (13,4%) and traffic accidents (10,8%) are following reasons of calls. The other reasons are differentiating among people like the situation of KOAH, high blood pressure, poisoning. Apart from these, the importance and the real emergency situations of calling ambulances should be taught to public because the unconsciousness leads to call 112 for needlessly. To illustrate, somebody has a phobia of ambulances and he/she wants to get rid of this situation. This purpose let to call 112 and it results of the intensity in emergency line.

The heart attack is serious emergency situation. Time is very important for not damaging heart. However, most people do not aware that they have symptoms of it and they do not visit the emergency departments. The deaths from heart attacks are always high in every time like in survey results and maybe the rate is higher than in survey but people do not realize it. 109 people (40%) are generally codify easily to the 112-emergency line. Emergency call center is good as infrastructure. However, 40 people (14%) who could not codify the line easily. The reason could be the intensive calls or the telecommunication system of their location. It could also be due to the insufficient of an emergency call center. Deficiencies must be eliminated. Everyone should be able to get help quickly in case of emergency.

79 people (29%) are satisfied with the arrival time of the ambulance. 52 people (19%) are not satisfied with the arrival time of the ambulance. The arrival time of the ambulance varies. This period should be the same everywhere. For the patient who has an emergency, the time has same importance everywhere. Thus, it's not enough for the ambulance to arrive quickly sometimes. Because human life is sacred. The ambulance must always come fast.

84 people (30%) are satisfied with the knowledge of the ambulance team. They think that the ambulance team have had a good training. Education system may also be

a good elimination for the ambulance team. Also there are people who are not satisfied. This ratio is 14% (40 people). The reason could be lack of information. If the patient's condition is learned more clearly before arrival, this ratio will decrease.

101 people (37%) know that ambulances are called unnecessarily. This is a serious problem. The number of ambulances usually does not match the number of cases. An emergency patient can wait for an ambulance because of unnecessary callings. Waiting could lead to the end of his or her life. People often don't know the seriousness of their situation. So, they call an ambulance without knowing their health situation. Generally, people are not sick but they feel bad. But there is no emergency situation. If they learn about their health situation, they will not call an ambulance unnecessarily.

The night doctors and general doctor practitioners work in hospitals' emergency departments. Sometimes, the responsible of emergency medicine doctors are found. The situation leads to the public to distrust them. They think that they do not have enough experience of various health cases, they are newly graduate. Also, the intensity in emergency departments cause the deterioration in connections of patients and healthcare personnel. Thus, the time per patient is decreasing.

The graph shows that approximately 40% of people are neutral of the knowledge and interest to patients of healthcare personnel. However, the residual parts which are the total rate of dissatisfaction and satisfaction are close to each other. Number of hospital beds are correlated with the capacity of hospitals. However, the elderly people in need of care are taken to the hospital by their family due to the financial impossibility and morale. Also, public could not arrange the hospital equipment to their homes and the maintenance is not as same as hospital. This reason induces the insufficient hospital beds in general.

Here, approximately 39% of people are neutral of the number of beds in hospitals. They think that the beds are enough in some cases. However, the great majority thinks that they are insufficient and so they are dissatisfied. The complaints are increased day by day.

Table 1 is the table summarizing the concluded needs of public for Docto'Ring'.

4 Application Execution

The information of patients like their location address, symptoms, personal details is not shared with anybody. All of the information is kept safely. The commitment is established among private ambulance companies and government.

Docto'Ring' could be downloaded free from Play and App Store and after usage, patients could make interpretation. There are 3 main sections of the application which are classified as below:

Home Page: It is first page of the application that gives an information about Docto'Ring' which includes ambulances and Telehealth for heart attack situation and if necessary, ambulance press button could be seen. Here, patients can login or create an account.

Registration Page: If patients do not have any account, they can create from here with entering his/her personal information. After reading and approving consumer agreement, the new account is established. Then, home page is opened again.

Table 1. Concluded needs for docto'ring'

Concluded Needs of Docto'Ring'
The application is free and work without Internet.
The privacy, data security are provided.
The waiting period of responses is short.
Malpractise is reduced with artificial intelligence and remote diagnosis.
No waste of time to answer questions due to the feature of voice comment.
Doctors and other health materials are sufficient.
It is accesible and come back to memory of people in emergency situations.
The state guarantee is taken.

Telehealth Page: Patients who have a pain in their chest and think that they have a heart attack situation use Telehealth segment. Here, there are 8 questions to determine the situation of them. If patients have that symptoms they will mark otherwise they will not mark. Every question can be listened with speaker. If patients feel panic and cannot mark the answers, they can speak via microphone and their voice is recorded.

Then, the guidance of Docto'Ring' is starting. Patients can hear all guidance from speakers and record their voice with microphone again. All guidance should be done and checks should be marked.

After completing all questions and guidance, the situation is evaluated. If there is an emergency situation, the ambulance calling button is appeared. They can press it and in the period of waiting time, they can save their lives with the guidance. All of guidance can be heard from speakers again. Also, thanks to the location future of Docto'Ring', their location is sent to the ambulances when they allow to share it with ambulances Thus, ambulances can reach right patient at the right time. If there is not any need of emergency, Docto'Ring' tells them.

4.1 Testing Process

To be able to determine the application's efficiency, the testing process and inspection will be done by government and Canadian software engineer. There are too much health legal regulations and laws are strict. Also, we do not have any permission from Health Ministry of Turkey for applying to the patients so the testing process could be implemented by only Turkish government.

5 Conclusion

All in all, this project aims to help decreasing unreliable callings to 112 and unnecessary intensity in emergency departments by developing mobile health application for heart attack situation. Turkey is still developing country in all areas. The basic problems in health industry have not be solved yet. The social awareness about usage of ambulances and emergency departments could not be established well. Turkish government tried to establish new health system and proposed to the private ambulance companies but nobody interested it with. Thus, it was decided that the project will be sold to the government which results to increase in reliability of Docto'Ring' and with marketing tools such as public service announcements, the social awareness was aimed to increase. Canadian Telehealth was taken as a basis to establish and develop Docto'Ring'. Our application differentiated into the scope; it was only for heart attack situation.

During this study, first, the literature review was conducted to determine and analyze world, Canadian and Turkey healthcare system. Then, the global, Canadian and Turkish mobile health applications were observed. The survey was prepared with 271 people to determine needs. At the same time, the interviews with paramedics, private ambulance company and doctors were done. The common remarks helped us to understand main needs of public. The top 9 reasons of calling ambulances were aligned and with Analytic Hierarchy Process, the most important reason which leads to looking for ambulances mostly was determined. It was a heart attack situation which could be severe for people if the interference could not be done, it may lead to deaths. Therefore, Docto'Ring' was prepared for heart attack situation to reduce deaths and provide timely intervention.

The feasibility analysis which includes technical, market, commercial analysis and overall risk assessment was done to achieve the aim. Meanwhile, stakeholder analysis, project management plan and quality planning were studied due to the situation of selling the Docto'Ring'.

Lastly, the application design with privacy policy and the application interface design were conducted step by step. The audible, written questions and microphone took a part of Docto'Ring' according to the target segment needs.

For further research, it is highly recommended that taking all permissions from Ministry of Health and reach more people's health data which leads to more accurate results.

References

A Strategy to Reduce Emergency Department Wait Times in Newfoundland and Labrador 2012 (2012). https://www.health.gov.nl.ca/health/wait_times/pdf/emergency_department_strategy.pdf

Akdur, R.: Sağlik Sektörü: Temel Kavramlar, Türkiye ve Avrupa Birliği'nde Durum ve Türkiye'nin Birliğe Uyumu, vol. 2. Ankara Üniversitesi Basimevi, Yayin (2006)

Aringhieri, R., Bruni, M., Khodaparasti, S., van Essen, T.: Emergency medical services and beyond: addressing new challenges through a wide literature review. Comput. Oper. Res. **78**, 3 (2016). https://www.researchgate.net/publication/308575332_Emergency_Medical_Services_and_beyond_A

Aslan, Ş., Güzel, Ş.: Pre-hospital emergency health services in Turkey. J. Soc. Human. Sci. Res. **5**(31), 4997 (2018). https://www.researchgate.net/publication/330363391_TURKIYEDEKI_HASTANE_ONCESI_ACIL_SAGLIK_HIZMET

Ball, D.: The future of risk assessment? (2012). https://davidjball.com/tag/risk-matrices/

Bickenbach, J.: WHO's Definition of health: philosophical analysis. In: Schramme, T., Edwards, S. (eds.) Handbook of the Philosophy of Medicine, p. 1. Springer, Dordrecht (2015)

Canadian Cancer Statistics Advisory Committee. Canadian Cancer Statistics 2019. Canadian Cancer Society, Toronto (2019). cancer.ca/Canadian-Cancer-Statistics-2019-EN

Canadian Institute for Health Information. National Health Expenditure Trends, 1975 to 2018. CIHI, Ottawa (2018). https://secure.cihi.ca/free_products/NHEX-trends-narrative-report-2018-en-web.pdf

Chongsuvivatwong, V., Phua, K., Yap, M., Pocock, N., Hashim, J., Chhem, R., et al.: Health and health-care systems in southeast Asia: diversity and transitions. Lancet **377**(9763), 429–437 (2011). https://doi.org/10.1016/s0140-6736(10)61507-3

Çelikay, F., Gümüş, E.: TÜRKİYE'DE SAĞLIK HİZMETLERİ VE FİNANSMANI. Eskişehir Osmangazi Üniversitesi Sosyal Bilimler Dergisi (1), 184–186 (2000). https://dergipark.org.tr/tr/download/article-file/113275

Deshpande, A., Khoja, S., McKibbon, A., Jadad, A.R.: Overview of assessments of real-time (synchronous) and asynchronous telehealth [Technology overview number 38]. Canadian Agency for Drugs and Technologies in Health, Ottawa (2008)

Donev, D., Kovacic, L., Laaser, U.: Health: Systems – Lifestyles – Policies, 2nd ed., p. 7. Jacobs Verlag, Lage (2013). https://www.researchgate.net/publication/257830385_The_Role_and_Organization_of_Health_Care_Systems

Fact Sheet: Telehealth (2019). https://www.aha.org/system/files/2019-02/fact-sheet-telehealth-2-4-19.pdf

GELİR İDARESİ BAŞKANLIĞI BÜYÜK MÜKELLEFLER VERGİ DAİRESİ BAŞKANLIĞI Mükellef Hizmetleri Grup Müdürlüğü. Yurtdişindan alinacak bilgisayar yazilimlari ve bilgi transferi bilgi transferi hizmetleri ve bunlara yönelik yurt içinde ve yurt dişinda eğitim, danişmanlik, teknik destek, güncelleme ve bakim hizmetleri nedeniyle yapilacak ödemelerden e tevkifat, Özelge (2013). https://www.gib.gov.tr/node/95544

Gürsoy, K.: Sağlikta Dönüşüm Programi Sonrasi Türk Sağlik Sistemine Genel Bir Bakiş: Başarilar, Performans Değerlendirmesi, Muhtemel Zorluklar ve Çözüm Önerileri. Sosyal Güvence Dergisi (7), 84–101 (2015). https://www.researchgate.net/publication/304241347_An_Overview_of_Turkish_Healthcare_System_a

Eray, O.: Türkiye'de Acil Servise Başvuran Hastaların Profilleri. Kardiyovasküler Akademi Derneği (2014). http://www.kvakademi.org/giris/KPDData/userfiles/file/OktayBulten14.pdf

Health Ministry of Turkey. Annual Health Report (2018). https://sgb.saglik.gov.tr/Dkmanlar/TC%20Sa%C4%9Fl%C4%B1k%20Bakanl%C4%B1%C4%9F%C4%B1%20Faaliyet%20Raporu%202018.pdf

Health Ministry of Turkey. Annual Health Statistics 2018 Newsletter. General Directorate of Public Health (2019)

International Pharmaceutical Federation (FIP). mHealth—Use of mobile health tools in pharmacy practice. The Hague, International Pharmaceutical Federation (2019)

Irvine, B., Ferguson, S., Cackett, B.: Healthcare Systems, Canada (2013). http://www.civitas.org.uk/content/files/canada.pdf

Martin, D., Miller, A., Quesnel-Vallée, A., Caron, N., Vissandjée, B., Marchildon, G.: Canada's universal health-care system: achieving its potential. Lancet **391**(10131), 1718–1735 (2018). https://doi.org/10.1016/s0140-6736(18)30181-8

Mcgregor, S.L.T.: Neoliberalism and health care. Int. J. Consum. Stud. Special Ed. Consum. Health **25**(2), 82–89 (2001)

Milliye. TÜRKİYE'NİN KALP HARİTASI (2013). http://www.milliyet.com.tr/cadde/turkiye-nin-kalp-haritasi-1781078

Mutlu, A., Işik, A.K.: Sağlik Ekonomi-sine Giriş. Ekin Yayinlari, Bursa (2012)

OECD. Health resources - Health spending - OECD Data (2019). https://data.oecd.org/healthres/health-spending.htm

PROJE HİBE DESTEKLERİ – ARALIK 2019 (2019). Accessed Dec 2019. http://danismend.com/kategori/altkategori/proje-hibe-destekleri/

Quick Stats | CIHI (2018). https://www.cihi.ca/en/quick-stats?xQSType=Pre-formatted%252520Table&amp%253bxTopic=Hospital%252520Care&amp%253bpageNumber=2&amp%

Razzak, J., Kellermann, A.: Emergency medical care in developing countries: is it worthwhile?, pp. 90–91 (2002). https://www.who.int/bulletin/archives/80(11)900.pdf

Research Spotlight on Telehealth (2019). https://www.pcori.org/sites/default/files/PCORI-Research-Spotlight-Telehealth.pdf

Saaty, T.L.: The Analytic Hierarchy Process. McGraw-Hill, New York (1980)

Sadiku, M., Shadare, A., Musa, S.: Mobile health. Int. J. Eng. Res. **6**(11), 450 (2017). https://www.researchgate.net/publication/320453153_Mobile_Health

Allin, S., Rudoler, D.: The Canadian health care system. In: Mossialos, E., et al. (eds.) International Profiles of Health Care Systems, (Commonwealth Fund, May 2017), pp. 21–30 (2017)

Sargutan, A.E.: Sağlik Sektörü ve Sağlik Sistemlerinin Yapisi. Hacettepe Sağlik İdaresi Dergisi **8**(3), 400–428 (2005)

Servicios Sociales e Igualdad, M.: Health care systems in the European Union countries. Presentation (2019)

Sözmen, K., Ergör, G., Ünal, B.: Hipertansiyon sikliği, farkindaliği, tedavi alma ve kan basinci kontrolünü etkileyen etmenler. Dicle Med. J./Dicle Tip Dergisi **42**(2), 199 (2015). https://doi.org/10.5798/diclemedj.0921.2015.02.0558

Şimşek, P., Günaydin, M., Gündüz, A.: Hastane Öncesi Acil Sağlik Hizmetleri. Türkiye Örneği. **8**(1), 120–124 (2019). https://www.researchgate.net/publication/332058317_Hastane_Oncesi_Acil_Saglik_Hizmetleri_Turkiye_Ornegi_Pre-Hospital_Emergency_Health_Services_The_Case_of_Turkiye

Tatar, M.: Sağlik Hizmetlerinin Finansman Modelleri: Sosyal Sağlik Sigortasinin Türkiye'de Gelişimi. SOSYAL GÜVENLİK DERGİSİ (1), 126 (2011). http://www.acarindex.com/dosyalar/makale/acarindex-1423912001.pdf

The Percentage of Individuals' Status of Smoking Tobacco Products by Sex and Age group. Turkey Health Interview Survey (2016). http://tuik.gov.tr/PreTablo.do?alt_id=1095

The Prevalence of Hypertension in gender and age groups. Turkish Society of Hypertension and Renal Diseases (2012). http://www.turkhipertansiyon.org/

Tuik.gov.tr. Türkiye İstatistik Kurumu, Ölüm Nedeni İstatistikleri 2018 (2019). http://www.tuik.gov.tr/PreHaberBultenleri.do?id=30626

Turkey Population. Accessed 10 June 2019. http://worldpopulationreview.com/countries/turkey/

Turkey Smoking Rate 2000–2019 (2019). https://www.macrotrends.net/countries/TUR/turkey/smoking-rate-statistics

TurkStat. Health Statistics. http://tuik.gov.tr/PreTablo.do?alt_id=1095

Türk Kardiyoloji Derneği 2016 raporu (2016). https://www.archivestsc.com/tr/jvi.aspx?volume=44&issue=4
Türkiye İstatistik Kurumu (TUİK). Nüfus Projeksiyonlari, pp. 2013–2075 (2017). http://www.tuik.gov.tr/
Türkiye İstatistik Kurumu. Sağlik Harcamalari İstatistikleri 2018 (2019). http://www.tuik.gov.tr/PreHaberBultenleri.do?id=30624
ORG Haber. Kalp Krizi Haritasi Çikartildi! Siz hangi bölgedesiniz? (2015). http://www.orghaber.com.tr/saglik/kalp-krizi-haritasi-cikartildi-siz-hangi-bolgedesiniz-h7114.html
Öksüz, H.: En fazla bu hastaliklar öldürüyor! Habernediyor (2019). https://www.habernediyor.com/saglik/en-fazla-bu-hastaliklar-olduruyor-h6971.html
World Bank. Good Practices in Health Financing, Washington (2008)
Who.int. Cardiovascular diseases (CVDs) (2017). https://www.who.int/news-room/fact-sheets/detail/cardiovascular-diseases-(cvds)
Yildirim, S.: Sağlik Hizmetlerinde Harcama ve Maliyet Analizi, Sosyal Sektörler ve Koordinasyon Genel Müdürlüğü Uzmanlik Tezi (1994)
Yüksel, E: Türkiye'de ve Dünyada Obezite(2018). https://www.dogrulukpayi.com/bulten/turkiye-de-ve-dunya-da-obezite

Parametric Design and Modeling of 3D Printed Prosthetic Finger

Yunus Emre Kahraman[1], Mert Madan[2], Binnur Sağbaş[1(✉)], and Numan M. Durakbasa[3]

[1] Department of Mechanical Engineering, Yıldız Technical University, Istanbul, Turkey
yunusemrekahraman17@gmail.com, bsagbas@gmail.com
[2] Department of Mechanical Engineering, Boğaziçi University, Istanbul, Turkey
mert.madan@gmail.com
[3] Research Group of Industrial Metrology and Adaptronic Systems, TU Wien (Vienna University of Technology), Vienna, Austria
durakbasa@ift.tuwien.ac.at

Abstract. Three dimensional (3D) printed prostheses have emerging trends in recent years. They become popular with their cost-effective and easy to build structures which provided by the developments of the additive manufacturing technology. By these advancements, many open-source prosthetic hand models started to appear in the market. However, some of them are not suitable for all kinds of patients yet. So, a good deal of designs has failed to meet all of the user's physical needs which might cause the device not being used by the patients. In this paper, it is tried to do a parametric design and produce a mechanical prosthesis with a simple mechanical design that can be adapted easily to each patient. For this purpose, the information about a planar four-link mechanism, design, and analysis steps of the mechanism and how it can be improved in the future has been discussed.

Keywords: Prosthetic hand · Additive manufacturing · 3D production · Upper limb · Low-cost production

1 Introduction

Amputation changes the body structure, affects daily activities, and decreases the quality of life. The most common causes of amputations are vascular diseases, infections, and traumatic injuries. The amputation of a finger reduces the functionality of the hand and creates psychological and social damages. It is often the latter ones that have greater impacts compared to the functionality. The finger amputation is one of the most common upper region amputations and it causes traumatic consequences in general thus requires severe rehabilitation. To reduce all these phenomena, much work has been done and progress has been made in the field of prosthetics [1], however it has still not reached a satisfactory level.

Currently, there are many prostheses on the market with their goal as mimicking all the hand and arm movements. There are some multifunctional prostheses introduced in

N. M. Durakbasa and M. G. Gençyılmaz (Eds.): ISPR 2020, LNME, pp. 80–90, 2021.
https://doi.org/10.1007/978-3-030-62784-3_7

the market such as the Cyborg beast low-cost prosthetic [2], the I-limb hand from Touch Bionics [3], the Espire from Steeper [4], the Bebionic hand from Ottobock [5], the VINCENTevolotion 3 from Vincent Systems [6], the Hero Arm from Openbionics [7]. These prosthetic devices are primarily designed for above the wrist. All of the devices are working with electric power according to the myoelectric principle [8]. With the realization of single powered fingers, prosthetic devices have started to change. Some researchers worked on the differences of multi-joint myoelectric prosthesis I-LIMB and single-joint Dynamic Mode Control (DMC) plus hand. They have concluded that the functional scores of the myoelectric hand were equal or even lower than the dynamic controlled one which is interesting because all of the technical developments they have made for the I-LIMB [9]. Unlike, other prostheses the Cyborg Beast is designed in a way that works body-powered. While working on the project of the cyborg beast, the researchers have tried to create a prosthesis without physically contacting patients. They had 11 participants, nine of them were treated in a laboratory and two of them were treated distantly. Researchers compared the participant's upper limb measurements taken from a picture with the ones taken directly. The result of this study shows that there is no significant difference between the measurements taken by a picture or taken directly [2]. 3D-printing is an additive manufacturing method which generates the 3D geometries by layer-by-layer from a computer-aided model. It provides an opportunity to generate partial hand prosthesis by open-source concepts [10]. The first 3D-prosthesis is designed by Ivan Owen and used by Liasm. After completing their project Owen has decided to upload all the files online and his efforts led to the creation of e-NABLE. As a result of this, rapid improvements in the prosthesis area have been achieved [11]. Although this method has a lot of merits such as, it can be built quickly and no-assembly is needed, it also has drawbacks such as limited material range. Despite the disadvantages, this method has lowered the cost of prostheses [12].

In this study, it is aimed to design and model a prosthetic finger for 3D printing applications that can fit any person and can be designed from any distance. In order to create an optimal body-powered prosthesis for all kinds of patients, an average length and width were used for all the fingers [13, 14].

2 Design of the Model

2.1 Mechanical Design

Quality management systems of the medical devices have been explained in ISO 13485 [15] while upper limb prosthetic parts are described in ISO13405-3 [16]. Mechanical design of the model generated in accordance with the related standards. Most upper limb amputation patients have partial finger loss and current prosthesis options cannot provide the required solution to these patients. The starting point of the whole design process was a single finger. The human hand was visually studied while it was grasping or holding many different objects. It was logically concluded that the hand consists of four similar fingers and that the finger design could potentially be repeated three times. This meant that when the size and space requirements to activate a single finger proved to be too large, another method would be needed. The finger design process began by determining the required movements for each finger. The human hand simply appeared to grasp a variety of household items, such as a cup or a pen which can frequently be found in a person's daily routine. The locations of the various joints were measured and translated into drawings [17]. The human finger is a superb engineering product consisting of three articulated joints that can be moved individually with muscle tendons. The four main fingers can spread out sideways and reach an impressive amount of total degree of freedom. Fortunately, in order to achieve most tasks of gripping, only a small amount of motion is required. The human finger achieves a suitable adaptive grip by bending the joints while holding an object. The main goal of this design is to create a lightweight and easy to use a body-powered prosthesis that can provide the required movements for the patient. After a considerable amount of research, the four-bar linkage mechanism is found to be able to transfer the movement of the finger to the fingertips [2, 11, 18, 19]. Haulin et al. [20] reported the mathematical way to use prosthetic fingers without a dead point and this design uses these linkage ratios to create a suitable prosthesis for all individuals.

Figure 1 describes the four-link mechanism which was stated by Haulin et al. [20]. R4 is a fixed link attached to the palm. MA is the driving link. TP1, TP2, TP3 respectively angles to P1, P2, P3 and selected from the Fig. (1). P1 is proximal phalange, P2 is intermediate phalange and P3 is distal phalange of a finger. P3' is the effective length of distal phalange where the force P applied from and point B is the IP point with an eccentricity "e" of 2 mm to obtain a proper force at the fingertip. The eccentricity reduces the stretching of the glove at the external side of the Intermediate Phalange (IP) joint and the necessary driver couple [21]. The mechanism has only one degree of freedom (DOF). With applying the Gruebler's formula [22] for bar mechanisms. DOF = 3(N − 1) - 2P1–2P2. N = number of bars, P1 = number of joints with 1 degree of freedom, P2 = number of joints with 2 degrees of freedom.

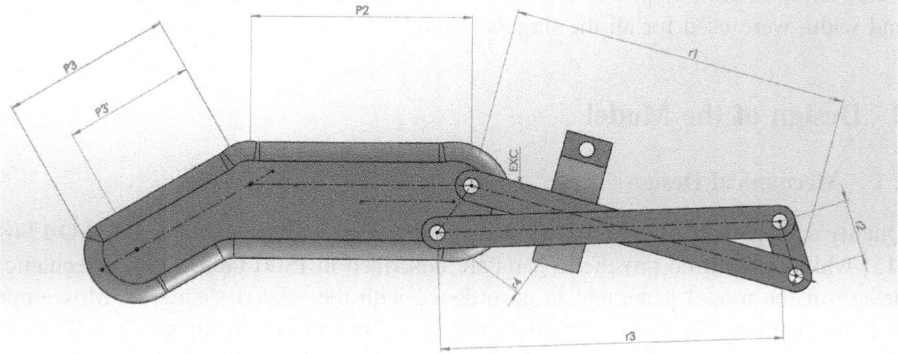

Fig. 1. A four-link mechanism [18]

The basic equational solution for the system;

$$\overrightarrow{R}_1 + \overrightarrow{R_2} + \vec{R}_3 + \vec{R}_4 = 0 \quad \text{With} \quad R_{xi} = r_x e^{j\theta xi} = r_x(\cos\theta_{xi} + j\,\sin\theta_{xi}) \tag{1}$$

As a result,

$$r_1 \cos\theta_{1i} + r_2 \cos\theta_{2i} + r_3 \cos\theta_{3i} - r_4 \cos\theta_{4i} = 0, \tag{2}$$

$$r_1 \sin\theta_{1i} + r_2 \sin\theta_{2i} + r_3 \sin\theta_{3i} - r_4 \sin\theta_{4i} = 0. \tag{3}$$

Eliminating θ_{3i} by combining Eq. 2 to get connection between θ_{1i} for driving link and θ_{2i} for finger-tip with using Haulin et al. [20] finger length ratios;

$$\frac{r_4}{r_2}\cos(\theta_{1i} - \theta_4) + \frac{r_4}{r_1}\cos(\theta_{2i} - \theta_4) + \frac{r_1^2 + r_2^2 - r_3^2 + r_4^2}{2r_1 r_2} = \cos(\theta_{1i} - \theta_{2i})$$

After defining the mechanical equation for mechanism, angles must be calculated, From fig $\theta_{1i} = TP_{1i} + \theta_B$ and $\theta_B = -arctan\left(\frac{EXC}{P_1}\right)$

EXC = Eccentricity "e"

$$\theta_{2i} = \theta_A + (TP2_i + TP1_i) = \theta_{2ir}(r = required),$$
$$L = P_1 + P_2 + P_3'$$
$$LP_1 = \tfrac{44.5}{91}L, \quad P_2 = \tfrac{29.2}{91}L, \quad P_3 = \tfrac{17.3}{91}$$
$$r_1 = \sqrt{P_1^2 + EXC^2}$$

Reference finger path for the mechanism was based on the equations provided by Guo for two grips (pinching and holding) of a regular finger [18], θ_1, θ_2, θ_3 symbolize the Metacarpal Phalange (MCP) joint angle, Distal Phalange (DIP) joint angle and PIP joint angle.

For pinching ($5^\circ \le \theta_1 \le 42.5^\circ$)

$$\theta_2 = 2.45\theta_1$$
$$\theta_3 = 3.35\theta_1$$
$$x_s = P_1 \cos\theta_1 + P_2 \cos\theta_2 + P_3' \cos\theta_3$$
$$y_s = P_1 \sin\theta_1 + P_2 \sin\theta_2 + P_3' \sin\theta_3$$
$$\theta_3 = 30^\circ$$
$$315^\circ \le \theta_4 \le 360^\circ$$

For cylindrical holding ($42.5^{\circ} \leq \theta_1 \leq 80^{\circ}$)

$$\theta_2 = 2.45\theta_1 + 19.12^{\circ}$$
$$\theta_3 = 3.35\theta_1$$
$$x_s = P_1 \cos\theta_1 + P_2 \cos\theta_2 + P_3' \cos\theta_3$$
$$y_s = P_1 \sin\theta_1 + P_2 \sin\theta_2 + P_3' \sin\theta_3$$
$$\theta_3 = 30^{\circ}$$
$$315^{\circ} \leq \theta_4 \leq 360^{\circ}$$

And for the bars [18]

$$r_1 = \frac{44.5}{91}L, \quad r_2 = \frac{7.58}{44.54}r_1, \quad r_3 = \frac{45.3}{44.54}r_1, \quad r_4 = \frac{7.42}{44.54}r_1.$$

By carefully examining the fingertip, it can be seen that there is only a slight rotation in the last joint. Therefore, it can be concluded that for simplification purposes the finger joint can be treated as a fixed link, instead of two separate connections that are joined by a finger joint.

An assembly was created that allowed the movement of the two joints in SolidWorks. When viewed from the side, the human finger is suitable for movement about 90° from full extension to full closure. Similarly, the second joint also rotates about 90°. When the hand is slowly brought to the fully closed position from the fully opened one, approximately two joints with a 1:1 ratio are observed to act. In this embodiment, the base finger allows easy rotation with a simple connection.

2.2 Parametric Design

Parametric modeling used for creating a basic model, by changing a few things the model can be adapted for different individuals. This modeling type requires mathematical approaches such as equations, lengths, etc. With this approach, models can be unique for all. The width and finger length were used for the parameterizing operation. In the design, the length related components are distal phalanges, intermediate phalanges, proximal phalanges, driving linkage, coupler linkage and MCP connection parts with sets of the percentage of finger length. The width related components are width and radius of the finger. With those parameters, equations are created for fingers and mate between them such as distances and angles.

Angles described in Fig. 2 such as teta_f1, teta_f2 are pinching angles for the finger from 5° to 42.5° and cylindrical holding for 42.5° to 80°. These degrees and mechanical constraints are applied to avoid dead centers. All the locking positions have been expressed as functions of the design variables of the mechanism.

The middle finger in Fig. 3 was used for a base model for all and all the equations come from that part with changing a few things. This parameterizing method used for all individuals who have different finger lengths and different finger widths. If this mechanism will be used for patients it has to be adaptable to all different statuses for all patients.

```
"L"= 91mm'fingerlength
"P1"= 44.5 * "L" / 91'PP
"P2"= 29.2 * "L" / 91'IP
"P3'"= 17.3 * "L" / 91'DP
"r1"= 44.54 * "L" / 91'linkage_r1
"r4"= 7.42 * "r1" / 44.54'linkage_r4
"r3"= 45.3 * "r1" / 44.54'linkage_r3
"P3"= 22.5 * "L" / 91'P3
"r2"= 7.58 * "r1" / 44.54
"D1@Sketch2"="P2"
"D2@Sketch2"="P3'"''''''''''''''''
"D1@Sketch4"="r2"
"W"= 15mm'Width
"D1@Surface-Extrude1"="W"
"D1@Fillet1"=0.2 * "W"
"D4@Sketch2"="P3"
"teta_f1"= 5deg'finger_grip_min
"teta_f2"= 1deg * 42.5'finger_grip_max
"teta_h1"= "teta_f2"'holding_grip_min
"teta_h2"= 80deg'holding_grip_max
"TP2_f"= "D1@finger grip" * 1.45
"EXC"= 2mm
"TP2_c"= "D1@finger grip" + ( 1deg * 19.125 )
```

Fig. 2. Equations for mechanism

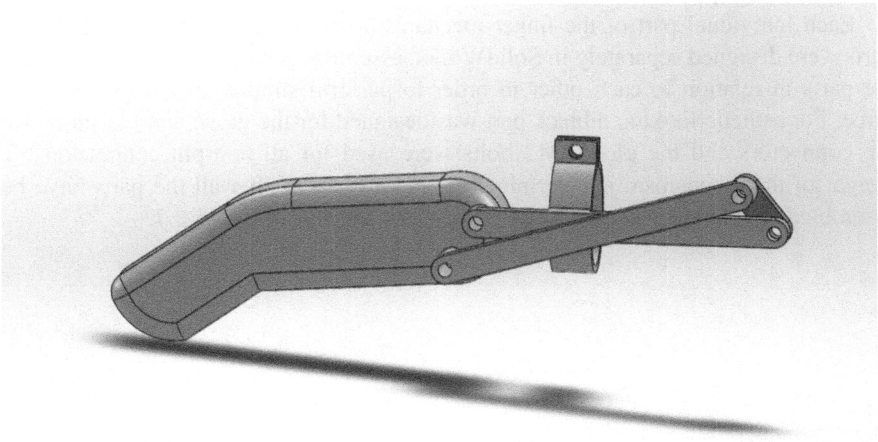

Fig. 3. Middle finger

To change parameters for different statuses, a text folder is created in SolidWorks which is shown on the Fig. 2. In that folder changing parameters such as finger length "L" changes the solid model on the program and creates the way to assembling the mechanisms which have the same parts such as this mechanism. After importing text folder to SolidWorks, it is shown like Fig. 4 and will be ready to use as parameters on the program.

Global Variables				
"L"	= 91mm	91mm	☑	fingerlength
"P1"	= 44.5 * "L" / 91	44.5mm	☑	PP
"P2"	= 29.2 * "L" / 91	29.2mm	☑	IP
"P3""	= 17.3 * "L" / 91	17.3mm	☑	DP
"r1"	= 44.54 * "L" / 91	44.54mm	☑	linkage_r1
"r4"	= 7.42 * "r1" / 44.54	7.42mm	☑	linkage_r4
"r3"	= 45.3 * "r1" / 44.54	45.3mm	☑	linkage_r3
"P3"	= 22.5 * "L" / 91	22.5mm	☑	P3
"r2"	= 7.58 * "r1" / 44.54	7.58mm	☑	
"W"	= 15mm	15mm	☑	Width
"teta_f1"	= 5deg	5deg	☑	finger_grip_min
"teta_f2"	= 1deg * 42.5	42.5deg	☑	finger_grip_max
"teta_h1"	= "teta_f2"	42.5mm	☑	holding_grip_min
"teta_h2"	= 80deg	80deg	☑	holding_grip_max
"TP2_f"	= "D1@finger grip" * 1.45	7.25deg	☑	
"EXC"	= 2mm	2mm	☑	
"TP2_c"	= "D1@finger grip" + (1deg * 19.125)	24.13degdeg	☑	

Fig. 4. Equations on solidworks

2.3 Assembling Mechanism

Assembling process was made on SolidWorks. For connections, pins were designed to hold parts together in the virtual environment and in real-life too. Pins are shown below in the Fig. 5. The gap, at the end of the pin, allows it to enter into the part, while the taper at the end prevents the pin from coming back out of the part.

Each individual part of the finger mechanism can be seen in Fig. 6. After all the parts were designed separately in SolidWorks, assemble mates were used to positioning the parts in relation to each other in order to perform simulations of movement and force. For esthetic looking a black part was designed for the glove and placed between the connectors and the glove. M2 bolts were used for all non-pin connections. The weight of the mechanism is determined as 125.12 grams after all the parts have been assembled. Exploded view of the assembly can be seen in the Fig. 7.

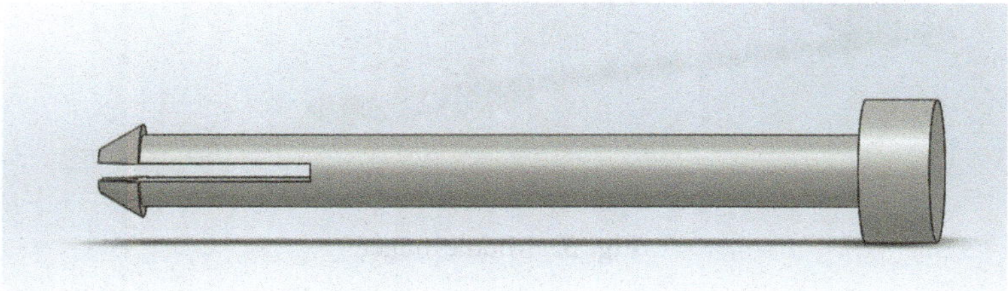

Fig. 5. Pins for assembling mechanism

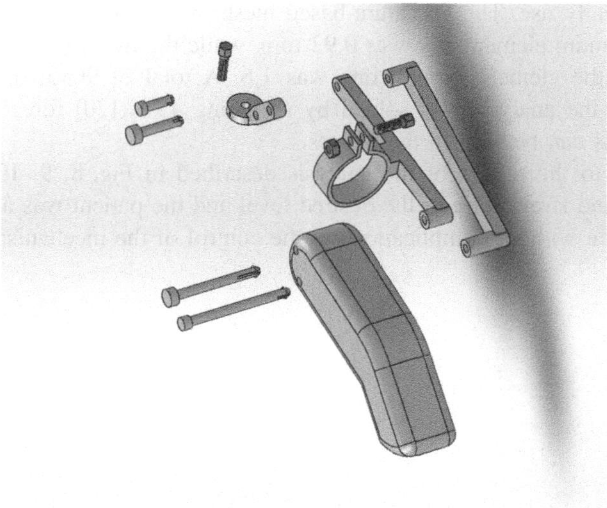

Fig. 6. Exploded view of finger mechanism

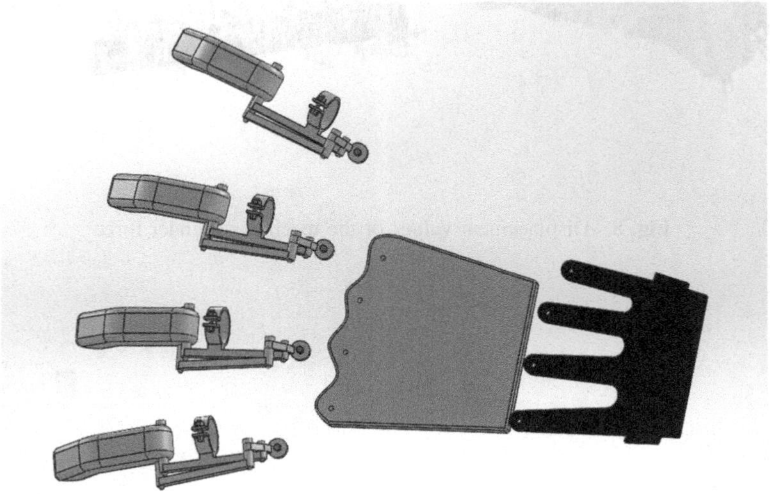

Fig. 7. Exploded view of assembly

3 Analysis and Results

In the analysis, the parts were simplified, bolts and nuts were removed from the analysis piece for easy analysis and relief of mesh parameters. The glove part was fixed to the hand where it was connected, each pin was set as a hinge or support so that the mechanism could be rotated and all connections were bonded so that they should be connected in daily use. The curvature-based mesh method was used during the analysis. The minimum element size was 0.92 mm, while the maximum element size was 4.64 mm and the element growth rate was 1.6. A total of 90856 mesh cells were assigned. And the analysis was solved by assigning 45 N [20] force to each finger. Analysis results can be seen in the figures.

According to the results of the analysis described in Fig. 8, 9, 10 the values of displacement and stress were at the desired level and the patient was able to continue his/her daily life without complications in the control of the mechanism.

Fig. 8. Displacement values of the mechanism under force

Fig. 9. Stress values of the mechanism under a force

Fig. 10. Strain values of the mechanism under a force

4 Conclusion

In this study, finger loss which is the most common upper extremity limb loss was examined. A mechanical finger prosthesis design has been developed and according to the design a final product has been produced. The information about the production and design has been given.

The equation-based parametric design method was used. The reason for using this method was to provide the patient with the fastest possible prosthesis and providing help for each different individual without changing the design but simply implementing the finger sizes of each patient. Afterward, this product has been tested for its suitability for use, and information on future studies has been given. The most important future studies are the addition of a thumb, the addition of the DIP joint and control using myoelectric sensors to maximize human-prosthesis compliance.

References

1. Mio, R., Ccorimanya, L., Flores, K., Salazar, G., Elias, D.: Design of a 3D-printable powered prosthetic hand for transmetacarpal amputees. In: Chang, I., Baca, J., Moreno, H. A., Carrera, I.G., Cardona, M.N. (eds.) Advances in Automation and Robotics Research in Latin America, vol. 13, pp. 83–96. Springer, Heidelberg (2017)
2. Zuniga, J., Katsavelis, D., Peck, J., Stollberg, J., Petrykowski, M., Carson, A., Fernandez, C.: Cyborg beast: low-cost 3D-printed prosthetic hand for children with upper-limb differences. BMC Res. Notes **8**, 1–9 (2015)
3. Össur website. https://www.ossur.com/en-us/prosthetics/arms/i-limb-access. Accessed 15 May 2020

4. Steeper Group website. https://www.steepergroup.com/prosthetics/upper-limb/wrists/espire-hybrid-elbow/
5. Ottobock website. https://www.ottobockus.com/prosthetics/upper-limb-prosthetics/solution-overview/bebionic-hand/. Accessed 15 May 2020
6. Vincent Systems GmbH website. https://vincentsystems.de/en/prosthetics/vincent-evolution-3/. Accessed 15 May 2020
7. Open Bionics website. https://openbionics.com/hero-arm/. Accessed 15 May 2020
8. Uellendahl, J.E., Uellendahl, E.N.: Experience fitting partial hand prostheses with externally powered fingers. Grasping Future Adv. Powered Upper Limb Prosthet., 15–27 (2012)
9. Niet, O.V., Reinders-Messelink, H.A., Bongers, R.M., Bouwsema, H., Van Der Sluis, C.K.: The i-LIMB hand and the DMC plus hand compared: a case report. Prosthet. Orthot. Int. **34**, 216–220 (2010)
10. Mio, R., Bustamante, M., Salazar, G., Elias, D.A.: A 3D-printed prosthetic hand with modular reconfigurable fingers. In: Kecskeméthy, A., Geu Flores, F., Carrera, E., Elias, D. (eds.) Interdisciplinary Applications of Kinematics, Mechanisms and Machine Science, vol. 71, pp. 93–102. Springer, Switzerland (2019)
11. Kate, J., Smit, G., Breedveld, P.: 3D-printed upper limb prostheses: a review. Disabil. Rehabil. Assist. Technol. **12**, 300–314 (2017)
12. Young, K.J., Pierce, J.E., Zuniga, J.M.: Assessment of body-powered 3D printed partial finger prostheses: a case study. 3D Print. Med. **5**(1), 1–8 (2019)
13. Varas, C.E.G., Thompson, T.J.U.: Metric dimensions of the proximal phalanges of the human hand and their relationship to side, position, and asymmetry. HOMO J. Comp. Hum. Biol. **62**, 126–143 (2011)
14. Alexander, B., Kotuik, V.: Proportions of hand segments. Intern. J. Morphol. **28**, 755–758 (2010)
15. ISO International Organization for Standardization Medical devices - Quality management systems (ISO Standard No: 13485:2016) (2016)
16. ISO International Organization for Standardization Prosthetics and Orthotics – Classification and Description of Prosthetic Components- Part 3: Description of Upper limb Prosthetics Components (ISO Standard No: 13405-3:2015) (2015)
17. Ventimiglia, P.: Design of a Human Hand Prosthesis. Bachelor Thesis, Worcester Polytechnic Institute, USA (2012)
18. Lim, D., Georgiou, T., Bhardwaj, A., O'Connell, G.D., Agogino, A.M.: Customization of a 3D printed prosthetic finger using parametric modeling. In: Proceedings of the ASME 2018 International Design Engineering Technical Conferences & Computers and Information in Engineering Conference (IDETC/CIE), Quebec (2018)
19. Graczyk, E.L., Resnik, L., Schiefer, M.A., Schmitt, M.S., Tyler, D.J.: Home use of a neural-connected sensory prosthesis provides the functional and psychosocial experience of having a hand again. Sci. Rep. **8**, 1–17 (2018)
20. Haulin, E.N., Lakis, A.A., Vinet, R.: Optimal synthesis of a planar four-link mechanism used in a hand prosthesis. Mech. Mach. Theory **36**, 1203–1214 (2001)
21. Vinet, R., Lozac'h, Y., Beaudry, N., Drouin, G.: Design methodology for a multifunctional hand prosthesis. J. Rehabil. Res. Dev. **32**, 316–324 (1995)
22. Norton, R.: Design of a Machinery: An Introduction to the Synthesis and Analysis of Mechanism and Machines. McGraw-Hill, New York (1992)

Industrial Applications

Conceptual Design and Fluid Structure Interaction Analysis of a Solar Powered High-Altitude Pseudo-Satellite (HAPS) UAV Wing Model

Mohammad Sakib Hasan[1(✉)], Jelena Svorcan[1], Dragoljub Tanovic[1], Gökçen Baş[2], and Numan M. Durakbasa[2]

[1] Faculty of Mechanical Engineering, University of Belgrade, Belgrade, Serbia
sakibhasan89@yahoo.com,
{jsvorcan, dtanovic}@mas.bg.ac.rs
[2] Vienna University of Technology, Vienna, Austria
{goekcen.bas, numan.durakbasa}@tuwien.ac.at

Abstract. High altitude platforms or Pseudo-satellites (HAPS) are unmanned aerial vehicles that can fly above 17 km from sea level. It can take advantage of weak stratospheric winds and solar energy to operate without interfering with current commercial aviation and with enough endurance to provide long-term services as satellites do. The technological innovations and the growing urgency to expand the availability of broadband led to the development of HAPS. Besides that, Earth observation, positioning, astronomy, and science are the main applications of High altitude platforms, or Pseudo-satellites (HAPS). In this paper, the conceptual design of the novel HAPS UAV based on the initial requirements regarding cruising height of 20 km, the cruising velocity of 25 m/s, and payload of 15 kg were performed and described. The HAPS wing was defined and aerodynamically studied in detail. The computed nominal load was used as input parameter for structural analysis of the wing's inner structure comprising outer shell, main spars, and ribs made of composite and plastic materials. All computations were performed using commercial software package ANSYS. The obtained results are discussed and graphically presented by computed stress and deformation fields.

Keywords: HAPS · UAV · Wing · Structural analysis · ANSYS

1 Introduction

High Altitude Pseudo–Satellite is a type of unmanned aerial vehicle (UAV) that operates at stratospheric altitude (17–25 km altitude) for a very long period. It may be capable of remaining in the sky for an entire year. For their ability to fly for an extended period at a higher altitude, these types of drones are described as high altitude long endurance (HALE) UAV. After the first flight of Qinetiq's Zephyr 7 in the stratosphere for two weeks, HALE UAVs have received much attention and are being globally developed for use as low earth orbit satellite with the advantage of being less

N. M. Durakbasa and M. G. Gençyılmaz (Eds.): ISPR 2020, LNME, pp. 93–105, 2021.
https://doi.org/10.1007/978-3-030-62784-3_8

expensive, closer to the sea level and being able to execute tasks that offer greater flexibility [1, 2]. Besides that, these types of UAVs also have the potential to be used in the delivery of communications networks including 5G, as well as provide other services such as surveillance, weather research, and forecast, environmental monitoring, etc. At present, it is only possible to reach ambitious goals like extended endurance or continuous flight by using solar cells. Photovoltaic cells would collect the energy from the sun during the day time which would power the motor and the electronics and also store enough energy for flying in darkness or under clouds [3].

Several types of high altitude solar UAVs have already been designed. NASA started its Environmental Research Aircraft and Sensor Technology (ERAST) project at the end of 1993. AeroVironment, Inc. developed Pathfinder, Centurion, and Helios drones under NASA's ERAST program which were powered by solar cells [4]. In July 1997, Pathfinder achieved a world altitude record of 21.8 km for propeller-driven aircraft [5]. Three HALE (High Altitude Long Endurance) configurations are being evaluated under the EC Project "CAPECON: Civil UAV Applications & Economic Effectivity of Potential Configuration Solutions" [6]. After buying the Zephyr system from the QinetiQ in 2013, Airbus re-flown the Zephyr as part of the High Altitude Pseudo-Satellite (HAPS). On 25 July 2018, Zephyr S achieved a flight endurance record for flying more than 14 days without refuelling [7, 8]. In less than two years BAE SYSTEM has designed a Persistent High Altitude Solar Aircraft called PHASA-35 collaborating with Prismatic Ltd. which made his first flight in 2020. Phasa-35 has the potential to fly continuously for a year [8].

HAPS UAVs usually fly for a long period of time. To be able to fly for months, UAVs need to have wings with high-aspect ratios, but at the same time high-aspect ratio wings also easily generate deflection and deformation during the flight. Fluid Structure Interaction (FSI) analysis helps to predict the interaction between the aerodynamic loads generated from the fluid flow and the structural components of the wing [9]. Aeroelastic analysis of the wing based only on the rigid structure would lead to the erroneous results, therefore, it is recommended to use an FSI analysis that uses the coupling effect of both aerodynamic load and structural deformation. Aeroelastic properties of a long-endurance UAV (EAV-2) were simulated by the FSI method in [10]. Authors in reference [11] investigated the influence of different materials on wing structure for high velocity environment by FSI analysis method. The performance of wing structure integrated on the HAPS require Quality assessment and control in order to obtain high performance and increased reliability. The concept of modelling and testing together with implementation of high precision metrology is an inevitable strategy of every next-generation device [12]. Therefore, this study focuses on the precise conceptual design phase with detailed analysis of the geometries. The novel technologies with better capabilities face challenges such as high production costs. This study takes into consideration the challenges of design and production by aiming to obtain long endurance and lower operating cost device modelled with high precision [13]. During the research and development phase, the international standards provide the guidelines. The innovative device is designed considering the requirements of quality objectives and planning to achieve them based on the international standards [14].

In this paper, a one-way coupling method for fluid structure interaction analysis is presented to discuss the deformation and stress caused by the aerodynamic forces on the wing structure. The main characteristics of the wing were obtained from the conceptual design phase where 24 wings were chosen by varying airfoils and taper ratio to analyse in XFLR5. Selected wing was then modelled by using CAD program CATIA V5 which was then moved to ANSYS-Workbench to simulate the FSI analysis to calculate the stress and deformation caused by the aerodynamic pressure. Composite and plastic materials were used in structural analysis as it has higher modulus of elasticity and lower weight.

Section 2 gives the competitive study of existing solar-powered aircraft from which design weight of the UAV was obtained. Section 3 deals with wing loading which is a very important factor in the conceptual design phase. Section 4 shows the detailed analysis of the airfoils and wings geometry. Wing design in CATIA and FSI analysis is presented in Sect. 5. Conclusion of the paper are discussed in the Sect. 6.

2 Competitive Study

Several solar powered manned and unmanned aircraft have flown in the last three decades. The conceptual design phase began with the competitive study to get an insight into existing HALE drones and also to get an idea if there is a possibility for improvement. The competitive study of historical data from existing airplanes helps to predict the weight of the aircraft. This method is known as the statistical weight estimation method [15]. After the competitive study, the authors agreed to design a drone with a total weight of 150 kg. Initial values of geometric features such as aspect ratio, wing area, and taper ratio can also be derived from the realized UAV structures. Table 1 shows a few high-altitude long endurance solar-powered aircraft that have been flown to this date.

Table 1. Competitive study of solar powered aircraft

	Wingspan [m]	Aspect ratio	Wing surface [m^2]	Total mass [kg]
Pathfinder	30.00	12.40	72.58	252
Pathfinder plus	36.30	15.13	87.12	315
Centurion	61.80	25.80	148.32	861
Helios	75.30	30.40	186.60	930
Zephyr S	22.50	11.60	43.64	50
Zephyr T	33.00	NA	NA	145
PHASA-35	35.00	NA	NA	150

3 Wing Loading Estimation

Wing loading is one of the very first tasks for designing an aircraft. It is defined as the ratio between the total weight of the aircraft and the reference area of the wing. This parameter has effects on some very important characteristics like the stalling speed, rate of climb, runway distance, drag coefficient, etc. Stalling speed is the minimum speed at which the lift coefficient is maximum. Equation 3 shows the relation between the maximum lift coefficient which been derived from the steady level flight equation [16] :

$$W = L \tag{1}$$

$$W = mg = C_L \frac{\rho}{2} S v^2 \tag{2}$$

$$C_{Lmax} \cdot q = \frac{W}{S} \tag{3}$$

Where C_{Lmax} is the maximum lift coefficient, q is the dynamic pressure and *W/S* is the wing loading.

The great solar flight diagram with the weight prediction model of Rizzo been shown in [3]. There Rizzo uses the following equation to predict the wing loading of the aircraft:

$$W/S = 63.24 W^{-0.524} AR^{0.992} \tag{4}$$

As the weight of the aircraft (1487 N) has already been estimated from the historical data in the previous section, Eq. 4 will be useful for finding the wing loading (W/S) for various aspect ratios. Table 2 shows wing loading (W/S) and surface area (S) of the wing for different aspect ratios.

Table 2. Wing loading and surface area estimation for different aspect ratio

Aspect ratio, AR	Wing loading, W/S [N/m^2]	Surface area. S [m^2]
20	27.02	54.46
21	28.36	51.88
22	29.70	49.54
23	31.04	47.41
24	32.38	45.45
25	33.72	43.64
26	35.05	41.98

NASA's technical paper about design consideration of Solar-powered aircraft stated that the wing loading should be as low as possible (15 [N/m^2]) [17]. Before concluding on the final aspect ratio and wing loading, the authors decided to investigate wing geometry for different airfoils by varying aspect ratio and taper ratio of the wing.

4 Airfoil and Wing Geometry Selection

4.1 Airfoil Selection

Airfoil selection is very important as it affects the entire aerodynamics of the aircraft. Three low Reynold's number airfoils (Eppler 387, S1210, and s1223) were examined for five different Reynolds numbers ranging from 150000 to 300000 with an increment of 50000 by using XFLR5 [18]. Figure 1 shows all three airfoils polars for different Reynolds no.

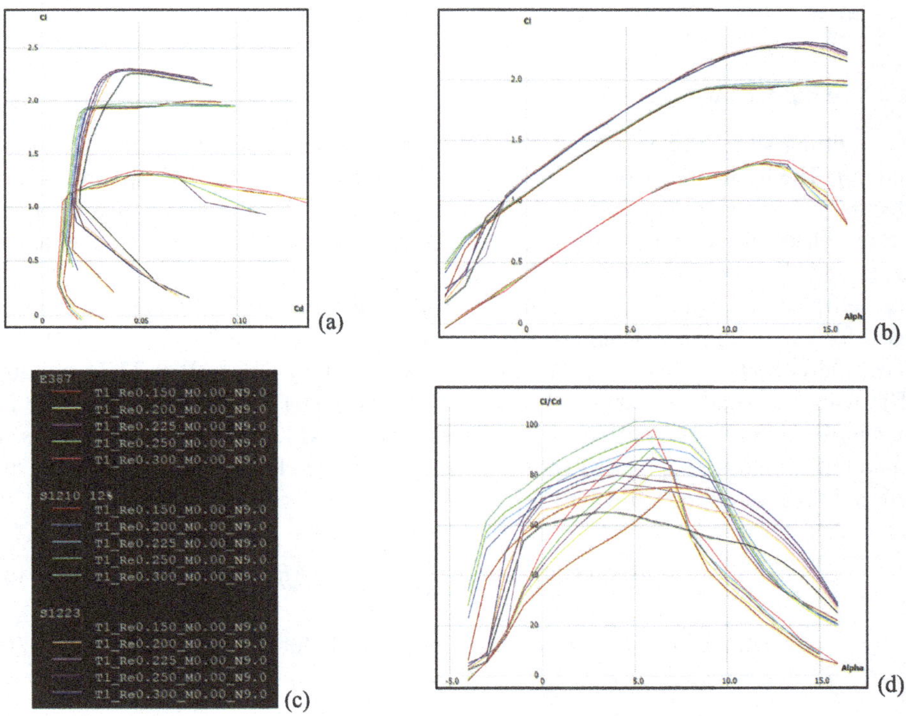

Fig. 1. Airfoil polars (a). Cl – Cd, (b) Cl – alfa, (c) Legend of Reynolds no, (d) Cl/Cd – alfa

Reynolds no can be derived from the following equation. Where υ is the velocity, c is the mean aerodynamic chord and ν is kinematic viscosity at 20 km from sea level.

$$\text{Reynolds no, } Re = \frac{\upsilon c}{\nu}$$

Table 3 shows all the important parameters of the airfoils. The design lift coefficient of the HALE drones varies from 0.8 to 1.75 [19].

Table 3. Airfoil Properties at Cruise altitude

Re = 250000								
Airfoil	Max thickness	C_{lmax}	$(C_l/C_d)_{max}$	Stall angle (α)	C_{di} @C_{lmax}	α @$(C_l/C_d)_{max}$	Cd @C_l = 1	α @C_l = 1
e387	9.10%	1.32	91	12	0.052	6	0.0110	5.58
s1210	12%	1.96	95	14	0.068	6	0.0134	−0.41
s1223	12.10%	2.29	80	14	0.04	4	0.0170	−1.08

4.2 Wing Geometry

Main lifting surface of an aircraft is its wing as it produces more than 90% lift for an aircraft. The basic dimensions of the wing (span and chord at root and tip) are usually determined by wing area, taper ratio, and aspect ratio as governing variables at the design phase of the wing. Following formulations were used to calculate the required wingspan (b), root chord (C_r), tip chord (C_t) and mean aerodynamic chord of a simple tapered planform wing [15]. Airfoil e387 was used for the first 8 wings, s1210 was used for wings from 9–16 and lastly s1223 was used for the last 8 wings (17–24). The taper ratio of 0.6, 0.8 and 1 was used for each wing aspect ratio (AR = 22,24 and 26). The wingspan of all 24 wings is between 33.01 and 33.04. Table 4 depicts all the 24 wings that have been considered for 3D analysis in XFLR5 and their properties at operating altitudes. Program XFLR5 considered the vortex lattice method to predict the lift and drag values of the wing [18].

$$\text{Aspect ratio}, AR = \frac{b^2}{S} \Rightarrow b = \sqrt{AR.S} \tag{5}$$

$$\text{Taper ratio, } \lambda = \textit{tip chord}/\textit{root chord};\ [\lambda = 1, \text{ for rectangular wing}] \tag{6}$$

$$\text{Root chord (RC)}, C_r = \frac{2b}{(1+\lambda)AR} \tag{7}$$

$$\text{Tip chord (TC)}, C_t = \lambda C_r \tag{8}$$

Table 4. Basic dimensions of the wings and their properties at operating altitude

Wing	AR	RC [m]	TC [m]	MAC [m]	C_{lmax}	$\alpha @ C_l = 1$	C_d @$C_l = 1$	$(L/D)_{max}$
Wing 1	22	1.88	1.13	1.53	1.2	6.8	0.0266	38.9
Wing 2	22	1.67	1.33	1.51	1.19	6.99	0.0275	37.9
Wing 3	22	1.50	1.50	1.50	1.18	7	0.0279	37.6
Wing 4	24	1.72	1.03	1.40	1.21	6.78	0.0261	39
Wing 5	24	1.53	1.22	1.38	1.2	6.82	0.0265	38.4
Wing 6	24	1.38	1.38	1.37	1.19	6.93	0.0272	37.78
Wing 7	26	1.41	1.13	1.28	1.22	6.75	0.0262	38.35
Wing 8	26	1.27	1.27	1.27	1.2	6.8	0.0268	37.63
Wing 9	22	1.88	1.13	1.53	1.9	−0.4	0.0295	34.5
Wing 10	22	1.67	1.33	1.51	1.88	−0.3	0.0303	33.6
Wing 11	**22**	**1.50**	**1.50**	**1.50**	**1.86**	**−0.25**	**0.0312**	**32.7**
Wing 12	24	1.72	1.03	1.40	1.91	−0.48	0.0288	34.9
Wing 13	24	1.53	1.22	1.38	1.89	−0.4	0.0296	34
Wing 14	24	1.38	1.38	1.37	1.88	−0.32	0.0303	33.3
Wing 15	26	1.41	1.13	1.28	1.91	−0.45	0.0291	34.3
Wing 16	26	1.27	1.27	1.27	1.89	−0.4	0.0298	33.7
Wing 17	22	1.88	1.13	1.53	2.205	−2.7	0.035	30
Wing 18	22	1.67	1.33	1.51	2.19	−2.6	0.036	29.2
Wing 19	22	1.50	1.50	1.50	2.17	−2.5	0.0361	28.4
Wing 20	24	1.72	1.03	1.40	2.23	−2.8	0.0348	31
Wing 21	24	1.53	1.22	1.38	2.21	−2.7	0.0356	30.1
Wing 22	24	1.38	1.38	1.37	2.19	−2.65	0.0365	29.1
Wing 23	26	1.41	1.13	1.28	2.23	−2.7	0.0355	30.8
Wing 24	26	1.27	1.27	1.27	2.221	−2.7	0.0363	29.7

$$\text{Mean aerodynamic chord}, MAC = (2/3)C_r\left(\frac{1+\lambda+\lambda^2}{1+\lambda}\right) \tag{9}$$

After both airfoils and wings analysis in XFLR5, a wing with maximum glide ratio $(L/D)_{max}$ has been selected. It has already been established that a wing with a taper ratio similar to the elliptical wing (taper ratio = 0.57) gives the best glide ratio and also a minimum induced drag ratio. On the other hand, they have bad stall characteristics and harder to manufacture. As a result, a rectangular wing with an aspect ratio of 22 and an airfoil s1210 was selected for the conceptual design phase which is Wing 11. Even

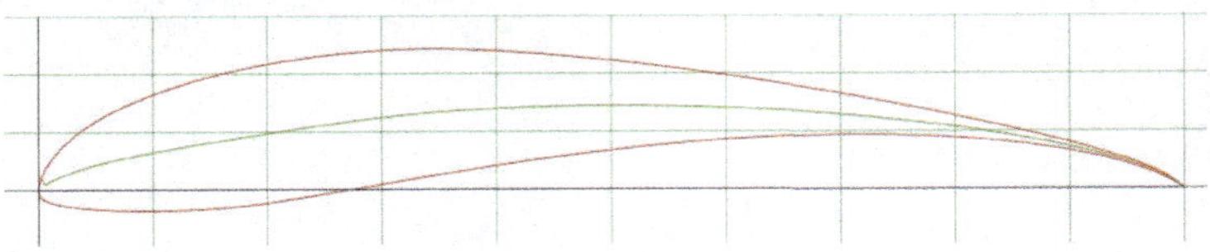

Fig. 2. Airfoil s1210 [20]

though wing with airfoil e387 gives better glide ratio than the s1210, its design angle-of-attack (AoA) corresponding to the design lift coefficient is dangerously close to its critical AoA. On the other hand, airfoil s1223 has very high lift coefficient but it comes with poor glide ratio (Fig. 2).

Specification of **wing 11**:
Wingspan, b = 33.01 [m]; Taper ratio, λ = 1; Chord, C = 1.50, Airfoil = s1210
Value of design lift coefficient (C_l = 1.08) at operating altitude (20 km) for cruising speed (25 m/s) and surface area (S = 49.50 m^2) was derived from the horizontal flight equation (Eq. 2).

5 Fluid Structure Interaction Analysis

5.1 Wing Model

FSI procedure begins with the modelling of the wing geometry using CATIA V5. Wing is constructed with upper surface, lower surface, spars and ribs. Spars are situated at 0.25C (Chord length = C) and 0.65C from the leading edge of the wing. Spacing between the ribs are 0.5C which is 750 [mm]. Figure 3 shows the model of the wing.

Grid was generated by using ANSYS Meshing. A semi-cylinder domain was created around the wing where radius of the arc was 2b from the leading edge of the wing. Outlet of the wing is at 100 m from the leading edge. It is recommended to create outer boundary as far as possible. Inflation of 20 layers with growth rate of 1.2 was

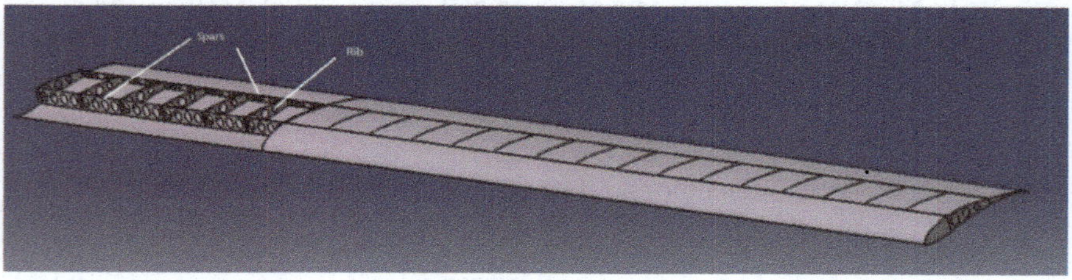

Fig. 3. Wing model

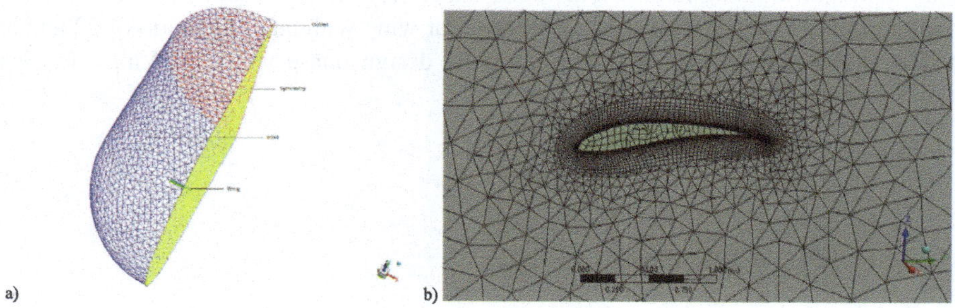

Fig. 4. (a) 3D domain grid model, (b) meshing around wing

used to locally refine the mesh around the wing surface. Different edge sizing was used for root and tip airfoils and for the leading and trailing edges of the wing. Eventually, a mesh was created with 944000 elements. Although this is a medium fine mesh but it can be considered adequate for conceptual design stage. Figure 4 shows the domain grid model and details of the mesh around the wing.

5.2 Simulation and Analysis of the Wing Flow Field

After checking the grid, pressure-based steady state solver is selected as it is more suitable for incompressible flow. *k-ω* SST and transition SST turbulence models were used to check which model gives more realistic results. Both these models have been well proven and designed for aerodynamic applications with the differences that transition SST model solves two additional equations compared to *k-ω* SST and that it assumes partly laminar flow that transforms to transitional and ultimately turbulent. It is well suited for small speed flows around laminar airfoils. SIMPLE pressure velocity coupling method was performed for *k-ω* SST whereas SIMPLEC method was implemented for transition SST model and both the simulations were carried out with 2nd order of accuracy.

Physical properties of fluid and boundary conditions are considered at operating altitude (h = 20 km). Simulations were performed to get the design lift coefficient (C_l = 1.08) for given angle of attack (AoA) as this C_l value corresponds to the desired

Table 5. Aerodynamic coefficients of wing 11 at operating conditions from ANSYS fluent and XFLR5 for different AOA

Program	Model	AOA [°]	C_l	C_d	C_l/C_d	Force [N]	Mass [kg]
ANSYS	Transition SST	2.8	1.08	0.043	25.11	1480	150.86
	Transition SST	3.3	1.13	0.045	24.74	1538	156.75
	k-ω SST	3.2	1.08	0.050	21.60	1482	151.07
	k-ω SST	2.3	0.999	0.045	22.20	1360	138.60
XFLR5	Ring Vortex (VLM2)	2.8	1.288	0.043	29.95	1471	150.00

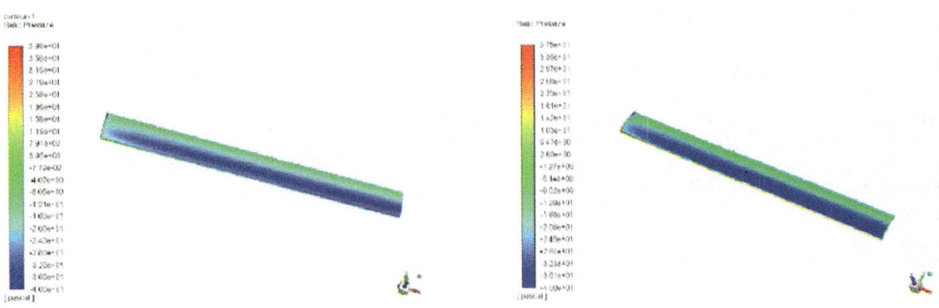

(a) (b)

Fig. 5. Pressure contours of (a) SST k-w and (b) transition SST MODEL

UAV mass of 150 kg. The inlet velocity was kept at 25 m/s. Both models were analysed for 2000 iterations.

After the simulations, the pressure contours were obtained from the post-processing stage. Figure 5 shows the pressure contours on the wing surface for *k-ω* SST and transition SST model. Table 6 shows the C_l values for different AoA and different turbulent models.

As demonstrated in Fig. 5, pressure distributions along the upper wing surface obtained by the two turbulence models are similar. Blue zones of negative gauge pressure take up most of the wing leading part, with small losses appearing near the wing tips. Due to the high aspect ratio of the wing, induced drag is small and span-wise pressure distribution is almost uniform.

On the other hand, even though differences in pressure distributions are almost imperceptible, the values of lift-to-drag ratio given in Table 5 prove that the assumption of partially laminar flow improves computed aerodynamic characteristics by nearly 15%. The high value of $C_l/C_d > 25$ obtained by transition SST is most satisfactory since it reduces the necessary thrust (and consequently power) that must be generated (and spent) by the electrically-driven propellers. It is also quite close to the value obtained by XFLR5 that uses the simplest numerical model, assumes most flow simplifications and does not take into account all the viscous drag that appears in reality.

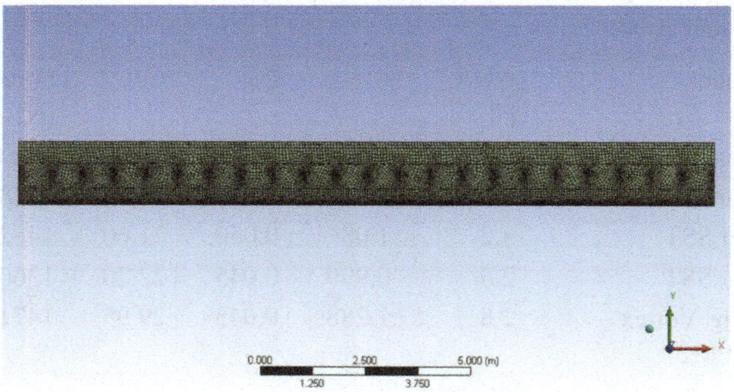

Fig. 6. Wing structural mesh

Table 6. Properties of the materials used in wing structure

Material	Young's modulus [Pa]	Poisson's ratio	Density [kg/m^3]	Shear modulus [Pa]
Epoxy Carbon UD	2.09E +11	0.33	1540	5.50E+09
PLA	3.50E+09	0.33	1300	1.32E+09

5.3 Structural Analysis

The analysis of the wing structure was performed by finite element method in ANSYS Structural. The assumed geometries of wing skin, spars and ribs are discretised into nearly 50000 plate-type elements illustrated in Fig. 6.

Used materials are unidirectional carbon-epoxy prepreg and polylactide (PLA). Their mechanical properties along the main load bearing direction are given in Table 6. Table 7 provides details on the thicknesses, materials and masses of individual wing component groups. One of the important factors while designing a HAPS UAV is the structural weight of the aircraft. It is recommended that the weight of the structures be about 35% of the total weight of the aircraft [3, 21]. In our case, the structural weight of the wing of 60 kg is about 40% of the total weight of the aircraft. It is considered adequate for the initial design phase of the UAV.

Wing structure is clamped on one end (lying in symmetry plane *yz*). Non-uniform pressure distribution, obtained from the fluid flow computation, is applied along the upper and lower surfaces of the wing, while gravitational (mass) loads are applied to the complete wing structure.

Computed displacement and stress fields along the wing surface are illustrated in Fig. 7. As expected, the greatest displacement appears at wing tips and it is less than 3 m. Given the great half-span of the wing (nearly 16 m), and its high aspect ratio, this is an acceptable value that results in the greatest total deformation smaller than 20%. Shape of the deformed wing can be additionally controlled by introducing dihedral angles to wing's outer parts which will be the topic of future studies.

The greatest stress values appear near the wing root in the zone between the two spars. The fact that the obtained stress levels are significantly lower than the stress limits of carbon plies validates the assumed wing structure (that weighs approximately 60 kg in total) and proves that the structure will remain operational even if exposed to

Table 7. Weight of different elements of wing structure

Element	No	Material	Thickness [mm]	Mass[kg]
Spar	2	Epoxy Carbon UD	1.5	8.9
Skin	2	Epoxy Carbon UD	0.28	20.5
Ribs	23	PLA	0.6	0.713
Total				30.11

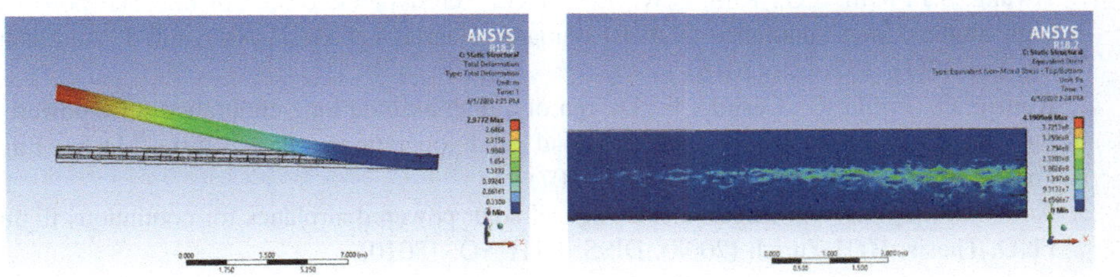

(a) (b)

Fig. 7. Displacement and stress along the wing

greater load factors (that might appear while performing different flight manoeuvres). Some future structural analyses might involve optimization studies of structural parameters to further decrease the mass of the wing.

6 Conclusions

- The paper covers several initial but crucial phases of conceptual design regarding the starting estimations of mass, aerodynamic performances and inner structure of high-altitude pseudo-satellite (HAPS) UAV wing.
- The assessment of wing mass is performed for a very specific aircraft category using a relatively novel relation provided by Rizzo [3].
- Aerodynamic computations are validated through comparison of two numerical methods that differ in starting assumptions and complexity.
- Computed aerodynamic pressures are used as an input load for half-wing structural analysis. The assumed wing geometry and applied materials satisfy the imposed constraints regarding the total mass and structural safety.
- Near future work on the presented HAPS UAV will include two-way fluid-structure interaction studies to better understand the behaviour of the assumed thin wing, optimization of its geometric features to reduce negative aeroelastic effects, inclusion of dihedral angles to outer wing segments, optimization of structural components, selection of the remaining aircraft components, etc.
- Given that the final goal of the proposed concept is to produce and make operational HAPS UAV it is necessarily to fundamentally change or further develop the approach to design to ensure that the final product meets the functional requirements of its application. This difficult task can only be achieved through the development of improved and contemporary techniques in both design and production phases [22].

Acknowledgment. This research work was supported by the Ministry of Education, Science, and Technological Development of Republic of Serbia through contract no. 451-03-68/2020-14/200105.

References

1. Hwang, S.J., Kim, S.G., Kim, C.W., Lee, Y.G.: Aerodynamic design of the solar-powered high altitude long endurance (HALE) unmanned aerial vehicle (UAV). Int. J. Aeronaut. Sp. Sci. **17**(1), 132–138 (2016)
2. Romeo, G., Frulla, G., Cestino, E.: Design of a high-altitude long-endurance solar-powered unmanned air vehicle for multi-payload and operations. Proc. Inst. Mech. Eng. Part J. Aerosp. Eng. **221**(2), 199–216 (2007)
3. Noth, A., Siegwart, R., Engel, W.: Design of solar powered airplanes for continuous flight. Ph.D. Thesis, ETH Zurich (2008). DISS. ETH NO. 18010
4. Gibbs, Y.:NASA Dryden Fact Sheet - Helios Prototype (2015)

5. Flittie, K., Curtin, B.: Pathfinder solar-powered aircraft flight performance. In: 23rd Atmospheric Flight Mechanics Conference, p. 4446 (1998)
6. Cestino, E.: Design of solar high altitude long endurance aircraft for multi payload & operations. Aerosp. Sci. Technol. **10**(6), 541–550 (2006)
7. Zephyr - UAV – Airbus (2020). https://www.airbus.com/defence/uav/zephyr.html
8. Zephyr S set to break aircraft world endurance record - Defence – Airbus (2018). https://www.airbus.com/newsroom/press-releases/en/2018/07/Zephyr-S-set-to-break-aircraft-world-endurance-record.html
9. Ansys MFX. Advanced FSI Technology: ANSYS MFX Multi-fieldTM Solver, pp. 010–4456 (2006)
10. Son, S.H., Choi, B.L., Won, W.J., Lee, Y.G., Kim, C.W., Choi, D.H.: Wing design optimization for a long-endurance UAV using FSI analysis and the kriging method. Int. J. Aeronaut. Sp. Sci. **17**(3), 423–431 (2016). https://doi.org/10.5139/IJASS.2016.17.3.423
11. Zhang, L., Sun, C.: Simulation analysis of fluid-structure interaction of high velocity environment influence on aircraft wing materials under different mach numbers. Sensors **18** (4) (2018). https://doi.org/10.3390/s18041248
12. Durakbasa, M.N., Bas, G., Bauer, J.M., Poszvek, G.: Trends in precision manufacturing based on intelligent design and advanced metrology. Key Eng. Mater. **581**, 417–422 (2014)
13. Durakbasa, M.N., Bas, G.: From nano- to pico- to femto-: future challenges for cost-oriented manufacturing. Elsevier Sci. Direct IFAC-PapersOnLine **52**(25), S382–S386 (2019)
14. EN 9100:2018 Quality Management Systems – Requirements for Aviation, Space and Defence Organizations (2018)
15. Gudmundsson, S.: General Aviation Aircraft Design: Applied Methods and Procedures. Butterworth Heinemann, Oxford (2013)
16. Raymer, D.P., Design, A.: A Conceptual Approach. American Institute of Aeronautics and Astronautics, Washington, D.C. (1989)
17. Phillips, W.H.: Some design considerations for solar-powered aircraft (1980)
18. XFLR5. http://www.xflr5.tech/xflr5.htm. Accessed 30 Apr 2020
19. Mermer, E., Serkan, O.: Conceptual design of a hybrid (Turbofan/Solar) powered HALE UAV. Master Thesis, Middle East Technical University, Ankara, Turkey (2016)
20. S1210 12% (s1210-il) (2020). http://airfoiltools.com/airfoil/details?airfoil=s1210-il
21. Alsahlani, A.A.M.: Design of a Swept-Wing High-Altitude Long-Endurance Unmanned Air Vehicle (HALE UAV). Ph.D. Thesis, University of Salford Manchester (2017)
22. Durakbasa, N.M., Bauer, J.M., Kräuter, L., Bas, G., Poszvek, G.: Intelligent integrated management and advanced metrology for quality toward the factory of the future. In: AMP 2018, LNME, pp. 98–108. Springer, Heidelberg (2018)

Design and Development of Drum Granulator

Ozge Tufan[1], Pınar Demircioğlu[2(✉)], İsmail Böğrekçi[2], and Utku Köse[1]

[1] EYS Metal R&D Center, Aydin, Turkey
{otanrikulu,ukose}@e-y-s.com
[2] Mechanical Engineering Department, Aydın Adnan Menderes University, Aydin, Turkey
{pinar.demircioglu,ibogrekci}@adu.edu.tr

Abstract. The aim of this study is to design and develop of drum granulator used in compost fertilizer production sectors. The production capacity of the designed drum granulator is approximately 10 tons/h and the drum speed is 11 rpm. With the designed nozzle system of the drum granulator, humidification process is carried out by adding water to the material fed to the granulator. The total length of the drum is 8 m and the drum diameter is 2 m. The drum is gear driven and has a tilt angle of 2–3°. 2–4 mm granules with a moisture content of about 60–65% (for fertilizer) are obtained in the drum granulator. The trial process of the granulator was completed and R&D studies were carried out. In development studies; PVC conveyor belt assembly preventing the material from sticking to the drum surface, drum inlet design preventing overflow problems at 2.5° drum inclination angle, outlet design preventing the material to be clogged when leaving the drum. Furthermore, a system adjusting the amount of water given according to the weight of the material fed has been made to reduce the rejection rate. This system has made a significant contribution to the granule production process.

Keywords: Granule · Granulator design · Tumbling granulator · Wet granulation

1 Introduction

Granulation is a size enlargement process that commonly used in many industry such as agrochemical, food, detergent, pharmaceutical, manure management, mineral processing and chemical industries [1]. The granulation is particle design and the properties obtained by granulation depend on the physical and chemical properties of the binder [2].

The granulation method provides to get the homogen mixture without any material segregation during the process, improve the material structure, flow ability and electrostatic properties of powders [3]. There are many different types of equipment which enable agglomeration of powder particles to form larger masses and the process when carried out a mixing device is usually described as granulation [4]. There are many factors affecting granule quality such as size distribution, porosity, attritional ability, dispersibility, dissolution rate, strength, shape, color and homogeneity [4].

N. M. Durakbasa and M. G. Gençyılmaz (Eds.): ISPR 2020, LNME, pp. 106–114, 2021.
https://doi.org/10.1007/978-3-030-62784-3_9

There are many techniques are enabling to use to form a granule such as formation of solid bridges, binding, sintering, chemical reaction, crystallization or deposition of colloidal particles. To achieve desired properties of the granule in the process for the granulation of the main particles depends on control of the adhesional forces between particles. Also, if the distance between particles is reduced, van der waals forces and electrostatic forces can interact to create bond between particles. These forces are used in dry granulation methods. In wet granulation method, there are cohesive forces that creates the liquid bridges between particles. Also, electric forces contribute to keep particles together [5]. Porosity, binder content and composition are independent properties of granules and significantly affect the granule growth process and excludes studies which assume that the granule size is the only independent granule property that controls coalescence [6].

Drum granulators are widely used in fertilizer and metallurgical industries to get high production capacity applications. In drum granulator designs, many parameters such as drum length, drum diameter, inclination angle, drum speed are important in determining the degree of agglomeration and the physical properties of the granule to be obtained and some of these parameters affect the total time (also called retention time) spent by the input materials in the drum. In the drum granulator, the material coming into the granulator (including reworked material) is fed to the inlet of the drum. At the inlet of the drum, due to natural segregation, the powdered particles condense at the bottom of the drum, larger particles condense on the surface of the drum [7]. Therefore, during the study, every parameter in drum granulator design was examined and studies were carried out to obtain maximum granule formation [8].

2 Material and Method

2.1 Material

St37 steel was generally used in drum granulator design. The fact that St37 structural steel is cheap and easily accessible and its mechanical properties are sufficient is a valid reason for its use in chassis manufacturing (Tensile strength of St37: 360 MPa, Yield strength of St37: 240 MPa).

Drum was produced by cylindrical bending of sheet metal. The length of the drum is eight meters and it is very difficult to make an eight meter long cylindrical bend. Therefore, four sheets of two meters in length are cylindrically bent and welded together. Technical drawing of the drum part is shown in Fig. 1. Sectional view of the drum layer (a) and Final state of the drum design (b) are given in Fig. 2.

2.2 Method

Autodesk Inventor software was used in the design and 3D modeling process of the granulator and ANSYS software was used in the analysis process. The method section is briefly explained as granulator design, production methods, bearing calculations, strength analysis, gearbox selection and irrigation system design. In finite element

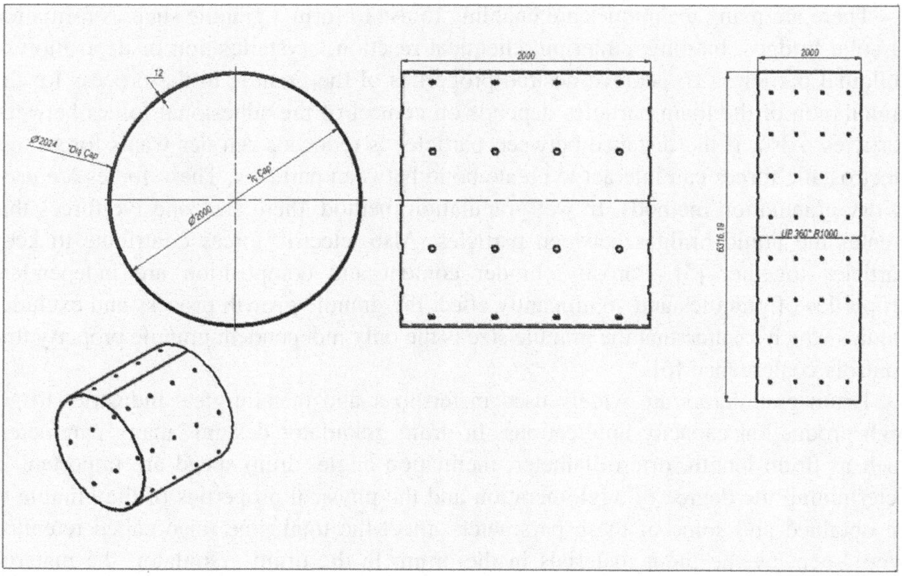

Fig. 1. Technical drawing of the drum part [8]

(a) (b)

Fig. 2. Sectional view of the drum layer (a) and final state of the drum design (b) [8]

analysis, there are strength analyses for chassis, drum and resistance analysis of the drum against torsion.

Static structural analysis of chassis is given in Fig. 3. The support point A shown in the image is fixed. Bearing point shown in the visual B is considered as a moving bearing, adopted free on X axis but fixed on the Y and Z axes. In Fig. 4, placement of forces on the geometry is shown.

Three different forces were applied to the chassis; one is the sum of the weights of the motor, reducer and pinion gear in the drive section. The other two forces are calculated from the weight of the drum delivered to the chassis via the carrier wheels and the weight of the material contained in the drum. Machine vibration was also effective in the calculation of force applied to the chassis.

Assuming that the vibration on the machine is not more than 5 m/s^2, masses on chassis multiplied by 15 instead of the gravity acceleration (9.82 m/s^2).

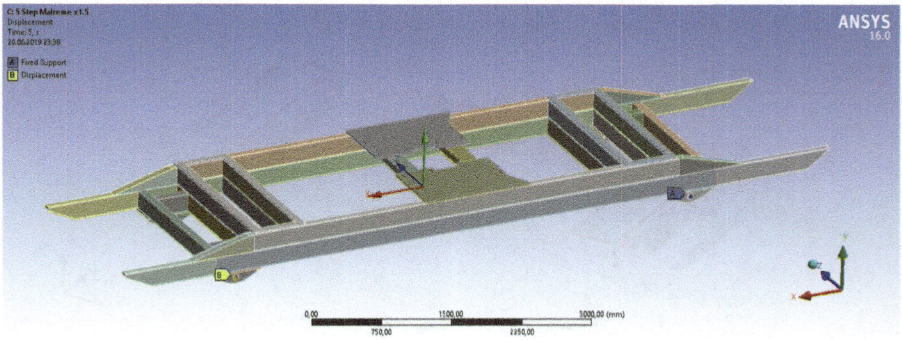

Fig. 3. Fixing points [8]

For example; the reason that force number three is 11745 N, weight of the drive group is 783 kg.

F = m.a
F = 783 × 9.81 + 783 × 5.
F = 11796.23 N.

Approximate total acceleration;

atotal = 9.81 + 5 = 14.81 m/s2 ≈ 15 m/s^2.
F3 ≈ 783 × 15 = 11745 N.

The reason that force number one and two are 65250 N; drum weight is 8000 kg and instant total weight of the compost and water in the drum is 700 kg.

Mtotal = 8000 + 700 = 8700 kg.
atotal = 15 m/s^2.
Ftotal = 8700 × 15 = 130500 N.

Total weight is equally distributed to both sides;
F1 = F2 = 130500/2 = 65250 N.

In Fig. 5, surface C is actually the surface on which the gear is located. Therefore, the torque from the gear unit is applied to the drum via this surface. The maximum torque of the gear unit is 1794 Nm. Also, the pinion gear 20 teeth, drum gear 140 teeth are designed. This gives a gear ratio of 140/20 = 7.

Therefore, the torque applied to the drum from the surface "C" calculated as;

$$M_{gearbox} \times i_{total} = M_{bowl} 1794 \times 7 = 12558 \text{ Nm}$$

A: Fixed support, B: Displacement, C: Earth Gravity, D: Force, E: Force 2, F: Force 3, G: Force 4, H: Acceleration are loads shown in Fig. 5. The reason why the forces on the analysis are 2000 N; It is assumed that 800 kg compost was in the drum. So, the mass of 800 kg was simulated as 8000 N force. The force of 8000 N is evenly distributed to the four part drum.

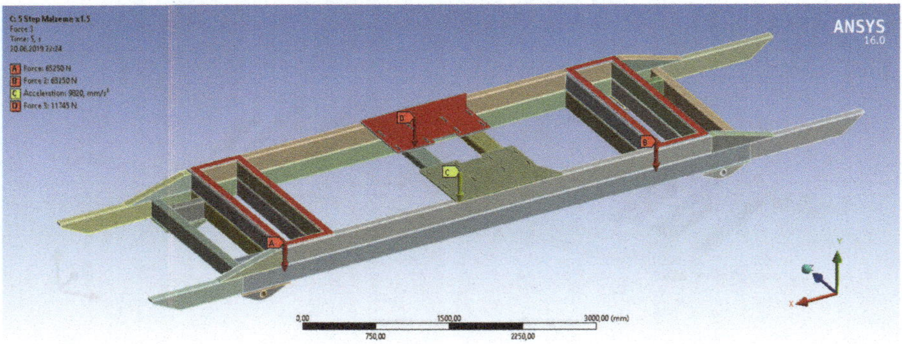

Fig. 4. Placement of forces on the geometry [8]

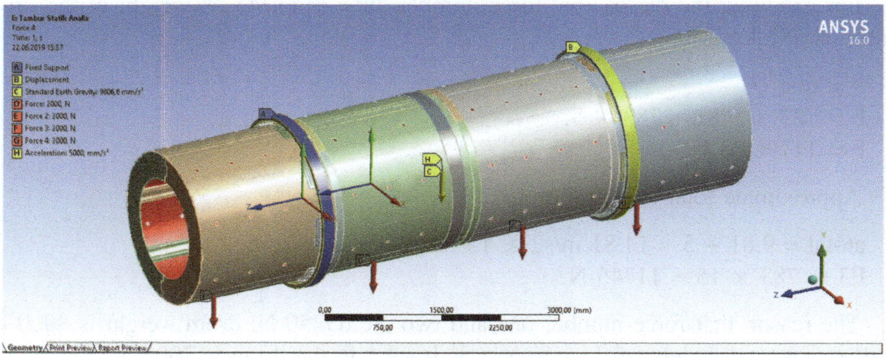

Fig. 5. Loads of the drum [8]

2.3 Testing of the Drum Granulator

After the manufacturing process is completed, the drum granulator becomes ready for machine performance tests. The granulator is placed at the designated location in the plant. In the first trials of granule production, the reject material ratio was quite high (Fig. 6). This is mainly due to the fact that the main binding element, chemical, is ammonia, but only water is used to produce granules. According to the results obtained in trials using only water, the production of small granules is quite difficult. The way to obtain smooth granules is to produce larger granules by increasing the water flow rate.

In order to produce granules of 2–4 mm diameter with granule sizes suitable for packaging, ammonia, the main binder chemical, must be added in the recipe. In addition, granules that do not contain chemicals are easily disintegrated into powder. After several days of experiments, it was found that for the formation of granules without chemical use, adding a 2.5° drum angle and adding 1 m^3/h water for 1 t/h compost was the best result. When ammonia was added to the compost entering the machine, the participation rates changed. The way to add ammonia into the compost is to mix by adding urea.

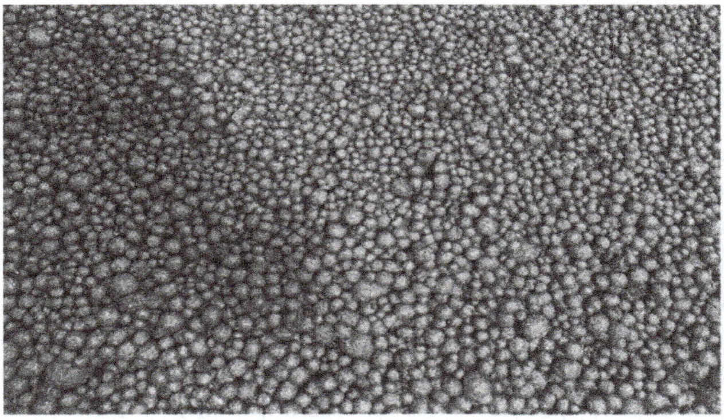

Fig. 6. (Oversized) reject granules [8]

340 kg of compost was added to 40 kg of urea and blended, then 1 t/h speed was given to the granulator.

In the experiment with urea added material, it was seen that giving 0.65 m^3 of water to 1 ton of material is enough to produce granules of 2–4 mm diameter (Fig. 7).

Fig. 7. Granules with a diameter of 2–4 mm (desired granule size) [8]

Ammonia was added to granulate fertilizer, then dried and passed through the screen was found to be 27% reject material. As the chemicals such as potassium, phosphorus and nitrogen are added to the recipe, the ratio of reject material decreases further. It was also determined that 15 kW motor value is very suitable for drum granulator. In the measurements, it was determined that the drum granulator uses 13.5 A electric current.

3 Results and Discussion

3.1 Design Process Results

During the design process, drive system design, gearbox selection, drum design and chassis design was carried out. Drum length is 8 m and drum diameter is 2 m. However, since the cylindrical folding of the material in the drum size is not possible during manufacturing, 4 pieces of 2 m lengths are cylindrical curled and welded. The drum is made of St37 material with a wall thickness of 12 mm. According to the researches, drum granulators with 10t/h production capacity are generally operated effectively at 11 rpm. For this reason, the drum granulator was designed to have a drum speed of 11 rpm, engine speed of 1450 rpm, and a total gear ratio of (i) 131.8.

The frame is designed to allow the drum to operate on different slopes. In order to make angle changes at any time, four supported pivoting feet are designed. In addition, 4 drum wheels are designed to support the drum's rotational movement. During the inclined operation of the drum, the axial fixing wheels are designed to limit the drum. In the design process, the forces to axial fixation wheels were calculated. The force applied by the drum to the wheel is 5600 N and the force applied by the bearing is 31500 N. With respect to the positions of the force vectors specified in Fig. 8; the force applied by the bearing multiplied by the distance to the bearing point, must be greater than the drum force multiplied by the distance from the bearing point.

3.2 Analysis Process Results

In the structural analysis, it can be said that a healthy result was obtained because there were 1,003,911 nodes and 170,837 elements. The minimum and maximum values of total deformation and Equivalent Stress values of the results required for structural strength are given in Table 1.

Table 1. Results of static structural analysis of chassis

Type	Total Deformation	Equivalent (von-Mises) Stress
Minimum	0.10649 mm	28.344 MPa
Maximum	1.3349 mm	285.83 MPa

According to the total deformation results; maximum deformation is seen as 1.33 mm. Most deformation is seen on the sheet on which the transmission is mounted. The yield stress of St37 structural steel is 240 N/mm^2 (MPa). But, this stress value is due to mesh errors found in the analysis and is a point value. As can be seen from Fig. 8, the maximum stress value is around 87 MPa. This result shows that the tension of the structural steel is below the safe tension.

$$\sigma\text{yield} \geq \sigma\text{maxin other words}, 240\ \text{MPa} \geq\ 87\ \text{MPa}$$

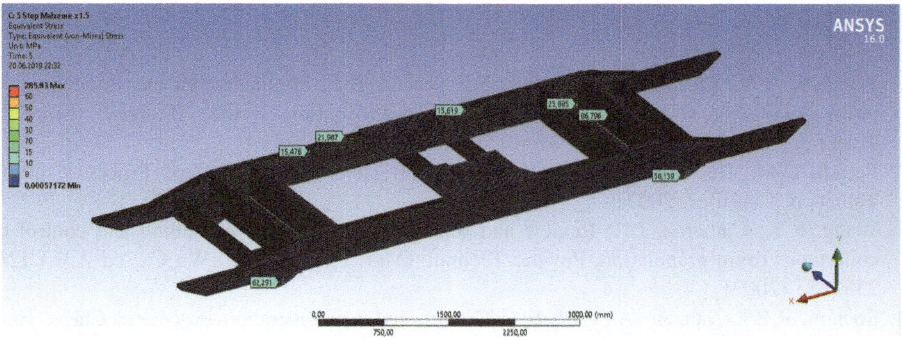

Fig. 8. Granules with a diameter of 2–4 mm (desired granule size) [8]

For this reason, stresses on the chassis are within safe limits. Optimisation of machining [9] in the context of Industry 4.0 is essential by following the required standards.

4 Conclusion

The following results were obtained in this study;

- The subject drum granulator has a drum length of 8 m and a drum diameter of 2m. The drum granulator operates with 11 rpm drum speed and 15 kW motor power and has 10 t/h production capacity. In the above study, 7.5 t/h production capacity was achieved at 11 kW, while the granulator were produced 10t/h product by operating the drum designed in larger dimensions at 15 kW. The drum has a foot design that can work horizontally at 2–3°. It is particularly suitable for working at 4°.
- After the production of the drum granulator subject of the thesis, trial studies were carried out. Even in the first trials, although the desired granule size was 2–4 mm, (As the desired granule size decreases, process control becomes more difficult and recycle rate increases.) a recycle rate of 27% was achieved. With the development studies carried out, a system was designed to feed water according to the weight of the incoming material. Sensor technologies are used in system design. This study made an important contribution to the reduction of recycle rate by feeding the granulator itself optimally.
- In development studies; PVC conveyor belt assembly which prevents the material from sticking to the drum surface, drum inlet design which prevents overflow problems at 2.5° drum inclination angle, outlet design which prevents the material to be clogged when leaving the drum. Furthermore, in order to reduce the rejection rate, a system which adjusts the amount of water given according to the weight of the material fed has been made. This system has made a significant contribution to the granule production process.

References

1. Adetayo, A.A., Litster, J.D., Pratsinis, S.E., Ennis, B.J.: Population balance modelling of drum granulation of materials with wide size distribution. Powder Technol. (Mort P., Teleki A., Wu C., Yu A.B.) **82**, 37–49 (1993)
2. Parikh, D.M.: Recent Advantages in Granulation Theory, Modeling and Process Control. Tablets & Capsules (2007)
3. Wang, F.Y., Cameron, I.T.: Review and future directions in the modelling and control of continuous drum granulation. Powder Technol. (Mort P., Teleki A., Wu C., Yu A.B.) **124**, 238–253 (2002)
4. Sochon, R.P.J., Salman, A.D.: Particle Growth and Agglomeration Processes. Chem. Eng. Chem. Process Technol. (Pohorecki, R., Bridgwater, J., Molzahn, M., Gani, R., Gallegos, C.) **2**, 299–317 (2010)
5. Parikh, D.M.: Introduction. In: Handbook of Pharmaceutical Granulation Technology (James Swarbrick), pp. 1–7. Taylor & Francis Group (2005)
6. Iveson, S.M.: Limitations of one-dimensional population balance models of wet granulation processes. Powder Technol. (Mort, P., Teleki, A., Wu, C., Yu, A.B.) **124**, 219–229 (2002)
7. Walker, G.M.: Drum granulation processes. In: Handbook of Powder Technology Book (Salman, A.D., Hounslow, M.J., Seville, J.P.K) **11**, 219–254 (2007)
8. Tanrikulu, O.: Design and Development of Drum Granulator. Unpublished M.Sc. Thesis (2019-M.Sc.-074), Aydın Adnan Menderes University, Turkey (2019)
9. Matras A., Zębala W.: Optimisation of Machining in the Context of Industry 4.0 – Case Study. In: Durakbasa, N., Gencyilmaz, M. (eds.) Proceedings of the International Symposium for Production Research 2018. ISPR 2018. Springer, Cham (2019)

Forecasting Electricity Generation and Shares by Energy Resources by Time Series Analysis: A Case-Study of Turkey

Aziz Kemal Konyalıoğlu[1(✉)] and Nuri Çelik[2]

[1] Department of Management Engineering, Istanbul Technical University, Istanbul, Turkey
konyalioglua@itu.edu.tr

[2] Department of Mathematics, Gebze Technical University, Kocaeli, Turkey
cnuricelik@gmail.com

Abstract. Electricity generation has always been a debatable topic to investigate since industrialization grows up day by day. The accurate planning and directing of the system of supplying electricity depends on reliable modelling and forecasting. Better forecasting and modelling provide more efficient planning, more suitable investments of time, cost and performance and more satisfied customers and citizens. A suitable forecast model for electricity generation is a difficult process to perform, since it involves many parameters such as climate conditions, population, industrial tendency and habitualness of each country or region. In this paper, we tried to forecast electricity generation and shares by energy resources in Turkey by using Time Series Analysis in R software. The study provides not only an effective forecasting for electricity consumption in order to meet the demand in Turkey, but also an efficient segmentation according to the shares by energy resources since the usage of electricity differs from resources. The data, which correspond to the period 1970–2018, are used to forecast the electricity generation in Turkey.

Keywords: Electricity forecasting · Electricity generation · Time series analysis · Box-Jenkins models

1 Introduction

Growth in population and the latest developments in technology and industrialization begin to increase the need of energy usage while energy sources are scarce. Furthermore, the decrease of energy resources becomes an important issue to think about [1]. In that case, electricity usage sharply grows because of its wide usage in the world and electricity demand forecasting can be assessed as one of the crucial topics for energy management in the economic perspective (Singh et al., 2013) [13]. Especially, in developing countries, electricity generation and demand forecasting has been a very difficult topic to investigate since many factors that affect forecasting. In Turkey there is an increase in electricity consumption, as well as in urbanization and economic activity [2]. Turkey is a rapidly developing country with increasing electricity usage parallel to industrialization. As such the country has begun to focus on new energy policies which

N. M. Durakbasa and M. G. Gençyılmaz (Eds.): ISPR 2020, LNME, pp. 115–120, 2021.
https://doi.org/10.1007/978-3-030-62784-3_10

are supposed to change energy industry radically due to the implementation of new regulations for producers and consumers [3].

In recent years, Turkey's energy demand has grown, which has made Turkey a net importer since 1997 [4]. Because of unpredictable conditions and sharp fluctuations, Turkey's energy consumption, generation and distribution cannot be easily projected.

In this paper, we forecast electricity generation and shares by energy resources in Turkey by using Box-Jenkins model which is also a model of time series analysis. The paper includes four main parts, which are: introduction in order to give a general perspective about the study, a literature review for investigating the latest studies about electricity generation and demand forecasting, a modelling and methodology part to introduce the Box-Jenkins model, which are also called ARMA models, and finally a conclusion part to discuss the results and make suggestions for future studies.

2 Literature Review

Several studies that use different methods for different scenarios of electricity generation and forecasting exist. Bianco et al. (2009) used a linear regression model in order to forecast electricity consumption in Italy between the years 1970 and 2007 by considering gross domestic product (GDP) and gross domestic product per capita based on a stationary data [15]. Ju and Hong (2013) tried to see electricity demand behavior by using chaotic gravitational search algorithm, which is a hybrid algorithm [5]. A different approach, named as triple seasonal methods, is used by Taylor (2009) for extending the accommodation of the data of Britain and France by comparing double seasonal methods and triple seasonal methods [6]. Hyndman and Fan (2010) assessed a temperature simulation in mixed form in order to estimate the probability distribution of South Australia's peak point of electricity demand [7]. El-Telbany and El-Karmi (2008) assessed a particle swarm optimization approach for short-term forecasting of electricity demand in Jordan comparing two methods which are back-propagation and autoregressive moving average methods [8]. Vilar et al. (2012) applied nonparametric functional method in order to forecast next-day electricity demand and compared the results with these methods integrating with a naïve approach [9].

Chang et al. (2010) proposed a fuzzy neural network approach for electricity demand forecasting including the affecting factors in Taiwan and compared this model with other methods in the literature [1]. Küçükali and Barış (2010) applied fuzzy logic methodology in order to forecast short-term gross annual demand of Turkey's electricity with several parameters including GDP [10]. On the other hand, Halicioglu (2007) assessed residential electricity dynamics by using Granger causality analysis, especially elasticity, and they have shown that elasticities in the long-run for income and price are greater than the elasticities in the short-run [3]. Erdoğdu (2007) used ARIMA and cointegration modelling for electricity demand analysis including forecasting and estimation by comparing the other projections performed [11]. Ozturk et al. (2005) tried two different non-linear estimation model by the aid of Genetic Algorithm in order to predict the electricity consumption between the years 2002–2025[12]. Another study for electricity demand for industrial and total sector done by Ozturk and

Ceylan (2005) has shown that genetic algorithm electricity demand can be used for real projection for electricity usage forecasting in Turkey [4].

3 Methodology and Modelling

Our aim was to forecast the electricity generation of Turkey. For this reason, we use the electricity generation data from 1970 to 2018 annually. Figure 1 shows the line graph of the data. It can be easily seen from the graph that there is an increasing trend from year to year.

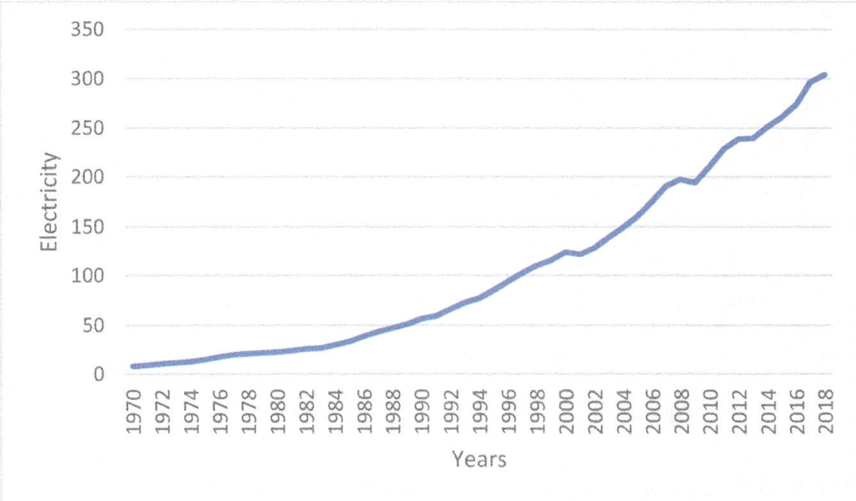

Fig. 1. Line graph of the electricity generation in turkey (1970-2018)[14]

In order to forecast the electricity generation, we use the Box-Jenkins models, which have been widely used in time-series analysis. The Box-Jenkins models, which are also called ARMA models, are very flexible, since they include both autoregressive and moving average terms. The first step in modeling is to detect stationary. The stationary means the constant mean and variance based on time. It is also a primary assumption for Box-Jenkins models. For this reason, we perform Dickey-Fuller test. Dickey-Fuller unit root test is the most acknowledged stationary test in literature. After performing the test, it is concluded that there is a stationary on data set with 0.05 significance level (p-value = 0.00).

The next step is to estimate the order (i.e. the p and q) of the autoregressive and moving average terms. The main tool to determine the orders is to use autocorrelation (ACF) and partial autocorrelation (PACF) plots. If there is an exponential decay to zero on ACF plot, use the PACF plot in order to detect the order of AR models and for MA models one can use the same plots with inverse thinking, Figs. 2 and 3 show the ACF and PACF plots for the electricity generation data set.

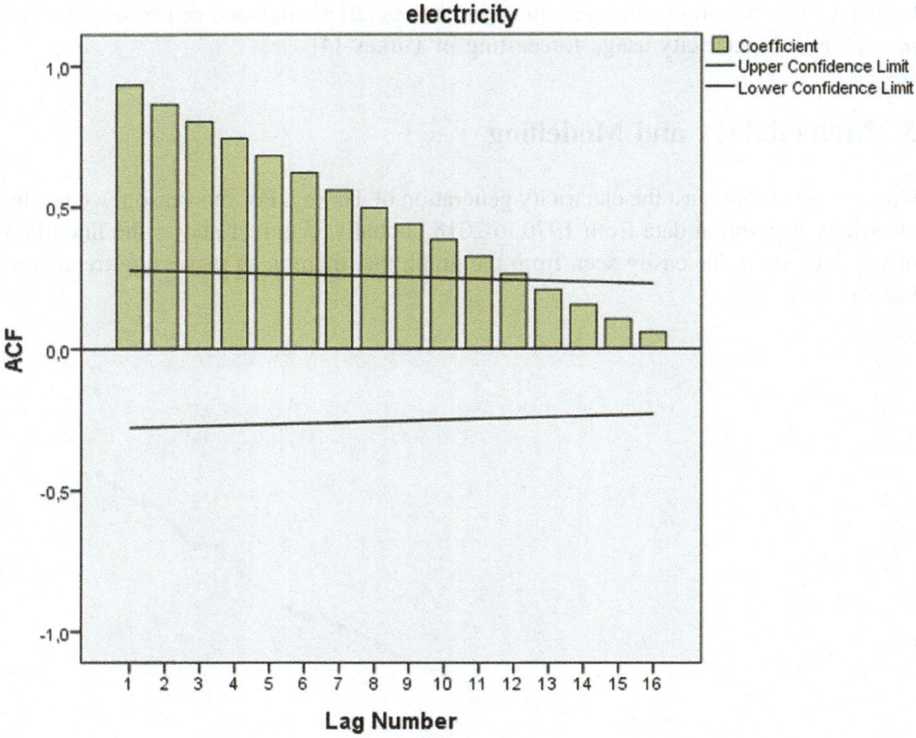

Fig. 2. ACF plot for the electricity generation data set

It can easily be detected from Figs. 2 and 3 that the data set can be modelled with AR(1) model. The term autoregression indicates that it is a regression of the variable against itself. Thus an AR(1) model can be written as

$$Y_t = \phi_0 + \phi_1 Y_{t-1} + \varepsilon_t \tag{1}$$

where ε_t is white noise, ϕ_0 and ϕ_1 are coefficients of the model. Therefore, we estimate the unknown model parameters based on the electricity generation data of Turkey as,

$$Y_t = 10.424 + 0.985 Y_{t-1} + \varepsilon_t \tag{2}$$

Figure 4 shows the electricity generations of Turkey between the years 1978-2018 and 5 years forecasted values. The forecasted values are obtained by using the Eq. (2) as 310.65, 361.41, 322.09, 327.69 and 333.19 for the years 2019 to 2023.

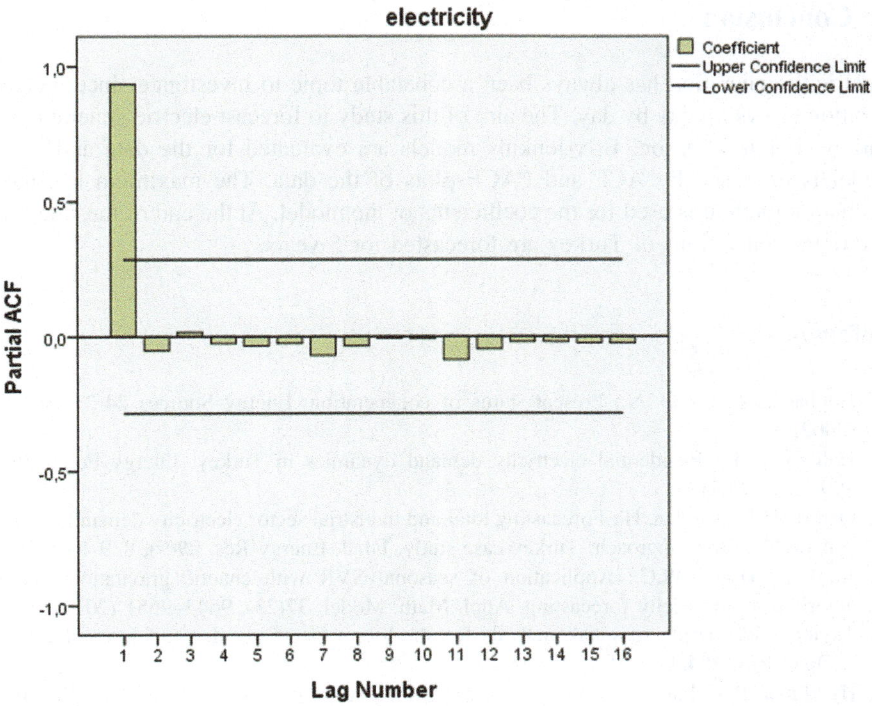

Fig. 3. Pacf plot for the electricity generation data set

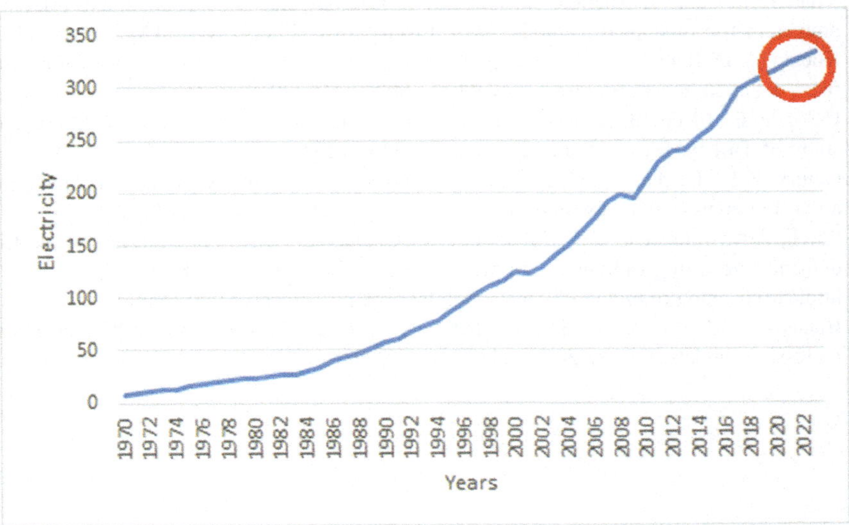

Fig. 4. Line graph of the electricity generation of turkey (projection until 2023)

4 Conclusion

Electricity generation has always been a debatable topic to investigate since industrialization grows up day by day. The aim of this study to forecast electric generations of Turkey. For this reason, Box-Jenkins models are evaluated for the data and AR(1) model is proposed by ACF and PACF plots of the data. The maximum likelihood estimation method is used for the coefficients of the model. At the end of the study, the electricity generations of Turkey are forecasted for 5 years.

References

1. Hepbasli, A., Ozalp, N.: Present status of cogeneration. Energy Sources **24**(2), 169–177 (2002)
2. Halicioglu, F.: Residential electricity demand dynamics in Turkey. Energy Econ. **29**(2), 199–210 (2007)
3. Ozturk, H.K., Ceylan, H.: Forecasting total and industrial sector electricity demand based on genetic algorithm approach: Turkey case study. Int. J. Energy Res. **29**(9), 829–840 (2005)
4. Ju, F.Y., Hong, W.C.: Application of seasonal SVR with chaotic gravitational search algorithm in electricity forecasting. Appl. Math. Model. **37**(23), 9643–9651 (2013)
5. Taylor, J.W.: Triple seasonal methods for short-term electricity demand forecasting. Eur. J. Oper. Res. **204**(1), 139–152 (2010)
6. Hyndman, R.J., Fan, S.: Density forecasting for long-term peak electricity demand. IEEE Trans. Power Syst. **25**(2), 1142–1153 (2009)
7. El-Telbany, M., El-Karmi, F.: Short-term forecasting of Jordanian electricity demand using particle swarm optimization. Electr. Power Syst. Res. **78**(3), 425–433 (2008)
8. Vilar, J.M., Cao, R., Aneiros, G.: Forecasting next-day electricity demand and price using nonparametric functional methods. Int. J. Electr. Power Energy Syst. **39**(1), 48–55 (2012)
9. Kucukali, S., Baris, K.: Turkey's short-term gross annual electricity demand forecast by fuzzy logic approach. Energy policy **38**(5), 2438–2445 (2010)
10. Erdogdu, E.: Electricity demand analysis using cointegration and ARIMA modelling: a case study of Turkey. Energy Policy **35**(2), 1129–1146 (2007)
11. Ozturk, H.K., Ceylan, H., Canyurt, O.E., Hepbasli, A.: Electricity estimation using genetic algorithm approach: a case study of Turkey. Energy **30**(7), 1003–1012 (2005)
12. Singh, A.K., Ibraheem, S.K., Muazzam, M., Chaturvedi, D.K.: An overview of electricity demand forecasting techniques. Network Complex Syst. **3**(3), 38–48 (2013)
13. http://www.emo.org.tr/ekler/89bb9e665db0b5a_ek.pdf. Accessed 20 May 2020
14. Bianco, V., Manca, O., Nardini, S.: Electricity consumption forecasting in Italy using linear regression models. Energy **34**(9), 1413–1421 (2009)

Fused Filament Fabrication of Ceramic Components for Home Use

Günther Poszvek[1(✉)], Clemens Wiedermann[1], Erich Markl[1], Jorge M. Bauer[2], Rolf Seemann[1], Numan M. Durakbasa[3], and Maximilian Lackner[1]

[1] Department Industrial Engineering, UAS Technikum Wien, Vienna, Austria
poszvek@technikum-wien.at
[2] UTN-FRBA Industrial, Buenos Aires, Argentina
[3] Department of Industrial Metrology and Adaptronic Systems, TU Wien, Vienna, Austria

Abstract. Ceramic materials offer a variety of desirable material properties, but due to being particularly hard and brittle, are challenging to machine in subtractive processes. Additive manufacturing of ceramics parts requires costly machines and raw materials, limiting the use of additively produced ceramic parts to the professional realm. Fused Filament Fabrication (FFF) has become a widespread additive manufacturing technology, due to low cost of machines and materials (typically polylactic acid (PLA) and ABS)). This paper addresses the suitability of the FFF process to manufacture ceramics parts cost-effectively.

There are many "effect" filaments for FFF use, e.g. with metallic powder content (glitter effect), wood powder (biobased raw materials) or inorganic filler (stone effect). The purpose of this research is to study FFF-derived ceramic parts, which do not only contain a certain fraction of inorganic particles in a polymer matrix, but which have been debindered and sintered to yield "true" ceramic parts comparable to those from a conventional ceramics manufacturing process.

The experiments have focused on FFF-printing available filaments, debindering (using solvents and heat) and sintering (using heat). Commercially available ceramic filaments were identified, all of them having a polymer matrix. For comparison, industrial grade machines for all common additive manufacturing processes for ceramics are listed, with prices that partly exceed € 250,000. During the practical experiments, the filament, containing 60% by volume of zirconium silicate, was successfully printed. The printed specimens were subsequently debindered and sintered with success, but the formation of pockets of trapped air could not be avoided completely during the debindering process. This paper shows that basically, FFF-made ceramic components. Semi-professional "makers" can use ceramic filaments on FFF printers, and apply post-processing in the steps of debindering (chemically with solvents or thermally) and final sintering to obtain parts.

Keywords: Ceramic materials · Fused Filament Fabrication (FFF) · Debindering · Cost-effectively

N. M. Durakbasa and M. G. Gençyılmaz (Eds.): ISPR 2020, LNME, pp. 121–139, 2021.
https://doi.org/10.1007/978-3-030-62784-3_11

1 Introduction

Additive manufacturing, or the expression 3D printing [1], which is often used interchangeably today, has become an integral part of product design and prototype development [2]. In addition to selective laser sintering (SLS) and stereolithography (SLA), the melt layer process (fused filament fabrication - FFF), in particular, has found widespread use due to the relatively low cost, good strength and wide choice of materials [3]. Most often, polymers such as PLA and ABS are printed, but furthermore solutions for metals, ceramics and hybrid materials are available.

As shown by Bauer and Durakbasa [68], cost effective printing is essential for the proliferation of an additive manufacturing technology, as a broad user base is required.

In the well-established FFF method, a filament of a thermoplastic is brought into a semi-liquid state in a heated extruder and pressed through a nozzle. Layer by layer the desired product is shaped [3, 4]. The melting of new layers onto already existing and now significantly cooled layers leads to strongly anisotropic material properties [5].

The physical properties of the printed part are variable through the choice of material and extrusion parameters when creating the tool path. Nevertheless, properties such as tensile strength and stiffness in the pure FFF process are limited by the material properties of the plastic used and, depending on the direction of loading, only achieve up to 70% of the strength of a comparative sample produced by injection molding [5] in the x-y-plane. In the z-direction, i.e. perpendicular to the print, the mechanical properties were found to be approx. half. This implies that the values in technical data sheets of filaments need to be treated with care. In order to increase their strength and stiffness, plastics are often used in the form of fiber composite materials. The plastic forms the matrix in which the fibers, often carbon fibers, glass fibers or aramid fibers, are embedded. A rough distinction is made between short fiber composite and continuous fiber composite, both approaches are also used in 3D printing. In addition to pure and fiber-reinforced plastics, there are also a number of others materials that can be printed in FFF based processes, including some metals [6] and ceramic materials [7]. Printing of endless fibre reinforced materials has been demonstrated, too [69, 70].

Additive manufacturing is particularly interesting for the production of ceramic components because ceramics are very difficult to process subtractively due to their high hardness and brittleness. With selective laser sintering (SLS), stereolithography (SLA) and fused filament fabrication (FFF), ceramics, in addition to other processes, can be produced using these three most common 3D printing processes [8]. Depending on the process, intermediate steps and subsequent processes such as debinding and sintering are required.

Compared to the FFF printing of plastics such as PLA or ABS, the production of ceramic parts in this process is not yet very well established, although procedures have already been described in 1996 [9]. In addition to the parameters already known from FFF plastic printing, such as layer thickness, speed of movement and mechanical stability of the printer, the print quality that can be achieved is affected by many additional factors that complicate the process. The filament used for printing contains up to 60% ceramic powder in addition to the binder polymer (the matrix) and any additives [10]. After printing, depending on the binder used, that matrix must be removed by solvents or by pyrolysis before the part is sintered [11]. When the green part is sintered directly, the huge inorganic fraction of the material would decompose too fast and damage/destroy the part. Although there are already a few filaments containing ceramics powders that can be printed in conventional FFF printers, there is hardly any information available about suitable printing parameters, any modifications to the printer, as well as what physical properties can be achieved with these filaments [12–15]. The purpose of this paper is to bring more clarity here.

Ceramics are a material class that is becoming more and more interesting to engineers, due to low weight, high hardness and abrasion resistance, high melting point and low thermal and electrical conductivity.

1.1 Objective and Research Question

The aim of this work is to provide an overview of the current state of the art in 3D printing of ceramics in the FFF process (as of May 2020). To describe the process and its most important parameters, to address existing problem areas in the process, to list the available range of filaments and printers for FFF printing of ceramics and to make a comparison to other additive manufacturing methods.

What is the state of the art in FFF ceramic printing and what are the offers on the market for this?

Under what conditions can the available filaments be reliably printed, and what dimensional tolerances and physical properties can be achieved with them?

The focus is to clarify whether, and if so, how far, the spread of FFF printers made manufacturing of ceramic components accessible to small companies, and private makers. The application and processes for printing ceramics for professional/industrial, the medical and dental applications are not considered in this work.

1.2 Relevance of the Question

There is a large amount of research on FFF printing in general and also specifically on the printing of ceramics and metals [9, 11], and [16–19]. For plastics, 3D printing in the FFF process is already being used very widely, right down to easy-to-use printers for a few hundred euros, which can already be found in many private households today [20]. In [9, 11], and [16–19] some promising approaches have already been described and tried out experimentally. The FFF process has advantages over other 3D printing processes such as good scalability of the machine size and cost-effective construction of the machines [17]. This work is intended to investigate what offers for ceramic FFF printing are now available on the market and how reliably the available filaments can be printed on conventional FFF printers. Answering these questions is not only relevant for end customers, who are to be given easier access to a new material class, but also to support industrial suppliers of ceramic components in deciding whether FFF printing is currently an appropriate option for their intended purpose.

1.3 Methods

This paper is initiated by a comprehensive literature review of the current state of the art in the FFF printing of ceramics based on available scientific publications. The key influencing factors for the process are identified and described, comparisons are made with non-filament-based extrusion methods, and the differences and properties of other processes for additive manufacturing of ceramics are discussed. Furthermore, it is determined which providers are currently active in this field and in which price range they have placed their machines and materials. The research part of the work is concluded with an overview of the problems that can occur with FFF printing of ceramics. In the experimental part of the work, predefined specimens made from purchased filament with a ceramic component are printed, debindered and sintered independently. The samples are measured immediately after printing and again after sintering, the optical quality is assessed and the hardness of the samples is determined. These test series are repeated several times, whereby individual factors are changed. The factors examined are the extrusion temperature, feed rate and diameter of the extruder nozzle. If the adhesion of the filament to the printing surface turns out to be problematic, a suitable material should be found that offers sufficient adhesion.

2 Fused Filament Fabrication

Fused filament fabrication was invented in 1989 by S. Scott Crump, who later co-founded Stratasys, and is a patented process for additive manufacturing [10]. Since the expiry of the patent in 2009, the process has spread rapidly due to the relatively simple mode of operation and inexpensive mechanical engineering and has provided the basis for a very active open source community and a large number of manufacturers who sell the relevant machines, often simply called 3D printers today [21]. Thermoplastic materials such as ABS, PLA, PC and PA are mainly used as materials [10], but applications for metals [22], ceramics [22], concrete [23] have also been used for the process, as well as - in a slightly modified form - foods such as chocolate [24].

2.1 Process

During production using the FFF process, the material to be extruded is heated to just above its melting point and pressed through a nozzle while it is moving on a planar plane parallel to the base plate [17]. The extruded material should solidify quickly after leaving the nozzle; this is often supported by additional air cooling [25]. Through the choice of the movement path and the interruption of the extrusion in places, a layer of the component to be produced is generated. After the layer has been completely applied, the plane of movement of the nozzle is removed from the base plate by a further layer thickness, either by lowering the base plate or by lifting the nozzle [17, 26].

2.2 Extrusion Methods

In additive manufacturing by means of extrusion, MEAM (Material Extrusion Additive Manufacturing), there are several processes which, however, have similarities [17]. In all of these processes, the layers are built up by planar movement of the nozzle relative to the building level and selective extrusion. The type of material feed can be roughly divided into piston, screw and filament-based extruders [17]. Piston-based extrusion is similar to the robocasting process, with the difference that water is used as a binder in robocasting, while thermoplastics, mostly as binders for metallic fillers, are used here [17, 27]. Screw-based extruders for direct use in additive manufacturing, offer theoretically advantages due to a higher filler content compared to filament extruders, but pose difficulties in generating a constant volume flow of the extruded material, as this differs greatly from one another [28]. Filament based extrusion is the most common one. In addition to the easy availability, a variety of materials and the comparatively inexpensive construction of the extruder is mainly due to the very simple and safe handling of the materials [17].

2.3 Ceramics

Ceramic components are used in a wide variety of applications from biotechnology to space travel due to their high hardness, strength and thermal and chemical resistance. The parts are made of ceramic powders and molded into the desired shape using injection molding, casting or molding [29, 30]. However, these methods are limited in the ability to produce corresponding shapes. They are time-consuming and expensive. It is also only possible to a limited extent, due to the high hardness and brittleness, to machine of ceramic components, since in addition to high tool wear, cracking, poor surface quality and poor dimensional accuracy often occurs [10]. Most of the ceramics production processes rely on an indirect manufacturing process, whereby a so-called "green body" is first produced [10, 19, 26]. The finished green body already has the optical shape of the component to be manufactured, but here the ceramic filling is only kept in shape by the binder, which limits the physical properties of the green body to those of the binder used [10, 26]. In order to produce a purely ceramic part with the appropriate properties from the green body, debindering and sintering of the part must be carried out.

2.4 Binder

FFF printing of ceramics and metals uses thermoplastic binders with a high proportion of filler because they cannot be extruded in their pure form due to their high melting points. Waxes are also possible. Different information can be found on the maximum percentage of filler. According to S. Cano et al. this is limited to about 50 vol.% [11], while Bose et al. describe materials with up to 65 vol.% of ceramics content [26]. The choice of binder mix plays a very central role in the overall process. The filament produced with it must be flexible and stable enough to be rolled up or unrolled from the spool. At the same time, it has to be stiff enough not to kink in the extruder during printing and to have the lowest possible viscosity when heated in order to keep the dynamic pressure in the extruder low, which is particularly problematic when the filler content is high because the filler tends to agglomerate particles. [10, 11, 17]. The adhesion on the construction level must be strong enough that the part does not come loose during printing, but weak enough so that it can be removed from the level at the end of the printer without damaging the level or the part [26, 31]. After printing, the binder must be removable from the product produced without damaging it, which, depending on the binder, can be done thermally, chemically or in a combination of both [10, 11]. Essentially three groups of components are used to achieve these properties: 1) The main component of the binder, usually a wax or thermoplastic polymer, with a share of 50–90% by volume. This component is removed in the first step of debinding [17].

2) The so-called backbone, which maintains the shape of the part after removing the main component [17].

3) Additives such as dispersants, compatibilizers and stabilizers for avoiding phase separation and agglomeration [17].

2.5 Debinding

Before sintering, the binder must be removed as completely as possible without damaging the part [32]. The main reason for the anisotropic shrinkage that occurs is the release of viscoelastic tensions in the material caused by pressure differences [33]. Compared to the shrinkage that occurs during later sintering, this is low when debinding [17]. The exact course of the debinding process depends on the binder system chosen and can either be purely thermal [19], supported by capillary action by embedding it in a powder bed when separating the binder [34], or by a multi-stage process in which the main component of the binder is dissolved using a solvent such as cyclohexane while the backbone of the binder is only subsequently removed thermally. [11, 16, 17].

2.6 Sintering

In the final processing step, the previously debindered parts, also called "brown body" at this stage, are sintered. During sintering, the material solidifies and shrinks at a high temperature, generally between 70–90% of the melting temperature of the material to be sintered [17, 30]. Although the melting point of the main part of the material composition is not reached during sintering, small amounts of liquid phases can form, sometimes promoted by additives. The structure of the material, which was originally very porous after washing out or burning out the binder, has a very high surface energy. At a correspondingly high temperature, the material tends to reduce this surface energy through solid connections along the particle boundaries, whereby the material shrinks by up to 20% by reducing the pores [17]. In addition to plastic flow, diffusion also plays an important role in sintering [30].

2.7 Parameters

When using a purchased filament with a ceramic filling, the same influencing factors apply to the printing of the green part as to the other materials used for FFF printing. The quality, accuracy and speed of the print depend, among other things, on the travel speed, the extruder temperature, the bed temperature, the layer thickness, the nozzle diameter and the orientation of the part on the printing surface [35].

The exact procedure for debinding and the relevant process parameters depend on the filament or binder system used. In addition to chemical and thermal debinding, a catalytic process is also described in which the main component of the binder system, polyoxymethylene, is dissolved out of nitric acid or oxalic acid at a relatively low and therefore thermally unproblematic temperature of 120 °C [36]. Another variant of debinding, which however still requires a subsequent thermal debinding step, is chemical debinding using a solvent [11]. The main component of the binder should be known for the choice of a suitable solvent. The binder must also contain a component which is insoluble for the solvent and which maintains the shape of the printed part after the main binder component has been removed [11]. The advantage of this combined method compared to purely thermal debinding lies in the porosity that is present after the main binder component has been removed, as a result of which the thermal debinding of the remaining backbone component can be done much more quickly and with less risk of defects than with exclusively thermal debinding [17].

Depending on the binder system, purely thermal debinding can either be done entirely by pyrolysis, or by a component with a low melting point that melts out at a lower temperature before the rest is evaporated and removed at temperatures up to 800 °C by pyrolysis [22, 37].

2.8 Known Problem Areas in the Process

In addition to the influencing factors already described here for additive manufacturing in the FFF process, the problems that can occur must be taken into account. The printing of ceramic-filled filaments and the necessary debinding and sintering processes involve some additional pitfalls, which affect the selection of the filament and suitable process parameters [38]. Depending on the nozzle diameter, extruder temperature and filler content in the filament, nozzle blockages can occur due to particle agglomeration [11, 39]. With thermal debinding it comes to rapid increase in temperature to form cracks and bubbles, as pyrolysis products released inside the part cannot diffuse out quickly enough [11]. Unless countermeasures have been taken by additives in the filament, a high proportion of filler leads to a stiff filament that tends to become brittle and can sometimes only be rolled up and unrolled at elevated temperatures [39]. Anisotropic shrinkage occurs during both debinding and sintering, which must be taken into account when creating the print program or the raw part model in order to obtain a true-to-size end product [19]. Due to the air inclusions and flaws in the printed layers that are almost unavoidable in the FFF process, the density of fully sintered parts made by FFF is lower than that of the ceramic solid material, sometimes only 75–90% [40].

3 Market Situation

The market situation shown below does not claim to be complete. There is an overview of ceramic-filled 3D printing filaments available at the time of writing this work, obtained via a search engine query in German and English. Further offers of some professional suppliers of additively manufactured ceramic components and a simple cost comparison are presented.

3.1 Filaments

In the filaments on offer, a distinction must be made between materials whose ceramic portion only serves to achieve certain optical and haptic properties, and those that are advertised with the possibility of a sintering process and are therefore also interesting for technical applications. To estimate the material properties that can be achieved with the filament, information is required about which ceramic filler is contained in which proportion, and in order to simplify successful debinding, the binder system used should be known.

As can be seen in Table 1, the range of ceramic-filled filaments is rather manageable. Some of the offered filaments advertise their product name with a ceramic component, but it is not specified what filler it is. Some sinterable filaments are offered under the ZetaMix® brand, which are filled with aluminum oxide, zirconium oxide and zirconium oxide reinforced aluminum oxide with technically usable high-performance ceramics. Although no specific information is given about the binder system used for the ZetaMix® filaments, the manufacturer also offers a solvent debinding device and a tube furnace for sintering. However, these filaments were not continuously available during the period of this work [41–43].

A filament with 60% zirconium silicate filling was purchased and used for all subsequent experiments, because of all available filaments only for this filament exact information about filler, filler content and binder was available. Zirconium silicate is an interesting ceramics material with corrosion resistance against alkaline materials. The zirconium silicate filament is not explicitly advertised on the manufacturer's website as sinterable, but it is stated that the binder can be completely removed by a firing process [44]. Information on sinterability and the relevant process parameters are available in other research work [45].

Table 1. Commercially available ceramic filaments. NA = not available. ZTA = Zirconia Toughened Alumina. PLA = polylactic acid.

Product name	Filler	Filler content (weight%)	Binder	Sinter able	Price/kg	Source
3D filament 1.75 mm Ceramic	NA	NA	NA	NA	€ 30	[46]
LAYCeramic Ceramic Filament	NA	NA	NA	yes	€ 185	[47]
China ceramics filament high quality	NA	NA	NA	NA	€ 20	[48]
Lay-Brick (Ceramic)	Sand stone	NA	NA	NA	€ 23	[49]

(continued)

Table 1. (*continued*)

Product name	Filler	Filler content (weight%)	Binder	Sinter able	Price/kg	Source
ZetaMix Filament	Al_2O_3	NA	NA	yes	€ 600	[41]
ZetaMix Filament	ZTA	NA	NA	yes	€ 600	[42]
ZetaMix Filament	ZrO_2	NA	NA	yes	€ 600	[43]
Zirconium Silicate (Zircopax®) Ceramic Filamet	$ZrSiO_4$	60%	PLA	yes	€ 130	[44]
Amaco X-23 Ceramic Clay Filamet	Clay	40%	PLA	yes	€ 34	[50]
Amaco 46-D Ceramic Clay Filamet	Clay	40%	PLA	yes	€ 34	[51]
Amaco White 25-D Ceramic Clay Filamet	Clay	40%	PLA	yes	€ 34	[52]
Markforged Ceramics support material	NA	NA	NA	NA	€ 125	[53]

3.2 Industrial Providers

A list of some suppliers and machines for additive manufacturing of ceramic parts grouped by process is listed in Table 2. At the lower end of the price range, the Ceramo One from Kwambio is a printer in the binder jet process, which is primarily aimed at prototype development and for which no information is available regarding the debinding and sinterability of the material used. Also in the more economical area of industrial offers there are two devices for additive manufacturing using the FFF process. Both suppliers chose screw extruders as the extrusion method, which have the advantage that commercially available material granules, as they are common for metal injection molding and ceramic injection molding, so-called feedstock, can be used [54, 55]. With machine prices over $ 50,000, however, these machines are also aimed at companies that use them as an extension to their existing metal or ceramic injection molding systems.

In terms of level of detail and achievable surface quality, the SLA devices from Lithoz, 3DCeram, Prodways and Admatec are clearly in the upper class of additive manufacturing of ceramics. For example, with the current generation of SLA printers from Lithoz such as the CeraFab S25, a lateral resolution of 25 µm and a layer thickness of only 10 µm is possible, which is why they are used in the dental sector, products for microchip production or for the production of complex cast cores for turbine blades. As can be seen in Table 2, SLA systems with at least $ 125,000 for the Admaflex 130 are also priced in the top segment.

Since the question of this work focuses on the FFF process and specifically the suitability of currently available filaments for use by small businesses and private makers, the offers for the industrial area are not discussed in detail

Table 2. Industrial suppliers. FFF = fused filament fabrication. BJ = binder jetting. SLS = selective laser sintering. SLA = stereolithography.

Technology	Provider	Printer	Price [$]	Building space [mm]	Source
FFF	3D-figo	FFD 150H	>50,000	150 × 150 × 120	[56]
FFF	AIM3D	ExAM 255	>50,000	255 × 255 × 255	[57]
BJ	ExOne	Innovent	>100,000	160 × 65 × 65	[58]
BJ	Kwambio	Ceramo One	>25,000	350 × 350 × 380	[59]
BJ	Voxeljet	VX4000	>250,000	4000 × 2000 × 1000	[60]
SLS	3D Systems	ProX DMP 200	>250,000	140 × 140 × 100	[61]
SLA	Lithoz	CeraFab 7500	>250,000	76 × 43 × 150	[62]
SLA	3DCeram	Ceramaker	>250,000	300 × 300 × 110	[63]
SLA	Admatec	ADMAFLEX 130	>125,000	96 × 54 × 120	[64]
SLA	Prodways	ProMaker V6000	>250,000	120 × 500 × 150	[65]

4 Experimental

In the empirical part of this work, a filament suitable for the FFF printing of ceramics was procured and according to the manufacturer's instructions a number of tests were carried out with it. The first attempts were aimed at optimizing the most important printing parameters such as extruder temperature and feed speed with regard to dimensional accuracy, optical quality and reliability of the printing process. As there was no clear information about the binder system used, even after asking the manufacturer, tests were carried out for both chemical and thermal debinding. According to the manufacturer, no sintering tests were carried out by the manufacturer with the purchased filament, which is why no information was available on this either. In order to check the sinterability and the changes in the material properties that occurred, some of the samples were also subjected to a sintering test.

4.1 Print

To carry out the tests, the filament Zirconium Silicate Filamet™ was purchased from the US manufacturer The Virtual Foundry. The choice fell on this filament because it was the only product that could be found, for which precise information about a filler known as sinterable was available and that was available for order. Information on the binder system used was not available on the manufacturer's website, on request via email it was communicated that PLA was used as a binder and that the manufacturer had made no attempts to debinder using solvents, but the material was successfully thermally debindered. In addition, the manufacturer's website contains recommendations for preheating the filament, the pressure nozzle diameter to be used, the printing temperature, feed speed and also a temperature curve for the thermal debinding process.

The Chinese manufacturer Qidi Tech used the Qidi X-Plus as a 3D printer for all printing tests. The printer offers a completely enclosed installation space of 270 × 200 × 200 mm, a heated print bed and is supplied with a second print head which is suitable for high-temperature materials such as polycarbonate or nylon. The print head, the extruder system and the motors for all axis movements are standard components as they are used in most other 3D printers for home and semi-professional applications, which is why the results of the print tests can also be applied to many other commercially available 3D printers. The filament, the temperature and the printing speed were varied.

4.2 Debinding

First, chemical debinding was done using different solvents. In the course of the experiment, two identical specimens with the external dimensions 20 mm x 20 mm x 5 mm were weighed. After their weight was known, they were immersed in 100 ml solvent for 8 h at room temperature. Cyclohexane and dichloromethane were purchased as solvents for this experiment. The cyclohexane was chosen because it has been used in several research projects as a solvent for corresponding binder systems [11, 22]. Dichloromethane was used as the second solvent because PLA is soluble in it and is less dangerous to handle, and is easier to obtain than trichloromethane, which is also suitable for dissolving PLA [66]. If possible, the samples were removed from the solvent after the end of the test period, were air-dried for 3 h at room temperature and then weighed again in order to be able to measure any fraction that may have been removed. Dichloromethane: The dichloromethane showed drastic effects on the sample placed in it during the test period. Removing the PLA from the filament, the binder system of which, according to the manufacturer, only consists of PLA, leads, as expected, to a complete loss of the dimensional stability of the printed component. After only 3 h in dichloromethane, delamination of the printed layers and the breaking out and disintegration of large pieces of the sample were encountered. Cyclohexane: No changes in weight or dimensions were ascertained in the sample placed in cyclohexane within the defined test period, only the color of the sample showed a slight lightening. In order to recognize a possible long-term effect of the cyclohexane, the sample was put back into the solvent after drying and weighing in the course of the experiment and left there for a period of about 4 weeks. After these 4 weeks and a few additional days in the air to dry the sample, the sample showed a significant swelling. The sample was not only significantly lighter than an untreated comparative sample, but had grown to 21.5 mm × 21.5 mm × 5.6 mm in its outer dimensions, which corresponds to a volume increase of 29.5%. However, the weight of the sample remained unchanged. The swelling of the sample is an effect that has already been described by Cano et al. has been described, but for a different binder system [11]. The fact that this change in volume remains after the sample has dried, without the sample mass having changed, means a change in density, and could indicate a resulting porosity that would favor subsequent thermal debinding. However, since this effect was only noticed after several weeks and thus after the end of the test series for thermal debinding, a possible influence on the change in density on the thermal debinding was not tested.

Second, thermal debinding was done using different temperature profiles. The thermal debinding test was carried out twice. The MR170E hardening furnace from Heraeus was used for both tests. The oven has a usable interior of 150 mm × 100 mm × 250 mm and an output of 3000 watts, and can therefore heat up to a nominal temperature of 1000 °C. All samples intended for later sintering were placed individually in stainless steel cups in both experiments and, according to the filament manufacturer's recommendation, were completely embedded in aluminum oxide powder purchased for this purpose. Only one sample was placed in the cup without the Al_2O_3 powder to observe the effects. During the debinding process, the aluminum oxide serves as a support material for the embedded parts, which pass through a phase before the binder escapes in gaseous form through pyrolysis, in which the melting temperature of the plastic still present has been exceeded and the material is again as soft as at the time of extrusion. On the one hand, the flat samples with 20 mm x 20 mm x 5 mm produced in the printing tests were used as samples for the tests. In addition, a sample geometry was constructed in which three square bars, each 6 mm × 6 mm × 20 mm, are connected orthogonally to each other. Directly in the 3D model, a label was created for the three spatial axes, which can be used to read the orientation of the part even after debinding and sintering, so that any anisotropic properties that may occur can be assigned to an axis.

The two tests carried out in this way differed only in the temperature profile over time. In the first experiment, the procedure was largely based on the manufacturer's recommendation, with the maximum temperature being chosen arbitrarily since it was not possible to reach the sintering temperature in the oven used for debinding. After viewing the results of the thermal debinding test according to the manufacturer's instructions, the test was repeated with a changed temperature curve. The temperature curve of the second experiment was designed in such a way that the temperature range in which the pyrolysis of the binder was expected was stretched over a considerably longer period in order to give the resulting gas more time to diffuse out.

4.3 Sintering

The sintering temperature of zirconium silicate is given in the literature between 1500 °C and 1600 °C [67]. These temperatures are far beyond the possibilities of the hardening furnace used for debinding. However, there was the opportunity to place a few samples with a dental technician who sintered the zirconium oxide, which is why the temperature curve used and shown in was determined by the sintering process of the zirconium oxide. An AUSTROMAT μSiC from Dekema was used as the sintering furnace.

4.4 Results

Although the sintering temperature used was in the lowest limit of the sintering temperatures described in the literature and the holding time of 3 h is shorter than in experiments documented in other research work, the parts were successfully sintered in the sense that a further significant shrinkage occurred and that after the debinding only hardened samples that are extremely weak after sintering [67]. The outer dimensions of the four sintered flat samples were 14.8 mm × 14.8 mm × 3.65 mm and differed from each other by only a maximum of 0.05 mm at the points that were not deformed by bubbles. The shrinkage in the axial direction from the print to the sintered part was therefore mathematically 26% in the print plane and 27% normal to the print plane. Due to the small number of samples and the relatively high measurement inaccuracy in relation to the total height of the sample, no clear statement about the isotropy of the shrinkage could be derived from this. The weight of the samples remained between the debindered and the sintered state unchanged, was still 2.4 g. Due to the further reduction of the sample volume to 799.5 mm^3, the relative density of the sample has increased to 63.87%. Of the six samples that were originally printed to measure the isotropy of the shrinkage, only one survived the debinding and the transport to the sintering furnace without damage. The measured lengths of the square bars arranged in the three spatial axis directions were 15.1 mm in the X direction, 15.2 mm in the Y direction and 15.1 mm in the Z direction. Based on the originally printed length, this resulted in a shrinkage of between 24% and 24.5%, and here, too, any anisotropy was not detectable, see Fig. 1.

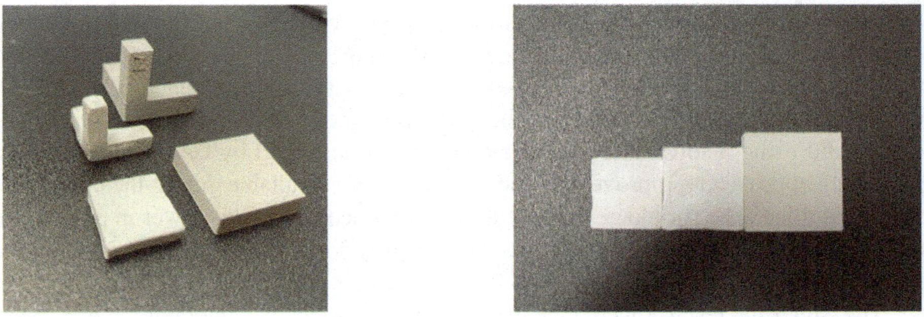

Fig. 1. a. Samples after sintering (left) and samples after printing (green body, right). b. Samples after sintering (left), after debindering (middle) and after printing (right). Dimensions approx. 5 cm × 5 cm.

5 Discussion

Obtaining a filament suitable for FFF ceramic printing was surprisingly difficult. Most of the results found in a German or English-language search engine are filaments that advertise ceramics in their product name, but do not provide any information as to which ceramic it is and how printed parts can be debindered or sintered. Such materials have very interesting applications due to their optical and haptic properties, but they are not suitable for the production of technical ceramic components. The company Zetamix is the only one of the offers found to have a filament in its program that includes a chemical debinding step using solvents, but this was not available and would also be relatively expensive at € 600 per kilogram. The purchased filament from The Virtual Foundry was therefore promising not only because of the precise information on the composition, but also because of the lack of other offers that met the criteria.

The filament printing was surprisingly unproblematic. The material could be printed in a very wide range of temperatures and speeds, and only showed weaknesses compared to pure PLA when bridging open areas without underlying support material. There were two misprints during the printing of the various samples and the nozzle clogged. Possible explanations for this would be particle agglomeration or contamination in the filament [10].

The causes of the low influence of the extrusion temperature on the print quality for overhangs and bridges can only be speculated without further investigation. However, a connection with the 60 vol.% zirconium silicate would be obvious, which influences the connection between temperature and viscosity, as well as the heat capacity and thus the cooling rate. The slightly higher surface roughness when printing samples at 175 °C compared to the comparison samples at 210 °C indicates an irregular extrusion volume flow.

Debinding was the most difficult of all the steps carried out. Apart from sintering, it is also the most time-consuming activity (although green body production is still slower than by conventional method, that is typically not the time-critical step.

A purely thermal debinding process is also described in the literature as very time-consuming [11], however, the tests did not succeed either by using the temperature curve recommended by the manufacturer or by using an adapted curve that extended the critical phase over a much longer period of time to achieve a satisfactory result with regard to the technical usability of the parts produced. A filament that is partially debindered using a solvent such as cyclohexane, as described by Cano et al. can be a promising approach to improve the process, but was not available during the period this work was done. The shrinkage that occurred during debinding was lower in percentage than during the subsequent sintering, which corresponds to the information in the literature [17].

The sintering test showed that the material is basically sinterable. For a more precise analysis of the properties that can be achieved with this, however, further tests are necessary in which process parameters such as temperature and holding time are varied. In addition, such tests require a suitable test procedure that would make the results quantifiable.

The potential of the FFF printing of ceramic materials lies primarily in making this material accessible to customer groups for whom professional devices for additive manufacturing of ceramics are financially out of the question. For example, small companies whose main area of business is not the industrial production of ceramic parts, but for whom ceramic components would be useful in some applications. Due to the very inexpensive availability of 3D printers with which the corresponding filament can be used, as well as the availability of suitable sintering furnaces that are already available for less than 2000 €, the manufacture of ceramic parts already reaches a price range that for individuals, for example becomes affordable in the craft sector or in model making.

Although the chosen material zirconium silicate does have significance, there is not special reason why that ceramics substrate was chosen for small scale 3D production other than ease of availability. Al_2O_3, as highest-volume technical ceramics material, would be very interested to print as well in future.

6 Conclusions

In the course of this work, the entire process of creating a ceramic component using additive manufacturing using the fused filament fabrication process was first worked up theoretically and then practically reproduced. An overview of the corresponding processes was given, the FFF process explained in detail and the available range of filaments for FFF printing as well as the most common professional suppliers and devices listed. The influence of the most important printing parameters for the purchased filament was examined and the conditions under which the filament can be successfully printed were explained. Tests for chemical and thermal debinding were carried out and the results discussed. Finally, the sinterability of the material was confirmed in the experiment.

With the tests carried out in this work, and also in other research work referred to in this work, it could be shown that the FFF process can be a technically feasible and economically sensible process for the production of ceramic components under certain conditions. However, this process represents a niche in the market in which very few providers participate.

The purchased filament could be reliably printed in a relatively wide temperature range from 175 °C to 210 °C. A print speed between 1000 and 2000 mm/min gave very good results in most printing situations. After sintering, the dimensional accuracy of the printed parts showed deviations of less than 0.1 mm at points that were not affected by gas inclusions, which suggests that the process can be easily repeated, but some parts of the samples were deformed by gas inclusions. The achievable physical properties of the material after sintering could not be determined apart from the relative density due to the lack of suitable test methods.

Meanwhile, there are providers on the market that offer ceramics prints made from FFF, e.g. Hage 3D [71] or HotEndWorks [72]. Their performance was not compared in this work, however, is expected to be similar.

References

1. Weller, C., Kleer, R., Piller, F.T.: Economic implications of 3D printing: market structure models in light of additive manufacturing revisited. Int. J. Prod. Econ. **164**, 43–56 (2015). https://doi.org/10.1016/j.ijpe.2015.02.020
2. Mathias, D., Snider, C., Hicks, B., Ranscombe, C.: Accelerating product prototyping through hybrid methods: coupling 3D printing and LEGO. Des. Stud. **62**, 68–99 (2019). https://doi.org/10.1016/j.destud.2019.04.003
3. Wang, X., Jiang, M., Zhou, Z., Gou, J., Hui, D.: 3D printing of polymer matrix composites: a review and prospective. Compos. Part B Eng. **110**, 442–458 (2017). https://doi.org/10.1016/j.compositesb.2016.11.034
4. Schirmeister, C.G., Hees, T., Licht, E.H., Mülhaupt, R.: 3D printing of high density polyethylene by fused filament fabrication. Addit. Manuf. **28**, 152–159 (2019). https://doi.org/10.1016/j.addma.2019.05.003
5. Koch, C., Van Hulle, L., Rudolph, N.: Investigation of mechanical anisotropy of the fused filament fabrication process via customized tool path generation. Addit. Manuf. **16**, 138–145 (2017). https://doi.org/10.1016/j.addma.2017.06.003

6. Thompson, Y., Gonzalez-Gutierrez, J., Kukla, C., Felfer, P.: Fused filament fabrication, debinding and sintering as a low cost additive manufacturing method of 316L stainless steel. Addit. Manuf. **30**, 100861 (2019). https://doi.org/10.1016/j.addma.2019.100861
7. Hensen, T.J.: Additive manufacturing of ceramic nanopowder by direct coagulation printing. Addit. Manuf. **23**, 140–150 (2018). https://doi.org/10.1016/j.addma.2018.07.010
8. Deckers, J.: Additive manufacturing of ceramics: a review. J. Ceram. Sci. Tech. **04** (2014). https://doi.org/10.4416/jcst2014-00032
9. Agarwala, M.K., Weeren, R.V., Bandyopadhyay, A., Safari, A., Danforth, S.C.: Filament feed materials for fused deposition processing of ceramics and metals. Int. Solid Free. Fabr. Symp., 8 (1996). http://hdl.handle.net/2152/70277
10. Chen, Z.: 3D printing of ceramics: a review. J. Eur. Ceram. Soc., **399**(4), 661–687 (2019). https://doi.org/10.1016/j.jeurceramsoc.2018.11.013
11. Cano, S.: Additive manufacturing of zirconia parts by fused filament fabrication and solvent debinding: selection of binder formulation. Addit. Manuf. **26**, 117–128 (2019). https://doi.org/10.1016/j.addma.2019.01.001
12. Amaco White 25-D Ceramic Clay FilametTM. The Virtual Foundry. https://shop.thevirtualfoundry.com/products/amaco-white-25-d-ceramic-clay-filamet. Accessed 26 Jan 2020
13. Amaco X-23 Ceramic Clay FilametTM: The Virtual Foundry. https://shop.thevirtualfoundry.com/products/amaco-x-23-ceramic-clay-filamet. Accessed 26 Jan 2020)
14. LAYCeramic Ceramic Filament - 3.00 mm (1 kg). MatterHackers. https://www.matterhackers.com/store/3d-printer-filament/layceramic-3.00mm. Accessed 26 Jan 2020
15. New Ceramic 3d Printing Filament 1.75 mm Ceramic Texture Impressora 3d Plastic Filament (1 kg). https://www.alibaba.com/product-detail/New-Ceramic-3d-printing-filament-1_60736064964.html. Accessed 26 Jan 2020
16. Abel, J.: Fused Filament Fabrication (FFF) of Metal-Ceramic Components. J. Vis. Exp. **143**, 57693 (2019). https://doi.org/10.3791/57693
17. Gonzalez-Gutierrez, J., Cano, S., Schuschnigg, S., Kukla, C., Sapkota, J., Holzer, C.: Additive manufacturing of metallic and ceramic components by the material extrusion of highly-filled polymers: a review and future perspectives. Materials **11**(5), 840 (2018). https://doi.org/10.3390/ma11050840
18. Gorjan, L., Tonello, R., Sebastian, T., Colombo, P., Clemens, F.: Fused deposition modeling of mullite structures from a preceramic polymer and γ-alumina. J. Eur. Ceram. Soc. **39**(7), 2463–2471 (2019). https://doi.org/10.1016/j.jeurceramsoc.2019.02.032
19. Travitzky, N.: Additive manufacturing of ceramic-based materials. Adv. Eng. Mater. **16**(6), 729–754 (2014). https://doi.org/10.1002/adem.201400097
20. Umfrage: 3D-Drucker zieht ein in deutsche Privathaushalte: reichelt.de. https://www.reichelt.de/magazin/studien/umfrage-3d-drucker/. Accessed 28 Jan 2020
21. Rundle, G.: A Revolution in the Making. Affirm Press, Melbourne (2014)
22. Lengauer, W.: Fabrication and properties of extrusion-based 3D-printed hardmetal and cermet components. Int. J. Refract. Met. Hard Mater **82**, 141–149 (2019). https://doi.org/10.1016/j.ijrmhm.2019.04.011
23. Manikandan, K., Wi, K., Zhang, X., Wang, K., Qin, H.: Characterizing cement mixtures for concrete 3D printing. Manuf. Lett. **24**, 33–37 (2020). https://doi.org/10.1016/j.mfglet.2020.03.002
24. Larano, M.: 3D printing complex chocolate objects: platform design, optimization and evaluation. J. Food Eng. **215**, 13–22 (2017). https://doi.org/10.1016/j.jfoodeng.2017.06.029
25. Lee, C.Y., Liu, C.Y.: The influence of forced-air cooling on a 3D printed part manufactured by fused filament fabrication. Addit. Manuf. **25**, 196–203 (2019). https://doi.org/10.1016/j.addma.2018.11.012

26. Bose, S.: Additive manufacturing of ceramics. Addit. Manuf. (2015). https://www.taylorfrancis.com/. Accessed 19 Dec 2019
27. Cesarano, J.: A review of robocasting technology. MRS Proc. **542**, 133 (1998). https://doi.org/10.1557/proc-542-133
28. Bellini, A., Shor, L., Guceri, S.I.: New developments in fused deposition modeling of ceramics. Rapid Prototyp. J. **11**(4), 214–220 (2005). https://doi.org/10.1108/13552540510612901
29. Bengisu, M.: Engineering Ceramics. Springer Science & Business Media (2013)
30. Hornbogen, E., Eggeler, G., Werner, E.: Werkstoffe: Aufbau und Eigenschaften von Keramik-, Metall-, Polymer- und Verbundwerkstoffen. Springer, Heidelberg (2019)
31. Spoerk, M., Gonzalez-Gutierrez, J., Sapkota, J., Schuschnigg, S., Holzer, C.: Effect of the printing bed temperature on the adhesion of parts produced by fused filament fabrication. Plast. Rubber Compos. **47**(1), 17–24 (2018). https://doi.org/10.1080/14658011.2017.1399531
32. Lewis, J.A., Smay, J.E., Stuecker, J., Cesarano, J.: Direct ink writing of three-dimensional ceramic structures. J. Am. Ceram. Soc. **89**(12), 3599–3609 (2006). https://doi.org/10.1111/j.1551-2916.2006.01382.x
33. Bellini, A.: Fused deposition of ceramics: a comprehensive experimental, analytical and computational study of material behavior, fabrication process and equipment design. ProQuest Information and Learning Company (2002)
34. Rahaman, M., Rahaman, M.N.: Ceramic Processing. CRC Press, Boca Raton (2006)
35. Peterson, A.M.: Review of acrylonitrile butadiene styrene in fused filament fabrication: a plastics engineering-focused perspective. Addit. Manuf. **27**, 363–371 (2019). https://doi.org/10.1016/j.addma.2019.03.030
36. Todd, I., Sidambe, A.T.: Developments in metal injection moulding (MIM). Adv. Powder Metall, 109–146 (2013), https://doi.org/10.1533/9780857098900.1.109
37. Onagoruwa, S., Bose, S., Bandyopadhyay, A.: Fused Deposition of Ceramics (FDC) and Composites. Int. Solid Free. Fabr. Symp., p. 8 (2001). http://dx.doi.org/10.26153/tsw/3267
38. Balani, S.B., Chabert, F., Nassiet, V., Cantarel, A.: Influence of printing parameters on the stability of deposited beads in fused filament fabrication of poly(lactic) acid. Addit. Manuf. **25**, 112–121 (2019). https://doi.org/10.1016/j.addma.2018.10.012
39. Rangarajan, S.: The Role of Materials Processing Variables in the FDC Process
40. Danforth, S.: Fused deposition of ceramics: a new technique for the rapid fabrication of ceramic components. Mater. Technol. **10**(7–8), 144–146 (1995). https://doi.org/10.1080/10667857.1995.11752614
41. ZetaMix Filament: ZetaMix. http://zetamix.fr/en/filaments/31-79-zetamix-filament.html. Accessed 26 Apr 2020
42. ZetaMix Filament: ZetaMix. http://zetamix.fr/en/filaments/31-81-zetamix-filament.html. Accessed 26 Apr 2020
43. ZetaMix Filament: ZetaMix. http://zetamix.fr/en/filaments/31-85-zetamix-filament.html. Accessed 26 Apr 2020
44. Zirconium Silicate (Zircopax®) Ceramic FilametTM: The Virtual Foundry. https://shop.thevirtualfoundry.com/products/zirconium-silicate-filamet-1. Accessed 26 Apr 2020
45. Suárez, G., Acevedo, S., Rendtorff, N.M., Garrido, L.B., Aglietti, E.F.: Colloidal processing, sintering and mechanical properties of zircon (ZrSiO4). Ceram. Int. **41**(1), 1015–1021 (2015). https://doi.org/10.1016/j.ceramint.2014.09.024
46. 3D filament 1,75 mm Ceramic Keramik 1000 g 1 kg: printmania.at. https://printmania.at/spezialmaterialien/3500-3d-filament-175-mm-ceramic-keramik-1000g.html. Accessed 26 Apr 2020

47. LAYCeramic Ceramic Filament - 2.85 mm (1 kg): MatterHackers. https://www.matterhackers.com/store/l/layceramic-ceramic-filament-285mm-1kg/sk/M7RE5J6P. Accessed 26 Apr 2020
48. China keramik filament hohe qualität 3d drucker filament (1 kg): aliexpress.com. https://de.aliexpress.com/item/4000890880279.html?src=ibdm_d03p0558e02r02&sk=&aff_platform=&aff_trace_key=&af=&cv=&cn=&dp. Accessed 26 Apr 2020
49. Lay-Brick (Ceramic): Filament2Print. https://filament2print.com/gb/wood-ceramic/587-lay-brick.html. Accessed 26 Apr 2020
50. Amaco X-23 Ceramic Clay FilametTM: The Virtual Foundry. https://shop.thevirtualfoundry.com/products/amaco-x-23-ceramic-clay-filamet. Accessed 26 Apr 2020
51. Amaco 46-D Ceramic Clay FilametTM: The Virtual Foundry. https://shop.thevirtualfoundry.com/products/amaco-46-d-ceramic-clay-filamet. Accessed 26 Apr 2020
52. Amaco White 25-D Ceramic Clay FilametTM: The Virtual Foundry. https://shop.thevirtualfoundry.com/products/amaco-white-25-d-ceramic-clay-filamet. Accessed 26 Apr 2020
53. Markforged Keramik Stützmaterial – 200 cm3 Rolle: Mark3D. https://www.mark3d.com/de/produkt/markforged-stuetzmaterial-keramik/. Accessed 26 Apr 2020
54. Werkstoffe – Additive Fertigung für besondere Ansprüche: 3d-figo. https://3d-figo.de/werkstoffe/. Accessed 01 May 2020
55. Keramik - Verdrucken Sie CIM-Keramiken mit AIM3D: AIM3D. https://www.aim3d.de/materialien/keramik/. Accessed 01 May 2020
56. 3d-figo FFD 150H review - professional ceramic 3D printer: Aniwaa. https://www.aniwaa.com/product/3d-printers/3d-figo-ffd-150h/. Accessed 01 May 2020
57. AIM3D ExAM 255 review - industrial 3D printer (pellet extruder): Aniwaa. https://www.aniwaa.com/product/3d-printers/aim3d-exam-255/. Accessed 01 May 2020
58. ExOne Innovent review - industrial 3D printer (sand, metal, and ceramics). Aniwaa. https://www.aniwaa.com/product/3d-printers/exone-innovent/. Accessed 01 May 2020
59. Kwambio Ceramo One review - ceramic additive manufacturing system. Aniwaa. https://www.aniwaa.com/product/3d-printers/kwambio-ceramo-one/. Accessed 01 May 2020
60. Voxeljet VX4000 review - industrial 3D printer (large build volume). Aniwaa. https://www.aniwaa.com/product/3d-printers/voxeljet-vx4000/. Accessed 01 May 2020
61. 3D Systems ProX DMP 200 Dental review - professional 3D printer: Aniwaa. https://www.aniwaa.com/product/3d-printers/3d-systems-prox-200-dental/. Accessed 01 May 2020
62. Lithoz CeraFab 7500 review - industrial ceramic 3D printer: Aniwaa. https://www.aniwaa.com/product/3d-printers/lithoz-cerafab-7500/. Accessed 01 May 2020
63. 3DCeram Ceramaker review - industrial ceramic 3D printer: Aniwaa. https://www.aniwaa.com/product/3d-printers/3dceram-ceramaker/. Accessed 01 May 2020
64. Admatec ADMAFLEX 130 review - industrial ceramic 3D printer (DLP):, Aniwaa. https://www.aniwaa.com/product/3d-printers/admatec-admaflex-130/. Accessed 01 May 2020
65. Prodways ProMaker V6000 review - industrial ceramic 3D printer: Aniwaa. https://www.aniwaa.com/product/3d-printers/prodways-promaker-v6000/. Accessed 01 May 2020
66. Informationen zum PLA Filament: material4print. https://www.material4print.de/de/infos-zum-filament-pla__138/. Accessed 03 May 2020
67. Kaiser, A., Lobert, M., Telle, R.: Thermal stability of zircon (ZrSiO4). J. Eur. Ceram. Soc. **28**(11), 2199–2211 (2008). https://doi.org/10.1016/j.jeurceramsoc.2007.12.040
68. Bauer, J., Durakbasa, N.: Application of multi-technology in the manufacture of parts by 3d printing, ESIAM19, Trondheim, Norway, 9–11 September 2019

69. Gonzalez-Gutierrez, J., Treitler, M., Spoerk, M., Arbeiter, F., Schuschnigg, S., Lammer, H., Lackner, M., Aburaia, M., Poszvek, G., Zhang, H., Sapkota, J., Holzer, C.: Carbon fiber reinforced thermoplastics for material extrusion additive manufacturing. In: Conference Proceedings of 35th International Conference of the Polymer Processing Society (2019)
70. Aburaia, M., Lackner, M., Grünbichler, H., Engelhardt-Nowitzki, C., Markl, E., Lammer, H., Haiguang, Z., Wang, J., Sapotka, J., Janics, T., Hailberger, M.: Freeform-FDM process development using natural fibre reinforced biopolymers. In: 2nd International Conference on 3D Printing Technology and Innovation March 19–20, London, UK (2018)
71. Hage 3D. http://hage3d.com/index.php/hage3d-3ddrucker-84l/. Accessed 24 Aug 2020
72. HotEndWorks. https://www.aniwaa.com/product/3d-printers/hotend-works-hdfab-advanced-material-3d-printer/. Accessed 24 Aug 2020

Fused Filament Fabrication of Metallic Components for Semi-professional and Home Use

Günther Poszvek(✉), Gerald Stattler, Erich Markl, Rolf Seemann, and Maximilian Lackner

Department Industrial Engineering, USA Technikum Wien, Vienna, Austria
poszvek@technikum-wien.at

Abstract. State-of-the-art 3D-printing with metal filaments based on the Fused Filament Fabrication (FFF) process is discussed in this paper, by a literature review and own printing tests. Commercially available metal filaments (composites with metal powder content) are flexible enough to be processed on a low budget 3D-printer and, by debinding and sintering the produced objects, it is possible to reach 80% and more of the mechanical strength of machined metal counterparts. A market research for available metal filaments from different sellers was conducted. The filament in question is compared by price and its printing and sintering properties, and particularities have been outlined. The prices for these metal filaments range from about €100/kg to €500/kg, with one exception: a filament which contains titanium is being offered for nearly €1500/kg. Objects have been successfully printed from a purchased, sinterable metal filament ("Ultrafuse® 316L" filament from BASF containing steel (DIN 1.4404, X 2 CrNiMo 17 13 2)) with the use of a low-budget 3D-printer to verify its feasibility. The testing model for printing was a "benchy", which is a common test object in the hobby area of 3D-printing. Physical adjustments which had to be made to the printer itself and the printing parameters, as well as the difficulties faced while printing, are described in detail. This paper should serve hobbyist makers in doing the transition from plastics FFF to metal FFF to allow them to produce more durable parts.

Keywords: 3D-printing · Fused Filament Fabrication (FFF) · Metal filaments · Transition from plastics to metal

1 Introduction

In this work, the central topic of 3D printing is the melt layer process, which is also known as "Fused Filament Fabrication" (FFF) or also Filament Deposition Modeling (FDM) [1]. The focus is on a special aspect: the printing of metal parts using the FFF process. The first attempts to manufacture metal parts by this process were made in 1996 by the "Center for Ceramic Research" at Rutgers University in New Jersey [2]. This area was then continuously researched [2–4] and recently this technology has also made it into the business world, including the FFF printers that are manufactured and

N. M. Durakbasa and M. G. Gençyılmaz (Eds.): ISPR 2020, LNME, pp. 140–149, 2021.
https://doi.org/10.1007/978-3-030-62784-3_12

sold specifically for these filaments, including the necessary accessories (debinding and sintering units).

Despite the rapidly growing market for 3D metal printing in recent years, only a few suppliers for filaments and special 3D printers could be found. In addition to the largest representative in this niche market, Markforged [5], competitors hardly seem to advertise their products, and information about both competitors and their products is difficult to find. A strong increase can be seen especially in the market for 3D printers for the home desktop area [6].

The cost-effective 3D printing of plastic objects is becoming more and more accessible to people. The technology of 3D printing with metal filaments, which has only been used commercially for a relatively short time, allows printing due to the very high proportion of metal powder, such as ANSI 316 L, as is the case with the "Ultrafuse® 316L" filament from BASF of so-called green compacts [7]. The material is a composite containing steel (DIN 1.4404, X 2 CrNiMo 17 13 2). These green compacts can then be sintered into stable and resistant metal objects by washing to remove most of the binder and then heating them in a sintering furnace [8–10].

Through the FFF process, these filaments enable metal objects to be produced at lower entry costs when compared to other metal 3D printing technologies such as laser sintering or "metal injection molding" (MIM for short) [11, 12].

This makes printed metal objects more affordable, and the timely production of prototypes from metal is possible with comparatively little effort.

The printing of metal prototypes is important in that the strength of plastics parts cannot compete with the mechanical properties that can be achieved by objects printed from metal filament after sintering [13]. This is also due to the fact that the sintered objects can withstand almost the same loads as objects made of the same full metal, [4] which clearly underlines the suitability of the process for prototypes and small series production.

The market for 3D metal printing is growing steadily, [5] which is probably due to the advantages of the FFF process and the inexpensive production option, especially of prototypes and small series [12]. It is all the more important for companies that want to invest in this technology. The market is rather confusing in this regard, because despite the few providers, their offer varies greatly. There are companies that not only offer their filament, but an entire system, such as Markforged. [14], Other companies such as BASF, which entered into a partnership with EVO-tech, also sell their filament independently, and advertise their product with its compatibility with conventional 3D printing devices. The offered filament itself plays the greatest role for the customer, but it is difficult to compare filaments on the basis of the sometimes sparse company information, which is why these would first have to be verified in order to make a good purchase decision.

Since research is constantly evolving and for potential buyers a comparison of the devices available on the free market that can print metal objects with the FFF process is necessary, the following research questions arise: What do the current market offers look like? How far does what is on offer match the current state of research and do the materials of the metal filaments keep what they promise?

2 Methods

In order to obtain an overview of the current state of research, extensive research was first carried out on 3D printing with metals using the FFF process. For this purpose, eJournals were primarily used on various platforms such as ScienceDirect and IEEE Xplore. A market research on the current providers for this market was drawn up in a contrasting form. For this purpose, market prices should be obtained from the metal filaments and special 3D printers available on the market. Various factors, such as the price per kilogram, should ensure comparability. Recommendations for the printing of the metal filaments found, such as the printing temperature, or information about a necessary preheating of the filament are also clearly presented. Any information on post-processing (debinding and sintering) should also be described. Another part of the work deals with the verification of the recommended print settings of purchased filaments. For this purpose, a spool was sourced from a manufacturer with metal filament, with which the usability is tested on a simple hobby 3D printer by printing several test objects. Finally, the results of the scientific work found should be compared with the manufacturer's information on the materials and own experience.

3 Theory

In 3D printing with the FFF process, the desired object is cut into thin, two-dimensional slices, which are called layers. Stacked on top of each other, they result in the original 3D model. In these layers, the printing material is applied layer by layer through the print head, first directly on the printing plate, then onto each other (extruded), which creates the three-dimensional print [15]. The material is extruded by pressing the material through a heated nozzle. The contour of the individual layers is first generated by planar movements of the print head (or the print bed, depending on the structure of the 3D printing device) and then this contour is filled. This filling of the contour, as well as the thickness of the contour, can be variably adjusted, so that mechanically less stressed component sections can be printed with less filling in order to reduce weight and thus the volume of printing material required for this. This also reduces the associated printing time. For more complex shapes that contain overhangs or sub-objects that would have no connection with the main object or only a very small connection when stacked on top of each other, a support structure is required. The supports are comparatively thin columns, which are built under the overhangs with the actual model to support them, and thus enable the printing of independent sub-objects [15]. With the FFF process, a prototype can be produced easily without having to rely on special manual skills, expensive industrial machines and equipment. In addition, one can quickly receive test objects and thus accelerate the development process [15]. A weakness of 3D printing is the usually very different strength in the various printing directions. This results from a poorer connection of the individual, stacked layers with the layers below and above them than the connection that is present within a print layer itself. However, this effect occurs in the sub-area of metal printing that we focus on after sintering only to a minor extent [4].

3D printing of metals through the FFF process places high demands on the metal filament: it must have a high degree of rigidity, and at the same time be flexible to be wound on spools without breaking the filament to the hot end (heated part of the print head). The filament should also keep a constant, uniform diameter over the entire process in order to maintain a constant delivery rate for a good printing result. In addition, the material to be processed should have an even distribution of the binder, which should be easy to remove from the metal powder. Either by "washing" or burning/evaporating this binder [2, 16]. That debinding is done prior to sintering.

Fine metal powder is required to manufacture metal filaments. A uniform, as small as possible size of the metal particles is important in order to obtain a uniform filament, to produce uniform prints and reproducible sintering results. As a rule, the grain sizes range between around 2 to 44 μm. The average grain size of the metal powder was 17.7 μm in the current research [4]. The material is first treated with an agent to reduce the interaction forces between the particles and to lubricate the powder for subsequent mixing with the binder. Cyclohexane or stearic acid is mostly used for this [2, 16].

The metal powder is mixed with the polymer binder, and the metal powder is gradually added. The uniform distribution of the components can be estimated on the basis of the viscosity of the mixed material and therefore on the basis of the torque of the driving motor of the mixing device [2, 16],

The filament is usually manufactured using a single or twin screw extruder [2, 4, 16].

Due to the high metal content, a fine nozzle (the exchangeable extrusion nozzle) would tend to clog, which is why filament manufacturers recommend starting the first prints with a nozzle diameter of at least 0.6 mm [17]. Since the material is also very abrasive, it is recommended to use harder nozzles than brass nozzles, which are standard with many printers [18]. To minimize the risk of breakage of filament during printing, some manufacturers recommend preheating the filament. The polymer portion should soften to such an extent that it can compensate for the inflexibility of the solid metal powder due to the slight preheating [19]. Before the printed object can be sintered, a large part of the polymer portion must be removed. There are two approaches to this. The amount of binding agent can be removed from the printed green parts by "washing" using a suitable solvent or by thermal treatment of the green body. To remove the binder sufficiently, it may also be necessary to carry out both a washing and a thermal treatment [2–4, 16].

Depending on the material selected, sintering in a vacuum furnace or an inert furnace flooded with protective gas can be used for sintering, for example to avoid oxidation of the metal particles, as is also to be expected with steel at higher temperatures [20].

4 Empirical Market Research

A market research was carried out. Materials from suppliers who do not state whether their material can be sintered, but whose material data sheets allow the assumption of this fact, are also considered here.

The Virtual Foundry: The company "The Virtual Foundry" is a young company (founded in 2014) [21], which sells its filaments filled with metal, glass and ceramics under the name "Filamet" and should also have filaments according to customer requirements [22]. The company already offers the filaments in its web shop from a spool weight of 0.25 kg to 0.5 kg, depending on the material [23].

BASF: BASF sells its Ultrafuse 316L filament in 3 kg packaging units via third party webshops [18].

PT + A: PT + A (Powder Technologies and Additives) has the largest product range with a range of eleven metal filaments and also offers customers new developments if they provide materials [24].

Markforged: Markforged offers a complete system for purchase and advertises less with the printing properties of its filaments, but with the end results, the sintered components. The metal filaments are available on spools of 200 cm^3 filament volume and are designed for their proprietary 3D printer, which in the set, along with a washing station and a sintering furnace sells for 195,000 € [25].

ColorFabb: The manufacturer ColorFabb does not provide any information on the sinterability of the offered filaments. However, the information in the material data sheets of "approximately 80%" indicates a sufficient amount of metal.

Form Futura: The supplier Form Futura also has two filaments on offer, which record a metal content that is sufficiently high for sintering in the data sheets.

Desktop Metal: Unlike the filaments of the competition, "Desktop Metal" does not offer its filament in the form of spools, but in cartridges with rods made of metal filaments. The stainless steels 17-4PH and 316L as well as H13 tool steel, copper and Inconel 625 are offered in the form of bars [26]. In order to be able to process these filament rods, Desktop Metal offers a 3D printer designed for this purpose. The printer is offered as a set, with a delivery station and a sintering furnace, for € 198,000 [27, 28].

Aim3D: Aim3D also sells its materials in a form other than filament. The products sold in the form of pellets range from hard metal, stainless steel and non-ferrous metals to ceramics and conventional plastics such as ABS and PLA [29]. A special 3D printer is also required to print these materials [30].

5 Experimental

In order to verify the usability of the materials available on the free market on commercially available 3D printers and to gain experience with the printing of filaments with a high metal content, a 0.5 kg spool was purchased from The Virtual Foundry. The aim was to test the material for printing on an inexpensive 3D printer using more complex objects with overhangs and bridges.

An Anycubic i3 Mega-S was used to print the filament with 66 mass% 316L stainless steel. The device, which costs around € 250 [31], was in its original condition at the time of the tests, see Fig. 1 left.

Only the 0.4 mm brass nozzle was exchanged for a 0.6 mm stainless steel nozzle. Ultimaker's Cura 4.3 was used as slicing software. The printing result is shown in Fig. 1 right.

Fig. 1. Left: The printer used for this work, an Anycubic i3 Mega-S; Right: "Benchy" made from Ultrafuse® 316L" filament (green body before debinding and sintering)

The filament is fed to the nozzle in the 3D printer used via a Bowden tube, which leads from the extruder (component that drives the filament and thereby presses through the nozzle) to the print head and moves with it. As a result, the arc diameter of the tube changes continuously during the printing process and the filament guided therein is thus bent along with it. The purchased metal filament is very brittle when cold (compared to PLA and ABS), which is why it would break through the pressure movements without adapting the pressure conditions. Such a break can have little effect in the form of a reduced print quality or, in an unfavorable case, can mean a complete failure of the print. The risk of breakage can be minimized by preheating the filament to the softening temperature. Such preheating was achieved by housing the entire 3D printer and using the printing plate as a heating element.

A special device that was designed to preheat filament could also have been used. Such a device is called a "filament warmer" [32, 33]. In this work, no filament warmer was used because the filament in the Bowden tube can cool down again, which means that the risk of breakage during operation cannot be ruled out and is just as great at the start of printing as without this device, since one can then use a completely cooled filament to start.

The risk of breakage of the filament on the way from the spool to the extruder due to excessive tensile load was reduced by storing the filament spool on smooth-running ball bearings in order to ensure smooth unspooling.

Before each start of a print, it was noted that the installation space temperature was raised enough by the pressure plate and the flexibility of the print material was given to such an extent that breakage of the material due to the bend in the Bowden hose was reduced to a minimum.

In order to minimize the risk of the filament breaking initially, the printing speed was set to a very low 10 mm/ s for the first attempts. A value of 215 °C was initially set as the printing temperature and the material flow was set to 100%.

After leveling the new nozzle, the feed and flow were adjusted to the new material. For this purpose, 6 mm long, 10 mm wide and 8 mm high cuboids 8 mm apart were created in one print. This test print was repeated several times, with slowly increasing feed values. It was found that a 7 mm draw-in was necessary to prevent material accumulation on the component corners due to an unwanted filament exit.

The change in feed from 0 mm to 7 mm already led to a significant improvement in the fill density of the edges of the printed test objects. A significant increase in the material flow to 125% was necessary to close all visible gaps. A further increase to 130% had the result that a significant excess of material began to collect on the top of the object and on the nozzle. A value of 128% produced good surfaces and little to no unwanted deposits, which is why this value was selected for the following prints.

Over the course of these first prints, the printing speed was increased from 10 mm/s to up to 30 mm/s. It was decided to set the print speed to 15 mm/s for the larger test prints and therefore also for all further attempts at print settings.

Previous personal experience with PLA showed that a slightly lower material flow can be helpful for the clean printing of bridges, which is why this was attempted on a small model. The model produced had a length of 24 mm, a width of 8 mm and a height of 6.3 mm. The inclination from one side made it possible to evaluate the bridges formed across different distances in one print.

A tower consisting of several layers of the model previously used should provide information about the optimal printing temperature, and a second tower with a lower printing speed was created in order to be able to observe the effect of the printing speed on the printing of bridges.

The printing program was adjusted via the slicer, so that the bottom part started with a printing temperature of 235 °C, and this was reduced by 10 °C after each "floor" until it was only 205 °C in the last one. There is a clear difference in bridging due to the different printing speeds. The tower, which was printed at 6 mm/s, has a significantly poorer bridge formation throughout. The best settings were a printing speed of 15 mm/s, at a printing temperature of 225 °C. Table 1 shows the mechanical properties after debinding and sintering.

Table 1. Mechanical properties of Ultrafuse 316L, BASF Polymer and 316L Stainless Steel Composite Filament (previously named "Innofil3D"). Source: [34]

	[1]Specimen shape form E2x6x25 according to DIN 50125 [2]Undersized impact test specimen according to DIN EN ISO 148-1		
Print direction	Standard	XY	ZX
Orientation	-	Flat	Upright
Tensile strength	DIN EN ISO 6892-1[1]	561 MPa	521 MPa
Yield strength, Rp 0.2	DIN EN ISO 6892-1[1]	251 MPa	234 MPa
Elongation at break	DIN EN ISO 6892-1[1]	53 %	36 %
Impact strength charpy (notched)	DIN EN ISO 148:2017-05[2] (2mm V-notch)	111 J/cm²	-
Vickers hardness	DIN EN ISO 6507-1	128 HV10	128 HV10

6 Discussion

The wide range of different materials and the options offered by some companies to be able to develop further filaments filled with metal on customer request was surprising, as it currently seems to be a niche market. This is probably also due to the ignorance that many companies are not yet aware of the possibility of being able to manufacture components in this way, or that they may be sceptical of this technology, which is new to them.

The successful printing of the benchy models shows that metal printing on inexpensive 3D printers can be successful, and that bridging over at least 20 mm is possible, but in practice this bridging will rarely be relevant, because for high strengths, a high component density is crucial, and this cannot be given for the roof of the benchy. This is because when the first layer is printed, only loose lines are drawn, which give in when the next layer is printed on them, resulting in a lower contact pressure of the new layer against the old one. This lower pressure presumably exists over several layers and thus causes a low component density. This is why one can assume a lower component strength in the benchy top than in the rest of the component. This work has used the filament "Ultrafuse® 316L", which is a composite containing steel (DIN 1.4404, X 2 CrNiMo 17 13 2). By debinding and sintering, parts that have comparable mechanical properties as conventionally manufactured ones can be obtained. The focus of this work was on printing parameter optimization. To conclude, hobbyist makers can do the transition from plastics FFF to metal FFF to produce more durable parts.

References

1. Legal Information. Stratasys. https://www.stratasys.com/legal/legal-information. Accessed 28 Jan 2020
2. Agarwala, M.K.: Filament feed materials for fused deposition processing of ceramics and metals. In: 1996 International Solid Freeform Fabrication Symposium. Accessed 25 Jan 2020. http://hdl.handle.net/2152/70277
3. Wu, G., Langrana, N.A., Sadanji, R., Danforth, S.: Solid freeform fabrication of metal components using fused deposition of metals. Mater. Des. **23**(1), 97–105 (2002). https://doi.org/10.1016/S0261-3069(01)00079-6
4. Thompson, Y., Gonzalez-Gutierrez, J., Kukla, C., Felfer, P.: Fused filament fabrication, debinding and sintering as a low cost additive manufacturing method of 316L stainless steel. Additive Manuf. **30**, 100861 (2019). https://doi.org/10.1016/j.addma.2019.100861
5. David: Starkes Wachstum im industriellen 3D-Druck. 3Druck.com - The Independent AM Magazine. https://3druck.com/marktbericht/starkes-wachstum-im-industriellen-3d-druck-4287713/. Accessed 26 Jan 2020
6. Kappner, B., Kahlenborn, W., Richter, S., Jetzke, T., Lessmann, A., Bovenschulte, M.: Die Zukunft im Blick: 3D-Druck, p. 64, May 2018
7. Ultrafuse 316L|BASF Polymer and 316L stainless steel composite filament. https://www.basf.com/global/en/who-we-are/organization/locations/europe/german-companies/basf-3d-printing-solutions-gmbh/metal-solutions/Ultrafuse_316L.html. Accessed 04 May 2020

8. Biggs, J.: The Markforged Metal X prints real metal parts. TechCrunch. http://social.techcrunch.com/2017/01/06/the-markforged-metal-x-prints-real-metal-parts/. Accessed 26 Jan 2020
9. BASF launches Ultrafuse 316LX for fused filament fabrication of metal parts. Metal Additive Manufacturing, 03 July 2017. https://www.metal-am.com/basf-launches-ultrafuse-316lx-fused-filament-fabrication-metal-parts/. Accessed 25 Jan 2020
10. Ultrafuse 316L Metal Filament - 1.75 mm - 3 kg. Ultrafuse FFF. https://www.ultrafusefff.com/product/ultrafuse-316l-metal-filament-1-75mm-3kg/. Accessed 19 Dec 2019
11. FDM vs. SLA vs. SLS: 3D-Drucktechnologie im Vergleich. Formlabs. https://formlabs.com/de/blog/3D-drucktechnologie-vergleich-fdm-sla-sls/. Accessed 19 Dec 2019
12. DEO, The Economics of MIM Versus 3D Metal Printing. 3DEO. https://news.3deo.co/metal-injection-molding-vs-3d-printing. Accessed 26 Jan 2020
13. Rösler, J., Harders, H., Bäker, M.: Mechanisches Verhalten der Werkstoffe, 5, aktualisierte und erweiterte Auflage. Springer Vieweg, Wiesbaden (2016)
14. The Metal X metal 3D printer. https://markforged.com/metal-x/. Accessed 26 Jan 2020
15. Fastermann, P.: 3D-Druck/Rapid Prototyping: eine Zukunftstechnologie - kompakt erklärt. Springer Vieweg, Berlin (2012)
16. Wu, G., et al.: Fabrication of metal components using FDMet: Fused Deposition of Metals. In: SFF Symposium Preceedings, University of Texas (1999).http://utw10945.utweb.utexas.edu/Manuscripts/1999/1999-090-Wu.pdf
17. Stainless Steel 316L FilametTM. The Virtual Foundry. https://shop.thevirtualfoundry.com/products/stainless-steel-316l. Accessed 02 May 2020
18. BASF Ultrafuse 316L Metal 3D Printing Filament - 1.75 mm (3 kg). MatterHackers. https://www.matterhackers.com/store/l/basf-ultrafuse-316l-metal-composite-3d-printing-filament-175mm/sk/MRDKJRRS. Accessed 18 Dec 2019
19. Use Instructions - Metal Filaments. Metal Filament|3D Printing|The Virtual Foundry, LLC. https://www.thevirtualfoundry.com/help. Accessed 04 May 2020
20. Debinding and sintering of metal injection molding (MIM) components. Handbook of Metal Injection Molding, pp. 129–171, January 2019. https://doi.org/10.1016/b978-0-08-102152-1.00009-x
21. The Virtual Foundry|LinkedIn. https://www.linkedin.com/company/the-virual-foundry-llc. Accessed 03 May 2020
22. The Virtual Foundry: Complete 3D Printing Materials. Metal Filament|3D Printing|The Virtual Foundry, LLC. https://www.thevirtualfoundry.com. Accessed 07 May 2020
23. Metal 3D Printing Filaments - Additive Manufacturing. Metal Filament|3D Printing|The Virtual Foundry, LLC. https://www.thevirtualfoundry.com/about-us. Accessed 03 May 2020
24. Pischang, Karsten: Bachelorarbeit zu Metall-3D-Druck, 05 May 2020
25. Markforged_Broschüre_2019.pdf. https://www.bibus.at/fileadmin/editors/countries/biaus/Mechatronik/Medien/MArkforged/Markforged_Brosch%C3%BCre_2019.pdf
26. Materials|Desktop Metal. https://www.desktopmetal.com/materials. Accessed 03 May 2020
27. Preisuebersicht (Desktop Metal Studio System inkl. Fleet).pdf. https://shop.solidpro.de/media/products/Preisuebersicht%20(Desktop%20Metal%20Studio%20System%20inkl.%20Fleet).pdf. Accessed 03 May 2020
28. Meiser, W.: Desktop Metal Studio System+, 3D CAD ProMachine. https://shop.solidpro.de/de/desktop-metal-studio-system.html. Accessed 05 May 2020
29. Materialien - 3D-Druck mit kostengünstigem Spritzgussgranulat. https://www.aim3d.de/materialien/. Accessed 05 May 2020
30. ExAM 255 - Ein 3D Drucker der Metallspritzgussgranulat verarbeiten kann. AIM3D. https://www.aim3d.de/produkte/exam255/. Accessed 05 May 2020

31. [Pre-order]i3 Mega S: ANYCUBIC 3D Printing. https://www.anycubic.com/products/anycubic-i3-mega-s. Accessed 05 May 2020
32. Der Mohawk Filamenthalter und Wärmer für den T850P, T850 und T1250. Tractus3D. https://tractus3d.com/de/unsere-produkte/mohawk/. Accessed 07 May 2020
33. FilaWarmer for Fialment, iMakr. https://www.imakr.com/eu/filamet-metal-/fila-warmer-for-filament, Accessed 3 May 2020
34. https://www.igo3d.com/mediafiles/Sonstiges/BASF/Ultrafuse/Broschuere/Ultrafuse-316L-Product-Brochure.pdf. Accessed 04 May 2020

Increasing of Stud Welding Efficiency in the Body Shop

Marian Králik(✉) and Vladimír Jerz

Institute of Manufacturing Systems, Environmental Technology and Quality Management, Faculty of Mechanical Engineering, Slovak University of Technology in Bratislava, Bratislava, Slovak Republic
{marian.kralik, vladimir.jerz}@stuba.sk

Abstract. The article describes possibilities of increasing of stud welding efficiency for future SUV projects proposed by the Volkswagen concern. The present state simulation and the proposal of welding line optimization of stud welding with the use of the Process Simulate software are presented in the article. Furthermore, the description of stud welding technologies is involved. The basic concepts of stud welding are analyzed in the chapters below. The last chapter aims at the description of the model creating process, simulation and the proposal of welding line optimization for stud welding with the use of the Process Simulate software.

Keywords: Stud welding · Efficiency · Robot · Process simulate · Simulation · Welding line

1 Description of the Used Stud Welding Technologies

The technology of stud welding was discovered before World War II in order to simplify the construction of huge military vessels. The complicated and time-consuming drilling of holes in the steel deck boards was eliminated, which significantly saved time.

Compared to other joining technologies, e.g. resistance welding, manual welding, drilling and threading, stud welding is a faster and easier method. The welded stud provides a high-quality connection joint that is resistant to refracting [1, 2]. The stud material and base material must be compatible and suitable for the welding process. Aluminum, mild steel and stainless steel studs are used for arc welding methods. The carbon content of steels should not exceed 0.3%. When welding steel studs, a small amount of flux is required, which is fixed to the stud head. It is instrumental to stabilize the burning of the arc [1–3].

This technology is currently very widely used. In the automotive industry, they are used to fasten mounting components, e.g. power steering, dashboard, insulation, exhaust system, upholstery, wiring (Fig. 1) [1, 2].

As the number of studs increases, so does the number of robots used to weld the bolts to the body shell. Studs of different types and sizes are used for each model. For example, both the steel and aluminum types are used for the latest Audi Q7. Some differ only in the surface finish of the material (Fig. 2).

N. M. Durakbasa and M. G. Gençyılmaz (Eds.): ISPR 2020, LNME, pp. 150–161, 2021.
https://doi.org/10.1007/978-3-030-62784-3_13

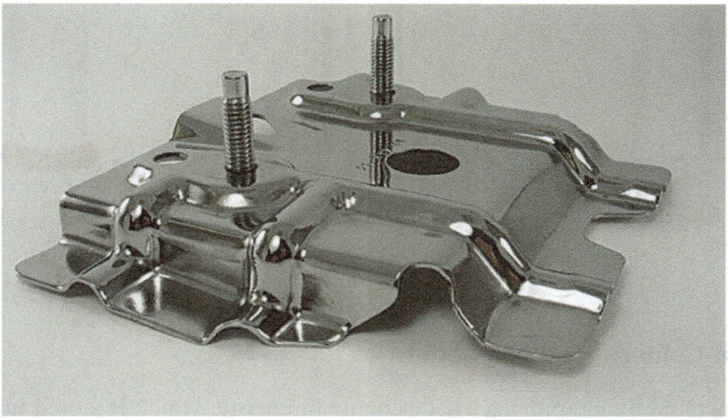

Fig. 1. Studs welded to bodyshell part

Steel studs			Aluminum studs			
T5x16 FB	M6x23F13	12,5x21HM	T5x15,5F7,5	M6x18,5F13.4,3	14,5x27	18,5x33,5 HM

Fig. 2. Studs used on the Audi Q7 NF

1.1 Arc Welding of Bolts with Stroke

The welding with stroke method is intended for welding connecting elements (studs) from 3 to 30 mm. Due to the greater depth of melting of the base material (1–3 mm), this method of welding is used for sheets with a thickness of more than 2 mm. This method is suitable for pins made of aluminum alloys or stainless steels due to the use of a protective atmosphere. With this technology, ceramic rings are used to form the weld. The advantage of this technology is especially high strength, possible diameter of bolts from 3 to 30 mm, etc.

The basic precondition for the start of the welding process is the contact of the stud with the base material. First, the tip (face) of the stud touches the surface of the metal sheet. Then the current is switched on and the bolt is lifted from the sheet metal surface by means of a linear motor. The resulting gap between the bolt and the sheet initiates the formation of an electric arc, which begins to melt the surface of the stud and the surface of the sheet. When the arc forms a melt on the surface of both the bolt and the sheet, the opposite movement occurs and the bolt is pushed into the melt by a controlled movement. Subsequently, the current is switched off, the molten metal begins to harden and a weld is formed (Fig. 3).

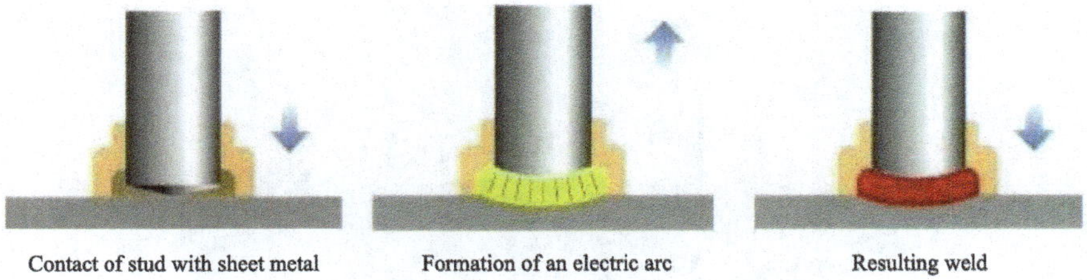

Fig. 3. Arc stroke welding of studs

1.2 Point-Contact Welding of Studs

Heat is generated by a rapid electric discharge of energy, which is stored in a series of capacitors. Discharge is performed in an extremely short time - approximately 0.001 to 0.003 s. This type of welding is suitable for steel, aluminum and brass elements with a diameter of 3 to 8 mm. Since the cycle time is extremely short, the heat output is small compared to conventional arc welding and thus the heat affected zone is very small. The cooling rate is high, which results in a heat-affected area with high hardness and also brittleness, which can cause problems e.g. under cyclic loading of the component [1, 3]. The advantage is the possibility to weld the studs to very thin sheets (from 0.5 mm) without the risk of remelting the entire sheet thickness. The basic precondition for the start of the welding process is the contact of the stud with the base material. The principle of this technology is shown in Fig. 4.

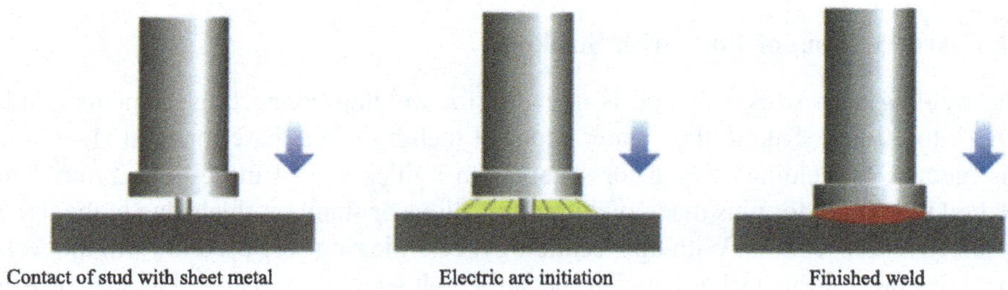

Fig. 4. Point-contact welding of studs

1.3 Welding of Studs in a Magnetic Field

The patented method of welding studs in a magnetic field (Magnetic field stud welding-SRM) is the most modern way of welding studs. It expands the use and increases the quality and efficiency of work in both manual and automated welding. This type of welding is suitable for steel, aluminum, brass and stainless steel elements with a diameter of up to 16 mm. The advantage is the reduction of the stud scorching. The studs have a thread along their entire length. First, the tip of the stud comes into contact with the sheet metal, and then the stud is removed and a welding arc is created. Figure 5 shows the procedure. The arc is guided by a magnetic field. Both the stud and the metal sheet are melted. The stud is immersed in a shallow zone of molten material and a non-detachable joint is formed.

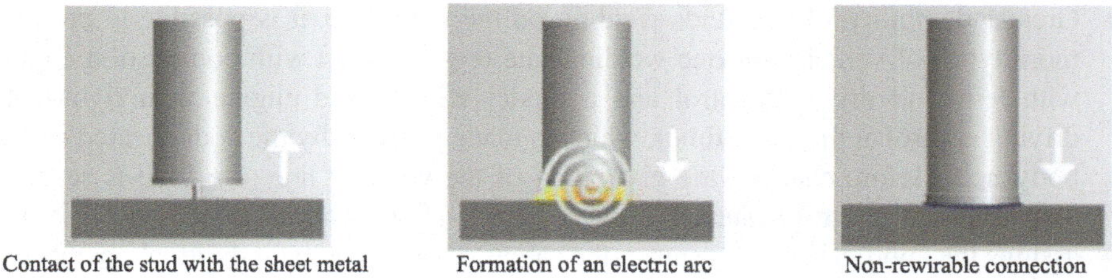

Fig. 5. Welding of studs in a magnetic field

2 Automation and Robotization of the Process

Two basic concepts of stud welding are used:

- concept of studs welding on the main welding line
- concept of studs welding on subgroup lines (line for welding of, e.g. front part of bodyshell floor, etc.)

The first way is to weld the studs on the main welding line. In this way, the bodyshell moves on a roller conveyor on a technological pallet ("skid").

The second way is the welding of studs on individual lines of bodyshell subgroups. (e.g. front floor welding line). A robot with a gripping effector is used to manipulate bodyshell parts. On the subassembly welding lines, the stud welding heads are mounted on static structures (Fig. 6).

Fig. 6. Stud welding on the main line and stud welding on the line of subgroups

In an SUV body builder manufacture, most of the studs are welded on the main line and some on the subgroups. In the C-SUV body shop, the concept of welding on subgroup lines was used in particular. The information obtained shows that as the most suitable concept of stud welding, it is appropriate to use the concept of stud welding only on the main line, which would increase the use of robots in the technological process, as used in the SUV body shop.

The advantages of the proposed solution include:

- Greater flexibility: When studs welding on the main line, it is possible to produce more types of vehicles on one welding line (e.g. a model with combustion engine with left-hand driving control and a model with hybrid engine with right-hand driving control at the same time), which is made possible by the code located on the bodyshell. Before entering the given part of the welding line, the code is scanned and the robot control system receives a signal about what type of body it is and assigns the appropriate programs to the robots performing the individual operations.
- Higher use of robots. The analysis showed that using this concept, it is possible to achieve a higher use of robots. In the NSF body shop, the level of utilization of robots is almost 60%.
- Lower failure rate: When welding steel studs, the failure rate of the equipment used for welding is 0.8% and for aluminum studs 1.6%. These data are statistically processed and regularly updated in the internal systems of Volkswagen Slovakia, Inc.

Disadvantages include:

- Investment costs for the swiveling table: To weld the studs from the lower part of the bodyshell, for access it is necessary to use a positioning device that rotates the body 180 degrees and also a support system for robots that are elevated approximately 2.5 meters above the welding shop floor.

3 Solution of a Part of the Welding Line Using the Process Simulate Software

Based on the analysis of the use of welding robots and the proposed recommendations for increasing the load of welding robots, two welding stations AFO3830 and AFO3840 were selected for welding steel studs on the main welding line U2A7. Process Simulate software from Siemens, which is part of the Tecnomatix software package, will be used to verify the possibility of increasing efficiency of the robots.

3.1 Description of the Welding Line

The given welding line consists of two welding cells (AFO3830 and AFO3840) and is used only for welding steel studs. The AFO1830 station consists of four KUKA angular robots and the AFO1840 station consists of three KUKA angular robots with six degrees of freedom. It is concerned the KR210R3100 model. The robot's designation indicates the permissible load - 210 kg. The robots have a range of 3100 mm and positioning accuracy is ± 0.06 mm.

The bodyshell (Fig. 7) goes along the roller conveyor on the technological pallet ("skid") first to the station 3830. In this station, the body is clamped on a swiveling table (Fig. 8), where it is geometrically precisely fastened, thanks to RPS (Reference

Point System). It is a system of openings and surfaces by means of which the body is guyed with a tolerance of ±0,1 mm. Subsequently, the bodyshell rotates 180 degrees and the robots perform individual operations on the lower part of the body. After the welding of the studs is completed, the body is rotated again by 180 degrees and then passes on the technological pallet to the station 3840. Here the studs from the upper part of the bodyshell are welded.

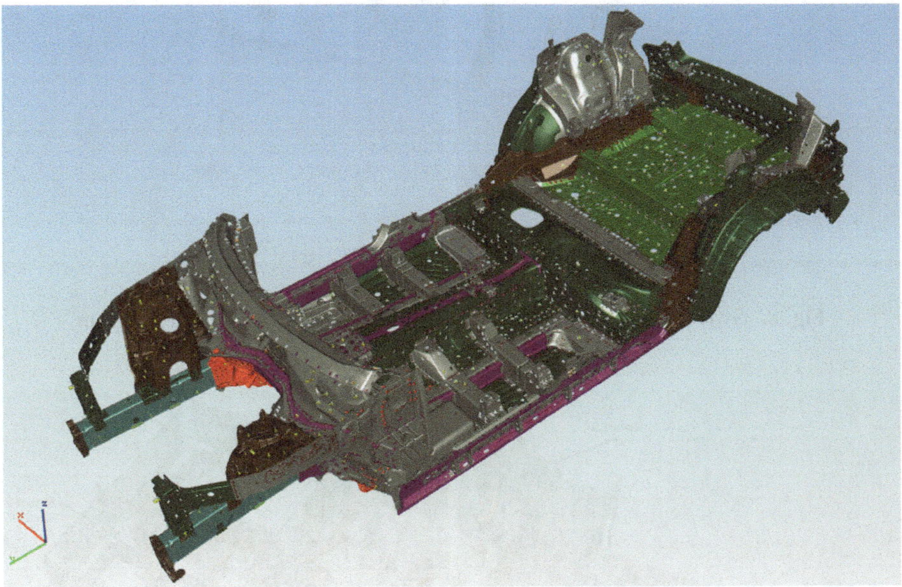

Fig. 7. Body model of the new generation Audi Q7

3.2 Dynamic Simulation

Volkswagen Slovakia, Inc. uses digital factory tools, specifically Process Designer (PD). Using the library of parts of welding workplaces and body parts, it was possible to model a welding line U2A7 (stations 3830 and 3840) in the Process Simulate program (hereinafter also PS - Fig. 9).

The model of line consists of several basic parts:

Parts: Models of parts and assemblies (e.g. bodywork) that are used in operations, parts are read from the library.

Resources: Welding line equipment models - robots, welding equipment, clamping tables, etc.

Manufacturing Features: In Process Simulate, this data is abbreviated MfG and is located in a separate library. The coordinates of spot welds, trajectories after continuous welding (Continuous Features), cutting, paintwork, etc. are considered such data.

Operations: As part of the rough planning, sub-operations are prepared for the workstations and their estimated times, which are then combined into the final simulation.

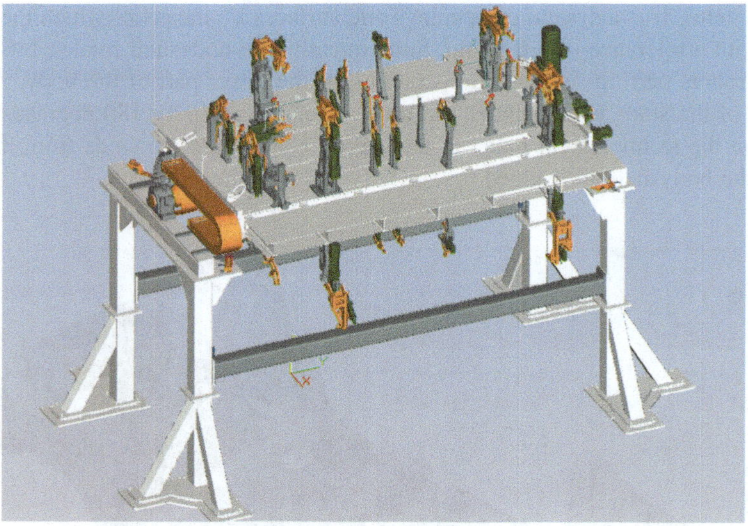

Fig. 8. Swiveling table model with fixtures in the Process Simulate program

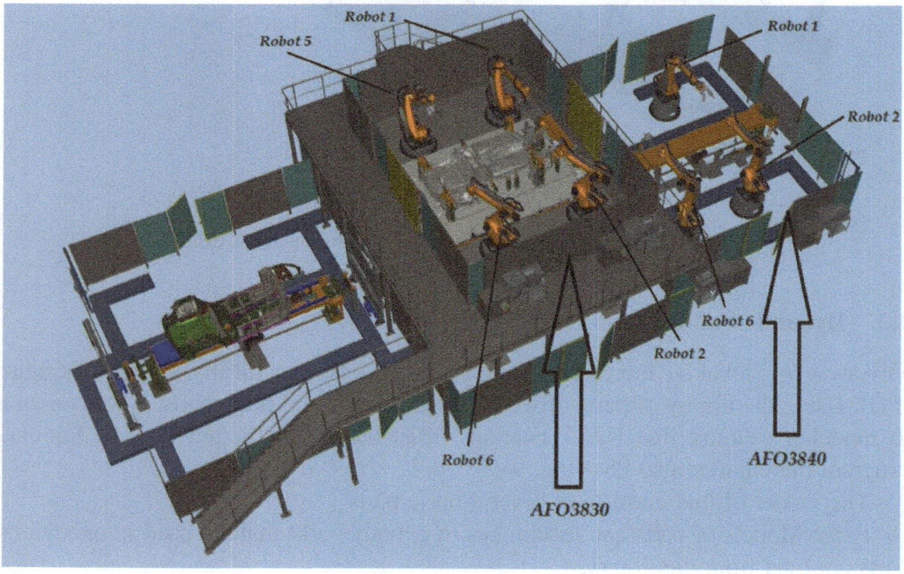

Fig. 9. U2A7 welding line model in Process Simulate

Creating a dynamic simulation in Process Simulate is divided into several steps:

1. Conversion of all models of lines to jt format. Setting kinematic chains, positions and coordinate systems for models of lines.

2. Creation of welding paths by projection of welding points on individual bodyshell parts.
3. Testing of simulation of partial operations and stations.
4. Integration of stations simulations into a complete welding line simulation.

3.3 Setting the Kinematics of the Line Parts

When defining kinematic relationships, the first step is to define the individual members of the chain. That is, which elements will form a whole from the point of view of kinematics and together with other members will be connected by joints. The joint is formed by the bond of two members of the kinematic chain, which defines the properties and conditions of movement. It is possible to realize rotary and sliding binding.

The Kinematics Editor tool is used in the PS to set the kinematics. The tool window shows a schematic layout representing the kinematic chain, where the individual color-coded blocks correspond to the individual members of the kinematic chain. The arrows show the individual bonds, with the blue color corresponding to the sliding bond and the black to the rotary bond (Fig. 10).

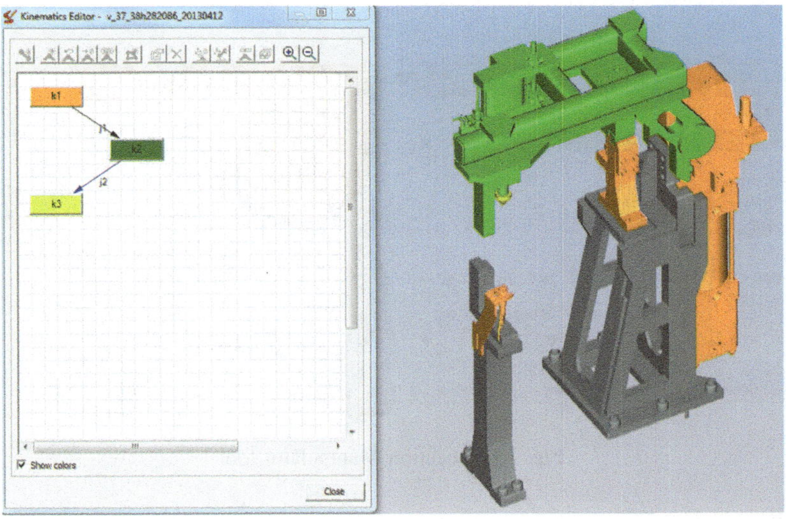

Fig. 10. Kinematic chain of a part of the clamping table

To perform the simulation of welding operations, it was necessary to read welding points from the MfG library (Fig. 11). These points have coordinates synchronized with the product coordinate system. Subsequently, they were assigned to a particular connecting parts (using the Project Weld Points tool). This created the position of the welding point (Fig. 12) and created a location containing a coordinate system that determines the direction of access. The welding head points to the point using the TCP (Tool Center Point) coordinate system.

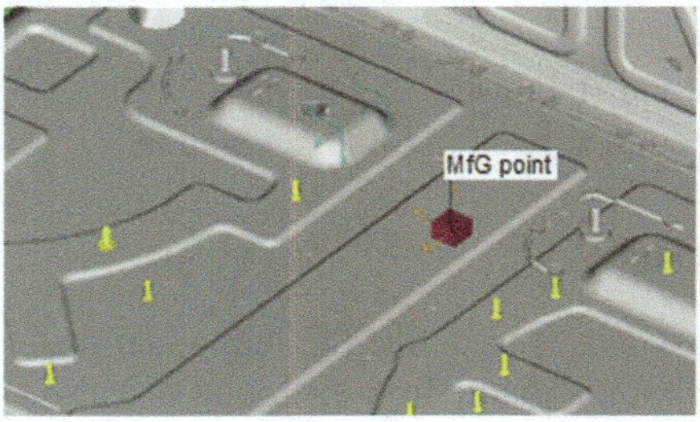

Fig. 11. Visualization of the MfG point

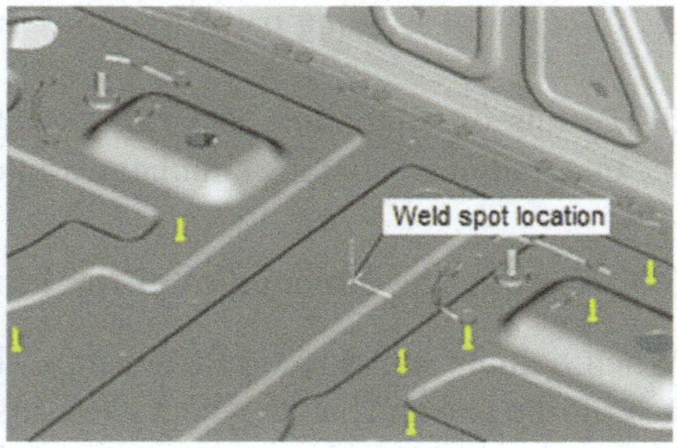

Fig. 12. Visualization of the created weld spot location

The initial rotation of the welding spot may not be suitable for a collision-free state to perform the operation and it is necessary to test the given points. The Process Simulate (PS) program allows several ways to rotate the welding spot. These are the Location Pie Chart and Rotate Locations tools, respectively Multiple Rotate Locations, which allow direct manual rotation of the welding head at a given location (Fig. 13).

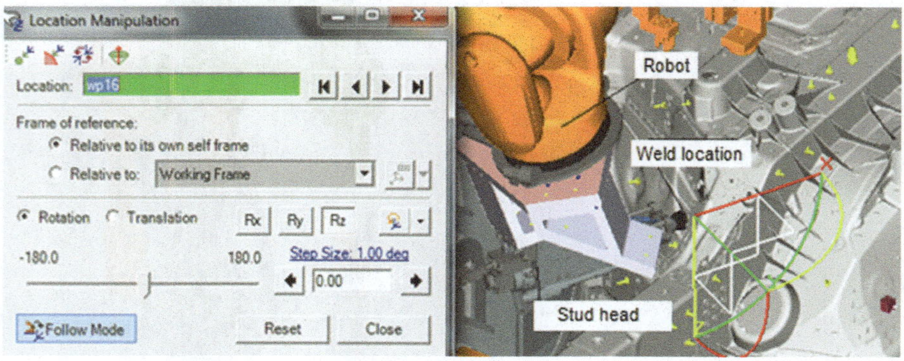

Fig. 13. Location Manipulator tool

3.4 Solution Simulation

After defining the welding points and individual points of the trajectory, the next step is to proceed to the simulation. The Object Flow operation was used to simulate the transport of the bodyshell between workplaces. The Device operation was used to open and close the fixtures, to move the body to the fixtures on the lifting device. The Weld operation was used to perform the stud welding.

By combining individual operations, a Time-based simulation was created during one cycle. A Gant diagram (Fig. 14) is used to show the sequence of operations and time consuming of operations, from which it is clear which operations take place simultaneously and how they are linked to each other.

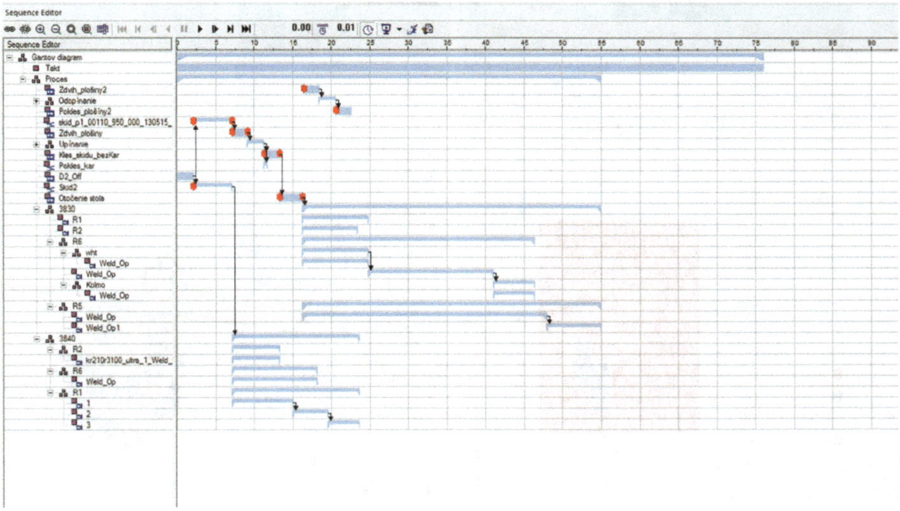

Fig. 14. Gantt diagram

The final phase of the simulation can be the generation of programs for a specific robots (offline programming method) in Process Simulate and their subsequent upload to the real control system.

3.5 Design of Welding Line Optimization

The optimization of the U2A7 welding line is aimed at increasing the load on robots. The simulation of the current state showed that the robots on the U2A7 line are not sufficiently used. The aim is to select the studs welded on the subgroup lines, to verify the possibilities of their welding on the main line and to quantify the potential savings in removing robots on the subgroup lines. The manner of this optimization can be used in the design of a new welding line for future models of the Volkswagen Group built on the MLB EVO platform (the next generation Porsche Cayenne, Volkswagen Touareg and others).

Another powerful tool that can be used is Smart Placement. After assigning welding points for a given robot, it is possible to optimize the location of the robot in terms of reach (Fig. 15).

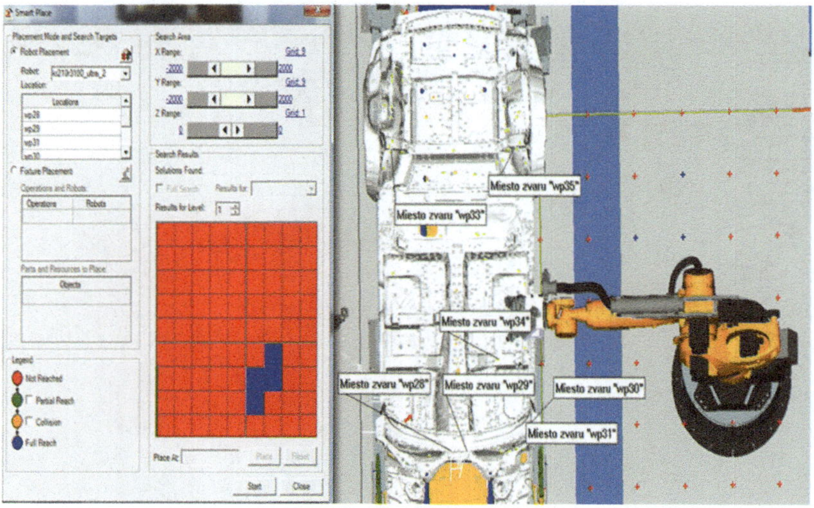

Fig. 15. Smart Place tool

4 Conclusions

The aim of the solution on the U2A7 welding line is to improve the usability of robots for welding steel studs, by reassigning a part of the studs welded on the lines of subgroups, whereby the number of robots will be reduced. Studs have been selected that can be reassigned to the main welding line. Subsequently, in the Process Simulate program, the possibilities of welding selected studs were verified using the Smart Place and Location Manipulator tools, whether the given weld points are within reach of the robot and there are no collisions of the robot or stud welding head with the bodyshell. The solution can reduce the number of robots on the line and thus reduce considerable financial costs for its realization.

Acknowledgment. This paper was created with support of the Cultural and Educational Grant Agency of the Ministry of Education, Youth and Sports of the Slovak Republic under the KEGA 027STU-4/2019 contract.

References

1. NELSON Stud Welding: Principles of stud welding. https://nelsonstud.com/pdf/. Accessed 02 Feb 2020
2. Hobson, B.: Welding: Arc Stud Welding. In Assembly Magazine 2001,. http://www.assemblymag.com/articles/83719-welding-arc-stud-welding. Accessed 23 Apr 2020
3. Stud Welding Products, Inc.: Capacitor Discharge. http://studweldprod.com/downloads/cat64.pdf. Accessed 23 Apr 2020

4. Heinz Soyer Bolzenschweißtechnik GmbH: Magnetfeld-Bolzenschweißen. http://www.gunmak.com.tr/tr/baglanti/assets/soyer_katalog_2015_lr.pdf. Accessed 23 Apr 2020
5. GoWelding: The History of Welding. http://www.gowelding.org/articles/history-of-welding. Accessed 23 Apr 2020

Measurements of a Finished Garment Using 3D Laser Image Processing

Semih Dönmezer[1,2(✉)]

[1] Interchangeable Manufacturing and Industrial Metrology, Vienna University of Technology, Vienna, Austria
semihdonmeze@anadolu.edu.tr
[2] Anadolu University EMYO, Eskişehir, Turkey

Abstract. This paper presents a determine defect detection system, measurement error or gather problem for the garment industry. The measurement of the completed suit and the followed evaluation of sewing quality and the correct joining of clothes is an important part in the industrial garment industry. However, the main difficulty in using visual inspection to evaluate the finished garment quality is the time taken and labor wasted. In order to solve these problems, if we consider the structure of the dress, a three-dimensional measurement is required. The trouble with respect to vision-based system is the robustness of image processing algorithms. Safety and stability of finished goods are the key importance when dealing with an especially textile and automotive project. The stability of the finished products largely depends on their overall geometry, and the results of the analysis depend on the certainty that this geometry can be measured practically. A three-dimensional laser scan will be required to determine whether a finished industrial product is manufactured within nominal scales. it helps to find different points of based on 3D-point measurement, and to allows determining the lines of connections point, straightness, flatness, roundness, cylindrical shape, symmetry, co-axiality, parallelism, perpendicularity, inclination, uncertainty and position in the structure. This contribution gives a critical appraisal of the process used in the preliminary research phase: gathering the point clouds, 3D-modelling, structural analysis calculating from the normative design to the real design and the tolerance requirements that are obtained. The focus on the added value of the methodology is applicability of product specification, verification, cost-effectiveness, overall advantages and drawbacks.

Keywords: Edge detection · 3D shape laser measurement · Image processing

1 Introduction

The purpose of this study is to measure non-solid objects and techniques used for measurements. It is extremely important to know that size of a product is suitable for manufacturing and fit for human body. Considering the companies producing textile, it is necessary to know the suitability of the produced jackets to the drawings or projected designs. Because designs for clothing are usually drawn in two dimensions and used by people in three dimensions.

N. M. Durakbasa and M. G. Gençyılmaz (Eds.): ISPR 2020, LNME, pp. 162–174, 2021.
https://doi.org/10.1007/978-3-030-62784-3_14

Furthermore, it should be known whether there is a difference between the nominal value of the measured product and the actual value in the application, and it should be checked whether this difference is significant enough to prevent production. It is not possible to produce a quality product without knowing the difference between the drawing (nominal) value and the actual value. Quality is often an expensive approach. Measurement is a beginning to produce quality products, and we can only understand quality by measuring them. A clothing design is papered in two sizes, but used in three sizes. A dress is designed in two sizes and manufactured in three dimensions to fit the person perfectly.

2 Related Works

Gray code is the basic and most widely used coding scheme in the field of light, structured for its simplicity and robustness (Sansoni 2000; 49(3)). If a binary Gray code pattern with a code word length of n is to be used, an image sequence consisting of n + 1 binary stripe patterns must be reflected in order to encode the scene. With the decoding procedure, the scene image can be divided into 2n sub-regions, each with a unique code word ranging from 1 to 2n. To achieve sub pixel measurement accuracy, sinusoidal phase shifting layouts are often used in collaboration with Gray code layouts (Song 2017). Vision-based measurement has gained wide applications in instrumentation and industrial inspections (Shirmohammadi 2014). Compared with traditional 2D image-based measuring techniques, more recent research works are focused on the 3D measuring technologies, such as stereo vision laser 3D scanning and structured light sensing (e.g. (Sun 2015), (Liu 2016), (Zhang 2010)). With the system calibration parameters, 3D coordinates of the surface points with coding information can be precisely calculated via triangulation principle (Hartley 2003). In computer vision and patter recognition jargon, the term features is of- ten used to refer to persistent elements of a shape (such as corners or sharp edges), which capture most of the relevant information, and based on which one is able to perform object analysis. In the last decade, feature-based methods have become a standard and broadly used paradigm in various applications, including retrieval and matching (e.g. for multi-view geometry reconstruction), due to their relative simplicity, flexibility and excellent performance in practice. An important milestone was the introduction of scale invariant feature transform (SIFT- Scale-İnvariant Feature Transform) (Lowe 2004). In computer vision, it is common to work with an image of a physical object, representing both its geometric and photometric properties. Such a representation simplifies the task of shape analysis by reducing it to simple image processing operations, at the cost of loosing information about the object's 3D structure, which cannot be unambiguously captured in a 2D image. In computer graphics and geometry processing, it is assumed that the 3D geometry of the object is given. Depending on application, the geometric representation of the object can differ significantly. A similar trend is emerging in 3D shape analysis in a variety of areas including non-rigid shape retrieval and shape matching, where fundamental differences between 2D and 3D shapes make it difficult to apply existing techniques from computer vision (Bronstein 2010).

3 Measurement Strategy

The measurement strategy contains specifications for the selection of the measuring equipment, the auxiliary equipment and the type of measurement. The number and distribution of the measuring points determines how representative the measuring points are for mapping the work piece surface. In addition to a sufficiently high number of probing points, special attention must be paid to the best possible distribution of the probing points on the work piece surface or on the geometric elements. This can have a decisive influence on the measurement uncertainty. The measuring method defines the measuring principle and the method for a measurement. The measuring principle determines the physical basis used, e.g. B. Optical measurement using the triangulation method. The method defines the procedure for the measurement, e.g. B. whether relative or absolute is measured. The evaluation strategy defines the association procedure used for the evaluation. The choice of a suitable evaluation strategy is an essential criterion for the determination of functional measurement results. This choice can have a decisive influence on the measurement uncertainty (Keferstein 2017).

The influences on the measurement result consist of a large number of components and can be divided into:

- Systematic deviations and
- Random deviations

Known systematic deviations are constant and determinable in terms of size and sign. In order to determine a systematic deviation, a standard with a known correct value is required. The correction of the measured value by the known systematic measurement deviation through the correction K leads to the corrected value. If systematic deviations are known, they should always be used for correction (Keferstein 2017).

3.1 Key Aspects of Quality for the Customer (https://www.inspection-for-industry.com/finished-goods-inspection.html) (2020)

- Aesthetics- Looks and style
- Performance + can go it faster
- Reliability + will it be available when needed
- Good functionality- it does the job well
- Reliable- acceptable level of breakdowns or failure
- Consistency
- Durability- lasts as long as it should
- Value for money.

3.2 Surface Form

The past two decades have seen the growing importance of computer aided manufacturing methodology as a way to control industrial production. Testing the technical product with high accuracy and improving the quality of all kinds of products.

Therefore, an advanced manufacturing technique should be considered as a crucial requirement for the production of industrial goods with controlled and optimized quality. It is common in both consumption and capital goods, free from surfaces. Electrical appliances and home appliances, tribune wings are a few examples.

Unlike the geometry element, freeform surfaces are difficult to define mathematically as circles, cylinders and cones. Therefore, an increasing CAD (Computer-Aided Design) application in construction is used in many areas today. Thanks to the increasing CAD application, the nominal geometry of the work pieces will be defined and available in the data storage environment of the CAD system.

Non-technical surface is such as free from surfaces, which can be not described closed analytical. Therefore, especially the automotive industry developed already prematurely characteristic mathematical approximation and interpolation methods in order to be able to describe of the diverse body form.

- Coons-representations
- Bezier-ply nominal and
- Basis-polynomial.

The polynomial degree is depending on the amount of the polynomial fulcrums. For the data exchange between different Computer-aided design (CAD)/Computer Aided Manufacturing (CAM)/Computer-aided quality assurance (CAQ)systems has achieved the free from surface interface specified by the association of the automobile industry (VDA- German Association of the Automotive Industry) and is supported by almost all system manufactures (Manikanta 2003).

4 Edge Detection

Using edge detection and edge detection. First, edges are corresponded to object bounders where pixels image brightness changes significantly. Usually edge detection method calculated from image function behavior in the pixel as a vector variable.

Edges is abrupt in the intensity of pixels, discontinuity in images brightness or contrast. Usually edges occur on the boundary of two regions. Process of identifying edges in an image to be used as fundamentals asset in image analysis. Reduce unnecessary information in the image while preserving the structure of the image. Extract important features of an image includes corners, lines and curves. For the recognize the objects it is necessary to find boundaries and segmentation. That means computer vision and recognition. There are various edge types, such as step edge, ramp edge, ridge and roof (Fig. 1).

The main idea of edge detection, first classical gradient edge detection. Sobel operator. The Sobel filter is used for edge detection. It works by calculating the gradient of image intensity at each pixel within the image. The result shows how abruptly or smoothly the image changes at each pixel, and therefore how likely it is that that pixel represents an edge. The Prewitt Edge filter is used to detect edges based applying a horizontal and vertical filter in sequence. Both filters are applied to the image and summed to form the final result. The two filters are basic convolution filters of the form: Horizontal Filter.

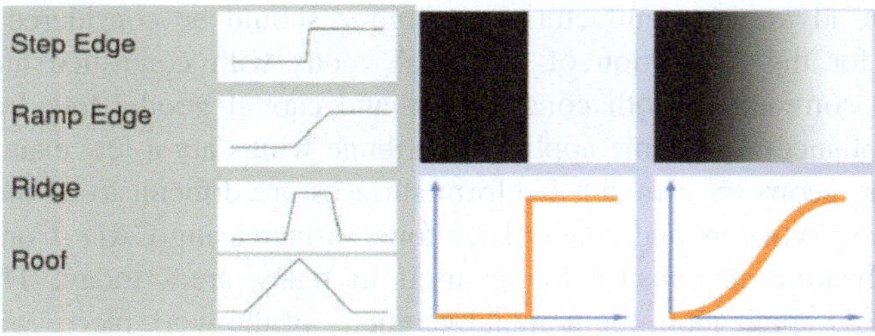

Fig. 1. The Types of edges

4.1 Main Steps in Edge Detection

- Smoothing which suppress as much noise as possible to avoid the destroying the edges,
- Enhancement, differentiation to enhance the quality of edges by using sharpening,
- Using thresholding to determinate which edge pixels should be discarded as noise and which pixel should be retained.
- Localization is determinate the exact edge location. Edge thinning and linking are usually required in this step (Manikanta 2020) (Fig. 2).

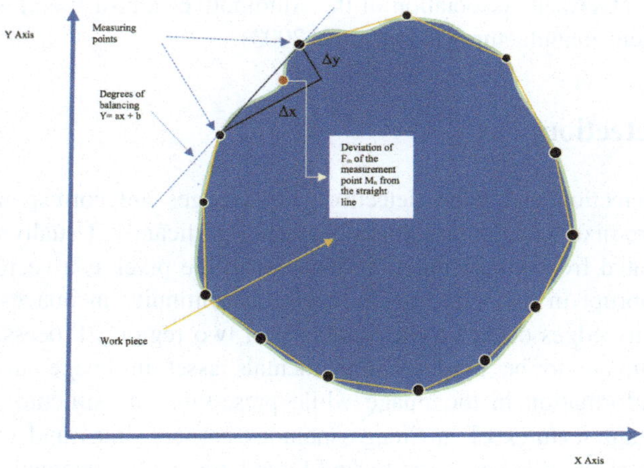

Fig. 2. Calculation of the geometry element straight according to Gauss

If work piece surfaces are touched with a larger number of measuring points than corresponds to the mathematical minimum number of the geometric element to be assigned, this is mathematically over determined. A compensation element is then calculated using the regression calculation. There are various approaches to this. Compensation calculation according to Gauss, example straight line: A straight line is sought and its slope (a) and its intercept (b) are calculated so that the sum of the squared deviations between this straight line and the probed coordinates of the work piece surface will be minimal (Keferstein 2017).

Gradient operators; Prewitt, Sobel, Laplacian of Gaussian (Marr-Hildreth) and Gradient of Gaussian (Canny). The Goal of gradient operators to identify changes of an image in manufacturing as surface discontinuities, characterization and sizing of discontinuations and perpendicular to the surface. Intuitively, most sematic and shape information from the image can be encoded in the edges and sometimes more compact than pixels. As a ideal edge of image, artists line drawing, but artist is also using-level knowledge. This would be carry us a slightly different dimension.

4.2 Image Smoothing

Expect for the pixels to be like their neighbors, they contain relatively few reflectance changes and are generally expected to be pixel-to-pixel independent, as smoothing the noise suppresses the noise.

Gaussian Smoothing is as σ increases, more pixel are involved in average and image become more blurred (Figs. 3, 4, 5).

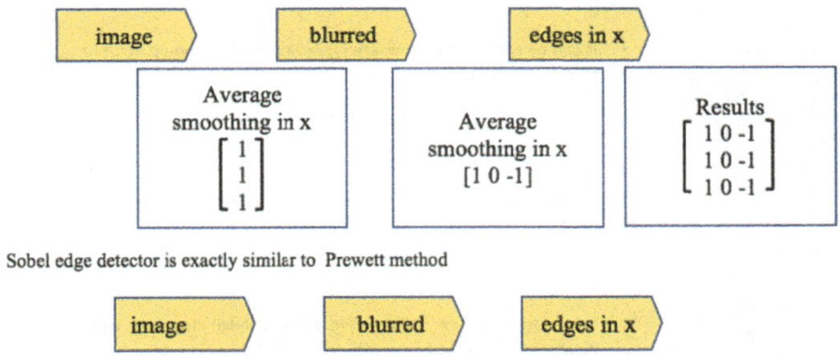

Fig. 3. Prewitt edge Detector

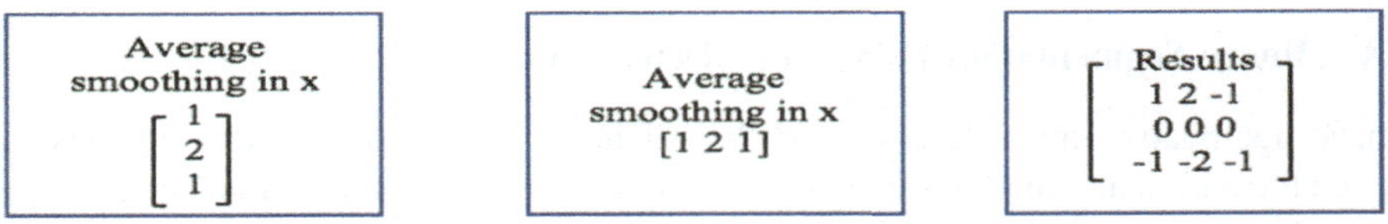

Fig. 4. Quality of an edge

For Canny edge detector, first criterion is good detection that optimal detector must minimize the probability of false positives as well as false negative. Second criterion should be good localization that the edges detected must be as close as possible to the true edges. The detector only needs to rotate one point for each edge point, which is usually necessary in these steps. (Shah 2012).

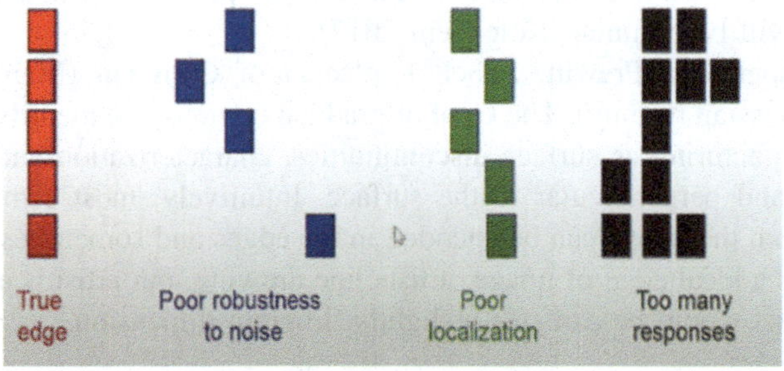

Fig. 5. Canny edge detector

4.3 Edges and Boundaries

Edges include local intensity discontinuities, points and dependent on models. Boundaries consist extensive and composed of many points (Shah 2012) (Fig. 6).

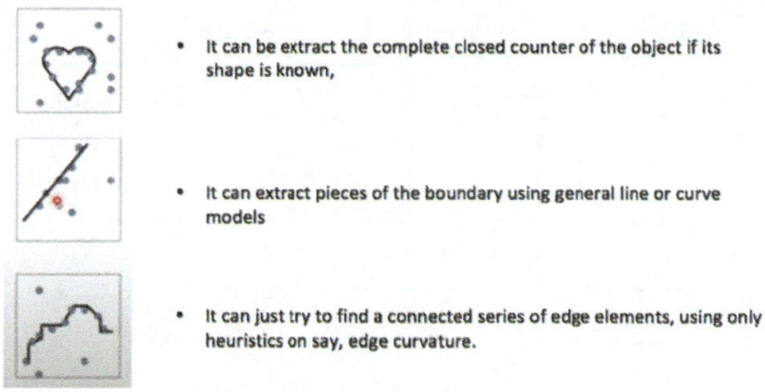

Fig. 6. Knowledge about boundary

4.4 Image Segmentation Using Edge Detection

Image aggregation algorithms classically fall into one of two classes: methods that define the region and methods that define the edge. Region-based methods typically use clustering to create regions based on the characteristics of the region being scanned, as well as using region enlargement or bins and merging methods. In contrast, edge-based methods use edge detection to find the edges in the image, and then use a method to link the edges to create boundaries that define regions. This method covers my field of study. Both classes of segmentation methods have their own advantages and disadvantages, so research is also being studied on hybrid methods that combine zone and edge-based segmentation.

Hybrid method should be used to image segmentation based on the topology of calculating the scanned regions. Suppose we have a set of points X = {x1,... is defined by some edge detection algorithms in xn}. For any positive radius α, we can consider the set of all α discs of radius at the X centers. As A grows, larger areas of the image are covered and discs begin to overlap. The centers of the overlapping discs can be combined to form edges and triangles. In my measurements, ellipses are used close to this method (Gevers 1997).

The proposed segmentation algorithm is a hybrid split and merge segmentation algorithm that uses computational join topology and permanent homology for image split and region feature characteristics for region join. The algorithm first uses some edge detection algorithm, such as the Canny algorithm (Canny 1986) or the wavelet-based detector (Mallat 1992) we are using, to determine the set of edge points (X) in the image. The scan is performed over X to create an initial segmentation with three types of regions: p-permanent regions, p-temporary regions, and d-triangles. The topological division is controlled by two parameters, α and p, where α defines the radius of the discs and p indicates the permanence. Informally, the kth Betti number refers to the number of k-dimensional holes on a topological surface. The first few Betti numbers have the following definitions for 0-dimensional, 1-dimensional, and 2-dimensional simplicial complexes: b0 is the number of connected components (Fritts 2007) (Fig. 7).

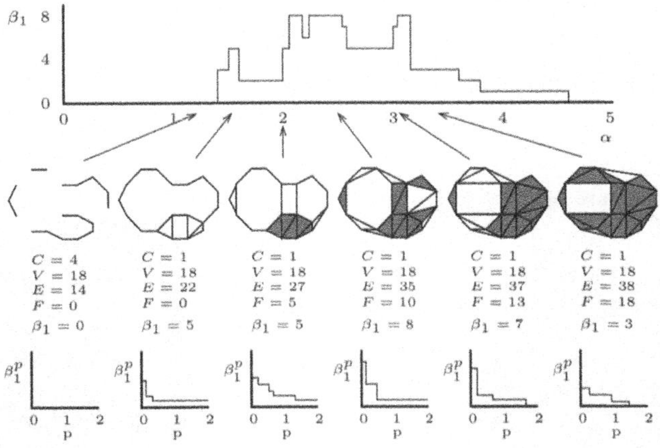

Fig. 7. Various complexes in a filtration, including their Betti numbers

4.5 Using the Hough Transform

For an image, it makes sense to connect a particular set of pixels only if we know they are part of the boundary of a meaningful region. Generally, we have to work with unstructured environments where everything we have is an edge image and there is no information about where interesting objects can be. In such cases, all pixels are candidates for connection and therefore should or should not be accepted according to predefined global features. An approach based on whether pixel sets are on the curves of the specified shape is required. Once detected, these curves will form the edges or region boundaries of interest (Gonzalez 2000).

4.6 Segmentation Methods

Segmentation methods that are based on detecting sharp, local changes in intensity. The three types of image features in which we are interested are isolated points, lines, and edges. Edge pixels are pixels at which the intensity of an image function changes abruptly, and edges (or edge segments) are sets of connected edge pixels (Fig. 8).

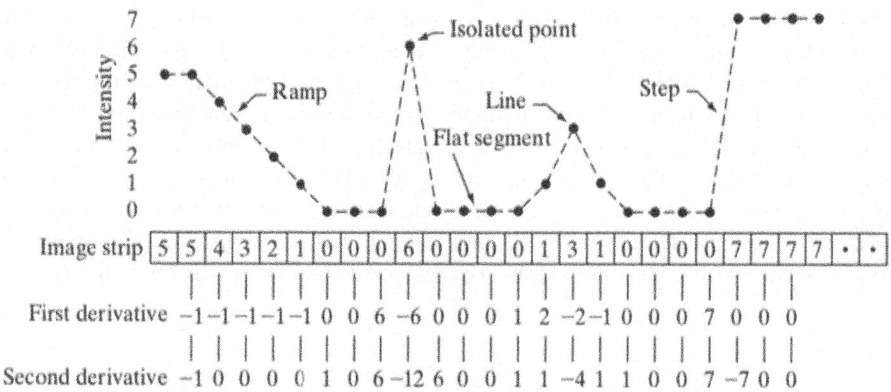

Fig. 8. Simplified profile (the points are joined by dashes for clarity)

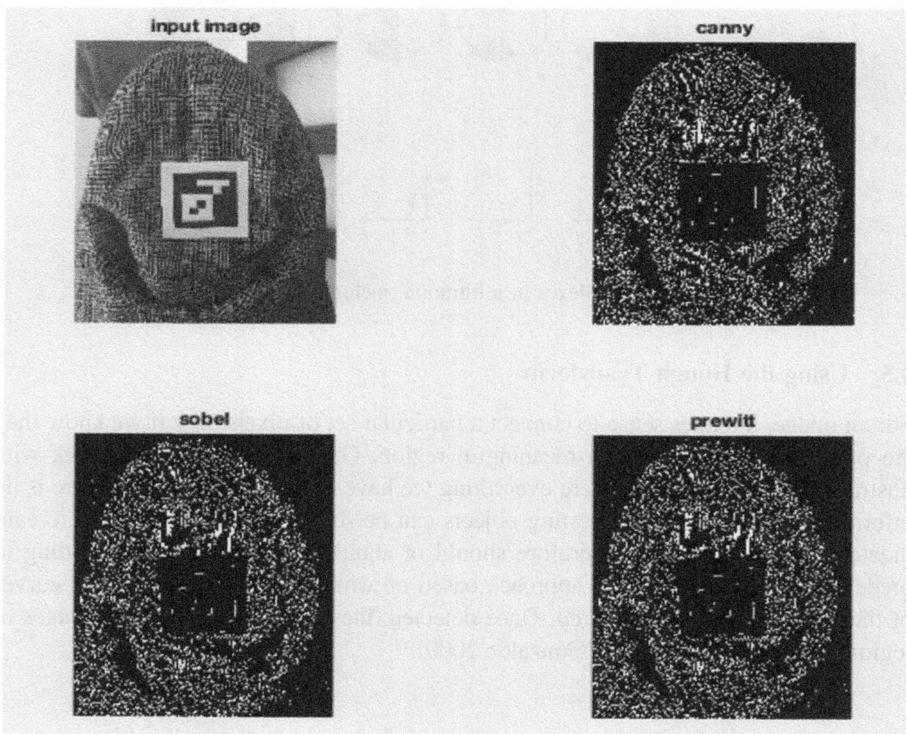

Fig. 9. Junction of jacket sleeve with four different methods by using MATLAB

The image strip corresponds to the intensity profile, and the numbers in the boxes are the intensity values of the dots shown in the profile. The derivatives were obtained using Eqs. (10.2-1) and (10.2-2) (Gonzalez 2000) (Fig. 9, 10, 11).

Algorithm Method	Output Type	Output at Edge Port
All	Binary image	Boolean matrix where the zero elements of this correspond to the edge pixels and the zero elements correspond to the background pixels.
Sobel, Prewitt	Gradient component	Gradient components that correspond to the horizontal and vertical edge responses at the Gh and Gv ports.
Roberts	Gradient components	Gradient components that correspond to the 45- and 135-degree edge responses at the G45 and G135 ports.
All	Binary image and gradient components	

Fig. 10. Specify the format of the Edge output (MatWorks 2020)

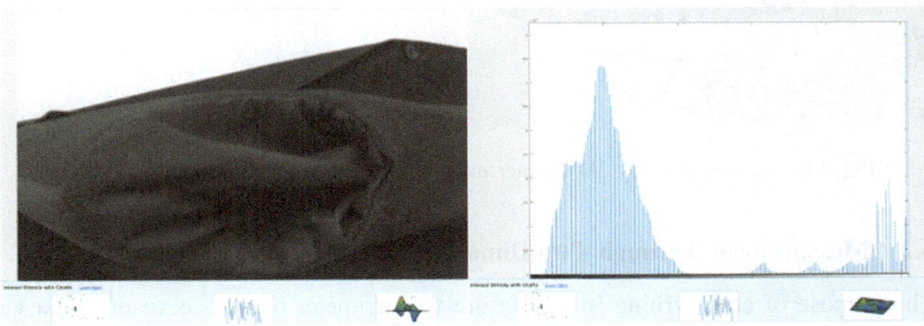

Fig. 11. Right figure; black and white (gray) form of the photo

Size = 2448 3264 3, gray pixel, 2448 × 3264 uint8-Numbers between 0 and 255, the pixel value of each color, more black colors dominate the display. 250 white 0 black value In 8-bit color level, left figure; The color distribution for gray tones as Intensity profile.

5 Measurement of Jacket and Conclusion

5.1 Measurement by Using 3D Laser Scanner-Non-contact Blue Laser

An operator must carefully mark about 50 points to determine the junction of a jacket and its sleeve with a three-dimensional laser scanner. This scan can be linked with

Auto-Cad and statistical calculations are also possible. The sample measurements were made by Faro-laser scanner with non-contact scanning and this scanner provides the highest accuracy point cloud data with unique non-contact measurement capabilities at high speed using blue laser technology. It captures the radius between the reference points commonly given in this scan. Here, when we give the reference points of the arm circumference, it finds the right curve passing through the recess, and this curve (picture on the left) gives the length of the ellipse. Due to the folds formed on the jacket in the picture on the right, it causes the reference points to be taken more frequently. This increases the probability of making errors in calculating the curve. As a result, the curve can be calculated very precisely, but it is the attention and precision that the operator will show for the measurement to be accurate. In addition, the operator should scan the scanner until the blue colors disappear (Fig. 12).

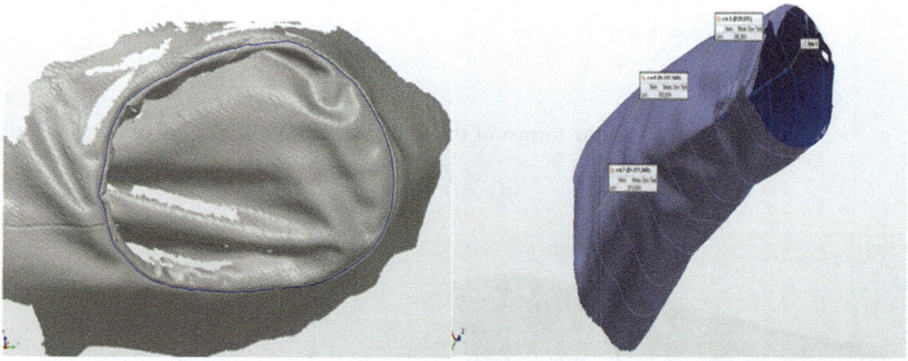

Fig. 12. Measurement by 3D scanner image the junction of a jacket and its sleeve

5.2 Measurement Through Two-Dimensional Image

The purpose of code writing is to measure the perimeter of the sleeve of the jacket through a photograph. As the jacket sleeve exhibits a geometrical similarity to the ellipse shape, but it is not a perfect ellipse. For this reason, when a jacket sleeve matched with a single ellipse, very high error rate comes out. In order to avoid from this problem, it was collected by calculating the arc length of the regions where the ellipse matches the jacket seam lines by using more than one ellipse. To avoid this problem, the arc length of the regions where the ellipse matches the jacket seam lines was calculated by using more than one ellipse. The calculated arc length is a value in pixels. A marker was used to convert this value obtained in pixels into a metric system. An edge length is determined by the corner points of known brands. Subsequently afterwards, the value of the edge of the marker in pixels is found. In this way, the transition was made to the metric system in pixels. The unit of length in cm entered in the "Measured Value" area in the system interface, the tape measure of the jacket, etc. it is the real value measured with a measuring instrument. The error rate was determined by comparing the resultant value obtained after the developed method with the real value (Fig. 13).

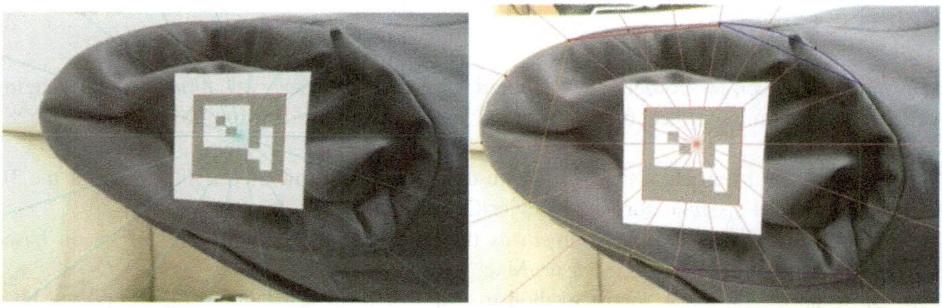

Fig. 13. Measurement of marker and jacket sleeve combination with two-dimensional image

6 Conclusion

In this article, the importance of machine vision systems is increasing and there are many approaches related to textile and clothing production. Textile manufacturers mean to increase the quality of the product by reacting to the applications that occur during the production process. These systems are Robustness of real-time vision-based systems, such as visual servo and largely depends on the methods used for feature extraction. Three characteristics of robust real-time vision is speed, accuracy, and reliability (Janabi 2002).

For all these solutions, they expand their visions with the smart factories and the tools they use. it is also needed high investments to detect the flaw and debug systems and it can be assumed that it will be easier to monitor the quality by measuring the relevant parameters and products (Gries 2018). With the information gained in line with the understanding of customer needs, the right yarn, fabric and colors are selected and designed for the new dress. Preparation of prototype garments is also time consuming. Companies like JCP Penny start design in advance, which means that it is about 40 weeks of preparation and design time before the product goes on sale. Companies like Zara start designing approximately 17 weeks in advance. For all these periods, modern technologies should be used in design, such as computer-aided (CAD). The design process begins in two dimensions, and by making the designed garment in 3-dimensional, it is looked how the customer will stand. This is now an important step to increase internet sales (Singh 2018).

Virtual development is an integral part of the development process for new products. Virtual Reality (VR) - a sub-area of virtual development - offers the possibility of quickly visualizing development stages and interacting with them in order to detect errors early on or to support ergonomics and installation studies or design inventions. In this way - by taking a holistic view of the product - costly optimization loops can be reduced at a later point in time. Practice shows that virtual models are not sufficient to answer all the necessary questions. But technology-oriented further development of VR has taken place in the past without examining the feasibility of specific work tasks from the user's perspective (Rademacher 2014).

References

Sansoni, G., Carocci, M., Rodella, R.: Calibration and performance evaluation of a 3-D imaging sensor based on the projection of structured light. IEEE Trans. Instrum. Measur. **49**(3), 628–636 (2000)

Song, Z., Jiang, H., Lin, H., Tang, S.: A high dynamic range structured light means for the 3D measurement of specular surface. Elsevier-Optics Lasers Eng. **95**, 8–16 (2017)

Shirmohammadi, S., Ferrero, A.: Camera as the instrument: the rising trend of vision based measurement. IEEE Instrum. Measur. Mag. **17**(3), 41–47 (2014)

Sun, Q., Chen, J., Li, C.: A robust method to extract a laser stripe centre based on grey level moment. Opt. Lasers Eng. **67**, 122–127 (2015)

Liu, Z., Yin, Y., Wu, Q., Li, X., Zhang, G.: On-site calibration method for outdoor binocular stereo vision sensors. Opt. Lasers Eng. **86**, pp. 75–82 (2016)

Zhang, S.: Recent progresses on real-time 3D shape measurement using digital fringe projection techniques. Opt. Lasers Eng. **48**(2), 149–158 (2010)

Hartley, R., Zisserman, A. : Multiple View Geometry in Computer Vision. Cambridge University Press, Cambridge (2003)

Lowe, D.G.: Distinctive image features from scale-invariant keypoints. Int. J. Comput. Vis. **60** (2), 91–110 (2004)

Bronstein, A., Bronstein, M., Ovsjanikov, M.: 3D features, surface descriptors, and object descriptors. 3D Imaging, Analysis, and Applications, 1–27 (2010)

Keferstein, C.P., Marxer, M., Bach, C.: Fertigungsmesstechnik: Alles zu Messunsicherheit, konventioneller Messtechnik und Multisensorik. Springer (2017)

(2020) https://www.inspection-for-industry.com/finished-goods-inspection.html

Durakbasa, N.M.: Geometrical Product Specification and Verification for the Analytical Description of Technical and Non-Technical Structures. Austauschbau und Messtechnik, TU Wien-AUM, pp. 137–138 (2003)

Manikanta, V.: (2020) https://www.slideshare.net/vaddimanikanta/edge-detection-algorithm-and-code

Shah, M.: UCF Trustee Chair (2012). https://www.youtube.com/watch?v=7mEiTU-XgCo

Gevers, T., Smeulders, A.W.N.: Combining region splitting and edge detection through guided Delaunay image subdivision. In: Proceedings of IEEE Computer Society Conference on Computer Vision and Pattern Recognition, pp. 1021–1026. IEEE (1997)

Canny, J.: A computational approach to edge detection. IEEE Trans. Pattern Anal. Mach. Intell. **6**, 679–698 (1986)

Mallat, S., Zhong, S.: Characterization of signals from multiscale edges. IEEE Trans. Pattern Anal. Mach. Intell. **14**(7), 710–732 (1992)

Letscher, D., Fritts, J.: Image segmentation using topological persistence, computer analysis of images and patterns. In: 12th International Conference, Proceedings, pp. 27–29 (2007)

Gonzalez, R.C., Woods, R.E.: Processamento de imagens digitais. Editora Blucher (2000)

MatWorks: Edge Detection (2020)

Janabi, S.F.: Real-Time Feature Extraction. Ryerson University Toronto, Ontorio (2002)

Gries, T., Lutz, V., Niebel, V., Saggiomo, M., Simonis, K: Automation in Quality Monitoring in Fabrics and Garment Seams. Woodhead Publishing, Cambridge (2018)

Singh, A., Nijhar, K.: Recent Devolopments in the Garment Supply Chain, pp. 377–396. Woodhead Publishing, Cambridge (2018)

Rademacher, M.H.: Virtual Reality in der Productentwicklung. Vieweg+Teubner Verlag (2014)

Porosity Measurement by X – Ray Computed Tomography: Different Porosity Analysis Application

Michaela Kritikos(✉)

Faculty of Materials Science and Technology in Trnava,
Slovak University of Technology, Bratislava, Slovakia
michaela.kritikos@stuba.sk

Abstract. Nowadays, defectoscopy is very popular for evaluation of the parts in different kind of industry. It is very important to know internal structure of the used parts. The most applications of porosity evaluation can be find in the plastic and aluminium parts control. There are still many questions about accuracy of the non- destructive porosity evaluation according to many possibilities for part′ s scanning and evaluation methods. Experimental investigation of this paper is dealing with differences in porosity evaluation of the plastic part scans with different voxel sizes. Two different porosity analysis were used for data achievement. There is still not enough knowledge about the influence of the voxel size (3 dimensional pixel) on the evaluated porosity.

Keywords: Defectoscopy · X – Ray computed tomography · Porosity evaluation · Voxel size

1 Introduction

X – Ray computed tomography is used for dimensional control and defectoscopy. Many scientists and academicians are dealing with accuracy and uncertainty of the application of this non – destructive method in the industry to substitute optical measurements and destructive methods.

X-ray CT machines for industrial applications are primarily equipped with an X-ray source tube and a detector, between which is a specimen mounted on a rotatory table that rotates 360° while the detector records radiographic images at different angular positions. The general workflow for CT data acquisition, including evaluation of dimensional measurements, is illustrated in Fig. 1. The most important algorithms for reconstruction from radiographic projections are Fourier-based methods, and algebraic and statistical approaches [4–7]. However, the algebraic and statistical algorithms are typically disregarded in practice because of the longer reconstruction times they need, which hinders their use in industrial environments. Two-dimensional Fourier-based reconstruction methods are more computationally efficient, and therefore more widely used. The measuring process is described in the Fig. 1 [1].

The result of the CT scan evaluation might be influenced by many different causes which can be described by Ishikawa diagram (Fig. 2).

N. M. Durakbasa and M. G. Gençyılmaz (Eds.): ISPR 2020, LNME, pp. 175–185, 2021.
https://doi.org/10.1007/978-3-030-62784-3_15

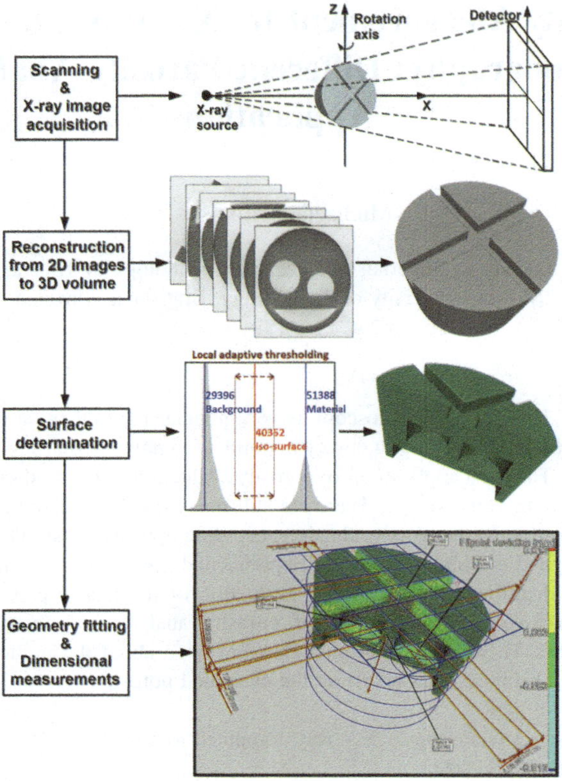

Fig. 1. Workflow chart for the full process of X-ray CT that includes evaluation of dimensional measurements. The 3D images here are generated with VGStudio MAX [1]

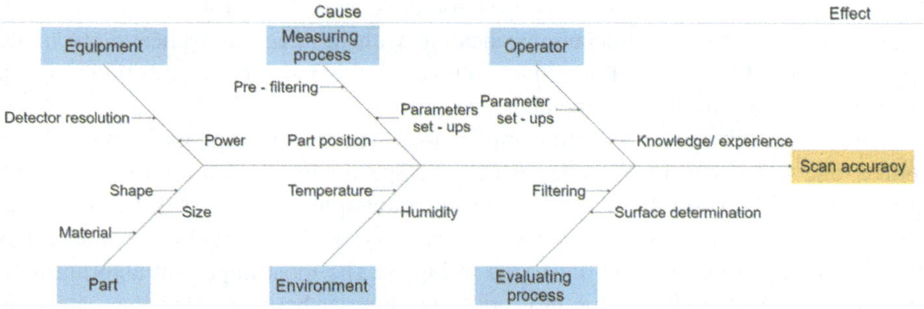

Fig. 2. Ishikawa diagram – effect of the different causes on the scan accuracy

However, there are also another causes influencing accuracy of GPS (geometrical product specifications) measurement or porosity evaluation. That is why computed tomography device was used as a metrology tool in many different papers. Kruth et al.

[2] used it for dimensional quality control, for traceable measurement of dimensions of technical (mechanical) components and for tolerance verification of such components.

The conclusion of their research was:

- CT metrology has a high potential for dimensional quality control of components with internal cavities that are not accessible with other measuring devices.
- CT is the only inspection process that allows combining dimensional and material quality control.
- The number of applications and number of installed systems is growing rapidly.
- Full benefit of CT will in future be achieved by the advent of multi-sensor systems applying data fusion techniques.
- CT might substitute other measurement and testing systems (e.g. optical systems) to a certain extent.

Another authors in their studies compare dimensional measurements on artifacts containing internal structures and mechanically compliant features using current state-of-the-art CT and coordinate measure machine (CMM) techniques. To illustrate the challenges of CMM and the potential of X-ray CT for reliable dimensional inspection, two different problems are presented: 1) characterization of internal geometry in a metallic artifact that has features inaccessible to tactile or vision-based measurement techniques; and 2) evaluation of dimensional measurement of diameters, form, and relative distances in the components of a flexible object with stems that bend due to mechanical contact by tactile instruments, e.g., CMM [1]. For dimensions of geometric features ranging from 0.6 mm to 65 mm, a comparison between CT and CMM measurements typically resulted in differences of approximately 5 μm or less for most of the measurements, while expanded uncertainties computed for the CT measurements ranged from 1 to 20 μm. CT measurements on the internal geometries resulted in uncertainties comparable to CMM data that were obtained with the metallic artifact disassembled. However, in the particular case of measuring compliant structures, the conventional CMM strategies led to large systematic errors in the measurements. It was demonstrated that using extrapolation methods, measurements from these two systems converged to within ± 2 μm.

Alloisi and Carmignato studied in their research influence of surface roughness on X-ray computed tomography dimensional measurements of additive manufactured parts [8]. The influence of surface roughness on CT dimensional measurements and the relation between tactile CMM and CT measurements were studied. Experimental results confirmed that surface roughness causes a systematic error between CMM and CT measurements. External diameters were always smaller than the corresponding CMM reference values of approximately Rz/2, while internal diameters were always bigger than CMM values of approximately Rz/2.

Accurate identification and measurement of internal voids and porosity is also an important step towards improvement of production processes to obtain high quality materials and products.

Recently, the importance of knowing the exact size, shape, volume and location of defects has become even higher as tighter requirements and new standards have been introduced in industry. There are several well-established methods for defects evaluation based on various principles (both destructive and non-destructive).

However, all conventional methods have various deficiencies and the information about internal voids/porosity that can be extracted is limited. Most of these drawbacks can be overcome by using X-ray computed tomography (CT). Unlike other methods, CT provides full three-dimensional information about shape, size and distribution of internal voids and porosity; however, the accuracy of measurements is still under investigation.

Hence, further evaluations on CT porosity measurements must be performed in order to consider X-ray computed tomography a reliable instrument for the assessment and detection of internal defects [3].

Hermanek and Carmignato [3] in their experimental investigation evaluate the accuracy of porosity measurements by CT. The reference object was manufactured by ultra-precision micro-milling. The object contains dismountable components with embedded internal hemispherical features that simulate internal porosity. The artificial porosity was micro-milled on top surfaces of dismountable cylindrical inserts. The hemispherical calottes were thereafter calibrated by traceable coordinate measuring systems and calibrated values were compared to actual values measured by a CT system. The accuracy of CT porosity measurements was then evaluated based on results obtained on various measurements, using different software tools and measuring procedures, comparing real scans to numerical simulations and investigating the influence of CT system parameters settings on measurement results.

The study on the influence of voltage and current on porosity measurement errors has proven that these factors are important, particularly for small defects. The variation caused by voltage and current is below 4 μm in the case of diameter and depth measurements and below 3% in case of volume measurements. The relationship between tube voltage and current, and deviations in measurements has been determined as follows: the higher the values of tube voltage and current, the lower the observed deviations from reference values.

The measurement repeatability was evaluated based on results of 12 different scans, repeated under the same "optimal" conditions. Mean deviations and standard deviations were calculated for the results of the 12 scans. The outputs of this study show that the mean error of diameter and depth measurements can be below 5 μm with standard deviation ranging from 11 μm for the smallest defects to less than 1 μm for the largest ones. As for volume measurements, the relative mean deviation is below 10% for defects below 200 μm in diameter, while it is below 5% for defects larger than 200 μm in diameter, with standard deviations up to 6% in the case of smaller defects.

Stoessel et al. (2011) [9] investigated application of computed tomography for 3D porosity evaluation in carbon fibre reinforced plastics (CFRP). Their research was based on the parameters that might have an influence on the evaluated degree. The CFRP reference sample and the investigated part were measured at the same time with varying testing parameters. The three main fields taken into account are the parameters of the CT system, the reconstruction settings and software parameters for evaluation. According to their paper the potential porosity influencing factors can be split into three main sections (Fig. 3).

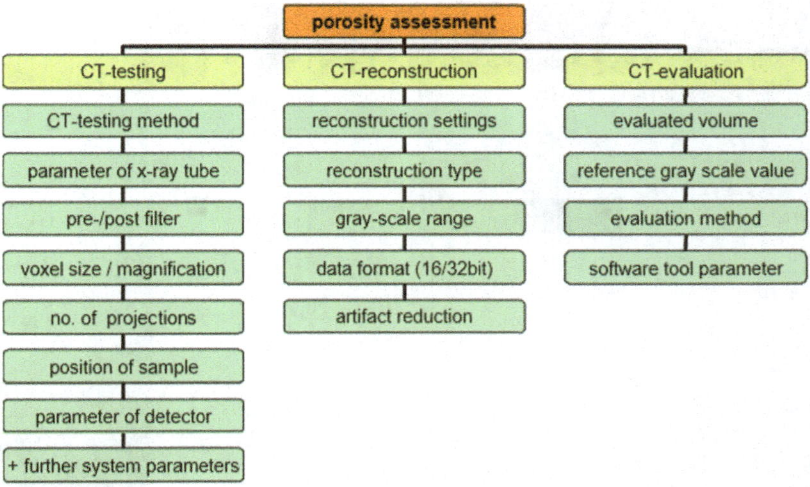

Fig. 3. Overview of potential influencing factors on the porosity assessment of CFRP with μ-CT [9]

This research leads to the conclusions:

- The parameters of the CT reconstructions effects were minor,
- The pre-filtering by placing a pre-filter in front of X - ray tube can be changed and reduced beam hardening, but with increasing thickness and atomic number the image noise level is increasing.
- With decreasing of magnification and increasing of the voxel size the evaluated porosity value is increasing.

Likewise, there is a publication from Chiffre et. al. [10] which gives an overview of the industrial applications of X – ray computed tomography in the manufacturing industry as well as in other industries. The paper contains a survey of application examples from the manufacturing industry as well as from other industries, e.g. electrical and electronic devices, inhomogeneous materials and the food industry.

2 Experimental Investigation

Nowadays, porosity evaluation is very necessary in many different fields of industry. Many automotive companies have porosity as a criterion for part's quality classification. The limit for defects can be volume of pores in % in whole part volume or the K. O. criterion can be size of the biggest pore. It is very important the accuracy of the method of getting data and porosity evaluation analysis.

The experimental investigation of this paper was based on the use of the computed tomography device for scanning of plastic part and the porosity of this part was evaluated in different positions and by different porosity evaluation methods. METROTOM 1500 from Zeiss company (Fig. 4) was used. It using leads to achievement of three dimensional scans. Software used for scanning was METROTOM OS 2.8.

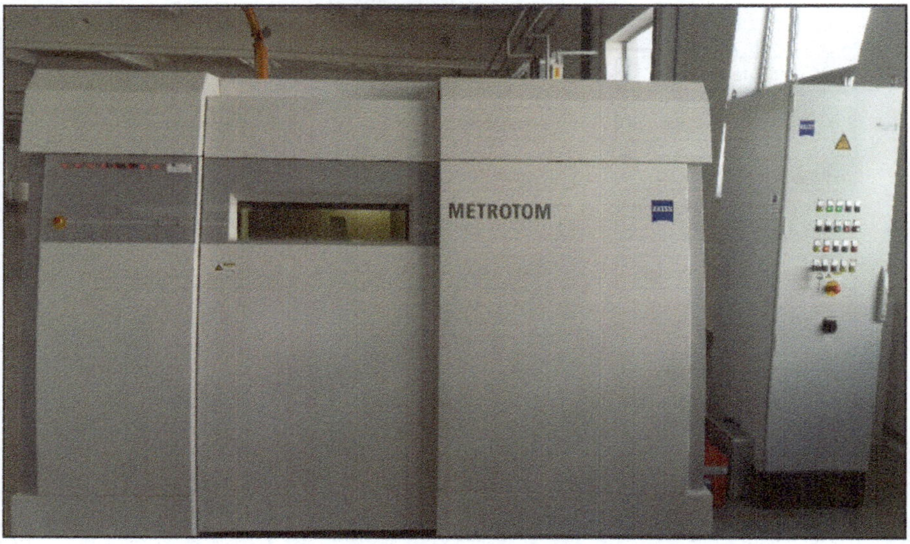

Fig. 4. CT device - METROTOM 1500

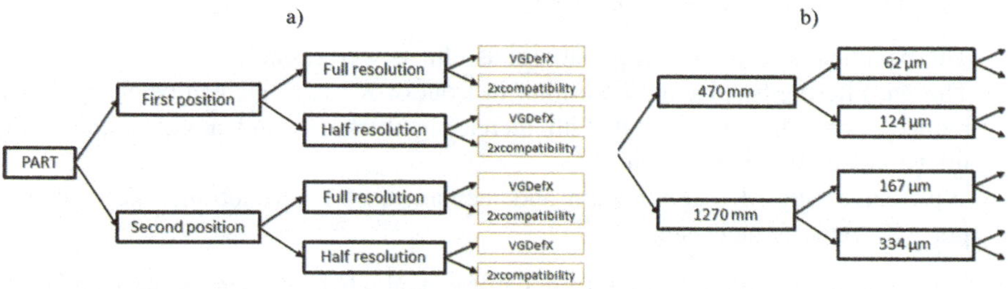

Fig. 5. a) Scheme of getting data for experimental investigation, b) Distance of the part from X – RAY lamp while scanning and achieved voxel values with full and half resolution

The plastic part was scanned in two different positions from X-Ray lamp. First distance was 470 mm and the second distance was 1270 mm. This device can use two detector resolutions: full (2048 × 2048 px) and half (1024 × 1024 px). Half resolution use causes double scan voxel size. Therefore, the outputs of this scanning were 4 different scans with different voxel sizes. After that two different porosity analysis were applied (Fig. 5a). The achieved voxel values are in the part b) of the same figure.

The settings of CT machine were done according to this data:

- the voltage – 120 kV,
- the current – 500 µA,
- integration time – 1000 ms,
- number of projections – 1000 – 2000 (according to detector resolution).

Calibration of the machine was done before the part's scanning process. It includes detector calibration, geometrical calibration and axis qualification.

This experimental investigation leads to comparison of the different results gotten from 2 different porosity analysis application and also comparison of the pores volume in the scan with different voxel size.

It was used VGStudio MAX for scans evaluation. There is possibility to use different analysis for porosity evaluation – VGDefX analysis and 2 × compatibility. Their definitions are not really clear and there is not described exact application. This experiment can lead to declaration there is a big difference in porosity evaluation when it is used different porosity analysis even if it is investigated the same scan in both cases.

First comparison (Fig. 6) can be done for VGDefX porosity analysis between scans with the smallest voxel size (full resolution, first position) and the biggest voxel size (half resolution, second position).

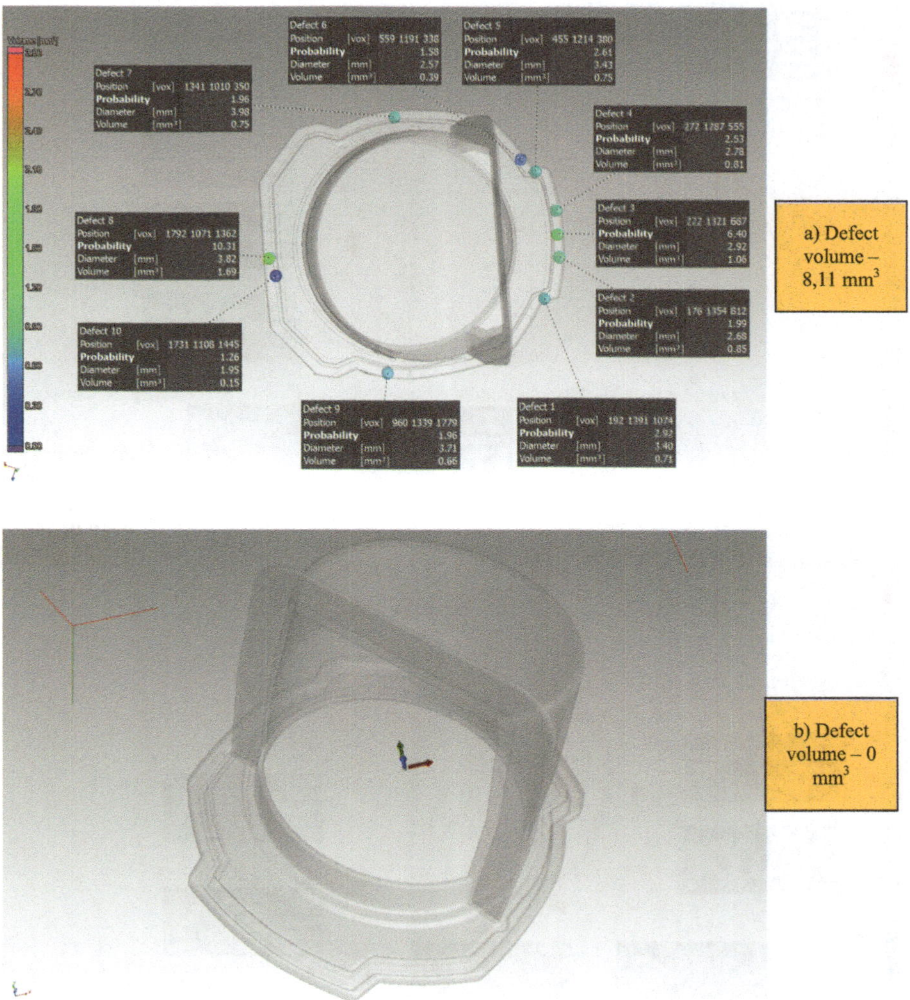

Fig. 6. Comparison of the porosity evaluation when apllicated VGDefX analysis on the scan with a) the smallest voxel size and b) the biggest woxel size.

As it is obvious from the Fig. 6, the application of the VGDefX analysis method on the scan with the smallest voxel size achieved by full resolution of the detector evaluated defect volume of 8,11 mm^3. In the scan with the biggest voxel size after application of the VGDefX analysis were not determined any defects. It can be assumed that this scan is not enough precise for the porosity evaluation for application of this kind of porosity analysis. Some operators who have not enough knowledge could make mistake and to evaluate this part that it is a part without any pores. But very important information in porosity evaluation is the probability of the pore occurrence. As it is seen from the previous figure, probability evaluation is between 1,26–10,31 when it was used VGDefX analysis method. The numbers are very small and the probability of the pores occurrence is not really sure.

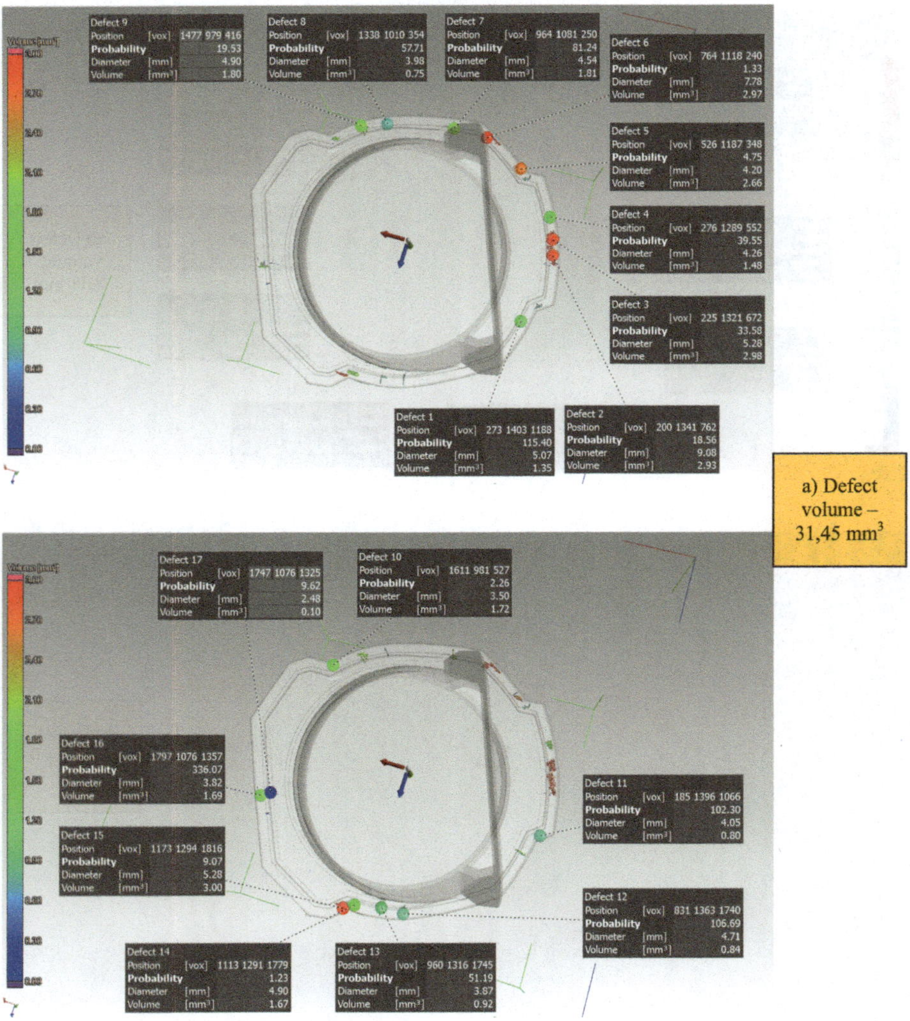

Fig. 7. Comparison of the porosity evaluation when apllicated analysis 2 × compatibility on the scan with a) the smallest voxel size and b) the biggest voxel size

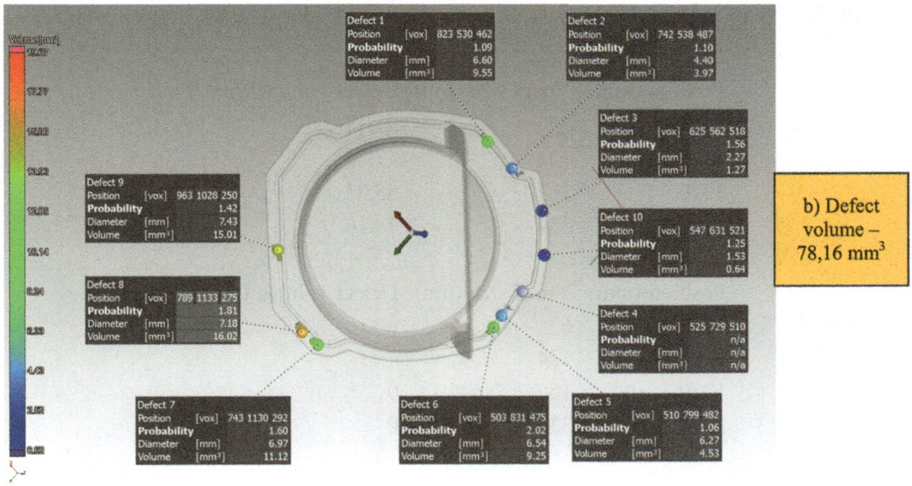

Fig. 7. (*continued*)

On the other hand, the probability values of the pores occurrence for 2 × compatibility analysis method application it was between 1,23–336,07 what are much higher values (Fig. 7). On this figure is the same comparison between scans with the smallest voxel size (full resolution, first position) and the biggest voxel size (half resolution, second position) but it is done for 2 × compatibility porosity analysis. The pores determined on the smallest voxel size scan are divided into two pictures because the highest number of the flags in the software is 10.

This comparison leads to the declaration that voxel size has a big influence on the evaluation of the porosity of plastic parts. There are differences in the defect volume even when it is used the same porosity evaluation method. The difference is 31,45 mm^3. The smaller voxel size of the scan demonstrated less defect volume and in the scan with bigger voxel size was evaluated defect volume of 78,16 mm^3. As it is seen on the figure, there are also found different defect on each scan.

There is big difference between these two methods for porosity evaluation. The scan with the biggest voxel size was evaluated as it has no pores by VGDefX porosity analysis application and in the same scan evaluated by 2 × compatibility method was determined defect volume of 78,16 mm^3.

The porosity evaluation of the scan with the smallest voxel size shows also differences in achieved results. While application of the VGDefX analysis leaded to the defect volume of 8,11 mm^3 in the part, the 2 × compatibility method determined 31,45 mm^3 of defect volume, what is almost 4 times higher value.

The result of the parameters combination for scanning is four scans and they are evaluated by 2 porosity analysis. Porosity defect volumes gotten from evaluation of all scans are illustrated in the Table 1.

Table 1. Defect volumes evaluation for 2 × compatibility and VGDefX porosity analysis

2 × compatibility		
Position (mm)	Voxel size (μm)	Defect volume (mm^3)
470	62	31,45
470	124	47,74
1270	167	55,03
1270	334	78,16
VGDefX		
Position (mm)	Voxel size (μm)	Defect volume (mm^3)
470	62	8,11
470	124	4,11
1270	167	0
1270	334	0

Accordingly, it is also important to identify how big influence can be brought into the scanning process by the recognition of the detector. In this experimental investigation the half (1024 × 1024 px) and full (2048 × 2048 px) detector resolution were used. Advantage of the use of half resolution is that it is half time consuming in comparison to full resolution. On the other hand, the disadvantage is the voxel size has every time double value. If it will be identified that there is not a notable difference between the pores volumes detected in the scan achieved by different detector resolution, it will have a significant sense for companies using this non – destructive method for porosity evaluation because their scanning process can be half time consuming. As it is evaluated in the Table 1, full resolution for 2 × compatibilty analysis method leaded to detection of about 30% less defect volume, what is not negligible. For VGDefX method application were result different – for first position it was half size of defect volume when it was used half resolution. There was not a difference for second position, when both defect volume values were 0 mm^3.

In this case, it can be deduced that the evaluated defect volume is increasing voxel size when using 2 × compatibility porosity analysis. VGDefX analysis application determine different effect – the defect volume is decreasing with voxel size growing. This conclusion is very important for future non – destructive porosity research and for operators of CT devices applicated in industry.

3 Conclusions

Findings of the experimental investigation of this paper are very important for scientist, academicians and operators using CT devices for porosity evaluation. It was researched that the porosity evaluation methods are not freely exchangeable (replaceable). There is a big influence of the porosity analysis algorithm on achieved defect volume.

In this paper were applied two different porosity analysis method on the scans of plastic part – VGDefX and 2 × compatibilty. The influence of the 2 × compatibility method was that defect volume is increasing with voxel size and the application of the VGDefX analysis leads to decreasing of defect volume with increasing of voxel size.

Thus, the effect of the voxel size was also examined (for plastic part scanning). In some case it seems the scanned part is ideal according to internal defects (so there are none) and in some case it was determined that the same part has 78,16 mm^3 defect volume.

However, this experimental investigation does not give an answer on the question which of those two porosity methods are closer to reality representation. Whereupon, there is possibility for future research with comparison of achieved porosity values to nominal (reference) values and to evaluate the accuracy of the used porosity analysis method.

Acknowledgment. This article was written within the project VEGA 1/0747/19: "Optimization of geometry of cutting tools produced by foundry technology and powder metallurgy to increase durability".

References

1. Villarraga-Gómez, H., Lee, C., Smith, S.T.: Dimensional metrology with X-ray CT: A comparison with CMM measurements on internal features and compliant structures. In: Precision Engineering, vol. 51, pp. 291– 307. Elsevier (2018)
2. Kruth, J.P., Bartscher, M., Carmignato, S., Schmitt, R., De Chiffre, L., Weckenmann, A.: Computed tomography for dimensional metrology. Manuf. Technol. **60**, 821 – 842 (2011)
3. Hermanek, P., Carmignato, S.: Porosity measurements by X – Ray computed tomography. Precis. Eng. **49**, pp. 377–387 (2017)
4. Buzug, T.M.: Computed Tomography from Photon Statistics to Modern Cone-Beam CT. Springer, Leipzig (2010)
5. Hsieh, J.: Computed tomography: principles, design, artifacts, and recent advances. In: SPIE –The International Society for Optical Engineering, 3rd edn., Bellingham, WA, USA (2015)
6. Kak, A.C., Slaney, M.: Principles of Computerized Tomographic Imaging, IEEE Press, New York
7. Herman, G.T.: Fundamentals of Computerized Tomography: Image Reconstruction from Projections. Springer, London
8. Valentina, A., Carmignato, S: Influence of surface roughness on X-ray computed tomography dimensional measurements of additive manufactured parts. In: Case Studies in Nondestructive Testing and Evaluation 6, pp. 104 – 110. Elsevier (2016)
9. Rainer, S., Denis, K., Reinhold, O., Bjorn, D., Llorenc, L.P.: µ-Computed tomograpfy for 3D porosity evaluation in carbon fibre reinforced plastics (CFRP). In: International Symposium on Digital Industrial Radiology and Computed Tomography – Tu.3.1 (2011)
10. De Chiffre, L, Carmignato, S., Kruth, J.P., Schmitt, R., Weckenmann: Industrial applications of computed tomography. Manuf. Technol. **63**, 655 – 677 (2014)

Industrial Robotics in Production Systems and Industrial Engineering

Accuracy Improvement and Process Flow Adaption for Robot Machining

Martin Kainrath(✉), Mohamed Aburaia, Kemajl Stuja, Maximilian Lackner, and Erich Markl

Competence Center Digital Manufacturing & Robotics, Department Industrial Engineering, UAS Technikum, Vienna, Austria
{mrl8m008, buraia, stuja, lacknerm, markl}@technikum-wien.at

Abstract. Complex structures and a high level of customization in architecture and construction industries require the usage of industrial robots for milling or additive manufacturing applications. Due to economic aspects and the intensive research on robot machining, conventional machine tools are increasingly superseded for specific tasks by industrial robots. However, assembling errors during the manufacturing process and low structural stiffness of industrial robots are still main reasons for inaccuracies of robotic based machining. This paper presents a comparison between industrial robots and cnc machine tools and evaluates their strength and weaknesses. Furthermore, recently published works considering robot calibration and offline compensation strategies in order to increase the absolute and machining accuracy are presented. Finally, the adaption of an existing robot machining process chain with additional components underlines the potential of robots for machining tasks.

Keywords: Robot machining · CNC · Calibration · Offline compensation · Process chain

1 Introduction

Robot machining is a combination of manufacturing engineering and robotics. Production industry uses numeric controlled machine tools to perform manufacturing processes with great reliability and accuracy. However, industrial robots are widely used for handling applications by accomplishing ordinary pick and place tasks. In a modern production line, various machining processes such as deburring, grinding and polishing are necessary for product finishing. Due to the low process forces, industrial robots are able to perform these processes easily. Based on the large workspace, high flexibility and low accuracy requirements, industrial robots from many manufacturers are suitable for these applications. The use of modern industrial robots in the field of machining is still controversial due to the lack of process stability. The rapid increase of additionally installed robots every year is a consequence of their economic benefits compared to a regular computer numerical controlled (CNC) machine tool. Another important aspect is the extensive research work in terms of robot machining during the

N. M. Durakbasa and M. G. Gençyılmaz (Eds.): ISPR 2020, LNME, pp. 189–200, 2021.
https://doi.org/10.1007/978-3-030-62784-3_16

last decade. Fig. 1 displays the total number of publications concerning robot machining from 1987 to 2018.

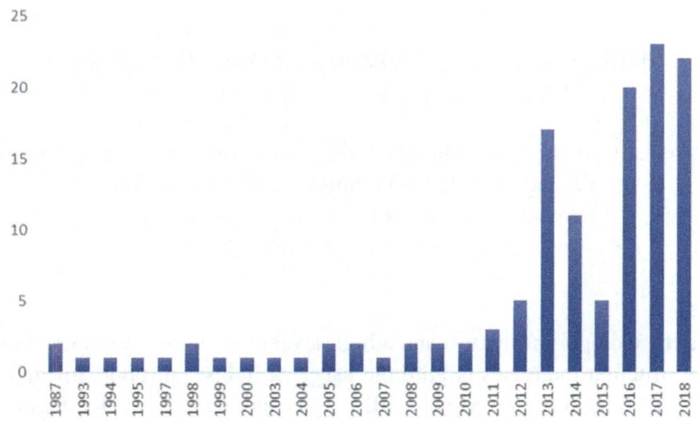

Fig. 1. Number of publications with emphasis on robot machining from 1987 to 2018 [1]

More than 75% of all publications have been reported from 2013 to 2018 and most of the studies are from Germany, China, France, and Canada. The investigation of various parameters influencing the structural stiffness of the serial kinematics of an articulated robot in case of machining aluminium components was the main subject of recent research. Due to the interaction between the endmill and the workpiece, a considerable cutting force arises, which results in static and dynamic effects for the entire system. The low manufacturing tolerances for aluminium components are technically not feasible without compensating end effector deviations. Researchers worldwide are trying to analyse the static and dynamic properties of industrial robots for machining applications and develop an appropriate compensation and process control strategy. In this paper, the second section introduces different types of industrial robots and their capability for robot machining, compared to CNC machine tools. The third section focuses on actual calibration methods and their usage for increasing the absolute accuracy of industrial robots. To overcome occurring end effector deviations, currently used offline compensation methods including static and dynamic stiffness modelling are presented as well. Finally, additional hard- and software components for the improvement of the automated process chain of an existing robot milling station at the UAS Technikum Wien will be proposed in section four.

2 Industrial Robots and CNC Machine Tools

In order to select an appropriate production system for manufacturing aluminium components, certain factors and constraints need to be considered. Basically, robot machining systems can be defined by flexibility, available workspace, structural stiffness, and positioning accuracy. Figure 2 illustrates the most important systems for robot machining and their strengths and weaknesses.

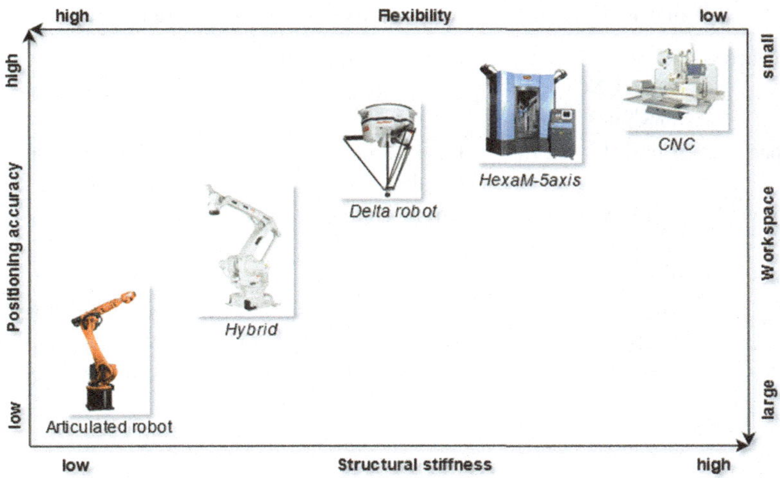

Fig. 2. Overview of important parameter and systems for robotic machining

Articulated robots provide the largest workspace and are unique in their flexibility. Milling hard materials (i.e. Aluminium, Titanium, Steel) causes enormous process forces and requires a small production tolerance, which is not consistent with their low structural stiffness. Due to their reinforced design, hybrid and delta robots reach a better positioning accuracy and are able to follow predefined paths more precisely. In terms of batch production and manufacturing of parts with small dimensions (≤ 1 m), modern CNC machine tools produce huge numbers in a short amount of time. The enormous weight of these machines results in high structural stiffness and enables precise production in combination with best dynamic properties. In case of processing large dimensions (>1 m), the heavy CNC machining centres grow with the raw parts due to the lack of flexibility. An additional external axis enables the extension of the workspace of an articulated or hybrid industrial robot easily. Table 1 shows a comparison of conventional CNC machine tools and robotic based machining regarding to five key performance factors.

In aircraft and aerospace industry, massive multi axis CNC machines are still used to produce components with large dimensions such as wing parts, fuselages, and alighting gears in an inefficient and energy wasting way. The low structural stiffness of a serial kinematic industrial robot leads to a significant deviation between the desired and the real path of the tool. Due to intensive research, industrial robots are already able to perform the complex production processes of large-scale components by using modern compensation methods for minimization of tool centre point (TCP) deviations. Complex calibration methods followed by offline or online compensation techniques provide a higher accuracy with additional technical effort.

Table 1. Comparison of Cnc and robotic – based machining [2]

Performance factor	Conventional XYZ CNC-machines	Robotic – based machining	
		Serial robots	Parallel robots
Workspace	Limited	Large (limited by link lengths)	Limited (by parallel architecture)
Flexibility	Limited number of operations	Any operation in the workspace	
Dynamics	High	Limited	High
Accuracy	High	Depends on the stiffness, link weights and payload	
Stiffness	High	Moderate, compliance is accumulated along the chain	High, separate chain stiffness is aggregated

3 Calibration and Offline Compensation

3.1 Calibration

The demand on Robots with optimized repeatability and good volumetric accuracy has increased steadily over the past decade. Today many manufacturers of industrial robots provide the service of calibration based on three or six-dimensional measurement equipment. However, little information about the actual accuracy of the robot can be found in data sheets, nevertheless the arising inaccuracies during the manufacturing and assembling process of industrial robots are not considered. Andrew Liou et al. [3] and Karan et al. [4] summarized the causes of inaccuracies in robots in the following five factors: environment, parameter, measurement, calculation and application.

Basically, the calibration process can be classified in three levels. Level one is also called mastering and only considers the deviations between the actual and the specified joint displacements and only refers to the home position of the robot. Level two (kinematic calibration) contains only cinematic parameters and affects the entire geometric robot calibration, including joint length and angular misalignment. The Level three calibration also takes non – geometric errors such as rigidity, friction and compliance into account for the modelling process [5]. The modified Denavit-Hartenberg (DHM) notation is used by many authors for the mathematical representation of the modelling of robot kinematics [5–7]. Brutscheck et al. [8] presented a detailed description of the approach and creation of a DHM table for a kinematic structure. Based on a DHM table, homogenous transformation matrices enable a relation between the end effector and the robot base coordinate system. By adding additional, non-geometric error parameters, a level three calibration can be performed. For parameter identification, several methods were proposed from many different authors. The least square optimization method [9], or statistical optimization such as extended Kalman filter [10, 11], generic algorithm [12] or particle filter [11].

Ginani et al. [13] performed a level two calibration on an ABB IRB 2000 industrial robot. An IGT ROMER measuring arm was used to measure a single point along the sixth robot axis in various robot configurations. The authors achieved a mean/maximum positioning error of 0.3 mm/1.4 mm. Another example for level two calibration was presented by Ye et al. [14]. Instead of a measuring arm, a widely used Laser Tracker system was utilized for precise measurement of the coordinates of a single point at the centre of the flange. The result was a mean/maximum positioning accuracy of 0.470 mm/0.640 mm. Lightcap et al. [15] investigated a level three calibration method on a Mitsubishi PA 10-6CE based on a stiffness model for non-geometric parameter consideration. The deflections caused by a payload of 44 N were considered by a torsional spring model which represents the flexibility of a harmonic drive. The presented model achieved a reduction of the mean/maximum positioning accuracy from 1,800 mm/2,450 mm to 0,330 mm/0,710 mm. Nubiola et al. [5] improved the absolute accuracy of an ABB industrial robot with a 29 parameter calibration model. In comparison to other presented works, the authors performed 1000 measurements (robot configurations) in the entire joint space for accuracy validation. After calibration, the mean/maximum positioning accuracy was reduced to 0.346 mm to 0.696 mm. Due to the errors listed in Table 2, the deflection between an ideal robot and a real manipulator can be up to 10 mm. Therefore, ABB provides a service for absolute accuracy. This calibration concept promises an absolute accuracy of the TCP of at least ±0.5 mm over the entire workspace of the robot [16]. According to ABB, the mean/maximum positioning accuracy ca be reduced to 0.3 mm/0.7 mm for an IRB 4400.

Table 2. Main Reasons for Inaccuracies in Robot Machining [5]

Error type	Description
Environment	Temperature fluctuations, warm up process
Parameter	Adaption of kinematic parameters due to manufacturing and assembling tolerances, influence of dynamic parameters, nonlinearities such as backlash and friction
Measurement	Resolution, nonlinearities from sensors /encoders mounted on the joints
Calculation	Round-off, steady state controller offset
Application	Installation errors

3.2 Offline Compensation

The models used for level two calibration are based on the kinematics of the robot. Advanced approaches (level three calibration) also consider non-geometric parameters. However, the mechanical processing of aluminium causes static and dynamic effects on the manipulator and results in tool deviation. With offline compensation, these effects are considered for the modelling process.

1) *Milling force modelling:* Milling forces arise during the interaction between the tool and the workpiece. The quality of the compensation method is heavily dependent on the process force modelling. Therefore, it is important to choose an appropriate

approach. Altintas and Ber [17] developed a mechanistic milling force model for the infinitesimal tangential, radial and axial force at a specific immersion angle ϕ with respect to the ith milling cutter tooth. After integration and transformation to a fixed coordinate system, the feed, transverse, and axial forces of the milling process can be predicted. Another modelling method was presented by Wang et al. [18]. The proposed model includes the depth of cut, width of cut, feed, process gain and coefficients describing nonlinearities between milling forces and model input values. For alternative modelling methods see [19–21].

2) *Joint stiffness modelling:* Joint stiffness of an industrial robot can be modelled statically and dynamically. The main difference is the consideration of inertia and damping in case of dynamic modelling. Static models are characterized by the displacement prediction based on time invariant and static loads. Although highly dynamic process forces arise during milling and turning of aluminium, numerous researchers used static stiffness models for robot configuration optimization [22–25] . Dynamic models can be developed utilizing joint space transformation and state space interpolation methods. Cartesian modal properties are predicted by the joint space transformation of the end effector by identifying the robot's joint and link dynamics followed by a transformation in cartesian space. State space modelling allows the description of a dynamic transfer function with a first order differential equation and an output equation. The parameterization of the state model is limited due to the linearity of the system using polynomial functions [26]. After determination of the joint stiffness matrix, the conservative congruence transformation (CCT) is widely used for transformation to the cartesian stiffness matrix. The transformation can be executed by means of the Jacobi matrix which is obtained from the kinematic model of the manipulator. The deflection of the end effector can be estimated in cartesian space with known joint stiffness matrix and milling force in any pose [27].

The implementation of a dynamic stiffness model is based on a second order inhomogeneous differential equation. The external milling force excites the system and causes chatter and a periodic deflection of the end effector. In comparison to static modelling, all influencing factors are time varying. Figure 3 shows the sequential chart of a dynamic chatter avoidance model, based on a milling force and robot stiffness model.

The complementary stiffness matrix K_c considers a change in the robot configuration under external load. It can be neglected if the external force is very small in comparison to the handling capacity [27]. In order to solve the cartesian robotic milling system model, the CCT transforms the joint stiffness matrix to the cartesian stiffness matrix first. By following the standard modal analysis procedure, the stability of the system is dependent on the eigenvalues of the resulting matrix [19]. The experimental results show a reduction in the average resultant force of about 50%.

Tyapin et al. [28] estimated the process forces based on a static force model. The authors achieved satisfying results and the accuracy requirements could be realized by the used modelling method. The cutting forces are heavily dependent on the cutting

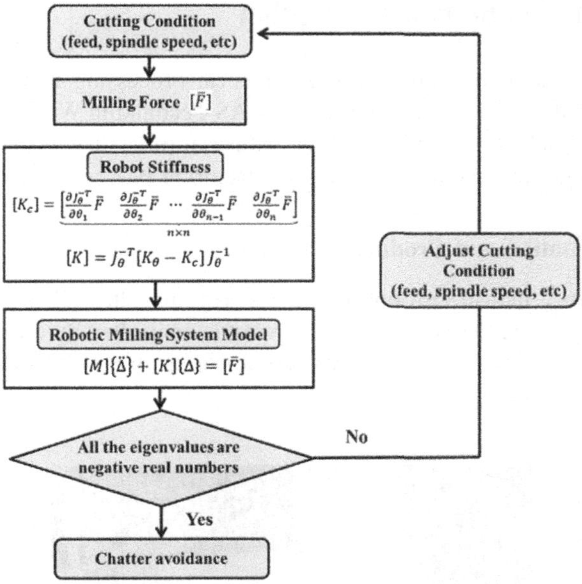

Fig. 3. Sequential chart of vibration suppression based on a dynamic robot model [19]

parameters, nevertheless with a cutting depth of 1 mm and a cutting width of 8 mm, the deviation of the TCP could be reduced from 0.092 mm to 0.007 mm. Cordes and Hintze [27] developed a method for path deviation at TCP based on reversal error and joint compliance. Both can be predicted in joint space relatively easy by utilizing the existing location invariance. A Laser Tracker enables Parameter identification for joint stiffness and reversal error. For validation of the implemented model, a circular contour was milled into an aluminium block using an ABB IRB 6660-205 and a 12 mm cutter. The accuracy of the proposed method depends on the applied force model, therefore the maximum deviation along the circular path was reduced from 0.119 mm to 0.055 mm. Schneider et al. [29] have developed an offline error compensation method based on a detailed investigation of various error sources. During the experimental milling tests and free movement in the workspace, they identified the joint compliance and gear backlash are the main causes of errors. Considering static and dynamic friction, gear backlash, and nonlinearities in joint stiffness could reduce the deviation of the TCP for milling a rectangular pocket in aluminium from 0.5 mm/0.24 mm to 0.07 mm/ 0.05 mm. Klimchik et al. [30] analysed the dynamic milling forces in time domain and derived the corresponding robot configurations for minimizing the compliance along the desired trajectory. With this method, the maximum deviation of the end effector could be reduced from 0.0632 mm to 0.0047 mm. The authors neglected high frequency vibration modes.

4 Hardware and Software Adaption

In order to integrate a fully automated mechanical processing item into the process chain of the digital factory, the team from the UAS Technikum Wien started to develop a station for mechanical production with an industrial robot. There is ongoing research to improve the system structure and the manufacturing process.

4.1 Existing Station and Production Process

The Kuka KR 16 2 industrial robot, impresses with high flexibility and a small footprint within the large-scale workspace. The robot is controlled by a Kuka KR C4 with open industry standards. Figure 4 shows the actual setup with all additional components required for robot machining.

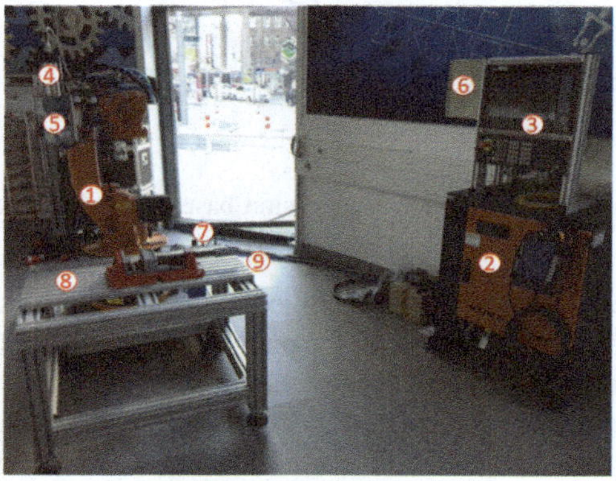

Fig. 4. Existing Robot Machining Station. 1 Kuka KR 16 2; 2 KR C4 Controller; 3 Sinumerik 840D; 4 Mechatron Milling Spindle; 5 Tool Change Adapter; 6 Frequency Inverter; 7 Lubrication system; 8 Working Table with Mechanical Vice; 9 Tool Magazine

With the possibility of using industrial robots as a manufacturing component in mechanical production, new requirements occurred for the programming of industrial robots. The calculation of transformation algorithms, avoidance of singularities and collisions as well as the pre- and post processing of manufacturing processes such as position and orientation of the workpiece or measurement of the final product can be realised by the robot controller (RC). The RC is parameterized machine specifically, therefore the controller is equipped with additional CNC functions for the machining process. Path planning and the creation of the trajectory are done by the CNC. Kuka offers the option of an integrated NC kernel (KUKA.CNC). Another possibility is the combination of CNC and RC using a Siemens 840D controller. RC – CNC combinations support the use of G-code programming and reduce the aversion to industrial robots in the manufacturing industry. If these modifications are not available, the processing quality depends on the postprocessor of the CAM software. The CAM supported path planning is based on a CAD file of the desired workpiece. After the corresponding tool path has been generated from individual machining steps, the program can be exported in standardized (G-code) or machine – specific programming language (Kuka Robot Language KRL for Kuka). Figure 5 illustrates the actual flow chart for the manufacturing process.

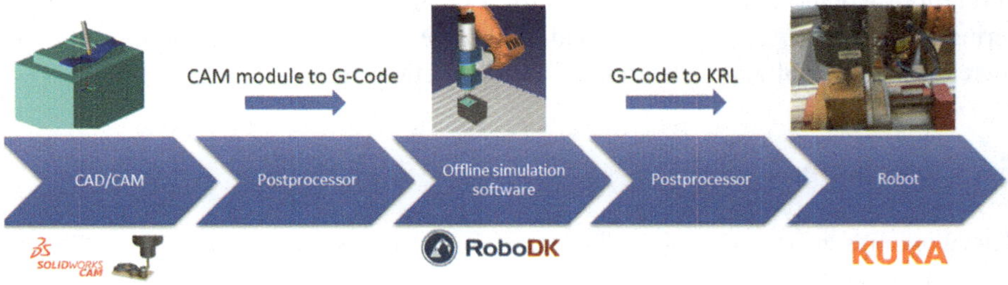

Fig. 5. Actual process chain for manufacturing of a desired workpiece

4.2 Enhancements for Automated Process Environment

The design of the workpiece and the following processing steps for the production process at the existing station are done using SolidWorks and SolidWorks CAM. With integrated postprocessors, a standardised G-code can be generated based on predefined processing steps. Due to the limited compatibility between SolidWorks CAM and RoboDK, even simple contours lead to misinterpretations of the given tool paths. The conversion to G-code and retranslation into KRL results in additional effort and causes unnecessary errors. Therefore, the actual process flow should be revised. Figure 6 proposes an adapted process chain for the manufacturing procedure.

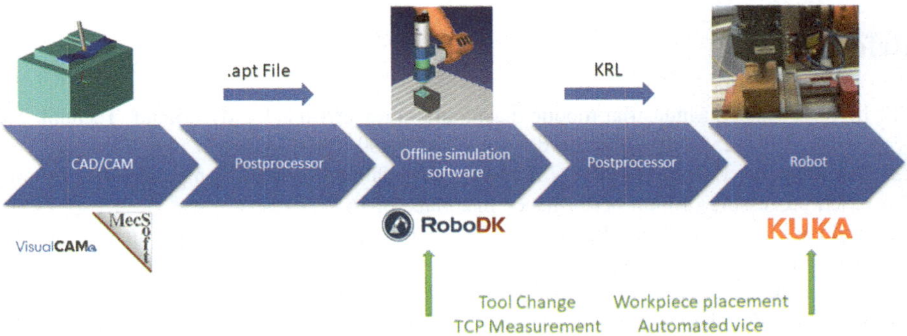

Fig. 6. Adapted process chain for a simplified workflow

With the exchange of the manufacturing software, a better compatibility between CAM and the simulation software is achieved. Visual CAM provides a suitable postprocessor for a six-axis industrial robot. The predefined processing steps are sent to the simulation software by a generated .apt file. RoboDK is able to convert the prepared machining commands to KRL commands and enables a simple communication between planning and production. Figure 6 shows planned and partially implemented hardware expansions for efficiency improvement of the station. The already integrated tool change system reduces downtimes and saves equipment costs for additional tools. An automated measurement of new tools using a forked light barrier and a 6D measuring tool would be an important step for a rapid integration of additional manufacturing options. An Automated vice and workpiece placement by the robot itself could increase the level of automatization of the production process.

5 Conclusions

This work presented an overview of the current state of research and compared different types of robotic machining possibilities. Good programmability, adaptability, and flexibility of articulated robots result in significantly lower investment costs compared to CNC machine tools with the same workload. Studies reveals a total cost reduction of up to 30% by using industrial robots [31]. Therefore, a lot of research focuses on the improvement of accuracy in robot machining. Despite the detailed modelling of advance calibration methods, the accuracy during milling operations cannot be improved because external forces (milling forces) are not considered in the modelling process. Only the deviation between the ideal robot in the simulation software (RoboDK) and the real robot can be reduced to a minimum. The mean positioning accuracy determined experimentally by all presented authors was about 0.3 mm. The integration of milling forces and cutting parameters in the modelling process enables offline compensation methods a significant reduction of TCP deviation. Depending on the predefined parameters and other boundary conditions, the presented strategies reported a maximum deviation between 4.7 μm and 55 μm.

References

1. Ji, W., Wang, L.: Industrial robotic machining: a review. Int. J. Adv. Manuf. Technol. **103** (1–4), 1239–1255 (2019). https://doi.org/10.1007/s00170-019-03403-z
2. Klimchik, A.: Enhanced stiffness modeling of serial and parallel manipulators for robotic-based processing of high performance materials, p. 161 (2012)
3. Andrew Liou, Y.H., Lin, P.P., Lindeke, R.R., Chiang, H.D.: Tolerance specification of robot kinematic parameters using an experimental design technique—the Taguchi method. Robot. Comput. Integr. Manuf. **10**(3), 199–207 (1993). https://doi.org/10.1016/0736-5845(93)90055-o
4. Karan, B., Vukobratović, M.: Calibration and accuracy of manipulation robot models. Mech. Mach. Theory **29**, 479–500 (1994). https://doi.org/10.1016/0094-114X(94)90130-9
5. Nubiola, A., Bonev, I.A.: Absolute calibration of an ABB IRB 1600 robot using a laser tracker. Robot. Comput. Integr. Manuf. **29**(1), 236–245 (2013). https://doi.org/10.1016/j.rcim.2012.06.004
6. Nguyen, H.-N., Le, P.-N., Kang, H.-J.: A new calibration method for enhancing robot position accuracy by combining a robot model–based identification approach and an artificial neural network–based error compensation technique. Adv. Mech. Eng. **11**(1), 16878140 1882293 (2019). https://doi.org/10.1177/1687814018822935

7. Newman, W.S., Birkhimer, C.E., Horning, R.J., Wilkey, A.T.: Calibration of a Motoman P8 robot based on laser tracking. In: Proceedings 2000 ICRA Millennium Conference on IEEE International Conference on Robotics and Automation. Symposia Proceedings (Cat. No.00CH37065), vol. 4, pp. 3597–3602, April 2000. Bd.4, https://doi.org/10.1109/robot.2000.845292
8. Brutscheck, T., Bücker, M.: Modellierung von Roboterkinematiken nach Denavit und Hartenberg. In: Bandow, G., Holzmüller, H.H. (eds.) Das ist gar kein Modell!: Unterschiedliche Modelle und Modellierungen in Betriebswirtschaftslehre und Ingenieurwissenschaften, pp. 149–165. Gabler, Wiesbaden (2009)
9. Mooring, B., Roth, Z.S., Driels, M.R.: Fundamentals of Manipulator Calibration. Wiley, New York (1991)
10. Martinelli, A., Tomatis, N., Tapus, A., Siegwart, R.: Simultaneous localization and odometry calibration for mobile robot. In: Proceedings 2003 IEEE/RSJ International Conference on Intelligent Robots and Systems (IROS 2003) (Cat. No.03CH37453), vol. 2, pp. 1499–1504, Okt. 2003. https://doi.org/10.1109/iros.2003.1248856
11. Jiang, Z., Zhou, W., Li, H., Mo, Y., Ni, W., Huang, Q.: A New kind of accurate calibration method for robotic kinematic parameters based on the extended Kalman and particle filter algorithm. IEEE Trans. Industr. Electron. **65**(4), 3337–3345 (2018). https://doi.org/10.1109/TIE.2017.2748058
12. Song, Y., Zhang, J., Lian, B., Sun, T.: Kinematic calibration of a 5-DoF parallel kinematic machine. Precision Eng. **45**, 242–261 (2016) https://doi.org/10.1016/j.precisioneng.2016.03.002
13. Ginani, L., Motta, J.: Theoretical and practical aspects of robot calibration with experimental verification. J. Brazilian Soc. Mech. Sci. Eng. **33**, 15–21 (2011). https://doi.org/10.1590/s1678-58782011000100003
14. Ye, S.H., Wang, Y., Ren, Y.J., Li, D.K.: Robot calibration using iteration and differential kinematics. J. Phys. Conf. Ser. **48**, 1–6 (2006). https://doi.org/10.1088/1742-6596/48/1/001
15. Lightcap, C., Hamner, S., Schmitz, T., Banks, S.: Improved positioning accuracy of the PA10-6CE robot with geometric and flexibility calibration. IEEE Trans. Rob. **24**(2), 452–456 (2008). https://doi.org/10.1109/TRO.2007.914003
16. ABB. Absolute Accuracy, February 2005. https://library.e.abb.com/public/0f879113235a0e1dc1257b130056d133/Absolute%20Accuracy%20EN_R4%20US%2002_05.pdf (zugegriffen Mai 18, 2020)
17. Altintas, Y., Ber, A.A.: Manufacturing automation: metal cutting mechanics, machine tool vibrations, and CNC design. Appl. Mech. Rev. **54**(5), B84–B84 (2001). https://doi.org/10.1115/1.1399383
18. Wang, J., Zhang, H., Pan, Z.: Machining with Flexible Manipulators: Critical Issues and Solutions (2006)
19. Cen, L., Melkote, S.N.: CCT-based mode coupling chatter avoidance in robotic milling. J. Manuf. Process. **29**, 50–61 (2017). https://doi.org/10.1016/j.jmapro.2017.06.010
20. Klimchik, A., Ambiehl, A., Garnier, S., Furet, B., Pashkevich, A.: Efficiency evaluation of robots in machining applications using industrial performance measure. Robot. Comput. Integr. Manuf. **48**,. 12–29 (2017). https://doi.org/10.1016/j.rcim.2016.12.005
21. Engin, S., Altintas, Y.: Mechanics and dynamics of general milling cutters. Part I: helical end mills, p. 18
22. Diaz Posada, R.J., Schneider, U., Sridhar, A., Verl, A.: Automatic motion generation for robotic milling optimizing stiffness with sample-based planning. Machines **5**(1) (2017). https://doi.org/10.3390/machines5010003

23. Diaz Posada, J.R., Schneider, U., Pidan, S., Geravand, M., Stelzer, P., Verl, A.: High accurate robotic drilling with external sensor and compliance model-based compensation. In: 2016 IEEE International Conference on Robotics and Automation (ICRA), pp. 3901–3907, Mai 2016. https://doi.org/10.1109/icra.2016.7487579
24. Xie, H., Li, W., Yin, Z.: Posture optimization based on both joint parameter error and stiffness for robotic milling. In: Intelligent Robotics and Applications, pp. 277–286. Cham (2018)
25. Lin, Y., Zhao, H., Ding, H.: Spindle configuration analysis and optimization considering the deformation in robotic machining applications. Robot. Comput. Integr. Manuf. **54**, 83–95 (2018). https://doi.org/10.1016/j.rcim.2018.05.005
26. Cvitanic, T., Nguyen, V., Melkote, S.N.: Pose optimization in robotic machining using static and dynamic stiffness models. Robot. Comput. Integr. Manuf. **66**, 101992 (2020). https://doi.org/10.1016/j.rcim.2020.101992
27. Cordes, M., Hintze, W.: Offline simulation of path deviation due to joint compliance and hysteresis for robot machining. Int. J. Adv. Manuf. Technol. **90**, April 2017. https://doi.org/10.1007/s00170-016-9461-z
28. Tyapin, I., Kaldestad, K.B., Hovland, G.: Off-line path correction of robotic face milling using static tool force and robot stiffness. In: 2015 IEEE/RSJ International Conference on Intelligent Robots and Systems (IROS), pp. 5506–5511, Okt. 2015. https://doi.org/10.1109/iros.2015.7354157
29. Schneider, U.: Improving robotic machining accuracy through experimental error investigation and modular compensation. Int. J. Adv. Manuf. Technol. **85**, 1–13 (2014). https://doi.org/10.1007/s00170-014-6021-2
30. Klimchik, A., Bondarenko, D., Pashkevich, A., Briot, S., Furet, B.: Compliance error compensation in robotic-based milling. Inf. Control Autom. Robot. **283**, September 2014. https://doi.org/10.1007/978-3-319-03500-0_13
31. Caro, S., Dumas, C., Garnier, S., Furet, B.: Workpiece placement optimization for machining operations with a KUKA KR270-2 robot. In: 2013 IEEE International Conference on Robotics and Automation, pp. 2921–2926, Mai 2013. https://doi.org/10.1109/icra.2013.6630982

Artificial Neural Networks Based Place Categorization

Andreas Kriegler[1(✉)], Wilfried Wöber[2], and Mohamed Aburaia[2]

[1] Mechatronics/Robotics, UAS Technikum Vienna, Vienna, Austria
krieglerandreas@gmail.com
[2] Digital Manufacturing and Robotics, UAS Technikum Vienna, Vienna, Austria
{wilfried.woeber,mohamed.aburaia}@technikum-wien.at

Abstract. Localization using images is a fundamental problem in computer vision for autonomous systems. The information gained enables a high-level understanding of the environment for the mobile robot, which is much more in line with a person's understanding of their surroundings. It is therefore a critical task for human-robot collaboration and enables robots to find places and identify objects applying this a priori information using semantic maps. Customarily, deep learning models such as convolutional neural networks can be trained for location classification from video streams. However, to achieve acceptable place classification results, previous trained neural nets must be adapted to the specific environment. This work tackles this adaptation of an existing system for place categorization including an extension in terms of possible places. The extension is based on classical machine learning models which were trained based on extracted features of a pre-trained neural network. The developed system outperforms the initial system by correctly classifying 90% of images, including places which were unknown to the neural network.

Keywords: Place categorization · Convolutional Neural Networks · Support vector machine · Machine learning · Mobile robotics

1 Introduction

With rapid advancements in technology and science, the level of autonomy in mobile robotics is ever increasing [1, 2]. The exponential increase in computing power in graphics processing units (GPU) has paved the way for the field of deep learning (DL) to emerge [3, 4] – a new approach to solve statistical problems regarding autonomous agents or systems [5]. Convolutional Neural Networks (CNN), introduced in [6], are at the forefront of deep learning when large amounts of visual data need to be processed in a way that enables the overarching system to learn with this data [7– 13]. This has made CNNs a natural addition to mobile robotics systems that are often equipped with multiple cameras to obtain images of their surroundings. The analysis of visual data using CNNs enables mobile robots to extract information of their environment. Based on this information, robots can then answer questions such as Where am I? [14], but also What is the place I am in like? [15]. Images or samples drawn from videos or datasets can be classified based on a known set of existing classes – in this

N. M. Durakbasa and M. G. Gençyılmaz (Eds.): ISPR 2020, LNME, pp. 201–209, 2021.
https://doi.org/10.1007/978-3-030-62784-3_17

case different semantic place labels such as *hallway* or *kitchen*. Combining place categorization with a map of the robot's environment yields semantic maps [16, 17]. This work focuses on the place categorization aspect, using techniques of supervised learning that shall now be discussed in greater detail.

2 Supervised Learning

Supervised learning is a general term used to describe several machine learning (ML) algorithms that learn a function which maps a given input to an output based on previously learned input-output pairs [18]. A set of n training samples of the form $\{(\boldsymbol{x}_1, y_1), ..., (\boldsymbol{x}_n, y_n)\}$ are given as input to train the algorithm, where $\boldsymbol{x}_i$ is the feature vector of the i–th sample and y_i is its label (or class). The algorithm then approximates a function $g : X \rightarrow Y$ where X is the input space and Y is the output space. These features are often expressed in vectorized form; hence one sample consists of a vector of features and the corresponding class label. All supervised learning tasks therefore require many labelled training examples, which are often difficult and expensive to obtain, requiring either intensive human labelling or accurate measurements. The goal is to obtain a trained model, which correctly determines the class labels for unseen instances, i.e. images that are new to the system. This requires the model to generalize properly [19] – the model must not overfit [20] on the training samples and can classify new examples with high accuracy.

3 Convolutional Neural Networks

Artificial Neural Networks (ANN) are learning algorithms used to generate the aforementioned input-output mapping functions while making use of the vast computing power available with GPUs. ANNs approximate the mapping-function with a multitude of layers while employing a correction method to tune its weights during training. Every input value is assigned an input neuron $x_1 ... x_k$ in the input layer forming the input vector $\boldsymbol{x} = (x_1, ..., x_k)^T$. The values are then propagated through all the nodes and combined with a weighting vector $\boldsymbol{w}$ in a feed forward manner. The activation function is classically the sigmoid function acting as a smooth threshold to map the input to the output following Eq. (1).

$$f(\boldsymbol{x}) = sig(\boldsymbol{x}) = \frac{1}{1 + e^{-x}} \tag{1}$$

Weights influence the steepness of the sigmoid and bias neurons translate the curve along the input axis. The output of one neuron therefore follows Eq. (2).

$$y = sig(\boldsymbol{w} \cdot \boldsymbol{x} + b) \tag{2}$$

A convolutional neural network is a class of deep ANNs that has become very popular for extracting all kinds of features from images. The hidden layers in a CNN consist of further convolutional, pooling, normalization and fully connected layers. Figure 1 shows the basic layout of a CNN [21].

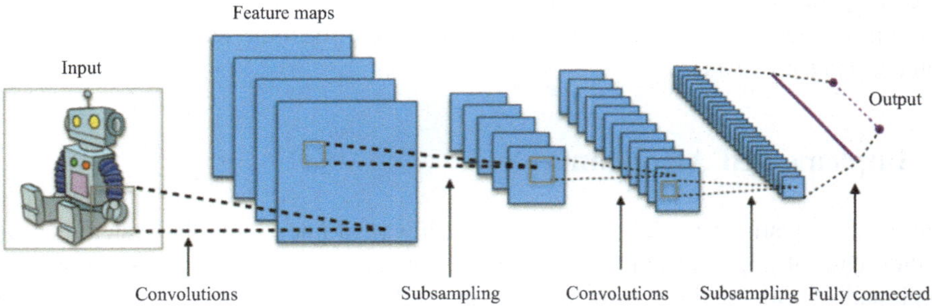

Fig. 1. *CNN architecture.* The basic layout and connectivity of the layers in a convolutional neural network [21]

The convolutional layer's parameters consist of a set of learnable filters or kernels with a small receptive field that extends through the full depth of the input volume. During the forward pass, for every filter the dot product between the filter and the input is computed, resulting in a stack of two-dimensional activation maps that form the full output volume [22]. This stack is then passed through pooling and batch-normalization layers before being fed into another convolutional layer. The end marks a fully connected layer that transforms the two-dimensional activation maps into a single value for every neuron in the output layer, making up the output vector with classification likelihood values [23].

A CNN was used in [15] for place categorization by using a set of images and place labels. Sünderhauf et al. [15] trained a CNN on the Places205 dataset [24] – a dataset featuring images from 205 different place categories including images from places such as *kitchen* and *auditorium* but also *desert* and *igloo*. From the original Places dataset with 476 classes, Places205 is a subset of all 205 classes which contain more than 5000 images per category. Since their network models were trained on such a vast number of different places the classification accuracy can be improved via fine-tuning on a dataset only containing a further subset of these 205 classes. The environment that the mobile robot was to be evaluated in, was expected to include e.g. a kitchen and a hallway, but not a desert, hence the system should be fine-tuned for this area of operations. On the other hand, the mobile robot is expected to encounter new places during its lifespan that might not be part of the 205 classes, therefore new classes should be easily included and trained on to allow classification amongst those as well. To this end, three different machine learning algorithms were trained using the output vector of the CNN as input. The architecture of the Places205-CNN is akin to the popular AlexNet-architecture [6] with a few minor modifications; mainly the last fully connected layer relates to 205 neurons, one for every place class. The last layer after that is a softmax layer [4] used to normalize the classification values of the 205 output neurons to sum up to 1; they can then be interpreted with probabilistic reasoning. While in the work of Sünderhauf et al.

[15] this vector is seen as the final result, with the highest value corresponding to the predicted class, in this work this 205 × 1 vector was instead continuously extracted to create the design matrix $\boldsymbol{m}_{205} \in \mathbb{R}^{u\times 205}$ and corresponding target vector $\boldsymbol{t}_{205} \in \mathbb{R}^{u\times 1}$ where u is the number of analysed frames obtained from the CNN given a specific video as input. This way a new dataset was created, which was then split 75–25% into training and test data respectively. Afterwards, three machine learning models were trained using this dataset.

4 Implemented Algorithms

The machine learning models were chosen as a way of fine-tuning the system because of their ease of implementation and to show that classical ML algorithms can work in combination with modern deep learning methods to great success, especially if the problem, such as classification among six classes, is not overly complex. This section discusses these implemented algorithms, namely the Naïve Bayes filter, multinomial logistic regression and a support vector machine.

4.1 Naïve Bayes Filter

Naïve Bayes filters (NBF) are probabilistic classifiers that are based on the Bayes' rule which relates the conditional probability $p(x\,|\,y)$ to its inverse $p(y\,|\,x)$. The probability density function for every feature for each class is first estimated using the Gaussian distribution. This gives a matrix of kernels $m_{205} \in \mathbb{R}^{K\times 205}$ where K is the number of classes that are expected to be encountered during operation of the mobile robot. The posteriors are then modelled according to the extended form of the Bayes' rule which can be observed in Eq. (3). This step is done for every class C_k with $k \in \{1, ..., K\}$.

$$\hat{p}(C_k\,|\,\boldsymbol{x}_1, ..., \boldsymbol{x}_n) = \frac{\pi(C_k)\prod_{i=1}^{n} p(\boldsymbol{x}_i\,|\,C_k)}{\sum_{k=1}^{K}\pi(C_k)\prod_{i=1}^{n} p(\boldsymbol{x}_i\,|\,C_k)} \tag{3}$$

It gives the probability that the current image belongs to class C_k, given the input vector $\boldsymbol{x}$, based on the conditional probabilities $p(\boldsymbol{x}_i\,|\,C_k)$, where $\boldsymbol{x}_1, ..., \boldsymbol{x}_n$ are the random features of an observation and $\pi(C_k)$ is the prior point-probability that a class index is k. Finally, a decision rule is used that picks the most probable hypothesis; also known as the maximum a posteriori (MAP) rule. The classifier is therefore the function that assigns label $\hat{y} = k$ for some k according to Eq. (4).

$$\hat{y} = \underbrace{argmax}_{k\in\{1,...,K\}}\, \hat{p}(C_k\,|\,\boldsymbol{x}_1, ..., \boldsymbol{x}_n) \tag{4}$$

4.2 Multinomial Logistic Regression

Multinomial logistic regression (MLR) is an extension of logistic regression for multiclass problems [25]. The log-odds of the probability of a sample belonging to a certain class is a linear combination of independent variables x. In multiclass logit the dependent variable y can take one of multiple discrete outcomes. If the model is parameterized by $\boldsymbol{\theta}$, with N being the number of drawn samples and N^{-1} therefore being a normalization factor, the model can be described with Eq. (5).

$$\begin{aligned} &\lim_{N\to+\infty} N^{-1}\sum_{i=1}^{N}\log p(y_i\,|\,\boldsymbol{x}_i;\theta) \\ = \sum_{x\in X}\sum_{y\in Y} p(\boldsymbol{x},y)&\left(-\log\frac{p(Y=y\,|\,X=\boldsymbol{x})}{p(Y=y\,|\,X=\boldsymbol{x};\theta)}+\log p(y\,|\,\boldsymbol{x})\right) \\ &= -D_{KL}(Y\,||\,Y_\theta)-H(Y\,|\,X) \end{aligned} \tag{5}$$

$H(Y\,|\,X)$ is the conditional entropy and quantifies the amount of information needed to describe the outcome of Y given that the value of X is known. $D_{KL}(Y\,||\,Y_{\boldsymbol{\theta}})$ is the Kullback-Leibler divergence, which measures the information loss using the learned model and quantifies this information loss when the distribution from parameterized model Y_θ is used instead of the ground truth distribution Y. By adjusting the parameters $\boldsymbol{\theta}$ to maximize the log-likelihood of the model (via iterative schemes such as the Newton-Raphson method), the KL divergence is minimized, thus the model which makes the least number of assumptions in its parameters is found; it is closest to the ground-truth.

4.3 Support Vector Machines

Support vector machines (SVM) try to classify data by separating the data points with a line in the case of two-dimensional data, or a $(p-1)$-dimensional hyperplane for the case of a p-dimensional feature vector. The Hessian normal distance from the group of data-points to the maximum-margin hyperplane is maximized by minimizing a loss function (the Hinge loss) that can be observed in Eq. (6) [26].

$$\left[\frac{1}{n}\sum_{i=1}^{n}\max(0,1-y_i(\boldsymbol{w}\cdot\boldsymbol{x}_i-\mathrm{b}))\right]+\lambda||\boldsymbol{w}||^2 \tag{6}$$

λ determines the trade-off between increasing the margin-size and ensuring that $\boldsymbol{x}_i$ lies on the correct side of the margin with n being the number of data vectors. SVMs use the *kernel trick* to generate the nonlinear decision boundary. Kernel functions are similarity functions in the original input space that correspond to the dot products of the individual data points in some expanded feature space. In this expanded space the data points are more easily separated with some linear boundary. Figure 2 gives a visual explanation of the kernel trick [27].

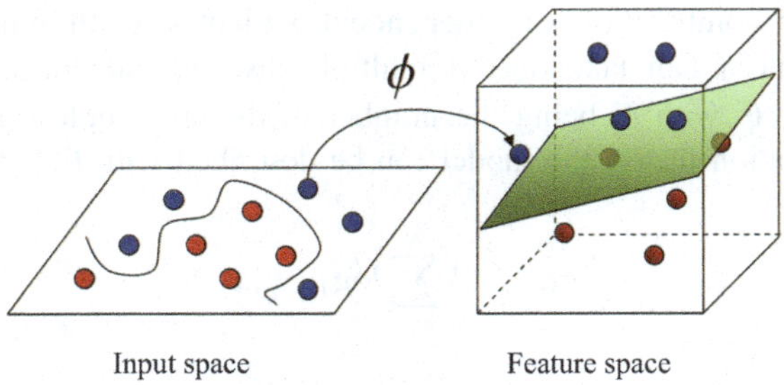

Fig. 2. *Kernel trick.* While data points are not linearly separable in the input space, they are in a higher-dimensional feature space [28]

5 Results and Discussion

The system was evaluated on 5 of the possible 12 observable places on university campus: the digital factory resembling an industrial setting as the Places205 class *assembly_line*, a big lecture theater as *auditorium*, a *corridor*, a *kitchen* and an *office*. To show the capabilities of the system to extend the set of classes, data of an additional place was gathered not known to the CNN: a *door*. The videos were captured using the ROS wrapper for Intel Real Sense devices [28] and a BlasterX Senz3D [29] camera. The SVM and other models were tuned on the distributions of the features oftentimes showing correlations between objects that were not common in the original Places205 dataset. This is especially obvious for the frames captured in the digital factory, which resembles partially an office environment as well as an industrial production line, a combination of features that the CNN was not trained on. Table 1 shows the final classification results. Since the number of frames in every video recording were nearly equal, no further weighting had to be done. With the models having correctly labelled 100% of the samples from the *door* class, it is shown, that the classification can easily be extended to environments not part of the 205 places.

Table 1. Classification results: top-1 classification accuracy of the CNN, limited to 5 out of 205 classes and the ML models trained with the inclusion of the *door* class.

Classification results (recall)				
Environment	CNN-5	MLR	SVM	NBF
Assembly_line	54%	**99%**	97%	92%
Auditorium	62%	88%	**89%**	81%
Corridor	**100%**	93%	93%	95%
Door	–	**100%**	**100%**	**100%**
Kitchen	**100%**	89%	87%	83%
Office	75%	63%	73%	**79%**
Total average	*78.2%*	*88.7%*	***90.2%***	*88.3%*

Figure 3 gives the normalized confusion matrix of the best performing model, the SVM – it shows per-class predictions against the ground truth.

	assembly_line	auditorium	corridor	door	kitchen	office
assembly_line	0,97	0,00	0,00	0,00	0,00	0,03
auditorium	0,09	0,89	0,00	0,01	0,00	0,00
corridor	0,00	0,00	0,93	0,07	0,00	0,00
door	0,00	0,00	0,00	1,00	0,00	0,00
kitchen	0,13	0,00	0,00	0,00	0,87	0,00
office	0,03	0,00	0,00	0,23	0,01	0,73

Fig. 3. *SVM confusion matrix.* Normalized confusion matrix for the SVM-model. Rows show the true label vs. Predicted labels in the columns

The recording of the *door* and the *office* were suboptimal, in the sense that the *door* carried large similarities to a closet that was part of the *office*, meaning a fair number of visually similar images were part of two separate classes. This led to misclassifications with the ML-models, while the CNN was not influenced. Most errors therefore come from images from the *office* class being classified as elements of the *door* class (see Fig. 3).

As a last experiment, an SVM was trained on a subset of the dataset featuring only three instead of the five classes: an *auditorium*, a *corridor* and a *kitchen*. For this problem the SVM boasted an accuracy of 99.54% showing the ambiguous *office* vs. *Door* recordings were the primary cause for misclassifications in the five-class environment; although as the number of classes increases, the prediction accuracy is expected to decrease naturally. The method for classification was One-Against-One [30] – 13, 16 and 9 support vectors were used to create the binary classifiers amongst the three classes *auditorium*, a *corridor* and a *kitchen* respectively.

6 Conclusion

This work introduced an extension for a place categorization system, based on a CNN, using machine learning. These models included a Naïve Bayes filter, a multinomial logistic regression model and a support vector machine. All three models yielded strong classification results between 88% and 99%. These three ML models are a perfect light-weight addition to the CNN for fine-tuning. The system, compared to an exclusively CNN-based approach, further makes no strong hardware- or complex software-demands, using a webcam and open-source software. The semantic information obtained is an important enabler of more advanced robotics tasks, especially human-robot collaboration, since humans describe places, rooms and goals not with coordinates but place labels. To this end, the system was successfully tested in an industrial- and office-environment, proving its usability for manufacturers and the commercial service industry alike.

References

1. Siegwart, R., Nourbakhsh, I.R., Scaramuzza, D.: Introduction to Autonomous Mobile Robots. The MIT Press, Cambridge (2011)
2. Amazone. "BoniRob", *BoniRob – Forschung zur Feldrobotik*: https://www.amazone.at/1857.asp
3. Hof, R.D.: Deep Learning. https://www.technologyreview.com/s/513696/deep-learning/
4. Deng, L., Yu, D.: Deep learning: methods and applications. Found. Trends Signal Process. **7** (3–4), 197–387 (2014)
5. Goodfellow, I., Bengio, Y., Courville, A.: Deep Learning. The MIT Press, Cambridge (2016)
6. LeCun, Y., Boser, B., Denker, J.S., Henderson, D., Howard, R.E., Hubbard, W., Jackel, L. D.: Backpropagation applied to handwritten zip code recognition. Neural Comput. **1**, 541–551 (1989)
7. Krizhevsky, A., Sutskever, I., Hinton, G.E.: ImageNet classification with deep convolutional neural networks. In: Advances in Neural Information Processing Systems 25 (NIPS 2012), pp. 1097–1105 (2012)
8. Szegedy, C., Liu, W., Jia, Y., Sermanet, P., Reed, S., Anguelov, D., Erhan, D., Vanhoucke, V., Rabinovich, A.: Going deeper with convolutions. In: Proceedings of the IEEE International Conference on Computer Vision and Pattern Recognition (CVPR 2015), pp. 1–9 (2015)
9. He, K., Zhang, X., Ren, S., Sun, J.: Deep residual learning for image recognition. In: Proceedings of the IEEE International Conference on Computer Vision and Pattern Recognition (CVPR 2016), pp. 770–778 (2016)
10. Girshick, R., Donahue, J., Darrell, T., Malik, J.: Rich feature hierarchies for accurate object detection and semantic segmentation. In: Proceedings of the IEEE International Conference on Computer Vision and Pattern Recognition (CVPR 2014), pp. 580–587 (2014)
11. Ren, S., He, K., Girshick, R., Sun, J.: Faster R-CNN: towards real-time object detection with region proposal networks. In: Advances in Neural Information Processing Systems, vol. 1, pp. 91–99 (2015)
12. Redmon, J., Farhadi, A.: YOLOv3: An incremental improvement, 1st edn. (2018). https://arxiv.org/abs/1804.02767
13. Singh, B., Li, H., Sharma, A., Davis, L.S.: R-FCN-3000 at 30fps: decoupling detection and classification. In: Proceedings of the IEEE International Conference on Computer Vision and Pattern Recognition (CVPR 2018), pp. 1081–1090 (2018)
14. Borenstein, J., Everett, H.R., Feng, L.: Where am I? Sensors and methods for mobile robot positioning, 1st edn. April 1996. https://citeseerx.ist.psu.edu/viewdoc/download?doi=10.1.1.444.5014&rep=rep1&type=pdf
15. Sünderhauf, N., Dayoub, F., McMahon, S., Talbot, B., Schulz, R., Corke, P., Wyeth, G., Upcroft, B., Milford, M.: Place categorization and semantic mapping on a mobile robot. In: Proceedings of the IEEE International Conference on Robotics and Automation (ICRA 2016), pp. 5729–5736 (2016)

16. Pronobis, A., Jensfelt, P.: Large-scale semantic mapping and reasoning with heterogeneous modalities. In: Proceedings of the IEEE International Conference on Robotics and Automation (ICRA 2012), pp. 3515–3522 (2012)
17. Hemachandra, S., Walter, M.R., Tellex, S., Teller, S.: Learning spatial-semantic representations from natural language descriptions and scene classifications. In: Proceedings of the IEEE International Conference on Robotics and Automation (ICRA 2014), pp. 2623–2630 (2014)
18. Russell, S.J., Norvig, P.: Artificial Intelligence: A Modern Approach. Pearson Education Limited, Malaysia (2016)
19. Mitchell, T.M.: The need for biases in learning generalizations. Department of Computer Science, Laboratory for Computer Science Research, Rutgers Univ., New Jersey (1980)
20. Geman, S., Bienenstock, E., Doursat, R.: Neural networks and the bias/variance dilemma. Neural Comput. **4**(1), 1–58 (1992)
21. Aphex34: typical CNN architecture. https://upload.wikimedia.org/wikipedia/commons/6/63/Typical_cnn.png
22. ujjwalkarn: An Intuitive Explanation of Convolutional Neural Networks. https://ujjwalkarn.me/2016/08/11/intuitive-explanation-convnets/
23. Li, F.-F., Johnson, J., Yeung, S.: CS231n. University Course, Topic: "Convolutional Neural Networks for Visual Recognition." University of Stanford. https://cs231n.github.io/convolutional-networks/
24. Zhou, B., Lapedriza, A., Xiao, J., Torralba, A., Oliva, A.: Learning deep features for scene recognition using places database. In: Advances in Neural Information Processing Systems 27 (NIPS 2014), pp. 487–495 (2014)
25. Greene, W.H.: Econometric Analysis. Pearson Education India, Chennai (2003)
26. Ben-Hur, A., Horn, D., Siegelmann, H.T., Vapnik, V.: Support vector clustering. J. Mach. Learn. Res. **2**, 125–137 (2001)
27. dante: Kernel trick explanation. https://datascience.stackexchange.com/questions/17536/kernel-trick-explanation
28. Intel Corporation: ROS Wrapper for Intel RealSense Devices. https://github.com/intel-ros/realsense
29. Creative. "BlasterX Senz3D." https://us.creative.com/p/web-cameras/blasterx-senz3d
30. Bishop, C.M.: Pattern Recognition and Machine Learning. Springer, New York (2006)

Logic Control of an Industrial Automation Cell with ROS as Part of SMEs and Industry 4.0 Interconnection

Erich Markl[1], Jorge M. Bauer[1,2,3], David Riepl[2](✉), and Thomas Steiner[1]

[1] UAS Technikum Wien, Vienna, Austria
erich.markl@technikum-wien.at,
doktorjbauer@yahoo.com.ar, steiner.thomas@gmx.at
[2] AuM, TU Wien, Vienna, Austria
david.riepl@aon.at
[3] Robotic, Depto Industrial, UTN-FRBA Argentina, Buenos Aires, Argentina

Abstract. The global market for industrial robotics is controlled by a small number of leading producers. Especially for SMEs that face the challenges of the Industry 4.0 paradigm shift, not only the acquisition costs of robotic equipment but also the cost of opting into the producer's proprietary software ecosystem impedes cost-efficient operations. Inevitable this creates a vendor lock-in for the customer, which causes more secondary expenses like training costs and unpredictable migration costs in case of a vendor change. As a large fraction of the work necessary to implement an industrial automation cell is dedicated to process control tasks, the required logic usually has to be implemented using the software ecosystem and proprietary programming language provided by the vendor. There is a major trend within SMEs to reduce the vendor lock-in for process control logic by introducing open solutions and open communication standards, rather than relying on closed and proprietary software components and protocols. The implementation of open technologies not only helps SMEs to reduce software license costs, but also enables them to increase the overall system flexibility, either by incorporating multiple types of industrial robots (articulated, SCARA, cartesian coordinate, etc.) or by defining and consuming open interfaces in order to facilitate communication with partner systems. The goal of this work is to evaluate the use of open standards and open-source solutions regarding their applicability for modern production lines, taking into account the need for open communication standards, which are demanded the by Industry 4.0 revolution. In addition, a simulation of an specific automation task common for an SME environment is developed within the scope of this work in order to show the practical feasibility and flexibility of an open-source process control solution.

Keywords: Industry 4.0 · Industrial robotics · Autonomation · Retrofitting robots

N. M. Durakbasa and M. G. Gençyılmaz (Eds.): ISPR 2020, LNME, pp. 210–221, 2021.
https://doi.org/10.1007/978-3-030-62784-3_18

1 Introduction

1.1 State of the Art

The process control logic is the brain of any industrial automation cell. The leading vendors of robotic equipment also offer robotic controllers and proprietary software ecosystems that are equal to the task and provide a technologically advanced solution. Although, these solutions come with financial disadvantages due to the high costs as well as they force the customer into a vendor lock-in situation since OEM controllers often have their own unique language and therefore training requirements, which requires the customer to heavily invest into the vendors software ecosystem.

A solution based on a programmable logic controller is a well-established alternative to proprietary solutions. PLC-based robotic controllers are more flexible regarding the integration of different components within an automation cell (robotic components, camera system, sensors, pneumatic actuators, etc.) and are usually programmed using a standardized programming language[1]. Despite their advantages, an automation cell controlled by a PLC-based process logic still comes with considerable costs, which poses a competitive disadvantage especially for SME's. The cost factors are not only related to the necessary hardware equipment, since PLC-vendors also earn a considerable amount from of software licensing.

Proprietary software pricing models can also be very complex because vendors offer multi-tier license models and use feature or user limitation to maximize revenue, which makes the overall cost non-transparent and difficult to compare. Furthermore, software vendors protect their software from being copied and used without the necessary license agreement by relying on methods like activation codes, license servers or hardware dongles. SME's often work on low-cost projects and special use cases where it is often not feasible to commit into a proprietary software ecosystem with large investments, an open robotics framework is a possible alternative. There are a number of such frameworks publicly available of which the Robot Operating System (ROS) is one of the most advanced solutions. The use of an open framework to implement the process control tasks on an automation cell reduces the software costs to a minimum. Although training costs still exists since engineers need time to familiarize themselves with the ecosystem, an SME can benefit from the open community behind such a project, which provides and openly shares solutions for common robotic problems (collision modeling, interfaces to actuators, etc.) Furthermore, an open framework provides an extended flexibility to integrate with existing automation setup as well as reduce implementation costs since no proprietary software licenses are required.

This paper shows that the use of open frameworks, like ROS, is feasible for implementing an industrial automation cell consisting of a multi-axis robot with up to 6 degrees of freedom. Section 2 introduces the used open source controller framework (ROS) and explains the main concepts for modeling simulation and coordinate transformation. Section 3 describes the industrial automation cell used for this evaluation and explains the process of modeling and simulation within the open framework.

[1] see IEC 61131.

Section 4 discusses the results of the evaluation and draws a final conclusion regarding the feasibility of the use of open frameworks for the particular use case.

1.2 Related Work

The concepts of the industry 4.0 paradigm shift, including design principles like interoperability, service orientation and modularization is summarized by Herman, Pentek and Otto [3] in their working paper. Furthermore, Kagermann et al. [4] state the importance of SMEs the Industry 4.0 revolution due to their ability to adapt more quickly adapt to a digital transformation process.

In a prior work Bauer, Durakbasa and Bas [5, 6] present the concept of a Multi-function Intelligent Measure Robot (MFI-MR), which is a use case for open source robotic frameworks. The Robot Operating System (ROS) was introduced by Quigley et al. [1] as a "structured communications layer above the host operating systems of a heterogeneous compute cluster".

1.3 Problem Definition

Industry 4.0 refers to the technological evolution from closed embedded systems to cyber-physical systems (CPS), which combine computation technology with production processes and, at the same time, are open to integration with a large system (plant floor level, enterprise, cloud). According to Lee [7] on of major problems of embedded system is the use of low-level software and specialized computer architectures needed to solve automation problems. The goal of the CPS's is to implement a robust abstraction hierarchy where each layer of abstraction aims to be predictable.

The key to achieve a robust abstraction of software components is the access to open interfaces and communication standards. This poses also one major opportunity for SME's that engage with open technologies to solve their automation problems, since open networking standards and open source programming are publicly available.

Without open alternatives, SME's are forced to buy into the OEM's software ecosystem which not only can be a competitive disadvantage against larger companies because of the needed investment, but also causes a vendor lock-in that, depending on the degree of process integration, makes the SME dependent on the vendor. Furthermore, proprietary software usually prevents the user from extending or adopting its functionality either with technical (e.g. closed source code base) or legal methods (terms of the software license). Due to this fact, the integration with a partner system becomes more difficult to implement, to test and to maintain.

Furthermore, open source programming would have the benefit to react easier to the needs of a smart manufacturing production system. Such a modern system has to adapt quickly on internal and external changes. With the access of open interfaces and communication standards it would be easier to implement new technology's, as an example a deep learning software, combined with a camera system. Such a system would be highly adaptive. There is almost no limitation of possible extensions.

2 Open Source Process Logic

For this evaluation we use the ROS, an open-source Robot Operating System introduced by Quigley et al. [1], to show a feasible way to develop and implement process control logic for a robotic automation system. ROS provides a modular software ecosystem comprised of components that solve problems like simulation and visualization, motion control planning and collision control. The main reason ROS was chosen is that the core development concepts are aligned with the main design goals of the Industry 4.0 paradigm shift summarized by Herman, Pentek and Otto [3], which are interoperability, service orientation and modularization. Although, other open source robotic frameworks are available[2], due to its architecture, ROS is a promising candidate for industry application in production within an SME's context.

2.1 Basic Concepts

The ROS framework is built around nodes. Such ROS Nodes are independent software components that implement a specific functionality and are able to communicate with other nodes, sensors or actuators. There is no limitation of how many nodes can simultaneously run and the system scales because the nodes can run on different hosts within the network. The communication is handled by a messaging layer named roscore, which implements messaging patterns like subscriber, publisher, topics and services.

Each message from a ROS node contains a certain key which includes the message type description. The message type descriptions allow to parse and reconstruct the message. There are several different message types available. The "std_message" consist of an 64 Bit Stream (Float 64). The defined structure of the message helps the subscriber to translate the message [9].

A topic allows to receive published messages from other nodes. Both publisher and subscriber are anonymous and decoupled, which means that a subscriber does not know which node has provide generated the message and a publisher does not know which subscriber will process its message. A topic is allowed to have several publishers and subscribers. The coordination of the communication between the publisher and the subscriber is ordered by the roscore node. Due to its central coordination role, roscore has to be launched before the nodes do. After that publisher nodes can be started, which will advertise its topic and start sending messages with a certain frequency. Other nodes can now subscribe to the topic and start receiving data. A node can be both a subscriber and a publisher. For example, a node receives a value of a sensor (via a sensor topic) and provide the data for an actuator node by publishing to the actuator's topic.

As an alternative to the publisher/subscriber communication mode, the ROS system also provides a service concept for node interaction. Since for the publisher/subscriber model there is no feedback that a message has been received, for some use cases, a

[2] see CARMEN (http://carmen.sourceforge.net/), OROCOS (https://www.orocos.org/) or XARP (https://www.yarp.it/).

remote procedure call (RPC) is required. ROS services are able to build an RPC request/reply communication structure (Fig. 1).

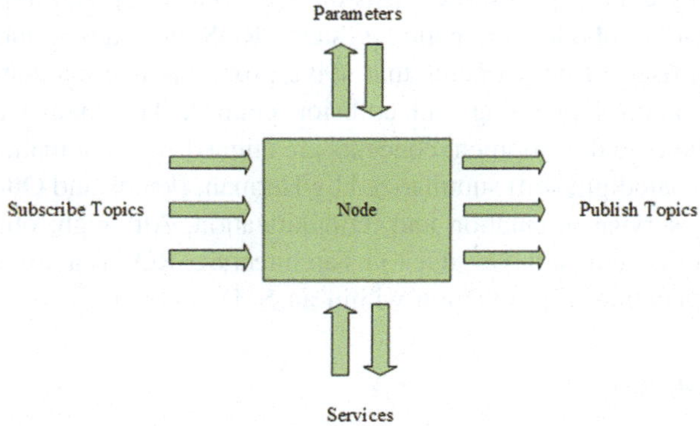

Fig. 1. Concept of an ROS node

Every ROS System also have a configuration services implemented by a so-called parameter server. The nodes can store and retrieve public and private parameters via a API.

2.2 Robot Modeling

In order to develop a realistic dynamic simulation of a robot, the robot must have a detailed description. According to Pan et al. [2], the robotic simulation is import tool to incorporate online robotic programming techniques and provides a fast way to test the control logic without the physical robot being available. In ROS, this simulation model is divided into four categories; namely the *dynamic model*, the *collision model*, the *visual model* and the *kinematic model*. Those models will be described in an XML-based format called Unified Robot Description Format (URDF). Furthermore, the simulation environment (gazebo) implements their own description format named Simulation Description Format (SDF). The SDF format meets the needs of a dynamic robotic simulation using the gazebo software. In the end, the SDF File can also be translated into a URDF Format.

The *visual model* is used to display the model graphical. The description file includes specification of color, reflectivity and transparency.

```
< visual name = " visual" >
        <geometry>
                <box>
                        <size> 2 2 4 </size>
                </box>
        </geometry>
</visual>
```

The collision model defines the “skin” of the object. Through the collision model it is possible to compute intersections between solids. The simulation is able to react to points of contacts. The model should be kept as easy as possible, to remove the computational overhead. In order to do that, it is recommend to avoid cylinders or spheres to describe a model and to use rectangular prisms or triangles instead [8]. For models with simple shapes, it is possible that the collision model and the visual model are identical. For complex shapes the visual model is a more accurate description of the robot.

```
< link name = “link1”>
        <collision>
                <origin xyz= ” 0 0 0.5” rpy= ” 0 0 0” />
                <geometry>
                        <box size “ 0.2 0.2 0.7” />
                </geometry>
        </collision>
</link>
```

The *dynamic model* describes the center of mass properties of every Link. The tag <inertial> defines the mass, the center of mass and the rotational inertia of the link. The rotational inertia is represented by a 3×3 matrix.

```
< inertial>
        <origin xyz= ” 0 0 0.5” rpy= ” 0 0 0” />
        <mass value =”1”>
        <inertia
                ixx=”1.0” ixy=“0.0” ixz=”0.0”
                iyy=”1.0” iyz=”0.0”
                izz=”1.0” />
</inertial>
```

The *kinematic model* describes the relationships between links trough joints. Each link of the model has a single parent link, but one link may have several child links. A link has only one reference frame. The link and the frame are hard linked to each other.

```
<link name=”link1” />
<link name=”link2” />

< joint name=”joint1” type=”continuous”>
        <parent link=”link1” />
        <child link=”link2”>
        <origin xyz= ” 0 0 0.5” rpy= ” 0 0 0” />
        <axies  xyz= ” 0 1 0”  />
</joint>
```

Every joint has a unique name and a type and there are several types available. In this example the type continuous is used. The type “continuous” describes a joint which rotate along the defined axis, with upper and lower limits. The tags, parent and child link describe the relationship of the joint. The transformation xyz and rpg (roll, pitch, yaw) between the parent and the child frame are specified through the origin tag. The <axies> value defines the rotation of the axis in the child frame.

2.3 Motion Planning

In order to plan and execute robot movement a ROS module named Moveit is used. Moveit provides several ROS nodes to simulate, control and manipulate different kind of robots (Fig. 2).

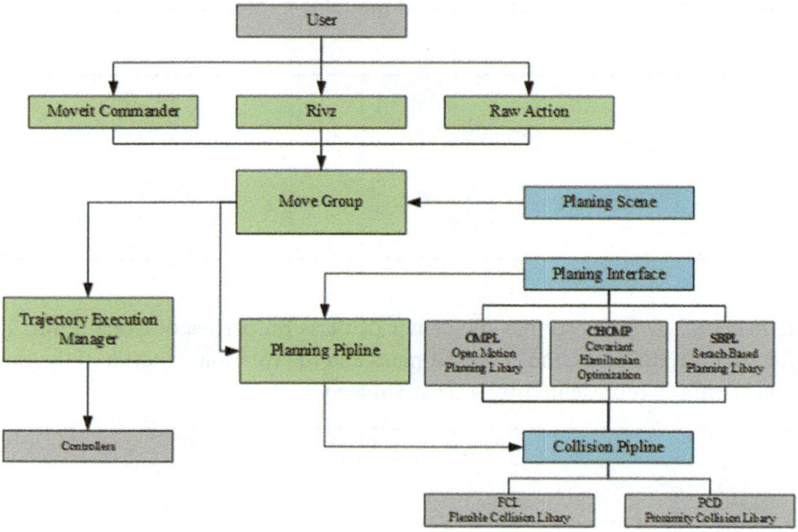

Fig. 2. Concept of motion planning with Moveit

The primary node collects components, like the URDF File of a robot together and provide ROS actions and services to use. Those actions and services can be accessed in several ways (p.e. through the available C++ and Python interface libraries). Furthermore, is possible to reach the actions and services by other GUI's such as Rviz[3], which is a 3D visualization tool used in ROS.

The "move_group" nodes talks to the robot through ROS actions and topics. It collects the current state information, like the position of the joints, point cloud of 3D sensors or data from other sensor sources to communicate with the controller of the robot. The motion planning of moveit works with a plugin interface. This plugin interface allows the user to integrate different motion planer to moveit. By default, moveit using the OMPL (Open Motion Planning Library) motion planer.

Like for motion planning, moevit uses a plugin interfaces also for solving of the kinematic model of the robot. This allows the user to implement alternative inverse kinematic algorithms. As default inverse kinematic solver, moveit uses the KDL numerical jacobian-based algorithm. Alternatively, one can also use the C++ based IKFast solver.

[3] http://wiki.ros.org/rviz.

3 Automation Cell

3.1 Components of Process Control

The idea is to create an adaptive and flexible robot control system. To provide such a flexibility model, we try to separate the software and hardware part of a robot. Those separated parts of the model communicate through a specified interface. Depending on the model the software and hardware has to be adapted. There should be also a possibility to react on values outside of the system, like cameras and sensors (Fig. 3).

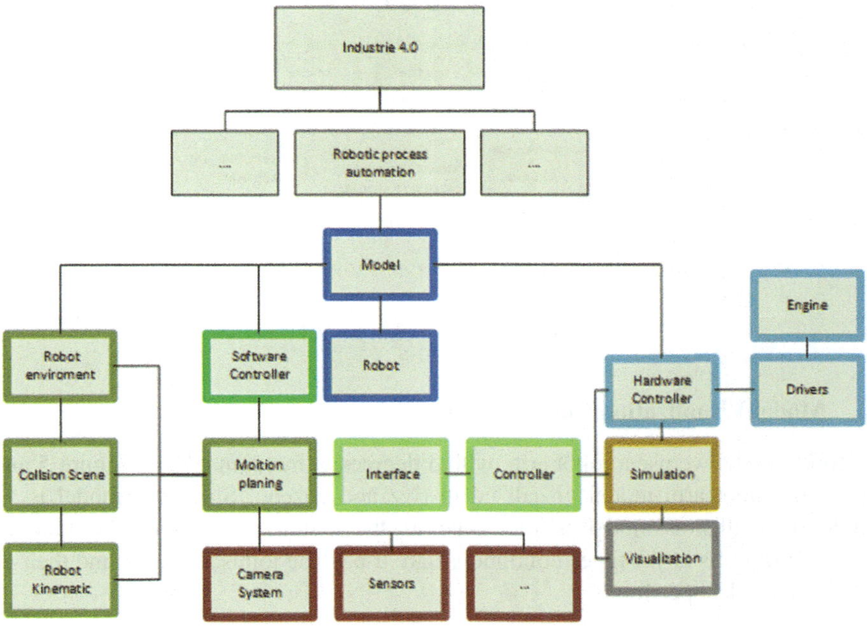

Fig. 3. Flexible robot control system model

3.2 Robot Arm System

Figure 4 shows how our Robot Arm Model looks like. The ROS environment provides a lot of tools to simplify the handling of robots. We adapt our base module to the ROS environment. In addition to the interface, an additional controller is needed to translate to calculate movement of the motion planning algorithm to angle values for the use engines.

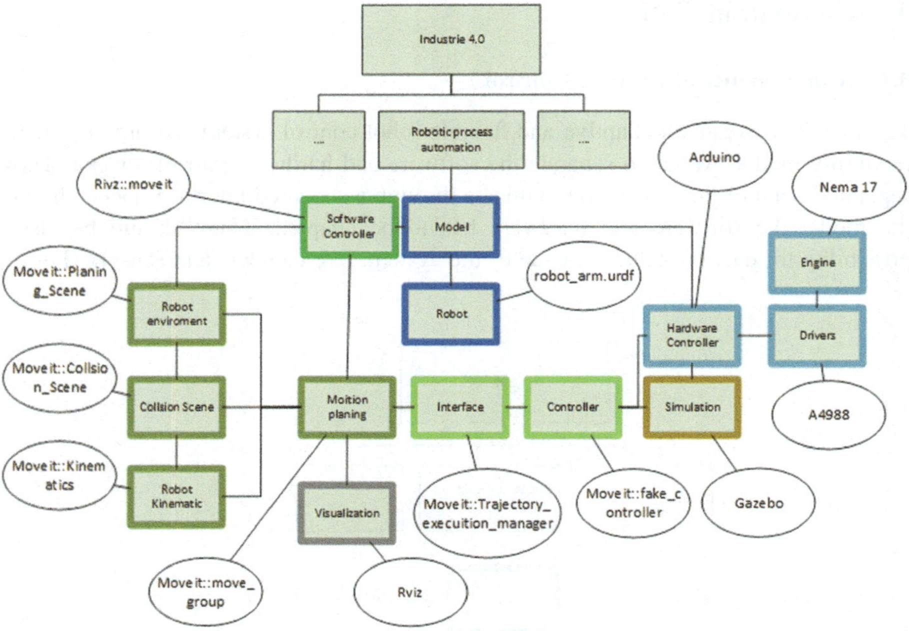

Fig. 4. Robot Arm Model

3.3 Model/Visualization/Control GUI

As Robot model we use a robot arm, with 3 degrees of freedom (DOF). Figure 5 and 6 shows the robot arm model visualized in rivz and gazebo. Since our model is very simple the collision model is the same as the visual model. For the kinematic descriptions joints of the type “continuous” are used. The joints rotate around their axis with lower and upper limits.

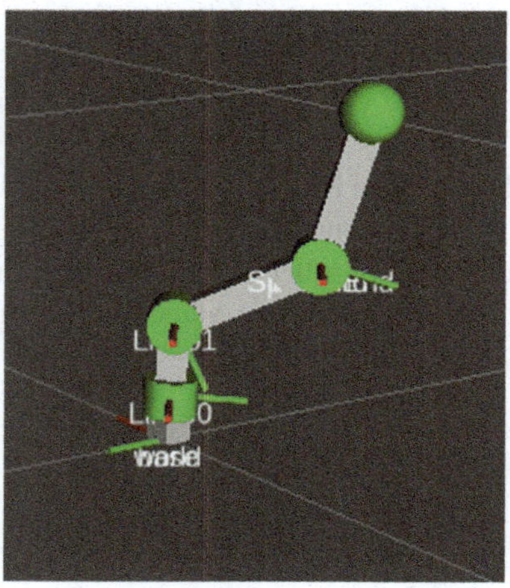

Fig. 5. Robot arm visualized in Rivz

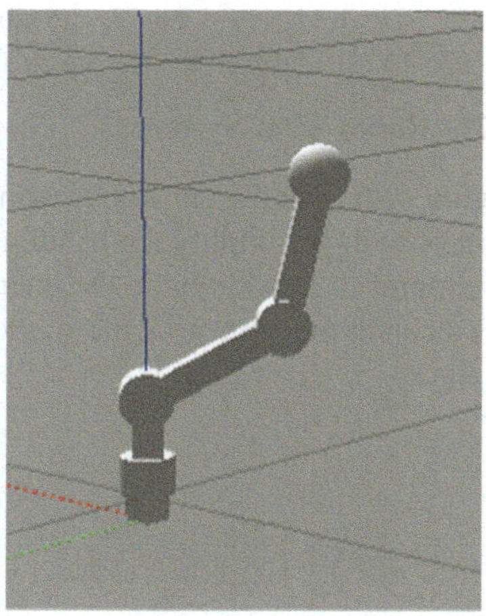

Fig. 6. Robot arm visualized in Gazebo

As already mentioned, ROS provides a lot of tools to simply the handling of robots. Two of the tools are the visualization software rivz and the robot control tool moveit. In this case moveit runs as a plugin for rivz. It is also possible to interact with moveit through provided c++ and python interfaces.

Moveit is responsible for the control and the planning of our robot arm. The robot model will be described through a URDF xml file. To get RIVz with moveit started, there has to be a special folder hierarchy created. The folders include launch files and description files. To simplify the implementation of such environment moveit provides a setup assistant. The setup assistant is an graphical user interface to configure a robot out of an URDF xml File. It generates necessary configuration files for the movit control.

3.4 Software Controller/Simulation/Robot Controller

To control the simulated or the real robot a translator between the "moveit::Trajectory_execution_manager" and the actual movement of the robot is required, which is a controller that converts the calculated movement of the moveit::move_group positions, to robot euler values. Moveit provides several different controllers. Among them are also controllers that are called moveit::fake_controller. This moveit::fake_controller are usually used for simulation reasons. Because the moveit::fake_controller is a node, other nodes are able to subscribe to fake_controller_joint_states topic to get the angle value of the joints. It is also possible to write his one controller. But for the time being we will stick with the moveit::fake_controller.

Gazebo is an robot simulation toolbox. The model of the robot will be again loaded though an URDF description file. Figure 7 shows the robot arm in the gazebo environment. The moveit::fake_controller provides the position data for the joints in the gazebo simulation. Furthermore the moveit::fake_controller controls the robot in the simulation.

```
header:
  seq: 1202857
  stamp:
    secs: 25346
    nsecs:  54000000
  frame_id: ''
name: [Link00_Link01_joint, Link01_linkEnd_joint, base_Link00_joint]
position: [-1.1045908390462014, 0.8609969184927122, 2.3998133512310282]
velocity: [-0.029379674014089488, -0.00438956575854876, -0.00015948785110820782]
effort: [0.0, 0.0, 0.0]
```

Fig. 7. Controller output

The ROS environment provides also an Arduino programming IDE, called rosserial. Rosserial is a serialized message protocol in ROS, which allows to communicate with topic and services. On the Arduino board runs a small program which subscribe to the fake_controller_joint_states topic. The Arduino program parse the received data and prepare them to control the robot.

As an alternative there is the possibility to program a dedicated node which subscribes to the joint state topic. This node would have the task to prepare the data and send the sorted data to the Arduino board. The Arduino board would work as a pipeline between the controller and the engines. As engines we use stepper motors from type NEMA17. The associated motor drivers are from type A4988.

4 Conclusions

The principal conclusion of this work is that the presented open source robotic framework provides a feasible way to implement the process control logic of an industrial automation application. The distributed system architecture of ROS implements concepts like service orientation, decoupling of components and modularization trough ROS packages, which meets some important design goals of Industry 4.0.

Our practical evaluation has shown that because of the used model language URDF, ROS can be adapted for many different industrial and mobile robotic applications. Furthermore, it is possible to freely share and reuse models via the online community, which becomes a great benefit for research purposes and SME sector.

The component model proposed in Sect. 3 highlights that a ROS system is composed of modular and interchangeable components, which can be implement by software written in higher programming language (C++, Python, etc.) or even functional programming languages (LISP). Due to this flexibility it is possible to implement a controller against any predefined interface, which makes it possible to incorporate already existing components like robots' sensors or PLC's. Due to this fact, ROS proves itself as a viable option for the implementation of process control logic in industrial applications with the benefit of reducing a vendor lock-in situation and the software license related costs.

5 Next Steps

The next important milestone is to implement the developed POC (proof of concept) into a real physical process. For this purpose, different environments will be used in the university pilot factories AuM-TU-Wien, UAS Technikum Wien, Austria & Robotic-Lab, UTN-FRBA Argentina.

First stage will be to complete the control of the robot-shield in the Jidoka automation environment built at AuM-TU-Wien by logically and physically linking automatic production equipment (3D printers & CNC milling machine) with coordinate metrology machine and other quality control units. Within the framework of the established inter-university cooperation, this experience is systematically projected and linked to actions in the UAS Technikum Wien, Austria & Robotic-Lab, UTN-FRBA Argentina nodes. At the same time, different SMEs are already integrated into the concrete experience.

References

1. Quigley, M., Conley, K., Gerkey, B., Faust, J., Foote, T., Leibs, J., Berger, E., Wheeler, R., Ng, A.Y.: ROS: an open-source robot operating system. In: ICRA Workshop on Open Source Software, vol. 3, no. 3.2, p. 5, May 2009
2. Pan, Z., Polden, J., Larkin, N., Duin, S.V., Norrish, J.: Recent progress on programming methods for industrial robots. In: ISR 2010 (41st International Symposium on Robotics) and ROBOTIK 2010 (6th German Conference on Robotics), Munich, Germany, pp. 1–8 (2010)
3. Hermann, M., Pentek, T., Otto, B.: Design principles for industrie 4.0 scenarios. In: 2016 49th Hawaii International Conference on System Sciences (HICSS), pp. 3928–3937. IEEE, January 2016
4. Kagermann, H., Wahlster, W., Helbig, J.: Securing the future of German manufacturing industry: recommendations for implementing the strategic initiative INDUSTRIE 4.0. Final report of the Industrie, 4(0) (2013)
5. Bauer, J.M., Durakbasa, M.N., Bas, G.: Multi functional intelligent measurement robot, as a part of a global quality management system for collaboration between small, medium and large scale enterprises. In: Quality-Access to Success, vol. 13, no. S5, November 2012 (2012). ISSN: 1582-2559, S. 67–72
6. Durakbasa, M.N., Bauer, J.M., Kräuter, L., Bas, G.: Novel developments in advanced manufacturing and multi functional intelligent factories (mfif) towards production in the future - challenges of autonomation in industry 4.0. In: Hauptvortrag: 2016 International Conference on Production Research Regional Conference Africa, Europe and the Middle East and 4th International Conference on Quality and Innovation in Engineering and Management, Cluj Napoca, Rumänien (eingeladen), 25 June 2016–30 July 2016, Proceedings Capturing Dynamics of Production Networks: Combining Agility, Innovation and Quality, D. Popescu et al. (Hrg.), Technical University Of Cluj-Napoca, 1/Cluj-Napoca, Romania (2016). ISBN: 978-606-737-180-2, S. 388–394
7. Lee, E.A.: Cyber physical systems: design challenges. In: 2008 11th IEEE International Symposium on Object and Component-Oriented Real-Time Distributed Computing (ISORC), pp. 363–369. IEEE, May 2008
8. Newman, W.: A Systematic Approach to Learning Robot Programming with ROS. CRC Press, Boca Raton (2017)
9. Open Source Robotics Foundation, Inc.: ROS Online Documentation. http://wiki.ros.org/. Accessed 12 June 2020

Robot Retrofitting by Using LinuxCNC Complemented with Arduino/RaspberryPI

Peter Kopacek[1], Jorge M. Bauer[1](✉), Miguel Llaja[3], Santiago Fauda[3], Jorge M. Acosta[2], Rodrigo Algoberro[3], and Félix Cuello[2]

[1] TUWien, Technische Universitaet Wien, Vienna, Austria
doktorjbauer@yahoo.com.ar
[2] Universidad de Palermo, Buenos Aires, Argentina
felix.cuello@gmail.com
[3] UTN-FRBA, Universidad Tecnológica Nacional, Buenos Aires, Argentina

Abstract. The use of robots for any manufacturing process is present in almost every industry nowadays. Nevertheless, automating the manufacturing process has always been a challenge for every one of them due to costs and complexity of adding this kind of automation. Acquiring robotic equipment in a concentrated market where the vendors have locked-in their clients can be expensive. In addition to this, the development of intern-factory made solutions has a very steep learning curve and it is not easily achievable. Due to these barriers small and medium size players can't easily jump into this revolution.

In this paper we analyze and develop tools in order to retrofit robotic components, legacy computer hardware with a combination of inexpensive hardware to help SME's to reduce costs and increase their productivity. Also, we demonstrate the feasibility of testing the tools in a virtual environment before moving into the real world. This virtual environment also remains a powerful tool for training and process simulation as the retrofitted robot is operated in production.

We started working with an old pen plotter with two degrees of freedom as proof of concept (POC) and then developed and investigated the tools to generalize this concept to control a 6 degrees of freedom robotic arm (first step PUMA560) and then project it to a concept Industry 4.0 cell with several controllers that interact with each other.

The goal of this paper is to evaluate the use of widely available open source technologies such as Linux Operating System and LinuxCNC and, also, the use of inexpensive hardware like Arduino-Raspberry PI and legacy hardware for motor-divers to create controllers that can be used in a variety of different robot types in the virtual world and also laying the groundwork to control a real robotic arm.

Keywords: Retrofitting robots · Industry 4.0 · Linux · LinuxCNC · Arduino · Raspberry PI

N. M. Durakbasa and M. G. Gençyılmaz (Eds.): ISPR 2020, LNME, pp. 222–235, 2021.
https://doi.org/10.1007/978-3-030-62784-3_19

1 Introduction

The general idea of this paper is to establish a cornerstone to demonstrate that we can control two robots at the same time; on one side a real-world robot that could perform operations in the real world and on the other side a virtual robot which will behave exactly the same as the real one. Achieving this goal could let us demonstrate that we can create any kind of proof of concept (POC) in a virtual controlled environment to make tests and debugs, and then plug the same type of real robot and control it as well.

We have achieved this for a 2 degrees of freedom plotter, but we also founded the bases to extend this for a 6 degrees of freedom robot.

This work also has a very specific boundary in terms of expenses because another target set is to achieve this work by using free open source software, inexpensive hardware and, therefore, spending just a small fraction of what it could take to buy a controller available in the market.

1.1 The Industrial Automation Need, Yesterday, Today, Tomorrow

The evolution of automation - autonomation and its influence on forms of production in particular and on life in society as a whole is a factor that any no one denies and that are analyzed from the most diverse disciplines. For our part, we participate and consider helpful the transdisciplinary approaches. Many authors have already catalogued this stage of robotic automation-autonomation and the productive information network as the fifth or even sixth industrial revolution (+). The importance of Small and Medium Enterprises (SME) in the globalized productive process, due to their dynamic capacity to create value and the number of people they employ, is also an accepted fact in which all specialists agree that their evolution in the coming years will not way disappear or have a decreasing evolution. If we look for a link between automation-automation and SMEs, we can say that automation-autonomation-robotics is a capital-intensive topic, a situation that especially makes SMEs in the global village in general and in emerging countries more difficult to adapt to that process. We can also say, agreeing with all the experts, that the different SMEs of the global village cannot leave automation-autonomation-robotics apart, since. It is very probable that if they do not work permanently in creative technological reconversion, they will lose their qualitative and price competitiveness in the hard national and international competitions. To this situation, which we have been observing for years, we have added in recent months the awareness that it is essential to protect ourselves against pandemics of viruses such as Covid19. In the face of this "new pandemic" the only method known by humanity for centuries (until specific remedies and vaccines are found) is the social isolation, better known as quarantine. The only new thing that humanity these days has had to learn in this hard way is how interrelated and small the global village has become, as contagion, disease and death are no longer limited to towns, cities, countries or continents, not do they make any distinction in terms of skin color or economic resources, but rather they touch all of humanity, requiring changes in habits in the ways of living, socializing and producing.

To contribute with some ideas and solutions proposal to the descript aspects, especially to the problem of SMEs [6] to automate-autonomate-robotize with restricted capital', is the technical content that we want to present here.

1.2 Yesterday, the Puma560 Milestone

The PUMA (Programmable Universal Machine for Assembly, or Programmable Universal Manipulation Arm) is an industrial robotic arm developed by Victor Scheinman at pioneering robot company Unimation [9]. Initially developed for General Motors, the PUMA was based on earlier designs Scheinman invented while at Stanford University. Unimation produced PUMAs for years until being purchased by Westinghouse (ca. 1980), and later by Swiss company Stäubli (1988). Nokia Robotics manufactured about 1500 PUMA robots during the 1980s, the Puma-650 being their most popular model with customers. Some own Nokia Robotics products were also designed, like Nokia NS-16 Industrial Robot or NRS-15 [1]. Nokia sold their Robotics division in 1990.

Reference to the PUMA560 as a real industrial robot used is a milestone. Seeing in that case how the academy, the technical universities continued to play and surely will continue to play a key role in the technological development of automation- autonomation - robotic in particular and in the transformation INDUSTRY 4.0 is a fact that surely no one will argue.

1.3 The "Engelberger Robotics Award"

Named after Joseph F. Engelberger, known throughout the world as the founding force behind industrial robotics, the Engelberger Robotics Award is the world's most prestigious robotics honor. The awards are presented to individuals for excellence in technology development, application, education, and leadership in the robotics industry. The Leadership Award is given each year, while the other categories are rotated by year [7, 8].

For all of us who have been working in automation - automation - robotics for years, the prize in general and especially the persons that received the prize are beacons that illuminate and guide us, not as big static towers but as examples to learn from and imitate in terms of the famous "live-learning".

1.4 Today Industrie 4.0

Within the framework call from experts as the "NEW" industrial revolution, the society of the global village as a whole, and especially for the European Industrial Company, have set the conceptual goal with the name "INDUSTRIE 4.0". It is built on the intensive use of Computer-Aided x (CAx) technologies and intensive data networking through the "Internet data highway". INDUSTRY 4.0 is, in this sense, an new milestone of CIM (*) (Computer Integrated Manufacturing, also call in the 70 years of the previous century as the factory in the dark or also factory without human workers), IMS (+) (Intelligent Manufacturing System) or the concept definition of MFIF (Multifunction Integrated Factories) as already formulated in the 80 years [5].

A new qualitative leap is also felt in more divulgation and use of the concept of autonomation, as a term for autonomous production = intelligent automation with integrated measuring units directly in the production cell, for quality assurance and automatic protocol generation.

Making low-cost systems with a high technological level and transferring that knowledge within the framework of "open-source" and "open-engineering" is the challenge to which we want to contribute as a university and academic centers.

1.5 Tomorrow, Automation Robotic Autonomation

From rational engineering the way to project technology into the future is to understand past milestones, then the today's milestones and last to propose a criterion for evolution. In this sense, the evolution shared by all the experts and the general public who "automation - autonomation – robotics" will still, is see as many years of exponential development that will surely surprise us on many occasions in the next time.

2 LinuxCNC

2.1 LinuxCNC History and Strong Current Use

LinuxCNC (formerly **Enhanced Machine Controller** or **EMC2**) is a free, open-source GNU/Linux software system that implements numerical control capability using general purpose computers to control CNC machines. Designed by various volunteer developers at linuxcnc.org, it is typically bundled as an ISO file with a modified version of 32-bit Ubuntu Linux which provides the required real-time kernel.

LinuxCNC is not today the only open-source software that allows real-time robotic automation control, but it is almost certainly the one that has gathered more experience over the years, keeping it updated with an open conception, being therefore a benchmark.

2.2 LinuxCNC Main Idea and Possibilities Foreseen in the Original Concept

Regardless of the "open-source" philosophy in which LinuxCNC was always developed and expanded, which allows you to change and adapt any aspect of the controller software, there are two aspects that should be emphasized.

a) It is a parametric concept, where customization does not necessarily imply writing new code. In almost all cases, working with change of parameters, we can adapt to specific to the computer hardware in particular, to the electronic hardware including the specific mechanical hardware of the machine to be automated
b) It was designed to handle up to nine degrees of freedom. Although the greatest experience is in CNC machine tools that generally work with three degrees of freedom, sometimes with four and rarely with 5, the LinuxCNC concept supports up to 9 independently controlled axes and interpolation linking. In that sense, although the experience in that field is not much, it is suitable for anthropomorphic

robots of 6 degrees of freedom like those that are common in the present industrial cells. Even in a cooperative robotics scheme with multiple anthropomorphic robots dialoguing among themselves, in a distributed control of 6, 12, 18, 24 and sometimes 30 degrees of freedom synchronized for a specific action.

Its Open-Source structure and experience on many real CNC machines, allows it to project as an important retrofitting [3] tool and to project as a "low cost" [4] controller especially for SMEs.

2.3 Components

The LinuxCNC has dozens of tools and each tool has some degree of complexity, and by adding components it is possible to control very complex machines. On this particular project we have used a subset of these components and we will explain in detail each of them.

We have divided the LinuxCNC components in a way that it was easier to understand for our project, but potentially you can divide them in many other ways. For the purpose of this project we have the **LinuxCNC** user interface, the Hardware Abstraction Layer (and its internal controllers) and external tools (both graphical tools and command line tools). So, let's start reviewing the tool top to bottom.

2.4 User Interface

LinuxCNC comes with different user interfaces depending on the type of work or robot you want to use. For example, by selecting a PUMA560 we get a very simple user interface which allows you to manage joints separately.

Depending on the chosen type of this robot, a user interface will appear on the screen. This interface can be changed or redesigned. However the ultimate goal is to work with a 6 degrees of freedom robot like a PUMA560, so opening that type of robot you get an interface that let you move joints separately, can let you move the tool and the LinuxCNC calculates the kinematic automatically, or just enter G-CODE and execute it (Fig. 1).

Fig. 1. A. PUMA NASA (1985). B. LinuxCNC showing the interface for a PUMA560. C. PUMA560 at emulation/simulation environment V-REP.2020

LinuxCNC **depending** on the kinematics chosen will automatically calculate the volume the robot can do and restrict the movements before doing it.

2.5 Hardware Abstraction Layer

Hardware Abstraction Layer (HAL) [1] function in some way resembles a protoboard because it's based on the same principles that are used to design hardware circuits and systems. At the highest level it is simply a way to define the logic of your machine by adding "building blocks" such as drivers, controllers, functions, cycles, logic operations, pins, cables which permits you to build a complex system. Later on, we will talk about the HAL in detail since it's the most important part.

Hardware Abstraction Layer
The LinuxCNC manual states that "*The designer of a hardware system not only selects the parts, he also decides how these parts will be interconnected.*". Therefore, the biggest amount of work we have done was to investigate the hardware abstraction layer works, what are available components and how we can create a new HAL or modify an existing one to control a real-world robot.

Before moving on with the details we need to understand the HAL concepts. HAL incorporates a big set of concepts, but we are going to use just a subset of them:

2.6 HAL Concepts (LinuxCNC, 2020)

Component: Is a piece of software with inputs and outputs with a defined set of inputs and outputs. Equivalent to a piece of hardware.
Parameters: Are the input, output or input/output that offers each component in the hal.
Pin: Are hardware terminals to interconnect.
Physical Pin: Are terminals in the real world. For example: **a parallel port pin**.
Signal: The type of information used to interconnect the components.
Type: Typifies the information for example it can be a **bit, float, unsigned int 32** (4,294,967,295) **or signed int 32** (−2,147,483,647 to 2,147,483,647)
Function: Are blocks of code that permits us to do some actions in a given order.
Threads: Is a list of functions that runs in a given moment. Each thread will run in a loop at a particular speed cycle.

2.7 HAL Tools

These tools are external but let us to watch the different components inside the HAL. And even though they are mandatory to use, but they are very useful in terms of debugging purposes:

Halcmd: Is a command line tool configuration and tuning. Very useful to run commands like a **repl** [2] in a programming language.
Halmeter: It has a similar function like a multímeter for the signals in the real world.
Halscope: It is an oscilloscope in the physical world.
+Building a HAL

Even though our first goal was to create a 2 degrees of freedom controller for a legacy plotter that required a new controller, the most interesting part relies on the fact that we can extend this to control a 6 degrees of freedom robot. So afterwards we will focus on a general view of a 6 degrees of robot and the most important aspects of its HAL.

2.8 Examining the HAL

An empty HAL does not have any components beside the **halcmd** which is the tool that is connected to the HAL to send commands directly to it. For this purpose we will use the **show** command, which is a very simple but powerful tool because it permits us to inspect the HAL as a whole or use wildcards to inspect what is loaded, the type of signals and if the different parameters or pins are input or output, and many other things (Fig. 2).

```
halcmd: show comp
Loaded HAL Components:
ID      Type  Name                    PID   State
2       User  halcmd31495             31495 ready
halcmd:
```

Fig. 2. Empty HAL

2.9 Loading Components

LinuxCNC includes a plethora of components to perform all kinds of tasks. As I said we are going to focus on a 6 degrees of freedom robot, but this can be easily narrowed to a 2 degrees of freedom robot just by changing a few of these modules (Fig. 3).

```
loadrt genserkins
loadrt motmod servo_period_nsec=1000000 traj_period_nsec=10000000 num_joints=6
loadrt stepgen step_type=0,0,0,0,0,0
loadrt hal_parport cfg="0x378 out"
```

Fig. 3. Loading modules in the HAL.

genserkins: Is a serial link manipulator up to 6 angular joints.
motmod: Is a motion planner that calculates the blended intermediate points given a set of waypoints.
stepgen: is a controller for stepper motors
hal_parport: Is the module in charge of interacting with the parallel port.

2.10 Robot Geometry: GENSERKINS

The **genserkins** module permits you to establish the geometry of the robot by specifying modified Denavit-Hartenberg parameters to the links. Once this module is loaded a number of pins are created specifically to define the geometry of the linked chain.

Particularly in this case we are going to use the default configuration because we still don't have a real-world robot and we can configure our robot in the virtual world with the same parameters. Here's an example about how the LinuxCNC specifies the modified Denavit-Hartenberg parameters for our robot (Fig. 4):

```
setp genserkins.A-0 0
setp genserkins.A-1 0
setp genserkins.A-2 17
setp genserkins.A-3 0
setp genserkins.A-4 0
setp genserkins.A-5 0

setp genserkins.ALPHA-0 0
setp genserkins.ALPHA-1 1.570796326
setp genserkins.ALPHA-2 0
setp genserkins.ALPHA-3 1.570796326
setp genserkins.ALPHA-4 -1.570796326
setp genserkins.ALPHA-5 1.570796326

setp genserkins.D-0 26.45
setp genserkins.D-1 -5.5
setp genserkins.D-2 0
setp genserkins.D-3 17.05
setp genserkins.D-4 0
setp genserkins.D-5 2.2
```

Fig. 4. Defining the geometry.

2.11 Planning Movement: MOTMOD

After defining the geometry module then we can load the motion module so we can get a blended set of points. This module takes several arguments, but we are using just 3 of them which are:

servo_period_nsec: Is the period of update for the servo thread. I am going to use this thread later to attach other components to it, for updating our servo motors.
traj_period_nsec: Is the period of update for the trajectory planner.
num_joints: This parameter specifies the number of joints that we are going to use which has to be in coincidence with the kinematics module used.

2.12 Controlling the Step Motors: STEPGEN

By loading this module we are able to control the step motors. This module has more than a dozen types of ways to control the step motors. However, for the purpose of this paper we want to use 6 step motors that are the same type and have the same scale for all of them. We should point out that the stepgen module gives you the possibility to control different kinds of motors and scales for each of them (Fig. 5).

```
loadrt stepgen step_type=0,0,0,0,0,0

setp stepgen.0.enable 1
setp stepgen.1.enable 1
setp stepgen.2.enable 1
setp stepgen.3.enable 1
setp stepgen.4.enable 1
setp stepgen.5.enable 1

setp stepgen.0.position-scale 10
setp stepgen.1.position-scale 10
setp stepgen.2.position-scale 10
setp stepgen.3.position-scale 10
setp stepgen.4.position-scale 10
setp stepgen.5.position-scale 10

addf stepgen.update-freq servo-thread
addf stepgen.make-pulses servo-thread
```

Fig. 5. Enabling the step generator for each step motor.

The first line defines how many motors you have and what type of motor they are in terms of the signals that we are going to send to them. The type 0 specifies that the motor is controlled by two different signals which specify **steps** (one pulse per step) and **rotation movement** (clockwise or counter-clockwise).

2.13 Bringing Signals to the Real World: HAL_PARPORT

To use this as a real-world robot controller we somehow need to bring the pulses to the real world. For this purpose we loaded the **hal_parport** module which is in charge of interacting with the computer parallel port. For the purpose of controlling a 6 degrees of freedom robot we are on the limits of the parallel port because we need to use 12 pins which is about the limit of the amount of the output pins we could specify inside the HAL, as follows (Fig. 6):

```
loadrt hal_parport cfg="0x378 out"
addf parport.0.write servo-thread

net joint0_lpt1_0_dir  parport.0.pin-01-out stepgen.0.dir
net joint0_lpt1_0_step parport.0.pin-02-out stepgen.0.step
net joint0_lpt1_1_dir  parport.0.pin-03-out stepgen.1.dir
net joint0_lpt1_1_step parport.0.pin-04-out stepgen.1.step
net joint0_lpt1_2_dir  parport.0.pin-05-out stepgen.2.dir
net joint0_lpt1_2_step parport.0.pin-06-out stepgen.2.step
net joint0_lpt1_3_dir  parport.0.pin-07-out stepgen.3.dir
net joint0_lpt1_3_step parport.0.pin-08-out stepgen.3.step
net joint0_lpt1_4_dir  parport.0.pin-09-out stepgen.4.dir
net joint0_lpt1_4_step parport.0.pin-14-out stepgen.4.step
net joint0_lpt1_5_dir  parport.0.pin-16-out stepgen.5.dir
net joint0_lpt1_5_step parport.0.pin-17-out stepgen.5.step
```

Fig. 6. Enabling the output signals for each motor in the parallel port.

With this final configuration in the HAL we are able to control a 6 degrees of freedom robot

2.14 Other Component, DDT

Another useful component that we used for some analysis was the ddt module which computes the derivative of a function. It is often used to get the velocity and acceleration given a position over time.

The usage is very simple (Fig. 7):

```
loadrt ddt count=12

net J0pos => ddt.0.in
net J0vel <= ddt.0.out
```

Fig. 7. Getting the joint velocity given the joint position over time.

2.15 A Real World Controller

We have proven in real life that these tools permit us to control a 2 degrees of axis machine by using the exact same methods descriptive here with slight differences. And this paper explains the theory of how we can proceed and extend this into a 6 degrees of freedom robot.

However the first step in this work is to control both a virtual AND a real robot, and to do that we had to split the information that we send from the parallel port from LinuxCNC as we described, and send both signals to the real motors and the virtual motors (Figs. 8 and 9).

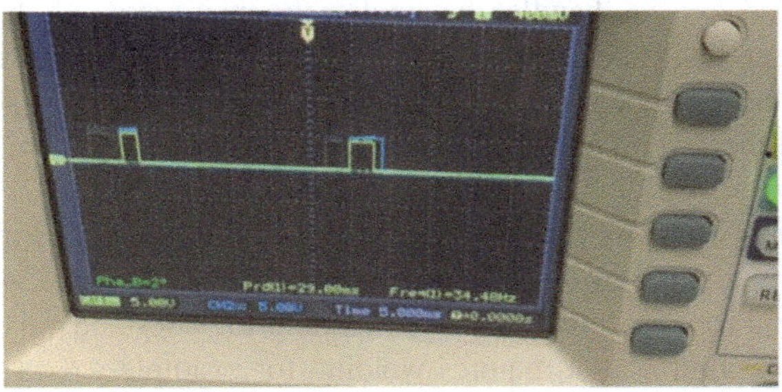

Fig. 8. Analyzing signals to be sent to the virtual plotter and the real plotter.

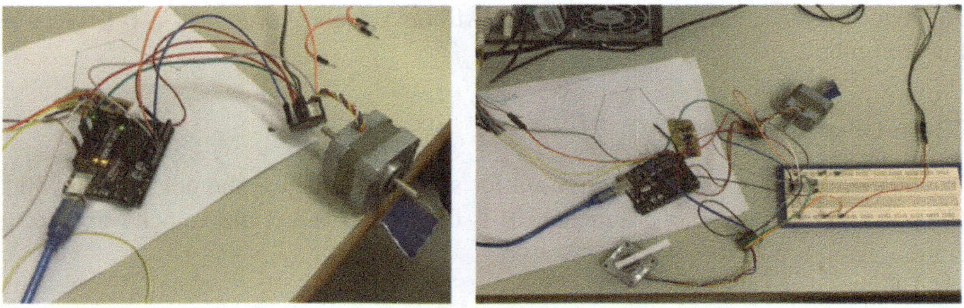

Fig. 9. A. Controlling one step motor by over LPT1 using LinuxCNC + Arduino. B. Controlling two step motors with LinuxCNC over LPT1 + Arduino.

3 Arduino/Raspberry PI, (and Other Similar Hardware)

ARDUINO/RASPBERRY PI are two popular and inexpensive plates (low cost hardware). They are not the unique but probably the most popular in the global village for simple tasks in automation.

One performance comparison for the specific task far exceeds the scope of this paper. We would just like to make a brief comparison of these two plates according to the task we assigned to them within the framework of the project. In this project with LinuxCNC through a parallel port of a classic PC, information is generated for engine control. Two signals per motor, pulses and direction of rotation as on-off information.

In the project, the task of the Arduino and Raspberry PI Board is to capture the emitted pulses, redirecting them to:

a. real motors via drivers
b. virtual motors in the V-Rep emulator environment (customized)

For more pulses we can handle, the smaller the micro-steps of the real and virtual robot's movements will be, resulting in greater position and trajectory control. LinuxCNC and the parallel port have a much higher capacity to discriminate pulses, and this is one of the several elements that the LinuxCNC can also handle in parametric form.

Arduino in that sense, its pins and logic of interruption works with a micro that in that sense has a relatively low top, adequate for robots in the virtual world and didactic robots or industrial robotic equipment of palletizing-logistics and very limited in terms of some trajectory control as it requires welding for example.

Raspberry Pi shows a more powerful micro, its way of programming and multitasking environment leaves actually open even for us its real behavior in the task of cutting many pulses per second and replicating them by a digital pin to the real motor driver at the same time that it transforms them in encoder position for the robotic control of the emulated robot.

4 Modern Motors, Design and Control of Special Brushless Motors

The POC proof-of-concept tests were carried out with stepper motors. Even stepper motors can be used in open loop and in modern times improved to closed loop with real position feedback.

For the POC of this project an old pen plotter with X-Y movement was used which was built with stepper motors. Didactic scale type robots built for teaching engineering for students are also simpler to construct with stepper motors. We can add opto-coupled light as encoders, or a resistive potentiometer and the best where one all absolute encoders by Hall effect to have a better control (Fig. 10).

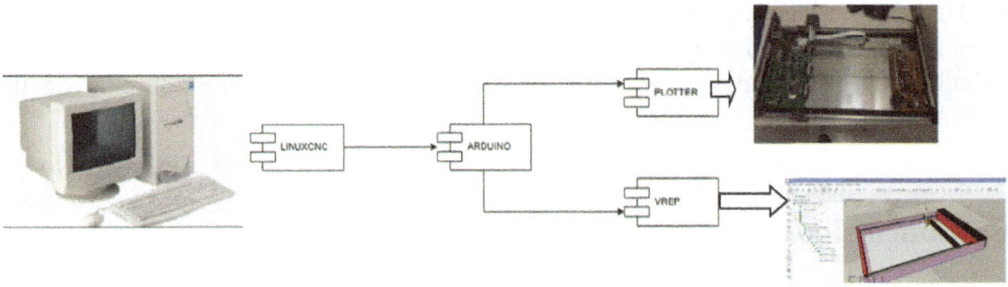

Fig. 10. Controlling Plotter (real/virtual) by over LPT1 using LinuxCNC + Arduino.

For retrofitting of real industrial robots, there is the need to adapt to existing motor drivers, generally servo motors with mechanical gearboxes of harmonic-drive or servo-drive type.

For more modern equipment that we are developing and in which we also foresee logical control with LinuxCNC we implement special brushless motors, with high torque. The development and simulation of outrunner-brushless is part of the mechatronic development of our team. To be able to simulate-emulate in our environment the operation of robots in terms of accelerations and efforts, inertial behaviors is a fundamental feedback to choose to design the most appropriate motor for each task.

Within the framework of our task for retrofitting robots in SMEs, having our own technology of brushless motors (low-cost) is a fundamental element in the transdisciplinary environment we are going through.

5 Advantages of Quickly Switching from: The Real World to the Virtual World (Emulator)

The possibility of a rapid change between the real and virtual world offers many advantages that we would like to briefly highlight here.

The LinuxCNC robot controller software always runs on a specific processor (usually a PC that has already completed a life cycle in the office, i.e. from the

Windows XP, NT and parallel port). It is important to note that the KUKA robot we have at the university, donated by VW-Argentina, works perfectly with its original controller a PC with Windows XP operating system and can control a path process at speed with a maximum deviation of 0.2 mm.

The real control system on PC with LinuxCNC of an anthropomorphic robot, which switches to real equipment and simulated environment without any internal change, allows us:

a. Optimize software especially when it is for retrofitting of SME equipment that cannot interrupt the work of their obsolete robots for many days
b. To generate a training environment for the personnel of the SME in which not only movement but also the results of inertia, collisions and interaction with all types of the robotic arm with elements, machines and/or equipment are seen
c. To generate learning environments for the university academic world, with interaction in the virtual world. We have technology to walk in a FULL INDUSTRY 4.0 cell and see in the virtual reality glasses the result of all programming on the controllers

6 Conclusions

We have developed a tool, or rather a set of tools that in a low-cost scheme allows SMEs to actively participate in automation - autonomation - robotics either by retrofitting their existing equipment or by incorporating used equipment whose mechanics are adjusted with a replacement of standard bearings and whose control system is adapted to the modern requirements needed in the conception INDUSTRY 4.0 to compete with products in cost and universal quality in the global village.

It's not a closed system in any aspect. It is an Opensource & Open Hardware system that adapts to different needs, expandable in different aspects. The LinuxCNC controller for anthropomorphic robots with six degrees of freedom interacts dynamically with ROS technology equipment (open source as well) or technology such as, kuka, fanuc, abb and all types of PLC (Figs. 11 and 12).

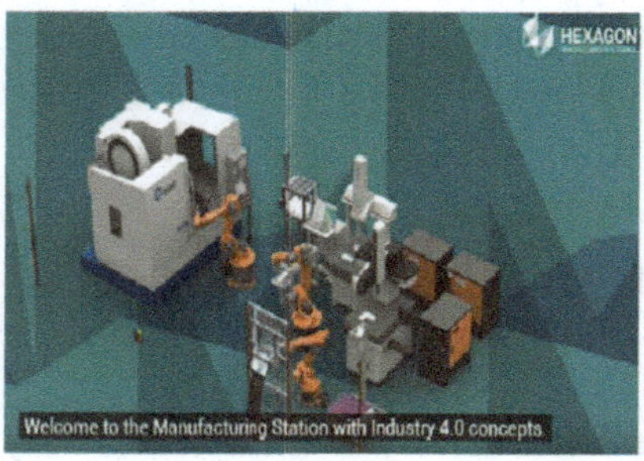

Fig. 11. INDUSTRY 4.0 as real cell Hexagon-Brazil.

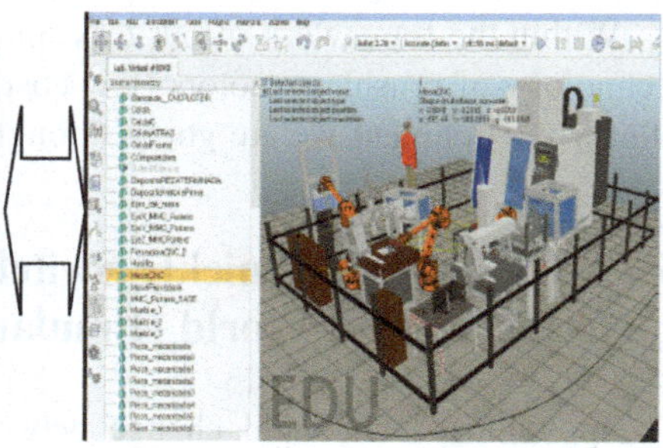

Fig. 12. INDUSTRY 4.0 as emulation environment (virtual)-with real linuxCNC controllers.

7 Next Steps

The next important milestone is to make the 1200 kg "Kawasaki Robot" received as a donation with a good mechanic but with a destroyed controller, transforming it into a modern anthropomorphic control robot, part of UMCARI (Mobile Robotics Training Unit), a tool designed to bring technical schools and motivate his student, the next generations, with real industrial robotics (Fig. 13).

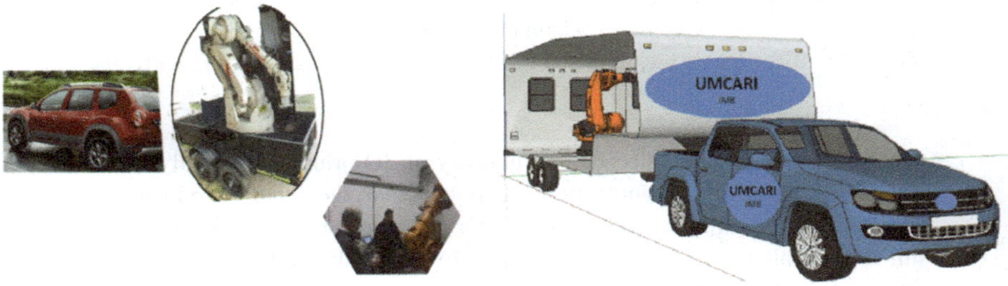

Fig. 13. Kawasaki Robot using LinuxCNC as part from UMCARI.

References

1. http://linuxcnc.org/docs/html/hal/intro.html
2. https://en.wikipedia.org/wiki/Read–eval–print_loop
3. Alvares, A.J., Toquica, J.S., Bomfim, M.H.S.: Retrofitting of ASEA IRB2-S6 industrial robot using numeric control technologies based on LinuxCNC and MACH3-MatLab. In: IEEE International Conference 2017 Engineering, Computer Science. http://doi.org/10.1109/ROBIO.2017.8324737. Corpus ID: 4309019
4. Bauer, J.M., Kopacek, P., dApolito, F., Kräuter, L., Dorna, I.: Cost Oriented (humanoid) robots (2011)
5. Durakbasa, N., Bauer, J.: Multi functional intelligent measurement robot, as a part of a global quality management system for collaboration between small, medium and large scale (2012)
6. Osanna, P.H., Durakbasa, M.N., Bauer, J.M., Kräuter, L.: Global international cooperation of collaborating small and medium sized enterprises to achieve total quality management. In: Vortrag: Proceedings Fourth International Working Conference, Belgrad, 27–30 May 2007, Total Quality Management - Advanced and Intelligent Approaches, V. Majstorovic (Hrg.) (2007). ISBN: 978-86-7083-594-8, S. 1–6
7. https://apps.derstandard.at/privacywall/story/2458131/hoechster-roboter-preis-fuer-peter-kopacek
8. https://de.wikipedia.org/wiki/Joseph_Engelberger
9. https://en.wikipedia.org/wiki/Unimation

Parameter Optimization for the 3D Print of Thermo-Plastic Pellets with an Industrial Robot

Stefan Otto Felber(✉), Mohamed Aburaia, Wilfried Wöber, and Maximilian Lackner

Competence Center Digital Manufacturing and Robotics, UAS Technikum Wien, Vienna, Austria
{mr18m009, aburaia, woeber, lacknerm}@technikum-wien.at

Abstract. Since the last decade, the industry of 3D printing is sharply on the rise. A new trend in the manufacturing of polymers deals with the production of upscaled products that exceed an object size of one cubic meter. This trend has still some challenges to overcome that aim to reduce the costs and the process time in order to enable a more economical production. A method called Fused Granular Fabrication (FGF) uses thermoplastic pellets as a working material. FGF faces the challenges of upscaled 3D printing. Plastic pellets come to a friction of the cost of filament rolls for typical 3D printing application. Furthermore, they offer a greater variety of materials and can be refilled constantly during the printing process, which guarantees a production of almost unlimited big 3D prints. This paper deals with an industrial robot application of the UAS Technikum Wien that is equipped with a simple pellet extruder. The process parameters for good 3D prints were evaluated in a series of tests. All the tests were performed with Ingeo 4034D PLA pellets. A developed cooling system could speed up the process time by 17% for qualitative 3D prints. This work shows the potential of the FGF method and focuses on the build of an upscaled vase with a height of one meter. It was found out that the system still has an insufficient print speed for larger plastic objects, because of the slow feed rate of the used pellet extruder. As a result, a new research started for an alternative pellet extruder. Some of the potential extruders were good substitutes to produce upscaled 3D prints but they come to a high cost. As a self-made solution a new concept for a pellet extruder was designed. The introduced concept should achieve a material flowrate above 2 kg/h, to enable the 3D print of upscaled plastic products within a single day. As industrial robots are already an essential part of our modern industry 4.0, this technology allows it to easily produce individual products with existing systems. The designed pellet extruder can be mounted and exchanged smoothly to make the modern production more efficient.

Keywords: 3D printing · FGF · Industrial robot · Pellet extruder · Print parameter

N. M. Durakbasa and M. G. Gençyılmaz (Eds.): ISPR 2020, LNME, pp. 236–247, 2021.
https://doi.org/10.1007/978-3-030-62784-3_20

1 Introduction

Since the first introduction of additive manufacturing in 1986 by Stratasys, the whole concept of our modern industry has changed a lot. This new technology of 3D printing enables the cost-effective and resource-saving production of various component geometries from a wide range of materials. Another advantage of additive manufacturing is, that it can be easily adapted to modern processes of the current industry 4.0. In the last decade, the 3D printing industry has become more and more prominent, which is manifested in the strong growth of its market value. The current market value of the entire 3D printing industry is estimated to be $22.4 billion US dollars, and it is expected to double up to $44 billion in the next five years [1]. Half of the current market value is the sole share of thermoplastic polymer 3D printing, which is expected to be worth $11.7 billion in 2020 and is still on the rise with the whole industry of additive manufacturing [2]. A new rising 3D printing trend is the production of upscaled products that exceed an object size of one cubic meter [3]. To enable an economical mass production with this technology, the process itself has to be accelerated [4]. Nowadays 3D printers are either large cartesian machines or industrial robot arms that print the upscaled objects in a layer-by-layer process. Cartesian machines are already well researched and offer some good solutions in a variety of applications. This work specializes on industrial robots, their various challenges, and their potentials in terms of the use in our modern industry 4.0. In the course of this project, a system solution for an industrial robot arm was designed, which is capable of printing upscaled objects out of thermoplastic polymers. The system was further evaluated to find good process parameters for the Fused Granular Fabrication (FGF) method.

2 Problem Description

The most popular 3D printing technology for processing polymers is Fused Filament Fabrication (FFF), which has a worldwide distribution of 46% by 2018 [5]. This cost-effective method builds any object in a layer-by-layer process. Current challenges of this method are the production of upscaled products exceeding a component size of one cubic meter, the further reduction of manufacturing costs and the reduction of production time [6]. A decisive disadvantage of the FFF process are the filament rolls as working material that cause a high cost. A new manufacturing method that counteracts this problem is the FGF method. This process uses thermoplastic pellets instead of filament as printing material. It is available for a fraction of the filament cost, allows unlimited 3D printing by continuously feeding plastic pellets and offers a wider range of possible combinations for thermoplastics. With FGF systems for material extrusion of plastic thermoplastics, material flow rates can be set 200 times higher than with a conventional FFF system [7]. Another advantage is, that this method can make use of recycled polymers, which enables an environmentally friendly production [8].

Cartesian 3D printing machines offer a huge build volume that has several dozen cubic meters in size [9]. The biggest printer for thermoplastics has a construction space of 144 m^3. It was especially designed for the construction of a bridge and can print up to 8 kg of plastic per hour [10]. The BAAM (Big Area Additive Manufacturing) by the

Oak Ridge National Laboratory (ORNL) can even reach a material flow of 21.3 kg/h through a 5 mm nozzle for materials like ABS reinforced with glass fibres. Another professional System by BLB Industries can print up to 30 kg/h of plastic with a 14 mm nozzle as hot end [11]. The bigger the nozzle diameter gets the more material can be printed in the same amount of time, but this also results in a decrease in the print quality of the product. Pellet extruders that reach such a high material flow of plastic always have a high mass. The extruder mass is between 10 and 200 kg, which require a special transport facility. The big advantage of cartesian 3D printing machines is, that they can easily bear the extruder weight. Most of these systems are limited to three degrees of freedom, that only allows movements in x, y and z direction.

Industrial robots in comparison provide a good solution for a versatile production in a large workspace with several degrees of freedom. Recent projects show industrial robots for the use-case of 3D printing. These robot systems are already used to build houses [7], bridge structures [12] or even furniture [13] out of a wide selection of building materials. There are only a few examples for industrial robots that print with thermoplastic pellets. Some of the systems make use of the FFF method with special filament extruders [14]. Those extruders are feed by several filaments rolls simultaneous to reach a higher material flow. Other systems have self-build pellet extruders with a low material flow of molten plastic between 0.05 and 2 kg per hour with a maximum nozzle size of 5 mm in diameter [13, 15, 16]. This low flow of plastic comes due to the limited load capacity of the robot flange. On the one side industrial robots offer a versatile production and on the other side they require much more print time in comparison to a cartesian 3D printing system. In addition, industrial robots come to a high cost and require a lot of know-how to prepare them for the process of 3D printing, especially when implementing all the digital interfaces for the print signals. One the one hand implementing digital interfaces to control the process can be challenging, but on the other hand a well connection enables a flexible and individual production with industrial robots in our modern industry. The production of upscaled objects with an industrial robot and the FGF method remains a big challenge. This scientific work concentrates on the conception of a suitable station to enable the FGF method with an industrial robot and on the optimization of the print process.

3 Materials and Methods

The basis manipulator of this work was an ABB IRB 2400/16 6-axis robot. This robot has a spherical workspace and can bear a payload of 16 kg on its flange. As a building platform for the 3D print, a 900 × 600 mm^2 plate of glass was used. Within these boundaries the robot can print up to a height of 1.5 m, which covers a big construction space in comparison to a regular desktop 3D printer. The glass plate can be heated up to 60 °C, which is a reliable temperature for materials like PLA to stick properly to the plate during process of printing. To enable the upscaled 3D print of plastic objects a pellet extruder was chosen. The Noztek Pro Filament Extruder was mounted on the flange of the ABB robot to continuously feed the extruder through a tube from a pellet storage above the ABB robot. This pellet extruder is typically used to create PLA or ABS filaments for desktop 3D printers, but in this work, it was transformed to a direct pellet extruder. The extruder has standardized nozzle diameters of 1.75 mm and 3 mm that create filament with a speed of up to 2.5 m/s or with a material flowrate of up to 0.5 kg/h [17]. For this work a nozzle with 3 mm in diameter was chosen together with Ingeo 4043D PLA pellets [18] as working materials. The cooling system for the printing process was designed out of 3D printed parts and uses a forced air stream out of a pneumatic compressor. This cools the extruded PLA down to its glass transition temperature of 65 °C to 70 °C to get solid again. The airstream can be accelerated up to 5 m/s without negatively interfering the 3D print. The printing technique of this work was surface printing of 3D models. This means that the printed objects do not have any infill structure. The parts are only built out of surface walls of the CAD model, which lead to faster print times. This technique could be achieved with the special mode "spiralize outer contours" of the slicer Cura by Ultimaker [19]. To transform the generated GCode of the slicer to a RAPID robot language, the translating software RoboDK was used [20]. This software serves as a digital interface which transforms the GCode and communicates it to the robot in small packages to guarantee fluent movements during the printing process.

4 Practical Realization

To prepare the used pellet extruder for the upscaled 3D printing, the best parameters for the 3D print had to be evaluated in the course of this work. For this a test object was chosen to analyze the print results of the used setup. This object was a hollow cuboid with the measurements of $100 \times 100 \times 50$ mm^3. The PLA pellets showed their best flow characteristics at a nozzle temperature of 210 °C and had their best first-layer bonding at a bed temperature of 60 °C. The walls of the cuboid where created by a single layer thickness of the extruded material in the "spiralize outer contours" mode of Cura [21]. The only two parameters that had changed in this entire setup were the layer height and the printing speed of the process. The layer height was alternated in steps of 0.01 mm and the speed in steps of 0.25 mm/s. A variety of test objects were created to examine the quality of each print. In the beginning all the prints were created without the use of a cooling system to create a reference part for the results of the test object with a cooling system.

In the next step the same reference part was printed with the cooling system. An air stream was forced to the top end of the nozzle. The cooling jet is designed in a circular form of 210°, so that the liquid material out of the nozzle gets cooled from a variety of directions (Fig. 1). This special form reduces irregularities, because a too strong air flow from a single direction could relocate the molten plastic from the nozzle. The nozzle of the extruder is in the centre of the half-circle. The air is forced from the backside into the tunnel system. The strongest airstream flows from the middle of the cooling jet and is gradually decreasing to the front ends of the 210° ring tunnel-system (Fig. 1). The airstream can be accelerated up to a velocity of 5 m/s without negatively effecting the temperature of the nozzle or the relocation of the material. This robot setup was the basis to evaluate the 3D print of upscaled products. This cooling system was then used for to 3D print of the same test object again, to compare the differences (Fig. 2).

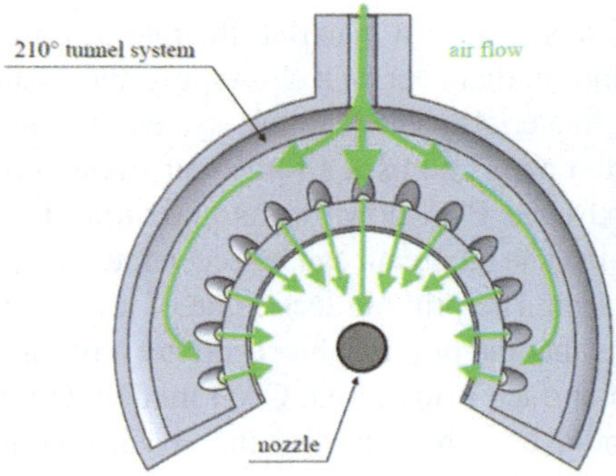

Fig. 1. Cooling jet of the printing system

Fig. 2. 3D print of test object

5 Print Results

The print evaluations of the test objects were based on optical observations. First, it was essential to check if the product quality was consistent. For this all the layers were checked if they have the same height and are located vertically to each other. Selecting the wrong layer height led to mistakes like the "elephant foot". This mistake occurred, when the first few layers were pressed on each other while they were still flexible, so that the wall lines are no longer vertical to each other. This happens when the layer height is too low (Fig. 3). If the layer height was selected too high, the layers did not longer stick to each other and lost their bounding properties in z-direction. The other evaluation was based on the print speed of the process. A too slow print speed led to long building times. Furthermore, the wall of a product got much thicker and the object gained a lot of weight. This fact also led to a line displacement between the layers in z direction. Once the print speed was selected too high, the material did not stick to the layer properly. The material got uncontrollable. It either felt down to the side of the layer below or made holes within the surface walls (Fig. 4).

The parameter optimization of this work showed the best results with the listed parameters (Table 1). The layer height was set to 1.2 mm for the 3 mm wide nozzle of the pellet extruder. The temperatures of the nozzle and of the bed were set to 210 °C and 60 °C. Without a cooling system the system showed good results with a print speed of 9 mm/s. With the cooling system the robot translation speed of 10.5 mm/s revealed the best print quality together with an airflow speed of 2 m/s out of the cooling jet. The test object without a cooling system had a weight of 54 g and was printed in 25 min, which results in a material flowrate of 0.13 kg/h. The same system with a cooling system could increase the print speed from 9 mm/s up to 10.5 mm/s, which is an improvement of about 17%. The test object with a cooling system was printed in 20.5 min and had an object mass of 44 g, which is a mass reduction of 22% with a similar quality without a cooling system. The resulting flowrate of the plastic is again at 0.13 kg/h, which shows that the flow of the Noztek Pro Filament Extruder is reasonably constant.

Fig. 3. Layer height was too low

Fig. 4. Parameter optimized test object

The next step of this project was the 3D print of an upscaled product. In order to achieve this task, a large vase was printed with the optimized setup. Green coloured pigments were added to the PLA pellets to colourize the product. The entire printing process had a duration of 235 h until the product was finished. The setup of the industrial robot with the pellet extruder was constantly working for 10 days to create a one meter high PLA object with the weight of 30.2 kg (Fig. 5). This long process time was very time consuming and not cost-effective to produce any further objects with this setup. According to the datasheet of Noztek the material flow should extrude the pellets at a rate of 0.5 kg/h but after all the tests with PLA pellets a constant flow of 0.13 kg/h was measured. This flow rate is only 26% the value according to the setpoint from the datasheet.

In conclusion to this long printing times a new solution had to be found in the continued task of this work. The main goal was to research for an alternative pellet extruder or to design a new self-made extruder system, that enables a higher material flowrate and faster print times.

6 Alternative Pellet Extruders

Due to the long printing times of the Noztek Pro Filament Extruder an extensive search for an alternative extruder started. All the commercially available pellet extruders are ranked after the flowrate of the molten material in kg/h. They are all specially designed to attach them easily to any robot flange. With the rising flowrates the prices for the robot pellet extruder units are also sharply rising. The results of this search are shown in Table 2.

On the first place of this table is the Notzek Pro Filament Extruder of this project. The 0.5 kg/h flowrate of this extruder is the worst in comparison to other available pellet extruders. The Notzek extruder is followed by two other pellet extruders to produce filaments, the FileLab Pro 350 [22] and the FelFil Evo [23]. Those two extruders have a bad material flow, similar to the MDPH2, which is really designed for an industrial robot use-case [24]. The Noztek Xcalibur is the improved version of the reference product but exceeds the maximum payload of the ABB robot [25]. The Pulsar pellet extruder with 5 times the flow rate of the reference product is not commercially

Table 1. Parameter optimization for the Noztek Pro Filament Extruder

Print parameters	
Nozzle diameter	3 mm
Layer height	1.2 mm
Nozzle temperature	210 °C
Bed temperature	60 °C
Cooling jet airflow	2 m/s
Print speed	10.5 mm/s

Fig. 5. 3D print of an upscaled vase

available yet to the time of this project [26]. Three alternative extruders, the MDPE10 [27], Exon 10 [28] and CEAD pellet extruder [29] have a good material flowrate for upscaled products between 4.5 and 12 kg/h but their weight exceeds the 16 kg of the robot flange. The only two possible pellet extruders for an acquisition of the Noztek Extruder would be the GMEK-600-ROB [30] and the Cosine Extruder [31], as they weigh less than 16 kg and achieve an adequate material flow of 7.3 kg/h to 11.3 kg/h of plastic for upscaled 3D printing applications. These extruders also come to a considerable price tag of about 10,000 to 17,000 €. The search for an alternative product resulted in some possible substitutes, but none of these products dominated the others at a cost-efficient price.

Table 2. Commercially available pellet extruders

Pellet extruder	Flowrate	Mass	Cost
[Name]	[kg/h]	[kg]	[€]
Noztek Pro Filament Extruder	0.5	6	1,180.37
FileLab Pro 350	0.8	12	1,536.65
FelFil Evo	0.85	5.5	599.00
MDPH2 - Pellet Head Extruder	0.9	8.39	4,567.80
Noztek Xcalibur	2	65	9,484.47
Pulsar Pellet Extruder	2.5	2.2	N/A
MDPE10	4.5	18	11,949.98
Dohle/Herz Exon 10	6	19	18,634.50
Munsch GMEK-600-ROB	7.3	10.7	16,357.20
Cosine Pellet Feed Extruder	11.3	15.9	9,253.30
CEAD Pellet Extruder	12	29	24,900.00

7 Design of New Pellet Extruder

In addition to the research a new concept for a pellet extruder was created. To produce upscaled 3D prints with a volume size of several cubic meters an adequate material flowrate is needed. Alternative systems with an industrial robot already show good and fast results with a flowrate slightly under 2 kg/h [13–16]. The new flowrate of thermoplastic material should be set at above 2 kg/h to print big objects in an acceptable time. With this technology a large object like the demonstrated vase in Fig. 5 or a regular chair could be printed easily within the duration of a single day. The requirements for the new extruder were the right amount of material flow and a maximum mass of 16 kg, so that it does not exceed the highest payload of the ABB robot.

There are several types of pellets extruders. The screw extrusion with a single screw is widely used in the polymer processing sector due to its excellent attributes. There is also an extruder type that uses two screws, which can transport a higher amount of material [32]. This version also generates a higher overall mass therefore it is unsuitable for this use-case. The plastic melts when moved along due to the rotation of the screw. The length of the screw ensures that the plastic continuously liquefies during its rotation. The normal length of a screw in the plastics industry is between 200 and 500 mm. The design of a plastic screw has typical three section (Fig. 6). They are used for feeding, compression and metering of the plastic pellets. The core diameter of the screw only changes in the compression zone. This increase in diameter is important to create a backpressure as the pellets melt along the screw. The backpressure forces unwanted air outside of the screw barrel in the opposite direction of the plastic flow. The L/D ratio is the compression ratio along the screw. It is defined by the overall screw length divided by the screw diameter. Typical ratios are 24:1, 30:1 or 32:1, the ratio is determined according to the application. Due to the small design, a ratio of 10:1 was selected for the industrial robot use-case.

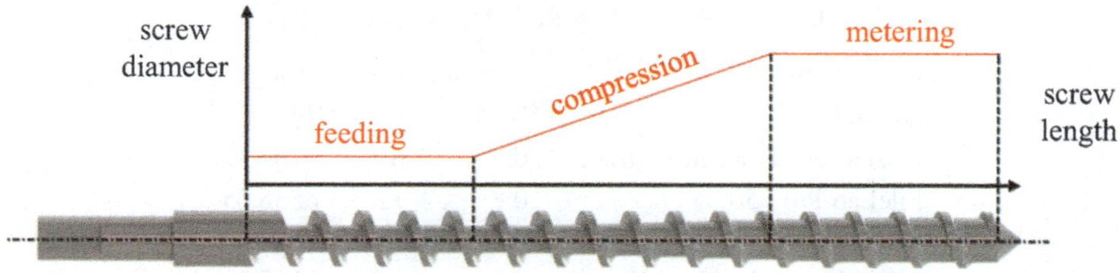

Fig. 6. Three zones of an extruder screw

The density of PLA is typically between 1.21 and 1.45 g/cm^3. In order to achieve a material flowrate of at least 2 kg/h PLA an amount of 1350 to 1650 cm^3/h has to flow through the nozzle at the end of the screw. For the use-case an amount of 2000 cm^3/h was aimed, which equals 555.56 mm^3 per second. The print speed through the nozzle is defined by the material flow through the cross-sectional area of the nozzle. This results in a constant print speed of 28.29 mm/s out of the nozzle. The robot transition speed has to be synchronized this extruder velocity.

$$v = \frac{\dot{V}}{A} = \frac{555.56\,\mathrm{mm}^2/\mathrm{s}}{5^2\,\mathrm{mm}\cdot\frac{\pi}{4}} = 28.29\,\mathrm{mm/s}$$

In addition to the calculation a first concept of the extruder was designed as a CAD model. Three heat zones supposed to melt the plastic pellets along the screw. According to some practical references the heat zones need different temperatures to guarantee the best print results [16, 33]. The temperature increases along the nozzle. If the temperature is too hot in the feeding section the pellets are likely to clog in the hopper, so that no further material can enter the screw. The first design of a new pellet extruder is shown in Fig. 7. This design is the first prototype for further research in this project. The developed extruder design is easy to mount and exchange on a robot flange. It enables upscaled 3D printing processes for industrial robots in our modern industry.

The next step of the extruder design is the dimension of the torque and the gearbox of the stepper motor that drives the extruder screw. After this the pellet extruder could be built and tested in the industrial robot setup. These upcoming steps will be conducted in further works.

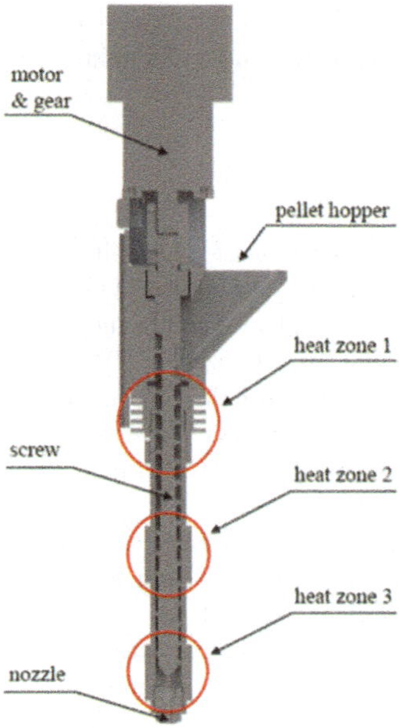

Fig. 7. CAD concept of the pellet extruder

8 Conclusion

In this project an industrial robot was equipped with a pellet extruder to research in the sector of upscaled 3D printing with thermoplastic pellets. A parameter optimization had been conducted for the evaluation of a good setup for the 3D print of upscaled products, that exceed a volume size of one cubic meter. An ABB IRB 2400/16 was fit to a Noztek Pro Filament Extruder. All the experiments were based on Ingeo 4043D PLA pellets. First, test objects were printed to evaluate the process parameters. The studies of the work showed that a designed cooling system could increase the print speed by 17%. It was found out that the used pellet extruder of the setup was still too slow, to enable an appropriate upscaled production with the setup. The 3D print duration of a one-meter high plastic vase took an overall time of 10 days. The measured material flowrate of 0.13 kg/h out of the pellet extruder is way too slow to produce bigger thermoplastic objects. This material flowrate was only a quarter of the setpoint value of the Noztek extruder. A research started for alternative pellet extruders. Commercially available products have material flowrate from 0.5 up to 12 kg/h but most of them exceed the payload of the robot flange. Only a couple of the commercially available extruders could substitute the Noztek Pro Filament Extruder for better results. As those extrudes come to a high price, a new concept for a pellet extruder was designed. The presented extruder should achieve material flowrate above 2 kg/h to enable the 3D print of bigger products like the demonstrated vase within a day. The next step of this project is the detailed dimension of the pellet extruder. The new extruder should open the gateway for a new research in the upscaled 3D printing with an industrial robot. As industrial robots are part of our modern industry 4.0, essential steps can be done to further improve and individualize processes.

References

1. Geschätzter Umsatz der 3D-Druckindustrie weltweit (2019). https://de.statista.com/statistik/daten/studie/388698/umfrage/umsatz-der-3d-druckindustrie-weltweit/
2. SmarTech Analysis: Polymer Additive Manufacturing Markets and Applications: 2020–2029. SmarTech Markets Publishing, Crozet (2020)
3. Buchanan, C., Gardner, L.: Metal 3D printing in construction: a review of methods, research, applications, opportunities and challenges. Eng. Struct. **180**, 332–348 (2019)
4. Fastermann, P.: 3D-Drucken - wie generative Fertigungstechnik funktioniert, 2. aktualisierte Auflage. Springer-Verlag, Heidelberg (2016)
5. 3D-Druck – Umfrage zu den meistgenutzten 3D-Druck-Technologien 2018 (2018). https://de.statista.com/statistik/daten/studie/760408/umfrage/meistgenutzte-3d-druck-technologie/
6. Roschli, A., Gaul, K.T., Boulger, A.M., Post, B.K., Chesser, P.C., Love, L.J., Blue, F., Borish, M.: Designing for big area additive manufacturing. Addit. Manuf. **25**, 275–285 (2019)
7. Paolini, A., Kollmannsberger, S., Rank, E.: Additive manufacturing in construction: a review on processes, applications, and digital planning methods. Addit. Manuf. **30**, 100894 (2019)
8. Reich, M.J., Woern, A.L., Tanikella, N.G., Pearce, J.M.: Mechanical properties and applications of recycled polycarbonate particle material extrusion-based additive manufacturing. Materials **12**(10), 1642 (2019)
9. World Economic Forum - Demonstrating the Viability of 3D Printing at Construction Scale (2016). https://futureofconstruction.org/case/winsun/download/
10. Tyson, M.: Worlds largest single 3D printed pedestrian bridge (2019). https://www.3dprintingsolutions.com.au/News/Australia/worlds-largest-single-3d-printed-pedestrian-bridge
11. BLB Industries - The Box Large (2020). https://blbindustries.se/wp-content/uploads/2019/11/techsheet_BOXlarge_ENG.pdf
12. Bolderen, G.V.: Exploration of stability of 3D-printed steel members: a study to buckling behaviour of wire and arc additively manufactured stainless steel tubular columns. Ph.D. thesis, Delft University of Technology, Delft, The Netherlands (2017)
13. Kooji, D.V.: Collection (2020). https://www.dirkvanderkooij.com/collections/collection/
14. Leonidou, L.: Engineering the AiCell (2019). https://medium.com/ai-build-techblog/engineering-the-aicell-e65a4deb0f3c
15. Valkenaers, H., Vogeler, F., Ferraris, E., Voet, A., Kruth, J.: A novel approach to additive manufacturing: screw extrusion 3D-printing. In: 2013 Proceedings of the 10th International Conference on Multi-material Micro Manufacture (4M13), San Sebastian, Spain, 8–10 October 2013, pp. 235–580 (2013)
16. Magnoni, P., Rebaioli, L., Fassi, I., Pedrocchi, N., Tosatti, L.M.: Robotic AM system for plastic materials: tuning and online adjustment of process parameters. Procedia Manuf. **11**, 346–354 (2017)
17. Noztek Pro Filament Extruder (2020). https://www.noztek.com/product/noztek-pro/
18. Ingeo 4043D (2020). http://www.eumakers.com/file_download/pla/Polymer_DataSheet_4043D.pdf/
19. Ultimaker Cura (2020). https://ultimaker.com/software/ultimaker-cura/
20. RoboDK - Simulate Robot Applications (2020). https://robodk.com/

21. Ultimaker - Experimental (2020). https://ultimaker.com/en/resources/20426-experimental/
22. FilaFab PRO 350 (2020). http://d3dinnovations.com/filafab/product/filafab-pro-350/
23. FelFil Evo (2020). https://felfil.com/Felfil_Evo_technicals.pdf?v=fa868488740a/
24. MDPH2 - Pellet Head Extruder (2020). https://massivedimension.com/products/mdphe-v1-pellet-head-extruder-system/
25. Noztek Xcalibur (2020). https://www.noztek.com/product/noztek-xcalibur/
26. Pulsar Pellet Extruder (2020). https://dyzedesign.com/pulsar-pellet-extruder/#tech-specs/
27. MDPE10 - Massive Dimension Direct Print Particle Extruder (2020). https://massivedimension.com/collections/extruders/products/mdpe10-massive-dimension-direct-print-particle-extruder/
28. ExOn 10 (2020). https://www.herz-gmbh.com/index.php?id=288&svartpin=VJVXmwXsGPx6wEes&action=3/
29. Pellet extruder for composite thermoplastic 3D printing (2020). https://robotextruder.com/
30. Rob-Extruder (2020). https://munsch-kunststoff-schweisstechnik.de/en/product-detail/rob-extruder/
31. Extruder Options: Pellet Feed Extruder (2020). https://www.cosineadditive.com/en/extruders/
32. Liu, X., Chi, B., Jiao, Z., Tan, J., Liu, F., Yang, W.: A large-scale double-stage-screw 3D printer for fused deposition of plastic pellets. J. Appl. Polym. Sci. **134**(31), 1–9 (2017)
33. Woern, A.L., Byard, D.J., Oakley, R.B., Fiedler, M.J., Snabes, S.L., Pearce, J.M.: Fused particle fabrication 3-D printing: recycled materials' optimization and mechanical properties. Materials **11**(8), 1143 (2018)

Industry 4.0 Applications

A Comparative Sectoral Analysis of Industry 4.0 Readiness Levels of Turkish SMEs

Oğuz Emir[1](✉), Zeynep Gergin[1], Fadime Üney-Yüksektepe[1], Uğurcan Dündar[2], Güneş M. Gençyılmaz[3], and Ali Ihsan Çavdarlı[4]

[1] Department of Industrial Engineering, İstanbul Kültür University, İstanbul, Turkey
{o.emir, z.gergin, f.yuksektepe}@iku.edu.tr

[2] Institute for Interactive Marketing and Social Media, Vienna University of Economics and Business, Vienna, Austria
ugurcan.duendar@wu.ac.at

[3] Department of Industrial Engineering, İstanbul Aydın University, İstanbul, Turkey
gunes.gencyilmaz@aydin.edu.tr

[4] KOSGEB Small and Medium Sized Enterprises Development Organization, Ankara, Turkey
ali.cavdarli@kosgeb.gov.tr

Abstract. The concept of Industry 4.0 brings new standards and has an essential impact on business models, work organizations, and processes. Nowadays, companies are enforced to perform such challenges as meeting personal customer requests, flexibility, responsiveness, customer-oriented solutions, and delivering continuous value to survive in the competitive market. The most innovative companies are resourceful to integrate new digital tools into their business models by following initial trends. For Small and Medium-Sized Enterprises (SMEs), adopting these new industrial challenges, and responding to them quickly is vital when it is considered that they have 99.8% of the enterprise share of Turkey. To do so, an intensive effort is needed to integrate Industry 4.0 applications within the enterprises. It has been suggested in the literature that manufacturing organizations should begin with understanding their current level of maturity by defining their strengths and weaknesses. Further, a significant effort is required with the collaboration of government, academia, and industry leaders to guide enterprises for improving their capabilities in an objective and standardized manner. Despite the diversity of approaches directed to Industry 4.0 concepts, it is not easy to find a study that measures how the manufacturing sector is adopting Industry 4.0 and compares manufacturing sectors with each other. To this end, broad research is initiated in March 2018 to identify the current situation of SMEs in Turkey. In this paper, it is intended to extend the study by providing a comparative sector analysis for manufacturing SMEs in Turkey in terms of Industry 4.0 transformation. Possible action plan suggestions are presented according to the results obtained at the end of the paper.

Keywords: Industry 4.0 · Small and medium enterprises · Turkish industry · Sectoral analysis · Assessment models

N. M. Durakbasa and M. G. Gençyılmaz (Eds.): ISPR 2020, LNME, pp. 251–263, 2021.
https://doi.org/10.1007/978-3-030-62784-3_21

1 Introduction

Industry 4.0 is a concept that has been emerged in Germany for the first time in 2011, and the definitions of the terminology can be found extremely varied in the literature. The concept is initiated with the objective of integrating digitalized technological solutions into manufacturing processes to control physical systems [1]. Industry 4.0 appeals to companies and provides competitive advantages in terms of eliminating unnecessary production costs, increasing throughput, reducing cycle times, and improving product quality and manufacturing environment [7]. The rapid pace of new technological developments pushes the companies in terms of accelerated adaptation and implementation. Regardless of developed or undeveloped countries, there is a global movement all over the world that is embarking on its transformation called the 4th Industrial Revolution [7]. Reference [8] states that this revolution is not coming out from an energy source but from the potential of a new technological fact called digitalization.

Industry 4.0 brings new standards and have an important impact on business models and processes. Companies are enforced to perform such challenges as meeting personal customer requests, flexibility, responsiveness, customer-oriented solutions, and delivering continuous value to survive in the competitive market. Industry 4.0 applications are becoming attractive for organizations. However, it can be complicated to implement them, and for this reason, there is an essential need for assisting companies in the transition phase.

SMEs are independent companies with a limited number of employees. This number varies according to the national files of the different countries. In Turkey, SMEs are defined as the enterprises that have less than 250 employees and less than 125 million Turkish Liras (TL) endorsement or balance [9]. According to the Turkish Statistical Institute (TÜİK), SMEs have 99.8% of the enterprises share of Turkey [13]. SMEs play a crucial role in countries' economy, and that is defined as the driving force of many manufacturing economies. Unlike big corporate companies, SMEs have to confront different challenges and barriers. Large companies are more capable of integrating information technology solutions into their production plants [10]. In order to protect SMEs against these conditions and prevent SMEs from becoming victims of this industrial revolution, it is required to fully grasp the organizational, human, and technological challenges of Industry 4.0 [2, 8].

Industry 4.0 is becoming widely used, and there is a growing number of academic studies. The literature review is conducted in Google Scholar by searching the following keywords and terms: "Industry 4.0", "Industry 4.0 Assessment Models", "SME", "Industry 4.0 Enablers" and "Industry 4.0 Barriers". After reviewed the titles of resulting papers from the initial search, in a first iteration, only 17 potential papers were scanned. However, after thoroughly reading the papers in a second iteration, seven papers relevant to this research were determined. The limited number of relevant papers points out the novelty of this trending research area.

Gergin et al. (2019) performed a study by measuring the Industry 4.0 scores of SMEs in the most industrialized cities in Turkey to find out their level of awareness and readiness for these cities. A survey is applied to 623 SMEs and the companies are evaluated with a scale of 1.0 to 4.0 level, and the overall score of these SMEs is calculated as 2.69 [5]. In parallel with this project, papers [2, 3] present a study by applying different data classification algorithms to determine the best method for the Turkish SMEs Industry 4.0 score prediction. It is desired to use the developed approach by creating a decision support tool with the classification algorithm behind to determine the current situation of SME. Gökalp, Şener and Eren (2017) analysed 7 existing studies about assessment/maturity models from the literature and reached an opinion that none of these models satisfies all expected criteria. For this purpose, it is proposed a structured Industry 4.0 assessment model in order to guide companies in terms of achieving a higher level of maturity stages [6]. Mittal et al. (2018) clarified specific requirements of maturity models and provided some useful insights that help towards developing an accurate model for SME's in the presence of industrial realities. When these insights are taken into consideration, it is expected that the enterprises can complete the transition with the least risk by defining their industry 4.0 vision and roadmaps clearly [10]. Schumacher, Erol and Sihn (2016) proposed an empirically new model for assessing the Industry 4.0 maturity levels of manufacturing enterprises. The objective of the study was to extend the recently developed models in the literature. After validation of the model, it is concluded that it can be easy to use and applicable to the real production environment [12]. Branco et al. (2019) presented a study by measuring the Industry 4.0 readiness for each country at the European Union (EU). Nine variables are selected for measuring different aspects of Industry 4.0, and 22 members of European Union countries are included in the study. Factor and cluster analyses were applied in the methodology part of the work to measure and interpret the levels of Industry 4.0 across the EU countries [1]. Sarı and Yiğitol (2020) discussed the level of awareness and readiness of Turkish manufacturing firms by surveying with managers of 427 firms. At the end of the study, it is deduced that majority of the participant firms know the concept of Industry 4.0 and try to be attuned to the new industrial revolution. Sectors that have the highest ratio of implementing Industry 4.0 principles are automotive, electronic, and machinery manufacturing. On the other hand, sectors that have the lowest ratio are textile and clothing, food, and beverage manufacturing [11]. Bibby and Dehe (2018) established an assessment model to measure the Industry 4.0 levels of implementation by bearing in mind some dimensions as a factory of the future, people, strategy, and culture [7].

Since there is not a specifically designed measurement tool that is accepted in the literature, assessing how the manufacturing firm is adopting Industry 4.0 is a challenging issue. Also, no studies that indicate a comparative sectoral analysis of Industry 4.0 scores for SMEs have yet been presented. Thus, this paper is initiated to fill this gap in the literature by applying a study to Turkish SMEs of manufacturing sectors. The paper is structured as follows. Section 2 presents the methodology of this study. Followed by Sect. 3 where industrialization levels of selected sectors are discussed and compared. Section 4 concludes the paper with the main findings and future research directions to researchers.

2 Methodology

This study is a continuum of a project launched in 2018 with a purpose to identify the current situation of SMEs in Turkey evaluating their various implementations for Industry 4.0 transformation. The questionnaire designed for this project is still collecting responses from SMEs, and the number of companies surveyed has increased to 1040. Consequently, at this stage of the project, the analysis has been carried out for these companies for this current study. Differently, in this current work, comparative sector analysis for SMEs is incorporated according to the survey results conducted.

The questionnaire consists of 3 sections and includes 28 questions in total that are prepared by academicians taking into consideration the relevant literature studies, technical reports, and comments of KOSGEB (Small and Medium-Sized Enterprises Development Organization), and SME representatives. Table 1 shows the design of the questionnaire. The aim of this question set is to gather some information from SMEs to measure their industrialization level and observe their awareness and readiness for Industry 4.0 transformation.

Each question has different options in the designed questionnaire and, an industrialization level is assigned ranging from 1.0 to 4 0 to the manufacturing organization following the completion of the questionnaire. Figure 1 explains the definitions and requirements of each industrialization level, considering all industrial revolutions with this perspective. In the previous papers [4, 5], it is mentioned that a Delphi study is applied to test the validity and reliability of the questionnaire and it is concluded that the questionnaire is consistent and reliable.

Table 1. Questionnaire design

Main section(s)	Number of questions	Question areas
First section	6 questions	General company information
Second section	12 questions	Industrialization level
Third section	10 questions	Barriers and Enablers for adopting Industry 4.0

3 Implementation and Results

As it is mentioned in the introduction part, it is intended to make a comparative sector analysis for Turkish SMEs in this current study. After conducting the e-survey to companies, responses are collected with the support of KOSGEB. This study is implemented with the analysis of 1040 SMEs. Almost all manufacturing sectors have participated in e-survey except for beverages and tobacco products manufacturing sectors. Furthermore, it is noticed that there is not enough participation in some sectors (see Table 2) that will represent the sector mean. Therefore, some sectors are excluded in the detailed analysis due to this insufficient participation. Manufacture of fabricated metal products, food products, rubber and plastic materials, base metal, machinery and equipment, and other productions are selected for the scope of this study in this regard.

Industrialization Level	Definitions and requirements of levels
Industry 1.0 Level	The operations are realized manually. There is no electrical system introduced to the processes.
Industry 2.0 Level	Some systematic implementations are introduced to the system. Electricity is used instead of manpower. Personal computers, simple software and Internet are present in the company, but not integrated with each other and not for managing the processes.
Industry 3.0 Level	All operations are realized in a systematic manner. Automation and sensors are used in the operations, but they are not in collaboration with each other. There are systems for data collection, but the data are not used for online decisions. The infrastructure of the company is ready for the full transformation.
Industry 4.0 Level	The company is in the adaptation phase. The internal operations and those with the stakeholders are realized in an integrated manner via digitalization. Most of the technological drivers such as big data, augmented reality, horizontal and vertical integration etc. are implemented in the system.

Fig. 1. Definitions and requirements of levels [5]

Manufacture of sports equipment, glassware products, musical instruments, jewellery products, games, toys, and other categories are categorized under other productions sector.

It is seen that the average industrialization level for the 1040 companies is 2.43. This score is consistent with the results obtained in the previous studies of the broad project. It is seen that the SMEs are still not achieving Industry 4.0 level, and yet they are behind 3.0 level. When the sectors are analyzed individually, the difference between the industrialization levels is quite close to each other. Hence, we cannot conclude a pioneer sector for industrialization. Companies operating in "Manufacturing Basic Pharmaceutics and Pharmaceutics Materials" sector has the highest score with 2.72, whereas the lowest score is observed for "Manufacturing leather and its bi-products" sector with 2.22. The two sectors, namely "Manufacturing beverages" and "Manufacturing tobacco products" could not be analyzed due to the fact that there are no respondents to the questionnaire from the SMEs operating for this sector.

For the continuing analysis of the study, sectors that have a representative amount of data are merged under clusters. Consequently, seven clusters are analyzed and compared from the perspective of their industrialization levels and Industry 4.0 adoptions.

Sector clusters are analyzed for the industrialization status of their different processes. Table 3 shows the scores of 12 different processes. Based on this comparative analysis, "Manufacture of food products" sector has the lowest industrialization level with 2.27 on average, and other sectors levels are close to each other. When the average scores of the process are investigated, the use of *online traceability* and *mobile applications* stand out with the highest averages. *Warehouse management* and *order receiving* positioned far behind other processes with 1.53 and 1.91 levels. Industry leaders should prioritize these processes in their action plans.

It is seen that *research and development* is highly adopted in "Machinery & Equipment" sector, while "Food" sector is not implementing *R&D* that much. There is a significant difference between the scores for these process as 3.07 for "Machinery & Equipment" and 1.85 for "Food" sector company averages. Similarly, "Food" sector is considerably behind the general average and the other sector averages regarding the Industry 4.0 implementations for *work order* processes.

Table 2. Survey participation numbers of sectors and industrialization levels

Manufacturing sectors	Number of participant SME	Average industrialization level
Manufacturing basic pharmaceutics and pharmaceutics materials	13	2.72
Manufacture of other transport vehicles	5	2.66
Manufacturing computers, electronics and optic materials	25	2.64
Printing and copying recorded media materials	5	2.62
Manufacture of motor vehicles, trailers and semi-trailers	22	2.54
Manufacturing rubber and plastic materials	94	2.52
Manufacturing chemicals and chemical products	47	2.48
Base metal industry	86	2.47
Manufacture of machinery and equipment not elsewhere classified	78	2.47
Manufacturing non-metalic minerals	16	2.47
Other productions	148	2.46
Manufacture of fabricated metal products, except machinery and equipment	109	2.46
Manufacturing electronic equipments	46	2.45
Outfit manufacturing	17	2.45
Manufacture of textile products	96	2.44
Manufacture of paper and paper products	22	2.44
Furniture manufacturing	53	2.42
Installation and repair of machinery and equipment	36	2.42
Manufacture of wood, wood and cork products (except furniture);	14	2.28
Manufacture of food products	97	2.27
Manufacturing leather and its bi-products	11	2.22
Manufacturing beverages	0	–
Manufacturing tobacco products	0	–
Grand total	**1040**	**Average - 2.43**

Another analysis is done for individual industrialization levels of the respondent companies. When SME industrialization levels are individually checked, there are three primary group levels as 2.0, 3.0, and 4.0. To obtain that evaluation, the industrialization score measured for each company is rounded to the nearest integer. For instance, if the score is between 1.50 to 2.49, the company's score is assumed at 2.0 level. Considering this evaluation, SMEs are divided into these levels for each sector, as shown in Fig. 2. Except for "Food manufacturing" sector, the number of level 3.0 companies are more than the number of level 2.0 companies. Notably, this situation shows double in the "rubber and plastic manufacturing" sector. (Level 2.0: 31 companies; Level 3.0: 62

Table 3. Overall results for industrialization level

Process	Average	Fabricated metal	Food	Textile	Rubber & plastic	Base metal	Machinery & equipment	Other productions
Order receiving	1.90	2.10	1.74	1.74	2.00	2.01	1.83	1.94
R&D	2.58	2.81	1.85	2.26	2.79	2.83	3.07	2.54
Work order	2.39	2.37	1.96	2.50	2.48	2.47	2.34	2.43
Online traceability	3.59	3.66	3.52	3.66	3.56	3.53	3.65	3.59
Purchasing	2.61	2.54	2.54	2.56	2.63	2.61	2.64	2.61
Production technology	3.11	3.17	3.02	3.05	3.14	3.17	3.16	3.11
Software	2.94	2.96	2.86	2.94	2.89	2.98	2.96	2.98
Quality control	2.57	2.61	2.55	2.57	2.62	2.57	2.54	2.56
Inventory management	2.46	2.42	2.27	2.54	2.48	2.43	2.56	2.46
Warehouse management	1.54	1.48	1.40	1.74	1.66	1.42	1.49	1.50
Maintenance management	2.08	1.99	2.00	2.05	2.13	2.03	2.07	2.19
Mobile applications	3.59	3.57	3.58	3.62	3.63	3.60	3.60	3.59
Average	**2.43**	**2.46**	**2.27**	**2.44**	**2.52**	**2.47**	**2.47**	**2.46**

companies in rubber&plastic). Regarding the selected 7 cluster sectors, there are only a few companies that show Industry level 4.0 implementations among these sectors. Based on the analysis, 302 companies show Industry level 2.0 while 400 companies show Industry level 3.0.

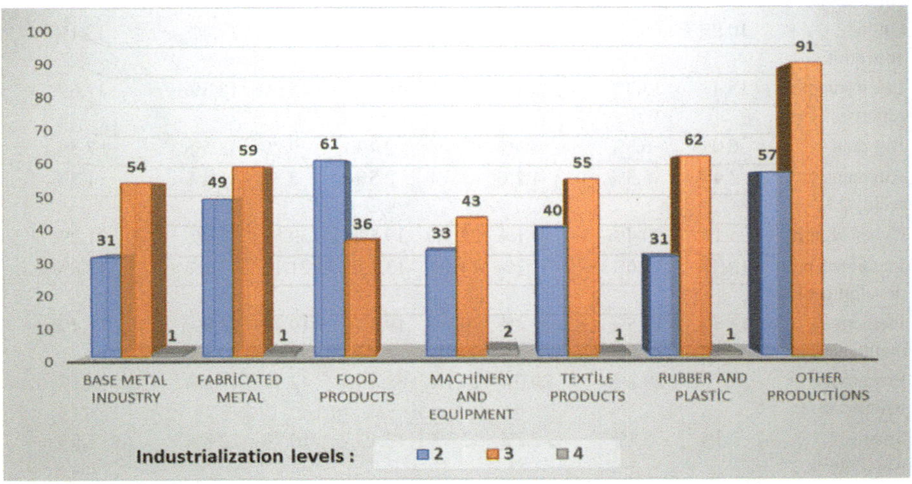

Fig. 2. Overall industrialization scores

When the answers for the third section of the questionnaire that investigate the awareness, barriers, and enablers for adopting Industry 4.0 are examined, the results are shown in Tables 4, 5, and 6 respectively. *Preemptive maintenance, automation, and sensory systems* are mostly preferred Industry 4.0 tools by every sector according to the data provided in Table 4. The usage rates of these three technologies are over 30% on average. On the other hand, *augmented reality, virtual reality, and smart warehouses* have an awareness rate below 5% on average for all sectors, displaying the low usage of these technologies. *Preemptive maintenance* is currently used in "machine & equipment" manufacturing sector as 47.2% that is the highest rate from existing sectors. When sectoral comparisons are made, the higher average rates of the "Machinery & Equipment" sector draws attention. It has the highest awareness rate in 8 of 19 Industry 4.0 applications. However, most of the sectors have lower awareness rates that

Table 4. Awareness of technologies (currently in use)

	Average	Fabricated metal	Food	Textile	Rubber & plastic	Base metal	Machinery & equipment	Other productions
3D Printing	11.9%	9.9%	7.9%	13.5%	18.4%	13.9%	15.3%	7.5%
Sensor systems	31.3%	32.7%	29.7%	31.0%	33.7%	39.0%	29.7%	31.2%
Virtual reality	3.5%	4.0%	3.5%	2.3%	0.0%	5.1%	2.8%	3.0%
Cloud technologies	20.1%	17.0%	12.5%	13.8%	8.4%	26.0%	37.5%	23.4%
Internet of things	5.5%	6.1%	4.7%	5.1%	2.4%	4.0%	7.1%	5.4%
Additive (3D) manufacturing	7.4%	13.3%	3.5%	7.4%	8.3%	13.0%	6.9%	7.5%
Improved robot technologies	7.2%	12.0%	3.5%	0.0%	17.2%	15.6%	8.5%	4.6%
Automation	33.2%	35.0%	30.4%	22.1%	39.6%	43.8%	43.8%	28.5%
Modeling and simulation	23.8%	32.3%	3.5%	15.7%	35.7%	35.0%	41.7%	18.0%
Online education	10.8%	8.2%	3.5%	2.5%	6.0%	17.1%	17.4%	13.1%
Electronic perf. reports	15.4%	13.3%	9.1%	17.7%	18.3%	15.6%	18.3%	17.6%
Big data	6.0%	6.3%	4.6%	5.1%	2.4%	9.2%	8.5%	7.5%
Augmented reality	2.4%	5.3%	1.2%	1.3%	2.5%	5.3%	1.4%	1.5%
Cyber security	23.1%	20.4%	9.1%	22.5%	19.5%	39.5%	22.9%	21.2%
Integrated prod development	13.9%	12.6%	4.7%	14.1%	13.3%	21.1%	17.1%	13.8%
Predictive maintenance	13.5%	13.5%	2.3%	9.1%	19.8%	19.2%	23.9%	11.4%
Preemptive maintenance	33.7%	29.3%	20.0%	24.1%	38.4%	42.9%	47.2%	34.1%
Smart warehouses	3.5%	3.1%	2.2%	2.4%	2.4%	10.7%	2.8%	1.5%
Mobile systems	16.1%	13.3%	20.2%	13.3%	10.8%	17.3%	13.9%	19.5%

are under 30%. The use of *cloud technologies, internet of things and big data* are deficient in "Rubber and plastic" manufacturers when compared with the general average and the other sectors. On the contrary, *improved robot technology* adoption is considerably high in this sector in comparison to the others. *Robot technologies* are not seen in the "Textile" sector, and *virtual reality* is not implemented in any company operating in "Rubber and plastic" sector. Rest of the Industry 4.0 technologies are all observed in all sectors.

Table 5 displays the opinions of surveyed companies in terms of barriers in Industry 4.0. The great majority of the companies consider that cost-related issues and lack of support are the essential barriers. *Financial resource shortage and the high cost of the new technologies* reside as the top barriers, especially for "Fabricated metal" and "Base metal" sector. Similarly, *high costs* set a barrier to these sectors more than the others. *High cost or lack of consultancy and lack of support from universities/ government* follows as the second important barriers for all surveyed SMEs. "Fabricated metal" sector companies complain more on *lack of education of workers in production* and technical *personnel's insufficient knowledge.* The rest of the answers do not display a significant difference between the barriers for different sectors.

Table 6 indicates what extent the information sources are considered as necessary for sectors. All companies believe that *websites* are the most crucial sources comparing to others with 83.9%. Manufacturing sectors think that articles on *search engines, trade fairs, and government foundations* are found significant after the websites. *Papers in journals* are the last preferred information source for companies with 49.7%. When all answers are analysed, the responses for information resources do not display a significant difference between different sectors.

Table 5. Barriers of Industry 4.0

Barrier	Average	Fabricated metal	Food	Textile	Rubber & plastic	Base metal	Machinery & equipment	Other
Current hardware is not suitable	58.4%	62.9%	59.6%	65.4%	60.0%	51.9%	56.8%	51.5%
Current software is not suitable	57.8%	61.9%	59.6%	62.5%	59.3%	50.0%	55.4%	53.9%
Financial resource shortage	**82.3%**	90.2%	73.9%	80.2%	83.0%	89.9%	81.8%	74.3%
High cost of the new technologies	**81.5%**	87.0%	80.9%	73.8%	81.9%	87.3%	81.3%	75.2%
Weak technical support from suppliers	52.6%	51.5%	57.3%	47.4%	56.5%	50.0%	47.3%	42.6%
Complexity of new technologies	30.4%	33.3%	35.2%	38.0%	33.3%	28.9%	24.0%	30.6%
Mismatch of work flow in the facility	41.4%	42.7%	44.9%	45.5%	37.9%	32.9%	45.3%	42.4%
Lack of knowledge on new technologies	52.5%	56.6%	59.6%	64.6%	49.4%	49.3%	47.4%	46.0%

(continued)

Table 5. (*continued*)

Barrier	Average	Fabricated metal	Food	Textile	Rubber & plastic	Base metal	Machinery & equipment	Other
Lack of education of workers in production	66.1%	78.0%	70.8%	68.4%	65.9%	65.3%	68.0%	62.4%
Technical personnel's insufficient knowledge	59.8%	70.7%	64.8%	62.3%	57.0%	53.2%	58.1%	56.3%
Resistance of employees to change	45.1%	57.3%	36.8%	56.4%	39.3%	44.7%	44.7%	44.0%
Resistance of top management to change	25.5%	24.7%	27.3%	34.2%	19.8%	35.1%	23.7%	24.6%
Lack of support from universities/government	**71.3%**	77.6%	67.0%	69.2%	69.3%	73.1%	62.7%	71.9%
High cost or lack of consultancy	**72.6%**	75.5%	74.2%	82.3%	72.1%	75.3%	68.4%	68.5%
Low demand	50.9%	43.4%	54.5%	63.8%	45.2%	56.6%	47.9%	53.6%
Lack of support of the big clients	47.0%	50.0%	52.3%	46.8%	50.0%	43.4%	37.8%	44.7%
Uncertainty of return on investment	67.4%	66.3%	65.5%	68.8%	67.4%	77.6%	65.8%	63.2%
Hesitation on data security and privacy	38.4%	35.1%	34.8%	42.1%	44.0%	38.2%	32.9%	38.7%

Table 6. Information sources for Industry 4.0

Information source	Average	Fabricated metal	Food	Textile	Rubber & plastic	Base metal	Machinery & equipment	Other
Websites	**83.9%**	83.8%	77.2%	81.8%	73.6%	86.4%	81.6%	85.3%
Articles on search engines	**76.8%**	72.6%	68.2%	74.4%	70.6%	79.2%	75.7%	80.3%
Trade fairs	**76.4%**	71.4%	82.0%	72.5%	75.9%	72.5%	78.4%	73.2%
Direct mails	60.6%	60.0%	61.8%	62.2%	53.6%	51.9%	46.7%	59.2%
Papers in journals	49.7%	51.0%	41.4%	40.0%	45.8%	50.0%	50.0%	53.6%
Trade unions	69.3%	69.1%	72.7%	71.4%	68.6%	72.4%	56.8%	63.4%
Presentations in congress	69.3%	72.2%	69.0%	63.5%	65.5%	68.0%	66.7%	67.2%
Consumers	65.5%	69.4%	60.5%	60.3%	61.0%	63.5%	60.5%	64.1%
Holding company	65.3%	69.1%	59.3%	58.6%	65.9%	62.2%	71.2%	63.3%
Equipment vendors	65.4%	71.1%	64.7%	54.8%	53.6%	66.7%	74.0%	60.5%
Consultants	69.3%	66.3%	65.2%	62.5%	62.7%	65.4%	74.0%	69.8%
Universities	69.6%	74.2%	55.8%	72.7%	60.0%	66.7%	73.3%	67.7%
Government foundations	**75.2%**	74.7%	76.1%	73.1%	70.6%	72.4%	71.2%	70.8%

4 Conclusions

In 2011, the concept of Industry 4.0 has emerged with the objective of integrating digitalized technological solutions into manufacturing processes to control physical systems. This modern concept brings new standards and has an important impact on business models, work organizations, and processes. Innovative companies take rapid actions by integrating Industry 4.0 tools like the internet of things, cloud computing, autonomous robots, augmented reality into their internal processes. Thus, they play an active role in the sector by meeting personal customer requests and delivering continuous value to the customer. On the other hand, it is observed that SMEs are usually financially constrained and lack of knowledge to follow innovations. Since SMEs have a high enterprise share of 99.8% in Turkey, an improvement will definitely contribute not only to the local economy but also to the global competition strength of the country. Therefore, a significant effort is required with the collaboration of government, academia, and industry leaders to guide enterprises for improving their capabilities. With that motivation, this is the continuum of the broad research that is initiated in March 2018, with a purpose to identify the current situation of SMEs in Turkey. In this paper, a comparative sector analysis is given for manufacturing SMEs in Turkey.

Except for companies from beverages and tobacco manufacturing sector, 1040 companies participated in total to the conducted e-survey. Some companies are excluded in the detailed comparative analysis due to lower participation, and in this context, 7 clusters are analyzed and compared. The average industrialization level is calculated as 2.43, and it is observed that each manufacturing sector has close levels around the average score except for food manufacturing with 2.27. Industry leaders of the food manufacturing sector should focus on their transformation to Industry 4.0 in that case. It is also realized that the readiness levels are consistent with the previous studies of the broad project.

When industrialization levels for different processes are investigated, sectors displayed higher levels on mobile applications and online traceability. However, warehouse management and order receiving processes have lower levels with the scores of 1.53 and 1.91 respectively. Since these two processes have lower scores than level 2.0, it can be concluded that a prioritized action plan for these processes is highly needed.

When the sectoral analysis is examined, readiness levels of each manufacturing sector are still behind the 3.0 level. To achieve higher levels, there is a lot to be done, and absolute cooperation is needed. There are only a few companies that show the readiness level of 4 among all SMEs.

The average of using Industry 4.0 technological implementations by each sector demonstrates quite low rates except for preemptive maintenance, automation, and sensory system adoptions. According to results, most companies do not prefer to use augmented reality, virtual reality, and smart warehouses implementations. For this situation, increasing incentive programs and expanding budgets will possibly increase awareness in each sector.

When barriers for adopting Industry 4.0 to are examined, it is concluded that SMEs are financially constrained and lack support for integration of Industry 4.0 applications. Because, companies are claimed that financial resource shortage and the high cost of the new technologies are the top barriers for Industry 4.0 transformation. Financial resource shortage and the high cost of the new technologies reside as the top barriers, especially for "Fabricated metal" and "Base metal" sector due to their costly production processes. High cost or lack of consultancy and lack of support from universities/government are considered as the second important barriers, according to surveyed SMEs. As repeatedly mentioned, financial aid programs, digital investments, education, and training policies can be provided to develop alternative solutions with the collaboration of stakeholders.

Lastly, websites are considered as the most preferred information sources with 83.9% sector average. Search engines, trade fairs, and government foundations follow as preferred information sources respectively. However, the least preferred information source for companies is papers in journals with 49.7%. The response for information resources do not display a significant difference between different sectors.

Overall, Turkish SMEs should focus on the adoption of Industry 4.0 by using this specifically designed readiness model. This model presents significant insights and results for surveyed companies. Companies can revise their long-term business plan after identifying their readiness level and create a roadmap for higher levels. All stakeholders should actively get involved in the transition phase and complete their responsibilities in this process. Governments should increase incentive and financial aid programs, organize trainings and encourage digitalization investments. Also, academia should work together with manufacturing companies in a coordinated way and fully support SMEs to increase Industry 4.0 awareness.

For future studies, pilot companies can be determined and follow carefully to observe the development of their industrialization level. The significant results obtained at the end of the study with the pilot group can be shared with other SMEs as guidance.

Acknowledgment. This research is made possible with the support of KOSGEB. The researchers express their gratitude to KOSGEB Management for their support on this study and for the ongoing research to cover all regions of Turkey.

References

1. Castelo-Branco, I., Cruz-Jesus, F., Oliveira, T.: Assessing Industry 4.0 readiness in manufacturing: evidence for the European Union. Comput. Ind. **107**, 22–32 (2019)
2. Dundar, U., Üney-Yüksektepe, F., Gergin, Z., İlhan, D.A., Gençyılmaz, M.G., Çavdarlı, A. İ.: Industry 4.0 score prediction of Turkish SMEs via data classification. In: International Symposium of Production Research 2019, Wien (2019)
3. Dundar, U., Üney-Yüksektepe, F., Gergin, Z., İlhan, D.A., Gençyılmaz, M.G., Çavdarlı, A. İ.: A decision support tool for classification of Turkish SMEs Industry 4.0 score levels. In: International Symposium of Production Research 2020, Antalya (2020). (accepted for publication)
4. Gergin, Z., Yüksektepe, F., Gençyılmaz, M.G., Aktin, A.T., Gülen, K.G., İlhan, D.A., Dündar, U., Cebeci, Ö., Çavdarlı, A.İ.: Industry 4.0 scorecard of Turkish SMEs. In: Proceedings of the International Symposium for Production Research, pp. 426–437 Wien (2018)

5. Gergin, Z., İlhan, D.A., Üney-Yüksektepe, F., Gençyılmaz, M.G., Dundar, U., Çavdarlı, A. İ.: Comparative analysis of the most industrialized cities in Turkey from the perspective of Industry 4.0. In: International Symposium of Production Research 2019, Wien (2019)
6. Gökalp, E., Şener, U., Eren, P.E.: Development of an assessment model for Industry 4.0: Industry 4.0-MM. In: International Conference on Software Process Improvement and Capability Determination, pp. 128–142. Springer, Cham, October 2017
7. Bibby, L., Dehe, B.: Defining and assessing Industry 4.0 maturity levels – case of the defence sector. Prod. Plan. Control **29**(12), 1030–1043 (2018)
8. Luco, J., Mestre, S., Henry, L., Tamayo, S., Fontane, F.: Industry 4.0 in SMEs: a sectorial analysis. In: Ameri, F., Stecke, K., von Cieminski, G., Kiritsis, D. (eds.) Advances in Production Management Systems. Production Management for the Factory of the Future. APMS 2019. IFIP Advances in Information and Communication Technology, vol 566. Springer, Cham (2019)
9. Ministry of Science, Industry and Technology: Kosgeb, Kobi Stratejisi ve Eylem Planı (KSEP) 2015–2018 (2016)
10. Mittal, S., Khan, M.A., Romero, D., Wuest, T.: A critical review of smart manufacturing & Industry 4.0 maturity models: implications for small and medium-sized enterprises (SMEs). J. Manuf. Syst. **49**, 194–214 (2018). http://www.sciencedirect.com/science/article/pii/S0278612518301341
11. Sari, T., Gules, H.K., Yiğitol, B.: Awareness and readiness of Industry 4.0: the case of Turkish manufacturing industry. Adv. Prod. Eng. Manag. **15**, 57–68 (2020). https://doi.org/10.14743/apem2020.1.349
12. Schumacher, A., Erol, S., Sihn, W.: A maturity model for assessing Industry 4.0 readiness and maturity of manufacturing enterprises. Procedia CIRP **52**, 161–166 (2016)
13. Turkish Statistical Institute (TÜİK): Small and Medium Sized Enterprises Statistics (2016). Press release no. 21540, November 2017

A Decision Support Tool for Classification of Turkish SMEs' Industry 4.0 Score Levels

Uğurcan Dündar[1(✉)], Fadime Üney-Yüksektepe[2], Zeynep Gergin[2], Oğuz Emir[2], Güneş M. Gençyılmaz[3], and Ali İhsan Çavdarlı[4]

[1] Institute for Interactive Marketing and Social Media, Vienna University of Economics and Business, Vienna, Austria
ugurcan.duendar@wu.ac.at

[2] Department of Industrial Engineering, Istanbul Kültür University, Istanbul, Turkey
{f.yuksektepe,z.gergin,o.emir}@iku.edu.tr

[3] Department of Industrial Engineering, İstanbul Aydın University, İstanbul, Turkey
gunes.gencyilmaz@aydin.edu.tr

[4] KOSGEB Small and Medium Sized Enterprises Development Organization, Ankara, Turkey
ali.cavdarli@kosgeb.gov.tr

Abstract. The concept of industry 4.0 aims to enhance the employment of digitalization with the help of burst in computer usage after the third major revolution in industry. Hence, to keep up with the new trends in industry, companies should transform their processes into digital platforms. In Turkey, 99.8% of the companies are SMEs, so in order to be in the global economic competition Turkey should enable their SMEs to make the process shifts to digitalization. In this manner, in 2018 a project is initiated by KOSGEB to evaluate the readiness of SMEs and help them for Industry 4.0. As the last step of this project, a decision support tool is created with classification algorithm behind. In this paper, chosen instruments are used to predict the Industry 4.0 score of SMEs that are calculated with previously published algorithm in the scope of the project. In terms of classification, three algorithms are compared with ROC metric. CatBoost algorithm which is specifically created for categorical classification is compared with Support Vector Machine algorithms (SVM and μ-SVM) that performed best in previous research. CatBoost outperformed other algorithms and is used as base classification method in decision support tool. This decision support tool will be used in decision making process of KOSGEB to which companies to help.

Keywords: Industry 4.0 · Data classification · CatBoost · Support vector machines · Decision support tool · Small and medium enterprises

N. M. Durakbasa and M. G. Gençyılmaz (Eds.): ISPR 2020, LNME, pp. 264–276, 2021.
https://doi.org/10.1007/978-3-030-62784-3_22

1 Introduction

Technological developments are one of the main drivers of trends in life. This statement is also applicable to industry. With the enhancements, trends change and that leads industry to reshape, or so-called to revolutionize. Until the last decade, the industry had three major revolutions. The first one was employing steam-powered mechanizations in the late 1700s. The usage of electricity as a main power source of mechanics and mass production led to second revolution in the beginning of 20th century. The third revolution occurred in the end of 1970 s with the invention of computers. It changed the whole industry from top to bottom as it did in the daily life of the human beings. Lately, we are experiencing a new revolution which is called Industry 4.0. Roughly, Industry 4.0 aims to increase digitization level of systems while integrating them, with the help of burst in computer usage with the third revolution [1]. However, there are various explanations for Industry 4.0 in the literature [2, 3]. Researchers, mostly, agree that followings are the drivers of the new concept: Big Data and Analytics, Autonomous Robots, Simulation, Horizontal and Vertical System Integration, Internet of Things, Cyber Security, Cloud Systems, Additive Manufacturing and Augmented Reality [4].

As it is declared by Mocan [5], it is crucial for developing countries, such as Turkey [6], to keep up with digitalization era. In order to not to fall behind, Turkey should transform its industry. This is only possible with the transformation of Small and Medium Enterprises (SMEs) which constitutes 99.8% of the whole economic environment [7]. According to Ministry of Science, Industry and Technology [8], SMEs are the organizing units with lower than 250 employees and lower than 40 million Turkish Liras endorsement.

In the light of these information, in 2018, we started a project with alignment of KOSGEB, to take the most efficient steps on towards Industry 4.0. The idea of the project to see the current situation of SMEs in this era, and help them to catch up with this new trend. Thus, this project consists of following phases: Measuring the industry 4.0 level of SMEs in Turkey, analyzing the data and deciding which companies, sectors etc. to work on. In this manner, a survey distributed and answers are analysed. The survey design, and the results of these analysis' have been published in [9, 10] and [11] with different number of attendees and results. In [12], first data classification study is been done on the data and first outcomes have been published.

In this paper the number of survey attendees are increased to 864 with distributing it throughout the Turkey instead of four most industrialized cities. These 864 companies' -from different sectors and places, industry 4.0 scorecards are prepared according to [9] and usage of some technological instruments are collected. This usage is used to predict the industry 4.0 level of the company -as High or Low. The threshold value is assigned as the median of the whole dataset which is 2.59. In classification algorithm the best one from [12] -Support Vector Machine, and Catboost algorithm is compared. Catboost algorithm is used because all of the attributes in the dataset are categorical and this algorithm is designed to apply boosting algorithm specifically on categorical dense data. Then, with the winning algorithm, the user-friendly decision support tool is developed.

The organization of the rest of the paper as follows. In Sect. 2, the theoretical foundation of employed methods is defined briefly. In Sect. 3, the analyzed data is explained step by step as procedure flows from preprocessing to algorithm application and finally to decision support tool. At the end, the conclusion and further study opportunities are discussed in Sect. 4.

2 Methodology

The flowchart of the methodology used in this study is shown in Fig. 1. The data is obtained by using the survey conducted for Turkish SME's by the help of KOSGEB. The raw data obtained is preprocessed as it includes some empty cells and some logical controls are performed. The popular data classification methods Catboost and Support Vector Machines (SVM) are selected to be studied. Those methods include some parameters that must be input. Hence, parameter tuning is performed by analyzing a number of runs whose training and test sets are randomly selected. By using these optimized parameters, 10-fold cross validation is performed by using CatBoost, SVM and μ-SVM methods. Among them, the one with the highest performance is selected. Finally, a decision support tool is developed by coding in Python. As a result, either KOSGEB or the company can predict their Industry 4.0 level automatically.

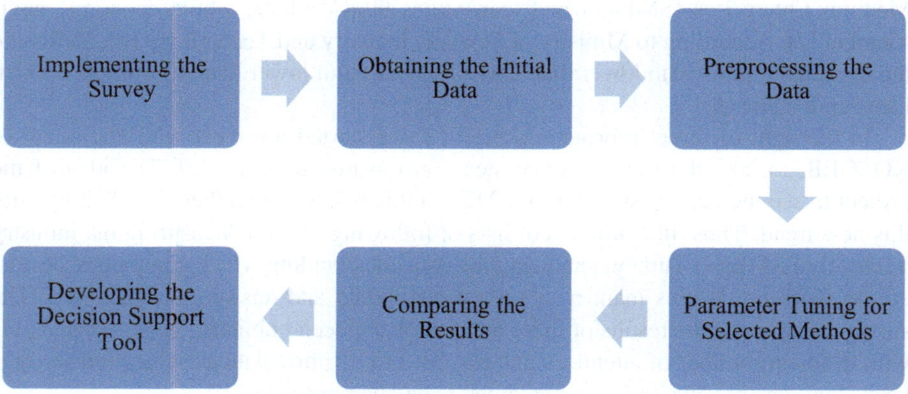

Fig. 1. Flowchart of the methodology.

2.1 Data Classification

Data classification is a supervised data mining algorithm which has wide range of application areas. The aim of data classification is to develop a prediction algorithm for an output characteristic by using some input characteristics. In order to achieve that goal, a training set which includes a number of instances with known input and output attributes are used. By working on these instances, a prediction tool will be developed, and it will be tested on a new instance to find out the value of output attribute [13]. For instance, by using previous breast cancer patient data with many different patient information, a new patient's breast cancer status can be predicted by the help of a data classification algorithm.

A number of different classification algorithms exist in the literature. The popular and widely used ones are selected for this study and explained in the following subsections.

1) *CatBoost:* CatBoost is a recently developed powerful data classification algorithm which based on gradient boosting. It can handle not only the numerical values but also the categorical values at the same time. While developing a number of decision trees new approach is proposed to select the tree structure and leaf nodes. By the help of that, overfitting is reduced, and performance is improved [14]. This algorithm includes same parameters to be set such as depth of the tree, learning rate, L2 leaf reg and number of iterations.
2) *Support Vector Machines (SVM):* Support Vector Machines (SVM) are also considered as popular classification algorithms. During the experiments, a hyperplane that maximized the distance to the closest instances of each class is constructed to separate the classes from each other. While a new instance is evaluated, its position with respect to the hyperplane is used to do the prediction. For the cases where the classes are not linearly separable, attributes are transformed into a higher dimension where they will become linearly separable. If the number of classes is more than two, then for each class a hyperplane is obtained by considering the remaining classes together as one class. Although SVMs have high accuracies for two class problems, their efficiency is affected in the multi-class data classification problems [15]. In SVMs, soft margin regularization parameter C, kernel function selection and gamma parameter should be determined and tuned before applying the algorithm.
3) *μ-Support Vector Machines (μ-SVM):* This algorithm is a special case of the SVM where μ, upper bound for margin errors and lower bound for fraction of support vectors, should be minimized [15].

2.2 Performance Evaluation

In this study, area under the Receiver Operation Characteristic (ROC) Curve is deployed. This metric is used to evaluate the performance of the classification algorithms over a threshold value which is 0.5 initially [16]. Authors in [17] shows that area under ROC curve is the probability of correct distinguish in binary classification cases. To understand how this metric works, a use of confusion matrix should be revisited.

Confusion matrix given in Table 1 is one of the most primitive metrics to evaluate the performance of a classification algorithm. This representative example is for a two-class classification problem. In the training set, actual class labels of the algorithms are known. By using the developed algorithm, a class prediction is performed. Then, actual and predicted classes are compared and confusion matrix is filled to measure the accuracy of the solution. In this table, if the predicted class and actual class are the same (TP & TN), then it is a correct prediction. On the other hand, if they are different (FP & FN), then is a misclassification.

Table 1. Confusion matrix example.

Confusion matrix		Predicted class	
		Class 1	Class 2
Actual class	Class 1	True Positive (TP)	False Negative (FN)
	Class 2	False Positive (FP)	True Negative (TN)

As it can be seen in Fig. 2, in ROC curve, the relation of True Positive Rate (TPR), Eq. 1, and False Positive Rate (FPR), Eq. 2, is plotted. The area under this curve is the indicator of the capability of classification algorithm, and it can take values in between 0 and 1. Higher the area means higher capability. Sensitivity (TPR) and Specificity (FPR) is inversely related. So, if we decrease the threshold value the we get more positive values which result higher sensitivity and lower specificity. And vice versa for the situation where we increase the threshold. This area under ROC curve is called score of classification algorithm from this point on.

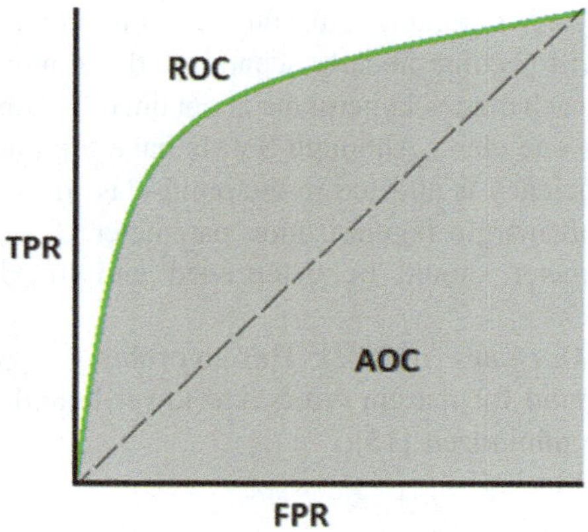

Fig. 2. Example of roc curve.

$$TPR(Sensitivity) = \frac{TP}{TP + FN} \tag{1}$$

$$FPR(Specificity) = \frac{TN}{FP + TN} \tag{2}$$

2.3 Estimation Methodologies

In data classification, in order to get rid of the bias resulted in using a training and test set, 10-fold cross validation is popular nowadays. In this estimation methodology, whole data set is randomly divided into 10 mutually exclusive subsets with equal number of instances. 10 distinct runs are performed by using the 9 subsets as training set and remaining one subset as test set. Hence, each time different training and test sets are used in order to minimize the bias [18]. The overall accuracy is calculated by taking the average of these 10 distinct runs.

3 Implementation and Results

In this section the dataset, applied procedures on dataset with Python's scikit-learn library and decision support tool which is coded by Tkinter library of Python is explained and results of procedures are discussed.

3.1 Data Description and Preprocessing

Data used in this study is gathered by the survey that is applied to SMEs in Turkey. In total 1042 SMEs attended this survey that consists of 28 question that inspects Industry 4.0 score of SMEs, recognition of Industry 4.0 concept in SMEs and usage of some certain technological instruments. In this study, we tried to predict Industry 4.0 score levels of SMEs with their usage status of these instruments.

The questioned instruments in the survey are: (i) 3D Printing, (ii) Sensor Systems, (iii) Artificial Intelligence, (iv) Cloud Technologies, (v) Internet of Things, (vi) Additive Manufacturing, (vii) Advanced Robot Technologies, (viii) Automation Systems, (ix) Modelling and Simulation, (x) Virtual Education, (xi) e-Performance Reports, (xii) Big Data, (xiii) Augmented Reality, (xiv) Cyber Security, (xv) Integrated Product Development, (xvi) Predictive Maintenance, (xvii) Productive Maintenance, (xviii) Smart Stocks, and (xix) Mobile Systems.

Each SME took part in the survey, state the usage status of instruments as: (i) no recognition of the instrument, (ii) not using the instrument with recognition, (iii) planning the use and (iv) using. Therefore, there is 19 categorical attributes which are the usage status of instruments and 1 continuous variable which is the Industry score of SMEs.

In the preprocessing phase, in order to switch the problem into a classification problem, the median of the scores is calculated as 2.57, and scores are converted to two categories as either High – if higher than median, or Low – if lower than median. Secondly, number of categories are decreased from four to three with merging the "No recognition of the instrument" and "Not using the instrument with recognition" into only "Not using". So that, we have three possible categoric values an SME can have for each of the 19 attributes.

While dealing with missing data, we first erased answers that have higher than three "NaN", and used "NaN" as 4th category since there were only 104 answers left with "NaN" in the total of 889 answers. But it's seen that it decreases the score. Hence, all the answers that includes "NaN" are erased from the dataset and no overfitting or underfitting is observed with this version. At the end, we have 794 answers with 19 categorical attributes that can have 3 possible values and 1 class attribute that can either be High or Low. There are 395 High and 399 Low score leveled SMEs in the dataset.

The dataset is divided into two as training set and test set with 80%–20%.

In the Fig. 3, a small sample of the data is shown.

ID	3D Printing	Sensor Systems	Artificial Intelligence	Cloud Technologies	Internet of Things	Additive Manufacturing	Advanced Robot Technologies	Automation Systems	Modelling and Simulation	Virtual Education	e-Performance Reports	Big Data	Augmented Reality	Cyber Security	Integrated Product Development	Predictive Maintenance	Productive Maintenance	Smart Stocks	Mobile Systems	Level
1	Not Using	Not Using	Not Using	Not Using	Not Using	Not Using	Not Using	Using	Not Using	Not Using	Not Using	Not Using	Not Using	Using	Using	Not Using	Using	Not Using	Using	High
2	Not Using	Not Using	Not Using	Not Using	NaN	Not Using	Not Using	Not Using	Not Using	NaN	Not Using	Not Using	Not Using	Not Using	Not Using	Not Using	Not Using	Not Using	Not Using	Low
.	.	.	.	.	.	.	.	.	.	.	.	.	.	.	.	.	.	.	.	.
.	.	.	.	.	.	.	.	.	.	.	.	.	.	.	.	.	.	.	.	.
.	.	.	.	.	.	.	.	.	.	.	.	.	.	.	.	.	.	.	.	.

Fig. 3. Sample of the data.

3.2 Hyperparameter Tuning

All three of classification algorithms are working based on specific parameters that should be optimized in order to gather the maximum utility. In this phase, we used grid search with cross validation. Each algorithm is run with 3-fold cross validation and fitted with training set. The outputs are recorded and the parameter set with the best score is selected for further analysis. Optimized parameters are shown in Table 2, and the change in score in grid search is shown with graphics in Fig. 4a, b and c.

Table 2. Parameters and optimized values of algorithms.

	Parameters	Optimized values (in order)
SVM	C, gamma and kernel	5, 0.01, sigmoid
μ-SVM	nu, gamma and kernel	0.8, 0.01, sigmoid
Catboost	depth, learning rate, L2 leaf reg and iterations	2, 0.009, 9, 500

As it can be seen from the table gamma which is a kernel coefficient and kernel parameters are the same with μ-SVM and SVM, the only difference is C and μ which are already different measures. C is regularization parameter that is inversely proportional to the strength of regularization while μ is upper bound for margin errors and lower bound for fraction of support vectors. For CatBoost tuned parameters are chosen to control overfitting. Depth specifies the depth of the tree, learning rate states the gradient step, L2 leaf reg is the coefficient of the L2 regularization term of the cost function and iterations as it can be derived by its name it controls the number of trees that can be built.

The parameter set shifts can be seen from the Figs. 4a, b and c. When shift occurs, drastic change is observed in the score. During grid search, CatBoost performs the least variation while μ-SVM variates the most. And gathered scores on training set can be ranked from top to bottom as CatBoost, μ-SVM and SVM. Even though, μ-SVM beats SVM in terms of train score, since it variates much SVM can be considered as more reliable algorithm.

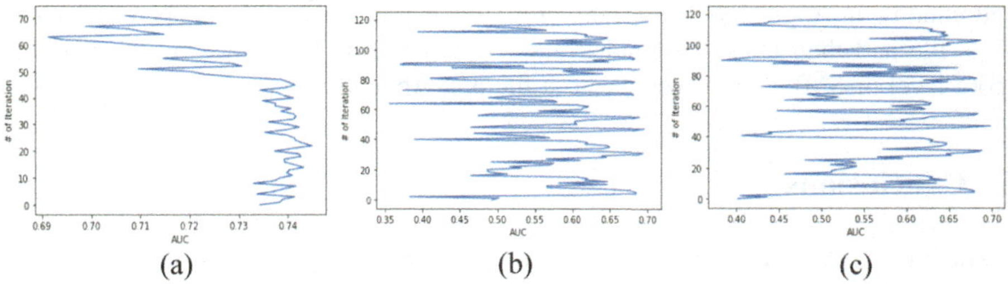

Fig. 4. a. Grid search performance of Catboost. b. Grid search performance of μ-SVM. c. Grid search performance of SVM

4 Results

After tuning the parameters, all three algorithms are fitted with training sets with the optimized parameters. Then prediction procedure is done by 10-fold cross validation with the test set and means of the scores are calculated. The scores, as it is stated earlier, are measured with ROC's area under curve. In Table 3, gathered scores and standard deviations are shown.

Table 3. Scores of classification algorithm.

	Mean score	Std deviation
SVM	68.6%	8.2%
μ-SVM	65.5%	8.3%
Catboost	78%	15.4%

As it can be seen from the Table 3, the observed variation in the score is changed drastically, in both SVM and especially in μ-SVM. Even though SVM and μ-SVM have lower variations, CatBoost outperforms both algorithms in terms of mean score. So, Catboost is chosen to be used in decision support tool.

4.1 Decision Support Tool

As the last step of the project, the decision support tool is developed based on CatBoost algorithm with Graphical User Interface (GUI). With the help of the tool, decision maker is able to input usage status of either one SME or a list of SMEs to system, and tool will provide the predicted Industry 4.0 score level of them. The opening window of the tool is shown in Appendix-I. In this main window, a user can choose if one or multiple SMEs will be entered to the system. If user chooses one SME, the window shown in Appendix-II appears and user can input the usage status choosing from the dropdown list and then clicks submit. After the submission of the statuses the score level of the SME is calculated and result is shown to user in the window that is shown in Appendix-III. If user wants to input multiple SMEs, "Enter list of SMEs" is clicked and window in Appendix-IV appears. From this window, user selects open and loads a CSV file and then clicks run and triggered the system. Then, system calculates the score level of SMEs and saves it into a file and Appendix-V appears.

5 Conclusions

Competition environment in industry is always shaped by the trends -revolutions, that are mainly driven by technological developments. So, in order to keep themselves in the race, companies should not only produce good product or service, they also should keep themselves evolve with respect to changes in trends. The velocity of the change is increased in a lot especially in the last decade. So, that while employing these changes, companies should also be quick in that. The importance of this employment cannot be ignored especially in developing countries such as Turkey.

In this project, in order to adapt Industry 4.0 in Turkey SMEs are targeted. The project first evaluates the current situation of SMEs and decides which sector, or company to help to make them catch up with the new era in industry. In this manner, the scope of the survey is increased and in total 1049 SME attended the survey. Their readiness is scored, and analyzed. Then, a decision support tool is built with proper classification algorithm behind.

While building the support tool, the first step was deciding the classification algorithm. Therefore, an algorithm to predict Industry 4.0 levels of companies via the usage status of technological instruments is formed. Firstly, in preprocessing phase, number of values that 19 categoric attributes can take is reduced from 4 to 3. Then, all instances that have "NaN" values are erased from the dataset, Score column is converted into High or Low score levels based on the median of 2.57 of scores. As a last step of preprocessing, the dataset is divided into 80–20% train and test sets.

Second step was the decide best algorithm with optimized parameters. Hence, the prior work's [19Data] best performing algorithm -SVM, is selected and two versions of SVM as SVM and μ-SVM is compared with CatBoost which is relatively new and one of the best suited algorithms for categoric variables on ROC metric. It is seen that Catboost outperforms other two SVM algorithms with 78% score against 65.5% and 68.6% in μ-SVM and SVM although their standard deviations are lower. In general, standard deviations are relatively high but since the number of total instances small, this is expected.

After choosing CatBoost as a classification algorithm, the decision support tool is coded as GUI in Python's Tkinter library to decide which companies to help, and windows are shown in Appendix.

For further research, number of attendees to survey might be increased and algorithm with lower standard deviation that performs relatively high score might be formed.

Acknowledgment. This research is made possible with the support of KOSGEB. The researchers express their gratitude to KOSGEB Management for their support on this study and for the ongoing research to cover all regions of Turkey.

Appendix-I

Appendix-II

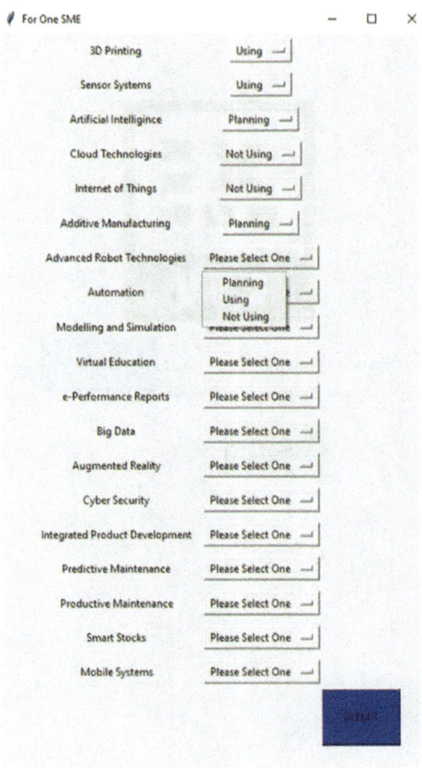

Appendix-III

Appendix-IV

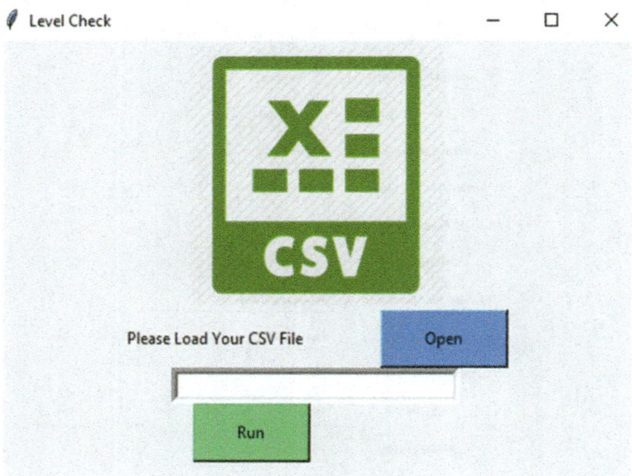

Appendix-V

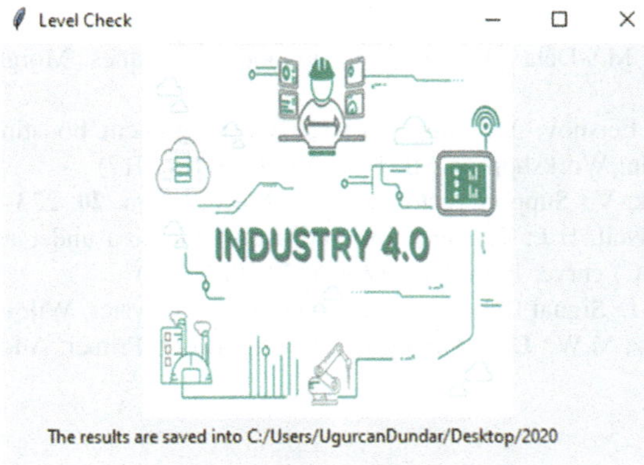

References

1. Erboz, G.: How to define Industry4.0: the main pillars of Industry 4.0. In: 7th International Conference on Management (ICoM 2017), Nitra, Slovakia, pp. 761–767 (2017)
2. Drath, R., Horch, A.: Industrie 4.0: hit or hype? IEEE Ind. Electron. Mag. **8**(2), 56–58 (2014)
3. Hermann, M., Pentek, T., Otto, B.: Design principles for Industrie 4.0. In: 49th Hawaii International Conference on System Sciences (2016)
4. Turkish Industry & Business Association (TÜSİAD): Industry 4.0 in Turkey as an Imperative for Global Competitiveness: An Emerging Market Perspective, TÜSİAD-T/2016-03/576
5. Mocan gelecek
6. United Nations, State of Commodity Dependence 2016, United Nations Conference on Trade and Development 2017
7. Turkish Statistical Institute (TÜİK): Small and Medium Sized Enterprises Statistics (2016). Press release, No:21540, 11/2017
8. Ministry of Science, Industry and Technology: Kosgeb, Kobi Stratejisi ve Eylem Planı (KSEP) 2015–2018
9. Gergin, Z., Üney-Yüksektepe, F., Gençyılmaz, M.G., Aktin, A.T., Gülen, K.G., İlhan, D.A., Dündar, U., Cebeci, Ö., Çavdarlı, A.İ.: Industry 4.0 scorecard of Turkish SMEs. In: Proceedings of the International Symposium for Production Research 2018 (ISPR 2018), Vienna, pp. 426–437 (2018)
10. Gergin, Z., Ilhan, D.A., Üney-Yüksektepe, F., Gençyilmaz, M.G., Dündar, U., Çavdarli, A. I.: Comparative analysis of the most industrialized cities in Turkey from the perspective of Industry 4.0. In: Proceedings of the International Symposium for Production Research 2019 (ISPR 2019), Vienna, pp. 263–277 (2019)
11. Emir, O., Gergin, Z., Üney-Yüksektepe, F., Gençyılmaz, M.G., Dündar, U., Çavdarlı, A.İ.: Comparative analysis of the most industrialized cities in Turkey from the perspective of Industry 4.0. In: Proceedings of the International Symposium for Production Research 2020 (ISPR 2020), Antalya. (2020). (accepted for publication)

12. Dündar, U., Gergin, Z., Üney-Yüksektepe, F., İlhan, D.A., Gençyılmaz, M.G., Çavdarlı, A. İ.: Industry 4.0 score prediction of Turkish SMEs via data classification. In: Proceedings of the International Symposium for Production Research 2019 (ISPR 2019), Vienna, pp. 25–35 (2019)
13. Han, J., Kamber, M.: Data Mining: Concepts and Techniques. Morgan Kaufmann, San Francisco (2001)
14. Dorogush, A.V., Eershov, V., Gulin, A.: CatBoost: gradient boosting with categorical features support. In: Workshop on ML Systems at NIPS (2017)
15. Cortes, C., Vapnik, V.: Support vector network. Mach. Learn. **20**, 273–297 (1995)
16. Hanley, J.A., McNeil, B.J.: The meaning and use of the area under a receiver operating characteristic (ROC) curve. Radiology **143**(1), 29–36 (1982)
17. Green, D., Swets, J.: Signal Detection Theory and Psychophysics. Wiley, New York (1966)
18. Roiger, R.J., Geatz, M.W.: Data Mining - A Tutorial Based Primer. Addison Wesley Press, Boston (2003)

Adaptation of CNC Machine Tools in Educational Center That Matches the Concept of Industry 4.0

Wojciech Zębala(✉), Grzegorz Struzikiewicz, and Emilia Franczyk

Production Engineering Institute, Cracow University of Technology, Kraków, Poland
{zebala, struzikiewicz}@mech.pk.edu.pl,
franczykemilia@gmail.com

Abstract. The paper presents an analysis of the possibilities of extension, adaptation and development of the existing machine park, used in educational and training centers, to the Industry 4.0 concept. The concept of the adaptation and extension of the CNC machine tools to the production nest in accordance with the Industry 4.0 concept was discussed. An analysis of the development possibilities of the educational and training centers in the areas of manufacturing technology, robotics, measuring techniques and IT was carried out. An algorithm, allowing the optimization of the functional extension process of numerically controlled machine tools was proposed. A case study was shown for the extension of the MiniMill2 milling center used in the Haas Technical Educational Center at the Cracow University of Technology.

Keywords: Industry 4.0 · Development · CNC machine tools

1 Introduction

Nowadays, almost every machine tool has the ability to collect, transfer and exchange data with humans and external devices available in a common technology platform. This is the result of continuous development and is contained in the Industry 4.0 idea, which is based on the concept of IoT (Internet of Things) [1]. Recent studies on the use of IoT systems in business show that the manufacturing industry accounts for about 32% of this market. Shares at around 20% refer to energy companies, the mobile industry and smart cities, while healthcare, suppliers and the services reach levels about ten percent. A rapid increase in the use of IT platforms supporting IoT systems by global production companies is expected [2]. According to some data, up to 60% of companies in the world will start using them during a year, and IoT platforms will generate 30% of their revenues. The increase in revenue will result from the ability to connect producers, suppliers, customers and users through the IT platform. Such systems are characterized by open architecture, simple access and high flexibility. IoT platforms can be extended with various supporting technologies that may be necessary for the implementation of the production process or various types of projects. In critical situations, conducting advanced analysis of obtained data they can be supported by

N. M. Durakbasa and M. G. Gençyılmaz (Eds.): ISPR 2020, LNME, pp. 277–288, 2021.
https://doi.org/10.1007/978-3-030-62784-3_23

artificial intelligence and machine learning. When monitoring or supervising production processes and transferring large amounts of data, digital platforms allow the usage of edge or cloud computing, etc. [3, 4]. Due to the huge functionality of the systems and the wide range of available technologies and IT tools, proper analysis should be carried out before the implementation of IoT software. The requirements for the platform should be taken into account, as well as the tool stock, the number of available technologies, and the costs of its implementation [5].

An important factor in the development of the Industry 4.0 concept, in addition to economic and technological analysis of its implementation, is the flexible integration of people, machines and processes. Constant development requires appropriate knowledge from current and future employees, engineers or users of IT platforms. In order to meet these demands, it is necessary to introduce changes in the education of future engineers and to train current employees of manufacturing companies. Educational units face the need to properly adapt their curricula and to introduce new or to modify existing teaching aids. For many educational units, a desirable solution would be to upgrade existing laboratories equipped with various machine parks, machine tools and measuring devices. Therefore, the authors attempted to analyze the possibilities of developing and adapting a CNC machine tool to educate students in accordance with the idea of Industry 4.0. The article presents a case study on the extension of the Mini Mill 2 CNC machine tool used at the Haas Technical Educational Center at the Cracow University of Technology.

2 Problem Analysis

In order to develop and adapt the educational stand to compliance with the Industry 4.0 concept, an analysis of the existing laboratory infrastructure was carried out and main areas of development and student education opportunities were selected. General algorithm for optimizing the process of adapting technological education center to the Industry 4.0 idea is shown in Fig. 1.

Accepted concept of training center development covered four main areas: manufacturing technology, robotics, measurement and IT. The main pillars of the technological education center are shown in Fig. 2. The above-mentioned areas were selected due to the type of classes currently taking place in the educational center for students of various fields and specialties available at the Faculty of Mechanical Engineering of the Cracow University of Technology. Current classes mainly focus on production methods and techniques and programming of numerically controlled machine tools for the faculties of Mechanical Engineering as well as Automation and Robotics.

A comparison between the current didactic structure and the structure enabling education according to the idea of Industry 4.0 is schematically shown in Fig. 3.

Analysis of the currently used methods of educating students showed thematic separateness of individual elements of the curriculum and the lack of interconnection between them (Fig. 3a). The subjects of education mainly included issues related to:

- construction and operation of CNC machine tools,

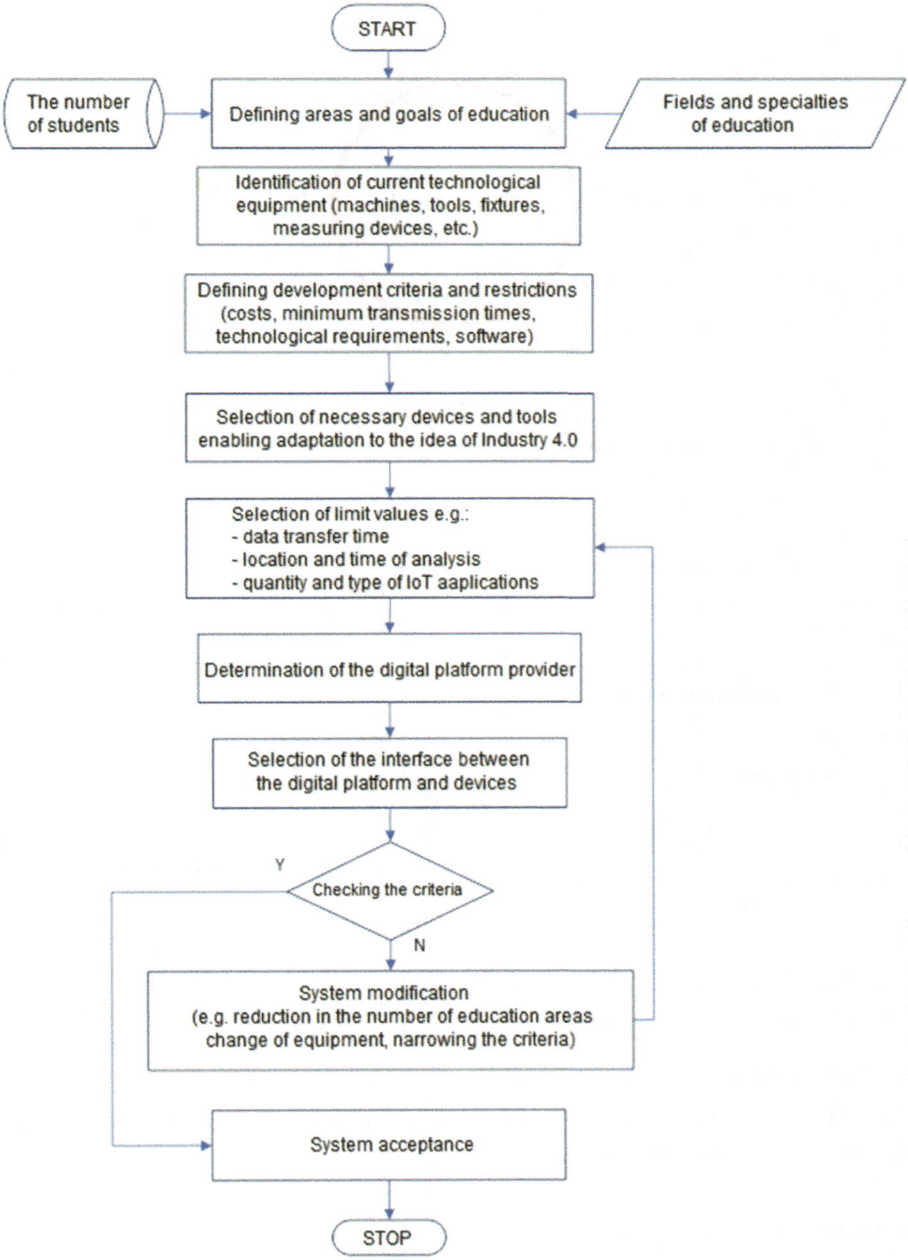

Fig. 1. General algorithm for the development of the technological machining center

- programming of machine tools by workshop and computer-aided methods CAD/CAM/CAE,
- methods of selection of tools and cutting parameters,

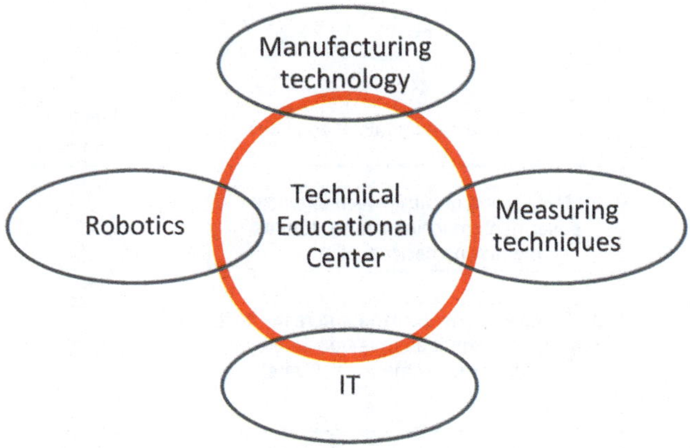

Fig. 2. Pillars of technological development of the HTEC education center

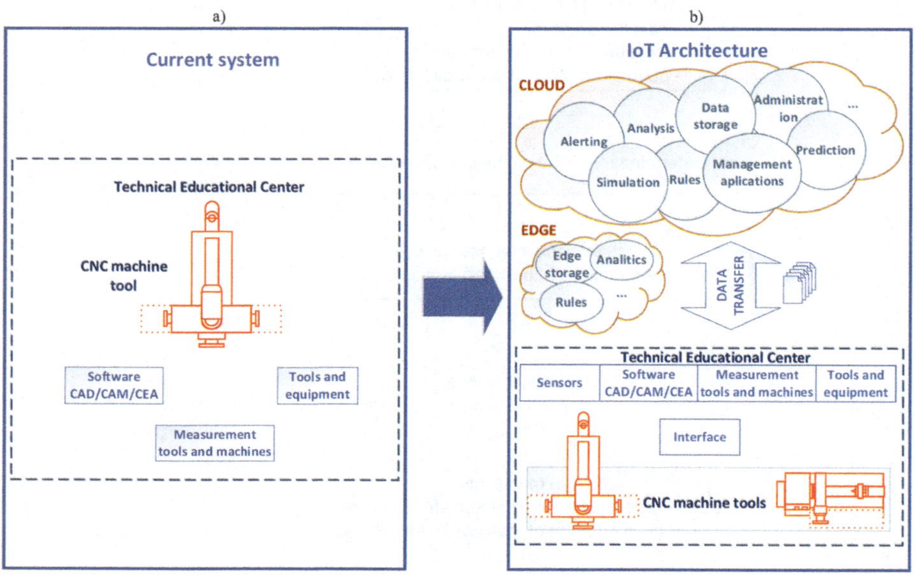

Fig. 3. Diagram of the structure of education in the technological training center a) current state, b) adapted to the concept of Industry 4.0

– measuring methods for cutting tools and workpieces.

Based on the analysis, a modified, flexible didactic structure was designed. The goal was to integrate and combine previously independent elements and to adapt to the teaching requirements in accordance with the concept of Industry 4.0. The structure (Fig. 3b) includes an autonomous and robotized machining center connected to an IT platform and having access to tools and applications of the IoT cloud system. The educational stand constructed in this way will allow students to assimilate, use and configure modern tools and solutions used in practices compliant with Industry 4.0 and i-Factory. The structure also includes the Edge Computing system working in the field of monitoring and supervising machines and machining processes as well as data transfer operations. Such designed structure of education requires the reconstruction and expansion of the educational stand and the extension and integration of didactic contents provided during classes. They mainly concern areas related to robotics, engineering and manufacturing, logistics, IT and measuring technologies. Each of these areas needs adaptation to the requirements of modern trends related to the idea of Industry 4.0. A general diagram showing the development areas and the system of their interconnection for the analyzed technical education center is presented in Fig. 4.

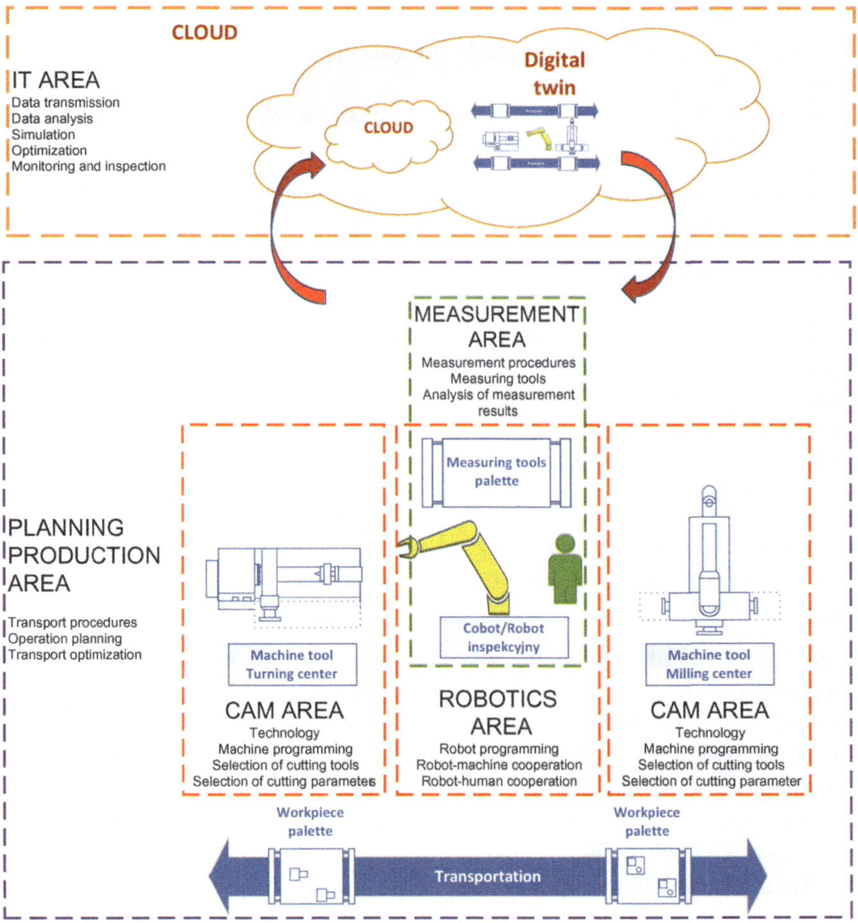

Fig. 4. Main areas of HTEC development

Separate teaching areas enable the transmission of particular content to students in the following scope.

a). Robotization of machining centers in a technological education center includes:

- cooperation between turning and milling centers and an industrial robot,
- programming industrial robots to implement technological processes,
- programming processes of loading and unloading workpieces from pallets and from/to the machine tool,
- programming activities related to removing the workpiece from the spindle and reattaching it in order to complete the machining,
- programming robots used to perform measurement activities using various tools and procedures (e.g. measurement of the workpiece or the tool using a touch probe; measurement of surface roughness, measurement of dimensions and surface quality on the basis of recognition the images received from vision systems etc.),
- cobot programming and their use in machine-human-cobot system,
- developing procedures and methods for data transmission from robot measuring sensors to the cloud,
- configuring, controlling and programming alerts as well as analyzing transmitted data using IoT tools.

b). Informatization of machining centers in technological education center and their adaptation to didactics in accordance with the concept of Industry 4.0 includes:

- developing methods of data transmission from machine tool sensors, robots, measuring tools and other technical devices to the cloud,
- performing selection and analysis of data in the cloud using appropriate applications,
- creating a so-called 'Digital twin' and simulation of the machining cell,
- monitoring, supervising and diagnosing devices using applications available in the cloud,
- creating mobile applications that support engineering work.

c). Engineering and manufacturing technology in production planning, machining processes and programming methods for numerically controlled machine tools includes:

- computer aided selection of cutting tools and cutting parameters,
- methods of manual and computer-aided programming of machining processes,
- methods and the use of computer simulations of manufacturing processes available in CAD/CAM, CAE programs and applications available in the cloud,
- operating and programming the machining cell,
- planning, optimizing and programming workpiece flow before and after production - transport and palletizing,
- production planning and optimization.

d). Measuring techniques include:

- measuring methods, tools and procedures,
- analyzing the results of measurements and reporting them using applications available in the cloud,
- technological cooperation of robots and measuring devices.

For the above-mentioned content, it is necessary to specify the objectives required to optimize the process of functional expansion of the technology center, in particular numerically controlled machine tools. Figure 5 schematically presents the main objectives for robotization of machining cells in a technological machining center.

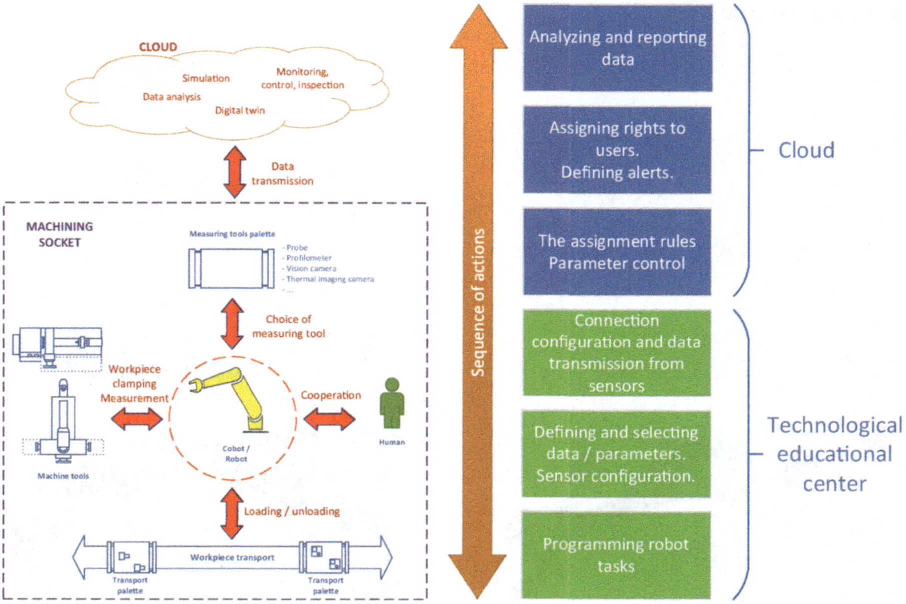

Fig. 5. The main objectives of robotization of the technological machining center

3 Case Study

A case study was carried out for didactic activities in a modified HAAS (HTEC) technology center based on a Mini Mill 2 milling machine. Analyzes were conducted for students of Automatic Control and Robotics. It has been assumed that the functionality of MindSphere cloud-based IoT operating system from Siemens will be used to implement teaching processes. Three main didactic areas were identified in order to develop the scenario of classes:

1. programming machine tools and industrial robots using CAD/CAM systems,
2. construction, applications and configuration of sensors used in measuring systems,
3. application and configuration of IoT applications used in cloud solutions.

Ad. 1. In the CAD/CAM area, basic tasks carried out by the students were:

- 2D and 3D modeling and preparation of technical documentation needed to produce chosen part (e.g. CAD module of the NX Siemens application),
- defining and computer-aided selection of tools and cutting parameters (e.g. Sandvik Coromant's Tool Guide application),

- planning technological operations and choosing a machining strategy with a use of selected CAD/CAM application (e.g. NX Siemens),
- simulation of the machining process and generation of the NC code for the machine tool.

2D and 3D drawings of the workpiece along with strategy and simulation of pocket machining are shown in Fig. 6.

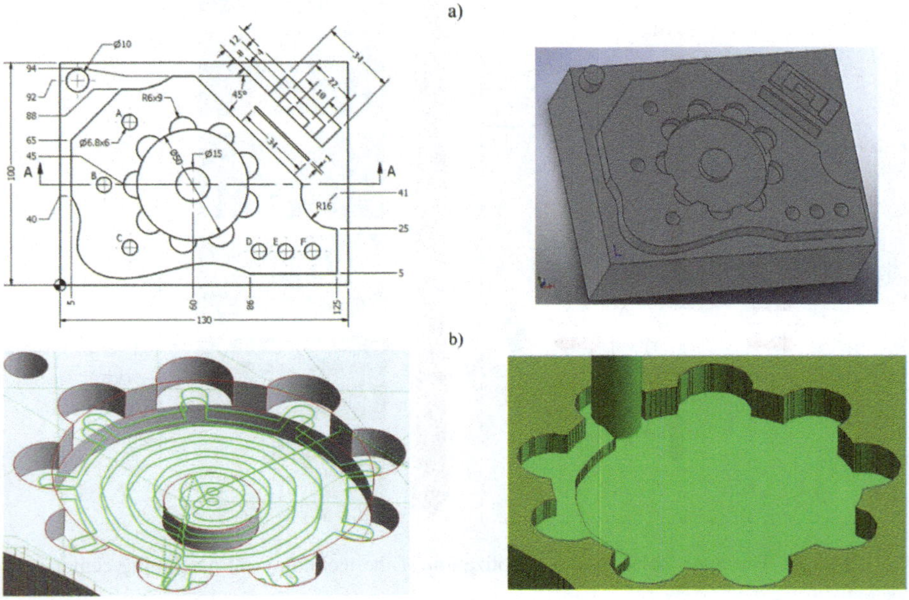

Fig. 6. Examples of workpiece: a) 2D and 3D model and b) simulation of machining of selected surfaces

Ad. 2. Basic tasks carried out by students in the area related to the machine tool included:

- configuration and assembly of the Kistler cutting force measuring path,
- configuration and assembly of a cutting temperature measuring system (Flir SC 620 infrared camera),
- configuration and assembly of a visual system for monitoring chip flow direction with use of a Phantom 5.2 high-speed camera,
- programming working movements of the Kawasaki FS06L industrial robot (e.g. workpiece loading/unloading operations, etc.). Robot specifications: 6 controlled axes, load capacity 6 kg, rotation approx. 320 horizontal reach 1550 mm, vertical reach approx. 1880 mm, accuracy in the repetition 0.1 mm.
- configuration and connection of measuring sensors with the cloud via an appropriate interface (e.g. Siemens MindSphere - The Cloud-based, Open IoT Operating System).

Figure 7 presents a view of a robotized educational stand with measuring devices, he main element of which is the Mini Mill 2 – numerically controlled milling machine from HAAS.

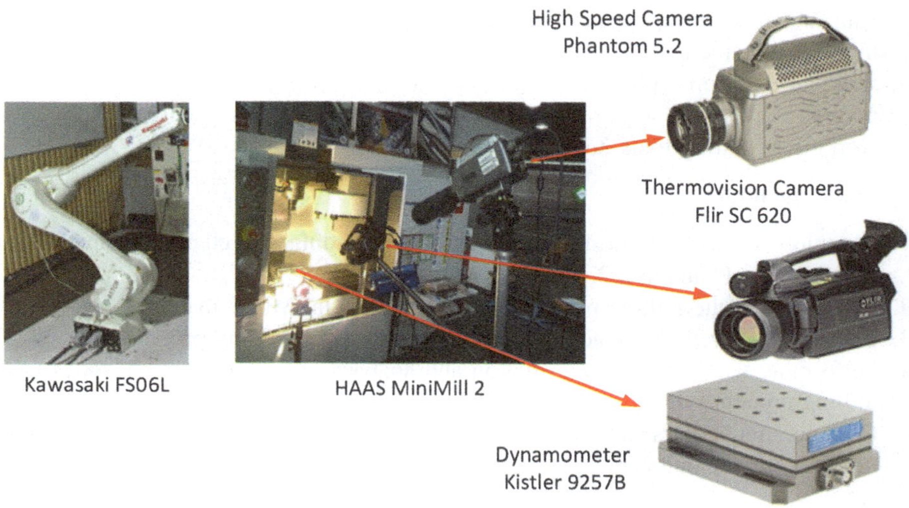

Fig. 7. A view of the educational stand based on the HAAS Mini Mill 2 machine tool, Kawasaki FS06L robot and measuring equipment available at the Cracow University of Technology.

Ad. 3. The following issues have been identified in the area related to cloud services:

- configuration of data transmission from measurement sensors to the cloud,
- assigning user rights and defining restrictions (e.g. transmission time, analysis time, maximum value of cutting forces, maximum temperature in the cutting zone, chip form),
- selection of IoT applications and tools available on the digital platform along with their configuration,
- selection and application of tools for analyzing measurement results (e.g. specialized applications, artificial intelligence) [6],
- defining system alerts and generating reports,
- optimization of the cloud system structure in terms of the amount of transmitted data, i.e. considering issues related to BigData and the choice between Cloud Computing and Egde Computing. The authors [7, 8] presented an analysis of the transmission time for data obtained from a dynamometer, a thermal imaging camera and a high-speed camera.

Exemplary class scenario at the Technological Educational Center:

Lesson objective: Determination of data transfer time in Cloud Computing and Edge Computing structures.

Requirements for students: Basic knowledge of computer science, production technology, programming of CNC machine tools and automation.

Duration of classes: 2 h

Teaching aids, implements and measuring equipment used:

- educational stand equipped with a Mini Mill 2 machine tool and a Phantom 5.2 high-speed camera,
- a digital platform with the structure of Cloud and Egde Computing and Internet access,
- workpiece, cutting tools, tooling.

Class agenda:

1. Assembly of the measuring track used for recording fast-changing images and based on the Phantom 5.2 camera.
2. Programming the milling process on the HAAS Mini Mill 2 machine tool.
3. Carrying out preparatory activities and starting the machining program.
4. Performing the process of registration and analysis of fast-changing images of the milling process.
5. Configuration and transmission of data from the process of recording fast-changing images in the Cloud and Egde Computing structure.

Calculation of the total data transmission time in the Cloud and Edge Computing structure on the basis of the following formulas:

- T_{totalC} – total time for *Cloud Computing* (1);
- T_{totalE} – total time for *Edge Computing* (2).

$$Ttotal_C = (Tr + Ta + Tt_i) \cdot f(\mathrm{s}) \tag{1}$$

$$Ttotal_E = (Tr + Ta) \cdot f + Tt_{iR}(\mathrm{s}) \tag{2}$$

Where:

- Tr – time of measurement/recording (for example: it was assumed that the measurement time was $Tr = 0.084275$ s).
- Ta – time of results analyzing of the measurement data, (for example $Ta = Treaction = 0.050525$ s).
- Tt_i – transmission time of measurement data for a specific medium (for example transmission speed range VTt was 100 MB/s):
- Tt_{iR} – transmission time of the report (for example assuming the size of the report file 1 MB).
- f – number of film files (sequences), necessary for supervising the whole machining process (in this case $f = 10$). The size of the file in MB obtained for the time of measurement Tr was $Wf = 200$ MB.

Performing an analysis of the results obtained and drawing conclusions. A comparative graph with examples of T_{total} calculations for Cloud and Edge Computing is shown in Fig. 8.

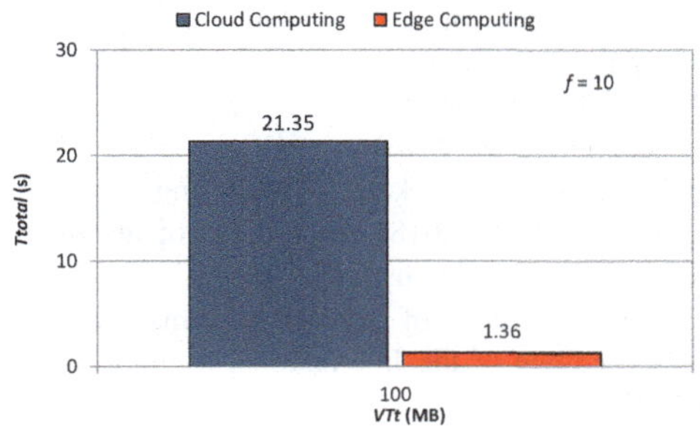

Fig. 8. Total time of data transmission from high-speed camera in Cloud and Edge Computing architecture.

4 Conclusions

The proposed structure of the technological machining center corresponds to those that operate on production lines in plants operating according to i-Factory. The purpose of learning in the laboratory corresponding to the Industry 4.0 idea is to combine elements of robotics, automation, mechatronics, applied computer science, measuring technology and machine design. Students can work with hardware and software used in the manufacturing industry. Proposed educational stand is equipped with infrastructure reflecting modern, robotic production centers. The scope of classes may include following topics: programming of machines and robots, monitoring and supervision of machining processes, communication protocols between individual sensors and the cloud, defining and managing alerts, logging, analyzing and reporting data results, etc.

The experience gained by students in the education cycle will be a valuable preparation for work in modern industrial production and related industries.

References

1. Zheng, P., Wang, H., Sang, Z., Zhong, R.Y., Liu, Y., Liu, C., Mubarok, K., Yu, S., Xu, X.: Smart manufacturing systems for Industry 4.0: conceptual framework, scenarios, and future perspectives. Front. Mech. Eng. **13**(2), 137–150 (2018)
2. Zhang, L., Luo, Y., Tao, F.: Cloud manufacturing: a new manufacturing paradigm. Enterprise Information Systems **8**(2), 167–187 (2014)
3. Khan, W.Z., Ahmed, E., Hakak, S., Yaqoob, I., Ahmed, A.: Edge computing: a survey. Future Gener. Comput. Syst. **97**, 219–235 (2019)
4. Sittón-Candanedo, I., Alonso, R.S., Corchado, J.M., Rodríguez-González, S., Casado-Vara, R.: A review of edge computing reference architectures and a new global edge proposal. Future Gener. Comput. Syst. **99**, 278–294 (2019)
5. Shi, W., Cao, J., Zhang, Q., Li, Y., Xu, L.: Edge computing: vision and challenge. Internet Things J. **3**(5), 637–647 (2016)

6. Struzikiewicz, G., Zębala, W., Rumian, K.: Application of Taguchi method to optimization of cutting force and temperature during turning of difficult to cut materials. In: PRECISION MACHINING VIII Book Series: Key Engineering Materials, Conference Proceedings, vol. 686, pp. 114–118 (2016)
7. Zębala, W., Struzikiewicz, G., Franczyk, E.: Monitoring of machining in the context of Industry 4.0 – case study. In: ISPR 2018, Proceedings of the International Symposium for Production Research, pp. 660–673, August 2018
8. Zębala, W., Struzikiewicz, G.: Aspects of monitoring of machining in the context of Industry 4.0. In: ISPR 2019, Proceedings of the International Symposium for Production Research, pp. 233–242, August 2019

Digital Maturity Assessment Model for Smart Agriculture

Aysu Melis Büyük[1], Gizem Ateş[1], Salim Burghli[2], Deniz Yılmaz[2], Gül Tekin Temur[2], and Çağlar Sivri[1(✉)]

[1] Management Engineering, Bahcesehir University, Istanbul, Turkey
aysumelis.buyuk@bahcesehir.edu.tr,
{gizem.ates1, caglar.sivri}@eng.bau.edu.tr
[2] Industrial Engineering, Bahcesehir University, Istanbul, Turkey
{salim.burghli, deniz.yilmaz3,
gul.temur}@eng.bau.edu.tr

Abstract. The main goal of this study to create a digital maturity assessment model for companies which want to prepare a future policy by evaluating their competencies in Smart Agriculture. Producers in the agricultural sector can learn their digital maturity levels by using the proposed evaluation model and analyze how successful they are in terms of Industry 4.0 requirements. By the help of the obtained strategic road maps, they can take advantage of cost and time by completing digital adaptation with more accurate decisions. For the digital maturity assessment model, Industry 4.0 requirements will be revealed both technically and administratively in the agricultural sector. For this purpose, the relevant literature is reviewed and a sustainable list of requirements for sector-environment-human dimensions are created. These requirements are grouped as main and sub-criteria. All criteria created for the digital maturity evaluation model for the agricultural sector are evaluated by using Best-Worst Method (BWM). After applying BWM, the weights of each criteria are figured out and they are used to evaluate the maturity level of each company by applying simple weighted average method. Also, a user friendly web site is designed to let companies easily measure their levels and have opportunity to identify their strong and weak points in the Smart Agriculture process.

Keywords: Smart agriculture · Digital transformation · Multi-Criteria Decision Making · Maturity model · Digital competency measurement

1 Introduction

Agriculture is evaluated as the basis of the life of human, since it is the main source of food products and other raw materials. It has an important role in the growth of the country's economy, especially in developing countries, it is one of the main sources of employment. It can be said that the country development in agriculture-based countries mostly depends on the development of agriculture. According to World Bank's statistics, the share of agriculture in GDP in 2018 is 5.8% for Turkey (while the world was determined to be 3.5%), as well as our country is the country with the highest production in terms of products, such as certain fruits, vegetables and nuts [1].

N. M. Durakbasa and M. G. Gençyılmaz (Eds.): ISPR 2020, LNME, pp. 289–301, 2021.
https://doi.org/10.1007/978-3-030-62784-3_24

With the impact of Industry 4.0, the agricultural sector has come a long way in terms of digitalization. According to a study by Start-up Genome [2], it has been emphasized that the number of entrepreneurship within the scope of deep learning technology has increased rapidly and one of the sectors where this situation has experienced the most is Agricultural Technologies. Besides that, agricultural sector is mostly based on manpower, in the cases of where advanced technology is not used, consequences such as loss of harvests and waste of farm animals become inevitable. While these factors revealed the need for improvement in agriculture, they showed that a more optimized system was needed and required human errors to be prevented. The fact that the needs are revealed in order to make agriculture "smart" in the digital world and the necessity to work on solutions by creating a platform facing the problems in the sector can be considered as the main starting point of this study.

In the current system, the efforts of agricultural companies and farm owners to manage the digitalization process with a limited vision using only high technology; ignoring human and environmental dimensions, not being able to handle it briefly in a short way, and the lack of an auxiliary tool to evaluate 'digitalization processes' objectively, was seen as a deficiency to be eliminated. The digital maturity assessment model for the agricultural sector within the scope of this study, and the deficiency in the existing system was seen as a source of motivation. In the literature review, the absence of a model designed for this purpose and purpose was also considered as a deficiency in the field of research and was regarded as an academic motivation source.

In this study, a digital maturity model that takes into account economic, environmental and social dimensions is created for firms which has tried to adapt digitalization. The model in question is a digital maturity assessment model for manufacturers that have operations in both agriculture and farming. With this model, answers to the following research questions can be created:

- What are the criteria for integrated agricultural institutions producing in both agriculture and livestock sectors to be considered "smart"?
- What are the significance levels of the Smart Agriculture criteria determined in the data analysis stage?
- What will be the outcomes that firms can obtain ideas about their maturity as a result of using this model?
- How will the availability of the proposed model be increased among the relevant manufacturers and stakeholders?
- What is the sectoral and economic benefit of the study, what are the areas of future development?

The rest of this study is organized as follows. Section 2 explains the smart agriculture and studies in Turkey. Section 3 mentions about the methodology used in the study, explains determination of criteria and weights and then shares an application made through a company. Section 4 concludes the paper and offer some future directions.

2 Smart Agriculture

In terms of definitions, Agriculture 4.0, also known as Smart Agriculture, represents the integrated internal and external network of agricultural activities, similar to Industry 4.0. This can be explained as all farm sectors and process can reach information in digital form. For example, some activities like communication with suppliers and customers etc. can be done in an electronic way and data transmission and data analysis are mostly became automated [3]. Smart Agriculture aims to prevent waste of resources with the use of advanced information and control systems, to increase the gross return of the product and to minimize environmental pollution caused by production. In Turkey, the private sector, unions, public and universities together with cooperatives involved this process and, by Turk Telekom group 'Animal Shelter Monitoring and Tracking Solutions' was initiated, and by the cooperation of Vodafone and Tabita, 'Smart Village' was established, by collaboration of Gebze Technical University and Hektaş, R&D studies on plant protection, plant nutrition, smart agriculture practices and seed breeding have been carried out [4].

3 Methodology

In this study the Best-Worst Method (BWM), which is one of the most up-to-date multi-criteria decision analysis methods, is used. The results of previous researches and the theses and the existing articles are used to determine the criteria that make agricultural practices "smart". Afterwards, questionnaires including a comparative evaluation of the criteria will be applied to obtain the necessary inputs for the use of BWM by meeting with experts. This method is based on binary comparison of criteria to determine the weight of the decision criteria [5]. The BWM method was developed by Jafar Rezaei (Delft University of Technology) in 2015 as an alternative to the analytical hierarchy process (AHP), one of the most popular multi-criteria decision-making methods.

The outstanding features of the proposed BWM method as compared to the existing MCDM methods are: (1) requires less comparison data; (2) leads to more consistent comparisons, meaning it produces more reliable results [6].

The stages of the BWM method can be explained as follows: [7]

Step 1: Decision criteria are determined.

Step 2: The best criterion (the most important criterion -B) and the worst criterion (the poorest criterion -W) are determined by experts.

Step 3: The best decision criterion (B) is determined to be superior to other determined criteria. A 9-point scale is used for this procedure. Numbers 1 to 9 are used on this scale, and 1: B is equally important for j; 9: B is much more important than j. As a result of the process, the best case (BO) vector is obtained according to other criteria. This vector is defined as follows;

$$A_B = (a_{B1}, a_{B2}, \ldots a_{Bn}),$$
$$(a_{Bj}, \text{ indicates the superiority of B over j, and } a_{BB} \text{ is equal to 1.})$$

Step 4: The superiority of all decision criteria over the worst decision (W) criteria is determined. A 9-point scale is used for this procedure. This scale uses numbers 1 to 9 and 1: W is equally important for j; 9: W is much more important than j. At the end of the process, the status (OW) vector of other criteria is obtained according to the worst criteria. This vector is defined as follows;

$$A_W = ((a_{1W}, a_{2W}, \ldots a_{nW}))^{\mathrm{T}}$$
$$(a_{jW}, \text{ indicates the superiority of j in W, and } a_{ww}, \text{ equals 1.})$$

Step 5: Criteria have optimum weights (w^*1, w^*2, …, w^*n).

There must be maximum absolute diameters to find optimum weights $\{|w_B - a_{Bj}w_j|, |w_j - a_{jW}\, w_W|\}$ minimized for all js.

$$\begin{gathered} min\, max_j\{|w_B - a_{Bj}\, w_j|, |w_j - a_{jW}\, w_W|\} \\ \text{s.t.} \\ \textstyle\sum_j w_j = 1 \\ w_j \geq 0, \text{ for all j's.} \end{gathered}$$

The equation expressed above is equal to the linear problem below;

$$\begin{gathered} min\, \xi^{\mathrm{L}} \\ \text{s.t.} \\ |w_B - a_{Bj}\, w_j| \leq \xi^{\mathrm{L}}, \text{ for all j's.} \\ |w_j - a_{jW}\, w_W| \leq \xi^{\mathrm{L}}, \text{ for all j's.} \\ \textstyle\sum_j w_j = 1 \\ w_j \geq 0, \text{ for all j's.} \end{gathered}$$

By solving this problem, the optimum objective function value ξ^{L*} and optimum weights (w^*1, w^*2,…, w^*n) are obtained. ξ^{L*}, indicates the consistency index. The fact that this value is close to zero indicates the consistency of decision makers in comparison.

3.1 Determination of Criteria Set

As mentioned earlier, a digital maturity assessment model prepared specifically for the agricultural sector has not been found in the literature. In light of the work done in other sectors and frequently cited in this area (e.g. [8–11]), by discussing with experts and developments in the field of Smart Agriculture and based on the industrial policies, the set of criteria shown in Fig. 1 was created. Detailed explanation of the criteria is given below.

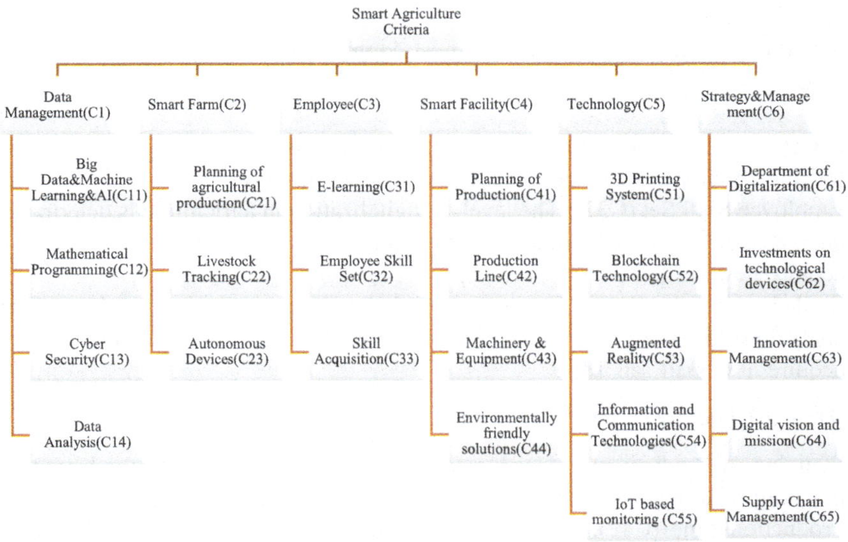

Fig. 1. Smart agriculture criteria set

1) *Data Management (C1)*:

- Big Data & Machine Learning & AI (C11): Big data technologies have come together to create new opportunities for machine learning (ML) to solve, measure and understand intensive data processes in agricultural operating environments [12].
- Mathematical Programming (C12): It was used to determine the most suitable sustainable practices for agriculture. Linear programming models are used to formulate the lowest cost feed rates for livestock, the most suitable crop models for individual farms, the most appropriate marketing plans to sell farm outputs and numerous other applications. [13].
- Cyber Security (C13): The use of new information systems and services will be increasingly used to maintain and improve operations, competitiveness and profitability in the agricultural sector. Therefore, it will be very essential to user Cyber Security systems to keep the system agains digital attacks.
- Data Analysis (C14): Crop production and yield, applications for land forecasting, mapping, weather forecasting and food safety need a huge amount of data and this makes data analysis necessary [14].

2) *Smart Farm (C2):*

- Planning of agricultural production (C21): It refers to the creation of the most effective process, taking into account the data of agricultural production (sowing, harvesting, fertilizing, etc.) together with the sensor technology.
- Livestock Tracking (C22): Digital devices such as wearable devices gain ground in animal tracking and provide real-time information about cows, ensuring health, breastfeeding or reproductive follow-up [15].

- Autonomous Devices (C23): It refers to the autonomous vehicles used in the farm, which can operate unmanned. The most common use is autonomous tractor.

3) *Employee (C3):*

- E-learning (C31): It refers to the digitalization of the employee development systematically and to be made continuous through platforms.
- Employee Skill Set (C32): The level of digitalization in agriculture is determined by analyzing the skills of employees in various fields and the company's efforts to acquire new skill sets [8].
- Skill Acquisitions (C33): The level of digitalization in agriculture is determined by analyzing the skills of employees in various fields and the company's efforts to acquire new skill sets [8].

4) *Smart Facility (C4):*

- Planning of Production (C41): It refers to the digitality of the method used in the planning of production at the facility (Enterprise Resource Planning, etc.).
- Production Line (C42): It refers to the automation of the line established in the production process and the need for minimum manpower.
- Machinery&Equipment (C43): It refers to the digitalization of the machines used and the presence of communication from the machine to the machine.
- Environmentally friendly Solutions (C44): It refers to the adaptation of the energy use at the facility to the digital world (Solar Energy Usage etc.).

5) *Technology (C5):*

- 3D Printing Systems (C51): It can reduce costs by providing substitutions for both special equipment and mass production products in the agricultural sector [16].
- Blockchain Technology (C52): In agriculture, Block Chain is applied to improve food safety and processing times.
- Augmented Reality (C53): Augmented reality can be used to assist farmers in insect identification and pest management [17].
- Information and Communication Technology (C54): By feeding information to research and development initiatives; by spreading information to farmers, connecting producers and consumers; It collects and shares timely and accurate information about weather, inputs, markets and prices [18].
- Iot based Monitoring (C55): Monitoring environmental parameters is the main factor to improve the yield of productive crops. Monitoring includes the monitoring of temperature and humidity in farm, in summary, by means of sensors [19].

6) *Strategy&Management (C6):*

- Department of Digitalization (C61): It expresses the existence of an autonomous department that will direct the digitalization process on a company basis.
- Investments on technological Devices (C62): It expresses how much of the revenue is returned to technological devices on a company basis.

- Innovation Management (C63): It refers to the company's support to its employees and external resources for innovation.
- Digital Vision and mission (C64): It refers to the emphasis on digitality in the company's future plans and strategies.
- Supply Chain Management (C65): It refers to the digitalization and minimum human support in the process starting with the purchasing process and extending to logistics.

3.2 Determination of Weights

The criteria chart created for the digital maturity assessment model for the agricultural sector was evaluated by 3 experts in the field at the international level. These experts have a farm and business in Houston, USA which engage with smart technologies and livestock tracking. Also, IoT based monitoring and other data analysis applications have been implemented by them during agriculture. Founder and the employees of Burghli Cattle Ranch are the experts who know how smart systems work in agriculture and which criteria are more important to make agriculture smart. The evaluation was made through a questionnaire in accordance with the method created and weights were calculated accordingly. Within the scope of the questionnaire, firstly the main criteria were handled and the experts were asked which were the best and the worst criteria that affect digital maturation in smart agriculture, and the other criteria were placed relative to each other considering the importance level. As a result of this process, the importance order of the main criteria was obtained. The same process was repeated for the sub-criteria under each main criterion title and the importance order of the sub-criteria was obtained for each main criterion. In line with this output, the weight of the criteria was calculated by applying the Best-Worst method. The evaluation was also evaluated in terms of consistency rate and it was checked whether this value was below 0.25 in accordance with the Best Worst method. In this regard, it was observed that all evaluations were satisfactory in terms of consistency.

As given in Table 1, the most important criteria affecting digital maturity in smart agriculture are Data Management (C1) and Technology (C5), while the Employee factor (C3) is the least effective factor in digital maturation. Within the Data Management heading, the most important sub-factor was observed as Data Analysis (C14). Accordingly, it can be seen that the most important benefit of digitalization is to minimize the loss. Technology (C5), on the other hand, has been the most important criterion because it enables smart agriculture applications to be carried out more controlled and faster, and it has ensured that the processes based on manpower are minimized. For this reason, Blockchain Technologies (C52) have been observed as important criteria for process control. In the digital maturity process, it has been stated that the smartness of the farm and the facility is effective in following the Data Management and technology criteria and in digitalization. Within the scope of the smart farm (C2), especially the importance of productive planning, that is, making accurate predictions using sensor technology and planning the process based on this, can be realized in smart agriculture. On the company basis, the digitalization of strategy and management is a criterion that affects digital maturity less than other factors. It can be seen from this data that companies prefer to invest in technology and data that will provide practical benefits in the digitalization process, and the development of the farm and facility has been prioritized in the follow-up.

Table 1. Final criteria weights

Main criteria	Local weights	Sub-criteria	Local weight of subcriteria	Global weight of subcriteria
C1	0.3125	C11	0.0636	0.0199
		C12	0.10909091	0.0341
		C13	0.19090909	0.0597
		C14	0.63636364	0.1989
C2	0.1875	C21	0.714286	0.1339
		C22	0.071429	0.0134
		C23	0.214286	0.0402
C3	0.03125	C31	0.3333	0.0104
		C32	0.3333	0.0104
		C33	0.3333	0.0104
C4	0.09375	C41	0.588235	0.0551
		C42	0.176471	0.0165
		C43	0.176471	0.0165
		C44	0.058824	0.0055
C5	0.3125	C51	0.089686	0.0280
		C52	0.523169	0.1635
		C53	0.209268	0.0654
		C54	0.125561	0.0392
		C55	0.052317	0.0163
C6	0.0625	C61	0.089686	0.0057
		C62	0.523169	0.0067
		C63	0.209268	0.0033
		C64	0.125561	0.0134
		C65	0.052317	0.0334

As a next step, simple weighted average method is used to determine the maturity levels between 0-5 (0=Beginner and 5=Leader). To provide a user friendly solution, a web site (https://akillitarim.wixsite.com/smagri) is designed to allow firms have a quiz and give answer to the questions related with the all sub criteria. After completing the quiz, the system gives reponders a score between 0-5 that is calculated with simple weighted average (multiplication of the obtained weights by answers of the responder). Firms' evaluations will be made according to the following scale (Table 2):

Table 2. Digital maturity levels

Levels	Beginner	Intermediate	Experienced	Expert	Leader
Scores	0–1	+1–2	+2–3	+3–4	+4–5

Beginner level companies refer to companies that are in the first step towards maturation in digitalization. This level shows the companies in the agricultural sector that have just met the concept of Smart Agriculture or want to develop in this field. Companies' suggestions are briefly summarized as follows:

- The correct understanding of the concept of Smart Agriculture.
- Recruitment of experts required for the implementation of the concept.
- Planning the beginning of the process of smart agriculture and creating awareness among employees.

Intermediate level companies refer to companies that are familiar with the Smart Agriculture concept and started to implement, compared to the beginner level. Mid-level companies have started to learn smart farming practices and understand the benefits of the concept. In the field of digitalization, innovations are given importance and the importance of technology and data usage is understood. Companies' suggestions are briefly summarized as follows:

- Improving Smart Agriculture concept perception.
- Integrating smart farming practices into the strategy process.
- Long-term planning and implementation of Smart Agriculture process.

Experienced level companies refer to experienced companies using Smart Agriculture processes. While the process is managed more professionally, current 'Smart Agriculture' technologies are followed. Usually, a team or department has been established to manage the process, and progress has started in the field of data. The digitalization of employees and strategies at this level has also been brought to the fore and steps have been taken. Companies' suggestions are briefly summarized as follows:

- Providing horizontal integration of the Smart Agriculture concept.
- Planning smart agriculture strategies and following the progress in digital maturity.
- Identification of non-digitalized points and improvement works.

Expert level companies refer to the companies that have been implementing the Smart Agriculture processes for a while and where other companies can take models in this process. These companies planned the digitalization process in detail and formed a team or department for this. IT infrastructures were established with digitalization in mind and data analysis gained importance. The farm and facility used have transformed digitally and smart solutions such as monitoring, and VR technologies have been adapted to the process. Companies' suggestions are briefly summarized as follows:

- Providing vertical integration of Smart Agriculture applications.
- Close monitoring of the practices of leading companies in the industry.
- Continuous training and studies for the adaptation of employees to digitalization.

Leader level companies refer to the best performers and the companies that successfully implement all activities. The vision of smart agriculture and the strategies leading to this vision were determined, and the company, farm, employees and management were shaped accordingly. Companies' suggestions are briefly summarized as follows:

- Aim to fully digitalize processes.
- Sustaining Smart Agriculture methods, searching for new resources and trial of new studies.

3.3 Application

To apply the model, a company dealing with agriculture and producing organic products was chosen. The company answered all questions through the website and saw its corresponding level and recommendations. According to Table 3, the company's score is between 2 and 3, which means that it is in the category of experienced level companies and some suggestions made through its level. This score is calculated by multiplying data that is entered by the firm and weights of criteria calculated using BWM. The sum of these multiplications gave us the firm's score, which equals 2.5. As a result of this application, the company can learn its strengths such as Data analysis usage, digitalization level on the planning of agricultural production and machine&equipment. Meanwhile, they can observe that they lack information about Big Data and its application, 3D printing systems, blockchain technologies, and ICT. They also clarified that they do not have any department or specialized group responsible for digitalization. Other factors can be classified as in progress. When the data entrance is examined, as a medium-scale firm, the company might be prioritized the elements that can be easily digitalized, such as planning of agricultural production. As an experienced level company, the company might be considered one of the most significant candidates to become a smart agriculture leader. They have many advantages in this process, and it is possible to say that they understood the concept and built construction for digital transformation. However, it is a fact that smart agriculture is a developing topic that does not have many examples or companies that are fully adapted smart applications and that is the reason of this company may be considered as one of the successful ones.

Table 3. Application

Question: As a company, how would you evaluate your…	Data entrance by firm (0–5)	Weights	Result
Level of usage Big Data & AI and ML	1	0.0199	0.0199
Level of usage Mathematical Programming	3	0.0341	0.1023
Level of usage Cyber Security	2	0.0597	0.1194
Level of usage Data Analysis	4	0.1989	0.7956
Level of digitalization on Planning of Agricultural Production	4	0.1339	0.5356
Level of usage Livestock Tracking	3	0.0134	0.0402
Level of usage Autonomous devices	2	0.0402	0.0804
Level of usage E-learning	3	0.0104	0.0312
Level of digitalization on Employee Skill Set	3	0.0104	0.0312
Level of digitalization on Skill Acquisition	1	0.0104	0.0104
Level of digitalization on Planning of Production	3	0.0551	0.1653
Level of digitalization on Production Line	3	0.0165	0.0495
Level of digitalization on Machine & Equipment	4	0.0165	0.066
Level of usage Environmentally friendly solutions	3	0.0055	0.0165
Level of usage 3D Printing Systems	1	0.0280	0.028
Level of usage Blockchain Technology	1	0.1635	0.1635
Level of usage Augmented Reality	1	0.0654	0.0654
Level of usage Information and Communication Technologies	1	0.0392	0.0392
Level of usage IoT based Monitoring	3	0.0163	0.0489
Level of digitalization in department based	1	0.0057	0.0057
Level of investment on technological devices	2	0.0067	0.0134
Level of innovation management	2	0.0033	0.0066
Level of usage digital vision and mission	2	0.0134	0.0268
Level of digitalization on Supply chain Management	2	0.0334	0.0668
Total			***2.5278***

4 Conclusions

This study provides a digital maturity assessment model using a novel MCDM method titled Best-Worst Method (BWM). In this assessment model, 5 different levels were determined, and 0-5 Likert scale was used. The process followed in this direction can be expressed as follows: A detailed literature review has been made in the field of smart agriculture and digital maturity requirements have been acquired and these findings have been defined as criteria. The established criteria were evaluated by experts and their importance levels were revealed by using Best-Worst Method. Using these findings, the website was designed, a quiz was created based on the criteria and weights within the site, where companies can evaluate themselves. In this study, the main aim is to make the infrastructure of the assessment model ready for use and to provide an interface in which it becomes easier to renew the findings by rerunning the model as the world-wide expertise on the subject increases. To achieve this, this model should be delivered to a large number of agricultural producers with the support of public or R&D centers, and a systematic approach should be followed in analyzing the data held. In this way, by keeping the pulse of the adaptation ability of the sector in digital transformation, sectoral tendency, superiority and weaknesses will be observed, the necessary moves will be planned more effectively.

For further studies, the uncertainties created by the subjective evaluations in the model can be addressed through MCDM models based on the fuzzy set theory and the results can be compared. At the same time, the effects of the changes that will occur in the evaluations of the experts can be observed by conducting sensitivity analyzes.

References

1. Republic of Turkey Ministry of Industry and Technology: Strategy of Industry and Technology. Republic of Turkey Ministry of Industry and Technology, Ankara (2019)
2. Startup Genome: Global Startup Ecosystem Report (2019)
3. CEMA: Digital Farming: what does it really mean? European Agricultural Machinery Industry Association, Brussels (2017)
4. Republic of Turkey Ministry of Agriculture and Forestry: Report of Current Status of Smart Agriculture in Turkey. Republic of Turkey Ministry of Agriculture and Forestry, Ankara (2019)
5. Rezaei, J.: Best-worst multi-criteria decision-making method: some properties and a linear model. Omega **64**, 126–130 (2016)
6. Rezaei, J.: Best-worst multi-criteria decision-making method. Omega **53**, 49–57 (2015)
7. Rezaei, J., Salimi, N.: Measuring efficiency of university-industry Ph.D. projects using best worst method. Scientometrics **109**, 1911–1938 (2016)
8. Lichtblau, K., Bertenrath, R., Stich, V., Blum, M., Bleider, M., Millack, A., Schmitt, K., Schmitz, E., Schröter, M.: IMPULS-Industry 4.0 Readiness. Impuls-Stiftung des VDMA, Aachen-Kölb (2015)
9. Tonelli, F., Demartini, M., Loleo, A., Testa, C.: A novel methodology for manufacturing firms value modeling and mapping to improve operational performance in the Industry 4.0. Procedia CIRP **54**, 122–127 (2016)
10. Schumacher, A., Erol, S., Sihn, W.: A maturity model for assessing Industry 4.0 readiness and maturity of manufacturing enterprises. Procedia CIRP **52**, 161–166 (2016)
11. Akdil, K., Ustundag, A., Cevikcan, E.: Maturity and readiness model for Industry 4.0 strategy. In: Industry 4.0: Managing The Digital Transformation, pp. 61–94 (2018)
12. Liakos, K., Busato, P., Moshou, D., Pearson, S., Bochtis, D.: Machine learning in agriculture: a review. Sensors **18**, 2674 (2018)
13. Kaiser, H.M., Messer, K.D.: Research application: crop farm model. In: Mathematical Programming for Agricultural, Environmental, and Resource Economics, pp. 158–164. Wiley (2012)

14. Kamilaris, A., Kartakoullis, A., Prenafeta-Boldú, F.X.: A review on the practice of big data analysis in agriculture. Comput. Electron. Agric. **143**, 23–37 (2017)
15. Deloitte: Smart Livestock Farming. Deloitte (2017)
16. Pearce, J.: Applications of open source 3-D printing on small. Org. Farming Librello **1**, 19–35 (2013)
17. Nigam, A., Kabra, P., Doke, P.: Augmented reality in agriculture. In: IEEE 7th International Conference on Wireless and Mobile Computing, Networking and Communications, Wuhan (2011)
18. FAO: Information and Communication Technology in Agriculture. Food and Agriculture Organization of the United Nations, Roma (2017)
19. Prathibha, S., Hongal, A., Jyothi, M.: IOT based monitoring system in smart agriculture. In: International Conference on Recent Advances in Electronics and Communication Technology, Bangalore (2017)

Do We Need Synchronization of the Human and Robotics to Make Industry 5.0 a Success Story?

Mary Doyle Kent[1(✉)] and Peter Kopacek[2]

[1] Department of Engineering Technology, Waterford Institute of Technology, Waterford, Ireland
mdkent@wit.ie
[2] Institute for Mechanics and Mechatronics, IHRT, TUWien, Vienna, Austria
kopacek@ihrt.tuwien.ac.at

Abstract. Industry 4.0 has brought about a revolution in how the manufacturing industry operates, from the product concept generation to the automated, high-tech factories that fabricate the products. Manufacturing systems are fully integrated, multi-robot and collaborative. The key elements of Industry 4.0 are information and communications technology (ICT), connectedness, sensor technology, robotics and innovative production systems. Smart manufacturing systems producing smart products. The result of this highly automated environment of Industry 4.0 is that the humans are at risk of no longer playing a valuable role. This paper asks the following questions: Can we ensure that humans have a place in the highly automated workplace of the future (Industry 5.0) by optimizing human capital? Can the traditional educational provider supply the skills required to educate this modern worker or do we require an innovative educational system?

Keywords: Technology · Robotics · Social impact · Industry 5.0 · Education and human centered systems

1 Introduction

The personalization of products and services is now a reality. Our cities have become more connected and smarter. All around us the division of work between machines and humans is changing, and is continually being redefined. Everyday examples of technology transforming how we live our lives are to be found routinely. For example, in the areas of buildings, manufacturing, transportation, healthcare, logistics and many other industries.

Before the hyper-connectivity of manufacturing, systems often operated in a standalone manner. Machines and people used paper, manual and semiautomated systems to run machines with little or no feedback and optimisation. The fact that Industry 4.0 machines are connected to the Internet of Things (IOT) means that all the data that is being generated by connected sensors can be captured, processed and used to optimise processes and factories. This data can be retained and used to benchmark

N. M. Durakbasa and M. G. Gençyılmaz (Eds.): ISPR 2020, LNME, pp. 302–311, 2021.
https://doi.org/10.1007/978-3-030-62784-3_25

best practice. One of the main goals of these developments is a "smart factory" being able to produce economically "Lot Size One".

This new smart way of manufacturing has become a revolution by the increasing levels of automation, cyber physical systems, digital twins and the intensive use of data analytics. It is driven by modern industrial and societal challenges and evolutions, and the integration of information and operational technology.

The modern manufacturing environment will require both robots and humans to collaborate and work together harmoniously, each bringing their positive attributes to the table. Anecdotal evidence suggests that industrialists are finding it more and more difficult to find the appropriate staff for this high-tech working environment. When staff are hired and trained a significant investment is made. It is then a high priority of the company to keep the staff employed for the longer term. Favourable working conditions and opportunities and career development are essential to keeping staff engaged and motivated.

Multi Robot Systems or Multi Agent Systems (MAS) consist of cooperative, industrial, mobile and humanoid intelligent robots which will work together as a team to solve common tasks. This team of robots will have to be cooperative and will require the replacement of the central control computer by a distributed, networked computer system [1].

Interesting opportunities are now opening in the field of automation between collaborative robots (cobots) and robocollaborators (coboters). This paper looks at how this relationship can be optimised and enhanced for both the company and the employee. It focuses on small to medium size businesses (SMEs) where the introduction of automation and robotics is traditionally more of a challenge. Theory based practical guidelines and a framework, which as aimed at enabling businesses to embrace human centered systems, is outlined.

2 Background

From research undertaken in 2019 Doyle-Kent et al. [2] questions arose in several areas in relation to the future of human contributions to Industry 5.0. Some of the questions are outlined here.

1. "*Education and Skills* •How can we ensure that humans have a place in the highly automated workplace of the future by optimizing human capital? •Can the traditional education provider supply these skills?
2. *Working Environment* •How can we ensure that the work of the employee becomes more interesting and fulfilling into the future? •Can the workspace of the employee be safe and comfortable with the new technologies surrounding them?
3. *Ensuring the Best Technologies are used without making Humans redundant* •How can we ensure humans are not made redundant in this new environment? •How can decisions be made to ensure the newest and most optimised technologies are used in manufacturing?"

The role of humans in industry, specifically in manufacturing, has been highlighted from the very first industrial revolution all the way through to the present day. From

Taylorism in the USA in the early 1900's [3] to Tavisstock's Social Technical Design in Liverpool in the 1950's [4] to Mike Cooley's Human Centered Systems in London in the 1980's [5] to Enid Mumford's Redesigning Human Systems in 2000's, [3] is, to name but a few examples. Cyber physical systems form an important pillar of Industry 4.0 today. When a smaller company is faced with the challenge of moving into the highly automated world of Industry 4.0 they face many additional challenges as compared to the larger, more resource rich companies. How do we justify the investment? How to we train our small workforce? How can we compete with the larger multinationals? How can we give our customers what they want whilst remaining flexible, sustainable and competitive? However, if the SMEs are to remain in business, they must leverage the advantages of automation in a world that is rapidly evolving. Could academics help SMEs in this challenge? Could recent technical advances in robotics level the odds?

2.1 Automation of Work

As previously stated, there are many concerns for SMEs when moving into the world of automation. How can we afford automation both in terms of the capital and human investment? Even when the companies have decided to take the leap into the robotics and automation arena how do they know what processes to automate? Questions around the social impact when introducing automation often arise. Will staff lose their jobs, will their work become less interesting and will incomes be reduced? Manufacturing is highly competitive and will become even more so into the future.

The ability of a company to move into the Industry 4.0 space depends on multiple factors, and because of the high level of complexity they may require a third party to help them understand their strengths and weaknesses. A smart factory is one in which systems are fully integrated and collaborative. The customer drives the demand and the system can adjust and react in real time to the order. The supply chain spans from the customer to the supply chain network, to the factory floor and all the way through to delivery to the customer.

Industry 4.0 operates is a very different way to the conventional automated pyramid system. One of the major challenges is to understand how to transition older traditional, less flexible, technologies to the newer flexible ones. At each level and stage in manufacturing, the focus changes. Table 1 summarises these changes of focus.

Lalanda et al. in 2017 discussed how Industry 4.0 builds on the connectivity of all systems. "Field devices and control systems are horizontally networked and vertically connected to supporting software". As a result of the integration, specific orders can be placed and managed individually. There is scope to look at alternative ways to achieve manufacturing objectives. This allows "the systematic oversight and improvement of production activities. Enhanced connectivity brings great benefits but also raises formidable challenges, calling for new architectures and techniques [6].

Greater digitisation will shift production to smart factories. They use IoT (use of sensors, actuators and data communication technology to build physical products) not just to automate production but communicate and share information, thus optimising the value chain.

Table 1. Comparing the traditional manufacturing to smart manufacturing

Traditional	Industry 4.0
From a component level the focus is on precision	From a component level the focus is on precision and comparison to a digital twin
From machine level, the focus is on quality and productivity	From a machine level, it's about a self-aware or self-compare and self-predict
From a production system level, the focus is on efficiency	From a production level, it's about a self-configuration and self-optimisation, eventually achieving an automated, optimised, self-regulating production facility

According to Hirsch-Kreinsen et al. in 2019 "Further investigation and application of existing and new technologies are fundamental to the ongoing development of Industry 4.0. It is apparent from the first initial steps into Industry 4.0 that it is no longer only developments in production, information and communication technologies that will form the vision of networked production. Instead, the future will be marked by transformation via flexible, modular production systems to autonomous, learning systems. These systems will self-program, self-organize, set and adapt new requirements for themselves and self-optimize" [7].

Following on from this is plausible that flexible, modular autonomous processes can be integrated into SMEs using low cost, highly flexible Collaborative robotics together with Robotcollaborators to form human centered manufacturing systems for Industry 5.0.

2.2 Collaborative Robots (Cobots)

A Collaborative robot is a subset of the robotics family that have been specifically designed to work alongside and support humans in the working environment. They are not designed to replace humans but to assistant as a helping, or third hand. Cobots were developed in 2012 to play an ever-increasing role in the multi-robot, highly automated manufacturing plants. They are low cost and easy to use, hence reducing the company's reliance on external or internal high cost robotics experts. Staff can be easily trained to program and operate them with a very fast turnaround time. Indeed, a leading manufacturer of cobots states that their product is so straightforward that no previous experience of robotics is required. The time needed to program and operate the robot from taking it 'out of the box' is typically half a working day.

Another advantage is cobots are so affordable that it is a worthwhile investment for any company, regardless of the size of the company. The total cost of the product and training is very low, and it has a typical return on investment of between 6–12 months, not only because of the reasonable cost of the robot, but also they can work without any safety devices and save space on the shop floor. The low total cost of ownership means that these products are within reach of all companies and educational training facilities. The disadvantages of Cobots are that currently they have a limited payload, and a limited speed because of the 7 axes.

Cobots are extremely versatile and can be used for Packaging and Palletizing, Machine Tending, Industrial Assembly, Pick and Place, Quality Inspection, Injection Molding, CNC Tending, Assembly, Polishing, Screw driving, Gluing, Dispensing and Welding. They can be easily changed over from one operation to another is a very short time. At the moment they are being used in the larger manufacturing companies but there is an opportunity to introduce this technology into SMEs.

2.3 Robocollaborators (Coboters)

A Robotcollobrator is a human who works in collaborative manner, with a Cobot. This new relationship in manufacturing plants is at the early stage of development, and, as a result, there is very little information in the current body of literature. As stated by Kopacek et al. in 2020 [8] the education of the Coboter "is currently more or less an only partially solved problem". This new type of operator is situated between a classical operator and one with some mechatronic knowledge. Furthermore, they should be trained annually because of the tremendous progress in this field". It is true to say that currently it is only the manufacturers of the Cobots that train the Coboters and this is a combination of free online training together with classroom-based training. This training is limited and there are opportunities to develop more highly technical training courses that can open wider possibilities of using this technology in small to medium enterprises (SMEs).

3 Systems and Theories

Change in organisations is difficult and can lead to social breakdown if not introduced in a cohesive manner. Many design problems are very complex. When bringing change to an organisation Mumford in 2006 [3] states that "the rights and needs of the employees must be given as high a priority as those of the non-human system". "The world of socio-technical design is democratic, humanistic and provides both freedom and knowledge to those who are part of it".

3.1 Socio-Technical Methodology

Mumford describes methodology called 'socio-technical' which was originally pioneered in the early 1950's by the London Travistock Institute. It meant that the technology (defined as machines and their associated work organization) should be equal but not superior to high quality and satisfying work environment for employees when new work systems and methodologies were being designed and implemented. Mumford states that "Socio-technical design also had an important democratic component: employees who used the new systems should be involved in determining the required quality of working-life improvements".

Skills that are required to develop successful systems are: **Capability** this is "the power, capacity and knowledge to achieve the desired objectives". **Competence** this is "the problem-solving ability of the individuals" or the "know-how". **Coordination** this is "the ability to work closely, democratically and creatively with other groups". These can be broken down into more detailed tasks.

Other requirements for complex socio technical designs are knowledge, teamwork, vigilance, entrepreneurialism and innovation. Problems need to be solved as they occur in real time. Consensus with the wider groups can lead to better problem solving. Knowledge of the nature and scope of the problem, together with the means to solving this problem are the essential elements required. "This knowledge will assist the development of an ethical position, a set of values, and a series of practical designs steps seen as achieving the desired result".

There are four stages in the problem-solving process:

1. **Seeing the total picture** - seeing the complex systems design as a unified holistic process including technical, economic, organisational and social issues at every stage of the design process.
2. **Developing strategies** - they need to be developed at each stage of the design process including technical, economic, organisational and social issues. In volatile situations these strategies may need to be continually adjusted and reviewed.
3. **Taking action** - A mission statement can start the process. A risk analysis is required.
4. Use a **design methodology** to successfully implement the changes. This facilitates the team to move from problem analysis to the successful implementation of the solution.

Company will need to consider its business strategies when looking at automation. In some cases, the automation of certain tasks and processes will give a high return on investment through improved performance, but this is not always the case. To ensure that humans have a value-added role in Industry 5.0 it is necessary to look at current job specifications and then break the job down into tasks that could be automated and tasks that cannot. The type of automation required is also part of this process.

3.2 A Design Method: Reinventing Jobs

Jesuthasan and Boudreau in 2018 [9] addressed the problem of automating processes in a book called "Reinventing Jobs". They describe how jobs can be deconstructed into work tasks and then the compatibility of these tasks for automation can be assessed. There are three elements that determine the compatibility of the task with its automation. The simple classification system is:

- Repetitive Vs Variable
- Independent Vs Interactive
- Physical Vs Mental

A task that is repetitive, independent and physical is the most readily automated. If the task is variable, interactive and/or mental it becomes more challenging to convert from manual to automated. Each job can be deconstructed into tasks and there are on-line programmes and databases available to help with is process. The Occupational Information Network (O*NET) database from the USA [10] is a free database which can be used for this. The O*NET database contains a rich set of variables that describe work and worker characteristics, including skill requirements. It displays a list of tasks associated with the selected O*NET-SOC occupation. Table 2 on the following page shows an example of a packaging and filling machine operator job being reinvented [10].

Table 2. Example of a packaging and filling machine operator job being reinvented [10]

Job & associated tasks	Characteristics of the work element			Return on improved performance	Role of automation	Type of automation
Example: 51-2092.00 team assemblers:	Repetitive vs variable	Independent vs interactive	Physical vs mental	Negative constant incremental exponential	Replaces augments creates	Robotic process Cognitive Social Or Collaborative
Perform quality checks on products and parts	Variable	Interactive	Physical	Incremental	Augment	Cobot
Package finished products and prepare them for shipment	Repetitive	Independent	Physical	Constant	Replaces	Cobot
Rotate through all the tasks required in a particular production process	Repetitive	Independent	Physical	Constant	Augment	Cobot
Review work orders and blueprints to ensure work is performed according to specifications	Variable	Interactive	Mental	Constant	Replaces	Cognitive

When a job has been deconstructed into tasks the organisation can clearly identify how optimise the application of automation and understand how to transform the activities. By using this procedure, it is noted that automation can shift elements of work into other roles and augment, eliminate and create new activities. Automation and reinventing jobs are critical to connecting work to the strategic goals of an organisation. "What started with simple remote-controlled machines to improve operating control and reduce variance has evolved to encompass integrating work with sensors, automated analytics, AI enabled machines that adapt to changing conditions. The work must similarly evolve and be reinvented" [9].

Finally, the tasks that are difficult to automate will be carried out by humans. This ensures that these new reinvented human roles are no longer the monotonous, dangerous or heavy-duty tasks that are more suited to automation. When Cobots are introduced into the workplace the Coboter will plays a key role in setting up and working side by side with the Cobots. They are involved in redistributing tasks and redefining their roles. Thus robotics, automation and humans have a valuable role in the manufacturing plant which can be adjusted over time and optimised.

3.3 Synchronising Cobots and Coboters

To define the new higher value roles of the Coboters a methodology such as the ETHICS can be used. [4] This facilitates groups unused to being involved in participative system design by giving simple design tools to assist logical thinking. This 11 step by step approach is used with the design group which are the direct users of the system. The steps are:

Step1. Why change?
Step 2. Definition of system boundaries
Step 3. Description of the existing system
Step 4. Definition of key objectives
Step 5. Efficiency needs analysis of variances
Step 6. Analysis of job satisfaction needs
Step 7. Future analysis
Step 8. Setting the system objectives
Step 9. Organisational, educational and technical design
Step 10. Implementation
Step 11. Evaluation

4 An Industrial Proposal

To test this proposal, it will be necessary to do a trial run in SME settings. It is important step to and will require collaboration with both robotics companies and manufacturing plants. The important elements of the strategy are Organisation (management), Technology (Cobots), People (Coboter) and the Tasks (job specification). Each type of industry and individual manufacturing company will have its own hierarchy of needs and priorities and the methodology will be bespoke for each application.

- *The first step will be to discuss the socio-techno methodology with management and put together a team with the core skills of capability, competence and co-ordination.*
- *The next step is to put together a plan, looking at the total picture, developing strategies, taking action and using a design methodology.*

- *The design methodology is the deconstructing of jobs into their work elements, looking at the return of improved performance, the role of automation for each task and then choosing the best type of automation for each task.*
- *Finally, the focus will be on redesigning and upskilling the roboters. These workers will have higher value and higher skilled jobs that are more comfortable safer in the modern industry 5.0 factory.*

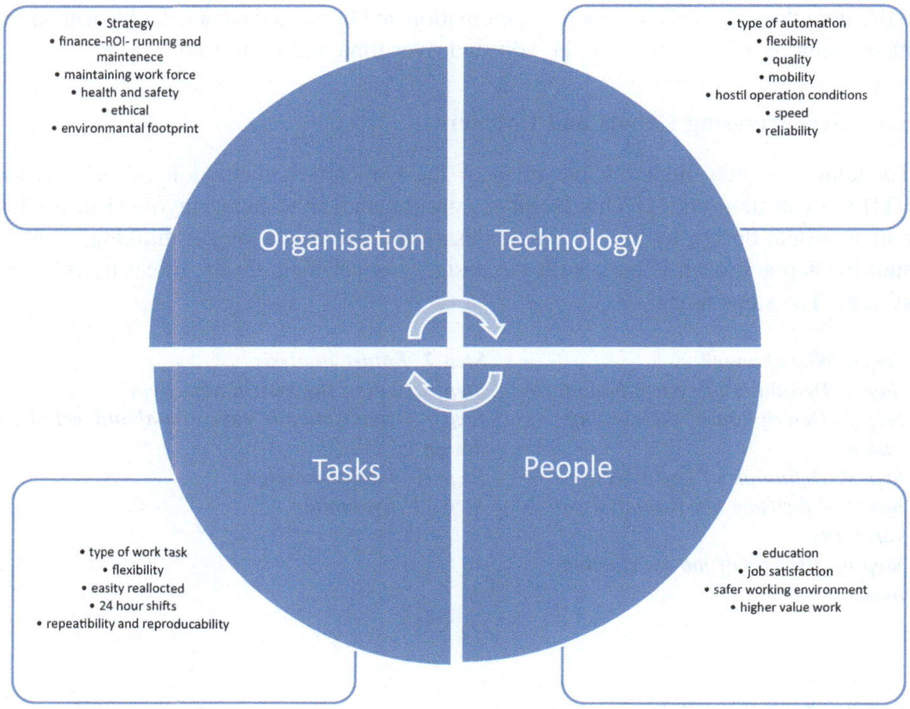

Fig. 1. The key stakeholder for industry 5.0

Figure 1 above shows the key stakeholder is when design an optimised Industry 5.0 framework for SMEs. Each SME will have a bespoke framework based on its specific requirement.

5 Conclusions

This paper looks at how the relationship between humans and robotics can be optimised and enhanced in an Industry 5.0 setting. It focuses on SMEs where the introduction of automation and robotics is traditionally more of a challenge for various reasons.

By systemically analysing the work tasks it is possible to find the optimum compromise between humans and robotics, building on the positives of both. In addition, by focusing on and enhancing the work and skills of the operators, this will ensure a long-term engagement of staff which are a critical resource in the modern factory.

Theory based practical guidelines and a framework have been proposed which is aimed at enabling businesses to embrace human centered systems. The next step is to apply the proposals to practical applications in SMEs and to assess the outcomes and to think about Production 6.0.

References

1. Kopacek, P.: Development trends in cost oriented production automation. IFAC-PapersOnLine **51**(30), 39–43 (2018). https://doi.org/10.1016/j.ifacol.2018.11.242
2. Doyle-Kent, M., Kopacek, P.: Industry 5.0: is the manufacturing industry on the cusp of a new revolution? In: Proceedings of the International Symposium for Production Research 2019, pp. 432–441. Springer, Cham, August 2019. https://doi.org/10.1007/978-3-030-31343-2_38
3. Taylor, F.W.: Scientific Management. Routledge (2004). ISBN 0-415-27982-8
4. Mumford, E.: The story of socio-technical design: Reflections on its successes, failures and potential. Inf. Syst. J. **16**(4), 317–342 (2006). https://doi.org/10.1111/j.1365-2575.2006.00221
5. Cooley, M.: Architect or bee? p. 66. Langley Technical Services, Slough (1980). ISBN 0-7012-0769-8
6. Lalanda, P., Morand, D., Chollet, S.: Autonomic mediation middleware for smart manufacturing. IEEE Internet Comput. **21**(1), 32–39 (2017). https://doi.org/10.1109/MIC.2017.18
7. Hirsch-Kreinsen, H., Kubach, U., Stark, R., von Wichert, G., Hornung, S., Hubrecht, L., Sedlmeir, J., Steglich, S.: Key themes of Industrie 4.0: research and development needs for successful implementation of Industrie 4.0. platform Industry 4.0 research council (2019). https://en.acatech.de/publication/key-themes-of-industrie-4-0/
8. Kopacek, P., Doyle-Kent, M.: "Social and Ethical Aspects of Automation" awaiting publication. In: Proceedings of the International Symposium for Production Research 2020. Springer (2020)
9. Jesuthasan, R., Boudreau, J.: Reinventing Jobs: A 4-Step Approach for Applying Automation to Work. Harvard Business Press (2018). ISBN: 9781633694071
10. Ontcenter homepage (2020). https://www.onetcenter.org/database.html

Evaluation of the Challenges of Companies in Industry 4.0 Transformation by GRA Method

Hilmi Yüksel(✉), Gülper Basmacı, and Serkan Genç

Dokuz Eylül University, İzmir, Turkey
{hilmi.yuksel,gulper.basmaci,serkan.genc}@deu.edu.tr

Abstract. In recent years, Industry 4.0 is met with increasing interest in the academic and sectoral fields. Despite this, issues are encountered in this process as a result of the challenges of companies in the transformation to industry 4.0. When the literature is analyzed, the challenges faced by the companies during the Industry 4.0 transformation process are mentioned, but it is not revealed which ones are more important in these studies. Considering all these, this study aims to contribute by revealing the importance order of the challenges encountered in the process of transformation to Industry 4.0 both in sectoral and academic terms. In the study, Grey Relational Analysis (GRA), which is a multi-criteria decision making (MCDM) method, was preferred. It is known that Gray Relational Analysis gives good results even with a small number of data. In the study, 5 expert opinions (EO) were consulted. It is thought that the order of importance obtained in our study will guide policy makers and sectoral practitioners.

Keywords: Industry 4.0 · Challenges · Multi criteria decision making · MCDM · GRA

1 Introduction

The new paradigm of the production sector is explained by the concept of Industry 4.0 [1]. Industry 4.0 refers the Fourth Industrial Revolution initiated by the integration of the Internet of Things (IoT) and the Internet of Services (IoS) in the manufacturing process [2]. Industry 4.0 is defined as "a paradigm that includes a new approach to production and changes in traditional, centralized control structures in favor of decentralized structures" according to Prause and Weigand [3]. The main purpose of Industry 4.0 is to create changes in production systems and provide "adaptability, resource efficiency as well as the improved integration of supply and demand processes between the factories" [4]. Industry 4.0 may present many opportunities for increasing productivity, efficiency and flexibility, and for decreasing costs. However, companies are facing unique challenges in the era of the Fourth Industrial Revolution. Adopting Industry 4.0 is not just a matter of new technologies and/or tools and/or production methods, but implies changes in all management aspects, and involves all actors of the ecosystem in which the company operates [5]. Businesses need to develop new

N. M. Durakbasa and M. G. Gençyılmaz (Eds.): ISPR 2020, LNME, pp. 312–323, 2021.
https://doi.org/10.1007/978-3-030-62784-3_26

strategies and define new business models and processes that will enable them to take advantage of the opportunities brought by digitalization in order not to fall behind the age. Businesses that fail to catch these trends on time will face premature demise in Industry 4.0 era. From the perspective of Turkey, identifying potential challenges faced by companies and knowing the importance order of them will result in sustainable competitive advantages. Otherwise, they are likely to encounter the mentioned premature demise. In addition, it is foreseen that increasing the awareness of this information throughout the country will facilitate and accelerate Industry 4.0 transformation of companies. Therefore; the aim of this paper is evaluation of challenges encountered in Industry 4.0 transformation of companies in context of Turkey. The challenges, which were used in the implementation of this empirical study, were obtained by literature search. The most frequently mentioned challenges of Industry 4.0 transformation in companies are "cultural barriers, employment disruptions, high implementation cost, lack of clear comprehension about IoT benefits, lack of clearly defined economic benefit, lack of Internet coverage and IT facilities, lack of knowledge management systems, lack of standards and reference architecture, lack of support by top management, lack of understanding of the strategic importance of Industry 4.0, lack of understanding the interplay between technology and human, legal and contractual uncertainty, more focus on operation at the expense of developing the company, need for enhanced skills, need for new business models, need to find suitable research partner, organizational and process changes, preferred autonomy, regulatory compliance issues, requires continued education of employees, seamless integration and compatibility issues, security and privacy issues, technology immaturity, too few financial resources, too few human resources, uncertainty about the reliability of the systems" [1, 6–12]. Kamble et al. (2018) defined 12 barriers to adaptation of Industry 4.0, based on a literature review and interviews with experts in industry and university [6]. Luthra and Mangla (2018), using an extensive literature review, identified 18 key challenges to Industry 4.0 initiatives for developing supply chain sustainability, classified into four main dimensions, which in turn were ranked using AHP [10].The method of our study was chosen as GRA, which is a multi-criteria decision making method. Multi-criteria decision making (MCDM) methods allow optimal decisions to be taken by simultaneously considering both objective and subjective decision criteria's that are in conflict with each other with small amount of data [13, 14]. In this study, the opinions of 5 experts were taken. As a result, it is ensured that challenges are placed in order of importance to guide policy makers and sectoral practitioners.

2 Challenges of Companies in Industry 4.0 Transformation

The challenges that companies have encountered in the industry 4.0 transformation, obtained through the literature review, are included in Table 1 with their explanations. The literature review was carried out using databases such as Google Scholar, Scopus, Elsevier, Science Direct and Web of Science. Challenges of a similar nature are combined under one name, resulting in twenty-seven challenges.

Table 1. Challenges of companies in industry 4.0 transformation

Cultural barriers (C1)	Organizational culture needs to be open to innovation and makes employees part of this project	[9]
Employment disruptions (C2)	The disruptions in employment as a result of developing technologies and automation. Automation of jobs in the manufacturing sector leads to job losses	[6]
High implementation cost (C3)	Refers to the capital expenditures that companies must bear to develop the industry 4.0 infrastructure in their organizations	[1, 6, 8–10]
Lack of clear comprehension about IoT benefits (C4)	Another technological innovation that is affected by the changes in today's technologies is IoT technology. The fact that this technology cannot be fully understood within the organizational framework with its conceptual dimension also restricts the use and efficiency of this technology	[6]
Lack of clearly defined economic benefit (C5)	Uncertainty prevails over what benefits the financial aspects of Industry 4.0 will create. In addition to this uncertainty, the emphasis on their economic negative effects is more common and this creates bias	[8, 10]
Lack of Internet coverage and IT facilities (C6)	Internet coverage and IT facilities that can support technological innovations, especially IoT, are needed to implement Industry 4.0 applications in a healthy way. However, many facilities do not have such an investment	[6, 8, 11]
Lack of knowledge management systems (C7)	One of the most important concepts brought by Industry 4.0 is big data. In order to process such a big data type, knowledge management systems with adequate competence are required. The current information management systems are not capable to handle real-time data flow	[6, 8]
Lack of standards and reference architecture (C8)	The concept of Industry 4.0 has been spoken as an industrial revolution only since 2011. As a result, it does not have established standards and reference architecture. Thus, this makes it difficult to apply and spread	[6–9, 11, 12]

(*continued*)

Table 1. (*continued*)

Lack of support by top management (C9)	Support of top management is an important variable in the implementation and success of industrial innovations. Top management may be biased against this new paradigm called Industry 4.0. This results in, the top management not supporting the Industry 4.0 transformation	[1, 8, 10, 11]
Lack of understanding of the strategic importance of Industry 4.0 (C10)	The awareness of how strategically important Industry 4.0 applications are for the company is necessary for the success of the applications. This lack of awareness is a barrier encountered in the adoption of Industry 4.0	[7, 11, 12]
Lack of understanding the interplay between technology and human (C11)	With Industry 4.0, there will be an increase in human-machine interaction, and appropriate skill and computational infrastructures must be considered. User friendly interfaces should come to the fore. Lack of understanding the interplay between technology and human is an obstacle to the transition to Industry 4.0	[7]
Legal uncertainty (C12)	Legal regulations that are valid today are always based on legal and natural persons. With Industry 4.0, businesses will be managed with artificial intelligence. The issue of how to deal with artificial intelligence is lacking in today's legal arrangements	[6, 8, 10, 11]
Contractual uncertainty (C13)	Contractual regulations that are valid today are always based on legal and natural persons. With Industry 4.0, businesses will be managed with artificial intelligence. The issue of how to deal with artificial intelligence is lacking in today's contractual arrangements	[6, 8, 10, 11]
Need for enhanced skills (C14)	Industry 4.0 technologies create a need for more qualified personnel. The majority of the workforce trained today does not have these competencies enough. Companies need to reorganize their Human Resource Management departments to master the issues of selecting and training the competent workforce	[6, 7]

(*continued*)

Table 1. (*continued*)

Need for new business models (C15)	In the literature, a clear framework for an enterprise to benefit from transition to Industry 4.0 has not been identified. This situation disrupts Industry 4.0 studies of businesses that are to be established or are currently operating	[8]
Need to find suitable research partner (C16)	Companies need to find suitable research partners to speed up and facilitate the transition to Industry 4.0. Limited cooperation in industry, university and research center is an issue in this sense	[8]
Organizational and process changes (C17)	With the proliferation of smart factories, vertical structures that have a certain hierarchy leave their place to decentralized process structures. This situation requires review of time-consuming and costly processes. Smart factories cannot be functional unless necessary changes are made	[6]
Preferred autonomy (C18)	Companies prefer the freedom to make their own decisions rather than being dependent on other companies' decisions. The technology that evolves with Industry 4.0 fully integrates the supply chain, which reduces the autonomy of decisions. Due to this bias, companies seeking autonomy do not prefer industry 4.0 transformation	[6, 8]
Regulatory compliance issues (C19)	Compliance with the legislation means that companies operate by taking into consideration the laws, policies and regulations in force. However, the issue of how sensitive artificial intelligence will be on these issues is uncertain	[6]
Requires continued education of employees (C20)	Industry 4.0 technologies require the continuous development of the workforce and lifelong learning. This new understanding of education is an obstacle to transition to Industry 4.0	[7, 12]
Seamless integration and compatibility issues (C21)	It is a problem encountered when upgrading existing machinery and equipment. Ensuring seamless integration and interoperability of different technologies and network structures is an important obstacle to the transition to Industry 4.0	[6]

(*continued*)

Table 1. (*continued*)

Security and privacy issues (C22)	Security and privacy issues are also on the agenda with the widespread use of smart factories and the Internet of Things. Companies become vulnerable to cyber-attacks, and with these attacks, companies face problems such as their trade secrets exposed and their operations stopped	[6–8, 10, 11]
Technology immaturity (C23)	Most of the technologies that are planned to be used within the scope of Industry 4.0 are still in their early stages. This causes the technologies to be used in Industry 4.0 to have both relatively low efficiency and high costs	[11]
Too few financial resources (C24)	Since Industry 4.0 technologies are in their early stages, the implementation costs for companies are relatively high. Especially the financial powers of small and medium-sized companies are not enough to do these practices	[7, 8, 10]
Too few human resources (C25)	As Industry 4.0 technologies are new, human resources competent for these technologies are few. In this respect, it is difficult to implement industry 4.0 by finding suitable human resources	[7, 11, 12]
Uncertainty about the reliability of the systems (C26)	There is not enough information about the stability of Industry 4.0 systems. This causes uncertainty and results in companies avoiding these new technologies	[8]
Lack of service delivery of Industry 4.0 technologies providers (C27)	The problem of lack of service providers faced by all other industry revolutions while living in its early ages also applies to industry 4.0. This causes a lack of know-how circulating in the market and slows down the development of Industry 4.0	[15]

3 Methodology

Multi-criteria decision-making (MCDM) methods are an important set of tools that enable the decision-maker to make better decisions on problems that are difficult to decide because it offers appropriate tools in situations where uncertainty, complexity and conflicting goals are present [16]. In MCDM problems, decision makers make decisions based on the decision variable, namely evaluation criteria and alternative set.

Decision makers sort or choose alternatives based on multiple and conflicting criteria values. In other words, alternatives that can be evaluated by MCDM methods can be sorted, grouped or selected from among them [17]. In multi-criteria decision-making problems, each criterion can be of different significance. Accordingly, determining the appropriate weight for each criterion becomes a necessary condition in terms of revealing the importance level of that criterion [18].

In this study, GRA, which is a multi-criteria decision making method, was used to weight the criteria obtained by literature review. The opinions of 5 experts were asked for the criterion weights and accordingly the group decision was used. Experts were selected from the companies that are "pioneers in Industry 4.0 transformation in our country, which are also serve as business model for other companies that will perform Industry 4.0 transformation and provide consultancy services for this purpose". The selected experts are the managers of Industry 4.0 department in these companies. Experts were asked to score the significance levels of the criteria between 1 and 9. The reason for choosing the GRA is that despite the small number of experts' opinions, it yielded successful results. Decision making becomes difficult when more than one criterion is made in the decision process, and the uncertainty inherent in decision making complicates the decision making process even more. Gray Relational Analysis method is one of the alternative and effective approaches that can be used to operate the decision making process more effectively in cases where there is little or incomplete information or where there is a lot of data and uncertainty [19].

Gray Relational Analysis is a member of the Gray System Theory launched by Deng Joung in 1982. Recently, GRA has been used in decision making for selection/evaluation/alternative rankings in many areas [20]. The main procedure of GRA is firstly converted into a comparable performance of all its alternatives. This step is called gray relational production. These sequences are defined based on an ideal target sequence. Then the grey relational coefficients between all comparable strings and sample sequences are calculated. Finally, the grey relational degree between the sample sequence and each comparable series based on the gray relational coefficients is calculated [21]. The implementation steps of GRA are as follows [22];

Step 1: Preparation of data set and creation of decision matrix

$$\begin{bmatrix} x_1(1) & \cdots & x_1(n) \\ \vdots & \ddots & \vdots \\ x_m(1) & \cdots & x_m(n) \end{bmatrix}, 1 \leq i \leq m,\ 1 \leq j \leq n$$

Step 2: Creation of reference series and comparison matrix: The reference value is the best value of the j. criteria within normalized values. The reference series is added to the first line of the decision matrix, resulting in a comparison matrix.

$$x_o = (x_o(j)), 1 \leq j \leq n$$

Step 3: Normalizing the decision matrix and creating the normalization matrix: There are three different situations for normalization. These are benefits, costs and optimal conditions. Benefit status shows that the "higher is the better", cost status is "lower is the better", and the optimal situation is the value determined by the decision maker.

$$\text{Benefit}: x_i^*(j) = \frac{x_i(j) - min_{jx_i}(j)}{max_{jx_i}(j) - min_{jx_i}(j)}$$

$$\text{Cost}: x_i^*(j) = \frac{max_{jx_i}(j) - x_i(j)}{max_{jx_i} - min_{jx_i}(j)}$$

$$\text{Optimal}: x_i^*(j) = \frac{|x_i(j) - x_{0b}(j)|}{max_{jx_i}(j) - x_{0b}(j)}$$

Step 4: Creation of the absolute value table:

$$\Delta_{0i}^*(j) = \left|x_o^*(j) - x_i^*(j)\right| = \begin{bmatrix} \Delta_{01}^*(1) & \cdots & \Delta_{01}^*(n) \\ \vdots & \ddots & \vdots \\ \Delta_{0m}^*(1) & \cdots & \Delta_{0m}^*(n) \end{bmatrix}$$

Step 5: Creating the grey relational coefficient matrix:

$$\gamma_{0i}(j) = \frac{\Delta_{min} + \Delta_{max}}{\Delta_{0i}(j) + \Delta_{max}}$$

Step 6: Calculation of gray relational grades:

$$\Gamma_{0i} = \sum_{j=1}^{n} [w_i(j) x \gamma_{0i}(j)]$$

4 Data Analysis

Twenty-seven challenges were identified from literature review. In the study, five experts' opinions were taken. While taking the opinions of experts, 1–9 scale was used (1 is not important at all, 9 is extremely important). The decision matrix and its normalized form, created with the opinions of experts, are shown in Table 2.

The values obtained after calculating deviation sequences and grey relational coefficients are shown in Table 3. Gray relational grade (GRG) values were used to determine the order of importance of the challenges encountered at Industry 4.0 transformation. While making the calculations, the challenges were considered as cost-based. The reason it is accepted in this way is that the challenges are negative situations.

While evaluating the gray relational grades (GRG), the lowest value was accepted as the most important by taking the Luthra et al.'s article as reference and the order of importance was made accordingly [23].

Table 2. Decision matrix and normalized decision matrix for challenges of companies in industry 4.0 transformation

	EO1	EO2	EO3	EO4	EO5
C1	9	4	3	5	8
C2	6	4	6	9	7
C3	6	9	7	9	8
C4	7	5	7	9	8
C5	8	6	8	9	7
C6	3	8	4	8	7
C7	5	8	6	8	6
C8	6	9	5	9	7
C9	9	8	8	9	9
C10	7	9	8	9	9
C11	6	9	8	6	8
C12	3	4	3	6	6
C13	3	5	4	6	6
C14	8	9	7	9	9
C15	7	7	7	9	7
C16	6	9	6	7	7
C17	7	7	5	7	8
C18	7	5	6	6	4
C19	4	3	4	6	4
C20	6	4	6	6	6
C21	5	7	8	8	9
C22	7	7	6	7	7
C23	7	8	5	9	7
C24	5	9	8	9	9
C25	7	8	9	9	8
C26	6	6	7	6	7
C27	6	8	6	5	7

	EO1	EO2	EO3	EO4	EO5
C1	0	0,8333333	1	1	0,2
C2	0,5	0,8333333	0,5	0	0,4
C3	0,5	0	0,3333333	0	0,2
C4	0,3333333	0,6666667	0,3333333	0	0,2
C5	0,1666667	0,5	0,1666667	0	0,4
C6	1	0,1666667	0,8333333	0,25	0,4
C7	0,6666667	0,1666667	0,5	0,25	0,6
C8	0,5	0	0,6666667	0	0,4
C9	0	0,1666667	0,1666667	0	0
C10	0,3333333	0	0,1666667	0	0
C11	0,5	0	0,1666667	0,75	0,2
C12	1	0,8333333	1	0,75	0,6
C13	1	0,6666667	0,8333333	0,75	0,6
C14	0,1666667	0	0,3333333	0	0
C15	0,3333333	0,3333333	0,3333333	0	0,4
C16	0,5	0	0,5	0,5	0,4
C17	0,3333333	0,3333333	0,6666667	0,5	0,2
C18	0,3333333	0,6666667	0,5	0,75	1
C19	0,8333333	1	0,8333333	0,75	1
C20	0,5	0,8333333	0,5	0,75	0,6
C21	0,6666667	0,3333333	0,1666667	0,25	0
C22	0,3333333	0,3333333	0,5	0,5	0,4
C23	0,3333333	0,1666667	0,6666667	0	0,4
C24	0,6666667	0	0,1666667	0	0
C25	0,3333333	0,1666667	0	0	0,2
C26	0,5	0,5	0,3333333	0,75	0,4
C27	0,5	0,1666667	0,5	1	0,4

Table 3. Deviation sequences and grey relational coefficients

	EO1	EO2	EO3	EO4	EO5
C1	1	0,1666667	0	0	0,8
C2	0,5	0,1666667	0,5	1	0,6
C3	0,5	1	0,6666667	1	0,8
C4	0,6666667	0,3333333	0,6666667	1	0,8
C5	0,8333333	0,5	0,8333333	1	0,6
C6	0	0,8333333	0,1666667	0,75	0,6
C7	0,3333333	0,8333333	0,5	0,75	0,4
C8	0,5	1	0,3333333	1	0,6
C9	1	0,8333333	0,8333333	1	1
C10	0,6666667	1	0,8333333	1	1
C11	0,5	1	0,8333333	0,25	0,8
C12	0	0,1666667	0	0,25	0,4
C13	0	0,3333333	0,1666667	0,25	0,4
C14	0,8333333	1	0,6666667	1	1
C15	0,6666667	0,6666667	0,6666667	1	0,6
C16	0,5	1	0,5	0,5	0,6
C17	0,6666667	0,6666667	0,3333333	0,5	0,8
C18	0,6666667	0,3333333	0,5	0,25	0
C19	0,1666667	0	0,1666667	0,25	0
C20	0,5	0,1666667	0,5	0,25	0,4
C21	0,3333333	0,6666667	0,8333333	0,75	1
C22	0,6666667	0,6666667	0,5	0,5	0,6
C23	0,6666667	0,8333333	0,3333333	1	0,6
C24	0,3333333	1	0,8333333	1	1
C25	0,6666667	0,8333333	1	1	0,8
C26	0,5	0,5	0,6666667	0,25	0,6
C27	0,5	0,8333333	0,5	0	0,6

	EO1	EO2	EO3	EO4	EO5	GRG	Rank
C1	0,3333333	0,75	1	1	0,3846154	0,6935897	23
C2	0,5	0,75	0,5	0,3333333	0,4545455	0,5075758	17
C3	0,5	0,3333333	0,4285714	0,3333333	0,3846154	0,3959707	5
C4	0,4285714	0,6	0,4285714	0,3333333	0,3846154	0,4350183	9
C5	0,375	0,5	0,375	0,3333333	0,4545455	0,4075758	6
C6	1	0,375	0,75	0,4	0,4545455	0,5959091	21
C7	0,6	0,375	0,5	0,4	0,5555556	0,4861111	16
C8	0,5	0,3333333	0,6	0,3333333	0,4545455	0,4442424	11
C9	0,3333333	0,375	0,375	0,3333333	0,3333333	0,35	1
C10	0,4285714	0,3333333	0,375	0,3333333	0,3333333	0,3607143	2
C11	0,5	0,3333333	0,375	0,6666667	0,3846154	0,4519231	12
C12	1	0,75	1	0,6666667	0,5555556	0,7944444	25
C13	1	0,6	0,75	0,6666667	0,5555556	0,7144444	24
C14	0,375	0,3333333	0,4285714	0,3333333	0,3333333	0,3607143	2
C15	0,4285714	0,4285714	0,4285714	0,3333333	0,4545455	0,4147186	7
C16	0,5	0,3333333	0,5	0,5	0,4545455	0,4575758	13
C17	0,4285714	0,4285714	0,6	0,5	0,3846154	0,4683516	15
C18	0,4285714	0,6	0,5	0,6666667	1	0,6390476	22
C19	0,75	1	0,75	0,6666667	1	0,8333333	26
C20	0,5	0,75	0,5	0,6666667	0,5555556	0,5944444	20
C21	0,6	0,4285714	0,375	0,4	0,3333333	0,427381	8
C22	0,4285714	0,4285714	0,5	0,5	0,4545455	0,4623377	14
C23	0,4285714	0,375	0,6	0,3333333	0,4545455	0,43829	10
C24	0,6	0,3333333	0,375	0,3333333	0,3333333	0,395	4
C25	0,4285714	0,375	0,3333333	0,3333333	0,3846154	0,3709707	3
C26	0,5	0,5	0,4285714	0,6666667	0,4545455	0,5099567	18
C27	0,5	0,375	0,5	1	0,4545455	0,5659091	19

As a result of the analysis, according to the 5 experts whose opinions are consulted, the most important challenge is "Lack of support by top management (C9)" with an importance value of 0.350. Considering this information, it is seen that there are important problems in the adoption of Industry 4.0 at the organizational level. This challenge is followed by "Lack of understanding of the strategic importance of Industry 4.0 (C10)" and "Need for enhanced skills (C14)" challenges, which have the same importance value (0,360). While one of these challenges is at the organizational level, similar to the most important criterion, it is seen that the other challenge belongs to the labor market. Small and medium sized companies constitute a significant part of the industrial enterprises. The financial opportunities of these companies are not sufficient to perform Industry 4.0 applications. For this reason, it is striking that financial difficulties are an important challenge in the transformation to Industry 4.0, and this is seen in the study with "Too few financial resources (C24)" and "High implementation cost (C3)", which are the next challenges with the importance value of 0,395. The "Lack of clearly defined economic benefit (C5)" challenge which is following these challenges is also related to the financial dimension. This challenge with 0,407 importance value also shows that companies are not aware of the financial benefits of Industry 4.0. Considering the challenges that constitute an obstacle to the industry 4.0 transformation of companies, it is seen that all three challenges with the lowest importance value are all related to legal regulations. These difficulties are "Contractual uncertainty (C13)", "Legal uncertainty (C12)", "Regulatory compliance issues (C19)" and their significance values are "0,714", "0,794", "0,833" respectively.

5 Conclusions

The world is still at an early stage in Industry 4.0. There are also deficiencies in understanding and evaluating the challenges encountered by this early period. For this reason, in this study, the difficulties encountered in the transition to Industry 4.0 were analyzed and the results are shown in the "Data Analysis" section.

According to the data analysis of the study, the problems at the organizational level are the main difficulties encountered in the industry 4.0 transformation. This is followed by difficulties regarding the labor market. Other important challenges are financial and monetary problems. It is recommended that the problems are handled by policy makers and sectoral practitioners in this order of importance.

Luthra and Mangla analyzed key challenges to Industry 4.0 and prioritized these challenges with AHP method. According to their study, organizational challenges were determined as the most important dimension [10]. Sevinç et al. analyzed the difficulties faced by SMEs in Industry 4.0 processes and applications with AHP and ANP method. The weight of organizational criteria has been determined as the highest according to AHP and AHP method [1]. The results of our study conformance the results of the researches in the literature. Industry 4.0 transformation should not be evaluated only as a technological investment. For the success of Industry 4.0 transformation, organizational structure and the strategic viewpoint of the organization and support of top management are also very important factors.

References

1. Sevinç, A., Gür, Ş., Eren, T.: Analysis of the difficulties of SMEs in Industry 4.0 applications by analytical hierarchy process and analytical network process. Processes **6**(12), 264 (2018)
2. Kagermann, H., Wahlster, W., Helbig, J.: Recommendations for implementing the strategic initiative Indusrie 4.0. Berlin Industrie 4.0 Working Group of Acatech (2013)
3. Prause, M., Weigand, J.: Industry 4.0 and object-oriented development: incremental and architectural change. J. Technol. Manag. Innov. **11**(2), 104–110 (2016)
4. Varghese, A., Tandur, D.: Wireless requirements and challenges in Industry 4.0. In: 2014 International Conference on Contemporary Computing and Informatics (IC3I), pp. 634–638. IEEE (2014)
5. Piccarozzi, M., Aqilani, B., Gatti, C.: Industry 4.0 in management studies: a systematic literature review. Sustainability **10**, 3821 (2018)
6. Kamble, S.S., Gunasekaran, A., Sharma, R.: Analysis of the driving and dependence power of barriers to adopt Industry 4.0 in Indian manufacturing industry. Comput. Ind. **101**, 107–119 (2018)
7. Stentoft, J., Jensen, K.W., Philipsen, K., Haug, A.: Drivers and barriers for industry 4.0 readiness and practice: a SME perspective with empirical evidence. In: Proceedings of the 52nd Hawaii International Conference on System Sciences, pp. 5155–5164 (2019)
8. Orzes, G., Rauch, E., Bednar, S., Poklemba, R.: Industry 4.0 implementation barriers in small and medium sized enterprises: a focus group study. In: 2018 IEEE International Conference on Industrial Engineering and Engineering Management (IEEM), pp. 1348–1352. IEEE (2018)
9. Shinohara, A.C., da Silva, E.H.D.R., de Lima, E.P., Deschamps, F., da Costa, S.E.G.: Critical success factors for digital manufacturing implementation in the context of Industry 4.0. In: IIE Annual Conference, Proceedings, pp. 199–204. Institute of Industrial and Systems Engineers (IISE) (2017)
10. Luthra, S., Mathra, S.C.: Evaluating challenges to industry 4.0 initiatives for supply chain sustainability in emerging economies. Process Saf. Environ. Prot. **117**, 168–179 (2018)
11. Glass, R., Meissner, A., Gebauer, C., Stürmer, S., Metternich, J.: Identifying the barriers to industrie 4.0. In: 51st CIRP Conference on Manufacturing System, pp. 985–988 (2018)
12. Türkes, M.C., Oncioiu, I., Aslam, H.D., Marin-Pantelescu, A., Topor, D.I., Capusneanu, S.: Drivers and barriers in using industry 4.0: a perspective of SMEs in Romania. Processes **7** (3), 1–20 (2019)
13. Gavade, R.K.: Multi-criteria decision making: an overview of different selection problems and methods. Int. J. Comput. Sci. Inf. Technol. **5**(4), 5643–5646 (2014)
14. Ho, C.: Measuring bank operations performance: an approach based on grey relation analysis. J. Oper. Res. Soc. **57**(4), 337–349 (2006)
15. Müller, J.M., Kiel, D., Voigt, K.I.: What drives the implementation of Industry 4.0? The role of opportunities and challenges in the context of sustainability. Sustainability **10**(1), 247 (2018)
16. Salah, S.B., Yahia, W.B., Ayadi, O., Masmoudi, F.: Definition and classification of collaborative network: MCDM approaches for partner selection problem. In: International Conference Design and Modeling of Mechanical Systems, pp. 733–744. Springer, Cham (2017)
17. Kumar, A., Sah, B., Singh, A.R., Deng, Y., He, X., Kumar, P., Bansal, R.C.: A review of multi criteria decision making (MCDM) towards sustainable renewable energy development. Renew. Sustain. Energy Rev. **69**, 596–609 (2017)

18. Štirbanović, Z., Stanujkić, D., Miljanović, I., Milanović, D.: Application of MCDM methods for flotation machine selection. Miner. Eng. **137**, 140–146 (2019)
19. Zhai, L.Y., Khoo, L.P., Zhong, Z.W.: Design concept evaluation in product development using rough sets and grey relation analysis. Expert Syst. Appl. **36**(3), 7072–7079 (2009)
20. Chan, J.W.: Product end-of-life options selection: grey relational analysis approach. Int. J. Prod. Res. **46**(11), 2889–2912 (2008)
21. Kuo, Y., Yang, T., Huang, G.W.: The use of grey relational analysis in solving multiple attribute decision-making problems. Comput. Ind. Eng. **55**(1), 80–93 (2008)
22. Wu, H.H.: A comparative study of using grey relational analysis in multiple attribute decision making problems. Qual. Eng. **15**(2), 209–217 (2002)
23. Luthra, S., Garg, D., Mangla, S.K., Berwal, Y.P.S.: Analyzing challenges to Internet of Things (IoT) adoption and diffusion: an Indian context. Procedia Comput. Sci. **125**, 733–739 (2018)

Industry 4.0 in Educational Process

Jan Vetiska, Michal Holub(✉), Petr Blecha, Frantisek Bradac, Jakub Brazina, Vaclav Stanek, Jiri Kroupa, and Zdenek Tuma

Department of Production Machines, Systems and Robotics, Faculty of Mechanical Engineering, Brno University of Technology, Brno, Czech Republic
{vetiska,blecha,bradac,kroupa,tuma}@fme.vutbr.cz, Michal.Holub@vutbr.cz, {jakub.brazina,vaclav.stanek}@vut.cz

Abstract. The presented paper describes the individual elements of industry 4.0 already introduced into the educational process of students of bachelor and master degree studies at the workplace of the Faculty of Mechanical Engineering, Brno University of Technology. For the purposes of education, approaches were taken into account with the requirements of Industry 4.0 in the field of sensor systems, OPC UA data collection, data processing, metrology, simulation, automation, robotics, M2M communication, Virtual Reality, and digital commissioning. Individual technologies are presented to students in the form of partial tasks that are mutually related. The result is an overall overview of the elements of industry 4.0 and their implementation into production systems.

Keywords: Education · Industry 4.0 · Digitalization · Production systems · Virtual reality

1 Introduction

This paper is focused on implementing the elements of Industry 4.0 into the process of education at the Faculty of Mechanical Engineering, Brno University of Technology (BUT) in the field of manufacturing machines, systems and robotics. The growing interest in the deployment of Industry 4.0 elements in industrial applications requires qualified personnel with sufficient knowledge in the field. An essential step in recruiting such qualified staff is to implement Industry 4.0 principles into the educational system. At BUT, the issues of Industry 4.0 in the areas of automation, robotics, digitization, quality prediction and failure states of production systems have been addressed for a long time. Research and development activities in this area were presented in the framework of conference contributions, symposia, workshops, and international trade fairs.

One of the areas of visualization of real-time data and processed data is the deployment of virtual reality. Use of virtual reality as a tool for a faster and more comprehensible understanding of production systems and their configured processes is described in [1].

N. M. Durakbasa and M. G. Gençyılmaz (Eds.): ISPR 2020, LNME, pp. 324–332, 2021.
https://doi.org/10.1007/978-3-030-62784-3_27

The preparation of a virtual laboratory with the aim to increase the effectiveness of teaching by working with an interactive model of machine tools is presented in [2]. The interactivity of the model lies in the possibility of assembly and disassembly of the basic parts of the machine tool to the basic components. Oculus head mounted displays and controls were used to visualize and interact with the machine tool.

The deployment potential of Industry 4.0-based applications is described in [3]. These include the collection of data from control systems, implementation of sensor systems in the production process, development of knowledge for efficient decision making for machine operators or maintenance, but also the risks associated with remote access in communication.

Another element falling within the field of manufacturing system design is the use of virtual twin manufacturing systems. The publication [4] presents a virtual twin headstock of the Multi-spindle Lathe model-based design in MCS Adams and co-simulation in Matlab-Simulink.

A proposal to upgrade a production cell in the context of Industry 4.0 is described in [5] where different approaches to production cell digitization using virtual reality environment are presented. Particular emphasis was placed on visualizing the information about the machine's surroundings, geometrical accuracy, and workplace safety.

With the advent and development of the INDUSTRY 4.0 initiative, a rapid growth of requirements for industrial applications started in the fields of digitization, advanced automation of robotics, etc. These applications represent a high share in automotive and aerospace industry where there is an enormous demand for high quality production. Implementation of these applications in small series production is very rare because new innovative technologies are expensive and have a long - time cost recovery.

Thanks to the support of different initiatives covered by government, the elements of the INDUSTRY 4.0 are introduced into small series production in small and medium - sized companies. As part of the implementation of these technologies, there is often a lack of qualified employees, which results in rejecting these innovations.

At present, requirements of industries for the knowledge and skills of future graduates of technical institutions are growing. A response of BUT to industrial companies' requirements is the implementation of production work cell into the teaching for 4th and 5th grade students as a complex unit with elements of mechanical design, sensors, electrical design, virtual reality, robotics, metrology, data processing, etc. The implementation of INDUSTRY 4.0 into the teaching is further described in this paper.

2 Industry Requirements

A lot of papers deal with implementation of INDUSTRY 4.0 elements. It is a multi-disciplinary topic with different approaches depending on the solution to the problems of particular disciplines. The publication [6] presents qualification and skill requirements for the IT work profile in industrial applications - knowledge of IT (Informatic Specialist), PLC programming (PLC programmer), robot programming (Robot Programmer), software engineering knowledge (Software Engineer), data analysis (Data

Analyst), and cyber security. These qualification and skill requirements are also presented in the work profile of production systems.

2.1 Requirements for the Area of Production Systems

The publication [7] describes a logical derivation of basic competences for INDUSTRY 4.0 employees. The main emerging challenges were first analyzed and then the competencies were derived according to the following challenges – economic challenges, social challenges, technical challenges, environmental challenges, and political and legal challenges.

The requirements for future employees of Industry 4.0, or graduates focused on production systems in general, are today very closely connected with the definition of the INDUSTRY 4.0 philosophy itself. In the past, the industrial requirements could be divided into mechanical design, robotics, automation, and quality. With a very rapid advent of INDUSTRY 4.0 elements, there is also a dynamic reorientation of the educational approach itself. The main intention is to educate the graduate who will be able to understand the issues of INDUSTRY 4.0, but also to make a multidisciplinary extension of this field. It can be stated that the requirements of the industry for the multidisciplinary education include sub-fields such as design, robotics, automation, and quality.

The requirements for production systems from the point of view of the implementation of INDUSTRY 4.0 elements can be shown on the example of CPPS (Cyber-physical Production Systems) presented by Stock [8]. These requirements can be seen in Fig. 1.

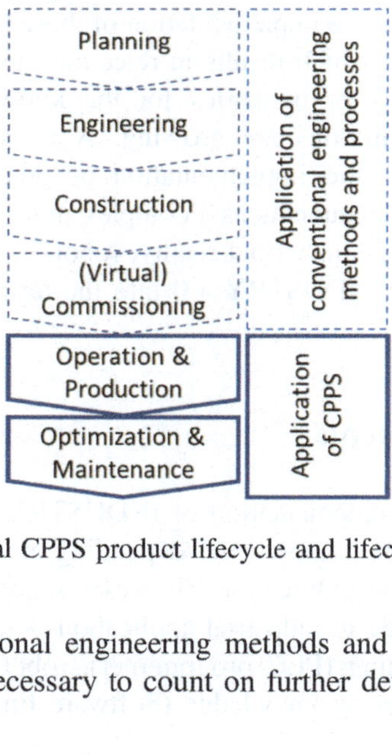

Fig. 1. General CPPS product lifecycle and lifecycle phases [8]

In addition to conventional engineering methods and processes in the design of production systems, it is necessary to count on further development according to the above-mentioned CPPS.

2.2 Qualification and Skills for Industry

Within the study at technical educational institutions focused on production systems, a necessary prerequisite is to complete such courses where the qualification and skill requirements for the given field are met. Table 1 describes a set of qualification and skill requirements for employment in the field of design and operation of production systems connected to INDUSTRY 4.0.

Table 1. Production systems job profiles

	Qualification	Skills
Design	Mechanical designer, production planning	CAD, MBS, FEM, Production systems modelling
Robotics	Design of robotic workstation	On-line, off-line programming, virtual commissioning, real robotic workstation commissioning and validation of the digital model
Automations	Design of automated systems, commissioning of automated systems	Sensors, PLC programming, electrical/electronics, pneumatics/hydraulics
Quality	"Safety engineer", Metrology	Basics of risk management, Statistics, CMM, mobile measuring devices
IT	"Programmer"	MATLAB, databases, networks, Virtual reality

As part of the study, the student will gain knowledge and skills in the fields of machine, machine parts, and equipment design. This study requires the basic knowledge of CAD systems, the ability to work independently in the field of machine parts and units design, including the processing of technical drawings. The stress-strain analysis (FEM), dynamics analysis (MBS), and the ability to implement machines and devices into production units (production systems) are the essential parts and today's requirements for mechanical design works.

Another area of study are activities associated with the acquisition of knowledge and skills with industrial robots and manipulators. In this phase, students work with programs for online and offline programming, virtual commissioning of the robotic workstations and workstation commissioning in the real workshop with the final validation of the digital model.

Sensor systems and advanced control systems ensuring the functionality and safety are necessary components for production systems design and operation. In this part, students acquire knowledge and skills of basic sensor systems, work with PLCs and other media. This knowledge is verified on basic teaching stands up to the programming of three-axis milling production machines.

A domain of quality is focused on strengthening the knowledge and skills especially in metrology and control of the production process, as well as on the safety of production systems. Within these activities, students solve the tasks of basic metrology up to the evaluation of production process. They are also involved in solving the tasks focused on the risk analysis and functional safety, which are closely connected to the above-mentioned areas.

The last area is the acquisition of sufficient knowledge and skills in the field of IT for production systems. In this part, the students solve the tasks related to the operation of the production network and data processing from the production system and, last but not least, the use of VR elements to work with advanced visualizations and to display production systems information in VR.

3 Implementation Phase

Regarding to the changing knowledge requirements for graduates from the industrial sector, it is necessary to accept the changing profile of the future graduate of a technical school with a focus on production systems already in the educational phase.

3.1 Education in Production Systems

Quality teaching of production systems with I4.0 elements is currently based on the synthesis of knowledge and skills. The production system must be viewed as a complex unit consisting of mechanics, electrical engineering and electronics, automation, and computer science. All these engineering disciplines within the production system have to be applied together to achieve a mutual synergy, see the mechatronic approach [9], and the base design model [10]. All the above-mentioned engineering disciplines work with very advanced SW tools in the design phase. The software allows for the creation of partial digital models for various purposes and their mutual co-simulation. This places high demands on the professional readiness of teachers and good knowledge of students from the previous educational process.

At present, teaching of production systems at individual faculties/institutes takes place separately without a sufficient multidisciplinary overlap, which results in a very narrowly specialized graduate without multidisciplinary basics. The Institute of Production machines, Systems and Robotics of Faculty of Mechanical Engineering of BUT is trying to respond to this fact through teaching where the emphasis is placed on the necessary multidisciplinary approach, i.e. mechanical engineering knowledge and skills (CAD, CAM, FEM) supplemented by the knowledge of robotics, electrical engineering and electronics, automation, computer science, metrology, quality and safety. The following figure shows the concept of a production system represented by a production cell consisting of two CNC machine tools (lathe and milling machine), an industrial robot, a measuring station and safety elements necessary for operation in fully automatic mode. The concept of the production cell (Fig. 2) is based on the modularity of systems, where the elements of industry 4.0 are gradually implemented based on the requirements of industry and the financial capabilities of the workplace.

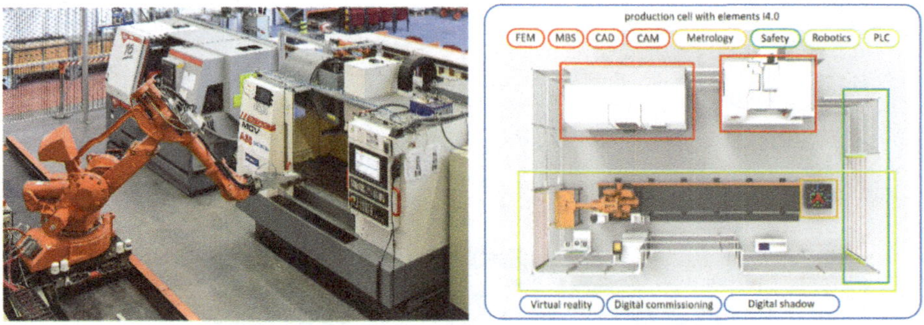

Fig. 2. Concept of production system for educational purposes

3.2 Educational Concept

The educational concept is based on the current requirements of industry and its pressure to implement I4.0 elements. These elements are applied within the school workshop at the development workplace shown in Fig. 3, on the left. This workplace lays down the hardware foundation for a series of consecutive subjects through which the students are introduced to the entire development process of the production system and its consecutive maintenance. These activities can be seen in Fig. 3, on the right.

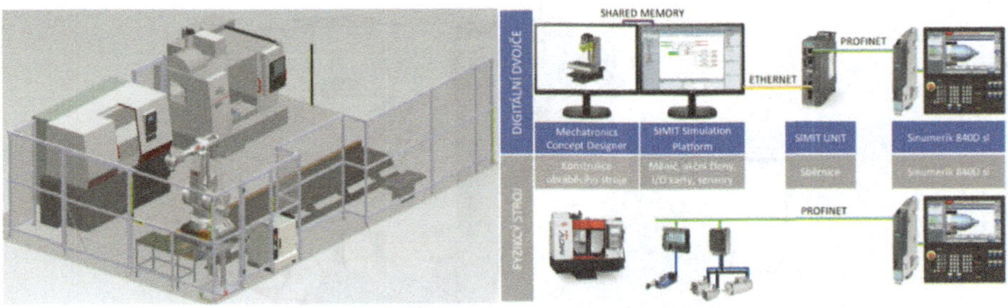

Fig. 3. Digital twin production system

Basic engineering disciplines, such as conceptual design of the production system, creation of CAD models, strength, kinematic, and dynamic analyses, creation of technical documentations are extended by automation and information technology. Figure 4 shows an example of defining a kinematic bond of the Hinge Joint type in the Siemens NX SW for the purpose of virtual commissioning of the machine axis Y of the milling CNC machine tool according to Fig. 3 (right).

The above-mentioned areas are taught using SIL and HIL, and the desired result is a complete digitally commissioned machine tool/production system where the programs for PLC, CNC machines, robots and their interconnection are tested. Then the students test the proposed SW solutions on the real production system. To apply the I4.0 elements, students propose how to collect relevant data from individual parts of the production system (robot, CNC milling machine, CNC lathe, Safety). Data are stored

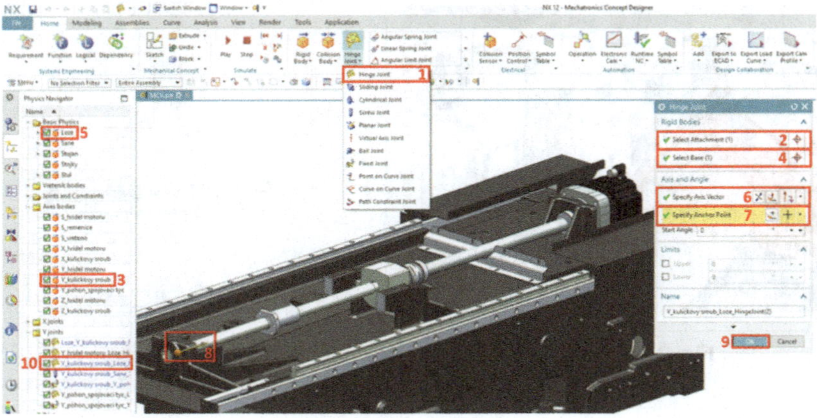

Fig. 4. Sample of kinematic constraint definition for virtual commissioning of Y axis of milling machine tool CAD model

and visualized using VR (digital shadow). Figure 5 shows an example of the application of workstation visualization using VR in online mode.

Fig. 5. Presentation of production system with usage of VR in online mode at International Engineering Fair 2019

Thanks to the validation of all partial tasks, students are able to gain the necessary overview of individual activities from designing to commissioning of the production system. This approach provides the students with the appropriate knowledge and skills for industrial applications focused on the concept of Industry 4.0.

4 Conclusions

Due to the gradual transformation of industry and its applications, the requirements for graduates of industrially oriented educational institutions are also changing. These changes bring about high investments associated not only with human resources, but also with the renewal or upgrade of machines. However, it is not possible to react flexibly within educational institutions focused on industrial applications, including production systems. Designing a comprehensive concept, production and commissioning of production system and also preparation of tasks for educational purposes can take several years. This is possible only in the case when funding is guaranteed. As part of universities financing, the necessary funding can be achieved in the form of grant programs, which can be also devoted to modernization of teaching.

In today's rapidly expanding fields such as robotics, automation, and metrology, it is not possible for educational institutions to keep pace with industry. One of the possible ways is to gradually build such tasks that are multidisciplinary and represent further integration of advanced technologies from a long-term perspective. This modular concept of a production cell is presented in this paper. The production cell is designed for multidisciplinary teaching in the field of production systems and is composed of machines and equipment acquired in the last 20 years. An important prerequisite for the use of such obsolete machines and equipment is to add more I/O signals, which can be used to control the entire production system at different controlling levels.

The current effort is to further expand the production system presented in this paper. More equipment will be added in the near future, such as a single-purpose measuring station, AI elements, etc. This will be also an immediate response to the industry demands. Based on the experience obtained from teaching, it seems to be a sufficient model to present both partial and complete elements of Industry 4.0.

Acknowledgment. These results were obtained with the financial support of the Faculty of Mechanical Engineering, Brno University of Technology (Grant No. FSI-S-20-6335).

References

1. Kovar, J., Mouralova, K., Ksica, F., Kroupa, J., Andrs, O., Hadas, Z.: Virtual reality in context of industry 4.0 proposed projects at Brno University of Technology. In: Proceedings of the 2016 17th International Conference on Mechatronics - Mechatronika, ME 2016 (2016). ISBN: 978-800105883-1
2. Kroupa, J., Tuma, Z., Kovar J., Singule, V.: Virtual laboratory for study of construction of machine tools. MM Sci. J. **2018**(2018), 2503–2506 (2018). https://doi.org/10.17973/mmsj.2018_11_2017100
3. Blecha, P., Durakbasa, N., Holub, M.: Digitized production – its potentials and hazards. In: Proceedings of the International Symposium Production Research 2018, pp. 402–411. Springer (2019). https://doi.org/10.1007/978-3-319-92267-6_35
4. Hadraba, P., Hadas, Z.: Virtual twin of the multi-spindle lathe for the chatter time-domain analysis. In: Proceedings of the 2018 18th International Conference on Mechatronics - Mechatronika, ME 2018 (2019). ISBN: 978-802145544-3

5. Holub, M., Tuma, Z., Kroupa, J., Kovar, J., Blecha, P.: Case study of digitization of the production cell. In: Lecture Notes in Mechanical Engineering, pp. 253–262. (2020). https://doi.org/10.1007/978-3-030-31343-2_22
6. Benesova, A., Tupa, J.: Requirements for education and qualification of people in industry 4.0. Procedia Manufact. **11**, 2195–2202 (2017). https://doi.org/10.1016/j.promfg.2017.07.366. ISSN 2351-9789
7. Hecklau, F., Galeitzke, M., Flachs, S., Kohl, H.: Holistic approach for human resource management in industry 4.0. Procedia CIRP **54**, 1–6 (2016). https://doi.org/10.1016/j.procir.2016.05.102. ISSN 2212-8271
8. Stock, D., Schel, D., Bauernhansl, T.: Middleware-based cyber-physical production system modeling for operators. Procedia Manuf. **42**, 111–118 (2020). https://doi.org/10.1016/j.promfg.2020.02.031. ISSN 2351-9789
9. Brezina, T., Vetiska, J., Hadas, Z., Brezina, L.: Simulation modelling and control of mechatronic systems with flexible parts. In: Mechatronics: Recent Technological and Scientific Advances, pp. 569–578 (2011). ISBN 9783642232435
10. Andrs, O., Hadas, Z., Hadas, Z., Kovar, J., Vetiska, J., Singule, V.: Model-based design of mobile platform with integrated actuator - design with respect to mechatronic education. In: Mechatronics 2013: Recent Technological and Scientific Advances, pp. 891–898 (2014). https://doi.org/10.1007/978-3-319-02294-9-113. ISBN 9783319022932

Industry 4.0 vs. Society 5.0

Leyla Polat[1(✉)] and Alptekin Erkollar[2]

[1] Necmettin Erbakan University, Konya, Turkey
lpolat@erbakan.edu.tr
[2] ETCOP, Klagenfurt, Austria
erkollar@etcop.com

Abstract. Our latest stop in the industrialization journey that starts with steam engines is the Industry 4.0 that commonly heard nowadays and that was first introduced in 2011 at the Hannover Fair in Germany. Industry 4.0 provides an autonomous system by using a large technology such as artificial intelligence that transfers the decision-making capability (mechanism) that belongs to human to electronic systems and machines. In this concept, although there is a work that provides prosperity of societies by thinking consumers in Industry 4.0, it ignores human factor when it is considered as individually. On the other hand, when it is mentioned about technology, Japan that is a technology giant comes with Germany together. However it falls behind of Germany for this super modern era of industrialization. Japan that doesn't want to stay in back, In April 2016, it emerges a philosophy that supports to name Society 5.0 thanks to Internet of Things and Artificial Intelligence instead of today's society that is called Information Society. This movement that deals with individual well-being, aims to use systems that work in harmony with each other in the interest of society, through the Internet of Things. Industry 4.0 trend that is also called digitalization in the industry with smart factories make various promises to businesses, on the other hand Society 5.0 trend that integrates cyber and physical systems for individual or common use make a lot of promises to humanity. Thus, these trends come face to face each other. In this study, Industry 4.0 and Society 5.0 trends are discussed conceptually in detail and a comparison is made by handling the positive and negative aspects of these concepts. As a result of the comparison, it is stated whether that both concepts have superiority over each other or not. Suggestions are made about the need to direct the future academic and sectoral studies with superior concept that one of the Industry 4.0 or Society 5.0. In this way, it is aimed to contribute to the literature, to raise awareness of societies and to enable them to deal with their development in different dimensions.

Keywords: Industry 4.0 · Society 5.0 · Smart businesses · Smart society · Smart factory

1 Introduction

There are various perceptions that technology is a great threat or a blessing to societies. With the artificial intelligence behind it, it is thought that smart devices where human abilities are transferred would create an inevitable danger for humans. From a different

N. M. Durakbasa and M. G. Gençyılmaz (Eds.): ISPR 2020, LNME, pp. 333–345, 2021.
https://doi.org/10.1007/978-3-030-62784-3_28

point of view, it is argued that these high intelligent objects with self-decision property and with communication property via internet of would benefit to the industry, economy, or society, thus prosperity will increase.

When it is looked at the literature, it is reported that with the concept of Industry 4.0, which is described as Germany's 4th Industrial Revolution at the Hannover Fair in 2011 [1] by advanced developments in technology would be rebuilt in production, manufacturing, consumption, retailing. Then, it is predicted that in 2012 by U.S.A., with the use of Internet of Things (IoT) technology in the industry would prevent costs in this item by eliminating only the use of cables and wires, and would provide full control within the enterprise thanks to the cloud system [2].

Evaluating the initiatives of Germany and America in a social context, Japan aims to create a smart society rather than smart machines. It presents this work as Society 5.0. According to advocated and conducted research, Japan aims to provide services to the individual in a special context with artificial intelligences that are developed with different algorithms by getting data from smart devices. Starting from this point of view, it is seen that they want all humanity to access the internet. E-Estonia project that encompasses the process of digitalizing Estonia's people could be given as an example of this study [3].

In this research study, Industry 4.0 of Germany that aims economic development by advocating that advanced technology should be used in industry is handled and Society 5.0 of Japan that aims social prosperity by considering that advanced technology should be used in favor of humanity are handled. Firstly, the common parameters between the two concepts have been determined. And then the usage purpose of these parameters in determined concepts and their returns are examined in detail. And finally, whether the two concepts have any superiority over each other is investigated.

2 Industry 4.0 and Society 5.0 Joint Parameters

When getting to the root of Industry 4.0 and Society 5.0, it is observed that they have a lot in common. In summary, in this study, for these two concepts, Big Data, Artificial Intelligence, Internet of Things (IoT), Cloud Computing and the human that affects to these systems or that is affected by these systems are identified as common parameters. Their effects are going to be examined under subheadings, respectively (see Fig. 1).

2.1 Big Data

There are countless sensors providing information such as position, movement, temperature, humidity that are located in each environment and each object, from cars to telephones, from household machines to clothings. When the data from all sensors are transferred to the computer, a large amount of data accumulation occurs. The amount of stored data (Volume), this accumulation that is operated at different speeds (Velocity) and that creates big variety (Variety) reveals the concept of Big Data [4, 5].

With the discovery of the Internet, many kinds of data are collected in virtual environment. These data were not considering as Big Data at the time. As a matter of

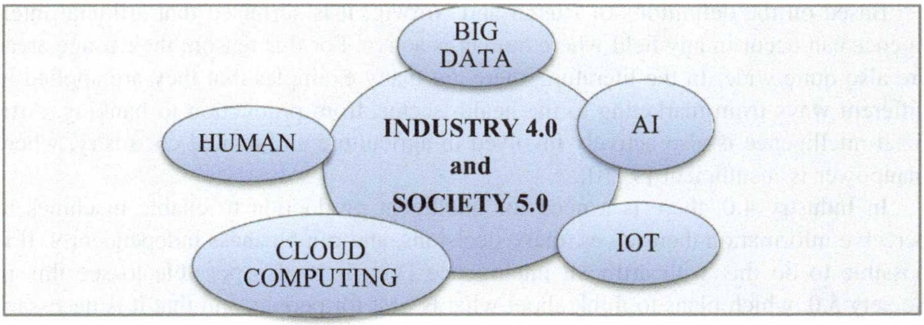

Fig. 1. Joint Parameters of Industry 4.0 and Society 5.0

fact, now, thanks to the following features Big Data concept is evaluated in a separate way.

- Cloud system is required for big data [6].
- Big data is scalable and takes part in the application section [6].
- Big data processing should proceed systematically [6].
- It is widely used in Production, Consumption and Mobile [6].

As mentioned above, Big Data, which is also present in the production area, plays a major role in the formation of self-sufficient smart factories thanks to cyber-physical systems. Once the necessary infrastructure has been established, all the data received from the machines and the necessary units constitutes the industrial big data. Subsequently, this industrial data that is processed and shaped thanks to the designed artificial intelligences is used for desired purposes [7]. This leads to a new generation of industrial revolution. As to the new generation of community revolution, this is realized through Society 5.0. Here, the physical characteristics of the people and the physical environment around them are taken by the equipment that they have into the virtual environment and the digital data are collected. This constitutes the social big data. Society 5.0 uses this data as individual or as communal thanks to the necessary artificial intelligences.

2.2 Artificial Intelligence

Artificial intelligence [7], which is defined in different forms in the literature, it is explained briefly stated that machines that are shaped as human thinking system can make decision, solve problems, take action by itself thanks to developed, designed computations, algorithms, in contrast to the machines that are executed according to the commands given by coding designed systems.

Russel and Norwig have gathered the definitions of artificial intelligence in a table and have divided them into 4 categories in their study. Accordingly, it is stated that electronic digital systems think, act, think realistically and act realistically like humans [8].

Based on the definitions of Russel and Norwig, it is surmised that artificial intelligence can occur in any field where human is active. For this reason, their usage areas are also quite wide. In the literature, there are many examples that they are applied in different ways from marketing to the health sector, from production to banking. Artificial intelligence is also actively involved in agriculture and related chemistry, where manpower is insufficient [9, 10].

In Industry 4.0, there is a need for intelligent production to enable machines to perceive information themselves, make decisions, and run business independently. It is possible to do this with artificial intelligence [11]. It is also possible to see this in Society 5.0, which plans to think about what is best for people, and that it is necessary to create different artificial intelligence that will work for the benefit of the community by using raw data in digital media through applications.

2.3 Internet of Things (IoT)

The Internet of Things (IoT), which is an inseparable parameter for both Industry 4.0 and Society 5.0, is used in three terms in the widest sense. The first of these is to create a global network to which each object is connected by means of advanced internet technologies and communication technologies and to ensure communication of every object on the network. The second is to supply the sensors, actuators, communication devices for communication and information, and thirdly to combine the necessary applications and services to create new business and market opportunities following the provision of the first two terms [12]. When we look at the new generation production, it is seen that innovative ideas are developed thanks to IOT and lean production is realized with different production styles.

With the transition of physical identities to the digital environment, objects with virtual personalities integrate seamlessly into the global internet network created. In this way, a self-shaping dynamic network infrastructure that can work together is occurred [13]. First informatics and then the production, communication, health, logistics, aviation, agriculture, entertainment, and food sectors level up by using this dynamic infrastructure of IoT. In this way, they broaden their horizons [14].

The IOT factor, which is known for its use in many fields and the convenience it brings to society, is determined as the common parameter of Industry 4.0 and Society 5.0 in this study. IOT, which forms the basis of smart factories, plays a major role in the development of smart, healthy societies too. For example, The Company Bosch has developed a device for people that have asthma disease. Thanks to the device and the sensors that are located in the device, asthma patients are monitored by the doctor in their daily lives, in their own environment and activities, and then the required dose is given after the patient's exact needs are determined.

2.4 Cloud Computing

Cloud Computing basically aims to make available Big Data to all users at any time with a non-stop system on the network and covers all the hardware and software that are required to store these data in a proper way. The stored central software and hardware in which data is located is called Cloud [15]. The providers of cloud services that collect data for a particular purpose call it Software as a Service (SAS). Only providers who want to promote their products differently use the terms Infrastructure as a Service (IaaS) or Platform as a Service (PaaS) [16].

For Industry 4.0, the importance of cloud computing is very large. Using a global internet infrastructure to provide intelligent production services, Industry 4.0 needs cloud computing to provide relevant data and service to every object, person, and unit in the factories. Thanks to easy integrations, Cloud Computing, which facilitates the continuation and collaboration of the entire value chain, provides innovations in the formation of new business processes and models for enterprises at the same time [17]. Society 5.0, which wants to provide the best services for communities and the environment, demands the active use of cloud computing. With its internet infrastructure aiming to reach every segment, Japan firstly aims to help people whenever and wherever they want, and to help in unexpected disasters, and to provide services to the community online thanks to cloud computing and subsequent smart applications. Again, it is possible to store digital data gathered in all areas throughout the country and waiting to be used when necessary, thanks to cloud computing software systems.

2.5 Human

Mankind, who played an active role in the development of technology, again has also been greatly affected by this development. It is also affected by Industry 4.0 and Society 5.0 at different rates. When considered a general user, the humanity is expected to be less affected by Industry 4.0 than by Society 5.0. Because when examined, this 4th Industrial Revolution covers production and consumption and addresses a more specific sector. Consumers in this concept can obtain products in less time, at a lower cost and with a high level of satisfaction. The individual logic technology at the core of Society 5.0 includes all people. It offers people an approach according to their needs, not monotype.

If the human parameter is considered by the producer, supplier and developer for Industry 4.0 and Society 5.0, it is seen that they need qualified personnel for both. Although unqualified or elderly employees draw the reaction to Industry 4.0, Tüsiad reports that qualified labor force will provide great added value to the country's economy and that their development levels will rise with this development [18]. For Society 5.0, it is observed that business models need to be changed as Ogawa states. It is said that necessary configurations should be made regarding employee diversity and continuity [19].

3 Industry 4.0

In addition to societies that do not have telephone communications in different parts of the world, or are surprised by predicting that this will be impossible when they hear such a technology, there are also developed societies that use objects, machines that act as an employee in their factories instead of human labor. So, what is the technology that astonishes people by enabling them to communicate and deliver faster production at less cost? That is the Industry 4.0 that was introduced by Germany in 2011 [1]. This idea, which emerged as a result of a multidisciplinary study, initially has had a great impact in developed countries and then it has been spreaded all over the world.

Although Germany, which is a developed country, does not have fast internet access network and does not include one of the leading companies in the IT sector, this country promises to use the information and communication technologies to communicate the objects in the factories via the Internet [20]. When it comes to the basis this, it is necessary to mention the socio economic structure of the country as well as it's the medium-sized enterprises and industrial working system. The aging of the population in the country, an inactive increase in the population, the provision of first-order needs for each citizen, and the existing industrial environment society provides an environment for being connected with industry and technology. For these reasons, it is possible to name Industry 4.0 and Germany together.

Germany has revealed smart factory concept by bringing together the artificial intelligence, the Internet of Things (IoT), cloud computing, such as many modern technologies in Industry 4.0 concept. On the other hand, these smart factories cover not only large, institutional firms but also small and medium sized enterprises [1]. This concept can be defined as all of the systems that allow the improvement of production processes with a flexible structure that can be dynamically adapted to changing customer requirements. In order to achieve this, the enterprise must change its entire structure from end to end to cover each working section. In this way, it has advanced software, hardware and mechanical supports; while on the other hand, it can reduce unnecessary labor costs by encouraging the recruitment of qualified personnel [21].

Professor Kagermann and the scientists at the National Academy of Science and Engineering (Acatech) in Germany has collected the promises of intelligent factories that come with Industry 4.0 in both production and consumption stages in a report. In addition to individually approaching customer demands instead of uniform production, there are promises to provide flexibility for each customer. It is also planned to improve resource use and decision-making processes for sustainability and efficiency. Apart from customer needs, motivation within the company is also considered. Accordingly, it is planned to create a balance between the social and business lives of the employees and to meet the needs according to the population distribution [22].

The report presented by Acatech employee's shows that Germany is still among the world's giants about technology, production and innovation thanks to smart factories by introducing the country's vision. In order to realize the promises made, it is necessary to create a cyber-physical system environment. Thanks to sensors, actuators, RFID codes that are embedded to the objects in the factory, motions, changes, situations on physical environment are transferred to the digital environment. With the aforementioned hardware, all types of data coming to the virtual space called Cloud accumulate as large amounts of information and these form the Big Data concept which is determined as the first parameter in our study [23]. These data is processed by data mining that is so popular in these days and data is used for different purposes thanks to various algorithms. In this way, machine learning develops and objects know that what their duties and responsibilities are. They also have the ability to make decisions thanks to the artificial intelligence behind them and inform their users about their maintenance time [24].

An enterprise aiming to realize Industry 4.0 applications by showing an end-to-end change will show a horizontal and in-depth growth that will achieve optimum productivity in all areas from stock management to supply chain management in the materials it uses in production. In addition to this, its qualified personnel specialized in its field and unqualified personnel will be affected [25].

In this context, either less manpower or more economic production and service affect human that is a main unit of community. When this interaction is examined, the human factor should be considered in two different categories that firstly is during production or service processes, and secondly is after production or service processes. In the first stage, human factor that is very important appears as how to raise or provide labor force in line with changing demands. In the second stage, the human factor emerges as the factor that will benefit from the product or service, and the changing demands and the role that the social economic structure will play.

4 Society 5.0

Is technology for society, or for society's interests, or for the individual? As stated in the OECD (2001) report, today, countries need to give importance to communication and information technologies in order to reach a certain level of prosperity and to get to the level of development [26]. When the results of these studies are considered, it is generally to make certain financial inferences and to strengthen the economy by creating high added value. Japan, one of the leading technology leaders in the world, has added a different perspective to this and stated that it aims to use all advanced technologies on behalf of social life of society, and it will take an individual approach [3, 27]. This is called Society 5.0.

The factors that exist in the system and the current economic and social order shape the lives of people. Based on this, it is seen that the people in the first ages have had to discover their environment and obtain their food by making primitive hunting tools from the objects around them and had to continue their lives in this way. People have brought a new dimension to their lives, with the discovery of the soil. They have provided their nutrition by planting food seeds, and then they have planted various seeds such as silk, flax etc. Economically, they have made progress as an agricultural society. The people who made production without using any industrial machinery with the materials in their hands entered the Industrial Era with the invention of steam machines in the 18th century and were called as industrial society. Together with the developing science in the 1970s, people have invented the machine called computer and thus stored the information they had acquired in this machine and processed it for the purpose they wanted. At the same time, by enabling digital communication, the transition of advanced societies to the information age has been provided [29].

Even though it is said that we are living in a digital age and that we all have easy access to the internet, is this situation true in reality? Japan argues that this is not the case, and states that people have limited access to information and technology due to a number of variables such as their physical condition and age etc. [30]. To this end, firstly it proposes to use the stack of information created by cyber physical systems, which transfer the physical world to virtual environments, in a cloud system for beneficial purposes for the public. And this country which would use this advanced technology in a positive way according to their own conditions to improve the quality of life, has evaluated this as Society 5.0 after the previous 4 types of society, and has confirmed the accuracy of this concept with the decision of the Council of Ministers on January 2016 [31].

The main objectives of Japan aiming to develop Society 5.0, based on their own community characteristics, such as scarcity of population, the increase in the number of older populations and the frequent occurrence of natural disasters in the country, are listed below [3, 32]:

- To be able to prevent decreasing population,
- To provide solutions to the increasing elderly population that make life easier and improve,
- To create a sustainable environment by taking technological measures against environmental pollution and natural disasters,
- Acting for the interests of society together with technology.

In line with the set goals, it is seen that with Society 5.0, Japan will form a human-centered (individually approaching) community structure. In this way, individuals will be able to lead a higher quality life by taking the necessary service regardless of their position, age, sex and languages [33].

The main parameters of Society 5.0 are the Big Data, which is accumulated in a virtual environment called Cloud, respectively, some cloud calculations that process this data, the Internet of Things (IoT) which enables the devices that are connected to the Internet to communicate, and the artificial intelligence that follows. Of course, the most important is the people who use it and the engineers who develop it after this technology.

5 Industry 4.0 vs. Society 5.0

Germany, which outperformed other major countries in the process of industrialization, contributes 23.4% to the gross national product according to the figures given in 2017 together with its industry. In addition, employment provided to employees as much as the number of population in a country emerges in the same study [34]. According to this information, it is not surprising that Germany is the pioneer of the 4th Industrial Era. Afterwards, 'The Industrial Internet' introduced by the United States, and 'The Industrie du futur' supported by France and 'Made in China 2025' concept for China's future show that it is necessary to transit to industrial change. Thanks to that, it is foreseen that a country would be dynamic and perminent [1].

The main objective of Industry 4.0, which is advocated and developed among the countries that are mentioned as a new perspective to the industry, is to realize a high value-added production and to improve their production processes by enabling the machines in the factories to communicate with each other. In this way, it is promised that there will be a certain decrease in cost and time in various departments in a factory and that the productivity will increase to top levels for both the consumer and producer sides. At this stage, German Federal Ministry of Education and Research In the project named Picasso, it finances the efficiency in different departments for a business. Accordingly, expected reductions in production, distribution processes and costs that will undergo change are given below [35]:

- Inventory costs - between 30% and 40%
- Production costs - between 10% and 30%
- Logistics costs - between 10% and 30%
- Complexity costs - between 60% to 70%
- Quality costs - between 10% and 20%
- Maintenance costs - between 20% and 30%

Industry 4.0, which covers production and its processes in general, also affects all employees who works and will work in these areas [36]. On this subject, Tüsiad and the Boston Group advocate that high value-added production would be occured by increasing the number of qualified personnel by decreasing unqualified workforce [18]. They show that with a cycle that is in their report. In this cycle, it is shown that an unqualified workforce will decrease thanks to the investments made in Industry 4.0, and thus, it will take place in international competition. As a result, it is illustrated that a high value-added production will be made with the new and highly qualified workforce and the annual growth of the enterprise will increase. This phase becomes a cycle thanks to the investments to be made in Industry 4.0 again [18].

In this study, the literature studies about Industry 4.0, which is described as the 4th Industrial Era, show that this concept does not affect the whole society, however it affects the large organizations that are active in the fields of industry, production, R & D, manufacturing. Moreover, it has an effect on each employee of small and medium enterprises from managers to workers. And for these, it foresees a positive change together with a sustainable economy. Indeed Promises of Kagermann et al.'s on employees that the company has, on relations between their competitors, and on communications between their customers confirm this state. In other words, it seems that it exists for a limited community of people and will serve for them. It is possible to mention that there is a limited usage and benefit in this topic.

The digitalization of the physical world and the existence of a cyber-physical system not only inspired Germany in its industry, but also inspired Japan that gives an importance on social values and environmental values. Japanese scientists believe that technology that cannot be stopped and that develops more and more each day should be for the benefit of society, and moreover, that it is accessible and usable for the individuals. They do not want shape the future of the world depending on just industry and production. They advocate that shaping the future depending on level of welfare of public by enhancing their social life.

In order to develop of this concept they support, through the Internet of Things, they process data of all types and sizes from the relevant unit, person, location, sensors, actuator, RFID or any device that can be used in this context. They adopt the creation of a new artificial intelligence and thus provide every segment of society with the services they want at any time. In this way, they guarantee that a sustainable environment can be maintained. To give an example, in the country where the number of elderly people has increased considerably, it is promised that necessary solutions will be made for elderly people. Required medical measures will be taken for this segment of the population and these individuals will be followed. In case of emergency or necessity, communication between the hospital and the elderly individual will be provided and relevant interventions will be made. The specified follow-up will be performed with smart mobile devices [37]. Similarly, Japan, which is a country that has suffered a lot from natural disasters due to its geographical location, states that they will solve this situation with the idea of Society 5.0. Artificial intelligence will be created with radar, sensor and satellite data placed in possible disaster areas, warning mechanism will be established for those places and major damages will be prevented from occurring. In the event of a disaster, location information is received by determined devices via GPS and status information of the damaged area is obtained via drones and terrestrial weather radars. The information received will be transmitted to the relevant health, public and relief teams nearby via smart mobile devices [38].

As it is seen, the concept of Society 5.0, that aims to create innovative, fast and efficient solutions by creating different artificial intelligence for every segment of the society, introduces a sustainable life opportunity by integrating human and nature in harmony. Society 5.0 emerges with smart societies as opposed to smart factories of Industry 4.0. Japan thus breaks the perception of technology as a threat to people and wants it to be considered a useful factor for people.

6 Conclusions and Related Work

In this study, the common parameters between Industry 4.0, which was introduced to the public under the leadership of Germany, which is one of the current trends, and Society 5.0 that is presented first by Japan, are determined and a broad comparison is made between them. As a result of the comparisons, it is seen that both concepts will rapidly become widespread among countries as a gain of globalization and will deeply affect people in the specified fields with the new generation production and social structure. However, it is concluded that the concept that will affect the way people live, society structure and social aspect of the country substantially is Society 5.0 (see Table 1).

Because Industry 4.0 has been introduced to the market before, it is more demanded by people by contrast with Society 5.0, and is better known to the public and governments. The desire of companies to move to new generation factories, rapid adaptation processes and sample applications in this field are taking the lead for this. When looking at from this perspective, Industry 4.0 is thought to provide a base for Society 5.0, whose with has many common points [39].

As emphasized in this study, Industry 4.0's most important fact is to increase efficiency and decrease the effect of human factor by using modern systems and infrastructure such as IoT, cloud computing, simulation, augmented reality, big data, etc. It has an effective role especially in the field of production. However, in Society 5.0, we see that the subject is directed towards social structuring and that the factors are analyzed and integrated in much more detail.

Industry 4.0 is one of the most important concepts today and is applied or will be applied all over the world. It should be noted that Industry 4.0, which is not only limited to production, but can be applied in different areas, should also take into

Table 1. Industry 4.0 vs. Society 5.0

	Industry 4.0	Society 5.0
Origin/Year	Germany/2011	Japan/2016
Master Domain	Production sector that includes customers and employees	All society and their life styles
Approach	Individual for customer Holistic for business	Individual
Purpose	Commercial	Public
COMPARISON FOR JOINT PARAMETERS		
Big Data	Mostly Involved	Mostly Involved
AI	Mostly Used	Mostly Used
IoT	Mostly Used	Mostly Used
Cloud Computing	Mostly Used	Mostly Used
Human	Partial Involved	Mostly Involved

consideration the implications and changes for both employees and the social community. In this work, the subject is discussed in general and in the ongoing studies, differences and solutions arising in the practices in different countries will be discussed. The same applies to Society 5.0, and it is planned for the future studies that it will be researched whether that they are implemented in different countries and the benefits of these practices.

It is our general aim to find differences and similarities between two concepts that bring a new era with numerical data by conducting detailed researches. In this study, which is planned as a beginning, conceptual analyzes are made between Industry 4.0 and Community 5.0 by determining the general lines and common parameters, and the differences between them are presented. Numerical analysis to be carried out for our purpose will be done in future studies and the results will be examined.

References

1. Rojko, A.: Industry 4.0 concept: background and overview. Int. J. Interact. Mob. Technol. (iJIM) **11**(5), 77–90 (2017)
2. Evans, P.C., Annunziata, M.: Industrial internet: pushing the boundaries. General Electric Reports, pp. 488–508 (2012)
3. Federation, K.J.B.: Toward realization of the new economy and society—reform of the economy and society by the deepening of "Society 5.0" (2016)
4. Lohr, S.: The age of big data. New York Times **11** (2012)
5. McAfee, A., Brynjolfsson, E., Davenport, T.H., Patil, D.J., Barton, D.: Big data: the management revolution. Harvard Bus. Rev. **90**(10), 60–68 (2012)
6. Labrinidis, A., Jagadish, H.V.: Challenges and opportunities with big data. Proc. VLDB Endowment **5**(12), 2032–2033 (2012)
7. Smith, T.: Artificial ıntelligence. Comput. Sci. A **165** (2005)

8. Russell, S., Norvig, P.: Artificial İntelligence: A Modern Approach Artificial Intelligence: A Modern Approach, pp. 122–145. Alan Apt, New Jersey (1995)
9. Lateef, Z.: 25 October 2018. www.edureka.com. https://www.edureka.co/blog/artificial-intelligence-applications/. Accessed: 15 Apr 2019
10. Lindsay, R., Buchanan, B., Feigenbaum, E., Lederberg, J.: Applications of artificial intelligence for organic chemistry (1980)
11. Qin, J., Liu, Y., Grosvenor, R.: A categorical framework of manufacturing for industry 4.0 and beyond. Procedia CİRP **52**, 173–178 (2016)
12. Miorandi, D., Sicari, S., De Pellegrini, F., Chlamtac, I.: Internet of things: vision, applications and research challenges. Ad Hoc Netw. **10**(7), 1497–1516 (2012)
13. Van Kranenburg, R.: The ınternet of things: a critique of ambient technology and the all-seeing network of RFID. Institute of Network Cultures (2008)
14. Polat, L., Karakuş, G.: Approach of medıum-sized ındustry enterprıses to ındustry 4.0 a research in Konya. In: The International Symposium for Production Research, pp. 345–354. Springer, Cham, August 2018
15. Fox, A., Griffith, R., Joseph, A., Katz, R., Konwinski, A., Lee, G., Stoica, I.: Above the clouds: a Berkeley view of cloud computing. Department Electrical Engineering and Computer Sciences, University of California, Berkeley, Report, UCB/EECS, 28(13) (2009)
16. Armbrust, M., Fox, A., Griffith, R., Joseph, A.D., Katz, R., Konwinski, A., Zaharia, M.: A view of cloud computing. Commun. ACM **53**(4), 50–58 (2010)
17. Schmidt, R., Möhring, M., Härting, R.C., Reichstein, C., Neumaier, P., Jozinović, P.: Industry 4.0-potentials for creating smart products: empirical research results. In: International Conference on Business Information Systems, pp. 16–27. Springer, Cham, June 2015
18. Tüsġad, R.: Türkiye ‘nin Küresel Rekabetçiliği için Bir Gereklilik Olarak Sanayi 4.0 Gelişmekte Olan Ekonomi Perspektifi (2016)
19. Keidanren: Society 5.0: C-o-creating the future Keidanren (2018). https://www.keidanren.or.jp/en/policy/2018/095_outline.pdf
20. Otte: BDI The Voice of German Industry, 10 February 2016. https://english.bdi.eu/article/news/industry-40-an-opportunity-for-germany/. Accessed 15 Apr 2019
21. Radziwon, A., Bilberg, A., Bogers, M., Madsen, E.S.: The smart factory: exploring adaptive and flexible manufacturing solutions. Procedia Eng. **69**, 1184–1190 (2014)
22. Kagermann, H., Helbig, J., Hellinger, A., Wahlster, W.: Recommendations for implementing the strategic initiative INDUSTRIE 4.0: securing the future of German manufacturing industry; final report of the Industrie 4.0 Working Group. Forschungsunion (2013)
23. Lee, J., Bagheri, B., Kao, H.A.: A cyber-physical systems architecture for industry 4.0-based manufacturing systems. Manuf. Lett. **3**, 18–23 (2015)
24. Lee, J., Kao, H.A., Yang, S.: Service innovation and smart analytics for industry 4.0 and big data environment. Procedia CİRP **16**(1), 3–8 (2014)
25. Gilchrist, A.: Industry 4.0: The İndustrial İnternet of Things. Apress (2016)
26. Clarke, R., Durand, M., Pilat, D., Torres, R.: The New Economy: Beyond the Hype. The OECD Growth Project. Economics. Organisation for Economic Cooperation and Development, Washington, 2001 L St., NW, Suite 650, Washington, DC 20036-4922 (2001)
27. Wang, F.Y., Yuan, Y., Wang, X., Qin, R.: Societies 5.0: a new paradigm for computational social systems research. IEEE Trans. Comput. Soc. Syst. **5**(1), 2–8 (2018)
28. Günay, D.: Sanayi ve sanayi tarihi. Mimar ve Mühendis Dergisi **31**, 8–14 (2002)
29. Webster, F.: Theories of the İnformation Society. Routledge (2014)
30. Cabinet Office. www8.cao.go.jp. https://www8.cao.go.jp/cstp/english/society5_0/index.html . Accessed 20 May 2019
31. Cabinet Office: Report on The 5th Science and Technology Basic Plan (2015)

32. Özdestici: 14 May 2017 https://webrazzi.com. https://webrazzi.com/2017/05/14/toplum-5-0/ . Accessed 21 May 2019
33. Harayama, Y.: Society 5.0: aiming for a new human-centered society. Hitachi Rev. **66**(6), 556–557 (2017)
34. Orth, M.: 24 August 2018 www.deutschland.de. https://www.deutschland.de/en/topic/business/germanys-industry-the-most-important-facts-and-figures. Accessed 21 May 2019
35. Bauernhansl, T.: WGP-Standpunkt Industrie 4.0. WGP, Wissenschaftliche Gesellschaft für Produktionstechnik (2016)
36. Erol, S., Jäger, A., Hold, P., Ott, K., Sihn, W.: Tangible Industry 4.0: a scenario-based approach to learning for the future of production. Procedia CİRP **54**(1), 13–18 (2016)
37. N. B. Design: Realizing Society 5.0, JapanGov
38. Cabinet Office www8.cao.go.jp. https://www8.cao.go.jp/cstp/english/society5_0/bosai_e.html. Accessed 22 May 2019
39. Skobelev, P.O., Borovik, S.Y.: On the way from Industry 4.0 to Industry 5.0: from digital manufacturing to digital society. Industry 4.0 **2**(6), 307–311 (2017)

Intelligent Design and Advanced Precision Metrology on the Geometrical Structure of Medical Needles Based on GPS and ISO Standards

Osman Bodur(✉), Günther Poszvek, Eva Maria Walcher, Numan M. Durakbasa, and Jorge M. Bauer

Institute for Production Engineering and Photonic Technologies, Research Group of Industrial Metrology and Adaptronic Systems, TU Wien, Getreidemarkt 9, 1060 Vienna, Austria
bodur.osman@hotmail.com, guenther@poszvek.net, walcher@ift.at, numan.durakbasa@tuwien.ac.at, doktorjbauer@yahoo.com.ar

Abstract. Medical products generally occupy a significant place in advanced manufacturing technology. Due to their sophisticated production systems, these productions are looking for high-precision measurement technology on a micro-scale within the high-quality functional requirements. Sophisticated metrology, as an important part of the production system in the context of Industry 4.0 reduces the uncertainties of measurement processes in the geometry of the manufactured parts and contributes to the preparation of critical solutions for high product and process quality through established standards and integrated management systems. In the context of Industry 4.0, manufacturing systems are updated to an intelligent level. The medical products shall design intelligently according to the functional requirements, and geometrical tolerances and form deviations adjusted the functional characteristics in production performance specifications. However, there is some deficiency of research on the form and geometrical tolerances, even if there are several articles that give information on the mathematical model of needle geometry. The purpose of this research is to indicate an advanced approach with precise measurement methods and intelligent measuring instruments in the production of medical devices to integrate the specifications of precision metrology and to encounter the sustained improvements towards more accurate geometrical structures in the manufacturing of workpieces. Considering this context, the congruent angles of the needles are specified, which affect the functional specifications of the medical needles. The form deviation and geometrical tolerances of the medical needle are implemented on the 2D and 3D technical drawings depending on the associated applicable GPS system and ISO standards. Besides, the surface structures of the medical needles are observed by using advanced metrology techniques through a digital microscope. Accordingly, the surface and multi-line roughness parameters of each surface formed in manufacturing steps by a ground operation, are evaluated.

N. M. Durakbasa and M. G. Gençyılmaz (Eds.): ISPR 2020, LNME, pp. 346–362, 2021.
https://doi.org/10.1007/978-3-030-62784-3_29

Keywords: Advanced metrology · Medical needle · Hypodermic · GPS · Geometrical tolerance · Form deviation · Surface roughness · Line roughness · Gauge · Pre-fillable syringes · BPMN

1 Introduction

Design inputs are the most significant parameter of product design controls in the medical device industry, like precision manufacturing industries. They might absorb almost 30 percent of the whole project time. Inputs have to be clear, practical and described precisely by confirmed through some technical methods, assessments, or examines. Also, they have to include measurable terms like geometrical tolerances, where it is possible at the technical drawings. The dimensions and tolerances should be clarified accurately and defined the acceptable limits for the functional characteristics in the product performance specification [1].

Hypodermic needles are used for injecting and retracting fluids for principally the human body. The assignment of needles shall be suitable, according to ISO 9626:2016 [2]. In ISO 7864:2016, min and max tolerances on the length of the needles are respectively −2 mm and +1 mm for up to 25 mm height. As viewing under 2,5x amplification, the needle points should show up keen and clear from the further edges, burrs, and hooks [3].

The gauge number system is utilized openly around the world for dimensioning medical instruments and equipment. Sterile Hypodermic Needles are diversified with different tubing gauge numbers for the manufacture of medical devices. International Standard provides the geometrical dimensions and the mechanical specifications of tubing of defined metric sizes between 10 (3,4 mm) Gauge - 34 (0,18 mm) Gauge. This research is carried out on one of them which is 27 Gauge medical needle. According to the table of dimensions of tubing, 27 Gauge number has 0,40 mm designated metric size, min. and max. outer diameter of sterile hypodermic needle respectively 0,40 mm and 0,42 mm [2]. Although the minimum inner diameters depending on types of the wall as explained in ISO 9626:2016, when the designated metric size of the needle is 0,4 mm, regular walled tubing of the needle is 0,184 mm and thin-walled tubing of the needle is 0,241 mm. Also, the needles generally are characterized by their two different bevel angles and the angle of the rotation for the second bevel angles. One of them is the primary bevel angle ξ and tip angle φ, which separates the bevel type of the needles [4]. The assignment of needles is used to define the dimensions and character of the needles are demonstrated in Fig. 1a:

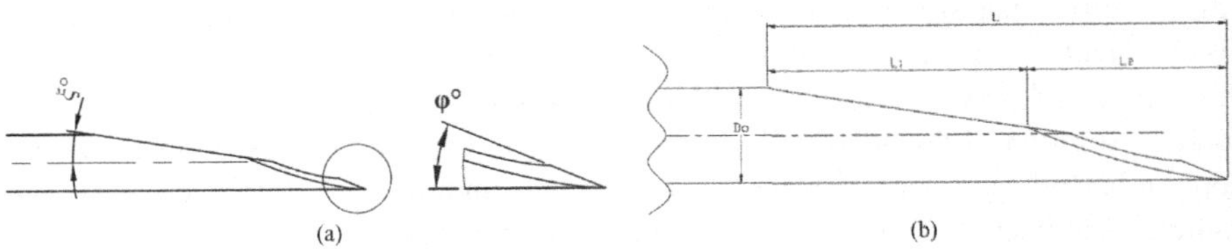

Fig. 1. (a) Primary bevel angle, tip angle and (b) the cut lengths of the needles

However, the cut lengths on the surface of the needle (L is cut length outcome of angle on bevel angle ξ, L_1 is primary length, L_2 is secondary length) specify depending on the angles of the needle in DIN 13097-4:2019 [5] and ISO 7864:2016 [3] as seen on the above Fig. 1b.

Peidong Han and Demeng Che have created a mathematic model for the included and tilt angles of cutting edges for some variations of needles in their work. They explain how the included and tilt angles, according to the bevel angles at the geometry of the needles, are essential for the percutaneous operations. That procedure affects the pain and trauma of the patients. The penetration force and placement errors would be reduced by the limitation of the angles [6]. A triple grinding process generates the medical needle. Yancheng Wang and his colleagues mentioned the significance of the needle insertion force depending on the tip angle (φ) and bevel angle (ξ) as Peidong Han's article [7].

Likewise, the inner cylinder of the needle that fluid flows without bream and roughness, and the sharpness of the needle affect the efficiency of penetrating and damaging the tissue and nuisance [4]. The higher roughness parameter increases the damaging of the tissue. That's why R_a (Surface roughness value) parameter on the critical surfaces has to be no more than 0,8 µm according to the ISO 80369-7 [8].

Based on all the above information, the motivation of that research is generally integrated with advanced precision metrology and the geometrical product specification on the needles, which has descriptive angles and diameters. It is possible the geometrical definition of the medical needle in the literature or the scientific article; conversely, it isn't mentioned the geometrical product specifications of the hypodermic needle. It is a general problem with components/"machine elements," which are standardized about their geometry, but whose tolerance (e.g., due to the age of the relevant standards) is not described in terms of ISO GPS. From experience with similar problems, and (approximate) "interpolation" is quite tricky and time-consuming (but not impossible).

2 Methods and Theoretical Parts

This section focuses on the functional specifications, the medical needle manufacturing process, the associated applicable GPS system that can be implemented on technical drawings, and validation through GPS concept and sophisticated measurement systems.

2.1 Purpose of Using Glass Pre-fillable Syringe with Needle G27

All we know that some of the medical solutions are injected into the body against diseases and for intervention or therapies through the needle or cannula. G27 is one of those types of needles and used within ready-to-use syringes containing enoxaparin sodium solution. Enoxaparin sodium is obtained by alkaline depolymerization of heparin benzyl ester derived from the porcine intestinal mucosa. However, enoxaparin sodium is indicated in the treatment of the following essential conditions:

- Especially those who undergo general surgery, including orthopedic surgery or cancer surgery and patients undergoing moderate and high-risk surgery
- patients with acute medical diseases and reduced mobility, including high-risk acute heart failure, respiratory failure, serious infections, and rheumatic diseases
- In patients with acute coronary syndrome, deep vein thrombosis, and thrombolytic therapy [9].

2.2 Injection Application Technique of Needle G27 of Glass Pre-fillable Syringe

The injection is preferably done while the patient is lying down. The application is performed alternately between the left and right anterolateral or posterolateral abdominal wall. The needle is inserted into the crimp of the skin vertically, without pressing it between the thumb and forefinger. This skinfold isn't released until the injection is completed [9].

2.3 Functional Requirements and Specifications

According to the section A and B and previous studies, the functional requirements shall be technically reducing the insertion force and friction force at the tissue, and increasing the position accuracy of the needle insertion, medically reducing the pain and trauma of the patients. Another essential functional requirement is tissue damage and conformity. Although the flawless results are obtained on the skin model[1], unfortunately, the needles aren't ideally manufactured as the same quality as their skin models. So, in line with the goals, depending on the functional specifications of the needle, the models should be limited by geometrical tolerances. Besides that, roughness parameters on the tip and surface of the needle shape the efficiency of the needle to indwell the tissue with less resistance and to entail the tissue damage. The following section shows the production method and the geometrical definition that affect functional specifications.

2.4 Production Process and Geometrical Definition

Glass pre-fillable syringes are manufactured from the tubing cone glass. That tubing cone glass is cut to the desired volume scales, which are variously available in 0.5 mL up to 20 mL volumes. Tubing glass is heated approximately 1500 °C to form the tip, which acts as the sliding bearing for the needle. The other side of tubing glass is also heated to give shape for the flange. The flange is cut as circle segments from each side of the flange with angles 90° and 270°. The barrel is printed on the surface of the cylinder form glass syringes to show the scale of volumes in ml. All of the solid volumes of glass syringe are heated once more to insert the needle into the tip, and the adhesive is dispensed to strengthen the needle also used as a sealing element. UV light is used to harden the adhesive and hold the needle in place. After inspection of all

[1] Skin model: geometric model is defined non-ideal features.

profiles, syringes are cleaned and used silicone chemicals to the internal part. The needle, which is produced by Stainless Steel Z6CN18/10[2], is covered by the needle shield, and indeed, the process of packaging continues [10, 11].

One of the most critical steps of syringes production is "manufacturing a medical needle" by grinding methods. In the beginning, stainless steel G27(outer diameter 0,40 mm) number metal pipe is cut as a predetermined length according to the standards. The shape of the needle tip eventuates with 4 step operation. At the first step, the tip of the needle is sharpened with ξ primary bevel angle by grinding wheel. At the second step, the primary bevel angle is increased to the tilt angle ϕ, which also is interpreted as the second bevel angle, according to the datum plane of the grinding wheel. The contact line of the grinding wheel on the needle tip creates a datum plane. In the third step, the needle is rotated around itself by an orientation angle described as β, and also, the grinding wheel is moved upwards q mm from its first position is defined "datum plane." The needle is sharpened the second time. At the last step, the needle is rotated to another direction with a 2β angle according to the first rotation around the centerline of the needle again, and the needle is ground third times. Thus, the second lancet is created [7]. Besides that, tip angle φ appears as a result of other angles parameters, which are tilt angle ϕ and rotating angle β. The triple grinding process is shown as a four-step operation in the following manufacturing flow charts in Fig. 2. One of the critical points is, all manufactured needles are inspected through the microscope after 4th steps, and the medical needles are separated according to availability. Needles that fail in the inspection are sent recycling. Otherwise, they get ready to montage with the syringes.

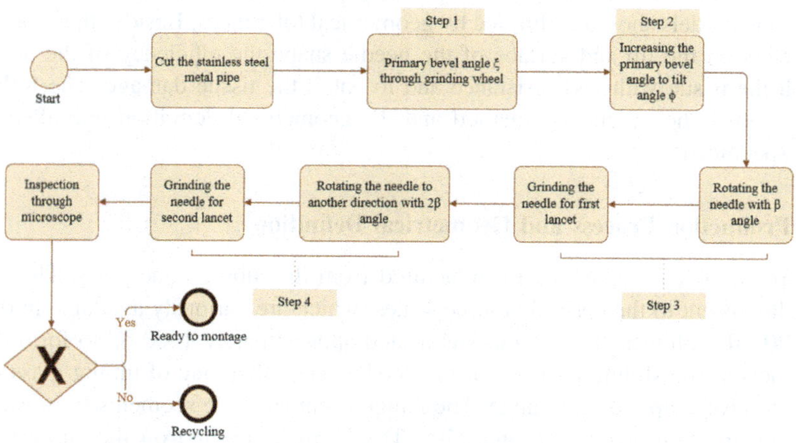

Fig. 2. The Manufacturing flow chart of the medical needle

After the first grinding process as explained on the flow chart, the orientation to the third step from the second step is demonstrated with the technical drawing as seen in

[2] Z6CN18/10 is also equivalent to 08KH18N10T, it is Corrosion-resistant high-temperature steel.

Fig. 3a to comprehend the second grinding process. The position of the medical needle after the 2nd step is depicted with the thick line and after being turned at the 3rd step is described with thin lines. The geometrical definition of the needle tip depends on the three angles ξ (primary bevel angle), ϕ (tilt or secondary bevel angle), β (rotation angle), and the geometrical dimension q is the significance for the lancets of a needle and tip angle of a needle. P is the needle lancet point, P' is the position of the needle lancet point after orientation, X' rotated axis by β angle, the parameter g is the distance between points P and P', and the parameter d is the distance between a datum plane and point P at the same direction with the X' axis. 3D geometry of the hypodermic needle in Fig. 3b is the consequence of the production. This figure can be utilized to better understand the geometrical sizes with the help of the planes. The planes are demonstrated as PP_d is perpendicular plane to the cutting direction, P_t is the cutting face plane of the needle tip, and PP_b is the normal plane of the P_t plane. Axis p is on the PP_d plane, axis b is on the PP_b plane, axis t is in the same direction with P_t plane, and axis d is the same direction with P_d datum plane. Between datum plane and tip plane occurs φ tip angle; between PP_d and PP_b occurs bevel angle ξ as same as between datum plane P_d and bevel plane Pb.

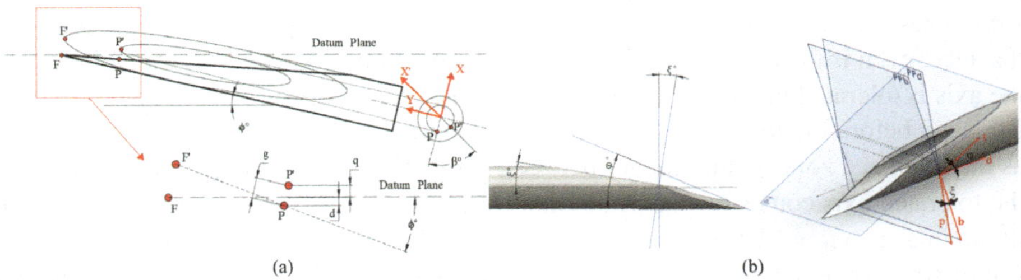

Fig. 3. (a) 3D view of the bevel and tip angles between planes and axis (b) The orientation to the third step from the second step in the manufacturing process.

2.5 Geometrical Product Specifications

The researcher has some difficulties in merging his/her concepts for the function of a fragment with mandatory orientation, form, and location tolerances. These tolerances provide to understand the geometric and functional specification clearly and precisely so, setting the fundamental tolerancing guidelines are required [12].

For the time being, the principle of independence is addressed again in connection with the disposable syringes. The principle of independence is laid down in ISO 8015 as an elementary principle of tolerance [13]. Therefore, it is essential to produce the entry in the drawing as clearly defined. The principle of independence always applies, even without a particular drawing entry. Because of the clarity and clarity, however, the indication "Tolerancing ISO 8015" should be noted on the drawing in or above the title block and the reference "Concepts, principles, rules: ISO 8015" would be even more appropriate. The general note, "This document is following ISO standards," should also be made at the beginning [14]. Independence also means that every tolerance is checked individually [13]. According to ISO DIN 11608-2, the outer surface structure of the tubing should be flat and flawless. Also, the importance of form deviations is mentioned. The size of the stylet, which is a slender medical probe to test the patency of lumen, is recommended as 0,15 mm for the needle of regular walled tubing and 0,19 mm for the needle of thin-walled tubing for G27 needles. It turns out that the stylet is also a parameter affecting the surface roughness of the flow area of the needle. One of the other vital points is geometrical and dimensional tolerances such as cylindricity, perpendicularity, concentricity, and envelope requirement[3] is highlighted in the section test gauge for needle [15].

In our research, we have to consider several types of tolerance. In general, we start from the ideal shape of the element. In this case, we take into account the shape tolerances (straightness, roundness, and cylindricity) and positional tolerances (inclination and Coaxiality/Concentricity). There is also the possibility to define and check the corresponding tolerances using running tolerances, but this is not relevant in our example. Straightness tolerance is used for axes and center lines (derived elements). This tolerance serves to ensure the overall straightness in a longitudinally or rotationally symmetrical geometric element. For clear identification, the tolerance arrow must be on the dimensional arrow that measures the shape element from which the axis or centerline is to be derived [16]. For the axis of a circular cylinder, only the straightness deviation of the axis is restricted, this does not apply to its surface line. For instance, even if the axis is straight, the cylinder could be crowned. The straightness of the axis is tolerated in all directions; this is a general requirement. The diameter symbol is placed before the tolerance value. It creates a circular cylinder with the diameter t_G[4] as the tolerance zone and no longer between parallel straight lines. The entire length of the tolerated axis comes to rest within this circular cylinder. What happens if you omit the diameter symbol before the tolerance value? Nothing at all due to the rotational symmetry, one would have to check the straightness in all directions. The diameter symbol is a prerequisite for uniqueness and should always be set [16, 17].

After functional requirements are defined, and functional specifications are determined according to them, geometrical tolerances should be assigned concerning functional specifications. Geometrical tolerances of the needle can be specified with tolerances of form, orientation, and location. Straightness of surface line and straightness of axis of the cylinder as form tolerances, roundness as form tolerance, cylindricity as form tolerance, angularity as orientation tolerance, Coaxiality (for axes) is one of the location tolerances. Straightness is one of the characteristic form tolerances; also, it doesn't need any datum element. Toleranced surface line and the axis of the needle can be an integral feature or a derived feature. The cylindrical surface line of the needle should be kept between two parallel planes apart. The tolerance zone for the axis of the cylinder is bounded by a cylinder of diameter ϕt. Roundness is another specific form of tolerance and contains between two concentric circles in any same cross-section of the surfaces, so these circles create the tolerance zone for roundness.

[3] This requirement means that the envelope of perfect form at the maximum material size is not to be violated. Used in special circumstances to ensure that mating features will always assemble.

[4] t_G: straightness tolerance zone.

Cylindricity is also one of the form tolerances and contains between two coaxial cylinders, which have different radius creates a limited tolerance zone. Angularity is one of the orientation tolerances which is needed datum elements. Datum elements also are chosen as planes or axis. The tolerance zone is limited by two parallel planes with distance t and inclined at the referred angle to the datum element. Coaxiality is a location tolerance which is limited by datum elements like axis or planes. In any cross-section, the tolerance zone of the centered two circles is limited as a circle of diameter ϕt, concentric with datum point A [17].

2.6 Micro and Macro Geometrical Assessment Through Measurement Device

Nowadays, measurement technologies are developed thoughtfully. Those developments reduce the measurement time, uncertainty, and increase precision. A digital microscope is one of the impressive devices used to record the visualization of the small details at high resolution. Also, Numerous measurements can be done out with only one device like such as residual dirt analysis, grain size analysis, or roughness measurements. It is able to be measured 3D and 2D roughness parameters both quickly and at high-resolution thanks to such an opportunity. Besides the surface structures, the geometrical parameters of the needle, such as angles, outer and inner diameter, are possible to measure through the new generation digital microscopes, as seen in Fig. 4.

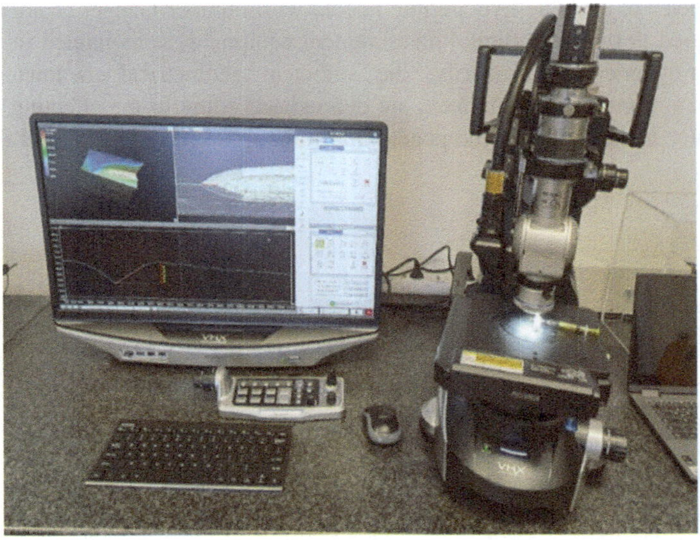

Fig. 4. Digital Microscope [18]

3 Results

The geometrical tolerances of needles used for pre-fillable syringes and hypodermic needles are defined and demonstrated in 2D and 3D technical drawings depending on the functional and geometrical specifications according to ISO 1101:2017 as seen in Fig. 5. The technical drawings are prepared through CAD programs. Besides, there are limits in the CAD programs: drawing as it is used in the construction field is not the same when drawing in the micro dimension [19]. According to ISO 7864:2016 and DIN-13097-4:2019 and some relevant scientific works explained in the second part of the paper, three central angles (ξ first bevel angle, φ tip angle, ϕ tilt angle or secondary bevel angle after rotated β angle) are limited by angularity tolerance zones to the datum elements which are surface line as specified A. The axis of the needle as specified B., The surface structure of the needle profile, is limited by form tolerance characteristics such as straightness, roundness, cylindricity. The tolerance zone of the inner diameter of the needle is defined with the axis of the outer diameter as a datum element. The tip area of needle structure is scaled by 20:1 in the details X and Y. The dimension parameters that define the profile geometry are D_0 is the outer diameter of the needle, D_i is the inner diameter of the needle in Fig. 5. The inner diameter of the needle is located by the Coaxiality tolerance zone of the centered inner and outer circles are limited through a circle of ϕ 0,05 mm with a middle point datum. Geometrical tolerances of medical needles are approximate results and possible to optimized by extensive researches. Besides the surface structure, the bevel angles, and the inner and outer diameter of a sterile hypodermic needle with number Gauge 27 are determined and measured through the digital measurement equipment as explained in the chapter of measuring method. Primarily, the sufficient geometrical parameters, surface roughness, and multi-line roughness are defined according to the effecting of the pain and nuisance of the patients, the penetration force, placement errors, and the flux.

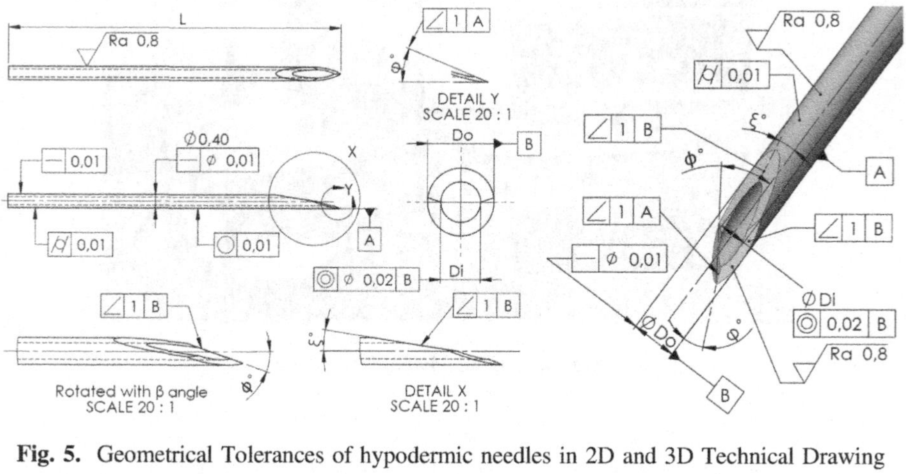

Fig. 5. Geometrical Tolerances of hypodermic needles in 2D and 3D Technical Drawing

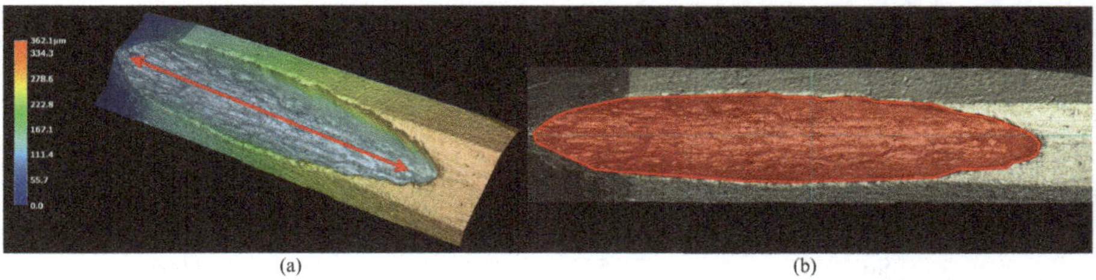

Fig. 6. 3D images of the defined (a) multi-line and (b) surface roughness profile – 1000x enlargement

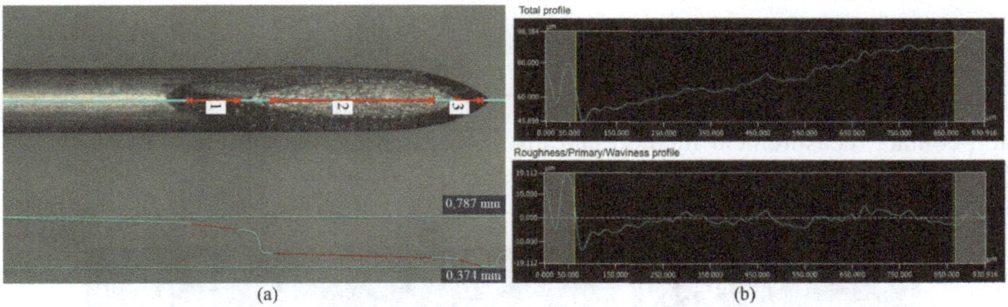

Fig. 7. (a) 3D images of tip and bevel angles – 40x enlargement (b) Multi-Line Total and Roughness Profile

All these measurements are realized for the defined area and multi-line through high resolution whose precision under 1000x (times) magnitude in Fig. 6, and the needle is enlarged 40 times in case of recognizing the bevel and tip angle in Fig. 7a. There are 3 described numbers between up and down surface lines in Fig. 7a; respectively number 1 is the primary bevel angle line, number 2 is the surface line of the inner cylinder and number 3 is the tip angle line. It is possible to see the distribution of height parameters in different colors between 0–362.1 μm in Fig. 6a. Colors clarified the distance of the microscope to the surface, and multi-line roughness parameters are measured and calculated according to ISO 25718:2012 and ISO 4288 as well as used filter type is Gaussian and measurement results, and the roughness parameters of the output profile line of injected fluid are shown in Fig. 7b. Roughness cut off parameter chooses as 0,8 μm according to ISO 4288 [20]. The defined profile line is limited from both sides. The standard deviation of the total profile is more than the standard deviation of the roughness profile. Furthermore, the measurement results are demonstrated in two categories as Ra (Arithmetical mean deviation of the assessed profile), Rz (Maximum peak to valley height of the profile) in Table 1. Besides, spatial, functions, and related parameters are characterized according to the ISO 25718-2 for the defined surface, as seen in Fig. 6b. L-filter (eliminates the largest scale elements from the surface) is chosen as 0,8 mm from the relevant standard [21]. Likewise, the measurement results of the surface roughness parameters are shown in Table 1. Average surface roughness (S_a) is 17.093 μm and the little burrs on the outer side of the inner cylinder (output surface of injected fluid) affects the surface roughness parameter.

Table 1. Measurement results of profile roughness parameters

No.	Measurement name	Average	Max.	Min.	Std. Dev.	Unit
1	Ra	1.969	2.494	1.716	0.270	μm
2	Rz	14.070	18.717	10.717	2.594	μm
3	Sa[a]	17.093				μm
4	Sz[b]	94.620				μm
5	Spc[c]	71.522				1/mm
6	Sdr[d]	0.062				

[a]Sa: Arithmetical mean height of the surface
[b]Sz: Maximum height of the surface
[c]Spc: Arithmetic mean peak curvature
[d]Sdr: Developed area ratio

The stated tip and bevel angles in ISO 7864 and DIN 13097-4 are measured by non-contact measurement method and also compared the measurement results of different needles of the glass pre-fillable syringes:

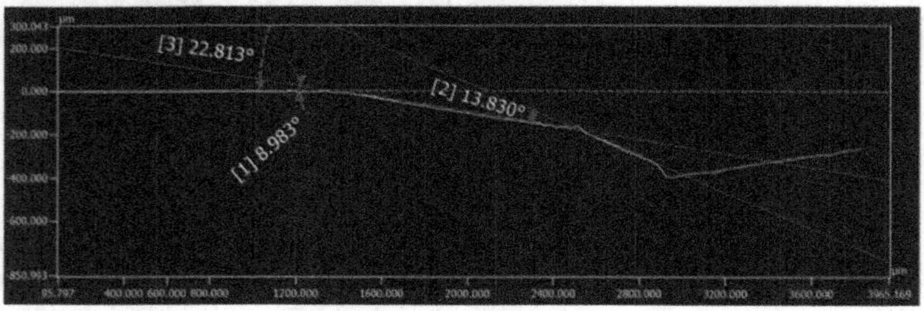

Fig. 8. Measurement the tip and bevel angles of the needle through Digital Microscope

The measurement results are founded out through the 3 different lines, which are respectively, no 1 – primary bevel angle between the surface line of the needle and bevel plane, no 2 – angle difference between bevel plane and tip plane, no 3 - tip angle between tip plane and surface line as seen in Fig. 8. As explained in the outer diameter of stainless-steel needles with Gauge number 27 in ISO 9626, the measurements are provided under appropriate conditions by the digital measurement device. According to measurement results, the measurement interval of outer diameter, tip angle, and bevel angles are extensive, and the controllability of geometric parameters is really low, as seen in Table 2. Designing and manufacturing of the needles based on geometrical tolerances bring closer results and minimum interval range.

Table 2. Digital microscope measurement results

Dimension name	1	2	3	4	5	Avg.	Unit
Outer Diameter Do	418.698	426.534	434.246	412.241	427.633	423,8704	μm
Tip Angle φ	22.813	21.440	22.010	21.845	22.507	22.123	°
Bevel Angle ξ	8.983	8.527	8.477	8.726	8.879	8,7184	°

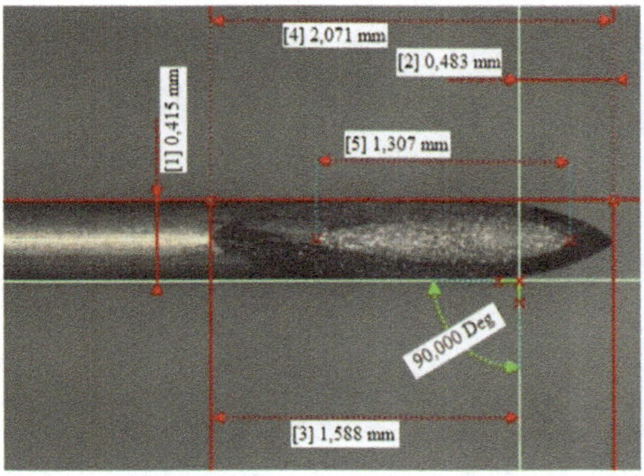

Measurement Number	Dimension Name	Measurement Results
1	Outer Diameter (D_o)	0,145 mm
2	Secondary Length(L_2)	0,483 mm
3	Primary Length (L_1)	1,588 mm
4	Cut Length (L)	2,071 mm
5	Cavity Length (L_i)	1,307 mm

Fig. 9. Measurement of the cut length of the needle – 40x enlargement and Measurement Results

The measurement of the cut length of the needle is done through VHX[5] Communication Software as seen in Fig. 9. According to the stated information about the cut length in chapter 4.11 of ISO 7864:2016 and chapter 4.2.3 of DIN 13097-4:2019; point length (from tip point to end of the cut length as seen measurement number 4 in Fig. 9), primary bevel length (it is the result of grinding with ξ bevel angle, also as seen measurement number 3 in Fig. 9), secondary bevel length(it is the result of grinding with ϕ tilt angle after rotating the needle β angle, as seen number 2 in Fig. 9) and also the outer diameter of needle (measurement number 1 in Fig. 9) are measured. However, the cavity length of the needle as vertical was measured (measurement number 5 in Fig. 9). As the measurement results compare with the dimension parameters in appendix B in DIN 13097-4:2019-06, the measurement cut length is less the calculated cut length. Even if the bevel angle is in the tolerance zone, the cut length depending on the bevel angle is out of the tolerance zone.

The used Digital Microscope has also 2000x enlargement specification. So, it is concentrated on surface and line roughness parameters on the lancet[6] surface, bevel surface, inner and outer cylinder surface as seen in Fig. 10. The multi-line and surface roughness parameters of 3 different needles of the pre-fillable glass are measured 3 times from each surface above. The results are shown in Table 3:

When the measurement results of the surfaces are compared, the lancet surface has more defects than the other surfaces because, after 3rd step in Fig. 2, lancet surface is exposed to grind in the direction of orientation angle after rotating, not like bevel plane which is formed after grinding in the same direction of bevel angle. Since the metal pipe prepared before manufacturing the needle tip has also been surface treated beforehand, the best roughness, compared to the other three surfaces, is on the outer surface.

[5] VHX: Video Highway Xtreme.

[6] Lancet surface: it is the surface area which occurs as conclusion of orientation and tilt angle.

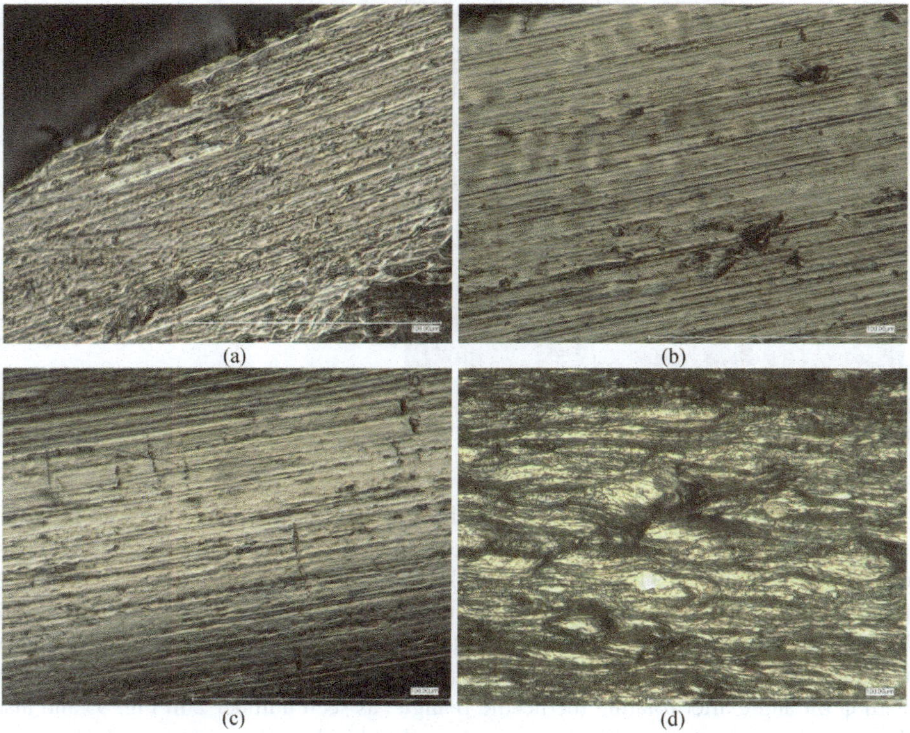

Fig. 10. 3D structure of (a) the lancet surface (b) the bevel surface (c) the outer cylinder surface (d) inner cylinder surface

The existing manufacturing flow chart, as seen in Fig. 2, is recreated by advanced precision metrology technics according to Industry 4.0 in Fig. 11. All manufactured needles are inspected through advanced metrology methods after each step, and the medical needles are separated according to availability as seen in Fig. 11. Needles that fail in the inspection are sent recycling and also given the failure information to the research and development department, so the solution is found against the failures. Otherwise, the process continues until the last step. After the last step, they get ready to montage with the syringes. Thus, quality assurance is implemented in the manufacturing process and the system runs faster and more reliably within the circumference of Industry 4.0.

Table 3. The line and surface roughness parameters of the lancet surface, bevel surface, the outer and inner cylinder surface.

Dimension name	1	2	3	Avg. (μm)	Unit
Lancet Surface (Line Rough.) Ra	8.19	8.92	6.78	7.83	μm
	7.80	8.20	7.78		μm
	7.11	8.22	7.42		μm
Lance Avg. Ra	7.70	8.45	7.33		μm
Lancet Surface (Surf. Rough.) Sa	5.65	7.57	3.82	5,45	μm
	5.60	6.35	4.78		μm
	5.00	7.08	3.17		μm
Lancet Avg. Sa	5,42	7,00	3,92		μm
Bevel Surface (Line Rough.) Ra	4.32	2.11	2.77	3,62	μm
	4.25	3.46	4.04		μm
	4.47	3.44	3.76		μm
Bevel Avg. Ra	4,35	3,00	3,52		μm
Bevel Surface (Surf. Rough.) Sa	1.65	3.57	2.03	2,7	μm
	1.92	3.52	2.54		μm
	1.90	4.05	3.14		μm
Bevel Avg. Sa	1,82	3,71	2,57		μm
Outer Surface (Line Rough.) Ra	2.04	2.58	2.80	2,43	μm
	1.79	2.62	2.79		μm
	1.77	2.80	2.64		μm
Outer Avg. Ra	1,87	2,67	2,74		μm
Outer Surface (Surf. Rough.) Sa	2.41	3.13	1.62	2,09	μm
	1.81	3.87	1.38		μm
	1.06	2.27	1.22		μm
Outer Avg. Sa	1,76	3,09	1,41		μm
Inner Surface (Line Rough.) Ra	3.03	3.31	2.86	2,93	μm
	3.79	2.17	2.79		μm
	3.24	2.12	3.06		μm
Inner Avg. Ra	3,35	2,53	2,90		μm
Inner Surface (Surf. Rough.) Sa	4,38	2.04	3,98	3,78	μm
	4,18	3.08	4.20		μm
	4,93	3.95	3.28		μm
Inner Avg. Sa	4,50	3,02	3,82		μm

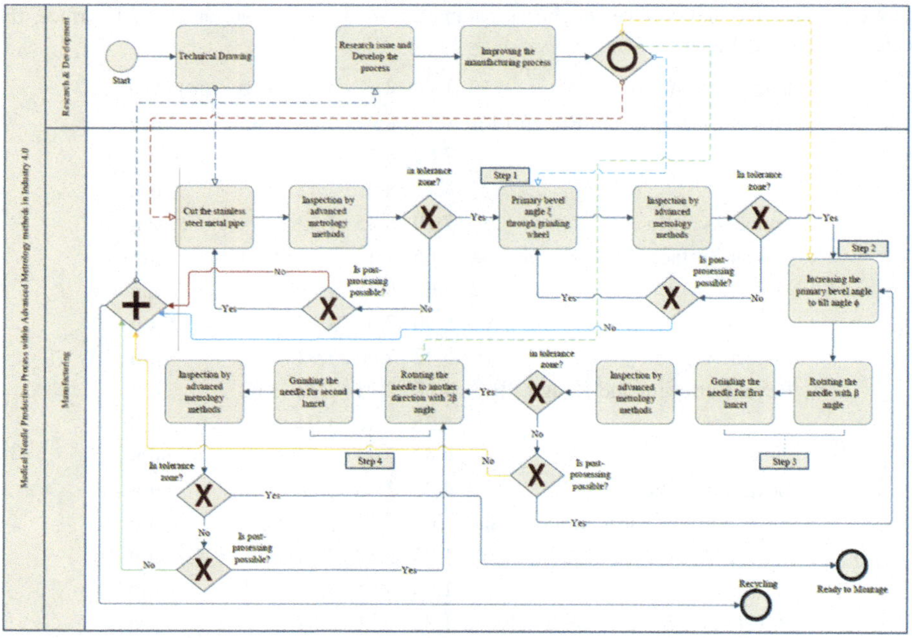

Fig. 11. The medical needle production process within advanced metrology methods in industry 4.0

4 Conclusion and Prospect

The improvements in high precision measurement technology render sophisticated medical productions with high-quality functions on micro-scale. In medical technology, in particular, it is important that the product and process quality are very high, which is why the medical products shall meet the geometric tolerances and form deviations. Especially in conjunction with Industry 4.0, the manufacturing systems should be brought to a technologically advanced level. It should be achieved a uniform product specification in the manufacture of medical devices using intelligent measuring instruments to increase the accuracy in production. Hypodermic needles and the needles of glass pre-fillable syringes have a significant place in the Medical Device Industry. Gauge numbers are diversified the needles types. G27 is one of them and used in acute heart failures and high-risk surgeries. Usage of those needles affects the functional needs at a high level. When the health conditions are considered, the essential functional requirement is tissue damage so, roughness parameters on the tip and surface of the needle and geometric dimensions depending on functional specifications are the points that need attention in terms of intelligent design. According to the previous studies and relevant ISO standards, roughness and geometrical tolerances are given on the defined geometrical parameters of the needles. Certainly, standardization of the needles of pre-fillable syringes and hypodermic needles by geometrical product specifications decrease the failures in functional specifications. Otherwise, it might have dramatic results, especially in the medical industry, which is one of the most critical areas of healthcare. Furthermore, high precision measurement technologies close the lack of evaluation and validation of sensitive high technological products like needles. The measurement results by digital microscopes are shown why the form, location, and orientation tolerances are noteworthy. Thus, the geometrical tolerances reduce the error rate that may occur during usage in medical operations and define the geometric structures in ISO standards and production technologies. Nevertheless, digital microscopes with 6000x enlargement are so carefully responded in microscale for roughness parameters to detect faults during production, to reduce surface defects, and to approach the skin models. By another perspective, explaining the manufacturing process of the hypodermic and glass pre-fillable needles by BPMN provide the opportunities to evaluate in terms of quality standards. As future work, the geometrical tolerances of the needles are made more precisely according to the realized surface, and line measurement results in standards.

References

1. Teixeira, M.B.: Design controls for the medical device industry. Taylor & Francis Group, LLC, 978-1-4665-0355-7 (2014)
2. ISO 9626:2016 – Stainless steel needle tubing for the manufacture of medical devices—Requirements and test methods", Edition, 15 December
3. ISO 7864:2016 – Sterile hypodermic needles for single use—Requirements and test methods, Edition, 1 August 2016
4. Kucklick, T.R.: The Medical Device R&D Handbook, 2nd edn. Taylor & Francis Group, LLC, 978-1-4398-1195-5 (2012)
5. DIN 13097-4 Hypodermic needles – Part 4: Point geometry, requirements, and testing, June 2019
6. Han, P., Che, D., Pallav, K., Ehmann, K.: Models of the cutting-edge geometry of medical needles with applications to needle design. Int. J. Mech. Sci. **65**, 157–167 (2012)
7. Wang, Y., Tai, B.L., Chen, R.K., Shih, A.J.: The needle with lan-cet point: geometry for needle tip grinding and tissue insertion force. J. Manufact. Sci. Eng. ASME **135**, 041010-1 (2013)
8. EN ISO 80369-7 – Small-bore connectors for liquids and gases in healthcare applications, Part 7: Connectors for intravascular or hypodermic applications, Edition, 15 October 2017
9. Clexane short product information, Turkey drugs, and medical devices agency. https://titck.gov.tr/storage/kubKtAttachments/vfhW7SacwkK6R.pdf
10. Sacha, G., Rogers, J.A., Miller, R.L.: Prefilled syringes: a review of the history, manufacturing, and challenges, Pharm. Dev. Technol. **20**(1), 1–11 (2015)
11. The Video of BD manufacturing of the pre-fillable syringes, source: https://drugdeliverysystems.bd.com/video-gallery?video=5837047753001
12. Durakbasa, N.M., Bodur, O., Poszvek, G., Demircioglu, P.: Additive miniaturized-manufactured gear parts validated by various measuring methods. IMEKOTC14 2019, LNME, pp. 276–290. Springer (2019)
13. ISO 8015:2011: Geometrical product specifications (GPS)—Fundamentals—Concepts, principles and rules (2011)
14. Läpple, V.: Dimensionelle und geometrische Tolerierung; Version 4-0-4, Steinbeis-Beratungszentrum Konstruktion. Werkstoffe. Normung (2019)

15. ISO/DIS 11608-2, "Needle-based injection systems for medical use—Requirements and test methods, Part 2: Double-ended needles" Voting terminates, 12 May 2020
16. Jorden, W., Schütte, W.: Form- und Lagetoleranzen Handbuch für Studium und Praxis; 7. Auflage; Carl Hanser Verlag München (2012)
17. ÖNORM EN ISO 1101:2017 08 01; Geometrische Produktspezifikation (GPS) - Geometrische Tolerierung - Tolerierung von Form, Richtung, Ort und Lauf (ISO 1101:2017)
18. Digital Microscope, Model VHX-7000, Keyence Corporation VHX-7000-KD-C3-DE (2019)
19. Durakbasa, N.M., Bauer, J.M., Bodur, O., Poszvek, G.: Challenges of miniaturizing a precision gear. In: ISPR 2018, Proceedings of the International Symposium for Production Research 2018, pp. 239–253 (2019)
20. ISO 4288:1996/COR 1:1998: Geometrical Product Specifications (GPS)—Surface texture: Profile method—Rules and procedures for the assessment of surface texture—Technical Corrigendum 1 (1998)
21. ISO 25718: Geometrical product specifications (GPS)—Surface texture: Areal—Part 2: Terms, definitions, and surface texture parameters (2012)

Social and Ethical Aspects of Automation

Mary Doyle Kent[1(✉)] and Peter Kopacek[2]

[1] Department of Engineering Technology, Waterford Institute of Technology, Waterford, Ireland
mdkent@wit.ie

[2] Institute for Mechanics and Mechatronics, IHRT, TUWien, Vienna, Austria
kopacek@ihrt.tuwien.ac.at

Abstract. This contribution deals with social and ethical aspects of automation. Because of the fast development of manufacturing automation, especially in the field of robotics, there is increasing emphasis on using modern robotics in Production 4 (5).0. Therefore, the main goal of this contribution is to give an overview about the state of the art and further development trends in this field. Special emphasis will be on new approaches like the new generation of collaborative robots (COBOTS). This yields to new questions in ethical and social sciences like the education of the operators (robocooperators). For the first time legal aspects of robots in manufacturing automation and their influences on ethics and social issues are introduced and shortly discussed with special emphasis to SME's.

Keywords: Automation · Social aspects of technology · Engineering ethics · Engineering education · Industry 4.0

1 Introduction

Modern engineering is the process of creating solutions that can simultaneously achieve the goals of the problem and remain within certain constraints. These goals and constraints are referred to as "design drivers". The three most common "design drivers" are: the cost of the solution, the schedule of the project and the features of the solution.

These drivers can act as either goals or constraints, depending on the situation. Ethical considerations can similarly be viewed as "design drivers", constraints or goals. Ethical behaviour is either something to be sought or something used to limit the design space. The technical features are analogous to the moral issues at stake in an ethical case. The main Manufacturing Systems "Evolution Drivers" are: [1]

Global growth & competition Knowledge Economy Environmental pressures	**Molecular manufacture Conflict over resources Ideology, & culture, ICT- ambient & networked**	**Global competition in services, Human need Physical Product**

These "Drivers "are influencing social and ethical aspects in a positive as well as negative manner.

N. M. Durakbasa and M. G. Gençyılmaz (Eds.): ISPR 2020, LNME, pp. 363–372, 2021.
https://doi.org/10.1007/978-3-030-62784-3_30

2 Automation

A manufacturing process consists of processing operations which transforms a raw material from one state of completion to a more advanced state that is closer to the final desired product by means of shaping operations, property enhancing operations, surface processing operations.

We are on the way to a 4th, 5th and 6th Industrial Revolution: Characterized by "everything and everybody is connected to everything and everybody, anywhere and at any time". The 4th Industrial Revolution is IT-driven. Connectivity and networking are its most characteristic features.

Autonomous technological systems enter the scene and they are capable of dealing within a *"human-like complexity"*. The central driver is the enhanced IT technology, mainly in the field of (distributed) artificial intelligence. Multi Agent Systems (MAS) are important architecture to deal with systems of distributed intelligence. They represent interacting, autonomous, heterogeneous agents and are inspired by social system and biological models. This is just the beginning. In the future the development of artificial intelligence may change all products, their design, their production processes and even infrastructure massively.

This yielded in the past, the development of Computer Integrated Manufacturing (CIM), Intelligent Manufacturing Systems (IMS) and Agile Manufacturing Systems (AMS). At that time (1999) the hardware and software possibilities were extremely limited, and therefore industrial applications of AMS were impossible.

2.1 Production 4.(5).0

In manufacturing automation, we are now in the era of Production 4.0. Production 5.0 is knocking on the door! Industry 5.0 focuses to validate creativity, high-quality custom-made products, and of the highest quality standard. It aims to increase the quality of life of humans. This can be called as the fifth revolution of mankind. According to Doyle-Kent et al. in 2019 the products manufactured by this new industrial era *"will not be just agile and lean, but automated, digital and data driven. Products will be of extremely high quality and available at more affordable prices. These products must be the best available and the supply chain optimised to support manufacturing"* [2]. Smart services are for both communication of humans and non-humans. The communication between humans and non-humans is the sharing of information and data regarding the "system". It is important to remember this when designing organizations, data-sharing platforms and technologies which support the "system" need to be optimally designed for this goal.

Now let's discuss the rise of robotic systems. Robots are entering into all areas of life. They are no longer restricted to certain "highly structured" environments but are right amongst us. Some decades ago the social aspects of robotics were discussed. Because of the results and the rapid development of this field ethical issues became more and more important Therefore the term "roboethics" was introduced in the literature [3]. Clearly this new automation philosophy requires highly cooperative industrial, mobile, and probably in the future, humanoid robots – a so called multi-robot system. A multi-robot system is a distributed system that consists of a collection of autonomous computers, connected through a network and distribution middleware which enables them to coordinate their activities and to share the resources of the system so that user perceives the system as a single, integrated computing facility. Nevertheless, multiple robot systems are different from any other distributed systems because of their implicit "real world" environment, which is more difficult to model.

In order to perform such collaborative tasks a new generation of robots - Collaborative robots (Cobots) were introduced in 2012. These robots are safe around humans by using sensors, force limiting and rounder geometries than traditional robots. They are lightweight (for them to be moved from task to task) and to be easy to implement and use without skills in programming. Above all, a collaborative robot is not a replacement robot; it assists workers rather than replaces them. Figure 1 shows a family of Collaborative robots. UR3 on the left has a payload of 0–3 kg, UR5 has a payload of 3–5 kg and the larger UR10 has a payload of 5–10 kg [4].

Fig. 1. Cobot examples from universal robots [4]

As detailed by the ISO10218 standard, robots can have four types of collaborative features. They are:

- Safety Monitored Stop
- Hand Guiding
- Speed and Separation Monitoring
- Power and Force Limiting.

The Safety Monitored Stop is implemented in environments where the robots operate mostly alone, with occasional human interference. The feature will cause the robot to pause (though not shutdown) when the safety zone is violated (i.e. a human enters its workspace).

The speed and separation monitoring feature are an extension of Safety Monitored Stop. Instead of adopting a single behaviour throughout the robot's entire workspace, the latter is graded into several safety zones (Fig. 2).

Hand Guiding enables the robot to move while the worker is in its workspace (as is possible with Speed and Separation Monitoring). Using an end-of-arm device capable of detecting applied forces, the robot can be guided by an operator for hand guiding and rapid path teaching.

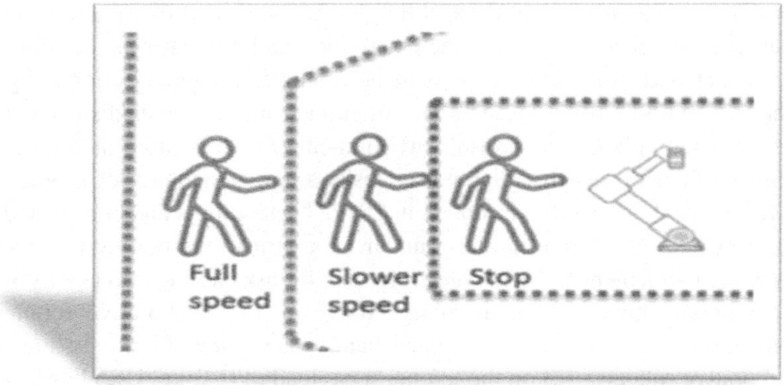

Fig. 2. Speed and separation monitoring

Power and Force Limiting feature is probably the safest of them all as this type of robots are purposefully designed to operate around humans rather than being adapted to do so. As the name suggests, the forces and energy applied to a human are limited so as to avoid any harm that might be inflicted.

Cobots are now introduced in Production 4.0 and will be also in Production 5.0 and 6.0. Teams of humans and robots collaborating efficiently can reduce the human idle time by up to 85% and the reasons for this are as follows:

- Easy to include in a network
- Minimum space on the floor
- Easy to program
- Price and safety

The International Federation of Robotics began together data in 2019. They produce forecasts for the collaborative industrial robots uptake, based on the sales figures of robot suppliers. They state that in 2018 approximately 14,000 cobots were installed out of approximately a total of 422,000 industrial robots. This is approximately 3.3% and is an increase of 23% over 2017 [5]. Cobots are an ever-increasing important subset of industrial robots in the manufacturing companies of Industry 4.0, 5.0 and 6.0.

2.2 Humans in Manufacturing Automation

The question arises *"what role do humans have in this highly automated environment?"* To have a role, humans must bring value. They must be suitably skilled to adapt to a highly automated, ever changing workplace. Educations plays a key role in this.

In October 2019 the German Research Council published a paper titled "Key themes of Industrie 4.0". In its opening paragraph Hirsch-Kreinsen comments on the social impact of this new era. He says that *"digitalization is enabling a fourth industrial revolution. This root-and-branch change in terms of new technologies, patterns of work and corporate organization, business and revenue models, value networks, right up to dynamic digital ecosystems, has the potential to have all-encompassing social impact which is, as yet, difficult to grasp in its entirety."* Cyber-physical systems play a strong part in this change [6]. The German research council noted the following key areas of opportunity:

- Value creation scenarios for Industry 4.0
- Prospective technological trends
- New methods and tools for Industry 4.0
- Work and society

It is accepted that robots are responsible for an increase in demand for workers that are more highly skilled. This, in turn, this means a positive impact on wages. The difficulty that remains is how to help middle-income earners in the lower-income range to upskill or retrain?

An important aspect in future manufacturing scenarios is defining the human-machine interaction. Changes in technology will continue to be dynamic in nature. How will human and robots communicate and interact together? Training and education will define the success, or otherwise, in the collaboration. Humans will be required to maintain the complex systems when they break down. Current technical skills need to be valued and elaborated. Workers roles will change from a machine operator to a strategic decision-maker and a flexible problem solver [7].

Going forward it will be especially important to establish learning environments that will engage the worker in a practical manner to facilitate upskilling. Technicians and operators will need to become digitally competent. They will use new assistance systems to make informed decisions. Tablets, smart phones and other devices will relay information to the humans facilitating real times decision making.

In a document from the World Bank Group [8] titled *"Trouble in the making, the future of manufacturing led development"* there is a call for the reform of education priorities and capabilities. For countries to take advantage of the new opportunities education and training policies need to be completely redesigned to ensure the employability of workers in a more complex manufacturing high tech environment. They say that jobs will be more *"non routine and cognitive"*. This will require greater investment in ICT related skills (software skills and programming). They suggest that these skills, and complementary skills in engineering, need to be *"offered in ways that are inclusive to ensure that women and men can benefit from these opportunities"*. There is growing pressure to ensure that diverse communities have opportunities to upskill in future manufacturing factories if the best talent is to be available [9–11].

Education needs also to be more responsive to industry demand. To facilitate this, private source of upskilling will become more important. In addition, due to the rapid unexpected changes in the global landscape a high priority should be given to "soft" skills. Adaptability, taking initiative, creativity and problem solving are premium.

3 Ethics

In the production scenarios of tomorrow, employees play a much larger role than today. For the company's future both, the know-how and the employee's qualifications are what really matter. Workers in manufacturing must be able to deal with an increasingly wide range of responsibilities and even greater tasks complexity, thus suitable education is becoming an important factor [6].

The workforce of the future will need skills different from those it has today. There will be less manual work and more planning, coordination and decision making. People won't vanish from factories because even the best computers and machines will never be able to relieve people of the most important thing they do - deciding on the best solution from a number of alternatives, managing unforeseeable events, or deriving new and creative ideas from experience [5].

Given that right decisions must be made based on enormous volumes of data, as the consequence of production digitization, and under severe time pressure, also management tasks will change. Moreover, most companies have increasingly complex value and supply chains, not to mention the demanding management of volatile markets and the increasing individual customer requirements. Many executives will have to give their organizations a new structure and introduce new leadership methods [6].

Industry 5.0 focuses to validate creativity, high-quality custom-made products and with life standard. It aims to increase the life quality of humans. It can be called as the fifth revolution of mankind. The communication will be between humans and non-humans. The standardization of architectures, data exchange formats and interfaces as well as concepts and vocabularies will be vital to the interoperability between the different technologies involved in a complex field like Industry 4.0 and 5.0 [2].

4 Social Aspects

Manufacturing conditions are much better with Industry 4.0. There is more occupational safety, through increased automation. The working conditions are better through ergonomically adapted workstations. Collaboration is increased through the availability of consistent data. Environmental protection is also better because of the optimized use of resources like more energy-efficient operation of machinery.

4.1 Robocollaborators

One of the most important changes is the increased innovative capability. Industry 4.0 has this main advantage since machines follow steps by themselves, so innovation is easy to develop. Cobots in the future can increase the *"job satisfaction"* because they relieve the human worker from tedious, non-ergonomic and vulnerable tasks.

Furthermore, they support the operator with precise placement of parts like positioning of chips, heavy and repetitive lifting and the loading and handling of hazardous materials. Cobots are like the "third-hand" for assembly. The close collaboration of humans and robots in the workplace increases the acceptance of robots not only because of the safe collaborative workspaces for close human robot interaction. This yields to social networking between "smart" operators and connect smart operators with smart parts of the smart factory. Probably "Robocollaborators" and their robots will become better friends with the time, helping humans instead of replacing them [4, 5].

4.2 Legal Questions

In 2017 in a study by Djordievic on the *"Laws of Robotics"* the author concludes that the Directive 85/374/EEC can cover only the damage caused by the robot as a result of manufacturing defects. They state that "the current legal framework would not be sufficient to cover the damage caused by the new generation of robots insofar as they can be equipped with adaptive and learning abilities entailing a certain degree of unpredictability in their behaviour since those robots would autonomously learn form their own variable experience and interact with their environment in a unique and unforeseeable manner therefore new legislation must be found" [12].

The cobot and the human worker work closely together. As precise as the robots are, it cannot be expected that no accident will ever happen. This kind of close collaboration raises number of radically new ethical and legal questions such as:

- Who would be responsible for the malfunctioning of the robot?
- Who would be responsible for the mistake made by a robot?
- What would happen if the worker got hurt by a robot?
- Should a robot be treated as a co-worker since it is doing part of the job?

In order to keep the technology developing, it is important to find the answers to these questions, either in the current legal framework or by creating new legal categories. However, these developments raise some ethical questions.

- Can human be attached to these robots, is it acceptable that people start communicating with robots and accept their recommendations and advices?
- Would they rely too much on them?
- In addition, what if something happens to a robot in this situation, how would a human react?

If there was no need for these kind of technologies in our lives, these robots would certainly never have been developed. Therefore, if there is a need for them, we need to find solutions to make them part of our lives, without jeopardizing anyone involved.

Law regulation in the field of technology influences innovation to a large degree [12]. Liability rules play a big role in the decision-making process of the manufacturer of certain products. If the risks of liability are too high it could make the manufacturer delay the launch of a product, which can decelerate the development of new technology. This is certainly not desirable. However, if the field of robotics is not well regulated, it would bring uncertainty, both to the manufacturer and the user.

A good example is a fully automatized car. The risk of the liability is completely transferred to the manufacturer. The driver is just a mere passenger in the car, with no obligation to act. Therefore, if an accident occurs, it is very unlikely that the driver could be held liable. In such cases the reason for the accident is usually the malfunction of the machine, so all the responsibility falls onto manufacturer [12, 13].

In a blog "The Conversation" Lecturer in Applied Ethics and Sociotechnical Studies at Griffith University David Tuffley takes the other extreme position. *"Let's say the car maker is successful in deflecting liability. In that case, the user becomes solely responsible whether or not they have a well-considered code of ethics that can deal with life-and-death situations"* [14].

In the case of a partially automated car, the human driver has the final responsibility. However, it is not the same as driving a regular car. When driving a non-automated car, the human driver is involved in driving the whole time, and he is not relying on the machine to do something instead of him. Driving a partially automated car, however, makes a driver more relaxed, and unprepared to react in every dangerous situation. Therefore, the risk of a manufacturer's liability is again higher than in the case of regular cars. There a currently discussions about a soft law (Table 1) and a hard law (Table 2).

Table 1. Advantages and disadvantages of a soft law [15]

Advantages and disadvantages of implementing a soft law	
Advantages	Disadvantages
The flexibility to adapt to the dynamics in technology development The recommendations and standards ensure the continuous adaptation of rules, without the need for statutory intervention The norms can be adopted on a voluntary basis in the member states They help building the legal environment for robotic technologies They involve more stakeholders in the decision-making process	Too general character of soft law norms could not cover all the aspects of technological innovation in detail The harmonization depends on a voluntary compliance, which makes the whole system uncertain The involvement of more stakeholders makes the process much more complicated and favors the needs of the more powerful ones Implementing soft law norms undermines procedural values, such as due process, accountability and transparency

Table 2. Advantages and disadvantages of a hard law [15]

Advantages and disadvantages of implementing a strong framework of rights and values	
Advantages	Disadvantages
The strong rules could ensure the higher level of safety in innovations They can determine more precisely the limits in the use of technology It could encourage innovation with implementing the rules that favour the technological development It brings more certainty to the legal system	It is not easily adapting to the fast-changing character of technological innovation Strong framework could have a chilling effect on technology It is expensive to implement new norms with every progress in the field of technology

One of the dilemmas is whether a soft law approach or a strong framework of rights and values should be implemented. A soft law approach is easily adaptable, which is very convenient for the extremely fast developing field of robotics. On the other hand, strong framework brings much more certainty into legal system. A strong legal framework is hard to change, and the field of robotics is constantly changing, making it hard for the legal system to catch up.

The ownership and use of data in Industry 4.0 will need a strong and well-defined framework. Data protection and data security will be *"central legal challenges*" [3]. Hirsch-Kreinsen states *"Machine data and the functionally relevant metadata arising with adaptive robotics or from the use of wearables increasingly make reference to individual people and their behavior and thus present new challenges to existing personal data protection requirements*". In addition, they stress that there is legal uncertainty with how current data protection legislation can be applied. How is Industry 4.0 defined legally? There is also a need for legal models for *"transboundary collaboration outside the borders of the EU"* [6].

In summary according to this report there is a research need to look at the training requirements and methods for handling actual and legal uncertainty in short technical innovation cycles. In addition, the criteria for lawful and legally compatible definitions of new technology needs to be established. And finally, new strategies for handling personal data and individual performance monitoring need to be constructed while maintaining data sovereignty and identifying ethical and legal boundaries [6].

The education is currently more or less an only partially solved problem. This new type of operators is situated between a classical operator and one with mechatronic knowledge. Furthermore, they should be trained annually because of the tremendous progress in this field.

We have now mostly education possibilities offered by companies producing components for these new technologies or independent from them but at a low-tech level. It is urgent to offer independent education for this new type of operator especially for SME's. The level needs to be at least the technical junior high school level [7].

5 Summary and Outlook

This contribution deals with the social and ethical aspects of automation, in particular, manufacturing automation (automation of discrete processes). Because of the progress in the field of robotics this area was chosen as an example.

After a short introduction in production 4.(5.).0 a new generation of collaborative robots (COBOTS) were described with special emphasis on this new production paradigm and the arising social and ethical questions. Initially the legal aspects of robots in manufacturing automation and their influences on ethics and social issues are introduced and discussed.

In the future technology development should definitely go hand in hand with an adequate legal framework. This framework should readily transform progressive nature of technology. It should also bring legal certainty, both for innovators and end users of the technology. However, it should not have a deceleration effect on the technology and should develop standards that would encourage progress.

Concerning the education of the robocollaborators adequate education possibilities should be offered. This includes both how they are educated and what knowledge they require in the ever-changing highly technical work environment. Companies and governments must work at investing in appropriate educations programmes to support the acquisition of skills necessary to secure and thrive in jobs that are being created or changed by the deployment of robots and automation.

References

1. Kopacek, P.: Trends in production automation. In: Proceedings of TECIS 2019, pp. 509–512. Elsevier B.V., Netherlands (2019). IFAC PapersOnLine, 52–25. https://doi.org/10.1016/j.ifacol.2019.12.595
2. Doyle-Kent, M., Kopacek, P.: Industry 5.0: is the manufacturing industry on the cusp of a new revolution? In: Proceedings of the International Symposium for Production Research 2019, pp. 432–441. Springer, Cham, August 2019. https://doi.org/10.1007/978-3-030-31343-2_38
3. Kopacek, P.: Robo-ethics a survey of developments in the field and their implications for social effects. In: Proceedings of TECIS 2019, pp. 131–135. Elsevier B.V. (2019). IFAC PapersOnLine 52–25. https://doi.org/10.1016/j.ifacol.2019.12.460
4. Universal Robots: Universal Robots website (2019). https://www.universal-robots.com/products/ur5-robot/
5. IFR: Demystifying Collaborative Industrial Robots (2019). https://ifr.org/downloads/papers/IFR_Demystifying_Collaborative_Robots_Update_2019.pdf
6. Hirsch-Kreinsen, H., Kubach, U., Stark, R., von Wichert, G., Hornung, S., Hubrecht, L., Sedlmeir, J., Steglich, S.: Key themes of Industrie 4.0: research and development needs for successful implementation of Industrie 4.0. Research Council (2019). https://en.acatech.de/publication/key-themes-of-industrie-4-0/
7. Acatech: Skills for Industrie 4.0. training requirements and solutions (2018). https://www.acatech.de/wp-content/uploads/2018/03/161205_KF_eng_POS_Kompetenzentwicklung_fin.pdf
8. Hallward-Driemeier, M., Nayyar, G.: Trouble in the Making?: The Future of Manufacturing-led Development. World Bank Publications (2017). ISBN 9781464811937
9. Doyle-Kent, M., Costello, O., Kopacek, P.: Where are all the Irish women engineers: a case study. IFAC-PapersOnLine **52**(25), 136–141 (2019). https://doi.org/10.1016/j.ifacol.2019.12.461
10. Doyle-Kent, M., Chowdhury, F.N., Costello, O., O'Neill, B., Organ, J., Kopacek, P., Larry Stapleton, L.: TECIS Inclusion and Diversity Working Group Vision. IFAC World Congress. Awaiting Publication, Berlin (2020)
11. Doyle-Kent, M., Kopacek, P.: Do we need synchronization of the human and robotics to make Industry 5.0 a success story? In: ISPR 2020. Awaiting Publication (2020)
12. Djordjevic, I.: Robotics Law. MSc thesis, TU Wien (2019). Unpublished
13. Karnouskos, S.: Self-driving car acceptance and the role of ethics. IEEE Trans. Eng. Manag. (2018). https://doi.org/10.1109/TEM.2018.2877307
14. Tuffley, D.: Selfdriving cars need -adjustable ethics- set by owners (2014). https://theconversation.com/self-driving-cars-need-adjustable-ethics-set-by-owners-30656
15. Leenes, R., Palmerini, E., Koops, B.J., Bertolini, A., Salvini, P., Lucivero, F.: Regulatory challenges of robotics: some guidelines for addressing legal and ethical issues. Law Innov. Technol. **9**(1), 1–44 (2017). https://doi.org/10.1080/17579961.2017.1304921

Social Transformation - Industry 4.0

Uğur Sütlü[1(✉)] and Fatma Serab Onursal[2]

[1] Business Master's Program, İstanbul Commerce University, İstanbul, Turkey
ugursutlu@outlook.com
[2] Department of International Logistics and Transportation, İstanbul Commerce University, İstanbul, Turkey
sonursal@ticaret.edu.tr

Abstract. The impact of advanced infrastructure and the rapidly rising production technologies on the industry have caused the desire for faster but less costly production, leading to Industry 4.0. This study addresses the general concepts in Industry 4.0, our country's position in the Industry 4.0 processes, changes to be caused by Industry 4.0 in the society and Project Management of these changes. Also, the data from the OECD and Dijital Dönüşüm Derneği (Digital Transformation Association) will be interpreted.

Keywords: Industry 4.0 · Digital transformation · Digital agility · Social transformation · Society 5.0

1 Industry 4.0

The first social change in human history was brought by agriculture. Settled life brought solutions to life's problems with permanent housing, regular nutrition and led to the first permanent revolution in human history.

Following agriculture, humanity has experienced other 3 big revolutions until journeying into Industry 4.0 and with each revolution, people have kept up with social changes and regulations. Learning to use the steam power in the First Industrial Revolution, people have changed the weaving, transportation and metallurgy industries with the steam engine. With the introduction of steel production, steam and steel came together and taking one of the important steps for navigation and railroad transportation.

With the Second Industrial Revolution, electricity entered into every part of life and oil started to be used in transportation. These developments put concepts such as production line and mass production into people's vocabularies, efficiency started having importance in production areas and humanity started to get socially stronger for the next revolutions.

The Third Revolution, technology, uncovered developments in computers, microchips and genetics. Innovative discoveries and trials that break down traditional discourses in medicine and the spread of electricity in the production and service sectors and the speed arising from the invention of the computer were other remarkable innovations. Computer technology has been improving day by day and humanity is

N. M. Durakbasa and M. G. Gençyılmaz (Eds.): ISPR 2020, LNME, pp. 373–382, 2021.
https://doi.org/10.1007/978-3-030-62784-3_31

now adding new terms such as information society to its vocabulary. The world is also undergoing a complete transformation along with humanity and sectors, institutions and all the techniques applied until now are now reconsidered with the changing world dynamics.

The Fourth Revolution, Industry 4.0, brought the people of the information society up against a whole new experience. Introduction of cyber-physical systems in production, generating billions of information per second with an Internet connection and robots working in collaboration carry humanity to a point far beyond the information society mentioned in the technology revolution.

Industry 4.0 is a term that emerged in the Hannover Fair in Germany in 2011 as a solution to Europe's falling behind countries such as China in terms of production and speed. With this approach, philosophies of having objects talk to each other by advancing Internet technologies far beyond, evaluating the collected information in meaningful data, meeting customer demands quickly by taking rapid actions against the changing world, and removing the human factor from production that will not create added value and cause efficiency issues are now spreading around the world and societies will experience radical changes. New professions, adaptation to the new order, speed, agility are beginning to become important and this new revolution will not only speed up production but also touch the lives of people in every sense.

2 The Door to Change

3D Printers, COBOT Technologies, Augmented Reality, Cybersecurity, Internet of Things and Big Data Technologies are among the technological innovations in the Industry 4.0 journey.

With the Internet of things, billions of devices will have an Internet connection and the speed gained by the 5G technology will increase data transfer to incredible sizes. Some major cybersecurity incidents and their impacts are shown in Table 1 as below.

Table 1. A cybersecurity incident example

Company	Industry	Country	Impact
T-Mobile	Telecom	Czech Republic	1.5 million records were seized.
Yahoo	Technology	USA	Over 1 billion customer account were seized.
Facebook	Social Media	USA	Profile information(birth date, name etc.) were seized.
Marriott	Hotel	USA	Customers' name, adress, passaport number etc. were seized.
Tech Bureau Corp	Technology	Japan	About 60 million dollars of digital money was stolen.

Ref.: (Digital Turkey & pwc - Digitalization Journey Turkey, 2019) [4]

Many companies who chose to work from home during the pandemic were exposed to attacks called phishing that try to penetrate cybersecurity. Also, worldwide cyber security company Kaspersky offers a service that we can monitor instant cyber-attacks on website which is "https://cybermap.kaspersky.com".

With three-dimensional printers taking more part in our lives, the costs of the spare parts manufacturing industry can be reduced to very low levels, and 24/7 production can be made with the support of an operator at the computer. Recent global health crises such as COVID-19 have also shown how robot-dominated enterprises could continue their production. Robotic technologies have become business continuity tools for the production and service sectors.

As with any change, this change process will also reveal new professional groups. For example, when software developed with artificial intelligence technologies and loaded with constitutional information is designed to be fed with big data, it can render a verdict by combining the lessons it has learned and the past records of the person judged. Even though it may seem like a threat to the profession of attorneyship, it can also open new doors for attorneys with which they can improve themselves. For this reason, each professional chamber and its members need to evaluate this transformation for their professions, present possible threats and opportunities and need to take actions to implement them.

Consumers are becoming more and more demanding and companies have no choice but to understand and get to know them better. Consumers, who have easy access to information and the product they want, started to take an interest in software that can renew itself instead of a product with fixed features. As can be seen in the Fig. 1, market shares of Tesla and Ford brands reveal that interest in electric vehicles and products equipped with software is increasing and Tesla has caught up with Ford in their market shares.

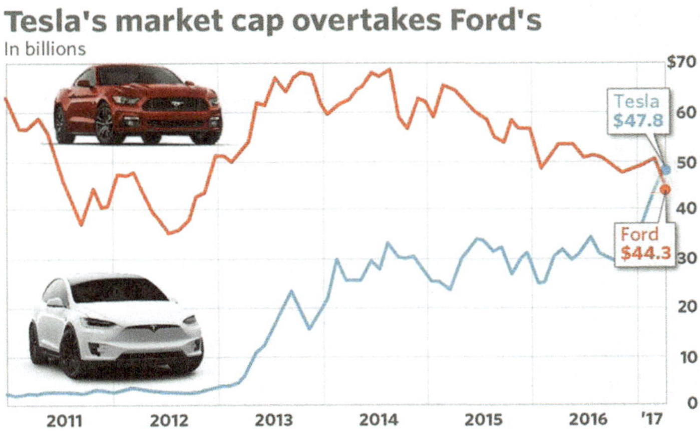

Fig. 1. Market shares of ford and tesla (MarketWatch, 2017) [1]

All these developments have also introduced a concept called service economy. Works performed without the content and the product to be sold are the ones with the highest turnover lately. Netflix, Spotify, BluTV and other platforms are the common examples of the service economy. In addition, Nike has established a membership model with a monthly subscription whose members can use their new shoes as the first consumer. The Nike membership system stands out as an important example both in terms of customer loyalty and the service economy.

ZAPPOS, a shoe company, sends your orders to wherever you are free of charge. Without the need for a fixed shipping address, your product is delivered to wherever you want. All these practices now highlight the quality of service rather than the product itself. For all these reasons, companies need to invest in the service economy and customer relations in order to achieve success in the Industry 4.0 investments in production.

3 How Prepared Are We for Change?

Even though Turkey is shown to be breaking its shell Digital Competitive Analyses, especially sectors with dominant foreign investment adapt to the change more easily and accelerate Turkey in this regard. Evaluations made in the production sector reveal that SMEs and lower levels of businesses that are positioned between Industries 2 and 3 pose the biggest challenge in this change.

For Internet use and access to the Internet per household, the OECD figures showing that our country has seen a significant upward trend in the last 20 years can be seen in the Fig. 2. However, "Digital Transformation Power" is not affected by the figures for access to the Internet alone.

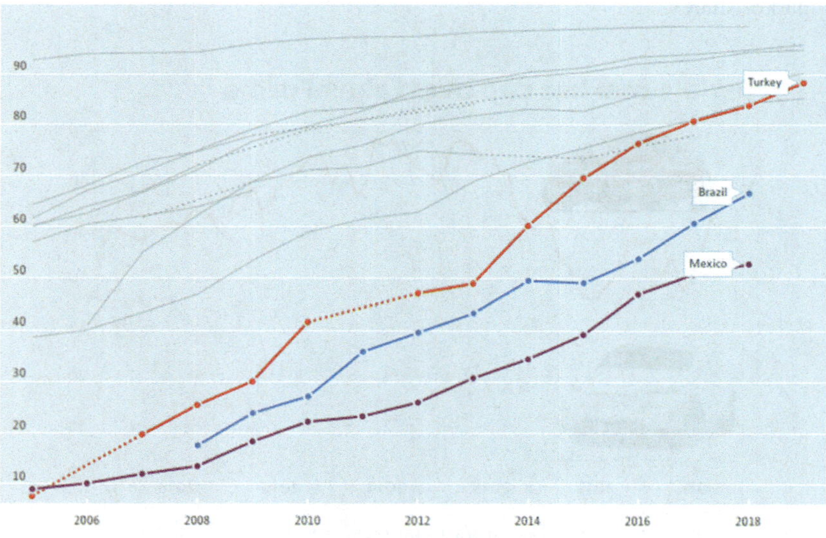

Fig. 2. İnternet use rates among OECD countries (OECD, 2020) [6]

Three main KPIs are considered in digital competitiveness analyses;

3.1 Technological Knowledge Level

How much research do countries do on new technologies, how much knowledge they have to understand and implement these new technologies are gaining importance. The indicators to measure this is the quality human resources, educational level and quality, and the level of scientific studies in the country.

3.2 Technological Development

Topics such as how advanced countries are in terms of technological development, whether they have the required environment for the digital technological development, the statuses of economic models, incentives and regulations are addressed under "Technological Development".

3.3 Digital Transformation Agility

The data of how prepared countries are for the digital future, how agile they are for digital transformation, how much place is given for digital technologies in the economic model is used to determine the country's "Digital Transformation Agility" score.

Turkey's position in World Digital Competitiveness rankings has not changed in the last 3 years. Turkey's 52nd place among 63 countries in IMD World Digital Competitiveness ranking (Table 2) is interpreted as Turkey's missing the digital transformation train (International Institute for Management Development(IMD) - Digital Competition) [5]

The current situation is analyzed in the report of Digital Türkiye Platformu (Digital Turkey Platform) as follows;

*"While only **34%** of individuals aged 16–74 have digital skills at or above the basic level, this average is **57%** in the European Union. Workforce rate with digital skills at or above the basic level is **46%** in Turkey and **65%** in the European Union.1 Turkey is seen to have fallen behind European countries in terms of digital skills. This poses a significant competition and growth issue for the Turkish economy and is also a major social risk. In addition, the effects of the fourth industrial revolution, in other words, Industry 4.0 are now beginning to be felt by companies and employees almost all around the world. Major changes in business models in most sectors cause the labor market to transform as well."*

And this change is anticipated to bring new occupations with it. Inadequate educational levels of Turkey's young population in recent studies seems to be the biggest threat in the way of adaptation to new occupations and the transformation. As shown in the chart below, 33% of the young people aged 20–24 in Turkey don't receive any kind of education. Sadly, this can be interpreted that we will not be easily integrated into new occupational groups that will emerge with Industry 4.0 and there will be an increase in unemployment numbers. (Figure 3) (Digital Turkey & pwc - Digitalization Journey Turkey, 2019) [4]

Table 2. World digital competitiveness rankings (Ref.: (International Institute for Management Development (IMD) - Digital Competition) [5])

2019	Country	2018	Change	2019	Country	2018	Change
1	USA	1	-	33	Poland	36	+3
2	Singapore	2	-	34	Portugal	32	-2
3	Sweden	3	-	35	Kazakhstan	38	+3
4	Denmark	4	-	36	Latvia	35	-1
5	Switzerland	5	-	37	Czech Republic	33	-4
6	Netherlands	9	+3	38	Russia	40	+2
7	Finland	7	-	39	Saudi Arabia	42	+3
8	Hong Kong SAR	11	+3	40	Thailand	39	-1
9	Norway	6	-3	41	Italy	41	-
10	Korea Rep.	14	+4	42	Chile	37	-5
11	Canada	8	-3	43	Hungary	46	+3
12	UAE	17	+5	44	India	48	+4
13	Taiwan, China	16	+3	45	Bulgaria	43	-2
14	Australia	13	-1	46	Romania	47	+1
15	United Kingdom	10	-5	47	Slovak Republic	50	+3
16	Israel	12	-4	48	South Africa	49	+1
17	Germany	18	+1	49	Mexico	51	+2
18	New Zealand	19	+1	50	Jordan	45	-5
19	Ireland	20	+1	51	Croatia	44	-7
20	Austria	15	-5	52	Turkey	52	-
21	Luxembourg	24	+3	53	Greece	53	-
22	China	30	+8	54	Cyprus	54	-
23	Japan	22	-1	55	Philippines	56	+1
24	France	26	+2	56	Indonesia	62	+6
25	Belgium	23	-2	57	Brazil	57	-
26	Malaysia	27	+1	58	Colombia	59	+1
27	Iceland	21	-6	59	Argentina	55	-4
28	Spain	31	+3	60	Ukraine	58	-2
29	Estonia	25	-4	61	Peru	60	-1
30	Lithuania	29	-1	62	Mongolia	61	-1
31	Qatar	28	-3	63	Venezuela	63	-
32	Slovenia	34	+2				

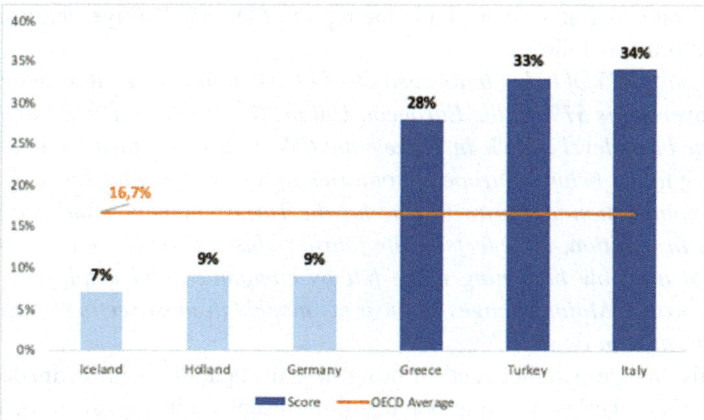

Fig. 3. Young population (20-24 Age) which not working and untrained (Accenture Turkey Digitalization Index, 2016) [2]

Qualified and knowledgeable citizens are a country's greatest power in the road to digital transformation and its most valuable capital that cannot be copied by anyone. Given that 200,000 software developers migrated other countries in 2019, we can say that Turkey needs to train a qualified and knowledgeable workforce to break its shell and keep up with this change.

In the evaluations regarding the sector-based digital transformation (Fig. 4) suitability and current status, it has been seen that the production-dominant sectors have not yet passed the digitalization bar but are not too far from the target.

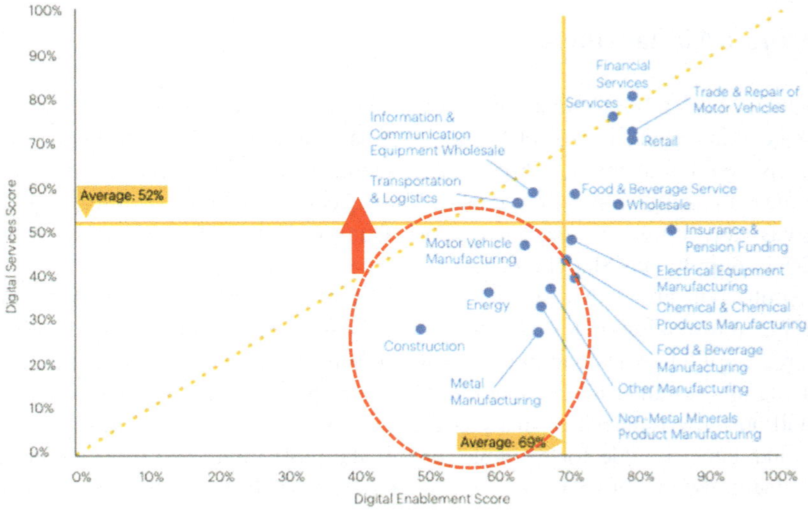

Fig. 4. Sector-Based Digital Transformation (Accenture Turkey Digitalization Index, 2016) [2]

As mentioned before, if production-dominant sectors position themselves at Industry 3.0 levels and apply computerized production technologies, transformation can be achieved more easily.

4 Society 5.0

There will be a capital need by integrating robots into the production sector and the change will be felt by the society, especially on the service economy, everyone will have smart devices at home and people will turn into devices with wearable technologies. These will also create development needs in the internet infrastructure. With these changes, it is estimated that human-robot interaction will rise to the highest levels, affecting societal relationships. In Japan, where Industry 4.0 is heavily discussed, the academic and social organizations started putting Society 5.0 (Human 5.0) into the agenda. (Prof. Dr. Nüket Saracel, Arş. Gör. Irmak Aksoy, 2020) [7]

Every change needs to be managed but this change affects not only the producers but customer demands and methods of addressing these demands in a society. Therefore, people will need to prepare action plans for the level of knowledge, welfare and culture that they want to reach. For example, what prevents the world's most famous surgeons from performing surgery on a person in another part of the world by connecting to a surgery robot like da Vinci? There is only one answer for it, Infrastructure!

You cannot entrust an infrastructure that will detect movements late due to the speed of the Internet with processes that need to be managed with 0 (zero) error, but the innovations that will arise by meeting these needs can completely change a society's behavior.

5 Project Management

A project is basically non-repetitive works that serve a specific purpose, carried out within a certain calendar. Project Management is a multidisciplinary profession 90% of which is communication. Project management activities (launching-planning-executing-monitoring-controlling-closing) are executed within 10 different information areas (Integration, Scope, Time, Cost, Quality, Human Resources, Communication, Risk, Supply and Shareholder).

A profession group that needs to give 90% of its energy to communication and check the information fields above is directly affected by the changing company structures, digital transformation moves, the policies of suppliers and human resources, and changing consumer needs. Traditional project management methodologies such as waterfall are now replaced in many companies by agile and hybrid models. In Cost, Scope and Time triangle for example, one of the most traditional methods in project management, we cannot keep the scope fixed these days. But the expected output needs to be in better quality than before. Needs and resources have changed, so Project Management needs to renew itself.

6 The Effects of Change on Project Management

In 2001, 17 software gurus published an Agile Manifesto. This manifesto consists of 12 articles and focuses on 4 main messages. Afterwards, this message starts to include not only the software world but also everyone involved in project management.

- **Individuals** rather than processes and tools **and interactions between them**
- **Running software** rather than comprehensive documentation
- **Cooperation with the customer** rather than contract negotiations
- **Responding to change** rather than sticking to one plan

There are several important steps to this change. The most important thing is to make processes lean; removing the processes that do not add value in terms of the customer and bringing value-added processes to the fore (AgileManifest, 2001) [3]

However, as the most important part of the traditional project management process is that the scope of the project has clear limit, while in the agile project management, the scope is locked within the sprints determined between 1 and 4 weeks, and after this sprint, the issues of the next sprint with meetings with the team and the product owner are prioritized. The most important gain of the claimant is flexibility. Agile project management philosophy decreases time to market periods for companies and aims to reflect outputs on the field quickly because projects that take months and years to get into the market can face problems too big to compensate in their market share. The flow of agile project management can be summarized as Fig. 5.

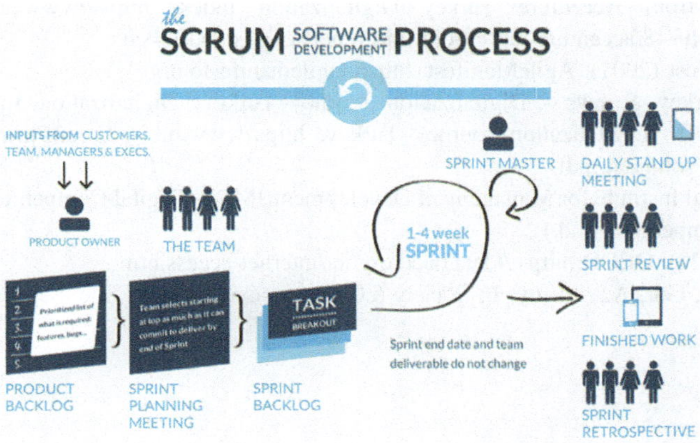

Fig. 5. Scrum framework (AgileManifest, 2001) [3]

Agile project management's one of the biggest goals is to give meaningful parts of the scope as partial but usable outputs iteratively, rather than giving the whole scope at once.

7 Conclusion and Evaluation

Even though the concept of Industry 4.0 seems to be a need for a transformation with a focus on the production sector due to its advantages, the conditions such as that customers are on the more demanding side and the best service they receive in the digital environment becomes a criterion for their next service greatly affect our social lives and the service sector. Attaching importance on education and increasing digital skills are very important steps to take in order not to miss the Fourth Industrial revolution.

Customers have become as digital and super smart profile with society 5.0 approach in all sector so companies should improve digital capabilities for these profiles. People who live in society 5.0, want to reach the high-quality products in short time so efficiency, lean management and customer loyalty have become the most important steps for all company on the digitalization roadmap.

It is very important to take the right steps for the new world order by doing further research on the education sector and shaping the digital society.

References

1. MarketWatch (2017): https://www.marketwatch.com/story/tesla-surpasses-ford-to-become-no-2-us-car-maker-2017-04-03
2. Accenture Turkey Digitalization Index: Accenture Turkey Digitalization Index (2016). Retrieved from Accenture Turkey Digitalization Index. https://www.accenture.com/_acnmedia/pdf-48/accenture-turkey-digi-ind-report-english.pdfla=en
3. AgileManifest (2001). AgileManifest. https://agilemanifesto.org
4. Digital Turkey & pwc - Digitalization Journey Turkey: Digitalization Journey Turkey, January 2019. Digitalization Journey Turkey. http://www.tubisad.org.tr/tr/images/pdf/dtp_dijitallesmeyolundatr.pdf
5. International Institute for Management Development(IMD) - Digital Competition: Analysis of Digital Competition (n.d.)
6. OECD (2020). OECD. https://data.oecd.org/ict/internet-access.htm
7. Saracel, N., Gör, A.,. Aksoy, I.: Society 5.0: super smart society. Soc. Sci. Res. J. (2020)

Strategic Model Proposal Related to IoT Applications in Disaster Management

Buket Karatop[1(✉)], Buşra Taşkan[2], and Cemalettin Kubat[3]

[1] Istanbul University Cerrahpasa, Istanbul, Turkey
buket.karatop@istanbul.edu.tr
[2] Muş Alparslan University, Muş, Turkey
b.taskan@alparslan.edu.tr
[3] Sakarya University, Sakarya, Turkey
kubat@sakarya.edu.tr

Abstract. In today's world, humanity was obliged to learn living with disasters. Although disasters are mostly accepted as events that occur outside the control of human beings, the natural disasters that have increased due to reasons such as the development of technology, the unconscious destruction the environment of human beings etc., have started to be a part of our daily life in today's world. Disaster Management discipline has been formed due to the needs such as managing disasters, minimizing damages caused by disasters and even developing methods for disaster prevention. Disaster management is a dynamic process with a very high risk coefficient. It is critical to establish strategies and to take advantage of technology correctly and adequately for taking precautionary warnings prior to disasters, providing effective intervention at the time of disaster and ensuring normalization after disaster. Based on the current importance of the subject, in this study a general strategic model was presented about IoT application areas related to the three phases of disaster management (before disaster, during disaster and after disaster).

Keywords: Disaster management · IoT · Strategy · Strategic model

1 Introduction

Today, natural and human related disasters have become a part of our lives. Human beings, who encountered only natural disasters in ancient times, are faced with a wide variety of human-induced disasters (technological reasons, diseases, wars, human migration, fires, etc.) in our age. For this reason, with each passing day it is seen that disaster occurrence periods are shortened and the severity of disasters increases. At the same time, no matter how much humanity advances in technology, it may become helpless in the face of disasters and suffer great losses. Pre-disaster preparedness and good planning, implementation of plans with qualified teams during disasters, providing rapid return to the new normal after the disaster are necessary to reduce the effects and risks of disasters. While rapid advances in technology may cause disasters on the one hand, they are also used to prevent disasters and reduce disaster damages and risks. One of these technologies is Internet of Things (IoT). Based on the current

N. M. Durakbasa and M. G. Gençyılmaz (Eds.): ISPR 2020, LNME, pp. 383–388, 2021.
https://doi.org/10.1007/978-3-030-62784-3_32

importance of the subject, in this study a general strategic model was presented about IoT application areas related to the three phases of disaster management (before disaster, during disaster and after disaster).

2 Disaster Management

Although disasters are mostly accepted as events that occur outside the control of human beings, the natural disasters that have increased due to reasons such as the development of technology, the unconscious destruction the environment of human beings etc., have started to be a part of our daily life in today's world. Disaster Management discipline has been formed due to the needs such as managing disasters, minimizing damages caused by disasters and even developing methods for disaster prevention. Disaster management is a dynamic process with a very high risk coefficient.

Disaster management aims to reduce the damage caused by disasters, provide quick and accurate assistance to the victims and normalize effectively (Sinha et al. 2017). Disaster management can be examined in 3 main sections: pre-disaster, during disaster and post-disaster. Pre-disaster preparedness or planning stage is the phase where strategies are determined to reduce risks, awareness trainings are given, and efforts are made to minimize the effects of possible disasters which be a threat. Determining risk reduction strategies in disaster management is also the activity of this phase. Post-disaster is the stage that enables to return to normal of life in the disaster area.

According to Karatop et al. (2020) "Natural disasters are events whose time, severity and impact are not certain. Although it is also the human beings most affected by disasters are own, now the majority of disasters stem from the human beings' own actions. Nowadays, the number of human-induced disasters have increased in a way which challenge natural disasters. Climate changes, wars, epidemics, migrations are almost normalized human-induced disasters that we are constantly experiencing. Although the number, type, severity and impact of disasters increase, the efficiency and effect of disaster prevention techniques and activities also increases. Disaster Management discipline has been created due to needs such as disaster management and minimizing the damage caused by disasters and developing disaster prevention methods. Disaster management is a dynamic process with very high risk coefficient".

Coordination of resource use and information sharing is of great importance in disaster management (Çağlayan et al. 2018). At the same time, it is a critical point to take preventive measures before a disaster, providing effective intervention at the time of disaster, to develop strategies and to benefit from technology for ensuring normalization after disaster.

3 IoT

Kevin Aston who first used the term Internet of Things in 1999 (Zhang 2011), has said the following about IoT in his article published in 2009.

"If we had computers that knew everything there was to know about things—using data they gathered without any help from us—we would be able to track and count everything, and greatly reduce waste, loss and cost. We would know when things needed replacing, repairing or recalling, and whether they were fresh or past their best".

We need to empower computers with their own means of gathering information, so they can see, hear and smell the world for themselves, in all its random glory. RFID and sensor technology enable computers to observe, identify and understand the world —without the limitations of human-entered data. .... The Internet of Things has the potential to change the world, just as the Internet did. Maybe even more so."

IoT has strategic importance in terms of disaster management (Karatop et al. 2020). Küçük et al. (2019) defines IoT as an ecosystem formed by devices which have perceiving and data processing capabilities, communicating with each others and the Internet.

4 Strategic Model Proposal to Disaster Management with IoT

Logistics and warehouse management are of strategic importance in disaster management (Karatop et al. 2020). The proposed model is based on this basis. At the same time, it is imperative that all warehouses have drones with optimum number and functions. 4 types of warehouse are suggested in the strategic model. These warehouses are assumed to be in optimized locations. Proposed warehouse types are as follows;

- Info-warehouse
- Main-warehouse
- Sub-warehouse
- Mobil-warehouse

As can be seen in Fig. 1, it can be said that the warehouses are in the form of clusters containing each other in terms of service areas and target audiences. At the same time, the things that info-warehouse, main-warehouse and sub-warehouse stock, are connected with each other but completely different. Info-warehouse can be defined as the control center where the data of entire system is stored and can access real-time data where immediate data is followed. This place is an information center and information warehouse where scenarios are studied with real data and simulations are made. In the model, Info-warehouse is in the position of decision maker and activator the warehouses according to these decisions. Reliability of data is important in decision making. In Information Logistics activities, with RFID technology all data in processes can be recorded and used as real-time. It provides ease of access to reliable data and helps to make the decision accurately and fastly (Karatop et al. 2020). While main warehouse's target audience and service area is international, national and regional, the target audience and service area of sub warehouses are national and local. Sub-warehouses are created according to the target audience and service area depending on the population density (province, town, village and neighborhood). While sub- warehouses deal with the immediate environment, main-warehouses have to be related to the big environment or the huge environment. Mobile warehouses are always ready to move. Because they are mobile, they deal with all places.

The proposed model consists of 3 stages (pre-disaster preparedness period, disaster period, post-disaster period);

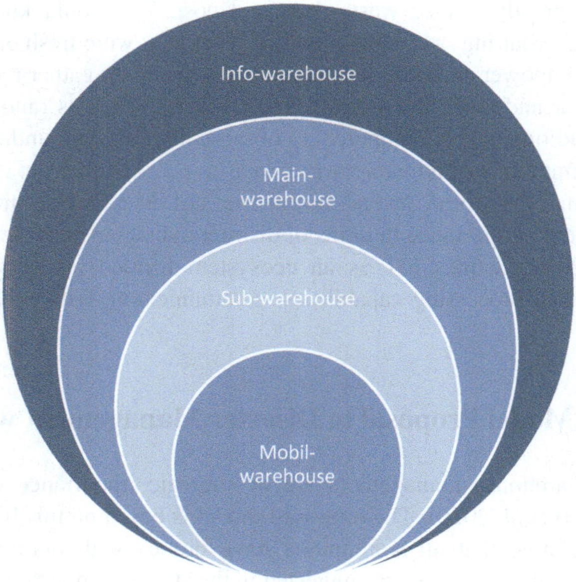

Fig. 1. The relationship of warehouses with each other

In the pre-disaster period, it is important to optimize the locations of the main warehouse, sub warehouse and mobile warehouses. These studies have to be done periodically. Optimization parameters may change according to changing conditions. Decision making criteria and optimization parameters must be updated. Each warehouse (main-warehouses, sub-warehouses, mobil-warehouses) is in real-time communication with each other and info-warehouse through IoT. Drones with optimized quality and number must be available for each warehouse. As Karatop et al. (2020) said "The lifeblood of disaster management is disaster logistics". For this reason, it is a very strategic decision to use drones in the first logistics activities and humanitarian aid during the disaster. In addition, every warehouse must have drones, but their number and characteristics must be optimized.

It is also an important issue that the status of the materials in the warehouses is up-to-date before the disaster. Warehouse supplies can be classified as food, health and hygiene, clothing and housing, fuel and other. During a disaster, all of these materials are desired to be at the desired quality and quantity. For this reason, it must be ensured that the current situation and the desired situation are the same. The current situation is the situation reported to info-warehouse from warehouses with IoT. The desired situation refers to the results obtained by optimizing according to parameters such as disaster risk, population, and construction status. Monitoring the current status of food, medicine and healthcare materials which have expiration dates, is essential for healthy products and materials during disaster. Before expiration dates, materials should be reported to info-warehouse via IoT and used for people waiting for help. At the same time, renewal is provided at the warehouses by transferring the durable materials that do not move at all and perhaps the their technologies are old, to the demand centers in a similar way. The proposed model is not static but is dynamic. New orders are placed to replace all materials whose expiration date is approaching and the technology is outdated. Order quantity is determined according to the periodically repeated optimization results.

In the event of a disaster, it is critical for info-warehouse to equip other warehouses with information and direct them.Info-warehouse ensures that the right timed and right amount of help is delivered to the right place in the minimum time with the flow of information. Here, the second critical object are drones. In drones, info-warehouse fulfills its related task by communicating with main-warehouse and sub-warehouse through IoT. It takes photographs, carries aid materials, reaches and transports the victim … etc. At the time of disaster, info-warehouse determines which mobile warehouses will move in which direction. Info-warehouse has to make fast and correct decision. It accelerates decision making with softwares supported by previously studied simulations and artificial-intelligence algorithms. The most important resource during disaster is time. Quick decision and action will ensure to be saved the maximum number of lives. Mobile warehouses take action according to the decisions made. It is known that the use of IoT in transportation and warehouse management affects efficiency positively. IoT is very successful in fleet management and determining the most suitable route (Karatop et al. 2020). In this context, mobile warehouses can quickly reach the disaster area using IOT.

Immediately after the disaster, the data where the IoT objects are located for damage detection, are collected at the center in info-warehouse. Küçük et al. (2019) proposed a similar system architecture in their work and named it "platform of post disaster management" for post-disaster management. After the disaster in order for life in the disaster area to return to normal, many activities are essential such as sending great aid to the region, removing the scrap and debris and improving the psychology of the victims.

After the disaster works that increase the effectiveness of disaster management such as maintenance type of machinery, equipment and vehicles, maintenance periods, tracking of things that are in the scrap pile and/ or will be in the scrap pile, can be done with IoT without the need for human monitoring. Thus, human-induced errors are also minimized (Karatop et al. 2020). It is certain that a disaster management system using IoT technologies will contribute to the reduction of post-disaster effects. With the use of IoT technologies, Info-warehouse obtains fast and accurate informations about the disaster area and the victims affected by the disaster. Thus, rescue personnel, equipment, vehicles, etc. resources can be managed more accurately, quickly, effectively and efficiently (Sakhardande et al. 2016; Küçük et al. 2019).

5 Conclusions

Advances in technology are increasing. With these developments, it increases both the type and number of disasters and reduces the disaster risks and damages. It is a critical point to put forward technology utilization strategies in disaster management. There are many benefits of using IoT in disaster management. At the same time, it accelerates decision makers and ensures efficient use of resources. In this context a general strategic model was presented about IoT application areas related to the three phases of disaster management (before disaster, during disaster and after disaster). The proposed model can provide the following benefits;

- Effective warehouse management is made before a disaster,
- Material loss is minimal,
- Disaster material information is available in real time,
- Damage assessment is made quickly after a disaster,
- Mobile warehouses are provided to reach the disaster site as quickly as possible,
- Chaos at the time of disaster is minimized,
- After the disaster, the normalization process is accelerated.

References

Ashton, K.: That 'Internet of Things' Thing. RFID J. (2009)

Çağlayan, N., Satoğlu, Ş.I., Kapukaya, E.N.: Afet Yönetiminde Büyük Veri Ve Veri Analitiği Uygulamaları: Literatür Araştırması, 7. Ulusal Lojistik ve Tedarik Zinciri Kongresi, ULTZK 2018 Bildiriler Kitabı 3–5 Mayıs 2018, Bursa (2018)

Karatop, B., Taşkan, B., Kubat, C.: Internet of things in disaster logistics productivity. In: Internet of Things (IoT) Applications for Enterprise Productivity, pp. 91–111. IGI Global (2020)

Küçük, K., Bayılmış, C., Sönmez, A.F., Kaçar, S.: IoT Teknolojilerini Kullanan Afet Sonrası Yönetim Sistemi. Akademik Platform Mühendislik ve Fen Bilimleri Dergisi **7**(2), 298–305 (2019)

Sakhardande, P., Hanagal, S., Kulkarni, S.: Design of disaster management system using IoT based interconnected network with smart city monitoring. In: 2016 International Conference on Internet of Things and Applications (IOTA), pp. 185–190. IEEE, January 2016

Sinha, A., Kumar, P., Rana, N.P., Islam, R., Dwivedi, Y.K.: Impact of internet of things (IoT) in disaster management: a task-technology fit perspective. Ann. Oper. Res. **283**, 759–794 (2017)

Zhang, Y.: Technology framework of the internet of things and its application. In: 2011 International Conference on Electrical and Control Engineering, pp. 4109–4112. IEEE, September 2011

Usage and Applications of the Swarm Robotics Concept at Industrial Level

Beritan Can(✉) and Şakir Esnaf

Istanbul University-Cerrahpasa, Istanbul, Turkey
beritaancan@gmail.com, sesnaf@istanbul.edu.tr

Abstract. In this study, it is seen as an effective method to ensure the flow of material/product between production units in lights-out factories, which currently have only a few but are expected to become widespread under Industry 4.0. In the current system, this technology can be used in a factory with human employees to prevent inefficiency caused by labor and time loss due to the use of the workforce to provide in-factory logistics. The study aims to reveal the benefits of integrating the concept of swarm robotics in the rapidly developing industrial era into the same rapidly changing production systems and implementing swarm robotics for a lights-out factory universe.

Keywords: Swarm robotics · Industry 4.0 · Lights-out factory

1 Introduction

Unmanned systems have ceased to be utopian in this century where vehicles can be sent to a point at billions kilometer away in deep space and data can be collected, images of black holes that have not been even known a hundred years ago and even sketches of life on other planets can be created. Small-scale developments, initiated and developed to meet the basic needs of human control and to make life easier, have evolved into larger dimensions over time and have become revolutionary from time to time and have reached the level to reveal human-like mechanical and intangible structures in the extent of intelligence and perception. Industry 4.0, which emerged as the last link of all these developments chain and expressed as the 4th Industrial Revolution, shows its effect in the fields of production, management, marketing, and even in the field of finance with the concept of blockchain. Industry 4.0 consists of nine building blocks: big data, cybersecurity, autonomous robots, simulation, internet of things, cloud, additive production (3D printers), augmented reality, and lights-out factories (Yıldız 2017). When Industry 4.0 enters human life, through above-mentioned components, objects can communicate each other as independently of human, machines can be controlled remotely for miles, millions of lines of data can be stored in a cloud environment, by designing the desired part, it can be produced at the home, office or anywhere through 3D printers. This means it is expected that there will be a reality in which many peculiar thoughts that have been the subject of only science fiction movies in the past decade have been brought to life.

N. M. Durakbasa and M. G. Gençyılmaz (Eds.): ISPR 2020, LNME, pp. 389–397, 2021.
https://doi.org/10.1007/978-3-030-62784-3_33

At the industrial level, the process that started with the use of hand tools in production continued with mechanization, and the developments have reached the magnitude of designing or even partially implementing unmanned production systems today. These unmanned production facilities which called lights-out factories are a concept with complex structures and smart robots customized for their tasks. These robots are mechanical structures that have the necessary software and hardware to make production. They can communicate with each other and, if necessary, with a central system. The ingredients of Industry 4.0 such as the internet, big data, and simulation of the above-mentioned objects must be used with advanced activity for these mechanisms to be used efficiently. Research by the McKinsey Global Institute has shown that only 78% of predictable physical work for production can be automated in a viable way (Chui et al. 2018).

2 Swarm Robotics

One of the concepts envisaged to be implemented under the title of Industry 4.0 is the swarm robot concept. The use of robots in non-factory areas such as social, military, health, and security has been made more complex to develop already complex robot mechanisms to adapt to multitasking and hard environmental conditions (Arvin et al. Arvin et al. 2009). As an alternative solution to these complex structures, the concept of "swarm robotics" is inspired by the concept of "swarm", which is the communities of animals or insects that come together by working together and systematically in complex and hard work and so the "Swarm Robotics" approach has emerged (Mohan and Pollanbalam 2009).

Swarm robot systems are systems that work for a specific and common purpose, with a large number of robots communicating and coordinating with each other and their environment, independent of any center. Thus, during the task, if one or more robots in swarm encounter a possible malfunction or are disabled, robust robots will continue to perform their tasks because of communication in the swarm, so there will be no disruption. As swarm robotic systems are made up of a large number of robots, robot structures are expected to be as simple mechanisms as possible, and each robot that makes up the swarm is expected to be manufactured at minimum cost.

The purpose of swarm robotic systems is basically to do a job that a single robot can do with many simple robots with the same or different features. Systems consisting of robots that have the same features are called homogeneous, and systems consisting of robots of different features are called heterogeneous swarm robot systems (Navarro and Matia 2013).

Swarm robotic systems do not need an inspection unless an extraordinary situation (malfunction, etc.) is involved. The speed to adapt to changing conditions is high. If a robot in the swarm is out of order for any reason, the rest of the swarm can continue to perform the task. The swarm size - the number of robots in the swarm - can be adjusted according to the job description. The advantages provided by these features increase the preferability of this technology.

In the swarm robotic scenarios, while the robot teams perform their pragmatic duties, in addition to the perception of the environment, they can also communicate with other robots belonging to the swarm (Arvin et al. 2009). Although the robots in the swarm change depending on the task expected to be performed, they are expected to recognize the environment they are in when necessary. They are expected to have the ability to follow a moving target, to influence objects as grasp them, and to influence each other by holding on to them. The swarm should behave at the minimum level by other robots around it and likewise to affect them to the minimum level and not to cause pollution (Turgut et al. 2007).

3 Definition of the Problem

When a lights-out factory universe containing all the components of Industry 4.0 is realized, it is a matter of curiosity how to provide logistics for the factory, which is one of the basic phenomena of a production facility. It is relatively easier to achieve large-sized transport operations, such as transporting a manufactured product with conveyors, compared to carrying out smaller and special tasks such as transporting a wrench, a screwdriver, 10 nuts, or 5 bolts. In this paper, the problem of how to carry small-sized transports that are not involved in a certain production process and which can come into view stochastically in a lights-out factory will be discussed.

4 Related Works

There are many studies within the scope of swarm robotics. Bodi et al. (2012) performed a batch cleaning scenario using the Beeclust algorithm with swarm robot systems and the robot named Jasmine has been shown the ability to find the shortest path between two target locations. Turgut et al. (2007), using a parallel programming algorithm, the behavior of acting as a swarm with an infrared communication system. Schmickl and Crailsheim (2007) used a new robot-inspired navigation principle based on a signal propagation technique inspired by the slime mold in the work of the previously mentioned batch cleaning scenarios. Arvin et al. (2009) talk about systems that allow multiple robots within a lot of robots to design an individual AMiR unit with simple detection and mobile capabilities. With this system, copying a large number of robots is easy and economically possible. Experimental results show the feasibility of using this robot to apply swarm robotics applications.

5 Method

Previous studies on swarm robotics have been studied. In this paper, a different method was used. The method is modeled on a lights-out factory universe where internal logistics is provided with a set of 2 robots that provide all the conditions of being swarm. This method was preferred because it is the most efficient method in terms of both cost and applicability in the current conditions. Working with a small number of robots has been an important factor in choosing this method as it provides a great advantage in terms of the simplicity of the system. The model is built on the following constraints:

1. Whether there is a load or not,
2. Proximity to the task (distance)
3. Percentage of charge

$i = 1, 2$ i = Number of robots
$j = 1, 2, 3, \ldots, m$
$k = 1, 2, 3, \ldots, n$ $m, n \in N$
w_i = load status of ith robot, $w_i \in (0,1)$, $\forall_i$
$A_i = y_i * D_i$, A_i = the distance that the ith robot can go and return with y_i charge
y_i = charge percentage of the ith robot
D_i = distance determination function
$D_i = \sqrt{(Lj - Lxi_j)^2 + (Lk - Lxi_k)^2}$, L is location
Lxi_j = The x coordinate of the location of the ith robot
Lxi_k = The y coordinate of the location of the ith robot
Lj = The x coordinate of the location of the task
Lk = The y coordinate of the location of the task

$$x_i = \begin{cases} 1, & \textit{assign to task} \\ 0, & \textit{not assign to the task} \end{cases}$$

$$\text{Min f} = \sum_{i=1}^{2} x_i \cdot y_i \cdot w_i \cdot D_i$$

Subject to,

$$\sum_{i=1}^{2} x_i = 1$$

$$y_i = \begin{cases} 1, & Di \leq Ai \\ 0, & \textit{otherwise} \end{cases}$$

$$w_i = \begin{cases} 1, & \textit{no load} \\ 0, & \textit{otherwise} \end{cases}$$

$$f > 0$$

In the model, the minimum function of f is written considering the status of the load, charge status, and distance to the task, and whether it is assigned to the task. The fact that only one robot will be assigned to a task is the key point of the study. All conditions will be evaluated and the most suitable robot will be assigned. The robot, which has no load, is at a level where the charging status can complete and return, and which is in the position closest to the task, fulfills all these conditions, and is assigned to the task.

6 Application

The application scenario of the study was realized with the model specified in the method section using 2 robots. In the study, the application has been made with swarm habitual robots which are produced by an entrepreneur company named Mikro-p and robots have the same name as the company. They have sensors that make it easy to exhibit swarm behavior on micro-ps and detect potential danger situations (obstacle, another robot), and in these cases, various LEDs and buzzers can be used as warning systems. In Fig. 1, technical details of micro-p robots are shared visually.

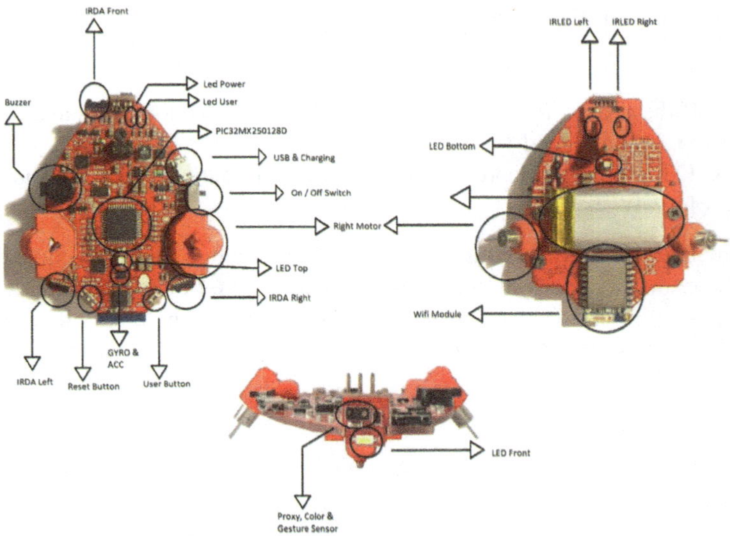

Fig. 1. Technical details of micro-p robots

Microprocessor: The microprocessor model used in MICRO-P is PIC32MX250F1 28D.

RGB LEDs: Mikro-P has two RGB LEDs, one on the top and one on the bottom. These LEDs produce a color depends on an 8-bit red, 8-bit green, and 8-bit blue value.

Led Headlights: It is a monochrome white LED.

PWMProxy, Gesture and Color Sensor: The TMG3992 color and distance detection sensor, which enables the detection of the amount of light, color and proximity of the object in the front of as the direction of movement, is located just above the Led Headlight on the front board of the MICRO-P.

IRDA: Mikro-P is equipped with three IRDA (TFBS4650 Infrared Transreceiver) for communicating with other Mikro-Ps using infrared rays. They are located at the left/right rear and at the front side.

WIFI Module: Mikro-P, which uses the ESP8266 wifi module, can be used as a wifi access point or in client mode connected to an access point. This module is located at the bottom surface, behind the battery.

As mentioned above, communication is provided by infrared transreceiver within the robots. Robots in the swarm are modeled with three constraints: charge status, load holding status, and distance from where the command will be performed. More than one robot cannot be assigned to the same job at the same time. The process of assigning a job to a robot works as follows:

- **Step 1:** The distance of the robots to the task is determined
- **Step 2:** Percentage of charge of the robot closest to the task is obtained. The charging status of the robot must be sufficient to complete the task and return.
- **Step 3:** If the charge status of the nearest robot is sufficient, the load holding status is checked.
- **Step 4:** If the nearest robot with sufficient charge status does not have loads at the same time, it is assigned for the task.
- **Step 5:** If it contains a load, go to Step 1(it will continue until the most suitable robot is found).

Considering the swarm abilities of robots, a universe that could belong to a unit of a factory where they can perform their tasks were designed. In this design, task positions are simulated by marking them in red and green colors. The area where the swarm operates is composed of 6 workstations. Design and realization of the operation are shown in Fig. 2:

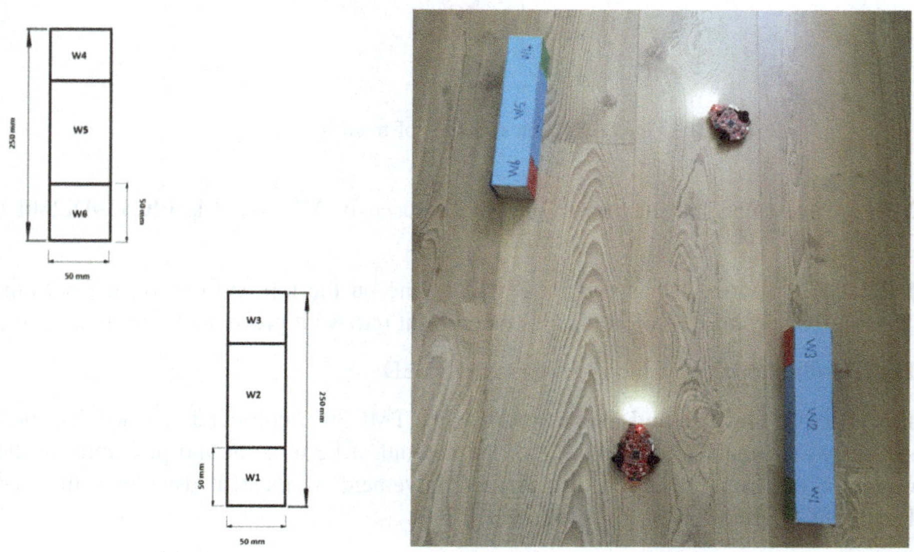

(a) *(b)*

Fig. 2. a) Design of the field swarm operates b) Realization of the field swarm operates

Each quadrangular region refers to a workstation. Workstations are marked with the letter W.

Robots are assigned a task that requires movement from the W1-coded workstation marked in green to the W3-coded workstation. A task has been assigned to the nearest robot in the current location. Robots are expected to perform their duties without hitting each other and other obstacles. The assigned robot first went to station W1. The assignment of the assigned robot to the W1 station, which is the task position, and the position of the other robot in the swarm at the same time are as shown in Fig. 3:

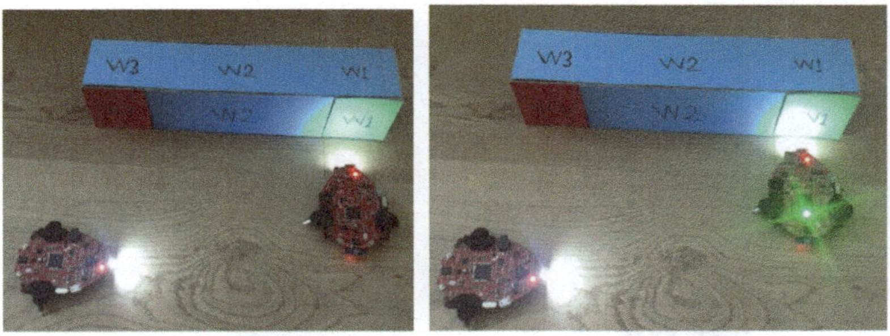

(a) *(b)*

Fig. 3. (a) Assigned robot at station W1 (b) Assigned robot completed the task

When the W1 station task is completed, the LED of the robot located on the top lights up green. Thus, the robot gives information that it has completed the task. After completing the part of his mission related to W1 station, the robot moved to the second part about W3 station. Meanwhile, there was no collision with the other robot in the same field from the swarm. The situation where the assigned robot arrives at the W3 station and the position of the other robot in the swarm at this time is shown in Fig. 4:

(a) *(b)*

Fig. 4. Robot during W3 task

After the robot has completed its defined task at the W3 station, it lights up the LED located on the top in red to give information about completing the task.

When the robots encounter an obstacle or go to a location other than the assigned task location due to a possible error, it stops briefly and the LED located on the top lights up blue.

In Fig. 5, the robot was left at a location other than W1 and W3 stations that are within the scope of the task. In this case, it was expected that the robot would be in a station outside its duty area and would illuminate the led light positioned at the top in blue color. The robot lit the upper led in blue as expected.

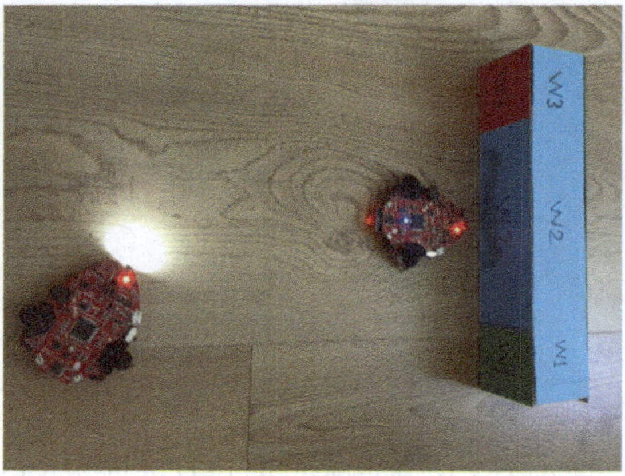

Fig. 5. The robot entered the wrong area

7 Conclusion

This study is a lot of robotic applications that can be used in the automation of the transportation activities in the existing factories and in the lights-out factories that can be implemented within the scope of Industry 4.0, which is expected to take place in the near future as the Fourth Industrial Revolution. In the study, the concept of swarm robotics, swarm behavior, and features that a swarm should possess are explained. Swarm robots' capabilities, availability, and advantages when used are described in terms of capabilities and cost.

In the application part of the study, a unit belonging to a factory is designed. In the factory universe, the swarm was assigned and it was observed whether it acted according to the expected characteristics. The swarm successfully completed the assigned tasks by going to the specified workstations.

Among the short-term targets of the swarm consisting of micro-p robots, it is included in the model that the function of releasing the load to the designated areas in the factory for the task priorities if the robot, which is the most suitable in terms of charge and distance, holds a load. Besides, a mobile application is developed and all the necessary information of the robots is captured instantly and the assignment of the most suitable robot from a center is among the planned future studies.

Acknowledgments. Thanks to Bürkan Törün of Mikro-P Electronics for his helpful suggestions and technical support on this paper. Mikro-P Electronics is a start-up company within ITU Cekirdek, İstanbul, TURKEY.

References

Yıldız, T.; Yaklaşan Dördüncü Sanayi Devrimi ve Türkiye'deki Mevcut Durum. ResearchGate (2017)

Chui, M., Manyika, J., Miremadi, M., Henke N., Chung, R., Nel, P., Malhotra, S.: Notes From The At Frontier Insıghts From Hundreds Of Use Cases. McKinsey Global Institute (2018)

Mohan, Y., Ponnambalam, S.G.: An extensive review of research in swarm robotics. IEEE (2009)

Navarro, I., Matía, F.: An Introduction to Swarm Robotics. Hindawi Publishing Corporation (2013)

Arvin, F., Samsudin, K., Ramli, A.R.: Development of a miniature robot for swarm robotic application. Int. J. Comput. Electr. Eng. **1**, 4 (2009)

Turgut, A.E., Gökçe, F., Celikkanat, H., Bayindir, L., Sahin, E.: Kobot: Suru robot çalismalari için tasarlanmis gezgin robot platformu. In: TOK 2007, p. 259 (2017)

Bodi, M., Thenius, R., Szopek, M., Schmickl, T., Crailsheim, K.: Interaction of robot swarms using the honeybee-inspired control algorithm BEECLUST. Math. Comput. Modell. Dyn. Syst. **18**, 1 (2012)

Schmickl, T., Crailsheim, K.: A Navigation Algorithm for Swarm Robotics Inspired by Slime Mold Aggregation, University of Graz (2007)

Information Management

Design of a Database Management System for Movie Recommendation Related to the History of Industrial Engineering for Courses

Damla Akın(✉) and Sakir Esnaf

Industrial Engineering, Istanbul University-Cerrahpasa, Istanbul, Turkey
damlakn@yahoo.com, sesnaf@istanbul.edu.tr

Abstract. This study aims to combine education and cinema to provide suitable movie content for students, academics, or users. At the first stage, the topics related to industrial engineering courses from each semester are determined by the development of the industry through industrial revolutions. The terms are then associated with the lessons and, matching film-semester-film is made as an appropriate auxiliary resource for the lessons. The database created in the SQLite program with the help of the keywords searched on the IMDb site is connected to C# and the database is used in the Windows form application. In the application presented to the user, three different pages are presented as lectures, semesters, and movies. It is possible to see which films are related to the selected course from the course list. This is a pioneer study in the literature suggesting a movie to the industrial engineering undergraduate students related to their courses. According to the best of our knowledge, an application-based project proposing films for courses has not been found in the literature.

Keywords: Database management system · Industrial engineering courses · Movie recommendation · Industrial revolutions

1 Introduction

The Industrial Revolution, in other words, Industry 1.0 started in England in the 1750s and continued until the 1850s. With the industrial revolution, a new economic system, production order, and new social classes have emerged [1].

In the workshop production process, not only the labor but also the thought, idea, and creativity of the final product were effective in the production. Taylorism was born with the labor and production processes in the system. In the Taylorist system, the production process is a production system where the work is divided into parts. The system determines how much work needs to be done and how long each part is going to take [1].

According to Taylor (1934), "U.S.A. It was written to point out the production losses encountered within the company, to convince the worker to operate and manage more effectively and efficiently, and finally to prove that the best management is scientific management and this is a management style suitable for all business styles".

N. M. Durakbasa and M. G. Gençyılmaz (Eds.): ISPR 2020, LNME, pp. 401–412, 2021.
https://doi.org/10.1007/978-3-030-62784-3_34

Taylorism aimed to increase the efficiency of the time it works, qualitatively and quantitatively, not to employ the worker more [1].

In the period from Industry 1.0 to Industry 3.0, we first see that muscle strength is important, and over time, this muscle force is replaced by mechanical strength. In Industry 4.0, we are now going through a period in which machines can communicate with each other, and concepts such as the Internet of Things enter our lives.

If we look at the history of cinema, it starts with Thomas Edison designing a camera to "do for the eyes what the phonograph does for the ears". Edison aimed to apply the phonograph system to the visual field and earn money there. Edison invented the Kinetoscope in 1891.

The kinetoscope allowed customers to watch the movie, which lasted several minutes through a binocular on the machine. Different films were shown on different kinetoscopes and these machines were working with coins. Edison invented the kinetophone, another device that combined image and sound after the kinetoscope, but the kinetoscope did not deliver the expected performance and did not work very well [2].

With the Internet age, we were able to move movies from videotapes to televisions, to small cinemas and computers that we set up with projections at home. In 2005, YouTube and many other online video websites started to be established. Publishing movies on YouTube and most video websites without permission from distributors has created a new copyright problem. With the introduction of platforms like Netflix and HBO into our lives, the viewing habit of the audience has completely changed. Today, most production companies have started to create their internet platform and publish their films here. The films started to be watched on television, computer and phones rather than movie theaters.

In 2016 and 2017, a script was written through an artificial intelligence called Benjamin, and the films of these scenarios were shot. In 2016, the movie "Sunspring" was written entirely by Benjamin. "It's No Game" film, shot in 2017, has been trained and co-authored with six other models. These changes over the years have shown that cinema is in a constant state of change [3].

Studies that use cinema in education were seen, but no interface source to assist at the university level was found. National Academy of Visual Education (1919), American Educational Film Association (1919), Visual Education Association (1920–1932), American National Education Association (1920–1932), American Visual Education Association (1922–1932), National Education Association Visual Education Department (1923–1946), National Education Association Audio-Visual Education Department (1947–1971) and Education and Technology Union, these associations were established to benefit from the educational opportunities of films in Europe and America. An institution called Innovations of Filmmaking has created the School Films Catalog [4].

The basis of this study will be formed with the terms determined from the mentioned industry periods, and movies will be associated with the course through these terms. The use of films in education is thought to be a useful auxiliary resource for learning difficult-to-understand topics, analyzing the subject, keeping the subject in mind, and often embodying situations that cannot be observed during education at school. Besides the uniformity of the education system, the information transferred with limited and classical methods, the participation of the cinema in the education will add to the vision as well as the rate of student learning. The inclusion of art in education in some way, this includes being accessible and continuous, not one-time, will be effective on the student. Domination of the course can be increased through films. Even if the student has nothing to do with the course, it will be of interest to be the source of this learning or learning, which is outside the classical methods, as a cinema. This study might be the first study in this manner.

2 Using Movies as Educational Tools

Cinema has been used as a tool for education in various ways, but in the literature searched, no application with a course-term-film relationship has been found, as was done in this research.

In a study on using cinema films in education as a socialization tool, it has been argued that the information gained from movies is more permanent in memories and can be a guiding and helpful resource in case studies, language development, and problem-solving. Studies are stating that the use of cinema as an educational tool especially in psychology is necessary for students to see the cases. Students were presented with a list of films and asked to select two films from this list [5]. Some questions have been prepared to develop critical thinking directed to the student by watching these films. These questions are:

1. What are the prominent features of psychological disorders that are portrayed in the characters in the movies you watch?
2. What kind of social or cultural attitude is reflected in the revived mental illness, treatment, and the role of the health personnel?
3. Describe your reaction to the characters in the movie.
4. Does your psychiatric treatment knowledge change your reaction to the movie or the characters in the movie?
5. How would you use this film to make a mental illness or its treatment better understood by patients, families, population, or medical staff? [5]

With these questions, it was tried to emphasize how useful cinema can be in education. Preschool USA National Science Foundation aimed to introduce students to the concepts of cartoon engineering and to meet science and cinema at an early age.

In another study, it was stated that cinema was frequently used in chemistry education. In this study, it has been analyzed how useful it is for students to spend their education period with the help of motion pictures. At the same time, the project aims to assist the literature as a movie archive in chemistry education [6].

"TeachWithMovies.com" shows classroom teachers how to teach with movies every day. TWM has more than 30 films on science, technology, aviation, space exploration, and medicine. It is a site that guides teachers for high school level courses [7].

There is a lot of work on making use of cinema especially in the fields of literature, chemistry, religion, and psychology. In the studies in the literature, the importance of using cinema in education was emphasized and suggestions were presented. The Teach

With Movies site went beyond the suggestions and offered a movie archive to high school level courses for teachers.

3 Database Management and Suggestion Systems Developed for Movies

With a study of 52 films and 113 participants, a project that created a new database of emotional movie clips without auditory content, based on the fact that movie clips were more active at subjective and physiological levels, was created. A psychophysiological evaluation was performed according to the skin conductivity level and heart rate of the participants. In the film selections, attention was paid to the protection of the context throughout the film, and films in various categories (horror, socially positive, social negative, landscape) were selected. The database movie clips described in the article cannot be used for commercial purposes, but are available only free of charge at EMDB@psi.uminho.pt [8].

Internet Movie Database (IMDb), which serves the user by using the database developed for movies, is one of the most used movie suggestion systems. Academy rewards (Oscars) are particularly highlighted on IMDb. The films are subjected to two scoring systems with separate scoring of users and cinema writers. Metascore scoring is the average of the scores given by film writers. IMDb also allows users to create their movie lists, watch lists. IMDb provides not only movies but also people from various business lines, such as behind-the-scenes and front-of-camera, in the industry. The dataset is a two-sided graphic where each node corresponds to an actor or a movie [9].

Many platforms such as Letterboxd, Blutv, and Netflix mainly use movie databases.

In another study, with the increase in the number of films, an approach is offered to the cinephiles and viewers that a suggestion system can be developed and this suggestion system can be realized based on how it is scored by people. Features describing the year of production, cast, director, genres, and average scoring are extracted and then used with a kNN classifier to decide how much someone will score any movie in the database. Based on this rating, many movies that have not yet been seen are selected and recommended [10].

4 Method

Since the project does not have an example in the literature, the method steps were tried for the first time, and since there was no database sample, films were selected by using the keywords from IMDB, and a database was first created with SQLite and then Windows Form application was written in C# programing language. The project initially focused on industrial engineering courses and it is aimed to be implemented for all undergraduate programs during the further development stages of the project.

4.1 A Database Management System that Associates Education Topics with Movies and Suggests a Course

Steps:

1. A database is created with the determined keywords and/or terms.
2. Associating the course, terms, and films by establishing a database
3. Connecting the database and C#
4. Establishment of the application structure

4.2 The Database Is Created with the Determined Keywords and/or Terms

Various terms have been identified by following the development of industrial history. These determined terms were later associated with industrial engineering undergraduate education courses. Following the history and literature of undergraduate education is very important in determining keywords and terms and makes the relationship between the courses and terms strong.

Various database systems can be used to create a database. The reason for using SQLite in this project is that it is open-source, compatible with C#, and small in size.

Separate tables are created for keywords and terms in the database. The created tables are related to the database used by writing a query or ER diagrams. Tables prepared in the SQLite database system are related to each other with queries. For association, there should be fields in common in two tables. This created database is a preliminary preparation for the database to be used in the application and is an auxiliary source.

4.3 Associating Courses, Terms, and Films by Establishing a Database

In the first step, the selected database and the keywords and terms determined as a result of the researches and the course that the movie will be recommended are selected. It is necessary to create three main tables, which are the course, terms, and films that form the main lines of the application. These tables must have common areas. Three new tables are created by associating the end of query tables to be made over common areas.

4.4 Connecting Database and C#

C# connects with the created database. This link is loaded with the libraries required to connect the database application used primarily. After the necessary libraries are loaded into the application, the database is called and its location is specified. At this point, specifying the location correctly is very important for the database to run smoothly. After calling the database without any problems, the necessary codes are entered to read the database.

4.5 Building the Application Structure

In the first stage of the application, the design and code structure of the homepage is created. Later, courses, films, and terms pages are created. By linking these pages with the main page, switching between pages should be made possible. Course detail, film detail, and term detail pages are created for the transition to detail pages. With these pages, course, movie, and term pages are connected and a transition is provided between each page. The course detail page contains the necessary information about the course, the films, and terms of the course. The film detail page contains the necessary information about the film and the list, of course, s it is linked to. The terms detail page lists the necessary information about the term and the movies it is attached to. Connecting C# and SQL correctly is very important at this stage of the project. SQL queries required for the vehicles on the course, film, terms, and detail pages, whose information will be printed, are printed and the necessary information is called from the database and printed on the vehicles. The project is completed by linking the pages and printing the necessary data.

In this paper, only industrial engineering courses have been studied, but the structure of the database management system and application is suitable for all undergraduate programs.

5 Application

The terms determined through the industrial revolutions will be associated with the industrial engineering courses, and tables will be created in the SQLite application. Later, by adding the film's table, a relationship will be established between the three tables and new tables will be created as course films, terms films, course terms. This database prepared in SQLite will be used in the C# program language and a decision support system will be prepared as a Windows Form desktop application.

First of all, the right model should be installed correctly and run with the right algorithms and the most correct result should be reached. After model management, data will be pulled from SQLite to C#. As a result of the application, when the user opens the desktop application, there will be transitions over three titles that can obtain information through films, terms, courses. Not only course suggestions about films but also necessary film analysis and information will be available. This project, which will be presented to the user as a Windows Form application, will be realized for the first time and the design, data, and application process are project-specific.

5.1 SQLite

SQLite is a SQL database engine developed with free and open-source C/C++ languages. The SQLite database, which is also compatible with the C# language to be used in the application, is also small in size and compatible with most operating systems. SQLite does not need a separate server operation [11].

In C#, data to be used in the application before the application passes, tables are created in SQLite and then data is entered into the tables. Three basic options will be presented to the user on the main page. Then, pages with lists of these options and detailed information will be organized. The tables created in Fig. 1 are included. To create a table from the Execute SQL button, necessary queries are written as in Fig. 2. A total of 6 tables have been created to provide a connection between three main titles and these titles.

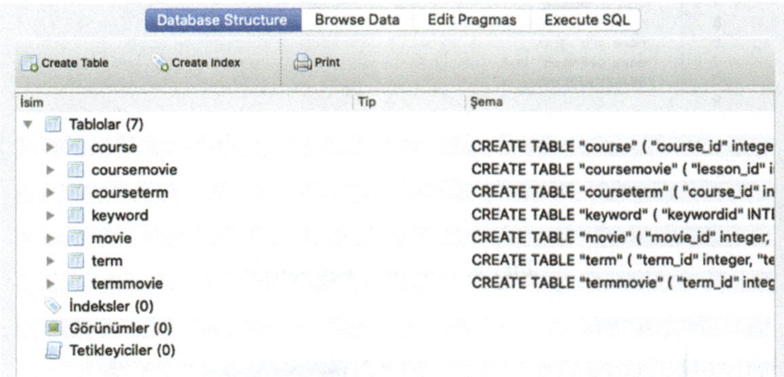

Fig. 1. SQLite table

The table in Fig. 3 is the course table, which is one of the three options offered to the user. The course table contains course_id, course_name, course_definition, and a column of notes to be taken from the user. In the term, film and course tables, the "Primary Key" column is created primarily and the relationship between the tables is created with these keys.

The output of the course names in the course table in Fig. 3 is as in Fig. 4 and after clicking on the course which movies are related to the courses in the course table, the new page appears.

The relationship between all tables is made with the primary key. These tables are created with the SQL query in Fig. 5.

In Fig. 6, there is the first page of the application. When the user enters the application, he/she selects which page he/she will go to in the application via three buttons. In Fig. 7, there is the third page of the course pages. After the users choose the course button on the home page, they first access the page in Fig. 4 (where the course list is located). After the course is selected, they access the film and course information about the course on the course detail page. It is also used in the flow term and movie pages on the lecture pages.

Figure 8 shows how the libraries needed to create the movie page are called. Since the database will be connected to the application with SQLite, the Data.SQLite library has been called.

After calling the necessary libraries, the database was connected to the application with the codes shown in Fig. 6, and then the movie names in the movie table were printed with the necessary queries as shown in Fig. 8. The same procedures have been applied for term and course pages (Fig. 9).

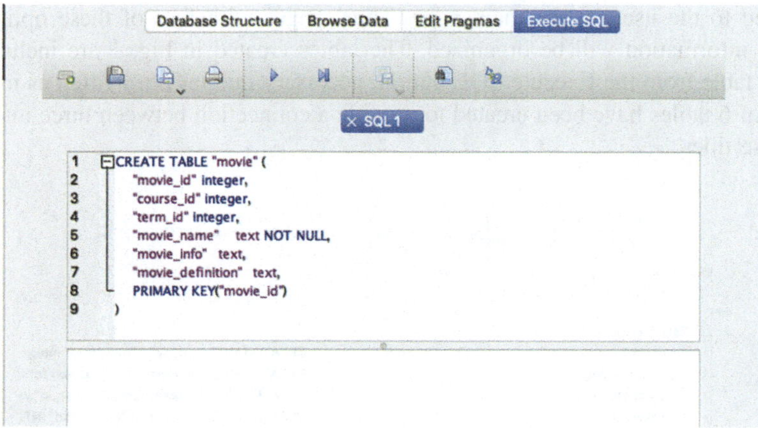

Fig. 2. Execute SQL query

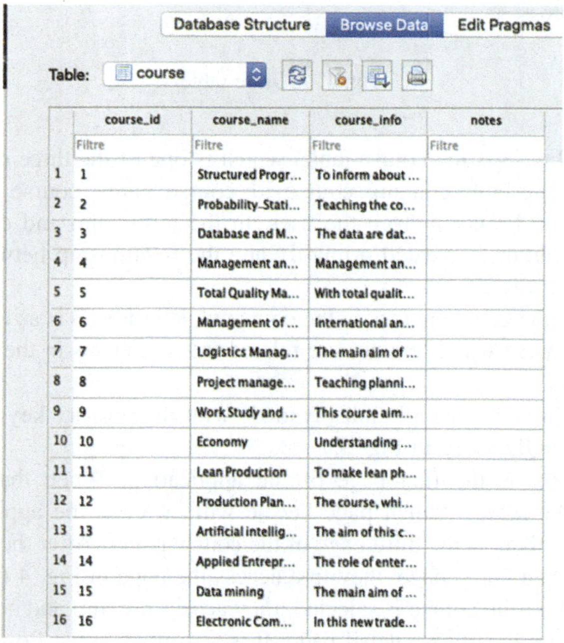

	course_id	course_name	course_info	notes
	Filtre	Filtre	Filtre	Filtre
1	1	Structured Progr...	To inform about ...	
2	2	Probability-Stati...	Teaching the co...	
3	3	Database and M...	The data are dat...	
4	4	Management an...	Management an...	
5	5	Total Quality Ma...	With total qualit...	
6	6	Management of ...	International an...	
7	7	Logistics Manag...	The main aim of ...	
8	8	Project manage...	Teaching planni...	
9	9	Work Study and ...	This course aim...	
10	10	Economy	Understanding ...	
11	11	Lean Production	To make lean ph...	
12	12	Production Plan...	The course, whi...	
13	13	Artificial intellig...	The aim of this c...	
14	14	Applied Entrepr...	The role of enter...	
15	15	Data mining	The main aim of ...	
16	16	Electronic Com...	In this new trade...	

Fig. 3. SQLite course table

6 Conclusions

In this study, the place of the cinema in education was based on the idea that the knowledge that students acquire through cinema will be more permanent. No study has been found to support undergraduate students in education through a movie recommendation. Since there were no examples in the literature, the method steps were tried

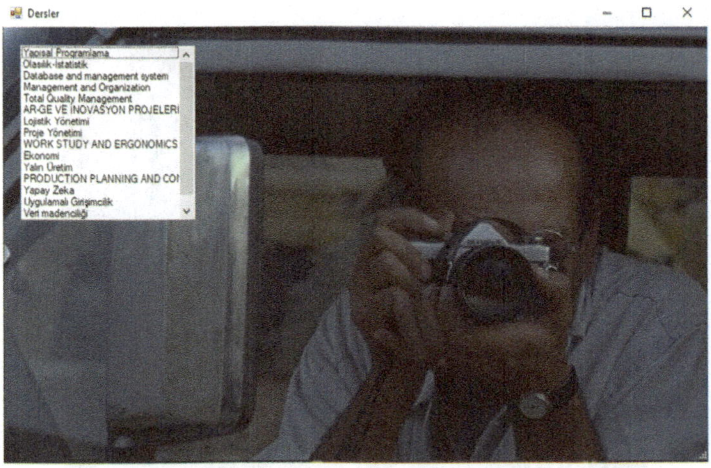

Fig. 4. Courses page (in Turkish)

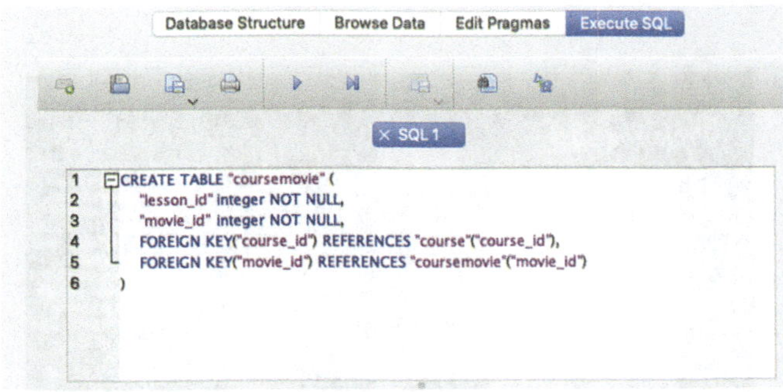

Fig. 5. SQLite query

for the first time and since there was no database sample, films suitable for the courses were selected by searching the IMDB for keywords. The project initially focused on industrial engineering courses and it is aimed to be implemented for all undergraduate programs during the development stages of the project. Since there is no similar study in the literature, so this makes the study original and unique.

Thanks to the application carried out in the project, users will be able to access film data for industrial engineering courses and make associations with the films. SQLite database is used in this application developed with C# Windows Form. The database is continuously updated and the application is open to development. It is suitable for the development of the project, by applying the application to all undergraduate programs, a larger database of users, and the contribution of cinema to education/learning. The application will likely be made an appropriate option not only for students but also for academic studies to grasp the subject.

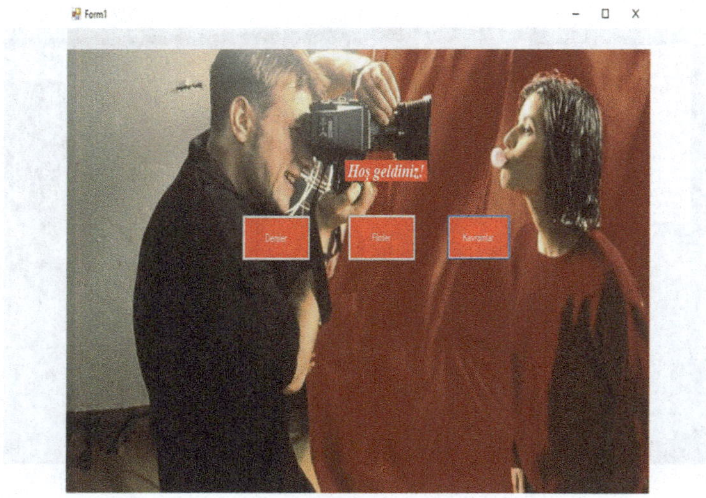

Fig. 6. Home page (in Turkish)

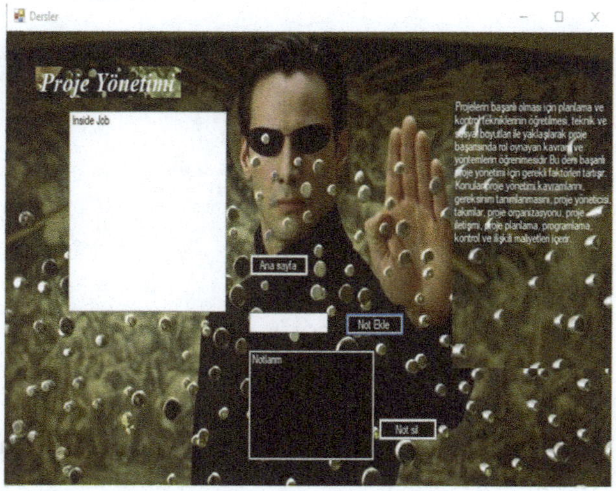

Fig. 7. Course detail page (in Turkish)

As a result, the undergraduate, graduate, and academic studies will be able to present film information to the user by using education and cinema together in the works that continue with the project originality. The project aims to provide support to students and academics through cinema in all undergraduate courses. With the development of the project, the application is made more comprehensive than Windows Form and aims to support not only students and academicians but also various platforms as an additional package. One of the objectives of the project is to develop a database system that universities can integrate into their systems.

```
using System;
using System.Collections.Generic;
using System.ComponentModel;
using System.Data;
using System.Drawing;
using System.Linq;
using System.Text;
using System.Threading.Tasks;
using System.Windows.Forms;
using System.Data.SQLite;
using System.Collections;
```

Fig. 8. C# library

```
namespace DamlaBitirme
{
    5 başvuru
    public partial class Filmler : Form
    {
        ListBox mylist = new ListBox();
        SQLiteConnection sqlite_conn;
        SQLiteCommand sqlite_cmd;
        SQLiteDataReader sqlite_datareader;
        private ArrayList filmler = new ArrayList();
        1 başvuru
        public Filmler()
        {
            InitializeComponent();
        }

        1 başvuru
        private void Filmler_Load(object sender, EventArgs e)
        {
            this.BackColor = Color.FromArgb(255, 232, 232);
            sqlite_conn = new SQLiteConnection("Data Source=C:/Users/damla/Downloads/deneme.
            sqlite_conn.Open();

            sqlite_cmd = sqlite_conn.CreateCommand();

            sqlite_cmd.CommandText = "SELECT * FROM film";


            sqlite_datareader = sqlite_cmd.ExecuteReader();
```

Fig. 9. C# movie code

References

1. Tezgel, O.: The birth of the contemporary factory system and the phases that passed up. Doctoral Thesis, Ankara University (2010). in Turkish
2. Özuyar, A.: Silent Period Turkish Cinema History. Yapi Kredi Publications, Istanbul (2017). in Turkish
3. Anadolu, B.: The impact of artificial intelligence on cinema in the context of digital storytelling: an analysis of sunspring and it's no game. Erciyes İletişim J. Spec. Iss. **1**, 39–56 (2019). in Turkish
4. YAKAR, H.G.İ.: Educational use of cinema films: a historical evaluation. Hasan Ali Yücel Fac. Educ. J. **19**(1), 21–36 (2013). in Turkish
5. Birkok, M.C.: Alternate media usage in education as a socialization tool: movies. Int. J. Hum. Sci. **5**(2), 7–8 (2008). in Turkish
6. Pekdağ, B., Le Maréchal, J.F.: Movies in chemistry education. Asia-Pac. Forum Sci. Learn. Teach. **11**(1), 1–19 (2010)
7. Frieden, J.A., Elliott, D.W.: Teach with Movies: Using the Storytelling Power of Movies to Motivate Students, Teacher Librarian **34**(3), 1 (2007)

8. Carvalho, S., Leite, J., Galdo-Álvarez, S., Gonçalves, Ó.F.: The Emotional Movie Database (EMDB): A Self-Report and Psychophysiological Study. Appl. Psychophysiol. Biofeedback **37**(4), 279–294 (2012)
9. Herr, B.W., Ke, W., Hardy, E., Borner, K.: Movies and actors: mapping the internet movie database. In: 2007 11th International Conference Information Visualization (IV 2007) (2007)
10. Frykowska, A., Zbieć, I., Kacperski, P., Vesely, P., Studenicova, A.: Movies recommendation system. Adv. Intell. Syst.Comput. 579–585 (2019)
11. Bhosale, S.T.: SQLİTE: lite database system. Int. J. Comput. Sci. Mob. Comput. **4**(4), 882–885 (2015)

Process Mining Research in Management Science and Engineering Fields: The Period of 2010–2019

Atik Kulakli[1(✉)] and Semra Birgun[2]

[1] College of Business Administration, American University of the Middle East, Egaila, Kuwait
atik.kulakli@aum.edu.kw
[2] Istanbul Gedik University, Istanbul, Turkey
semrabirgun@gmail.com

Abstract. Process mining is an analytic method to improve processes by extracting the information from various information systems. The interest has increased recently in scholarly publications. Authors aimed to explore the research outputs to identify and synthesize the results from studies on process mining, specifically in management science and engineering domains. The bibliometric approaches employed to analyze the papers published in highly ranked and cited journals which indexed and ranked in Web of Science Core Collection from 2010 to 2019. This pilot study concentrates on the management science and engineering field. Further, similar studies will be extended with different academic fields and publication sources. The data of 35 highly ranked journal articles found in SCI-Exp and SSCI indexed sources. The bibliometric study reveals the patterns of scientific areas of research domains.

Keywords: Process mining · Management science & engineering · Bibliometrics · Information management · WoS core collection

1 Introduction

Process mining has received substantial attention in management and engineering literature recently. Many studies from different fields highlighted and explained the importance of the domain. Process mining is an analytical discipline to detect, monitor, and improve real processes through an easy extraction of information from event logs in current information systems, and provides easy use for digitization. Specialized dataset mining algorithms used to identify trends, patterns, and descriptions of event logs reported by an information system during process mining. Process mining aims at improving process efficiency and process understanding. Due to the nature of process mining, there is no single approach to use for the application process [1]. Process mining has similar steps, like managing the projects where systems developed for compliance with each other. Process mining also has standard process methods such as concentration to current situations, data analysis for discovery, and performance evaluations like any other management and technological approaches [2].

N. M. Durakbasa and M. G. Gençyılmaz (Eds.): ISPR 2020, LNME, pp. 413–425, 2021.
https://doi.org/10.1007/978-3-030-62784-3_35

The process mining studies varied by different business applications with multi-dimensional techniques, principles, and content-specific industrial cases. Knoll et al. [3] developed a methodology [combined with multidimensional PM techniques, and principles of lean and VSM] to enable continuous recording, evaluation and waste analysis for internal logistics, and conducted a case study to evaluate the methodology in an automotive company. Jans and Hosseinpour [4] presented an approach combined with data mining and PM for the transaction verification level of continuous auditing. This approach includes a human expert as an auditor.

Authors applied bibliometric analysis with WoS Core Collection data in order to explore the academic level of process mining research. Significant findings in this study include: (a) the number of journals published in 2010 significantly increased by 2019; (b) there are three leading countries publishing research in the process mining field, including the USA, PR China, Belgium, and S. Korea; (c) there are three predominant sub-fields by subject category, including the Operations Research Management Science are the leading field with 51.43% articles, Computer Science Artificial Intelligence area with 34.29%, Business and Business Finance area with 31.43%, Management with 22.86% and Engineering Industrial with 20% respectively; (d) among journals the leading in publishing sources are Expert Systems with Applications with seven records (20%), and Decision Support System with five records (14.29%).

2 Literature Review

In their study, Roldan et al. [5], based on VR and PM, proposed a training system for assembly operators and simulated it to validate. They reported that this training system was better in terms of user evaluations, mental demand, perception, learning, results, and performance compared to traditional systems. Wu et al. [6] presented a 6-steps [Accimap] methodology to analyze the business process in different system levels and validated by a case study in the complaint handling process in a manufacturing company. In the research of He et al. [7], a conceptual model developed with PM for gas explosion accident emergency rescue in China coal mines.

Baader and Krcmar [8] developed a conceptual model "reduce false-positive numbers by combining red flag approach and process mining" [8], and they implemented to identify fraud within the purchase-to-pay business process. They obtained a lower positive false rate as % 37. Zerbino et al. [9] developed a 5-step model based on a PM framed expert system for information system audits and conducted a case study for validation of the model in a port process. They reported that the findings of the case study are useful for reengineering the export process, and for revising the constraints in the process flow of the Port Comunity System.

Baykasoglu et al. [10] presented a novel approach based on process mining and similarity analysis for evaluating the performance of students in a real-life online exam. They compared the assessment mechanism against the instructor to evaluate the performance of the assessment system and reported that they obtained perfect agreement between the automatic assessment system and instructor.

Park et al. [11] analyzed the characteristics of information navigation based on process mining techniques used for nuclear power plant event logs. Erdogan and

Tarhan [12] developed a goal-driven process evaluation method by using PM for improving the healthcare processes and conducted a case study of a surgery process to investigate the feasibility of this method. They stated that the results showed that the proposed method has the potential to initiate proposals.

Er et al. [13] demonstrated the use of process mining in modeling and analyzing at a process level in a real manufacturing company by using PM for production planning in an ERP environment. Van der Aalst [14] showed that spreadsheets could be used for PM and applied them to the order handling process in a telecom company. Cho et al. [15] proposed a framework contained with performance indicators to assess the life-cycle phase of the business process redesign. Werner [16] studied on providing accurate information on the accounting control flow and obtaining a lower complexity.

Rojas et al. [17] proposed a data reference model to guide the emergency room experts in conjunction with a detailed analysis using data and process mining techniques and tested this methodology in the emergency service of a university hospital. Greyling and Jooste [18] conducted a case study for effectively managing and improving the asset management process by using PM. Park, Jung, and Jung [19] prepared a feasibility study to improve the quality of work process in assessing for Human Reliability Analysis, and they stated that process mining techniques are useful for identifying the necessary information.

Lee et al. [20] developed a slippery genetic algorithm-based PM system to optimize fuzzy rules to facilitate a comprehensive quality assurance scheme in the garment industry. A case study is conducted in a garment manufacturing company to evaluate the sGAPMS. The results illustrate that better quality. They evaluated this system with a case study in a garment producer and stated that it obtained better quality assurance. Carrera, Lee, and Jung [21] proposed a probabilistic model to process discovery based on an extended hidden Markov chain and evaluated the model with real data gathered from a Facebook fan page. Rovani et al. [22] constituted a methodology and conducted a case study to show how to use PM techniques based on declarative models to analyze medical treatment processes. Cheng and Kumar [23] proposed a rule-based approach to remove noisy traces from the log and presented results from experiments with a real dataset.

Maita et al. [24] studied on a systematic literature review to evaluate how the use of artificial neural networks and support vector machines in process mining. Cho, Song, and Yoo [25] proposed a methodology to analyze outpatient processes based on process mining, and they show its validity with a case study. Claes and Poels [26] provide a technique that consists of a method and an algorithm for an inter-organizational process and tested with data sets to confirm its effectiveness and efficiency. Lee et al. [27] studied an RFID-based PM system as an intelligent system using fuzzy association rule mining with a recursive process mining algorithm in order to achieve quality assurance. They applied the proposed system in a garment factory and obtained improvements in defects, production time, and rework cost.

Yzquierdo-Herrera, Silverio-Castro, and Lazo-Cortes [28] conducted a systematic approach to estimate the lack of information in the traces used in processes mining, and they discussed their experimental results. Jans, Alles, and Vasarhelyi [29] presented a PM study to make a case whether internal and external auditors should leverage the capabilities of process mining. Caron, Vanthienen, and Baesens [30] proposed a comprehensive rule-based process mining approach with a two-dimensional business rule taxonomy.

Huang and Kumar [31] described new quality metrics and a novel quality-based algorithm for model extraction from noisy logs and reported. The results of the experiments with this algorithm on real and synthetic data. Liu et al. [32] proposed a generic approach of business process simulation for operational decision support and, they denoted this approach plays a vital role in process management and improvement by simulating for a credit card application. Jans et al. [33] conducted a case study at the procurement process for internal transaction fraud mitigation by using PM.

Ou-Yang and Winarjo [34] proposed a Petri Net integration approach to support multi-agent process mining by using the alpha algorithm. Ou-Yang and Juan [35] developed a Petri Net based evaluation approach consists of conceptual, system design, and implementation stages, and applied the third stage in an agent-based robotic assembly system. Edgington, Raghu, and Vinze [36] empirically analyzed the database used in the help desk process between a national US public agency and its global outsourcing provider and suggested the refinements in service level agreements. Li et al. [37] proposed Policy-based Process Mining as a novel approach to provide a solution to the information overload problem.

3 Methodology

3.1 Aim of the Study and Research Questions

The main objective of this study is to explore scientific publication patterns in research domains of "process mining" concerning management science and engineering domains. This pilot study reveals domain-specific contributions from various research outcomes. Therefore further studies on the process mining would be extended with different subject areas to compare the research patterns and trends further show the gaps. Research questions formed based on those objectives and scope and listed below;

RQ1 What are the characteristics (descriptive) of publication results? How many papers on "process mining" have published in 2010 and 2019?
RQ2 In which sources were the papers published most frequently?
RQ3 Who are the most productive authors/co-authors published in those periods? What are the citation results of authors?
RQ4 What are the most productive organizations and countries?

3.2 Bibliometric Study and Data Collection

Having conducted a bibliometric study requires a structured database to gather and analyze the data. There are various analysis methods for literature review, along with bibliometric tools [38, 39]. In this study, the authors employed detailed publication results, author/co-author productivity, in parallel organizations and country productivity, publication sources, and general citation results for the period of search terms. In this pilot study, a single database of Web of Science Core Collection (SCI-Exp and SSCI) selected. There are two reasons for conducting the data collection, the first structure of data for bibliometric design, and the second is for citation accuracy and ease of comparison. ISI and Scopus have separate citation counts according to indexes. Therefore for uniqueness purposes, this is vital to select a single database.

In consequence, the authors decided to work on highly ranked and impacted sources for initial research. Thus, publication data retrieved from the WoS database with below search strategy

Search criteria and strategy;

Database ISI Web of Science Core Collection (Clarivate Analytics)

Indexes = SCI-EXPANDED, SSCI

Timespan = All years of (2010–2019)

Step 1: General search without academic field restricted

TITLE: ("process mining*") 87 articles found in DOCUMENT TYPES of "Article and Proceedings Paper."

Step 2: Management Science and Engineering field selected (refined)

TITLE: ("process mining*") 35 articles found

Refined by: DOCUMENT TYPES: ("ARTICLE" OR "PROCEEDINGS PAPER") AND WEB OF SCIENCE CATEGORIES: ("ENGINEERING MULTIDISCIPLINARY" OR "OPERATIONS RESEARCH MANAGEMENT SCIENCE" OR "ENGINEERING MANUFACTURING" OR "MANAGEMENT" OR "BUSINESS" OR "ENGINEERING INDUSTRIAL")

Search criteria above conducted and retrieved data from the WoS database and downloaded as plain.txt, and.bib file formats for further analysis in Microsoft Excel and R Studio for descriptive and bibliometric analysis.

4 Results

4.1 Main Findings of Descriptive Publication Results

Total of (n = 35) articles found in Web of Science Core Collection database (SSCI and SCI indexes publications) for the period of 2010–2019. The only (n = 1) article is written in the Spanish language (2.86%) while (n = 34) articles are written in the English language (97.14%). There are 101 authors/co-authors from 19 countries leading the USA, followed by P.R. China, Belgium, South Korea, Netherlands, and Taiwan. Based on search criteria, 35 journal articles found (2 of them considered as proceedings). The keyword distribution found as Keyword Plus (ID) (n = 93) and Author's Keywords (DE) (n = 118).

Table 1 shows the top five Web of Science subject categories distribution of the retrieved data. In addition to the table, results with *(n = 6) Engineering Electrical Electronic, (n = 5) Computer Science Information Systems, Engineering Multidisciplinary, (n = 4) Computer Science Interdisciplinary Applications, Engineering Manufacturing, (n = 3) Chemistry Multidisciplinary, Materials Science Multidisciplinary, Physics Applied, (n = 1) Education Scientific Disciplines, Information Science Library Science, and Robotics.*

Table 1 Top five WoS subject areas by articles (2010–2019)

WoS subject categories	Record	Percent (%)
Operations Research Management Science	18	51.43
Computer Science Artificial Intelligence	12	34.29
Business and Business Finance	11	31.43
Management	8	22.86
Engineering Industrial	7	20.00

4.2 Distribution of Articles by Years

Figure 1 presents the publication records by years starting from 2010 through the end of the year 2019. Records started with (n = 3) in 2010 then slightly decreases to (n = 2) for 2011, 2012, and 2014 for each year. From 2013 to 2019, rises to (n = 5). The distribution of publications is not steady. 2016 and 2017 are below the average for the research period. The sharp increase can be seen in 2018 (n = 7), which is the highest year of overall comparable in the period. The publication trend shows the interest of publication and upward movement in the counts—the mean of the publication period as of (n = 3.5) yearly.

Fig. 1. Publication records by years

4.3 Journal and Source Frequency and Productivity

According to published data, there are 17 sources published process mining-related articles. Table 2 shows the publication frequency in the most relevant sources. Although there are no significant differences in publication history, the top contributed journals of 7 sources (n >= 2 records) listed in Table 2. The most productive publication sources sorted from the highest on top of the list with (n = 7; 20%) records are *Expert Systems with Applications*. Followed by (n = 5; 14.29%) *Decision Support Systems*, (n = 4; 11.43%) *International Journal of Accounting Information Systems*, (n = 3; 8.57%) *Applied Sciences Basel*, (n = 2; 5.71% each) three of them are as follows, *Business Process Management Journal, International Journal of Industrial Engineering Theory Applications and Practice,* and *Reliability Engineering System Safety.*

Table 2. Top publication Sources

Most relevant-top publication sources	Records	Percent (%)
Expert Systems with Applications	7	20.00
Decision Support Systems	5	14.29
International Journal of Accounting Information Systems	4	11.43
Applied Sciences Basel	3	8.57
Business Process Management Journal	2	5.71
Int. J. of Industrial Engineering Theory Applications and Practice	2	5.71
Reliability Engineering System Safety	2	5.71

Besides, data shows that there are ten Journals published as single articles in the following sources. Those journals are namely as *Computer Applications in Engineering Education, Information Systems and E-Business Management, Informs Journal on Computing, International Journal of Production Research, Journal of Enterprise Information Management, Journal of Systems Science and Systems Engineering, Revista Facultad de Ingenieria Universidad de Antioquia, Robotics and Computer Integrated Manufacturing, Safety Science, South African Journal of Industrial Engineering.*

4.4 Author's Productivity

Table 3 presents the author/co-author's detailed results according to published data. Two articles published as single-authored documents, while multi-authored documents stand as (n = 99) documents and (n = 101) authors with (n = 117) appearances. Documents per author ratio is (n = 0.35), while authors per document are (n = 2.89).

There are no significant differences in publication records among authors. The top two contributors have (n = 3; 8.57% each) articles namely Jans M and Jung JY followed by 13 authors (n = 2; 5.71% each) Cho M, Choy KL, He Z, Ho GTS, Kumar A, Lee CKH, Ou-Yang C, Park J, Song M, Wang HJ, Wen LI, Wu Q, and Yoo S. The rest of the 86 authors appeared in articles (n = 1; 2.86% each).

4.5 Organizations and Countries Productivity

Table 4 illustrates the most productive (top 7 with n >= 2 article) countries on the process mining research domain. The leading country is South Korea, followed by the

Table 3. Author/co-author descriptive statistics (2010–2019)

Descriptive results	Records counted
Authors	101
Author appearances	117
Authors of single-authored documents	2
Authors of multi-authored documents	99
Single-authored documents	2
Documents per author	0.35
Authors per document	2.89
Co-Authors per documents	3.34
Collaboration Index	3
Average citations per documents	9.74

USA, Belgium, China, Germany, Taiwan, and Turkey. The table shows two dimensions of collaboration patterns as Article Counts (n), Frequency, Single Country Publications (SCP) and Multiple Country Publications (MCP), and MCP Ratio (MCP/Articles).

Table 4. Country productivity by the corresponding author

Country	Articles (n)	Frequency	SCP	MCP	MCP_Ratio
S. Korea	5	0.1429	5	0	0
USA	5	0.1429	2	3	0.6
Belgium	4	0.1143	2	2	0.5
China	4	0.1143	4	0	0
Germany	2	0.0571	2	0	0
Taiwan	2	0.0571	2	0	0
Turkey	2	0.0571	2	0	0

On the other hand, country-specific distribution of publication (without corresponding author sorted) are the USA (n = 7; 20%), followed by P.R. China with (n = 6; 17.14%), Belgium, and S. Korea, with (n = 5; 17.14%), the Netherlands, and Taiwan with (n = 3; 8.57% each), Cuba, Germany, Italy, Spain, and Turkey with (n = 2; 5.71% each), Brazil, Chile, England, Estonia, Indonesia, New Zealand, Singapore, and South Africa (n = 1; 2.86% each).

Figure 2 represents organization affiliated and contributed to the Process Mining research (n >= 2). There are 55 organizations in total, 15 are more than (n >= 2) articles (shown in Fig. 2), remaining 40 of them have (n = 1) articles each.

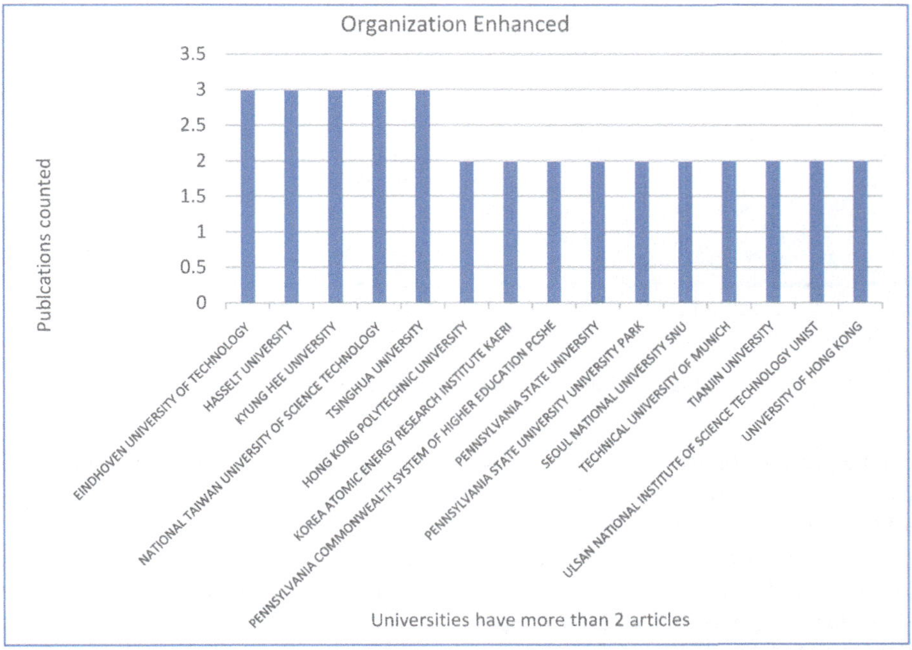

Fig. 2. Affiliated organizations

4.6 Citation Results

Citation report of 35 publications resulted from Web of Science Core Collection between 2010 and 2019 statistics show that h-index 11; average citations per item 9.74; the sum of times cited 317; without self-citations 302; citing articles 295; without self-citations 284 counted. Figure 3 illustrates the total citation distribution throughout the research timeframe. Although the citation figures started in 2010, it starts rising in 2016 with (n = 16) then steadily increase until 2017. Between the years 2017 and 2018, nearly doubled the citation counts and reached (n = 91) in the year 2019.

5 Discussions and Conclusions

The purpose of the study was to conduct a bibliometric analysis of all the scientific articles published by the journals, which indexed and ranked in Web of Science Core Collection the period of 2010 to 2019. Therefore the structure of this study prepared to explore and identify the main research domain of "process mining" in scholarly publications. Moreover, research questions have been formed to understand and find out the outputs in publication patterns.

According to the publication results, the majority of the papers presented in the following categories as methodological, conceptual, and literature review (20 papers);

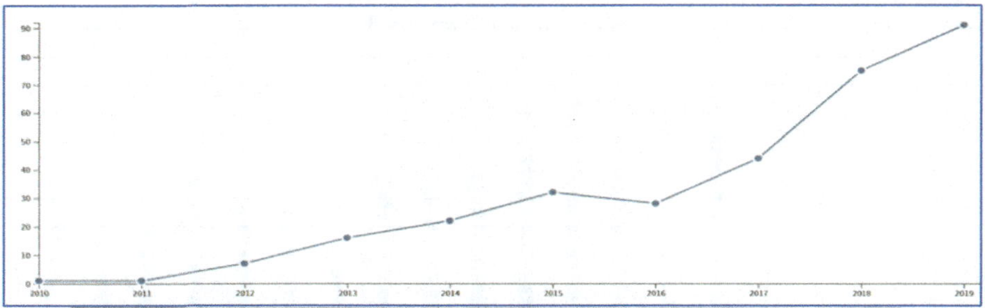

2010	2011	2012	2013	2014	2015	2016	2017	2018	2019
1	1	7	16	22	32	28	44	75	91

Fig. 3. Total citation counts per year (2010–2019)

empirical and applied studies (14 papers); case studies, and technical papers (12 papers). The majority of the papers represent multiple categories. Very few papers were a literature review or theoretical papers.

The articles published in the period has increased from 2015 to 2019 than the average publication (3.5 average articles per year, n = 35 articles in total). Results show that the leading countries for publication count as of the USA, P.R. China, Belgium, and South Korea. Moreover, the leading journals are Expert Systems with Applications, Decision Support Systems, International Journal of Accounting Information Systems, Applied Sciences Basel, and Business Process Management Journal, Int. J. of Industrial Engineering Theory Applications and Practice and Reliability Engineering System Safety in Seventeen among the journals.

Although the distribution of publications are different than the order of the organization rank (Fig. 2), Eindhoven University, Hassel University, Kyung Hee University, National Taiwan University, and Tsinghua University are the leading organizations among (n = 55) in total. Another interesting data can be seen in country productivity by corresponding author distribution. While South Korea has the highest, the article count with Single Country Publications (SCP, n = 5) interestingly has Multiple Country Publications (MCP, n = 0). Similarly, China, Germany, Taiwan, and Turkey have similar figures and no MCP count. On the other hand, the USA and Belgium have SCP and MCP in comparison to other leading countries. It means the country-level studies more common than international collaboration and co-operation.

Concerning the limitations of the study, it is a pilot study to show the research pattern in a specific research domain, with the size of the publication data and narrow domain field. Furthermore, results cannot be generalized for the entire process mining context. Consequently, the cross-sectional relationship among other variables, including various academic areas, and the link to process mining could be examined. In order to recognize and take account of the most basic data collection needs, researchers and practitioners should be able to understand the patterns and trends of existing publications in this field.

Process mining, which is an analytical discipline to improve real processes, would become widespread as one of the most effective techniques that managers who are trying to adapt to changing market conditions, including applications such as changing business models, adaptation of new technologies, learning machines, and IoT, in the digital transformation process.

References

1. Altan, Z., Birgun, S.: Using process mining approach for machining operations. In: Gençyılmaz, G., Durakbasa, N. (eds.) Proceedings of the International Symposium for Production Research 2019, LNME – Lecture Notes in Mechanical Engineering, pp. 452–464. Springer, Heidelberg (2020)
2. Shubina, I., Kulakli, A.: Pervasive learning and technology usage for creativity development in education. Int. J. Emerg. Technol. Learn. (iJET) **14**(1), 95–109 (2019)
3. Knoll, D., Reinhart, G., Prüglmeier, M.: Enabling value stream mapping for internal logistics using multidimensional process mining. Expert Syst. Appl. **124**, 130–142 (2019)
4. Jans, M., Hosseinpour, M.: How active learning and process mining can act as continuous auditing catalyst. Int. J. Account. Inf. Syst. **32**, 44–58 (2019)
5. Roldan, J.J., Crespo, E., Martin-Barrio, A., Pena-Tapia, E., Barrientos, A.: A training system for Industry 4.0 operators in complex assemblies based on virtual reality and process mining. Robot. Comput. Integr. Manuf. **59**, 305–316 (2019)
6. Wu, Q., He, Z., Wang, H.J., Wen, L.J., Yu, T.Z.: A business process analysis methodology based on process mining for complaint handling service processes. Appl. Sci. Basel **9**, 274–293 (2019)
7. He, Z., Wu, Q., Wen, L.J., Fu, G.: process mining approach to improve emergency rescue processes of fatal gas explosion accidents in Chinese coal mines. Saf. Sci. 111, 154–166 (2019)
8. Baader, G., Krcmar, H.: Reducing false positives in fraud detection: combining the red flag approach with process mining. Int. J. Account. Inf. Syst. **31**, 1–16 (2018)
9. Zerbino, P., Aloini, D., Dulmin, R., Mininno, V.: Process-mining-enabled audit of information systems: methodology and an application. Expert Syst. Appl. **110**, 80–92 (2018)
10. Baykasoğlu, A., Özbel, B.K., Dudaklı, N., Subulan, K., Senol, M.E.: Process mining based approach to performance evaluation in computer-aided examinations. Comput. Appl. Eng. Educ. **26**, 1841–1861 (2018)
11. Park, J., Jung, J.Y., Heo, G., Kim, Y., Kim, J., Cho, J.: Application of a process mining technique to identifying information navigation characteristics of human operators working in a digital main control room - feasibility study. Reliab. Eng. Syst. Saf. **175**, 38–50 (2018)
12. Erdogan, T.G., Tarhan, A.: A goal-driven evaluation method based on process mining for healthcare processes. Appl. Sci. Basel **8**, 274–293 (2018)
13. Er, M., Arsad, N., Astuti, H.M., Kusumawardani, R.P., Utami, R.A.: Analysis of production planning in a global manufacturing company with process mining. J. Enterp. Inf. Manage. **31**, 317–337 (2018)
14. van der Aalst, W.: Spreadsheets for business process management using process mining to deal with "events" rather than "numbers"? Bus. Process Manage. J. **24**, 105–127 (2018)
15. Cho, M., Song, M., Comuzzi, M., Yoo, S.: Evaluating the effect of best practices for business process redesign: an evidence-based approach based on process mining techniques. Decis. Support Syst. **104**, 92–103 (2017)
16. Werner, M.: Financial process mining - accounting data structure dependent control flow inference. Int. J. Account. Inf. Syst. **25**, 57–80 (2017)

17. Rojas, E., Sepulveda, M., Munoz-Gama, J., Capurro, D., Traver, V., Fernandez-Llatas, C.: Question-driven methodology for analyzing emergency room processes using process mining. Appl. Sci. Basel **7**, 274–293 (2017)
18. Greyling, B.T., Jooste, W.: The application of business process mining to improving a physical asset management process: a case study. S. Afr. J. Ind. Eng. 28, 120–132 (2017)
19. Park, J., Jung, J.Y., Jung, W.: The use of a process mining technique to characterize the work process of main control room crews: a feasibility study. Reliab. Eng. Syst. Saf. **154**, 31–41 (2016)
20. Lee, C.K.H., Choy, K.L., Ho, G.T.S., Lam, C.H.Y.: A slippery genetic algorithm-based process mining system for achieving better quality assurance in the garment industry. Expert Syst. Appl. 46, 236–248 (2016)
21. Carrera, B., Lee, J., Jung, J.Y.: Discovery of gatekeepers on information diffusion flows using process mining. Int. J. Ind. Eng. Theo. Appl. Pract. **23**, 253–269 (2016)
22. Rovani, M., Maggi, F.M., de Leoni, M., van der Aalst, W.M.P.: Declarative process mining in healthcare. Expert Syst. Appl. **42**, 9236–9251 (2015)
23. Cheng, H.J., Kumar, A.: Process mining on noisy logs - can log sanitization help to improve performance? Decis. Support Syst. **79**, 138–149 (2015)
24. Maita, A.R.C., Martins, L.C., Paz, C.R.L., Peres, S.M., Fantinato, M.: Process mining through artificial neural networks and support vector machines a systematic literature review. Bus. Process Manage. J. **21**, 1391–1415 (2015)
25. Cho, M., Song, M., Yoo, S.: A systematic methodology for outpatient process analysis based on process mining. Int. J. Ind. Eng. Theo. Appl. Pract. **22**, 480–493 (2015)
26. Claes, J., Poels, G.: Merging event logs for process mining: A rule based merging method and rule suggestion algorithm. Expert Syst. Appl. **41**, 7291–7306 (2014)
27. Lee, C.K.H., Ho, G.T.S., Choy, K.L., Pang, G.K.H.: A RFID-based recursive process mining system for quality assurance in the garment industry. Int. J. Prod. Res. 52, 4216–4238 (2014)
28. Yzquierdo-Herrera, R., Silverio-Castro, R., Lazo-Cortes, M.: Treatment of lack of information in process mining. Revista Facultad De Ingenieria-Universidad De Antioquia **41**, 67–68 (2013)
29. Jans, M., Alles, M., Vasarhelyi, M.: The case for process mining in auditing: sources of value added and areas of application. Int. J. Account. Inf. Syst. **14**, 1–20 (2013)
30. Caron, F., Vanthienen, J., Baesens, B.: Comprehensive rule-based compliance checking and risk management with process mining. Decis. Support Syst. **54**, 1357–1369 (2013)
31. Huang, Z., Kumar, A.: A study of quality and accuracy trade-offs in process mining. Inf. J. Comput. **24**, 311–327 (2012)
32. Liu, Y., Zhang, H., Li, C.P., Jiao, R.J.: Workflow simulation for operational decision support using event graph through process mining. Decis. Support Syst. **52**, 685–697 (2012)
33. Jans, M., van der Werf, J.M., Lybaert, N., Vanhoof, K.: A business process mining application for internal transaction fraud mitigation. Expert Syst. Appl. **38**, 13351–13359 (2011)
34. Ou-Yang, C., Winarjo, H.: Petri-net integration - an approach to support multi-agent process mining. Expert Syst. Appl. **38**, 4039–4051 (2011)
35. Ou-Yang, C., Juan, Y.C.: Applying process mining approach to support the verification of a multi-agent system. J. Syst. Sci. Syst. Eng. **19**, 131–149 (2010)
36. Edgington, T.M., Raghu, T.S., Vinze, A.S.: Using process mining to identify coordination patterns in IT service management. Decis. Support Syst. **49**, 175–186 (2010)
37. Li, J.X., Wang, H.J., Zhang, Z., Zhao, J.L.: A policy-based process mining framework: mining business policy texts for discovering process models. Inf. Syst. E-Bus. Manage. **8**, 169–188 (2010)

38. Kulakli, A., Shubina, I.: A bibliometric study on mobile applications for PTSD treatment: the period of 2010–2019. In: Proceedings of 6th International Conference on Information Management, IEEE (ICIM 2020), Imperial College London, UK, 27th–29th March 2020, pp. 319–323 (2020)
39. Kulakli, A., Osmanaj, V.: Global research on big data in relation with artificial intelligence (a bibliometric study: 2008–2019). Int. J. Online Biomed. Eng. (iJOE) **16**(2), 31–46 (2020)

Lean Production

Waiting as Waste in Lean Production Processes

Viktor Molnar(✉)

Institute of Manufacturing Science, University of Miskolc, Miskolc, Hungary
viktor.molnar@uni-miskolc.hu

Abstract. From the lean production point of view, waiting of components or products within a production process is a phase which creates no value to the customer. In addition to that, a plant uses resources in this time (e.g. storage) which produce unnecessary costs to the company. In the automotive industry mass production is often applied. If the waiting time is only a fragment of the whole production process, it could mean a great amount of loss in annual profit due to the large number of produced objects. At the same time, allocation of overhead costs is not always fair. In this study activity-based costing procedure is applied to demonstrate a fairer allocation when the waiting time reduction is built into the production process. With this procedure decision-makers can obtain more accurate information about the first costs of the products and the wastes of the production process. A case study is constructed to illustrate our model, which can serve as a benchmark for companies who have adapted lean philosophy.

Keywords: Lean production · Activity-based costing · Waiting loss

1 Introduction

In order to increase production efficiency several production and management concepts appeared. The lean production philosophy, which started to spread in the 1990s, was adapted and applied mainly by automotive industrial companies. The efficiency-increasing methods and techniques applied in this concept have continuously improved in the past decades and other sectors have also begun to apply them. The focal point of lean philosophy is the reduction of waste of processes to as high an extent as possible. In automotive industrial production, technology improvement was a determining factor in its progress for decades. Optimal choosing of technological data or developing new procedures contributed greatly to earning more profit in automotive industrial production, which is characterized by mass production [1–5]. Today, however, technology improvement results only in a relatively low extent of extra profit. This is particularly true if the current role of technology improvement is compared to the possibilities of production process organization. Lean production or management have introduced options for significantly reducing lead times in a whole production process [6, 7]. At the same time these improvement solutions require complex approaches. Therefore, success is not guaranteed in all cases. Practitioners meet several difficulties when applying such complicated organizing solutions, mainly due to the complex

N. M. Durakbasa and M. G. Gençyılmaz (Eds.): ISPR 2020, LNME, pp. 429–437, 2021.
https://doi.org/10.1007/978-3-030-62784-3_36

implementation of a solution or failing to take into consideration the effects of connected processes. In the state-of-the-art, when implementation success of the lean concept is analyzed, the work culture is frequently mentioned as a critical factor that is difficult to influence (time-consuming, complex). It has to be noted that lean management has strongly links to the six sigma concept and the two are often mentioned as the lean-six sigma (LSS) concept. This hybrid solution offers the advantages of the two basically different systems.

Beyond collecting the success factors of LSS solutions, specific problems have also been studied. In such studies, the success of process or system improvement is connected to exact indicators. Shafer and Moeller analyzed economic indicators through empirical research [8]. Shah and Ward [9] and Bortolotti et al. [10] analyzed production efficiency indicators based on empirical data. Shanmuganathan et al. analyzed damage indicators using experiments [11]. Erbiyika and Saru studied the success of six sigma implementation through defect costs in an automotive industrial supply chain [12]. Connected to the analysis of complex organization systems, the application of simulation methods has gained prominence in recent years. Aldowaisan et al. carried out a simulation for processes characterized by non-normal distributions, analyzing the cost of quality [13]. Meade et al. considered the first costs of products in a simulation process [14]. Gottmann et al. carried out a comprehensive literature review for various types of costs of production [15].

Several factors influence whether a company is able to adapt the lean concept successfully or not. Many researchers have attempted to summarize these success factors. There are success factors collected by empirical research [16–18]. Several authors have based their work on comprehensive literature reviews [19–21]. The most frequent success factors collected by these works are:

- managers' commitment,
- work culture,
- communication,
- continuous education and training,
- continuous and clear goal designation.

In the toolset of lean production and LSS several summarizing and typifying models can be found to describe waste types in a structured way. The reason for this is that one of the focal elements of the lean concept is the reduction or elimination of all types of waste of production or business processes. One of the most relevant types of waste is waiting. Waiting is considered as waste because during this period there is no alteration done in a product that could mean value creation for the customers. Inventory, material handling and other types of costs are connected to waiting. During waiting no value-creating quality is built into a product.

It is not easy to consider waiting in the cost structure of processes. Activity-based costing (ABC) is a suitable tool to gain relatively accurate information about the cost structure by building cost drivers (e.g. waiting) into the calculation. The basic idea of ABC is that the emerging costs of a company are always caused by activities or activity groups of certain processes. These activities can directly or indirectly be connected to products or services (Fig. 1).

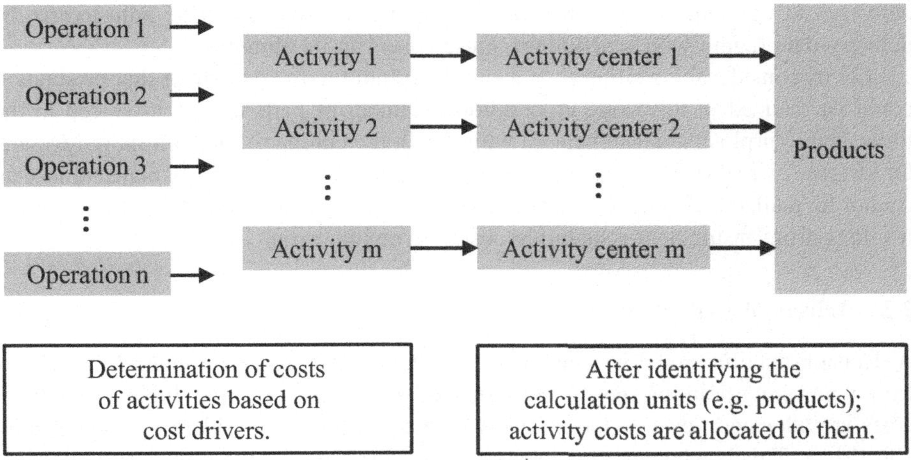

Fig. 1. Phases of activity-based costing [22]

The mechanism of this is that costs of activities are determined based on the cost drivers. After that the calculation units (products or services) are identified and the activity costs are allocated to them. This allocation can be done by division of costs based on logical considerations. Scaling is applied in the division, i.e. the division is made on the basis of proportions of the identified activity-based cost drivers [22–24].

2 Model Construction

2.1 Scheduling and Waiting

The most significant problem of scheduling is that several products of product families have to be machined or assembled on a line in a given period of time. Within this period customers can modify their needs, which would mean that a line has to be changed over to the production of another product within a relatively short time. Based on preliminary information (forecasts or historical data) production planners have to determine when and on which machine or work station the products have to be produced. This has to be done with a relatively steady and continuous capacity allocation. If this continuity is solved the number of periods without production – the machines, the work stations and the products wait, work-in-process is piled up – will be reduced. If the production is close to continuous, the reason for that mainly is that greater efforts were made in production scheduling (surplus work, higher cost of production scheduling). It is a fact that the continuity of production means that the number of changeovers also increases. Changeover of machines and equipment to the production of another product implies the increase of production support and servicing time and costs. The process can be more complicated: if the waiting time of a process is radically reduced or even eliminated by ensuring continuity, the finished product will wait for delivery. This increases the inventory costs. However, this depends on the date of delivery. At the same time, if the end of production process can be adjusted to the delivery date (again with scheduling), these costs can be eliminated.

Let us consider the waiting time as an opportunity cost. Assuming that production could be realized in this time, an additional number of parts can be produced in the time. Thus, surplus parts can be sold with which the company can earn more revenue. Of course, in reality the ordered quantity is fixed, so the profit calculated this way cannot be realized. However, it serves to inform the managers about what the waiting resulting from insufficient scheduling costs the organization.

2.2 Effects of Reduction of Waiting Time

A decision-making model was elaborated for how to manage these interrelated factors from controlling point of view. The essence of the model is that the direct and indirect costs are allocated between the products so that waiting is considered as opportunity cost driver. This is where the model differs from the classic ABC model. In Fig. 2 it is demonstrated how waiting as a cost driver affects the elements of ABC. Waiting is defined as a cost driver, that is, a ratio: the proportion of time without production due to waiting within the planned lead time. Waiting generates various changes. Reduction of waiting, which is the basic mechanism of the model, can be realized by increasing the cost of scheduling (activity pool). At the same time the number of changeovers (cost driver) increases because changeovers help in realizing the steady resource use. The reduced waiting time implies the production of extra parts. Increase in the number of changeovers requires extra production organizing work, therefore the cost of production organization (activity pool) increases too. The possibility for increasing the number of parts affects the change (positive or negative) in the other cost components included in the costing model. In the elaborated model the scheduling-related costs are

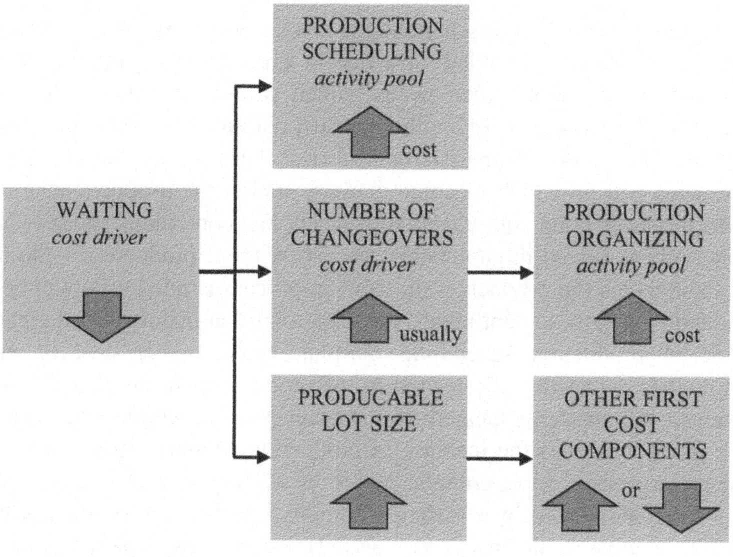

Fig. 2. Effects of reduction of waiting time [25]

collected by the scheduling activity pool and the changeover-related costs by the production organizing activity pool.

The allocation of costs of scheduling activity pool for products is carried out in proportion to the original values of waiting.

3 Discussion – Case Study

Seven different components are machined on a line of a plant producing automotive industrial components. The lot sizes of each product were determined based on the customers' forecast but the products are delivered in several batches. The customers determine how many parts will be required in the forthcoming days. That is why the products are not machined continuously and changeovers are necessary. Table 1 summarizes the number of produced parts and their prices.

Table 1. Number of produced parts and sales prices

	P_1	P_2	P_3	P_4	P_5	P_6	P_7
Number of produced parts	1060	3900	350	1250	1100	500	1050
Price (€)	21	16	9	12	10	8	14

The basic data of ABC (cost drivers) are summarized in Table 2. The allocation of cost drivers and activity pools is provided in Table 3. The components of first cost are the activity pools (indirect costs) and the direct costs. The proportional allocation of costs between the products within an activity pool is based on the rates of cost drivers, considering the above-mentioned allocation. The structure of first cost is summarized in Table 4.

Let the rate of waiting rates of the production processes be reduced to zero by scheduling. This can be realized by more work hours of scheduling. This results in the increase of the scheduling cost from €3,850 to €8,465. Meanwhile the number of changeovers increases. This implies the increase of production organization cost from €3,952 to €7,904 because changeover requires work station servicing (Fig. 3). The components of the new first costs are summarized in Table 5.

Table 2. Direct costs and cost drivers

Cost drivers	Products						
	P_1	P_2	P_3	P_4	P_5	P_6	P_7
Direct material cost (€)	6100	19500	820	4400	3500	520	4800
Direct wages (€)	3280	15050	640	3100	3480	180	3850
Lead time (day)	3	2	1	4	3	1	2
Number of changeovers	1	4	1	1	2	1	3
Waiting rate (within the lead time)	0.05	0.25	0.09	0.12	0.35	0.06	0.08

Table 3. Activity pools and cost drivers

Activity pool	Cost (€)	Cost driver
Logistics	8109	Direct material cost
Human resource administration	6822	Direct wages
Production support	4840	Lead time
Production organization	3952	Number of changeovers
Scheduling	3850	Waiting rate

Table 4. Structure of first costs (€/piece)

	Products						
	P_1	P_2	P_3	P_4	P_5	P_6	P_7
Direct material cost	5.75	5.00	2.34	3.52	4.00	8.80	4.19
Direct wages	3.09	3.86	1.83	2.48	3.16	0.36	3.67
Logistics	1.18	1.02	0.48	0.72	0.65	0.21	0.94
Human resource administration	0.71	0.89	0.42	0.57	0.73	0.08	0.85
Production support	0.86	0.16	0.86	0.97	0.83	0.61	0.58
Production organization	0.29	0.31	0.87	0.24	0.55	0.61	0.87
Scheduling	0.18	0.25	0.99	0.37	1.23	0.46	0.29
Total	12.06	11.49	7.80	8.87	11.15	11.13	11.38

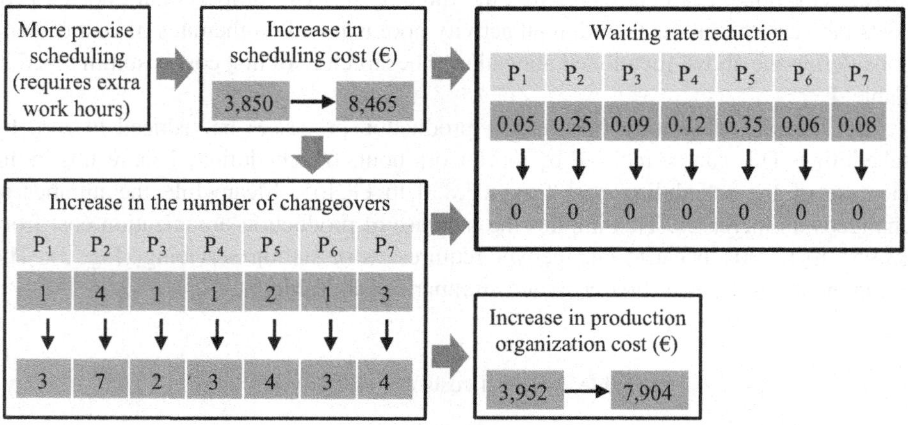

Fig. 3. Effects of the process improvement

Table 5. Structure of first costs after waiting time reduction costs (€/piece)

	Products						
	P_1	P_2	P_3	P_4	P_5	P_6	P_7
Direct material cost	5.47	3.75	2.14	3.10	2.07	0.98	4.21
Direct wages	2.94	2.89	1.67	2.18	2.06	0.34	3.37
Logistics	1.12	0.77	0.44	0.63	0.42	0.20	0.86
Human resource administration	0.68	0.67	0.38	0.50	0.47	0.08	0.78
Production support	0.81	0.12	0.79	0.85	0.54	0.57	0.53
Production organization	0.82	0.41	1.58	0.64	0.72	1.72	1.07
Scheduling	0.38	0.41	1.98	0.72	1.75	0.96	0.59
Total	12.22	9.01	8.98	8.63	8.03	4.84	11.41

During waiting time additional parts could be produced. The increase in this number can increase the revenue. The number can be calculated by Eq. 1.

$$n\prime = n + wn/(1 - w), \tag{1}$$

where n' is the increased number of parts, n is the original number of parts and w is the rate of waiting within the lead time of the whole series.

In case of the hypothetical surplus production within the time of waiting, the changes in the number of produced parts and in the first costs are summarized in Table 6.

Table 6. Change in number of parts and first costs after the process improvement

	Products						
	P_1	P_2	P_3	P_4	P_5	P_6	P_7
Change in number of produced parts (%)	+5.18	+33.33	+9.71	+13.6	+53.81	+6.20	+8.67
Change in total first cost (%)	+1.33	−21.58	+15.13	−2.71	−27.98	−56.51	+0.26

Table 7. Total cost, total revenue and profit of the analyzed products before and after the process improvement

	Before process improvement	After process improvement	Change [%]
Total cost (€)	101,173	105,360	+4.14
Total revenue (€)	132,510	164,253	+23.96
Profit (€)	31,337	58,893	+87.93

Applying the ABC model before the improvement of process loaded by waiting, the profit of the analyzed products was €31,337. After the process improvement this value increased to €58,893, which is an increase of about 88% (Table 7). However, this is rather a more theoretical value because of the next two reasons. On hand, the scheduling in practice does not result in absolutely 0 waiting rates. On the other hand, this profit could be realized only if the customer buys or intends to buy the extra number of parts in the analyzed period. However, in reality the values are more likely to provide information about what waiting-in-process, which creates no value for the customers, means to the company expressed in money.

In the demonstrated case we have neglected the fact that waiting can only rarely be totally eliminated in a profitable way. In most of the cases, mainly in processes in which several types of products are machined, waiting waste cannot be eliminated totally. The reason for this is that reducing waiting to zero could lead to unrealistically high costs due to the great scheduling effort. This could reduce the profit radically or make the process unprofitable. That is why the waiting rate that yields the maximal profit function should be found. This requires further development of the introduced model.

4 Summary and Conclusions

A further-developed activity-based costing model was introduced in this paper. The model proved to be suitable for including waiting of the production process as an opportunity cost in the first cost structure of the products. An illustrative case study was introduced to describe how the cost structure and profit of a production process can change. The basis of application of the method is one of the pillars of the lean concept: the reduction or elimination of not value-creating elements of production processes. The theory of opportunity cost was applied to express waiting waste as a cost type. The calculation method, of course, requires adaptation to the particular situation, but its central element – activity-based costing – ensures the possibility of general application. The introduced model highlighted the significance of scheduling efforts in production processes. Based on the analyzed components and cost structure, after decreasing the waiting rates within the lead times, the total cost of the components is increased by about 4% and the total revenue can increased by about 24%. The total profit, therefore can increase by 88%. This is a theoretical category which implies that the waiting loss has a significant effect in the studied process. In our further research it is intended to extend the model with an optimization step to determine the minimal waiting rate (connected to maximum profit) of the production process.

References

1. Mamalis, A.G., Kundrák, J., Horváth, M.: On a novel tool life relation for precision cutting tools. J. Manuf. Sci. Eng. Trans. ASME **127**(2), 328–332 (2005)
2. Felho, C., Kundrák, J.: Comparison of theoretical and real surface roughness in face milling with octagonal and circular inserts. Key Eng. Mater. **581**, 360–365 (2014)
3. Kundrák, J., Felho, C.: 3D roughness parameters of surfaces face milled by special tools. Manuf. Technol. **16**(3), 532–538 (2016)
4. Karpuschewski, B., Kundrák, J., Emmer, T., Borysenko, D.: A new strategy in face milling - inverse cutting technology. Solid State Phenom. **261**(1), 331–338 (2017)

5. Mathiselvan, G., Sundaravel, S., Chaitanya, S., Kaja Bantha Navas, R.: Application of taguchi method for optimization of process parameters in analyzing the cutting forces of lathe turning operation. Int. J. Pure Appl. Math. **109**, 129–136 (2016)
6. Szentesi, S., Tamas, P., Illes, B.: Process improvement methods for manufacturing system. In: Proceedings of COMEC 2016 IX International Scientific Conference of Mechanical Engineering, Santa Clara, pp. 1–6 (2016)
7. Banyai, T., Telek, P., Landschutzer, C.: Milkrun based in-plant supply – an automotive approach. In: Lecture Notes in Mechanical Engineering, pp. 170–185 (2018)
8. Shafer, S.M., Moeller, S.B.: The effects of six sigma on corporate performance: an empirical investigation. J. Oper. Manage. **30**, 521–532 (2012)
9. Shah, R., Ward, P.T.: Lean manufacturing: context, practice bundles, and performance. J. Oper. Manage. **21**, 129–149 (2003)
10. Bortolotti, T., Boscari, S., Danese, P.: Successful lean implementation: organizational culture and soft lean practices. Int. J. Prod. Econ. **160**, 182–201 (2015)
11. Shanmuganathan, V.K., Haran, A.P., Gayathri, N.: Condition monitoring maintenance of aero-engines through LUMS – a method for the implementation of lean tools. Measurement **73**, 226–230 (2015)
12. Erbiyika, H., Saru, M.: Six sigma implementations in supply chain: an application for an automotive subsidiary industry in Bursa in Turkey. Procedia Soc. Behav. Sci. **195**, 2556–2565 (2015)
13. Aldowaisan, T., Nourelfath, M., Hassan, J.: Six sigma performance for non-normal processes. Eur. J. Oper. Res. **247**, 968–977 (2015)
14. Meade, D.J., Kumar, S., Houshyar, A.: Financial analysis of a theoretical lean manufacturing implementation using hybrid simulation modeling. J. Manuf. Syst. **2**, 137–152 (2006)
15. Gottmann, J., Pfeffer, M., Sihn, W.: Process oriented production evaluation. Procedia CIRP **12**, 336–341 (2013)
16. Crute, V., Ward, Y., Brown, S., Graves, A.: Implementing lean in aerospace - challenging the assumptions and understanding the challenges. Technovation **23**, 917–928 (2003)
17. Scherrer-Rathje, M., Boyle, T.A., Deflorin, P.: Lean, take two! reflections from the second attempt at lean implementation. Bus. Horiz. **52**, 79–88 (2009)
18. Staats, B.R., Brunner, D.J., Upton, D.M.: Lean principles, learning, and knowledge work: evidence from a software services provider. J. Oper. Manage. **29**, 376–390 (2011)
19. Hoon Kwaka, Y., Anbari, F.T.: Benefits, obstacles, and future of six sigma approach. Technovation **26**, 708–715 (2006)
20. Brun, A.: Critical success factors of six sigma implementations in Italian companies. Int. J. Prod. Econ. **131**, 158–164 (2011)
21. Aziz, R.F., Hafez, S.M.: Applying lean thinking in construction and performance improvement. Alexandria Eng. J. **52**, 679–695 (2013)
22. Musinszki, Z.: Industry 4.0 – cost system 4.0? innovations in cost systems. In: Ipar 4.0 – költségrendszer 4.0? Innovációk a költségrendszerekben, Hungarian, Controller Info, vol. 3, pp. 2–9 (2016)
23. Musinszki, Z.: Cost allocation problems and solutions. In: Költségallokációs problémák és megoldások, Hungarian, Controller Info, vol. 4, pp. 2–10 (2015)
24. Kaplan, R.S., Cooper, R.: Cost & impact, integrated costing systems: the basics of profitable corporate operation. In: Költség, Hatás, Integrált költségszámítási rendszerek: az eredményes vállalati működés alapjai, Hungarian. IFUA Horvath & Partner, Panem (2011)
25. Molnar, V., Tumik, A.: Waiting loss costs of non-production processes in lean six sigma approach: an ABC-based decision support model. In: Várakozási veszteségből adódó költségek Lean Six Sigma megközelítésben: egy ABC-alapú döntési modell, Hungarian, Controller Info, vol. 1, pp. 35–40 (2017)

Logistics Management

A Critical Review of Quality Function Deployment (QFD) Tool Through Logistics Centre Development Projects

Rıza Gürhan Korkut[1(✉)], Metin Çelik[2], and Süleyman Özkaynak[3]

[1] Maritime Transportation Management Engineering, Piri Reis University, Istanbul, Turkey
rizagurhan.korkut@pru.edu.tr
[2] Maritime Faculty, Istanbul Technical University, Istanbul, Turkey
celikmet@itu.edu.tr
[3] Maritime Faculty, Piri Reis University, Istanbul, Turkey
sozkaynak@pirireis.edu.tr

Abstract. The early-stage design and development schemes for the Logistics Centre projects require an inclusive analysis on both qualitative and quantitative data. Many successful organizations listen to the Voice of Customer (VoC) and then incorporate that into the design and implementation of their products and services. Quality Function Deployment (QFD) is the widely used methodology committed to listening to the VoC for easily translating their needs into technical design specifications. In this study, QFD process is reviewed over its major stages and then it is further assessed in terms of its strengths and facilitating aspects applied of any problem for multi-criteria decision-making purposes. The House of Quality (HoQ) is the basic tool of the planning and communication practices of QFD. The HoQ is also a pathfinder's guide delivering the resources involving the interdisciplinary persons with diverse responsibilities and perfection. Proposing clues about improving benefits and outputs of a QFD process applied in any developing and design project applications is also one of the endeavours of this study.

Keywords: Voice of customer · Quality Function Deployment · House of Quality · Logistics centre · Decision-making

1 Introduction

1.1 What Is the Aim of This Symposium Paper?

Keeping the business investments continuously competitive in an industry is one of the crucial aims of the companies. If the investors are able to operate their businesses effectively and efficiently enough, and more distinguishable and governable over the supply chain while reducing operational costs and possessing and controlling the customer satisfaction at higher and acceptable levels then they have a resilience and energy to survive in the market that they exist.

N. M. Durakbasa and M. G. Gençyılmaz (Eds.): ISPR 2020, LNME, pp. 441–452, 2021.
https://doi.org/10.1007/978-3-030-62784-3_37

Customer satisfaction for logistics centres that looking for competitive advantage is very substantial since if they do not satisfy the expectations of the customers then their places will easily be replaced by the other companies whose activities are more concentrated on customer requirements. So, logistics service providers must ensure that every aspect of customer service-related process is covered and included in their design of any service module guaranteeing that client's needs are understood.

The objective of applying QFD in logistics service design is to assure that the ultimate service design meet customer requirements as indicated in reference [1]. The practical implication of QFD approach is that customer requirements to be transformed into customer value with regards to the logistics services.

This study proposes an approach that supporting and improving a project conceptualization and design through transforming customer requirements into integrated logistics service designs by criticizing a very well-known decision criteria decision making tool named QFD. If the readers of this paper do not have any experience on methodology and implications of QFD methodology, then this document would not be helpful for them since it is not a complete article of QFD methodology containing every aspects from alpha to omega whereas this is only a symposium paper recalling the basics of QFD while proposing a critical review about its applications.

1.2 How Do We Describe a Logistics Centre?

EUROPLATFORMS EEIG is the only European Association of Freight Villages and Logistics Centres and it is an European Economic Interest Grouping oriented in the development of Logistic Platforms as nodal transport infrastructure as declared in reference [2]. Europlatforms EEIG's (European Association of Freight Villages) general assembly in 2015 approved the definition of a Logistics Centre as in the following sentence: "Centre in a defined area within which all activities relating to the transport, logistics and distribution of goods, both for national and international transit, are carried out by various operators on a commercial basis".

1.3 What Is the Contribution of This Study in Regards with a Logistics Centre Development Project?

An investment project of Logistics Centre starting from scratch or aiming to implement development and/or maintenance objectives for an existing project shall be successful and sustainable if customer's requirements are listened to at earlier stages of the project for checking against the design requirements and technical constraints for attaining conformance to stakeholders' expectations at all.

An effective and sustainable investment performance and customer satisfaction of Logistics Centre are all dependent on the following aspects but not limited to: the selection of centre location, the level of development of the region, integration of transportation means and their related cargo planning and schedules, the dynamism and commercial understanding of the stakeholders and the organization of operational activities and the level of integration of those with the logistical requirements of the customers.

The successful realization of Logistics Centre design projects and their sustainability are highly likely dependent on the decisions done right at the first sight. This study is not organised only pursuing to answer such like that "finding the most appropriate location where the Logistics Centre is located". It is to reveal the customer requirements and expectations about a Logistics Centre and to rank and prioritize them in order of their importance as their fulfilment by the technical and design requirements and/or constraints.

2 Review and Assessment Framework for QFD Processes

In the first section, data gathering by listening to the voice of the customer is mentioned as the major task of any QFD processes. In the next section, reconciliation of the collected data is discussed due to relativity and connectivity integrity of the stakeholders' input. Afterwards, there is debate on how to apply a priority check to all the input of the process. Subsequently, some clues are explained for used to improving the QFD processes for higher success story cases. Later in the Sect. 2, process outputs and benefits of QFD methodology are discussed. Thereafter, conclusion and references sections exist respectively at the end of this study. By the last section of this part of the paper you may find application phases of QFD methodology realised through a logistics centre development project.

2.1 Data Gathering: Converting VoC into Input for QFD Processes

In the early design stages of product and service development projects that there is a crucial step where the customer requirements are collected and listed then all of them compared with the technical requirements. These tasks are all needed to be executed with the highest level of precision since any changes to be done in the late design phases have extremely high cost.

It is stated in reference [3] one of their recent studies that involvement effort of QFD into the early stages of any project brings effectivity and efficiency increase at all. Such an effort replaces the traditional approach of brand-centric value proposition with the customer-oriented process of value creation. It does also help to emphasize if there is any asymmetry in perceived thoughts between the customers and the service providers. If the communication strategy of a company for any development effort or project not having such a condition, then all the service designers and engineers of the same company indirectly interact with the customers and generally interpret customer's preferences subjectively into the technical requirements. To appropriately catch customer requirements at the early stages of product and/or service design and development is remarkable to avoid costly design changes later in the process.

It has great importance of what the customers' voice means rather than what them tell surveyors. It is emphasized in a study issued by reference [4] in regards with QFD processes that the agreed and accepted way to approach to the customer requirements is to communicate with the customers through listening to their voice. Authors of this article underlined the importance of what the customer means rather than what them to say, and this is maintained by individualised attention given to the customers by showing empathy and willingness to provide the swift action for satisfying their immediate requirements.

One another crucial necessity of QFD process in terms of VoC is that customer requirements needed to be organised into a hierarchical structure and interpreted in customer's own words not in the industry or investor's jargon.

"Listening to VoC" is a product/service-development technique that produces a detailed set of customer wants and needs, which are organized into a hierarchical structure, and then prioritized in terms of relative importance and satisfaction with current alternatives in [5]. There are 4 aspects of the VoC; "customer needs", "a hierarchical structure", "priorities", "and customer perceptions of performance". It is extremely important that these needs declared by the customers be stated in the customers' own words in order not to lose the meaning at all. The VoC is useful for structuring the customer needs into a hierarchy of strategic, tactical, and operational needs just prior to filling up the QFD charts through comparisons against technical requirements for further investigations.

As it is mentioned in the introduction that House of Quality (HoQ) is the basic tool of the planning and communication practices of QFD. It is required to understand that HoQ is a diagram or a chart consisting of data collected through technical and competitive benchmarking. This aspect of QFD has been studied by reference [6].

The House of Quality (HoQ) is described as a matrix of making plans for service and/or products during initiative stage of their design process. This matrix is rather a diagram depicting how customer expectations establish a direct connection with the companies' way of achieving requirements/needs of their customers. The aforesaid diagram can be created using data gathered through technical and competitive benchmarking. "Technical Benchmarking" indicate stating that the service provider and its competitors how to satisfy the customer needs and expectations in regards with the technical requirements. On the other hand, "Competitive Benchmarking" means check against one provider's performance versus a selection of superior performing competitors. For both two types of benchmarking data should execute a consistency providing if the score is high in one then the other also score high.

As we have already said that QFD is one of the best suiting methods for introducing current and the most probably the future expectations of customers in any of the design, development, and maintenance projects. The same methodology is also used to reflect all the customer expectations into new products and services as technical requirements allow as possible. Whereas customers do not have any idea about neither what they do not experience in their life nor what they do not dream. Therefore, those customers do not help the owners and/or their designers to render competitive and developing features into new products and services. They might refuse progressive ideas and even hinder and do not help to boost their inspiration.

Reference [7] proposed an output-based methodology for converting the ideas and expectations collected from the customers into innovative designs of new or already developed products and services. By means of such a methodology, it is made possible to design or develop Logistics Centres projects inclusive of reformist ideas. At the first glance aspects of new or needed to be renewed products and services are defined and the customers whose voice to be listen to for gathering their commentary in regards with those design developments. Data collected from the stakeholders are categorised into specific groups in terms of the requirements and then checked against whether they are clear enough and suited with correct formats and measurement units. If there are unnecessary or weak inputs that might distort the whole composition of the database, then they are all cleared out from compilation and re-sorted into the categories ready to upload in line with the QFD process steps. This advanced list of customers' expectations is re-communicated with the customers and asked their final response on it just prior to the introducing the charts of QFD process. This is the final list that gathers the requirements and expectations declared together characterised by regeneration.

As we have already mentioned above in this section that "customers' expectations" are the most crucial input for any QFD process. So, chasing up with the stakeholders' dynamic ideas whose expectations needed to be considered by the designer in such an environment involve too much effort and continuously exposed financial burdens.

There are also models substituting all types of costly data surveys which making the continuous interactions with the customers as redundant. Reference [8] proposed a model that estimating the customers' requirements in regards with their economical, technological, cultural, and environmental perspectives without continuously contacting with them. In this model, currently collected customer expectations used as "known parameters of a hidden Markov model" aiding to configure the time-dependent descriptions in a hierarchy of entities of the model. In the end of this model, hidden parameters are accepted as the necessary inputs for updating the House of Quality chart.

2.2 Relativeness and Connectivity Integrity of QFD Inputs

Inner dependencies of technical characteristics and engineering specifications with the customers' expectations are the most disputing elements of any QFD process to be carefully sorted out.

In their study, reference [9] proposed an advanced QFD model by underlying the significance of taking the interrelationship between the whole components of QFD into account while enhancing the practicability of QFD process. Those components of QFD are literally Customers' Requirements, Engineering Characteristics, and Investors' Demand and Expectations, and all of those demand or expectations shall be listed as "the cost of the services provided", "competency level of stakeholders for providing services", "stakeholder's ability to extend the available services" and "their inner dependencies". It is also proposed in a QFD process where development and/or maintenance of projects is aimed that the developers should define the current and the requested specifications to be used at a later time for evaluating their technical difficulties and assessing the level of bottlenecks prior to the execution.

It is also worth to mention the significance of customer's expectations and their relative influence on each other and overall impact on the total benefit or harm of the project here in this section.

Reference [10] underlined the importance of assigning attributes to each of the expectations and requirements communicated by the customers as per their interrelationship reflecting their characteristic impact on overall benefit in their study where they focused on "Conjoint Analysis". By means of the mentioned analysis, it is experimentally shown that the attributes affecting the competitiveness of an airport project can easily be ranked among each other and overall benefit of the major characteristics affecting the investment as per its attribute's basis be calculated. It is also revealed in this study that airport brand and the cost of landing charges are categorised as the most important attributes that affecting the decision of airport choice.

Nevertheless, in another study realised by reference [11] that it is discovered there are subjective assessments, priority checks and control dependencies to internal characteristics needed to be overcome during process of putting a decision or plan into effect.

If you want QFD charts and diagrams bring you at most success as per the expectations from the mentioned methodology then "design requirements" should be ranked among each other in terms of relative relationship just prior to the loading them into the charts.

In traditional calculations of QFD process, the priority ratings, and relative preferences of Design Requirements are considered based on important customer's needs with the support of only one additional factor such as cost. However, adding flexibility characteristic of Data Envelopment Analysis (DEA) on ratings to treat them as qualitative variables facilitates the calculations of the relative importance of decision requirements in QFD processes. In such a compounded method, the ratings of decision requirements with respect to "customer requirements" can be added regarding several factors simultaneously such as cost, environmental impact, ease of implementation. In their recent paper, reference [12] have suggested data envelopment analysis to compute the relative importance of decision requirements when several additional factors need to be considered.

2.3 Priority Check and Ranking the QFD Inputs

As it is already known that Dry Bulk segment of seaborne trade has 40% share in the world and with this percentage it is the largest group in terms of loaded commodities. Reference [13] published a study aiming to apply QFD for dry bulk commodities terminal improvements from the perspective of the customers and their expectations. A comprehensive survey is applied for collecting the expectations of the customers of eighteen dry bulk ports. The list of customer requirements prepared for this study as usual is imprecise. So that a "factor analysis" is used for grouping of those requirements prior to load the QFD process such as "loading efficiency", "infrastructure improvements", "time performance" and "operation improvements".

In this analysis, QFD team has identified the technical specifications in a hierarchical order for the sake of determining the engineering characteristics which are directly affecting the customer requirements and/or perceptions. It is proposed through the study realised in this article that size of the loading and unloading platforms significantly influence labour and machine utilization and overall operational efficiencies. Location of warehousing areas also affects the trucking efficiency inside the dry bulk terminal area whereas loading/unloading and trucking operations are the least important focus area compared to the others.

There is another question to be answered here in this section is that how to rank design characteristics while taking the cost of the services, stakeholders' priorities and their environmental effects and other design constraints.

Reference [14] as it is proposed in their study that QFD processes is one-dimensional in terms of systematic inputs and therefore directly focused on the customers' requirements and expectations approaching them through listening to the voice of customer. In their study Data Envelopment Analysis and Russell Graph Measure are combined together with QFD, and "cost of the services", "environmental effects of the stakeholders' decisions" and "the other design criteria" enlighten the whole analysis for providing a basis for criteria to the decision makers. They also added that if the technical requirements and design criterion are ranked as per "their costs" and "the ease of application" which brings consistency into the priority list having design requirements are ranked for improving customer satisfaction.

2.4 Improving the QFD Processes

Having doubts about success and precision of the decisions made by means of QFD processes is acceptable at some extent due to consistency of several tasks such as data gathering and preparing them for the process. It might be because of misleading compilation of customers' expectations and even erroneously completed uploading task into the charts of QFD.

Reference [15] mentioned a novel approach applying SERVQUAL (service quality scale) methodology for defining the strengths and weaknesses of the design characteristics and the services offered. After having listed technical aspects of the design, it is suggested to categorise them carefully by means of KANO Model for finding the most satisfactory customers' expectations to be uploaded into QFD charts. KANO Model is a useful tool for designers of QFD processes since it is capable of categorise the type of services offered in terms of technical requirements, relationship matrix and target values as per the customer preferences.

Those types of service preferences transferred by the customers to the designers are as the followings; "exciting", "nice to have or attractive quality attributes", "performance availability requirements" and "all other basic quality requirements" classified as significant attributes by the customers.

SERVQUAL is proposed to define the weaknesses and strengths of the services to be offered by designers against to the expectations declared by the customers. Those expectations of the customer shall be explained as "facilities and equipment" as also described in physical terms and potential of performing the designed services.

The fundamental source or basis underlying HoQ is to implement any attempt to assign relevant connections between Logistics Centre service operations and customer pleasure is not so simple to realize. If we can manage to interpret technical aspects as successful as expected and functional enough than the customer requirements section of HoQ become the technical design specifications of another HoQ primarily related to the details of service design.

Reference [16] indicated a linkage in their study that transferring the voice of customers indirectly to the stakeholders who make the service design decisions. Structuring a HoQ is not so difficult or impossible to understand and then identify the ingredients of it. But to grasp the details of its layout and discipline is crucial to solely deal with it.

QFD causes to arise effectiveness and efficiency in the decision process by bringing an exchange of informative environment into existence which provides the stakeholders who can optimize their decision-making paradigm. This phenomenon is proposed in [17].

QFD adds transparency, precision and stability to the multiple criteria decision-making process which is actually because of the service mapping ability and integrated procedural flow of QFD. It helps the stakeholders to develop their knowledge about their own processes and services which increase their maturity level in regards with the visible and easy to measure effectiveness and efficiency of their operational design decisions.

Shared information in QFD framework is very well structured so that it ultimately optimizes the consciousness on decision-making behind the decision initiatives. Hence, the decision-makers can build a business case together with their external parties, customers, planners, and designers.

2.5 Interpretation of QFD Process Outputs

As we have discussed above section that QFD is applicable in any design projects starting from scratch and in some other development or maintenance schemes for Logistics Centres and similar ventures. It is experienced in many occasions that if we carefully apply a survey for gathering customers' expectations asked for their satisfaction level then this set of information would be the decision criteria for a QFD process organised for future investments.

Reference [18] published an article which taking a customer focus that ranking the service-offering dimensions of Logistics Centres by bearing the customer requirements in mind. In this article, a survey is applied featuring QFD methodology to logistics service providers and it is used to categorize the customers' expectations and their related service characteristics. Value added services delivered by the service providers and all dimensions of the centre such as infrastructure, warehouse and intermodal were ranked at the higher priorities as per the customers' preferences. It is also added in the conclusion of the study that logistics centre investors took the advantage of using the results of this analysis applying the same methodology to assess customers' requirements while guiding their design of market offerings.

Sometimes earlier experience or inherited opinions from the former practices may influence the decision makers' judgements and bring their prejudice in to the front causing biased decisions. If you want to escape from such traps of biased decisions, then knowledge-centric approaches may be the best option instead of experience-dependent methodologies like TRIZ (Theory of Inventive Problem Solving).

Reference [19] proposed the idea that service designs are built typically based on the former practices and experiences in the field. The authors stated the usefulness of TRIZ in such an environment and underlined that the advantage of reliability of achieving systematic innovation from knowledge centric TRIZ approach. On the other hand, the advantage of transferring customer requirement into the charts of customer centric QFD methodology is also shown that has the capability of obtaining the service design requirements.

2.6 Maximizing the Benefits of QFD While Get Rid of the Means that Hinder the Ultimate Success

If we carefully understand the abilities, shortcomings, and features of QFD methodology then we may easily expect a definitive level of success from use of that power of QFD for benefitting through multi-criteria decision-making environment supplied by.

As it is suggested in [20] that there might be conflicts between several Technical/Design Requirements and large matrix of QFD which may lead to possible confusion and excessive details superseding the expected benefits and other prospects of QFD method. In the same article, authors underlined the benefits of QFD by categorising them into two major group that one is qualitative benefits and the other is quantitative benefits.

If we are able to understand QFD processes correctly as well as reducing its chart down to a manageable size as what the authors declared in their study to Logistics Centre projects, then we might get benefit by experiencing shorter stages at early design processes with fewer design changes. Any reduction in start-up costs and interdepartmental harmony due to multifunctional teamwork are also be attainable.

Reference [21] underlined the importance of competitive advantage of contemporary Supply Chains while stating that there are ten trends dominating the manufacturing in next two to three years. He also added keep remaining competitive means delivering more value to the customers than the competitors. This is also an objective of QFD methodology and its way of maintaining and/or developing the competitiveness of supply chain management projects comprehensively. While escaping from any financial war escalated by means of price leverages, the entrepreneurs may increase their competitiveness by utilizing new technology simplifying the Supply Chain that in turn delivering higher customer satisfaction level.

It is worth to mention that QFD methodology always takes care of "satisfying quality level" instead of "pleasing quality level". For the sake of understanding what the meaning of "customer satisfying quality level" is representing the desired quality level covering benchmarks derived out from the competitive analysis.

The major benefit of QFD is having realised higher quality-level-in-house. As it is underlined in [22] that QFD makes the employees to share their ideas in the direction towards the higher quality levels for both services and products. This is because of QFD brings all personnel together and improve the flow of information among them.

If multiple of designers and/or decision makers have doubts about the output of QFD analysis and especially if there is a disagreement concluded about the importance of any technical requirement then House of Quality (HoQ) process of QFD will eventually overcome this difficulty by going beyond the conceptual charting.

Sometimes the project analysts face to difficulties for considering a design of a Logistics Centre and struggle for introducing a framework for that design because of individuality and uniqueness of the industry. At such occasional times QFD methodology might be supported with some other robust methodologies such as Analytic Hierarchy Process (AHP) for formulating several operating and management strategies. Such kind of complementary initiative is suggested in [23] that brings us much more structured decision mapping which realise a systematic approach enabling the decision maker to consider both subjective as well as the objective factors at the same time affecting the design procedure.

2.7 How We Have Applied the Phases of QFD Methodology in a Logistics Centre Development Project?

QFD methodology needs at least 4 consecutive phases incorporating the operations during course of any service design and/or development project.

Prior to the first phase, involvement of a knowledgeable and qualified QFD design team is crucial and they contacted with the stakeholders through structured interviews and collected their expectations and requirements about a Logistics Centre. The team also included the results from a benchmarking study and comparisons among the service characteristics and design parameters of best of the competitor centres in the region.

In the 1st. phase, the tool of HoQ is used to transform the customer expectations and requirements into service design attributes through the help of a relationship matrix. In the matrix format, the relationship between the customer requirements and expectations with the technical and design requirements are identified and evaluated by a scale like "9 – 3 – 1 – 0" where 9 points indicate "strongest relation" and 1 point is enough to represent the "fair or weak relation". All the customer expectations and requirements are ranked with "communicated customer priority" and their relative weights are given as per comparison done, among others.

The 2nd. phase is dedicated for design development purposes and design and technical requirements are ranked as per their relative importance which is calculated in the previous phase. After that those ranked values are set as the expectations and requirements column of the second matrix of HoQ, and the service processes and their respective limits, constraints identified where all those systems components describe the building blocks of Logistics Centre such as services, processes, teams and their activities. For example, "total area size of centre", "expansion area size of it" and "currently developed area size" are all the attributes that reflecting the customers' expectations from the centre.

Any correlation between the critical service quality processes required by the customers are emphasized and their desirable quality levels are underlined in the 3rd. phase which is named as the process development phase of the complete QFD process. For example, "location of the logistics centre guaranteeing its proximity to gates and other logistics hubs" is less important than "terminal services offered in the centre" whereas "container handling capacity in the centre" has the utmost importance in the customers preferences.

In the 4th. phase standards of operations are defined which used for developing the quality targets for each of the design and technical requirements. By having this phase of QFD is done, all the customer expectations built into the service design and included in the services developed at the logistics centre. In this last phase of QFD process that logistics services platform is identified, and then specific quality target values and process parameters associated with this service platform is proposed.

3 Conclusions

Any user who practicing multiple criteria decision making efficacies of QFD method and its tool of the planning and communication for the project development purposes might face to the limitations of the mentioned method at some certain extent while pursuing the quantitative results from their initiative. They may also realise that the loss of relationship between the stakeholders' expectations and technical attributes. Nevertheless, there are enough characteristics in QFD method that help any researcher to overcome such limitations that might be faced to realise. So, it is worth to criticise conceptual soundness of QFD methodology for making it is used more thoroughly rewarding tool in an any design projects such as already done for their Logistics Centre Development Project by the authors of this symposium paper.

In this study, it is aimed to arise the effectiveness and efficiency of use of QFD tool by making it strengths more visible and also attainable while underlining its characteristics that might not be easily grasped by any project design team for a better result.

References

1. Lin, Y., Pekkarinen, S.: QFD-based modular logistics service design. J. Bus. Ind. Market. **26** (5), 344–356 (2011)
2. Europlatforms EEIG: The European Logistics Platforms Association. https://www.europlatforms.eu/
3. Stylidis, K., Rossi, M., Wickman, C., Söderberg, R.: The communication strategies and customer's requirements definition at the early design stages. Procedia CIRP **50**, 553–558 (2016)
4. Cudney, E.A., Elrod, C.C.: Quality Function Deployment in Continuous Improvement, Missouri University of Science and Technology, July 2011
5. Gaskin, S.P., Griffin, A., Hauser, J.R., Katz, G.M., Klein, R.L.: The Voice of the Customer. Wiley International Encyclopaedia of Marketing, February 1991
6. Revelle, J.B.: Quality Essentials: A Reference Guide from A to Z, April 2004
7. Ulwick, A.W.: Turn Customer Input into Innovation. Harvard Business Review, January 2002
8. Bolar, A.A., Tesfamariam, S., Sadiq, R.: Framework for prioritizing infrastructure user expectations using Quality Function Deployment (QFD). Int. J. Sustain. Built Environ. **6**, 16–29 (2017)
9. Liu, H.-T., Wang, C.-H.: An advanced QFD model using fuzzy analytic network process. Appl. Math. Modell. **34**, 3333–3351 (2010)
10. Chung, T.W., Lee, Y.J., Jang, H.M.: A comparative analysis of three major transfer airports in Northeast Asia focusing on Incheon international airport using a conjoint analysis. Asian J. Shipping Logistics **33**(4), 237–244 (2017)
11. Lee, Y.-T., Wu, W.-W., Tzeng, G.-H.: An effective decision-making method using a combined QFD and ANP Approach. WSEAS Trans. Bus. Econ. **12**(5), 541–551 (2008)
12. Ramanathan, R., Yunfeng, J.: Incorporating cost and environmental factors in quality function deployment using data envelopment analysis, Omega 37, 711–723 (2009). https://doi.org/10.1016/j.omega.2007.12.003

13. Ashlyzan Razik, M., Tahar, R.M., Hasrulnizzam, W., Mahmood, W., Mohd Rozar, N.: Integrated quality function deployment (QFD) model for dry bulk terminal improvements (DBTI) in Malaysian ports. J. Econ. Bus. Manage. **3**, 4 (2015)
14. Azadi, M., Farzipoor Saen, R.: A combination of QFD and imprecise DEA with enhanced Russell graph measure. Soc. Econ. Plann. Sci. **47**, 281–291 (2013)
15. Baki, B., Sahin Basfirinci, C., Cilingir, Z., Ilker Murat, A.R.: An application of integrating SERVQUAL and Kano's model into QFD for logistics services. Asia Pac. J. Mark. Logistics **21**(1), 106–126 (2009)
16. Hauser, J.R., Clausing, D.: The House of Quality. Harvard Business Review, May 1988
17. Dehe, B., Bamford, D.: Quality function deployment and operational design decisions – a healthcare infrastructure development case study. Prod. Plann. Control. (2017) https://doi.org/10.1080/09537287.2017.1350767
18. Ceren Altuntaş Vural and Okan Tuna: The prioritization of service dimensions in logistics centres: a fuzzy quality function deployment methodology. Int. J. Logistics Res. Appl. **19**(3), 159–180 (2016)
19. Wang, Y.-H., Lee, C.-H., Trappey, A.J.C.: Service design blueprint approach incorporating TRIZ and service QFD for a meal ordering system: a case study. Comput. Ind. Eng. **107**, 388–400 (2017)
20. Chan, L.-K., Wu, M.-L.: Quality function deployment: a comprehensive review of its concepts and methods. Qual. Eng. **15**(1), 23–35 (2002)
21. Martin Boggess: 10 Trends That Will Dominate Manufacturing in 2019. https://us.hitachi-solutions.com/blog/top-manufacturing-trends/ (2019)
22. Olewnik, A., Lewis, K.: Limitations of the house of quality to provide quantitative design information. Int. J. Qual. Reliab. Manage. **25**(2), 125–146 (2008)
23. Das, D., Mukherjee, K.: Development of an AHP-QFD framework for designing a tourism product. Int. J. Serv. Oper. Manage. **4**(3), 321–344 (2008)

Two-Phase Fuzzy C-Means and Genetic Algorithm for Food Distribution

Engin Baytürk[1(✉)], Sakir Esnaf[2], and Tarik Kucukdeniz[2]

[1] Engineering and Technology, American College of the Middle East, Egaila, Kuwait
Engin.Bayturk@acm.edu.kw

[2] Industrial Engineering, Istanbul Cerrahpasa University, Istanbul, Turkey
{sesnaf,tkdeniz}@istanbul.edu.tr

Abstract. Covid-19 is a pandemic that effects all around the world nowadays. Some countries decided to isolate areas to reduce the spreading of the virus. During this isolation, one of the most important things for people to get food and necessary items to survive during lockdown period. In this study, two-phase Fuzzy C-Means and Genetic Algorithms proposed for this problem. Proposed algorithm compared with genetic algorithm in a real case study in Kuwait. As a result, prosed algorithm gives better result according to total travelled distance by all trucks.

Keywords: Fuzzy C-Means · Genetic algorithm · Multiple travelling salesman problem · Covid-19 · Food distribution

1 Introduction

COVID-19 is a virus that has been reported by health officials in China on September 2019 as a first time. The first infected group of people are detected in Wuhan City, China [4]. Since then, thousands of people were infected and died because of the virus in China. The virus continued to spread across the world very quickly. Nowadays, more than 7 million of people affected by the virus in all around the world. World Health Organization (WHO) determined many precautionary plans to reduce spreading of the virus.

During Covid-19, some countries isolate the areas where number of cases count considerably a lot to reduce rate of virus spread. However, people who are in the lockdown areas need basic goods such as water, bread and so on to survive during this period. Kuwait, one of the Middle East countries fighting with Covid-19, provides basic goods to people who are in lockdown areas during this time period. While people collect their needs from trucks, there are a lot of people that build a line, and thus, they may increase the rate of virus spread.

Travelling salesman problem (TSP) is an NP-hard problem which aims to reach find minimum total travelled distances or minimum total time in a case that each city is visited only once [7]. Travelling salesman problem is a special version of multiple travelling salesman problem (mTSP) which the number of salesmen (m) is equal to one. Heuristics can be utilized for solving travelling salesman problem [1]. To solve

N. M. Durakbasa and M. G. Gençyılmaz (Eds.): ISPR 2020, LNME, pp. 453–460, 2021.
https://doi.org/10.1007/978-3-030-62784-3_38

multiple travelling salesman problem (mTSP) some heuristic approaches are provided by researches such as ant colony optimisation algorithm [5], particle swarm optimization [12] and genetic algorithm [8].

Fuzzy C-Means (FCM) algorithm first proposed by [6]. The FCM method is a fuzzy version of the k-means algorithm and the aim is to minimize the objective function [10]. After this study, researches provided different hybrid algorithms to improve performance of the FCM such as, Fuzzy C-Means and Centre of Gravity (FCM-COG) [9], Fuzzy C-Mean and Nelder-Mead (FCM-NM) [11].

In mTSP, there are researches that combine clustering algorithms and Genetic Algorithm in various studies. Xu et al. [13] introduce the two-phase heuristic algorithm for mTSP, in the first phase cities are grouped using K-means algorithm and in the second phase, route planning is provided to find minimum total travelled distance by genetic algorithm. Kumar et al. [2] provides a hybrid algorithm that combines a Fuzzy C-Means and Genetic Algorithm (FCM-GA) to solve travelling salesman problem (TSP) which number of travelling salesman is equal to 1. In conclusion, this algorithm gives effective solution to the TSP.

In this study, two phase Fuzzy C-Means and Genetic Algorithm is proposed as alternative way of food distribution to lockdown areas during Covid-19. Proposed algorithm provides a solution to Multiple Travelling Salesman Problem (mTSP). Multiple travelling salesman problem (mTSP) aims to find minimum total travelled distances or minimum total time by using m travelling salesmen. In this particular problem, each city needs to be visited once by one travelling salesman. With the proposed algorithm we strongly believe that there will be no long queues to collect necessary items since trucks will visit all streets one by one. The reason why, this way rate of the virus can be reduced. Proposed algorithm first using Fuzzy C-Means Algorithm, assign streets to trucks with a minimum total distance. In the second phase, problem is solved as a travelling salesman problem for each truck individually.

In this case study, since, the capacity of each truck is bigger than total demand, multiple travelling salesman problem (mTSP) is considered instead of vehicle routing problem (VRP).

According to best of our knowledge, no research has been done with Fuzzy C-Means and Genetic Algorithm regarding with food distribution during Covid-19.

The motivation study is the opinion that the vehicles are parked in square of areas during the food distribution, there are long queues in front of the vehicles and this can be effective in spreading the virus.

The rest of paper is organized as follows, in the second section problem definition is discussed and general formulation of multiple travelling salesman problem (mTSP) is given. In the third section, proposed algorithm is discussed. In the fourth section, result of proposed algorithm and comparison of algorithms are provided. In the last section of the paper contains discussion about conclusion and suggestions for the future researches.

2 Problem Definition

Multiple travelling salesman problem (mTSP) aims to find minimum total travelled distances or minimum total time by using n travelling salesmen. In this particular problem, each city needs to be visited once by one travelling salesman.

The mathematical model proposed by [3] is given below in Eq. (1)–(4):

$$\min Z = \sum_{t=0}^{m} \sum_{i=0}^{n} \sum_{j=0}^{n} X_{ijt} d_{ij} \quad (1)$$

Subject to:

$$\sum_{t=0}^{m} X_{ti} = 1, i = 1, \ldots, n \quad (2)$$

$$\sum_{i=0}^{n} X_{ijt} = y_{tj}, j = 1, \ldots, n \; \forall_t \quad (3)$$

$$\sum_{i=0}^{n} X_{ijt} = y_{ti}, i = 1, \ldots, n \; \forall_t \quad (4)$$

where,

i: cities should be visited by salesmen (i = i,….,n)
t: salesman (t = 1,…,m)
d_{ij}: distance between city i to city j
X_{ijt}: if the salesman visited city j right after city i $X_{ijt} = 1$, otherwise $X_{ijt} = 0$
y_{ti}: if the salesman t visited city I $y_{ti} = 1$, otherwise $y_{ti} = 0$

In here, Eq. 1 refers to objective function which calculates minimum total travelled distances by all travelling salesmen. Equation 2 ensures each city will be visited one time by one travelling salesman. Equation 3 and 4 provides that all salesmen's starting and end will be in from the same city.

3 Proposed Algorithm

In this part, Fuzz C-Means (FCM) and two-phase Fuzzy C-Means and Genetic Algorithm (FCM-GA) methods will be explained in detail.

1) Fuzzy C-Means Algorithm: Fuzzy C-Means algorithm first discussed by [6]. Fuzzy C-Means algorithm is fuzzied version of K-Means algorithm and aims to reach minimum objective function value [10].

Steps of Fuzzy C-Means algorithm is given in Eq. (5)–(7) [3]:

$$J_p(U,c) = \sum_{i=1}^{n} \sum_{j=1}^{k} w_i (u_{ij})^p a_i - c_j^2 \tag{5}$$

Step 1: Initially, $U^{(0)} = [u_{ij}]$ membership matrix is assigned randomly.
Step 2: Cluster centers are calculated by Eq. (2):

$$c_j = \frac{\sum_{i=1}^{n} (u_{ij})^p a_i}{\sum_{i=1}^{n} (u_{ij})^p} \tag{6}$$

Step 3: Membership values are updated by Eq. (3):

$$u_{ij} = \frac{1}{\sum_{j=1}^{k} \left(\frac{a_i - c_i}{a_i - c_j} \right)^{\frac{2}{p-1}}} \tag{7}$$

Step 4: Algorithm stops if the differences between total costs (J_p) that are calculated by consecutive membership matrixes U_k, U_{k+1} less than stopping criteria.

c: number of clusters
n: number of customers
p: coefficient of membership
c_j: coordinates of cluster centre j
$[u_{ij}]$: membership matrix
ai: coordinates of customer i
$J_p(U,c)$: total cost function
$\|a_i - c_j\|$: Euclidean distance between customer i and cluster j

2) *Two Phase Fuzzy C-Means and Genetic Algorithm:*
Two phase FCM-GA steps are as follows:

Phase 1: In this phase, FCM is used to assign customers to clusters.
Step 1: k number of customers is divided into (c_j) sub clusters. For this:
Step 1.1: Using FCM, k number of cluster centres are determined.
Step 1.2: Customers are assigned to closest cluster centre. Sub clusters are created by assigned customers.
Step 2: Objective function is calculated with membership matrix.
Step 3: Algorithm stops if the differences between total costs (J_p) that are calculated by consecutive membership matrixes U_k, U_{k+1} less than stopping criteria.

When the algorithm stops, final version of cluster centres is determined. So, n number of customers are assigned to k number of sub clusters with a minimum total cost.

After clustering with FCM algorithm, multiple travelling salesman problem is divided into k number of individual travelling salesman problem.

Phase 2: Genetic algorithm is employed to solve travelling salesman problem.

Step 1: Clusters are calculated by FCM algorithm are considered as k number of travelling salesman problem.

Step 2: Each travelling salesman problem is calculated by Genetic Algorithm and minimum total travelled distance are calculated.

Step 3: If difference between consecutive total costs is less than stopping criteria, algorithms stops and minimum total travelled distance of travelling salesman problem is found.

Step 4: When each minimum total travelled distance of n number of travelling salesman is found, algorithm sum the total travelled distances of each clusters.

4 Results

In this part of the paper, results obtained from two phase FCM-GA and GA are discussed and compared with each other. Data collected from one of the lockdown areas in Kuwait during Covid-19. In real case, 4 trucks are used to distribute food in this area. The reason why, 4 trucks will be used in both algorithms as a comparison. Each algorithm is run 1000 times and comparison is made by average total travelled distance.

One of the results obtained from MATLAB using Genetic algorithm that has been used as multiple travelling salesman problem is given in Fig. 1.

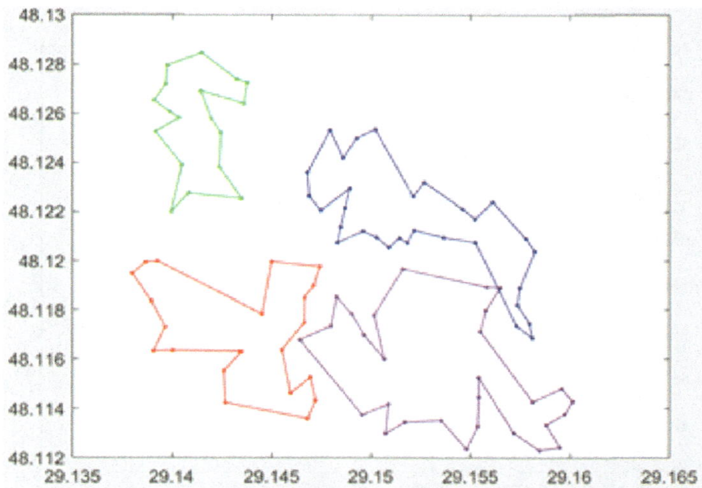

Fig. 1. Genetic Algorithm illustrative graphical solution

Clustering result obtained by MATLAB using Fuzzy C-Means is given in Fig. 2.

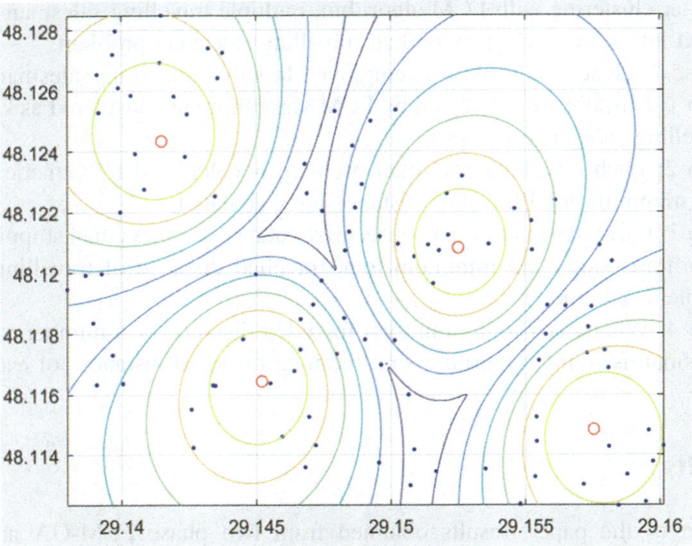

Fig. 2. Clustering result of Fuzzy C-Means Algorithm

After clusters are determined by Fuzzy C-Means algorithm, problem is held as four separate travelling salesman problem. Then, each total travelling distance are added to find total travelled distance of method.

Example of individual travelling salesman problem solution after FCM algorithm using by MATLAB is given Fig. 3.

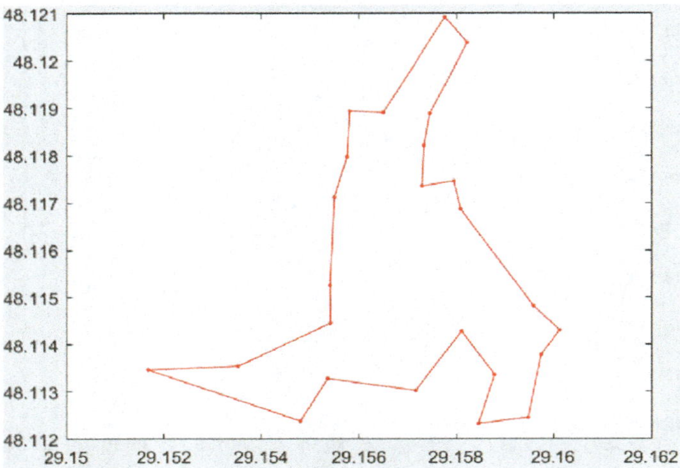

Fig. 3. Individual travelling salesman solution example

Algorithms are stopped after 1000 trials and averages of total travelled distances are calculated for both methods. Comparison of both methods is given in the Table 1 below.

Table 1. Result comparison

Number of demand points	Number of trucks	GA	Two phase FCM-GA
100	4	0.144	0.1423

Area that has been used to compare algorithms has 100 streets and now four trucks are used to deliver food to people who live in this area. As a result of average total travelled distances, two phase FCM-GA performs better than GA after 1000 trials.

5 Conclusion

In this study a two phase FCM-GA proposed and compared with GA using the real data case study for food distribution during Covid-19. As a result, proposed algorithm performs better than GA in terms of total travelled distances.

In the future studies, Fuzzy C-Means and Nelder-Mead method can be used to determine clusters of food distribution points. This method can be compared with GA and two phase FCM-GA in terms of total travelling cost. In addition to this, proposed algorithms can be compared using most common data sets in multiple travelling salesman problem according to total travelled distances and CPU time.

References

1. Kiraly, A., Abonyi, J.: Optimization of multipletraveling salesmen problem by a novel representation based genetic algorithm. Int. Comput. Eng. **366**, 241–269 (2011)
2. Kumar, A., Singh, P., Joshi, P.: Travelling salesman problem using genetic algorithm and fuzzy c-mean clustering algorithm. Int. J. Sci. Technol. Res. **8**, 1391–1397 (2019)
3. Balasko, B., Feil, B.: Fuzzy Clustering and Data Analysis Toolbox for Use with Matlab. Veszprem University, Hungary (2006)
4. Centers for Disease Control and Prevention (2020) Coronavirus Disease 2019 (COVID-19). https://www.cdc.gov/coronavirus/2019-ncov/index.htm
5. Singh, G., Mehta, R.: Implementation of travelling salesman problem using ant colony optimization. J. Eng. Res. Appl. **6**, 285–289 (2014)
6. Bezdek, J.B.: Pattern Recognition with Fuzzy Objective Function Algorithms. Plenum Press, New York (1981)
7. Garey, M.R., Johnson, D.S.: Computers and Intractability: A Guide to the Theory of NP-Completeness, vol. 24, pp. 90–91, New York (1979)
8. Avin, R., Agin, D.: Solving TSP using genetic algorithms-case of Kosovo. Adv. Comput. Sci. **1**, 256–260 (2012)
9. Esnaf, S., Kucukdeniz, T.: A fuzzy clustering-based hybrid method for a multi-facility location problem. Int. Intell. Manuf. **20**, 259–265 (2009)

10. Kenesei, T., Balasko, B., Abony, J.: A MATLAB toolbox and its web based variant for fuzzy cluster analysis. In: Proceedings of 7th international symposium on Hungarian researchers on computational intelligence, Hungary (2006)
11. Kucukdeniz, T., Esnaf, S.: Hybrid revised weighted fuzzy c-means clustering with Nelder-Mead simplex algorithm for generalized multisource Weber problem. J. Enterp. Inf. Manage. **31**, 908–924 (2018)
12. Yan, X.S., Zhang, C., Luo, W., Li, W., Chen, W., Liu, H.: Solve traveling salesman problem using particle swarm optimization algorithm. Int. J. Comput. Sci. **46**, 264–271 (2012)
13. Xu, X., Yuan, H., Liptrott, M., Trovati, M.: Two phase heuristic algorithm for the multiple-travelling salesman problem. Soft Comput. **22**, 6567–6581 (2017)

Optimization

Modelling and Optimization of Dissimilar Welding Between 304L and HSLA-X70 Using Response Surface Methodology

Mohamed Farid Benlamnouar[1], Tahar Saadi[1(✉)], Nabil Bensaid[1], Mokrane Gousmine[1], Youssef Touggui[2], and Mustapha Temmar[2]

[1] Research Center in Industrial Technologies CRTI, 16014 Cheraga, Algiers, Algeria
csc.farid@yahoo.fr, ta_saadi@enst.dz, nblbensaid@gmail.com, mokrane.goss@gmail.com
[2] Structures Research Laboratory (LS), University of Blida 1, 09000 Blida, Algeria
touggui youssef@gmail.com, temmarmustapha@yahoo.fr

Abstract. The industrial applications of TIG welding process are based on the choice of suitable input values for welding parameters like current, speed and gas flow, which determine mechanical properties of weld joint the like tensile strength. Welding parameters were measured according to 3^3 full factorial parameter designs with independent variables of TIG dissimilar welding process between austenitic stainless steel 304L and a high strength low alloy steel HSLA-X70. The response surface methodology (RSM) and analysis of variance (ANOVA) were used to check the validity of quadratic regression model and to determine the significant parameter affecting the output responses. The results show that tensile strength is influenced principally by welding speed. Finally, the ranges for best welding conditions are proposed for serial industrial production.

Keywords: TIG · Tensile test · ANOVA · RSM · Dissimilar welds

1 Introduction

Dissimilar welding attracted great interest to researchers in parallel to its multiple-use in industry. Dissimilar welding is used to combine the attractive properties of at least two materials; such properties could be e.g. tenacity, tensile strength and resistance of corrosion [1].

Therefore, the innovation in dissimilar welds can be done by proposing new technological solutions to deal with strongly perturbed properties of the welded materials [1, 2]. However, the feasibility of such combinations is often limited by the strong difference of the physico-chemical properties of materials intended for welding [2, 3]. This is because TIG welding involves a high speeds of cooling and consequently, a short thermal cycles that influence in particular the kinetics of the formation of metallic phases during solidification, the stabilization of these phases, the formation of carbide precipitations and the solubility rate of the alloying elements of the welded materials [4, 5].

N. M. Durakbasa and M. G. Gençyılmaz (Eds.): ISPR 2020, LNME, pp. 463–472, 2021.
https://doi.org/10.1007/978-3-030-62784-3_39

The extent of convergence among chemical composition of materials to be welded is main factor to get better mechanical characteristics of the weld joints [6]. Thus, it is necessary to optimize welding parameters in order to avoid spacing of chemical and mechanical properties in dissimilar weld [7].

Dissimilar welding has been subject of several investigations because of its top technical and economic potential [8]. Consequently, In several studies, dissimilar welds was combined the attractive properties of HSLA and SS steels such as high strength, highly resistant to corrosion cracking, to treat mechanical and thermal problems of dissimilar welds in some industrial applications [4, 9], in which numerical methods were proposed for optimising the welding process. In these methods, when searching to establish correlations between welding parameters and output responses, it is necessary to use a plan of experiments. Mostly, the experiments often rested on the adoption of the classic experimental tests, the identification of critical parameters and the search for optimal values allowing us to arrive at the target value of the weld characteristics. In all of these steps, we must produce a factorial matrix comprising very large number of experiments. But these experiments realization will be very expensive, from both a time and practical view. So, experimental approach of Response Surface Methodology (RSM) is used. According to this approach, to study the impact of several parameters we just need few experiments, which allow us to save time and have a better experimental cost/quality ratio [8].

RSM is a collection of mathematical and statistical techniques that are useful for the modelling and analysis of problems in which response of interest is influenced by several variables and the objective is to optimize the response [10]. In this approach, Benlamnouar et al. [9] obtained a strong correlation between welding parameters (current, speed and gas flow) and tensile strength of X70-304L dissimilar weld using Taguchi method. In another study [9, 11] it has been reported that current has highest effect on tensile strength. In this work, an experimental contribution that focuses on optimization and modelling process parameters in TIG welding of dissimilar weld.

This study focuses on the optimization of welding parameters by Response surface methodology using mechanical characteristics across the HSLA/SS dissimilar welds. Mechanical tests were achieved to analyse precisely the tensile strength. Mathematical models for tensile strength were developed using the response surface methodology (RSM) is presented. The ANOVA study involves the effects of welding parameters (current, speed, gas flow) on tensile strength.

2 Design of Experiments

A Chemical composition of materials is given in Table 1.

The base Materials is used in the form of rolled plates of 3 mm thickness and filler metal diameter of 1.5 mm, welding process was conducted by automatic tungsten arc welding (TIG) using in two-pass supported by Miller welding machine, without preheating, according to ASME Standard [12]. Argon gas flow is measured using LPM flowmeter.

Standard tensile specimens were tested using SATEC INSTRON tensile testing machine at a strain rate of 0.1/s, according to ASTM-E8 Standard (Fig. 1).

Table 1. Chemical composition of base materials and filler materials

Element (%)	C	Si	Mn	Cr	Ni	Mo	Cu	Nb	Ti
X70	0.08	0.7	2.39	0.04	0.02	0.1	0.01	0.4	0.09
304L	0.047	0.5	1.85	17.8	9.4	0.2	0.15	0.08	0.01
E2209	0.027	0.46	0.9	22.5	9.1	2.8	–	–	–

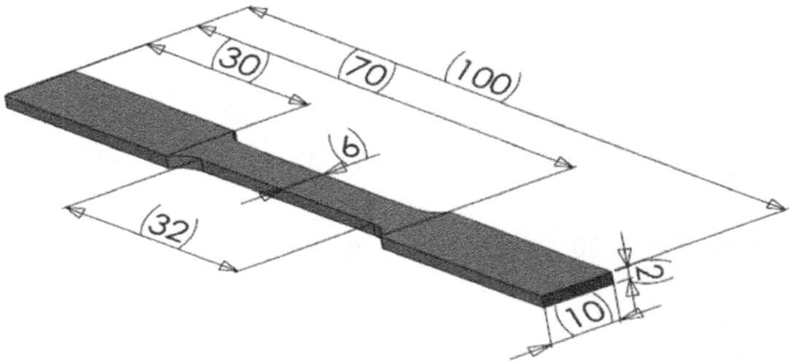

Fig. 1. Tensile specimen

The response surface methodology (RSM) has been proven to be a powerful tool for determining the effects of each factor and the interactions between them, which allows the optimization of the processes to be carried out effectively [8]. Response surface procedures involve experimental strategy, mathematical methods, and statistical inference, which, when combined, allow us to do effective empirical exploration of the phenomenon in which we are interested [9]. The experimental strategy allows us to explore the response surface with the same precision, in any direction. Following this investigation, the experimental design allows the analyst to explore the response surface in a systematic way in the direction that offers the most promise for improvement tensile test.

The experimental investigations are designed on the basis of RSM method. Three factors, 3 levels design consist of 27 sets of coded conditions (Table 2). The tensile strength of the TIG weld samples were analyzed by employing Response Surface Methodology and Analysis of variance (ANOVA) to determine the direct and interaction effects of the process parameters.

Table 2. Welding parameters and their levels

Parameter	Designation	Level		
		1	2	3
Current (A)	A	60	100	140
Speed ($mm.min^{-1}$)	B	30	50	70
Gas flow ($l.min^{-1}$)	C	6	10	14

Design-Expert software was used for optimizing welding parameters; the optimal welding conditions for the tensile strength dissimilar weld with the constraints of the welding parametric range are those corresponding to the highest values of tensile strength (Table 3).

Table 3. Experimental results for Tensile strength

Test number	Current (A)	Speed ($mm.min^{-1}$)	Gas flow ($l.min^{-1}$)	Tensile Strength (MPa)
1	60	30	6	182,5
2	60	30	10	187,7
3	60	30	14	217,3
4	60	50	6	314,5
5	60	50	10	321,5
6	60	50	14	348
7	60	70	6	463,5
8	60	70	10	424,5
9	60	70	14	468,2
10	100	30	6	173,5
11	100	30	10	193,8
12	100	30	14	201,7
13	100	50	6	295,3
14	100	50	10	328,5
15	100	50	14	342,5
16	100	70	6	456
17	100	70	10	456,8
18	100	70	14	467,5
19	140	30	6	182,4
20	140	30	10	178,5
21	140	30	14	306,3
22	140	50	6	308,8
23	140	50	10	311,5
24	140	50	14	326,5
25	140	70	6	401,5
26	140	70	10	419,3
27	140	70	14	436,5

3 Results and Discussion

3.1 Statistical Analysis

The Response Surface Methodology (RSM) comprised the following six major components:

- Defining the independent input variables and the desired output responses;
- Adopt an experimental design matrix using design of experiments;
- Perform regression analysis with the quadratic model of RSM;
- Analyze the results using ANOVA;
- Determine the situation of the quadratic model of RSM and decide whether the model of RSM needs screening variables or not;
- Optimize and conduct confirmation experiment and verify the predicted performance characteristics.

The Table 4 shows the ANOVA table for response surface 2FI model for tensile strength. The value of "Prob > F" in Table 4 for model is less than 0.05 which indicates that the model is significant, which is desirable as it indicates that the terms in the model have a significant effect on the response. It confirms that the speed is the most influential factor on welding because of his highest exponent. The second parameter, which influences the welding, is gas flow. Where as the third parameter, which influences less, is current.

Table 4. ANOVA for tensile strength

Source	DF	SC sq.	MS	F-value	Prob > F	Cont. %	Remarks
Model	6	223.62	37.27	87.04	< 0.0001		Significant
A	1	0.053	0.053	0.12	0.7292	0.00	Significant
B	1	212.95	212.95	497.31	< 0.0001	95.0	Significant
C	1	5.81	5.81	13.58	0.0015	3.00	Significant
A × B	1	1.90	1.90	4.45	0.0478	1.00	Significant
A × C	1	0.79	0.79	1.85	0.1885	0.00	Significant
B × C	1	2.11	2.11	4.92	0.0383	1.00	Significant
Error	20	8.56	0.43				
Total	26	232.19					

Mean = 17.72, Coefficient of variation (%) = 3.69, Predicted residual error of sum of squares (PRESS) = 19.15, R^2 = 0.9631, R^2 Adjusted = 0.9520, R^2 Predicted = 0.9175, Adequate precision = 25.552.

3.2 Regression Equations

The relationship between the factors and the performance measures was modelled by quadratic regression. The regression equations obtained were as follows. The tensile strength model is given below in Eq. (1).

$$\begin{aligned} \text{Sqrt (Tensile strength)} = {} & 4.340 + 7.469 * 10^{-3} * A + 0.274 * B + 0.243 \\ & * C - 4.980 * 10^{-4} * A * B + 1.607 * 10^{-3} * A * C \\ & - 5.236 * 10^{-3} * B * C \end{aligned} \tag{1}$$

3.3 Effect of Welding Parameters on Surface Response Factors

The correlation between the effective factors (current, voltage and gas flow) and the tensile strength were obtained by quadratic model (Eq. (1)); this correlation can be represented by 3d surface obtained by Design-Expert software. It investigates the influences of welding parameters on the tensile strength, response surface are drawn in Figs. 2 depicts the influence of welding speed, current and gas flow on the tensile strength.

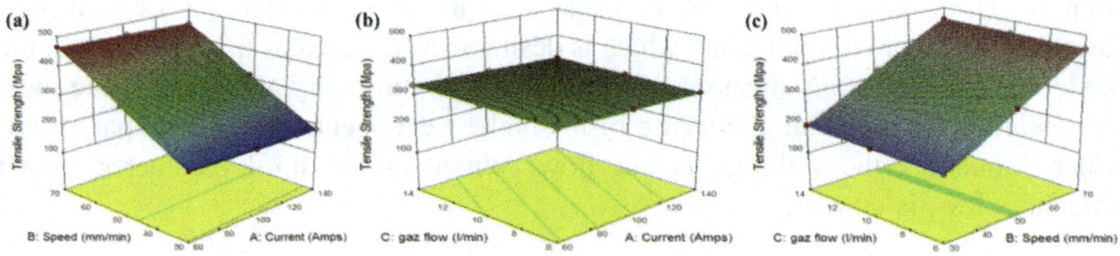

Fig. 2. Effect of welding parameters on tensile strength. (a) Speed and current, (b) Gas flow and current, (c) Gas flow and speed

The predicted values are compared with the corresponding experimental values. This is depicted in Fig. 3.

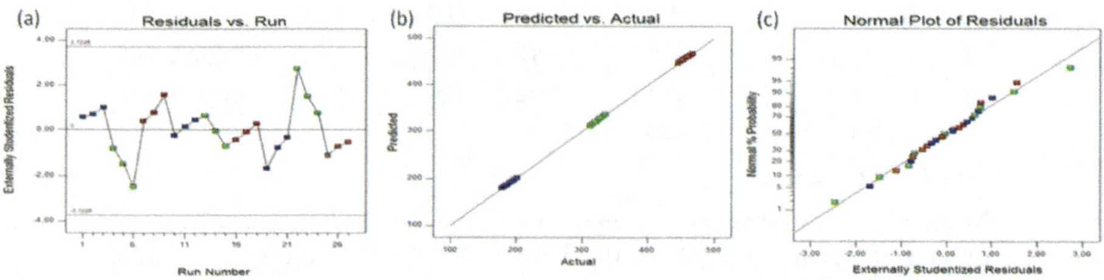

Fig. 3. Analyses: (a) Plot of residuals versus run numbers for tensile strength, (b) Comparison between measured and predicted value of tensile strength (c) Comparison of Predicted and Actual tensile strength

4 Optimization of Welding Conditions

In this study, a desirability function approach was used for single response parameters, namely tensile strength. During the optimization process, the aim was to find the optimal values of welding parameters. To solve this type of parameter design problem, an objective function F(x), is defined as follows [13]:

$$F(x) = -DF \tag{2}$$

$$DF = \left(\prod_{i=1}^{n} d_i^{wi}\right)^{\frac{1}{\sum_{j-1}^{n} wi}} \tag{3}$$

The purpose here is to find the best solution that maximizes a combined desirability function DF. The constraints used during the optimization process are summarized in Table 5.

Table 5. Constraints for optimization of welding conditions

Condition	Goal	Lower limit	Upper limit	Lower weight	Upper weight
Current	Is in range	60	140	1	1
Speed	Is in range	30	70	1	1
Gas flow	Is in range	6	14	1	1
Tensile strength	Maximize	178.9	468.1	1	1

The ramp plot showing the optimum values in the range of input parameters and the predicted output responses is given in Fig. 4.

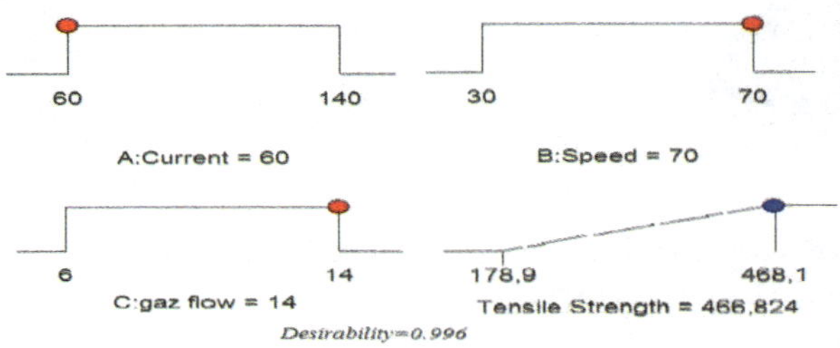

Fig. 4. Ramp plot for combined objective with desirability of 0.996

Figure 5 presents the bar graph of desirability for the welding conditions and the responses together with a combined desirability = 0.996.

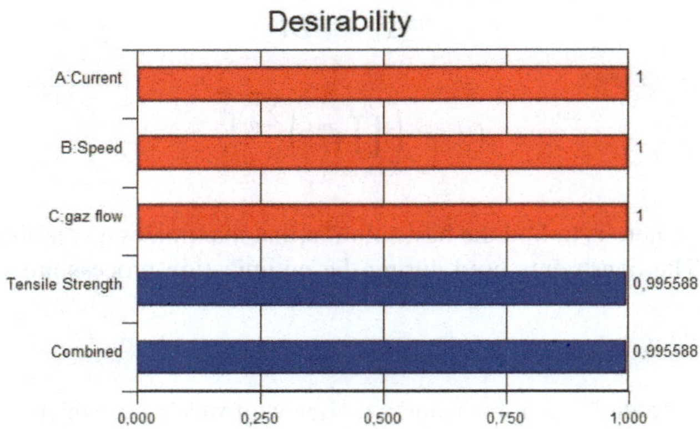

Fig. 5. Bar graph of desirability

In the present study, optimization is carried out with desirability approach. This approach generally used to find the optimum parametric combination for single and multiple objective optimizations [14].

The desirability functions during the optimization process are summarized in Fig. 6 and Fig. 7.

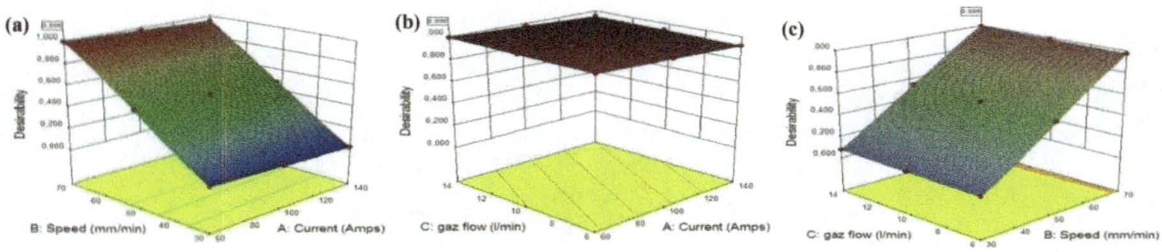

Fig. 6. 3D surface plots of welding parameters on desirability, (a) Speed and current, (b) Gas flow and current, (c) Gas flow and speed

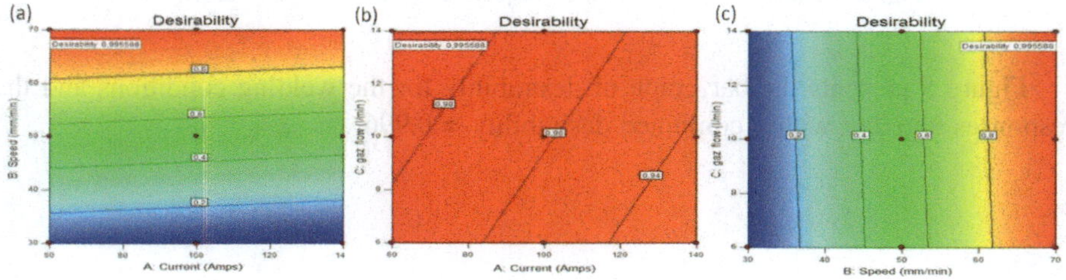

Fig. 7. Desirability Contours: (a) Contour plots of current and welding speed on tensile strength, (b) Contour plots of current and gas flow on tensile strength, (c) Contour plots of speed of welding and gas flow on tensile strength

Contour plots show distinctive circular shape indicative of possible independence of factors with response. A contour plot is produced to visually display the region of optimal factor settings. RSM is used to find the optimal set of process parameters that produce a maximum value of the tensile strength. Analyzing the contour graphs and by solving Eq. (1) it will be possible to attain the maximum tensile strength. Figure 6 presents three dimensional response surface plots for the desirability obtained from the regression model. The interaction effect between welding speed (Fig. 7(a)), current and gas flow (Fig. 7(b)), predicted optimum tensile strength obtained from the response surface and contour plots (Fig. 7(c)).

5 Conclusions

This research work presents the application of response surface methodology (RSM) models to the study of the influence of welding parameters on tensile strength. The relationship between the factors and the measured performance were modelled by quadratic regression. Three process variables, namely speed of welding, current and gas flow, were used for the development of the models. The actual models were found to satisfy the optimization of the welding parameters.

The foremost conclusions which can be drawn are as follows:

- The result of ANOVA proved that the quadratic mathematical models allow prediction of tensile strength parameter with a 95% confident interval.
- Comparison of experimental and predicted values of tensile strength shows that good agreement has been achieved between them.
- The welding speed has the highest statistical influence on tensile strength.
- The predictive model is found to be capable of better predictions of tensile strength.
- Verification of the experiments carried out shows that the empirical models developed can be used for SS 304L/X70 dissimilar weld.
- This work suggests the optimization of mechanical properties of dissimilar weld by RSM method using tensile strength.

References

1. Lomozik, M.: Effect of the welding thermal cycles on the structural changes in the heat affected zone and on its properties in joints welded in low alloy steels. Weld. Int. **14**(11), 845–8S0 (2000)
2. Kravanja, S., Zula, T.: Cost optimization of industrial steel building structures. Adv. Eng. Softw. **41**, 442–450 (2010)
3. Gousmine, M., Miroud, D., Benlamnouar, M.F., Demri, B., Younes, A.: Study of composite with metallic matrix WC/W2C–20 W–20Ni realized by spontaneous infiltration of the bronze alloy Cu–30Mn–3P. In: Applied Mechanics, Behavior of Materials, and Engineering Systems. Springer, pp. 365–374 (2015)
4. Ranjbarnodeh, E., Serajzadeh, S., Kokabi, A.H., Fischer, A.: Effect of welding parameters on residual stresses in dissimilar joint of stainless steel to carbon steel. J. Mater. Sci. **46**, 3225–3232 (2011)

5. Bensaid, N., Hadji, M., Badji, R., Benlamnouar, M.F., Saadi, T., Laib dit Laksir, Y., Senouci, S.: Microstructure and mechanical behavior of AISI 430 FSS welds produced with different elemental metal powder addition. Solid State Phenomena 297, 195–203 (2019)
6. Wang, J., Lu, M., Zhang, L., Chang, W., Xu, L., Hu, L.: Effect of welding process on the microstructure and properties of dissimilar weld joints between low alloy steel and duplex stainless steel. Int. J. Miner. Metall. Mater. **19**, 518 (2012)
7. Ragavendran, M., Chandrasekhar, N., Ravikumar, R., Saxena, R., Vasudevan, M., Bhaduri, A.K.: Optimization of hybrid laser-TIG welding of 316LN steel using response surface methodology (RSM). Opt. Lasers Eng. **94**, 27–36 (2017)
8. Devendranath Ramkumar, K., Mishra, D., Ganesh Raj, B., Vignesh, M.K., Thiruvengatam, G., Sudharshan, S.P., Arivazhagan, N., Sivashanmugam, N., Maximus Rabel, A.: Effect of optimal weld parameters in the microstructure and mechanical properties of autogeneous gas tungsten arc weldments of super-duplex stainless steel UNS S32750. Mater. Des. **66**, 356–365 (2015)
9. Benlamnouar, M.F., Hadji, M., Badji, R., Bensaid, N., Saadi, T., Laib dit Laksir, Y., Senouci, S.: Optimization of TIG welding process parameters for X70-304L dissimilar joint using taguchi method. Solid State Phenom. 297, 51-61 (2019)
10. Sun, X., Zhao, G., Zhang, C., Guan, Y., Gao, A.: Optimal design of second-step welding chamber for a condenser tube extrusion die based on the response surface method and the genetic algorithm. Mater. Manuf. Process. **28**, 823–834 (2013)
11. Ghani, S.J.A., Choudhury, I.A., Hassan, H.H.: Application of Taguchi method in the optimization of end milling parameters. J. Mater. Process. Technol. **145**, 84–92 (2004)
12. ASME: Qualification Standard for Welding and Brazing, an International Code of ASME Boiler and Pressure Vessel Committee on welding and Brazing, USA (2010)
13. Kumar, P., Chauhan, S.R.: Machinability study on finish turning of AISI H13 hot working die tool steel with cubic boron nitride (CBN) cutting tool inserts using response surface methodology (RSM). Arab. J. Sci. Eng. **40**, 1471–1485 (2015)
14. Myers, R.H., Montgomery, D.C.: Response Surface Methodology Process and Product Optimization Using Designed Experiments, 2nd edn. Wiley, New York (2002)

Proposing a Pre-emptive Resource Constrained Project Scheduling Problem (PRCPSP) Model to Optimize Manpower and Project Delivery Time (A Case Study)

Arman Saeidi(✉), Kamran Rezaie, Alireza Nazari, and Amir Hossein Ordibazar

School of Industrial and Systems Engineering, College of Engineering, University of Tehran, Tehran, Iran
{arman.saeidi,krezaie,alirezanazari.e, a.ordi.2012}@ut.ac.ir

Abstract. In this paper, a Pre-emptive Resource Constrained Project Scheduling Problem (PRCPSP) mathematical model is proposed to schedule a production project by considering uncertainty of time parameter for each activity. At last to evaluate the validity of the model, the mathematical model is implemented in a production factory and the results have been evaluated. In addition, some analyses are implemented to assess the behavior of the model and extract managerial insights.

Keywords: Fuzzy uncertainty · Resource management · Scheduling problem · Mathematical modelling · Project management

1 Introduction and Literature Review

Resource Constrained Project Scheduling Problem (RCPSP) is one of the applied issues in the field of operations research the and project management, so that one of the most important and challenging tasks in project management in levels of design and performance, time bandage and activities efficiency is resources limits. This study focuses on a specific development of RCPSP which is a Pre-emptive Resource Constrained Project Scheduling Problem (PRCPSP) with the cessation of activities by the aim of optimization of manpower and project delivery time. Limited PRCPSP assumption with the possibility of cessation of activities that cessation of activities in executive situations brings the model closer to real project scheduling issues. This assumption may be due to some factors. For instance, lack of resources and factors related to manpower or equipment failure.

Project scheduling is one of the most well-known issues in the field of optimization. In fact, the standard form of project scheduling with resource constraints can be defined as a combination optimization problem. Decision variables are defined as the integer numbers of times to start activities in a period of time, the objective function can be used to minimize in cases such as the maximum time to complete the activities, time delay in project implementation and so on. There are two types of constraints:

N. M. Durakbasa and M. G. Gençyılmaz (Eds.): ISPR 2020, LNME, pp. 473–480, 2021.
https://doi.org/10.1007/978-3-030-62784-3_40

1. Pre-emptive activities 2. Limitations of resources [1–3]. In addition, different uncertainty methods are used to confront the uncertainty of parameters like time duration, demand and so on [1, 4, 5]. In addition, some multi-objective approaches are used to solve mathematical modelling with more than one objective function [1, 6].

In the RCPSP and MRCPSP methods, it is assumed that each activity starts once and runs until it is completed. In the case of a project with limited resources, with the possibility of cessation of activities (PRCPSP), any activity can be broken at any time and resumed at no extra cost. In this paper, a PRCPSP is proposed and the fuzzy method is used to confront the parameters uncertainty and a case study will be solved by this approach. By reviewing the literature on the topics discussed, and identifying the research gaps, we distinguish our work from existing studies in the literature by bring up these contributions which we will outline below:

- The proposed model is applied in a real case study, and the applicability and the effectiveness of the model are verified.
- Pre-emptive Resource Constrained Project Scheduling Problem (PRCPSP) that is a specific development of Resource Constrained Project Scheduling Problem (RCPSP).
- This study is proposing a multi-objective fuzzy scheduling problem to deal with the inherent uncertainty of this case.

2 Problem Definition

In this paper, we proposed a model to optimize scheduling in an industrial unit that receives orders as a project and it is obliged to deliver the projects to the customer on time. What is important for this factory is to find a production plan that minimizes project completion time and uses the least possible manpower.

The conditions of this factory are as below:

- In this factory, there are n work stations.
- Activities at workstations do not require special expertise, and each operator is able to operate at each station.
- There is a prerequisite for the activities of different stations, so that the start of each activity in the new station must be after the completion of the activity in the previous station.
- If you stop an activity at a station without spending time and money, the activity can be resumed.
- Due to the lack of an information recording system, the instability of operators in each station and the seasonal conditions, there is no accurate estimate of the most important parameter of this issue, e.g. the time of each activity, and this parameter is estimated based on personal experience of the production manager.

We also need to add conditions to the above to formulate the problem. These conditions are as below:

- We have a lot size for this problem, its size is determined by the station that has the largest lot size. It is noteworthy that this condition is not unrealistic because if the lot size is smaller than this value, there is at least one station that operates with a capacity less than the potential capacity.
- Instead of assuming that the end of the activity on each piece is a prerequisite for starting the next activity on the piece, we set the end of the activity on each batch as a prerequisite. This assumption may take us a little further from the optimal point, but it makes the production program more practical and easier to implement.
- It is assumed that the activity time is an integer variable and does not take a decimal value. In fact, numbers with decimal places are round up to the nearest integer.

The indexes, parameters and variables of the mathematical model presented for the above problem is as follows:

i	Workstation Index (i = 1, 2, ., n)
j	Workstation Index (j = 1, 2, ., n)
t	Time index (t = 1, 2, ., t_{max})
n	Number of workstationsNumber of workstations
t_{max}	The longest possible time for project completion
C_{max}	Project completion time
X_{it}	Binary variable If the activity i in time t is running 1 otherwise zero
S_i	Activity i start time
C_i	Activity i finish time
Q_{ij}	A set of activities that activity i is prerequisite of activity j
d_i	Time required to perform the activity i
r_i	The number of manpower required to perform the activity i
R	Total manpower available

$$\min z_1 = C_{max} \quad (1)$$
$$\min z_2 = R \quad (2)$$
(1)

s.t.

$$\sum_{t=1}^{t_{max}} X_{it} \geq d_i \quad \forall i \quad (3)$$
$$S_i \leq t * X_{it} + (1 - X_{it}) * M \quad \forall i.t \quad (4)$$
$$C_i \geq t * X_{it} \quad \forall i.t \quad (5)$$
$$C_{max} \geq C_i \quad \forall i \quad (6)$$
$$S_j \geq C_i \quad \forall i.j \in Q \quad (7)$$
$$\sum_{i=1}^{n} r_i * X_{it} \leq R \quad \forall t \quad (8)$$
$$X_{it} = \{0.1\} \quad \forall i.t \quad (9)$$

Equation (1) and (2) are objective functions, the objective function (1) tries to minimize project delivery time. The objective function (2) is used to minimize the amount of manpower required to carry out the project.

Equation (3) to (9) represent constraints of the model. Constraint (3) is the minimum time required for each activity. In other words, this constraint selects the set of answers from points where the sum of the times performed at different times is at least equal to the time required. Constraints (4) and (5) show the start and end times of each activity, respectively. Constraint (6) indicates that the end time of the project should be greater than the end time of each activity. Constraint (7) indicates the prerequisite relationships between project activities. Constraint (8) states that the resources consumed per unit of time should not be greater than the maximum available resources. Constraint (9) indicates that the decision variable is binary.

As explained, because environmental conditions such as air temperature as well as operator conditions like the experience of doing work are not stable, there is no definitive estimate of time required to perform each activity, and this parameter will be considered uncertain by fuzzy method. Equation (3) changes to Eq. (1) as below:

$$\sum_{t=1}^{t_{max}} X_{it} \geq \tilde{d}_i \qquad \forall i \qquad (10)$$

To de-fuzzify the constraints, general fuzzy size (Me) will be used. The reason for this approach is that Me has the capability to adjust the parameter λ, therefore the problem may be solved pessimistic or optimistic.

The definitive form of the Eq. (10) is Eq. (11) as below:

$$\lambda * Poss\left(\tilde{d}_i \leq \sum_{t=1}^{t_{max}} X_{it}\right) + (1-\lambda) * Nec\left(\tilde{d}_i \leq \sum_{t=1}^{t_{max}} X_{it}\right) \geq \alpha \qquad \forall i \qquad (11)$$

If we assume that the fuzzy parameters of the problem are trapezoidal $(d_{i1}.d_{i2}.d_{i3}.d_{i4})$ and the parameter α value is greater than 0.5 and parameter λ value is less than 0.5 ($\lambda < 0.5.\alpha > 0.5$), which is almost actual conditions, then the Eq. (11) of the model turns to be as Eq. (12):

$$\sum_{t=1}^{t_{max}} X_{it} \geq \frac{(\alpha-\lambda)d_{i4} + (1-\alpha)d_{i3}}{(1-\lambda)} \qquad \forall i \qquad (12)$$

Now we come to the part where we have a two-objective deterministic problem. We will use fuzzy approaches to solve the problem.

The first step in using fuzzy approaches for solving multi-objective problems is calculating the ideal and anti-ideal answers.

To calculate the ideal answer, for the first objective function (z_1^+), we will solve the following model, Eqs. (13) to (20):

$$\min z_1^+ = C_{max} \tag{13}$$
$$\sum_{t=1}^{t_{max}} X_{it} \geq \frac{(\alpha-\lambda)d_{i4} + (1-\alpha)d_{i3}}{(1-\lambda)} \quad \forall i \tag{14}$$
$$\mathrm{S_i} \leq t * X_{it} + (1 - X_{it}) * M \quad \forall i.t \tag{15}$$
$$\mathrm{C_i} \geq t * X_{it} \quad \forall i.t \tag{16}$$
$$\mathrm{C_{max}} \geq \mathrm{C_i} \quad \forall i \tag{17}$$
$$\mathrm{S}_j \geq \mathrm{C_i} \quad \forall i.j \in Q \tag{18}$$
$$\sum_{i=1}^{n} r_i * X_{it} \leq R \quad \forall t \tag{19}$$
$$X_{it} = \{0.1\} \quad \forall i.t \tag{20}$$

Switching the second objective function with the Eq. (13) of the model, the ideal amount for second objective function (z_2^+) will be calculated.

In addition, the anti-ideal for the first objective function (z_2^-) will be calculated as Eqs. (21) to (29):

$$\min z_1^- = C_{max} \tag{21}$$
$$R \leq z_2^+ \tag{22}$$
$$\sum_{t=1}^{t_{max}} X_{it} \geq \frac{(\alpha-\lambda)d_{i4} + (1-\alpha)d_{i3}}{(1-\lambda)} \quad \forall i \tag{23}$$
$$\mathrm{S_i} \leq t * X_{it} + (1 - X_{it}) * M \quad \forall i.t \tag{24}$$
$$\mathrm{C_i} \geq t * X_{it} \quad \forall i.t \tag{25}$$
$$\mathrm{C_{max}} \geq \mathrm{C_i} \quad \forall i \tag{26}$$
$$\mathrm{S}_j \geq \mathrm{C_i} \quad \forall i.j \in Q \tag{27}$$
$$\sum_{i=1}^{n} r_i * X_{it} \leq R \quad \forall t \tag{28}$$
$$X_{it} = \{0.1\} \tag{29}$$

Similarly, anti-ideal answer for the second objective (z_2^-) function will be calculated.

Now, the membership for each of the objective functions according to the range of changes will be calculated as Eqs. (30) and (31).

$$\mu_{z_1}(\mathrm{C_{max}}) = \begin{cases} 1 & if \quad \mathrm{C_{max}} \leq z_1^+ \\ \frac{z_1^- - \mathrm{C_{max}}}{z_1^- - z_1^+} & if \quad z_1^+ \leq \mathrm{C_{max}} \leq z_1^- \\ 0 & if \quad z_1^- \leq \mathrm{C_{max}} \end{cases} \tag{30}$$

$$\mu_{z_2}(R) = \begin{cases} 1 & if \quad R \leq z_2^+ \\ \frac{z_2^- - R}{z_2^- - z_2^+} & if \quad z_1^+ \leq R \leq z_2^- \\ 0 & if \quad z_2^- \leq R \end{cases} \tag{31}$$

Now, We use the Torabi-Hosseini method (TH) method to solve the multi-objective problem. It will be as Eqs. (32) to (42).

$$\max\ \gamma\beta_1 + (1-\gamma)\left(w_1\mu_{z_1}(\mathrm{C_{max}}) + w_2\mu_{z_2}(\mathrm{R})\right) \quad (32)$$
$$\mu_{z_1}(\mathrm{C_{max}}) \geq \beta_1 \quad (33)$$
$$\mu_{z_2}(R) \geq \beta_1 \quad (34)$$
$$\sum_{t=1}^{t_{max}} X_{it} \geq \frac{(\alpha-\lambda)d_{i4} + (1-\alpha)d_{i3}}{(1-\lambda)} \quad \forall i \quad (35)$$
$$\mathrm{S_i} \leq t * X_{it} + (1 - X_{it}) * M \quad \forall i.t \quad (36)$$
$$\mathrm{C_i} \geq t * X_{it} \quad \forall i.t \quad (37)$$
$$\mathrm{C_{max}} \geq \mathrm{C_i} \quad \forall i \quad (38)$$
$$\mathrm{S}_j \geq \mathrm{C_i} \quad \forall i.j \in Q \quad (39)$$
$$\sum_{i=1}^{n} r_i * X_{it} \leq R \quad \forall t \quad (40)$$
$$X_{it} = \{0.1\} \quad \forall i.t \quad (41)$$
$$0 \leq \beta_1 \leq 1 \quad (42)$$

If the membership function is replaced, the final form of the problem will be as Eqs. (43) to (53).

$$\max\ \gamma\beta_1 + (1-\gamma)\left(w_1 * \frac{z_1^- - \mathrm{C_{max}}}{z_1^- - z_1^+} + w_2 * \frac{z_2^- - R}{z_2^- - z_2^+}\right) \quad (43)$$
$$\mathrm{C_{max}} \leq z_1^- - \beta_1 * (z_1^- - z_1^+) \quad (44)$$
$$\mathrm{R} \leq z_2^- - \beta_1 * (z_2^- - z_2^+) \quad (45)$$
$$\sum_{t=1}^{t_{max}} X_{it} \geq \frac{(\alpha-\lambda)d_{i4} + (1-\alpha)d_{i3}}{(1-\lambda)} \quad \forall i \quad (46)$$
$$\mathrm{S_i} \leq t * X_{it} + (1 - X_{it}) * M \quad \forall i.t \quad (47)$$
$$\mathrm{C_i} \geq t * X_{it} \quad \forall i.t \quad (48)$$
$$\mathrm{C_{max}} \geq \mathrm{C_i} \quad \forall i \quad (49)$$
$$\mathrm{S}_j \geq \mathrm{C_i} \quad \forall i.j \in Q \quad (50)$$
$$\sum_{i=1}^{n} r_i * X_{it} \leq R \quad \forall t \quad (51)$$
$$X_{it} = \{0, 1\} \quad \forall i.t \quad (52)$$
$$0 \leq \beta_1 \leq 1 \quad (53)$$

The PRCPSP mathematical model has described in this section. In addition, the fuzzy approach and the multi-objective approach has been declared in Eqs. (1) to (53). In, next section the case study will be discussed.

3 Numerical Results and Case Study

The case study for this problem is the MaroSanat Factory, which manufactures valves, springs and other related products for gas, oil and petrochemical industries. The problem, which is discussed in this paper is the factory production planning for a kind of springs. The amount of manufacturing by considering demands and offers is 1000 springs. In this factory, there are six workstations as described in Table 1. In Table 1, for each station, estimated time in a trapezoidal form is shown. It is calculated based on expert perspective. For two stations No. 4 & 5, the time is deterministic. Therefore, all parameters are the same. The required resources are used to consider manpower in work stations. Due to the size of batch, which is 200, so the number of batches of the problem is $\frac{1000}{200} = 5$.

Table 1. Stations description

No.	Station	Process time	Required resources
1	Cutting	$\tilde{d}_1(12.12.15.20)$	1
2	Furnace and press	$\tilde{d}_2(35.40.50.60)$	4
3	Quench	$\tilde{d}_3(20.25.30.35)$	1
4	Temper	$\tilde{d}_4(15.15.15.15)$	1
5	Shot Blast	$\tilde{d}_5(20.20.20.20)$	0
6	Packaging	$\tilde{d}_6(15.20.30.40)$	3

Using the criterion (Me) and parameters of the problem and proposed mathematical PRCPSP model, the model has been solved and the results were informed to managers and experts of the factory. The results show that the amount of parameter λ and α are equal to 0.2 and 0.8, respectively. It shows that the pessimistic perspective is selected. In addition, the amounts of process time duration for each station are 19, 58, 34, 15, 20 and 38, respectively. These results show that the pessimistic approach is correct, since most of the parameters are nearer to the worst states.

After solving the problem by GAMS software, ideal and anti-ideal answers are $z_1^+ = 416.z_1^- = 539.z_2^+ = 5\,and\,z_2^- = 8$. By putting these values in Eqs. (30) and (31), the following amounts will be calculated for $\mu_{z_1}(C_{max})$ and $\mu_{z_2}(R)$.

$$\mu_{z1}(C_{max}) = \begin{cases} 1 & C_{max} \leq 416 \\ \frac{539-C_{max}}{539-416} & 416 \leq C_{max} \leq 539 \\ 0 & 539 \leq C_{max} \end{cases}$$

$$\mu_{z_2}(R) = \begin{cases} 1 & R \leq 5 \\ \frac{8-R}{3} & 5 \leq R \leq 8 \\ 0 & 8 \leq R \end{cases}$$

The problem has been solved by TH method as it was shown in Eqs. (32) to (42). The compatibility coefficient between goals (γ) assumes 0.5, the weight of the objective functions $\omega_1\,and\,\omega_2$ are assumed 0.6 and 0.4 by the point view of experts.

The final answer to the problem after solving it in the TH method, and the above parameters were presented to managers and experts of the company. The minimum level of satisfaction of target functions is 0.33, the amount of the first objective function which is project completion time is 424 and number of required people is 7. These values show that by increasing the number of the personnel, the process completion time decrease.

In another test, by assuming, the weight of the objective functions ω_1 *and* ω_2 are assumed 0.4 and 0.6. The results show 473 for project completion time and 6 for required people. It shows that the objective functions have trade-off with each other.

4 Conclusions and Further Research

At last, the managers were informed about the results. They were satisfied to have personnel with multi-tasking abilities to enable their production planning more flexible. In addition, they were satisfied of pre-emptive resource constrained project scheduling approach because they became able to deal with their resources limits and scheduling and manage sudden offered projects from customers.

For future studies, it is recommended to use other uncertainty approaches like robust methods. In addition, personnel with different expertise may be modeled and evaluated with this model. And also, developing a multi-period model, considering time window of the model are good directions.in addition other recourses such as financial, equipment's and staff knowledge may be considered.

References

1. Tirkolaee, E.B., Goli, A., Hematian, M., Sangaiah, A.K., Han, T.: Multi-objective multi-mode resource constrained project scheduling problem using Pareto-based algorithms. Computing **101**(6), 547–570 (2019)
2. Pellerin, R., Perrier, N., Berthaut, F.: A survey of hybrid metaheuristics for the resource-constrained project scheduling problem. Eur. J. Oper. Res. **280**(2), 395–416 (2020)
3. Lin, J., Zhu, L., Gao, K.: A genetic programming hyper-heuristic approach for the multi-skill resource constrained project scheduling problem. Expert Syst. Appl. **140**, 112915 (2020)
4. Ma, W., Che, Y., Huang, H., Ke, H.: Resource-constrained project scheduling problem with uncertain durations and renewable resources. Int. J. Mach. Learn. Cybernet. **7**(4), 613–621 (2015). https://doi.org/10.1007/s13042-015-0444-4
5. Birjandi, A., Mousavi, S.M.: Fuzzy resource-constrained project scheduling with multiple routes: a heuristic solution. Autom. Constr. **100**, 84–102 (2019)
6. Torabi, S.A., Hassini, E.: An interactive possibilistic programming approach for multiple objective supply chain master planning. Fuzzy Sets Syst. **159**(2), 193–214 (2008)

Process Management

Improvements in Manufacturing Processes by Measurement and Evaluation Studies According to the Quality Management System Standard in Automotive Industry

Ali Mert Özcan[1], Anıl Akdoğan[1(✉)], and Numan M. Durakbasa[2]

[1] Mechanical Engineering Department, Yildiz Technical University, Istanbul, Turkey
mertaliozcan@gmail.com, nomak@yildiz.edu.tr
[2] Research Group Production Metrology and Adaptronic Systems, Institute for Production Engineering and Photonic Technologies, TU Wien, Vienna, Austria
numan.durakbasa@tuwien.ac.at

Abstract. The original equipment manufacturers and their suppliers have great importance in automotive manufacturing industry at all over the world. The industry has led the way by technological advancement, innovation and economic growth throughout the industrial history. The most important quality management standard known for the manufacturers of the automotive industry, which has such a great importance for the world, is the quality management system standard numbered 16949 prepared by IATF (International Automotive Task Force). The latest version of the standard in 2016 was created for the industry with the contribution of original equipment manufacturers, suppliers and interested parties. In this study; some basic quality techniques recommended by IATF 16949: 2016 were applied in one of the leading automotive supplier enterprises in Istanbul, Turkey. Some of these standard techniques like measurement systems analysis and data evaluation studies have been applied mainly on metrological activities and their effects on manufacturing processes have been observed. It has been firstly realized that the quality techniques helped to prevent the errors before they occur thus; it leads to increase the quality and the efficiency of the production. Additionally, the processes were better observed and better controlled. Moreover, the communication with the customers was more effective than before by mutually understanding. Finally, it became possible to identify the problems in measurements clearly by analyzing the measurement systems through scientific methods. The all helped to improve the manufacturing processes performance.

Keywords: IATF 16949 · Metrology · Measurement system analysis · Quality techniques · Automotive industry · Quality management · Manufacturing processes

N. M. Durakbasa and M. G. Gençyılmaz (Eds.): ISPR 2020, LNME, pp. 483–492, 2021.
https://doi.org/10.1007/978-3-030-62784-3_41

1 Introduction

The IATF 16949: 2016 standard was created to ensure the high-quality requirements in design to production processes in automotive industry to reach a prevented quality works published by the International Automotive Task Force (IATF: International Automotive Task Force). According to the IATF 16949: 2016, products can be manufactured all around the world in a common standard requirements and be produced using affordable qualifications from different countries. Quality improvement techniques are the basic tools of an effective quality management system. Among these improvement techniques; Advanced Product Quality Planning and Control Plan (APQP), Production Part Approval Process (PPAP), Error Mode and Effects Analysis (FMEA), Statistical Process Control (SPC) and Measurement Systems Analysis (MSA) are the requirements of the IATF 16949: 2016 standard [1]. Today, most automotive industry manufacturers and their suppliers require the use of one or more of these basic quality improvement techniques. Therefore; in such organizations which specialize in basic quality improvement techniques, processes can guarantee to do the industry's most qualified individual quality works and the requirements of the IATF 16949 which enables applicable quality reference.

In this study, it is aimed to approve the measurement accuracy of one of the automotive industry suppliers in Turkey by conducting some metrological and data evaluation studies. The metrological data collection and evaluation conducted by quality improvement tools. Therefore, the most important quality management standards' requirements are already concerned. Metrological processes, one of the most critical processes in manufacturing, were focused on and measurement systems in the shop floor were analyzed with MSA, one of the most important quality improvement tools.

2 Measurement Systems Analysis (MSA)

Measurement system analysis is defined as a statistical property of multiple measurements obtained from a measurement system operating under stable conditions. Comparison of one measuring device with another, ensuring the determination of a component necessary for calculating the process variation and the level of acceptability for a production process, determination of acceptance criteria of measurement equipment, comparison of operators performing the measurement, comparison of the instrument before and after repair and evaluation of a measuring instrument suspected of inadequate are the areas where measurement systems analysis is commonly used [2].

As it is a well-known metrological rule, the quality of the measurement results depends on two main factors; accuracy and precision. Accuracy is defined as the ability to accurately measure true value, while precision is defined by the closeness of measurement results to each other [3, 4]. Generally, measurements are assumed to be precise. The analysis and the results are often assumed to be based on this assumption.

One may not realize that there is a variation in the measurement system that affects measurements and then data-based decisions. But in fact, the measurement system error can be divided into five categories namely; Bias, Repeatability, Reproducibility, Stability and Linearity [5, 6]. In general, Gage RR (GRR) studies refer the evaluation of a measurement system. A basic gage RR study requires thee appraisers (Fig. 1). The total variances of a system is not limited to gauge but to all components of the measurement system like instruments and methods.

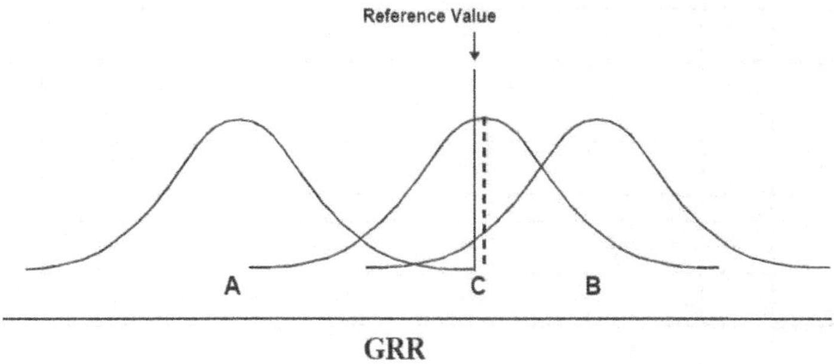

Fig. 1. GRR representation [2]

3 Application of Quantitative MSA

MSA and data evaluation studies were applied in an automotive supplier company with the collection and analysing the high quality data obtained mainly from the metrological activities of the company. The company is located in Istanbul-Turkey produces 500 different products in 17 main categories which are exported to more than 75 countries since 1979. The company also adds an average of 20 new products to its' product portfolio each year. With the help of having IATF 16949 and ISO 9001 Quality Management System Standard certificates, the company improved focus on performance evaluation studies [7, 8].

MSA applications were done for two different dimensional requirements in technical drawings. One of them is a hole diameter which is 5.1 ± 0.05 required. As the results of the repeated measurements done with a 1/1000 mm accurate calliper; the EV (Equipment Variation) value is 0.00331; the AV (Appraiser Variation) value is 0.00139; the PV (Part Variation) value is 0.01398 and the TV (Total Variation) is found 0.01444. The GRR value is calculated as 21,5% which is between 10% and 30%. In this study, a conditional acceptance decision is made. In subsequent measurement system analysis work the process can be closely monitored, and the system will continue to be improved continuously. The Number of distinct categories, ndc, is found as 5.49 and it is at the acceptance level since it is over 5 (Table 1).

Table 1. GRR study with 1/1000 mm caliper on a dimensional measurement

MEASUREMENT SYSTEM CAPABILITY

PART NAME Hinge
CHARACTERISTIC Diameter of pin hole
TOLERANCE 5.1 ±0,05
EQUIPMENT 0,001 mm Digital Caliper (016012)

APPRAISER		3		TRIAL		3		PART		10			
APPRAISER		PART										AVERAGE	
TRIAL		1	2	3	4	5	6	7	8	9	10		
APPRAISER A	1	5.105	5.100	5.130	5.090	5.100	5.090	5.080	5.090	5.120	5.100		5.101
2.	2	5.100	5.110	5.120	5.090	5.110	5.080	5.080	5.100	5.130	5.100		5.102
3.	3	5.110	5.100	5.120	5.100	5.100	5.080	5.080	5.090	5.120	5.100		5.100
4.	Average	5.105	5.103	5.123	5.093	5.103	5.083	5.080	5.093	5.123	5.100	Xa$_{av}$ =	5.1008
5.	Range	0.01	0.01	0.01	0.01	0.01	0.01	0.00	0.01	0.01	0.00	Ra$_{av}$ =	0.008
APPRAISER B	1	5.090	5.100	5.120	5.100	5.100	5.090	5.080	5.100	5.130	5.100		5.101
7.	2	5.090	5.110	5.120	5.100	5.110	5.090	5.080	5.100	5.120	5.100		5.102
8.	3	5.090	5.100	5.120	5.100	5.100	5.090	5.080	5.110	5.120	5.100		5.101
9.	Average	5.090	5.103	5.120	5.100	5.103	5.090	5.080	5.103	5.123	5.100	Xb$_{av}$ =	5.1013
10.	Range	0.00	0.01	0.00	0.00	0.01	0.00	0.00	0.01	0.01	0.00	Rb$_{av}$ =	0.004
APPRAISER C	1	5.114	5.110	5.100	5.100	5.100	5.090	5.080	5.110	5.130	5.100		5.103
12.	2	5.120	5.100	5.100	5.100	5.110	5.090	5.080	5.110	5.130	5.100		5.104
13.	3	5.120	5.098	5.100	5.090	5.110	5.090	5.090	5.110	5.130	5.100		5.104
14.	Average	5.118	5.103	5.100	5.097	5.107	5.090	5.083	5.110	5.130	5.100	Xc$_{av}$ =	5.1037
15.	Range	0.01	0.01	0.00	0.01	0.01	0.00	0.01	0.00	0.00	0.00	Rc$_{av}$ =	0.005
16. Part Average (X_p)		5.104	5.103	5.114	5.097	5.104	5.088	5.081	5.102	5.126	5.100	Xav$_{av}$ = / R_p =	5.1020 / 0.044
17.	(Raav = 0,008 + Rbav = 0.004 + Rcav = 0.005) / Appraiser = 3											Rav$_{av}$ =	0.0056
18.	(Max Xav = 5.1037) - Aax Xav = 5.1008) = Xdiffav							Trial	D_4	D_3 = 0			0.0029
19.	(Ravav = 0.0056) x (D_4 = 2.58) = UCL_R							2	3.27				0.0144
20.	(Ravav = 0.0056) x (D_3 = 0.00) = LCL_R							3	2.58				0.000

Measuring Equipment Analysis Tolerance : 0,100

% Total Variation (with reference to tolerance) (TA)

Repeatability - Equipment Variation (EV)

EV = Ravav x K1
= 0.0056 0.5908
= 0.00331

Trial	K1
2	0.8862
3	0.5908

% EV = 100 (EV / TA) 100 (EV/TV)
= 19.85 % 22.92 %

Reproducubility - Appraiser Variation (AV)

AV = √ (Xdiffav x K2)² - (EV² / nr)
= √ 0.000 - 0.00000
= 0.00139

Appraiser	K2
2	0.7071
3	0.5231

% AV = 100 (AV / TA) 100 (AV/TV)
= 8.3 % 9.64 %

Repeatability and Reproducubility (R&R)

GRR= √ (EV)² + (AV)²
= √ 0.000 + 0.000
= 0.00359

% GRR = 100 (GRR / TA) 100 (GRR/TV)
= 21.5 % 24.86 %

DECISION CONDITIONAL ACCEPTANCE

Part Variation (PV)

PV = R_p x K_3
= 0.044 x 0.31
= 0.01398

% PV = 100 (PV / TA) 100 (PV/TV)
= 83.9 % 96.86 %

Total Variation (TV)

TV = √ GRR² + PV²
= √ 0.000 + 0.000
= 0.01444

Part	K3
2	0.7071
3	0.5231
4	0.4467
5	0.4030
6	0.3742
7	0.3534
8	0.3375
9	0.3249
10	0.3146

ndc= 1.41* (PV / GRR)
= 5.49

DECISION : PART VARIATION - MEASUREMENT SYSTEM VARIATION IS ACCEPTABLE

The other dimensional requirement was 16 ± 0.05 sleeve diameter as nominal and tolerances. As a result of measurements made with a 1/1000 mm accurate caliper; the EV value is 0.00197; the AV value is 0.00038; the PV value is 0.01188 and the TV Value was found as 0.01205. The GRR value is calculated as 12.0% which is between 10% and 30%. In this study, a conditional acceptance decision has been made considering the costs of the measuring device like calibration and repair. In subsequent MSAs, the process can be closely monitored, and the system will continue to be improved continuously. The Number of distinct categories, ndc, is found as 8.36 and it is at the acceptance level since it is over 5 (Table 2).

Table 2. GRR study with 1/1000 mm caliper on a dimensional measurement

MEASUREMENT SYSTEM CAPABILITY

PART NAME Sleeve
CHARACTERISTIC Diameter
TOLERANCE 16±0,05
EQUIPMENT 0,001 mm Digital Caliper (015012)

APPRAISER	3		TRIAL		3		PART		10			
APPRAISER	PART										AVERAGE	
TRIAL	1	2	3	4	5	6	7	8	9	10		
1	16,010	15,970	16,000	16,000	16,010	16,010	15,990	16,000	16,010	16,020		16.002
2. 2	16,010	15,970	16,000	16,000	16,010	16,010	15,990	16,000	16,010	16,020		16.002
3. 3	16,010	15,960	16,000	16,000	16,010	16,010	15,990	16,000	16,000	16,010		15.999
4. Average	16.010	15.967	16.000	16.000	16.010	16.010	15.990	16.000	16.007	16.017	X_{aav} =	16.0010
5. Range	0.000	0.010	0.000	0.000	0.000	0.000	0.000	0.000	0.010	0.010	R_{aav} =	0.003
1	16,000	15,990	16,000	16,000	16,000	16,010	15,990	16,000	16,000	16,020		16.001
7. 2	16,000	15,990	15,990	16,000	16,000	16,010	16,000	16,000	16,000	16,020		16.001
8. 3	16,000	15,990	16,000	16,000	16,000	16,010	15,990	16,000	16,000	16,020		16.001
9. Average	16.000	15.990	15.997	16.000	16.000	16.010	15.993	16.000	16.000	16.020	X_{bav} =	16.0010
10. Range	0.000	0.000	0.010	0.000	0.000	0.000	0.010	0.000	0.000	0.000	R_{bav} =	0.002
1	15,990	15,990	16,000	16,000	16,000	16,010	15,990	16,000	15,990	16,020		15.999
12. 2	15,990	15,990	16,000	16,000	16,000	16,010	16,000	16,010	16,000	16,020		16.002
13. 3	15,990	15,980	15,990	16,000	16,000	16,010	15,990	16,010	16,000	16,020		15.999
14. Average	15.990	15.987	15.997	16.000	16.000	16.010	15.993	16.007	15.997	16.020	X_{cav} =	16.0000
15. Range	0.00	0.01	0.01	0.00	0.00	0.00	0.01	0.01	0.01	0.00	R_{cav} =	0.005
16. Part											X_{avav} =	16.0007
Average (X_p)	16.000	15.981	15.998	16.000	16.003	16.010	15.992	16.002	16.001	16.019	R_p =	0.038
17.	(R_{aav} = 0,003 + R_{bav} = 0,002 + R_{cav} = 0,005) / Appraiser = 3										R_{avav} =	0,0033
18.	(Max Xav = 16.0010) - (Min Xav = 16.0000) = Xdiffav							Trial	D_4	D_3 = 0		0.0010
19.	(R_{avav} = 0.0033) x (D_4 = 2.58) = UCL_R							2	3.27			0.0086
20.	(Ravav = 0.0033) x (D_3 = 0.00) = LCL_R							3	2.58			0.000

Measuring Equipment Analysis Tolerance : 0,100

% Total Variation (with reference to tolerance) (TA)

Repeatability - Equipment Variation (EV)

EV = R_{ortort} x K_1
= 0,0033 0.5908
= 0,00197

Trial	K_1
2	0.8862
3	0.5908

% EV = 100 (EV / TA) 100 (EV/TV)
= 11,82 % 16.34 %

Reproducibility - Appraiser Variation (AV)

AV = √ (Xdiff$_{av}$ x K_2)^2 - (EV^2 / nr)
= √ 0,000 - 0,00000
= 0,00038

Appraiser	K_2
2	0.7071
3	0.5231

% AV = 100 (AV / TA) 100 (AV/TV)
= 2,3 % 3,15 %

Repeatability and Reproducibility (R&R)

GRR= √ $(EV)^2$ + $(AV)^2$
= √ 0,000 + 0,000
= 0,00201

% GRR = 100 (GRR / TA) 100 (GRR/TV)
= 12,0 % 16,64 %

DECISION : CONDITIONAL ACCEPTANCE

Part Variation (PV)

PV = R_p x K_3
= 0,038 x 0,31
= 0,01188

Part	K_3
2	0.7071
3	0.5231
4	0.4467
5	0.4030
6	0.3742
7	0.3534
8	0.3375
9	0.3249
10	0.3146

% PV = 100 (PV / TA) 100 (PV/TV)
= 71,3 % 98.61 %

Total Variation (TV)

TV = √ GRR^2 + PV^2
= √ 0,000 + 0,000
= 0,01205

ndc= 1.41* (PV / GRR)
= 8.36

DECISION : PART VARIATION - MEASUREMENT SYSTEM VARIATION IS ACCEPTABLE

4 Application of Qualitative MSA

Qualitative analysis was performed on one of the dimensional measurements of a manufacturing part. Three different operators each are three times done inspections in this work. '1' input was used if the operators' decision is appropriate or ok, and '0' input was used if their decision is not-ok according to the inspections. As the same understanding, reference values were entered as '1' or '0'. As a result, 39 "ok" (positive) results and 11 "not-ok" (negative) results were obtained (Table 3).

Table 3. Qualitative MSA measurement data

Inspection Nr.	Operator Ü.K.			Operator C.B.			Operator B.B.			Reference	Result
	1st	2nd	3rd	1st	2nd	3rd	1st	2nd	3rd		
1	1	1	1	1	1	1	1	1	1	1	+
2	1	1	1	1	1	1	1	1	1	1	+
3	1	1	1	1	1	1	1	1	1	1	+
4	1	1	1	1	1	1	1	1	1	1	+
5	1	1	1	1	1	1	1	1	1	1	+
6	1	1	1	1	1	1	1	1	1	1	+
7	1	1	1	1	1	1	1	1	1	1	+
8	1	1	1	1	1	1	1	1	1	1	+
9	1	1	1	1	1	1	1	1	1	1	+
10	1	1	1	1	1	1	1	1	1	1	+
11	1	1	1	1	1	1	1	1	1	1	+
12	1	1	1	1	1	1	1	1	1	1	+
13	1	1	1	1	1	1	1	1	1	1	+
14	1	1	1	1	1	1	1	1	1	1	+
15	1	1	1	1	1	1	1	1	1	1	+
16	1	1	1	1	1	1	1	1	1	1	+
17	1	1	1	1	1	1	1	1	1	1	+
18	1	1	1	1	1	1	1	1	1	1	+
19	1	1	1	1	1	1	1	1	1	1	+
20	1	1	1	1	1	1	1	1	1	1	+
21	1	1	1	1	1	1	1	1	1	1	+
22	1	1	1	1	1	1	1	1	1	1	+
23	1	1	1	1	1	1	1	1	1	1	+
24	1	1	1	1	1	1	1	1	1	1	+
25	1	1	1	1	1	1	1	1	1	1	+
26	1	1	1	1	1	1	1	1	1	1	+
27	1	1	1	1	1	1	1	1	1	1	+
28	0	0	0	0	0	0	0	0	0	0	−
29	1	1	1	1	1	1	1	1	1	1	+
30	1	1	1	1	1	1	1	1	1	1	+
31	0	0	0	0	0	0	0	0	0	0	−
32	1	1	1	1	1	1	1	1	1	1	+
33	1	1	1	1	1	1	1	1	1	1	+
34	0	0	0	0	0	0	0	0	0	0	−

(*continued*)

Table 3. (*continued*)

Inspection Nr.	Operator Ü.K.			Operator C.B.			Operator B.B.			Reference	Result
	1st	2nd	3rd	1st	2nd	3rd	1st	2nd	3rd		
35	1	1	1	1	1	1	1	1	1	1	+
36	0	0	0	0	0	0	0	0	0	0	−
37	0	0	0	0	0	0	0	0	0	0	−
38	1	1	1	1	1	1	1	1	1	1	+
39	0	0	0	0	0	0	0	0	0	0	−
40	1	1	1	1	1	1	1	1	1	1	+
41	0	0	0	0	0	0	0	0	0	0	−
42	1	1	1	1	1	1	1	1	1	1	+
43	0	0	0	0	0	0	0	0	0	0	−
44	1	1	1	1	1	1	1	1	1	1	+
45	0	0	0	0	0	0	0	0	0	0	−
46	1	1	1	1	1	1	1	1	1	1	+
47	0	0	0	0	0	0	0	0	0	0	−
48	1	1	1	1	1	1	1	1	1	1	+
49	1	1	1	1	1	1	1	1	1	1	+
50	0	0	0	0	0	0	0	0	0	0	−

The three measurement operators were also compared to each other to determine the measurement abilities. The first operator namely Ü.K. and the second operator namely C.B. were subjected to the first comparison. The comparison results of the operators are shown in Table 4.

Table 4. MSA – Operator comparison-1

			Operator C.B.		Total
Operator Ü.K.	0	Count	33	3	36
		Expected count	8	28	36
	1	Count	0	117	114
		Expected count	25	91	114
Total		Count	33	117	150
		Expected count	33	117	150

The second operator namely C.B. and the third operator namely B.B. were subjected to the second comparison. The comparison results of the operators are shown in Table 5.

The first operator namely Ü.K. and the third operator namely B.B. were subjected to the third comparison. The comparison results of the operators are shown in Table 6.

Table 5. MSA – Operator comparison - 2

			Operator B.B.		Total
Operator C.B.	0	Count	33	0	33
		Expected count	7	26	33
	1	Count	0	117	117
		Expected count	26	91	117
Total		Count	33	117	150
		Expected count	33	117	150

Table 6. MSA – Operator comparison - 3

			Operator B.B.		Total
Operator Ü.K.	**0**	Count	33	0	36
		Expected count	8	28	36
	1	Count	0	114	114
		Expected count	25	89	114
Total		Count	33	117	150
		Expected count	33	117	150

The data obtained from the comparison were analyzed in the Kappa coefficient interpretation which is used to assess the inter-rater reliability or agreement. The results are shown in Table 7. The used formula is given in the Eq. (1) where P_0 is the sum of the observed proportions in the diagonal cells and P_c is the sum of the expected proportion in the diagonal cells.

$$kappa = \frac{P_0 - P_c}{1 - P_c} \tag{1}$$

The general rule in interpreting Kappa values is to evaluate the coefficient as 0.75 and above accepted as perfect agreement, between 0.40–0.75 as moderate-good, and below 0.40 as poor fit [2]. Perfect agreements found between the operators. The results we have obtained with reference here are shown in Table 8.

Table 7. MSA – Kappa results

Operators	Ü.K.	C.B	B.B
Ü.K.		0.94	0.94
C.B.	0.94		1
B.B.	0.94	1	

Table 8. MSA – Kappa agreement results

Comparison of operators	Kappa agreement
Ü.K. Vs. C.B.	Perfect
C.B. Vs. B.B.	Perfect
Ü.K. Vs. B.B.	Perfect

5 Conclusions

In this study, some of the quality techniques namely Measurement Systems Analysis and data evaluation studies recommended by the IATF 16949: 2016 quality management system were applied in a company work for the automotive industry.

As a result of the measurement and analyzing studies, it has been observed that the applied quality techniques increased the manufacturing performance and helped to prevent errors before they occur. Thus, it can increase the process quality and efficiency. That is of course, firstly because of the better observation of the processes and the better control capability than before. Additionally, the better communication supplied with the customer and enterprise is increased by mutually understanding. Finally, it becomes possible to identify the problems in the measurement system clearly by analyzing the measurement systems used by the company by scientific methods. As a small section is presented in this study, one of the biggest reasons of the companies' IATF 16949: 2016 certifications maintain successfully is the correct application of the quality improvement techniques and their sustainability.

References

1. IATF: IATF 16949: 2016 Automotive Quality Management System (2016)
2. M. Workgroup, MSA Manual (2010)
3. Pandiripalli, B.: Repeatability and reproducibility studies: a comparison of techniques (2010)
4. Corrigendum concerning standard, VDI/VDE 2600 Inspection process management Part 2: Determination of the measurement uncertainty of complex inspection processes (2019)
5. TSI, Statistical Quality Control Series of Official Statistics with Questions - 11, Turkey Statistics Institution Press (2011)
6. Adamczak, S., Świderski, J., Dobrowolski, T., Durakbasa, N.: Application of MSA analysis for the evaluation of instruments for measuring the surface texture. Mechanik **91**(5–6), 434–436 (2018)

7. Ozcan, M.: Quality improvement techniques based on international automotive task force 16949 quality management system and measurement system analysis, Master Science thesis, Graduate School of Science and Engineering, Yildiz Technical University, İstanbul, Turkey (2020)
8. Ozcan, M., Akdogan, A., Karakaya, C., Coban, E.: İmalat Proseslerinde IATF 16949 KYS Standardına Uygun Araçlarla Kalite İyileştirme ve Sertifikasyon, Balkan Zirvesi 2. Uluslararası Uygulamalı Bilimler Kongresi, Edirne, Türkiye, Mayıs (2020). ISBN 978-605-06766-2-4

Improving Business Processes of Human Resources Departments in Turkish Technic Inc.

Hande Nur Şavlı, Havva Burke Barot, Mirace Bengü Yüksel, Rumeysa Ekşi(✉), Bahadır Tunaboylu, and Selim Zaim

Department of Industrial Engineering, Istanbul Sehir University, Istanbul, Turkey
{handesavli,havvabarot,miraceyuksel,
rumeysaeksi}@std.sehir.edu.tr,
{btunaboylu,selimzaim}@sehir.edu.tr

Abstract. The aim of this project is to analyze and reorganize the current processes of the relevant company and to make improvements to increase the efficiency of the processes as a result of the studies and to manage the process flow of the Human Resources Departments in Turkish Technic INC. Current processes are prepared according to five thousand employes. During this period, the current number of employees of the company reached fifteen thousand employees. For this reason, these processes have been rearranged according to the current number of employees in the company. Process flow charts have been prepared by us using a program called VISIO. In the program, the swimlane diagram was used. With this diagram, it was shown which departments were doing what and which passive departments were in the flow. When using this program, the functional process flow chart has been chosen as a form in order to comply with processes. Possible risks were identified with the help of the prepared processes. Control situations to be done were determined. The main purpose is to reach ISO 9001 standards regarding the quality of existing processes. Project will consist of three stages. First, the processes were documented. Then, risks and directions to be developed based on ISO 9001 standards were determined. Improvements were applied in the last stage. As a result, the processes were reviewed and new forms prepared considering the risks were taken, and instruction lists were created for each process, and as a result of these instructions, the processes were adapted to a certain standard. As a result of the studies, all processes have been standardized.

Keywords: Process improvement · Process flow chart · Human resources · Process analysis · Quality standards · Functional flow chart · Swimlane diagram · Digitalization · Standardization

1 Introduction

The process is briefly expressed as "the process of converting existing inputs into outputs through some operations", as reference [1] shows. In order for management approaches to succeed, process management must be done flawlessly. For perfect process management, it is essential to analyze the process first. Process analysis

N. M. Durakbasa and M. G. Gençyılmaz (Eds.): ISPR 2020, LNME, pp. 493–503, 2021.
https://doi.org/10.1007/978-3-030-62784-3_42

determines the positive and negative aspects of the process and identifies the sections and activities that create or do not create added value in the processes. It provides great benefits in terms of time, cost and customer satisfaction, eliminating activities that do not add value. For these reasons; The work flow charts and processes created by Turkish Technic Inc. are inadequate today because current processes are prepared for 5000 employees since the company is a large company with 15000 employees and the process analysis and workflow diagrams are one of the biggest reasons. In a study (ref), [2], "The process is a transformation that adds value to the product, and the effectiveness of this transformation will give the efficiency of use of its resources". Therefore, the risks arising from the fact that the Human Resources processes, which have not been addressed for about 5 years, need to be improved and some processes no longer do business and most of the work is done by hand are added to the main reasons for the start of this study. In this study; Firstly, the project is to examine all processes in the Human Resources department of Turkish Technic Inc., to draw the current process flow with the Visio program and to make the necessary improvements. Considering the preliminary examination, one of the biggest deficiencies of the company is the manual progression of the processes. The aim of the project is to eliminate error rates and increase efficiency by manually converting these processes to digital. In the initial research conducted in the related company, 150 processes connected to the Human Resources Department were identified. The main objective of this project is to move all processes to a better location, to create workflow diagrams and to make improvements and test them to eliminate the risks that may occur in these diagrams. This test will determine your success rate and decide if you want to make a new improvement. The following processes and project objectives will be discussed in relation to the most important problems and their problem-solving objectives, depending on the preliminary work in the relevant company. Two processes to be improved by digitization will be discussed; Firstly, the annual process is one of the manual processes. The aim of this process is to digitize this process, which is progressing manually, and to integrate these processes with SAP. The second process is the Recruitment process. By digitizing this process, the recruitment department has significantly reduced the workload of employees and improved the overall recruitment process over time. Another factor is that all the personnel in the company are informed about the processes without the need of other personnel while they are drawing the work flow charts and they can perform the steps to be followed in the processes related to them (such as permission identification, absenteeism information). In addition, not only the existing staff but also the newly hired staff will be provided with detailed information about their work and will enable them to follow the work more efficiently in less time.

2 Lıterature Revıew

All employees of the organization, workers, engineers, managers, entrepreneurs and union members play a key role in increasing productivity. Each role has the direction of being appropriate and effective. Efficiency refers to the most effective way to fulfill the definition of roles. The role of people in ensuring efficiency is to provide effectiveness is an example of a book in [3]. This effectiveness is the performance of human input to achieve the desired result. Ensuring effectiveness is the ability of characteristics such as methods, techniques, personal skills, knowledge, behavior and ability to reflect the output. In a study (ref), [3], in order to ensure efficiency, training, workflow and rotation and placement of the right resources for the right job can be increased by job and process development and career planning. In a study (ref), [4], organizations take into account the human resources' knowledge, skills, capacity, flexibility and success in terms of the performance of the task. The three main factors determining the productivity of organizations are: 1. The quality level of labor in the process, 2. Time spent for the process, 3. Technology used in the process.

In order to ensure efficiency, organizations tend to undergo a major analysis and require great effort and time. Therefore, the method to be applied is important. In a study (ref), [5], "Human resources management has two main objectives. These can be listed as follows:- Increase productivity,- Improving the quality of work life."

Human Resources Management is of great importance in increasing productivity. A journal article in [6] stated that "In today's markets, an effective human resources power is needed to achieve the targets, produce high quality and provide more economic power and continuity". Therefore, many methods have been established and studies have been made on Human Resources Management and Development. Models on Strategic Human Resource Management; It is grouped under two main roofs: Twinning Strategic Human Resources Management Model and Resource Based Strategic Human Resources Management Model. The Twinning Strategic Human Resources Management Model is related to the matching of processes with the concepts of personnel selection, degree of success, personnel evaluation, reward and development, and this strategy and structure is influenced by external factors such as politics and economy. Therefore, in a study (ref), [7], by using SWOT analysis, the strategic importance of the external environment in the matching model is emphasized. The Resource Based Strategic Human Resources Management Model focuses on the basic workforce of the organization. Develops a management model for the determination of this value and the problems that arise in this regard. This model is defined as the collection of productive resources of the company and provides the resources of the organization in determining its capabilities. In a study (ref), [7], according to SWOT analysis, the resource-based model draws attention to the strategic importance of the strengths and weaknesses within the enterprise. Some sub-strategies are used in Strategic Human Resources Management. These: human resource planning, performance evaluation, organizational development, pricing and rewarding, in-service training and career development.

Among these, the main topics we will deal with the main theme of Human Resources Process Development are Strategies for Human Resources Planning and Strategies for Performance Evaluation. In order to ensure the highest efficiency of Human Resources Planning, the most important feature is keeping up to date with technology, age, changes and markets. It should not only keep up-to-date, but also take future-oriented measures, make possible market, market, labor force forecasts in the future, and take measures to meet the labor force with the right quality and number of resources and personnel in line with these forecasts. Organizations' plans for the operation are provided by hired personnel. If the right staff is not selected, the organization's plans unfortunately fail. Reference [8] shows, a strong and accurate Human

Resources Planning Organization is a very important activity in terms of efficiency. Strategies that ensure the conformity of organizations to their basic strategy, adaptation of knowledge and skills within the organization, measurement and evaluation of the level of workability are called Performance Evaluation Strategies. Thanks to these strategies, the gains gained as a result of the evaluations are taught to the personnel and guidance is provided for them to act towards the basic strategy. These gains have a significant impact on the determination of the business strategy. Therefore, these strategies should be considered by organizations and improvements should be made accordingly. Basic decisions such as workflows, training and development, personnel placement and performance evaluation are among the Human Resources management decisions and the strategy is determined accordingly. Among these, performance evaluation decisions are related to measurement and evaluation studies about how good they are in terms of accomplishing the task and accomplishing the task given to both employees, chiefs and the relevant department of the organization. Many enterprises/organizations have found different solutions for performance evaluation and have progressed in line with this in a study (ref), [9], these are grouped with complementary concepts and can be listed as follows:

- "Developing a special assessment system tailored to the characteristics and needs of various employee groups/using a uniform assessment system throughout the organization,
- Use evaluation results as a tool to improve employee performance/use evaluation as a control mechanism to extract low performances,
- Designing the evaluation system for various purposes (such as training, promotion, selection, career)/designing it for a limited purpose (e.g. remuneration),
- To design an evaluation system that envisages the active participation of various people and groups/to establish an evaluation system based solely on the opinion of the employee's top manager"

Workflow diagrams and methods are important in determining the most appropriate method for problem definition and objectives among these performance evaluation methods. A block diagram is used to specify some computational hardware items used to implement certain embodiments of the process. According to one embodiment of the process, the diagram showing an application of the workflow management system is called the message flow diagram. The diagrams showing the operation of the process are called work flow diagrams. Among these workflow diagrams, BPMN (Business Process Management Notation), which is the workflow modeling language, allows the modeling of corporate business processes on a common infrastructure. A process should have five basic features, as reference [10] shows;

- Identifiably: It is a feature that identifies the basic elements of the process.
- Measurability: It is the ability of the process to be monitored with performance criteria.
- Repeatability: It is the ability of the output formed as a result of the processing of the same or changing inputs that drive the process continuously to meet the needs and expectations of the customer

- Controllability: The ability of process managers to always know about the performance of the process and to take corrective actions when necessary.
- Value-added: The ability of the process to have a positive impact on the quality of the output and the satisfaction of the customer using the output.

Using this notation in a corporate firm, all employees involved in processes, including managers and staff, will be able to access all workflow data available in the firm and control these processes and analyze their results.

3 Methodology

3.1 Constraints

The main limitation of the project is the examination of the processes of the Human Resources department of Turkish Technic Inc. This greatly limits the project because Human Resources processes are interrelated or interfere with the processes of other departments. Since it is not possible to clearly understand the diagrams of the workflows of other processes, it is difficult to develop processes related to other departments of Human Resources. Another limitation is that it is not known which process in the section needs to be improved. Therefore, each process should be examined individually. Furthermore, the targeted completion date of the project is June 2020, the deadline for the evaluation of the project results. Therefore, by 2020 June, a flow diagram of all processes should be drawn, improvements should be integrated into processes, final processes should be printed and evaluated. The optimized process should be tested for at least 2 months to remove the results from the processes and evaluate the results. The third basic limitation is the Turkish Technic subsidiary, where all software processes are submitted for approval by the parent company, THY Inc. Turkish Technic cannot directly apply every decision it decides to make and implement. The approval process makes this difficult. Since the reliability of the company is more important than the price/performance ratio of the programs purchased for the processes mentioned above, there is very little choice to select the companies where the programs are purchased.

3.2 Analysis and Research

SWOT Analysis for Creating Process Flow Diagrams

SWOT analyzes of the determined process flows were made. Analyzes are made to facilitate the follow-up of the processes and to improve the quality of work. In a study (ref), [13], Performance criteria define effective methods and tools for determining the performance standard of the business, which is used to increase the strengths of the system and minimize its weaknesses. As mentioned in the quote, determining the KPIs of the processes has helped SWOT analysis in a way. Some information was obtained as a result of the SWOT analysis. Examples of weaknesses identified from these; Understanding and drawing all the processes in the department takes time, as the appropriate time is not formed with the people working in other departments related to the processes. In line with some information obtained from the processes drawn, the manual steps in the process remain manual, making the work more efficient. No software improvement can be made in such cases. Some opportunities identified are; It has facilitated the follow-up, execution and control of all processes drawn depending on the department.

Threats from the analysis results are; errors arising from the possibility of incorrect transfer of the drawn processes. For example; A process that will be drawn incorrectly may be misinterpreted at later stages and pose a greater risk.

1) *SWOT Analysis of Current Operation of Annual Leave:* With the Annual Leave Planning Program, which was prepared within the scope of the project, the current process has been digitized. Before the program, employees faced several strengths like lack of ability to manage annual leave plan changes systematically and demonstrating the necessary data for the conscious choice of leave. Although these strengths were solved with the program, some new problems emerged. One of them is lack of assignment and scoring algorithm to extend and complicate Software Development Process. Weakness of the program is if the employee does not provide the use of the accumulated leave with all the leave days he/she deserves in a year. Opportunity for the program is the company's rapidly growing staff. Lastly the threats of the current process like integrated structure of all steps from beginning to end of the process, frequency of site daily plan changes, changes in the number of personnel and personnel mobility and reorganizations between units and plan as if the data such as flight and training data are clear one year ago have eliminated with the improvements. Threat of the new program is that there is no restriction and follow-up by the employee on annual leave plan changes.
2) *Productivity Analysis of Annual Leave Planning Program:* In the improvement made within the scope of the annual leave planning process, the goal is to save man-hour. Annual Leave Planning is carried out through excel staff in the current process and the workload belongs to Personnel Relations Department. In order to prevent these situations and minimize the problems, the Annual Leave Planning process has been digitized. Responsible personnel determines the day quota of the company's 15,000 employees and permits for all departments. This process (since the year is divided into 4 periods) takes place 4 times a year. The personnel performing this process say that they spend 5 working days for each period. 5 working day, because of 1 working day is 8 h, means 40 h in total. As stated, this process be done 4 time, therefore 160 man-hour spent for this job in annual man-hour which is 2000 h. In this case; 1 employee spends 8% of the annual manhour for this job. This is quite a workload. Because of the improvements, the determination of quotas will be made from the SAP system, also 1 time in a year. The expectation was that this process would take 24 h and decrease the spent to 1.2% of annual man-hour. However, studies have shown that responsible personnel was able to finish the process in 38 h. This means that 1 worker requires 1.9% man hour per year to finish this process. As a result; 6.1% man-hour have been saved for annual man-hour.
3) *Criteria, Company Research and Analysis for Recruitment Software:* As a result of the work done, it was decided to switch the processes from manual to digital in order to shorten the recruitment process. With the improvement to be made, the recruitment process, which lasts 8 months in total, is expected to be shortened. With digitalization, it is aimed to reduce the recruitment process from 8 months to a

maximum of 3 months. The number of days with no waiting period corresponds to 125 working days and this has been reduced to 25 working days for the recruitment process. With budget research, it was determined that it is more appropriate to outsource the software. Following this decision, researches for suitable software companies were initiated and the criteria required for Recruitment Software were prepared to be presented to companies. These criteria were collected under 11 main headings. According to the feedback from the companies, it was decided to work with the company. In one study (ref) [12], transitioning to digitalization in the human resources department has been an effective way to reduce the time and money spent on organizing that department.

4) *Productivity Analysis of Recruitment Process Program:* One of the improvements made within the scope of the project is the improvement of the Recruitment Process. This improvement is mainly through the system of all processes from the applications of candidates (preparation, announcement, candidate application, exam, interview, paperwork completion, start-up, etc.) and recruitment processes in the General, Cabin, Overseas, Cockpit and Technician recruitment process. to provide a system that includes implementation, reporting and integration with other systems. Currently, the process is carried out through the current program. In general, the workload belongs to the Employment and Career Management department. An average of 6000 people are employed annually in the company. A total of 6 in-company personnel are involved with the people hired. The average duration of the staff working in the recruitment department is 5 h. With the program to be made, this period is reduced to 2 h. Finally, the average salary of the staff working in this department is 7000 TL. Thanks to the SAP program paid before the program by making the necessary calculations, the amount paid to the company employees for the total time they spent on this job is 216.000 TL. However, with this software program, this amount to be paid has decreased to 100.800 TL. As a result, the company made a 53.3% profit for this process thanks to the SAP program.
5) *Creating the Content and Structure of the Instructions:* Job descriptions were deemed necessary in order for the jobs to progress faster and employees to be more productive. In this, it was decided to prepare instructions. Written job descriptions enable employees to know what to do officially and require them to do so. *In a study (ref),* [13], through these guides, employees are informed about the behaviours, tasks and procedures expected from them. With instructions to be prepared, employees can see what is expected of them in writing and follow what they need to do step by step. In this way, it becomes a standard in work-led processes. It is easy to evaluate and follow. In this way, workflow speeds up and productivity increases.
6) *Implementation of 5S:* The processes that used is still the same since the company founded. In a study (ref), [13] shows that according to thermodynamics, Entropy is bigger than zero, which means that disorder is always greater than zero, so everything in the universe goes to disorder. The business environment is also prone to the disorder. In a study (ref), [14]; therefore, the existing processes are increasingly irregular and disrupted. Modern management means not only the quality management system based on the ISO series 9000: 2000 standards, but also pursuit to the continuous improvement. Processes should be reconsidered and improved to ensure continuous improvement. At this content, the philosophy of the

5S method adopted as a principle. 5S is the method for maintaining the improvement. In a study (ref), [14], 5S creates and maintains a well organized, clean, high effective, and high-quality workplace and safety of work. Although 5S is generally used for physical improvements in workplaces, in this project has been used to improve process flows and provide continuous improvement. In this content, "Sort, Set in order, Shine, Standardization, Sustain" steps which are the basis of the 5S method have been adapted to the project and process diagrams are treated as workplaces.

7) *Application of Kaizen Technique to All Improvement Processes:* The Kaizen philosophy means everyone working from the lowest level to the highest level to be found in the process and the improvement to be developed in this process. For this reason, everyone's role was determined in the processes, their contributions were laid out in detail and unnecessary workload or staff were eliminated accordingly. In Kaizen, instead of team leader and in-team competition, providing the best efficiency with the activity of the whole group is the main goal. It has a collective approach away from individuality. For this reason, it has been preferred to use this method in Human Resources because Human Resources is a human-oriented region. Process improvement is essential for better results in the Kaizen. So that, the processes are handled one by one and drawn to the finest detail in Visio, and meetings were held with the process owner team while process diagrams were being drawn. Reference [15] shows that Kaizen is focused on being better than the position where the process is located. That is "process improvement". With the effect of Kaizen, this study has adopted the method of dividing the sections into small parts, determining the processes clearly and moving these processes to the maximum position. In particular, Kaizen was applied in more detail in the Annual Leave Application and Recruitment processes. The aforementioned elements have been the basic principles that are emphasized in detail in addressing all processes. All processes were effectively analyzed and unnecessary steps were eliminated. The steps that created unnecessary handicaps in the number of personnel were changed. While some steps were taken care of with 3 staff, they were given to one staff. If the personnel with a connection point unnecessarily prolonged communication in a process, direct connection was provided. Unnecessary repetitions were prevented especially in approval processes and necessary preventive and supervisory measures were taken in this regard. Decision points were determined and processes around decision points were determined. The processes based on the simultaneous work of many staff and designed to be completed as soon as possible. It was aimed to complete the processes in the most efficient way with full team work. In some steps, since the work time of the staff created a lot of time gaps, technology was placed and the work was done with the least amount of staff as soon as possible. Process development principles were determined by considering the total quality control, suggestion system, automation, workplace discipline, total productive maintenance, zero error, small group activities, worker management cooperation, productivity improvement, new product development units under the umbrella of Kaizen philosophy.

3.3 Main Design

Functional flow chart selected while using the program which is called VISIO chart to show the processes. Actors were written in the charter and the duties assigned to the actors were entered under them. The pillars separated the tasks performed by the actors. Selected orange arrows make it easier to trace the path of the process. The events realized were represented with square boxes, the questioning/decision stages were represented with rectangles. The start and end are specified in oval shapes. Red triangular shapes are shown with identified risks and green circles represent quality control points. Notes are shown in a yellow square box. Calibri 11 was used as font size and type (Fig. 1).

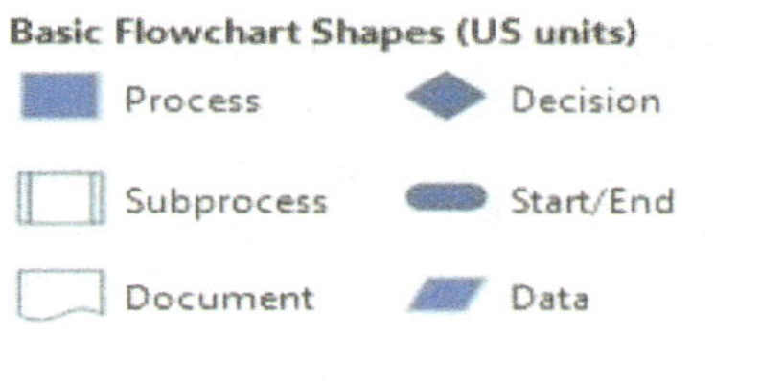

Fig. 1. Basic flow shapes

In the below image (Fig. 2) basic flowchart shapes are shown to decide which function to use depending on the job.

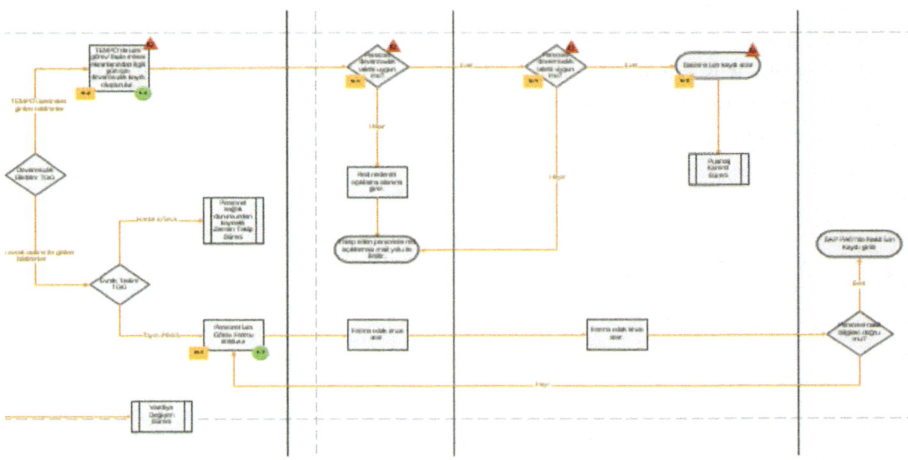

Fig. 2. Process flow chart

Process flow monitoring was provided by arrows. Process-related risks, quality criteria and notes are explained in detail under the Transaction Card title at the bottom of the diagram.

4 Conclusıons

This study was created through comprehensive Process Analysis, Human Resources Management, and reviewing all the studies in the literature that may or may be used in process analysis. As analysis and research methods, SWOT Analysis, Productivity Analysis, Key Success Factors, 5S and Kaizen were investigated and applied to improve the processes as much as possible. The position of the personnel has been determined, unnecessary personnel has been identified, the division of labor has been made more efficient, short contacts have been established between the personnel, the approval process between the manager and the personnel has been accelerated, and most importantly, what has been done exactly in the process has been achieved. For nearly 5 years, the risks arising from some processes no longer doing business and most work being done manually have been addressed. As a result of the studies, it was determined that the biggest problem for the company was the manual operation of the processes in line with the observations and experiences of the personnel. According to the new regulations here, work flow diagrams were drawn and instructions were prepared. The main purpose of this project was to move all processes to a better place, to create work flow diagrams and to make improvements and to eliminate the risks that may occur in these diagrams. As a result of the meetings and researches, the success rate of the improvements to be made was determined. When a result is obtained below this success rate determined when the improvements are applied, it is possible to make improvements again. However, as can be seen from the results of the above analysis, the success rate is satisfactory. Finally, the results realized within the scope of this project can be listed as follows;

1- Work flow charts of 150 processes affiliated to the Human Resources department were drawn,
2- A program was created for the monitoring and control of Annual Leave and the transition of the leave path to digital was achieved. Figure 3, an example of the program screens,

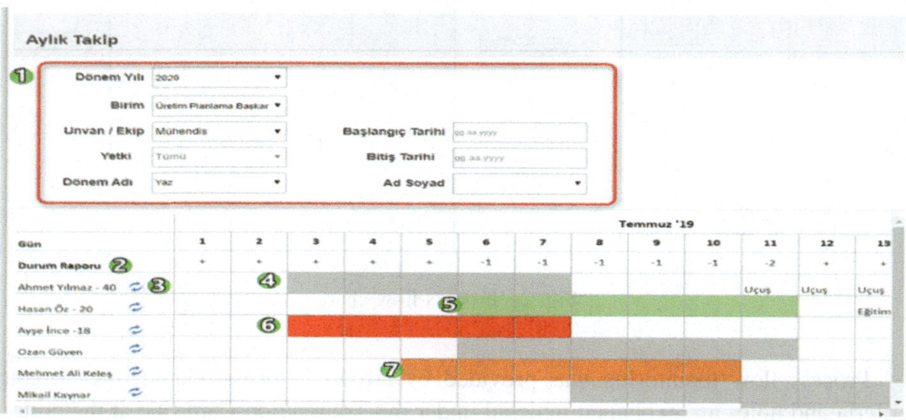

Fig. 3. Annual leave programme

3- The recruitment process has been changed from manual to digital, this process has been shortened.

References

1. Kurt, H.M.: Altı sigma ve istatistiksel uygulamaları (Doctoral dissertation, Fen Bilimleri Enstitüsü) (2008)
2. Soyateş, A.: Toplam Kalite ve İstatistiksel Süreç Kontrol, (Lisans Tezi, D.E.Ü.Müh.Fak. End.Müh.Böl.,İzmir) s. 11 (2003)
3. Prokopenko, J.: Verimlilik Yönetimi, Uygulamalı El Kitabı. Ankara: Milli Prodüktivite Merkezi Yayınları. Çevirenler: Olcay Baykal, Nevda Atalay, Erdemir Fidan (1995)
4. Edvinsson, L., Malone, M.: Intellectual Capital. Harper Collins Pub, New York (1997)
5. Kaynak, T., Adal, Z., Ataay, İ., Uyargil, C., Sadullah, Ö., Acar, A. C., Uluhan, R.: İnsan kaynakları yönetimi. İstanbul Üniversitesi İşletme Fakültesi Yayınları, pp. 278–289 (1998)
6. Filip, F.C., Marascu-Klein, V.: The 5S lean method as a tool of industrial management performances. IOP Conf. Ser. Mater. Sci. Eng. **95**(1), 012127 (2015)
7. Bilensoy, S.: Stratejik İnsan Kaynakları Yönetimi Yayınlanmamış Yüksek Lisans Tezi. Ankara Üniversitesi Sosyal Bilimler Enstitüsü İnsan Kaynakları Ve Kariyer Danışmanlığı Tezsiz Yüksek Lisans Programı (2004)
8. Palmer, M., Winters, K.T.: İnsan Kaynakları. İngilizceden Çeviren: Doğan Şahiner. Rota Yayınları, İstanbul (1995)
9. Barutçugil, İ.: Stratejik İnsan Kaynakları Yönetimi. Kariyer Yayıncılık, İstanbul (2004)
10. ÖZAY, H.E.: Süreçlerle Yönetim Sistemi ve Orta Ölçekli Bir İşletmede Uygulanması, D.E. Ü. Sosyal Bilimler Enstitüsü Yayımlanmamış Yüksek Lisans Tezi, İzmir (2003)
11. Türkan, Ö.U.: Üretimde yalın dönüşümün temel performans kriterleri. Balıkesir Üniversitesi, Fen Bilimleri Enstitüsü Dergisi **12**(2), 28–41 (2010)
12. Güler, E.Ç.: İŞLETMELERİN E-İNSAN KAYNAKLARI YÖNETİMİ VE E-İŞE ALIM SÜREÇLERİNDEKİ GELİŞMELER. Ege Akademik Bakış Dergisi **6**(1), 17–23 (2006)
13. Grote, D.: Performance evaluations: is it time for a makeover? HR Focus **77**(11), 6–7 (2000)
14. Rifkin, J.: Entropy: a new world view. [social and political implications of the Second Law of Thermodynamics]. United States (1980)
15. Michalska, J., Szewieczek, D.: The 5S methodology as a tool for improving the organization. J. Achievements Mater. Manufact. Eng. **24**(2), 211–214 (2007)
16. Soydan, S.: Süreç yönetimi ve iyileştirilmesi üzerine bir uygulama (Doctoral dissertation, DEÜ Sosyal Bilimleri Enstitüsü) (2006)

Improving Manufacturing Process of Mobile Recorder

Pınar Demircioğlu[1(✉)], İsmail Böğrekçi[1], Berk Sevinç[2], and Numan M. Durakbasa[3]

[1] Mechanical Engineering Department, Aydın Adnan Menderes University, Aydın, Turkey
{pinar.demircioglu,ibogrekci}@adu.edu.tr
[2] R&D Department, Beylerbeyi General Security Systems LLC., Aydın, Turkey
sevinc.berk@beylerbeyiguvenlik.com.tr
[3] Department of Industrial Metrology and Adaptronic Systems, Vienna University of Technology (TU Vienna), Vienna, Austria
numan.durakbasa@tuwien.ac.at

Abstract. Security vulnerabilities, failures and efficiency losses arose from errors due to the manufacturing processes and design of cases. For maintenance purposes, it is so vital to dismount cases and remove components in a short time. In classical cases, mobile recorders, time delay relay contacts, hard drives and power converters, poe card, recorder cards are mounted to the body of case by opening holes to the front-back panel and inside of the case and screwing cards to the cases. This study mainly focuses on developing a case without any screws and soldering. Components and cards can be mounted to the case with a click system. The projected case prevents from soldering errors, the jamming during removal of the hard disk, easing for both remote monitoring and re-opening the case by the end user in case of the insertion of a sim card for the internet. In this study, two molds are designed and manufactured using CAD software and classical subtractive manufacturing techniques. Two molds were designed and manufactured for using in extrusion process of aluminum and plastics materials. Aluminum and plastic parts were produced with extrusion machine. The cards were mounted to the case by clicking. This study improved production processes, reduced both soldering amounts in developing electronic cards and the cost & time delay. It also increased the efficiency, reduced product loss and idle/die change times, provided user convenience, and shortened production time.

Keywords: Mobile recorder · Mold · Taylor-made electronic card design · Power converters · Time delay relay module · Poe card

1 Introduction

Mobile Network Video Recorders (NVRs) are produced by soldering the hard disk, inverter, Poe card, and NVR card through cables and then mounted to the case. In case of defects caused by soldering errors, removal of the hard disk, remote monitoring, and insertion of a sim card for the internet, there are security gaps, failures, and efficiency

N. M. Durakbasa and M. G. Gençyılmaz (Eds.): ISPR 2020, LNME, pp. 504–512, 2021.
https://doi.org/10.1007/978-3-030-62784-3_43

losses such as opening the cases again by the end-user. The main purpose of the study is to improve production processes, reduce soldering processes by designing electronic cards, localize and incorporate inverter and term contact, reduce cost, increase efficiency, reduce product loss, ease user, shorten production time and reduce mold change times. Thus, it is aimed to develop ergonomic and technical changes in the interior and exterior design of the case, and a system for customer expectations.

Mobile recorders are used in passenger vehicles. Passenger vehicles, such as a school bus, can present a very dangerous environment for its passengers. Generally, a single bus driver is placed in charge of all the students riding the bus. The bus driver is not only burdened with the job of safely driving the bus, but also managing students riding to and from school. Because school bus drivers cannot continuously watch the students, they frequently act inappropriately without the bus driver taking notice or without the ability to do anything about it. For example, students often place themselves in danger by standing up while the bus is in motion, throwing various items, etc. Concomitantly, students are exposed to the danger of passing traffic every time they board and exit a school bus [1].

With a mobile and web application designed to solve this problem, the system has helped parents, schools, and buses communicate automatically and easily through the app to detect children's arrival time. In this system, it takes advantage of RFID and GPS technologies and connects to a remote server over WiFi using an ESP8266 microcontroller. An Ublox 6M GPS module is used to find the current geographic coordinates of the vehicle's location and the speed at which it travels. An MFRC522 RFID reader identifies each student while boarding or getting off the vehicle by reading the ID from the RFID tags [2].

All processes done manually are made orderly. For this purpose, the school service and student tracking system using HM-11 Bluetooth low energy module pointer devices, Firebase cloud platform, which is a database that can operate in real-time, MIT App Inventor Android mobile application and web application that can access pointer devices are presented [3].

A GPS module is used to track the location, and an RFID card built into the system is used to identify the child. When a child gets on the bus, the RFID tag on the ID card is detected by the reader on the bus, and the system will identify the child and sends a text message to parents with the current location and time [4].

Companies that convert cars, buses (school shuttle vehicles, passenger vehicles, etc.), new production vehicles in a heavy industry into school shuttle vehicles, currently working school service owners, all companies such as tour, bus companies engaged in logistics and transportation business, public vehicles, car rental companies, personal vehicle owners, vehicles of public officials, law enforcement vehicles, companies with construction equipment, etc. companies producing vehicle types; they use the mobile recording device, which is the output of the study for monitoring and recording purposes.

2 Material and Method

2.1 Design of the Interior and Exterior Case

Vehicle tracking system and the router are located inside the mobile recorder. In the new case, there is an electronic board plus invertor and term contact. In the previous design, there was an external dependency. It needs more soldering process. An ergonomic appearance will be obtained in the exterior design of the new case. The hard disk is removable and pluggable. In this way, vulnerabilities can be prevented. Fault detection and repair time may decrease. With the advancement of new design, for the sim card layout, ease of usage is reached.

The mobile recorder is a system where the cameras attached to the vehicles can be recorded by taking the internal-external images of the vehicles and can be remotely monitored at the centers via the internet. In conventional process, mobile NVRs are produced by soldering the hard disk, inverter, Poe card, and NVR card through cables and then assembling them into the case [5].

Within the scope of the study, firstly, literature and benchmarking studies are conducted, and the card designs used in mobile recorders are examined through patents and articles. Also, the current situation is analyzed by benchmarking. Conceptual design, electronic design, and analysis are carried out in the third phase while raw material procurement and purchases continue. Then virtual prototype work is done and the three-dimensional design of the case used in mobile recorders is carried out at this stage.

In order to compare the process time for building up classical case and projected case, following the completion of the three-dimensional design, the project planning and commissioning package is initiated, at this stage, workflow and time plans are created and are made ready for prototype production according to the standards [6, 7]. Then the design and manufacturing of test systems are carried out. After the mechanical and electronic designs are completed, test simulations are made and then the manufacturing of test systems is performed. Prototype production of the card design used in mobile recorders with the three-dimensional design is realized. Finally, the functionality is examined by performing function tests on the prototype in the test and verification stages.

2.2 Steps for the Production of Conventional Mobile Recorder

The conventional mobile NVR chassis consists of 4 main parts. These are top, bottom, front and back panels. The top panel acts as a cover. No action is taken on the top panel. In the lower panel, the necessary materials are fixed for the device. The front and back panels are used for input and output operations. In the production of classical mobile recorders, some necessary drilling operations are performed on the bottom panel. These drilling operations are made to fix the parts inside the device. Three types of cutting tools are used in drilling operations. These are 5 mm, 3.5 mm, and 2.5 mm in the hole diameters. For the front panel, 4 green leds and socket are mounted on the "LED indicator pcb" with soldering method. Power-Register-Error PCB in the same way is soldered within 4 min.

Video and audio output cables are combined with soldering method to BNC. Then, the heat shrink tubing is used to prevent the soldering areas from short-circuiting. 5 mm fireproof orange tubing is passed over all the cables. This process takes about 2 min. Vehicle tracking cable; the same operations on the video and audio output cables are repeated on the BNC. This process takes about 2 min. The assembly of all prepared parts to the front panel is completed in about 3 min. Preparation of 4.5G BNC cable is carried out in the same way as other cable soldering processes. Alarm cable preparation is carried out in the same way as other cable soldering processes. The assembly of all prepared parts to the back panel is completed in approximately 9 min. Preparation of the term contact takes approximately 2 min. Preparation of the inverter takes approximately 2 min. Term contact, Inverter and PoE card are mounted to bottom panel. Mounting time takes approximately 6 min. Preparation of the battery PCB (Printed Circuit Board) sockets and the KL-MNVR board takes about 20 min. Then it is mounted to the bottom panel. All the parts of Mobile NVR are mounted to the bottom panel within 23 min (Fig. 1).

Fig. 1. Production status of conventional mobile recorder (production time: 23 min)

2.3 Production of Planned Mobile Recorder

The schematic view of new Mobile Recorder is given in Fig. 2. After this scheme is designed, research is carried on the components to be used and purchases are made in line with these components.

Router is required for the recorder to connect to the internet. The model determined as a result of the researches is taken into consideration when determining the device model. The desired features of the model are as follows: compliance with industrial conditions, low power consumption, compliance with standards in Turkey (800, 900, 1800, 2100, 2600 MHz) and fast data flow.

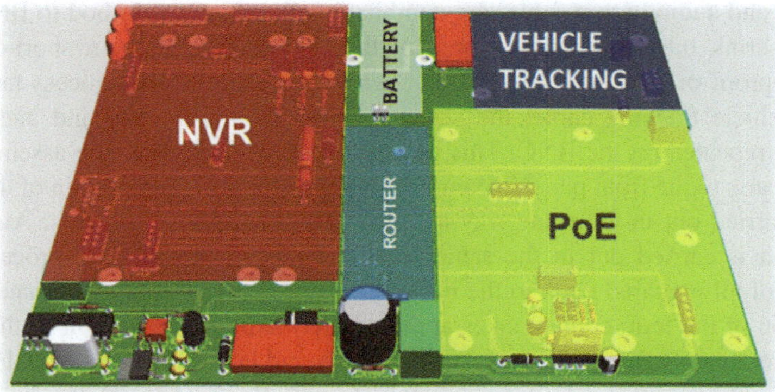

Fig. 2. Schematic view of the new production state of the mobile recorder (except for the case)

PoE Switch (data transfer device) is used to provide communication between the camera and the recorder, the criteria is taken into consideration when determining the model and the features of the model. Desired features of the PoE Switch are low power consumption, 10/100 MBPS speed, fast data flow and support 8 ports.

NVR board named "KL M 8M PBRT G" enables 8-channel IP and HD camera recording and monitoring produced within Beylerbeyi General Security Systems. The "VT-600" model is used for vehicle tracking. VT600 is a GPS/GPRS based tracking device specially developed and designed for the vehicle's real-time tracking and fleet management. VT600 has a built-in GPS module to obtain accurate location data. This device uses GSM to send location data to a specific mobile phone or server base for tracking and fleet management. The VT600 has the ability to store GPS coordinates in its internal memory when there is no GPRS connection or at a specified interval requested by the user. Also, the device has remote fuel disconnection, motion alarm, low battery alarm, GPS blind area alarm, power cut alarm, GPS antenna disconnect alarm and Geofence features.

3 Results and Discussion

The mobile recorder is a system where the cameras attached to the vehicles can be recorded by taking the internal-external images of the vehicles [8] and can be remotely monitored at the centers via the internet. The projected case prevents from soldering errors, the jamming during removal of the hard disk, easing for both remote monitoring and re-opening the case by the end user in case of the insertion of a sim card for the internet. In this study, the production and assembly of the 3D case design are carried out.

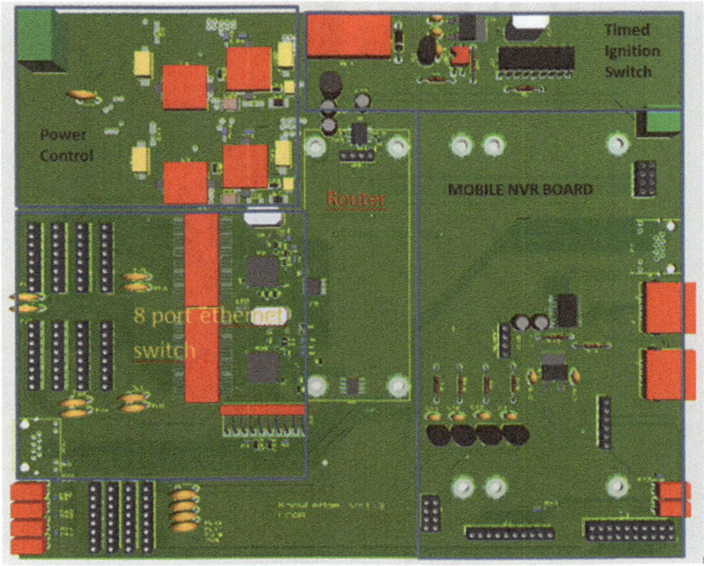

Fig. 3. Card design

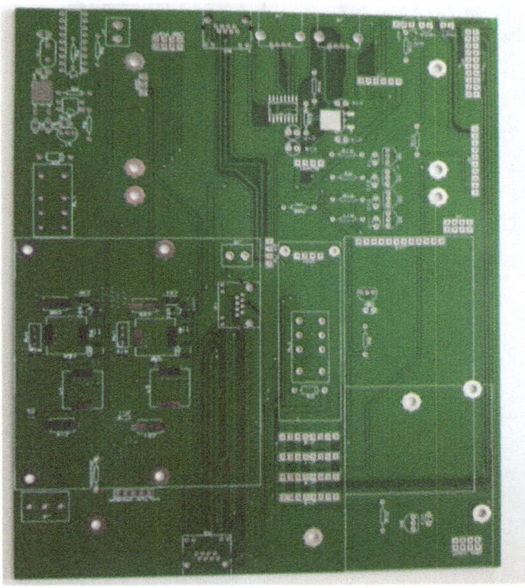

Fig. 4. Printed Circuit Board (PCB) of the designed card

The designed electronic card (Fig. 3), Printed Circuit Board (PCB) of the designed card (Fig. 4), final produced card (Fig. 5) and cards mounted to the new case (Fig. 6) are depicted.

Fig. 5. Produced card

Fig. 6. Cards mounted to new case

Virtual prototyping are conducted using Autodesk Inventor shown in Fig. 7. Considering the disruptions in production, increasing the production speed and customer demands, the need for a new mobile recorder case design was emerged.

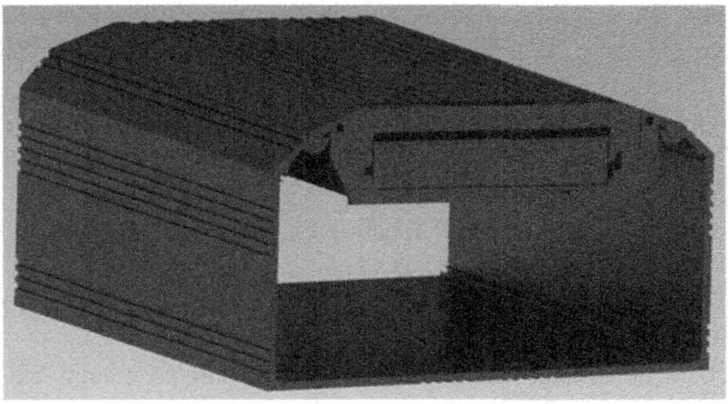

Fig. 7. Isometric 3D model of new case

4 Conclusions

In the new case, there is an electronic board plus invertor and term contact. In the previous design, there was an external dependency. It needs more soldering process. An ergonomic appearance will be obtained in the exterior design of the new case. The hard disk is removable and pluggable. In this way, vulnerabilities can be prevented. Fault detection and repair time may decrease. With the advancement of new design, for the sim card layout, ease of usage is reached.

The main purpose of the study is to improve production processes, reduce soldering processes by designing electronic cards, localize and incorporate inverter and term contact, reduce cost, increase efficiency, reduce product loss, ease user, shorten production time and reduce mold change times.

In this study, the production and assembly of the 3D case design completed. Two molds are designed and manufactured using CAD software and classical subtractive manufacturing techniques. Two molds were designed and manufactured for using in extrusion process of aluminum and plastics materials. Aluminum and plastic parts were produced with an extruder. The cards were mounted to the case by clicking. This study improved production processes, reduced both soldering amounts in developing electronic cards and the cost & time delay. It also increased the efficiency, reduced product loss and idle/die change times, provided user convenience, and shortened production time.

Acknowledgement. This study was supported and funded as KOSGEB under R&D and Innovation Program (Project ID: 277247).

References

1. Schmidt, W.P.: Video Recording System for Passenger Vehicle, School Bus Video Camera Catches Passbys in the Act, School Transportation News, August 1994
2. Raj, J.T., Sankar, J.: IoT based smart school bus monitoring and notification system. In: IEEE Region 10 Humanitarian Technology Conference (R10-HTC), Dhaka, pp. 89–92 (2017)

3. Küçük, K., Çelik, B., Bayılmış, C.: Nesnelerin İnterneti Teknolojileri ile Gerçek Zamanlı Okul Servisi ve Öğrenci Takip Sistemi Tasarımı. Düzce Üniversitesi Bilim ve Teknoloji Dergisi **6**(4), 1211–1223 (2018)
4. Shah, S., Singh, B.: RFID based school bus tracking and security system. In: 2016 International Conference on Communication and Signal Processing (ICCSP), Melmaruvathur, pp. 1481–1485 (2016)
5. Horowitz, P., Winfield, H.: The Art of Electronics. Cambridge University Press, Cambridge (1989)
6. BS EN 50155:2017. Railway applications. Rolling stock. Electronic equipment (2017). ISBN 978 0 580 82911 6
7. IEC 61508-2:2010. Functional safety of electrical/electronic/programmable electronic safety-related systems - Part 2: Requirements for electrical/electronic/programmable electronic safety-related systems (2010)
8. Pucher, E., Gruber, A., Eidmann, A., Spitzwieser, C.: Periodical technical emission control by using an IoT universal short-test procedure. In: Durakbasa, N., Gençyılmaz, M. (eds.) Proceedings of the International Symposium for Production Research 2019. ISPR 2019. Lecture Notes in Mechanical Engineering. Springer, Cham (2020)

Manufacturing Process Development for Thin Film Filaments as a New Product

Bahadır Tunaboylu[1], Biset Toprak[2(✉)], Ahmet Korhan Binark[2], Osman Öztürk[3], and Selim Zaim[2]

[1] Marmara University, Istanbul, Turkey
bahadir.tunaboylu@marmara.edu.tr
[2] Istanbul Sabahattin Zaim University, Istanbul, Turkey
{biset.toprak, ahmet.binark, selim.zaim}@izu.edu.tr
[3] Gebze Technical University, Gebze, Kocaeli, Turkey
osmanozturk@gtu.edu.tr

Abstract. Thin films of rare-earth hexaborides are promising for wear and corrosion protective, decorative, and thermionic coatings. Among many hexaborides available, due to its unique properties lanthanum hexaboride (LaB_6) is used as a bright and long life thermionic electron source mostly in electron microscopes. Unlike single-crystal LaB_6 cathodes used today, LaB_6 thin-film coated metal filaments can be advantageous due to their simple design, easy installation, cost-efficient manufacturing, and less energy consumption. In the study, LaB_6 thin films were grown onto tantalum, molybdenum, and tungsten wire-substrates using magnetron sputtering. Tape-test was used on the film surface to determine the adhesion of the LaB_6 films to the substrates. The morphology, structure, and stoichiometry of tape-tested films were investigated by stereo microscopy, scanning electron microscopy (SEM) and X-ray photoelectron spectroscopy (XPS). Stereo microscopy and SEM analysis showed that the films were grown by magnetron sputtering have dense, fine columnar structure without peeling or flaking. In the XPS analysis of LaB_6/W film it was observed that the film was oxidized and there were no peaks of any element other than lanthanum, boron, and oxygen found. The analysis shows that the composition of the film was LaB_x coated over the tungsten (W) substrate at the deposition conditions.

Keywords: Thermionic coatings · Lanthanum hexaboride (LaB_6) thin film filament · Magnetron sputtering (MS) · Scanning electron microscope (SEM) · Transmission electron microscope (TEM)

1 Introduction

Among many hexaborides available, lanthanum hexaboride has been accepted as the best electron emitter due to the combination of high chemical stability, low evaporation rate at high temperature, and low work function [1–5]. After the investigation of electron emitter properties of lanthanum hexaboride by Lafferty [5], LaB_6 rod cathode was first designed by Broers [5, 6]. It was suggested that LaB_6 cathodes are superior to tungsten cathodes in terms of brightness, service life and mechanical stability [6, 7].

N. M. Durakbasa and M. G. Gençyılmaz (Eds.): ISPR 2020, LNME, pp. 513–522, 2021.
https://doi.org/10.1007/978-3-030-62784-3_44

Currently, LaB_6 and CeB_6 single-crystal cathodes are used in transmission electron microscopy (TEM), scanning electron microscopy (SEM), electron beam lithography, and X-Ray microanalysis, which offer high brightness and service life [1, 6, 8–10] Fig. 1 presents the commercially available tungsten and single-crystal LaB_6 cathodes. As can be seen from the figure, tungsten cathodes consist of hairpin filament when LaB_6 cathodes consist of hairpin filament as well as a single crystal tip. From Table 1, it can be noted that LaB_6 cathodes are superior to tungsten cathodes in terms of brightness and service life. Since operating temperature is one of the factors that affect service life, LaB_6 cathodes are preferable due to their lower operating temperature [11]. However, LaB_6 cathodes have a complicated design with various tip shapes (round, flat, sharp) and cone angles which increases the cost compared to tungsten cathodes. The cathodes are typically the main consumables in the electron microscopy operational costs. The tungsten, LaB_6 and CeB_6 cathodes are imported and their prices are 429 EUR (4.391 TL), 1.592 EUR (16.300 TL), and 4.600 EUR (47.111 TL), respectively. Turkey that has the largest boron resources in the world (73%) provides a motivation for development of high value-added boron products with innovative approach. For this purpose, LaB_6 thin-film coated metal filaments are investigated here in this application since they are promising due to their potential in simple design, easy usage, and low operating energy and low cost with respect to the single crystal cathodes [11–13]. In this study, we aimed to grow lanthanum hexaboride thin films on tantalum, molybdenum, and tungsten wires to develop a manufacturing process for thin-film filaments. In order to analyse the adhesion of the LaB_6 films to the substrates, tape-test was used on the film surfaces. Then, the morphology, structure, and stoichiometry of tape-tested films were investigated in terms of stereo microscopy, scanning electron microscopy, and X-ray photoelectron spectroscopy.

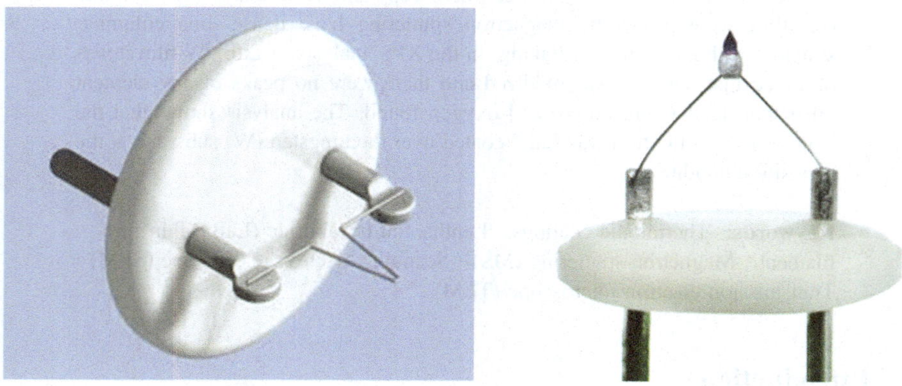

(a) (b)

Fig. 1. Image of commercially available (a) tungsten cathode and (b) LaB_6 cathode

Although LaB_6 thin film cathodes appeared to be advantageous when they are used as thin film-based products, the studies have mostly focused on electron emission

Table 1. Comparison of LaB_6 CeB_6 and tungsten cathodes [10, 14]

	LaB_6	CeB_6	Tungsten
Brightness (A/m^2-sr)	10^3	10^3	10^2
Typical service life (hr)	1000+	1500 +	30–100
Work function (eV)	2.70	2.65	4.5
Operating vacuum (torr)	10^{-7}	10^{-7}	10^{-5}
Operating Temperature (K)	1800	1800	2800
Current density at Operating Temperature (A/m^2)	20×10^{-4}–50×10^{-4}	20×10^{-4}–50×10^{-4}	3×10^{-4} (at 2800 K)
Evaporation rate (g/m^2-sec)	2.9×10^{-13}	2.1×10^{-13}	NA

properties of single crystal, pressed or sintered rod LaB_6 cathodes [5–7, 15–22]. In literature, lanthanum hexaboride thin films were grown by electrophoretic deposition [1, 11, 23], the evaporation [24], the magnetron sputtering [2, 3, 13, 25, 26], the electron beam deposition [27] and the pulsed laser deposition [8, 9, 28] in order to investigate their electron emission properties. Waldhauser found that among the rare earth elements, LaB_6 and CeB_6 are the most promising ones for utilization as coated cathodes but the highest electron emission was yielded from LaB_6 coated filament [26]. In addition, LaB_6 is very attractive for decorative hard coatings applied to luxury accessories due to its purple-red color spectrum and high hardness [12, 29, 30].

The first study to investigate the utilization of coated cathodes in electron microscopes was conducted by Khairnar and Mahajan [11]. In their study, LaB_6 was deposited on tantalum filament by electrophoretic deposition. The work functions were found between 2.6–2.8 eV. Before the deposition, the tantalum wire in hairpin shape was carburized to prevent diffusion of boron into the wire. It was found that LaB_6 coated tantalum filaments can be operated at 1873 K while uncoated tungsten filaments need a higher temperature than 3020 K at constant emission density. The lifetime of the filaments was found 55 h. Since the service life of the cathodes is affected by operating temperature, coated filaments were suggested. It was found that the service life of filaments is primarily determined by the adhesion of LaB_6 to the substrate.

However, the coatings produced by electrophoretic deposition were porous and poor in adhesion that results in fragile and undurable coating to thermal shocks. Besides, the preparation of that coatings is difficult because heat treatments are required before and after the deposition [13, 26].

Among the techniques mentioned above, the magnetron sputtering which is one of the physical vapor deposition (PVD) method, is used to deposit more dense, adhesive, and smoother films at low temperatures with ease of controlling the parameters [13, 31, 32]. The method is capable to produce thin films having the same functionality as thick films produced by other coating techniques [33–35]. In MS, the material called target, usually prepared by sintering or hot pressing, is deposited on the substrate as atomic-sized particles to form a thin film in a gaseous plasma [36].

Since the adhesion of the LaB_6 thin films is one of the main factors that affect service life, Mroczkowski used radio frequency (RF) magnetron sputtering to deposit adhesive LaB_6 thin films on tungsten. As a result, adherent, and stress-free thin films with 1 μm thickness were obtained. Work function and electron emission values were found to be consistent with results obtained from bulk LaB_6. The lifetime measurements were carried out at 1200 K and measured as more than 1000 h. The work function of the films was between 2.4–2.6 eV. The surface of the films contained lanthanum oxide (La_2O_3) in addition to LaB_6 [13].

Kinbara et al. deposited LaB_6 films direct current magnetron sputtering (DCMS) on glass substrates to investigate the impact of argon pressure on the composition of the films. It was found that (100) plane films, which are preferred as electron emitters due to their low work function, can be obtained under 1 Pa argon pressure [26]. In the study conducted by Waldhauser et al., deposition parameters were optimized by using DCMS to coat the molybdenum and tungsten sheets. After determining the optimum parameters, the work function was found as 2.8 eV [3]. Waldhauser et al. growed LaB_6 thermionic coatings on molybdenum and tungsten substrates by MS. It was found that films were over stoichiometric. However, with increase in the argon pressure the ratio was approximated to stoichiometry [26]. Craciun et al. obtained boron deficiency, which may be caused by the mass difference between lanthanum and boron atoms [9]. It can be concluded that determining optimal coating parameters have a crucial impact on the morphology and structure of the films [37]. It was found that the service life of filaments is primarily determined by the adhesion of LaB_6 to the wire surface rather than the evaporation rate of lanthanum and boron [11]. Late et al. stated that good adhesion is required for durable electron emitters [28]. Thus, in this study, the adhesion of the LaB_6 films, which is deposited by MS determined using tape-test on the film surfaces. Then, the morphology, structure, and stoichiometry of tape-tested films were investigated in terms of stereo microscopy, scanning electron microscopy, and X-ray photoelectron spectroscopy.

2 Experimental Method

The films were sputtered on tungsten, molybdenum, and tantalum wire-based substrates by magnetron sputtering (MS) system existing in Gebze Technical University, Surface Physics Laboratory with using commercially available disc target composed of 99.9% LaB_6 (Diameter: 76 mm and Thickness: 10 mm). The MS system consists of BesTech deposition chamber with 6 Guns: 3 direct current (DC), 2 radio frequency (RF), and 1 DC-Pulse. In the study, the RF gun was used with 25 W to deposit the tungsten, molybdenum, and tantalum wires with a diameter of 1.1 mm, 1 mm and 0.7 mm, respectively and 1.2 mm long. Then, the substrates were subjected to the wire shaping process by cutting into two pieces transverse for obtaining a flat surface. After the wire shaping process, all substrates were placed into the holder (see Fig. 2) to form thin films under the same conditions simultaneously. Then the holder was placed into the magnetron sputtering system as the distance between the substrates and target material was 50 mm. Before the deposition, substrate surfaces were cleaned with 100 W in a vacuum environment to remove contaminations and oxides. The films were deposited directly, without using any intermediate layer, on the substrates that were kept at room temperature. Sputtering was carried out under the 2×10^{-3} mbar pressure and

20 W sputtering power with maintaining high purity argon at a flow rate of 3 sccm. Since we have 3-in. guns and the thickness verifying was controlled by Quartz Crystal Microbalance (QCM) in situ, also the deposition ratio of guns calibrated by using XPS. Therefore, the thickness of the films was well-controlled magnetron sputtering and deposition was continued until 2.5 nm thickness was obtained.

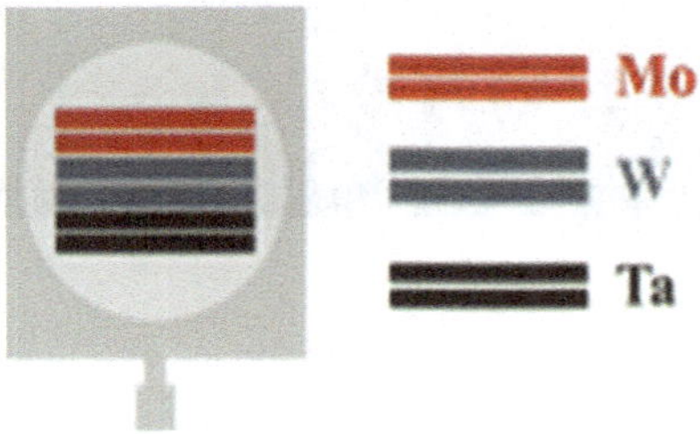

Fig. 2. Schematic illustration of the holder developed

3 Results

In order to analyse the adhesion of the LaB_6 films to the substrates, tape-test was applied on the film surfaces. Then, the morphology, structure and stoichiometry of tape-tested films were investigated in terms of stereo microscopy, SEM and XPS. The images obtained from the stereo microscopy are presented in Fig. 3. As shown in the figure below, there is no peeling or flaking observed after the tape test on the surface of the LaB_6 coated molybdenum, tantalum and tungsten samples.

(a) (b) (c)

Fig. 3. Stereo microscopy image of tape tested LaB_6 on (a) molybdenum (b) tantalum and (c) tungsten with 45X magnification factor

The morphology of the films was analysed by Phenom XL scanning electron microscopy. Figure 4 presents scanning electron micrograph of uncoated molybdenum, tantalum and tungsten with 3000X magnification factor. From the figure, it is apparent that films of LaB_6/W is smoother than those of LaB_6/Mo and LaB_6/Ta. Since the

Fig. 4. Scanning electron micrograph of uncoated surface of (a) molybdenum, (b) tantalum and (c) tungsten with 3000X magnification factor

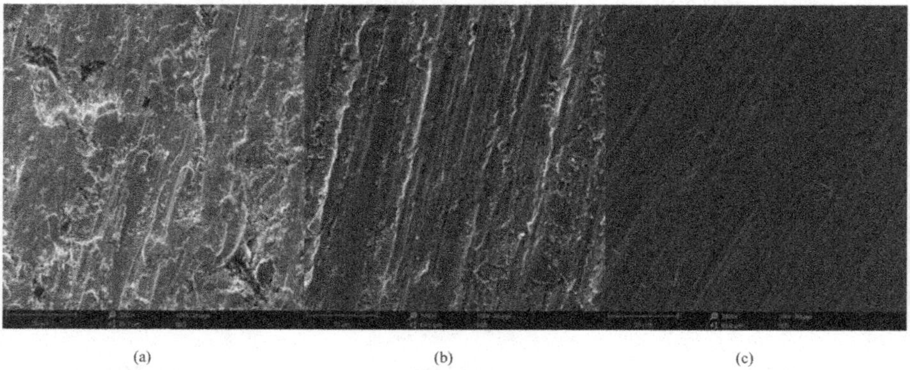

Fig. 5. Scanning electron micrograph of tape tested LaB_6 on (a) molybdenum, (b) tantalum and (c) tungsten with 3000X magnification factor

tungsten is the hardest among tantalum and molybdenum, this result is expected because of the wire shaping process. Figure 5 shows SEM image of LaB_6 film deposited onto molybdenum, tantalum and tungsten, respectively. Film growth seems to be dense and shows a fine-columnar structure. It can be noted from the SEM analysis films produced by magnetron sputtering are dense and strongly adhered to the wires that will be used as a filament. Since the adhesion depends on the substrate characteristics as well as film properties, good adhesion of the films was partly provided by the cleaning process done before deposition [31].

The surface composition and elemental ratios of the LaB_6/W film was determined by X-ray photoelectron spectroscopy measurements were performed by using SPECA Phoibos 150 specs charged particle analyzer (Charge analyzer: 9 MDC- charge detectors) including x-ray gun XR-50 (Anode:Al Kα- 350 W), which exist in GTU Surface Physics Laboratory. The attenuation depth was 6 nm for the XPS. It can be seen from the XPS scan in Fig. 6 the film was oxidized. No peaks of any element other than lanthanum, boron, and oxygen have been observed on the spectrum.

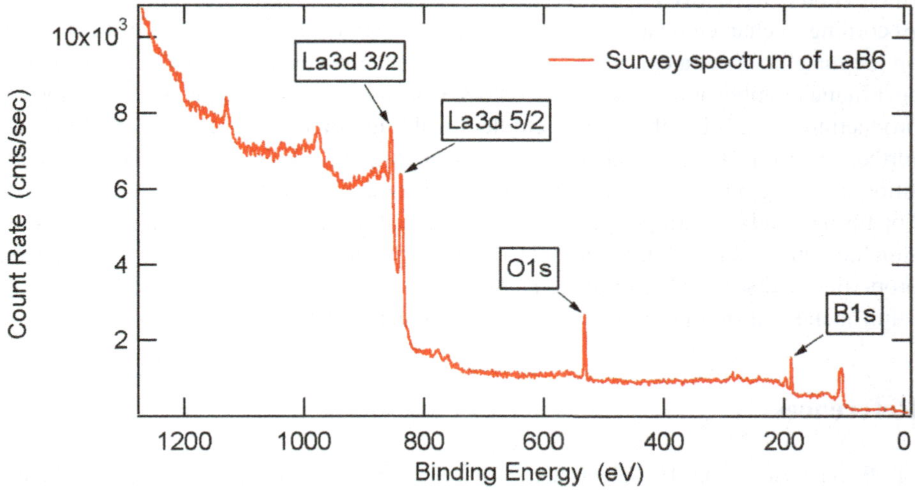

Fig. 6. X-ray photoelectron spectra of the LaB_6/W film

XPS was performed for survey and elemental windows to calculate elemental stoichiometry of the LaB_6/W film. The elemental ratios obtained from the XPS analysis are presented in Table 2. But the numbers on the table is taken from window scans of the elements. Also, we had three scans for one survey (with nine charge detectors), so it was enough for statistical calculation. From the X-ray photoelectron spectra, surface appearance of the LaB_6/W film was found to be LaB_x.

Table 2. Elemental ratio calculation of LaB_6 films

Peak Name	Peak Area (A)	ASF	A/ASF	%
O1s	1068.3	0.711	1502,53	28,55
B1s	482.8	0.159	3036,48	57,70
La3d	6602.9	9.122	723,84	13,75

4 Conclusion

The study aimed to perform the deposition of LaB_6 thin films on molybdenum, tantalum, and tungsten wires and investigation of their adhesion, morphology, and elemental ratio. In order to determine the adhesion of the LaB_6 films, tape-test was used on the film surfaces. Then, the morphology, structure, and stoichiometry of tape-tested films were investigated by stereo microscopy, SEM, and XPS. Stereo microscopy and SEM analyses showed that the films were grown by magnetron sputtering have dense and fine columnar structure without peeling or flaking. In the XPS analysis of LaB_6/W film, it was observed that the film was oxidized and there were no peaks of any element other than lanthanum, boron, and oxygen found. The analyses show that the composition of the film was LaB_x coated over W substrate at the deposition conditions. According to characterization results, the deposition and characterization of LaB_6 films on the metallic substrates were completed. These are encouraging results for the lower cost filament application. This study serves as a base of future studies for the domestic production of thin-film thermionic emitters with high quality and low price. However, further experiments are required to produce high-quality films in the desired composition by changing the target composition and magnetron sputtering system parameters. For further studies, LaB_6 target will be segmented to include a boron content to have stoichiometric films. After determining the optimal parameters, electron emission properties, and service life measurements of the samples will be investigated to develop a full manufacturing process for thin-film filaments for electron microscopes.

References

1. Buckingham, J.D.: Thermionic emission properties of a lanthanum hexaboride/rhenium cathode. Br. J. Appl. Phys. **16**, 1821–1832 (1965)
2. Nakano, T., Baba, S., Kobayashi, A., Kinbara, A., Kajiwara, T., Watanabe, W.: Structure modification of radio frequency sputtered LaB_6 thin films by internal stress. J. Vac. Sci. Technol. A **9**, 547–549 (1991)
3. Waldhauser, W., Mitterer, C., Laimer, J., Störi, H.: Structure and electron emission characteristics of sputtered lanthanum hexaboride films. Surf. Coat. Technol. **74–75**, 890–896 (1995)
4. Craciun, V., Cristea, D., Socol, G., Lambers, E., Trusca, R., Fairchild, S., Back, T., Gruen, G., Craciun, D.: Characteristics of LaB_6 thin films grown by pulsed laser deposition. J. Vac. Sci. Technol. A **34**, 547–549 (2016a)
5. Lafferty, J.M.: Boride cathodes. J. Appl. Phys. **22**, 299–309 (1951)
6. Broers, A.N.: Electron gun using long life lanthanum hexaboride cathode. J. Appl. Phys. **38**, 1991–1992 (1967)
7. Broers, A.N.: Some experimental and estimated characteristics of the lanthanum hexaboride rod cathode electron gun. J. Sci. Instrum. (J. Phys. E) **2**, 273–276 (1969)
8. Late, D.J., Date, K.S., More, M.A., Misra, P., Singh, B.N., Kukreja, L.M., Dharmadhikari, C.V., Joag, D.S.: Some aspects of pulsed laser deposited nanocrystalline LaB_6 film: atomic force microscopy, constant force current imaging and field emission investigations. Nanotechnology **19**, 1–6 (2008)
9. Craciun, V., Cristea, D., Socol, G., Lambers, E., Trusca, R., Fairchild, S., Back, T., Gruen, G., Craciun, D.: Characteristics of LaB_6 thin films grown by pulsed laser deposition. J. Vac. Sci. Technol. A Vac. Surf. Films **34**, 051509 (2016b)
10. https://www.emsdiasum.com/microscopy/products/brochures/Cathodes.pdf
11. Khairnar, R.S., Mahajan, P.W., Joag, D.S., Nigavekar, A.S., Kanitkar, P.L.: Processing and characterization of LaB_6 coated hairpin cathodes. J. Vac. Sci. Technol. A **3**, 398–402 (1985)
12. Mitterer, C.: Borides in thin film technology. J. Solid State Chem. **133**, 279–291 (1997)
13. Mroczkowski, S.J.: Electron Emission Characteristics of Sputtered Lanthanum Hexaboride. J. Vac. Sci. Technol. **9**, 586–590 (1991)

14. https://blog.phenom-world.com/thermionic-electron-sources-sem-filaments-emission-stability
15. Broers, A.N.: High resolution thermionic cathode scanning transmission electron microscope. Appl. Phys. Lett. **22**, 610–612 (1973)
16. Schmidt, P.H., Joy, D.C., Longinotti, L.D., Leamy, H.J., Ferris, S.D., Fisk, Z.: Anisotropy of thermionic electron emission values for LaB_6 single-crystal emitter cathodes. Appl. Phys. Lett. **29**, 7 (1976)
17. Schmidt, P.H., Longinotti, L.D., Joy, D.C., Ferris, S.D., Leamy, H.J., Fisk, Z.: Design and optimization of directly heated LaB_6 cathode assemblies for electron beam instruments. J. Vac. Sci. Technol. **15**, 1554–1560 (1978)
18. Swanson, L.W., Dickinson, T.: Single-crystal work-function and evaporation measurements of LaB_6. Appl. Phys. Lett. **28**, 578–580 (1976)
19. Oshima, C., Bannai, E., Tanaka, T., Kawai, S.: Thermionic work function of LaB_6 single crystals and their surfaces. J. Appl. Phys. **48**, 3925–3927 (1977)
20. Jacobson, D.L., Storms, E.K.: Work function measurement of lanthanum - boron compounds. IEEE Trans. Plasma Sci. **6**, 191–199 (1978)
21. Storms, E.K., Mueller, B.A.: A study of surface stoichiometry and thermionic emission using LaB_6. J. Appl. Phys. **50**, 3691–3698 (1979)
22. Goebel, D.M., Hirooka, Y., Sketchley, T.A.: Large-area lanthanum hexaboride electron emitter. Rev. Sci. Instrum. **56**, 1717–1722 (1985)
23. Favreau, L.J.: Cataphoretic coating lanthanum boride on rhenium filaments. Rev. Sci. Instrum. **36**, 856–857 (1965)
24. Oshima, C., Horiuchi, S., Kawai, S.: Thin film cathodes of lanthanum hexaboride (LaB_6). Jpn. J. Appl. Phys. **13**, 281–284 (1974)
25. Kinbara, A., Nakano, T., Kobayashi, A., Baba, S., Kajiwara, T.: LaB_x thin films prepared by magnetron sputtering. Appl. Surf. Sci. **70**(71), 742–745 (1993)
26. Waldhauser, W., Mitterer, C., Laimer, J., Stori, H.: Sputtered thermionic hexaboride coatings. Surf. Coat. Technol. **98**, 1315–1323 (1998)
27. Yutani, A., Kobayashi, A., Kinbara, A.: Work functions of thin LaB_6 films. Appl. Surf. Sci. **70**(71), 737–741 (1993)
28. Late, D.J., More, M.A., Sinha, S., Dasgupta, K., Misra, P., Singh, B.N., Kukreja, L.M., Bhoraskar, S.V., Joag, D.S.: Synthesis and characterization of LaB_6 thin films on tungsten, rhenium, silicon and other substrates and their investigations as field emitters. Appl. Phys. A **104**, 677–685 (2011)
29. Mitterer, C., Komenda-Stallmaier, J., Losbichler, P., Schmölz, P., Werner, W.S.M., Störi, H.: Sputter deposition of decorative boride coatings. Vacuum **46**, 1281–1294 (1995)
30. Mitterer, C., Komenda-Stallmaier, J., Losbichler, P., Schmölz, P., Störi, H.: Decorative boride coatings based on LaB_6. Surf. Coat. Technol. **74–75**, 1020–1027 (1995)
31. Baptista, A., Silva, F.J.G., Porteiro, J., Mígueza, J.L., Pintoa, G., Fernandes, L.: On the physical vapour deposition (PVD): evolution of magnetron sputtering processes for industrial applications. In: 28th International Conference on Flexible Automation and Intelligent Manufacturing (FAIM2018). Global Integration of Intelligent Manufacturing and Smart Industry for Good of Humanity, Columbus, OH, USA, 11–14 June 2018, pp. 746–757 (2018)
32. Swann, S.: Magnetron sputtering. Phys. Technol. **19**, 67–75 (1988)
33. Ray, A.K.: Hard nitride coatings by DC magnetron sputtering. Mater. Sci. Forum **246**, 61–90 (1997)
34. Constantin, D.G., Apreutesei, M., Arvinte, R., Marin, A., Andrei, O.C., Munteanu, D.: Magnetron sputtering technique used for coatings deposition; technologies and applications. Recent **12**(1), 29–33 (2011)

35. Abusuilik, S.B., Inoue, K.: Effects of intermediate surface treatments on corrosion resistance of cathodic arc PVD hard coatings. Surface Coatings Technology **237**, 421–428 (2013)
36. Pessoa, R.S., Fraga, M.A., Santos, L.V., Galva, N.K.A.M., Maciel, H.S., Massi, M.: Plasma-assisted techniques for growing hard nanostructured coatings: an overview. In: Anti-Abrasive Nanocoatings, pp. 455–479. Elsevier (2015)
37. Ivashchenko, V.I., Turchi, P.E.A., Shevchenko, V.I., Medukh, N.R., Leszczynski, J., Gorb, L.: Electronic, thermodynamics and mechanical properties of LaB_6 from first-principles. Phys. B Condens. Matter **531**, 216–222 (2018)

Modelling of the High-Chromium Cast Iron Surface Roughness

Borislav Savković[1], Pavel Kovač[1], Branko Štrbac[1], Marcela Pokusova[2], and Nenad Kulundžić[1(✉)]

[1] Faculty of Technical Sciences, University of Novi Sad, Novi Sad, Serbia
{savkovic, pkovac, strbacb, kulundzic}@uns.ac.rs
[2] Faculty of Mechanical Engineering, Slovak University of Technology in Bratislava, Bratislava, Slovakia
marcela.pokusova@stuba.sk

Abstract. The surface roughness is one of the functions of material machinability. In order to obtain the better results in the surface roughness, it must strive for the more ideal cutting parameters. In addition to the cutting parameters on quality of surface roughness influenced by a number of factors. This paper examines the influence of cutting parameters on surface roughness when machining high-chromium cast iron. Modelling of the machining process was performed using a multifactorial model. The experiments have been carried out in accordance with a design of experiments (DOE) to determine adequate model and significant parameters on surface roughness parameters. MarSurf 1 was used to measure specimens surface roughness.

Keywords: Surface roughness · Modelling · Machinability · High-chromium cast iron

1 Introduction

Milling machining is one of the most conventional machining methods used in the metal processing industry. The machining process itself is accompanied by many phenomena, which affect the output performance of the machining process. One of the phenomena is the quality of the machining surface.

The goal of production is high productivity with as little machining cost as possible. Over time, experimental methods have been developed that have the task of demonstrating the influence of machining regimes on the workability of materials. The basic parameters of control of the cutting process are the elements of the machining conditions. They have a direct impact on the quality of the workpiece surface, and therefore on the roughness.

The roughness of the machined surface, in addition to the cutting force, the stability of the cutting tool, the cutting temperature and the shape of the shavings, is a function of the workability of the material. Knowledge of material workability when designing cutting technology is very important. Accordingly, it seeks to create a database of workability and optimization of cutting modes.

N. M. Durakbasa and M. G. Gençyılmaz (Eds.): ISPR 2020, LNME, pp. 523–534, 2021.
https://doi.org/10.1007/978-3-030-62784-3_45

The modern mechanical engineering industry generally has high demands regarding the quality of the machined surface. Therefore, the roughness of the machined surface is of great importance. It strives for more economical production, but also increases product performance. Constant parameter control is required to improve the quality of the final product [1].

The process of face milling is very dynamic, so the final surface roughness is influenced by a number of factors. These factors can be divided into controlled ones, which include: cutting speed, feed and depth of cut. In addition to these three controls, there are a number of uncontrollable factors, such as: vibration, tool wear, movement errors, in homogeneity of tool material and workpiece, etc. [2].

The constant development of industrial materials also creates the need to test their machinability during the material removal process. Today, in the industrial sector, materials are demanded to have the best mechanical characteristic, and on the other hand, it is a challenge for their quality and cost-effective processing [3]. Cast iron, with certain admixtures of other alloying elements, certainly found its place in the application of new materials. Thus, we have found that high-chromium cast irons are known as materials that have a particularly high abrasion resistance, are very brittle and have low fracture toughness, which is why their use in the past has been limited. Many years of experience in the development and production of machining castings have shown that significant improvements in mechanical properties such as toughness and fracture strength can be significantly improved by special alloying. This allowed high-chromium cast iron to become promising materials with many unique and specific properties.

The results showed that for the wider use of these materials, several problems must be solved, such as: the effect of microalloying and the inoculation of molten metal [4–6], the possibility of changing certain properties by thermal treatment [4, 7], and the process of hardening the iron matrix at the local plastic deformation [4, 8].

The wear resistance and mechanical properties of high-chromium cast iron depend on the type, shape and amount of carbides contained in their microstructure, the matrices within which the carbides are housed, and the bond strength between the carbide and the matrix [4, 6, 7, 9, 10]. By increasing the carbon and chromium content, the amount of carbide increases. This improves the hardness and wear resistance of the castings but reduces the toughness as well as the fracture resistance.

The model shows the essential characteristics of an object, process or system. The mathematical model itself consists of a system of equations, states, and algorithmic rules. When setting up a mathematical model of an object, process, or system, basic sources of information are used.

The basic goal of mathematical modelling is:

- study and analysis of an object, process or system in order to obtain new and complete knowledge and legality of the object, process or system under study;
- discovering the mechanisms of interaction within the studied process;
- verification of hypotheses about the laws and mechanisms of internal interactions in an object or process;
- forecasting the state and behaviour of processes, systems and phenomena;
- optimization based on the set optimization criteria;
- managing of objects, processes or systems in space in time [11].

A. *A Second-Order Three-Factor Model*

The mathematical model in real coordinates takes the following form:

$$Y = b_0 + b_1x_1 + b_2x_2 + b_3x_3 + b_{11}x_1^2 + b_{22}x_2^2 + b_{33}x_3^2 + b_{12}x_1x_2 + b_{13}x_1x_3 + b_{23}x_2x_3 \tag{1}$$

Resolution of this model is possible by regression analysis by determining the regression coefficients using the least squares method based on the matrix formula:

$$B = (X'X)^{-1}X'Y \tag{2}$$

where is: B - matrix of coefficients b_i, b_{ij}, b_{ii} to be determined

X - experiment plan matrix
X′ - transposed matrix X
$(X'X)^{-1}$ - product matrix inverse X′X
Y - matrix of logarithmic values of measured size

This process is quite time-consuming, so using the appropriate software is implemented on the same principle. In this paper, MiniTab was used as a tool to accelerate the implementation of the proposed model.

2 The Experiment

The aim of this paper is to investigate the influence of machining elements on the roughness parameters of the machined surface at the front milling of high-chrome cast iron, hardness 44 ÷ 46 HRC. Experimental milling machining tests were carried out at the Department of Manufacturing Engineering in Novi Sad. The conditions under which the experiment was carried out relate to the machine tool, cutting tool, workpiece material and machining mode parameters.

A. Conditions During Experimental Testing
The machine tool on which the milling operation was performed is the vertical milling machine "PRVOMAJSKA" FSS-GVK-3.

he test used a milling head "JUGOALAT" G. 037 ϕ100 [mm], with mechanically attached cutting pads (Fig. 1), the following characteristics: number of teeth t = 5, tool cutting edge angle κ = 75, rake angle γ = 0°.

As a cutting material, hard metal coated inserts obtained by the CVD (Chemical Vapor Deposition) process at temperatures of 700 ÷ 900 °C were used. The inserts were designed for machining gray cast iron (ISO code K), steel (ISO code P) and hardened materials (ISO code H). The experiment was performed with a single-tooth tool, i.e. with one cutting pad (t = 1). The dimensions and geometry of the cutting insert are shown in Fig. 1.

1) Workpiece material:
   The workpiece material in the course of the experiment was precisely high-chromium cast iron, with a hardness of 44 ÷ 46 HRC and the chemical composition given below (Table 1).

Table 1. Chemical composition of high-chromium cast iron

Material label	C	Si	Mn	Cr	Mo	Ni
17680	3,95	0,5	0,98	21,25	1,7	5,7

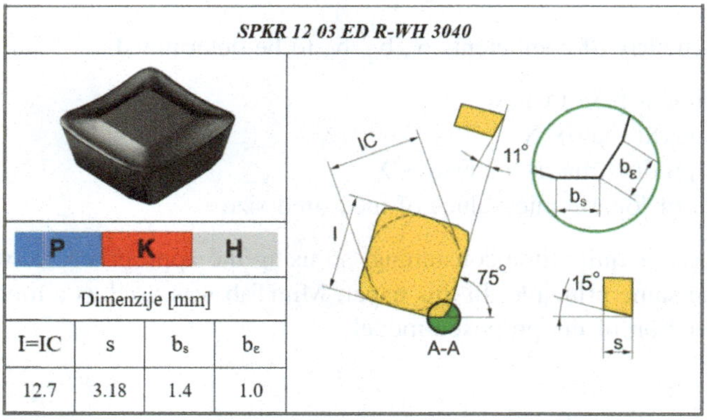

Fig. 1. Geometry of the Used HM tile cutters

Based on research on high-chromium cast irons conducted at the Faculty of Mechanical Engineering STU in Slovakia, the following conclusions were reached: with increasing carbon content (C) and chromium (Cr), the resistance to cast iron abrasion increases almost linearly. When the limit of 3% C is exceeded, brittleness is significantly increased, whereas an increase in chromium (Cr) over 20% had no greater effect on abrasion resistance. The study showed that cast iron containing about 20% Cr can have extremely high toughness when alloyed with 1.7% Mo + 5% Ni + 2% Mn. The high toughness of austenite allows the carbon content to be increased above the over-eutectic region (up to 5% C), Fig. 2. At the upper carbon content limit (about 5% C), the material in the cast state showed very high abrasion resistance and ductility of about 1.5%. At the lower carbon limit (about 2.5% C), cast iron has obviously lower abrasion resistance and hardness of about 38 HRC but also a particularly high ductility.

During heat treatment, the material received a smaller increase in hardness by (5 ÷ 10 HRC). After heat treatment, austenitization at 900 °C and release at 450 °C, a large number of relatively large eutectic carbides can be observed in addition to the primary M_7C_3 carbides embedded in the matrix, which maintained a high level of austenite.

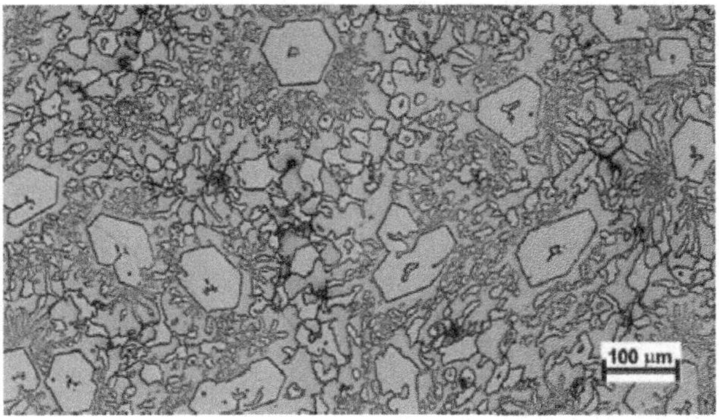

Fig. 2. Microstructure of high-chromium cast iron with 4.28% C in the cast state [12]

Today, high-chromium cast iron is considered a promising category of metallic materials. The high-chromium cast iron described above illustrates what exploitation properties a cast-iron carbide with a metastable austenitic matrix can achieve. The extremely high toughness of austenite allows for a significant increase in the carbon content up to the over-eutectic region, where a large amount of carbide provides high abrasion resistance, while at the same time, the iron has sufficient toughness and high fracture resistance. High-chromium cast iron belongs to hard-working materials due to its high hardness and abrasion [13, 14].

The hardness of the workpiece material was measured in four different places. The measured hardness on the HP 250 device installed in the Material Testing Laboratory is given in Table 2.

Table 2. Hardness of workpiece material

High-chromium cast iron				
No. Measurements	1	2	3	4
Rockwell hardness (HRc)	44	46	45	46

Values of machining regime elements were adopted based on literature recommendations. Factor variation was performed at 5 levels of values, so that any mean between two adjacent levels is the geometric mean of those values (Table 3). Cutting modes, adapted to the workpiece material and the design of the experiment, are shown in Table 4. The repeatability of this experiment is reflected through rows 9, 10 11 and 12, where it can be seen that the magnitude measured in these rims ranges in a tolerance of approximately 0.4 mm. It should be noted here once again that this high-chromium material is difficult to process and the tools are exposed to wear. That is, after each experimental row, a new cutting insert is used. Machining was performed without coolant and lubricant agent. The output parameter in this experiment was the roughness of the machined surface.

Table 3. Experimental parameter levels

Levels (Affiliation functions)	Cutting speed [m/s] on the diameter D = 100 mm	The feed per tooth [mm/t]	Cutting depth [mm]	Number of rotations [o/min]
Highest +2	0,586	0.443	0.5	112
Big +1	0,471	0.3739	0.4	90
Medium 0	0,371	0.3157	0.32	71
Small −1	0,293	0.2665	0.25	56
Minimum −2	0,235	0.225	0.2	45

Table 4. Measured and calculated values for mean arithmetic roughness (R_A)

No.	x_1	x_2	x_3	v (m/s)	f_z (mm/t)	a (mm)	R_a (μm)
1	−1	−1	−1	0.293	0.2665	0.25	1.063
2	1	−1	−1	0.471	0.2665	0.25	1.608
3	−1	1	−1	0.293	0.3739	0.25	1.841
4	1	1	−1	0.471	0.3739	0.25	4.361
5	−1	−1	1	0.293	0.2665	0.4	1.305
6	1	−1	1	0.471	0.2665	0.4	2.299
7	−1	1	1	0.293	0.3739	0.4	2.763
8	1	1	1	0.471	0.3739	0.4	3.663
9	0	0	0	0.371	0.3157	0.32	2.959
10	0	0	0	0.371	0.3157	0.32	3.056
11	0	0	0	0.371	0.3157	0.32	2.658
12	0	0	0	0.371	0.3157	0.32	2.702
13	−2	0	0	0.235	0.3157	0.32	1.088
14	2	0	0	0.586	0.3157	0.32	2.767
15	0	−2	0	0.371	0.225	0.32	1.521
16	0	2	0	0.371	0.443	0.32	3.278
17	0	0	−2	0.371	0.3157	0.2	1.834
18	0	0	2	0.371	0.3157	0.5	2.446
19	−2	0	0	0.235	0.3157	0.32	1.058
20	2	0	0	0.586	0.3157	0.32	2.940
21	0	−2	0	0.371	0.225	0.32	1.664
22	0	2	0	0.371	0.443	0.32	3.376
23	0	0	−2	0.371	0.3157	0.2	2.245
24	0	0	2	0.371	0.3157	0.5	3.444

For the purposes of this research, i.e. surface roughness measurements, a US-based MarSurf PS1 device from Mahr Metrology was used. The mean arithmetic roughness Ra was chosen as the output parameter [15, 16].

3 Model Implementation and Discussion

Data processing according to the three-factor plan of the second-order experiment (performed in the Minitab software package includes regression analysis and dispersion analysis as well as evaluation of the accuracy of the mathematical model.

To obtain the most relevant data for surface roughness, two models were applied. The first was a linear model with two forms presented: no interaction of the studied parameters and interaction of the input variables. After that, a square model with the interaction of input variables is presented. The resulting models of these two and three types of equation are shown below (3), (4) and (5).

$$R_a = -3,73 + 5,23v + 10,14f_z + 2,74a \tag{3}$$

$$R_a = -2,04 - 5,6v - 3,9f_z + 18,5a + 57,4vf_z - 22,7va - 22,9f_za \tag{4}$$

$$\begin{aligned} R_a = & -16,03 + 25,9v + 32,8f_z + 31,2a + 57,4vf_z - 36,66v^2 - 52,7f_z^2 \\ & - 18a^2 + 53,1vf_z - 23,4va - 23,7f_za \end{aligned} \tag{5}$$

Based on the analysis of variance, the following values for model adequacy were obtained, Table 5.

Table 5. Adequacy assessment of the proposed models

The type of equation	Lack-of-Fit (p-value)	Decision
Linear without interaction	0,010	The model is not adequate
Linear with iteration	0,008	The model is not adequate
Square with interaction	**0,106**	The model is adequate

When comparing the obtained P-values for the corresponding models with the reference P-value of 0.05, it is concluded that for both cases the linear equations should be rejected [17]. Accordingly, a quadratic model with an interaction was chosen as a representative model and the results obtained from it are presented below.

Figure 3 shows the influence of each individual parameter on the arithmetic mean surface roughness. It can be concluded that the speed and depth of cut at lower levels certainly have a positive effect on the achievement of the target value, i.e. minimum surface roughness. Then, by further increasing their values, they reach the most unfavourable parameters in terms of surface roughness. While at the highest levels they have a positive tendency, that is, irregularities on the machined surface are reduced, i.e. roughness is improved. On the other hand, the tooth per the constantly has a negative effect on the roughness of the machined surface.

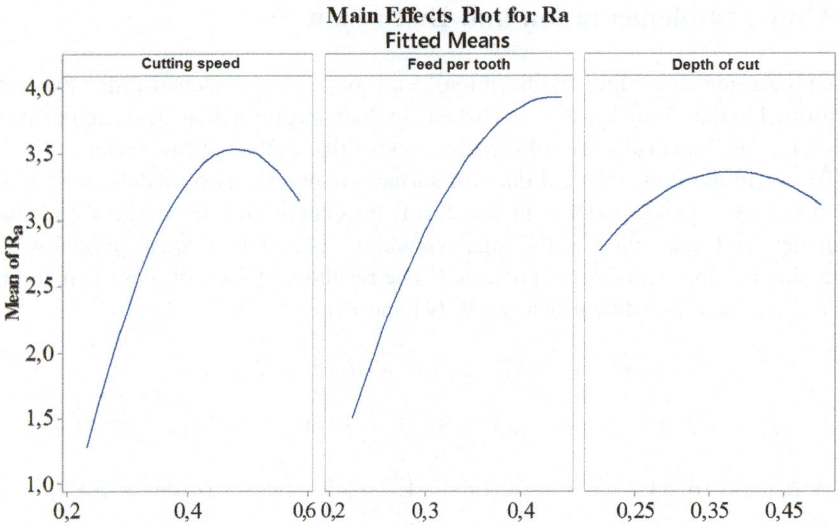

Fig. 3. Effect of the variables examined changing

In Fig. 4, a diagram is shown the interaction of the parameters studied. From Fig. 4a can be seen that there is an interaction of cutting speed and feed per tooth at the middle and higher levels of feed. On the other hand, there is no interaction at the lowest level of feed per tooth. Looking at Fig. 4b, can be seen that cutting speed and cutting depth at all levels have an interaction. Finally, the interaction feed per tooth and cutting depth is present in most cases except at the interval of change of cutting depth in interval from 0.2 mm to 0.35 mm per tooth.

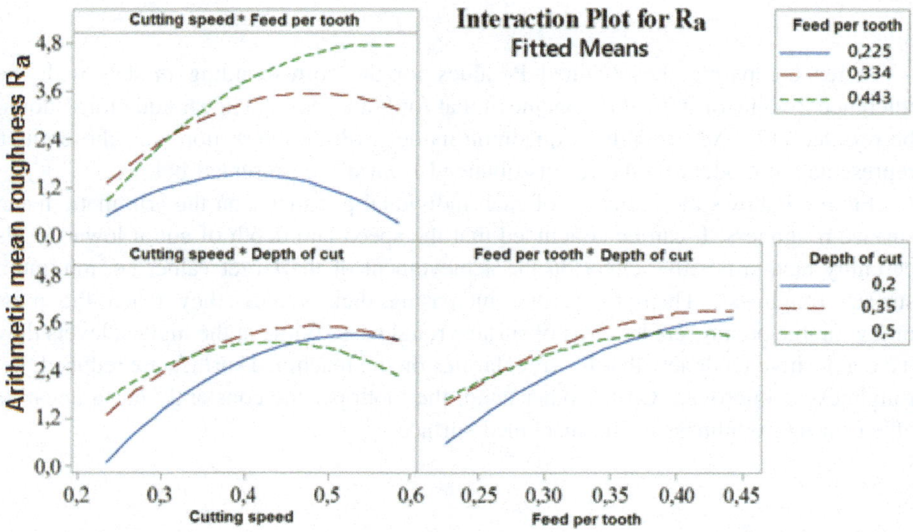

Fig. 4. Interaction of examined variables

From the look at the analysis of variance, it is concluded that the depth of cut has the least impact. The highest influence has the feed per the tooth on the roughness of the machined surface. The view of the influence of the interaction of individual parameters, the proposed equation should retain the square degree of cutting speed and feed per tooth, while the other parameters should be removed (Table 6).

Table 6. Analysis of Variance - quadratic model with interactions

Source	DF	Adj SS	Adj MS	F-Value	P-Value
Model	9	17,0748	1,89720	15,15	0,000
Linear	3	6,7019	2,23398	17,84	0,000
Cutting	1	3,2708	3,27076	26,12	0,000
Feed	1	4,7338	4,73375	37,81	0,000
Depth	1	0,0950	0,09497	0,76	0,398
Square	3	2,7130	0,90435	7,22	0,004
Cutting * Cutting	1	2,6253	2,62525	20,97	0,000
Feed * Feed	1	0,8117	0,81173	6,48	0,023
Depth * Depth	1	0,3497	0,34975	2,79	0,117
2-Way Interaction	3	0,7941	0,26468	2,11	0,144
Cutting * Feed	1	0,5216	0,52155	4,17	0,061
Cutting * Depth	1	0,1975	0,19747	1,58	0,230
Feed * Depth	1	0,0731	0,07315	0,58	0,457

The reliability of this model is reflected by the correlation coefficient R^2. The coefficient of multiple correlation R^2, which represents the fraction of explained variability in total variability, is 0.91. The adjusted coefficient of correlation R^2, which does not depend on the number of variables added to the model, is 0.85. R^2 for prediction in our experiment amounts to 0.63.

Another confirmation of the accuracy of the obtained model and the experimentally obtained data is also reflected in Fig. 5. It can be seen that the obtained residuals follow the centre line represented by the linear equation. That the experimental data collected is well collected can be seen from the other diagrams in Fig. 5.

Figure 6 shows a 3D diagram of the two most influential variables for the response surface, namely the arithmetic mean roughness. It can be seen that at the slightest feed per the tooth, get the best surface roughness. As regards the cutting speed, when processing cast iron with high chromium content, it is necessary to avoid a speed range of 0.4 to 0.5 mm/s.

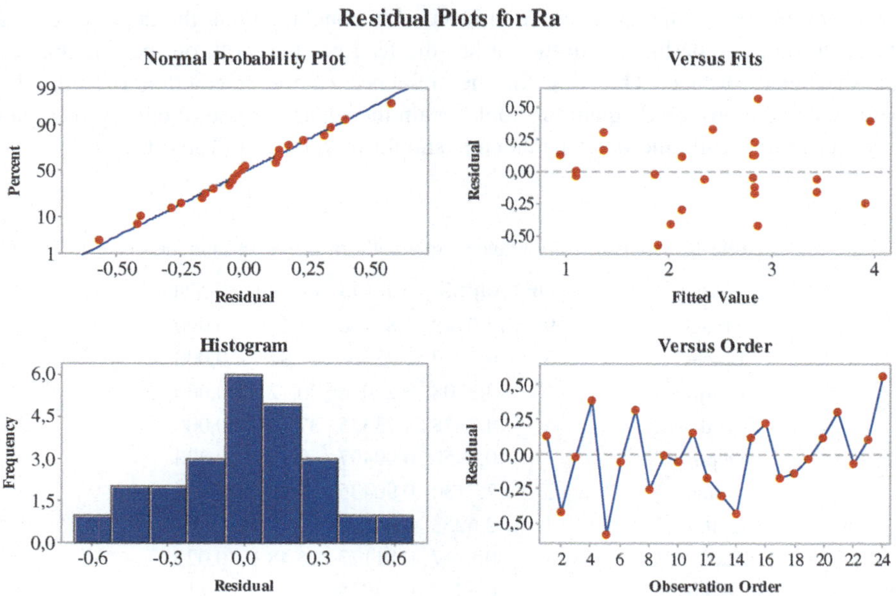

Fig. 5. The residual diagrams

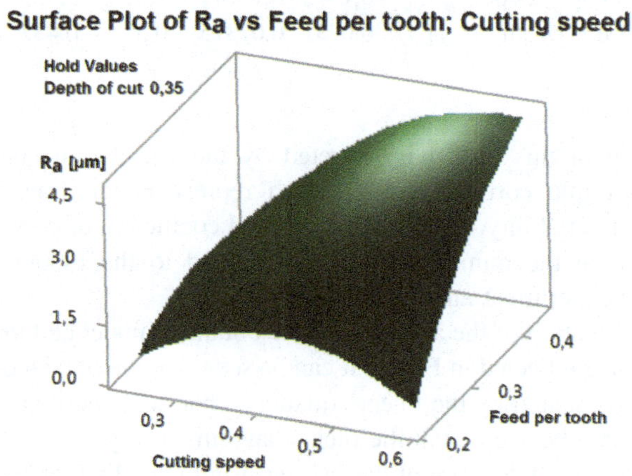

Fig. 6. 3D diagram of the two most influential variables, cutting speed and feed per tooth

4 Conclusion

Modern production needs to improve the quality of the machined area daily, increase productivity, but also reduce production costs. These requirements are largely realized by applying modern modelling techniques and simulating the output characteristics of the machining process. Production without a computer is unthinkable today.

The accuracy of the dimensions, shape and quality of the surface to be machined must be fulfilled. The operation and control of the cutting conditions tends to produce the desired results. There are also factors that cannot be completely predicted. The tendency is to minimize these factors. Cutting itself is a very dynamic process with frequent vibrations.

The conclusions drawn from this research must be taken into account in the further analysis of this topic, and they are presented in order:

Workpiece material, high-chromium cast iron has low machinability and requires precise selection of both cutting regimes and machining tools, specifically cutting inserts that can successfully process this material.

The input data in the experiment itself are measured values of the surface roughness parameters. These data depend, first of all, on the measuring instrument itself and on the overall measurement conditions.

Many software engineering and mathematical processing software are present in engineering practice. Their use is of great importance both in terms of accuracy of results and considerable time savings.

One of the conclusions is the impact of tool wear that has been somewhat neglected. Using the same cutting pad until it breaks and loses its function leads to the assumption that this fact has greatly influenced the results. With a better-quality cutting tool, its more frequent replacement, and a newer machine tool, the influence of some factors would be reduced, which would overall lead to an improvement in the quality of the machined surface and, consequently, an improvement of the finished product.

When conducting such tests, can be create a database, both internal and external for access to others, to know as closely as possible what affects the quality of the machined surface, so as not to wander and waste time processing some of the still unknown by experience materials with machining condition.

Finally, it should be noted that this experiment served to determine a more favorable roughness in relation to the examined processing parameters. Certainly, if the industrial sector would require even more favorable roughness, other factors affecting the cultivated area would have to be introduced. This primarily refers to the geometry of the tool.

Acknowledgment. The paper is the result of the research within the project TR 35015 financed by the ministry of science and technological development of the Republic of Serbia.

References

1. Savković, B.: Modeliranje funkcija obradivosti pri procesu obrade glodanjem, Doktorska disertacija, Fakultet tehničkih nauka (2015)
2. Ćirković, N., Hadžistević, M., Jurković, Z.: Analiza uticaja režima obrade na aritmetičku hrapavost obrađene površine pri čeonom glodanju. Zbornik radova Fakulteta tehničkih nauka **26**(6), 1311–1314 (2011)
3. Savkovic, B., Kovac, P., Dudic, B., Gregus, M., Rodic, D., Strbac, B., Ducic, N.: Comparative characteristics of ductile iron and austempered ductile iron modeled by neural network. Materials **12**(18), 1–16 (2019)

4. Laird, M., Gundlach, J., Roehring, K.: Abrasion-Resistant Cast Iron Handbook. American Foundry Society, Schaumburg (2000)
5. Kopyciński, D.: Inoculation of chromium white cast iron. Arch. Foundry Eng. **9**(1), 191–194 (2009)
6. Studnicki, A.: Effect of boron carbide on primary crystallization of chromium cast iron. Arch. Foundry Eng. **8**(1), 173–176 (2008)
7. Kopyciński, D., Guzik, E., Siekaniec, D., Szczęsny, A.: Analysis of the high chromium cast iron microstructure after the heat treatment. Arch. Foundry Eng. **14**(3), 43–46 (2014)
8. Hann, S.K., Gates, J.D.: A transformation toughening white cast iron. J. Mater. Sci. **32**(5), 1249–1259 (1997)
9. Kopyciński, D., Piasny, S., Kawalec, M., Madizhanova, A.: The abrasive wear resistance of chromium cast iron. Arch. Foundry Eng. **14**(1), 63–66 (2014)
10. Suchoń, J., Studnicki, A., Przybył, M.: Stereology of carbide phase in modified hypereutectic chromium cast iron. Arch. Foundry Eng. **10**(2), 169–174 (2010)
11. Kovač, P.: Modeliranje procesa obrade-Faktorni planovi eksperimenta. FTN, Novi Sad (2006)
12. Pokusova, M., Brúsilová, A., Šooš, L'., Berta, I.: Abrasion wear behavior of high-chromium cast iron. Arch. Foundry Eng. **16**(2), 69–74 (2016)
13. Ravi, A.M., Murigendrappa, S.M., Mukunda, P.G.: Experimental investigation on thermally enhanced machining of high-chrome white cast iron and to study its machinability characteristics using Taguchi method and artificial neural network. Int. J. Adv. Manuf. Technol. **72**, 1439–1454 (2014)
14. Zhou, J.M., Andersson, M.: Machinability of abrasion resistance cast iron with pCBN cutting tools. J. Mater. Manuf. Processes **23**(5), 506–512 (2008)
15. Balos, S., Rajnovic, D., Sidjanin, L., Savkovic, B., Kovac, P., Janjatovic, P.: Tensile and fatigue properties, machinability and machined surface roughness of Al-Si-Cu alloys. Materia-Rio de Janeiro **24**(3), 1–13 (2019)
16. Kovac, P., Savkovic, B., Rodic, D., Aleksic, A., Gostimirovic, M., Sekulic, M., Kulundzic, N.: Modelling and optimization of surface roughness parameters of stainless steel by artificial intelligence methods. In: Proceedings of the International Symposium for Production Research, Vienna, Austria, pp. 3–12 (2019)
17. Simunovic, K., Simunovic, G., Saric, T.: Single and multiple goal optimization of structural steel face milling process considering different methods of cooling/lubricating. J. Clean. Prod. **94**, 321–329 (2015)

Productivity and Performance Management

A System Dynamics Model for Long-Term Performance Monitoring of Projects Related to Oil Production in Iran

Fatemeh Masoudian[1(✉)] and Kamran Rezaie[2]

[1] Monitoring and Evaluation, TÜV NORD Iran, Tehran, Iran
fmasoudian@tuvnordiran.com
[2] Faculty of Industrial Engineering, University of Tehran, Tehran, Iran
krezaie@ut.ac.ir

Abstract. Investment is needed in all societies to sustain economic growth and capital. In fact, investment is the country's economic engine. It is necessary to assess the profitability of the investment. The assessment may include the technical, financial, economic, environmental, social, and political aspects. Projects in the oil industry are complex and have wide effects on society and environment. Therefore after careful financial studies, other aspects of their implementation should be evaluated. However, environmental impact and social consequences resulting from the implementation of these projects are seemingly non-economic factors, but ultimately they have economic effects.

System Dynamics (SD) is a proper and useful tool for analysing things that happen in the oil industry. A simple model that can simulate economic and social context in which the project is implemented, can analyse long-term results of ignoring environmental and social aspects.

Based on the result of simulation with the SD model in this study, ignoring environmental and social issues has a negative impact on Iran's GDP.

Keywords: Performance monitoring · System Dynamics · Economic assessment · Social assessment · Environmental impact assessment

1 Introduction and Literature Review

Iran has the world's second largest gas reserves and 4th largest oil reserves. This should lead to sustainable development, with significant economic growth, social welfare and high quality of life; but this is not the case.

If the economic and social context in which oil projects are implemented can be simply modelled, it could be possible to analyze the long-term performance of oil industry related projects implementation. A comprehensive and long-term view of oil industry projects across the country and the study of their economic, social and environmental impacts will enable the country to move towards a sustainable development.

N. M. Durakbasa and M. G. Gençyılmaz (Eds.): ISPR 2020, LNME, pp. 537–545, 2021.
https://doi.org/10.1007/978-3-030-62784-3_46

System Dynamics
System dynamics is an approach to understand the nonlinear behavior of complex closed systems over time and the behavior of the system is due to the interaction of its components. The characteristics and behavior of a system is not equal to summation of the behavior of its components, but it is the result of the interaction of its components.

Thus, it can be said that this approach is one of the best approaches to study the behavior of complex systems over time. Complex systems are like economic and social systems and subject to the existing limitations, SD is a very useful approach.

The SD methodology, which is adopted in this research, is a modeling and simulation technique specifically designed for long-term, chronic, dynamic management problems [1]. It focuses on understanding how the physical processes, information flows and managerial policies interact so as to create the dynamics of the variables of interest. The totality of the relationships between these components defines the "structure" of the system. Hence, it is said that the "structure" of the system, operating over time, generates its "dynamic behavior patterns". It is most crucial in SD that the model structure provides a valid description of the real processes. The typical purpose of a SD study is to understand how and why the dynamics of concern are generated and then search for policies to further improve the system performance.

2 Problem

As mentioned, evaluating oil industry projects in Iran, solely based on financial aspects and ignoring other aspects does not lead to a sustainable development. Since oil projects in Iran, have a great impact on country's economic, social and environmental situation and also a huge volume of projects in the country is related to the oil industry, it is necessary to pay attention to the long-term performance of the project, not just from the perspective of the project institution, which is always profitable as of low cost and oil related market price.

A comprehensive and long-term view of oil industry projects across the country and the study of their economic, social and environmental impacts will enable the country to move towards a sustainable development.

3 Model

The model assumes that Iran is an economic entity and GDP is considered as a measure of its economic performance. Ignoring recent sanctions, according to the Central Bank, the share of oil in Iran's GDP is high, until 2017, it was around 25%. Each year, a percentage of GDP is assigned to invest in the oil industry. Given the trend of government investment in the oil industry in recent years, it is assumed that 7% of GDP will be assigned to invest in oil projects. In order to see the impact of changes in investment policies on GDP, this ratio has been considered as a separate variable.

Investment in the oil industry, while having an impact on GDP growth, also affects the quality of the environment, and on the other hand, the Human Development Index.

The three variables of the model, GDP, EPI and HDI:

GDP = Gross Domestic Product
EPI = Environmental Performance Index
HDI = Human Development Index

are considered as economic, environmental and social indicators, respectively.

3.1 Variables and Reference Trends

1) *Gross Domestic Product*
Gross Domestic Product is the total value of goods produced and services provided in a country during one year. Among the macroeconomic indicators, GDP is of particular importance, because not only is it used as the most important indicator of economic performance of a country, many other macroeconomic items are also its by-products. If we consider the whole country as a company and an economic entity, GDP can be a measure to evaluate its economic performance.
2) *Environmental Performance Index*
The University of Yale's Center for Environmental Laws and Policies with the collaboration of other groups, including the International Center for Geology at Columbia University and the World Economic Forum, annually study the environmental performance of countries [3]. The EPI ranks countries in terms of important environmental issues. Protecting human health and protecting the ecosystem are the two main issues for measuring EPI.

With the focus on these two issues and in nine sub-categories (1 - health, 2 - air quality, 3 - water and sewage, 4 - water resources, 5 - agriculture, 6 - forestry, 7 - fisheries, 8 - indigenous and biological diversity, 9 - climate and energy), totally 20 indicators are studied in each country and finally one number is announced as the environmental performance index for each country. As the best possible existing databased, this index has been chosen to evaluate environmental performance of Iran over years.

Many studies have been conducted on the relationship between environmental quality and economic growth [4, 5]. It has been proven that there is a correlation between these issues, but what effect and how much is unclear.

The Environment Kuznets Curve expresses a hypothesis about the relationship between the environment and economic development [6]. According to this hypothesis, economic development first has negative effect on the quality of the environment and then acts in reverse and improves the situation (Fig. 1).

In this model, based on this hypothesis, the effect of GDP growth on EPI is assumed first in the unfavorable direction and then in the desired direction. The Kuznets curve is very similar to the function sin(x) and this function is considered as

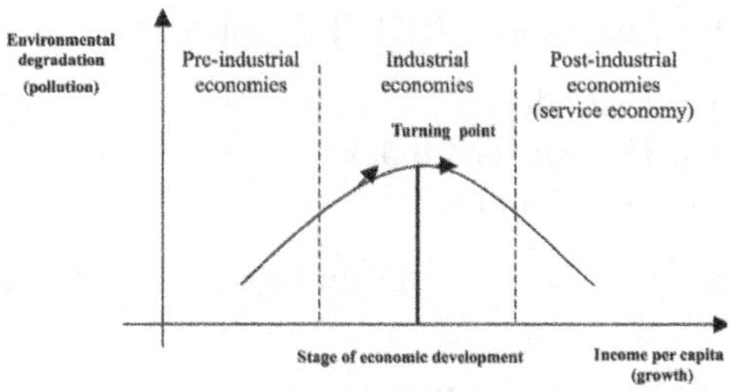

Fig. 1. Environment Kuznets curve

the reference trend in the model. Obviously, the interaction of EPI on GDP is the inverse function.

$$\begin{aligned} &\text{effect GDP on EPI} = \sin(x) \\ &\text{effect EPI on GDP} = \sin^{-1}(x) = \arcsin(x) \end{aligned}$$

3) *Human Development Index*
The Human Development Index is considered as an indicator to show the progress in the social field. The United Nations Development Program conducts a complete assessment for all countries and published the data annually [7]. This index is the geometric mean of the normalized indicators of life expectancy, education and income per capita. A country with a higher life expectancy, a higher income level and higher education, receives higher scores. It is calculated as follows [7]:

$$\text{Life Expectancy Index (LEI)} = \frac{LE - 20}{85 - 20}$$

If the life expectancy at birth is 85, this index is 1, and if it is 20, it is zero.

$$\text{Education Index (EI)} = \frac{MYSI + EYSI}{2}$$

$$\text{MYSI} = \frac{MYS}{15}$$

$$\text{EYSI} = \frac{EYS}{18}$$

The average age of people in school is MYS, with a maximum of 15 years until 2025.

The years that people are expected to spend in school are EYS, and 18 is the number of years that are expected to be spent to get a master's degree.

$$\text{Income Index (II)} = \frac{\ln(GNI) - \ln(100)}{\ln(75000) - \ln(100)}$$

If the income per capita is $ 75,000, this ratio is equal to one, and if it is $ 100, it is equal to zero.

Finally:

$$\text{HDI} = \sqrt[3]{LEI.EI.II}$$

There is a significant relationship between HDI and GDP. According to the calculation formula of this index, it can be concluded that the effect of GDP on HDI is similar to the function $\sqrt[3]{x}$ and this function is considered as the reference trend in the model. Obviously, the interaction of HDI on GDP is its reverse function.

$$\text{effect GDP on HDI} = \sqrt[3]{x}$$

$$\text{effect HDI on GDP} = x^3$$

3.2 Stock-Flow Diagram

See Fig. 2.

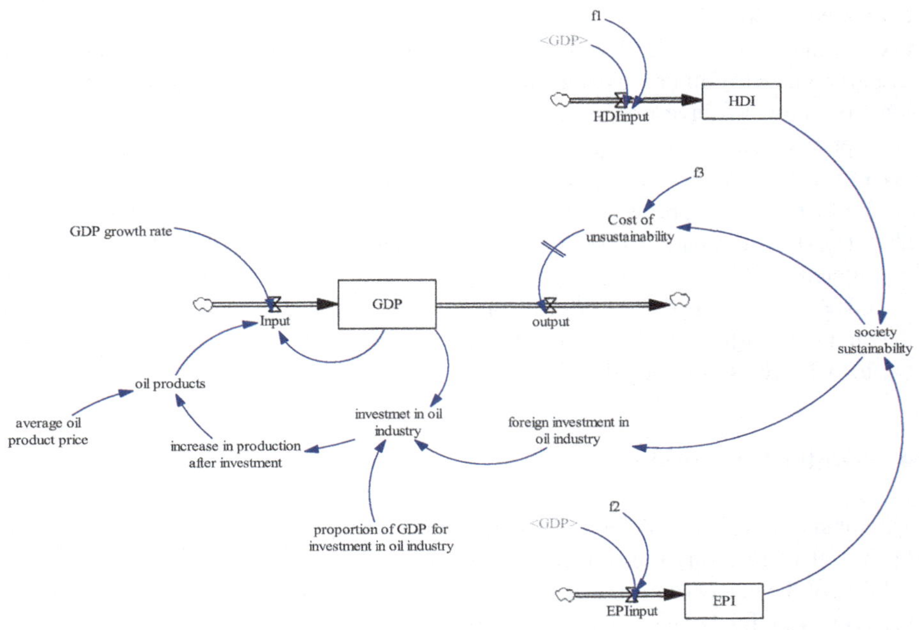

Fig. 2. Stock-flow diagram

Legend:

F1 … The effect of GDP on HDI
F2 … The effect of GDP on EPI
F3 … The effect of sustainability on Instability costs

With these three indicators, Iran can be evaluated as an investment platform from three economic, social and environmental aspects.

Naturally increase in GDP, increases the possibility of investment in oil industry projects. In addition to domestic investments on oil projects, foreign investment is also available and jointly determines the volume of investment in the oil industry. The resulting investment will increase oil production and affect GDP. On the other hand, the two variables HDI and EPI change based on GDP changes [2].

The relationship between GDP and EPI is reversed in the current situation in Iran. This means that GDP growth will inevitably lead to a weakening of the EPI. HDI and EPI affect "society sustainability". Because of direct effect of GDP on both variable, the direct link between GDP and the society sustainability has been discarded. HDI and EPI will have the same effect on society sustainability, and their algebraic averages determine society sustainability.

A stable society is a developed and naturally safe society. Such societies will be the destination of foreign investors. Therefore, there is a direct relationship between the society sustainability and foreign investment in the oil industry, the greater the society sustainability, the higher the annual foreign investment.

On the other hand, instability in society, which can be caused by social or environmental problems, imposes costs on the country. The medical costs incurred to compensate for human losses caused by unhealthy environments (unhealthy climate, unhealthy agriculture, unhealthy fisheries, etc.) and the costs of increasing crime, delinquency, poverty and illiteracy together, account for the costs of unsustainability and of course affects GDP with a delay. However, this delay and also delayed the negative impact of future investments on the environment can be ignorable, because a log from previous investments is existing.

Optimistically, it is assumed that economic growth rate is fixed and positive, because the recent trend in the country, shows negative GDP growth.

Furthermore, according to energy balance sheets, $ 1 billion investment is equivalent to 30 million barrels of crude oil per year and the price per barrel of crude oil is considered $ 40.

GDP output is the cost of unsustainability. The f3 function represents the change in the status of costs imposed on society by unsustainability. Average costs imposed in the next 15 years considered as a constant factor in the model.

4 Model Validation

The main criterion for SD models validation is structure validity, which is the validity of the set of relations used in the model, as compared with the real processes. The validity of the behavior is also important, but less than in forecasting models since first, behavior validity is meaningful only after the structure validity is established and secondly, a point-by-point match between the model behavior and the real behavior is neither possible nor as important as it is in classical forecasting modelling [1, 8]. Therefore, we conducted structural validation tests, which confirmed that the model structure yields meaningful behavior under extreme parameter values.

5 Result and Conclusion

Changes in GDP are observed in Fig. 3, while other variables affecting it are constant. Figure 4 shows the GDP situation if there was no output and a net growth rate of 4% has been applied. There is a huge difference between the two charts over the next 15 years.

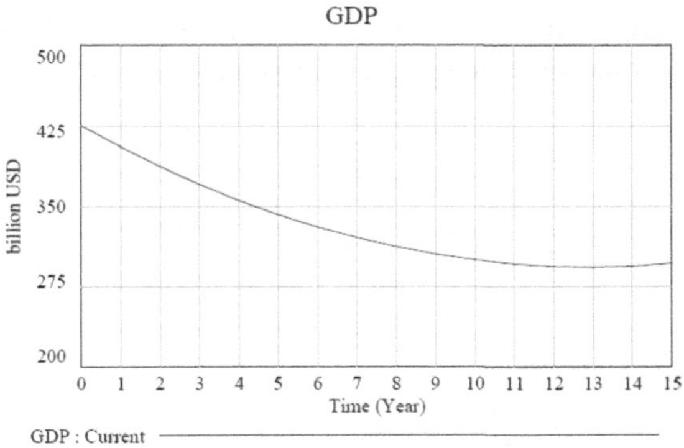

Fig. 3. GDP changes in 15 years

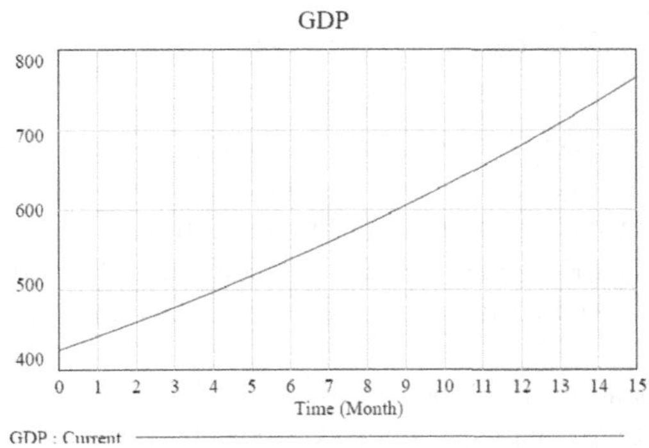

Fig. 4. GDP changes in 15 years with 4% growth

The Human Development Index is directly related to GDP. As long as GDP is a positive number, annual human development should grow at a slow pace. This index is currently 0.77 for Iran and as it can be seen in Fig. 5, diagram slope is decreasing as GDP declines, but the uptrend is still in the index.

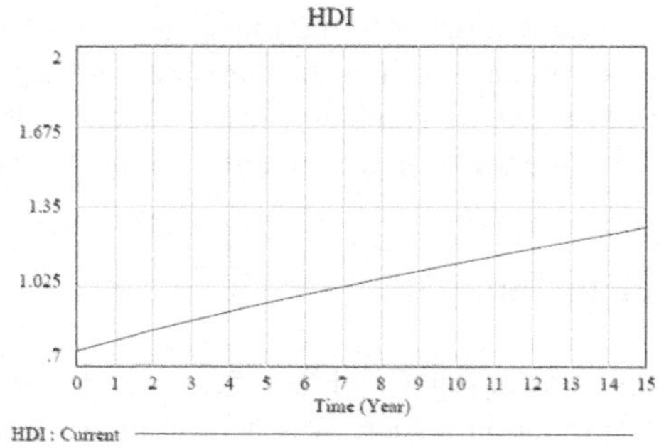

Fig. 5. HDI changes in 15 years

As investments are made annually in the oil sector, and these investments will necessarily be accompanied by an increase in the amount of pollutants, it is therefore obvious that the environmental index rate will be declining (Fig. 6).

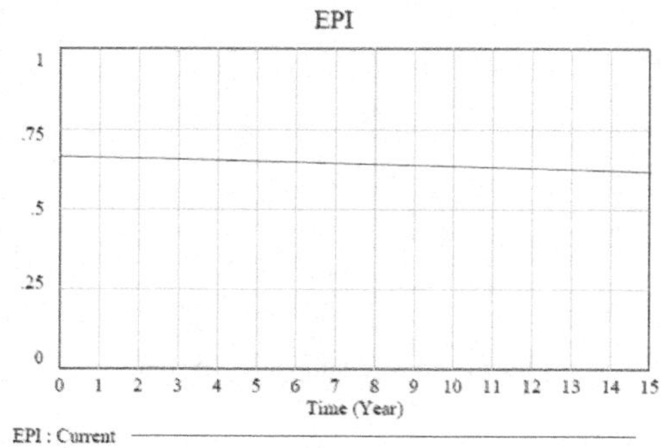

Fig. 6. EPI changes in 15 years

Another variable which is stimulated via this SD model is stability in society. Although Fig. 7 shows increase for this variable, its growth rate is very low. After 15 years, it has risen from 0.71 to 0.94. And of course, this increase is due to the growth that the human development index has had. It seems that if different statistics on human development were available that were less relevant to GDP, this index might have been lower.

As previously mentioned, oil industry projects have diverse effects, therefore there is a need to focus on environmental impacts and social consequences of implementing these projects after careful financial and economic studies. If such studies were conducted, the analyzed data and reports were available in the Ministry of Petroleum. Analyzing the long term performance of oil related projects in Iran, shows although optimistically GDP growth rate has been considered positive for the next 15 years, EPI decreased and HDI increased with a slow rate. Decision makers should really focus on

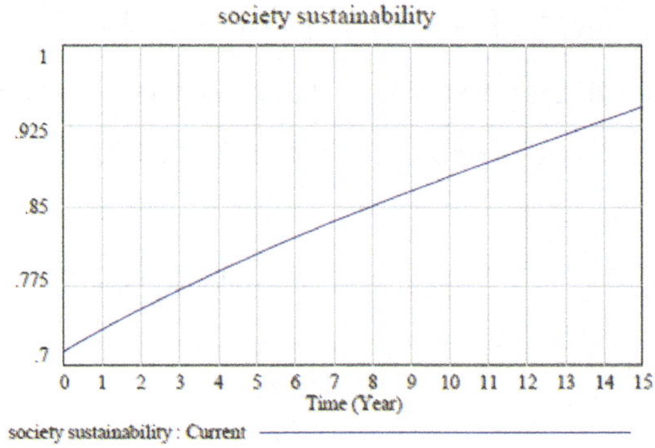

Fig. 7. Stability in society changes in 15 years

performance monitoring of implementing projects and this focus will lead to absorb more foreign investment. With all natural resources available in Iran, stability in society could grow faster and easier.

References

1. Forrester, J.W., Senge, P.M.: Tests for building confidence in system dynamics models. TIMS Stud. Manag. Sci. **14**, 209–228 (1980)
2. Acemoglu, D., Johnson, S., Robinson, J.A.: Institutions as a fundamental cause of long-run growth. In: Handbook of Economic Growth, vol. 1, pp. 385–472 (2005)
3. Index EP. Environmental Performance Index: Global metrics for the environment: Ranking country performance on high-priority environmental issues, vol. 2. Yale Center for Environmental Law & Policy, Yale University, Center for International Earth Science Information Network, Columbia University in collaboration with the World Economic Forum. https://epi.envirocenter.yale.edu/downloads/epi2018policymakerssummaryv01.pdf. Accessed July 2018
4. Piñeiro Chousa, J., Tamazian, A., Vadlamannati, K.C.: Does higher economic and financial development lead to environmental degradation: evidence from BRIC countries. Energy Policy **37**(1), 2009 (2017)
5. Persson, T.A., Azar, C., Lindgren, K.: Allocation of CO2 emission permits—economic incentives for emission reductions in developing countries. Energy Policy **34**(14), 1889–1899 (2006)
6. Stern, D.I.: The environmental Kuznets curve. In: Modelling in Ecological Economics, pp. 173–202. Edward Elgar, Cheltenham (2004)
7. Ross-Larson, B., Coquereaumont, M., Trott, C.: Human Development Report. United Nations Development Programme, 522146, pp. 0–19 (2006)
8. Barlas, Y.: Formal aspects of model validity and validation in system dynamics. Syst. Dyn. Rev. **12**(3), 183–210 (2000)

An SME Examination on the Effect of Transition to Automation Systems on Production Performance

Güzide Karakuş[1(✉)], Saliha Öztürk[2], and Ayşenur Güldoğan[2]

[1] Aviation and Space Sciences Department, Necmettin Erbakan University, Konya, Turkey
gkarakus@erbakan.edu.tr

[2] Industrial Engineering Department, Necmettin Erbakan University, Konya, Turkey
salihaozturk11@gmail.com, guldoganaysenur8@gmail.com

Abstract. Automation technologies are being exposed to negative criticism, as their high costs and will cause high unemployment rates in the future, yet their contributes to production systems are accepted by everyone. Automation systems, which are generally used to reduce production costs and increase product quality, can also create an important competitive power as it gives the work speed and flexibility. The advantages it provides in safety and ergonomics especially in heavy and dangerous works are among the important reasons for the transition to these systems. In other words, the transition to automation systems leads to a significant increase in production efficiency.

In this study, the effect of the transition to automation systems on production performance was investigated. Within the scope of the research, case analysis method consisting of observation, interview and document review stages was used. The process of transition to automation systems and the problems encountered in this process were examined through an interview that carried out with a pioneer company in Mobile Hydraulic Lifting Equipment in Turkey. In line with the production data of the automation systems before and after the transition, it has seen that the transition to automation systems provided advantages such as increasing process and product quality, decreasing error rates, decreasing lost times, eliminating bottlenecks, decreasing wastage, improving traceability etc. In addition to its advantages, there are disadvantageous aspects such as high investment cost, long investment return time, requiring a qualified workforce, long set-up times and employee resistance caused by the fear of losing their jobs. The study is aimed to benefit all experts and practitioners working in the field.

Keywords: Operational management · Manufacturing · Automation systems · Operational performance · Productivity

N. M. Durakbasa and M. G. Gençyılmaz (Eds.): ISPR 2020, LNME, pp. 546–557, 2021.
https://doi.org/10.1007/978-3-030-62784-3_47

1 Introduction

In today's competitive environment, businesses survival depends on having an efficient production management system and sufficient production technologies. Production management is the process of combining an enterprise's current materials, machinery and human resources to produce a certain amount of products at the targeted quality, at the desired time and the lowest cost [2]. To make accurate assessments and make decisions in this process, it is important to establish performance criteria that will provide accurate and apprehensible data in cost, quality, production process and logistics processes [10].

Neely [11] remarked that the importance given to performance measurement both in business life and in academic studies has increased substantially. She listed the reasons for this changing and developing interest in business performance as follows; the changing nature of the business, increased competition, specific improvement initiatives, national and international awards, change in organizational roles, changing external demands and the power of information technology. This change and the need for performance evaluation, which the author has portrayed clearly at the end of the 90s, has become more important with the rapid change in information technologies and the transition to automation systems in the 2000s.

Automation systems are mainly preferred in enterprises to increase production efficiency and thus improve their competitiveness. But the transition to automation systems that require large budgets is not easy. Cooperating with suppliers who can determine the needs of the business correctly and having a qualified human resources structure is very important in terms of the effectiveness of the system. Although it is easier to meet these requirements for large enterprises, for SME companies that have to work with scarce resources, this situation can sometimes cause significant losses.

The relationship between automation systems and production performance is an up-to-date subject that can provide businesses with foresight before such a major investment. In this direction, this study aims to reveal the effect of the transition to automation systems on production performance. In this context, this study has two research questions;

a. What kind of problems do businesses have in the transition to automation systems?
b. What is the effect of the transition to automation systems on production performance?

Within the scope of the research, the case analysis method consisting of observation, interview and document review stages was used. Interviews were carried out in Turkey's pioneer company in the crane manufacturing, the transition to automation systems and problems encountered in this process have been investigated. In line with the production data of the automation systems before and after the transition, the effects of the transition to automation systems on the production performance have been revealed.

2 Production Performance

Performance assessment plays a significant role in setting the targets, achieving the targets and setting the future action plans in line with the findings obtained [14]. In this way, the enterprise will be able to identify its shortcomings by evaluating its past activities, work to eliminate them, identify the factors that affect performance, and use its resources in this direction, and reach more realistic targets in a timely and more efficient way [6]. So performance measurement is a vital issue for businesses. Although many methods have been implemented in the literature about production performance, which has an important place in business performance, businesses do not have enough knowledge about this issue.

Considering the history of performance evaluation; In the 1970s, it is seen that performance evaluation was made with the evaluation of past accounting data and the inadequacy of this evaluation is known. Multidimensional evaluation methods were revealed in the 1980–1990s. Owing to these methods, it is seen that non-financial data are evaluated with external and future-oriented criteria. Then, new methods that different processes can also be taken into account in performance measurements were revealed [13]. In the age of globalization, businesses are increasingly concerned about the importance of efficiency and performance measurement [4].

To determine the appropriate, valid, reliable, open, easily measurable, accurate performance criteria for production processes is necessary for effective performance management [10]. However, it is not possible to develop a reliable and valid scale as in the fields of operational management, organizational behaviour, psychology or business administration. Although this situation prevented the development of a common language, the main requirement is; establishment of a company-specific performance evaluation system for production management [9].

Productivity, which is one of the most important concepts within the scope of production performance, is the ratio between the output (goods and services) obtained in line with the activities of the enterprise and the input used for these activities (labour, energy, materials, money). Many criteria which are seen as a simple fraction or ratio, such as safety, lost times, personnel circulation, attitude and behaviour, have a significant impact on productivity [8].

3 Automation System

The automation system is defined as implementing the works, that are implemented or controlled by using manpower, by machines through automatic systems. In the system, if machine power is used more than manpower (e.g. petrochemical plants), full automation can be mentioned, and if manpower is used more than machine power (e.g. use of robot welding), partial automation can be mentioned. Automation technologies are classified as software and network technologies, industrial automation, building automation and machine automation. The most important benefits of the automation system to the business are; strict quality control, improvement in repeatability, reduction of wastage, decrease in labour intensity and increase in productivity [12].

Although the high investment cost in automation systems is the most important obstacle for this transition, the advantages it provides in the long term are the reasons why businesses prefer these systems. Even if they do not have full automation systems, they often prefer the use of partial automation in their business processes. Especially in hard, dangerous or repetitive work, the use of automation creates positive effects on productivity and this improves performance. The effects of automation systems on efficiency are stated below [12];

- Increasing production by preventing manual delays,
- Increasing productivity by ensuring the optimum efficiency of the machine,
- Increasing productivity by eliminating the need for reprocessing,
- Increasing productivity by reducing personnel-related errors and increasing quality,
- Contributing to increased productivity by ensuring traceability.

4 Implementation

Many studies have been done in the national and international literature on production performance and efficiency measurement. However, academic studies are not compatible with practice. Production performance in enterprises is generally evaluated via partial productivity measurements or personnel performance. Felix and Riggs [8] stated that this complicatedness problem can be overcome by combining all the important efficiency criteria of operation in an interrelated format with the Objectives Matrix which they developed.

Balkan [5] investigated the reasons for the decrease in production capacity through efficiency parameters in a company operating in the textile sector using the Ramsey Model. Basumerda, Rahmi and Sulistio [4] conducted an efficiency analysis of a business that provides storage services using the project function or Riggs model, which is also suitable for use in the service industry.

While Beamon [7] conducted a performance measurement research in material transportation, Savaş [12] investigated the effect of automation use on productivity in a textile enterprise in her master's thesis. At the end of the study, she concluded that the system increases productivity with the advantages of increasing quality and reducing labour costs.

Kabadayı [10] and Bayyurt [6] conducted studies on performance indicators by mentioning the importance of performance evaluation. Akdeniz and Durmaz [1] examined the efficiency performance of a mould enterprise using the RIGGS Model. However, as mentioned in the previous section, production performance is not an area that can be measured on a certain standard. Basic formulas and models need to be adapted to the production system and process structure of the business, in other words, a performance measurement system suitable for the business must be designed. Sharing the results by applying such business-specific studies to each sector and different sized enterprises will make important contributions to the development of new performance variables. In this context, this study has two research questions;

a. What kind of problems do businesses have in the transition to automation systems?
b. What is the effect of the transition to automation systems on production performance?

In the direction of the specified questions, the case analysis method consisting of observation, interview and document review stages was used. The process of transition to automation systems and the problems encountered in this process were examined through interviews with the authorities of the Supply Chain Management in Hidrokon Hydraulic Machine Industry and Trade Inc. [15] which is Turkey's pioneer company in the crane manufacturing. The reasons for choosing the business examined in the application of the research are; crane manufacturing is a labour-intensive area, the transition to automation systems in crane manufacturing is very low in Turkey and these SME-level enterprises are biased against new technologies.

In the implementation section, the effects of the transition to automation systems on the production performance are revealed in line with the production data of the automation systems before and after the transition.

4.1 Problems Experienced in Transition to Automation Systems

In this section, a face-to-face interview was conducted by using the semi-structured interview technique with the authorities of the Supply Chain Management of Turkey's leading crane manufacturer. Before the interview, semi-structured questions were prepared in line with the information obtained in the literature research and were asked during the interview. During the interview, the company officials provided candid answers to our questions with document presentations and examinations at the production area. Findings related to the interview are as follows;

First of all, information about the products and production processes of the company was presented by the company officials. Hidrokon Hydraulic Machine Industry and Trade Inc. is one of the pioneers in the business sector in Turkey which produces mobile (on-vehicle) lifting equipment in different length and tonnages. The company has made continuous investments in new technologies and systems to improve its production performance in line with customer demands and increasing its competitiveness in the market.

- The overall process flow of hydraulic lifting equipment is as follows:
- Sheet metal material cutting,
- Bending the sheet metals which has cut in line with the project,
- Tacking and welding of the parts which has bent,
- Performing the necessary drilling operations on the parts by using the machining method in line with the project,
- Assembling of the chassis and boom groups (placing the chassis on the vehicle, mounting and installing the foot extension and pressing booms, mounting the ring gear and upright boom to the chassis, mounting the cylinders of the main boom and extension booms with the upright boom),
- Adding hydraulic cylinders to this assembly and adding hydraulic installation system,
- Painting and testing of the machine.

Some process steps (bending, cylinder manufacturing, etc.) in the production process are carried out with the procurement of services in line with the need to concentrate on more value-added works in the machine. The company has made significant investments to use automation systems for the stages that have an important place in the processes and some cases causing bottlenecks. Accordingly, the company invested in welding robots to improve both the quality of the welding process and to minimize losses due to standby. First, a robot was purchased for the main boom assembly, then two more were purchased and included in the system.

Firstly, the company official was asked, "what processes are used in the automation systems existing in the enterprise?". The response given by the relevant officers is "Over the years, partial automation systems have been installed in our business in line with the needs (increasing the quality of the process, reducing time losses, minimizing errors, eliminating bottlenecks, etc.). Two more robots were added to the first welding robot installed for a welding process within five years, and it was made useable in three processes. Similarly, partial automation systems were come in to use in machining processes, drilling and subassembly processes."

In response to "In this study, we want to examine the effect of using welding robots which are used in the main boom manufacturing process on production efficiency. Can you tell us the approximate value of the investment cost of welding robots in your company?" question, the company official said, "It cost approximately $1,5 million".

In response to "Could you give information about how decisions are made in the transition to automation systems?" question, "In the process of purchasing welding robots, first of all, which processes have key importance on our product quality has been analyzed. Along with the welding process has a significant intensity in our production processes, it also has a key role in the quality of our products. At the same time, the intensity of the welding process was causing a bottleneck in the production flow. In this context, it was decided to purchase a welding robot at first. Over time, the transition to automation systems has been made by similarly evaluating the processes, and it will be done in a similar way" reply was received.

In response to "Did automation systems affect the traceability and efficiency performance of your business?" question, "In automated processes, the processing times of the previous operations, waiting times, lost times, operator information, error rates, etc. are recording. Automatic recording of these records, which were kept by the operator before automation systems, prevented the occurrence of errors and contributed significantly to the traceability. Similarly, the robot's ability to repeat the same work at any time and without any errors has contributed significantly to our production efficiency" reply was given by the Supply Chain Executives.

In response to "Has your personnel costs decreased due to the purchase of a welding robot?" question, "The transition to the welding robot and the other partial automation systems caused a decrease in our staff number. However, our personnel costs have increased, not decreased. Because staff procurement was required to use these automation systems, and operators such as to this qualification were working at least twice the salary of the existing operators." reply was received.

Finally, "What were the advantages and disadvantages provided by the automation systems established in the business?" question is posed and the Supply Chain Officers gave the following response: "All automation systems included in the system, primarily welding robots; provided advantages such as increasing process and product quality, decreasing error rates, decreasing lost times, eliminating bottlenecks, decreasing waste, improve traceability etc. This has led to significant improvements in internal customer (other process owners whom a process offers the process semi-finished products to) satisfaction. In addition to its advantages, there are disadvantageous aspects such as high investment cost, long investment return period, requiring a qualified workforce, long set-up times and employee resistance caused by the fear of losing their jobs".

4.2 Effect of Automation System on Production Performance

After the transition to the automation system and the perceptions of the operating authorities regarding the automation systems were understood, the effect of the system on the production performance was examined and accordingly, the necessary data before and after the use of welding robot which is an automation system, was requested from the company. Employees of the Supply Chain Management of the company presented the necessary data in line with their records.

Within the scope of the evaluation of production performance, the quality, cost and delivery sub-dimensions defined by Amoako-Gyampah and Acquaah [3] are addressed by adding time and labour dimensions used by Akdeniz and Durmaz [1]. The performance indicators of each sub-dimension, the indicators used by Akdeniz and Durmaz [1] are shown in Table 1 by making them suitable for the production structure of the company we examined.

Table 1. Formulas used in performance indicators

Performance indicator	Formulation
Time Performance Indicators	
Make-Ready Time Ratio (Decreasing this ratio will increase the machine capacity utilization rate)	$Make - Ready\,Time\,Ratio = \frac{Make-Ready\,Time}{Production\,Time\,of\,One\,Part} \times 100$
Machine Efficiency Ratio (This ratio shows the efficiency of the machine)	$Machine\,Efficiency\,Ratio = \frac{Quantity\,of\,Parts\,Produced\,Daily}{Daily\,Total\,Actual\,Machine\,Working\,Time} \times 100$ $Daily\,Total\,Actual\,Machine\,Working\,Time = \frac{Daily\,Working\,Time}{Production\,Time\,of\,One\,Part}$
Machine Operating Performance (This ratio is an indicator of how effective the machines work. Approaching this ratio to 100% is an indication that the idle times, downtimes and preparation times of the machines are reduced)	$Machine\,Operating\,Performance = \frac{Malfunction\,Time}{Total\,Production\,Time} \times 100$ $Total\,Production\,Time = Daily\,Working\,Time \times Number\,of\,the\,Days\,Worked\,per\,Week$

(continued)

Table 1. *(continued)*

Performance indicator	Formulation
Compliance with Planned Production Time (This ratio is an effectiveness indicator that evaluates the business performance by clearing the effects of the factors beyond the control of the business. It is ideal that this ratio to approach 1)	$Compliance\ with\ Planned\ Production\ Time = \frac{Actual\ Production\ Time}{Planned\ Production\ Time} \times 100$
Labour Performance Indicators	
Production Employee Performance	$\text{Production Employee Performance} = \frac{Quantity\ of\ Parts\ Produced\ Daily}{\text{Number of Production Employee}} \times 100$
Delivery Performance Indicator	
Delivery Success	$Delivery\ Success = \frac{Actual\ Delivery\ Amount}{Planned\ Delivery\ Amount} \times 100$
Quality Performance Indicators	
Manufacturing Error Ratio (Low error rates are indicating a reduction in error correction costs (such as labour, materials, energy costs) along with indicating the success of the production unit)	$Manufacturing\ Error\ Ratio = \frac{Reprocessed\ Amount}{Total\ Manufactured\ Quantity} \times 100$
Average Number of Errors	$Average\ Number\ of\ Errors = \frac{Total\ Number\ of\ Errors}{Number\ of\ Products\ Controlled\ by\ Quality\ Control} \times 100$
Cost Performance Indicators	
Production Cost	$Production\ Cost = \frac{Total\ Cost}{Quantity\ of\ Parts\ Produced\ Daily} \times 100$
Unit Labour Cost	$Unit\ Labour\ Cost = \frac{Total\ Labour\ Cost}{Number\ of\ Production\ Employee} \times 100$

1) *Performance Inputs:* The inputs of the company to be used for performance measurement are presented in Table 2.

Table 2. Performance inputs

Inputs	Data before the automation	Data after the automation	Amount of deviation (%)
Number of Production Employee	2 persons	1 person	−50%
Make-Ready Time (Minutes/Day)	45 min	20 min	−55,5%
Production Time of One Part (Hours/Day)	7 h	4,5 h	−35,7%
Number of Parts Produced in One Day (Parts/Day)	1,29 parts	2 parts	−55%
Malfunction Time (Hours/Week)	4 h	1 h	−75%
Daily Working Time	9 h	9 h	–
Total Production Time	45 h	45 h	–
Monthly Cost of an Employee	4.500,00 TL	6.000,00 TL	+33,3%
Unit Part Expense	3.000,00 TL	3.000,00 TL	–

Number of Production Employee: While the number of employees before the automation was two, after automation, one-person labour force was sufficient to do the same job. In other words, the number of production employee decreased by 50%.

Make-Ready Time: Since this period is seen as a lost time for businesses, reduction of this period is gain for the enterprise. While calculating the make-ready time, the duration in minutes of the preparation made in manufacturing in one day at the company was taken as the base. While the make-ready time before the automation was 45 min, after automation it was observed that this period lasted 20 min. In other words, make-ready time was reduced by 25 min, resulted in a gain of 55,5%.

Production Time of One Part: Successful businesses aim to reduce the production time of one part without affecting efficiency. As seen in the study, while the manufacturing time of a part before the automation lasted 7 h, after automation it decreased to 4,5 h and it decreased by 35,7%.

Number of Parts Produced in One Day: Before the automation, while 1 part was completed in 7 h in the welding line that mentioned, after, the processing time of the same part decreased to 4,5 h. The daily working time in the enterprise is 9 h.

Before the welding robot;	After the welding robot;
$Number\ of\ Parts\ Produced\ Daily = \frac{9}{7} = 1{,}29$	$Number\ of\ Parts\ Produced\ Daily = \frac{9}{4{,}5} = 2$

It has been observed that the number of parts produced in one day increased by 55%. In other words, as the manufacturing time decreases, the number of parts produced in one day has increased.

Malfunction Time: Discussed weekly and in hours. While 4 h of loss was experienced before the automation, after the automation, this period decreased to 1 h. This decrease in the malfunction time provided a 75% time-saving.

Daily Working Time: The daily working time of the company is 9 h. This remained the same, without any change before and after the automation.

Total Production Time: The total production time was calculated weekly. The company works 5 days a week. When the "*Total Production Time* = Daily Working Time × *Working Days per Week*" formula was applied; the total production time did not change before and after welding robot use, and it was preserved as 9 × 5 = 45 h.

Monthly Cost of an Employee: While the monthly cost of welding personnel was 4.500 TL before automation, it was 6.000 TL afterwards. When it's calculated as a percentage, this increase corresponds to 33,3%. The reason for this increase is that the automation system requires qualified personnel. Nevertheless, based on the fact that 2 people worked before automation, the labour cost of welding line that mentioned decreased from 2 × 4,500 = 9,000 TL to 6,000 TL afterwards. In this context, while an increase in the monthly cost of an employee was observed, a decrease in the total labour cost of the process was observed.

Unit Part Expense: It is calculated daily. It refers to all kinds of expenses (raw materials, electricity, etc.) per part, excluding labour costs. This expense is 3.000 TL before and after the automation.

2) *Performance Measurement Results:* The outputs obtained for performance measurement results for before and after the robot manufacturing automation use of the main boom welding process of the crane manufacturing company are presented in Table 3.

Table 3. Performance outputs

Outputs	Data before the automation	Data after the automation	Amount of deviation (%)
Make-Ready Time Ratio	(45/420) × 100 = %10,71	(20/270) × 100 = %7.41	−3,3%
Machine Efficiency Ratio	(1,29/9) × 100 = %14,33	(2/9) × 100 = %22,22	+7,89%
Machine Operating Performance (Week)	(4/45) × 100 = %8,89	(1/45) × 100 = %2,22	−6,67%
Compliance with Planned Production Time (Week)	(39/45) × 100 = %86,67	(42,5/45) × 100 = %94,44	+7,77%
Production Employee Performance (Day)	(1,29/2) × 100 = %65	(2/1) × 100 = %200	+135%
Non-conforming Part Amount	4 parts	1 part	−75%
Delivery Success (Week)	[(6,45 − 4)/6,45] × 100 = %38	[(10 − 1)/10] × 100 = %90	+52%
Manufacturing Error Ratio (Month)	(4/25,8) × 100 = %15,10	(1/40) × 100 = %2,50	−12,6%
Average Number of Errors (Month)	(4/25,8) × 100 = %15,10	(1/40) × 100 = %2,50	−12,6%
Production Cost (Day)	4,170,00/1,29 = 3.233,00 TL	6,200,00/2 = 3.100,00 TL	−4%
Unit Labour Cost	300,00/2 = 150,00 TL	200,00/1 = 200,00 TL	+33,3%

5 Conclusion and Recommendations

It is vital for companies to transition to automation systems in production systems and to make this transition at the lowest cost. Because the return times of these systems, which require high investment budgets, are quite long. In this context, it is necessary to switch to automation systems with the right decision-making processes. Making the right decisions in production management is possible by making the right analyses. For this, accurate and timely production data are needed. In this study, the process of purchasing a welding robot to the main boom welding process in Hidrokon Hydraulic Machine Industry and Trade Inc. which is Turkey's pioneer company in the crane manufacturing was examined. The reason for choosing this enterprise is that crane manufacturing is a labour-intensive and heavy-duty sector, companies in the sector are predominantly SMEs, and the rate of the transition to automation is low. Within the scope of the study, it is aimed to investigate what problems businesses experience in transition to automation systems and the effect of the transition to automation systems on production performance.

First of all, the advantages and disadvantages of using welding robots in the enterprise have been revealed as a result of their observations and evaluations by discussing with the company officials. The reason that drives the business to supply welding robots is to increase product quality and to eliminate bottlenecks in the production process. The processing officials stated that all automation systems included in the system, especially welding robots, provide advantages such as increasing process and product quality, decreasing error rates, decreasing lost times, eliminating bottlenecks, decreasing wastage, and improving traceability etc., and this situation provides significant improvements in internal customer satisfaction. They also mentioned that they have disadvantageous aspects such as high investment cost, long investment return time, requiring a qualified workforce, long set-up times and employee resistance caused by the fear of losing their jobs.

It has been seen in the analysis that after the use of the welding robot, the machine efficiency ratio increased by 7,9%, while the machine operating performance decreased by 6,7%. While the compliance with planned production time increased by 7,8%, the production employee performance increased at a high rate of 135%. Although the number of personnel working in the process decreased from two to one, unit labour cost increased by 33% due to the need for qualified personnel.

After the use of the welding robot, the non-conforming part amount decreased by 75%, the manufacturing error ratio by 12,6%, and the average number of errors by 12,6%. In other words, the expectation of quality performance has been met. Although the company did not expect reducing costs, the production cost has decreased by 4% after the use of welding robots. With the use of the welding robot, the total manufactured quantity, in other words, the planned delivery amount has increased. But delivery success has increased by 52% due to the fact that the reprocessed amount has decreased more than the total manufactured quantity, in other words, the actual delivery amount has increased more than planned.

The company officials found the automation systems to be useful in general, and the analyses prove this benefit. In addition to this, it is important to apply the evaluation over the single process, which is the most important limitation of the study, in other processes in order to see the total benefit.

Acknowledgements. The authors would like to thank Hidrokon Hydraulic Machine Industry and Trade Inc. for information sharing and support, and especially Messrs. Ahmet YILMAZ and Fatih Mehmet Yılmaz for their contributions, interest and sparing their time.

References

1. Akdeniz, H.A., Durmaz, F.: Verimliliğin genel performans üzerindeki yansımalarının uygulaması, Dokuz Eylül Üniversitesi İktisadi ve İdari Bilimler Fakültesi Dergisi, vol. 13, no. 2 (1998)
2. Alpay, Ş., Yüzügüllü, N.: Eskişehir ve çevresindeki üretim işletmelerinde üretim yönetimi teknolojilerinin uygulanma düzeyleri. In: V. Ulusal Üretim Arastırmaları Sempozyumu. İstanbul Technical University, 25–27 October 2005
3. Amoako-Gyampah, K., Acquaah, M.: Manufacturing strategy, competitive strategy and firm performance: an empirical study in a developing economy environment. Int. J. Prod. Econ. **111**(2), 575–592 (2008)

4. Basumerda, C., Rahmi, U., Sulistio, J.: Warehouse server productivity analysis with objective matrix (OMAX) method in passenger boarding bridge enterprise. In: IOP Conference Series: Materials Science and Engineering, vol. 673, no. 1, p. 012106. IOP Publishing, December 2019
5. Balkan, D.: Tekstil sektöründe verimlilik ölçümü ve bir uygulama. Tekstil ve Mühendis **26**(113), 79–85 (2019)
6. Bayyurt, N.: İşletmelerde performans değerlendirmenin önemi ve performans göstergeleri arasındaki ilişkiler. Sosyal Siyaset Konferansları Dergisi **53**, 577–592 (2007)
7. Beamon, B.M.: Performance, reliability, and performability of material handling systems. Int. J. Prod. Res. **36**(2), 377–393 (1998)
8. Felix, G.H., Riggs, J.L.: Productivity measurement by objectives. Natl. Prod. Rev. **2**(4), 386–393 (1993)
9. Hensley, R.L.: A review of operations management studies using scale development techniques. J. Oper. Manag. **17**(3), 343–358 (1999)
10. Kabadayı, E.T.: İşletmelerdeki üretim performans ölçütlerinin gelişimi, özellikleri ve sürekli iyileştirme ile ilişkisi. Doğuş Üniversitesi Dergisi **3**(2) (2002)
11. Neely, A.: The performance measurement revolution: why now and what next? Int. J. Oper. Prod. Manag. **19**, 205–228 (1999)
12. Savaş, S.: Bir dokuma işletmesinde otomasyon kullanımı ile verimliliğin arttırılması. Master's thesis, Toros University, Mersin, Turkey (2018)
13. Türker, M.F.: Üretim performansı ölçümü için bir yöntem. Master's thesis, İstanbul Technical University, İstanbul, Turkey (2007)
14. Zaied, A.N., Mansour, M., Mostafa, M.: Evaluating the performance of order fulfillment process in supply chain. Egypt. Int. J. Eng. Sci. Technol. **20**, 38–48 (2016)
15. https://www.hidrokon.com/en/default.asp. Accessed 01 June 2020

Performance Evaluation in the Firms of Turkish Textile Sector

Aylin Duman Altan(✉) and Kemal Güven Gülen

Industrial Engineering Department, Corlu Engineering Faculty,
Tekirdag Namık Kemal University, Tekirdağ, Turkey
{aduman, kggulen}@nku.edu.tr

Abstract. In today's global competitive environment, firms are carried out various activities in order to increase their effectiveness. Performance measurement is a critical assessment tool to improve the business effectiveness. Malmquist total factor productivity calculation is an important method used to determine the success levels of firms. Various output and input data are used in this method. Accordingly, the evaluating the performances of the firms in the textile sector, which are among the top first and second 500 companies declared yearly by the Istanbul Chamber of Industry (ISO), have been calculated through the Malmquist total factor productivity index. Data from seventeen firms are used in the study based on from 2014 to 2018 years. Net assets, equity capital and employee numbers are selected as input factors. And also, net sales, gross value added, pre-tax profit and export data are selected as output factors.

Keywords: Performance evaluation · Productivity · Efficiency · Malmquist index · Textile sector

1 Introduction

Malmquist total factor productivity index method can be used to analyze whether the resources of enterprises are used effectively on a yearly basis. On the other hand, Data Envelopment Analysis (DEA) method enables us to evaluate whether enterprises use resources efficiently depending on input and output factors. Since the Malmquist Index contains the time dimension, the Malmquist Total Factor Productivity Method is one of the alternatives to eliminate this deficiency of the basic DEA method. Malmquist method is one of the alternatives that can be used in decision-making more effectively.

The change in the total factor productivity of the textile firms in the first and second 500 largest firms announced annually by ISO will be examined using the Malmquist total factor productivity index in this study. Our study is organized as the literature review, the data and the methodological framework, the empirical findings obtained from our analysis and the conclusion, respectively.

Malmquist total factor productivity index is a method that is used for measuring the changes in total factor productivity of firms' by time. More technically, Malmquist Total Factor Productivity index is defined as it measures the TFP change between two data points by calculating the ratio of the distances of each data point relative to a common industrial data. The distance function is a measurement that defines the

N. M. Durakbasa and M. G. Gençyılmaz (Eds.): ISPR 2020, LNME, pp. 558–568, 2021.
https://doi.org/10.1007/978-3-030-62784-3_48

production technologies for multiple inputs and multiple outputs without any need for optimizing the objectives. Input distance function defines the production technology according to the most contracted input vector when the output vector is given. Output distance function also defines the production technology according to the most expanded input vector when the input vector is given.

2 Literature Review

Many studies examining efficacy and total factor productivity in various sectors are included in the literature. In this study, the literature review focuses on the analysis of input and output parameters used in both different sectors and studies. As a result of the research, the information related to the studies performed is presented in the Table 1.

Table 1. Summary of literature

Refe rence	Year	Sector	Input	Output
[16]	2018	Manufacturing firms in Bangladesh	Total annual cost Total annual cost of raw materials Total annual cost of electricity Permanent full time workers	Total annual sales
[8]	2017	Automobile/Automobile parts firms in Pakistan	Total assets Operation costs Selling Administrative expenditures	Net sales
[17]	2017	Taiwan's information and communication technology industry	Number of employees Fixed assets Operating expenses R&D expenses	Net revenue
[4]	2017	A textile firm operating in the Aegean Region	Number of employees Total amount of raw materials Number of machines	Number of products The number of customers
[6]	2017	Turkish automotive industry	Net assets Equity capital Employee numbers	Net sales Pre-tax profit Export
[10]	2017	Province-level regions of China Industry	Employment Fixed-assets'capital stock R&D cost	Gross output Export
[3]	2016	Turkish SMEs	Short-term liabilities Long-term liabilities Equity	The sales revenue Net profit
[2]	2016	Defense industries of 21 Country	GDP Expenditures Import Employment Logistic performance index	Total sales Export Profit
[15]	2016	Turkish casting sector	Number of employees Total liabilities Equity T otal expense Fixed assets	Total sales Export Profit
[14]	2016	The sectoral based grouped first 500 and second 500 companies published by ISO.	Sales Revenue/Total Assets Capital/Total Assets Financing Expenses/Total Liabilities Liabilities/Equity Salary and Fees/Sales Revenue Sales Revenue/Current Assets Number of firms	Return on equity Return on assets before depreciation Return on assets Sales profitability

(*continued*)

Table 1. *(continued)*

Refe rence	Year	Sector	Input	Output
[1]	2015	Turkish Paper and Paper Products industry	Current rate Total foreign resources/total assets Equity/total assets Equity/total foreign resources Net sales/equity	Net profit/equity Net profit/total assets
[19]	2012	Turkish ISO 500 firms in weaving and clothing industry	Total net assets Equity capital Number of employees	Net sales Export Gross value added
[7]	2012	Greek manufacturing sectors	Total assets Equity Administrative, distribution and selling expenses	Net profit margin The return on equity The return on assets
[11]	2010	Turkish automotive industry	Total assets Equity Administrative, distribution and selling expenses	Net profit margin The return on equity The return on assets
[9]	2006	SMEs in Korea	Capital Number of staff Fixed assets Supporting funds	The sales amount Profit

3 Material and Method

Malmquist index measurement, first proposed by Malmquist [12], is a method for assessing the relative productivity across different regions or time periods. For each DMU (decision making unit) it is necessary to identify the same inputs and same outputs in DEA. The input and output variables used in the efficiency analyses of firms' have been examined in the literature and some variables in the various studies are shown in Table 1. Depending on the literature, net assets, equity capital and employee numbers are used as input factors. And also, net sales, gross value added, pre-tax profit and export data are used as output factors. Data of seventeen firms are obtained from 2014 to 2018 years based on ISO.

Although there are more firms operating in the periods examined, the data of some firms could not be reached and data is normalized since some companies' pre-tax profits are negative. In these circumstances, the study is limited to 17 firms. It can be seen that 17 DMUs are suitable regarding the input-output numbers used in the study.

The Malmquist productivity index can be interpreted as a measure of total factor productivity change (TFPC). Following Fare et al. [5], this paper adopts the Malmquist productivity change index, referring the emphasis on the equi-proportionate increase of outputs, within the context of a given level of input.

The TFPC is defined as follows in:

$$\frac{d_0^{t+1}(x^{t+1},y^{t+1})}{d_0^t(x^t,y^t)} \times \left[\frac{d_0^t(x^{t+1},y^{t+1})}{d_0^{t+1}(x^{t+1},y^{t+1})} \times \frac{d_0^t(x^t,y^t)}{d_0^{t+1}(x^t,y^t)}\right]^{1/2}$$

where the ratio outside the brackets measures the change in relative efficiency (i.e., the change in how far observed production is from maximum potential production) between years t and t + 1. The geometric mean of the two ratios inside the brackets captures the shift in technology between the two periods evaluated at xt and xt + M, that is,

Technical Efficiency Change (TEC): $\frac{d_0^{t+1}(x^{t+1},y^{t+1})}{d_0^t(x^t,y^t)}$

Technological Change (TC): $\left[\frac{d_0^t(x^{t+1},y^{t+1})}{d_0^{t+1}(x^{t+1},y^{t+1})} \times \frac{d_0^t(x^t,y^t)}{d_0^{t+1}(x^t,y^t)}\right]^{1/2}$

Finally, the TFPC could be due to either change in technical efficiency or change in the technology – technological progress in the industry– or both. The total factor productivity change is the product of technical efficiency change and technological change. In addition, technical efficiency change can be further decomposed into pure technical efficiency change and scale efficiency change.

The computer software DEAP2.1 [20] is used to calculate these indices. The software can measure relative efficiencies based upon DEA which is for input or output and constant and variable returns to the scale. It has also been applied to calculate the TPFC for measuring the influence of time shift in this study.

4 Results and Discussion

The Malmquist and its decomposition index within 17 textile firms during 2014–2018 is calculated by using DEAP2.1. The technical, technological, pure technique, scale efficiency and total factor productivity changes for the 2014–2018 periods are calculated for the 17 firms included in the analysis and the results are shown in Table 2, Table 3, Table 4, Table 5 and Table 6.

The TEC values are shown as in Table 2. While TEC is larger than 1, it indicates the improvement in the firm in technical terms, whereas it is less than 1 indicates that the firm is technically declining. In addition, the change in SEC (scale efficiency change), which is one of the components of TEC, and the change in PEC (pure technical efficiency change) shows whether the company uses managerial effectiveness and whether it uses the appropriate scale in production [11].

As indicated in Table 2, the analysis of the change in technical efficiencies shows that it has slight change during the period 2014–2018. With the exception of the year 2014–2015 (decline in efficiency, which is 4%), the textile industry has reported TECs in the study period (efficiency rose of 0,7%, 2% and 0,3% in the years 2015–2016, 2016–2017 and 2017–2018 respectively). For example, while the average TEC decreased by 4% in the 2014-2015 period; the biggest increase belongs to firm 16 (30.2%) and the biggest decrease belongs to firm 10 (−25,4%). Based on mean TEC data, it can be said that the technical efficiency performance is stable in textile sector.

Table 2. Period-based TEC values of firms

Firm	Period			
	2014–2015	2015–2016	2016–2017	2017–2018
1	1.000	1.000	1.000	0.958
2	0.888	0.950	1.001	1.047
3	1.000	1.000	1.000	1.000
4	1.000	1.000	1.000	1.000
5	0.923	**0.868**	**1.192**	1.065
6	0.943	0.962	**0.958**	**1.093**
7	1.000	1.000	1.000	1.000
8	0.825	**1.406**	1.000	0.954
9	1.000	1.000	1.000	1.000
10	**0.746**	1.059	0.998	1.002
11	1.000	1.000	1.000	1.000
12	0.923	1.052	1.030	1.000
13	1.000	0.952	1.051	1.000
14	0.876	0.981	1.163	1.000
15	1.000	**0.868**	0.987	**0.858**
16	**1.302**	1.119	0.994	1.091
17	1.000	1.000	1.000	1.000
mean	0.960	1.007	1.020	1.003

While TC is larger than 1, it indicates the improvement in the firm in technological terms, whereas it is less than 1 indicates that the firm is technologically declining. The TC values are shown as in Table 3. As indicated in Table 3, the analysis of the TCs show that it has just slight decline in efficiency (which is 1.1%) during the period 2015-2016. With the exception of the year 2015–2016, the textile industry has reported TC's in the study period (efficiency rose of 1,8%, 10% and 19.2% in the years 2014–2015, 2016–2017 and 2017–2018 respectively). For example, while the average TC decreased by 1,8% in the 2014–2015 period; the biggest increase belongs to firm 9 (33.7%) and the biggest decrease belongs to firm 4 (16%).

According to the use of Malmquist productivity index in this paper, therefore, if this amount is equal to 1, it signifies no change in performance, if bigger than 1 it shows performance advancement, and in case it is less than 1 it signifies corporation performance getting worse [13].

Table 3. Period-based TC values of firms

Firm	Period			
	2014–2015	2015–2016	2016–2017	2017–2018
1	1.116	**1.247**	1.087	1.075
2	1.071	1.006	1.110	1.120
3	0.989	0.935	1.127	1.355
4	**0.840**	0.948	1.009	1.204
5	1.059	1.156	1.006	1.203
6	0.956	0.886	1.188	1.086
7	0.987	1.088	**0.900**	1.150
8	1.150	0.926	**1.285**	**0.887**
9	**1.337**	0.915	1.111	**1.701**
10	1.010	1.086	1.179	1.266
11	1.047	1.001	1.105	1.097
12	0.965	1.033	1.129	1.093
13	1.001	0.829	1.101	1.132
14	0.965	1.059	1.207	1.255
15	0.946	1.118	1.078	1.295
16	0.859	0.887	1.069	1.413
17	1.106	**0.812**	1.070	1.121
Mean	1.018	0.989	1.100	1.192

The TFPC's are shown as in Table 4. The analysis of the change in efficiencies (Malmquist indices) shows that productivity has been increasing during the last two period. With the exception of the year 2014–2015 and 2015–2016 (slight decline in productivity, which is 2.3 and 0,4%) the textile industry has reported productivity progress in the study period (productivity rose of 12.3%, 19.5 in the years 2016–2017 and 2017–2018 respectively). For example, while the average TFPC decreased by 2.3% in the 2014–2015 period; the biggest increase belongs to firm 9 (33.7%) and it is apparent that the main source growth for the TFPC is attributed to the technological efficiency change (see Table 2). On the other hand, in the same period, the biggest decrease belongs to firm 10 (24,7%) and it is apparent that the main source downsizing for the TFPC is attributed to the technical efficiency change (see Table 2). With this comparison using Table 2 and Table 3, the source of the change in firm activities is determined on an annual basis.

Table 4. Period-based TFPC values of firms

Firm	Period			
	2014–2015	2015–2016	2016–2017	2017–2018
1	1.116	**1.247**	1.087	1.030
2	0.951	0.955	1.112	1.172
3	0.989	0.935	1.127	1.355
4	0.840	0.948	1.009	1.204
5	0.978	1.003	1.199	1.281
6	0.902	0.852	1.137	1.187
7	0.987	1.088	**0.900**	1.150
8	0.948	1.301	1.285	**0.847**
9	**1.337**	0.915	1.111	**1.701**
10	**0.753**	1.150	1.177	1.268
11	1.047	1.001	1.105	1.097
12	0.891	1.087	1.163	1.093
13	1.001	**0.789**	1.157	1.132
14	0.846	1.040	**1.404**	1.255
15	0.946	0.970	1.065	1.111
16	1.118	0.992	1.063	1.542
17	1.106	0.812	1.070	1.121
Mean	0.977	0.996	1.123	1.195

The Malmquist Index summary of firm means are shown as in Table 4. By comparing these data, the source of the change of average firm productivity is determined rather than the source of the annual change of firm productivity. In this scope, although there is a decrease in the technical efficiency and scale efficiency, an increase occurred in the technology and pure efficiency. When considered on the basis of firm, in 16 (94%) of a total of 17 firms, it is identified that total factor productivities increased. The number 9 firm is whose total factor productivity increases the most with the increase of 23.3%, as seen in the Table 5. It is apparent that the main source growth for the TFPCs are attributed to the technological efficiency change. The number 4 firm is whose total factor productivity decreases the most with the decrease of 23.3%. It is apparent that the main source downsizing for the TFPC is attributed to the technological efficiency change. In addition, the number 16 firm is whose technical efficiency increases the most with the increase of 12.1%. It is apparent that the main source growth for the TEC is attributed to the pure efficiency change. The following comments can be made for some firms whose situation is different. For example, it is apparent that the main source growth for the TEC of firm 16 is attributed to the PEC. On the other hand, it is apparent that the main source downsizing for the TEC of firm 6 is attributed to the SEC.

The PEC measures the ability of decision-making units to convert their inputs to outputs. In addition, the change in PEC and the change in SEC, which is one of the components of TEC, is greater than 1, which means that the firm is successful in managerial efficiency and production in appropriate scale. In this context, the firm 6 should increase the success of production on the appropriate scale.

The Malmquist Index summary of annual mean values are shown as in Table 5. As indicated in Table 5, the analysis of the change in efficiencies (Malmquist indices) shows that productivity has been increasing during the last two period. The Malmquist productivity change experienced by the textile industry as a whole has averaged 6,9% per year and suggest improvement in performance of TFPCs from 2014-2018. As the result showed over the sample period, the average annual rate of technical efficiency change is −0,3% while the rate of technological change is 7,2%. As indicated in Table 5, with the exception of the year 2014–2015 and 2015–2016 (slight decline in productivity, which is 2,3 and 0,04%) the textile industry has reported productivity progress in the study period (productivity rose of 12.3%, 19.5 in the years 2016–2017 and 2017–2018 respectively).

Table 5. Malmquist index summary of firm means

Firm					
	TEC	TC	PEC	SEC	TFPC
1	0.989	1.129	1.000	0.989	1.117
2	0.969	1.076	1.000	0.969	1.043
3	1.000	1.090	1.000	1.000	1.090
4	1.000	**0.992**	1.000	1.000	**0.992**
5	1.004	1.103	1.000	**1.004**	1.108
6	0.987	1.022	1.000	0.987	1.009
7	1.000	1.027	1.000	1.000	1.027
8	1.026	1.050	1.023	1.003	1.076
9	1.000	**1.233**	1.000	1.000	**1.233**
10	0.943	1.131	1.000	**0.943**	1.066
11	1.000	1.062	1.000	1.000	1.062
12	1.000	1.053	1.000	1.000	1.053
13	1.000	1.009	1.000	1.000	1.009
14	1.000	1.116	1.000	1.000	1.116
15	**0.926**	1.103	**0.930**	0.995	1.021
16	**1.121**	1.036	**1.121**	1.000	1.161
17	1.000	1.019	1.000	1.000	1.019
mean	0.997	1.072	1.004	0.993	1.069
<1	6	1	1	6	1
1	9	0	14	9	0
>1	2	16	2	2	16

By decomposing the Malmquist index, it is possible to determine the sources of productivity growth. As explained previously, technical efficiency change (TEC) and technological change (TC) are the efficiency changes (movement of micro finances towards the frontier - catching up) and technological changes (frontier shift) respectively. In this regard, the sources of growth or decline in textile industry are due to TEC, TC, or both. However, from Table 6, it is apparent that the main source of mean TFPC growth is attributed to the mean technological efficiency change (7,2% increase). In addition, it is apparent that the main source downsizing for the TFPC of firms are attributed to the SEC. It shows that firms in textile sector firstly should increase the success of production at appropriate scale for increase the TPFC depends on TEC.

Table 6. Malmquist index summary of annual means

Periods	TEC	TC	PEC	SEC	TFPC
2014–2015	0.960	1.018	0.994	0.966	0.977
2015–2016	1.007	0.989	1.025	0.982	0.996
2016–2017	1.020	1.100	0.993	1.028	1.123
2017–2018	1.003	1.192	1.003	0.999	1.195
Mean	0.997	1.072	1.004	0.993	1.069

5 Conclusion

In this study, DEA based Malmquist Total Factor Productivity analysis method is used to determine the efficiency of the firms in the textile sector among the first and second 500 firms registered to Istanbul Chamber of Industry. The TFPC values of the firms' are examined as a priority in the scope of this aim and the changes in the TEC together with the changes in the TC are examined. According to the results of index analysis, total factor productivity of Turkish textile industry increased by 6,9%. In this scope, it is identified that total factor productivity of 94% of the firms considered in the analyses increased, while maximum increase actualized in the firm 9 (23.3%). The total factor efficiency increases by 12,3% and 19,5%, respectively, in the years of 2016–2017 and 2017–2018. Subsequently, the average annual rate of technical efficiency change is −0,3% while the rate of technological change is 7,2%. There are firms that need to improve both the TC value and the TEC value. Generally, the results show that there is stagnation in TE value in textile sector. In this context, the firms should work towards increasing its managerial effectiveness and the success of production at the appropriate scale. Moreover, the rate of technological change is (7,2%) not satisfactory. The reason of inefficiencies of some firms' should be discovered and take action to improve them.

The results of the analysis are limited by available data, the selected variables to the firm and the period determined. Different variables and firms entering or leaving the analysis will change the results. In order to increase the efficiency of the textile sector in Turkey, firms should give importance to the technological development. Because technological developments are an important factor in increasing the productivity of both labor and capital factor.

As a conclusion, the relative performances of firms' are explored in this study. These findings may have important implications for evalotion of crisis management success. Therefore, the study should be repeated by adding the data of 2020 to evaluate the industry performance in the Covid-19 process.

References

1. Akyüz, C.K., Yıldırım, İ., Balaban, Y.: Kağıt Sektöründe Yer Alan Firmaların Veri Zarflama Analizi Yardımıyla Etkinliklerinin Ölçümü. Int. J. Econ. Adm. Stud. **14**, 23–38 (2015)
2. Bayrak, R., Bakırcı, F., Sarıkaya, M.: Savunma Sanayinde VZA Yöntemiyle Etkinlik Analizi. J. Entrepreneurship Dev. **10**(2), 26–50 (2016)
3. Büyükkeklik, A., Dumlu, H., Evci, S.: Measuring the efficiency of Turkish SMEs: a data envelopment analysis approach. Int. J. Econ. Finance **8**(6), 190–200 (2016)
4. Doğan, Ö.N., Ersoy, Y.: Etkinlik Ölçümü: Tekstil Sektöründen Bir İşletme Örneği. Hitit Univ. J. Soc. Sci. Inst. **10**(1), 35–44 (2017)
5. Fare, R., Grosskopf, S., Norris, M., Zhang, Z.: Productivity growth, technical progress, and efficiency change in industrialized countries. Am. Econ. Rev. **84**(1), 66–83 (1994)
6. Gedik, A., Koçarslan, H., Karaer, M.: ISO 500 İşletmelerinin Malmquist-TFV Endeksi ile Etkinliğinin Ölçülmesi: Otomotiv Sektörü Örneği. J. Soc. Humanit. Sci. Res. **4**(15), 1917–1923 (2017)
7. Halkos, E.G., Tzeremes, G.N.: Industry performance evaluation with the use of financial ratios: an application of bootstrapped DEA. Expert Syst. Appl. **39**(5), 5872–5880 (2012)
8. Hussain, H., Md-Rus, R., Al-Jaifi, A.A.H.: Technical and scale efficiency of automobile firms in Pakistan. Asian J. Multidiscip. Stud. **5**(12), 25–29 (2017)
9. Yang, J.-C.: The efficiency of SMEs ın the global market: measuring the Korean performance. J. Policy Model. **28**, 861–876 (2006)
10. Lan-Bing, L., Bing-Lian, L., Wei-Lin, L., Yung-Ho, C.: Efficiency evaluation of the regional high-tech industry in China: a new framework based on meta-frontier dynamic DEA analysis. Socio-Econ. Plan. Sci. **60**, 24–33 (2017)
11. Lorcu, F.: Malmquist Toplam Faktör Verimlilik Endeksi: Türk Otomotiv Sanayi Uygulaması. Istanbul Univ. J. Sch. Bus. Adm. **39**(2), 276–289 (2010)
12. Malmquist, S.: Index Numbers and Indifference Surfaces. Trabajos de Estadistica Y de Invest. Operativa **4**, 209–242 (1953)
13. Mohammadi, M., Karimizadeh, R., Abdipou, M.: Evaluation of drought tolerance in bread wheat genotypes under dryland and supplemental irrigation conditions. Aust. J. Crop Sci. **5** (4), 487–493 (2011)
14. Oralhan, B., Sümerli Sarıgül, S.: ISO Birinci 500 ve İkinci 500 Şirketlerin Veri Zarflama Analizi Yaklaşımıyla Sektör Etkinliklerinin Ölçülmesi. Hitit Univ. J. Soc. Sci. Inst. **9**(2), 771–792 (2016)
15. Sevinç, A., Eren, T.: Kobilerin Yönetim ve Organizasyon Sorunları ve Kırıkkale Kobi'lerinde Bir Uygulama. International Congress of Management, Economy and Policy, p. 86 (2016)
16. Smriti, T.N., Khan, Md.H.R.: Efficiency analysis of manufacturing firms using data envelopment analysis technique. J. Data Sci. **16**(1), 69–78 (2018)
17. Tzu-Chiang, C., Pei-Yun, C., Fang-Yie, L.: Prediction of technical efficiency and financial crisis of Taiwan's information and communication technology industry with decision tree and DEA. Soft. Comput. **21**(18), 5341–5353 (2017)

18. Wang, D.D.: Performance assessment of major global cities by DEA and malmquist index analysis. Comput. Environ. Urban Syst. **77**, 101365 (2019)
19. Yayar, R., Çoban, N.M.: ISO 500 Firmalarının Etkinliklerinin Ölçülmesinde Veri Zarflama Analizi Yaklaşımı: Dokuma ve Giyim Eşya Sanayi. Niğde University İİBF Dergisi **5**(2), 165–180 (2012)
20. Coelli, T.J.: "A Guide to Deap Version 2.1: A Data Envelopment Analysis (Computer) Program" Cepa Working Paper Series, No: 96/08 (1996). https://www.owlnet.rice.edu/~econ380/DEAP.PDF

Some Universities Performance Evaluation of Entrepreneurship and Innovation in Turkey with Multiple Criteria Decision Making Methods

Mehmet Karahan(✉) and Sariye Peşmen

Firat University, Elazığ, Turkey
mkarahan@msn.com, pesmensariye@gmail.com

Abstract. The concept of "Entrepreneurial and Innovative University" is frequently encountered in recent years. Innovation is becoming increasingly important for universities as well as for the regions they live in as in all areas of life. In this context, entrepreneurial and innovative university-based enterprises have become an important tool in the transfer of information and technology and in solving social problems. Student and academician mobility in universities, cross-border investments between countries and widening the scope of competition have made the change inevitable for universities to survive. This change forces universities to turn into an entrepreneurial and value-creating structure by abandoning their known education, training, and research-based structure. In the face of developments, many universities in developed countries are trying to adapt to this change and they have started to practice "entrepreneurial universities". This change has become inevitable for the universities of our country that feel the pressure of all the global, national and institutional effects and this has started to be used as a competitive tool. In this way, it will be possible for our country to come up with a suitable model and to transform the universities of our country into efficient, innovative, transparent and quality oriented universities. In this study, firstly, the phases of change developed by considering the historical process of universities and the concepts of third generation university and entrepreneur university are explained. Identifying needs to be done to increase the universities' entrepreneurship and innovation properties in Turkey, it is necessary to create the infrastructure for it. The entrepreneurial university paradigm to study how it works and the universities in Turkey to be able to produce some results that can be assessed how measured in this direction, that can bring a fundamental paradigm shift in the focus of university entrepreneurship makes it extremely important research.

In the study, Entrepreneurial and Innovative University Index using the data of 2018, the entrepreneurship and innovation performances of the universities were analyzed with Promethee GAIA, which is one of the MCDM methods. As a result of the analysis, the criteria with the highest weight and the performance ranking of alternative universities were determined. Comparison of the top ten universities and regional universities has been made and necessary suggestions have been developed to overcome the identified positive or negative differences.

N. M. Durakbasa and M. G. Gençyılmaz (Eds.): ISPR 2020, LNME, pp. 569–583, 2021.
https://doi.org/10.1007/978-3-030-62784-3_49

Keywords: Entrepreneurial and innovative university · Third generation university · Performance evaluation · Promethee GAIA · MCDM

1 Introduction

As a result of the globalization process, major changes in mobilization, information and communication technologies have deeply affected the universities. Accordingly, expectations from higher education have also increased in raising the educated people needed by the information society. The role of universities has increased greatly, both nationally and internationally. Universities have been affected by the social and economic changes caused by globalization and their functions, areas of responsibility, organizational structures, and funding skills have also changed. Thanks to this momentum created by globalization, new university types such as entrepreneurial university, corporate university, and virtual university have begun to emerge.

The increasing competition and mass education demands in higher education have turned universities into an entrepreneurial quality by pushing them to search for new resources. As a result of the reflections of the neo-liberal education approach in higher education, universities have become institutions that are student-oriented, aiming education suitable for global development, producing entrepreneurial and creative human resources and developing appropriate educational programs. In today's process, universities have adopted direct contribution to economic and social development. In the 21st century, higher education institutions have not only become an idea market, but have also turned into an economic market that competes with students, tutors, resources and respectability. The market in question has often created a management and entrepreneurial culture that conflicts with traditional and academic culture, making higher education institutions more competitive and more integrated with the industry [1].

In the changing and rapidly developing world, universities have taken their share and have entered a great change process. In the historical process, universities have transformed from science-oriented first-generation universities to education and research-oriented second-generation universities, and recently to third-generation (3K) universities integrated with entrepreneurs and society. In order to realize this transformation successfully, universities have to cooperate with the industry, to be in touch with the society, to create the best academic staff and to take the most successful students. Universities that will stand out in this competitive environment are dynamic in education and research and active in the information market. In this new model, universities; In order to adapt to the demands of the outside world and the demands of the market, teaching, research and service to the society have turned into organizations whose functions are loaded [2].

Entrepreneurship is the process of taking risks within the activities of a new or existing organization, performing economic activities or developing by combining creative and innovations with a sound management system. In general, the entrepreneurial university; is a university that promotes both education and training, and research and entrepreneurship. According to the most general definition, the entrepreneurial university is a university that creates value for the society but can create an innovative perspective, be different, and find innovative systems, people, rules, methods. Entrepreneurial universities are defined as institutions that develop networks of relations with their stakeholders, cooperate, break down strict hierarchical rules between disciplines and departments, and create new disciplines by building new structures while breaking down, and generate values [3].

The concept of entrepreneurship ranks first in the agenda of all businesses and operators in recent years. Entrepreneurship is a concept that expresses the individual's ability to put the idea into action. This concept includes creativity, innovation and risk taking, as well as the ability to plan and manage projects to achieve goals. In other words, entrepreneurship is a concept that includes many disciplines and helps employees and individuals in daily life, working life and society to see opportunities that may arise while doing social and commercial activities [4].

The entrepreneurial university in general; It is defined as the university that aims to conduct economic researches, to produce education and services through interventional activities. Entrepreneurial universities are redefining the "traditional" roles of universities in society as knowledge creators, information transfer providers, and promoting actors of innovative and economic development through basic and applied research. Entrepreneurial universities are reshaping the academic view by transforming knowledge into intellectual property and innovation. As the university is more involved in the establishment of new companies and technology transfer activities, it gains a new "entrepreneurial" identity [5].

Entrepreneurial universities have undertaken the tasks of technology transfer, innovation and contribution to society as well as classical academic activities. Entrepreneurial universities raise students who can start their own businesses with an entrepreneurial culture. It also facilitates students to find a job by establishing good relations with technology parks and business world. Public and university-industry cooperation is also very important in the entrepreneurial university understanding. These cooperation efforts have gained an institutional structure through technology development regions. With the concept of 3rd generation university, universities are encouraged to engage in entrepreneurial activities other than education and scientific research activities [6].

Rapidly changing internal and external environment variables affect universities' function and strategic thinking intensely and push them to become entrepreneurial universities. As information became an important part of innovation, the university, as an organization that produces and disseminates information, has started to play a bigger role in the creation and dissemination of innovation, inventiveness, especially industry circles. Universities become the main element of the industrial innovation system in the knowledge-based society, both as a human capital server and by sowing seeds in the establishment of new companies. In order for this change process to turn into an entrepreneurial structure, the organizational structure and management elements should change as well as the change of cultural infrastructure. In the Western world, these changes have turned towards the concept of "entrepreneurial university" and it is known that significant transformations have been made to fulfil its requirements. It is very clear that our country also needs to make these structural changes and renewal process. However, it is imperative to be successful in finding a model that is suitable for our own culture and social structure while making the big change in higher education in many countries in our country's universities [7].

In order to help increase entrepreneurship activities in our country, TÜBİTAK published the first index values for universities in 2012. This index was created within the scope of the decision of "Development of policy tools for the purpose of triggering innovation and entrepreneurship in the university" taken at the 23rd meeting of the High Council of Science and Technology and the index was called "Entrepreneurial and Innovative University Index". In 2016, TÜBİTAK published the Entrepreneurial and Innovative University Index, in order to encourage universities to entrepreneurship and innovation activities, to increase cooperation with industrialists and to produce quality information that turns into economic and social benefits. TUBITAK, 152 universities in Turkey have been included in the present study and the ranking of the top 50 public universities. In the weighed method of TÜBİTAK, the variables within themselves were brought to the fore and contributed to the observation of the interactions between the relevant variables.

In the study, firstly, a summary of the literature about the theoretical information and the methods used in the applications, which investigates the main factors and causes that form and develop the entrepreneurial university model, is given, then the explanations about the Promethee method are made and the performance of the entrepreneurial and innovative universities is evaluated in the last section. Made expected from this study to increase the amount of information generated from GYU index, demonstrate the feasibility of the entrepreneurial university model for Turkey and especially the remaining university in the lower level of the index ranking it is to show they can rise to the top positions by developing what matters.

2 Literature Research

In this study by Er and Yildiz [8]; published by TUBITAK 2016 and by 2017 Turkey entrepreneurial and innovative universities index values were examined ORESTE and by factor analysis method, according to the subsequent analysis results are described changes occurring in university rankings. In the results of working; some universities climbed higher in the ranking, while others dropped lower. In 2016, the biggest increase was in Bursa Technical University with 10 rows of increase, while the biggest decrease was in Cukurova University with 5 rows of decrease. In 2017; the biggest increase was in Hasan Kalyoncu University with 13 rows, while the biggest decrease was in Kocaeli university with 11 rows.

In this study by Omurbek and Karatas [3], entrepreneurs and innovators 50 universities in Turkey's 2016 performance was analysed with MCDM method. In the study, the performance evaluations of universities were made by MAUT and SAW methods according to scientific and technological research competence, intellectual property right, cooperation and interaction, entrepreneurship and innovation culture, economic contribution and commercialization criteria. According to the results of the study; it is stated that the criterion with the highest weight value is the intellectual property right, Sabanci University is in the first place in the performance evaluation ranking and the Middle East Technical University is in the second place.

Daglar [6] conducted by this research, entrepreneurship in the strategic plans of state and private universities in Turkey have investigated how they deal with. The research covers the strategic plans of 102 public universities and 30 foundation universities on their web pages and how entrepreneurship is included in these plans. In the results of working; It has been proposed to increase the number of entrepreneurship and innovation themed courses and the number of students taking these courses, to open entrepreneurship training programs, to provide entrepreneurship certificate training, to increase the number of entrepreneurship projects to be carried out together with seminar, consultancy services and techno park companies.

Ozdemir [9] in his study; entrepreneurship training given by the universities in Turkey were evaluated with a questionnaire application. As a result of the study; According to the data obtained from 93 universities that make up the research group, he suggested that entrepreneurship education at universities has accelerated in the last five years and structural arrangements are being made to support this education. In addition, the entrepreneurship education in recent years in line with the growing interest worldwide, also expressed interest in Turkey show parallelism.

In the study conducted by Sungur [5]; As a result of the change in regional development theory and practices, the increasing importance and changing roles of universities in regional development were investigated. According to the results obtained from the study, the type of university needed today; It has been suggested that it should be a dynamic university that can meet the needs of the working environment and that universities should be "entrepreneurs" in order to achieve regional (and national) economic development.

In a study by Korkmaz [10], it was tried to determine whether the students of Bulent Ecevit University business administration department have entrepreneurial personality and what psychological, demographic and family factors that are effective in entrepreneurship tendencies. As a result of the statistical analysis, it was determined that there were significant relationships between the psychological, demographic and family characteristics that were effective in students' seeing themselves as an entrepreneurial personality and their desire to start a business in the future. According to this result, it has been suggested that the students of Bulent Ecevit University Business Administration Department have an entrepreneurial personality.

In this study, conducted by Sakinc and Bursalioglu [1], the main factors and causes that make up the entrepreneurial university model and its development were investigated extensively through the related literature. In the study, it was tried to establish a framework that theoretically defines the dynamic processes of the entrepreneurial university model and the applicability of the entrepreneurial university model is discussed. In the results of working; It is stated that universities can play a more active role in the effectiveness and sustainability of entrepreneurship policies by developing individual and institutional capacities and interaction in the university-industry-state spiral.

Ozden et al. [11] in their study; Overview of the concept of entrepreneurship in Turkey and Kyrgyzstan university students' entrepreneurial intentions, have investigated the factors affecting the choice whether to become entrepreneurs. In this study conducted in line with this purpose; 2007–2008 spring semester, Erzurum Ataturk University and the Kyrgyzstan-Turkey Manas University Faculty of Economics and

Administrative Sciences conducted a survey application to final year students, it is made of statistical analysis and evaluation. According to the results obtained from the study; it was determined that entrepreneurship levels of both faculty students were low compared to expectations. It was suggested that students should be motivated and guided in terms of entrepreneurship and be given more theoretical and practical training starting from the first term in both faculties in order to overcome this deficiency.

In the study conducted by Koc and Mente [12]; Within the scope of university-industry-state cooperation, the importance of the concept of innovation was emphasized, a detailed literature review was made, and the "Triple Helix Model", which is extremely important in terms of university-industry-state cooperation, was examined. According to the results of the study, the researchers stated that there should be a supportive, active participatory role and a state structure that integrates science and production wings, and universities should turn into an industry-oriented, entrepreneurial, more emphasis on R&D, production and commercialization. In addition, university-industry-state cooperation should be implemented. It was also determined that the "Triple Helix Model" examined in the study is a very useful alternative.

3 Purpose of the Research

The aim of the study is to analyse the entrepreneurship identity and practices that the university phenomenon, which has come up by loading different missions, has undertaken in the globalization process. Entrepreneurial and innovative universities to increase the amount of information generated from the index, to demonstrate the feasibility of the entrepreneurial university model for Turkey, and in particular the index staying in lower levels of the university rankings is to show they can rise to the top positions by developing what matters.

For this purpose, since 2012 by TUBITAK annually published Entrepreneurial and Innovative University Index (GYUE) when using data for 2018 in this study, the performance of entrepreneurial and innovative universities in Turkey will be evaluated, sorting will be done in terms of entrepreneurship and innovation aspects of the university. Afterwards, the top five and the bottom five universities were compared with the regional universities (Firat University) and the necessary suggestions were developed to eliminate the positive or negative differences identified. Also in this study; The GYUE index ranking created by TÜBITAK has been handled from different perspectives and it is aimed to increase the amount of information produced from the index.

3.1 Methodology of the Study

In the Promethee method, which is one of the newest MCDM methods used today, decision making alternatives are compared in pairs. Decision-making is done by selecting one of the six fixed preference functions specified for each criterion in this comparison. There are also additional threshold values that must be determined in each preference function. The GAIA geometric plane is a graphical representation on which Promethee results are built. This chart offers decision makers and researchers a quick, simple and understandable view. Also, showing the Promethee results in a GAIA geometric plane on a two-dimensional plane makes it possible to analyse the results with a different approach.

The Promethee method is one of the most recent methods among MCDM methods, but due to its many superior features, it is one of the most used methods in recent years. The use of the method in both research and real life is increasing. The fact that this method has an extension, such as GAIA geometric representation, makes it easier for the decision-makers to present the image related to the analysis in a short time, making the decision process even faster.

3.2 Data Used in the Research

In Turkey, by TUBITAK being published annually since 2012, there are four sizes of GYUE. With the help of this index, both scientific activities of universities and collaborations with industry can be handled together, incentives to increase collaborations and produce qualified information that turn into social benefit. Thus, the successes of universities in innovation and entrepreneurship activities are increased and their future targets are determined more clearly.

GYUE consists of four main dimensions and a total of 19 indicators, each with a different number. Data of these 19 indicators; Provided by TUBITAK, YOK, Ministry of Industry and Technology, TURKPATENT, KOSGEB, European Union Education and Youth Programs Centre, TTGV, TUBA and Universities. During the determination of these variables, variables previously used in similar indices, expert opinions, focus group studies and data finding opportunities were taken into consideration. While determining the sub-variables, it is also taken into consideration that the index can be refreshed regularly and the developments achieved by the universities can be measured regularly over the years, and the data published at the university level in periods are included in the calculations [13].

In this study, the data of the four main dimensions in the GYUE were used, and this data used in the study belongs to 2018 and was published on the website by TUBITAK in 2019. For the analysis to be made, after determining the alternatives and criteria, the step of editing the decision matrix for the application of the Promethee method was started. For this, the starting table was created with the data obtained from the relevant sources.

3.3 Evaluation of GYU Performance with Promethee GAIA Method

Each of the four main criteria determining GYU performance includes a different number of sub-variables, with a total of 19 indicators. From the index data of these four main criteria, the Promethee GAIA geometric representation initial table has been created and shown in Table 1 below.

As can be seen in Table 1, the data of Firat University, where the comparison will be made with the top five and the bottom five universities in the GYU index, in the initial table. The data in the starting table were entered into the Promethee-GAIA package program and the performance analysis process was started. According to GYUE 2018 data, the performances of universities were evaluated according to

Table 1. Alternatives and criterion values (initial table)

Sn	University	Total	Scientific and Tech. Resear. Competence (BTAY)	Intellectual Property Pool (FMH)	Cooperation and Interaction (IE)	Economic Contribution and Commer. (ECT)
1	METU	93,16	23,63	16,24	28,54	24,75
2	ITU	90,16	21,93	15,89	27,59	24,75
3	Sabancı University	85,49	18,08	15,16	27,90	24,36
4	Bilkent University	84,42	21,30	14,02	26,05	23,05
5	Bogazici Universit	83,33	20,55	15,46	28,06	19,26
6	Firat University	40,52	10,94	5,38	9,43	14,77
7	Baskent University	32,59	6,18	4,92	7,68	13,80
8	Duzce University	32,23	7,06	5,46	12,54	7,17
9	Canakkale 18 Mart Un.	32,15	9,56	2,67	13,03	6,88
10	Tekirdag Namık Kemal	31,69	7,14	3,18	8,49	12,88
11	Istanbul Sehir Univer	29,63	8,20	4,82	13,60	3,00

Promethee I partial ranking and Promethee II full ranking and university performances were ranked according to positive and negative Phi values. In Table 2, where the Promethee partial ranking and full ranking values are shown together, the performance ranking of the universities was made according to the net superiority score size (net Phi).

As can be seen in Table 2, the performance ranking of GY universities is made according to the net superiority (Phi) score obtained by subtracting the negative superiority value (Phi−) from the positive superiority value (Phi+). Accordingly, METU's score (Phi = 0.9712) among the entrepreneurial and innovative universities included in the study is higher than all other alternatives and ranks first. The net weight of the university of Baskent (Phi = −0.6575) is in the lowest place, since it is both negative and smallest.

Table 2. Promethee II ranking table

SN	GY Universities	Phi	Phi+	Phi−
1	Orta Doğu Teknik University	0,9712	0,9712	0,0000
2	İstanbul Teknik University	0,7137	0,8425	0,1288
3	Bogazici University	0,4950	0,7475	0,2525
4	Sabancı University	0,4675	0,7338	0,2662
5	Ihsan Doğramacı Bilkent Ünv.	0,3525	0,6763	0,3237
6	Firat Üniversitesi	−0,2100	0,3950	0,6050
7	Duzce University	−0,4775	0,2612	0,7388
8	Istanbul Sehir University	−0,4950	0,2525	0,7475
9	Canakkale Onsekiz Mart Unv.	−0,5225	0,2387	0,7612
10	Tekirdag Namık Kemal Unv.	−0,6375	0,1812	0,8187
11	Baskent University	−0,6575	0,1713	0,8287

Based on the above ranking results obtained, the locations of the eleven alternative universities in the performance rankings were determined, but in the next step, the Promethee GAIA geometric analysis was decided to be performed in order to minimize the potential mistakes of the decision makers.

As seen in Fig. 1 below, this analysis made on the geometric plane has given a different visual richness to the evaluation. Thus, decision-makers were given the opportunity to see healthier and consistent interpretations by seeing the relationships between alternatives and criteria on a geometric plane.

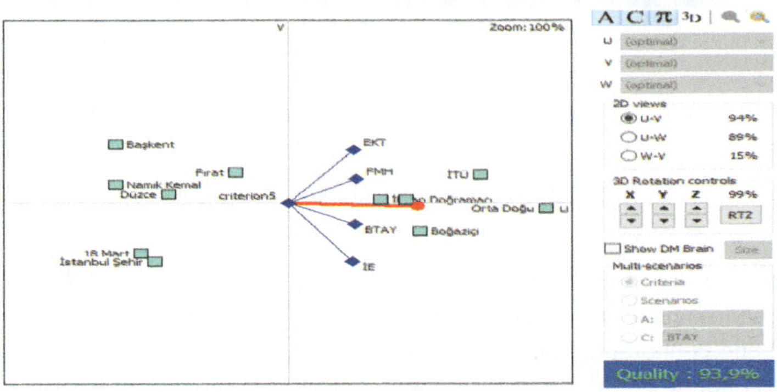

Fig. 1. GAIA graphical display of GYU performance

As seen in the GIA geometric plane in Fig. 1, alternatives are shown in the form of green squares, criteria in the form of vectors with blue stripes, and the decision bar showing the optimum status is a vector with a circle at the beginning of the thick red line. The value that shows the quality of the Promethee measurement as a percentage is opposite the quality expression in the bottom right of Fig. 1 and calculated as 93.9%.

This value calculated by the program indicates that if it is lower than 70%, an unhealthy result is obtained, and if it is greater than 70%, a healthy result [14] is achieved. Accordingly, it can be concluded that the quality of measurement of competitiveness performance calculations calculated for the study is very good.

As shown in Fig. 1, the vector (red bar) close to the horizontal axis but without any criteria written indicates the optimum situation. As seen in Fig. 1, the ordering of the alternatives close to the optimality line can be evaluated according to the 4 criteria discussed. Accordingly, the alternatives that can be evaluated in terms of their competitiveness in terms of their proximity to the four criteria are; Sabancı, Ihsan Dogramacı, Bogazici, ITU, METU universities.

As seen in Fig. 1, the alternatives that are close to the criteria indicated by thin blue lines are universities that have advantageous and high performance in terms of the criteria discussed. For example, in terms of FMH criterion, Ihsan Dogramacı and Sabancı University can be stated to be in a superior position since they are closest to this criterion. In terms of BTAY criteria, it can be interpreted that Bogazici University, which is the closest to the criteria, has a better performance.

1). *Comparison of Universities with Similar Profiles:* In this part of the analysis, METU and ITU alternatives are examined comparatively (Fig. 2, 3) to show that the profiles of the universities that are close to each other on the GIA geometric plane are similar (Fig. 2, 3) and the necessary interpretations are made below.

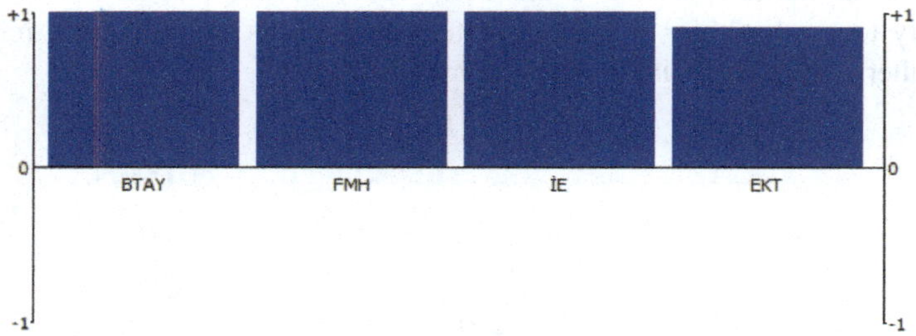

Fig. 2. Profile of METU alternative

As seen in Fig. 2, although the positive contribution levels of METU's scientific and technological research competence (BTAY), intellectual property pool (FMH) and cooperation and interaction (IE) criteria are quite high, the contribution of the economic contribution and commercialization (ECT) criterion is relatively is less.

The profile features of the ITU alternative, which is one of the two universities close to each other in Fig. 3, are shown below and the necessary interpretations are given below.

As seen in Fig. 3, the scientific and technological research competence (BTAY), intellectual property pool (FMH), economic contribution and commercialization (ECT) criterion of the ITU alternative is higher and the contribution of the cooperation and interaction (IE) criterion is lower. When Fig. 2, 3 is compared

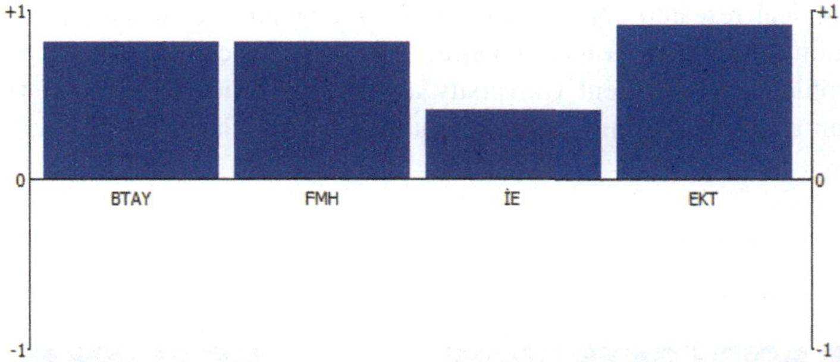

Fig. 3. Profile of ITU alternative

comparatively; Although the METU and ITU alternative profiles generally show some difference, it can be stated that they have a great similarity since all criteria contribute positively. However, it should be noted that the ODTU alternative is relatively weak in terms of ECT criteria and the ITU alternative is relatively weak in terms of IE criteria.

2). *Comparison of Alternatives in Terms of Criteria:* In this part of the analysis, the comparison of two universities (Bogazici and Baskent), which are far or away from the Firat university alternative, on the Promethee GIA geometric plane in terms of entrepreneurship and innovation criteria was made.
In Fig. 4 below, the profiles of Bogazici University, which is far from the positive direction of Fırat University, according to the entrepreneurship and innovation criteria are shown comparatively.

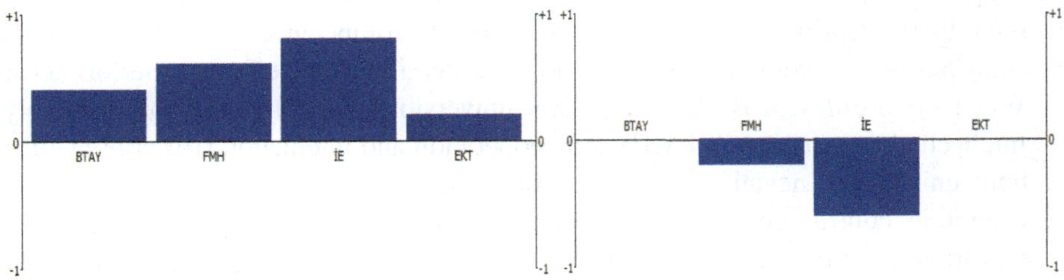

Fig. 4. Comparison of Bogazici and Firat University Profiles

Figure 4 as seen in the graphic on the left, Bogazici University alternative; Scientific and technological research competence (BTAY), intellectual property pool (FMH), cooperation and interaction (IE) and economic contribution and commercialization (ECT) criteria are positive for performance. Accordingly, it can be interpreted that it has effects that increase competitiveness.
Figure 4 shows the Firat University's intellectual property pool (FMH) and collaboration and interaction (IE) criterion (decreases performance) negatively to the performance in the graph on the right. In addition, it is seen that scientific and

technological research competence (BTAY), economic contribution and commercialization (ECT) criteria have no effect on performance.

The profile of the Baskent University alternative, which is far from the negative direction of the Firat University alternative, is shown in Fig. 5 below.

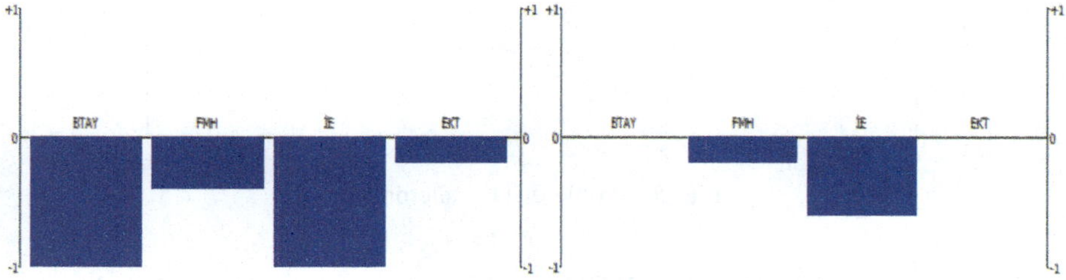

Fig. 5. Comparison of Baskent and Firat University Profiles

Figure 5 As seen in the graphic on the left, when looking at the profile of the Baskent University alternative, it is seen that all the criteria contribute to the performance (decrease performance) from the negative aspect. It is seen that especially technological research competence (BTAY) and collaboration and interaction (IE) criteria affect performance more and negatively than others. The criterion that affects Baskent University least negatively is the economic contribution and commercialization (ECT) criterion.

As can be seen in the graphic on the right side of Fig. 5, Fırat University contributes negatively to the performance in terms of intellectual property pool (FMH) and cooperation and interaction (IE) criterion (decreasing performance). In addition, it is seen that scientific and technological research competence (BTAY), economic contribution and commercialization (ECT) criteria have no effect on performance. When the profiles of Baskent and Firat universities are compared comparatively; Intellectual property pool (FMH) and cooperation and interaction (IE) criteria affect both universities negatively, but technological research competence (BTAY) and economic contribution and commercialization (ECT) criteria do not affect the performance of Firat University and negatively affect Baskent University. Accordingly, it can be said that the profile of Firat University is more preferable than Baskent University and it is in a better condition in terms of all criteria.

4 Conclusions and Suggestions

Today, as a result of the pressures created by rapidly changing conditions, strategic planning of the universities and the strategic transformation change that constitutes its cornerstone are also compulsory. In the Western world, it is known that this quest is directed towards the concept of "entrepreneurial and innovative university" and that important transformations and changes are made in order to fulfil its requirements. There is a need for re-structural changes and renewal processes in our country.

Therefore, it is accepted that the strategic transformation towards the concept and implementation of the entrepreneurial university is also very necessary for the universities of our country. However, while realizing this transformation, the necessity of finding a model suitable for its culture and social structure should not be forgotten.

In the study, firstly, a literature review was conducted on the theoretical information and methods used in applications that investigate the main factors and causes that form and develop the entrepreneurial university model. Then, to demonstrate the applicability of the ADL model for Turkey and especially the ADL index ranking lower levels in the remaining universities of which are development issues by making could rise to the top row shows the purpose of an application working with Promethean to-GAIA of the MCDA method has been applied. Then, the universities in the top five and the bottom five in the performance ranking and Firat University were analysed comparatively and necessary suggestions were developed to eliminate the positive or negative differences identified.

According to the results obtained from the study; According to Promethee I partial ranking, METU seems to be more dominant than other universities and has partial priority. Likewise, it is possible to say that Firat University has higher priority than Duzce University, and Istanbul Sehir University has more priority than 18 Mart University. According to Promethee II full rank, the universities that have net superiority have positive superiority and consequently have the highest net flow (Phi) value, respectively; METU, ITU, Bosphorus, Sabancı, İhsan Doğramacı and Bilkent universities are determined. The ones that provided negative superiority and as a result net value (Phi) value were negative respectively; Firat, Duzce, Istanbul Sehir, Canakkale 18 March, Tekirdag and Baskent are universities.

According to the results obtained from the study; Competition ranking was made between Firat University and 10 other universities that were compared. However, a different visual richness has been added to the problem solution by making use of GAIA geometric plane visual analysis in order to enable decision makers to identify potential problems more clearly. Thus, it has been ensured that the alternatives and criteria are visible on the geometric plane and the comments to be made are easier and more understandable. The accuracy rate (quality value) of the results obtained from the analyses was calculated as 93.9% and it was decided that the analyses were reliable accordingly.

According to the criteria discussed in the study, the universities that are close to the optimal line shown in the GAIA plane are respectively; Sabanci, Ihsan Dogramacı, Bogazici, ITU and METU universities. In addition, it was determined that the criteria of EKT, FMH and BTAY and IE were similar in terms of the four criteria considered in this plane. Accordingly, ITU, Ihsan Doğramacı and Sabancı universities are superior in terms of ECT and FMH criteria. Blue vectors close to the optimal line (red line), according to this plane; It can be said that the rate of ECT and IE criteria affecting the competition is higher. According to the GAIA plane, universities that are located in the same direction and close to each other are also considered to have similar profiles. In the opposite case, it is interpreted that the difference of criterion values between universities increases and their similarities decrease. Accordingly, universities whose profiles are close to each other; Ihsan Dogramacı and Sabancı are 18 March and

Istanbul Sehir universities. Those who are distant from each other; Baskent and Bogazici are the universities of Firat and METU, ITU and Istanbul Sehir.

Some suggestions that can be put forward within the framework of the results obtained from the study can be listed as follows.

It is encouraged by the university that the academic staff, one of the internal stakeholders of the universities, to establish their own companies, become partners with existing companies and become information producers by producing the information that forms the core of innovation.

Depending on the fact that entrepreneurship can be learned and taught, education and training needs to be restructured in a way that will allow students to grow up as technology entrepreneurs in the future.

Entrepreneurship education in universities in recent years; It shows a trend from the production of scientific knowledge about entrepreneurship to the training of entrepreneurs. In this regard, educators are required to organize activities to adopt and spread the entrepreneurial culture.

In the field of entrepreneurship and innovation, the companies that students will establish in order to realize their dreams and develop their skills should be supported and students should be given the opportunity to transfer information to companies in their close circle while they are doing their internship. Thus, these students can be made aware of competitive technologies and innovations that can make a difference in their environment, especially in the business world.

Following this study, the studies that are suggested to the researchers in future studies and which are thought to count the increase of knowledge on the subject are as follows.

TUBITAK's only the first 50 universities in addition to the explanation of the value of the index of other universities in Turkey to be more meaningful analysis of the results is very important. It will be more realistic to make action plans that universities will put forward to advance their rankings with the help of an explanation of all results and analysis. In future studies, comparing the national index with international indices will open the way for the universities in the country to increase their competitiveness at the international level.

References

1. Sakınc, S., Bursalıoglu, S.A.: A global change in higher education: entrepreneurial university model. J. High. Educ. Sci. **2**(2), 92–99 (2012)
2. Ciftci, M.: Girişimci Üniversite ve Üçüncü Kuşak Üniversiteler, Dumlupınar Üniversitesi Sosyal Bilimler Dergisi **27**, 341–348 (2010)
3. Omurbek, N., Karatas, T.: Girişimci ve Yenilikçi Üniversitelerin Performanslarının Çok Kriterli Karar Verme Teknikleri ile Değerlendirilmesi. Mehmet Akif Ersoy Üniversitesi Sosyal Bilimler Enstitüsü Dergisi. **10**(24), 176–198 (2018)
4. Er, F., Yıldız, E.: Türkiye Girişimci ve Yenilikçi Üniversite Endeksi 2016 ve 2017 Sonuçlarının ORESTE ve Faktör Analizi ile İncelenmesi. J. Oper. Res. Stat. Econ. Manage. Inf. Syst. **6**(2), 293–310 (2018)

5. Sungur, O.: Üniversitelerin Bölgesel Kalkınmada -Değişen- Rolü ve Girişimci Üniversite Kavramı. AİBÜ Sosyal Bilimler Enstitüsü Dergisi **15**(4), 35–61 (2015)
6. Dağlar, H.: Üniversitelerin Girişimciliğe Bakışı: Devlet ve Vakıf Üniversitelerinin Stratejik Planları Üzerine Bir İnceleme. In: ICOMEP 2018-Spring, International Congress of Management, Economy and Policy, 28–29 April 2018, pp. 47–55. Bildiriler Kitabı, İstanbul (2018)
7. Odabası, Y.: Değişimin ve Dönüşümün Aracı Olarak Girişimci Üniversite. Girişimcilik ve Kalkınma Dergisi **1**(1), 87–104 (2006)
8. Er, F., Yildiz, E.: Türkiye Girişimci ve Yenilikçi Üniversite Endeksi 2016 ve 2017 Sonuçlarının ORESTE ve Faktör Analizi ile İncelenmesi. Alphanum. J. **6**(2), 293–310 (2018)
9. Özdemir, P.: Girişimcilik Eğitimi ve Üniversitelerimiz. ÇÖMÜ Dergileri, Girişimcilik ve Kalkınma Dergisi **11**(1), 224–240 (2016)
10. Korkmaz, O.: Üniversite Öğrencilerinin Girişimcilik Eğilimlerini Belirlemeye Yönelik Bir Araştırma: Bülent Ecevit Üniversitesi Örneği", Afyon Kocatepe Üniversitesi, İİBF Dergisi. **14**(2), 209–226 (2012)
11. Özden, K., Temurlenk, M.S., Basar, S.: Girişimcilik Eğilimi: Kırgızistan-Türkiye Manas Üniversitesi ve Atatürk Üniversitesi Öğrencileri üzerine bir Araştırma. Rev. Soc. Econ. Bus. Stud. **11/12**, 1–20 (2008)
12. Koc, K., Mente, A.: İnovasyon Kavramı ve Üniversite-Sanayi-Devlet İşbirliğinde Üçlü Sarmal Modeli, Hacettepe Üniversitesi Sosyolojik Araştırmalar Dergisi (2007)
13. TUBİTAK: Girişimci ve Yenilikçi Üniversiteler (2019). https://www.tubitak.gov.tr/sites/default/files/289/gyue_ilk50.pdf
14. Bülbül, E.S., Köse, A.: Türk Sigorta Sektörünün PROMETHEE Yöntemi ile Finansal Performans Analizi. Marmara Üniversitesi İktisadi ve İdari Bilimler Fakültesi Dergisi **38**(1), 187–210 (2016)

Turkey and Some EU Countries' Economic Performance Analysis with Multi-criteria Decision Making Methods: Promethee GAIA Application

Mehmet Karahan[1(✉)], Fatma Çetintaş[1], and Merve Sinem Karahan[2]

[1] Department of Business Administration, Firat University, Elâzığ, Turkey
mkarahan@msn.com, cetintasfatoss@gmail.com
[2] Department of Computer Engineering, Beykent University, İstanbul, Turkey
msinemkarahan@gmail.com

Abstract. The economic performance of countries can be compared with various methods and the results obtained from these comparisons are accepted as the economic levels of the countries covering certain periods. Since the ranking or comparison of countries according to their economic performances can be made according to many criteria, mathematical methods are needed in which all criteria are taken into consideration in such performance measurements. Multi-criteria decision-making (MCDM) methods have the ability to alternate sorting, to choose and classify alternatives with the results frequently used and obtained in such comparisons. They are widely used in economic comparisons, in decision problems where many alternatives and criteria are evaluated. Promethee method, which is one of the newest MCDM methods developed to select the best alternative within the framework of determined criteria, is one of the most effective and easiest methods for solving such problems.

In the study, 10 EU countries, Turkey and the intended method of measuring the economic performance of Promethee 2018, five key economic criteria were used in this measurement. Determining the performance level of the countries covered in the study, Turkey's uncovering of inadequacy or superiority of some in comparison with EU countries is very important in the sense. According to the Promethee I partial ranking results obtained from the study; Luxembourg has a more dominant and partial priority than other countries. According to the Slovakia-Ireland Likewise, it is more prevalent than Turkey and the Netherlands. According to the ranking of the provinces that showed clear superiority in the full rankings of Promethee II; While the weight of Luxembourg is positive, the weight of Slovakia is negative. Luxembourg ranks first among the countries covered by the analysis, having the highest net Phi value. These countries respectively, Slovakia, Ireland, Belgium, Lithuania, Slovenia, Hungary, Estonia, Turkey, the Netherlands and Latvia are to follow. The Quality value, which represents the quality of the Promethee GAIA measurement, was calculated as 86% for this study.

Keywords: Performance management · Economic performance · MCDM · Promethee · GAIA

N. M. Durakbasa and M. G. Gençyılmaz (Eds.): ISPR 2020, LNME, pp. 584–597, 2021.
https://doi.org/10.1007/978-3-030-62784-3_50

1 Introduction

The main purpose of performance evaluation is managerial and future oriented personnel development. Managerial goals; To prepare the environment that will enable the measurement of the success levels from various units of the business to the lowest unit, to obtain reliable information about the workforce and management potential of the business, to gather information about the general success and problems of the business and to predict future events. Future personnel development objectives; Performance evaluations are carried out for purposes such as providing employees with information about their success levels, enabling their superiors to know what they think about themselves and what they expect, determining personal goals, recognizing the achievements achieved, and therefore meeting the individual's need for success, increasing job satisfaction and motivation [1]. Until the 1990s, economic performance evaluation was evaluated with three basic macroeconomic criteria such as unemployment, inflation and economic growth. In the "World Economic Outlook", which was first published in 1987, OECD created a new macroeconomic evaluation index, called "magic diamond". This index, which was composed of 4 variables in its first use, was obtained by adding the value of 100 to the total weight of 5 variables over time and has taken its place in the literature as the macroeconomic performance index [2].

The ranking of the countries according to their economic performances can be made by various methods and the results obtained from the comparisons are accepted as the economic performances of the countries covering certain periods. Multi-criteria decision-making (MCDM) methods are capable of alternative sorting, selection and classification of alternatives with the results used and obtained in such comparisons. This kind of mathematical methods are needed when it is necessary to use a multi-criteria structure in comparisons, where many alternatives and criteria can be evaluated for economic comparisons, and to analyse by taking into account a large number of independent and different factors. Although Turkey's economy showed a rapid development in recent years, the EU accession process, some issues related to the country's economic performance is brought to the agenda from time to time. For this reason, Turkey's economic performance of EU member and candidate countries with a comparative examination, shed light on the agenda of future discussions on the issue and will give way to take forward the measures for Turkey. As a result of the evaluations, it will be tried to determine whether the MCDM methods give close results in determining the economic performance ranking and the comparison performance of the Promethee GAIA method, which is one of the newest MCDM techniques commonly used in the literature, will be tested. In the study, the 2018 economic performance of Turkey and 10 EU countries will be analysed according to five key economic data. Decision criteria used in the research; unemployment rate, gross domestic product, inflation rate, exports and imports of goods and services. The countries discussed in the study will be compared in terms of these indicators. In particular, Turkey's determination of some sort between EU countries and are intended to reveal the lack of comparison with other EU countries and superiority.

The study consists of three main chapters, and in the first part, a literature survey of national and international studies conducted with the MCDM methods is included. In the next section, the theoretical foundations of the Promethee method used in the study are examined. In the third part of the study, information about the data set used was given and the results of the analysis related to the Promethee method for measuring the economic performance of the countries were evaluated and general evaluations were made on the results obtained.

2 Literature Research

Asaleye, et al. [3] investigated the effect of oil shock in Nigeria on macroeconomic performance using structural vector autoregression and normalized equations. In the study, gross domestic product, exchange rate, employment, consumer price, oil prices and interest rates were used to calculate macroeconomic performance. As a result of the study, it has been suggested that dependence on the oil sector does not support the formation of employment over time and an alternative way should be found to ensure sustainable growth and development.

Salas-Velasco [4] in his study; He examined the macroeconomic performance of OECD countries in terms of production efficiency, applied the stochastic econometric boundary model in measuring efficiency and predicted the determinants of efficiency. In the model used in the study, capital, labour, human resources, property rights and technology data were used as variables. As a result of the study, it was determined that OECD countries should increase their total output by 23% without consuming more resources in order to operate more efficiently. In addition, a methodology has been developed to determine the inefficiency problems of countries in the study and it has been determined that countries generally experience inefficiencies due to wrong resource allocation, skill mismatch, etc.

Lim et al. [5] in their study; Using the data obtained from the World Values Survey and the European Values Survey, they examined the impact of trust between countries, which focus on improving the social capital measure, on income and investment. As a result of the study, a strong positive relationship was found between the per capita confidence level predicted by the increase in investments and the real GDP. In addition, the impact of the predicted trust on macroeconomic variables was determined upwards.

Araujo et al. [6] in their study, analysed the relationship between money markets in Brazil and macroeconomic performance theoretically and empirically. In the results of working; It has been determined that the GDP and growth rates decrease, the increase in public debts, the appreciation of the currency in the domestic market and inflation are not kept below the target. According to these results, it has been suggested that serious revisions should be made in Brazilian money markets in order to improve the results in terms of both actual inflation and economic performance.

Masca [7] in his study; he analysed the macroeconomic performance evaluation of 28 EU countries with the TOPSIS method. Six macroeconomic indicators (long-term interest rates, general government deficit, GDP, general government gross debt, inflation rate, gross fixed capital formation and unemployment rate) data were used in the study. As a result of the study, it has been suggested that Sweden has the best and Greece has the worst economic performance.

Guney [8] in his study, analysed the sectors in stock market Istanbul through financial ratios and by using Promethee method. With the help of analysis, it is tried to determine which sectors are promising for investors in the study. Proportional data of the sectors, Reuters Limited and Matriks Bilgi Dagitim Hizmetleri A.S. provided by. According to the results obtained from the study, respectively; It has been suggested that the Financial Sector, Industry Sector, Service Sector and Technology Sector are promising as priority sectors.

Eyuboglu [9] in his study; He compared the macroeconomic performances of Azerbaijan, Kazakhstan, Kyrgyzstan, Tajikistan, Turkmenistan and Uzbekistan, among the countries of the Turkic world in the 10-year period between 2004–2013. In the performance measurement performed in the study; growth rate, inflation rate, unemployment rate and current account balance/GDP criteria were used, AHP and TOPSIS methods were applied. All the data related to the criteria used in the study were obtained from the World Bank and IMF sources. As a result of the analyses carried out in the study, Azerbaijan is the country with the highest macroeconomic performance.

Welsch and Kuhling [10] in the study in which OECD member countries examined macroeconomic performance; using the data of the thirty OECD countries'national income growth, unemployment and inflation on the welfare, the countries' macroeconomic performance indices were calculated and compared. These calculated index values cover the period of 1990–2009, and according to the general evaluation made in the study, it was concluded that the economic performance of the majority of OECD countries improved in the relevant period and international integration also increased the welfare of the member countries.

Sanlısoy and Cetin [11] in their study, using TOPSIS method were compared to the 1950–2015 period, economic performance of Turkey's economy. In determining the economic performances in the study; growth, unemployment, inflation, interest rate, public foreign debt stock, current account deficit and budget deficit criteria were used. This data was obtained from the data published by the TUIK electronic data distribution system, the statistics published on the website of the Ministry of Development, and the data published by the Ministry of Finance General Directorate of Public Accounts. According to the findings obtained from the study, it has been suggested that the country's economy exhibited a higher performance in the review period, however, the transition period to a foreign economy was a more successful period.

Uzun and Kazan [12] in their study; In the solution of the main machine selection problem suitable for the NB 25 Wartsila fishing ship project, 12 criteria were determined and according to these criteria, they tried to select the most suitable machine among 7 main machine alternatives using AHP, TOPSIS and Promethee methods. As a result of the study, it has been determined that AHP and Promethee methods give close results and AHP and Promethee results are consistent with the real situation in the selection of the host.

Karaatli [13] in his study, Turkey's performance between 2003 to 2014 years (tourism income, number of visitors, number of facilities, number of rooms, number of beds, the total occupancy rate, average length of stay, travel agent number, GDP of tourism revenue share, tourism income to exports, the place of tourism income in closing the foreign trade deficit, the number of domestic passengers, the number of international passengers, the number of cruise ships, the number of passengers arriving with cruise ships, the average expenditure amount, the US dollar buying average, the Euro buying average, the inflation rate) According to 19 criteria, it was compared with Entropy method, which is one of the MCDM methods. In conclusion, Turkey has concluded that the achieved success in the tourism sector for the relevant period.

Onder et al. [14] evaluated the economic performance models of F13 countries in the period of 2001–2013. F5 countries (Brazil, Turkey, India, Indonesia and South Africa) to identify vulnerabilities that occur in times of recession and its aftermath, ANP and TOPSIS methods were used. In the study, 11 macroeconomic parameters determined for F5 countries, Turkey's largest recession period (2008–2009), although the most fragile economies, the economic performance in the wake of the crisis has been suggested to be relatively high.

Dogan and Gencan [15] in their study; They aimed to measure the relative effectiveness of 26 public hospitals operating in Ankara. In the study, 5 output variables produced by 4 input variables used by hospitals to produce healthcare services were determined and two different DEA models were created with these variables. It has been suggested that the results obtained from the study can be used as a guide by decision-makers and hospital administrators and employees who determine the country's health policy.

Ardoss et al., [16] in his study; The main thoughts of the multi-criteria decision making theory and practice were examined through a questionnaire study. The aim of the study is to define various applications and approaches and to propose the most robust and effective approaches for the best alternative. This questionnaire is about banking, where uncertainty that occurs frequently in decision making occurs. In addition, the survey study tries to identify the problems encountered in fuzzy multi-criteria decision making techniques.

Urfalıoglu and Genc [17] in their study, the determination of Turkey's economic situation in the EU process and MCDA methods were intended to demonstrate the feasibility and economic data. The research, which 27 EU member states and accession candidates to complete Turkey, Macedonia, Iceland, Montenegro and Croatia were included, six of macro-economic analysis required by the study was conducted using data from 2010. According to the results obtained from the study; Economic ranking, Turkey ranks 31st with Electra method. It is 13th with the TOPSIS method. It is 32th with the Promethee method.

Akkaya et al. [18] in their study; The profitability and effectiveness of the Turkish tourism industry for three years were evaluated. The financial data of 8 tourism companies from 2009, 2010 and 2011 used in the study were obtained from Borsa Istanbul and the data were analysed by Visual Promethee Academic. As a result of the study, the best alternative was chosen based on the information determined by Promethee II.

Genc [19] in the study where Promethee and Electre methods examine the classification characteristics; classified some selected G7 countries (Germany, Japan, UK and Canada) according to 2010 macroeconomic criteria (governments' net debt/GNP ratio, unemployment, inflation). As a result of the study, the economic performances of G7 countries selected using the Electre and Flowsort method were determined and their classifications were made according to the performance.

Ozden [20] in his study; According to EU member states and Turkey's economic performance ranking of the Maastricht criteria, he has made the VIKOR method using 2010 data. In conclusion, made by the EU member states and Turkey's performance rankings relevant criteria and comparisons have been made with Turkey.

Dincer [21] in his study; In 2008, it analysed the economic activities of EU member and candidate countries according to five macroeconomic criteria using TOPSIS and WSA methods. As a result of the analysis; It is determined that Luxembourg, Holland and Denmark are in the first three places, and Macedonia, Latvia and Bulgaria are in the last three.

Ozden [22] in his study; According to 2009 data of the six macroeconomic criteria (public debt/GDP, unemployment rate, budget deficit/GDP, export/import, GDP/population and inflation) of EU member and candidate countries, their performances were measured by TOPSIS method. According to the results obtained from the study, the EU member and candidate countries are ranked in accordance with the relevant criteria and Turkey in this sort of place, missing or surplus is determined.

Mareschal and Smet [23] in his work, he investigated how visual representations can be developed both through the preference modelling process and the analysis of a multi-criteria decision problem using Promethee-GAIA methods. The factors and criteria (price, power, fuel, space and comfort) determined as a result of the study are shown in the form of GAIA Brain, Diamond display, Network representation.

Eleren and Karagul [24] in their study; seven macro variables using data from the 1986–2006 period of Turkey's economy, the country's economic performance has been calculated by the level of TOPSIS method. According to the performance scores obtained from the study, the best year is 1986, followed by 1990, 1987 and 1993, respectively. It was suggested that the worst years in terms of performance were 1999, 2001, 2006 and 2000.

Kılıc [25] in his study; Using the 7 basic macroeconomic criteria of 2004, it has classified the 25 member countries of the EU and 4 candidate countries with the MCDM approach. In the study, the extent to which candidate countries are ready for economic integration with the EU was analysed by the Electre method and as a result of the study, it was suggested that many countries that became full members of the EU in 2004 were not ready for economic integration with the EU in terms of economic performance.

Pohekar and Ramachandran [26] in their work; he scanned the publications on sustainable energy management and concluded that MCDM methods are more applicable in changing socio-economic conditions. In the study, more than 90 articles were examined and a classification was made regarding the application areas and application year in order to emphasize the trends of the publications. In MCDM applications; Multipurpose optimization, decision support system and multi criteria decision making methods have been used and it has been suggested that AHP, Promethee and Electre are the most popular techniques.

Yurdakul and Ic [27] in their study, Turkey operating in the automotive industry and ISE also been found with five large-scale automotive company, the period from 1998 to 2001 were analysed with TOPSIS method according to the balance sheet and financial ratios. The performances of the companies discussed in the study were determined according to 6 criteria (current rates, stock turnover rate, short-term debts/total debt, gross sales profit/net sales, main operating profit/net sales and net sales per person) and as a result of the study, their successes are ranked according to their performance scores.

3 Methodology of the Research

MCDM methods allow the decision maker to reach a solution through various methods. This solution takes place by sorting, grouping alternatives or choosing among them. Each MCDM method has its own approach and specific information to be known by the decision maker when using this method. These approaches constitute the methodology of MCDM methods.

In the Promethee method, which is one of the newest MCDM methods used today, it compares the decision making alternatives in pairs. In these comparisons, the decision making process is made by choosing one of the 6 fixed preference functions determined for each criterion. There are also additional threshold values that must be determined in each preference function. The GAIA plane is a graphical representation built on Promethee results. The plane provides decision making and researchers a quick, simple and understandable perspective. In addition, the GAIA plane displaying Promethee results on two-dimensional plane creates the opportunity to examine the results with a different approach.

3.1 Purpose of the Research

The purpose of this study; Turkey in 2018, with some EU countries with Promethean to-GAIA method is to compare the economic performance. In addition, the Promethee-GAIA method, which is quite new in the literature, is introduced to increase the recognition and usability of the researchers and decision-makers in our country.

An important advantage of the method; Having an extension such as GAIA geometric representation is that it visually facilitates the presentation of the decision-making picture to decision-makers in a very short time and speeds up the decision-making process.

3.2 Data and Method

In the study, of the 10 EU countries and Turkey, the economic performance (unemployment rate, GDP, inflation rate, exports and imports of goods and services% of

GDP) was compared with the data for 2018. The data used in the study were obtained from the OECD database for 2018 [28, 29]. The criteria used in the evaluation are the ones that are widely used as economic performance criteria as a result of the examination made in the related literature [2], [17], [30, 31] and those that can be accessed from these data were selected. These decision criteria are selected among the most basic economic variables that determine the economic performance of countries.

After determining the alternatives and criteria, the decision matrix was created for the application of Promethee methods and the economic data obtained from the sources are shown in Table 1 below.

3.3 Evaluation of Economic Performance with PROMETHEE Method

EU member or candidate countries and 2018 data pertaining to economic performance indicators of Turkey, the entry would be made to the Promethean program, it is designed as alternatives and criteria. A starting table from these arrangements was created and presented in Table 1 below.

Table 1. Alternatives and criterion values (initial table)

Countries acronym	Unemployment	GDP	Inflation	Export (% of GDP)	Imports (% of GDP)
BEL	5,950	51490,80	2,053	82,576	82,739
HUN	3,708	31080,93	2,850	84,939	80,565
IRL	5,739	83081,13	0,488	122,326	89,185
LUX	5,583	113136,7	1,528	211,559	175,544
NLD	3,832	56325,70	1,703	84,322	73,331
SVK	6,536	33742,63	2,514	96,090	94,065
TUR	10,887	28383,57	16,332	29,530	30,627
EST	5,374	36022,43	3,436	74,257	70,747
SVN	5,110	37964,63	1,739	85,379	77,098
LVA	7,415	30309,51	2,535	61,304	61,489
LTU	6,152	35309,40	2,698	75,621	73,679

Source: [28, 29].

As seen in Table 1; place in the ranking of some EU countries in the scope of Turkey's top research, while others are under. Thus, more realistic results when comparing economic performance of Turkey and the EU were intended to achieve. In the continuation of the initial table, Promethee II net current values are achieved by following the steps of the Promethee method. In this part of the study, comments on the GAIA geometric plane of the results obtained from Promethee analyses will be made.

As shown in Fig. 1 below, the Promethee II chart shows which option to choose. Accordingly, Promethee II results are shown in the figure below.

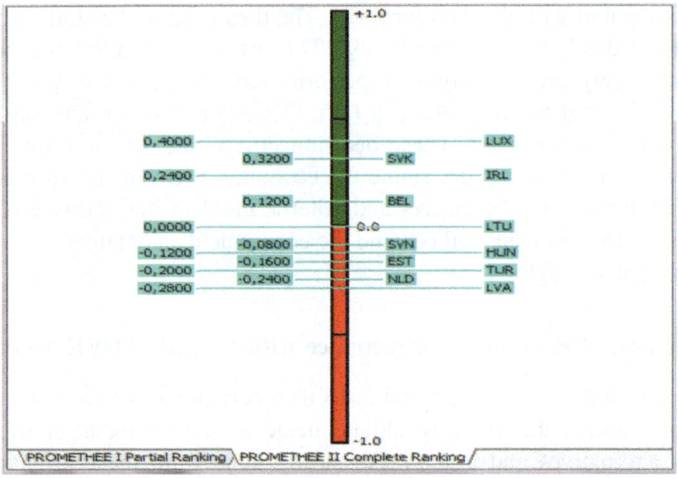

Fig. 1. Promethee II (Full Sequencing)

As can be seen in Fig. 1, with the results of Promethee II, both negative superiorities and positive superiorities will be determined and a decision will be made on which country to choose. Accordingly, the economic performance of the countries between 0 and +1 is high. In Promethee II, a weight-based decision emerges. Accordingly, while the weight of the Luxembourg country is positive, the weight of Slovakia is negative. Luxembourg ranked first with the highest net Phi value among the countries covered by the analysis. This country, respectively; Slovakia, Ireland, Belgium, Lithuania, Slovenia, Hungary, Estonia, Turkey, is followed by the Netherlands and Latvia.

In Table 2 below, Promethee II ranking table is given according to Phi score ranking.

Table 2. PROMETHEE II ranking table

Rank	Action	Phi	Phi+	Phi-
1	LUX	0,400	0,700	0,300
2	SVK	0,320	0,660	0,340
3	IRL	0,240	0,620	0,380
4	BEL	0,120	0,560	0,440
5	LTU	0,000	0,500	0,500
6	SVN	−0,080	0,460	0,540
7	HUN	−0,120	0,440	0,560
8	EST	−0,160	0,420	0,580
9	TUR	−0,200	0,400	0,600
10	NLD	−0,240	0,380	0,620
11	LVA	−0,280		

The net superiority value (Phi) obtained by subtracting the negative superiority value (Phi−) from the positive superiority value (Phi+) seen in Table 2 is shown. Accordingly, the positive superiority of Luxembourg (Phi = 0.4000) among the alternatives included in the study is greater than all other countries.

According to these results, economic performance rankings of the countries have been determined. In addition, the GAIA geometrical representation, which allows for more detailed and richer interpretations of this ranking, was provided by the program and shown on the plane of the decision below (Fig. 2). Thus, interpretation of the results has been made easier.

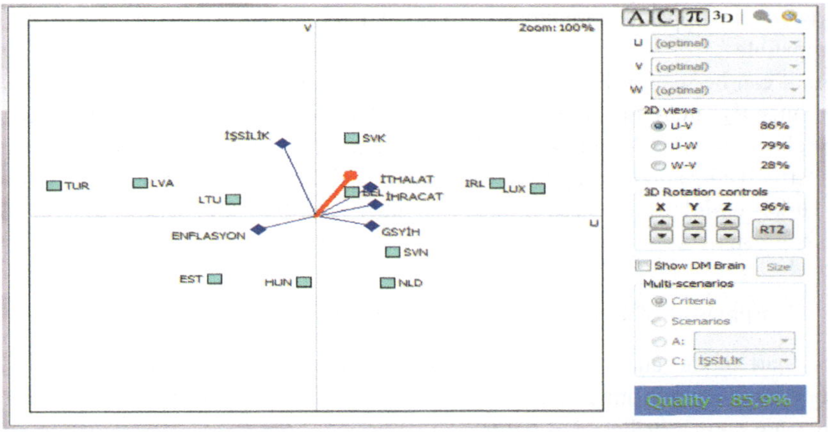

Fig. 2. Economic performance of EU countries GAIA graphic display

As can be seen in Fig. 2, alternatives are shown in the GAIA plane as square, and the decision bar showing the optimum status as a blue striped vector is shown in red. The Quality value in the lower right part of the graph, which indicates the quality of the Promethee measurement, has been determined as 85.9%.

As this determined value is approached to 100%, the accuracy of the analysis increases. If the value is below 60%, the analysis and calculations of the problem should be made again and more care should be taken. If this value obtained in the study is not below 60%, analysis can be continued [32].

As seen in Fig. 2, the red line close to the horizontal axis (criteria and alternatives) shows the optimum situation. The values close to the optimal line (red line) shown in Fig. 2 show the countries with high economic performance (Belgium and Slovakia) according to the relevant criteria (import, export and unemployment).

As seen in Fig. 2, the countries that are close to the criteria indicated by thin blue lines are the countries that should be preferred primarily according to the criteria discussed. Accordingly, if it is desired to make a choice in terms of import and export criteria, it can be interpreted that the Belgium country, which is the closest point to these criteria, shown in blue line, is in a better condition in terms of these criteria.

As shown in Fig. 2, if the blue-line vectors showing the axes of competition are long (for example, the unemployment vector), it is interpreted that these criteria have differentiating features in the alternative determination and the importance of influencing the decision bar. If the criteria have low parsing power, as shown in Fig. 2 (for example GDP), the vector length of the criterion is shorter.

In the GAIA geometric plane; The vectors showing the same direction are interpreted with the criteria bars that show similar characteristics and different directions that conflict with each other. Accordingly, it is possible to say that the unemployment criterion seen in Fig. 2 is more important in influencing competition. Since the import and export criteria are in line with each other, it can be stated that these two criteria are compatible with each other. On the contrary, since the inflation and unemployment criteria are in opposite directions, it can be interpreted that these two criteria are incompatible.

As it can be seen in Fig. 2, it is possible to say that their profiles are close to each other since the alternatives of Ireland and Luxembourg are in the same directions and relatively close to each other on the GAIA plane. This is the opposite of the situation, the alternative locations away from each other (with Slovakia and Estonia Luxembourg and Turkey) while differences between countries can be expressed by the value they have is great.

3.4 Comparison of Economic Performance of Countries

In this part of the analysis; The profiles of alternative close together on the GAIA geometric plane, indicating that similar, Turkey and Latvia were chosen alternative and profiles are shown in Fig. 3–4 below.

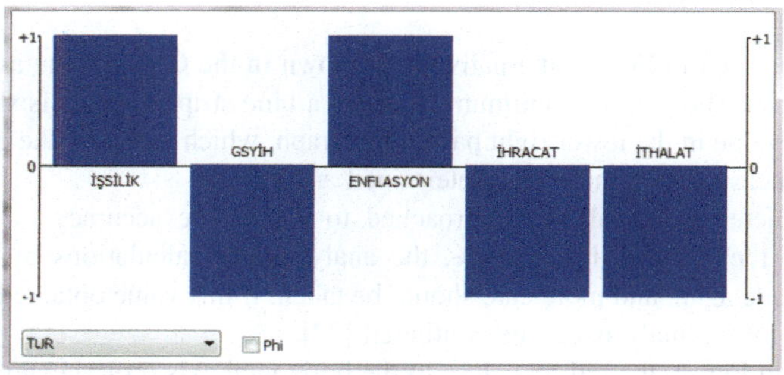

Fig. 3. Alternative profile of Turkey

As seen in Fig. 3; Turkey has a positive rule in the context of unemployment and inflation criteria, GDP, exports and imports across the board criteria are also negative lead.

The profile of the Latvian alternative, which is close to each other on the GAIA geometric plane and whose profiles are similar, is shown in Fig. 4 below.

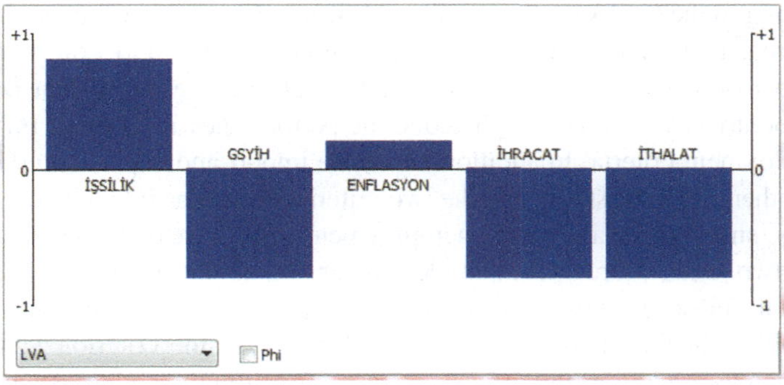

Fig. 4. Profile of the Latvian alternative

As seen in Fig. 4; In a similar manner in the context of Latvia's unemployment and inflation criteria on the Turkish position it has a positive advantage. It has a negative advantage in terms of GDP, export and import criteria. However, when Figs. 3 and 4 is analysed comparatively, it is possible to say that these advantages differ somewhat in terms of their level.

As seen in Figs. 3 and 4; when Turkey and Latvia's profiles examined closely, it is seen that unemployment and inflation criteria in the context of ever closer together, and in a positive direction. It is possible to say that there is a negative similarity in the context of GDP, import and export criteria.

4 Conclusion and Comments

According to Promethee II ranking, countries between 0 and +1 have high economic performance. According to this; In the Promethee II full rankings, the countries that showed clear superiority were ranked and the weight of Luxembourg was determined as positive and Slovakia was determined as negative. Luxembourg, which has the highest net Phi value among the countries covered by the analysis, ranks first. These countries, as with Slovakia, Ireland, Belgium, Lithuania, Slovenia, Hungary, Estonia, Turkey, the Netherlands, Latvia, respectively. Thus it was determined the economic performance ranking of Turkey and some selected EU countries. However, in order to take advantage of GAIA geometric plane visual analysis in order to reduce the possible mistakes of decision makers, a different visual richness was provided to the problem solution, alternatives and criteria were made visible on the geometric plane, and the comments to be made were made easier to understand. The accuracy rate of the results obtained from the analyses was calculated as 85.9%. According to this value, it was decided that the analysis is likely to be correct.

According to the five criteria discussed in the study, the countries that are close to the optimal line shown in the GAIA geometric plane are respectively; Belgium, Slovakia, Ireland and Luxembourg are countries. In addition, according to the evaluation made in the context of proximity to the five criteria considered in the said plane; Belgium is closer to import and export criteria, Slovenia to GDP criteria, Estonia to Inflation criteria, and Latvia to unemployment criteria.

According to the GAIA geometric plane shown in Fig. 2 (blue vectors that are close to the optimal line - red vector); the ratio of import and export criteria to affecting economic performance is higher. Accordingly, it can be interpreted that Belgium and Slovakia countries, which have high economic performance in terms of import, export and unemployment criteria. In addition, since the import and export criteria are in line with each other, it can be stated that the two criteria are compatible with each other. On the contrary, since the inflation and unemployment criteria are opposite, it is interpreted that these two criteria are incompatible. According to the GAIA geometric plane, countries with close positions and in the same direction are also considered to have similar profiles. In the opposite cases, the difference in the criterion value between countries increases and the similarities decrease. Accordingly, countries whose profiles are close to each other; The profiles of Ireland, Luxembourg, Slovenia and the Netherlands are similar. The remote countries together with Slovakia, Estonia and Turkey, while the Luxembourg country profiles are different.

References

1. Eraslan, E., Algun, O.: İdeal Performans Değerlendirme Formu Tasarımında Analitik Hiyerarşi Yöntemi Yaklaşımı, Gazi Üniversitesi, Müh. Mimarlık Fakültesi Dergisi **20**(1), 95–106 (2005)
2. Dasbası, B., Barak, D., Celik, T.: Türkiye İçin Makroekonomik Performans Endeksinin Analizi (1990–2017): Yapay Sinir Ağı Yaklaşımı, Bingöl Üniversitesi İktisadi ve İdari Bilimler Fakültesi Dergisi, Yıl: 3, Cilt: 3, Sayı: 1, pp. 93–112 (2019)
3. Asaleye, A.J., Charity, A., Adedoyin, I.L., Adeyemi, A., Ogundipe, H.I., Olabisi, P., Adewara, S., Olusegun, B.O.: Oil price shock and macroeconomic performance in Nigeria: implication on employment. Int. J. Energy Econ. Policy **9**(5), 451–457 (2019)
4. Salas-Velasco, M.: Resource misallocation and production inefficiency Estimating cross-country differences in macroeconomic performance. J. Econ. Stud. **45**(6), 1272–1287 (2018)
5. Lim, S., Morshed, A.M., Khun, C.: Trust and macroeconomic performance: a two-step approach. Econ. Model. **68**, 293–305 (2018)
6. Araujo, E, Araújo, E., Filho, F.F.: Macroeconomic performance in Brazil under the inflation targeting regime. Invest. Economica **77**(304), 72–101 (2018)
7. Masca, M.: Economic Performance Evaluation of European Union Countries by TOPSIS Method. North Econ. Rev. **1**(1), 83–94 (2017)
8. Güney, C.: Visual Promethee İle Yatırımcılar Açısından Sektörlerin Değerlendirilmesi. TURAN-SAM Uluslararası Bilimsel Dergisi **9**(34), 177–187 (2017)
9. Eyüboğlu, K.: Türk Dünyasında Yer Alan Ülkelerin Makro Performanslarının Karşılaştırılması. Bilig **83**, 331–350 (2017)
10. Welsch, H., Kühling, J.: Macroeconomic performance and institutional change: evidence from subjective well-being data. J. Appl. Econ. **19**, 193–217 (2016)
11. Sanlısoy, S., Cetin, M.: Türkiye Ekonomisinde Hükümetlerin Ekonomik Performanslarının TOPSIS Yöntemi ile Ölçülmesi. Sosyoekonomik **24**(28), 65–83 (2016)
12. Uzun, S., Kazan, H.: Çok Kriterli Karar Verme Yöntemlerinden AHP TOPSIS ve PROMETHEE Karşılaştırılması: Gemi İnşada Ana Makine Seçimi Uygulaması. J. Transp. Logistics **1**(1), 99–113 (2016)
13. Karaatlı, M.: Entropi-Gri İlişkisel Analiz Yöntemleri ile Bütünleşik Bir Yaklaşım: Turizm Sektöründe Uygulama. Süleyman Demirel Üniversitesi İktisadi ve İdari Bilimler Fakültesi Dergisi **21**(1), 63–77 (2016)

14. Onder, E.: Economic performance evaluation of fragile 5 countries after the great recession of 2008–2009 using analytic network process and TOPSIS methods. J. Appl. Finan. Bank. **5** (1), 1–17 (2015)
15. Dogan, N.O., Gencan, S.: VZA/AHP Bütünleşik Yöntemi ile Performans Ölçümü: Ankara'daki Kamu Hastaneleri Üzerine Bir Uygulama. Gazi Üniversitesi İktisadi ve İdari Bilimler Fakültesi Dergisi **16**(2), 88–112 (2014)
16. Aruldoss, M., Lakshmi, T.M., Venkatesan, V.P.: A survey on multi criteria decision making methods and its applications. Am. J. Inform. Syst. **1**(1), 31–43 (2013)
17. Urfalıoğlu, F., Genc, T.: Çok Kriterli Karar Verme Teknikleri İle Türkiye'nin Ekonomik Performansının Avrupa Birliği Üye Ülkeleri İle Karşılaştırılması. Marmara Üniversitesi, İ.İ. B. Dergisi **35**(2), 329–360 (2013)
18. Akkaya, G.C., Uzar, C.: The usage of multiple-criteria decision making techniques on profitability and efficiency: an application of promethee. Int. J. Econ. Finan. Stud. **5**(1), 149–156 (2013)
19. Genç, T.: PROMETHEE Yöntemi ve GAIA Düzlemi. Afyon Kocatepe Üniversitesi, İİBF Dergisi **15**(1), 133–154 (2013)
20. Ozden, U.H.: AB'ye Üye Ülkelerin ve Türkiye'nin Ekonomik Performanslarına Göre VIKOR Yöntemi ile Sıralanması. İstanbul Ticaret Üniversitesi Sosyal Bilimler Dergisi **11** (21), 455–468 (2012)
21. Dincer, S.E.: Multi-criteria analysis of economic activity for European union member states and candidate countries: TOPSIS and WSA applications. Eur. J. Soc. Sci. **21**(4), 563–572 (2011)
22. Özden, U.H.: TOPSIS Yöntemi ile Avrupa Birliğine Üye ve Aday Ülkelerin Ekonomik Göstergelere Göre Sıralanması. Trakya Üniversitesi, Sosyal Bilimler Dergisi **13**(2), 215–236 (2011)
23. Mareschal, B., Smet, Y.: Visual PROMETHEE: developments of the PROMETHEE and GAIA multi criteria decision aid methods. In: 2009 IEEE International Conference on Industrial Engineering and Engineering Management, pp. 8–11, December 2009
24. Eleren, A., Karagul, M.: 1986–2006 Türkiye Ekonomisinin Performans Değerlendirmesi. Yönetim ve Ekonomi **15**(1), 1–14 (2008)
25. Kılıç, S.B.: Avrupa Birliğine Üye ve Aday Ülkelerin Bazı Temel Makro Ekonomik Kriterlere Göre Sınıflandırılması: Çok Kriterli Karar Alma Analizine Dayalı Bir Modelin Tahmini. Ç.Ü. Sosyal Bilimler Enstitüsü Dergisi **14**(2), 339–352 (2005)
26. Pohekar, S.D., Ramachandran, M.: Application of multi-criteria decision making to sustainable energy planning a review. Renew. Sustain. Energy Rev. **8**, 365–381 (2003)
27. Yurdakul, M., I, Y.: Türk Otomotiv Firmalarının Performans Ölçümü ve Analizine Yönelik TOPSIS Yöntemini Kullanan Bir Örnek Çalışma. Gazi Üniversitesi Müh. Mimarlık. Fakültesi Dergisi **18**(1), 1–18 (2003)
28. OECD: website (2019). https://data.oecd.org/
29. TÜİK: website (2019). https://databank.worldbank.org/
30. Ela, M., Dogan, A., Ucar, O.: Avrupa Birliği Ülkeleri ve Türkiye'nin Makroekonomik Performanslarının TOPSİS Yöntemi İle Karşılaştırılması. Osmaniye Korkut Ata Üniversitesi İktisadi ve İdari Bilimler Fakültesi Dergisi **2**(2), 129–143 (2018)
31. Karabulut, K.: Avrupa Birliği Ülkeleri ve Türkiye'nin Ekonomik Performanslarının Karşılaştırılması: Veri Zarflama Analizi. Atatürk Üniversitesi, İktisadi ve İdari Bilimler Dergisi **22**(1), 1–11 (2008)
32. Bülbül, S.E., Köse, A.: Türk Sigorta Sektörünün PROMETHEE Yöntem ile Finansal Performans Analizi. Marmara Üniversitesi İktisadi ve İdari Bilimler Fakültesi Dergisi **38**(1), 187–210 (2016)

Warehouse Planning for Maximum Area Efficiency in the White Goods Sector

Gizem Berkdemir, Gizem Atak, and Ferhan Çebi(✉)

Department of Management Engineering, Istanbul Technical University, İstanbul, Turkey
gizemberkdemir@gmail.com, {atakl6, cebife}@itu.edu.tr

Abstract. In today's intense competitive environment, businesses focus on improving their systems. Warehouse management is one of the most important areas of supply chain management. The placement of storage areas; directly affects the processes such as receiving, picking and shipment and therefore plays an important role in the elimination of idle areas. In this study, it is aimed to develop a model that maximizes the storage area utilization efficiency for 3 different product groups of an enterprise operating in the white goods sector. The model is solved by using goal programming.

Keywords: Warehouse efficiency · Idle area minimization · Stocking · Supply chain management · Storage management

1 Introduction

Stocking; It has an undeniable importance as a result of the modern competitive environment in cost minimization, supply chain management and customer satisfaction.

In addition to the storage process, storage of manufactured products and semi-products, it also plays an important role in ensuring the uninterrupted flow of materials used in production and thus eliminating losses and costs in production [1].

Good recognition of storage systems, making the best design of the warehouse, determining the appropriate storage strategy and methods, planning which products and how will be stored will maximize the benefit obtained from the storage processes [2].

On the other hand, the white goods industry has a great importance within the Turkish manufacturing sector with its technology developed by producing, product designs, increasing production volume and export capacity. In addition, the white goods sector in our country has a wide range of products, from the large household appliances sector to the small household appliances sector. For this reason, our country's white goods sector also attracts the attention of the consumers in the western market. Studies in this area will be a guide for all companies operating in the white goods sector.

N. M. Durakbasa and M. G. Gençyılmaz (Eds.): ISPR 2020, LNME, pp. 598–603, 2021.
https://doi.org/10.1007/978-3-030-62784-3_51

2 Literature Review

In this section, studies on warehouse management and location are examined in detail. As mentioned before; storage has an important place in supply chain management. For this reason, there are many studies in this field from past to present. For example, in a study conducted in 1963, Kuehn and Hamburger (1963) worked on an intuitive program to determine the optimal location of warehouses [3].

When it comes to today, it is very important to be able to stock effectively and effectively in order to meet changing customer demands. For example; In 2016, Gül et al. worked on warehouse storage maximization. In their study, the areas to be allocated in the warehouse were calculated according to the frame sizes and volumes, so as to provide the maximum stocking area and volume in the product warehouse in order to ensure efficient storage despite the customer demands that change over time. In the light of these goals, mathematical models were developed to solve different versions of the problem [4].

In the supply chain area, customers have static and dynamic storage demands. This creates an uncertainty factor for warehouses. In a study conducted in 2015, a model was developed that targets this uncertainty. It has introduced a variable inventory ratio coefficient to deal with the uncertainty in storage space requirements. This reveals the original value of the study. Using the Robust optimization method, stocking costs have been reduced and the financial benefits of all interested parties have been increased [5].

A similar study was encountered in 2005 in the same area. In the study, it is aimed to place the products in an optimum way by taking into consideration the dimensions of the products in order to provide maximum space and volume in the warehouses. Problems in the study; high inventory costs, determination of the place where the product will be placed (shelf environment/no shelf environment), transportation times and transportation costs, shelf volume suitable for products of different sizes, etc. Especially materials that can be stacked with shelves cause a considerable storage cost when placed in any warehouse without shelves. In addition, the materials accumulated in front of the shelves create an extra cost item for the company. In the study, a mathematical model is developed that prevents this situation [6].

On the other hand, efficiency studies in warehouses are not only due to changes in customer demands. Not all product groups have the same frequency of movement in warehouses. Some products move more in the warehouse. Here is another study that considers this situation, in 2016, on the effective placement of the raw material warehouse. In this study consisting of two stages, it is aimed to improve the placement of the chemical raw material warehouse in a chemical company. Firstly, in the first step, the most moving raw material items were determined with the help of multi-criteria ABC analysis. Later on, a mathematical model was developed and the raw materials were placed in a way that minimizes the transportation according to the ABC analysis results. As a result of the study, a new settlement was proposed and it was observed that the improvement achieved with this settlement was revealed [7].

Good recognition of the storage systems, the best design of the warehouse and the selection of suitable methods for this design and planning how and how the products will be placed in the warehouse increase the efficiency to be obtained from the storage process. Therefore, it is important to classify the materials in this article. In a study conducted in 2012, a solution was sought for the problems related to material classification and lift placement. In the company discussed in the study, products are classified according to their prices, but products without price information can be grouped incorrectly. As a result, there are errors during counting and some materials are recounted unnecessarily, which has led to an increase in costs for the company. Especially, frequent counting caused delays in product delivery. This situation was reflected to the company as a cost. For all these reasons, a model that enables the rearrangement of the warehouse was developed and costs were avoided [1].

3 Application

In the study, ABC analysis is implemented. Firstly, the stock quantities of 2018 were examined for each product group. The data are in Table 1.

Table 1. Average production quantities according to 2018 product groups

		Item Groups		
	Period	**A**	**B**	**C**
Number of Units	January 18	29.120	79.546	55.505
	February 18	26.800	100.402	63.420
	March 18	40.656	128.671	81.561
	April 18	45.211	126.917	80.429
	May 18	43.054	126.247	78.932
	June 18	44.609	101.279	60.919
	July 18	28.628	57.640	39.997
	August 18	34.670	50.868	41.143
	September 18	33.732	50.138	35.712
	October 18	21.434	40.713	37.039
	November 18	20.102	35.608	31.884
	December 18	26.979	55.105	34.808
	Total	394.996	953.134	641.348

Secondly, products' suitable locations, where the products can be placed, are analyzed. Location capacities according to product groups are given in Table 2:

Table 2. Address capacities according to product groups

Address type	Address dims (cm)	A	B	C
LOCA1	238 × 250	48	48	48
LOCA2	238 × 275	48	60	48
LOCA3	238 × 300	60	60	60
LOCA4	238 × 350	84	84	84
LOCA5	238 × 475	96	96	96
LOCA6	238 × 750	168	192	168
LOCA7	238 × 1200	300	300	300
LOCA8	238 × 525	120	120	120
LOCA9	238 × 675	156	168	160
LOCA10	238 × 1000	240	252	240
LOCA11	238 × 1500	384	384	384
LOCA12	238 × 825	204	204	192
LOCA13	238 × 1125	276	276	276
LOCA14	238 × 775	192	192	192

There are 14 location groups in the warehouse. The size of each address type and how much of which product group it can be placed in these locations are also indicated in the table. For example: the size of the LOCA1 address type is 238 × 250 = 59500 cm2 = 5.95 m2. In this address type, only 48 of the A group can fit and if a product of the A product group is placed in an address of the address type, the product from the other two product groups cannot be placed. In other words, more than one product group cannot be placed in an address at the same time. The products of a product group can be placed at other addresses of the same address type.

Model:

In the study. the maximum period of the stock amount of each product group are taken into account by analyzing the historical stock data:

Maximum number of inventories of product group A in 2018 on a monthly basis = 45.211
Maximum number of inventories of B group in 2018 on monthly basis = 128.671
Maximum number of inventory of product group C in 2018 on a monthly basis = 81.561

Objective Function:

$$Min\ Z = \mathrm{d}+ \quad (1)$$

Wi = i. size of the area occupied by the address type
Ai, Bi, Ci = i. number of addresses required for address type
n: Number of address types

Nij = i. address type j. capacity including product group
NiA = {48, 48, 60, 84, 96, 168, 300, 120, 156, 240, 384, 204, 276, 192}
NiB = {48, 60, 60, 84, 96, 192, 300, 120, 168, 252, 384, 204, 276, 192}
NiC = {48, 48, 60, 84, 96, 168, 300, 120, 160, 240, 384, 204, 276, 192}
d− = Negative deviation
d+ = Positive deviation

Constraints:

Constraint 1: The stock amount of this year for product group A should be at least the amount of stock in 2018.

$$A \leq N_{iA} \sum\nolimits_{i=1}^{14} (A_i) \tag{2}$$

Constraint 2: For the product group B. this year's stock amount should be at least as much as in 2018.

$$B \leq N_{iB} \sum\nolimits_{i=2}^{14} (B_i) \tag{3}$$

Constraint 3: The stock amount this year for C product group should be at least as much as the stock amount in 2018.

$$C \leq N_{iC} \sum\nolimits_{i=1}^{14} (C_i) \tag{4}$$

Constraint 4: No product in product group B is assigned to address types LOCA12. LOCA13. LOCA14.

$$\sum\nolimits_{i=11}^{14} (B_i) = 0 \tag{5}$$

Target Constraint: There is a total area of 22000 m2.

$$(W_i) \sum\nolimits_{i=1}^{n} (A_i + B_i + C_i) + d - -d + = 22000 \tag{6}$$

$$Wi, Ai,\ Bi,\ Ci,\ Nij \geq 0$$

4 Results

The created mathematical model is solved by using Lindo. With this model. it has been calculated which address group of the company is assigned by which product group and how many addresses are required for each of them. According to the results;

- A total of 119 addresses must be created for group A products and these addresses cover an area of approximately 4204.87 m2 and fit in 45216 units of A-type products to the warehouse.

- A total of 670 addresses must be created for group B products. and these addresses cover an area of approximately 11963.67 m2 and fit into 128676 units of B type products from the warehouse.
- A total of 213 addresses must be created for group C products and these addresses cover approximately 7584.49 m2 and fit 81568 units of A-type products into the warehouse.
- The number of addresses created for each address type is 956 in total and the total area covered by these addresses is 23753.03 m2. Since the usable target area in the warehouse is determined as 22.000 m2. 1753.03 m2 area is needed.

5 Future Studies and Suggestions

The area of warehouse management. which is an important part of the logistics industry in recent years. Is being studied more and more and the number of studies in this area is increasing rapidly. The scope of the literature review can be extended in future studies. In particular, heuristic methods can be developed by applying up-to-date methods such as big data and machine learning in this field. For example, with the learning algorithms of warehouse management software. the product flow/movement frequency from historical data can be calculated and placement and collection algorithms can be written accordingly. Thus. idle area minimization can be achieved in warehouses.

References

1. Özyörük, B., Ak, S.: Etkin Depo Yerleşiminin Düzenlenmesi İçin Bir Model: Elektronik Firmada Uygulanmasi. Tübav Bilim Dergisi **5**(1), 21–29 (2012)
2. Le-Duc, T.: Design and Control of Efficient Order Picking Processes (2005)
3. Kuehn, A.A., Hamburger, M.J.: A heuristic program for locating warehouses. Manage. Sci. **9**, 643–666 (1963)
4. Gül, G., Erol, B., Öngelen, G., Eser, S., Çetinkaya, Ç., Özmutlu, H. C., Erhuy, C.G.: Ambar Depolama Maksimizasyonu. J. Ind. Eng. (Turkish Chamber Mech. Eng.) **27**(4), 26–38 (2016)
5. Luo, Y., Chen, Z.Y., Chen, C.X.: Robust optimization in warehouse space allocation of pledges in supply chain financing. Adv. Transp. Stud. (2015)
6. Heragu, S.S., Du, L., Mantel, R.J., Schuur, P.: Mathematical model for warehouse design and product allocation. Int. J. Prod. Res. **43**(2), 327–338 (2005)
7. Çolak, M., Keskin, G.A., Günel, G., Akkaya, D.: Bir Cam İşletmesinde Kimyasal Hammadde Deposunun Etkin Yerleşimi İçin Bir Model Önerisi. Beykent Üniversitesi Fen Ve Mühendislik Bilimleri Dergisi **9**(2) (2016)

Quality Management

Geometric Tolerancing Analysis of Multi-material Components (MMCs) Obtained by Industrial X-Ray Computed Tomography

Mario Sokac[1(✉)], Zeljko Santosi[1], Djordje Vukelic[1], Marko Katic[2], Numan M. Durakbasa[3], and Igor Budak[1]

[1] Faculty of Technical Sciences, University of Novi Sad, Novi Sad, Serbia
{marios, zeljkos, vukelic, budaki}@uns.ac.rs
[2] Faculty of Mechanical Engineering and Naval Architecture, University of Zagreb, Zagreb, Croatia
marko.katic@fsb.hr
[3] TU Wien, Vienna, Austria
numan.durakbasa@tuwien.ac.at

Abstract. The rising demand in recent years for the application of X-ray computed tomography (CT) has been influenced by its non-destructive abilities and accuracy. The demand for application of X-ray CT on Multi-Material components (MMCs) has also increased. However, due to the presence of two or more materials during X-ray CT scanning, the consequent analysis of MMCs can have a negative impact on the results. This research aims to investigate the application of advanced algorithms and their influence on the accuracy of the X-ray CT images for performing geometric dimensioning and tolerancing (GD&T). This paper presents a case study where the application of such method was investigated in two case studies containing MMCs. Results show that the newly developed methods, such as the one shown in this paper, can improve borders of segmented MMCs on X-ray CT images which will result in more accurate results regarding GD&T.

Keywords: Geometric dimensioning and tolerancing (GD&T) · Computed tomography · X-ray CT · MMC

1 Introduction

High demands in the industry for fast and accurate product measurement is opening the doors for the application of X-ray computed tomography (CT) systems due to their non-destructive abilities and high accuracy [1]. But when it comes to scanning of Multi-Material components (MMCs), X-ray CT systems are prone to different types of artefacts [2, 3], as they have a big influence on the quality of generated CT data [4]. This allows accurate visualization and application of different types of dimensional measurements to be performed. Artefacts present artificial structures that are found in the resulting X-ray CT datasets which don't correspond to the real data [5]. Therefore, their presence can deeply affect the acquisition of reliable measurement data. They may appear as variations in shades of grey, or as a deviation in the reconstructed data. The

N. M. Durakbasa and M. G. Gençyılmaz (Eds.): ISPR 2020, LNME, pp. 607–613, 2021.
https://doi.org/10.1007/978-3-030-62784-3_52

presence of these artefacts can have a great influence on the obtained results, i.e. reconstructed 3D models, which can make them unreliable for any further analysis [5].

2 Literature Review

When it comes to CT dimensional measurements there are numerous factors such as surface quality, material, acquisition parameters, CT machine setup, etc. [6] that can influence the result. MMCs are subjected to artefacts generation due to the presence of two or more materials. Therefore, their visualization can be challenging, and many different methods have been developed to cope with this challenge [5–9]. Visualization for exploration and analysis of geometric tolerances has been previously discussed in [10] for uncertainty estimation in CT metrology when measuring geometric tolerances such as straightness, circularity, or flatness. Authors in [11] investigated different measurement and evaluation methods of the rotational symmetrical workpiece where they used both Coordinate Measuring Machine (CMM) and X-ray CT systems. Tolerance verification of plastic housing using X-ray CT has been also discussed in [12].

Many authors also investigated and discussed industrial X-ray CT accuracy and measurement uncertainty as well in [13]. Currently, the situation is progressing towards developments of new methods, which will aid in the superior analysis of CT datasets acquired by X-ray computed tomography.

3 Materials and Methods

MMCs, as noted earlier, in which all elements are adequately reconstructed, present a great challenge in generating geometrically accurate 3D models due to the presence of two or more materials of different specific densities (mostly metal and polymer).

Our previous work, first investigated as a preliminary study [14], and later as a validated method [15] presented a novel method for surface extraction of MMCs from X-ray CT datasets which was used in this study. The presented method consists of implementing two methods to extract more information from the X-ray CT datasets with improved processing of generated X-ray CT images which consequently enables surface reconstruction of dimensionally more accurate surface 3D models. The proposed method was compared against two commercially available software, namely GOM Inspect (GOM GMBH, Germany) and VGStudio Max (Volume Graphics GmbH, Germany). The workflow for geometric tolerancing analysis is shown on Fig. 1, while Fig. 2 shows the two objects used in this analysis.

Reference measurements were collected on a three-axis coordinate measuring machine Contura G2 (CARL ZEISS) shown in Fig. 3a. Basic measurements were carried out using a contact measurement stylus with 1 mm diameter, and L = 50 mm length in a measurement laboratory with controlled microclimatic conditions (20 ± 0.5 °C). Reference measurements were collected on the basis of defining basic geometrical features, whose analysis, as well as extraction of reference measurements, was performed in CALYPSO v4.8 software. X-ray CT datasets were obtained using the NIKON XT H 225 industrial CT system (Fig. 3b).

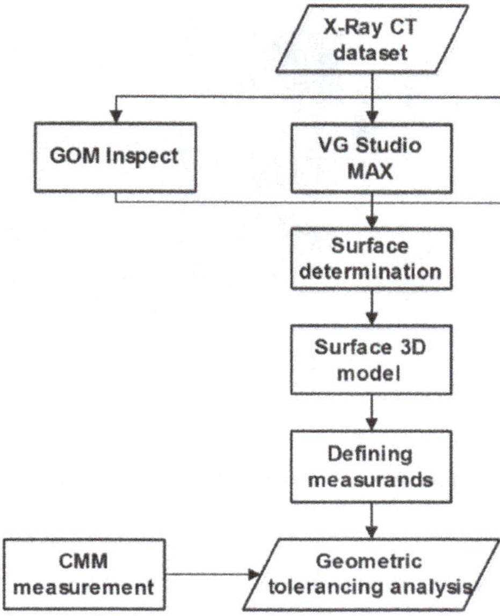

Fig. 1. Workflow procedure for geometric tolerancing analysis

Fig. 2. Two objects used in this study [15]

4 Results and Discussions

To test the effect of surface reconstruction on the geometrical shape of objects, an inspection of the geometric tolerancing of the shape was carried out on metal components for both case studies. The measurement was performed on the CMM shown previously in Fig. 3a. Cylindricity and flatness were selected as geometric characteristics of the form inspection for the first case study (steel screw). In addition to the cylindricity and straightness as geometric characteristics of the shape, orientation

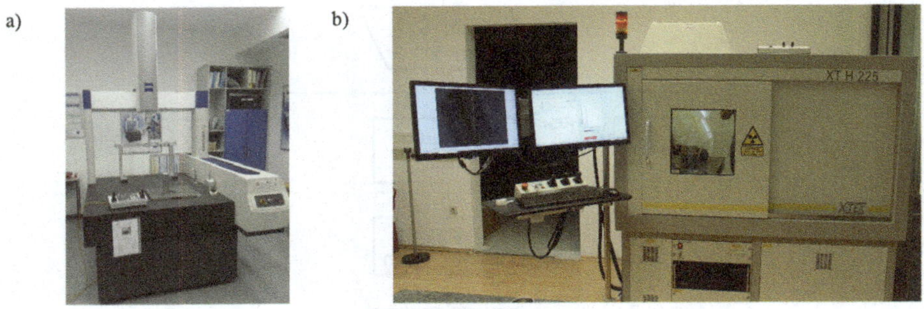

Fig. 3. Showing a) CMM CONTURA G2 by Carl Zeiss and b) Industrial X-RAY CT System NIKON XT H 225

tolerance analysis was also performed for the case study 2 (metal component with pins of DB-9M connector), namely the parallelism and position of pins C1 and C2 in reference to the plane P1.

Figure 4 shows the analysed characteristics for both case studies.

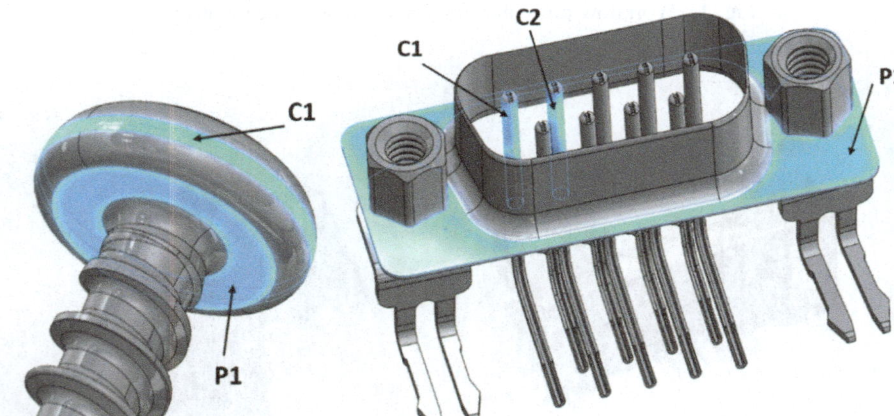

Fig. 4. Analysed characteristics C1, C2 and P1 for both case studies

Measurements for both case studies were performed on CMM with sampled points defined at an equidistant distance of 0.1 mm. A geometric feature extraction procedure was performed for surface 3D models generated using all three software (GOM Inspect, VGStudio Max, and the method proposed in [15]), to compare the shape deviation with the values obtained from CMM.

Figure 5 shows an analysis of the geometric tolerancing for both case studies. Geometric tolerancing analysis was performed in GOM Inspect v2018 software.

Table 1 shows the results of the analysis for both case studies comparing the results collected from surface 3D models from all three compared software and the results obtained using CMM.

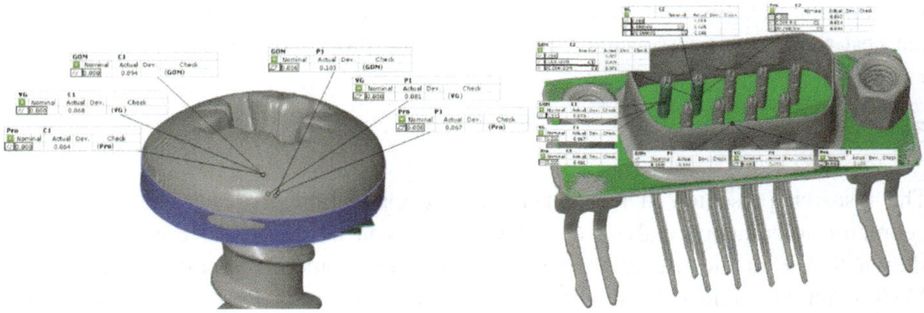

Fig. 5 Geometric tolerancing analysis for both case studies [16]

Table 1. Numerical values obtained by using CMM and three software GOM inspect, VGSTUDIO max and proposed method [15] for both case studies

	Geometric characteristic	CMM [mm]	GOM inspect [mm]	VGStudio max [mm]	Proposed [15] [mm]
Case study 1	Cylindricity (C1)	0.062	0.094	0.068	0.064
	Flatness (P1)	0.018	0.103	0.081	0.067
Case study 2	Cylindricity (C1)	0.072	0.073	0.067	0.066
	Cylindricity (C2)	0.066	0.072	0.061	0.060
	Flatness (P1)	0.111	0.182	0.153	0.139
	Parallelism (C1–C2)	0.038	0.035	0.025	0.024
	Position (C1–C2)	0.041	0.076	0.046	0.044

For case study 1 regarding the cylindricity analysis, when comparing the results obtained using CMM with the results obtained from all three analysed software, it can be observed that the surface 3D model generated by GOM Inspect software shows the largest deviation from the cylindrical shape compared to the results obtained at CMM, while the proposed method in [15] shows the least deviations, which are also very close to the measured reference value obtained using CMM. For flatness inspection, the best results are also obtained by the proposed method, while GOM Inspect showed the largest deviations from the results from CMM. However, the analysis of the flatness of the lower surface of the screw showed larger discrepancies for all three surface 3D models compared to the results obtained on CMM.

For case study 2, it can be observed that very similar results were obtained with the VGStudio Max and the proposed method [15] for cylindricity for pins C1 and C2. When analysing the flatness tolerance, the values obtained in all three compared software showed larger deviations compared to the results obtained using CMM. The largest discrepancies, compared to CMM, were obtained using GOM Inspect software. VGStudio Max software showed minor deviations from GOM Inspect, while the closest results to the CMM results were obtained using the method proposed in [15].

Examination of C1–C2 pin parallelism and the position of the C1–C2 pins showed that the best results were also achieved by the method proposed in [15].

5 Conclusions

The research presented in this paper aims at application of developed methods for geometric tolerancing analysis of MMCs. Based on the results of this analysis, it is shown that with the use of modern methods the geometric tolerancing analysis of MMCs can be conducted on the bases of X-ray CT datasets. Presented method shows that it can be useful for advanced extraction of information from MMCs, which will in return assure high quality of the consequent analyses of X-ray CT data.

Acknowledgements. This paper presents the partial results achieved within the PhD thesis in [16]. This research has also been supported by the Ministry of Education, Science and Technological Development through the project no. 451-03-68/2020-14/200156: "Innovative scientific and artistic research from the Faculty of Technical Sciences (activity) domain". Project IKARUS also supported parts of presented research (European Regional Development Fund, MIS: RC.2.2.08-0042, Faculty of Mechanical Engineering and Naval Architecture, University of Zagreb).

References

1. Akdogan, A.S., Vanli, A.S., Durakbasa, N.: Re-engineering of manufacturing parts by computed tomography data. In: Proceedings of the 12th International Conference on Measurement and Quality Control - Cyber Physical Issue IMEKO TC 14 2019, pp. 114–121 (2019)
2. ISO 15708–1:2017: Non-destructive testing – Radiation methods for computed tomography – Part 1: Terminology (2018)
3. ISO 15708-2:2017: Non-destructive testing – Radiation methods for computed tomography – Part 2: Principles, equipment and samples (2018)
4. Aloisi, V., Carmignato, S.: Influence of surface roughness on X-ray computed tomography dimensional measurements of additive manufactured parts. Case Stud. Nondestr. Test. Eval. **6**, 104–110 (2016)
5. Amirkhanov, C., Heinzl, M., Reiter, J., Kastner, J., Gröller, M.E.: Projection-based metal-artifact reduction for industrial 3D X-ray computed tomography. IEEE Trans. Vis. Comput. Graph. **17**(12), 2193–2202 (2011)
6. Baer, M., Hammer, M., Knaup, M., Schmidt, I.: Scatter correction methods in dimensional CT. iCT **14**(1), 41–46 (2012)
7. Shammaa, M.H., Ohtake, Y., Suzuki, H.: Segmentation of multi-material CT data of mechanical parts for extracting boundary surfaces. Comput. Aided Des. **42**(2), 118–128 (2010)
8. Nagai, Y., Ohtake, Y., Suzuki, H.: SegMo: CT volume segmentation using a multi-level Morse complex. Comput. Aided Des. **107**, 23–36 (2019)
9. Yagüe-Fabra, J.A., Ontiveros, S., Jiménez, R., Chitchian, S., Tosello, G., Carmignato, S.: A 3D edge detection technique for surface extraction in computed tomography for dimensional metrology applications. CIRP Ann. Manuf. Technol. **62**(1), 531–534 (2013)

10. Amirkhanov, A., Fröhler, B., Reiter, M., Kastner, J., Gr, M.E., Heinzl, C.: Uncertainty in CT metrology: visualizations for exploration and analysis of geometric tolerances. iCT **3**, 189–195 (2014)
11. Dovica, M., Végh, A.: Comparison of the cylindricity deviation using different evaluation methods. Am. J. Mech. Eng. **1**(7), 339–342 (2013)
12. Müller, P., Cantatore, A., Andreasen, J.L., Hiller, J., De Chiffre, L.: Computed tomography as a tool for tolerance verification of industrial parts. Procedia CIRP **10**, 125–132 (2013)
13. Dewulf, W., Kiekens, K., Tan, Y., Welkenhuyzen, F., Kruth, J.: Uncertainty determination and quantification for dimensional measurements with industrial computed tomography. CIRP Ann. Manuf. Technol. **62**(1), 535–538 (2013)
14. Sokac, M., Santosi, Z., Vukelic, D., Katic, M., Durakbasa, M.N., Budak, I.: Enhancement of images from industrial X-Ray computed tomography systems by hybrid approach. In: Proceedings of the International Symposium for Production Research 2019, vol. 58, pp. 1285–1301 (2020)
15. Sokac, M., Budak, I., Katic, M., Jakovljevic, Z., Santosi, Z., Vukelic, D.: Improved surface extraction of multi-material components for single-source industrial X-ray computed tomography. Measurement **153**, 107438 (2020)
16. Šokac, M.: Hibridni model za segmentaciju snimaka generisanih primenom kompjuterizovane tomografije. Doctoral Thesis, University of Novi Sad, Faculty of Technical Sciences (2019)

Intelligent Test Automation for Improved Software Quality Assurance

Gamze Uğur-Tuncer[1,2](✉), Can Davutoglu[2], and Numan M. Durakbasa[1]

[1] Industrial Metrology and Adaptronic Systems, Institute of Production Engineering and Photonic Technology, TU Wien, Vienna, Austria
numan.durakbasa@tuwien.ac.at

[2] QiTASC GmbH, Vienna, Austria
{gamze.ugurtuncer, can.davutoglu}@qitasc.com

Abstract. In the current fast digital transformation projects, with tight timelines, it becomes prudent to increase the quality of digital services and products swiftly. The business partners or customers request further abilities promptly with high quality. Hence, the quality of software is determinant on the market placement.

Latest World Quality Reports [1, 2] have stressed that new trends in QA and testing are also formed by this overall rapidity, that is by the concept of so called 'Quality at Speed'. Intelligent testing with increasing levels of automation is essential to achieve high quality with reduced time to market.

QiTASC's philosophy is rooted in the passion for creating high-quality testing pipes that help businesses to move away from tedious and time consuming QA activities. This allows the organizations to spend more time optimizing their products and preparing them for the market while their testing is taken care intelligently and efficiently.

Keywords: Software testing · Quality assurance · Intelligent test automation · Software life cycle testing · End to end testing

1 Introduction: Quality Assurance and Testing

QA and testing are not supportive roles in an organization, but they are critical to business outcomes and customer satisfaction. 'Customer focused' digital transformation brings up 'Quality at Speed' phenomenon. These indicate that QA and testing activities should be carried out efficiently and rapidly. Automation is the key to achieving the goals of increasing quality and better time to market, and even also with reduced costs.

Agile and DevOps are being adapted to greater extents in software development projects parallel to the need for speeding up. On the other hand, the organizations cannot fully benefit from these phenomena mainly due to the low levels of automation that causes inefficiencies in QA and testing activities. Improved level of automation is pointed out to be a key priority for QA, as argued in both recent WQR reports and scientific researches. Recently, the aim of automation has changed to a better coverage rather than just shortening of testing times.

N. M. Durakbasa and M. G. Gençyılmaz (Eds.): ISPR 2020, LNME, pp. 614–624, 2021.
https://doi.org/10.1007/978-3-030-62784-3_53

2 Software Testing

Software testing is a software quality assurance instrument carried out by a specialized testing teams utilized to check the software product's quality before it goes live. It is an activity to check whether the actual results match the expected results and to ensure that the software system is defect free. It involves the execution of a software component or system to evaluate its properties of interest.

Software Testing simply means the verification of Application Under Test (AUT) manually or using automated tools. Previously it was mostly performed at the end of development stage. But together with the realisation of importance of early detection of errors, early and often testing came into the attention of SQA professionals.

Software testing objectives are categorized [7] as follows.

Direct objectives:

- Spotting as many errors as possible.
- Performing the essential tests efficiently and effectively, considering budgetary and scheduling limitations.
- After the correction of spotted errors, carrying the software to an adequate level of quality.

Indirect objective:

- Compiling a log of software errors for future use in error prevention with corrective and preventive actions.

The benefits of using software testing can be counted as follows [6].

- Cost-Effectiveness: Testing any IT project on time helps to save resources for the long term. The costs of fixing the bugs will be less if they are caught in the earlier stage of software testing.
- Product quality: is significant for any software product. Testing ensures a quality product is provided.
- Security: Business is very sensitive to trust, so removing risks and problems earlier is essential. A less tested, defective product can cause not only money but also reputation loss.
- Customer Satisfaction: is key for the business outcomes to thrive. UI/UX Testing ensures the best user experience.

Software testing is undoubtedly the largest consumer of software quality assurance resources. According to WQR (2019–2020) [2] in the past seven years, the proportion of total IT budget allocated to QA and testing (costs of testing processes, tools and resource) is above a quarter, on average 27%. After 2015 where the highest proportion was observed with 35%, there is a slight downward trend which is mostly affiliated with the cost-effectiveness of the new tools and techniques introduced. Especially Automation significantly improves ROI on the new tools.

Companies are competing for customers with shrinking profit margins and tighter deadlines. Therefore, it is essential for them to find possibilities to reduce expenditures while sustaining high quality in their products and services. "Improving testing efficiency through automation" is the strategic key to achieve such goals [6].

3 Automation: Intelligent Testing

The companies, which adopt intelligent test automation approaches, elevate their ability to overcome time to market pressure together with the optimum level of quality to secure their business outcomes. With test automation the companies can make their projects transparent, so that they can take the right decisions, fulfil their projects faster and cheaper, reduce timelines due to parallel tasks and continuously growing repositories, verify features and services any time and in required accuracy which means reduced risk [3].

Driving forces for automation in the ICT industry:

- Agile Projects require automation
- Complex, mashed applications can no longer be manually tested. (Various technologies, less time, high complexity.)
- Projects ask for transparency (Automation provides an immediate snapshot of a project at any time.)
- Innovations and new approaches facilitate automation
- Increasing costs create pressure in software development, automated testing helps reducing costs.

Benefits of test automation:

1. Bugs can be detected during the development phase itself, which leads to shorter product development lifecycles.
2. Automated testing saves time by eliminating the need to write extensive test scripts.
3. It can also be used alongside manual testing procedures to cross check the test results.
4. Once created, automated tests can easily be repeated reducing the running time from days to hours. Time savings translates directly into cost savings.
5. The tests can be run 7/24 without the need for manual input, thus ensuring thorough quality checks.
6. Automated software testing can increase the depth and scope of tests, extended to perform tasks impossible with manual testing. They can even be run on the entire network or multiple devices simultaneously, also with different configurations.
7. Thousands of different complex test cases can easily be executed during every test run providing a wide coverage that is impossible with manual tests. Automated tests can look inside an application and see memory contents, data tables, file contents, and internal program states to determine if the product is behaving as expected.
8. The test scripts for automation tools can be reused in similar projects.

9. Automated testing also includes testing of web applications on all the browsers. Thousands of virtual users can be simulated interacting with a network, software and web applications.
10. Automated testing is perfect for user environment simulation because graphical user interface (GUI) testing is very time consuming and consists of repetitive actions for the software tester.
11. The reports that are generated are extensive and give the software tester details such as scripts executed, scheduled, bugs found and fixed, etc.
12. Testers freed from repetitive manual tests have more time to create new automated software tests and deal with complex features.
13. Whenever source code changes, tests can run automatically, the team or the developers are notified if they fail. Features like these save developers time and increase their confidence. Team members improve their skill sets and confidence, and in turn, pass those gains on to their organization.

4 Current Challenges in Test Automation

WQR 2019–2020 [2] indicates that there is an upward trend in automation in various phases of testing. Yet, the main existing challenges are sorted from highest to lowest as follows:

1. Big changes in the applications to be tested with every release
2. Challenges with test data environment and stability
3. Lack of skilled and experienced automation resources
4. Lack of having the right tool to automate
5. Having various automation tools
6. Having manual steps within the test process
7. Having started too late with testing and test automation
8. Having poorly defined project requirements

The highest challenges, continuously changing applications in each release and stability of test data environment underlines the importance of model based testing, test data environment and stability, together with end to end test automation. Robotics automation is also an important step which automates certain manual business processes, reducing cost and increasing speed and quality.

The rest of challenges are mainly related to "level of automation". Therefore, in order to better understand how automated testing supports high-quality, it is important to focus on level of automation that interfere with or prevent effective testing [6].

Which QA activities should be automated and which ones should be kept for manual testing, at least for a certain time? Besides automating the test execution; test case creation and requirement implementation; reporting services, data analytics and data exporting; scheduling tests are important automation areas.

The 'big bang approach' which is going from no automation to complete or a high level of automation at once, results in inadequate results. With this approach it is difficult to recognize sources of failure, as everything is integrated into the test scope at once. Correcting a fault, then requires “consideration of the possible effects of the correction on several modules at once and the same time” [7]. This creates uncertainty for budgeting procedures and testing-fixing.

On the other hand, an incremental approach integrates certain features or test modules in multiple phases, from a bottom-up or a top-down manner, which follows a hierarchical view of the software modules or products being tested [7].

5 QiTASC Test Automation: Smart Testing

5.1 Why intaQt Automation?

Recently, automation is used in many areas of software development life cycle (SDLC) as shown in Fig. 1 and the ultimate aim is to support all the life cycle of software development.

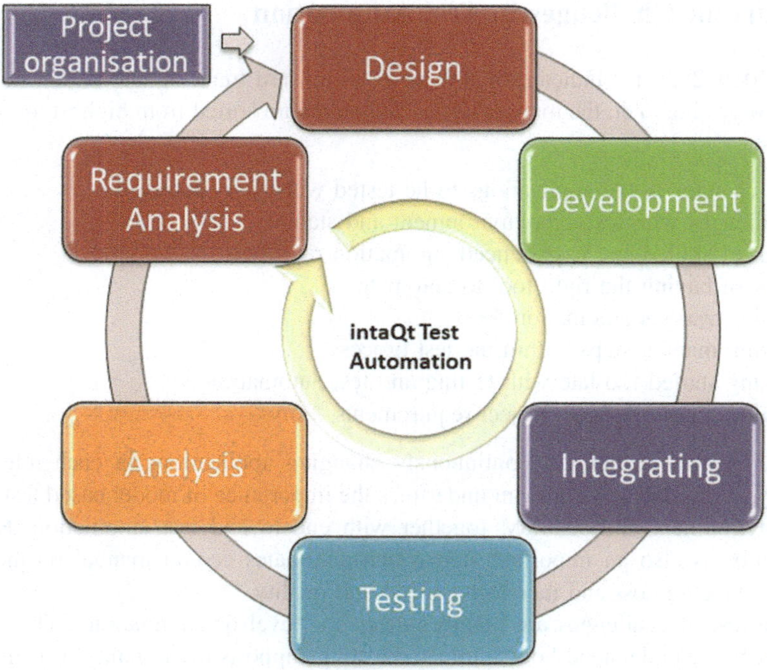

Fig. 1. Automation in various phases software development life cycle

Involving test automation in various phases of SDLC significantly optimizes testing efforts resulting in superior test outputs and significantly improved quality with improved times to market.

Test software intaQt®is an INtelligent Test Automation & Quality-assurance Tool. It is used to create, manage and execute end-to-end automated tests that reflect real-life scenarios, customer experiences and complex interactions with backend systems, different domains and technologies in a single test case [3].

It is a unique solution for an important need of performing full-stack automation where rule-based validation capability (such as Call Detail Record Verification/XML-Verification/Trace Verification) and reporting is possible within the same framework [3].

QiTASC's view of test automation in this process is as follows.

Requirement Analysis:

The data derived from the electronically readable requirements form are used in the phases of SDLC to specify and enrich the test cases.

Design:

The UI models required for automation are generated from the design model and from the automated user interaction.

Test cases are derived from design documents such as BPMN, IDEF, UML models, flow charts, etc.

Development:

Test data from unit tests are analyzed for change requests and corrections to be submitted to development units.

The use of a continuous integration system provides immediate feedback.

Integration:

Integration is accelerated through automated "Sanity Check Tests".

Satellite systems and other integration systems are integrated into the automation

Test:

Tests are designed to be executable in the following environments:

- in unit tests with mocks
- in the LAB in mixed operation with simulators and mocks
- in acceptance with live components and simulators
- in operation only with live systems

Analysis:

Analysis phase utilizes test cases and test automation tools to carry out "active probing" and to simulate problems.

To manage business risk alongside code risk, Continuous Testing provides delivering bug-free software at the speed the business requires. QiTASC believes in testing early and often where test driven development & continuous integration are the cornerstones.

Continuous Testing is the evolution of test automation in a DevOps and Continuous Deployment environment. The purpose of Continuous Testing is to authenticate and give feedback to all phases of the software delivery pipeline - from design to production. The main phases, such as functional testing, performance and security tests are fully automated.

intaQt is compatible with SDLC and intelligent test automation concepts, by bringing various automation steps together in one tool and orchestrating them as shown in Fig. 2.

This concept incorporates planning and modelling the test framework, configuring all elements and systems which takes part in the scenario, automated test execution, data collection and rule-based verification and reporting [8].

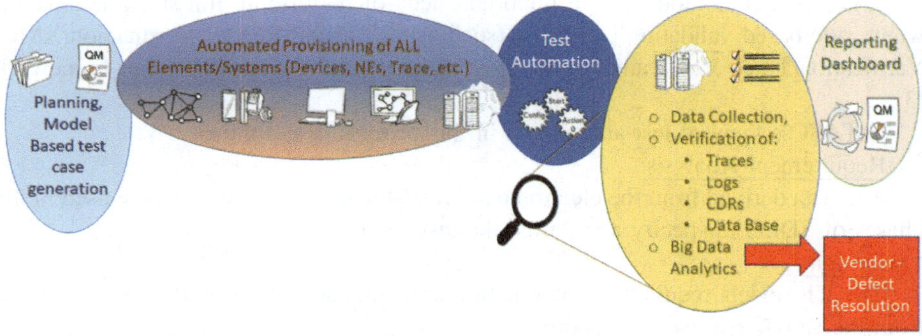

Fig. 2. Interpretation of test automation framework with iNTAQT

5.2 Combining QiTASC Products to Achieve High Levels of Automation

QiTASC provides a suite of tools that encourage incremental automation in test execution, reporting and bug tracking. While intaQt's flexible configurations and reusable modules encourage the incremental automation of test cases themselves, QiTASC's automated reporting service, conQlude, automates the collection and interpretation of the massive amount of data output generated by upwards of thousands of test executions per day. conQlude is also compatible with most project management systems (PMS), which gives users additional flexibility with post-hoc data analysis. In terms of device management, QiTASC's sQedule automates test scheduling which provides opportunities for eliminating inefficient time use by assigning test cases to Agents or Bots that execute test cases. Automated or "intelligent" scheduling is especially useful for tests that must be run outside of working hours or are very long. It is also a highly productive approach to running regression tests or tests that must be re-run multiple times. Finally, QiTASC's command line interface, intaQt Client, enables test projects to be executed via continuous integration services such as TeamCity and Jenkins. To validate evidence QiTASC provides also a tool called cheQ, cheQ compares traces in an intelligent way.

To further promote incremental automation, intaQt Studio provides an integrated Git client, which is a version control tool that tracks changes made to test cases and other artifacts within a project [6]. This lets users within a project maintain and access changed files, and even fall back to previous versions if needed. Additionally, intaQt's Built-ins facilitate the automation of external functionality, not normally included within the test scope, including protocols such as HTTP, SCP and SSH, languages such as SQL and XML, formats such as CSV and JSON, as well as other important utilities including date and time handling, data manipulation and multimedia recognition. For even greater levels of automation, intaQt Client allows tests projects to be run as builds via the continuous integration servers TeamCity and Jenkins, leaving the user with more time to analyze test output and correct errors. Finally, even test analysis activities can be automated via QiTASC's conQlude reporting service, which collects all test metadata from a project's text executions and sorts it into a centralized database for interpretation or further data manipulation [6].

One of the most complicated and time-consuming parts of a test project is all the work required to create and maintain test cases. This is because test planning maintenance has largely remained a manual activity within the scope of test automation [9].

To solve these productivity and accuracy issues, QiTASC works on a model-based testing (MBT) tool that is compatible with intaQt and all other QiTASC test automation suit. This tool will turn models of Flows into suites of test cases, which can then be executed via intaQt Studio or intaQt Client. MBT models cover not only the steps of a test case, but configurations, input parameters and decision points to ensure the optimal combination and number of test cases are created [9].

Using the same testing tools across all testing phases is an efficient, cost-effective approach. Therefore, using a testing solution that can be flexibly used in all phases—from development, to field acceptance and to live environments—and that should be given high priority when choosing test automation software [6].

Figure 3 shows some open sources integrated to intaQt® that enable automation of various interfaces [4]:

- Appium, Selendroid for App automation
- SOAP, REST, LDAP, ssh, sftp, etc. for Backend automation
- Selenium, sikuli for UI Test Automation

With this approach, there is no need to spend time on learning proprietary languages - our tools work out of the box with its own language.

Some possibilities that are a part of intaQt® test automation are as follows:

- Web-UI Automation
- Protocol & Network Simulation. CAP, Diameter, SIP, etc.
- APP iPhone Automation
- Mobile Device Automation
- PC-Client/MacOS/Windows Automation
- APP-Tablet Automation
- APP-Android Automation
- Trace Integration & Automation
- Attenuator Automation
- VoIP-Phone Automation
- Remote Access Automation
- Roaming Automation
- IoT Automation
- Robotics automation
- Backend, IF, Automation/Emulation, SOAP, REST, Ldap, SQL, ssh, sftp, etc.
- Automated Data Correlation & Verification
- Automated Reporting (HP-ALM, Jira, etc.)

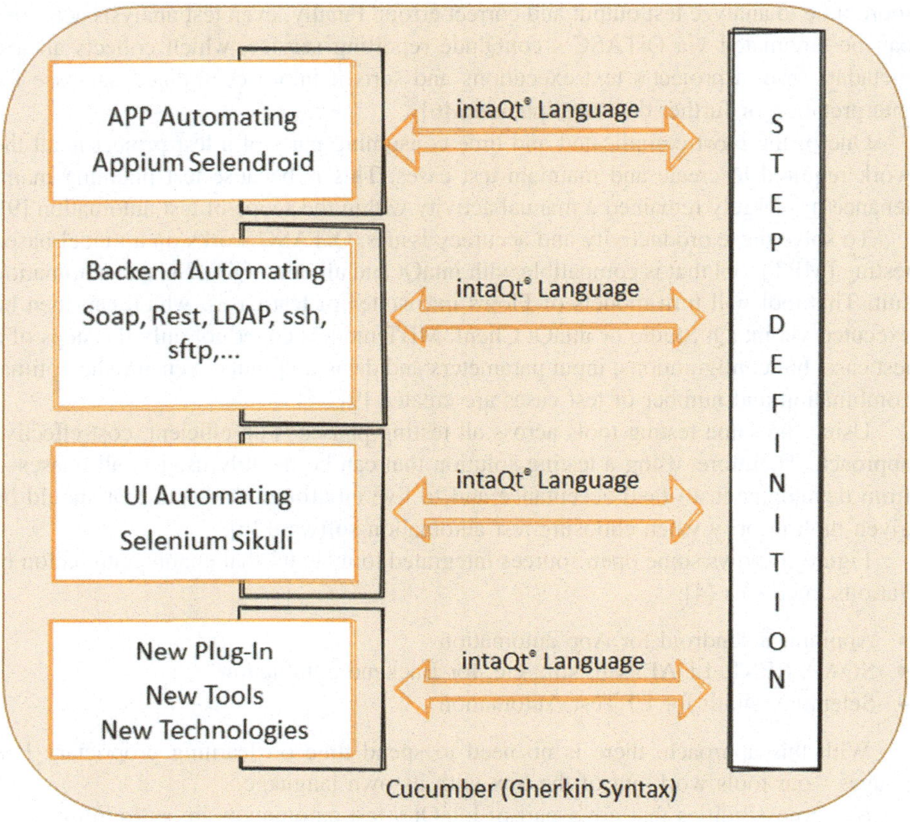

Fig. 3. Integration and iNTAQT® language

5.3 Complex and Multi-context Systems Can Easily Be Automated by intaQt

The emphasis on end-user satisfaction underlines the importance of end-to-end testing. Furthermore, together with swift technological, developments the systems to be tested and assured becomes more and more complicated. Hence, different technologies can be integrated in one use case.

End-to-end testing is a methodology to test the functionality and the performance of an application under product-like circumstances and data to replicate live settings. Applications are built on layers and entire networks of sub-systems, including UI and API layers, external databases, networks, and even third-party integrations. When one fails, so does the entire product, making the stability of each component vital to the success of an application. This means there is a clear need to test the entire application, from start to finish - at both the API and UI layers [6].

With intaQt it is possible to simulate what a real user scenario looks like from start to end, involving testing of a complete application environment in a situation that mimics real-world use, such as interacting with a database, using network communications, or interacting with other hardware, applications, or systems if appropriate. The completion of this testing is not only to validate the system under test, but to also ensure that its sub-systems work and behave as expected.

5.4 Further Automation Possibilities in the Agenda

The potential of test automation can be leveraged if it is complemented with predictive maintenance through AI-ML analysis to process the test data and see what patterns emerge. Algorithms for defect analysis can easily learn from past trends and go through the standard defect analysis process in the least amount of time and make sure that every bug that causes a defect is caught at its root.

Through data science, test data can be analyzed as they start to emerge (defect analytics- analyzing the log files on the code while it is being created), this data can be modeled to predict the patterns and steer to the root cause of the defects to give feedbacks to the developer and prohibit the possible errors before they arise supporting the implementation of Agile and DevOps. Utilizing AI-ML to understand the root cause of errors increase efficiency in testing and therefore minimize defect life cycle.

6 Conclusion

Leveling-up software test automation processes is key to a successful digital transformation. As systems grow increasingly complex and businesses become more reliant on them, ensuring reliability and recoverability are key business mandates. By extending test automation practices to not only to detect issues but also to remediate them, organizations reduce recovery time during production failures. Traditional approaches to software quality assurance typically fail to plan for remediation in production, increasing business risks.

intaQt‘s robust, reusable framework supports an incremental approach to automation by allowing different modules to be added and expanded upon as a project progresses. By integrating version control tools, continuous integration compatibility with intaQt Client, the conQlude reporting service, cheQ protocol verification capabilities and the capability to verify conformity of standards, finally with sQedule‘s intelligent resource management, QiTASC delivers a holistic test solution that helps stakeholders maintain high quality services while finding and correcting errors faster and reducing the time-to-market.

References

1. World Quality Report 2018–19 Tenth Edition. https://www.capgemini.com/resources/world-quality-report-2018-19-australia-new-zealand/
2. World Quality Report 2019-20 Eleventh Edition. https://www.capgemini.com/research/world-quality-report-2019/

3. Ugur-Tuncer, G., Davutoglu, C., Durakbasa, N.M.: Automated quality assurance applications in the rise of IoT. In: Proceedings of the International Symposium for Production Research 2019, pp. 361–368. Springer, August 2019
4. Davutoglu, C., Durakbasa, N.M., Bas, G., Guclu, E.: Improved quality and reliability in telecommunication network testing by means of a remote online test system. iJOE **10**(4), 66–69 (2014)
5. Bauer, A., Bas, G., Durakbasa, N.M., Kräuter, L., Ugur, G.: Tuncer measurement technology & quality & justicia in Industry 4.0. In: ISPR 2018: Proceedings of the International Symposium for Production Research 2018, pp. 438–450, August 2018
6. Courtney, L., Zehender, M.: The Future of Testing, Improving Levels of Automation and Test Coverage in Telecommunications. https://www.qitasc.com/whitepapers/
7. Galin, D.: Software Quality Assurance, From Theory to Implementation. Pearson Education India (2004)
8. Zehender, M., Davutoglu, C.: intaQt® (Intelligent Test Automation and Quality-assurance Tool) Handbook, Austria 2012–2020. https://docs.qitasc.com/
9. QiTASC website: Introducing disQover - Model Your Test Scope and Discover How We Automated Test Case Maintenance. https://www.qitasc.com/articles/disqover-model-based-testing

The Role of Product Perceived Quality in Building Customer Behavioral Loyalty Across Retail Channels

Florian Teleaba[1(✉)] and Sorin Popescu[2]

[1] T.U. Cluj-Napoca, Cluj-Napoca, Romania
florian.teleaba@gmail.com
[2] Design Engineering and Robotics Department, T.U. Cluj-Napoca, Cluj-Napoca, Romania
sorin.popescu@muri.utcluj.ro

Abstract. Retail today is increasingly omnichannel. Retailers must drive customer behavioral loyalty (repeat patronage) across channels, so they must do it optimally – no company can be the best at everything. Many other factors, beyond product quality, are required to maximize customer behavioral loyalty across channels. Academia and business seem inconclusive though in explaining which are those winning factors. When the end user of the product is (often) not the buyer, such as in the toys industry, matters complicate further. This study examines these gaps based on an online survey with 785 respondents in an Eastern European country. In offline retail perceived quality beats price, assortment and customer experience in maximizing behavioral loyalty, yet only once price is perceived fair. Being the best in quality and above average in price is the optimal differentiation strategy. Despite everyone talking about customer experience today, it appears to be the least effective factor; in fact, differentiating only in experience appears more damaging to customer loyalty than not differentiating at all. In online retail, being the best in assortment and above average in experience appears to be the optimal strategy; price not only matters the least but differentiating only in price is more damaging than not differentiating at all. Product quality does not appear to play a decisive role for differentiating and building customer behavioral loyalty in online, yet it is a must.

Keywords: Perceived quality · Product quality · Customer loyalty · Customer experience · Omnichannel · Toys retail

1 Introduction

Customer loyalty is without doubt a priority for any firm. In retail it becomes even more important today. As recent research shows [1], the world of retailing has changed dramatically over the last decade and is moving from a multi-channel to an omnichannel space. Cross-channel integration in this new space can increase a retailer's sales and brings benefits in terms of online and offline customer loyalty [2], even if

F. Teleaba—Manager in Kearney (the global consulting firm).

N. M. Durakbasa and M. G. Gençyılmaz (Eds.): ISPR 2020, LNME, pp. 625–640, 2021.
https://doi.org/10.1007/978-3-030-62784-3_54

each channel moderates in a different way these effects [3]. Retailers understand this opportunity and already more and more of them today offer for instance customer loyalty programs across channels [4].

There are many elements that are required to maximize customer loyalty in retail, beyond offering product of the right and required quality. Based on the paramount research work on customer experience creation in retail by [5], recent extensive work on retail loyalty management and omnichannel customer experience [6], and the consulting work in retail of one of the authors, three additional factors are required: customer experience, competitive pricing, and optimal assortment. Understanding which of these four overarching factors drives most the customer behavioral loyalty (repeat patronage), and how they interact in this new omnichannel space, is of paramount importance. The subtlety is that, while in theory a retailer may wish to maximize (be the best in) all factors, in practice it is rarely possible (i.e. unlikely to offer best customer experience, best assortment and quality of products, and the best/cheapest price). Additionally, there is no reason to think that what is important in brick & mortar retail is equally important in e-commerce, therefore different channel strategies are required. For multi-channel retailers the difficulty to balance these factors grows which each channel, therefore.

However, not many research papers have analyzed these factors together, not across channels, and not recently in the omnichannel space. Some research papers either (a) investigate loyalty levels across channels, but with less focus on specific loyalty factors others than channel-specific ones, and therefore not if those factors differ between channels, or (b) investigate if one or few factors have an influence on customer loyalty, but most often focus on one channel only (stores or online).

We did not identify though, to the best of our knowledge and research efforts, research papers or business reports that address (1) more holistically, across channels, the four principal factors of customer behavioral loyalty: customer experience, price, assortment and quality, (2) which combinations of these factors increase behavioral loyalty most, (3) if these factors or 'winning combinations' are the same between channels, and (4) if these factors change for different customer channel-buying behaviors, for instance between single-channel and omnichannel buyers. (We did our research across time periods, types of journals and types of retail sector).

This study addresses these four areas of scientific (and business) gap, areas of high practical and theoretical relevance. The overarching questions, unexplored by scientific research so far, are: at which factors should a retailer aim to excel in order to maximize customer behavioral loyalty, and are these 'winning' factors or combinations of factors the same across channels? What is the role of product quality and perceived quality in building customer behavioral loyalty?

This paper also addresses an amazing sector insufficiently explored so far: toys retail. In retailing of children products such as toys, unlike in most other sectors, the buyer is (often) not the end user of the product, and this affects significantly the consumer decision making process. Parent-child joint decision making is extremely important in children's products, and retailers that address the needs of both parent and child reach higher customer purchase likelihood [7]. But customer loyalty per-se in toys retail has benefited from little research so far. One of the very few examples exploring factors that affect children's store loyalty is provided by Williams & Burns,

2001. It looks at impersonal and interpersonal communication, experience, and perceptions of affective and functional qualities of store image, but it omits to consider the effects of other factors that are arguably highly important for customer loyalty, such as assortment, price, or products quality. It also looks at the loyalty of the child, not of the buyer (the parent). Additionally, these studies do not focus on toys retail, but on general retail. There is therefore a major opportunity, unexplored yet (to the best of our knowledge and research efforts), to investigate customer loyalty in toys retail.

This paper provides first a background on what customer loyalty and behavioral loyalty are, and few remarks about the role of each factor studied (price, quality, assortment and customer experience) on customer loyalty. Secondly, the paper explains an original framework and methodology to determine the optimal differentiation strategy in the market in order to maximize customer behavioral loyalty. This can be easily applied in practice and overcomes the major issues with customer research in real life, namely that asking customers directly what their preferences or choice factors are does not produce reliable or relevant results (there is a big difference between 'stated' and 'true' importance of factors in their purchasing decisions). Thirdly, we describe in detail how the research was done, the method (detailed online survey) and our data (of 785 complete responses). Fourth, we show the results of our analyses, key findings and implications for scientific research and retailers, as well as suggestions for future research.

2 Theoretical Background on Customer Loyalty, Role of Product Quality and Other Factors

2.1 Customer Loyalty in Omnichannel Retail

When we refer to customer loyalty, we must first understand what loyalty really is, and differentiate between different types of loyalty. Loyalty needs to be separated by attitude and repeat purchasing, concept existing since the 60s [8]. Dick & Basu, 1994, separate relative attitude toward an entity (brand, company) from repeat patronage behavior, arguing that relative attitude translates into repeat patronage. Similarly, Oliver, 1999 defines four loyalty phases: cognitive loyalty (belief or conviction that a brand should be preferred), affective loyalty (favorable attitude towards a brand), conative loyalty (behavioral intention influenced by repeated episodes of positive affect toward the brand), and action loyalty (which translates buying intention into actual buying decision/behavior). In this study we focus on behavioral loyalty – that is, repeat patronage.

In today's increasing omnichannel retail, customer experience plays a major role in maximizing customer loyalty [5]. As indicated by the authors of this paramount work, and as recent extensive work on retail loyalty management and omnichannel customer experience [6] shows, as well as based on the experience of one of the authors of this paper in retail consulting, next to customer experience also optimal pricing, assortment and product quality are highly important (and all should be balanced). But as described in the introduction, the scientific (and business) literature in retail seems to be entangled, and there is no consensus on which factors are most important, to either attitudinal or behavioral loyalty, and even less across channels. Research until now has either (a) investigated loyalty levels across channels, but with less focus on specific loyalty factors others than channel-specific ones, and therefore not if those factors differ between channels, or (b) investigated if one or few factors have an influence, but most often focus on one channel (stores or online).

On loyalty across channels, several studies investigate important aspects, such as the impact of multi-channel integration (see meta-analysis of Frasquet & Miquel, 2017), if multiple channel retail strategies increase customer loyalty [9], the effects of satisfaction with channel performance on behavioral intentions [10], influential factors of channel loyalty such as channel familiarity, perceived risk, hassle cost, or channel consistency [11], the role of multi-channel service quality [12], even the roles of brand and culture [13] or trust [14] on customer loyalty. Such studies however are focused more, as we can see, on the influence of channels' integration or performance on customer loyalty. Even very recent research on customer journey management across retail channels is focused on the role of technology and social or cultural factors, and how these drive cognitive, affective and behavioral consumer responses [15], or it is focused on the determinants of cross-channel journeys [16], or at most it is focused on general loyalty antecedents across different customer journey segments [17]. There is less systematic focus on the specific principal factors of loyalty: price, quality, assortment, and customer experience (and their roles across the different channels).

On factors of loyalty specifically, very little research touched upon customer experience, price, assortment and product quality all together. For example, one paper looked at how perceived value influences purchase attitude and behavior through four dimensions: emotional, social, quality/performance and value for money [18]. Yet, it left aside assortment which clearly represents an important factor for customer loyalty in retail. Another research example addressed several loyalty factors: price, quality, selection and staff (and found quality as most important attribute) [19]. Still, customer experience was only partially considered, "staff" being only one component of customer experience. The most recent research that addresses how different factors affect loyalty across offline and online channels [20] focuses on perceived value, quality and experience but does not address the role of assortment. Perhaps the most recent comprehensive research of multiple factors affecting customer loyalty in retail is the meta-analysis of Blut et al. 2018: all relevant factors discussed here (price, assortment, quality and experience) and many others (24 factors in total) have been analyzed on how they influence customer satisfaction, repeat patronage and attitudinal loyalty. Even in this research however there is no attention given to multiple channels, and if the factors of loyalty investigated play the same or different roles between channels. In fact the paper even states that the recent emphasis of research on omni- and multi-channel retailing has not significantly considered the phenomenon of cross-channel patronage and its drivers; to the best of our knowledge and research efforts that also did not happen since 2018 until now.

2.2 Product Quality and Loyalty

Quality is a key component of competitive advantage [21]. It is the place to start building customer satisfaction and loyalty [22], and its effect on loyalty becomes stronger over time [23]. How product quality is perceived matters also. Perceived product quality is one of the eight attributes of overall product quality by Garvin, 1987. Yet it is a delicate matter: it can increase customer satisfaction and product sales [24], even fully transmit the impact of brand innovativeness on customer loyalty [25], but there is proof that perceived quality can be also a consequence of satisfaction [26], therefore comes also as an outcome not only precedent. Perceived product quality depends on both intrinsic and extrinsic quality attributes of the product [27]. Product quality perception relates to brand as well. Consumer evaluate more favorably a brand when it is sold in a reputable retailer [28].

Service quality is also a determinant of customer loyalty, sometimes even more important than product quality [29] (and can even influence merchandise quality perception [30]). Service quality can be however highly correlated with customer experience, and previous research [31] showed that service quality significantly influences customer experience quality (which, further, influences customer loyalty).

We expect overall that product quality plays a decisive role in customer behavioral loyalty, especially once price is considered 'fair'.

3 Research Methodology

3.1 Research Framework and Questions

Our main research questions are:

R1. Which combination of factors of differentiation in the market brings the highest customer behavioral loyalty? (among the four overarching factors of loyalty in retail: product quality, price, assortment and customer experience).
R2. Is this similar between offline and online channels (stores vs. e-commerce)?

We propose a simple novel framework to identify the 'true' factors of customer behavioral loyalty and 'winning' combinations, as per Fig. 1 below.

For each factor we look at how customers perceive several brands in the market (each, individually, on a scale from 1 – low, to 10 – high), and isolating the one considered 'preferred' brand. Then, we compare the scores obtained by the preferred brand with the average score of the other brands in the market, and with the highest score obtained by any other brand in the market.

Quantifying perceptions in each factor is done on a scale from 1 to 10, where 1 is the lowest score (ex. Highest price, or lowest quality, assortment or customer experience) and 10 is the best score. For simplicity for respondents, note that the scale for price was inversed, i.e. they rated as 10 the most expensive brand, not the cheapest (i.e. "best" in price); then, during the comparative analysis of factors' scores, the scale in this case was inversed, i.e. score of 10 (most expensive) would become score of 1 (worst in terms of price).

To quantify customer loyalty, we used a simple yet effective question: asking them to think of purchases made in the last year and choose between (a) I buy from different retailers, (b) I have a small number of retailers I typically buy from, (c) I usually do all the purchases from a single retailer – three mutually exclusive collectively exhaustive situations, easy to choose from. Quantifying customer loyalty can be made through the customer's perception of loyalty (Odin et al. 2001), therefore we remain confident that this approach is satisficing enough.

The goal was to identify which combination of higher scores of preferred brand versus other brands, along the four factors, maximizes behavioral loyalty. While there are many more combinations possible, we looked at few important ones for answering our main research questions:

- Preferred brand is above market average (score above the average score of the other brands) in each factor, in each combination of two factors, in each combination of three factors, and in all four factors (15 combinations)
- Preferred brand is best in market (highest score) in each factor, in each combination of two factors, in each combination of three factors, and in all four factors (15 combinations in total)
- Preferred brand is best in market in each factor combined with above market average in each one of the other three factors (12 combinations in total)

Highest customer behavioral loyalty is found in the combination where there is the highest share of respondents that choose option (c) to the above question, i.e. they usually do all their purchases from only one retailer. We call these "loyals". For testing the statistical significance of the results, we compare the share obtained in each combination with the share of loyals in the entire sample as a baseline, i.e. the 'natural share' of loyal customers in that market.

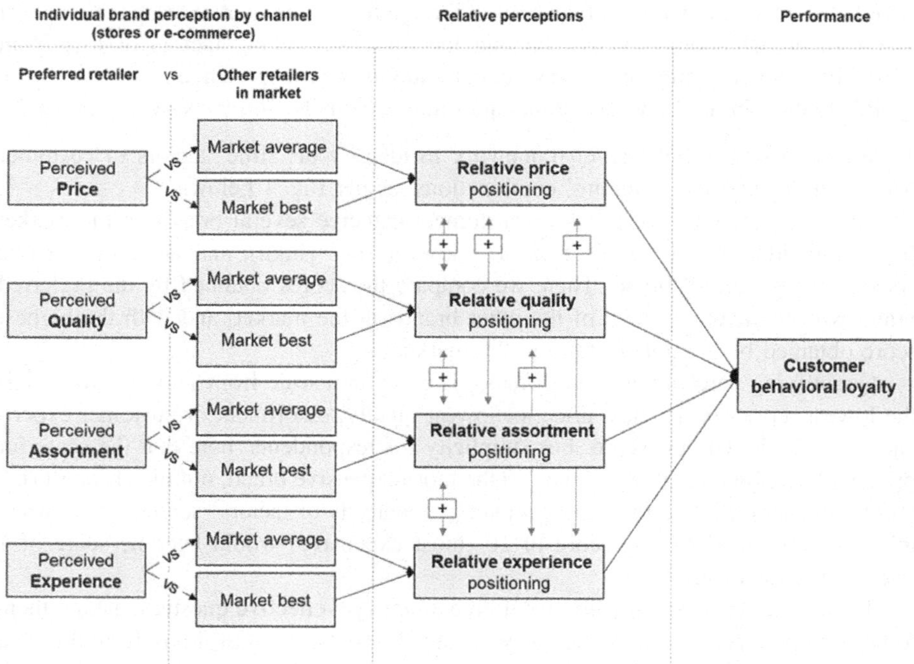

Fig. 1. Framework for evaluating customer behavioral loyalty factors in the competitive environment

3.2 Research Method and Data

This exercise was part of a consulting project led in the last year by one of the authors. An online customer survey was deployed through a specific professional online surveying tool, based on a random sampling approach, targeting adult (18+) population in the respective Eastern European country. Data collection was monitored frequently, based on age and income quotas, to ensure representativity of the sample for the population. The survey covered several questions relevant for the mentioned project, and only the ones relevant for this paper are described in Table 1 below.

After data cleansing and validation, 785 complete responses were obtained, split in two similar samples:

- a sample of 684 responses for offline channel behavior (evaluating offline retailer brands)
- a sample of 252 responses for online channel behavior (evaluating online retailer brands)

The collected samples can be considered statistically representative for the adult population in the respective country with a 95% confidence internal and ~4% error for the offline behavior, ~7% error for the online behavior. Notice that some respondents provided valid and complete responses to questions referring to both offline and online retailer brands, having purchased in the last year from both channels, therefore appear in both samples. The validated complete responses do not include those that selected "None" regarding which is their preferred retailer brand, therefore all responses indicate a preferred brand for either offline or online retail (where applicable) and are thus suitable for including in the analysis.

Table 1. Questions in the survey

Type of question	Question	Remarks/options to choose from
Filtering (continue survey or stop)	"Did you buy toys and games for children in the last 3 months?"	Yes or No question. Purpose of question is to filter and include in research only those that have bought toys and games (for children specifically) in the last 3 months (to more reliable opinions: recent, knowledgeable)
Main research question	(behavioral loyalty measure) "When buying toys and games, which statement best characterizes you?"	Options to choose from: "I buy from different retailers", "I have a small number of retailers I typically buy from", "I usually do all the purchases from a single retailer"

(*continued*)

Table 1. (*continued*)

Type of question	Question	Remarks/options to choose from
Main research question	"From which of the following offline retailer brands did you make purchases in the last year?" *Question repeated for online retailer brands*	Purpose of these two questions (one for offline brands, one for online brands) is to determine which retailer brands to ask for evaluation in the next questions. Several brands were displayed (all important ones in the market), and respondents had to mention "Yes" or "No" Respondents also had as possible option "None"
Main research question	"And which is your preferred offline retailer brand?" *Question repeated for online*	This question was addressed for each brand mentioned as purchased from in the last year (previous questions). Respondents also as possible option "None"
Main research question	"How would you evaluate the prices of retailer brand X?" *Question repeated for assortment range, quality of products sold, overall customer experience offered by each retailer*	These four questions (for each of the four factors) were addressed for each brand mentioned as purchased from in the last year (previous questions) Evaluation on a scale from 1 to 10, where 1 is very low and 10 is very high (for making it simple for respondents, prices were asked to be evaluated on same scale, i.e. 1 was cheapest, 10 most expensive, and scores inversed during analysis)

4 Results

From the 684 respondents related to offline retailer brands, 148, or 21.64%, chose they usually do all purchases from a single retailer; most of them, 353 (51.61%), have a small number of retailers to choose from, and 183 (26.75%) typically buy from different retailers. Therefore, our baseline for comparison onward is 21.64%, i.e. the share of loyal customers, and we aim to see in which combination of factors (price, assortment, quality, experience, of preferred brand relative to others) this percentage is maximized.

From the 252 respondents related to online retailers and loyalty behavior, 29 only, 11.51%, chose they usually do all purchases from a single retailer; most respondents, 140 (55.55%), typically buy from different retailers, and 83 (32.94%) have a small number of retailers to choose from. We already observe that, compared to offline channel, in online channel the customer behavioral loyalty is much lower, and most customers tend to buy from different retailers versus only from a small number of retailers in offline.

Table 2 below summarizes our results in terms of share of customers loyal to their preferred brand for different combinations of differentiating across factors in the market. Next to each percentage, in parentheses, we report its t-stat value (calculated by comparing it to the respective baseline). For simplicity we consider any t-stat higher than 1.8 (or lower than −1.8) as statistically significant at the 5% level. (Note that a t critical value of 1.796, for a 5% significant level, is the t critical value for a sample size of only 12 observations, which is significantly less than each of our individual sub-samples).

4.1 Factors of Loyalty in Offline Channel

Across offline results, we see that when price of preferred brand is not above average or best in market, the share of loyals is not high, indicating that price is the most important factor of preserving repeat purchase (behavioral loyalty) among customers. This is in line with previous research which indicates the importance of price, and price fairness, for customer loyalty [32, 33]. Highest percentage of loyals, 39.71%, is achieved when the preferred retailer is best in quality and above average in price (i.e. below average in actual price levels). This percentage is even higher than when being best in price and only above average in quality (38.2%). These results are statistically significant at the 5% level based on the t-test. This brings additional important insights regarding the importance of price for customer loyalty, and its relative importance compared to other factors. Previous studies, even very recent studies like [33] or [34], despite indicating price as a key determinant for loyalty, do not consider any trade-offs with, or its importance relative to, other factors such as product quality. This is highly important because price and quality go hand in hand, both in terms of economics (i.e. higher quality typically means higher price) and perception (price is often a cue for quality perception [35, 36]). This result improves understanding of the relationship between price and quality and their importance for customer loyalty, indicating the quality is more important, as long as price is considered 'fair' – in other words, price is a must, and quality is the way to differentiate while the 'musts' are realized. Interestingly, share of 'loyals' when the retailer is best in quality but there is no differentiation in price is much smaller, 23.5% (even if not statistically significantly different versus baseline). However, being the best in price without any differentiation in quality shows a share of loyals of 32.1% (statistically significantly different versus baseline this time). This is not the highest level, but still higher than the baseline, indicating again the importance of price.

Table 2. Results (shares of loyal customers in each channel)

Preferred retailer relative to others in the market		Loyals (% customers buying only from their preferred retailer in last 12 months) (t-stat)	
Best in:	Above average in:	Offline (stores)	Online (e-commerce)
Baseline (no differentiation)		**21.6%**	**11.5%**
	Price	24.0% (1.107)	10.3% (−0.462)
	Assortment	18.2% (−1.909)	13.2% (0.672)
	Quality	19.2% (1.29)	13.2% (0.631)
	Experience	18.1% (−1.973)	12.8% (0.476)
Price		32.1% (2.938)	6.9% (−1.247)
Assortment		19.5% (0.306)	17.1% (1.529)
Quality		23.5% (0.53)	13.7% (0.542)
Experience		20.0% (−0.548)	12.8% (0.363)
Price	Assortment	35.0% (3.245)	8.3% (−0.689)
Price	Quality	38.2% (3.795)	6.7% (−1.018)
Price	Experience	30.0% (2.13)	8.3% (−0.689)
Quality	Price	39.7% (3.618)	6.7% (−1.018)
Quality	Assortment	24.6% (0.84)	14.6% (0.706)
Quality	Experience	23.5% (0.515)	13.7% (0.496)
Assortment	Price	30.7% (2.06)	14.6% (0.706)
Assortment	Quality	20.6% (−0.297)	17.0% (1.249)
Assortment	Experience	19.2% (−0.722)	20.3% (2.126)
Experience	Price	29.4% (1.907)	9.3% (−0.518)
Experience	Assortment	20.2% (−0.44)	12.3% (0.202)
Experience	Quality	20.4% (−0.392)	13.6% (0.494)

Overall, we see that assortment and experience alone are not highly important for customer's behavioral loyalty, with no major uplift in the share of loyals. In fact, when being above average in assortment only, or in experience only, the share of loyals appears to be smaller than baseline, ~ 18% (both statistically significant results).

In case of assortment, higher variety of choice can sometimes have a negative impact on customer perception and behavior [37, 38]; sometimes customers may avoid stores with very large assortment as it becomes hard to make your purchases (cognitive effort) [39]. Therefore, despite the intuitive appeal of offering wider, bigger assortments, even with improved perception of products quality [40] or operational benefits [41], this study rather shows positive results might be obtained only for the short term sales, however for customers' repeat purchase behavior the size of the assortment is not really an effective differentiator.

In case of experience, although much discussed in business today (and in literature) as being of top importance for customer loyalty [5, 6], there can be different explanations why differentiating only in customer experience can decrease loyalty. Most likely people which value 'experience' a lot, or for which the preferred brand is one that offers an above average or exceptional experience, may be prone to more often switching brands in search for the best experience. This is because experience itself is not that easy to quantify (like price, or quality, or even assortment) and because perception of experience, like taste or general preference in consumer behavior, is not consistent over time – one time you appreciate a certain type of experience, another time you may appreciate or look for a different type of experience. As reference dependence changes over time (i.e. what ideal customer experience is, as preference), that can bring the need/desire to constantly try something new, increasing churn rates [42].

4.2 Factors of Loyalty in Online Channel

Being the best in assortment without any other differentiation in other factors is enough to uplift behavioral loyalty to 17.11%. When assortment is perceived as best in the market, and experience above average, highest share of behavioral loyalty is achieved – 20.34% (double the baseline of 11.51%), result being statistically significant. We draw attention to the high importance of this insight. There is a vast literature addressing online retail and factors of loyalty, but it does not address systematically how the trade-offs between different factors affect customer behavioral loyalty. This study does so. Also, much of the recent research of customer loyalty in e-commerce has been focused on the role of online customer experience; there is extensive research on the components of online customer experience in the last decade (from [43] to recent [44], as only two selected examples out of many more). Overall, it seems as if customer experience is the ultimate differentiator in online retail, and literature and business often favor this idea [6]. However, it appears that assortment is in fact the most effective way to differentiate, with customer experience being only the second most important factor. Being above average in quality next to being best in assortment and above average in experience, the behavioral loyalty is also among highest, 18.75%, indicating quality as the next important factor after assortment and experience.

4.3 Summary of Results and Managerial Implications Concerning Product Quality Role in Building Behavioral Loyalty

As easily seen in Table 2, there is a staggering difference between the combination of factors that maximizes customer behavioral loyalty in offline channel and the one that does so in online channel. In fact, the combination that maximizes behavioral loyalty in offline (best in quality, above average in price – 39.7%) appears to lead to very low loyalty levels in online (6.7%), even lower than the baseline; and the combination that maximizes behavioral loyalty in online (best in assortment, above average in experience – 20.3%) appears to lead to very low loyalty levels in offline (19.2%), even lower than the baseline there.

Table 3 below summarizes the results in simple words. It becomes clear that customer behavioral loyalty factors differ between channels even for the same customer, and that product quality (and perceived quality) plays different roles for customer behavioral loyalty across channels.

In offline retail, there are few important takeaways and implications for retailers. Price matters most for customer behavioral loyalty, but up to a point. Once the price is considered fair by customers (meaning better than market average), quality is what

Table 3. Summary of research results

Research question	Findings in		Additional remarks/observations
	Offline behavior (stores)	Online behavior (e-commerce)	
Factors in order of importance for building customer behavioral loyalty	Price, then Quality, then Assortment, then Customer Experience	(likely*) Assortment, then (likely) Customer Experience, then (likely) Quality, then (likely) Price	Factors are listed in order of importance For exact magnitude of effects see Table 2
Move effective combination of being best in the market in one factor and above average in a second one	Best in Quality, and Above average in Price	Best in Assortment, and Above average in Customer Experience	For exact magnitude of effects see Table 2

* likely' denotes that results were not statistically significant (at the 5% level) however there are relevant qualitative insights about the effects (and their direction)

matters most. This is in line with recent research on the role of price fairness [33] and brings new insights on the trade-offs between price and quality that must be made to maximize customer loyalty. Assortment matters less than price and quality. Customer experience appears to be the least important. Despite significant attention and importance given to it today, differentiating only in customer experience without attention to price, quality or assortment may influence negatively the customer behavioral loyalty, even if some positive influence on attitudinal loyalty may still exist. Note however that most of our respondents evaluated their preferred brand as having better than market customer experience, and that proved still insufficient for repeat purchase behavior – therefore, customer experience is rather a must, a 'hygiene' factor, not an effective differentiation opportunity.

In online (e-commerce), assortment appears to be the most important factor for customer behavioral loyalty, followed by customer experience. Price appears to be the least important, and even damaging if used as the only differentiating factor. Note however that product quality is a must, a 'hygiene' factor, even if not an effective differentiation opportunity like assortment or experience.

For omnichannel retailers, the most important lesson is that different differentiation strategies work in different channels, because customer behavior is different between channels. The mantra of omnichannel retail has long been to offer the 'same experience' (same prices, same assortment, same customer experience) across channels. However, this study puts under question and scrutiny if that really is the most effective way to maximize behavioral loyalty. The results demonstrate that approaching each channel with a different value proposition should be preferred.

5 Conclusions and Discussion

This study contributes in two major ways to the scientific research and theoretical understanding of customer loyalty and the role of product quality (and perceived quality) in customer loyalty across channels.

First, even the most recent research shows it takes a eta-analysis to analyze comprehensively factors of customer behavioral loyalty [34], and even that can be insufficient to explore systematically the factors across channels. It takes considerable effort and complexity to study behavioral loyalty factors across channels [20] often requiring focus on only a handful of factors. Current research is insufficient and inconclusive on how exactly to study factors of loyalty across channels in a specific setup (and not based for example on meta-analyses). This study presents an original and novel framework, simple to deploy and tailor, for analyzing the effects of the four principal factors of customer loyalty (price, quality, assortment and customer experience) in both offline and online channels.

Second, this study brings new insights but also theoretical questions on some of the critical factors of customer behavioral loyalty previously studied. For instance, on quality, it confirms what intuition and research shows, that it should be the place to start building customer loyalty [22], but goes beyond saying "quality is important" into showing how quality has different roles between channels: in offline retail it is a differentiator for maximizing customer behavioral loyalty, in online retail it is rather a must with optimal differentiators being other factors. Customer experience has been recently argued extensively [5], [6] to be a differentiator for customer loyalty, especially in omnichannel retail context [1]. This study finds this to be only partially true: in online retail it is important, yet only second to assortment, while in offline retail it is merely a must-have but insufficient as differentiator to maximize repeat purchase. In fact, in offline it may be the least important factor and even damaging to customer behavioral loyalty if used as the main lever. On price, this study confirms recent proof that price fairness is key to behavioral loyalty and consumer trust [33], and further explains why and how: up to a certain point it is the primary factor of loyalty, but once considered fair, its importance decreases leaving room for product quality as key differentiator in offline retail or assortment and customer experience in online retail. On assortment, there is either vast research suggesting wider assortment is beneficial for loyalty, vast research on choice overload indicating too much assortment can hurt, and research that tries to find the middle ground between these two apparently contradicting directions of thought [45]. This study brings a much clearer explanation as to why both wider and small assortment are beneficial, showing its influence on loyalty depends on channel and how it is perceived relative to price, quality and customer experience. Lastly: much of the research on behavioral loyalty factors across channels focuses on what are the effects of different factors in each channel and if they differ across channels [20], but not also on what are the required trade-offs to be made between those factors, to obtain an optimal combined result (across all channels). This study provides insights on how to manage such trade-offs.

In terms of limitations of this study, there is potential to increase the sample size and thus the statistical significance of results on the online behavior side especially. Bigger sample size can also give the possibility to analyze in depth the different subgroups/strata.

In terms of future research recommendations, we believe there are three main directions (horizontal, vertical and in-depth) to building a deeper understanding on how customers' perceptions of price, assortment, experience or quality can influence their behavioral loyalty across channels. Horizontally, replication of this study across few industries and product categories fundamentally different than toys retail could offer interesting insights; also, studying these effects between economical versus luxury product categories, and under different marketplace scenarios, might reveal important learnings, as recent research reveals (Lim et al. 2020). Vertically, looking at the attitudinal loyalty and satisfaction, their factors, and their potential mediating role on the relationship between product quality and behavioral loyalty, is an opportunity. In depth, the four factors studied in this paper can be broken down further into their components; product quality can be for example further split in Garvin's eight attributes of overall product quality. Lastly, there is a considerable amount of recent research showing that behavioral loyalty factors are different between consumers of various socioeconomic statuses [46], therefore future research should consider this aspect as well.

References

1. Verhoef, P.C., Kannan, P., Inman, J.J.: From multi-channel retailing to omni-channel retailing: introduction to the special issue on multi-channel retailing. J. Retail. **91**(2), 174–181 (2015)
2. Frasquet, M., Miquel, M.J.: Do channel integration efforts pay-off in terms of online and offline customer loyalty? Int. J. Retail Distrib. Manag. **45**(7), 859–873 (2017)
3. Cao, L., Li, L.: The impact of cross-channel integration on retailers' sales growth. J. Retail. **91**(2), 198–216 (2015)
4. TotalRetail, TotalRetail's Top 100 omnichannel retailers, 2nd annual report (2018)
5. Verhoef, P.C., Lemon, K.N., Parasuraman, P.A., Roggeveen, A.L., Tsiros, M., Schlesinger, L.A.: Customer experience creation: determinants, dynamics and management strategies. J. Retail. **85**(1), 31–41 (2009)
6. Ziliani, C., Ieva, M.: Loyalty Management: From Loyalty Programs to Omnichannel Customer Experiences. Routledge, Abingdon (2020)
7. Darian, J.C.: Parent-child decision making in children's clothing stores. Int. J. Retail Distrib. Manag. **26**(11), 421–428 (1998)
8. Dau, G.S.: A two dimensional concept of brand loyalty. J. Advert. Res. **9**(3), 29–36 (1969)
9. Wallace, D.W., Giese, J.L.: Customer retailer loyalty in the context of multiple channel strategies. J. Retail. **80**(4), 249–263 (2004)
10. van Birgelen, M., de Jong, A., de Ruyter, K.: Multi-channel service retailing: the effects of channel performance. J. Retail. **82**(4), 367–377 (2006)
11. Xu, X., Jackson, J.E.: Investigating the influential factors of return channel loyalty in omni-channel retailing. Int. J. Prod. Econ. **216**, 118–132 (2019)

12. Lin, H.H.: The effect of multi-channel service quality on mobile customer loyalty in an online-and-mobile retail context. Serv. Ind. J. **32**(11), 1865–1882 (2012)
13. Frasquet, M., Descals, A.M., Ruiz-Molina, M.E.: Understanding loyalty in multichannel retailing: the role of brand trust and brand attachment. Int. J. Retail Distrib. Manag. **45**(6), 608–625 (2017)
14. Chen, Q., Griffith, D.A., Wan, F.: The behavioral implications of consumer trust across brick-and-mortar and online retail channels. J. Mark. Channels **11**(4), 61–87 (2004)
15. Grewal, D., Roggeveen, A.L.: Understanding retail experiences and customer journey management. J. Retail. **96**(1), 3–8 (2020)
16. Hu, T.-I., Tracogna, A.: Multichannel customer journeys and their determinants: evidence from motor insurance. J. Retail. Consum. Serv. **54** (2020)
17. Herhausen, D., Kleinlercher, K., Verhoef, P., Emrich, O., Rudolph, T.: Loyalty formation for different customer journey segments. J. Retail. **95**(3), 9–29 (2019)
18. Sweeney, J., Soutar, G.: Consumer perceived value: the development of a multiple item scale. J. Retail. **77**(2), 203–220 (2001)
19. Birtwistle, G., Clarke, I., Freathy, P.: Customer decision making in fashion retailing: a segmentation analysis. Int. J. Retail Distrib. Manag. **26**(4), 147–154 (1998)
20. Hult, T., Sharma, P.M.I.F., Zhang, Y.: Antecedents and consequences of customer satisfaction: do they differ across online and offline purchases? J. Retail. **95**(1), 10–23 (2019)
21. Flynn, B.B., Schroeder, R.G., Sakakibara, S.: The impact of quality management practices on performance and competitive advantage. Decis. Sci. **26**(5), 659–692 (1995)
22. Torres-Moraga, E., Vasques-Parraga, A.Z., Zamora, J.: Customer satisfaction and loyalty: Start with the product, culminate with the brand. J. Consum. Mark. **25**(5), 302–313 (2008)
23. Pan, Y., Sheng, S., Xie, F.: Antecedants of customer loyalty: an empirical synthesis and reexamination. J. Retail. Consum. Serv. **19**(1), 150–158 (2012)
24. Lynch, J.J.: Customer Loyalty and Success. Macmillan Press Ltd., New York (1995)
25. Pappu, R., Quester, P.: How does brand innovativeness affect brand loyalty? Eur. J. Mark. **50**(1/2), 2–28 (2016)
26. Beerli, A., Martin, J.D., Quintana, A.: A model of customer loyalty in the retail banking market. Eur. J. Mark. **38**(1/2), 253–275 (2004)
27. Fandos, C., Flavian, C.: Intrinsic and extrinsic quality attributes, loyalty and buying intention: an analysis for a PDO product. Brit. Food J. **108**(8), 646–662 (2006)
28. Zhu, Y., Chen, H.: A tale of two brands: the joint effect of manufacturer and retailer brands on consumers' product evaluation. J. Brand Manag. **24**(3), 284–306 (2017)
29. Lee, W.I., Chang, C.Y., Liu, Y.L.: Exploring customers' store loyalty using the means-end chain approach. J. Retail. Consum. Serv. **17**(5), 395–405 (2010)
30. Sirohi, N., McLaughin, E., Wittink, D.: A model of consumer perceptions and store loyalty intentions for a supermarket retailer. J. Retail. **74**(2), 223–245 (1998)
31. Kim, H., Beomjoon, C.: The influence of customer experience quality on customers' behavioral intentions. Serv. Mark. Q. **34**(4), 322–338 (2013)
32. Xia, L., Monroe, K.: The price is unfair: a conceptual framework of price fairness perceptions. J. Mark. **68**(4), 1–15 (2004)
33. Heeryung, K., Eunjoo, H.: Premiums paid for what you believe in: the interactive roles of price promotion and cause involvement on consumer response. J. Retail. **96**(2), 235–250 (2020)
34. Blut, M., Teller, C., Floh, A.: Testing retail marketing-mix effects on patronage: a meta-analysis. J. Retail. **94**(2), 113–135 (2018)
35. Dodds, W., Monroe, K.G.D.: Effects of price, brand and store information on buyer's product evaluations. J. Mark. Res. **28**(3), 307–319 (1991)

36. Dyallo, M., Coutelle-Brillet, P., Riviere, A., Zielke, S.: How do price perceptions of different brand types affect shopping value and store loyalty? Psychol. Mark. **32**(12), 1133–1147 (2015)
37. Chernev, A., Bockenholt, U., Goodman, J.: Choice overload: conceptual review and meta-analysis. J. Consum. Psychol. **25**(2), 333–358 (2015)
38. Iyengar, S.S., Lepper, M.R.: When choice is demotivating: can one desire too much of a good thing? J. Pers. Soc. Psychol. **79**(6), 995–1006 (2000)
39. Chernev, A., Hamilton, R.: Assortment size and option attractiveness in consumer choice among retailers. J. Mark. Res. **46**(3), 410–420 (2009)
40. Berger, J.A., Draganska, M., Simonson, I.: The influence of product variety on brand perception and choice. Mark. Sci. **26**(4), 460–472 (2007)
41. Chevreux, L.: Happy Accidents. The Robin Report (2015)
42. Teleaba, F., Popescu, S.: A behavioral economics perspective over lean versus 10× improvement in new product development. Acta Technica Napocensis **61**(4) (2018)
43. Srinivasana, S.S., Anderson, R., Ponnavolu, K.: Customer loyalty in e-commerce: an exploration of its antecedents and consequences. J. Retail. **78**, 41–50 (2002)
44. Mainardes, E.W., de Almeida, C.M., de-Oliveira, M.: e-Commerce: an analysis of the factors that antecede purchase intentions in an emerging market. J. Int. Consum. Mark. **31**(5), 447–468 (2019)
45. Gao, L., Simonson, I.: The positive effect of assortment size on purchase likelihood: themoderating influence of decision order. J. Consum. Psychol. **26**(4), 542–549 (2016)
46. White, C., Tong, E.: On linking socioeconomic status to consumer loyalty behaviour. J. Retail. Consum. Serv. **50**, 60–65 (2019)

R&D and Technology Management

Innovation Efficiency in Automotive Industry: The Case of Turkey

Gizem Günay, Gökhan Aldemir, and Ferhan Çebi(✉)

Management Engineering, Istanbul Technical University, Istanbul, Turkey
{gunayg,aldemirg,cebife}@itu.edu.tr

Abstract. Innovation activities refer to renewal efforts as products, services or processes and the commercialization of these efforts. In this study, the relationships between the innovation activities and company performance in the automotive industry are examined. The R&D expenditures, the number of employees working in the R&D department, the number of projects carried out, the number of patents, trademarks and design registrations are the variables used in connection with the innovation processes. Efficiency values are obtained by using firm data of research and development activities, which can be described as innovation indicators, and their efficiency is compared in terms of innovation with Data Envelopment Analysis (DEA). Accordingly, suggestions are made the companies to be efficient by data envelopment analysis.

Keywords: Innovation · Efficiency · Automotive industry · R&D · DEA

1 Introduction

With the developments of technology into our lives, every enterprise tries to differentiate and innovate in order to adapt to today's conditions in the globalizing world. Product, process and business models are being developed with new methods and systems day by day. On the other hand, enterprises that resist the steps taken towards innovation have difficulties to continue their activities. In the increasingly competitive environment, the increase in firm values is provided by assets such as patents, trademarks and licenses that focus on information and research and developments potential rather than tangible assets. These have a major impact on the results achieved by businesses at the end of the period.

According to Ruttan, it is argued that the concept of innovation defined by Schumpeter is formed as a result of entrepreneurial movements. Within the scope of innovation, new formations such as producing products, services and processes, discovering new markets, and structuring a new organization make the current economic order shaking the business, which is only possible by being open to new technologies. There is also a positive relationship between innovation and creativity, which is a part of technological change. At this point, the concept of creativity gains great importance. By using imagination and generating unusual ideas and adding originality, innovation is revealed and put on the market. Variables such as R&D expenditures, number of R&D studies, patents, publications and the number of employees working in research

N. M. Durakbasa and M. G. Gençyılmaz (Eds.): ISPR 2020, LNME, pp. 643–657, 2021.
https://doi.org/10.1007/978-3-030-62784-3_55

and development in processes related to innovation enable changes and developments in the field of technology [4].

In this study, activities of companies in the automotive sector in Turkey are analyzed in terms of innovation. In the analysis, CCR (scale-based fixed return assumption) and BCC (scale-based variable return assumption) models are obtained with the MaxDEA package program. Data envelopment analysis for the output is used since it is aimed to determine how much output combinations can be obtained using certain input combinations. The analysis is made for "Turkey's the highest 250 companies of R&D expenditures" which operate the main and sub-industry in the automotive sector. According to results of the analysis, the automotive industry can examine innovation activities on both firm and industry basis.

2 Literature Review

Innovation is a common term used in various areas and in different meanings. It is often linked to creativity, originality, value creation and economic development. As stated in the Oslo Guide, innovation is "the application of a new or significantly developed product or process, a new marketing method or a new organizational method in business practices within workplace organization and foreign relations" [2].

Çiçek and Onat investigate the impact of innovation-oriented activities on the firm. The companies in the information and technology sectors are subjected to analysis and the companies engaged in innovation activities are selected by allocating resources for R&D activities. Input-oriented data envelopment analysis is used and analysis results are obtained with the EMS (Efficiency Measurement System) package program. As an input variable, it is seen that only two firms need improvement for research and development expenses, while other firms have sufficient research and development expenses and are compatible with the scale of the company It is determined that 5 of the 9 companies subject to analysis (Alcatel, Ericom, Link, Netaş and Aselsan) are effective companies in terms of determined criteria. It can be said that the main feature of the effective companies is having relatively effective outputs compared to their own scales. When evaluated according to the potential improvement rates, it can be said that the efficiency levels of the companies that increase their sales by investing in their intangible assets and their R&D activities increase [1].

Habidin et al. use explanatory factor analysis on innovation performance measures and test convergence. Confirmatory factor analysis, the most comprehensive method, is used to test and analyse how accurate the data is compared to a multi-factor first-order confirmatory stage. In the next stage of the analysis, the innovation performance measurement model is tested based on multiple factors. Application is carried out on 229 Malaysian automotive companies producing details based on each innovation performance (environmental innovation, employee innovation, technology innovation). The result of three factor analysis of innovation performance measurements shows that the innovation performance measurement model is a valid and reliable model for the automotive industry in Malaysia [3].

Tan, Zhang, and Khodaverdi analyse service performance levels using data envelopment analysis and balance scorecard techniques in the automotive industry. Data envelopment analysis is used to measure the service performance of car dealers at different levels. Afterwards, which aspects of decision making units are inefficient are analysed with the balance scorecard approach. According to the results, the dealers are insufficient to learn about the growth of the customer and this helps the dealers to become efficient. In addition, this study focuses on various understandings about performance evaluation and provides some useful suggestions that can be applied in the future [5].

3 Methodology

While calculating the efficiency values of the main and sub-industry companies operating in the automotive industry in terms of innovation activities, the results of the data envelopment analysis are obtained according to both the CCR and BCC models. Output oriented data envelopment analysis is used because it is aimed to determine how much output components can be acquired by using certain input components. The main and sub-industry companies operating in the automotive industry, located in the "Turkishtime Economy and Business Culture Portal – R&D 250" study and ranked as "Turkey's top 250 companies in R&D expenditures" are examined.

In accordance with the company policies, companies whose data is fully accessible are coded from Company 1 to Company 18 and their efficiency according to the innovation is calculated. As a result necessary suggestions are made for the inefficient companies to become efficient.

In the study, the R&D expenditure (TL), the number of employees working in the R&D department and the number of projects carried out in the R&D center are used as an input, while the number of patents received in the R&D center, the number of brands received in the R&D center and the number of design registrations received in the R&D center are used as an output. Input and output variables used are shown below in Table 1.

Table 1. Model structure

Model Structure	
Input variables	**Output variables**
R&D expenditures (TL)	The number of patents received
The number of R&D employes	The number of brands received
The number of projects carried out	The number of design registrations received

Input and output variables used in the data envelopment analysis are determined and then input-output components are obtained for all companies. It should be noted that the number of the decision-making units selected in the practice should be at least twice the total number of inputs and outputs. In this study, the total number of inputs and outputs is 6 and the number of decision-making units is 18, so this condition is provided. The input and output variables for each company are shown below in Table 2 and Table 3.

Table 2. Available values of the input variables for each company

	Inputs		
Company	**R&D expenditures (TL)**	**The number of R&D employes**	**The number of projects carried out**
Company 1	21.592.983,0	106	54
Company 2	22.338.474,24	112	30
Company 3	28.634.824,0	186	30
Company 4	46.937.835,0	148	23
Company 5	245.812.509,26	721	39
Company 6	26.393.406,0	195	15
Company 7	594.899.116,49	1396	83
Company 8	43.644.000,00	501	28
Company 9	26.802.876,0	82	16
Company 10	75.080.427,0	102	72
Company 11	60.914.197,54	280	160
Company 12	27.811.398,36	74	118
Company 13	17.098.904,2	107	23
Company 14	17.031.031,94	69	46
Company 15	12.521.919,16	74	23
Company 16	6.263.162,33	39	156
Company 17	4.839.252,0	32	21
Company 18	4.690.526,27	45	23

Table 3. Available values of the output variables for each company

Company	Outputs The number of patents received	The number of brands received	The number of design registrations received
Company 1	205	41	19
Company 2	198	66	146
Company 3	5	1	8
Company 4	159	47	13
Company 5	39	44	13
Company 6	30	16	5
Company 7	35	12	2
Company 8	8	11	9
Company 9	4	2	4
Company 10	172	10	4
Company 11	28	3	15
Company 12	4	38	6
Company 13	6	15	9
Company 14	7	9	8
Company 15	23	7	14
Company 16	13	3	5
Company 17	8	13	11
Company 18	12	39	46

In the study, the solution of the output oriented CCR and BCC model is operated with MaxDEA package program. Total efficiency values are calculated with the constant returns to scale assumption by using the CCR model, and the technical efficiency values are calculated with the variable return to scale assumption by using the BCC model. Then, by dividing the efficiency results obtained from CCR into the results obtained from BCC, the scale efficiency is obtained for each decision making unit. As a result of obtaining technical efficiency and scale efficiency values, it is determined why total inefficient decision making units are not efficient. If the sum of the calculated weights for decision-making units is greater than 1, it has decreasing returns to scale, if it is less than 1, it has an increasing return to scale and if it is equal to 1, it has a constant returns to scale.

3.1 *CCR Model Results*

According to the results obtained with CCR model, four companies (Company 1, Company 2, Company 4 and Company 18) of the 18 companies operating in the automotive industry are found to be total efficient. Three companies that have the lowest efficiency value among the inefficient companies are Company 3, Company 9 and Company 11. As one of the results obtained with data envelopment analysis, it is shown that reference companies that can be taken as an example in terms of output oriented model and how many times these companies are used by inefficient companies as reference in the Table 4. Although Company 4 is an efficient company, it has been determined that it is not used as a reference.

Table 4. Output oriented CCR model results

Company	Efficiency (CCR)	CCR reference set	The number of references
Company 1	1	Company 1 (1,0)	4
Company 2	1	Company 2 (1,0)	12
Company 3	0,054795	Company 2 (1,0)	-
Company 4	1	Company 4 (1,0)	-
Company 5	0,512821	Company 2 (1,3)	-
Company 6	0,484848	Company 2 (0,5)	-
Company 7	0,065717	Company 2 (2,766667)	-
Company 8	0,178571	Company 2 (0,933333)	-
Company 9	0,056818	Company 2 (0,533333)	-
Company 10	0,871927	Company 1 (0,962264)	-
Company 11	0,055046	Company 2 (1,765) Company 1 (0,776)	-
Company 12	0,592516	Company 18 (1,644444)	-
Company 13	0,296592	Company 2 (0,76) Company 18 (0,002)	-
Company 14	0,162871	Company 2 (0,14) Company 18 (1,169)	-
Company 15	0,204481	Company 2 (0,454) Company 1 (0,109)	-
Company 16	0,225668	Company 2 (0,131) Company 1 (0,153)	-
Company 17	0,493966	Company 2 (0,045) Company 18 (0,59)	-
Company 18	1	Company 18 (1,0)	4

The values in parentheses in the table show the density values (λ) of the companies located in the reference set. For example, there are Company 2 (0.131746) and Company 1 (0.153761) in the reference set for Company 16. The potential improvement rates of the total inefficient firms in terms of output-oriented model, that is how much they should decrease their input variables and how much they should increase their output variables in order to increase their efficiency levels, are calculated and these results are indicated in Table 4. Inefficiency sources of 14 companies determined to be inefficient according to the CCR model and target values of input-output variables are shown in the Table 5, Table 6, Table 7 and Table 8.

Table 5. Input slack values obtained according to the CCR model

Company	Input Slack Values		
	R&D expenditures (TL)	The number of R&D employes	The number of projects carried out
Company 1	0	0	0
Company 2	0	0	0
Company 3	-6296350	-74	0
Company 4	0	0	0
Company 5	-216772492	-575	0
Company 6	-15224169	-139	0
Company 7	-533096004	-1086	0
Company 8	-22794757	-396	0
Company 9	-14889023	-22	0
Company 10	-54302274	0	-20
Company 11	-4717803	0	-65
Company 12	-20098088	0	-80
Company 13	0	-21	0
Company 14	-8279911	0	-15
Company 15	0	-11	-4
Company 16	0	-8	-144
Company 17	-1017779	0	-6
Company 18	0	0	0

Table 6. Output slack values obtained according to the CCR model

Company	Output Slack Values The number of patents received	The number of brands received	The number of design registrations received
Company 1	0	0	0
Company 2	0	0	0
Company 3	107	48	0
Company 4	0	0	0
Company 5	181	0	165
Company 6	37	0	63
Company 7	15	0	374
Company 8	140	0	86
Company 9	35	0	8
Company 10	0	28	14
Company 11	0	94	0
Company 12	13	0	66
Company 13	131	0	82
Company 14	0	0	26
Company 15	0	0	0
Company 16	0	2	0
Company 17	0	0	12
Company 18	0	0	0

When the companies with the lowest efficiency in the tables above are analyzed, the factors that cause Company 11 to be inefficient are the R&D expenditures, the number of projects carried out and the number of brands received in the R&D center. In order to be total efficient for the company with the assumption of output-oriented and constant returns to scale, the number of brands received in the R&D center should be 97, the number of projects carried out should be 95, and it should reduce R&D expenditures by approximately 8%. The reference companies are Company 1 and Company 2 in determining the efficiency limits of the Company 11. Company 9 has achieved efficiency in terms of the number of projects carried out and the number of brands received in the R&D center, on the other hand, it should increase the number of patents received

Table 7. Input target values required to be achieved according to the CCR model

Company	Target Values R&D expenditures (TL)	The number of R&D employes	The number of projects carried out
Company 1	21.592.983,0	106	54
Company 2	22.338.474,24	112	30
Company 3	22.338.474,00	112	30
Company 4	46.937.835,00	148	23
Company 5	29.040.017,26	146	39
Company 6	11.169.237,00	56	15
Company 7	61.803.112,49	310	83
Company 8	20.849.243,00	105	28
Company 9	11.913.853,00	60	16
Company 10	20.778.153,00	102	52
Company 11	56.196.394,54	280	95
Company 12	7.713.310,36	74	38
Company 13	17.098.904,20	86	23
Company 14	8.751.120,94	69	31
Company 15	12.521.919,16	63	19
Company 16	6.263.162,33	31	12
Company 17	3.821.473,00	32	15
Company 18	4.690.526,27	45	23

to 39, the number of design registrations received in the R&D center to 12, decrease the R&D expenditures by approximately 55% and the number of R&D employees by approximately 27%. Its reference company is Company 2. Company 3, on the other hand, should increase the number of patents received by 107 and the number of brands received in the R&D center by 48 in terms of output variables, also it should decrease the R&D expenditures by about 22% and the number of the R&D employees by about 40% in terms of input variables. In other words, output values should be increased and input values should be decreased. The company that is the reference in determining the efficiency limits of the company is Company 2.

Table 8. Output target values required to be achieved according to the CCR model

Company	Target Values The number of patents received	The number of brands received	The number of design registrations received
Company 1	205	41	19
Company 2	198	66	146
Company 3	112	49	8
Company 4	159	47	13
Company 5	220	44	178
Company 6	67	16	68
Company 7	50	12	376
Company 8	148	11	95
Company 9	39	2	12
Company 10	172	38	18
Company 11	28	97	15
Company 12	17	38	72
Company 13	137	15	91
Company 14	7	9	34
Company 15	23	7	14
Company 16	13	5	5
Company 17	8	13	23
Company 18	12	39	46

3.2 *BCC Model Results*

As a result of the calculations, according to the BCC model, 8 companies (Company 1, Company 2, Company 4, Company 6, Company 9, Company 16, Company 17 and Company 18) of the all companies operating in the automotive industry are determined to be technically efficient in the Table 6. Three companies that have the lowest efficiency value among technically inefficient companies are Company 3, Company 8 and Company 11. As one of the results obtained by the analysis, reference companies that inefficient companies can take as an example and how many times these reference companies are used by inefficient companies are shown in the table below. It can be said that Company 16 is the company with the highest performance among the efficient companies since it is the company that has been received the highest number of references by inefficient companies. It is determined that Company 1 and Company 16 are not used as the reference companies although they are efficient companies.

Table 9. Output oriented BBC model results

Company	Efficiency (BCC)	CCR reference set	The number of references
Company 1	1	Company 1 (1,0)	-
Company 2	1	Company 2 (1,0)	8
Company 3	0,054795	Company 2 (1,0)	-
Company 4	1	Company 4 (1,0)	2
Company 5	0,666667	Company 2 (1,0)	-
Company 6	1	Company 6 (1,0)	1
Company 7	0,181818	Company 2 (1,0)	-
Company 8	0,181604	Company 2 (0,71) Company 4 (0,28)	-
Company 9	1	Company 9 (1,0)	1
Company 10	0,884995	Company 1 (0,94) Company 17 (0,05)	-
Company 11	0,139904	Company 2 (0,69) Company 1 (0,30)	-
Company 12	0,749706	Company 2 (0,43) Company 18 (0,56)	-
Company 13	0,36276	Company 4 (0,29) Company 18 (0,71)	-
Company 14	0,184913	Company 2 (0,35) Company 18 (0,64)	-
Company 15	0,343147	Company 2 (0,29) Company 1 (0,11) Company 9 (0,004) Company 17 (0,58)	-
Company 16	0,225668	Company 16 (1,0)	-
Company 17	0,493966	Company 17 (1,0)	3
Company 18	1	Company 18 (1,0)	4

According to Table 9, Company 2 (0.298012), Company 6 (0,110078), Company 9 (0.004328) and Company 16 (0.587582) are found in the reference set of the Company 15, and these values in parentheses show the density values for each reference company. The slack values of the input-output variables of the ten companies determined to be inefficient according to the BCC model are shown in the Table 10. Potential improvement rates of the technically inefficient companies in terms of output oriented model are given in the Table 11, that is, how much they should decrease their input variables and how much they should increase their output variables in order to increase their efficiency levels.

When the available values, target values and slack values of input-output variables is examined in terms of the 3 companies with the lowest score in technical efficiency, factors that cause Company 3 to be inefficient are the R&D expenditures, the number of the R&D employees, the number of patents received and the number of brands received. In order to be total efficient with the assumption of output oriented and constant returns to scale, the company should increase the number of patents received to 112, the number of brands received to 49, while reducing R&D expenditures by 22%

Table 10. Input slack values obtained according to the BBC model

Company	Input Slack Values R&D expenditures (TL)	The number of R&D employes	The number of projects carried out
Company 1	0	0	0
Company 2	0	0	0
Company 3	-6296350	-74	0
Company 4	0	0	0
Company 5	-223474035	-609	-9
Company 6	0	0	0
Company 7	-572560642	-1284	-53
Company 8	-14277137	-379	0
Company 9	0	0	0
Company 10	-54393051	0	-20
Company 11	-38803382	-169	-123
Company 12	-15482208	0	-92
Company 13	0	-32	0
Company 14	-6018852	0	-20
Company 15	0	0	0
Company 16	0	0	-132
Company 17	0	0	0
Company 18	0	0	0

and the number of R&D employees by 39%. In other words, in addition to increasing its outputs to 112 and 49, it should make 22% and 39% reductions on the inputs. The reference company is Company 2 in determining its efficiency limits.

The scale efficiency is calculated for each company by dividing the scores obtained from the output-oriented CCR model into the scores obtained from the output-oriented BCC model (CCR/BCC) and as a result, the returns to scale is determined. In terms of scale efficiency, it is concluded that 5 companies are totally efficient according to the output oriented model. Considering Table 12, among all companies, five of them continue their activities according to constant returns to scale, five of them continue their activities according to decreasing returns to scale and eight of them continue their activities according to increasing returns to scale. Companies that have constant returns to scale are scale efficient companies. Companies that have decreasing returns to scale

Table 11. Output slack values obtained according to the BBC model

Company	Output Slack Values The number of patents received	The number of brands received	The number of design registrations received
Company 1	0	0	0
Company 2	0	0	0
Company 3	107	48	0
Company 4	0	0	0
Company 5	140	0	127
Company 6	0	0	0
Company 7	6	0	135
Company 8	143	0	58
Company 9	0	0	0
Company 10	0	28	14
Company 11	0	37	0
Company 12	87	0	81
Company 13	39	0	12
Company 14	41	0	39
Company 15	0	9	10
Company 16	0	11	4
Company 17	0	0	0
Company 18	0	0	0

should use their resources more efficiently by reducing their capacity in order to be efficient and as a result they should reach the same output level with less input. Companies that have increasing returns to scale should utilize their growth potential in order to be efficient and thus they should produce more output.

Table 12. The scale efficiency for companies

Company	Efficiency (CCR)	Σλ	Efficiency (BCC)	Returns to scale
Company 1	1	1	1	Constant
Company 2	1	1	1	Constant
Company 3	0,054795	1	0,054795	Constant
Company 4	1	1	1	Constant
Company 5	0,512821	1,3	0,666667	Decreasing
Company 6	0,484848	0,5	1	Increasing
Company 7	0,065717	2,76	0,181818	Decreasing
Company 8	0,178571	0,93	0,181604	Increasing
Company 9	0,056818	0,53	1	Increasing
Company 10	0,871927	0,96	0,884995	Increasing
Company 11	0,055046	2,54	0,139904	Decreasing
Company 12	0,749706	1,64	0,749706	Decreasing
Company 13	0,296592	0,76	0,36276	Increasing
Company 14	0,162871	1,31	0,184913	Decreasing
Company 15	0,204481	0,56	0,343147	Increasing
Company 16	0,225668	0,28	1	Increasing
Company 17	0,493966	0,64	1	Increasing
Company 18	1	1	1	Constant

4 Conclusions and Recommendations

In this study, eighteen companies operating in the automotive sector entered in the ranking of "Turkey's highest 250 R&D spending companies", their efficiency are evaluated with many input and output variables to discuss and compare with each other. In the analysis, CCR (scale-based fixed return assumption) and BCC (scale-based variable return assumption) models are obtained with MaxDEA package program. Since more firms are determined to be efficient in the BCC model, it is found to be more optimistic when compared to the CCR model. According to the results; Firm 1, Firm 2, Firm 4 and Firm 18 are firms that are efficient according to both CCR and BBC models. The main reason ineffectiveness is low output. It is observed that 28% of the 18 firms have fixed returns, 28% have decreased returns, and 44% have increased returns according to the scale. In addition to these six variables, if the data are reached, other variables can be added to the analysis while examining their effectiveness in terms of innovation. Although the results obtained from the study are favorable with the applied method, it is expected that more clear results will be obtained with the analysis of all the companies operating in the automotive sector. Similarly, in addition to the automotive industry, the other sectors can be analyzed and companies' efficiency in terms of innovation can be compared.

References

1. Çiçek, H., Osman, O.N.A.T.: Determination of innovation based activities on firm performance via data envelopment analysis; a research on ISE. Mehmet Akif Ersoy Üniversitesi Sosyal Bilimler Enstitüsü Dergisi **4**(7), 46–53 (2012)
2. Data, I.I.: Oslo Manual (2005)
3. Habidin, N.F., Hashim, S., Zainol, Z., Mustaffa, W.S.W., Ong, S.Y.Y., Hudin, N.S.: Measuring the innovation performance of Malaysian automotive industry. Geografia-Malaysian Journal of Society and Space, vol. 11, no. 11 (2017)
4. Ruttan, V.W.: Usher and schumpeter on invention, innovation, and technological change. The quarterly journal of economics, pp. 596–606 (1959)
5. Tan, Y., Zhang, Y., Khodaverdi, R.: Service performance evaluation using data envelopment analysis and balance scorecard approach: an application to automotive industry. Ann. Oper. Res. **248**(1-2), 449–470 (2017)
6. [Online]. Available: http://www.turkishtimedergi.com/wpcontent/uploads/2018/12/arge250_2017_low.pdf

Overview of Technology Commercialization Options in Romania

Diana Dragomir[1], Diana Pitic[2], Mihai Dragomir[1(✉)], and Daniela Popescu[1]

[1] Design Engineering and Robotics, Technical University of Cluj-Napoca, Cluj-Napoca, Romania
{diana.dragomir,mihai.dragomir, daniela.popescu}@muri.utcluj.ro
[2] Economic Sciences and Business Administration in German Language, Babeş-Bolyai University Cluj-Napoca, Cluj-Napoca, Romania
diana.pitic@econ.ubbcluj.ro

Abstract. This article discusses the possibilities of evolving technology transfer and knowledge transfer frameworks and entities in Romanian universities into improved and competitive technology commercialization outlets that will better serve the community and the taxpayers with concrete solutions to boost economic competitiveness and social development. A fours stage investigation is set up that identifies best practices, structures existing and potential technology commercialization initiatives and proposes innovative management solutions for higher education institutions to the most common challenges and barriers in the field. These proposals are discussed from the points of view of the necessary effort and the probability of success.

Keywords: Technology commercialization · Technology transfer · Knowledge transfer · PDPC · TRIZ

1 Introduction

For over 30 years universities are working diligently to redefine their role in society. The democratization of access to higher education has created a tremendous force for change but at the same time it took away from the universities' appeal as elite establishments. The forces of globalization and digitalization have expanded accessibility, creating large generations with increasing academic accomplishments [1], but at the same time universities have been left to search for their meaning in the world. Two of the possible solutions to this conundrum which have been gaining traction in the past decade are stimulating entrepreneurship and fostering technology transfer. These approaches have gained maturity and are ready to make their mark in world [2, 3] but they still have to prove themselves in relation to the classical model of higher education.

Romania, as a developing country and a transitional economy is a good position to act as a study framework for the success of such endeavours, more so than advanced economies where economic and social relations are well established for centuries.

N. M. Durakbasa and M. G. Gençyılmaz (Eds.): ISPR 2020, LNME, pp. 658–664, 2021.
https://doi.org/10.1007/978-3-030-62784-3_56

Technology transfer, which is the focus of the current work, is struggling with legislative and market challenges, while organisational culture in Romanian universities is slowly evolving to support the need for true technology commercialization.

2 Context and Methodology

The performance of Romania in innovation and higher education remains modest, and rapid and decisive change is required to produce palpable results. Many of the outputs of research projects, financed nationally or using European funds, does not go beyond academia and publishers and there is a growing interest of companies and people (i.e. tax payer) to see more concrete return for their investment in education. At the same time, success in this field could mean an important source of additional income for universities, which would allow them faster development in order to become more competitive internationally.

The goal of this paper is to investigate the possibilities of implementing methods and mechanisms to support the acceleration of the technology (and by extension, solutions and knowledge) commercialization efforts in Romanian institutions of higher education. For this, the authors have worked with a focus group of 12 researchers (from PhD candidates to full professors) from 5 universities to implement the research methodology presented Fig. 1.

Fig. 1. The research methodology employed to study technology commercialization

The first stage of the undertaking has involved the identification and analysis of best and worst practices in the field of technology and knowledge transfer, as it exists right now in the institutions from where the focus group members are coming from. For this purpose, open discussions were initiated and documented under the guise of free form narratives relating to actual events, projects and situations. Some structure was created under the form of mind maps to help solidifying conclusions and findings into concentrated actionable study items.

Next, based on the consensus of the group, obtained through the moderation of the authors, a tier list has been established which structures the existing and possible processes and methods to be used for technology commercialization. The list has four different levels, with most interviewees finding themselves and their institutions on the first two levels.

In the third stage, the PDPC (Process Decision Program Chart) method has been used to uncover the possible, and very probable, unwanted events or forces acting as challenges and difficulties that would prevent the Romanian institutions of higher education from attaining the top levels of the list developed previously. At the same time, solutions have been proposed based on experience.

Finally, in the fourth and final stage, the TRIZ (Theory of Inventive Problem Solving) has been applied on the most important solutions to detail their mechanisms. The goal is to facilitate the process of overcoming seemingly impossible barriers, especially form a cultural point of view, and to push the technology commercialization functions further along their development within the university system in the country.

The following chapter of the paper presents the most important results and contributions made while implementing the methodology, focusing on stages 2 to 4. Stage 1, which has a looser structure and forms the preamble of the research activity is not presented due to space constraints.

3 Implementation of the Investigation Methodology and Main Results

As stated, in the first stage, the researchers forming the focus group have discussed with the authors of the paper in various combination of individuals and institutions and free discussions have been encouraged to help them share their experience. Similarly, the authors have shared their own experiences, both good or bad, and mind maps oriented on 6 main directions have been used to structure all the material. These directions were: engagement of beneficiaries, institutional mechanisms, faculty engagement, economic/social impact, outside influences, integration into existing research outputs and. Some of the questions addressed during the dialogues included the following:

- What was your biggest success story concerning technology commercialization?
- What were the most common reasons for failure in the field?
- How would you rate your level of knowledge concerning intellectual property?
- Do you know of methods that are successful in your partner universities?
- In what sectors or industries do you see the best potential for technology commercialization?

Based on the results from these discussions, and factoring in the experience of all involved, participants and authors, the 4-tier list of approaches has been developed (see Table 1).

Table 1. Structured tiered list of methods

Tier	Technology commercialization methods	Comments on the current situation
4th	University funded/supported innovative companies University equity in private technology ventures	At the moment, only aspirational approaches

(*continued*)

Table 1. (*continued*)

Tier	Technology commercialization methods	Comments on the current situation
3rd	University supported spin-offs/start-ups Researcher led entrepreneurial endeavours Valorisation and exploitation of patent portfolios	Rare initiatives, involving determined and motivated small teams or individuals Mostly encountered in research intensive or specialized universities
2nd	Company funded research competitions Company supported research entities Partnerships in high level research grants Management and promotion of patent portfolios	Becoming common place in specialized fields (e.g. engineering, agriculture) Not used in all situations or institutions
1st	Customized direct research contracts Voluntary collaborations for projects Sporadic communications, advice, etc.	Starting to gain traction in most universities, common in the ones oriented towards research Usually dealing with simple tasks, reduced responsibilities, small contract sums

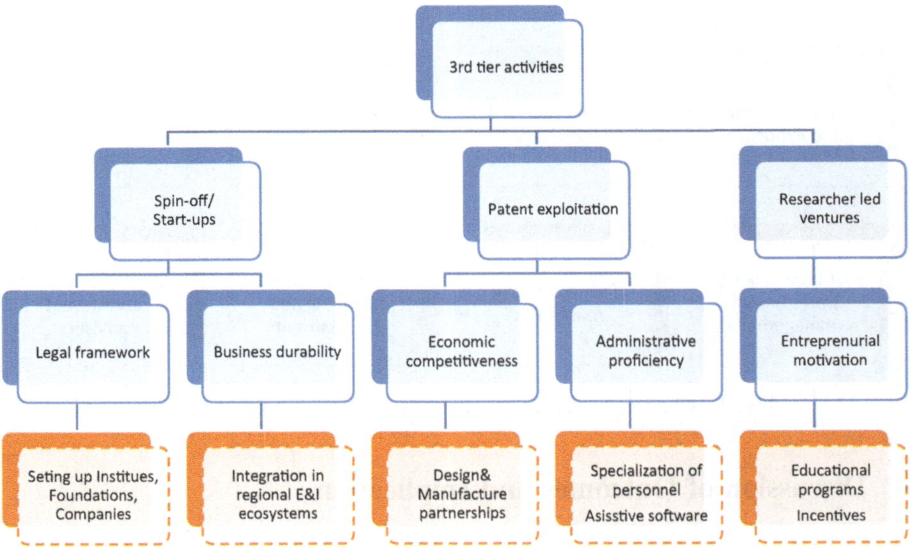

Fig. 2. Challenges and proposals for implementing tier 3 activities

Based on the identified actions and processes needed to implemented higher level technology commercialization functions, such as the ones in tier 3 and 4 above, the PDPC (Process Decision Program Chart) method has been used to guide the discussion towards identifying for each development axis 1 or 2 main challenges and proposing countermeasures for diminishing their effect. The match between the problems and the solutions is based on the current experience of the interviewees and the expertise of the authors, with detailed mechanisms for tackling contentious issues to be addressed in the next stage.

As it can be observed in Figs. 1 and 2, there are significant economic, legal and organizational aspects to deal with when implementing technology commercialization to its full extend. These must be considered in advance by universities in counter-balance with the advantages brought about by the image and funding source represented by becoming a market player (Fig. 3).

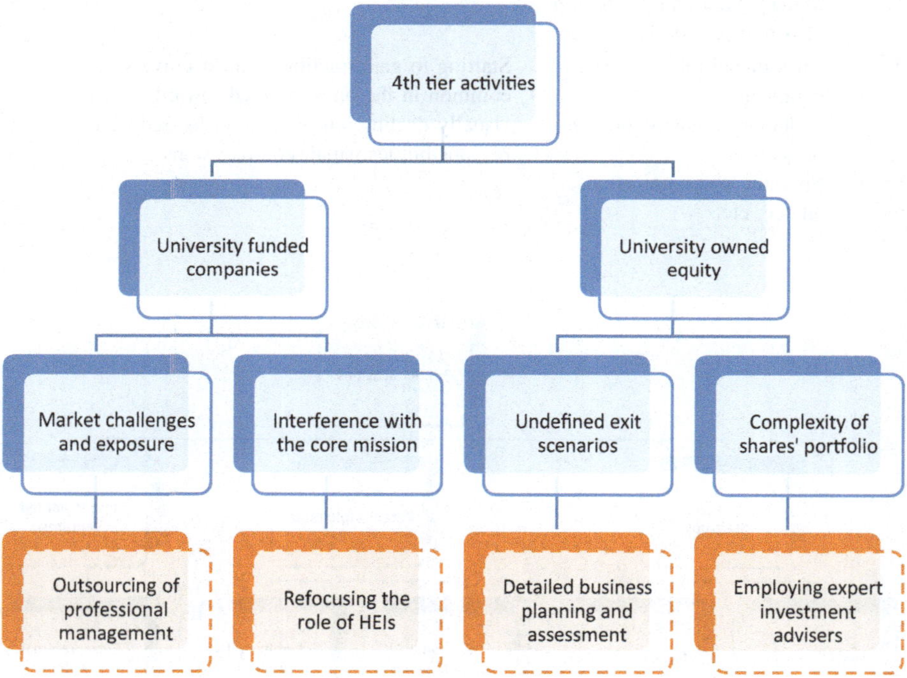

Fig. 3. Challenges and proposals for implementing tier 4 activities

4 Discussion of Outcomes and Implication

The final step of the methodology employed is used here to discuss the resulting changes needed to be implemented when detailing the 3 of the 9 solutions proposed in the previous chapter. By applying the TRIZ methodology with the help of the TRIZ40 platform by SolidCreativity [4], the investigation is focused on the bottlenecks that are expected to appear during their implementation, thus generating more applied success scenarios (see Table 2).

Table 2. Implementation details for the proposed solutions

Proposed solution	Main bottleneck	TRIZ conflict	TRIZ solutions	Implementation
Design & manufacture partnerships	Public accountability vs. commercial contracts	27 Reliability vs. 39 Productivity	1 Segmentation 35 Parameter changes 38 Strong oxidants	Create agile project-based teams that can be reconfigured by task Contribute with staff to operate company RDI infrastructure Stimulate competition-based cooperation in projects, structures and free associations
Incentive programs	Academic career vs. business career	10 Force vs. 9 Speed	13 The other way round 28 Mechanics substitution 15 Dynamics	Provide solutions to other problems (e.g. child care) to allow researchers to find business/academia balance Digitize and localize most work Include entrepreneurship as an evaluation criterion for faculty
Refocusing the role of HEIs	Tradition & legislation vs. competitiveness	13 Stability vs. 35 Adaptability	35 Parameter changes 30 Flexible shells and thin films 14 Spheroidality	Rewrite university charters by public consultation, including companies and the private sector Set-up structures by involvement in clusters, science parks, networks, associations, etc. Perform market research, customer discovery and matchmaking of requirements and capabilities

The implementation scenarios generated have a high degree of coherence and robustness and translating them to reality, especially in a coordinated and integrated manner, across the higher education system has the potential to increase the contribution of Romanian universities to economic development. Structured approaches could also be implemented at organizational level, taking into account the characteristics of each institution and adapting the overall plan to them.

5 Conclusions

In conclusion, we must underline the need to develop technology commercialization solutions in Romanian universities, which will include human components, organizational components and innovative implementation plans. The proposal made in this paper uses a straightforward and durable methodology to link the current situation with the visions of academics in the field. By deploying the results to the level of actionable items, the method demonstrates the possibility to effects significant changes in the higher education system by assuming and performing small changes of existing activities in line with a vision that integrates academia with the regional and national entrepreneurship and innovation ecosystems.

References

1. Bialik, K., Fry, R.: Millennial life: How young adulthood today compares with prior generations, Pew Research Center - Social & Demographic Trends, 14 February 2019. [Online]. Available: https://www.pewsocialtrends.org/essay/millennial-life-how-young-adulthood-today-compares-with-prior-generations. Accessed 18 June 2020
2. Good, M., Knockaert, M., Soppe, B., Wright, M.: The technology transfer ecosystem in academia. An organizational design perspective. Technovation **82–83**, 35–50 (2019)
3. Sutopo, W., Astuti, R.W., Suryandari, R.T.: Accelerating a technology commercialization; with a discussion on the relation between technology transfer efficiency and open innovation. J. Open Innov. Technol. Mark. Complex. **5**(4), 95 (2019)
4. PAO, Goguel, O.: TRIZ40 V2.1, SolidCreativity, 2014. http://www.triz40.com/TRIZ_GB.php. Accessed 20 June 2020

Risk Analysis and Management

Physical Asset Risk Management: A Case Study from an Asset-Intensive Organization

Mohsen Aghabegloo, Kamran Rezaie(✉), and S. Ali Torabi

School of Industrial Engineering, College of Engineering, University of Tehran, Tehran, Iran
{m.aghabegloo,krezaie,satorabi}@ut.ac.ir

Abstract. Asset-intensive organizations need to balance the costs and risks of their physical assets to increase their profitability. Therefore, these organizations should employ a proper procedure to manage the risks associated with physical assets. This study focuses on physical asset risk management and proposes a fuzzy risk management framework accordingly. To that end, all risks are identified based on the life-cycle of physical assets, and the critical ones are determined based on a fuzzy risk assessment process. In this regard, several factors influencing risk impact estimation are first determined and their relative importance are calculated using fuzzy best-worst method. Then, the identified risks are evaluated based on the risk assessment matrix to identify the critical risks. Finally, based on experts' opinions, risk treatment plans are proposed to handle critical risks. A case study for an asset-intensive organization is also provided to demonstrate the applicability of the model.

Keywords: Physical asset management · Risk management · Risk assessment matrix · Fuzzy sets theory · Multi-attribute decision making

1 Introduction

Today the competitive environment of the market forces operating and production enterprises to apply sound approaches to reduce capital, operating and support costs while increasing operational effectiveness and customer satisfaction [1]. In asset-intensive industries, the best way to optimize profitability is to manage physical assets effectively [2]. However, uncertainties involved in the business environment have negatively affected the management of physical assets. Accordingly, the ability of asset-intensive organizations to manage their physical assets while considering uncertainties in strategic and operational level can improve their competitive advantage due to the considerable reduction in operational and capital costs.

Physical asset management focuses on the value that can be realized from physical assets through balancing risk, cost, and performance level [3, 4]. In this way, every decision through the life-cycle of physical assets should be made based on the associated risks. Physical assets may face several risks with different severity and likelihood. Ranking these risks and identifying the critical ones is of great importance as handling all the risks is not practical due to resource limitations.

N. M. Durakbasa and M. G. Gençyılmaz (Eds.): ISPR 2020, LNME, pp. 667–678, 2021.
https://doi.org/10.1007/978-3-030-62784-3_57

Critical risks of physical assets are those risks with medium to high probability and medium to high impact on business objectives. To manage these critical risks, organizations should develop appropriate risk treatment plans taking the available resources into account. In this paper, physical asset risk management framework is proposed in which critical risks are identified and ranked in different stages of the asset life cycle. Then, different risk treatment plans are ranked and selected based on experts' opinions to handle critical risks.

2 Literature Review

PAS 55-1, [6] highlights managing asset risks as a key principle and attribute of asset management. D. Komljenovic et al. [7] proposed a risk-informed decision-making framework which integrates complexity and uncertainty factors in asset management decision making process. Mcgill et al. [8] analysed security risks of critical infrastructures through a quantitative risk management framework which considers different scenarios and their associated security vulnerabilities and threat likelihoods. Shi-Cho Cha et al. [9] proposed a risk management information system for managing asset risks considering both risk management and business continuity management disciplines. Thatshayini et al. [10] have classified the asset management risks in six categories including physical failure, operational failure, risks associated with natural environmental events, risks associated with factors outside of the organization's control, stakeholder related risks and asset life-cycle phases. Badurdeen et al. [11] developed a sustainable risk management framework for asset life-cycle considering triple bottom line of sustainability. Herder and Wijnia [12] presented a risk management process for infrastructure asset management in which stakeholders' opinions on the physical asset risks are considered. Masera et al. [13] proposed a risk governance structure for critical electricity infrastructure. Muhr [14], also studied risk management of electricity infrastructure in which ageing and degradation factors of electrical power components are combined with the asset and risk management to increase the reliability of assets. Nordg [15] assessed asset risks in electricity distribution companies applying Bayesian Network and bow-tie model. Nordgard and Solum [16] also developed a bow-tie model to support decision making in power distribution asset. Also, [17] reviewed different approaches for reducing risks through oil and gas field life cycle.

3 Physical Asset Risk Management Process

Reducing physical asset risks in asset-intensive organizations considerably increases the profitability of these organizations. For this, various physical asset plans should be implemented to manage risks in various stages of physical assets. However, there are not enough resources for mitigating all risks regarding physical assets. Therefore, a process within which critical physical assets are identified and risk treatment plans (RTPs) are determined is inevitable.

This paper focuses on the risk management of physical assets using a quantitative approach. The main steps of the process are: (1) identifying physical asset risks in different stages of asset life cycle, (2) determining critical physical asset risks, and (3) selecting RTPs. These stages are elaborated hereafter in this section.

3.1 Determination of Physical Assets' Risks

In this step, risks related to physical assets are identified based on experts' opinions. For this, experts from different departments of organization are asked to emphasize physical assets-related risks in different stages of physical asset life cycle.

3.2 Determination of Critical Physical Assets' Risks

To determine critical physical assets' risks, risk assessment matrix should be constructed. In this regard, impact of risks and their likelihood should be estimated. To estimate impact of physical assets' risks three factors of "Financial Factor (FF)", "Safety and Health Factor (SH)", and "Environmental Factor (EF)" can be considered. These factors have been widely considered in the literature of physical assets failures [18–20]. FF shows the effect of physical assets' risks on profitability of the organization. SH measures the effect of physical assets' risks on people injury (including personnel). EF emphasizes the effect of physical assets' risks on the surrounding environment.

The aforementioned factors have different influential weights regarding consequence of physical assets' risks. Hence, relative importance of these factors should be calculated to enhance the effectiveness of the consequence estimation. For this, a hybrid multi criteria decision analysis (MCDA) tool is applied in this study. Several MCDA techniques have been proposed in the literature. The best-worst method (BWM) developed by [21] is applied in this papers as it requires limited pairwise comparisons and presents consistent solutions [22]. However, BWM does not consider the uncertainty of human judgments [23]. To overcome this shortcoming, fuzzy BWM is applied in this paper. The steps of fuzzy BWM to obtain relative importance of consequence factors are as follows [24]:

1) *Determine the Most and the Least Important Criteria:* The decision makers are asked to determine the most and the least important factors.
2) *Execute the Fuzzy Reference Comparisons for the Most Important Factor and the Least Important Factor:* By considering the linguistic terms presented in Table 1, the fuzzy preference of the most important factor over all the selected factors are determined by the experts from the organization. Similarly, the fuzzy preference comparison of all factors over the least important factor are determined. Fuzzy vectors for most important factor to other factors and factors to the least important factors can be obtained as Eqs. (1) and (2) respectively.

$$\tilde{P}_B = (\tilde{p}_{B1}, \tilde{p}_{B2}, \ldots, \tilde{p}_{Bn}) \tag{1}$$

$$\tilde{P}_W = (\tilde{p}_{1W}, \tilde{p}_{2W}, \ldots, \tilde{p}_{nW})^T \tag{2}$$

where the preference of the most important factor over factor j and the preference of factor j over the least important factor are denoted by $\tilde{a}_{Bj}$ and $\tilde{a}_{jW}$ respectively.

Table 1. Linguistic terms and corresponding membership functions

Linguistic term	Membership function
Equally important	(1,1,1)
Weakly important	(2/3,1,3/2)
Fairly important	(3/2,2,5/2)
Very important	(5/2,3,7/2)
Absolutely important	(7/2,4,9/2)

3) *Calculate the Optimal Weights of Factors:* The optimal fuzzy weights of factors can be obtained by solving the following nonlinear optimization problem:

$$\min \tilde{k} \tag{3}$$

$$\left|\frac{\tilde{w}_B}{\tilde{w}_j} - \tilde{a}_{Bj}\right| \leq \tilde{k}$$

$$\left|\frac{\tilde{w}_j}{\tilde{w}_w} - \tilde{a}_{jw}\right| \leq \tilde{k}$$

$$\sum_{j=1}^{n} GMRI(\tilde{w}_j) = 1$$

$$w_j^l \leq w_j^m \leq w_j^u$$

$$w_j^l \geq 0, \text{ for all } j$$

where $\tilde{w}_j = \left(w_j^l, w_j^m, w_j^u\right)$ is triangular fuzzy number (TFN), and $GMRI\left(\tilde{w}_j\right)$ is the graded mean integration representation of $\tilde{w}_j$ which can be calculated by Eq. (4). Moreover, $\tilde{k}$ shows the consistency of the comparisons. Notably, the crisp values of the weights can be obtained by Eq. (4).

$$GMRI\left(\tilde{w}_j\right) = \frac{w_j^l + 4w_j^m + w_j^u}{6} \tag{4}$$

To construct the physical assets' risk assessment matrix, the committee of experts are asked to judge the impact and likelihood of each identified physical assets' risks based on the Table 2. Experts' judgments are presented as TFNs to consider the inherent uncertainty. Notably, as impact of physical assets' risks is estimated based on three different factors, it can be calculated as follows:

$$\tilde{I}_r = \sum_j \tilde{w}_j \otimes \widetilde{IF}_{rj} \tag{5}$$

where $\widetilde{IF}_{rj}$ denotes the fuzzy impact factors including FF, SH, and EF for physical asset risk r and $\tilde{I}_r$ is fuzzy estimated impact of physical assets' risk r.

Table 2. Linguistic terms and corresponding membership functions for impact factor and likelihood

Impact factor/likelihood	Membership function
Extremely high	(7,9,9)
High	(5,7,9)
Average	(3,5,7)
Low	(1,3,5)
Remote	(1,1,3)

To construct the risk assessment matrix, the fuzzy impact factor ($\tilde{I}_r$) and the fuzzy likelihood ($\tilde{L}_r$) are defuzzified by Eq. (4). Critical physical risks are determined based on the threshold identified by top manager of the organization. The threshold is often determined based on the risk appetite of the organization.

3.3 Select RTPs

Organizations should select appropriate RTPs to handle physical asset risks. Each RTP requires specified resources and can reduce the impact or likelihood of physical assets' risks up to a predetermined level. These plans can be identified according to experts' opinions. To choose appropriate RTPs, a fuzzy TOPSIS method is proposed in this paper. TOPSIS is a MCDA method proposed by Chen and Hwang [25]. To cope with the inherent uncertainty of experts' judgments, fuzzy TOPSIS is applied in this paper. Fuzzy TOPSIS has been widely studied in the literature [26–28]. The steps are as follows:

1) *Construct a comparison matrix:* Based on the linguistic terms presented in Table 3, a comparison matrix of RTPs with different criteria is constructed which is called a decision matrix. In this matrix, each value is shown as a TFN $\left(\tilde{S}_{ij} = \left(e_{ij}, f_{ij}, g_{ij}\right)\right)$. Notably, the criteria are determined based on the opinions of experts.

Table 3. Linguistic terms and corresponding membership functions in TOPSIS

Linguistic term	Membership function
None	(0,0,1)
Low	(1,1,3)
Moderate	(1,3,5)
Considerable	(3,5,7)
High	(5,7,9)
Extreme	(7,9,9)

2) *Create the Weighted Normalized Decision Matrix: The normalized decision matrix can be obtained as follows:*

$$\tilde{S}_{ij} = \left(\frac{e_{ij}}{c_j^*}, \frac{f_{ij}}{c_j^*}, \frac{g_{ij}}{c_j^*}\right) \quad ; \quad g_j^* = max_i\, g_{ij} \text{ ; Positive ideal solution} \tag{S5}$$

$$\tilde{S}_{ij} = \left(\frac{e_j^-}{g_{ij}}, \frac{e_j^-}{g_{ij}}, \frac{e_j^-}{g_{ij}}\right) \quad ; \quad e_j^- = min_i\, e_{ij} \text{ ; Negative ideal solution} \tag{S6}$$

Let ($\widetilde{wr}_{ij} = (\mathrm{l}_{ij}, m_{ij}, \mathrm{n}_{ij})$ be the weights assigned by decision maker to each criterion, the weighted normalized matrix can be calculated as follows:

$$\tilde{v}_{ij} = \tilde{s}_{ij}.\widetilde{wr}_{ij} \tag{S7}$$

3) *Determine the fuzzy positive ideal solution (FPIS) and the fuzzy negative ideal solution (FNIS):* FPIS and FNIS are calculated according to Eqs. (S8) and (S9).

$$A^* = \{\tilde{v}_1^+, \tilde{v}_2^+, \ldots, \tilde{v}_n^+\} = \{(\max v_{ij} | j \in J), (\min v_{ij} | j \in J')\} \tag{S8}$$

$$A^- = \{\tilde{v}_1^-, \tilde{v}_2^-, \ldots, \tilde{v}_n^-\} = \{(\min v_{ij} | j \in J), (\max v_{ij} | j \in J')\} \tag{S9}$$

where J and J' represent the positive and negative ideal solutions, respectively.

4) *Calculate the distance of each RTP from FPIS and FNIS: The distance of each alternative from FPIS and FNIS can be calculated as below:*

$$Dis_i^+ = \sum_{j=1}^{n} d(\tilde{v}_{ij}, \tilde{v}_j^+) \tag{S10}$$

$$Dis_i^- = \sum_{j=1}^{n} d(\tilde{v}_{ij}, \tilde{v}_j^-) \tag{S11}$$

where d is the distance function between two fuzzy numbers. Let $\tilde{P}_1 = (a_1, b_1, c_1)$ and $\tilde{P}_2 = (a_2, b_2, c_2)$ be two TFNs. The distance between these TFNs can be obtained by the following equation:

$$d(\tilde{P}_1, \tilde{P}_2) = \sqrt{\frac{1}{3}\left[(a_1 - a_2)^2 + (b_1 - b_2)^2 + (c_1 - c_2)^2\right]} \tag{S12}$$

5) *Ranking RTPs:* the closeness coefficient (CC_i) of each RTP as follows:

$$CC_i = \frac{Dis_i^-}{Dis_i^+ + Dis_i^-} \tag{S13}$$

The RTPs are finally ranked based on CC_i values in their descending order.

4 Case Study

In this section the proposed physical asset risk management process is applied in a petrochemical plant (hereafter called as PP) which produces Ethylene. In this study, two critical physical assets including the tower and reactor of the petrochemical plant are considered to identify physical assets' risks. The life-cycle of the physical assets is shown in Fig. 1.

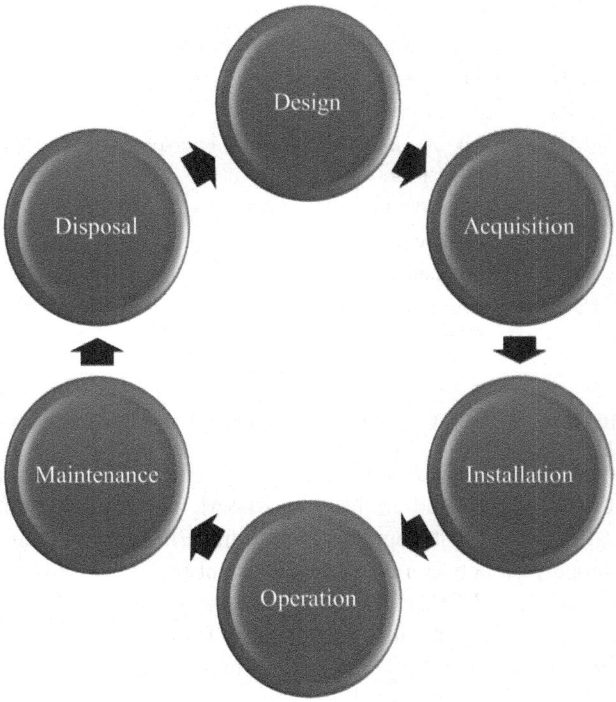

Fig. 1. Life-cycle of tower and reactor in PP

4.1 Determination of PP's Physical Assets' Risks

In this step, risks of physical assets regarding different stages of asset life cycle is determined through group meeting with committee of experts from process, asset management, risk management, and operation departments. These risks are presented in Table 4.

Table 4. Risks related to physical assets in PP

Life cycle stage	Risk ID	Risk description
Design	(D1)	Risk of fault in HAZID (Hazard Identification Study)
	(D2)	Risk of not considering licensor requirements
	(D3)	Risk of not having updated as build P&ID (Piping and Instrumentation Diagram)
Acquisition	(P1)	Purchasing poor quality equipment
	(P2)	Breaking contracts by supplier
Installation	(I1)	Defects in installation specifications and schedules
	(I2)	Changes in land specifications or insufficient consideration of the project site
Operations	(O1)	Cyber attack
	(O2)	Power outage
	(O3)	Operating the plant at above design rate
	(O4)	Failure of safeguards
Maintenance	(M1)	Lack of critical spare parts
	(M2)	Lack of qualified, experienced personnel dealing with maintenance and Inspection
	(M3)	Ineffective inspection plan
	(M4)	Equipment aging
Disposal	(Di1)	Compliance breaches
	(Di2)	Environmental pollution

4.2 Determination of PP's Critical Risks Related to Physical Assets

To determine critical risks related to physical assets, the relative importance of impact factors are first determined based on the fuzzy BWM. The committee of experts in PP has determined FF as the most important factor and EF as the least important factor. The pairwise comparison between the most important factor (FF) and other factors and the least important factor (EF) and the others are shown in Tables 5 and 6. Then, the weights are calculated whose results are shown in Table 7.

The committee of experts are asked to determine the status of the impact factors and likelihood of each identified risks by linguistic terms presented in Table 4. Then, the impact factor is estimated by Eq. (5) and the risk assessment matrix is constructed based on the crisp values of impact and likelihood of risks as Fig. 2. The risks with

Table 5. Preferences of the most important factor over other factors

Impact factors	FF	SH	EF
FF	Equally importance	Very important	Absolutely important

Table 6. Preferences of other factors over the least important factor

Impact factors	EF
FF	Absolutely important
SH	Fairly important
EF	Equally importance

Table 7. Fuzzy weights of impact factors derived from fuzzy BWM

Fuzzy weights of impact factors
$\tilde{w}_{FF} = (0.608, 0.619, 0.633)$
$\tilde{w}_{SH} = (0.198, 0.234, 0.282)$
$\tilde{w}_{EF} = (0.131, 0.142, 0.171)$

likelihood value more than 5.5 and impact value more than 5 Are considered as critical risks related to physical assets for which risk treatment plans should be implemented.

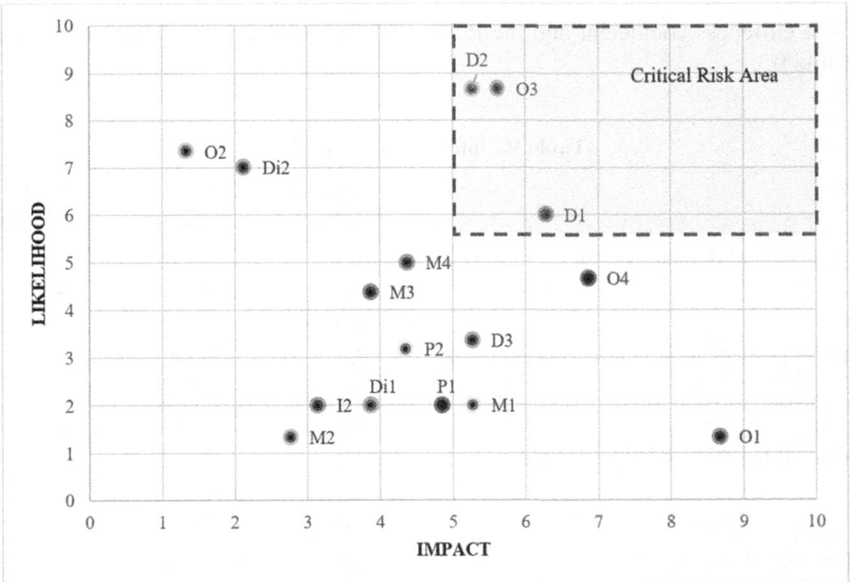

Fig. 2. Physical asset risk assessment matrix in PP

4.3 Select RTPs for PP's Physical Assets

To select appropriate RTPs, 3 criteria including: 1) the ability in likelihood reduction, 2) the ability in impact reduction, and 3) cost are determined by committee of experts in PP. moreover, four different RTPs are determined by the committee of experts. To rank these RTPs, the proposed fuzzy Topsis is applied. The decision matrix whose values represent the arithmetic mean of the committee of experts are shown in Table 8.

Table 8. Information about RTP

RTPs	Critical risks	Likelihood reduction	Impact reduction	Implementation cost
Review risk assessment records	D1	(3,5,7)	(1,3,5)	(2,4,6)
	D2	(1,3,5)	(3,5,7)	
	O3	(1,3,5)	(3,5,7)	
Approve changes in design by process manager	D1	(1,3,5)	(2,4,6)	(1,3,5)
	D2	(2,4,6)	(5,7,9)	
	O3	(5,7,9)	(5,7,9)	
Update all HAZOP (Hazard and Operability Study) and HAZID studies	D1	(7,7,9)	(1,3,5)	(3,5,7)
	D2	(1,3,5)	(5,7,9)	
	O3	(1,3,5)	(5,7,9)	
Reject requested changes that do not comply with licensor requirement	D1	(1,3,5)	(1,3,5)	(7,9,9)
	D2	(7,9,9)	(1,3,5)	
	O3	(7,9,9)	(1,3,5)	

The closeness coefficient and the RTP rankings for each critical risk are presented in Table 9.

Table 9. Information about RTP

RTPs	Critical risks	Closeness coefficient	RTP rankings
Review risk assessment records	D1	0.494	2
	D2	0.142	4
	O3	0.142	4
Approve changes in design by process manager	D1	0.23	4
	D2	0.285	3
	O3	0.561	2
Update all HAZOP and HAZID studies	D1	0.672	1
	D2	0.536	2
	O3	0.536	3
Reject requested changes that do not comply with licensor requirement	D1	0.413	3
	D2	0.715	1
	O3	0.715	1

Based on the ranking presented in Table 9, experts can choose appropriate RTP for each risk. Experts choose to update all HAZOP and HAZID studies to reduce the determined critical risks.

5 Conclusions

Managing the risks related to physical assets in an asset-intensive organization can considerably increase the competitive advantage of the organization by enhancing the availability and reliability of the physical assets. This paper thus develops a process for handling the risks related to physical asset management, where the critical risks are determined from the identified physical asset-related risks using the risk assessment matrix. For this, appropriate impact factors are considered and their weights are determined applying fuzzy BWM, then the risks are evaluated based on their likelihood and impact. Moreover, the fuzzy TOPSIS method is employed to rank possible RTPs and to select appropriate RTPs for each of the risk. A case study in a petrochemical plant was also provided to show the applicability of the process. Future research may focus on interdependency among physical asset risks.

References

1. Mitchell, J.S.: Physical Asset Management Handbook, 4th ed (2012). Reliabilityweb.com
2. Schuman, C.A., Brent, A.C.: Asset life cycle management: towards improving physical asset performance in the process industry. Int. J. Oper. Prod. Manag. **25**(6), 566–579 (2005). https://doi.org/10.1108/01443570510599728
3. Asset Management- an Anatomy: The Institute of Asset Management (2015). https://theiam.org/media/1781/iam_anatomy_ver3_web.pdf. Accessed 09 June 2020)
4. ISO: ISO 55000:2014 Asset management — overview, principles and terminology. Int. Organ. Stand. **2014** (2014)
5. Wijnia, Y.: Asset risk management: issues in the design and use of the risk matrix. In: Engineering Asset Management and Infrastructure Sustainability, pp. 1043–1059. Springer, London (2012)
6. British Standards Institution BSi: PAS 55–1: 2008, Asset Management, Part 1: Specification for the Optimized Management of Physical Assets. Standard (2008)
7. Komljenovic, D., Gaha, M., Abdul-nour, G., Langheit, C., Bourgeois, M.: Risks of extreme and rare events in asset management. Saf. Sci. **88**, 129–145 (2016). https://doi.org/10.1016/j.ssci.2016.05.004
8. Mcgill, W.L., Ayyub, B.M., Kaminskiy, M.: Risk analysis for critical asset protection, **27**(5) (2007). https://doi.org/10.1111/j.1539-6924.2007.00955.x
9. Cha, S.-C., Juo, P.-W., Liu, L.-T., Chen, W.-N.: RiskPatrol: a risk management system considering the integration risk management with business continuity processes. In: IEEE International Conference on Intelligence and Security Informatics, pp. 110–115 (2008). https://doi.org/10.1109/isi.2008.4565039
10. Thatshayini, P., Rajini, D., Weerainghe, A.: Risks associated with physical asset management: a literature review. In: The 7th World Construction Symposium 2018: Built Asset Sustainability: Rethinking Design, Construction and Operations, pp. 457–466 (2018)

11. Badurdeen, F., Shuaib, M., Liyanage, J.P.: Risk modeling and analysis for sustainable asset management. In: Mathew, J., Ma, L., Tan, A., Weijnen, M., Lee, J. (eds) Engineering Asset Management and Infrastructure Sustainability (2012). https://doi.org/10.1007/978-0-85729-493-7_6
12. Herder, P.M., Wijnia, Y.: A systems view on infrastructure asset management, pp. 31–46 (2012). https://doi.org/10.1007/978-94-007-2724-3
13. Masera, M., Wijnia, Y., De Vries, L., Kuenzi, C., Sajeva, M., Weijnen, M.: Chapter 5 Governing Risks in the European Critical Electricity Infrastructure, pp. 117–152 (2006)
14. Muhr, M.: Asset and risk management of electrical power equipment. In: International Symposium on Electrical Insulating Materials, pp. 861–864 (2005)
15. Nordg, D.E.: Dag Eirik Nordgård Risk Analysis for Decision Support in Electricity Distribution System Asset Management. Norwegian University of Science and Technology (2010)
16. Nordgard, D.E., Solum, G.: Experiences using quantitative risk assessment in distribution system asset management. In: 20th International Conference on Electricity Distribution Prague, 8–11 June 2009, no. 0171, Status among Norwegian distribution companies (2009)
17. Smalley, P.C., Begg, S.H., Naylor, M., Johnsen, S., Godi, A.: Handling risk and uncertainty in petroleum exploration and asset management: an overview. Am. Assoc. Pet. Geol. Bull. **92**(10), 1251–1261 (2008). https://doi.org/10.1306/06040808063
18. Theoharidou, M., Kotzanikolaou, P., Gritza-Lis, D.: Risk-based criticality analysis. In: International Conference on Critical Infrastructure Protection, vol. 1, pp. 35–49 (2009). https://link.springer.com/content/pdf/10.1007%2F978-3-642-04798-5_3.pdf
19. Santos, T., Silva, F.J.G., Ramos, S.F., Campilho, R.D.S.G., Ferreira, L.P.: Asset priority setting for maintenance management in the food industry. In: 29th International Conference on Flexible Automation and Intelligent Manufacturing (FAIM2019), vol. 38, no. 2019, pp. 1623–1633 (2019). https://doi.org/10.1016/j.promfg.2020.01.122
20. Jaderi, F., Ibrahim, Z.Z., Zahiri, M.R.: Criticality analysis of petrochemical assets using risk based maintenance and the fuzzy inference system. Process Saf. Environ. Prot. **121**, 312–325 (2019). https://doi.org/10.1016/j.psep.2018.11.005
21. Rezaei, J.: Best-worst multi-criteria decision-making method. Omega **53**, 49–57 (2015)
22. Mi, X., Tang, M., Liao, H., Shen, W., Lev, B.: The state-of-the-art survey on integrations and applications of the best worst method in decision making: why, what, what for and what' s next? R. Omega **87**, 205–225 (2019). https://doi.org/10.1016/j.omega.2019.01.009
23. Karimi, H., Sadeghi-dastaki, M., Javan, M.: A fully fuzzy best – worst multi attribute decision making method with triangular fuzzy number: a case study of maintenance assessment in the hospitals. Appl. Soft Comput. J. **86**, 105882 (2020). https://doi.org/10.1016/j.asoc.2019.105882
24. Gupta, H., Barua, M.K.: A novel hybrid multi-criteria method for supplier selection among SMEs on the basis of innovation ability. Int. J. Logist. Res. Appl. **21**(3), 201–223 (2018). https://doi.org/10.1080/13675567.2017.1382457
25. Chen, S.-J., Hwang, C.-L.: Fuzzy multiple attribute decision making methods. In: Lecture Notes in Economics and Mathematical Systems, vol. 375, pp. 289–486 (1992)
26. Amiri, M., Zandieh, M., Soltani, R., Vahdani, B.: A hybrid multi-criteria decision-making model for firms competence evaluation. Expert Syst. Appl. **36**, 12314–12322 (2009)
27. Ye, F.: An extended TOPSIS method with interval-valued intuitionistic fuzzy numbers for virtual enterprise partner selection. Expert Syst. Appl. **37**(10), 7050–7055 (2010)
28. Memari, A., Dargi, A., Akbari Jokar, M., Robiah, A., Abdul Rahim, A.R.: Sustainable supplier selection: a multi-criteria intuitionistic fuzzy TOPSIS method. J. Manuf. Syst. **50**, 9–24 (2019)
29. Jiménez, M., Arenas, M., Bilbao, A., Rodrı, M.V.: Linear programming with fuzzy parameters: an interactive method resolution. Eur. J. Oper. Res. **177**(3), 1599–1609 (2007)

Risk Analysis Application for a Medical Device Manufacturer

Ayşe İrem Pınar and Nükhet Tunçbilek(✉)

Industrial Engineering, İstanbul Doğuş University, İstanbul, Turkey
{20190303501,ntuncbilek}@dogus.edu.tr

Abstract. The concept of risk assessment became important with economic and technological developments in the 1920s. After World War II, damage became more important with the fact that potential accidents and all risks resulting in damage and injury could be controlled and managed. With this importance, organizations in developed countries have focused on risk assessment activities in their respective fields. Risk assessment is one of the most important and critical steps of risk management. While the concept of risk assessment is new in our lives, its content and methods are not new. Risk assessment and analysis methods have been applied around the world since the 1930s. The lack of practice in the companies that produce medical devices related to risk assessment reveals the importance of working on this issue. This study was conducted using Risk Assessment Method in a medium-sized factory operating in the medical device sector in Istanbul. This assessment was carried out in conjunction with the company's quality team. The results of the evaluation were examined and the risk outcomes were determined by taking into consideration the events, the returns of the patients and the possible risks.

Keywords: Medical device · Risk management · Risk Assessment Table · ISO 14971: 2012

1 Introduction and Purpose

Risk management can be applied in all areas of life. Although the same principles apply in different events, sectors and situations, they also differ in risk management practices. As a result of these differences, there are certain risk assessment requirements in ISO quality systems (ISO 22000, OHSAS 18001, ISO 14001, ISO 22716, API Q1, TS 16949 etc.). According to these requirements, each institution performs its analysis and evaluations in the application areas according to the ISO standard it is attached to. A company producing medical devices is subject to EN ISO 13485 Medical Devices Quality Management System Standard. Within the scope of this system, companies establish quality standards according to the field of activity of the company. EN ISO 14971: 2012, which this standard refers to, defines a process for manufacturers to identify risks that may arise in medical devices, to evaluate, control, and determine the effectiveness of these controls. The requirements of the standard can be applied to every stage of the medical device's life cycle.

N. M. Durakbasa and M. G. Gençyılmaz (Eds.): ISPR 2020, LNME, pp. 679–689, 2021.
https://doi.org/10.1007/978-3-030-62784-3_58

2 Basic Concepts

EN ISO 14971: 2012, as an international standard for application of risk management to medical devices, defines a process to assist manufacturers of medical devices to identify the hazards associated with the medical device, to estimate and evaluate the associated risks, to control these risks, and to monitor the effectiveness of the controls. The requirements enlisted in the standard are applicable to all phases of the life cycle of a medical device. The process described in the standard applies to risks associated with a medical device, such as risks related to biocompatibility, data and systems security, electricity, moving parts, radiation, and usability.

Harm: Physical injury or damage to the health of people, or damage to property or the environment [1].
Hazard: Source, situation, or act with a potential for harm in terms of human injury or ill health, or a combination of these [2].
Risk: Combination of the probability of occurrence of harm and the severity of that harm [1].
Risk Analysis: Systematic use of available information to identify hazards and to estimate the risk [1].
Risk Assessment: Process of evaluating the risk(s) arises from a hazard(s), taking into account the adequacy of any existing controls, and deciding whether or not the risk(s) is acceptable [2].
Acceptable Risk: Risk that has been reduced to a level that can be tolerated by the organization having regard to its legal obligations and its own OH&S (Occupational Health and Safety) policy [2].
Severity: Measure of the possible consequences of a hazard [1].
Probability: When something can happen.
Medical Device: Refers to all kinds of devices, tools, instruments, software, accessories or other materials including such software necessary for their intended use and which are manufactured by the manufacturer for diagnosis and/or treatment purpose in particular to be used use alone or in combination with others for the following purposes for human beings and which do not fulfill their primary function when used on human beings by means of pharmacological, immunological or metabolic impacts but somehow supported by them in the course of their operation:

1) Diagnosis, prevention, tracking, treatment or alleviation of diseases; or
2) Diagnosis, monitoring, alleviation or remedial of injury or disability; or
3) Research, change or replacement of an anatomic or physiological function; or
4) Birth control [3].

3 Risk Management Process

Risk management is a method that decision makers use to reduce or eliminate risk [4]. The risk management process, on the other hand, is a process that consists of determining, evaluating and controlling the events and situations that may occur beforehand while the company is trying to achieve its goals.

The Risk Management process consists of 4 basic elements given below:

1) Risk analysis,
2) Risk evaluation,
3) Risk control,
4) Production and post-production information [5].

Schematic representation of the Risk Management Process is shown in Fig. 1. This process is described at Annex B of the ISO 14971 standard text in detail [6]. An explanatory example is given to make sure that the risk management process steps are well understood. When the steps in Fig. 1 are followed one by one and the questions are answered, it is possible to determine whether the risk analysis is suitably done or not, and whether the steps are proceeding correctly for the first time risk analysis or not.

4 Risk Analysis Methods

Risk analysis methods can be analyzed in two main groups according to the probability of their realization and estimation of their possible effects. These are quantitative and qualitative methods. In qualitative methods, instead of mathematical risk assessment, verbal logic risk assessment is made, and the expert performing the application estimates risks and risk priority values based on their own experience and intuition. Descriptive values such as high, very high are used when calculating and expressing the estimated risk. This prediction is based entirely on subjective evaluations and often does not have a systematic behavior. In such methods, the intuition and reasoning abilities of the expert performing the assessment are important for the reliability of the method. Therefore, it is not proper to make a risk assessment only with qualitative methods for critical systems. Quantitative risk assessment methods use numerical methods when calculating risk. These numerical methods can be simple techniques such as probability and reliability analysis or complex techniques such as simulation models. Quantitative risk analysis gives numerical values to probability of a hazardous event occurring, the effect of the hazard, and these values are used to calculate the risk value. The basic formula of quantitative risk analysis is shown below:

$$\boldsymbol{Risk = Possibility\ of\ Threat * Effect\ of\ Threat}$$

The main purpose of the risk assessment is to answer these questions: “Are these risks acceptable?” and “What measures should be taken for unacceptable risks?” Calculating the probability and the severity of the risks do not eliminate the risks and risks may still occur. The important thing is to be able to take appropriate control measures against the identified risks [7].

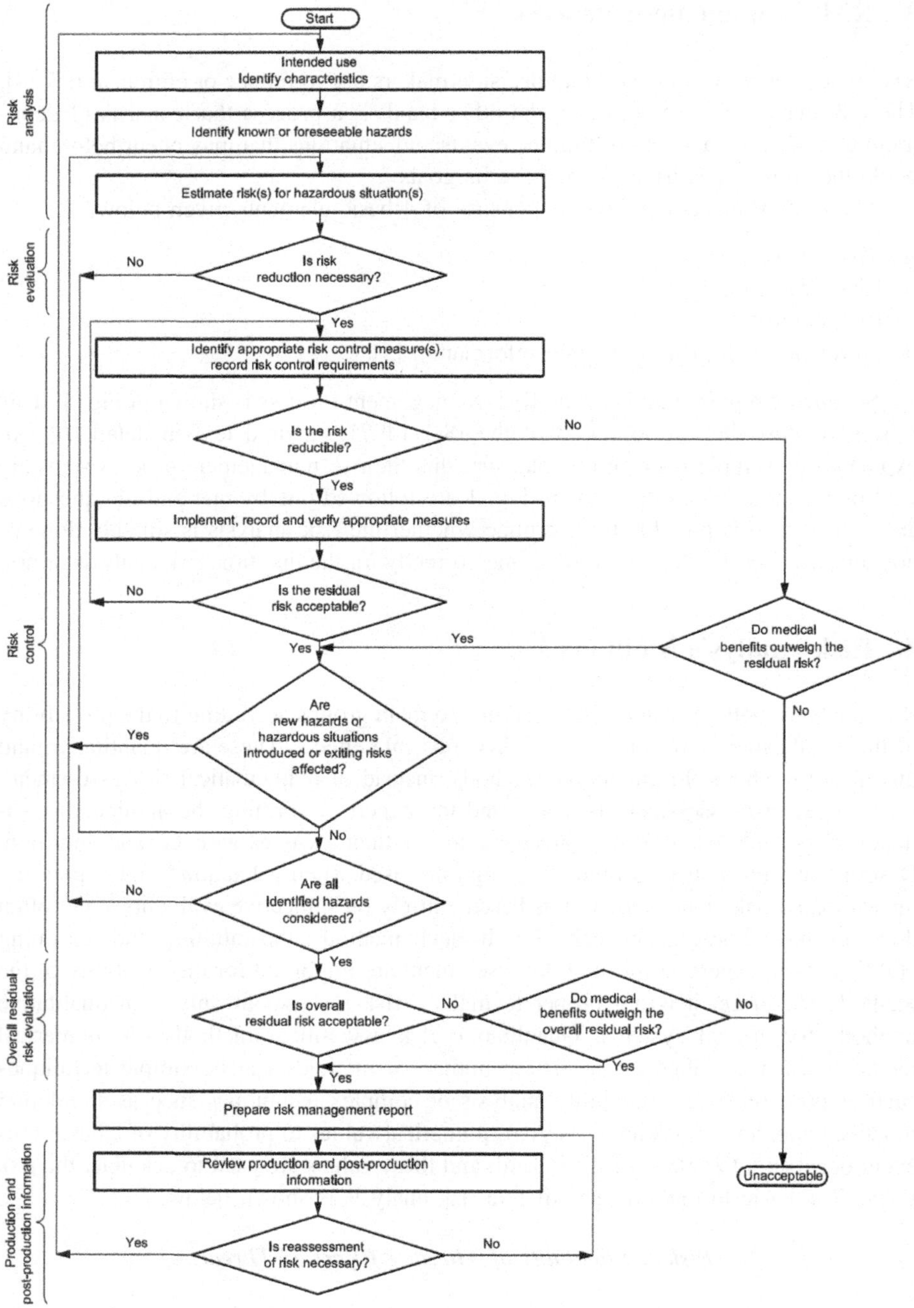

Fig. 1. Schematic representation of the risk management process [6]

There are more than 100 risk analysis methods and standards in application. Some of these methods are given below:

- Preliminary Hazard Analysis-PHA,
- Job Safety Analysis-JSA,
- What-If Analysis?
- Preliminary Risk Analysis (PRA) Using Checklists,
- Hazard and Operability Studies-HAZOP,
- Fault Tree Analysis-FTA,
- Failure Mode and Effects Analysis-FMEA,
- Event Tree Analysis-ETA,
- Cause and Effect Analysis,
- Mathematical Risk Evaluation Method,
- Risk Assessment Table-RAT
  a) L Type Matrix
  b) X Type Matrix

5 A Case Study on Risk Management Process

This case study was conducted using Risk Assessment Method in a medium-sized factory operating in the medical device sector in Istanbul, Turkey. This assessment was carried out in conjunction with the company's quality team. The results of the evaluation were examined and the risk outcomes were determined by taking into consideration the events, the returns of the patients and the possible risks.

I. Step: Risk Analysis:

- Stage 1: Identification and explanation of the product to be analyzed for risk; One of the products of the company is selected
- Stage 2: Identification of people who will perform risk analysis; The company's quality team
- Stage 3: All hazards that have occurred or are likely to occur are listed for the selected product by using historical data.
- Stage 4: Each item on the list prepared with the company's quality team are handled and evaluated individually.

II. Step: Risk Evaluation:

Risk Assessment Table (RAT) is a simple method used for qualitatively assessment of health, safety, environmental and other business risks. RAT is a technique used in assessing risks, prioritizing activities, and allocating resources based on the results of the assessment. Probability/Frequency Scores that are shown in Table 1 and Severity Scores that are shown in Table 2 are used to assess the level of risk.

Table 1. Probability/frequency scores

Frequency	Points	Rating steps for possibility of occurrence
Very Low	1	Improbable
Low	2	Remote
Medium	3	Occasional
High	4	Probable
Very High	5	Frequent (under normal operating conditions)

Table 2. Severity scores in case the event occurs

Severity	Points	Encountered case
Very Rare	1	Inconvenience or temporary discomfort
Rare	2	Junk-free, First-aid-free
Middle	3	Borderline situations, Injury
Serious	4	Important, Critical situations
Very Serious	5	Death, Extraordinary Situation, Disaster

To define the level of risk, Table 3 should be referred. The table is divided into colored areas as Dark Gray and Light Gray. 5×5 Risk Assessment Table was used for the identified risks. The risk here is calculated by the following formula:

Risk = Probability x Severity

Table 3. Risk level (probability x severity scores)

PROBABILITY \ SEVERITY	VERY SERIOUS 5	SERIOUS 4	MIDDLE 3	RARE 2	VERY RARE 1
VERY HIGH 5	HIGH 25	HIGH 20	HIGH 15	MEDIUM 10	LOW 5
HIGH 4	HIGH 20	HIGH 16	MEDIUM 12	MEDIUM 8	LOW 4
MEDIUM 3	HIGH 15	MEDIUM 12	MEDIUM 9	LOW 6	LOW 3
LOW 2	MEDIUM 10	MEDIUM 8	LOW 6	LOW 4	LOW 2
VERY LOW 1	LOW 5	LOW 4	LOW 3	LOW 2	LOW 1

When evaluating the risks, the severity and probability of the event caused by the hazard were taken into consideration. High risk values above 12 are considered as unacceptable and colored with dark gray. Low risk values below and equal to 12 are considered as acceptable and colored with light gray. This classification is shown in Table 4 including the activities necessary to take when such risks occur.

Table 4. Risk results

Evaluation	Risk Value	Activity
Unacceptable Risk	15,16,20,25	Production and shipment should be stopped until the determined risk is reduced to an acceptable level, and the Warning procedure should be put into effect.
Acceptable Risk	1,2,3,4,5,6,8,9,10,12	It may not be necessary to plan control processes and keep records of the activities to be performed to eliminate the identified risks.

The values given in Table 1 and Table 2 below are used for probability and severity values in the formula [7].

III. Step: Risk Control:

The purpose of this step is to reduce the risks to an acceptable level or to decrease the risk score from an acceptable level to more acceptable level. These steps are shown in Table 4. To reduce the risks to an acceptable level the necessary implementation steps are determined. The main purpose is to eliminate the risk completely. The aim is to reduce risks to the lowest possible level, as it cannot be completely eliminated. The process of reducing the risks to an acceptable level is possible only by changing the probability of occurrence. The severity value cannot be changed by taking precautions.

IV. Step: Production and Post-Production Information:

The manufacturer shall establish, document and maintain a system to collect and review information about the medical device or similar devices in the production and the post-production phases [8].

After all steps are completed, the company should issue a Risk Management Report. This report should at least guarantee the following:

- The risk management plan has been appropriately implemented;
- The overall residual risk is acceptable;
- Appropriate methods are in place to obtain relevant production and post-production information [9].

In this study conducted at the medical device manufacturing company a number of risks were detected. In Table 5, the risk known before the application are listed. After completing the risk assessment steps that are described in standard ISO 14971, a new list of risks are discovered. A subset of the actual information was shown in Table 6. Actual values and full list of risks were not shared due to the Company's Quality Assurance Policy. Table 6 below represents a sample.

Table 5. Risks ıdentified before the application and risk degrees

Danger	Potential	Damage	Risk before			Corrective actions remarks	Risk after			Assessment and monitoring of remaining risks
			Possibility	Severity	Risk		Possibility	Severity	Risk	
Failure to provide sufficient effect during use	Formulation error	Decrease of Product Performance	3	3	**9**	Production instructions and recipes are defined. Production is carried out by Competent Personnel	2	3	**6**	It will be followed up with customer complaints and non-compliance reports
	Incorrect raw material use	Decrease of Product Performance	3	3	**9**	Raw materials were determined during the design. All raw materials are kept in defined areas during storage. Initial quality controls are carried out	2	3	**6**	It will be followed up with customer complaints and non-compliance reports
	Usage error Using the product less than the specified dose	Decrease of Product Performance	4	2	**8**	The way of using the product is defined on the product label	2	2	**4**	It will be followed up with customer complaints and non-compliance reports
	Usage error Using the product more than the specified dose	Occurrence of side effects	3	3	**9**	The method of eliminating high dose use of the product is described on the product label	2	3	**6**	It will be followed up with customer complaints and non-compliance reports
Producing wrong	Resulting from measuring devices	Decrease of Product Performance	3	3	**9**	All critical measurement devices are calibrated within the calibration plan. Calibrations are renewed annually	2	3	**6**	It will be followed up with customer complaints and non-compliance reports

(*continued*)

Table 5. (*continued*)

Danger	Potential	Damage	Risk before			Corrective actions remarks	Risk after			Assessment and monitoring of remaining risks
			Possibility	Severity	Risk		Possibility	Severity	Risk	
Wrong use	Lack of labeling	Decrease of Product Performance	3	3	**9**	Product use is described in the package leaflet inside the product box	2	3	**6**	It will be followed up with customer complaints and non-compliance reports
use of the product for different purposes	Lack of labeling	Decrease of Product Performance	3	3	**9**	Product use is described in the operating instructions	2	3	**6**	It will be followed up with customer complaints and non-compliance reports
Quick exhaustion of the content of the product than planned	Incorrect storage	Decrease of Product Performance	5	1	**5**	The storage conditions of the product are indicated on the instructions for use	3	1	**3**	It will be followed up with customer complaints and non-compliance reports

All of risk management process steps described above were applied. Result of risk analysis and risk assessment are combined in Table 6. Initial risk analysis made and the list of risks is obtained. Upon completion of the initial risk assessment, there were 14 unacceptable risks.

After defining precautions and taking necessary actions, the risk assessment is repeated three months later. The precautious activities are taken and probability of all risks is re-evaluated. It is observed that all probability levels are decreased. When the new degrees of risks are calculated by using the new probability values and the severity values, there are no more unacceptable risks. The company knows how to analyze and evaluate risks in continuous manner and which actions to take in order to reduce probability of identified risks.

Table 6. Risks identified as a result of the application and risk degrees

RISK ANALYSIS FOR SAMPLE PRODUCT OF THE COMPANY									
		INITIAL EVALUATION				RE-EVALUATED AFTER PRECAUTIONS			
OPERATION STEPS	RISKS	PROBABILITY	SEVERITY	DEGREE OF RISK	PRECAUTIONS	PROBABILITY	SEVERITY	DEGREE OF RISK	IS ANOTHER PRECAUTION NECESSARY?
Storage/Shipment	1) Mixing of materials	4	3	12	Storage in separate areas, storage records	2	3	6	Customer complaints will be evaluated.
	2) Dirty materials	4	3	12	External cleaning control made at the entrance of the storage	1	3	3	
	3) Keeping the materials under improper conditions	4	4	16	Storage system procedure, storage records	1	4	4	
	4) Bacterial growth in storage conditions	4	4	16	Microbiological tests after each production	2	4	8	
	5) Deformed packaging	4	3	12	2nd check after production at the entrance to the warehouse	1	3	3	
	6) Dispatching the wrong product	4	3	12	Storage system procedure, storage records	1	3	3	
Manufacturing	1) Getting the wrong material for production	4	4	16	Visual label control, 2nd control, locked car	2	4	8	It will be evaluated according to the deviations and changes that opened on the subject.
	2) Exceeding the calibration dates of production equipment	4	4	16	Calibration label check, 2nd check	2	4	8	
	3) Use of wrong equipment	4	3	12	Trained personnel, usage procedures, 2nd control	1	3	3	
	4) Unsuitable product	4	5	20	Quality Control analysis at the end of production	2	5	10	
	5) Use of dirty containers / containers / bins	3	5	15	Cleaning procedure, clean / dirty status label, 2nd check	1	5	5	
	6) Production in dirty area	4	4	16	Cleaning procedure, clean / dirty status label, 2nd check	1	4	4	
	7) Use of faulty raw materials	4	5	20	Quality controls are carried out at the production entrance	2	5	10	
	8) Working with neglected equipment	4	4	16	Maintenance schedule, maintenance label	1	4	4	
Packaging	1) Wrong packaging material purchase for production	4	3	12	Visual label control, material codes, 2nd control, material list from the storage	1	3	3	It will be evaluated according to the deviations and changes that opened on the subject.
	2) Not reading the instruction book	3	4	12	Input control	1	4	4	
	3) Torn / hole in the box	3	3	9	Input control, visual control of the packaging part 3rd control	2	3	6	
	4) Dirty box	3	2	6	Input control, Packaging part visual inspection 3rd control	2	2	4	
Interaction of product components	1) Using the product at high dosage	3	5	15	The way of use of the product is defined on the product label.	1	5	5	It will be evaluated according to the deviations and changes that opened on the subject.
	2) Using the product at low dosage	4	2	8	The way of use of the product is defined on the product label.	2	2	4	
	3) Insufficient explanation of functions in the instruction book	3	3	9	Preparation by expert person, biocompatibility reports, ministry approval	2	3	6	
	4) Damage to the product as a result of damaged / faulty packaging	3	4	12	Packaging controls	1	4	4	
	5) Deterioration of products kept under extreme sun	3	5	15	Instruction book , storage conditions	2	5	10	
	6) Using the product for other purposes	4	4	16	Describes the use of the product in the leaflet.	2	4	8	
	7) Use of degraded products	3	5	15	Storage conditions, Product expiration date	2	5	10	
	8) Use of spoiled products	4	5	20	Storage conditions, Product expiration date, instruction book	2	5	10	

Acceptable Risk	1,2,3,4,5,6,8,9,10,12
Unacceptable Risk	15,16,20,25

6 Conclusion

In this study, within the scope of the ISO 14971:2012 standard for the application of risk management to medical devices, the process is applied at a medical device manufacturer in Istanbul, Turkey. The risks about one of the company's products were evaluated step by step by using the Risk Assessment Method together with the company's quality team. Prior to this study the company had a list of 14 risks. A subset of these risks are shown in Table 5. One of the 14 items was considered as unacceptable risk, one as acceptable risk and 12 as risks that should be kept under control.

A new risk assessment study is carried on by starting with a risk analysis meeting with Quality Assurance Manager, Quality Control Manager and General Manager. During this meeting 84 risks were identified after the initial evaluation. A subset of this list is shown in Table 5. 45 of the identified risks are at unacceptable risk and 39 risks are at acceptable levels. Control and precautionary actions are determined for the items that do not provide the acceptable level of risk. Risks that are above the acceptable level were reduced to an acceptable level after applying the determined control and precaution activities. Table 5 and Table 6 do not include all of the risks identified in the study due to the company's policy.

During this study the company has reevaluated the risks about one of their medical device products and discovered some more risks that were not known before. After defining precautions and applying them, risk level of many risks were decreased either by reducing probability or reducing severity. All 14 unacceptable risks were changed to acceptable level. This study resulted in a safer state by discovering and managing risks there were not known before and could not be prevented unless otherwise.

References

1. ISO 14971:2012 Medical devices - Standard for the Application of Risk Management to Medical Devices, pp. 1, 3, 4
2. TS 18001:2008 Occupational Health and Safety Management Systems; Conditions, pp. 4, 6
3. Medical Devices Regulation; Official Gazette of Turkish Republic: 9.1.2007 – no. 26398, p. 5
4. Özer, A.: Process, Performance and Risk Analysis/Method in Organizations, 2nd edn, p. 234. Adalet Publishing (2015)
5. ISO 14971:2012 Medical devices - Standard for the Application of Risk Management to Medical Devices, p. 5
6. ISO 14971:2012 Medical devices - Standard for the Application of Risk Management to Medical Devices- Annex B, p. 24
7. Ceylan, H., Başhelvacı, V.: Risk analysis with the risk assessment method: an application. Int. J. Eng. Res. Dev. **3**(2), 25–33 (2011)
8. ISO 14971:2012 Medical devices - Standard for the Application of Risk Management to Medical Devices, p. 13
9. ISO 14971:2012 Medical devices - Standard for the Application of Risk Management to Medical Devices, p. 12

Scheduling

Hybrid Flowshop Scheduling Problem at DrPaste

Ece Nural(✉), Ceren Sanisoğlu, Esennur Ermiş, Sinem Ünver, Zehra Gürçay, and Zehra Düzgit

Department of Industrial Engineering, İstanbul Bilgi University, İstanbul, Turkey
ece.nural@bilgi.edu.net, {ceren.sanisoglu, esen.ermis, sinem.unver, zehra.gurcay}@bilgiedu.net, zehra.duzgit@bilgi.edu.tr

Abstract. DrPaste is a manufacturer company of pastry materials and tools. In this study, we consider sugar paste production of this manufacturer. The company wants to minimize the makespan to deliver the sugar paste to its customers faster. There are 5 stages of sugar paste production: *i*) refining the sugar, *ii*) preparing the raw materials, *iii*) mixing by boilers, *iv*) packaging the sugar paste, *v*) controlling with a metal detector. In stage 3, there are 12 identical machines (boilers) for mixing. The boilers are not dedicated to a specific color. Each boiler has a capacity of 65 kg. In stage 4, there are 4 employees responsible for taking the sugar paste from the boilers and packing manually into different package sizes. We focus on the 3rd and 4th stages of sugar paste production, which can be handled as a two-stage hybrid flowshop environment. There are 17 colors of sugar paste. In stage 3, there is a clean-up process before mixing a different color. Cleaning the dark colors from a boiler such as brown is more difficult than light colors like pink. We have 4 types of color groups; *i*) white, *ii*) light colors, *iii*) midtone colors, *iv*) black. There are sequence-dependent setup times before switching to a different color group. In stage 4, employees need to separate the sugar paste into 0.5 kg, 1 kg, 2.5 kg, and 5 kg packages according to order specifications. Under these circumstances, the job (color) sequence on each machine/employee plays an important role in stages 3 and 4. Therefore, we want to find the best job sequence in mixing and packing stages. Since the problem is NP-hard, we employ Genetic Algorithm to solve the problem. We also develop a decision support system (user interface) to implement the proposed solution approach easily in the company.

Keywords: Scheduling · Hybrid flowshop · Sequence-dependent setup time · NP-hard · Meta-heuristic · Genetic algorithm · Makespan · User-interface design

1 Introduction

Scheduling is an important tool for manufacturing, where it can have a significant impact on the productivity of a process. According to [1], the purpose of scheduling is to minimize the production time and costs, by telling a production facility when to make, with which staff, and on which equipment. Production scheduling aims to

N. M. Durakbasa and M. G. Gençyılmaz (Eds.): ISPR 2020, LNME, pp. 693–707, 2021.
https://doi.org/10.1007/978-3-030-62784-3_59

maximize the efficiency of operations and reduce costs to achieve a specific purpose determined by the decision maker in specific environmental conditions. By environment, we mean the specifications of the servers such as the number or the speed, limitations etc. Since different environments require different approaches, it is crucial at this step to fully understand the types of production environment.

At the end of the 1970s, [2] introduced the 3-field notation $\alpha/\beta/\gamma$ where α represents the machine environment, β represents job characteristics and constraints, and γ field represents the objective function for the related problem.

In scheduling theory, there are mainly 4 types of environments; *i)* single machine, *ii)* parallel machine, *iii)* flowshop, *iv)* jobshop environment and in Graham's notation it corresponds to the α field. The details of the environments are as follows.

In a single machine environment, there is only one operation and each job are processed by the same machine and it can be considered as the most basic environment. In parallel machines, there are multiple machines which can process the same single operation. The speeds of the machines may differ. If the speeds are the same, the environment is said to be identical parallel machine environment. If the speed of the machines is different and the speed is not job-dependent, it is called uniform whereas it is called unrelated parallel machine environment if it is job-dependent.

Another environment is flowshop. Each job follows the same machine routing meaning, in other words each job first visits the first machine, then the second and the rest is similar. In addition to this, Hybrid Flow Shop (HFS) is a specific type of a flowshop. HFS consists of a series of production stages in the flowshop environment where each of these stages has several machines operating as a parallel machine environment. The machines in the stages can be identical, uniform or unrelated. Some stages may have only one machine, but at least one stage must have multiple machines. For each stage, there is at least one machine to process one operation and there are several operations to complete a job.

In the jobshop environment, each job may have a different machine routing.

In the β field, constraints are listed such as release time of the job, whether the preemption is allowed or not and if there are setup times etc. Release time is the time a job arrives to the system and it is available for processing. Another constraint may be setup time. In between two operations, machines may be arranged for the next job. In scheduling theory, setup times may depend on the sequence. If the next job depends on the previous job, it is called sequence-dependent setup time (SDST). Otherwise, it is called sequence-independent setup time. For instance, in an environment with sequence-dependent setup time, the setup time differs in between two cases; scheduling product A and product B, scheduling product B and then product A.

In the γ field, the objective is represented such as minimizing the total completion time, maximum lateness and maximum tardiness or makespan etc. Makespan (C_{max}) is the completion time of the last job that leaves the system. In scheduling theory, makespan should be reduced to deliver orders faster.

In this study, we examine a real-world scheduling problem of a company, namely DrPaste which mainly produces sugar paste. There are 5 sequential stages of sugar paste production: *i)* refining the sugar; *ii)* preparing the raw materials, *iii)* mixing in the boilers, *iv)* packing and *v)* controlling. We focus on 3rd and 4th stages of production.

For mixing, there are 12 identical machines with the same speed. For packing, there are 4 identical workers responsible from packages. Hence, the production environment under consideration can be said to be a two-stage hybrid flowshop. There are 17 different colors of sugar paste which can be categorized under 4 color groups: *i*) white, *ii*) light colors, *iii*) midtone colors and *iv*) black. In the mixing stage, a sequence-dependent setup (cleaning) time is required when switching between color groups. The company wants to deliver the orders to its customers as soon as possible. Therefore, the aim is to minimize the makespan. Therefore, our problem can be represented as HFS/SDST/C_{max}.

The study is organized as follows. In Sect. 2, NP-Hardness of the problem is explained. In Sect. 3, production environment of the company is described. In Sect. 4, genetic algorithm is introduced as the proposed solution approach. Section 5 contains the actual size problem data which is followed by small size problem definition in Sect. 6. The framework of parameter tuning is presented in Sect. 7. Design of user interface is given in Sect. 8. Conclusions are drawn in Sect. 9.

2 NP-Hardness of the Problem

HFS/SDST/C_{max} problem is shown to be NP-hard in scheduling literature. [3] develops a mathematical model upon fourteen different size problems which have different number of stages and different number of machines at related stages. They conclude that mixed integer programming (MIP) model does not work efficiently for large scale problems because of the running time. [4] concerns with two-stage manufacturing model where the first stage includes two parallel machines and the second stage has several parallel machines. MIP and Genetic Algorithm (GA) are proposed for two-stage HFS problem with dedicated machines and sequence-dependent setups. First, a mathematical model is developed and also GA is proposed because of the time complexity.

[5] develops a mathematical model for a HFS flowshop problem with sequence-dependent setup times to minimize the maximum completion time. However, the running time of the model is very large. Due to NP-hardness of the problem, they use a heuristic approach.

We expect that, in the light of these knowledge, running time of a mathematical model may be very large due to the time complexity of the problem. Therefore, we employ Genetic Algorithm to solve the problem at hand.

3 Production Environment

In this study, the main focus is sugar paste production. The production flowchart is represented in Fig. 1. Each color of sugar paste follows the same operations sequentially. For different colors, different amounts of food-colorings are put in the mixture, but other than that all the products follow the same route.

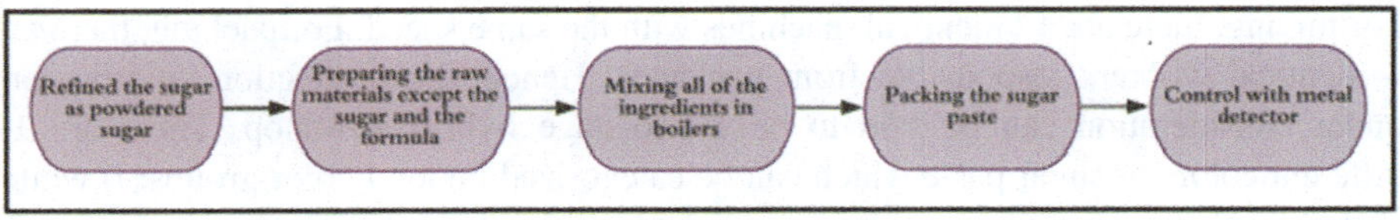

Fig. 1. Production steps of sugar paste

A detailed representation of the production environment is exhibited in Fig. 2. In order to shorten the overall makespan, we can reduce the makespan of a specific part of the production line. Hence, we decided to focus on the following processes:

i) mixing all of the ingredients in boilers,
ii) packing the sugar paste

which are represented in the red dashed line in Fig. 2. Then, the problem can be defined as a two-stage hybrid flow shop scheduling problem.

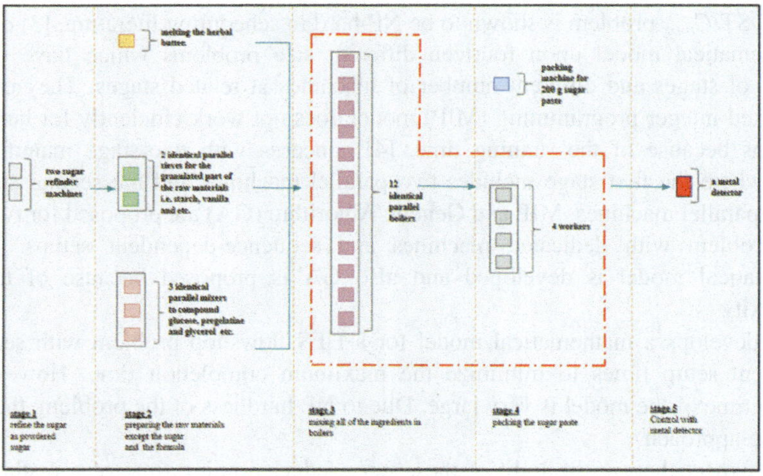

Fig. 2. The hybrid flowshop production environment

At the mixing stage, there are 12 identical machines (boilers). Each boiler mix with the same speed. The capacity of the machines at the mixing stage is 65 kg.

There are 4 types of color groups; white (W), light colors (L), midtone colors (M) and black (B). In total, we have 17 different colors (see Table 1). As it can be seen in Fig. 3, switching from W to W requires 10 min of setup time (cleaning time). If there is L before B, it takes 15 min. On the other hand, if there is M before L, it takes 20 min to clean. Lastly if there is B, then the cleaning time of it takes 30 min. It is important to remark that after producing black, it is not allowed to produce any other color.

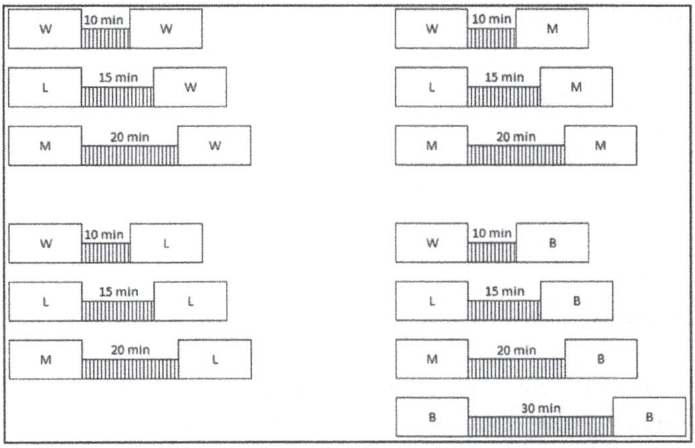

Fig. 3. Sequence-dependent setup times

Although the setup times are sequence-dependent, we always need 10 min for cleaning after W, 15 min after L, 20 min after M. Therefore, we embed setup times into processing times. The processing times (including setup times) per batch are given in Table 1.

Table 1. Colors and processing times at mixing stage

	Color ID	Color Name	Processing Time (min) / batch
White	1	White	30
Light Colors	2	Skin	35
	13	Baby pink	
	14	Baby Blue	
	7	Blue	
	9	Green	
	15	Silver	
	3	Yellow	
Midtone Colors	4	Orange	40
	6	Pink	
	5	Red	
	8	Dark Blue	
	12	Grey	
	16	Gold	
	10	Purple	
	11	Brown	
Black	17	Black	50

Company produces in batches. Each batch weighs 65 kg. After the mixing stage, the batches must be separated into 0.2 kg, 0.5 kg, 1 kg, 2.5 kg and 5 kg according to order specifications, which is done by 4 workers. Since 0.2 kg sugar paste is packed by a machine, it is out of our consideration. Therefore, we assume that the packing process of 0.2 kg is negligible. Since 17 colors have 4 different sizes, then there are 68 possible job types in the production system. All the servers (employees) are assumed to be identical in packaging stage. The packaging time required for each package size is given in Table 2. Note that, packing small sizes takes more time.

Table 2. Package sizes and processing times at the packaging stage

Package size	Number of packages	Time (min)
0.5 kg	130	30
1 kg	65	20
2.5 kg	26	15
5 kg	13	5

4 Genetic Algorithm

In GA, we first have to define two terms, 'individual' and 'population' which gained its inspiration from the theory of natural evolution from [6]. A population consists of a previously determined number of individuals. To define an individual, we use chromosomes and genes. Each gene represents a job and their sequence can be viewed as the chromosome. In each iteration, from the current population, the next generation is produced.

There are 4 stages in the GA: *i*) selection, *ii*) crossover, *iii*) mutation, and *iv*) reinsertion. The flowchart of the GA is demonstrated in Fig. 4. To briefly explain; in the selection phase, the 'fittest' individuals, which will be selected according to an algorithm defined by the decision maker, two parents are selected to produce offspring among the population. To calculate the fittest, several methods can be used.

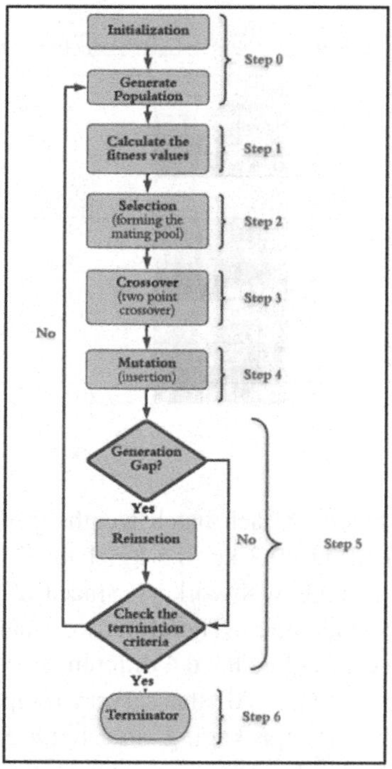

Fig. 4. Flowchart of genetic algorithm

Step 0: Initialization: The initial population is generated with a pre-determined population size by the decision maker. In this study, the population size is denoted by N_{pop}.
Step 1: Calculation of the fitness value: For each sequence, completion time of the last job is calculated for each step. Note that we need to maximize the fitness value whereas we need to minimize the makespan.
Step 2: Selection: In the selection phase, two parent chromosomes are selected with the help of the decision maker's method. To improve the next generation (offspring) the fittest ones are chosen as the parents. We implement a new method introduced by [7] where we distribute a probability of being selected to the individuals which will assure that the fittest (the one with the lowest makespan) having the highest probability of being selected. *P*(*k*) represents that probability with $P(\mathrm{k}) = 2k/Y(Y + 1)$ where Y is the number of individuals and we assign the value k to the individuals considering the makespans. The lowest makespan will be $k = 4$, the next lowest will be $k = 3$, next $k = 2$ and the highest makespan will be $k = 1$. Suppose we have a population with the following makespan values; 5, 4, 3, 2.

$$\text{Makespan} = 5;\ P(1) = 2(1)/4(4+1) = 2/20$$

$$\text{Makespan} = 4;\ P(2) = 2(2)/4(4+1) = 4/20$$

$$\text{Makespan} = 3;\ P(3) = 2(3)/4(4+1) = 6/20$$

$$\text{Makespan} = 2;\ P(4) = 2(4)/4(4+1) = 8/20$$

As a conclusion, lowest makespan 2 has a highest probability 8/20 of being selected.
Step 3: Crossover: In the crossover step, after the selection phase, parents may generate an offspring depending on the crossover probability. To perform the crossover step, a uniform random number is generated between (0,1) for each parent where we selected in the selection step. Then, if that random number is less than or equal to the crossover probability, crossover operation is performed. Otherwise, crossover operation is not performed for that parent. We employ a two-point crossover operator. The operator and the working procedure are explained in Fig. 5. Suppose that $N_{pop}= 4$ and we have completed the selection phase and selected parents are:

$$\text{First Parent} = 1 - 2 - 3 - 4$$

$$\text{Second Parent} = 3 - 4 - 1 - 2$$

Generated offspring by using the two-point crossover method are 1−3−2−4 and 3−1−4−2.

First parent = 1-2-3-4
Second parent = 3-4-1-2

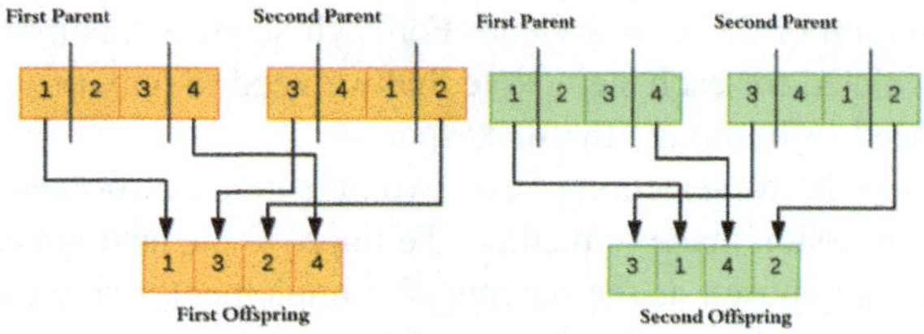

Fig. 5. Two-point crossover method

Now, assume that, due to crossover probability, crossover is not performed for the remaining two parents. In this case, we will end up having two chromosomes in the population, meaning that we will have a 'generation gap' in the population since the population size should be the same from beginning to the end. To fill that gap, at Step 5, we have the reinsertion step.

Step 4: Mutation: In this step, the generated offspring may or may not mutate which depends on the probability of mutation. A uniform random number is generated between (0,1) for each offspring. Then, if that random number is less than or equal to the mutation probability, mutation operation is performed. As it can be seen in Fig. 6, an insertion method is used for this study.

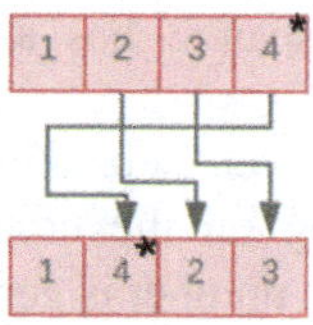

Fig. 6. Insertion method for mutation

Assume that mutation operation is not performed due to mutation probability. Then the offspring, which we generated at crossover step, continues as it is.

Step 5: Reinsertion: Population size should remain the same from beginning to the end. As we have described in the crossover step, there is a chance that the crossover may not occur, and a generation gap may be present. So, as a solution to the generation gap, we have a reinsertion stage which can be done with several methods depending on the decision maker to increase population size. In this study, we fill this gap by choosing the fittest ones in that generation, which is called as the elitist approach.

Step 6: Termination: The termination criterion is determined by the decision maker. For instance, the decision maker may decide to terminate when they reach a specific number of iterations or to stop when they reach a pre-determined time. In our case, we stop the algorithm when we reach a pre-determined number of iterations.

5 Actual Size Problem Data

The company generally makes 3-day schedules. Based on this information, we considered a 3-days' order list between December 9–11, 2019. The reason of considering a past data is to be able to make a comparison between the actual makespan and the makespan that we will obtain by the proposed solution methodology. Based on past orders, the company tells that 0.5 kg and 1 kg orders arrive in terms of quantity. For instance, the customer says that they need 50 boxes of 0.5 kg. On the other hand, for 2.5 and 5 kg, they tell the orders arrive in terms of kg. For instance, they say that they will buy 0.5 kg of 2.5 boxes. In Table 3, we convert all the customer orders into kilograms. As we mentioned before, in the production site, the mixers always mix 65 kg at a time. So, we need to convert each order to be a multiple of 65 kg. If it is not a multiple of 65, we choose to round it up. In total, we have 287 jobs.

Table 3. Customer order list

Colors	0.5 kg			1 kg			2.5 kg			5 kg		
	kg	Multiple of 65	Number of jobs	kg	Multiple of 65	Number of jobs	kg	Multiple of 65	Number of jobs	kg	Multiple of 65	Number of jobs
White	48	65	1	162	195	3	14080	14105	217	10480	10530	162
Red	24	65	1	72	130	2	70	130	2	1000	1040	16
Black	24	65	1	324	325	5	2500	2535	39	2500	2535	39
Pink	24	65	1	888	910	14	20	65	1			0
Blue	24	65	1	97	130	2	70	130	2	860	910	14
Yellow	24	65	1	474	520	8						0
Orange	12	65	1	132	195	3				100	130	2
Brown	12	65	1	312	325	5				600	650	10
Purple	12	65	1	150	195	3						
Dark blue	12	65	1	6	65	1	60	65	1			
Skin	12	65	1	276	325	5						
Green				438	455	7						
Gray				90	130	2						
Baby blue				762	780	12						
Baby pink												
Silver												
Gold												

6 Small Size Problem

Before solving our actual size problem, we generate a small size problem with two purposes: to check if the algorithm runs properly and to decide on the parameters of the algorithm via parameter tuning. To represent a small size problem, instead of 12 machines in the first (mixing)s stage and 4 machines in the second (packing) stage, we decided to have 3 machines in the first stage and 1 machine in the second stage. We constructed an order list which consists of 20 jobs and 12 orders which as seen in Table 4.

Table 4. Data of small size problem

Weight	Color	Size	Number of jobs	Multiple of 65
162	White	1	3	195
70	Red	2.5	2	130
24	Black	0.5	1	65
20	Pink	2.5	1	65
97	Blue	1	2	130
24	Yellow	0.5	1	65
100	Orange	5	2	130
12	Brown	0.5	1	65
150	Purple	1	3	195
60	Dark blue	2.5	1	65
12	Skin	0.5	1	65
90	Gray	1	2	130
			20	

By the help of [7], the Python model which was improved to run GA for hybrid flowshop environment, we have coded our problem and run with small size problem. It was observed that the algorithm works properly in terms of job-machine assignments.

7 Parameter Tuning

As we have mentioned before, there are four parameters to be determined by the decision maker before starting to employ GA: *i)* population size, *ii)* crossover probability, *iii)* mutation probability, and *iv)* termination criterion. Based on the related studies in literature, we determined some parameter intervals to find the best combination of parameters of GA as follows:

$$N_{pop} = [2, 4, 6, 8, 10]$$

$$P_c \text{ (Probability of crossover)} = [0.5, 0.6, 0.7, 0.8, 0.9, 1.0]$$

$$P_m \text{ (Probability of mutation)} = [0.01, 0.02, 0.03, 0.04, 0.05]$$

$$stopGen \text{ (Termination criterion in terms of number of generations)} = [100, 200]$$

There are several methods for parameter tuning. Grid search, which works by doing extensive research in a subset of a specific parameter, is a traditional way whereas random search works by doing random research. The best advantage of random search is decreasing running time. Therefore, random search may not guarantee to find the best parameters. In this study, we prefer grid search.

We run the algorithm 10 times and then calculate the average C_{max} since GA has some randomness. The randomness arises from the previously mentioned uniform random numbers drawn.

Since we have 5 candidates for N_{pop}, 6 candidates for P_c, 5 candidates for P_m and 2 candidates for *stopGen*, we have 300 different combinations in total. In Table 5, we report the best (smallest) 25 combinations. Hence, we select the best parameters as $N_{pop} = 4$; $P_c = 0.5$; $P_m = 0.04$; $stopGen = 100$ since this combination gives the best C_{max} value 420.

Table 5. Best 25 combinations of parameters, average Cmax, standard deviation and average running time

PARAMETERS	AVERAGE C_{max}	STANDARD DEVIATION	AVERAGE RUNNING TIME (sec)
N_{pop}:4; P_c: 0.5; P_m: 0.04; stopGen: 100	420.00	0.00	0.37
N_{pop}:4; P_c: 0.6; P_m: 0.03; stopGen:100	420.50	1.58	0.37
N_{pop}:2; P_c: 0.5; P_m: 0.05; stopGen: 100	421.00	2.11	0.20
N_{pop}:4; P_c: 0.5; P_m: 0.05; stopGen: 100	421.00	2.11	0.38
N_{pop}:4; P_c: 0.6; P_m: 0.02; stopGen: 100	421.00	2.11	0.38
N_{pop}:4; P_c: 0.9; P_m:0.03; stopGen: 100	421.00	2.11	0.38
N_{pop}:4; P_c:1.0; P_m: 0.03; stopGen: 100	421.00	2.11	0.39
N_{pop}:2; P_c:0.6; P_m:0.04, stopGen: 100	421.50	2.42	0.19
N_{pop}:2; P_c:0.7; P_m: 0.05; stopGen: 100	421.50	2.42	0.25
N_{pop}:4; P_c:1.0; P_m:0.02, stopGen: 100	421.50	2.42	0.38
N_{pop}: 2; P_c:1.0; P_m:0.01; stopGen: 100	422.00	3.50	0.20
N_{pop}:2; P_c:0.8; P_m: 0.04; stopGen: 100	422.00	3.50	0.20
N_{pop}:2; P_c:0.9; P_m:0.02, stopGen: 100	422.00	2.58	0.20
N_{pop}:2; P_c:0.6; P_m:0.03, stopGen: 100	422.00	2.58	0.21
N_{pop}:2; P_c:1.0; P_m:0.05, stopGen: 100	422.00	4.22	0.25
N_{pop}:4; P_c:0.6; P_m:0.04, stopGen: 100	422.00	3.50	0.38
N_{pop}:2; P_c:0.7; P_m:0.04, stopGen: 100	422.50	3.54	0.21
N_{pop}:2; P_c:0.7; P_m:0.01, stopGen: 100	422.50	4.86	0.22
N_{pop}:2; P_c:0.6; P_m:0.05, stopGen: 100	423.00	3.50	0.19
N_{pop}:2; P_c:0.8; P_m:0.05, stopGen: 100	423.00	2.58	0.20
N_{pop}:2; P_c:0.9; P_m: 0.01; stopGen: 100	423.00	3.50	0.20
N_{pop}:2; P_c:0.6; P_m:0.01, stopGen: 100	423.00	4.22	0.21
N_{pop}:2; P_c:0.8; P_m: 0.02; stopGen: 100	423.50	4.74	0.19
N_{pop}:2; P_c: 0.5; P_m:0.02; stopGen: 100	423.50	4.12	0.20
N_{pop}:2; P_c:0.8; P_m:0.03, stopGen: 100	423.50	4.12	0.23

8 Design of User Interface

Based on GA outputs, we can obtain the job-machine assignments, start and completion times of jobs. We also want to convert these outputs into a Gantt chart to follow the proposed schedule visually. With this objective, we designed a user interface. We have collected the outputs in an array called 'machine tracker'. Machine tracker holds all information as [start time, end time, stage index, machine index, job index]. In order to check the Gantt chart easily, we use our small size problem with 10 jobs and 2 stages (Fig. 7).

Fig. 7. User interface design

As it is seen in Fig. 8, the interface asks for the job data. The user can drop the excel file there but there is a certain format to be followed as in Table 6.

Table 6. Input format for input data of user interface

Weight	Color	Color Code	Size	Size Duration	Number of jobs	Multiple of 65
48	White	WHITE	0,5	30	1	65
24	Red	RED	0,5	30	1	65
12	Brown	BROWN	0,5	30	1	65

.
.
.

Population size, probability of mutation, probability of crossover and number of generations (termination criterion) are to be determined by the decision maker. The interface is not designed to work with constant parameters since these parameters may change depending on the problem. The Gantt chart of our small size problem, which is developed by the algorithm behind the user interface, is given in Fig. 8.

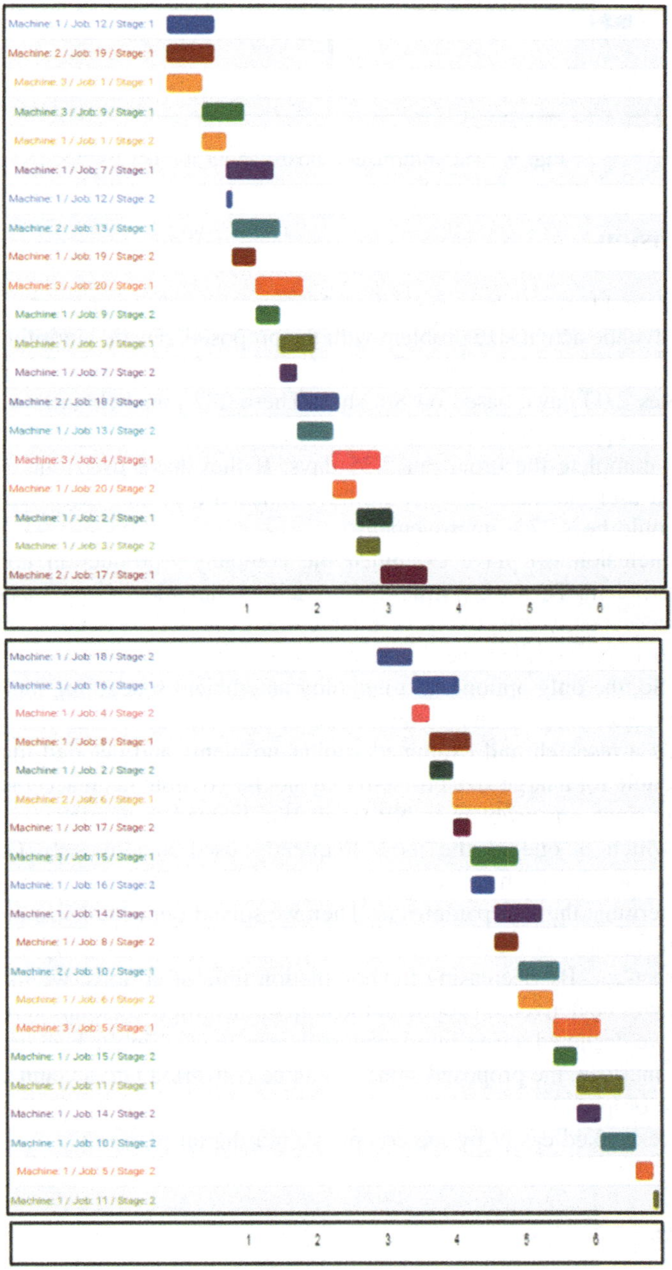

Fig. 8. Gantt chart of small size problem

When you click on a job, you can see the machine index, job index, stage index, start time, and end time of the related job as in Fig. 9.

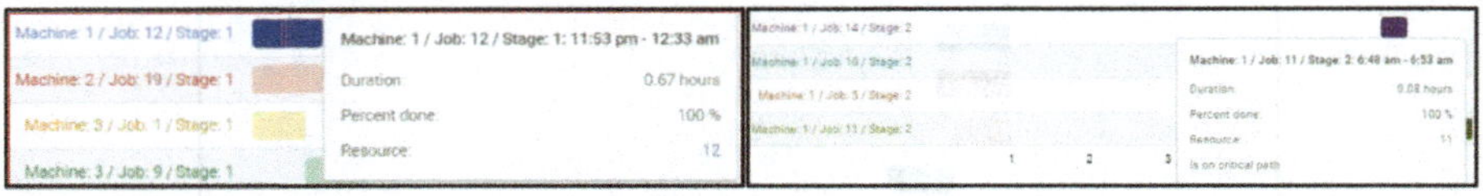

Fig. 9. The information boxes about job details

9 Conclusion

In the actual problem, we have a total of 287 jobs for a 3-day period (see Table 3). When we solve the actual size problem with the proposed genetic algorithm, the results show that C_{max} of these 287 jobs is 16 h and 15 min. The total processing time of the schedule takes 2.03 days, based on 8-h shift. These 287 jobs belong to a past 3 days' data, so the company knows their actual makespan for this old order list. They have managed to complete the orders in 2.25 days. If they have used our proposed GA approach and user interface, they would have finished it in 2.03 days. So, we can say that there would be 9.77% improvement.

As a conclusion, we have examined the company's production layout and the scheduling problem. They were struggling to deliver the orders as soon as possible and not using the capacity effectively in their current area. To extend their production facility or to buy new machines were not in their options because of their budget limitations. So, the only option was to employ an efficient scheduling method in order to use their capacity better and minimize the completion time of all jobs.

We made a research and examined similar problems and decided that finding an optimal schedule for a large size problem may not be possible in an acceptable range of running time since our problem is NP-Hard. We decided to employ a meta-heuristic algorithm, which is one of the most frequently used one, namely GA. We first implemented the algorithm on a small size problem and then performed parameter tuning to determine the best parameters. Then we solved our actual size problem using those parameters and compared with the actual makespan value. By using GA, we shortened the C_{max}. By decreasing the completion time of all jobs, we made sure that more orders will be delivered soon with the help of the proposed solution methodology.

Finally, we designed a user interface which uses the GA in the background. By the help of this interface, the proposed solution can be converted into a Gantt chart and the details of the proposed schedule (start time, end time, stage index, machine index, job index) can be tracked easily by the company on a digital platform.

References

1. Pinedo, M.L.: Planning and Scheduling in Manufacturing and Services, 2nd edn. Springer, New York (2009)
2. Graham, R.L., Kan, A.H.G.R.: Optimization and approximation in deterministic sequencing and scheduling: a survey. Annals of Discrete Mathematics. Erasmus University, Rotterdam (1979)

3. Kurz, M.E., Askin, R.G.: Scheduling flexible flow lines with sequence-dependent setup times. Eur. J. Oper. Res. **159**(1), 66–82 (2004). https://doi.org/10.1016/s0377-2217(03)00401-6
4. Harbaoui, H., Morineau, O.B., Khalfallah, S.: Scheduling a two-stage hybrid flow shop with dedicated machines, time lags and sequence-dependent family setup times. In: 2016 IEEE 46 International Conference on Systems, Man, and Cybernetics (SMC) (2016). https://doi.org/10.1109/smc.2016.7844695
5. Temizsoy, A., Göllü, İ., Kaplan, E., Özköse, S., Aydoğan, Z., Acar, T.: Hybrid flow shop scheduling. The Senior Design Project of The Industrial Engineering Students at Yaşar University (2018)
6. Mallawaarachchi, V.: Introduction to Genetic Algorithms - Including Example Code. https://towardsdatascience.com/introduction-to-genetic-algorithms-including-example-code-e396e98d8bf3. Accessed 20 Nov 2019
7. Suyunu: suyunu/Flow-Shop-Scheduling. https://github.com/suyunu/Flow-Shop-Scheduling. Accessed 4 July 2017

Service System Management

Effect of Gamification Applications on Individual Performance: Air Transportation Application

Samet Elitok and Hür Bersam Bolat(✉)

Management Engineering Department,
Istanbul Technical University, Istanbul, Turkey
samet.elitok@gmail.com, bolat@itu.edu.tr

Abstract. The game, which has been a part of human life since the birth, has found itself a new form with technology and has started to appear as gamification applications in businesses. Gamification, which was originally intended for educational purposes, has served many different purposes over time, such as health, production, performance improvement and communication enhancement. The spread of the domain over a wide scale requires confirmation of the validity and effectiveness of these applications. This study analyses the effect of gamification activities carried out in an internationally operating air transport company on the individual performance of employees. In the study, the independent and mediated effect of the factors that make up gamification is observed using structural equation modelling.

Keywords: Gamification · Performance · Air transportation · Structural equation modelling

1 Introduction

Today, the use of technology is spreading day by day and it has a place in every part of life. From the simplest item to use, to the most advanced electronic device; from our social life to our professional life, technology makes its impact felt. This affects not only individuals, but also the life cycles, method and management decisions of businesses. Businesses are becoming more and more technological structures day by day. New searches that will keep the business in a more profitable and sustainable form are shaped by the technology effect.

The game factor in human life since birth is affected by this search and takes its place in professional life with "gamification" practices. Various objectives such as increasing the amount of work output with the same inputs, increasing training efficiency, strengthening employee motivation and communication among employees trigger the use of gamification applications. It is assumed that the person sees the work or education he/she has received as an entertainment tool, not an essential action, thereby increasing the motivation of work and improving its performance. In these applications, the game factor manifests itself in various forms and turns into a new tool in achieving the chosen goal by using game familiarity.

N. M. Durakbasa and M. G. Gençyılmaz (Eds.): ISPR 2020, LNME, pp. 711–719, 2021.
https://doi.org/10.1007/978-3-030-62784-3_60

With this study, it is aimed to see whether gamification applications performed in enterprises have an impact on individual performance. It was carried out in an international enterprise operating in the air transportation sector, which is a sector that has to follow technology closely due to its nature of work. This study organized as follows. In the second part, an information about the gamification literature will be given. In the third part of the study, a structural equation model is proposed in accordance with the structure of gamification studies in the selected enterprise. In the third part of the study, a structural equation model is proposed in accordance with the structure of gamification studies in the selected enterprise. According to the model determined, the analysis results of the survey study conducted on 166 employees, working at three different strategic levels, will be mentioned. And last part is designed for recommendations and future studies.

2 Gamification Background

The gamification concept, which was first introduced by Nick Pelling, a game designer in 2002, reached its basic use in 2011. Generally speaking, gamification is the use of game design elements in non-game environments [1]. Although gamification is expressed in different definitions and scopes over time, almost all definitions are shaped to use game elements for common purposes such as making work enjoyable, increasing loyalty to work, increasing employee motivation, and increasing internal communication and training efficiency.

Each game is a structure consisting of several pieces. How to play the game, the scenarios that the player will encounter and how these scenarios will affect the game and the player are determined by the use of these components. Game elements are each of the components that make up this gaming experience. The most commonly used game elements are known as "The PBL Triad" [2]. Accordingly, points, badges and leader boards are the most basic and frequently preferred game elements. Apart from The PBL Triad, there are many inclusive and subset game elements. However, not every game element is required in every game. Game elements will be selected according to the structure of the enterprise where gamification studies are carried out, and the study will be created specifically for the enterprise.

The privatization factor in gamification practices has led to the diversification of the literature. Many ideal gamification model studies have been created based on component and design. Component-based models focus on the mechanical setup of the game, while design-based models consist of progressive installation models [2–4]. In gamification applications, it is seen that the game elements pyramid model is generally used in component selection. On the basis of design, it is seen that special structures are established for the enterprise.

According to the pyramid model of the game elements, gamification applications consist of three parts: dynamic, mechanical and components [2]. Dynamics are all elements that form the main structure of the game, such as constraints, rules and fiction, but are not directly included in the game. Mechanics is expressed as each of the actions that enable the player to be involved in the game. Components include smaller game elements such as badges, points and achievements.

Many studies have focused on how game mechanics can have an impact on individual behavioural outcomes, trying to explain how the psychological effect occurs [5, 6]. For this purpose, it is seen that support is received from many different disciplines. motivation and performance-based studies are largely based on Personal Satisfaction Theory [7] and Intrinsic and External Motivation [8] approaches. It is seen that many different disciplines are utilized such as Transteoric Model [9] for behaviour change, Universal Design [10] for educational studies and User Centered Design [11] for game designs. As seen in the reference [12, 13], many new approaches have emerged with the help of different fields, the effects of gamification elements on business factors and the orientation of the users involved in gamification can be measured.

This multidisciplinary structure, which manifests itself in theoretical approaches, has also been involved in gamification application integration studies in enterprises. For many businesses, especially Fortune 100 companies, the effect of gamification on performance was measured. It is seen that implementation studies are mostly focused on education, socialization and crowdsourcing, and it aims to encourage users to participate and direct them to the determined behaviour [14]. Although the gamification applications made in the aviation sector are high, there is no research variety in the sector. The studies generally focus on the subject of "Frequent Flyer Program". For example, reference [15] searched for an answer to the question of how the point-based and tier-based development of the Frequent Flyer Program, created for sustainability and profitability, will affect customer behaviour and satisfaction.

It is seen that a significant part of the studies carried out is disconnected between theoretical and practical studies, empirical verification of theoretical studies has not been done, and subsequent application studies have taken reference without confirming the practical validity of these studies [16]. For this reason, this study examines the effectiveness of gamification practices and under what factors it is shaped. In order to see the direct and indirect effects of the determined factors on each other, the analysis was done by structural equation modeling. As the research universe, the aviation sector, where gamification studies are still few, has been preferred.

3 Proposed Model and Results

In the proposed model, it is estimated that gamification will have an impact on the performance factor with the effect of business suitability, contribution to gamification and gamification design factors. In addition, it was estimated that the design of gamification applications will have an impact on performance through business suitability. While determining the factors, gamification literature and nature of the games were used. According to this; It is envisaged that a person involved in the game will be more active than a person who does not adopt the game, and this activity will be felt under the influence of gamification. In addition, past studies show that a gamification design that is not suitable for the enterprise will have a negative impact on achieving the expected results of the application. For this reason, the gamification studies are suitable for the company, the design is made specific to the company and the participants' adoption of this application will result in the final effect of gamification. The proposed model is shown in Fig. 1.

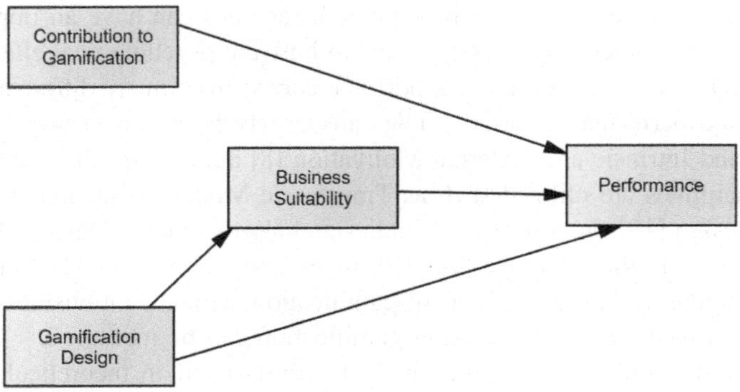

Fig. 1. Proposed model

The study was carried out on the employees of the operation unit in an internationally operating air transport company. The data for the model created was collected through a survey consisting of 18 expressions. The survey consists of three expressions for each gamification factor and nine expressions for the performance factor. For the measurement statements, the participants were asked to evaluate between 1 and 7 how much they participated in each statement. Before the main study, a pilot study was carried out, and the belonging of the expressions to the determined factors was checked with the Exploratory Factor Analysis conducted through the IBM SPSS 23 program. After the factor structures were confirmed, data collection was continued, 166 feedbacks were received from 280 employees to whom the survey was directed, and 162 feedbacks were included in the analysis. The data obtained was tested with Confirmatory Factor Analysis in IBM SPSS Amos 24 program. Maximum Likelihood approach was used in the analysis. The fit index values obtained as a result of the analysis are given in Table 1. When the fit indices are checked, it is seen that the values are appropriate.

Table 1. Fit index of model (CFA)

CMIN	*df*	P	CMIN/*df*	RMSEA	GFI	NFI	IFI	TLI	CFI
238,734	126	,000	1,895	,075	,870	,908	,955	,944	,954

When the results of the analysis were analysed, it was found that the regression weights, variance and covariance values were statistically significant in the 95% confidence interval. The Standardized Regression Weights obtained according to this are given in Table 2. Accordingly, it is seen that the lowest factor weight is 0,652 and it is acceptable.

Correlations among constructs, convergent validity, and discriminant validity are given in Table 3. The highest correlation appears to be between suitability and performance. Considering the fit indices and validity results, it is seen that the model represents the data well.

Table 2. Standardized regression weights

			Estimate
Contribution1	<—	Contribution	,923
Contribution2	<—	Contribution	,809
Contribution3	<—	Contribution	,842
Suitability1	<—	Suitability	,729
Suitability2	<—	Suitability	,900
Suitability3	<—	Suitability	,734
Design1	<—	Design	,652
Design2	<—	Design	,907
Design3	<—	Design	,842
Performance1	<—	Performance	,831
Performance2	<—	Performance	,808
Performance3	<—	Performance	,842
Performance4	<—	Performance	,837
Performance5	<—	Performance	,872
Performance6	<—	Performance	,946
Performance7	<—	Performance	,946
Performance8	<—	Performance	,756
Performance9	<—	Performance	,809

Table 3. Correlations among constructs, convergent validity and discriminant validity

	CR	AVE	Performance	Contribution	Suitability	Design
Performance	0,959	0,725	**0,852**			
Contribution	0,894	0,738	0,435	**0,859**		
Suitability	0,833	0,627	0,701	0,285	**0,792**	
Design	0,847	0,652	0,395	0,494	0,357	**0,808**

After Confirmatory Factor Analysis, the structural model was tested by going to the path analysis. The analysis was carried out through the IBM SPSS AMOS 24 program and Maximum Likelihood approach was used in the analysis. The fit indices obtained as a result of the analysis are given in Table 4.

Table 4. Fit index of structural model

CMIN	*df*	P	CMIN/*df*	RMSEA	GFI	NFI	IFI	TLI	CFI
240,736	127	,000	1,896	,075	,872	,908	,954	,944	,954

When the analysis results are examined, it is seen that the design factor has no significant effect on the performance factor in the 95% confidence interval. Contribution and suitability factors have a significant effect on performance. Standardized Path Estimates is given in Fig. 2.

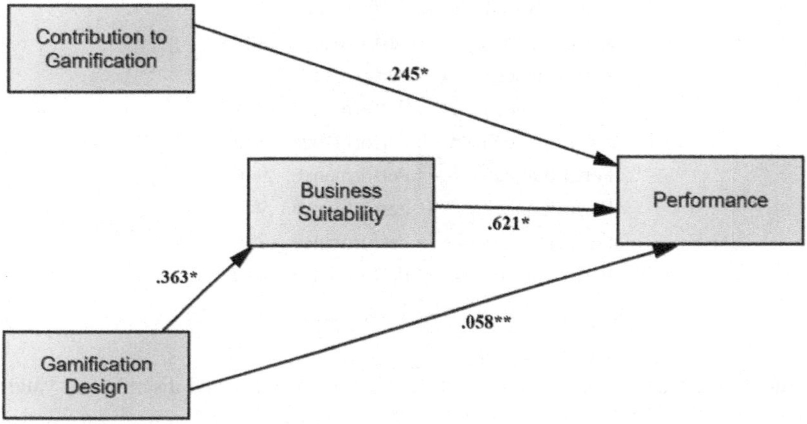

Fig. 2. Results of model with standardized path estimates. *P-VALUE < 0.05 **P-VALUE = .466

According to the results, the design factor does not have a direct effect on performance. However, when the values are analysed, it is seen that the suitability factor has a high effect on performance. In addition, it was determined that the design factor had a significant effect on the suitability factor. Accordingly, the design factor can have an indirect effect on the performance factor through suitability. In order to answer this question, mediation analysis was carried out through the SPSS Process 3.5 plugin created by Hayes [16]. In the analysis, 95% confidence interval and 10000 Bootstrap sampling were used. Analysis results are given in Table 5. Accordingly, although the design factor does not have a direct effect on performance, it is seen that it has a significant indirect effect through the suitability factor.

Table 5. Results of mediation analysis

Total Effect of Design on Performance							
Effect	se	t	p	LLCI	ULCI	C_ps	C_cs
,2570	,0743	3,4605	,0007	,1103	,4037	,1841	,2603
Direct Effect of Design on Performance							
Effect	se	t	p	LLCI	ULCI	C'_ps	C'_cs
,0987	,0640	1,5429	,1249	-,0277	,2251	,0707	,1000

Indirect effect(s) of Design on Performance				
	Effect	BootSE	BootLLCI	BootULCI
Suitability	,1583	,0539	,0603	,2731
Partially standardized indirect effect(s) of Design on Performance				
	Effect	BootSE	BootLLCI	BootULCI
Suitability	,1134	,0370	,0451	,1903
Completely standardized indirect effect(s) of Design on Performance				
	Effect	BootSE	BootLLCI	BootULCI
Suitability	,1603	,0521	,0629	,2653

4 Conclusions and Recommendations

This paper tests the impact of gamification applications on employee performance for an air transport company. It is important to test the validity of gamification applications due to the difficult integration process and high installation and fix costs. Since the performance outputs have a direct effect on revenue, the study was carried out on the operation department. With this work; gamification is accepted to be composed of gamification design, contribution to gamification and business suitability factors, and the effect of these factors on performance factor was observed. According to the results of the analysis, the following results were obtained:

- Contribution to gamification has a direct positive effect on performance.
- The suitability of gamification work to the enterprise has a direct positive effect on performance.
- The design of the gamification application does not have a direct impact on performance.
- The design of the gamification application has a direct positive effect on the suitability of the gamification work to the business.
- The design of the gamification application has an indirect positive effect on performance through the suitability of gamification to the business.

Accordingly, the results of the study show that in order for a gamification study to be successful, it is important that the gamification study is suitable for the business. A gamification application that is not suitable for the enterprise may not be successful and may result in waste of resources and time, as well as bring more harm than good. In addition, it is important that the gamification study is adopted by the participants. The results of the study show that although the gamification study is not as big as the business suitability, the contribution of the participants to the gamification work positively affects their performance. This contribution can both accelerate gamification integration processes, reduce resource and time costs, and help the integration to succeed.

The study was carried out in a full-service international air transport company. Other business models such as low-cost, luxury, cargo and charter in the sector were not reflected in the study. Since the aviation industry is a sector with high operating costs, the data collection phase was carried out in a very limited time. Due to this limitation, the study was carried out in only one department, the operation department, which is supposed to have a direct effect on revenue. Other departments such as sales, marketing and human resources were not included in the analysis and the study was limited within a single department. In addition, the questionnaire items of the proposed structure were kept in sufficient number for the analysis method. A questionnaire that would increase the detail of the study but lead to high costs in the business process was avoided.

Despite the limited data collection phase, the structural equation model allowed the limited data to be analysed in detail and to control the relationship between factors with different reliability methods. With structural equation modelling, the relationships between the items in the measurement tools and the errors of each item could be observed. In addition, not only direct effects, but also indirect effects and mediation relations could be examined in detail and quickly. In this way, the analysis results obtained were examined comprehensively and were reached with the control mechanisms provided by the selected analysis method and paved the way for not only a single prediction but also potentially different models. This contributed to the enrichment of the study.

For the future studies, the proposed model can be tested for airline companies with different business models such as low-cost, cargo and luxury, and its validity can be checked. Repeating the research on different business models will contribute to demonstration of industry-wide validity. As noted, a single enterprise and a single department were used. Repeating the analysis on different departments or including different businesses in the analysis will enrich the study. In addition, determining the sub-dimensions of gamification factor in detail and repeating the study with these sub-dimensions may also benefit the literature. In particular, taking into account the time factor and increasing the number of sub-items of the established structure will enrich the study and provide a more detailed validation check.

References

1. Deterding, S., Dixon, D., Khaled, R., Nacke, L.: From game design elements to gamefulness: defining "gamification". In: Proceedings of the 15th International Academic MindTrek Conference: Envisioning Future Media Environments, pp. 9–15. ACM, New York (2011). https://doi.org/10.1145/2181037.2181040
2. Werbach, K., Hunter, D.: For the Win: How Game Thinking Can Revolutionize Your Business. Wharton Digital Press, Philadelphia (2012)

3. Klevers, M., Sailer, M., Günthner, W.: Implementation model for the gamification of business processes: a study from the field of material handling (2016). https://doi.org/10.1007/978-981-10-0575-6_14
4. Chou, Y.K.: Actionable Gamification: Beyond Points, Badges, and Leaderboards, pp. 23–33. Octalysis Media, Fremont (2015)
5. van Roy, R., Zaman, B.: Unravelling the ambivalent motivational power of gamification: a basic psychological needs perspective. Int. J. Hum Comput Stud. **127**, 38–50 (2019). https://doi.org/10.1016/j.ijhcs.2018.04.009
6. Morschheuser, B., Hamari, J., Maedche, A.: Cooperation or competition – when do people contribute more? A field experiment on gamification of crowdsourcing. Int. J. Hum Comput Stud. **127**, 7–24 (2019). https://doi.org/10.1016/j.ijhcs.2018.10.001
7. Ryan, R.M., Deci, E.L.: Self-determination theory and the facilitation of intrinsic motivation, social development, and well-being. Am. Psychol. **55**(1), 68–78 (2000). https://doi.org/10.1037/0003-066x.55.1.68
8. Ryan, R.M., Deci, E.L.: Intrinsic and extrinsic motivations: classic definitions and new directions. Contemp. Educ. Psychol. **25**(1), 54–67 (2000). https://doi.org/10.1006/ceps.1999.1020
9. Prochaska, J.O., Marcus, B.H.: The transtheoretical model: applications to exercise. In: Dishman, R.K. (ed.) Advances in Exercise Adherence, pp. 161–180. Human Kinetics Publishers, Champaign (1994)
10. Rose, D.H., Meyer, A.: Teaching Every Student in the Digital Age: Universal Design for Learning. Association for Supervision and Curriculum Development, Alexandria (2002)
11. Norman, D.A.: The Design of Everyday Things, 2nd edn. Basic Books, New York (1988)
12. Nicholson, S.: A user-centered theoretical framework for meaningful gamification. In: Proceedings of Games Learning Society 8.0, Madison, WI (2012)
13. Rapp, A., Hopfgartner, F., Hamari, J., Linehan, C., Cena, F.: Strengthening gamification studies: current trends and future opportunities of gamification research. Int. J. Hum Comput Stud. **127**, 1–6 (2019). https://doi.org/10.1016/j.ijhcs.2018.11.007
14. Zuo, L., Xiong, S., Wang, Z., Iida, H.: An analysis of gamification effect of frequent-flyer program (2018). https://doi.org/10.1007/978-3-030-23712-7_8
15. Seaborn, K., Fels, D.I.: Gamification in theory and action: a survey. Int. J. Hum Comput Stud. **74**, 14–31 (2015). https://doi.org/10.1016/j.ijhcs.2014.09.006
16. Hayes, A.F.: Introduction to Mediation, Moderation, and Conditional Process Analysis: A Regression-Based Approach. Guilford, New York (2013)

Simulation and Modelling

A Brief General Industrial Virtual Reality Applications Overview Including a Simple Factory VR Simulation

Cem Yurci[1(✉)] and Numan M. Durakbasa[2]

[1] Department of Mechanical Engineering, Yildiz Technical University, Istanbul, Turkey
yurci_cem@yahoo.com
[2] Research Group Industrial Metrology and Adaptronic Systems, Institute for Production Engineering and Photonic Technologies, TU Wien, Getreidemarkt 9, Obj. 1, Vienna, Austria
numan.durakbasa@tuwien.ac.at

Abstract. Today, all enterprises are trying to shift their management and manufacturing processes to the new innovations according to the Revolution Industry 4.0. Among those are simulation and digitation which lead to digital twinning. For digital twinning, virtual, augmented, mixed and extended reality play the key role with the advancements in haptic devices and VR glasses. Those operations are used for many purposes and areas, for example by training or manufacturing organisation. Accordingly, a simple factory planning simulation is created in a VR environment with the physics engine Unity 3D. It can be accepted as a start point to reach to the importance conclusions of VR usage in industrial applications. So, it is aimed to emphasize the powerful and effective sides of VR today in technical applications through a simple instance for reaching the future conclusions and suggestions.

Keywords: Virtual Reality · Industry 4.0 · Digital twinning · Virtual Manufacturing · Education · Analysis · Factory planning

1 Introduction

The virtual world is defined by Damer [1] as *'a place described by words or projected through pictures which creates a space in the imagination real enough that you can feel you are inside of it'*. Its duration, started with visual books, theatre plays or movies, spread to digital computers and opened the doors of Virtual Reality applications in many fields. One of these fields is the Industry which is in the purposed scope of that study. Virtual Reality is a continuously developing area with the aids of technology which enables preventing of negative effects of VR as well.

1.1 Necessity of VR According to Industry 4.0

Today, companies have to adapt themselves to the novel strategic standards of the world to sustain their presence and are striving in that way. The most recent and valid

N. M. Durakbasa and M. G. Gençyılmaz (Eds.): ISPR 2020, LNME, pp. 723–735, 2021.
https://doi.org/10.1007/978-3-030-62784-3_61

industrial trend is Industry 4.0 which is leading to Industry 5.0 with following its own fundamentals. Intelligent manufacturing supports Industry 4.0 to achieve its goals such as raised flexibility in manufacturing, mass customization, higher quality, and improved productivity [2]. Basl [3] ranks 'extended reality', 'digitation' and 'simulation' in the sub-trends of Industry 4.0 for production, information and health technology. In addition, he emphasizes the importance of reaching to a maturity level in Industry 4.0 for a smart enterprise. Lasi et al. [4] are mentioning simulation, virtual and augmented reality as technological drivers. For example, today firms can use augmented reality to provide real-time information for workers efficiently by decision making and work procedures such as repairing [3]. During VR and AR implementations, 'visualisation' is the main object and method which lead to 'visual computing' with the support of 'artificial intelligence' [5]. VR also has changed the approach to 3D modelling and the way of scientists by perform mathematical simulations, data visualization and decision making by utilizing from computers [6]. That is valid especially for industrial engineers by VRP which is a requirement of market and its efficient contribution reduces the time to time market, innovation risks and increases the innovation head start and manufacturability [6]. For example, VR, AR and remote control have an AHP weight of 10% as a tool by providing the benefit 'rapid time to market' despite to the challenge of the need of government support and promotion, according to Erbay and Yıldırım [7]. Model-based manufacturing approach also uses virtual reality (VR) as advanced technology for 3D presentation of model structure, composition [8]. So, VR provides important potential in different engineering applications, such as product design, modelling, shop floor controls, process simulation, manufacturing planning, training, testing and verification through reducing physical try-outs [8].

The impacts of Industry 4.0 are inevitable for lean production systems as well with the aids of VR And AR prominently in human-machine interaction, as claimed in the study of Wagner et al. [9]. However, Götz and Jankowska [10] describe in their work the necessity and way of the integration of virtual and real production subsystems including their valid role-specific interaction. For instance, the opposition and diversity can be observed by the proposition of Beard-Gunter et al. [11] to emplace TQM (Total Quality Management) in gamification. The significance of games has been increasing in industrial application and education in a fast manner with a big amount. So, Contreras et al. [12] suggested using of serious games by the development of the competencies in industrial processes.

1.2 Digital Twinning for the Visualization Process

'Big Data' and 'Digital Twin' are two main elements of 'Smart Manufacturing' [13] in which Big Data interacts on digital twinning and assists it even by determining the role and efficiency of visual data transfer. Digital Twins creating technology is accepted as one of the most critical sub-trends in the digital transformation of the industry according to Luściński [14]. The prototyping of Digital Twins systems is being implemented for the concept of Digital Twins by utilizing from existing IT technologies. Even the Industry 4.0 applications are secured with Cyber-Ranges and Digital Twins to the provide the resilience and resist manipulation attempts and attacks to the general production systems [15]. For a Cyber-Physical System example, a digital-twin-simulation consisting a model equipped with sensor as the physical system has been carried out by Yang et al. [16]. Other digital twinning samples can be seen in the studies of Stojanovic et al. [17] and Umer [18] via the sematic enrichment of indoor point clouds and a data management system for digital twinning of power systems, respectively. Moreover, Masood and Edger [19] put forward that organisational issues are very important besides technological aspects for Industrial AR to implement challenges and to reach to the success factors of Industry 4.0.

2 A Brief Literature Review for Applications of VR

2.1 Industrial VR Applications

VR applications have started to cover an important place in many industry sectors. A significant part of a process beginning with logistics maintaining with production and ending with marketing and sell operations can be conducted by VR utilization. For instance, in the study of Moreno et al. [20] a virtual model of a sheet metal punching machine has been created. And that twin can connect to the virtual production line. Moreover, Mujber et al. [21] gave the virtual machine tool sample (Fig. 1) for VR applications in manufacturing process simulation. Besides, the research work of Teklemariam and Das [22] discussed VR and haptic feedback device technologies and their applications in various design areas with a virtual product design case of a phantom omni force feedback device [23].

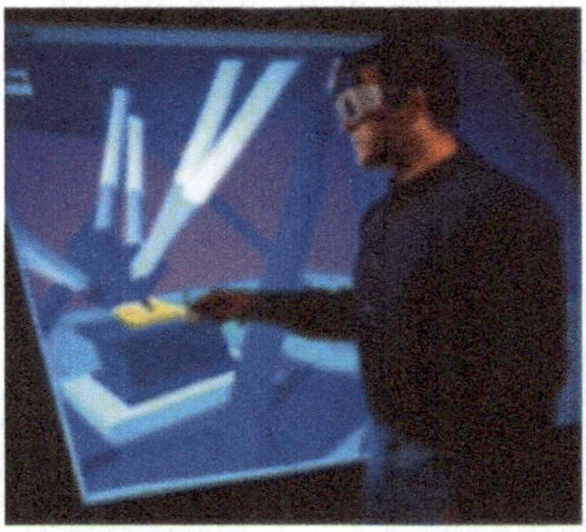

Fig. 1. VR for a machine tool [21]

Numerical FEM results can also be simulated and displayed in the virtual world. The study of Ingrassia and Cappello [24] presents a detailed research for the simultaneous combination of CAD, FEM and virtual environment by introducing a new virtual reality design approach. Such an instance can be met in the work of Zhou et al. [25] for a virtual injection molding machine. Berg and Vance [26] prepared a survey investigating the industrial use of VR in product design and manufacturing. For example, they gave place to an ergonomics Case New Holland VR Laboratories

(Fig. 2) by which VR is used to evaluate reachability of an exterior door handle within a vehicle.

Fig. 2. Checking out the seat place of a tractor through VR Visualisation [26]

According to one of the earlier studies, Shukla [27] counted Computer Aided Design, Computer Aided Manufacturing and Computer Aided Process Planning in the scope of Virtual Manufacturing while it is presenting an integrated, synthetic manufacturing environment. Also, Jayaram et al. [28] accepts CAD Environment as one of the first VR Applications with its all assembly design processes. Schelkle and Elsenhans [29] assert that such methods as load-path analysis, parametric geometry modelling, stochastic concept and prototype optimization will contribute to the concept design process of CAE for virtual vehicle development. Also, Carvajal et al. [30] created the autonomous and tele-operated navigation of an in CAD designed aerial manipulator robot (Fig. 3a) in digitalized virtual environments (Fig. 3b). The Unity 3D graphics engine has been used by them for robot-environment interaction.

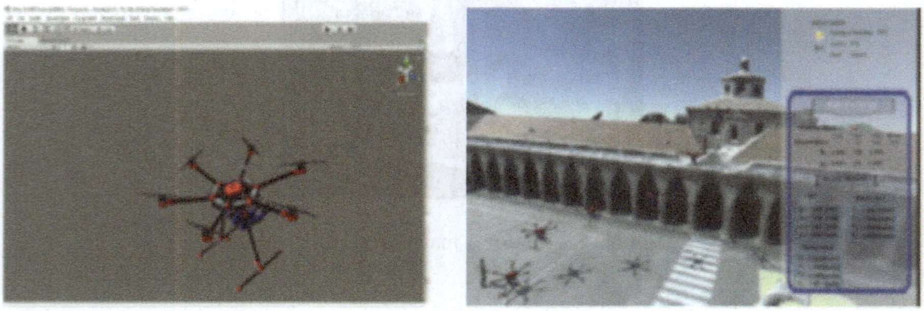

(a) (b)

Fig. 3. a) CAD model of the aerial manipulator imported to Unity 3D b) Autonomous control operation for positioning the aerial manipulator in the virtual environment [30].

2.2 VR Applications in Industrial Education

Training is a wide application area for VR. When it is connected with and included in Industry 4.0 it can be referred to Education 4.0. Many instances have been being observed for VR-AR-Mixed Reality training applications recently whose numbers are increasing rapidly with visual IT technological developments. Pan et al. [31] present in their study several virtual learning environments from war scenarios to storytelling (Fig. 4).

Fig. 4. An interactive VR storytelling environment [31]

The importance of gaming for Industry 4.0 is increasing to achieve more suitable qualifications for workers according to Industry 4.0 and new Trade World. Zarte [32] introduced a training concept related with ERP for non-IT students in the format of VR simulation game. Before that, a web-based interactive manufacturing engineering module including VR simulations and animations have been prepared by Ong and Mannan [33] to train students in NC programming and operations. So, the students will not need to work on physical-real NC machines in the laboratory. Additionally, there are many engineering training VR implementations, such as in the study of Richert et al. [34] investigating the human -robot interaction in the virtual world via 'Oculus Rift' glasses. Followingly, they mentioned about created virtual spaces or experimental setups including some scenarios. Also, a VR interactive training environment prototype has been formed by Goulding et al. [35] with its own procedure steps and scenario schemes. Another interactive VR based e-Education System have been instantiated by Chan and Lau [36] with the simulations of query crane operations and an air cargo terminal (Fig. 5).

Two scenarios, extracted from [37] and [38], are illustrated in Fig. 6. While the one (Fig. 6a) is for an accident simulation to train the process-industry operators, the other one (Fig. 6b) is from mining industry providing simulated exposure to real-world working conditions without the associated accident risks.

As it has been mentioned, serious games and gamification are important in Education 4.0. That is asserted by Almeida and Simoes [39] as well when it is especially connected with Industry 4.0 tools. An example for gaming in a different area can be seen in Fig. 7 illustrated in the work of Zyda [40]. It has been proved that soldiers, who played the rifle range segment of that game, achieved higher scores on the real-life rifle range.

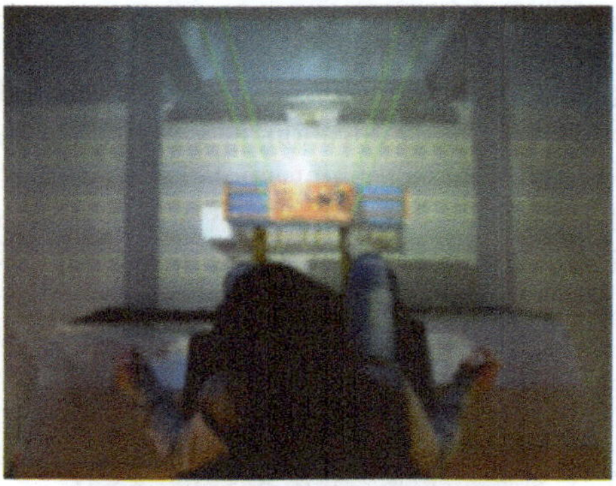

Fig. 5. A simulation example of query Crane operations: An operator using a crane simulator to perform container unloading operations [36]

(a) (b)

Fig. 6. Several industrial training VR applications [37, 38]

Fig. 7. A military training simulator [40]

During the design of AR simulations, some signs have to be created and inserted into them. They can be called as standard symbols for converting maintenance actions and a simple can be seen in Fig. 8. From the study of Scurati [41].

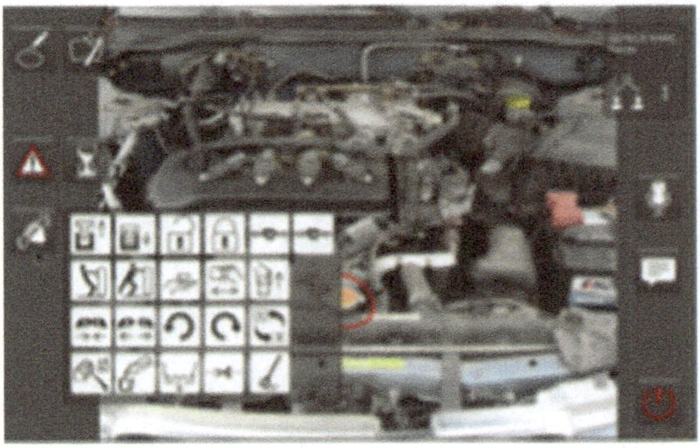

Fig. 8. VR simulation (application on server) with an alternative to select the symbol according to the desired action [41]

2.3 Other General VR Applications

VR and AR simulations are utilised for many sectors and fields with several objects. One of popular areas where VR is used is the medicine sector. A history of medical simulation is given in the research work of Rosen [42]. Grantcharov [43] shows an example for virtual reality simulation for laparoscopic skills training. As an advanced VR training application Vite et al. [44] composed a virtual simulation for neurosurgical training trough Brain Sylvian Fissure Exploration and Aneurysm Clipping. During the procedure, they used FEM and FFD (Free Form Deformation) techniques by surrounding the geometric model of the brain with a biomechanical space defined as a set of hexahedrons. After that, deformation tests of the brain model in the Sylvian fissure region are applied to simulate a microsurgical procedure in an interaction with a haptic device, as shown in Fig. 9. Even hospital patient rooms can be created by VR mock-up design [45] (Fig. 10).

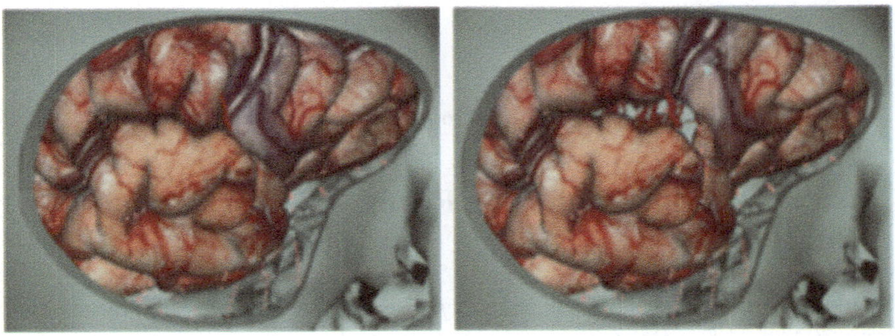

Fig. 9. Soft-tissue deformation Simulation of the brain: the Sylvian fissure at closed and opened states [44]

Fig. 10. The patient's view by immersive VR [45]

Guttentag [46] prepared a detailed review in his study of applications and implications for tourism informing from planning and management to heritage preservation. Even, Wang et al. [47] defined a virtual tourist community for tourism marketing in the travel industry trough Internet as a different and apart application. Besides, VR offers major, eventual application solutions for furnishing, home furnishing and textile industry, too [48] (Fig. 11). In addition to that, digital olympic museum and sports simulation can be created in virtual environment with its all components [49].

Fig. 11. Virtual room prototype [48]

3 Experimental Procedure and Results of VR Workshop Simulation

A workshop or small factory has been organised in the 3D software-physics engine 'Unity 2018.4.23.f1'. Related assets including machines, weld robot, workman and staff have been imported for the Asset Store. They have been placed rotated, positioned and scaled according to the design properties. A sparkle particle effect and audio have been added for the welded metal part with necessary adjustments. For VR attachment, related GvrViewerMain has been transferred to the hierarchy system. The whole simulation has been so organised, integrated and created. In Fig. 12, a creating phase has been illustrated. The simulation result and VR appearance have been achieved (Figs. 13 and 14).

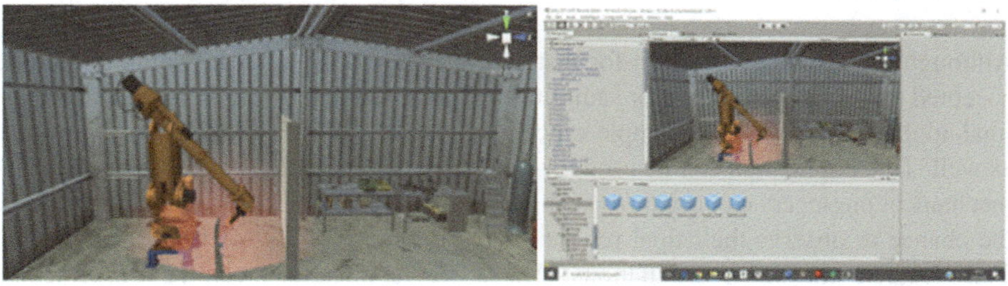

Fig. 12. Creation of the VR simulation in Unity 3D

Fig. 13. Several views of the designed workshop including imported assets

Fig. 14. VR application appearance for VR glasses applied and viewed in the Game option of Unity 3D

The achieved VR simulation has been transported to an Android application file with an extension '.apk'. For that, a Virtual Reality SDK (Cardboard) has been added in Player Settings in Unity and from minimum api levels 'Android 9.0 Pie (API Level 28) has been selected. Finally, the related '.apk' file has been built and run in the

Android device with a USB connection after making related developer settings in the Android device.

4 Conclusions

Although the virtual application for that study contains a simple simulation, it can be accepted as a good beginning for sampling and commenting VR implementations. That final application can be for example watched and navigated in itself through VR glasses as well. That application can be used for weld operator trainings. Besides industrial and mechanical engineers can organize such factories or workshops by themselves to have the chance to observe their final results lively. They also can import their self-created CAD objects into their simulations. Those VR applications can be implemented additionally by professional enterprises to obtain a foresight by factory planning with more realistic assets.

The importance and usage of VR applications will be increasing continuously. Those are leading today AR and extended reality with the rapid progress in haptics devices. With adding of digital twinning and usage of professional innovative glasses the operational and technical problems can be solved recently with such applications. The innovation in those areas and the sustaining of those applications in Industry 5.0 as well depends on fast internet connections, such as 5G. The industrial world is able to transform to a digital and visual world with AR simulations and their applications which will rise the economic profits by providing easiness and reduction of time consumption.

The future studies will depend on developing virtual simulations even by utilizing from FEM analyses. Other applications in several fields can be generated as well. The strong connection between digital and Industry 4.0 will be being researched.

References

1. Damer, B.: Meeting in the ether: a brief history of virtual worlds as a medium for user-created events. Artifact **2**, 94–107 (2009)
2. Zhong, R.Y., Xu, X., Klotz, E., Newman, S.T.: Intelligent manufacturing in the context of Industry 4.0: a review. Engineering **3**, 616–630 (2017)
3. Basl, J.: Companies on the way to Industry 4.0 and their readiness. J. Syst. Integr. **3**, 3–6 (2018)
4. Lasi, H., Kemper, H.-G., Fettke, P., Feld, T., Hoffmann, M.: Industrie 4.0. Wirtschtsinformatik, June 2014
5. Rüßmann, M., Lorenz, M., Gerbert, P., Waldner, M., Justus, J., Engel, P., Harnisch, M.: Industry 4.0: the future of productivity and growth in manufacturing industries, pp. 1–14, April 2015
6. Dönmezer, S.: Integrations management and product development for new product. In: Durakbasa, N.M., Gencyilmaz, M.G. (eds.) Proceedings of International Symposium for Production Research 2018, pp. 292–308. Springer, Cham (2019)

7. Erbay, H., Yıldırım, N.: Technology selection for digital transformation: a mixed decision making model of AHP and QFD. In: Durakbasa, N.M., Gencyilmaz, M.G. (eds.) Proceedings of International Symposium for Production Research 2018, pp. 480–493. Springer, Cham (2019)
8. Mandic, V.: Model-based manufacturing system supported by virtual technologies in an Industry 4.0 context. In: Wang, L., Majstorovic, V.D., Mourtzis, D., Carpanzano, E., Moroni, G., Galantucci, L.M. (eds.) Proceedings of the 5th International Conference on Industry 4.0 Model for Advanced Manufacturing, AMP 2020, pp. 215–226. Springer, Cham (2020)
9. Posada, J., Toro, C., Barandiaran, I., Oyarzun, D., Stricker, D., de Amicis, R., Pinto, E.B., Eisert, P., Döllner, J., Vallarino Jr., I.: Visual computing as a key enabling technology for Industrie 4.0 and industrial internet. IEEE Comput. Graph. Appl. **35**(2), 26–40 (2015)
10. Wagner, T., Herrmann, C., Thiede, S.: Industry 4.0 impacts on lean production systems. In: The50th CIRP Conference on Manufacturing Systems-Procedia CIRP, vol. 63, pp. 125–131 (2017)
11. Götz, M., Jankowska, B.: Clusters and Industry 4.0 – do they fit together? Eur. Plan. Stud. **25**, 1633–1653 (2017)
12. Gunter, A.B., Ellis, D.G., Found, P.A.: TQM, games design and the implications of integration in Industry 4.0 systems. Int. J. Qual. Serv. Sci. **11**(2), 235–247 (2019)
13. Contreras, M., Bonilla, D., Negrón, A.P.P.: Proposal of a model for the development of labor competencies based on serious games in the context of Industry 4.0. In: CIMPS 2018, pp. 80–87 (2019)
14. Qi, Q., Tao, F.: Digital twin and big data towards smart manufacturing and Industry 4.0: 360 degree comparison. IEEE Access **6**, 3585–3593 (2018)
15. Luściński, S.: Digital twinning for smart industry. In: MMS 2018 (2018)
16. Bécue, A., Savarit, A., Pouille, E., Fourastier, Y., Baron, C., Thomas, C., Praça, I., Gradussofs, B.: CyberFactory#1 – securing the Industry 4.0 with cyber-ranges and digital twins. IEEE (2018)
17. Yang, W., Tan, Y., Yoshida, K., Takakuwa, S.: Digital twin-driven simulation for a cyber-physical system in Industry 4.0. In: DAAAM International Scientific Book, pp. 227–234 (2017)
18. Stojanovic, V., Trapp, M., Richter, R.: Semantic enrichment of indoor point clouds an overview of progress towards digital twinning. In: Challenges - Big Data and Analytics, vol. 2, pp. 809–818
19. Umer, A.: Conception, development and evaluation of a data management system for digital twinning of power systems. M.Sc. thesis, Politecnico Milano, Milano, Italy
20. Masood, T., Egger, J.: Augmented reality in support of Industry 4.0—implementation challenges and success factors. Robot. Comput. Integr. Manuf. **58**, 181–195 (2019)
21. Moreno, A., Velez, G., Ardanza, A., Barandiaran, I., de Infante, A.R., Chopitea, R.: Virtualisation process of a sheet metal punching machine within the Industry 4.0 vision. Int. J. Interact. Des. Manuf. **11**(2), 365–373 (2016)
22. Mujber, T.S., Szecsi, T., Hashmi, M.S.J.: Virtual reality applications in manufacturing process simulation. J. Mater. Process. Technol. **155–156**, 1834–1838 (2004)
23. Teklemariam, H.G., Das, A.K.: A case study of phantom omni force feedback device for virtual product design. Int. J. Interact. Des. Manuf. (2015)
24. Ingrassia, T., Cappello, F.: VirDe: a new virtual reality design approach. Int. J. Interact. Des. Manuf. **3**, 1–11 (2019)
25. Zhou, H., Shi, S., Ma, B.: A virtual injection molding system based on numerical simulation. Int. J. Adv. Manuf. Technol. **40**, 297–306 (2009)

26. Berg, L.P., Vance, J.M.: Industry use of virtual reality in product design and manufacturing: a survey. Virtual Reality **21**(1), 1–17 (2016)
27. Shukla, C., Vazquez, M., Chen, F.F.: Virtual manufacturing: an overview. Comput. Ind. Eng. **31**, 79–82 (1996)
28. Jayaram, S., Connacher, H.I., Lyons, K.W.: Virtual assembly using virtual reality techniques. Comput. Aided Des. **29**, 575–584 (1997)
29. Schelkle, E., Elsenhans, H.: Virtual vehicle development in the concept stage -current status of CAE and outlook on the future-. In: 3RD MSC Worldwide Aerospace Conference & Technology Showcase, pp. 1–11 (2011)
30. Carvajal, C.P., Méndez, M.G., Torres, D.C., Terán, C., Arteaga, O.B., Andaluz, V.H.: Autonomous and tele-operated navigation of aerial manipulator robots in digitalized virtual environments. In: Paolis, L.T.D., Bourdot, P. (eds.) 5th International Conference, AVR 2018 Proceedings, Augmented Reality, Virtual Reality, and Computer Graphics, Part II, pp. 496–515. Springer, Cham (2018)
31. Pan, Z., Cheok, A.D., Yang, H., Zhu, J., Shi, J.: Virtual reality and mixed reality for virtual learning environments. Comput. Graph. **30**, 20–28 (2006)
32. Zarte, M., Pechmann, A.: Concept for introducing the vision of Industry 4.0 in a simulation game for non-IT students, pp. 512–517. IEEE (2017)
33. Ong, S.K., Mannan, M.A.: Virtual reality simulations and animations in a web-based interactive manufacturing engineering modüle. Comput. Educ. **43**, 361–382 (2004)
34. Richert, A., Shehadeh, M., Plumanns, L., Groß, K., Schuster, K., Jeschke, S.: Educating engineers for Industry 4.0: virtual worlds and human-robot-teams empirical studies towards a new educational age. In: 2016 IEEE Global Engineering Education Conference (EDUCON), pp. 142–149 (2016)
35. Goulding, J., Nadim, W., Petridis, P., Alshawi, M.: Construction industry offsite production: a virtual reality interactive training environment prototype. Adv. Eng. Inform. **26**, 103–116 (2012)
36. Chan, L.K.Y., Lau, H.: An Interactive Virtual Reality based e-Education System (2014)
37. Manca, D., Brambilla, S., Colombo, S.: Bridging between virtual reality and accident simulation for training of process-industry operators. Adv. Eng. Softw. **55**, 1–9 (2013)
38. van Wyk, E., de Villiers, R.: Virtual reality training applications for the mining industry, pp. 53–64. ACM (2009)
39. Simoes, F.A.J.: The role of serious games, gamification and Industry 4.0 tools in the Education 4.0 paradigm. Contemp. Educ. Technol. **10**, 120–136 (2019)
40. Zyda, M.: From visual simulation to virtual reality to games, pp. 25–32. IEEE Computer Society (2005)
41. Scurati, G.W., Gattullo, M., Fiorentino, M., Ferrise, F., Bordegoni, M., Uva, A.E.: Converting maintenance actions into standard symbols for Augmented Reality applications in Industry 4.0. Comput. Ind. **98**, 68–79 (2018)
42. Rosen, K.R.: The history of medical simulation. J. Crit. Care **23**, 157–166 (2008)
43. Grantcharov, T.P., Kristiansen, V.B., Bendix, J., Bardram, L., Rosenberg, J., Jensen, P.F.: Randomized clinical trial of virtual reality simulation for laparoscopic skills training. Br. J. Surg. **91**, 146–150 (2004)
44. Vite, S.T., Velasco, C.D., Valencia, A.F.H., Lomel, J.S.P., Casta'neda, M.A.P.: Virtual simulation of brain Sylvian fissure exploration and aneurysm clipping with haptic feedback for neurosurgical training. In: De Paolis, L.T., Bourdot, P. (eds.) 5th International Conference, AVR 2018 Proceedings, Augmented Reality, Virtual Reality, and Computer Graphics, Part II, pp. 230–238. Springer, Cham (2018)

45. Dunston, P.S., Arns, L.L., Mcglothlin, J.D., Lasker, G.C., Kushner, A.G.: An immersive virtual reality mock-up for design review of hospital patient rooms. In: Wang, X., Tsai, J.J.-H. (eds.) Collaborative Design in Virtual Environments, ISCA, vol. 48, pp. 167–176. Springer, Dordrecht (2011)
46. Guttentag, D.A.: Virtual reality: applications and implications for tourism. Tour. Manag. **31**, 637–651 (2010)
47. Wang, Y., Yu, Q., Fesenmaier, D.R.: Defining the virtual tourist community: implications for tourism marketing. Tour. Manag. **23**, 407–417 (2002)
48. Oh, H., Yoon, S.Y., Hawley, J.: What virtual reality can offer to the furniture industry. J. Text. Appar. Technol. Manag. **4**(1), 1–17 (2004)
49. Pan, Z., Yun, R.: Digital olympic museum and sports simulation. In: Ma, D., Gausemeier, J., Fan, X., Grafe, M. (eds.) Virtual Reality & Augmented Reality in Industry, pp. 121–132. Springer, Heidelberg (2009)

Investigation of the Efficiency of Vibration-Assisted Nano-Grinding with Molecular Dynamics

Nikolaos E. Karkalos[1], Angelos P. Markopoulos[1(✉)], and János Kundrák[2]

[1] School of Mechanical Engineering, Laboratory of Manufacturing Technology, National Technical University of Athens, Heroon Politechniou 9, 15780 Athens, Greece
{nkark, amark}@mail.ntua.gr

[2] Institute of Manufacturing Science, University of Miskolc, Miskolc, Miskolc-Egyetemváros 3515, Hungary
kundrak@uni-miskolc.hu

Abstract. The use of ultrasonic vibrations during grinding can lead to favorable results, such as easier processing of difficult-to-cut materials, improved surface quality, reduced workpiece damage and cutting forces. In order to investigate the efficiency of this technique to nanometric grinding, simulations are essential to be initially conducted. Thus, in the present work, a Molecular Dynamics model of vibration-assisted nano-grinding is introduced, with a view to determine the efficiency of this process under various conditions, by comparing its outcome to that of a conventional nano-grinding process via analyzing grinding forces, temperatures and subsurface damage of the workpiece.

Keywords: Vibration-assisted grinding · Nano-grinding · Molecular dynamics · Vibration frequency · Vibration amplitude

1 Introduction

Surface quality and integrity are among the most important requirements for mechanical components, as they are directly related to product life [1, 2]. For that reason, processes such as precision grinding and polishing are preferred to render high quality surfaces, especially when it is desired to machine difficult-to-cut materials. In the case of machining materials such as titanium alloys, ceramics or single crystal metals, requirements increase and excessive wear is observed, leading to a considerable increase of cost [3]. Several alternative solutions for the treatment of difficult-to-cut materials have been proposed, such as the use of more effective means of cooling or hybrid processes, such as laser-assisted or vibration-assisted machining, in order to facilitate cutting [4, 5]. In particular, vibration-assisted machining or ultrasonic machining possesses the advantages of reduction of cutting forces and burr, as well as increase of machining accuracy [5]. Vibration-assisted processes were first introduced

N. M. Durakbasa and M. G. Gençyılmaz (Eds.): ISPR 2020, LNME, pp. 736–747, 2021.
https://doi.org/10.1007/978-3-030-62784-3_62

during the 1960s [5] and are possible to be applied for abrasive cutting and micro/nano-machining processes, as well [5–8].

In the case of micro and nano-level material removal processes, there are inevitable difficulties, which render the experimental study and direct measurement of physical quantities of the process either impossible or considerably expensive. However, studies regarding nano-level machining process are important to be conducted, with a view to examine phenomena such as friction and deformation mechanisms in detail or assist in improving the efficiency of micromachining processes used in high-end industries such as semiconductor industry. When it is desired to investigate material removal in these scales, it is necessary to conduct numerical simulations with a method capable of representing the phenomena occurring in atomistic level. The Finite Element Method (FEM) is able to be applied until micro-scale level, but cannot account for interactions in lower scales, where matter cannot be treated as continuum. However, the Molecular Dynamics (MD) method has emerged as a viable alternative during the past three decades in the field of nano-manufacturing. Starting from various pioneering works, such as [9], regarding nano-grinding of copper, or the work conducted in [10] on the effect of large negative rake angle tools in nano-cutting, it has evolved into a more detailed means for nano-grinding simulations. An MD model with spherical abrasive grains and random geometry asperities was developed in [11] to examine the diamond polishing for silicon workpieces. Another study [12] was also presented on the phenomenon of material removal rate anisotropy during diamond polishing, with an MD model which enabled the determination of the correlation between material removal rate and the ratio of sp^2 to sp^3 phases on the substrate. Furthermore thorough work on the wear of ground surfaces during polishing, as well as the stability of abrasive grains, was conducted in [13, 14], by using models with randomly generated shape of the workpiece surface and detailed modeling of asperities and abrasive grains. Regarding vibration-assisted cutting, only a few works using MD models have been reported in the relevant literature, mainly on the field of general nano-cutting or nano-scratching models.

An MD simulation of vibration-assisted nano-cutting was first conducted in [15]. Various cases regarding the vibration-assisted cutting of an aluminum substrate were examined with different cutting speed, vibration frequency and amplitude values.

In that preliminary study, it was concluded that the vibrational motion is essential to reduce plastic deformation and cutting forces. In order to achieve a meaningful difference in process outcome, maximum vibrational speed should overcome the magnitude of nominal cutting speed and in that case, cutting force becomes lower with an increase in the vibration frequency. Later, two further studies were conducted by the same scientific group [16, 17] in order to determine the effect of various process parameters on the physical quantities of nano-cutting. In these studies, they found out that, when the product of vibration amplitude and frequency increases, the decrease of cutting forces and plastic flow becomes significant. Moreover, when the maximum vibrational speed exceeds three times the cutting speed, the beneficial effects of vibration are reduced.

The earliest study on elliptical vibration-assisted cutting of silicon substrates was later presented in [18]. After analysis of cutting forces and stress distribution was performed, they noted that cutting forces exhibit more variable behavior due to the vibrational motion and their values are lower than in conventional cutting. Moreover, another MD study on vibration-assisted machining focusing on the analysis of workpiece microstructure and cutting forces was reported in [19]. It was observed that localized plastic deformation occurred periodically and propagation and degeneration of shear bands occurred repeatedly, contributing also to the periodicity of force curve. An MD model was used in [20], in order to evaluate the efficiency of employing ultrasonic elliptical vibration for the cutting of ferrous metals. From the analysis of their results, they underlined the importance of the movement of cutting tool out of the workpiece, which can allow for heat dissipation, as well as reduction of tool wear in the form of diamond graphitization, due to lower cutting temperature. The effect of vibration frequency on the nano-scratching was investigated in [21]. Their model included a spherical indenter and a copper workpiece and three different values for the vibration period were selected. It was determined that, an increase in vibration frequency leads to a decrease of groove roughness, an increase of dislocations and facilitation of workpiece deformation, as well as a reduction of cutting force.

An MD model, regarding the effect of elliptical vibration on nano-cutting of single crystal silicon was employed in [22]. At first, they noticed that vibration-assisted cutting is able to change the cutting mode from brittle to ductile, as no crack formation was observed from atom flow. Furthermore, surface quality and material removal rate increased as well as temperature, but cutting force was reduced. Later, an MD model relevant to the elliptical vibration-assisted nano-cutting of iron was proposed in [23]. With the developed model, they were able to conduct a detailed comparison between conventional and vibration-assisted cutting regarding various quantities of the process. Their findings indicate that, there is a clear difference in the chip formation mechanism and the evolution of plasticity between the two cases. Furthermore, vibration-assisted cutting exhibits considerably lower cutting forces and friction, elimination of surface-initiated wear of the cutting tool, whereas cutting temperature was found to be higher. A study regarding the comparison of 1D vibration-assisted cutting, elliptical vibration-assisted cutting and conventional cutting of single crystal silicon was carried out in [24]. They concluded that, elliptical vibration-assisted cutting increases the material removal rate and surface quality of the workpiece and leads to an increased number of atoms with large displacements. Moreover, it can reduce the temperature on the cutting tool and the cutting force, whereas temperature in subsurface zone is generally higher than in conventional cutting. In another work [25] a more detailed study on elliptical vibration cutting of silicon was conducted, focusing on the effect of various characteristics of the vibrational motion, such as the vibration amplitude, frequency and phase difference.

Although a fundamental level of understanding has been achieved from the aforementioned studies on general nano-cutting, there are still no specific studies on abrasive processes such as nano-grinding with a number of abrasive grains. Thus, in the present study, it is aimed to extend the understanding of vibrational motion effect on machining and abrasive processes by an MD model regarding vibration-assisted nano-grinding. With this model, the effectiveness of vibration-assisted nano-grinding will be evaluated both qualitatively and quantitatively. Thus in the present study, an appropriate MD model is developed and presented and investigations on the effect of vibration parameters on grinding forces, temperature and subsurface damage of the workpiece are conducted, with a view to compare the findings with previous findings regarding nano-cutting and finally determine optimum conditions for achieving the desired process outcome.

2 Methodology

In the present study, an MD model for Vibration-Assisted Grinding (VAG) is created. The workpiece is an orthogonal parallelepiped single crystal copper (Cu) substrate, whereas grinding is performed by two cubic shaped, single crystal diamond (C) abrasive grains and the total number of atoms in the system is about 104,000. Interactions between copper atoms in the substrate are modeled by Embedded Atom Model (EAM) potential function, whereas interactions between Cu-C atoms are modeled by an appropriate Morse potential function [26]. The interactions between C-C atoms in the abrasive grains are neglected and the two abrasive grains are considered as rigid, because their hardness is sufficiently higher than the hardness of the workpiece material and thus no deformation can occur on them. It is important to note that, the choice of appropriate potential function and the proper representation of atomistic structures of both abrasive grains and workpiece are essential towards the validity of results. In the workpiece, three distinct atom zones are defined; namely the Newtonian, thermostat and boundary atom zones, as can be seen in Fig. 1. The atoms in the Newtonian zone, depicted in blue color in Fig. 1, constitute the deformable part of the workpiece, where nano-grinding will take place; their displacement is governed by forces due to their interaction with the atoms of the abrasive grains or other atoms in the workpiece and it is calculated from the Newton equation of motion. Furthermore, the atoms in the thermostat zone, depicted in yellow color in Fig. 1, are employed in order to remove artificially generated heat from the workpiece and the atoms in the boundary zone, depicted in red color in Fig. 1, are kept fixed in their initial positions in order to prevent rigid body motion of the workpiece and avoid boundary effects. Moreover, periodic boundary conditions are imposed on the two surfaces on the yz plane in order to simulate nano-grinding without the need of using a much larger workpiece.

The first abrasive grain is positioned at a distance of 0.4 nm away of the workpiece surface, whereas the second abrasive grain is positioned 4 nm away from the first and has a different protrusion height, in order to simulate more realistically the abrasive grains of a grinding tool which can have different protrusion height. The horizontal or nominal speed of the abrasive grains, denoted hereafter as V_0, is constant at 200 m/s for all simulations, whereas the speed in the vertical direction will be varied due to vibration. It is to be noted that the vibrations will be performed in the z-axis, perpendicular to the upper surface of the workpiece.

The choice of vibration parameters, namely vibration amplitude A and vibration frequency f is not totally arbitrary; according to the relevant literature [15], there exists a limit below which the vibrational motion does not provide a significant effect to the outcome of the material process. This limit is defined by the equation $V_{max} > V_0$ where V_{max} is the magnitude of the vibrational speed defined as $V_{max} = 2\pi fA$. More specifically, the case of 1D VAG will be considered in the present study with amplitudes of

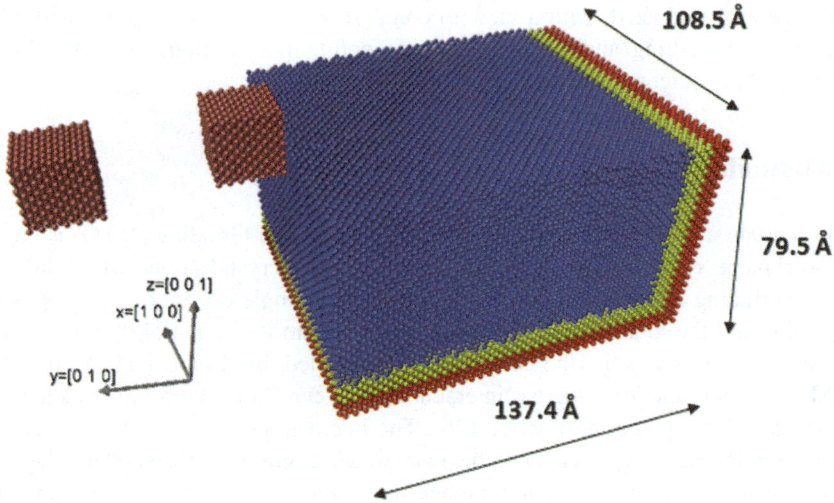

Fig. 1. Schematic of the MD model.

0.5 Å up to 2 Å and vibration frequency values between 79.5 and 795.7 GHz, as can be seen in Table 1.

Table 1. Characteristics of simulation cases

Case number	Vibration frequency (GHz)	Vibration amplitude (Å)
1	0	0
2	79.5	1
3	397.8	1
4	795.7	1
5	397.8	0.5
6	397.8	2

Two series of simulations will be carried out: in the first, three amplitude values, namely 0.5, 1 and 2 Å will be selected with a constant vibration frequency (397.8 GHz), whereas in the second three vibration frequency values, namely 79.5, 397.8 and 795.7 GHz, will be selected with a constant amplitude value (1 Å). These values are of the same order of magnitude with the values employed in the relevant literature for nano-cutting simulations [15, 23] and they were chosen in accordance with the cutting length and depth of cut. Total depth of cut will be 1 nm in any case, as the first abrasive grain removes material at 0.8 nm depth of cut and the second abrasive grain at an additional depth of 0.2 nm. Total cutting length is 12 nm, initial temperature of the workpiece is fixed at 293 K and numerical time-step is 1 fs. For the simulations, LAMMPS software is employed.

3 Results and Discussion

After the simulations were carried out, analysis of the results took place. Several snapshots from the nano-grinding simulations are depicted in Figs. 2(a) to (c).

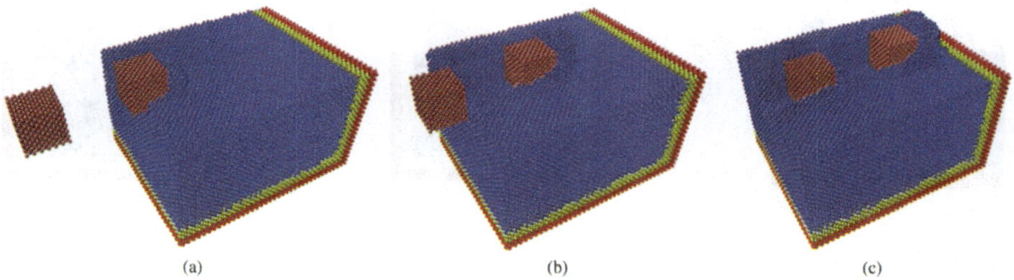

Fig. 2. Snapshots of nano-grinding at: (a) 10 ps, (b) 30 ps, (c) 50 ps

From these results, the characteristics of this process can be observed. When the first abrasive grain gets in contact with the workpiece, it removes a bulk of material and forms a nano-groove. During nano-grinding, the main part of the removed material is accumulated on the front side of the workpiece, whereas a small part of the removed material is forming a pile on each side of the groove, like the burr observed in various types of machining. When the second abrasive grain gets in contact with the workpiece, it removes a small amount of material and sometimes some of the misplaced atoms due to the action of the first grain. Contrary to conventional nano-grinding, in VAG the abrasive grains have a sinusoidal motion additional to the horizontal motion in order to simulate the vibration of the grinding wheel. Thus, according to the vibration frequency and amplitude values, the grains will move in a more complex way, which can affect material removal and workpiece deformation considerably. Depending on the values of the vibrational motion parameters, different chip morphology is expected to occur, especially in cases where V_{max} is greater than V_0.

Although in every case the same depth of cut and grinding speed are selected, the variation of vibration frequency and amplitude leads to different chip morphology, as can be seen in Figs. 3(a) to (f). It was noted that in the case of conventional grinding, the main bulk of removed atoms were concentrated in the front side of the abrasive grain, forming the main pile of atoms, whereas a smaller number of atoms were concentrated in the groove sides. When vibration frequency increased up to 397.5 GHz, the main pile of atoms was displaced towards the right side of the workpiece and the width of the pile of atoms on the groove side was larger, but at the largest frequency value, the main pile was again almost symmetrical and side piles were minimal. Furthermore, when vibration amplitude increased, it was observed that the height of the pile of atoms in front of the workpiece reduced in height and increased in width.

Regarding the grinding forces, the obtained results are depicted in Figs. 4(a) and (b). From these results, it can be seen that grinding forces are affected by both parameters of the vibrational motion. When vibration frequency is increasing, a slight

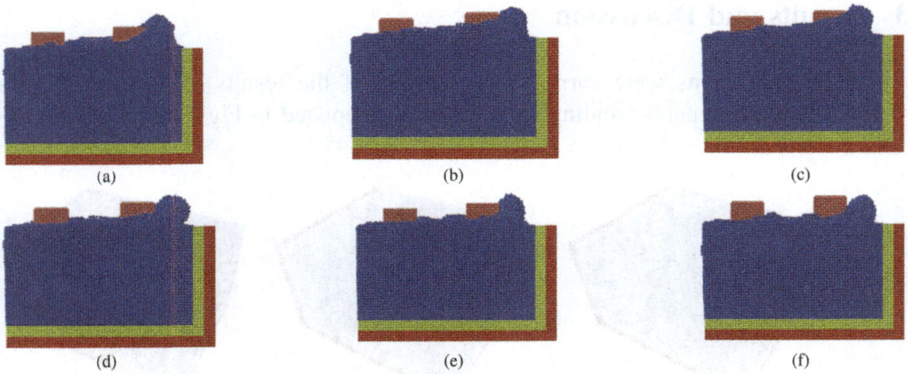

Fig. 3. Snapshots of nano-grinding at the final timestep. The snapshots (a)-(f) correspond to simulation cases 1–6, from Table 1

decrease of F_y force can be observed and then, the decrease becomes greater for larger values of vibration frequency. Moreover, vibrational motion with relatively low amplitude produces indistinguishable results from the case of conventional nano-grinding, but an increase of vibration amplitude can produce a considerable decrease of F_y grinding force. The results are absolutely consistent with the theory, as it is proven that vibration-assisted grinding with magnitude of vibrational speed lower than the nominal grinding speed almost does not affect the grinding forces, whereas starting with cases with V_{max} slightly over the theoretical limit, an observable difference occurs, indicating that grinding is considerably affected by the vibrational motion. Similar trends for the horizontal or tangential grinding force F_y, in respect to the parameters of vibrational motion have been reported in the relevant literature [16, 21–23], thus proving the validity of the present study. Furthermore, the present results lie within the same order of magnitude as similar studies, thus indicating their validity [16, 21].

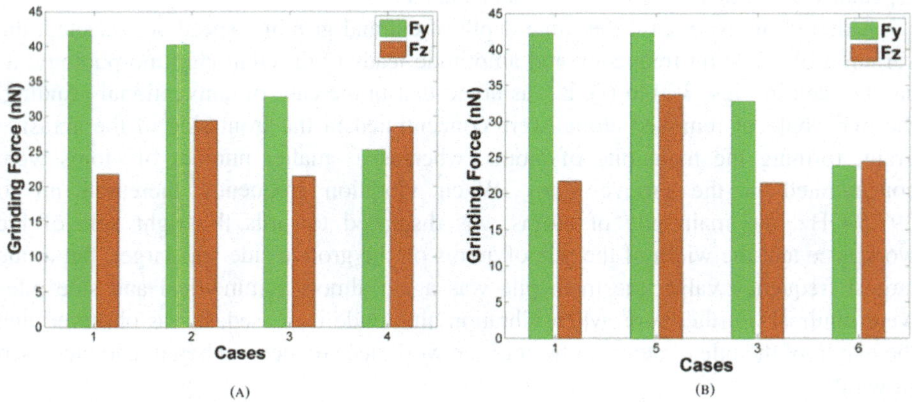

Fig. 4. Average grinding forces values for cases with: (a) different vibration frequency values, (b) different vibration amplitude values

As for the normal component of grinding force, F_z, the results are more affected by the intensity of the vibrational motion. The use of vibrations leads to an increase of F_z when $V_{max} < V_O$, as in case 2 of the simulations, but decreases slightly with increasing vibration frequency. However when the vibration frequency becomes larger, due to the more frequent impacts of abrasive grains and workpiece surface, the normal force values increase again. The same phenomenon can be observed in cases with varying vibration amplitude. When the condition $V_{max} > V_O$ is not satisfied, adverse effects are produced regarding F_z, whereas with an increase in vibration amplitude, the normal force decreases and increases again when the amplitude value becomes larger. These results indicate that apart from satisfying the condition $V_{max} > V_O$, the magnitudes of f and A should be chosen in such a way that the vertical impact of the abrasive grain is not very intensive. This is expected to occur when the amplitude is large and frequency is smaller so that the abrasive grain will remain more time out of contact with the workpiece.

Analysis of nano-grinding temperature is also essential, as material removal involves always thermo-mechanical phenomena and the determination of temperature is not just a complementary study, but it can also justify other findings. The average temperature of the Newtonian zone of atoms increases with an increase of vibrational frequency, as can be seen from the results of Figs. 5(a) and (b). In accordance with the trend observed in the case of grinding forces, an observable difference from the temperature value calculated for the case without vibrational motion, occurs only when the vibrational speed is near or above the theoretical limit. Thus, the temperature in case 3 is slightly higher and it becomes considerably higher in case 4.

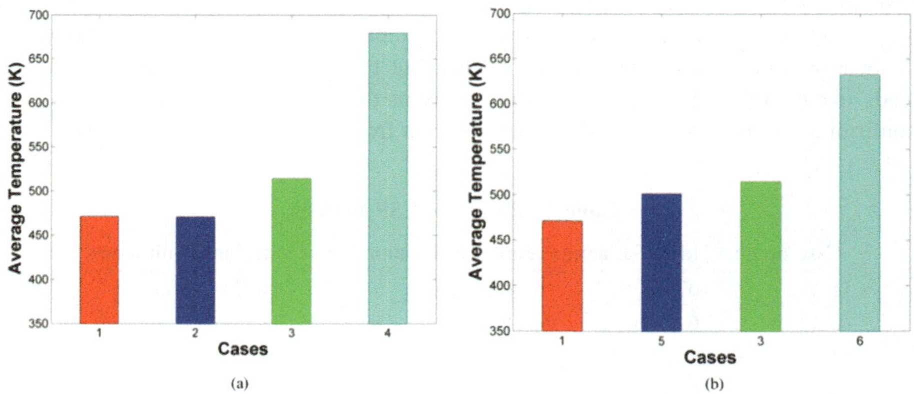

Fig. 5. Average Newtonian zone temperatures for cases with: (a) different vibration frequency values, (b) different vibration amplitude values

Moreover, the vibration amplitude affects also the temperature, as an increase of temperature is observed with an increase of vibration amplitude. In this series of simulation, it is shown that even in the case with $A = 0.5$ Å, a slight increase of temperature is observed although it is not related with a clear difference in grinding forces. The results regarding grinding temperature provide an explanation for the decreasing trend of grinding forces; it can be concluded that the increase of grinding temperature causes material softening and thus it contributes to the decrease of grinding forces, leading to an easier material removal in these cases. The temperature variation with vibrational parameters is generally attributed to the increased kinetic energy of the cutting tool which is transferred to the workpiece, leading to an increase of its temperature [22]. Moreover, the increase of temperature in cases with higher V_{max} is partially explained due to the more intense impact of the abrasive grains on the workpiece surface, as it was observed by the increase of normal grinding force, and the relatively small period of vibration, which does not allow for sufficient cooling of the workpiece when the abrasive grains are out of contact with it. The observed trends are generally in accordance with the trends observed in other simulations of nano-cutting or nano-scratching [22–25]. Furthermore, the present results regarding temperature lie within a reasonable range compared to similar studies [22].

Besides the analysis of grinding forces and temperature, it is considered important to conduct further analysis, regarding the deformation of the workpiece and the subsurface alterations after the completion of nano-grinding. In order to reveal the effect of process parameters on the subsurface alterations, it is essential to analyze the values of centrosymmetry parameter (CSP) for the workpiece. The value of CSP for each atom is calculated by the distance of each atom relative to its neighboring atoms in the lattice and thus, deviations from the ideal value indicate alterations in the atomistic structure of the workpiece. According to an MD study for copper (fcc) workpiece [27], the characteristic values of CSP were: $CSP < 3$ for an ideal FCC structure, $3 < CSP < 5$ for a partial dislocation, $5 < CSP < 8$ for a stacking fault, $8 < CSP < 21.5$ for surface atoms and $CSP > 21.5$ for surface edge atoms. From the presented results regarding CSP in Table 2, it can be seen that the introduction of the vibrational motion with $V_{max} < V_0$ leads to a slight reduction in the number of atoms pertinent to the partial dislocation and stacking fault atoms groups, but the increase of vibration frequency leads eventually to an increase of the atoms in these groups, especially in case 4, confirming the observed regarding the vibration frequency in the literature [21].

Table 2. Results of CSP analysis

Case number	Ideal fcc atoms	Partial dislocation atoms	Stacking fault atoms
1	67051	748	1455
2	67534	731	1233
3	67224	946	1275
4	66877	1198	1390
5	67181	920	1360
6	64510	2613	1988

The same phenomenon is noted for increasing values of vibration amplitude but the difference between the case with 1Å amplitude and 2Å amplitude is significantly large. Thus it can be concluded that increase of vibration frequency and amplitude leads to increase of partial dislocations in the workpiece, with an increase of amplitude exhibiting more significant effect than the increase of frequency.

Summarizing all the results stemming from the analysis of various physical quantities of VAG, it can be deduced that the vibrational motion has a considerable effect on the process, mainly by effectively reducing the tangential force component and facilitating material removal. The theoretical condition for an effective VAG was confirmed to be essential, as only when $V_{max} > V_0$ the tangential grinding force can be sufficiently reduced. However, in order to reduce the normal force component and maintain cutting temperature below a desired limit, a combination of higher vibration amplitude and lower frequency should be chosen in order to increase the period of no contact between abrasive grain and workpiece and reduce the intensity of the abrasive grain impact on the workpiece.

4 Conclusions

In the present study, an MD model for vibration-assisted nano-grinding was developed and applied for the investigation of the effect of vibration parameters on nano-grinding outcome, as well as the determination of optimal values of these parameters. From the obtained results various conclusions can be drawn.

The introduction of vibrational motion on the abrasive grains is advantageous for nano-grinding, as it can considerably reduce the tangential grinding force when the maximum vibrational speed exceeds the value of nominal speed of the grains. Although the normal force values cannot be so effectively reduced, the resultant force exhibits always decreasing trend.

At the same time, the vibrational motion was also shown to contribute to an increase of average temperature of the workpiece as vibration frequency and amplitude values increase, especially when the maximum vibrational speed is much higher than the nominal one, leading to material softening and lower force values. Moreover an increase of atoms pertinent to partial dislocations or stacking fault was attested, indicating the facilitation of material removal process.

Finally, it was suggested that in order to achieve lower cutting forces and maintain temperature and subsurface deformation within a desired limit, combinations of higher amplitude values and lower frequency values should be chosen in order to increase the time when the abrasive grain is out of contact with the workpiece.

References

1. Kundrák, J., Felhő, C.: 3D roughness parameters of surface face milled by special tools. Manuf. Technol. **16**(3), 532–538 (2016)
2. Kundrák, J., Gyáni, K., Felho, C., Deszpoth, I.: The effect of the shape of chip cross section on cutting force and roughness when increasing feed in face milling. Manuf. Technol. **17**(3), 335–342 (2017)
3. Mamalis, A.G., Kundrak, J., Gyani, K., Horvath, M.: On the precision grinding of advanced ceramics. Int. J. Adv. Manuf. Technol. **20**(4), 255–258 (2002)

4. Feucht, F., Ketelaer, J., Wolff, A., Mori, M., Fujishima, M.: Latest machining technologies of hard-to-cut materials by ultrasonic machine tool. Procedia CIRP **14**, 148–152 (2014)
5. Zhang, J., Cui, T., Ge, C., Sui, Y., Yang, H.: Review of micro/nano machining by utilizing elliptical vibration cutting. Int. J. Mach. Tools Manuf **106**, 109–126 (2016)
6. Thoe, T.B., Aspinwall, D.K., Wise, M.L.H.: Review on ultrasonic machining. Int. J. Mach. Tools Manuf. **38**(4), 239–255 (1998)
7. Brehl, D.E., Dow, T.A.: Review of vibration-assisted machining. Precis. Eng. **32**(3), 153–172 (2008)
8. Grabchenko, A., Fedorovich, V., Pyzhov, I., Kundrák, J.: 3D simulation of vibrating diamond grinding. Manuf. Technol. **14**(2), 153–160 (2014)
9. Rentsch, R., Inasaki, I.: Molecular dynamics simulation for abrasive processes. CIRP Ann. **43**(1), 327–330 (1994)
10. Komanduri, R., Chandrasekaran, N., Raff, L.M.: Some aspects of machining with negative-rake tools simulating grinding: a molecular dynamics simulation approach. Philos. Mag. B **79**(7), 955–968 (1999)
11. Agrawal, P.M., Narulkar, R., Bukkapatnam, S., Raff, L.M., Komanduri, R.: A phenomenologic model of polishing of silicon with diamond abrasive. Tribol. Int. **43**(1–2), 100–107 (2010)
12. Zong, W.J., Cheng, X., Zhang, J.J.: Atomistic origins of material removal rate anisotropy in mechanical polishing of diamond crystal. Carbon **99**, 186–194 (2016)
13. Eder, S.J., Bianchi, D., Cihak-Bayr, U., Vernesa, A., Betz, G.: An analysis method for atomistic abrasion simulations featuring rough surfaces and multiple abrasive particles. Comput. Phys. Commun. **185**, 2456–2466 (2014)
14. Eder, S.J., Cihak-Bayr, U., Bianchi, D.: Single-asperity contributions to multi-asperity wear simulated with molecular dynamics. IOP Conf. Ser. Mater. Sci. Eng. **119**, 021009 (2016)
15. Shimizu, J., Tanaka, H., Zhou, L., Eda, H.: Molecular dynamics of ultra high-acceleration and vibration cutting. Key Eng. Mater. **257–258**, 21–26 (2004)
16. Shimizu, J., Zhou, L., Eda, H.: Molecular dynamics simulation of vibration-assisted cutting: influences of vibration, acceleration and velocity. Int. J. Nanomanuf. **1**(1), 105–116 (2006)
17. Shimizu, J., Zhou, L., Eda, H.: Molecular dynamics simulation of vibration-assisted cutting: influences of vibration parameters. Int. J. Manuf. Technol. Manage. **9**(1–2), 120–129 (2006)
18. Liang, Y., Li, D., Bai, Q., Wang, S., Chen, M.: Molecular dynamics simulation of elliptical vibration cutting. In: Proceedings of the 1st IEEE International Conference on Nano/Micro Engineered and Molecular Systems, Zhuhai, China, 18–21 January 2006, pp. 635–638. IEEE (2006)
19. Zhou, X., Zhu, Z., Lin, J.: Evolution of workpiece microstructure and cutting force during ultraprecision vibration assisted machining. J. Comput. Theor. Nanosci. **10**(1), 78–85 (2013)
20. Huang, S., Liu, X., Chen, F.Z., Zheng, H.X., Yang, X.L., Wu, L.B., Song, J.L., Xu, W.J.: Diamond-cutting ferrous metals assisted by cold plasma and ultrasonic elliptical vibration. Int. J. Adv. Manuf. Technol. **85**(1-4), 673–681 (2015)
21. Zhu, B., Zhao, H., Zhao, D., Zhang, P., Yang, Y., Han, L., Kui, H.: Effects of vibration frequency on vibration-assisted nano-scratch process of monocrystalline copper via molecular dynamics simulation. AIP Adv. **6**(3), 035015 (2016)
22. Zhu, B., Zhao, D., Zhao, H., Guan, J., Hou, P., Wang, S., Qian, L.: A study on the surface quality and brittle-ductile transition during the elliptical vibration-assisted nanocutting process on monocrystalline silicon via molecular dynamic simulations. RSC Adv. **7**, 4179–4189 (2017)
23. Goel, S., Duarte Martinez, F., Zare Chavoshi, S., Khatri, N., Giusca, C.: Molecular dynamics simulation of the elliptical vibration-assisted machining of pure iron. J. Micromanuf. **1**(1), 6–19 (2018)

24. Dai, H., Chen, J., Liu, G.: A numerical study on subsurface quality and material removal during ultrasonic vibration assisted cutting of monocrystalline silicon by molecular dynamics simulation. Mater. Res. Express **6**, 065908 (2019)
25. Dai, H., Du, H., Chen, J., Chen, G.: Influence of elliptical vibration on the behavior of silicon during nanocutting. Int. J. Adv. Manuf. Technol. **102**(9-12), 3597–3612 (2019)
26. Fang, T.H., Weng, C.L.: Three-dimensional molecular dynamics analysis of processing using a pin tool on the atomic scale. Nanotechnology **11**(3), 148–153 (2000)
27. Tong, Z., Liang, Y., Jiang, X., Luo, X.: An atomistic investigation on the mechanism of machining nanostructures when using single tip and multi-tip diamond tools. Appl. Surf. Sci. **290**, 458–465 (2014)

Supply Chain Management

A New Approach to Location Routing Problem: Capacitated Periodic Location Routing Problem with Inventory

Esra Yaşar Boz[1(✉)], Ahmet Reha Botsalı[2], and Tuba Ulusoy[2]

[1] International Trade and Logistics, KTO Karatay University, Konya, Turkey
esrayasaarr@gmail.com

[2] Industrial Engineering, Necmettin Erbakan University, Konya, Turkey
rehabotsali@gmail.com, ulusoytuba@gmail.com

Abstract. Integration of the Facility Location Problem (FLP) and the Vehicle Routing Problem (VRP) is called the Location-Routing Problem (LRP). In this study, using the real data of a food processing company, the best warehouse locations and the customers who are served by these warehouses are tried to be determined by a new mixed-integer formulation in which inventory costs and time periods within the processes are taken into consideration. The problem is called Capacitated Periodic Location Routing Problem with Inventory (CPLRPI) which can be considered a new approach in LRP literature. The objective of this problem is to minimize the total cost of the system that includes warehouse location and vehicle routing. The new mixed-integer formulation is solved using GAMS, but an optimal solution cannot be obtained due to the size and complexity of the problem. Therefore, the problem is solved by Simulated Annealing (SA) meta-heuristic in Matlab. The parameter combination of the SA is generated by an orthogonal array.

Keywords: Facility location problem · Location-routing problem · Simulated annealing · Vehicle routing problem · Warehouse management

1 Introduction

Today, in conjunction with developments in production technology, supply chain management—also called logistics science- becomes more important for companies wishing to decrease their operational cost and to gain an advantage over their competitors. Therefore, to select their warehouse locations and to define a corresponding product distribution network are critical for companies. One way to ensure fast delivery to customers is to locate warehouses close to customers' locations. For this reason, conventionally, after warehouses are located in convenient places to minimize the total transportation cost of the product distribution system, the vehicle routes (optimal set of routes to be performed by a fleet of vehicles to serve a given set of customers) are determined. However, it is obvious that dealing with both warehouse location and vehicle routing simultaneously is more reasonable in order to minimize the cost associated with the distribution of products.

N. M. Durakbasa and M. G. Gençyılmaz (Eds.): ISPR 2020, LNME, pp. 751–766, 2021.
https://doi.org/10.1007/978-3-030-62784-3_63

In supply chain management literature, the integrated problem of the location of warehouses and the determination of transportation routes from these warehouses to customers is called the location-routing problem (LRP). The solution of LRP shapes the investment decision of companies regarding warehouse location and/or plant location selection based on customer orders. Generally, the solution is obtained in two stages: in the first stage, the warehouse site selection is performed and in the second stage, the vehicle routing problem is solved. LRP is an optimization problem that can make the following two decisions to be connected [1].

(i) Which facility or facilities should be used for specific purpose?
(ii) Which customers should be visited in which order?

In this study, we address an integrated warehouse location/vehicle routing problem that has not been previously studied. The problem is based on a case study of a food processing company. The structure of the addressed problem can be represented by a network which is illustrated in Fig. 1.

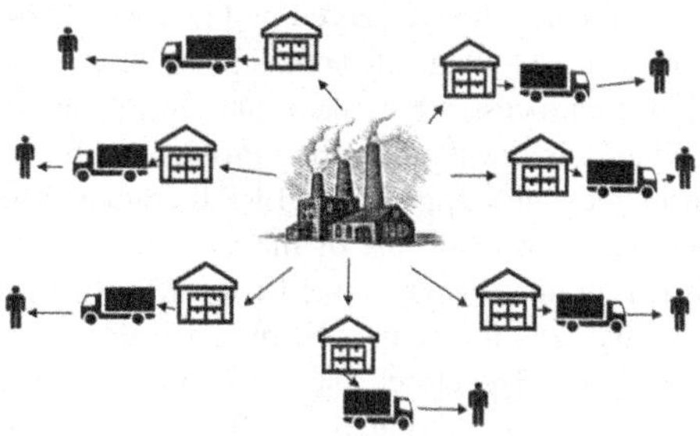

Fig. 1. The general representation of the problem's structure

In related literature, the most similar problem to our problem is the LRP. However, in a standard LRP, there is only one period, and the decision of location selection and vehicle routing are determined only considering this single period. Conversely, in our problem, the number of periods is not limited to one, so considering the opened warehouses, the routes are determined in each period according to the changing customer demand. In this respect, our problem can be referred to as a periodic LRP. In addition to this, we also consider the amount of inventory and inventory holding cost during the given time period. To the best knowledge of authors, in the related literature, no study has yet been carried out considering both multiple periods and the inventory holding cost in each period simultaneously. So, the problem is called Capacitated Periodic Location Routing Problem with Inventory (CPLRPI).

This problem is modelled as a mixed-integer mathematical model and is tried to be solved by using GAMS (The General Algebraic Modelling System), which is a software, that enables to model and to solve optimization problems. Since the problem is NP-Hard and the size of it is large, either a feasible or optimal solution is not obtained. So, in order to solve the problem, Simulated Annealing (SA) approach, which is an efficient meta-heuristic in solving LRPs, is coded in Matlab (MATrix LABoratory), which is a programming language and computing environment [2]. The parameter combination of the SA is generated by an orthogonal array based on Taguchi design to reduce the number of runs in Matlab.

The remainder of this paper includes the following sections: Problem Definition, Simulated Annealing for the CPRLPI, Discussion, Conclusions and Recommendations.

2 Literature

The Location-Routing Problem (LRP) is basically an integrated version of two primary decision problems. The first one is the selection of the facility locations and the second one is the vehicle routing problem. The LRP deals with determining the location of facilities and the routes of the vehicles for serving the customers under some constraints (i.e., facility and vehicle capacity, route lengths, level of budget) to satisfy demand of all customers and to minimize the total cost that includes routing costs, vehicle fixed costs, facility fixed costs, and facility operating costs [3]. The LRP was first introduced as a concept by Boventer in 1961 [4]. The first exact algorithm was developed by Laport and Nobert in 1981 [5]. After that time, the researchers have focused on different aspects of the problem by using both exact and heuristic algorithms.

Many survey studies which are devoted to LRP can be found in Supply Chain Management literature [1, 6–8]. Apart from these studies, there are many papers that classify LRP [9–12]. Lopes et al. (2013) initially group the LRP structurally as it can be seen in Fig. 2. After this grouping, they classify each group according to their solution methods and objective functions.

When it comes to the solution methods, the complexity and the size of the problem require problem specific approaches. As a result, various problem-specific exact and heuristic methods have been applied to solve LRPs. The non-deterministic polynomial-time version of LRP leads to a difficult solution process [13]. Therefore, exact solution methods may not be effective for large-scale problems [14]. In the literature, there are studies in which small-scale problems are solved by using exact approaches. Belenguer et al. solved Capacitated Location Problem (CLRP) with 5 depot and 50 customers using branch and cut method [15]. Ponboon et al. proposed an exact algorithm based on branch and price algorithm for LRP with Time Window (LRPTW) and solved the problem using many testing instances [16]. Capelle et al. developed an integer programming model for Location-Routing Problems with Pickup and Delivery (LRP-PDP), and they solved the problem with the method based on column generation by using the data found in the literature [17]. Since the problems with large-scale of data generally cannot be solved by exact solution methods, the solutions are searched via heuristic methods. Panicker et al. solved a two-stage LRP, which consists of multiple plants, warehouses, and customers by applying the Ant Colony Algorithm [18]. Asefi et al. developed a new hybrid method based on neighbourhood search and Simulated Annealing algorithm and solved the LRP of integrated solid waste management facility with this hybrid algorithm [19]. Moon et al. proposed a mixed-integer formulation for the Location Routing Problem with Multi-Compartment and Multi-Trip. In order to solve the problem, the constructive heuristic and hybrid genetic algorithms were

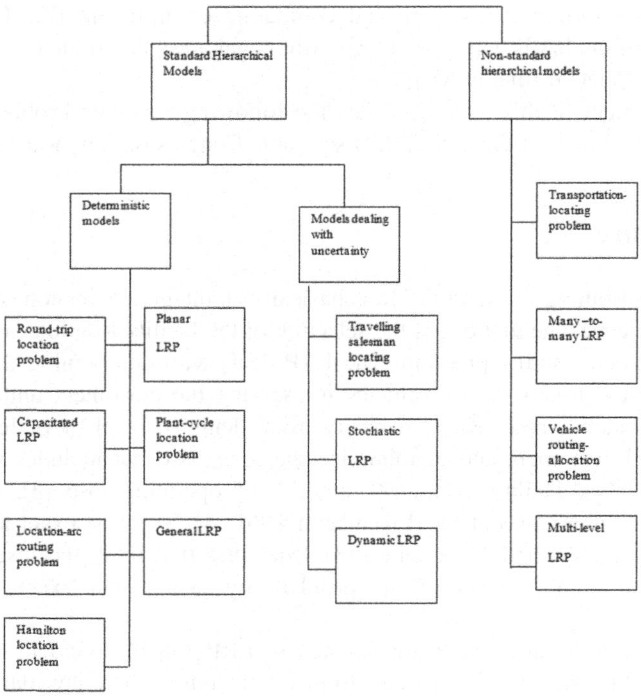

Fig. 2. Types of location-routing problems [12]

developed while for the small size instances both methods gave efficient results, for the large size instances hybrid genetic algorithm provided more promising results [20]. Fan et al. proposed a mixed-integer formulation for a two-echelon location routing problem with simultaneous pickup and delivery. The authors developed multistart hybrid heuristic with path relinking which composes of local search and a variable neighbourhood descent algorithm. This algorithm helped to reduce computational time [21]. Oudouar et al. developed a novel approach to solve a Capacitated Location Routing Problem (CLRP). This approach based on a heuristic and a neural network. These authors solved first part of the problem in which the depots to be opened were determined and the customers were assigned to depots with self-organizing map is unsupervised learning method and solved the second part of the problem which is routing problem with Clarke and Write technique. The results showed that this approach is efficiency [22].

In this study, we address the Periodic Location Routing Problem (PLRP). PLRP is basically a combination of two problems: Facility location and periodic vehicle routing problem. As a result of this problem, there is an answer to the questions about which warehouses will be opened at periods, which customers will be assigned to which warehouses and which customers use which routes. The PRLP was introduced by Prodhon and Prins in 2008 [23]. In the study, a memetic algorithm with population management was proposed for the solution of large-size instances. Pirkwieser and Raidl presented three very large neighbourhood search on integer linear programming and these authors showed that their approach is successful [24]. Prodhon proposed a meta-heuristic algorithm combination of a hybrid evolutionary algorithm and randomized extended Clarke and Wright algorithm to solve PRLP. The author showed that the algorithm outperforms the previous methods for the PRLP [25]. Albera-Sambola et al. studied a different variant of PLRP which is the Multi-Period Location Routing Problem with Decoupled Time Scales. The reason for being a different variant is that the decision to open the facilities is made in certain periods. The authors proposed two heuristic and approximation to avoid loss of capacity [26]. Hemmelmayr et al. defined a new variant of PLRP by adding various effects of certain costs and evolution of flexibility and proposed mix integer programming model and adaptive large neighbourhood search heuristic algorithm for the solution [27]. Mohammed et al. studied two-echelon stochastic multi-period capacitated location-routing problem. The first stage decision (location and capacity decisions) in the problem was solved as Benders master problem, the second stage decision (routing decision) was made by using a branch-cut-and-price algorithm [28]. Koç determined three variants of standard PLRP are the Heterogeneous PLRP with Time Windows, the Heterogeneous PLRP and the homogeneous PLRP with Time Windows. He developed Unified-Adaptive Large Neighbourhood Search meta-heuristic for the problems and the solutions showed that this algorithm outperforms the previous methods [29]. Prodhon proposed a meta-heuristic algorithm which is based on Randomized Extended Clarke and Wright Algorithm for the PLRP. The method implemented in three sets of instances and the results showed that the algorithm can provide efficient solutions [30]. Liu et al. proposed a hybrid algorithm integrated a large neighbourhood search and particle swarm optimization to solve the PRLP. The hybrid algorithm provided better solutions than the existing methods [31]. Hemmelmayr proposed a large neighbourhood search algorithm for the PRLP and the author used parallelization strategies. The result of this study shows that the algorithm can outperform previous solution methods [32]. Tunalıoğlu et al. proposed an adaptive large neighbourhood search algorithm for the solution of Multi-period LRP and the author applied this algorithm in the collection of olive oil mill wastewater [33]. Nasherahkami et al. proposed a new model for the PRLP. Its objective function minimized the lost demand costs and the results showed that this model is efficient to solve this problem [34]. Amiri et al. developed a mixed-integer nonlinear programming model for Maritime Fleet Sizing Mix Periodic Location-Routing Problem with Time Windows (MFSMPLRPTW). The authors solved this problem by using an exact two-phase solution approach for both small and medium cases. As they couldn't solve the large-sized instances with an exact two-phase solution approach, they used two meta-heuristic algorithms [35].

3 Problem Definition

The data of a company that integrates production and distribution in the food sector is used in this study to solve the Capacitated Periodic Location Routing Problem with Inventory (CPLRPI). The 55.000 customers who currently order from the company are classified according to their cities and since there are 77 cities, the number of customers is 77. These 77 customers' demands in 2014–2016 are classified considering their monthly demand on a date basis. Demand data is presented as a matrix of 77 customers in 28 periods. The distance between the cities for each customer is found via Google Earth and the shortest route between the two locations is taken into account. Since the return's distances are equal, the distance matrix is formed symmetrically.

The company has 11 main categories of products in total. Only the chocolate category of products is examined in this study. Demand is determined considering the average space requirement of the products in the chocolate category.

The warehouses which are likely to be opened are tried to be determined according to the orders of 28 periods.

Some assumptions are made on the problem within the context of the study. These assumptions can be listed as follows:

- Only the chocolate product categories are identified from the product categories of the company and the orders of the products with chocolate are summed up under the chocolate category.
- It is assumed that the number of vehicles is unlimited and the company would use the required number of vehicles obtained in the solution.
- The potential warehouse locations (7 locations in our case) are fixed and there is no limit on the number of warehouses to be opened. Each customer's demand is covered by only one warehouse.
- There are no constraints regarding the distance that the vehicles travel.
- The cost of a driver and/or a second driver is not considered because the time schedule is not created in the problem.
- Costs are considered with inflation-adjusted real values.
- Order to be delivered to the customer cannot be split.
- The order to be delivered to the customer cannot be more than the capacity of the vehicle.

This problem can be considered a management decision-making process in which there are existing customers and warehouses, also warehouses that are likely to be opened. The company wants to determine which warehouses should be opened by considering the vehicle routing cost. This cost emerges from the delivery of the orders to the customers which depends on the sequence and the warehouse from which the delivery is being performed. Therefore, the problem is a Location Routing Problem (LRP). However, it differs from conventional the LRP since time periods and inventory are taken into consideration in the problem. If we primarily deal with the selection process of warehouse locations, the problem can be described as follows:

Let G = (N, A) a network where $N = \{N_- \cup N_+\}$, in which N_- indicates the warehouse set and N_+ indicates customer set grouping with respect to city proximity (as discussed above). A indicates the arc node. The arc set is defined as $\{(i, j): i \in N_-, j \in N_+\ i \neq j\}$ Our main goal is to determine the warehouses that should be opened while considering minimizing the total system cost. The total cost includes warehouse opening costs, transportation cost from factory to warehouses and from warehouses to customers, and inventory holding cost at warehouses. In addition to that it is expected that the solution of the model helps to determine that the 77 customers are served from which warehouses. There are 7 potential warehouse locations. Each customer order must be met from only one warehouse and the sum of customer demands should not exceed the capacity of the factory. The notations of the problem is given in Table 1.

Table 1. Notations

Parameters	
d_{jt}	Demand of customer *j* at time *t*
h_i	Inventory in i^{th} warehouse
g_{ij}	Cost of shipping from i^{th} warehouse to customer
u_{ij}	Cost of shipping from i customer j customer
c_i	Product shipping cost going to the ith warehouse from the factory
A_i	Cost of opening the ith warehouse
OI_{it}	Inventory to be paid extra in the ith warehouse at time t
I_{mini}	Monthly amount of inventory for the ith warehouse (min)
h_i	Unit inventory holding cost in warehouse i
E	Number of customers
F	Number of vehicles
W	The capacity of the factory
Q	The capacity of the vehicle
R	The capacity of the warehouse
K	The set of the vehicles
T	The set of time periods
Variables	
Z_{ijt}	The amount of products sent to the j. customer from i. warehouse at period t
X_{it}	The amount of products sent to the i. warehouse from the factory at period t
I_{it}	Inventory that is held in i. warehouse during the time t
Y_i	Binary decision variable if i. warehouse is open
B_{ijkt}	Binary decision variable if k vehicle from goes to I customer from the j customer in t period

Objective Function

$$\min \sum_{i \in N-} A_i Y_i + \sum_{i,j \in N} \sum_{k \in K} \sum_{t \in T} B_{ijk} u_{ij} d_{jt} + \sum_{i \in N-} \sum_{t \in T} X_{it} c_i + \sum_{t \in T} \sum_{i \in N-} OI_{it} h_i \qquad (1)$$

Subject to

$$X_{it} \leq Y_i R \qquad (\forall i \in N-, \forall t \in T) \qquad (2)$$

$$d_{jt} = \sum_{i \in N-} Z_{ijt} \qquad (\forall t \in T, \forall j \in N+) \qquad (3)$$

$$\sum_{j \in N+} Z_{ijt} \leq Y_i R \qquad (\forall i \in N-, \forall t \in T) \tag{4}$$

$$I_{i0} = 0 \qquad (\forall i \in N-) \tag{5}$$

$$I_{it} = I_{i(t-1)} + X_{it} - \sum_{j \in N+} Z_{ijt} \qquad (\forall i \in N-, \forall t \in T) \tag{6}$$

$$OI_{it} \geq I_{it} - \mathrm{I}_{m\,ini} \qquad (\forall i \in N-, \forall t \in T) \tag{7}$$

$$\sum_{i \in N-} X_{it} \leq W \qquad (\forall t \in T) \tag{8}$$

$$\sum_{i \in N-} \sum_{j \in N+} \sum_{k \in K} B_{ijkt} \leq F \qquad (\forall t \in T) \tag{9}$$

$$\sum_{j \in N+} B_{ijkt} - \sum_{j \in N+} B_{jikt} = 0 \qquad (\forall t \in T, \forall k \in K, \forall i \in N) \tag{10}$$

$$\sum_{j \in N+} \sum_{k \in K} B_{ijkt} = 1 \qquad (\forall t \in T, \forall i \in N) \tag{11}$$

$$\sum_{i \in N+} \sum_{j \in N-} B_{ijkt} = 1 \qquad (\forall t \in T, \forall k \in K) \tag{12}$$

$$\sum_{i \in N-} \sum_{j \in N+} B_{ijkt} = 1 \qquad (\forall t \in T, \forall k \in K) \tag{13}$$

$$\sum_{i \in N-} \sum_{j \in N+} d_{jt} B_{ijkt} \leq Q \qquad (\forall t \in T, \forall k \in K) \tag{14}$$

$$Y_i \geq Y_j + 1 - N(1 - \sum_{k \in K} B_{ijkt}) \qquad (\forall i, j \in N, \forall t \in T) \tag{15}$$

$$Y_i \geq 0 \qquad (\forall i \in N-) \tag{16}$$

$$OI_{it} \geq 0 \qquad (\forall i \in N-, \forall t \in T) \tag{17}$$

$$I_{it} \geq 0 \qquad (\forall i \in N-, \forall t \in T) \tag{18}$$

$$X_{it} \geq 0 \qquad (\forall i \in N-, \forall t \in T) \tag{19}$$

$$Z_{ijt} \geq 0 \qquad (\forall i \in N-, \forall t \in T, \forall j \in N+) \tag{20}$$

$$Y_i \in \{0, 1\} \qquad (\forall i \in N-) \tag{21}$$

$$B_{ijkt} \in \{0,1\} \qquad (\forall i,j \in N, \forall t \in T, \forall k \in K) \tag{22}$$

The objective function is given by expression (1) above. The goal is to minimize the total cost which includes warehouse cost, routing cost (cost of product distribution from warehouses to customers, transportation cost (cost of product shipping from factory to warehouses), and inventory cost (the cost of excess inventory) in the objective function of the model. Constraint (2) ensures that the number of products shipped from the factory at time t does not exceed the warehouse capacity. Constraint (3) establishes that the amount of the product demanded by the customer equals the amount of product that goes to the customer. Constraint (4) ensures that if the warehouse is open, total number of products going to the customers does not exceed the warehouse capacity. Constraint (5) sets the value of the initial inventory in each warehouse to 0. Constraint (6) is used to determine the amount of inventory at time t in each warehouse. Constraint (7) is for determining the amount of extra inventory to be paid. Constraints (8) ensures total amount of products delivered to each warehouse does not exceed the capacity of factory at time t. Constraint (9) ensures that the total number of vehicles from all warehouses to customers equals F. Constraint (10) is called flow-balance constraint. Constraint (11) allows only one road to be used for reaching each customer. Constraints (12) and (13) allow vehicles to start the route from a warehouse and finish it at that warehouse. Constraint (14) is for ensuring that the capacity of the vehicle is not exceeded. Constraint (15) is a standard sub tour elimination constraint. Constraints (16)–(22) define the domains of the decision variables.

4 Simulated Annealing for the CPRLPI

Since solving NP-hard problems is very difficult by exact solution methods, meta-heuristic methods have been developed, improved and hybridized with heuristics and other meta-heuristics. Simulated Annealing (SA) is one of the most commonly used meta-heuristic for NP-hard problems. Thus this meta-heuristic is preferred to solve Location Routing Problems (LRPs) in the literature.

The SA is a local search process and it is based on the principles of a physical process which is called annealing. Annealing is a process in which a substance is heated and then slowly cooled to get a strong crystalline structure. The SA algorithm simulates the energy changes in a system which includes a cooling process that continues until it converges to an equilibrium state [36].

SA was introduced by Metropolis et al. and came to be known by Kirkpatrick et al. [37, 38]. While the objective function of the problem resembles the energy state of the system, a solution to the optimization problem is considered a system state. This algorithm includes probabilistic processes, thus, this feature helps to degrade the solution. The aim is avoiding the local minima or maxima by delaying the convergence. SA is an algorithm that does not have memory.

A random neighbour is generated at each iteration and an improved cost function is accepted [36, 37].

In this study, SA algorithm is used to solve the CPRLPI. This algorithm has been used in a standard way. As it is explained in Sect. 3, the size of the problem is quite large, therefore, we did not want to make the algorithm complicated. The following sets are defined in order to use in the problem:

The customer set is {1, 2, 3, …, 77}, potential depot set is {1, 2, 3, .., 7}, 28 time period set is {1, 2, 3, …, 28} and m vehicle set is {1, 2, 3, …, m}. The parameters are T_0, T_{min}, α, I_{max} and F. T_0 is the initial temperature which is really important. The algorithm does cool itself constantly at each iteration. The parameter T_{min} is the minimum temperature. This parameter is used as one of the stopping criteria in the study. A represents the factor of cooling procedure. In the cooling procedure, the temperature decreases gradually linear, geometric, logarithmic, nonmonotonic or adaptive [36]. In the SA algorithm, we prefer geometric decrease cooling procedure. I_{max} is the maximum number of iterations. It can be used as a stopping criterion. F represents the initial objective function. The objective function is calculated during the iterations but in order to compare with the first iteration of the algorithm, an initial objective function value is needed. The initial objective function value can be found by using any algorithm.

In the SA algorithm applied in this study, two stop criteria are used. The first is the temperature factor and the second is the number of iterations. At the end of the iterations, these two criteria are checked and the algorithm is terminated if at least one is met. Therefore, the algorithm is terminated when the current number of iterations is reached or if the current temperature value falls below the minimum temperature value determined beforehand.

In order to determine the best combination of the SA parameters, some experiments are needed to make. Taguchi method is preferred to determine the best combination of parameters by reducing the number of experiments that are needed to conduct [39]. This method is also a tool that helps to solve constrained optimization problems [40]. Defining problem, determining factors and levels, choosing most suitable orthogonal array, collecting appropriate data, conducting Taguchi-related analyses, identifying significant factors and interactions, finding the optimal combination of factor levels to minimize system response time and determining the benefits of optimal combination of factor levels to minimize system response time and, determining the benefits of optimal setting are the steps which are followed by applying Taguchi experimental design [41].

Based on the Taguchi design, the standard orthogonal array has been selected in order to reduce the number of runs in Matlab. It has 3 layers and 5 parameters, therefore we have 27 combinations of the parameters as it can be seen in Table 2.

Table 2. Taguchi orthogonal array with sımulated annealıng parameters

F	T_0	T_{min}	α	I_{max}
1	150	50	0,05	2000
2	150	50	0,05	2000
3	150	50	0,05	2000
1	125	25	0,01	2000
2	125	25	0,01	2000
3	125	25	0,01	2000
1	100	0	0,001	2000
2	100	0	0,001	2000
3	100	0	0,001	2000
1	125	0	0,05	1500
2	125	0	0,05	1500
3	125	0	0,05	1500
1	100	50	0,01	1500
2	100	50	0,01	1500
3	100	50	0,01	1500
1	150	25	0,001	1500
2	150	25	0,001	1500
3	150	25	0,001	1500
1	100	25	0,05	1000
2	100	25	0,05	1000
3	100	250	0,05	1000
1	150	0	0,01	1000
2	150	0	0,01	1000
3	150	0	0,01	1000
1	125	50	0,001	1000
2	125	50	0,001	1000
3	125	50	0,001	1000

The following SA algorithm, which was studied by Talbi in 2009 and by Osman in 1993 for the vehicle routing problem in the literature, is used to solve the CPRLPI [42].

Algorithm 1 Simulated annealing algorithm for the CPLRPI

1: **input:** $T_0, T_{min}, \alpha, I_{max}, F$
2: **let** $T=T_0, I=0, F=F^*$
3: **Assign a customer to a warehouse randomly**
4: **while** $(I \neq I_{max})$ **and** $(T > T_{min})$ **do**
5:     **if** $(F^* < F)$ **then**
6:         **set** $F^*=F$
7:     **else**
8:         **Generate a random number (R)**
9:     **else if** $(R < e^{-(F^*-F)/T})$ **then**
10:         **Accept and set** $F^*=F$
11:     **else if** $(T < T_{min}) \vee (I = I_{max})$ **then**
12:         **Stop**
13:     **else if** $(I = I + 1)$ **then**
14:         **Go to the Line 4**
15:     **end if**
16: **end while**

5 Discussion

The proposed mixed-integer formulation is coded using the GAMS modelling language but an optimal solution cannot be obtained by CPLEX solver due to the size and the complexity of the problem. In order to find a solution for the mixed-integer formulation, the simulated annealing algorithm coded in MATLAB is used.

The numerical solutions are showed in Table 3. In this table, F^* represents the results which the algorithm found as the best solution. I^* is the total number of iterations until the algorithm stops. T^* is the temperature when the algorithm stops.

According to the solutions given in Table 3, it can be said that the algorithm obtains the best solution at iteration 916. Hence, it is concluded that if one of the stop criteria of the algorithm is met after iteration 916, the best solution is found. If the iteration stops at any iteration before iteration 916, the best solution is not obtained. When these instances are used, the average running time of the algorithm is 53.50 s. As the number of iterations increases, the average running time of the algorithm gets longer. When the algorithm is terminated at iteration 22, the running time of the algorithm is 8.78 s and when the algorithm is terminated at iteration 2000, the running time of the algorithm is 95 s. These running times show that the algorithm can run fast and when the parameters are changed, the results can be taken dynamically (Fig. 3).

The best solution is found when the parameter combination is 7, 8, 9, 10, 11, 12, 16, 17, 18, 22, 23, 24, 25, 26 and 27 in Table 3. These combinations have the same value of objective function which is equal to 31.625. When these parameter combinations are run in the SA algorithm, warehouses 1, 3, and 5 are opened. However, we cannot ensure whether this result is the global optimum or the local optimum.

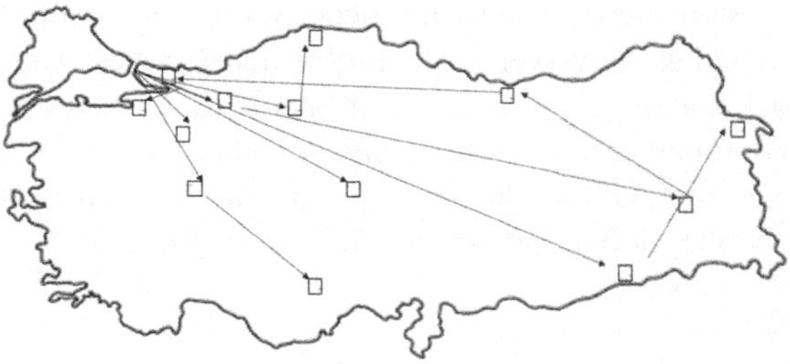

Fig. 3. Illustration of vehicle routes from an open warehouse

Table 3. Numerical solutions

Combination	I	α	F	T_0	T_{min}	I^*	T^*	F^*	Open warehouse
1	2000	0,05	90.000	150	50	22	48,53	35.590	1, 2, 3, 4, 5, 6, 7
2	2000	0,05	70.000	150	50	22	48,53	34.218	1, 2, 3, 4, 5, 6, 7
3	2000	0,05	40.000	150	50	22	48,53	34.122	1, 2, 3, 4, 5, 6, 7
4	2000	0,01	90.000	125	25	161	24,78	31.653	1, 2, 3, 4, 5
5	2000	0,01	70.000	125	25	161	24,78	31.703	1, 3, 4, 5, 7
6	2000	0,01	40.000	125	25	161	24,78	31.774	1, 2, 3, 4, 5, 7
7	2000	0,001	90.000	100	0	2000	13,52	31.625	1, 3, 5
8	2000	0,001	70.000	100	0	2000	13,52	31.625	1, 3, 5
9	2000	0,001	40.000	100	0	2000	13,52	31.625	1, 3, 5
10	1500	0,05	90.000	125	0	1500	4,81	31.625	1, 3, 5
11	1500	0,05	70.000	125	0	1500	4,81	31.625	1, 3, 5
12	1500	0,05	40.000	125	0	1500	4,81	31.625	1, 3, 5
13	1500	0,01	90.000	100	50	14	48,76	33.194	1, 2, 3, 4, 5, 7
14	1500	0,01	70.000	100	50	14	48,76	33.118	1, 2, 3, 4, 5, 7
15	1500	0,01	40.000	100	50	14	48,76	32.560	1, 2, 3, 4, 5, 6, 7
16	1500	0,001	90.000	150	25	1500	33,44	31.625	1, 3, 5
17	1500	0,001	70.000	150	25	1500	33,44	31.625	1, 3, 5
18	1500	0,001	40.000	150	25	1500	33,44	31.625	1, 3, 5
19	1000	0,05	90.000	100	25	28	23,78	33.673	1, 2, 3, 4, 5, 6, 7
20	1000	0,05	70.000	100	25	28	23,78	33.224	1, 2, 3, 4, 5, 6, 7
21	1000	0,05	40.000	100	25	28	23,78	33.261	1, 2, 3, 4, 5, 6, 7
22	1000	0,01	90.000	150	0	1000	0,0065	31.625	1, 3, 5
23	1000	0,01	70.000	150	0	1000	0,0065	31.625	1, 3, 5
24	1000	0,01	40.000	150	0	1000	0,0065	31.625	1, 3, 5
25	1000	0,001	90.000	125	50	916	49,99	31.625	1, 3, 5
26	1000	0,001	70.000	125	50	916	49,99	31.625	1, 3, 5
27	1000	0,001	40.000	125	50	916	49,99	31.625	1, 3, 5

For such costly strategic decisions, the decision-makers should use the proposed mixed-integer formulation. However, if the problem size is very large, a mathematical programming model cannot be solved in a reasonable time or an optimal or a feasible solution cannot be found. Hence, the decision-makers need to try meta-heuristic algorithms. However, it should not be forgotten that the solution found from the meta-heuristic may not be the global optimum or the local optimum.

6 Conclusion and Recommendations

In this paper, the LRP concepts, types, and solution methods are discussed in the literature review section to shed light on the topic. The remainder of the paper is devoted to a case study in which a solution is obtained for the CPLRPI of a company operating in the food sector, in Turkey. A mixed-integer formulation is used in order to model the problem by using the data of the company. The classical LRP is extended by adding the inventory and time periods.

This study has two main contributions to the literature. Firstly, the study deals with the real-world problem of a company in the food sector. The problem is handled as Capacitated Periodic Location Routing Problem with Inventory (CPLRPI) which can be considered a new approach in LRP literature. Secondly, a novel mixed-integer formulation is developed to solve the problem by considering inventory and time periods which are not negligible aspects.

Since some problems have been encountered in managerial decision making because of the lack of a program based on a scientific method to solve warehouse location and vehicle routing problems within the company, the solutions to these problems can help the company to make reasonable decisions.

The solutions obtained in this study are presented as feedback to the company. In addition, the firm is advised to use mathematical programming techniques to solve issues related to CPLRPI.

This study can be extended further by integrating the production planning problem in the factory to the original problem. Also, the effect of some of the assumptions we make (i.e., demand aggregation or availability of unlimited trucks) can be analysed on the solution quality in a future study.

Acknowledgment. This paper includes a part of the M.Sc thesis is written by Esra YAŞAR BOZ.

References

1. Drexl, M., Schneider, M.: A survey of variants and extensions of the location-routing problem. Eur. J. Oper. Res. **241**(2), 283–308 (2015)
2. Vincent, F.Y., Lin, S.W., Lee, W., Ting, C.J.: A simulated annealing heuristic for the capacitated location routing problem. Comput. Ind. Eng. **58**(2), 288–299 (2010)

3. Karaoğlan, I., Altiparmak, F.: A hybrid genetic algorithm for the location-routing problem with simultaneous pickup and delivery. In: The 40th International Conference on Computers & Industrial Engineering, pp. 1–6. IEEE (2010)
4. Von Boventer, E.: The relationship between transportation costs and location rent in transportation problems. J. Reg. Sci. **3**(2), 27–40 (1961)
5. Laporte, G., Nobert, Y.: An exact algorithm for minimizing routing and operating costs in depot location. Eur. J. Oper. Res. **6**(2), 224–226 (1981)
6. Prodhon, C., Prins, C.: A survey of recent research on location-routing problems. Eur. J. Oper. Res. **238**(1), 1–17 (2014)
7. Schiffer, M., Schneider, M., Walther, G., Laporte, G.: Vehicle routing and location routing with intermediate stops: a review. Transp. Sci. **53**(2), 319–343 (2019)
8. Schneider, M., Drexl, M.: A survey of the standard location-routing problem. Ann. Oper. Res. **259**(1-2), 389–414 (2017)
9. Laporte, G., Nobert, Y., Taillefer, S.: Solving a family of multi-depot vehicle routing and location-routing problems. Transp. Sci. **22**(3), 161–172 (1988)
10. Berman, O., Jaillet, P., Simchi-Levi, D.: Location-routing problems with uncertainty. In: Facility Location: A Survey of Applications and Methods, vol. 106, pp. 427–452 (1995)
11. Nagy, G., Salhi, S.: Location-routing: issues, models and methods. Eur. J. Oper. Res. **177**(2), 649–672 (2007)
12. Lopes, R.B., Ferreira, C., Santos, B.S., Barreto, S.: A taxonomical analysis, current methods and objectives on location-routing problems. Inter. Trans. Oper. Res. **20**(6), 795–822 (2013)
13. Tuzun, D., Burke, L.I.: A two-phase tabu search approach to the location routing problem. Eur. J. Oper. Res. **116**(1), 87–99 (1999)
14. Ghiani, G., Laporte, G.: Location-arc routing problems. Opsearch **38**(2), 151–159 (2001)
15. Belenguer, J.M., Benavent, E., Prins, C., Prodhon, C., Calvo, R.W.: A branch-and-cut method for the capacitated location-routing problem. Comput. Oper. Res. **38**(6), 931–941 (2011)
16. Ponboon, S., Qureshi, A.G., Taniguchi, E.: Branch-and-price algorithm for the location-routing problem with time windows. Transp. Res. Part E Log. Transp. Rev. **86**, 1–19 (2016)
17. Capelle, T., Cortés, C.E., Gendreau, M., Rey, P.A., Rousseau, L.M.: A column generation approach for location-routing problems with pickup and delivery. Eur. J. Oper. Res. **272**(1), 121–131 (2019)
18. Panicker, V.V., Reddy, M.V., Sridharan, R.: Development of an ant colony optimisation-based heuristic for a location-routing problem in a two-stage supply chain. Int. J. Value Chain Manag. **9**(1), 38–69 (2018)
19. Asefi, H., Lim, S., Maghrebi, M., Shahparvari, S.: Mathematical modelling and heuristic approaches to the location-routing problem of a cost-effective integrated solid waste management. Ann. Oper. Res. **273**(1-2), 75–110 (2018)
20. Moon, I., Salhi, S., Feng, X.: The location routing problem with multi-compartment and multi-trip: formulation and heuristic approaches. Transportmetrica A Transp. Sci. **1–35** (2020)
21. Fan, H., Wu, J., Li, X., Jiang, X.: Presenting a multi-start hybrid heuristic for solving the problem of two-echelon location-routing problem with simultaneous pickup and delivery (2E-LRPSPD). J. Adv. Transp. (2020)
22. Oudouar, F., Lazaar, M., El Miloud, Z.: A novel approach based on heuristics and a neural network to solve a capacitated location routing problem. Simul. Model. Pract. Theory **100**, 102064 (2020)
23. Prodhon, C., Prins, C.: A memetic algorithm with population management (ma—pm) for the periodic location-routing problem. In: International Workshop on Hybrid Metaheuristics, pp. 43–57. Springer (2008)

24. Pirkwieser, S., Raidl, G.R.: Variable neighbourhood search coupled with ILP-based very large neighbourhood searches for the (periodic) location-routing problem. In: International Workshop on Hybrid Metaheuristics, pp. 174–189. Springer (2010)
25. Prodhon, C.: A hybrid evolutionary algorithm for the periodic location-routing problem. Eur. J. Oper. Res. **210**(2), 204–212 (2011)
26. Albareda-Sambola, M., Fernández, E., Nickel, S.: Multi period location-routing with decoupled time scales. Eur. J. Oper. Res. **217**(2), 248–258 (2012)
27. Hemmelmayr, V., Smilowitz, K., de la Torre, L.: A periodic location routing problem for collaborative recycling. IISETrans. **49**(4), 414–428 (2017)
28. Mohamed, I.B., Klibi, W., Sadykov, R., Şen, H., Vanderbeck, F.: A benders decomposition approach for the two-echelon stochastic multi-period capacitated location-routing problem (2019)
29. Koç, Ç.: A unified-adaptive large neighbourhood search metaheuristic for periodic location-routing problems. Transp. Res. Part C Emerg. Technol. **68**, 265–284 (2016)
30. Prodhon, C.: A metaheuristic for the periodic location-routing problem. In: Operations Research Proceedings 2007, pp. 159–164. Springer (2008)
31. Liu, S.-C., Lu, M.-C., Chung, C.-H.: A hybrid heuristic method for the periodic inventory routing problem. Int. J. Adv. Manuf. Technol. **85**(9-12), 2345–2352 (2015)
32. Hemmelmayr, V.C.: Sequential and parallel large neighbourhood search algorithms for the periodic location routing problem. Eur. J. Oper. Res. **243**(1), 52–60 (2015)
33. Tunalıoğlu, R., Koç, Ç., Bektaş, T.: A multi period location-routing problem arising in the collection of olive oil mill waste water. J. Oper. Res. Soc. **67**(7), 1012–1024 (2016)
34. Nasherahkami, N., Bashiri, M., Bagherinejad, J.: Modeling of periodic location routing problem with time window and satisfaction dependent demands. In: 2015 IEEE International Conference on Industrial Engineering and Engineering Management (IEEM), pp. 1457–1461 (2015)
35. Amiri, M., Sadjadi, S., Tavakkoli-Moghaddam, R., Jabbarzadeh, A.: Optimal fleet composition and mix periodic location-routing problem with time windows in an offshore oil and gas industry: a case study of National Iranian Oil Company. Scientia Iranica. Trans. E Ind. Eng. **26**(1), 522–537 (2019)
36. Talbi, E.G.: Metaheuristics: From Design to Implementation, vol. 74. Wiley, Hoboken (2009)
37. Metropolis, N., Rosenbluth, A.W., Rosenbluth, M.N., Teller, A.H., Teller, E.: Equation of state calculations by fast computing machines. J. Chem. Phys. **21**(6), 1087–1092 (1953)
38. Kirkpatrick, S., Gelatt, C.D., Vecchi, M.P.: Optimization by simulated annealing. Science **220**(4598), 671–680 (1983)
39. Tang, S., Tan, Y., Sapuan, S., Sulaiman, S., Ismail, N., Samin, R.: The use of Taguchi method in the design of plasticinjection mould for reducing warpage. J. Mater. Process. Technol. **182**(1–3), 418–426 (2007)
40. Otto, K.N., Antonsson, E.K.: Extensions to the Taguchi method of product design (1993)
41. Kumar, A., Motwani, J., Otero, L.: An application of Taguchi's robust experimental design technique to improve service performance. Int. J. Qual. Reliab. Manag. **13**(4), 85–98 (1996)
42. Osman, I.H.: Meta strategy simulated annealing and tabu search algorithms for the vehicle routing problem. Ann. Oper. Res. **41**(4), 421–451 (1993)

Smart Manufacturing of Electric Vehicles

Büşra Alptekin[1(✉)], Bahadır Tunaboylu[2], Selim Zaim[3], and Pietro Perlo[4]

[1] Industrial Engineering, Istanbul Technical University, Istanbul, Turkey
alptekinb20@itu.edu.tr
[2] Metallurgical and Materials Engineering, Marmara University, Istanbul, Turkey
bahadir.tunaboylu@marmara.edu.tr
[3] Industrial Engineering, İstanbul Sabahattin Zaim University, Istanbul, Turkey
selim.zaim@izu.edu.tr
[4] Interactive Fully Electrical Vehicles (I-FEVS), La Loggia, Italy
pietro.perlo@ifevs.com

Abstract. The future of the manufacturing sector is evolving with the Industry 4.0, and its repercussions are felt in the factory, the business, product launch and also impacts customer aspects in the overall cycle. These technological innovations create opportunities for new disruptive entries to the market, for instance in automotive manufacturing. It is well known that the automotive industry is under the control of large OEMs since the very beginning. It is very difficult to compete as a newcomer because the production process of the Internal Combustion Engine (ICE) and the chassis require high technology, investment, and a well-defined supply chain that create a monopolistic environment for the sector. Therefore, we propose a microfactory concept as a novel manufacturing method that allows a quicker entry for new players in the automotive industry especially for electric vehicles (EVs).

The microfactory described in this paper, presents an alternative solution to overcome the existing monopolies with the electric motor, a tubular chassis allows for lean manufacturing, and a novel supply chain model. Digital transformation shows itself in production systems by Industry 4.0 applications and more recently in the blockchain-based supply chain applications are being investigated. Further, a detailed cost analysis to compare conventional car manufacturing and microfactory concept EV manufacturing are presented. In addition, the blockchain technology is proposed to improve the supply chain of EV manufacturing. This study indicates that the microfactory provides flexible and customized production of urban electric vehicles minimizing both ecological footprint and total investment.

Keywords: Advanced manufacturing technology · Electric vehicles · Flexible manufacturing · Sustainable production · Sustainable supply chain · Automotive industry

N. M. Durakbasa and M. G. Gençyılmaz (Eds.): ISPR 2020, LNME, pp. 767–773, 2021.
https://doi.org/10.1007/978-3-030-62784-3_64

1 Introduction

Almost a century later, electric vehicles are being discussed as a superior alternative in many ways to conventional ones, thanks to recent technological developments. The increasing trends in smart city concepts have also affected electromobility needs towards autonomous/electric vehicles (EV) with high security capabilities. Because, the mobility needs such as safety, lower emissions (CO_2, noxious gases, noise), congestion, energy security, employment and the resources can be met by safer, more efficient and lower footprint electric vehicles. With this motivation, the automotive industry is in the transformation process from Internal Combustion Engine (ICE) to electric motors to reduce global CO_2 emissions. Thus, the electrification, connected vehicles, changing ownership models and autonomous driving are important trends transforming the industry. The sales rate also verify that there is an increasing trend in EV sales year by year, for instance global stock of electric passenger cars reached 5.1 million units in 2018 [1] and global plug-in vehicle sales reached 2.2 million units for 2019 which was 2.1 million for 2018 and 64% higher than in 2017 [2]. China is still the largest growth contributor, with 1.2 million EV sales, which are 56% of all EV sales in 2018. For European market, plug-in vehicle sales -include all Battery Electric Vehicles (BEV) and Plug-in Hybrids (PHEV), passenger cars and light commercial vehicles reached 408 000 units in 2018, which is 33% higher than in 2017. Norway is the share leader, where 40% of new car sales were Plug-ins in 2018, Iceland and Sweden following closely. EV sales in USA increased by 79%, and Tesla Model-3 is the main contributor [1, 2].

1.1 Mobility Needs and Trends

Mobility as a service (MaaS) providers need safe, personalized and small EVs (four/three/two wheels) with affordable prices. To decrease the total cost of ownership, new manufacturing models should be developed for low investment, flexible and multipurpose platforms which have high convergence with renewable energies. EVs are claimed as the best alternative for environment, however, even though EVs have zero tailpipe emissions they are not GHG emissions-free when evaluated on a well-to-wheel basis. Nonetheless, increasing debate on electrification leads to some interesting future scenarios for 2030. Many reports indicate that the electromobility leads to a radical reduction of green gas emissions and use of oil [1, 3, 4, 5, 6]. For instance, a typical mid-size EV can generate up to 67% lower GHG emissions than a gasoline ICE car on a well-to-wheel basis [7, 8] which means they pollute much less than their combustion engine counterparts. As technology becomes widespread, it is likely to become more efficient and sustainable. Economies of scale will benefit EV manufacturing by offering better infrastructure, more efficient production techniques, recycling options and reducing the need for the mining of new materials.

However, to be competitive in the automotive sector, which consists of different sub-segments, using existing manufacturing infrastructure and products, it is difficult to penetrate and get an international market share for new comers. Because the current production systems heavily depend on conventional manufacturing where high capital cost and high technological knowledge required to operate, is mostly under the control of large OEMs. Indeed, the demand for higher quality, lower cost, and more customizable mobility systems leads transportation-related industries to invest more in advanced, smart, and sustainable technologies in the production of higher quality, affordable, energy-saving, competitive, and customizable vehicles. The microfactory can break this monopoly between large OEMs and Tier1 suppliers for new comers or emerging countries who want to manufacture their national car brands.

For the purpose, this paper represents the microfactory model as an alternative solution to automotive industry to produce EVs in a smart way to meet the urban mobility needs.

2 Microfactory Concept

The main drawback of conventional manufacturing is the complexity of conventional chassis of vehicle (Fig. 1). The huge investment is needed to produce body in white (BIW) because of the complexity of operations for moulds to shape metal sheets in a 3D geometry, followed by complexity of tooling for welding operations. Moreover, the conventional manufacturing needs large production volumes to reach ROI, which also leads to giga-scale production plants. The another challenge of this existing system is the lack of flexibility in production lines, for instance there is a great difference in operations between the chassis with one and two-doors on the same side.

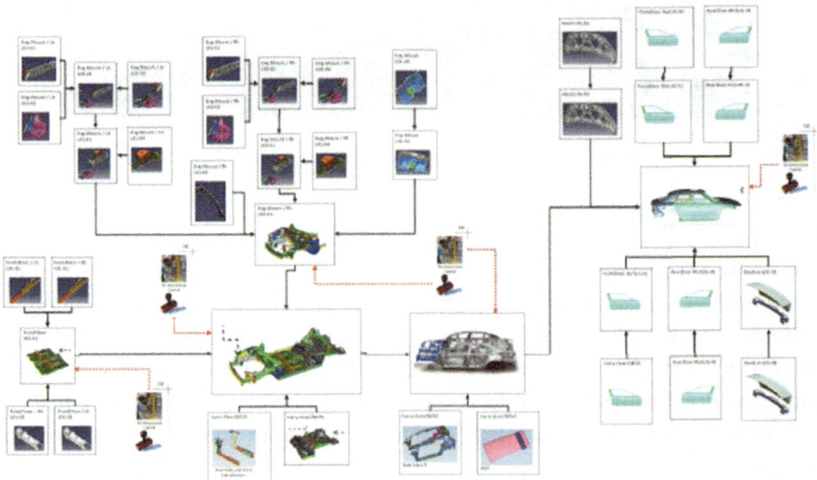

Fig. 1. The conventional assembly line

A remarkable novelty has been introduced in this study which involves a tubular-based chassis design to facilitate the coupling of body panels. The developed tubular chassis using a blend of super high strength steel (SHSS) with defined characteristics on the strength, the thickness, and sizes is manufactured with no moulds and simplified templates. Moulds are the major cost items in conventional manufacturing lines. The result is that very low investment is needed to meet the most severe EuroNCAP crash tests (full frontal, off-axis and side crash tests). Therefore, the microfactory described in this paper, differentiates itself as a concept, which can be defined as low investment manufacturing of safe, low footprint, secure and efficient urban electric vehicles using digital technologies. Within the microfactory, the production can cover all needs for urban mobility from e-bike, 3 wheel to 4 wheel vehicles with minimal additional efforts. Each microfactory is conceived to produce 50 vehicles/day over 2 shifts, all the production processes from laser cut, assembling of the chassis, assembling of the powertrain and the battery pack, painting, the integration of interior and external parts (panels, lights, sensors, seats, etc.) and testing are performed inside this factory. For the purpose, a 10,000 msq plant is enough (Fig. 2). When the demand exceeds what is possible in the third shift, another microfactory has to be established thus introducing parallel manufacturing concepts.

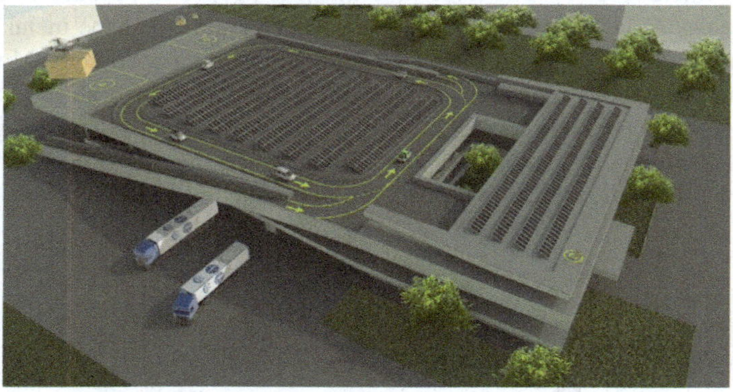

Fig. 2. The energy independent microfactory

The proposed design is also characterized by allowing the manufacturing of considerably different architectures with minimal changes in the production line. Figure 3 shows different architectures resulting from a flexible manufacturing assembly line: small changes to the chassis used to manufacture a pickup car, a food delivery and restaurant car allow the switch to passenger cars and taxi models. In summary, new investment is not required to produce six different platform architectures that can be further personalized to about 120 different aesthetic body panels within the microfactory.

Microfactories are highly integrated with Industry 4.0 technologies. The automation infrastructure of Microfactory is provided by automotive assembly line manufacturers such as Comau. The microfactory co-developed with Comau is capable of the batery assembly, e-motor assembly and body assembly, presented in Fig. 4. Microfactory provides a simplified manufacturing layout which allows minimal changes in the chassis to manufacture in a flexible way a variety of vehicles. The robust SHSS steel-based chassis eliminates moulds in production line with appropriately designed in 3D printed parts.

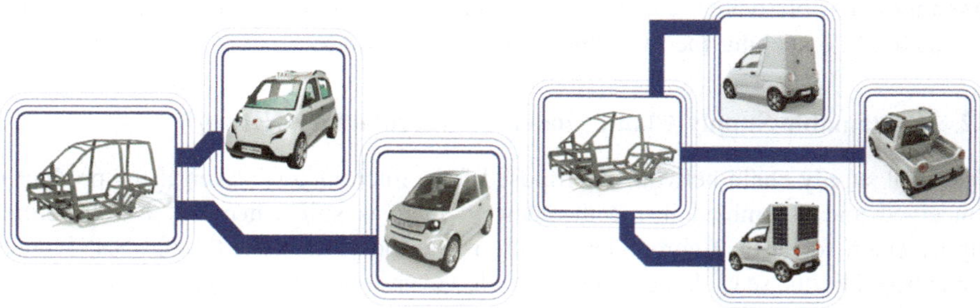

Fig. 3. Flexible manufacturing

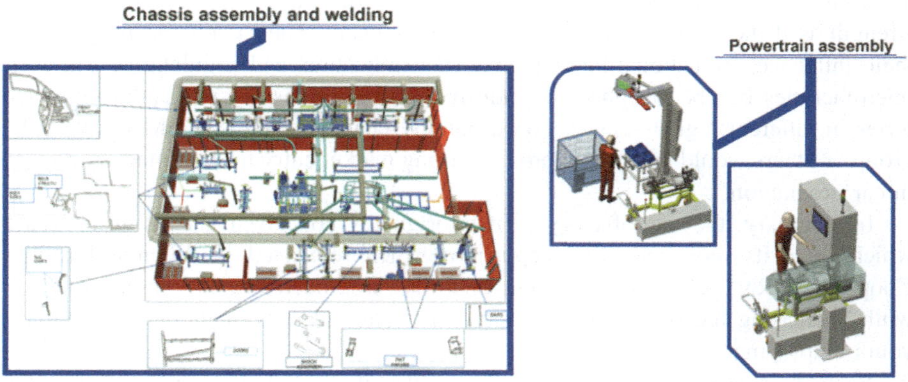

Fig. 4. The assembly line of microfactory

2.1 Cleaner Energy Management

Electric motors have zero tailpipe emissions when it compared to ICE vehicles. The source of energy is crucial to make EVs superior to conventional cars. The microfactory is designed as energy independent which the solar panels produce the necessary energy to run the factory (Fig. 2).

The technological developments in batteries triggered the launch of EVs in the industry today. The research in these two fields are supporting each other. These innovations lead to decrease in battery cell prices too, for instance, the price of battery cell for global automakers was around $700 in 2013, it is assumed that it will fall by 40–46% until 2025 and, it will almost catch to the Tesla's battery advantage for the industry [9]. Lithium-ion battery pack prices have fallen by 21% every year and remained at $176/kWh in 2018.

In line with the EV growth, different players along the battery supply chain are increasing the capacities of the module and the package production. For instance, Tesla has the 239 GWh Li-on capacity for the module and pack, Volkswagen has 206 GWh for a pack, and Daimler has 72 GWh for module and pack [10]. Within the microfactory approach, the battery tray should achieve high-level thermal control, allocated a maximum possible battery capacity, be ergonomic for driver and passengers and easy to install while keeping the investment as low as possible.

2.2 Sustainable Supply Chain Model for Microfactory Network

The local supply-chain network is proposed for a microfactory. Even if the product is conceived and assembled inside the factory, there is still a need of a well-defined supply chain for critical components such as the motor/inverter, tubular SHSS, electronic board lights, seats, battery cells, etc. However, low volume manufacturers cannot easily reach the large Tier1 automotive suppliers, in fact in most cases from a supplier point of view; it is not worth to supply a single microfactory. When microfactories located in different regions share the use of components/systems the interest of large Tier1 changes. The collaboration among microfactories transforms their weakness into strength as if they behave like a single large company. The same concept applies on their influence on national-international standardization and regulations. Since the microfactories can be designed for Industry 4.0 operations, by replicating microfactories in different regions automation capability enablers such as COMAU, the CLN group or others would potentially provide strong market entry for microfactory concept in car production.

In summary, the microfactory as it is originally conceived here is an approach, which finds its most attractive application when replicated in different locations. Cooperative R&D&I among different microfactories decreases the engineering cost while increasing the social sustainability from many aspects including promotion of youth employment.

Each microfactory must be able to produce its own brand with customization according to demand of its local market. The idea is that the collaboration with other microfactories will make local brand stronger. To sum up, this kind of supply-chain model for the microfactory concept allows cooperation of companies while they are economically independent entities.

For final note for microfactory concept, a blockchain integrated supply chain model can be designed for microfactory networks, since blockchain provides transactions among parties using collaboration and cryptography while eliminating intermediaries [11]. As a result, proposed microfactory model can have fast and secure supply chain model enabled with blockchain technology.

3 Conclusions

The transformation need of the automotive industry towards electrification creates a new market that has the opportunity to break the existing monopoly. In relation to this, connected vehicles, changing ownership models, and self-driving cars become new technologies to be investigated.

In this study, we presented the trends and opportunities towards electromobility and proposed a microfactory approach which provide flexible and customized production of urban electric vehicles minimizing both ecological footprint and the total investment.

For future studies, the blockchain integrated supply chain model can be studied in detail for microfactory network and more quantitative research can be obtained with the application of this proposed system.

References

1. Global EV Outlook: Scaling-up the transition to electric mobility. International Energy Agency, Technology Report, May 2019. https://www.iea.org/reports/global-ev-outlook-2019
2. Irle, R.: EV-volumes.com (2019). https://www.ev-volumes.com/country/total-world-plug-in-vehicle-volumes/. Accessed 07 Dec 2019
3. Europen Comission: EU Energy Trends to 2030 (2009). https://doi.org/10.2833/21664
4. Europen Roadmap: Electrification of Road Transport. ERTRAC, 2nd edn., 2–12 (2012)
5. Tsai, S., Champagne, J.M.: Climate Change and Financial Risk, The Growth of electric vehicles and their impact on oil. Publisher WWF-Hong Kong, December 2016
6. Cembalest, M., Morgan, J.P.: Eye on the Market. Annual Energy Paper, April 2018
7. Mackenzie: EVs up to 67% less emissions intensive than ICE cars (2018). https://www.woodmac.com/press-releases/evs-up-to-67-less-emissions-intensive-than-ice-cars/. Accessed 23 Dec 2019
8. Perlo, P., Ottella, M., Corino, N., Pitzalis, F., Brignone, M., Zanello, D., Ziggiotti, A.: Towards full electric mobility: energy and power systems. In: Barbaro, P., Bianchini, C. (eds.) Catalysis for Sustainable Energy Production, pp. 89–105. Wiley-VCH Verlag GmbH&Co. KGaA, Weinheim (2009)
9. Mc Kinsey: Recharging economies: the EV-battery manufacturing outlook for Europe (2019). https://www.mckinsey.com/industries/oil-and-gas/our-insights/recharging-economies-the-ev-battery-manufacturing-outlook-for-europe
10. Mc Kinsey: Making electric vehicles profitable (2019). https://www.mckinsey.com/industries/automotive-and-assembly/our-insights/making-electric-vehicles-profitable
11. Korpela, K., Hallikas, J., Dahlberg, T.: Digital supply chain transformation toward blockchain integration. In: Hawaii: 50th Hawaii International Conference on System Sciences (2017)

Capstone Projects

A Multi-sided and Multi-model Assembly Line Balancing Problem

Seda Gemici, Emine Otuzbir, İrem Almila Koçyiğit(✉),
Sinem Pekelli, Fethi Tüzmen, Hande Öztop, and Levent Kandiller

Department of Industrial Engineering, Yaşar University, Izmir, Turkey
seda.gemici.90@gmail.com, emine_otuzbir@hotmail.com,
iremm-almila@hotmail.com, sinempekelli@gmail.com,
fethituezmen@gmail.com, {hande.oztop,
levent.kandiller}@yasar.edu.tr

Abstract. In this paper, we study a real-life assembly line balancing problem (ALBP) of a cooler manufacturer brand in Manisa, Turkey. The aim of this study is to create an effective assembly line balancing tool for the company, which minimizes the number of stations and balances the total workloads of the stations while keeping the number of products produced the same. The studied ALBP is Type-1 ALBP that minimizes the number of workstations given a cycle time, which is determined by a bottleneck operation. However, different from the standard Type-1 ALBP, some of the stations are two-sided stations in the studied assembly line. There are special constraints in the studied ALBP, such as concurrent tasks, preemptive tasks, zone-restricted tasks, and parallel tasks. Due to these additional characteristics of the system, a novel mixed-integer linear programming (MILP) model is proposed for the studied ALBP to minimize the number of workstations. A secondary objective, which balances the workload of the stations, is also considered in the proposed MILP model using a lexicographic optimization. The computational experiments show that the proposed MILP model can obtain the optimal solution in reasonable computational time. When the model results are compared with the current system, there is a 44% improvement in the number of stations, on average. Furthermore, a sensitivity analysis is performed to analyze the trade-off between the number of stations and cycle time criteria, employing an ε-constraint method. Finally, a user-friendly DSS is developed by embedding the proposed MILP model.

Keywords: Two-sided assembly lines · Assembly line balancing problem · Parallel and preemptive tasks · Mixed integer programming · Decision support system

1 Introduction

In this paper, we study a real-life assembly line balancing problem at a cooler manufacturer in Manisa, Turkey. Based on our observations, we define the problem as an ALBP. The assembly line in the company is different from the classical ALBPs in the literature. The general features and assumptions of the problem are listed as follows:

- A single model or product is produced on the assembly line at a time.

N. M. Durakbasa and M. G. Gençyılmaz (Eds.): ISPR 2020, LNME, pp. 777–789, 2021.
https://doi.org/10.1007/978-3-030-62784-3_65

- Precedence relationships between the tasks are known.
- Operation times are deterministic.
- The workpiece is transported from one workstation to another on a belt conveyor at a constant speed.
- Parallel stations can be opened for tasks that have longer operation time than the cycle time (parallel tasks).
- Some tasks are done on either the left side or the right side of the workpiece (two-sided line).
- Some tasks are concurrently performed on both sides of the line (concurrent tasks).
- Some tasks must be assigned to specific stations (zone-restricted tasks).
- Some tasks can be done by more than one worker with different amounts (preemptive tasks).

When the assembly line of the company is examined, it is concluded that the cycle time cannot be changed because there is a bottleneck station containing chemical reactions, which determines the cycle time. Therefore, the aim here is not to minimize the cycle time, but to minimize the number of stations, which corresponds to a Type-1 ALBP. In addition, when the assembly line of the company is observed, workload imbalance has been identified in the line. For this reason, after the number of stations is minimized, to balance the workload of the existing stations is another important performance criterion in the project for increasing the line efficiency.

The special conditions of the assembly line must also be taken into consideration when the number of workers is to be minimized and distributed to all workers in a balanced manner. These special cases can be listed as follows; the tasks that take place on a certain machine/platform and therefore can only be performed in a particular zone of the line (zone-restricted tasks), the tasks that must be done by two workers at the same time (concurrent tasks), the tasks that more than one worker has to do in order to not exceed the cycle time (parallel tasks), and the tasks can be assigned to more than one worker with different amounts due to the long operation time (preemptive tasks).

Although there are many academic studies on the assembly line balancing problem in the literature as explained in Sect. 2, the traditional ALBP solution methods cannot be applied directly to the studied ALBP due to the special characteristics of the line such as the existence of two-sided stations, concurrent tasks, preemptive tasks, zone-restricted tasks, and parallel tasks. Hence, a novel mixed-integer linear programming (MILP) model is proposed for the studied ALBP to minimize the number of workstations. A secondary objective, which balances the workload of the stations, is also considered in the proposed MILP model using a lexicographic optimization. Furthermore, a sensitivity analysis is performed to analyze the trade-off between the number of stations and cycle time criteria, employing an ε-constraint method. The proposed model considers all the special characteristics of the line and the immediate precedence relationships of the tasks. Finally, a user-friendly decision support system is developed using the Excel-VBA interface by embedding the proposed MILP model.

The rest of the paper is organized as follows: A concise literature review is also provided in Sect. 2. The proposed MILP model formulation is presented in Sect. 3. The verification and validation of the developed solution methodology are explained in Sect. 4 as well as the sensitivity analysis. In Sect. 4, computational results indicating our contribution in terms of key performance measures are also stated, and the developed decision support system is briefly explained. Finally, the conclusions are made in Sect. 5.

2 Literature Review

Initially, the study of Boysen et al. [1] is examined, which presents a detailed classification for the assembly line balancing problem. The definition of the assembly line balancing problem, various objective functions, and constraints that may be encountered in the problem are mentioned [1]. Considering the literature examined, it was decided that our assembly line is a deterministic, two-sided, single model, and straight type. The aim is to minimize the number of workstations with the given cycle time. Research belongs to Gungor et al. [2], where they studied on a two-sided single-model assembly line balancing problem of a white goods manufacturer. Their aim is to reduce labor costs without changing the number of products produced by reducing the number of stations and balancing the total workload. They developed a mathematical model for the problem and solved the problem by supporting the outputs of their models with a heuristic method. Scholl's book [3] is also examined to build a mathematical model for our problem, which deals with two major decision issues that arise when flow line production systems are installed and operated. Balancing and sequencing problems are discussed, classified and arranged in the first part of the book. Then, precise and intuitive procedures are proposed to solve decision problems. Ayrım [4] also worked on the minimization of workloads. Workers must be allocated to jobs that have appropriate levels of workload for their capacities to ensure production efficiency.

Since the ε-constraint method is employed to perform a parametric analysis for the studied ALBP, a literature review on multi-objective optimization is also made. Mavrotas [5] studied on the effective implementation of the ε-constraint Method in Multi-Objective Mathematical Programming (MOMP) problems. This study discusses the effective application of the ε-constraint method to produce Pareto optimal solutions in a MOMP. He proposes a novel version of the ε-constraint method, i.e., an augmented ε-constraint method – AUGMECON, which avoids the production of weakly Pareto optimal solutions and accelerates the whole process by avoiding redundant iterations. Also, Bérubé et al. [6] studied on an exact ε-constraint method for bi-objective combinatorial optimization problems. This article describes an exact ε-constraint method for bi-objective combinatorial optimization problems with integer objective values. They deal with multi-objective optimization problems by solving a series of single objective subproblems, where all objectives, but except one objective, are converted to constraints. Parvizi et al. [7] also presented a hybrid method to solve the multi-objective optimization problems, which combines the ε-constraint method and a Cuckoo search algorithm. In their method, firstly, the multi-objective optimization problem is transferred to a single-objective optimization problem using the ε-constraint method; then, the Cuckoo search optimization algorithm optimizes the problem in each task. Finally, the optimized Pareto frontier is drawn. According to what they say, the advantage of this method is the high accuracy and distribution of the Pareto frontier.

As mentioned in Sect. 1, there are four special cases in the studied system: concurrent tasks, preemptive tasks, zone-restricted tasks, and parallel tasks. Since the machines of some stations are difficult and costly to move, the work done here is fixed by zone restrictions. Preemptive tasks can be allocated to different workers with different amounts regarding precedence relationships. Concurrent tasks are performed by two workers on both sides of the line together. Regarding these special characteristics of the system, solution procedures for similar problems were investigated in the literature. However, as seen in the review, there is no study in the literature that addresses the same problem with ours. Therefore, we proposed a novel mixed-integer linear programming model for the studied assembly line balancing problem with concurrent tasks, preemptive tasks, zone-restricted tasks, and parallel tasks.

3 Problem Formulatıon

The proposed MILP model for the studied ALBP is presented in this section. The sets, parameters and decision variables of the proposed model are listed below.

Indices

i: index of tasks, $i \in T$
j: index of workers, $i \in W$

Sets

T: Set of tasks,
T_A: Set of tasks with zone restriction,
T_P: Set of preemptive tasks,
T_C: Set of concurrent tasks,
A_i: Set of workers that tasks with zone restriction can be assigned, $i \in T_A$
I_i: Immediate predecessor of task i, $i \in T$
W: Set of workers,
W_P: Set of workers that preemptive tasks can be assigned,
W_C: Set of workers that concurrent tasks can be assigned

Parameters

t_i: Operation time of task i,
C: Cycle time,

Decision Variables

$$x_{ij} = \begin{cases} 1, & \text{if task } i \text{ is assigned to worker j} \\ 0, & \text{otherwise} \end{cases}$$

y_{ij} = amount of task i performed at worker j,

$$z_j = \begin{cases} 1, & \text{if any task is assigned to worker j} \\ 0, & \text{otherwise} \end{cases}$$

Mathematical Model

$$\text{Min} \sum_{j \in w} z_j \tag{1}$$

$$\sum_{j \in W} x_{ij} = 1, \forall i \in T \backslash \{T_P, T_C\} \tag{2}$$

$$\sum_{j \in W_c} x_{ij} = 2, \forall i \in T_C \tag{3}$$

$$x_{ij} = x_{i(j-1)} + x_{i(j+1)}, \forall i \in T_C, \forall j, (j-1), (j+1) \in A_i \tag{4}$$

$$lx_{kl} - 1 \leq (jx_{ij}) + M(1 - x_{ij}), \forall i \in T, \forall j \in W, k \in I_i \cap T_C \tag{5}$$

$$\sum_{j \in W \backslash W_C} x_{ij} = 0, \forall i \in T_C \tag{6}$$

$$\sum_{k \in A_i} x_{ik} = 1, \forall i \in T_A \tag{7}$$

$$\sum_{j \in W_P} y_{ij} = 1, \forall i \in T_P \tag{8}$$

$$y_{ij} \leq x_{ij}, \forall i \in T_P, \forall j \in W_P \tag{9}$$

$$\sum_{j \in W \backslash W_P} x_{ij} = 0, \forall i \in T_P \tag{10}$$

$$\sum_{j \in W \backslash W_P} y_{ij} = 0, \forall i \in T_P \tag{11}$$

$$x_{ij} - My_{ij} \leq 0, \forall i \in T_P, j \in W_P \tag{12}$$

$$\sum_{i \in T \backslash T_P} t_i x_{ij} + \sum_{i \in T_P} t_i y_{ij} \leq C z_j, \forall j \in W \tag{13}$$

$$lx_{kl} \leq (jx_{ij}) + M(1 - x_{ij}), \forall i \in T, \forall j \in W, k \in I_i \backslash T_C \tag{14}$$

$$x_{ij} \in \{0, 1\} \tag{15.1}$$

$$0 \leq y_{ij} \leq 1 \tag{15.2}$$

$$z_j \in \{0, 1\} \tag{15.3}$$

In the MILP model, we assume that each station is defined as a worker. In the primary objective function (1), it is aimed to minimize the number of workers along the line. Constraint set (2) ensures that tasks other than the concurrent and preemptive tasks are assigned to a single worker. Constraint set (3) ensures that concurrent tasks are assigned to two workers, and constraint set (4) provides that workers to be the consecutive workers to whom these tasks are assigned. Constraint set (5) regards the precedence relations between a task and its concurrent predecessor tasks. Constraint set (6) prevents the assignment of concurrent tasks to the workers that cannot be assigned. Constraint set (7) ensures that tasks with zone restrictions are assigned only to the workers at the specified locations. The assignment of preemptive tasks to specific workers, with the sum of one, is ensured by constraint set (8). Constraint set (9) states that if some amount of a preemptive task is assigned to a worker, then this preemptive task must be assigned to this worker. Constraint sets (10) and (11) prevent the assignment of preemptive tasks to ineligible workers. Constraint set (12) regards the balance between x_{ij} and y_{ij} decision variables. Constraint set (13) ensures that the total operation time of the assigned tasks to a worker does not exceed the cycle time. Constraint set (14) regards the precedence relations between a task and its nonconcurrent predecessor tasks. Constraint set (15.1) and (15.3) indicate that the decision variables x_{ij} and z_j can only take binary values of 0 and 1, while constraint set (15.2) indicates that the decision variable y_{ij} can take a decimal value between 0 and 1.

After minimizing the number of workers, a secondary objective function (16) has also been considered in the proposed MILP model for maintaining the workload balance between workers, while preserving the minimum number of workers. This secondary objective function is explained below with additional constraints and variables.

Parameters

S: Standard workload rate

Decision Variables

$O_j =$ Workload difference of worker j, which is over S

$U_j =$ Workload difference of worker j, which is under S

$$\text{Min} \sum\nolimits_{j \in W} O_j + U_j \tag{16}$$

$$\left(\sum\nolimits_{i \in T \setminus T_P} t_i x_{ij} + \sum\nolimits_{i \in T_P} t_i y_{ij}\right)/C - S = O_j - U_j, \forall j \in W \tag{17}$$

$$O_j, U_j \geq 0 \tag{18}$$

The secondary objective function (16) aims to balance the workload of workers, by keeping the number of workers at minimum based on the results of the above model with primary objective function (1). Additional parameter S and decision variables O_j and U_j are included in the above MILP model in order to keep the workload balance of workers at a desired standard value S. Constraint set (17) computes the deviations of the workers' workloads from the standard workload rate S. Constraint set (18) prevents the variables of O_j and U_j from taking negative values. The proposed MILP model is solved using lexicographic optimization by including these additional decision variables and constraints. Namely, the initial primary objective function is optimized, and then, the secondary objective function is optimized by keeping the primary objective function at its optimal value.

4 Computational Results and Decision Support System

4.1 Verification

The aim of this section is to verify our developed solution approach. Hence, a toy problem is prepared below to show how the proposed solution procedure works by using sample operation times that are measured by us in the facility. A set of tasks T = $\{T_1, T_2... T_n\}$ and a set of workers W = $\{W_1, W_2... W_m\}$ are defined. There are 12 tasks and 5 workers in the toy problem. The processing time and immediate predecessors of the tasks are shown in Table 1. The primary objective function, i.e., minimization of the number of workers is considered in this toy problem. In order to ensure that all the constraints are functioning, all special constraints are considered in the toy problem, which originates from the real problem such as zone restrictions, preemptive, parallel, and concurrent tasks. Here are the constraints that we considered: Task 10 must be assigned to Worker 4. Task 4 is a concurrent task and can be assigned to the worker pairs, either 1-2 or 2-3. Tasks 6 and 9 are preemptive tasks and can be assigned to Worker 3, 4, and 5. The model is run on IBM ILOG CPLEX Optimization Studio Version 12.10. Then, the task assignments are provided in Table 2; the total operation time of the tasks assigned to each worker by following the precedence rules does not exceed the cycle time. Also, Table 3 indicates the amounts of preemptive tasks that are assigned to the workers.

Table 1. Toy problem's data

Task no	Operation time	Immediate predecessors	Task no	Operation time	Immediate predecessors
1	20	–	7	10	3
2	20	–	8	5	3, 4
3	30	1	9	40	6, 7, 8
4	10	1, 2	10	10	5
5	20	2	11	15	9, 10
6	45	3	12	15	11

4.2 Validation and Improvements

In order to validate the proposed MILP model, we tried to observe the assembly processes of as many commercial cooler models as possible. We made observations on six different models. We have listed the number of tasks of these six different types of coolers based on the task categorization explained in Sect. 1. Thereby, Table 4 shows

Table 2. Worker-task assignments

Task\Worker	W_1	W_2	W_3	W_4	W_5
1	1	0	0	0	0
2	0	1	0	0	0
3	1	0	0	0	0
4	0	1	1	0	0
5	0	1	0	0	0
6	0	0	1	1	0
7	0	0	0	1	0
8	0	0	0	1	0
9	0	0	0	1	1
10	0	0	0	1	0
11	0	0	0	0	1
12	0	0	0	0	1

Table 3. Preemptive worker-task assignments

Task	W_1	W_2	W_3	W_4	W_5
6	0	0	0.88	0.12	0
9	0	0	0	0.50	0.50

Table 4. Real data

Model of cooler	Total # of tasks	# of concurrent tasks	# of preemptive tasks	# of zone-restricted tasks	# of parallel tasks
Model 1	94	4	8	10	16
Model 2	87	4	8	10	13
Model 3	93	4	8	10	16
Model 4	90	4	8	10	13
Model 5	96	5	8	10	16
Model 6	89	5	8	10	13

the total number of tasks and the number of special tasks in the refrigerant types that we have aimed to improve.

After the accuracy of the developed model is checked with the generated toy data, the MILP model is optimally solved for above six cooler models of the company, on IBM ILOG CPLEX Optimization Studio 12.10.0 in 22 s on average. Consequently, it can be said that the developed MILP model is valid. The optimal solution time of the MILP model for each cooler model can be seen in Table 5. The optimal solution time of the MILP model seems to vary according to the number and type of tasks in the cooler model.

Since the problem can be optimally solved in a very short time, heuristic methods are not required to solve the problem. The optimal solution time of the MILP model with respect to a different number of tasks is shown in Fig. 1. As shown in Fig. 1, when the number of tasks is increased, the solution time increases as expected.

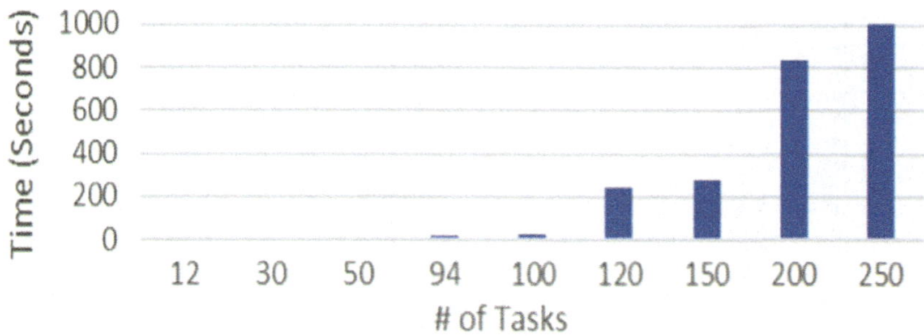

Fig. 1. Optimal solution time vs. number of tasks

The optimal solutions found by the proposed MILP model are compared with the current system. The improvement percentage of each model, as also shown in Table 5, the proposed MILP model improves the number of workers for each cooler model by more than 40%. Furthermore, the weighted improvement percentage of six different cooler models is 44% according to the annual demand rates of these models.

Table 5. Improvement percentages

Model of cooler	# of workers in current system	# of workers in improved system	Improvement percentage (%)	Optimal solution time (seconds)
Model 1	22	12	45	14.14
Model 2	22	12	45	17.32
Model 3	22	12	40	17.20
Model 4	22	13	40	23.12
Model 5	28	13	54	21.05
Model 6	28	10	64	36.70

4.3 Sensitivity Analysis

Moreover, the trade-off between the cycle time and the number of workers criteria is analyzed by employing a parametric analysis using the ε-constraint method. In the ε-constraint method, the problem is evaluated as a bi-objective optimization problem to analyze the trade-off between these two performance criteria. As a result of this analysis, the efficient frontier is created, as shown in Fig. 2, and the relationship between the cycle time and the number of workers criteria is investigated. As shown in Fig. 2, the number of workers decreases as the cycle time increases.

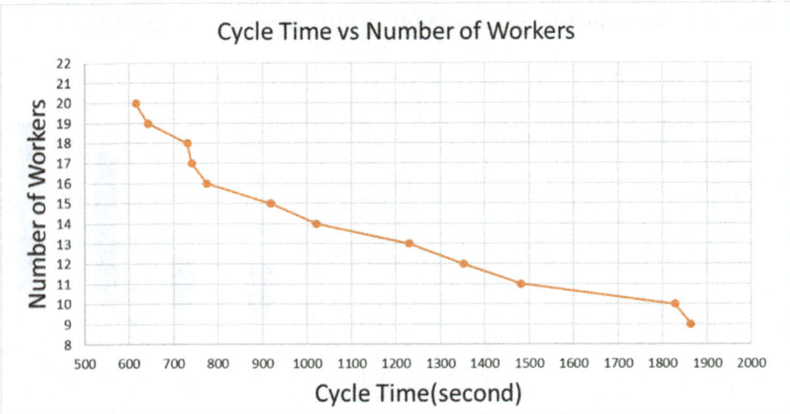

Fig. 2. Efficient frontier (# of workers vs. cycle time)

The effect of the secondary objective function on the workload distribution of workers as also shown in Fig. 3, there are significant changes in the workload distributions of some workers in the assembly line of Model 1, when the secondary objective is employed.

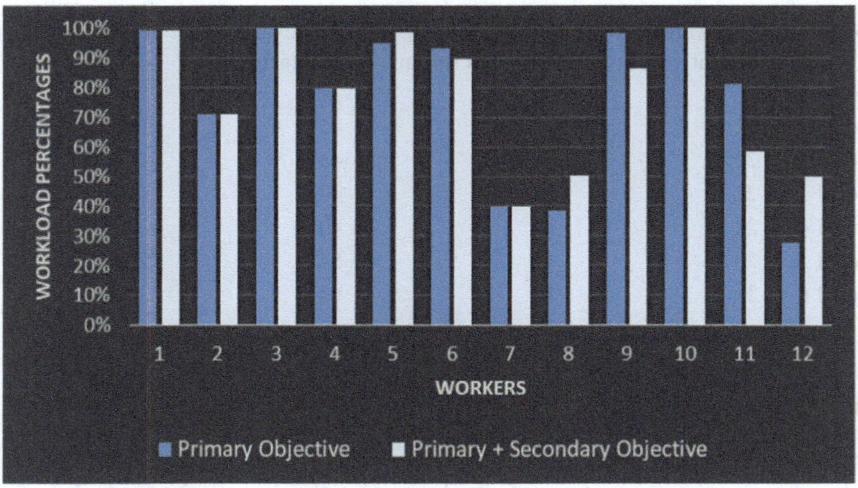

Fig. 3. Effect of secondary objective on workload distribution graph

4.4 Decision Support System

Further to that, a user-friendly decision support system (DSS) is developed using the Excel-VBA interface by embedding the proposed MILP model. The DSS is created by using an Excel VBA interface to increase the ease of use and comprehensibility of the developed work. In this DSS, there is a dynamic structure, which could be modified easily so that the user can edit the pre-existing model data, decode the model data, and add new model data. Moreover, when the company produces a new model, the necessary information such as the number of workers, number of tasks, and types of tasks, etc. is received from users and saved in a new workbook. The DSS creates a report table, which has four main titles as workers, workloads, total time, and assigned tasks. The optimal task-worker assignments are reported after the proposed MILP model is optimally solved. Then, the graphs, such as workload distribution and total work time of each worker, are reported to the user in detail. Next, the user is directed to the analysis screen, which is created in Microsoft Power BI: Data Visualization, as given in Fig. 4. This file has four pages, which are dashboard, workload, task assignment, the primary lead time. The dashboard includes a summary of the results, which are workload distribution of the workers, task assignment of the workers, production lead time of cooler between workers, and KPIs such as the total number of workers in the line. Additionally, these results are converted to a report as a PDF file and directed to the printer.

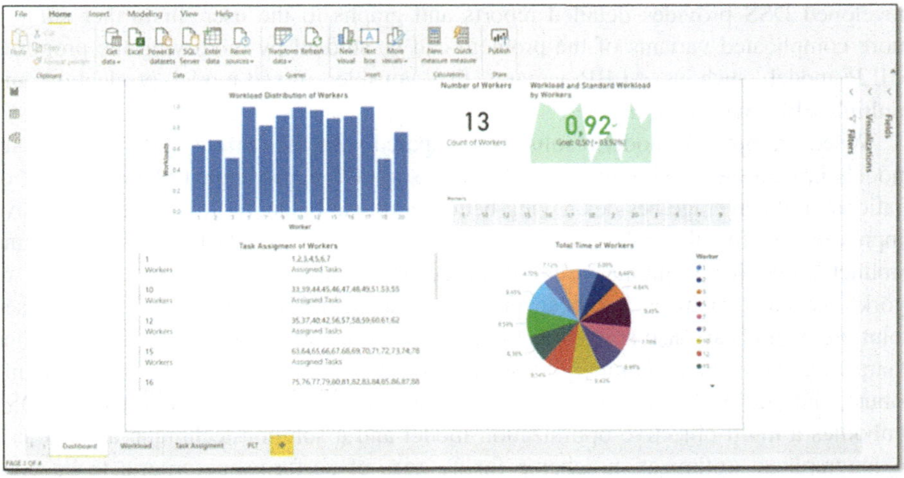

Fig. 4. DSS analysis screen

The decision support system developed within the scope of the project can be implemented on similar assembly lines by editing the database. The model we developed is at TRL 3. Their algorithms are at TRL 4, and the decision support system prototype is at TRL 6. In addition, it is possible to increase the technology readiness level above TRL 6 with the support of a software company.

5 Conclusion

The classic assembly line balancing literature is usually focused on the simple assembly line balancing problem. However, studies involving more realistic constraints have been increasing in recent years. In this project, we study a real-life ALBP of a cooler manufacturer brand in Manisa, Turkey.

The studied ALBP is Type-1 ALBP that minimizes the number of workstations given a cycle time, which is determined by a bottleneck operation. However, different from the standard Type-1 ALBP, some of the stations are two-sided stations in the studied assembly line. Furthermore, there are special constraints in the studied ALBP, such as concurrent tasks, preemptive tasks, zone-restricted tasks, and parallel tasks. Due to these additional characteristics of the system, a novel mixed-integer linear programming (MILP) model was proposed for the studied ALBP to minimize the number of workstations. A secondary objective, which balances the workload of the stations, is also considered in the proposed MILP model using a lexicographic optimization. By means of the secondary objective, the workload distribution of the workers was tried to be balanced according to a standard rate. The computational experiments show that the proposed MILP model is able to obtain the optimal solutions in reasonable computational time. Furthermore, a sensitivity analysis was performed to analyze the trade-off between the number of stations and cycle time criteria employing the ε-constraint method. Finally, a user-friendly decision support system was developed using the Excel-VBA interface by embedding the proposed MILP model. The developed DSS provides detailed reports and graphs to the user. In further studies, more complicated variants of the problem can be studied by modifying the proposed MILP model, such as ALBP variants that consider mixed-model production and multiple objective functions.

When the optimal model results are compared with the current system for the six models having the maximum demand, there is a 44% improvement in the number of stations on the average besides having better workload balances. It is expected to have improvements for all cooler models since other cooler models belong to the same product group have similar assembly processes. Thus, more efficient utilization of workforce and machines has been provided via the DSS developed. The proposed solution methods are also applicable to other production environments with similar line characteristics; even the number of tasks is beyond 250. The model formulation is quite robust, and our DSS reports the optimal solutions in a very reasonable time. Our DSS embodies a multi-objective optimization model and a solution technique as well as a legicographical optimization scheme for a series of performance criteria to be optimized. Thus, the DSS developed in this study help managers to take line design and balancing decisions effectively, even at the operational level. The DSS reduces the time that it takes these decisions to have an applicable setting by the operational planner, yet improving the efficiency of the line considerably.

Acknowledgments. This study is supported by TUBITAK (The Scientific and Technological Research Council of Turkey) in the program of "2209-B Undergraduate Thesis Support Program for Industrial Applications". This work cannot be completed without the assistance of Müge Özer and Buse Geniş. We are thankful for their contribution. Furthermore, we are grateful to the company for their co-operation.

References

1. Boysen, N., Fliedner, M., Scholl, A.: A classification of assembly line balancing problems. Eur. J. Oper. Res. **183**, 674–693 (2007)
2. Gungor, C., Kandiller, L., Kayserili, C., Matlar, A.G., Mutlum, E., Oztop, H., Sensu, C.: Montaj Hattı Dengeleme Problemi Whirlpool - Indesit Buzdolabı Fabrikası. In: Oner, A. (ed.) SİSTEM TASARIMI PROJE ÖZETLERİ, pp. 70–83 (2017)
3. Scholl, A.: Balancing and Sequencing of Assembly Lines. Physica-Verlag Heidelberg, Heidelberg (1999)
4. Ayrım, Y., Can, G.: İş Yükü Minimizasyonunu Hedefleyen Bir İşçi Atama Modeli Önerisi. Mühendislik Bilimleri ve Tasarım Dergisi **6**, 148–158 (2018). https://dergipark.org.tr/tr/pub/jesd/issue/41245/359201
5. Mavrotas, G.: Effective implementation of the e-constraint method in multi-objective mathematical programming problems. Appl. Math. Comput. **213**, 455–465 (2009)
6. Bérubé, J.F., Gendreau, M., Potvin, J.Y.: An exact ε-constraint method for bi-objective combinatorial optimization problems: application to the traveling salesman problem with profits. Eur. J. Oper. Res. **194**, 39–50 (2007)
7. Parvizi, M., Jahani, N., Shadkam, E.: A hybrid COA/ε-constraint method for solving multi-objective problems. Int. J. Found. Comput. Sci. Technol. **5**, 27–40 (2015)

A Parallel Machine Scheduling Problem for a Plastic Injection Company

Aytaç Ali Aslaner, Akdeniz Coşkun, Nur İlayda Tülemiş, Ece Özboyacı, Nil Ergün, Berk Bulut, Görkem Gülbent, Mert Paldrak, and Efthymia Staiou(✉)

Department of Industrial Engineering, Yasar University, Izmir, Turkey
aytac.aslaner@hotmail.com, akdenizcoskun@hotmail.com, ilaydatulemis@gmail.com, eceozboyaci@gmail.com, nilergun97@hotmail.com, berkkbuluttt@gmail.com, gorkemgulbent@gmail.com, {mert.paldrak,efthimia.staiou}@yasar.edu.tr

Abstract. In this study, a production planning and scheduling problem is carried out in a company operating in the plastic injection sector in Turkey. The scheduling problem is classified as unrelated parallel machine scheduling problem with sequence dependent setup times, including job-machine eligibility, common source and precedence constraints. The proposed solution methodology is based on weekly planning with order grouping and makespan minimization. As a result, a mixed integer programming model was formulated, and optimal results are obtained for a sample data set with real data received. A user-friendly Decision Support System (DSS) was created to help the company with the weekly production planning and scheduling process, applying a heuristic approach solution.

Keywords: Parallel machine scheduling · Sequence-dependent setup times · Precedence relations · Common molds · Machine eligibility · Heuristic approach · Best-fit algorithm

1 Introduction

Parallel machine scheduling has attracted attention from both academia and industry. Theoretically, parallel machine scheduling problems are a generalized version of single machine scheduling and a special case of flexible flow shop scheduling [1]. The techniques which are developed for parallel machines can also be applied to multi-state systems. In practice, resources are used in parallel and they can be identical, non-identical and unrelated. Several types of problems may arise in manufacturing systems such as scheduling, job assignment, tool loading and tool switching [2]. In this study, a parallel machine scheduling problem is examined, and a solution has been developed for a company operating in the plastic injection industry. Considering the minimum inventory policy of the company, an order-based approach, pull system, was deemed appropriate. The most important characteristics of the problem are determined as: eligibility of the products, molds and injection machines, precedence relations between

N. M. Durakbasa and M. G. Gençyılmaz (Eds.): ISPR 2020, LNME, pp. 790–803, 2021.
https://doi.org/10.1007/978-3-030-62784-3_66

the jobs, sequence and color dependent setup times, products which are using a common mold, difference between the number of cavities of the molds. In the suggested system, preemption is not allowed, in order to be able to schedule the jobs more systematically. In order to eliminate unnecessary setup times, orders which include same products are grouped and these orders are considered as one order. There are 19 injection machines and 3 duroplast injection machines in the factory. These machines are able to produce almost 600 different types of products. To produce some of the main products, sub-products must be produced as a predecessor.

The different stages of the production process in our study could be described as: plastic injection, duroplast press, intermediate assembly, outsourcing coating, outsourcing painting, outsourcing intermediate assembly and final assembly. The plastic injection sub-process is carried out by melting the plastic raw material at high temperature and injecting it into a mold. Duroplast press is the sub-process of shaping the raw material under high pressure. After pouring the duroplast material into the product molds, the product takes its main shape by processed in high pressure machines. Afterwards the workers remove the burrs which are on the product with the help of appropriate tools. After injection/press processes, some of the sub-products require an intermediate assembly process, were the small finished parts are assembled with each other. After this phase, the assembled parts continue to the next stages of the process. Since only the plastic injection operations are carried out in the factory, if the parts need to be painted or coated, the painting and coating processes are made by contracted manufacturers. When the sub-products are coated and painted, they proceed to the final assembly line. At this stage, all the finished components of a product are assembled. After the final assembly phase, all the final goods are ready to be packaged and shipped to the customers.

The paper continues with the detailed description of the problem definition. After defining the problem, the literature review section follows, presenting the studies examined during the developing phase of the solution methodology. Following the literature review, the problem formulation and the mathematical model developed is presented. In the solution methodology section, the developed heuristic approach is explained followed by the Decision Support System section which was developed to serve the weekly scheduling needs of the company applying the heuristic approach. Afterwards, the verification and validation of the proposed solution and results are presented. Finally, the conclusions of the study are summarized.

2 Problem Definition

2.1 Current System Observations and Symptoms

Currently, in the company, production planning engineers create the production schedule for a short-term period based on their experience. For instance, incoming orders are to be produced immediately according to the availability of the workers and the appropriate machine-mold pair. Because of these instantaneous decisions made and the cursory assignments, efficiency of the workers and the machines significantly decreases. Since priority is given to the overseas customers, production of the orders received from domestic customers are sometimes postponed. In addition, due date data for incoming orders are not properly kept by the sales department, which results in problems in the planning activities. Also, production planning is based on a make-to-order system for the majority of the products, and the company's policy for keeping minimum stocks, may result in inefficient production planning causing delays and deficiencies especially for the products which are not kept in stock.

2.2 Problem Definition

The main goal of this study is to create a production planning system that minimizes makespan in the production process. The problem is to determine the least time-consuming sequence of the jobs assigned to each identical machine. Therefore, the problem is defined as a parallel machine scheduling problem with identical machines.

There is only one unit of each mold type and each mold can produce only one type of a product. Some of the products have color options and although the products with different colors are considered as different products, these products are processed using the same mold. Also, each mold has different number of mold cavities. The factory has approximately 600 different types of molds and products. In addition, the molds can only be used with the technically appropriate machines.

When the orders are received, the raw materials and related molds are getting prepared for the injection process and after the production, deburring process is performed. Then, if an assembly process is necessary, the products are sent to the assembly line before the packaging process. If an assembly process is not required, the products are directly sent to the packaging line and prepared for shipment.

For solving the scheduling problem, a mixed integer programming model is formulated with the objective of minimizing the maximum completion time. In order to find an optimal solution for the problem, IBM ILOG CPLEX Optimization Studio 12.8.0 is employed to solve the mathematical model. Due to the size of the problem, CPLEX cannot solve the problem with optimality in a reasonable amount of computational time. Consequently, a heuristic algorithm is proposed to find a feasible solution in a shorter computational time. The proposed heuristic solution has been coded via Python and was integrated within the developed Decision Support System (DSS). Solving the problem with the mathematical model and the heuristic approach at the same time has been very useful in verifying and validating the results. Thus, the feasibility of the solution has been proven.

3 Literature Review

In the literature, there are numerous articles considering scheduling problems with minimum makespan. However, the literature on makespan is rather limited when the unrelated parallel machine scheduling problem with eligibility constraints are taken into consideration. Most studies in the literature focus on sequence independent setup times, which are assumed to be included in the total processing times [3, 4].

The problem of scheduling injection machines is a special case of parallel machine scheduling. Preparation times can be considerably longer in the production of plastic parts with injection machine, when the production of a light colored part after a dark colored part is scheduled, or when some specific raw materials have to be used after different types of raw materials. On the other hand, in order to produce a product in injection machines, the mold of this product must be connected to the appropriate machine. Tuğba Saraç and Aydın Sipahioğlu [1] have stated that although injection machines are identical in terms of production time, each machine is not an alternative to each other since all molds cannot be used on all machines. In their study, the mold usage of the machines takes into consideration preparation times depending on the sequence. They have proposed a mathematical model based on the knapsack problem in their solution.

Identical parallel machine scheduling problem has three important constraints: sequence-dependent preparation times, machine availability constraints and common resources [5].

During the literature review phase, different articles presenting different mathematical models and heuristic approaches to solve scheduling problems having similar characteristics to the problem of this study are reviewed. According to Cemalettin Öztürk and Arslan M. Ornek [6], a little modification in the proposed mixed integer programming model could be used to quote delivery times for customer orders in case due dates are not specified. This mixed integer programming model considers capacity constraints, operation sequences, processing times and due dates in a multi-machine, multi-order, multi-item environment where an item can be processed on a given set of eligible machines

Scheduling problems which are combinatorial optimization problems cannot be formulated as linear programs and there are no simple rules or algorithms that ensure optimal solutions in a limited amount of computational time [8]. It is difficult to find a solution to the problem by deterministic methods. Effective constructive meta-heuristic algorithms were presented to solve parallel machine scheduling problems [4]. Therefore, the genetic algorithm was presented to solve parallel machine scheduling problems in [7–10]. The experimental results of the genetic algorithm for the parallel machine scheduling problem with an objective of minimizing makespan and the number of tardy jobs, demonstrate that the genetic algorithm encoding method is performing successfully [8]. In addition, simulated annealing algorithm, which is a meta-heuristic algorithm, can be used to solve combinatorial optimization problems [11]. Simulated annealing algorithm is also used in several studies [12, 13].

Nevertheless, in this study, the unrelated parallel machine scheduling problem with sequence dependent setup times and eligibility constraints is studied with the objective of minimum makespan. A constructive heuristic algorithm is developed in order to solve the large instances in a reasonable amount of computational time. Furthermore, a user-friendly Decision Support System has been developed by integrating the heuristic algorithm.

4 Problem Formulation

In the current system, short-term plans are created instantaneously based on the incoming orders. The company prioritizes the orders according to customer importance and considering the economic returns and regulates the production accordingly. Consequently, changes in orders' priorities can occur, causing delays in their delivery and depending on the contract with the customer, may result in penalty costs and customer dissatisfaction.

In addition, when the production of a different job is about to start, the machine requires a mold changeover which is a very time-consuming process. Since there is not an optimal schedule for the machines, unnecessary setup times are very high between the production of different products. Each mold has a different structure, therefore the time of removing-placing the mold varies between the molds. Also, for some product groups color of the product may vary according to the customer's requirement. Therefore, the company cannot keep stock for these products and this situation can create delays in their delivery to the customer.

Another point that must be considered while dealing with the problem is that each product needs a single, specific mold to be produced. However, some molds can be used as a common mold while producing different products. The main reason for this situation is that different colors of a product are considered as different jobs. Thus, during the planning and scheduling stages, it is important not to assign these jobs in the same time frame.

Considering all the above constraints, a mathematical model for solving the unrelated parallel machine scheduling problem with sequence dependent setup times and eligibility constraints has been developed to achieve an optimal schedule for assigning jobs to machines with the goal of minimizing the total completion time (makespan). For the model formulation, we use the mathematical model developed by Cemalettin Öztürk and Arslan M. Ornek [6] which includes completion time, disjunctive, assignment, job-machine eligibility and conflicting constraints. In addition, we extend the model by adding common mold and precedence constraints that will be explained in detail later.

The proposed mathematical model, the sets and indices, the parameters and the decision variables together with the objective function and constraints of the model are explained in detail below:

Sets and Indices

N = number of jobs $i, j = \{1,2, \ldots |N|\}$

L = number of machines $l = \{1,2, \ldots |L|\}$

W = number of molds $w = \{1,2, \ldots |W|\}$

G_w = Set of jobs that can be processed with mold w

T = Set of common molds

Parameters

P_j = Cycle time of job j

E_{ij} = Total setup time between job i and job j considering the mold changes and color changeover

M = A big number

G_w = Set of jobs that can be processed with mold w

T = Set of common molds

$$F_{jl} = \begin{cases} 1 & \text{if job } j \text{ can be processed on machine } l \\ 0 & \text{otherwise} \end{cases}$$

$$A_{ij} = \begin{cases} 1 & \textit{if} \text{ job } i \text{ precedes job } j \\ 0 & \text{otherwise} \end{cases}$$

Decision Variables

c_j = Completion time of job j

$c_{\max}$ = Completion time of last job in schedule

s_j = Starting time of job j

$$X_{jil} = \begin{cases} 1 & \text{if job } j \text{ precedes job } i \text{ on machine } l \\ 0 & \text{otherwise} \end{cases}$$

$$Y_{wji} = \begin{cases} 1 & \text{if job } i \text{ and } j \text{ are both assigned to mold } w \\ 0 & \text{otherwise} \end{cases}$$

$$Z_{jl} = \begin{cases} 1 & \text{if job } j \text{ processed on machine } l \\ 0 & \text{otherwise} \end{cases}$$

Mathematical Model

$$\min Z = C_{max} \tag{1}$$

Subject to

$$s_j + \sum_{l \in L} (P_j) * Z_{jl} = C_j \qquad \forall j \in N \tag{2}$$

$$s_j \geq s_i + P_i + (E_{ij} * X_{ijl}) - M * (1 - X_{ijl}) \qquad \forall j, i \in N, \forall l \in L, j < i \tag{3}$$

$$s_i \geq s_j + P_j + (E_{ji} * X_{jil}) - M * (1 - X_{jil}) \qquad \forall j, i \in N, \forall l \in L, j < i \tag{4}$$

$$Z_{jl} + Z_{il} \geq 2 * (X_{jil} + X_{ijl}) \qquad \forall j, i \in N, \forall l \in L, j < i \tag{5}$$

$$C_{max} \geq C_j \qquad \forall j \in N \tag{6}$$

$$\sum_{l \in L} Z_{jl} = 1 \qquad \forall j \in N \tag{7}$$

$$F_{jl} \geq Z_{jl} \qquad \forall j \in N, \forall l \in L \tag{8}$$

$$s_i \geq s_j + \sum_{l \in L} (P_j + E_{ji}) * Z_{Jl} \qquad \forall j, i \in N, A_{ij} = 1 \tag{9}$$

$$C_j \leq s_i + M * Y_{wij} \qquad \forall j, i \in G_w, \forall w \in T, i \neq j \tag{10}$$

$$C_j \geq s_i + M * (1 - Y_{wij}) \qquad \forall j, i \in G_w, \forall w \in T, i \neq j \tag{11}$$

Sign Restrictions

$$X_{jil} \in \{0, 1\} \qquad \forall j, i \in N, \forall l \in L \tag{12}$$

$$Y_{wji} \in \{0, 1\} \qquad \forall i, j \in G_w, \forall w \in T \tag{13}$$

$$Z_{jl} \in \{0, 1\} \qquad \forall j \in N, \forall l \in L \tag{14}$$

$$C_j \geq 0 \qquad \forall j \in N \tag{15}$$

$$s_j \geq 0 \qquad \forall j \in N \tag{16}$$

$$C_{max} \geq 0 \qquad \forall j \in N \tag{17}$$

The boundaries of our solution are defined by the constraints presented above. In more detail, constraint (1) calculates the minimum completion time of all jobs. Constraint (2) calculates the completion time of a job. Constraints (3), (4) and (5) are disjunctive constraints, which provide that no two items can be processed on the same machine simultaneously if and only if both items are assigned to the same machine. Constraint (3) ensures that if job i is scheduled before job j on machine l, starting time of job j must be greater than or equal to the completion time of job i. Constraint (4) is the complementary of disjunctive constraint (3). Constraint (5) guarantee that if jobs i and j are scheduled on machine l, both items must have been assigned to that machine. Constraint (6) ensures that the completion time of any job is less than or equal to makespan (*Cmax*). Constraint (7) ensures that each job is assigned to one and only one machine. Constraint (8) guarantees that each job is assigned to only one machine which is eligible. Constraint (9), which has been created for the jobs that have precedence relationship, ensures that a predecessor job is processed before its successor job is completely processed. Constraint (10) and (11) ensures that jobs which are using common mold do not start at the same time. Constraints (12), (13), (14), (15), (16) and (17) are the sign restriction constraints which are used to specify whether the range of values that the decision variable can take is positive, negative, or free from the sign.

5 Solution Methodology

As it is described in detail in the problem definition section, this study focuses on finding a solution to a parallel machine scheduling problem. In this context, the suggested mathematical model has been coded via IBM ILOG CPLEX Optimization Studio 12.8.0 as the first solution alternative. A sample data set with real data received from the company has been used to test the model. While the first approach was developed, it was observed that this approach is not able to solve the problem in a reasonable amount of computational time and problems might occur while running the model with bigger data sets. Therefore, as the second alternative, a heuristic approach has been developed.

In addition, a preparation phase is created to group the orders and prepare the necessary input data. By grouping the orders received from different customers according to the product types, some of the unnecessary setup times will be eliminated. Thus, the total completion time of all jobs will be reduced. This grouping process can be done easily via the DSS by the developed "Add an Order" module. Then the DSS checks the related product BOMs for the orders defined in the system and calculates how many units must be produced for each job in total. If an order arrives during an already planned week, this new order will be postponed to the next planning period. When all the orders are added to the DSS, an Excel file which includes the input dataset such as sequence dependent setup times, precedence relationships, and cycle times will be created by using the "Get Matrices" and "Data for Python" modules. After this preparation phase is carried out, both approaches can be applied.

The heuristic approach was developed by integrating Python and Excel VBA. It is based on creating different combinations of job assignments by swapping the jobs. Firstly, the cluster of all jobs is mixed like a deck of cards at the beginning of the assignments. By using the Best-fit Algorithm, the jobs are assigned to the eligible machines by the order of the achieved mixed cluster. For the next job that will be assigned, eligible machines are checked, and the job is assigned to the machine which has the minimum completion time. In this way, the jobs can be assigned to different machines in every assignment. After assigning all the jobs, the jobs assigned to the machines are switching with each other by the help of the swap function on the machine basis. When the new schedules are obtained, the completion time of the final schedule is compared with the global best value which is current optimal. If any improvement on the current makespan value (global best) was made, the current makespan value is considered as the new global best. The algorithm makes the specified number of searches repetitively. Finally, it creates the output data of the best schedule obtained. The output of the Python algorithm is based on jobs assigned to machines, according to the eligibility matrix, but without taking into consideration the precedence and common mold constraints. Therefore, the Python solution, exported as an Excel file, is analyzed by the VBA code developed to check and finalize the schedule according to the precedence and common mold constraints. If there is a confliction between two jobs which have precedence relationship or use the same mold, one of these jobs is shifted by adding dummy time. Extra dummy times are added to the following jobs which will be produced in the same machine after the completion of the shifted job. Thus, the final schedule is prevented from new conflictions. Finally, the weekly production plan is decided, and the scheduling is presented in the form of jobs to machines assignments Gantt chart, a work orders file and machine-based scheduling reports.

The proposed heuristic solution procedure is presented in the flowchart in Fig. 1.

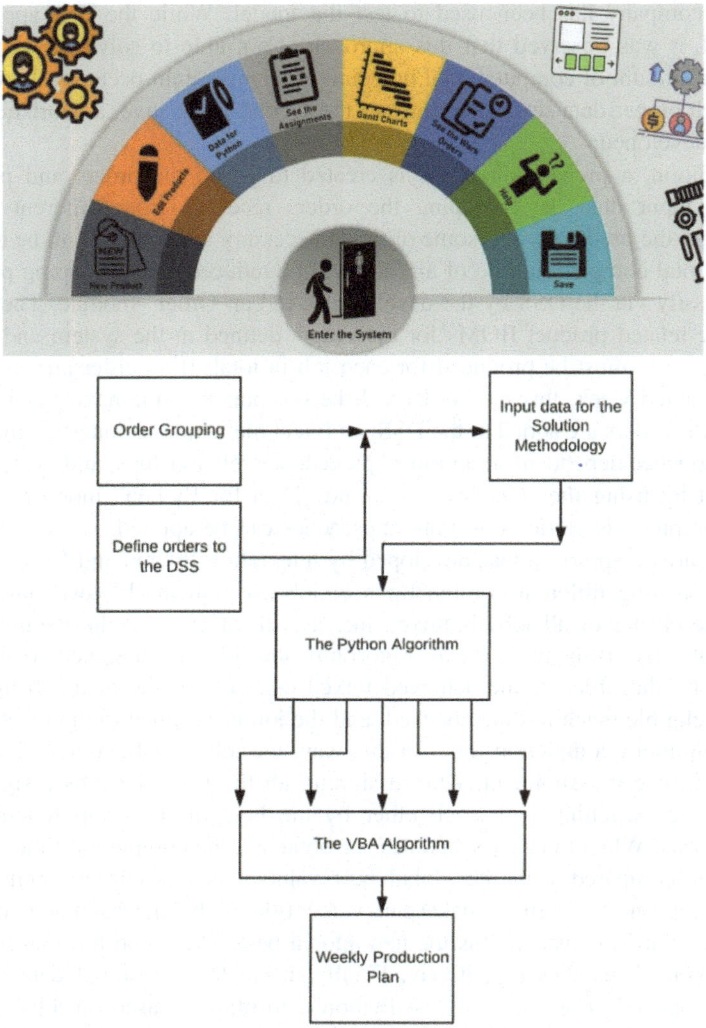

Fig. 1. The proposed solution procedure. The main page

6 Decision Support System

As a result of the study a Decision Support System (DSS) was developed to assist the company in their weekly production scheduling process. The DSS has been developed with a user-friendly interface using Microsoft Excel VBA and the heuristic solution which has been coded via Python. The main page of the DSS is presented in Fig. 1.

There are several modules in the DSS. The entrance module "Enter the System", contains the relevant tools to add the orders received from the customers. The user enters the product code and the order quantity information here and can see/reset the current orders to be scheduled. When all the orders are defined, the user uses the "Get

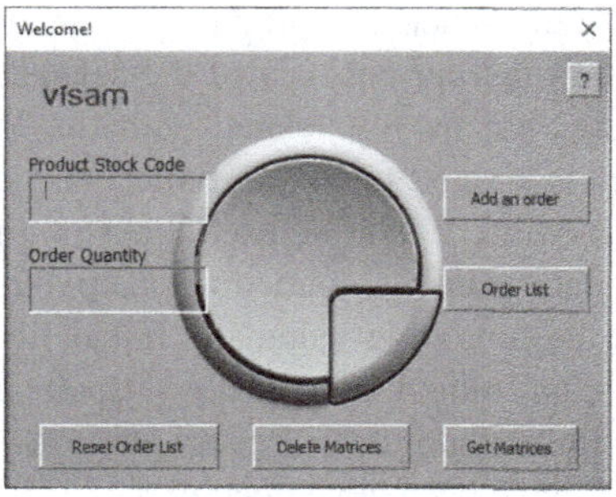

Fig. 2. Entrance module

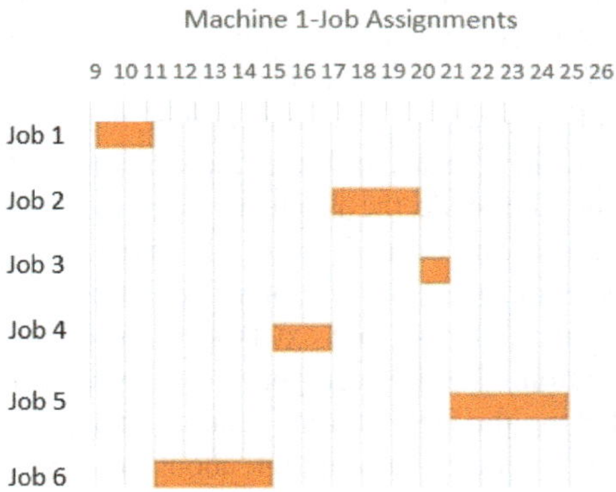

Fig. 3. Example of a gantt chart

Matrices" tool to create the input data matrices, which will be used by Python while generating the solution. The order grouping procedure will be applied on this phase. Via the "Delete Matrices" tool, the user can clear the input data from the system. the entrance module is shown in Fig. 2.

The DSS is created with the specific number of products existing today in the company. Therefore, when the user tries to define an order with a product not included in the systems database, the DSS delivers a message to the user to define the product first. Via the "New Product" module, the user adds the new product's code and the product's BOM by selecting the related Excel file. After the product BOM is defined, the user selects only the sub-components which will be produced, not the raw materials or color materials. When this phase is completed, the DSS checks whether the necessary information of these selected components exist. If the system needs the data for these components, it opens a new window to identify the new components. After all necessary information is provided by the user, the DSS creates a message box and says that the new product is completely identified and can be added, in the orders list now.

An "Edit Products" module allows the user to update any information of the predefined or post-defined products that needs to be changed in the system. This module lists all the products and components which were defined before, and the user can select the product to edit. When the product is selected, a new window opens, which allows the user to edit the current data of the selected item.

The rest of the modules are mostly related with the Python heuristics solution. "Data for Python" module is created to be used after all the necessary input data are obtained via the entrance module. When the user clicks on the module, an Excel file is created with all the input data matrices, named dataforpython <date>.xls. Afterwards, the user needs to run the Python code using the Excel file and get the results. The Python algorithm creates an output Excel file, which includes the machine-job assignments and the start-finish time data of every job.

Three different modules have been developed in the DSS to present the assignments to the user. In the "See the Assignments" module, the user can see all the machines in a window. By selecting the machine button, the user can see the jobs assigned to that machine. If the user wants to see the assignments in a graph with the jobs' starting and finishing times, the "Gantt Charts" module provides those graphs. Similarly, the user can click on a specific machine and see the relevant Gantt chart. A simple example of a Gantt chart is shown in Fig. 3 below. As a third option, the user can select the "See the Work Orders" module, which creates an Excel file containing the relevant assignment information, with the product code the product name, the assigned machine and the start and completion times. The created Excel file gives the user the opportunity to share these work orders with operators and other departments.

For assisting the users, a "Help" tool exists in each one of the user forms, noted with a question mark button. The user can get all necessary information about how to use each module by clicking the Help button. In addition, a general "Help" module has been created to give more detailed information about the DSS and the modules.

7 Verification and Validation

In order to validate the results, a sample problem has been created. In order for the results of this problem to be completely reliable and to test the solution methodology in every sense, weekly order data was received from the company. The weekly data includes the orders for 41 different products. These products have 75 sub-products in total. The Decision Support System grouped the orders according to the sub-products and provided all the necessary input data. After these input data, were embedded to the Python algorithm, the model was run considering the job-mold and mold-machine eligibility restrictions, and jobs were assigned to the machines accordingly. However, there were still two restrictions that must be met. To reshape the assignments according to the precedence and common mold constraints, the results have been sent to the DSS. By adding dummy time to some of the jobs' start/finish times, precedence and common mold restrictions have been assured. Finally, all the job-machine assignments have been completed. All the assignments have been checked to see if they are made according to machine-job eligibility, precedence, and common mold restrictions. These jobs have been assigned to an eligible machine. Precedence relations have been met for the jobs that a precedence job in needed. All the predecessors have been produced before their successor jobs. In addition, between the 4 jobs that use common molds, there is no confliction. As a special case, job 45 and job 46 use the same mold, but since they are being produced simultaneously, their start and finish time values are equal. In order to verify the solution methodology, the VBA and Python codes were checked step by step using the "Debugging" tools for each constraint. As a result, it was verified that each constraint is being properly executed. Consequently, the results and the solution methodology were verified and validated.

Also, in order to evaluate and compare results of the proposed heuristic solution approach with the mathematical model approach, the same problem as described above is solved with the mathematical model. As a result of the experiment which has been carried out using big dataset, the mathematical model solution has found the optimal schedule with the makespan of 4,759.34 min and the heuristic solution has found a feasible schedule with 5,098.1 min of total completion time. Considering the significant difference in the computational time needed to reach the final schedules by the two methods, the difference of 338.76 min, which results in 7% increase in the total jobs completion time, was considered acceptable by the company. The comparison of the two approaches are shown in Table 1. The same experiment is also conducted using a smaller data set and the results are shown in Table 2.

Table 1. Comparison of the two approaches using big data

	Mathematical solution approach	Heuristic solution approach
Input data	54 orders & 41 different products & 75 sub-products in total	
Computational time	603 min	18,45 min
Optimal makespan	4759 min/79,32 h	5098 min/84,96 h
Solution difference (%)	7%	

Table 2. Comparison of the two approaches using small data

	Mathematical solution approach	Heuristic solution approach
Input data	18 orders & 14 different products & 16 sub-products in total	
Computational time	92,65 min	4 min
Optimal makespan	1014 min/16,9 h	1056 min/17,6 h
Solution difference (%)	4%	

8 Conclusions

In this study, the parallel machine scheduling problem of a plastic injection company in Turkey is discussed. The current production planning process of the company is not based on a scientific approach, which results in inefficient usage the workforce and machines, delays in orders, penalty costs due to delays in delivery to customers. Taking all of these into consideration, the overall purpose of the project is to minimize the total production time of the orders while meeting customer requirements on time under the production restrictions. There are five main characteristics of the system which has been considered while creating the solution methodology. These characteristics can be listed as sequence dependent setup times, machine eligibilities, color variation, common molds, and precedence relationships between the jobs. In the first step of the project, literature review has been conducted for the parallel machine scheduling problem and according to these reviews, appropriate mathematical model to meet all the constraints has been formulated. The proposed mathematical model has been coded via CPLEX and operated with a sample of real weekly order data taken from the company. Due to the size of the data, CPLEX could not create the final schedule with optimality in a reasonable amount of computational time. Considering this problem, a heuristic approach is proposed to develop a fast and efficient scheduling process. This proposed approach has been coded by combining Python programming and Excel VBA. Furthermore, a Decision Support System has been developed to provide a useful tool to the company, using the heuristic approach solution. With the help of the DSS, the company can create the weekly production plan, presented in the Gantt charts, work orders file and machine-based scheduling report delivered by the system, reducing the total completion time of the production process.

Concluding, we can say that the developed DSS system, by improving the production planning process and decreasing the total completion time of jobs, will further assist the company to use more efficiently the existing workers and machines, increase customer satisfaction by on time deliveries and reduce penalty costs by not exceeding due dates; therefore providing extra time and income to the company.

Acknowledgment. This work could not be completed without the assistance of Asst. Prof. Dr. Mahmut Ali Gökçe, Dr. Cemalettin Öztürk and Dr. Kazım Erdoğdu. We are thankful for their contribution. Furthermore, we are grateful to the company for their co-operation.

References

1. Saraç, T., Sipahioğlu, A.: Plastik Enjeksiyon Makinalarının Çizelgelenmesi Problemi. MMO Endüstri Mühendisliği Dergisi **20**(2), 2–14 (2009)
2. Burak, G., Brahim, H., Selin, Ö.: Parallel machine scheduling with tool loading: a constraint programming approach. Int. J. Prod. Res. (2018). https://doi.org/10.1080/00207543.2017.1421781
3. Shen, L., Neufeld, J.S.: Solving the flexible job shop scheduling problem with sequence - dependent setup times. Eur. J. Oper. Res. **265**(2), 503–516 (2018)
4. Chen, C.L., Chen, C.L.: Hybrid metaheuristics for unrelated parallel machine scheduling with sequence- dependent setup times. Int. J. Adv. Manuf. Technol. **43**, 161 (2009)
5. Akyol, E., Saraç, T.: Paralel Makina Çizelgeleme Problemi için bir Karma Tamsayılı Programlama Modeli: Ortak Kaynak Kullanımı. GU J. Sci. Part C **5**(3), 109–126 (2017)
6. Cemalettin, Ö., Ornek, A.M.: Operational extended model formulations for advanced planning and scheduling systems. Izmir Univ. Econ. Dept. Ind. Eng. Appl. Math. Model. **38**(1), 181–195 (2014)
7. Ural, Ç.G., Kocamaz, M., Soyuer, H.: A developed encoding method for parallel machine scheduling with permutation genetic algorithm. Faculty of Economics and Administrative Sciences, Business Administrations, Ege University, Turkey (2009)
8. Ural, Ç.G.: A genetic algorithm to minimize makespan and number of tardy jobs in parallel machine scheduling problems. Ege University, IIBF, Business Administration Department (2016). https://doi.org/10.17671/btd.99806

9. Ramadan, S.Z., Süer, G.A.: A dual fitness function genetic algorithm: application on deterministic identical machine scheduling. World Acad. Sci. Eng. Technol. Int. J. Ind. Manuf. Eng. **7**(1), 163–170 (2013)
10. Liu, M., Cheng, W.: A genetic algorithm for minimizing the makespan in the case of scheduling identical parallel machines. Artif. Intell. Eng. **13**(4), 399–403 (1999)
11. Lee, J.H., Hoo, J.: Uniform parallel machine scheduling with dedicated machines, job splitting and setup resources. School of Business, Konkuk University, Seoul 05029, Korea, College of Global Business, Korea University, Sejong 30019, Korea (2019)
12. Senthilkumar, P., Narayanan, S.: Simulated annealing algorithm to minimize makespan in single machine scheduling problem with uniform parallel machines. VIT University, Vellore, India (2011)
13. Rajaee, A.A., Amindoust, A., Asadpour, M.: Applying simulated annealing algorithm for parallel machine tardiness problem subject to job splitting. Department of Industrial Engineering, Najafabad Branch, Islamic Azad University, Najafabad, Iran (2018)

A Single Machine Job Scheduling Problem with Sequence Dependent Setup Times

Aylin Elibol(✉), Selen Tosun, Emin Erbay, Egemen Orta, Öykü Gökşen, Asena Ceritoğlu, Çınar Arabacı, Hande Öztop, and Mustafa Arslan Örnek

Department of Industrial Engineering, Yaşar University, Izmir, Turkey
aylin_elibol@hotmail.com, selentosun1@gmail.com, eminerbay17@gmail.com, egemenorta17@gmail.com, oyku.goksen@hotmail.com, asenaceritoglu01@gmail.com, cinararabaci7@gmail.com, {hande.oztop,arslan.ornek}@yasar.edu.tr

Abstract. In this paper, a real-life single machine job scheduling problem with sequence-dependent setup times of a hood manufacturer company is addressed to minimize total weighted tardiness of the jobs with given due dates. Initially, a mixed-integer linear programming model is developed for the problem. Since the problem is NP-hard, heuristic algorithms are also proposed to solve larger instances. Namely, Apparent Tardiness Cost with Setups (ATCS), Earliest Due Date (EDD), Weighted Earliest Due Dates (WEDD), Shortest Processing Time (SPT) and Weighted Shortest Processing Time (WSPT) rule-based algorithms are developed for the problem. A swap move-based improvement is also employed in the proposed heuristic algorithms. To evaluate the effectiveness and efficiency of the proposed solution approaches, a comprehensive computational study is conducted by developing instances for the problem using the methodology from the related literature. Initially, optimal results are obtained for small instances by solving the mathematical model. Then, the performance of the proposed heuristic algorithms is evaluated by comparisons with the optimal results and time-limited model results. The computational results show that proposed ATCS rule-based heuristic is very effective to solve the problem. A user-friendly decision support system (DSS) is also developed to serve users with easy and efficient job scheduling.

Keywords: Single machine scheduling problem · Total weighted tardiness · Sequence-dependent setup times · Apparent Tardiness Cost with Setups · Heuristic · Decision support system

1 Introduction

This study considers a real-life single machine job scheduling problem with sequence-dependent setup times (SMSP-SDST) of a hood manufacturer company in Manisa, Turkey, where the objective is to minimize total weighted tardiness of the jobs in the manufacturing system with given due dates. In this study, we consider the new assembly line of the company for the hoods, which is installed at the hood and aspirator

N. M. Durakbasa and M. G. Gençyılmaz (Eds.): ISPR 2020, LNME, pp. 804–816, 2021.
https://doi.org/10.1007/978-3-030-62784-3_67

manufacturing plant in Manisa. In this assembly line, products such as engine groups, lamps, control groups, accessory groups, auxiliary products (chimney support, pipe adapter, bracket, deflector, etc.) and filters from the sub-industry are installed into the hoods. The main assembly line operations are: assembling the engine to body, assembling decorative materials, grouping to glass, lamp and bottom cover, last quality control and packaging. After the assembly process, the finished hoods are packaged and transported to the warehouse before being sent to the customer.

In this new assembly line, batches of various product groups (models) will be assembled, where the aim is to determine the sequence of these product batches on the assembly line. Therefore, there is a single machine job scheduling problem for this assembly line, where each batch of product group refers to a job and the assembly line refers to a single machine. Furthermore, there are sequence-dependent setup times in the system due to the model changes, and delivery dates of the jobs to the customers determine the due dates of the jobs in the assembly line. Additionally, jobs have different weights (penalties) for tardiness. Consequently, the studied problem is classified as SMSP-SDST, where the aim is to minimize the total weighted tardiness of the jobs. Note that, SMSP-SDST with total weighted tardiness criterion has already been proven as NP-hard [1]. This study aims to create effective solution methods and a user-friendly decision support system for the SMSP-SDST of the company, in order to determine the best schedule that minimizes total weighted tardiness of the jobs.

The remainder of the paper is organized as follows. The problem definition and a brief literature review are provided in Sect. 2. The mixed-integer linear programming (MILP) model of the problem is given in Sect. 3. The proposed heuristic algorithms are explained in Sect. 4. Detailed computational results are provided in Sect. 5. The developed user-friendly decision support system is explained in Sect. 6. Finally, conclusions and future work is provided in Sect. 7.

2 Problem Definition and Literature Review

As mentioned in Sect. 1, the company is planning to construct a new assembly line to produce traditional products, which are designed according to special customer requirements. In this assembly line, batches of various product groups (models) will be assembled and packaged based on a pre-determined schedule, to be delivered on time. In this assembly line, model-based orders of the customers are processed, and orders are produced as batches with specified due dates. Namely, a batch party order for a model of the customer refers to a job in the system. Thus, the processing time of a job is calculated by multiplying the unit processing time of the product and the number of products in that batch. Assembly of the next product batch (job) is not started on the assembly line (machine) until the previous product batch (job) is finished. Also, jobs (orders) cannot be interrupted, once they are started to be processed. Because of these features, the studied problem is addressed as a single machine scheduling problem, where the aim is to determine the best sequence for the jobs (model-based product batches) on the single machine (assembly line). Furthermore, when the production of a model is switched to the production of another model, the adjustment (setup operation) is made for the new model to begin. Since the setup times depend on the order of the jobs, the studied problem has also sequence-dependent setup times. Additionally, there are also weights of tardiness for customer orders. Then, the aim is to minimize the total weighted tardiness of the jobs.

Consequently, the problem is job scheduling of a new assembly line with sequence-dependent setup times, which consist of these following subproblems; deciding assembly sequence of the jobs according to given due dates and demand quantities, and determining the ready times of the subcomponents on the assembly line. Note that, once the sequence of the jobs is determined, the ready times of the subcomponents can also be determined accordingly based on the sequence of the jobs.

The SMSP with total weighted tardiness criterion has been widely studied in the literature. M'Hallah & Alhajraf [2] proposed a mathematical model and ant colony optimization algorithms for the single machine total weighted earliness tardiness scheduling problem. Tanaka & Araki [3] presented a mathematical model and an exact algorithm for the SMSP-SDST with total weighted tardiness objective. Lee et al. [4] also proposed a three-phase heuristic method for the SMSP-SDST with total weighted tardiness criterion, where the first phase computes some instance statistics; the second phase applies the Apparent Tardiness Cost with Setups (ATCS) dispatching rule to schedule the jobs; and the third phase applies an improvement procedure on the schedule obtained in the previous phase. Later, Valente and Alves [5] also proposed heuristic algorithms for the same problem based on a beam search technique.

Inspiring from the 3-phase heuristic algorithm proposed by Lee et al. [4], in this paper, we propose an ATCS rule-based algorithm for the SMSP-SDST with total weighted tardiness criterion, where we employ a swap move-based improvement heuristic in the third phase. Note that, the swap move-based improvement heuristic presented in this paper is different from the improvement procedure of Lee et al. [4]. We also present Earliest Due Date (EDD), Weighted Earliest Due Dates (WEDD), Shortest Processing Time (SPT) and Weighted Shortest Processing Time (WSPT) rule-based algorithms for the problem, by enhancing these rules with the swap move-based improvement heuristic. Additionally, we develop a user-friendly decision support system, which is integrated with the developed heuristic algorithms.

3 Problem Formulation

We use the MILP model that is proposed by Tanaka & Araki [3] for the SMSP-SDST with total weighted tardiness criterion. The sets, parameters and decision variables of the MILP model is listed in Table 1. Then, the MILP model is given as follows:

Table 1. Problem notation

Sets	
N :	Set of jobs $\{1,2,3,\ldots,n\}$
Parameters	
d_i :	Due date of job i
p_i :	Processing time of job i
w_i :	Weight of job i
M :	Very large number
s_{ij} :	Setup time if job j is immediately processed after job i
Decision Variables	
T_i :	Tardiness of job i
C_i :	Completion time of job i
y_{ij} :	$\begin{cases} 1 & \text{if job } j \text{ is immediately processed after job } i, \\ 0 & \text{otherwise} \end{cases}$

Minimize
$$\sum\nolimits_{1 \le i \le n} w_i T_i \tag{1}$$

Subject to:

$$T_i \ge C_i - d_i, \qquad \forall i \in N \tag{2}$$

$$\sum_{1 \le j \le n} y_{0j} = 1 \tag{3}$$

$$\sum_{1 \le i \le n} y_{i,n+1} = 1 \tag{4}$$

$$\sum_{\substack{0 \le i \le n \\ i \ne j}} y_{ij} = 1, \qquad \forall j \in N \tag{5}$$

$$\sum_{\substack{1 \le j \le n+1 \\ j \ne i}} y_{ij} = 1, \qquad \forall i \in N \tag{6}$$

$$C_j \ge (p_j + s_{0j})y_{0j} - M(1 - y_{0j}), \forall j \in N \tag{7}$$

$$C_j \ge C_i + (p_j + s_{ij})y_{ij} - M(1 - y_{ij}), 1 \le i,j \le n, i \ne j \tag{8}$$

$$y_{0j}, y_{i,n+1}, y_{ij} \in \{0,1\}, C_i \ge 0, T_i \ge 0 \tag{9}$$

The objective function (1) minimizes the total weighted tardiness of the jobs. Note that, dummy jobs 0 and n+1 are introduced into the model in order to define the first and last jobs. The constraint set (2) calculates the tardiness of each job j as the difference between its completion time and its due date. The constraint set (3) assigns the initial job in n jobs and the constraint set (4) does it for the last job in n jobs. Constraint sets (5) and (6) are job sequencing constraints. The constraint set (7) calculates the completion time of the first job and the constraint set (8) calculates the completion time of each job j regarding the sequence-dependent setup times. Finally, the constraint set (9) defines binary decision variables and the domain of other decision variables.

4 Heuristic Algorithms

As will be explained in Sect. 5, the above mathematical model can optimally solve only small instances due to the NP-hardness of the problem. However, in this study, we also aimed to solve the SMSP-SDST for larger number of jobs. As it is explained in Sect. 5, the mathematical model cannot obtain an optimal solution for the larger instances with more than 12 jobs. Therefore, we also develop heuristic methods to solve the problem for larger instances. After the literature review, the intuitive method that will lead to the fastest and effective solution has been determined as the ATCS rule. Inspiring from the 3-phase heuristic algorithm proposed by Lee et al. [4], in this paper, we propose an ATCS rule-based algorithm for the problem, where we employ a different swap move-based improvement heuristic in the third phase. We also present EDD, WEDD, SPT and WSPT rule-based algorithms for the problem, by enhancing these rules with the swap move-based improvement heuristic. The details of these algorithms are explained as follows.

4.1 ATCS Rule Based Heuristic Algorithm

A three-phase heuristic method that employs the ATCS rule is developed by Lee et al. [4]. Inspiring from the three-phase heuristic algorithm proposed by Lee et al. [4], in this paper, we propose an ATCS rule-based algorithm for the problem, where we employ a swap move-based improvement heuristic in the third phase. Note that, the swap move-based improvement heuristic presented in this paper is different from the improvement procedure of Lee et al. [4]. With the developed ATCS rule-based heuristic algorithm, it is intended to solve larger instances in reasonable computational times.

In this section, the developed ATCS rule-based heuristic algorithm is explained as follows. The ATCS rule-based heuristic algorithm consists of three stages. In the first phase, coefficients that are necessary for scaling parameters are computed, and makespan (C_{max}) is estimated to calculate these coefficients. The next phase is about determining scaling parameters for the priority index of the ATCS rule and applying the dispatching rule. Finally, in the last phase (improvement phase), the sequence, which is determined in the previous phase, is improved with a swap move-based improvement procedure in order to obtain a better schedule.

1) Phase 1: In this phase, Due Date Tightness factor τ, Due Date Range factor R and Setup Time Severity factor η are computed to characterize the instance. These coefficients are defined as in the following equations, where C_{max} (makespan) is the completion time of the last job in the system. τ coefficient represents the average tightness of the due dates, which is computed as below:

$$\tau = 1 - \bar{d}/C_{max}$$

where $\bar{d}$ is the average of the due dates. The R coefficient is computed as follows, which provides an indicator of the variability of the due dates regarding the total workload:

$$R = (d_{max} - d_{min})/C_{max}$$

where d_{max} and d_{min} represent the maximum and minimum values of due dates. Finally, the η coefficient is computed as below:

$$\eta = \bar{s}/\bar{p},$$

where $\bar{s}$ is the average of setup times and $\bar{p}$ is the average processing time. Since the studied problem depends on the sequence-dependent setup times, the calculation of C_{max} is difficult to determine before the jobs are scheduled. Therefore, C_{max} is estimated as proposed by [5] with the below formula, following the study of [4]:

$$\hat{C}_{max} = n(\bar{p} + \delta\bar{s}) \ with \ \delta \leq 1,$$

where n$\bar{p}$ represents the processing time part of the makespan. By defining $\delta \leq 1$, it is assumed that the setup time part of the makespan is generally much smaller than n $\bar{s}$, as a job with a smaller setup time is more likely to be chosen during the generation of a sequence. The value of δ is computed using the equation $\delta = 0.4 + 10/n^2 - \eta/7$. In the case of δ is greater than 1, a random number for C_{max} value is generated between C_{min} and C_{max} values, which are calculated as the sum of processing times and the minimum/maximum estimations of the setup times. After estimating C_{max} and computing the aforementioned coefficients based on C_{max}, it is proceeded to the next phase.

2) Phase 2: In the second phase, scaling parameters (k_1 and k_2) for the priority index of the ATCS rule are determined. The below equations are used for the selection of a proper value of k_1:

$$k_1 = 4.5 + R \ for \ R \leq 0.5$$

$$k_1 = 6.0 - 2R \ for \ R \geq 0.5$$

As it can be seen from the above equations, τ and η are not considered in the calculation of k_1. Hence, these factors are considered in the computation of k_2 as follows:

$$k_2 = \tau/2\sqrt{\eta}$$

After the computation of k_1 and k_2 values, which are important for applying the dispatching rule, the priority rule is expressed as follows:

$$I_j(t,l) = \frac{w_j}{p_j} exp\left[-\frac{max(d_j - p_j - t, 0)}{k_1\bar{p}}\right] exp\left[-\frac{s_{lj}}{k_2\bar{s}}\right],$$

where t represents the current time and l is the index of the job just finished. According to this rule, at each iteration, the job with the highest priority index is selected to be scheduled. Namely, after a job is scheduled, the priority indices of all the remaining jobs are computed and the job with the highest priority index is scheduled next. Then, the priority indices of all other remaining jobs are computed again according to updated values of t and l, and the job with the highest priority index is scheduled next. This rule is applied until there is no job left to be scheduled. After that stage, the initial job sequence is achieved.

3) Phase 3: In the previous phase, an initial sequence is determined for a given instance. In the third phase, this sequence is improved with a swap move-based improvement heuristic to obtain a better schedule. In the swap move, two jobs are selected and interchanged, while all other jobs are kept the same in the sequence. In the proposed improvement heuristic, this swap move is used for all job-pairs in the sequence. Furthermore, if there is an improvement, the swap local search procedure is restarted again. Figure 1 shows the pseudo-code of the swap move-based improvement heuristic. In the figure, p represents the current sequence, p' represents the new sequence after the swap operation and p(i) represents the job at position i of sequence p.

```
While improvement = True
  improvement = False
   For i = 1 To n
    For j = 1 To n
      p' = swap p(i) and p(j)
      If p'. tardiness < p. tardiness Then
      improvement = True
      p = p'
      End If
    Next j
  Next i
Wend
```

Fig. 1. The swap move-based improvement heuristic

To see the effect of the proposed swap move-based improvement heuristic (SWAP), the ATCS rule is run with and without improvement heuristic. Over 96 instances, the effect of the swap move-based improvement heuristic can be seen in Fig. 2. As shown in Fig. 2, in almost all instances, the proposed swap move-based heuristic improves the results of the ATCS rule. As can be understood from this graph, the contribution of the swap move-based improvement heuristic to the ATCS rule is significant. Note that, the RPD formula is given in Sect. 5.

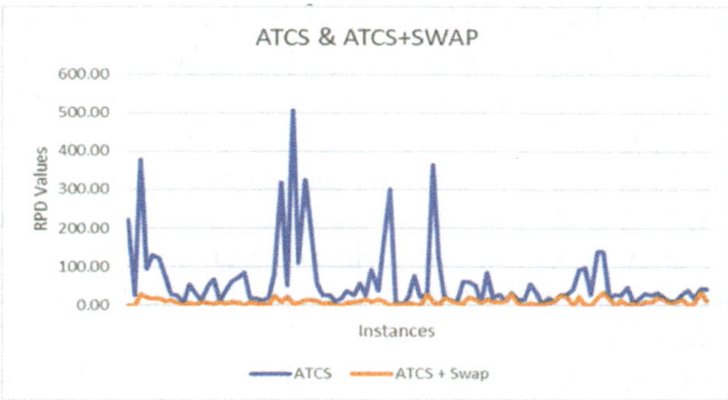

Fig. 2. Comparison of ATCS & ATCS with swap over 96 instances

4.2 EDD, WEDD, SPT and WSPT Rule based Heuristic Algorithms

In addition to the ATCS rule, well-known EDD, WEDD, SPT and WSPT priority rules are also coded with and without swap move-based improvement heuristic to compare their performances with the ATCS rule. To mention briefly, these simple rules are generally used for job scheduling. Firstly, EDD compares the due dates of jobs. The job with the earliest due date is processed first, followed by the second-earliest due date and so on. On the other hand, in the WEDD, weights of the jobs are also considered during the dispatching rule. The SPT rule orders the jobs in increasing order of their processing times. Differently, the WSPT schedules the next job with the highest ratio of weight over the processing time. Note that, the significant effect of swap move-based improvement heuristic is also observed for these rules as explained in Sect. 5.

5 Computational Experiments

The MILP model in Sect. 3 is coded using IBM ILOG CPLEX Optimization Studio 12.9. The heuristic algorithms in Sect. 4 are coded in Microsoft Excel VBA. The computational experiments are performed on an extensive set of instances with 8,10,12 and 14 jobs. These instances are randomly generated using a scheme similar to Valente and Alves [5] and Cicirello [6]. Each instance is characterized by the due date tightness factor, the due date range factor and the setup severity factor. For each job, an integer processing time is generated from the uniform distribution in the range [50,150] with an average processing time of 100, and an integer weight is generated from the uniform distribution in the range [1,10]. The integer setup times are generated from the uniform distribution in the range [0, $2\bar{s}$], where $\bar{s}$ is the average setup time. The average setup time is the multiplication of the average processing time and setup time severity factor. Two values are considered for the setup severity factor η as: 0.25 and 0.75.

The due date of a job is uniformly distributed over [$\bar{d}(1 - \mathrm{R})$, $\bar{d}$] with probability τ and uniformly distributed over [$\bar{d}$, $\bar{d} + (\hat{c}_{\max} - \bar{d})R$] with probability $1 - \tau$, where $\hat{c}_{\max}$ is a makespan estimate, τ is the due date tightness factor, R is the due date range factor and $\bar{d}$ is the average due date. The following values are considered for the τ and R: 0.3, 0.45, 0.6 and 0.1, 0.2, 0.25, 0.75 respectively. The makespan is estimated as $\hat{c}_{\max} = n(\bar{p} + \delta\bar{s})$ with $\delta \leq 1$ and the value of δ is calculated using the function $\delta = 0.4 + 10/n^2 - \eta/7$. There are $2 \times 3 \times 4 = 24$ combinations of (τ, η, R) in total.

The following values are considered for the number of jobs n: 8, 10, 12, 14. For each combination of (τ, η, R), 2 instances are generated for each setting of $n = 8$ and $n = 10$, resulting in 96 instances. For each combination of (τ, η, R), 1 instance is generated for each setting of $n = 12$ and $n = 14$, resulting in 48 instances. In total, 144 instances are generated as follows: 48 instances for $n = 8$, 48 instances for $n = 10$, 24 instances for $n = 12$ and 24 instances for $n = 14$. The generated instances are solved by mathematical model on the IBM ILOG CPLEX Optimization Studio 12.9 software. For the instances with 8 and 10 jobs, optimal results are obtained by the mathematical model. On the other hand, for the instances with 12 and 14 jobs, the feasible results obtained by the mathematical model under a 1-h time limit are reported.

5.1 Comparison of Algorithms with Respect to the Optimal Results

To evaluate performances of the 10 versions of the heuristic algorithms (SPT, SPT + SWAP, WSPT, WSPT + SWAP, EDD, EDD + SWAP, WEDD, WEDD + SWAP, ATCS AND ATCS + SWAP), the results of the heuristic algorithms are compared with the optimal results, for the instances with 8 and 10 jobs. Figure 3 shows the average RPDs (Relative Percent Deviation) of these algorithms for the instances with job 8 and 10 jobs. Relative Percent Deviation (RPD) is calculated for these instances as follows: $RPD = ((heuristic - optimal)/optimal) * 100$ where *heuristic* is the result of the heuristic algorithm and *optimal* is the optimal result obtained by the mathematical model.

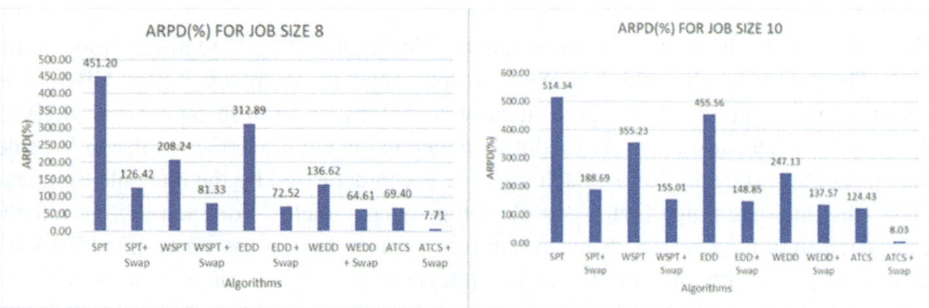

Fig. 3. The average RPDs (Relative Percent Deviation) of algorithms for job size 8 and 10

As depicted in the above graphs, ATCS rule performs much better than the other algorithms in both versions, i.e., with and without swap move-based improvement heuristic. While ARPDs of other algorithms without swap method are more than 100%, ATCS rule without swap method has an ARPD that equals to 69.40% for the instances with 8 jobs. Similarly, ATCS rule also performs better than the other algorithms in both versions for the instances with 10 jobs. Furthermore, it is obvious to see the effect of including the swap move-based improvement heuristic to the heuristic algorithms. As shown in Fig. 3, by including the swap move-based improvement heuristic, ARPD of ATCS rule is decreased from 69.40% to 7.71% and 124.43% to 8.03% for the instances with 8 and 10 jobs, respectively. From Fig. 3, it can also be observed that the second-best heuristic is WEDD with swap move-based improvement heuristic, since it has ARPD values of 64.61% and 137.57% for the instances with 8 and 10 jobs, respectively.

5.2 Comparison of Algorithms with Time-Limited CPLEX

Since the developed model cannot obtain the optimal results for the instances with 12 and 14 jobs under the 1-h time limit, the feasible results obtained by the mathematical model under the 1-h time limit are reported. Hence, comparisons are made using these feasible time-limited CPLEX results. Figure 4 demonstrates the average RPDs (Relative Percent Deviation) of the heuristic algorithms and time-limited CPLEX for the instances with job 12 and 14 jobs. Relative Percent Deviation (RPD) is calculated for these instances as follows: $RPD = ((heuristic - best)/best) * 100$ where *heuristic* is the result of the heuristic/time-limited CPLEX algorithm and *best* is the best result for that instance among the results of 10 heuristic algorithms and time-limited CPLEX. Since the model reports the feasible results for these instances under the 1-h time limit, ATCS rule has the potential to find a better solution than the mathematical model.

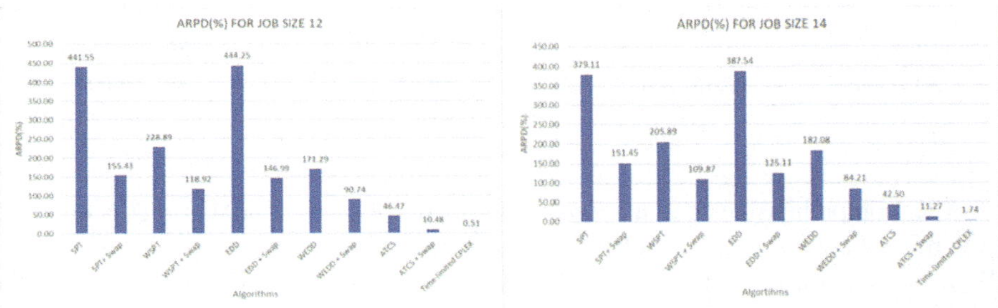

Fig. 4. The average RPDs (Relative Percent Deviation) of algorithms for job size 12 and 14

The above bar charts show that the ATCS without swap method performs better than the other algorithms with 46.47% and 42.50% ARPD values for the instances with 12 and 14 jobs, respectively. It can also be said that the ATCS rule performs better than the other algorithms in both versions (with/without swap) for the instances with 12 and 14 jobs. Furthermore, it is obvious to see the effect of including the swap move-based improvement heuristic to the heuristic algorithms. As shown in Fig. 4, by including the swap move-based improvement heuristic, ARPD of ATCS rule is decreased from 46.47% to 10.48% and 42.50% to 11.27% for the instances with 12 and 14 jobs, respectively. The ARPD values and the number of optimal/best solutions found by the ATCS with swap move-based improvement heuristic are also summarized in Table 2.

Table 2. The computational results of ATCS + SWAP for various job sizes

ATCS + Swap							
ARPD (%) from Optimal Results		# of Optimal Solutions Found (over 48 instances)		ARPD (%) from Best Results		# of Best Solutions Found (over 24 instances)	
$n = 8$	$n = 10$	$n = 8$	$n = 10$	$n = 12$	$n = 14$	$n = 12$	$n = 14$
7.71	8.03	15	14	10.48	11.27	3	5

Furthermore, the average RPD values for each R, τ and η value can be found in Table 3. From this parametric analysis, it is observed that the ATCS rule is sensitive to τ values. Note that, C_{max} estimation has a significant importance for τ values. Therefore, the estimation of C_{max} must be done effectively. Additionally, setup severity factor has no significant effect on the performance of ATCS rule.

Table 3. ATCS + SWAP sensitivity analysis

ATCS + SWAP	τ			η		R			
n	0.30	0.45	0.60	0.25	0.75	0.10	0.20	0.25	0.75
8	13.43	5.17	4.54	8.71	6.72	10.11	7.21	3.28	10.25
10	12.22	6.18	5.68	9.22	6.84	6.16	9.41	10.76	5.78
12	9.92	13.85	7.67	8.61	12.34	9.56	14.88	6.1	11.37
14	14.68	7.01	12.13	12.38	10.17	6.05	20.8	9.36	8.88
Average	**12.56**	**8.05**	**7.51**	**9.73**	**9.02**	**7.97**	**13.08**	**7.38**	**9.07**

After comprehensive computational experiments, it can be said that ATCS rule with swap move-based improvement heuristic outperforms the other heuristic algorithms for all job sizes as shown in Figs. 3 and 4. Also, from those charts, it can be seen that the second-best performing heuristic is WEDD with swap move-based improvement heuristic. Consequently, since the ARPD values of ATCS + SWAP are at acceptable levels, it is verified as our heuristic approach to solve the instances with 12 or more jobs with very low computational times. Moreover, we encoded WEDD + SWAP in the developed decision support system to improve the solution performance, since WEDD + SWAP has the lowest RPD values after the ATCS + SWAP. The details of the developed decision support system are explained in the following section.

6 Decision Support System

The proposed heuristic methods, which are ATCS + SWAP and WEDD + SWAP with minimum ARPD values, are coded using the Excel VBA program to create a user-friendly decision support system (DSS). While creating DDS, it is aimed that the developed DSS is to be easy to use, provides clear and visual solutions, to be dynamic, and to be integrated into many similar areas. The system takes delivery dates, product types, product numbers, order codes, and the weight of the orders as inputs for the orders that will be produced on the newly installed line. It incorporates with a database, which is created with the data collected from the factory. This database is changeable for applying the system to other assembly lines. Within DSS, both ATCS + SWAP and WEDD + SWAP algorithms run together, and the program takes the best result between them as the final output. Namely, the final schedule is created by choosing the minimum weighted tardiness value between the objective function values of the two heuristic algorithms.

Figure 5 illustrates the main page of the DSS. Adding a new job order, changing, or deleting orders can be done on the main page. Furthermore, the orders that are entered can be seen easily in that page. After the user completed the data entrance, with just one click to schedule, the output page will be opened, and users can access the detailed job schedule with a Gantt Chart. Furthermore, users can convert the detailed results to a PDF format. Figure 6 demonstrates the output page of the DSS system. Additionally, by pressing the Schedule Plan button, the user will access the Gantt Chart, which is shown in Fig. 7.

Furthermore, the developed DSS was verified by different scenarios to ensure the dynamism and operating properly. Also, the data collected from the company was entered to the DSS and results were examined. Owing to a successful result, it is concluded that the developed DSS can provide effective schedules to the company in very reasonable computational times.

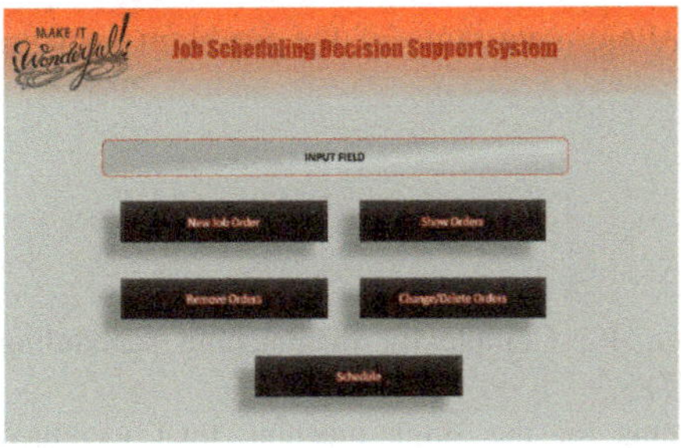

Fig. 5. Decision support system-main page

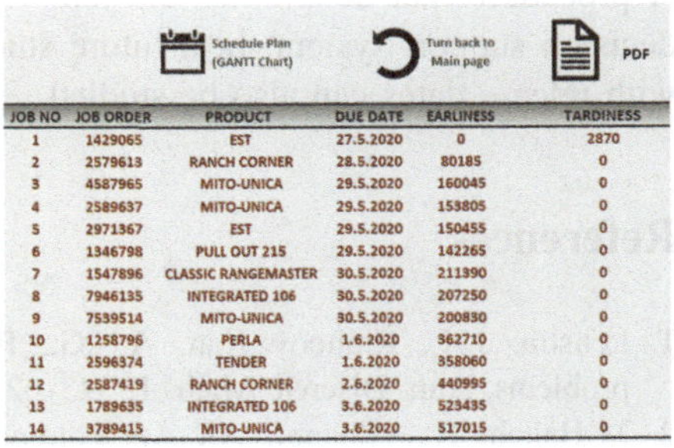

JOB NO	JOB ORDER	PRODUCT	DUE DATE	EARLINESS	TARDINESS
1	1429065	EST	27.5.2020	0	2870
2	2579613	RANCH CORNER	28.5.2020	80185	0
3	4587965	MITO-UNICA	29.5.2020	160045	0
4	2589637	MITO-UNICA	29.5.2020	153805	0
5	2971367	EST	29.5.2020	150455	0
6	1346798	PULL OUT 215	29.5.2020	142265	0
7	1547896	CLASSIC RANGEMASTER	30.5.2020	211390	0
8	7946135	INTEGRATED 106	30.5.2020	207250	0
9	7539514	MITO-UNICA	30.5.2020	200830	0
10	1258796	PERLA	1.6.2020	362710	0
11	2596374	TENDER	1.6.2020	358060	0
12	2587419	RANCH CORNER	2.6.2020	440995	0
13	1789635	INTEGRATED 106	3.6.2020	523435	0
14	3789415	MITO-UNICA	3.6.2020	517015	0

Fig. 6. Decision support system-output page

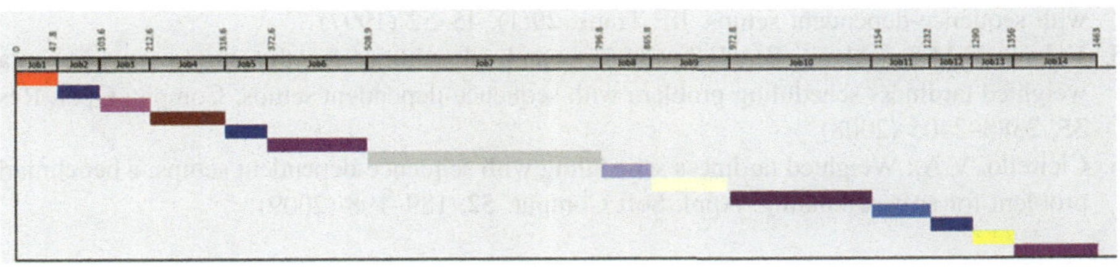

Fig. 7. Decision support system-gantt chart

7 Conclusion and Future Work

In this study, a single machine scheduling problem with sequence dependent setup time has been addressed to minimize the total weighted tardiness of the jobs. First, the mathematical model was developed and coded in IBM ILOG CPLEX software. It was observed that the run time of the model is at unacceptable levels for the instances with more than 12 jobs. Since the problem is NP-hard, heuristic algorithms were also proposed to solve larger instances. Namely, Apparent Tardiness Cost with Setups (ATCS), Earliest Due Date (EDD), Weighted Earliest Due Dates (WEDD), Shortest Processing Time (SPT), and Weighted Shortest Processing Time (WSPT) rule-based algorithms were developed for the problem. A swap move-based improvement heuristic is also employed in the proposed heuristic algorithms. To evaluate the effectiveness and efficiency of the proposed solution approaches, a comprehensive computational study was made by developing instances for the problem using the methodology from the literature. Initially, optimal results were obtained for small instances by solving the mathematical model. Then, the performance of the proposed heuristic algorithms was evaluated by comparisons with the optimal results and time-limited model results. The computational results show that the proposed ATCS rule-based heuristic is very effective to solve the problem.

Furthermore, a flexible and user-friendly decision support system has been developed for the company to enable the planning engineer to prepare quick and efficient scheduling plans for the new assembly line. Scheduling systems have high importance worldwide. So, this system can be implemented worldwide factories which have a single machine production environment and sequence dependent setup times. A procedure that determines the job weights can also be included in the developed decision support system. In a future study, different variants of the problem such as with release dates can also be studied.

References

1. Lenstra, J.K., Rinnooy Kan, A.H.G., Brucker, P.: Complexity of machine scheduling problems. Ann. Discrete Math. **1**, 343–62 (1977)
2. M'Hallah, R., Alhajraf, A.: Ant colony systems for the single-machine total weighted earliness tardiness scheduling problem. J. Sched. **19**, 191–205 (2015)
3. Tanaka, S., Araki, M.: An exact algorithm for the single-machine total weighted tardiness problem with sequence-dependent setup times. Comput. Oper. Res. **40**, 344–352 (2012)
4. Lee, Y.H., Bhaskaran, K., Pinedo, M.: A heuristic to minimize the total weighted tardiness with sequence-dependent setups. IIE Trans. **29**(1), 45–52 (1997)
5. Valente, J.M.S., Alves, R.A.F.S.: Beam search algorithms for the single machine total weighted tardiness scheduling problem with sequence-dependent setups. Comput. Oper. Res. **35**, 2388–2405 (2008)
6. Cicirello, V.A.: Weighted tardiness scheduling with sequence-dependent setups: a benchmark problem for soft computing. Appl. Soft Comput. **52**, 189–198 (2009)

An Implementation of Flexible Job Shop Scheduling Problem in a Metal Processing Company

Ali Ozan Özkul, Gamze Küçük, Ceren Çelik, Nevzatcem Öztuna, Mert Demirkan, Ezgi Çağlar Nizam, Yıldırımşan Büyükmertoğlu, Damla Yüksel(✉), and Ayhan Özgür Toy

Department of Industrial Engineering, Yaşar University, İzmir, Turkey
aliozanozkul96@gmail.com, ezgicaglarnizam@gmail.com, gamzekucuk_@hotmail.com, cerenceliik@hotmail.com, yildirim.san@hotmail.com, oztuna.nevzatcem@outlook.com, demirkan.mert@outlook.com, {damla.yuksel, ozgur.toy}@yasar.edu.tr

Abstract. In this study, we consider the flexible job-shop scheduling problem at a local metal sheet processing company. We aim to develop a model and an algorithm to generate a weekly production plan for the company. The objective is to minimize the makespan while meeting the demands of products for a given planning horizon. First, we provide an LP formulation of this problem. The computational complexity of the problem is NP-hard, hence the input data prohibits obtaining the optimal solution in a reasonable time. Therefore, we implement a metaheuristic and several rule-based heuristics. These are Genetic Algorithm, Giffler and Thompson's Algorithm, and three other Rule-Based Heuristic Algorithms that we developed. We first test our model and heuristics over a set of sample instances, then we solve for the real data. Our experimental study indicates that one of the rule-based heuristics we developed outperforms others in most of the instances.

Keywords: Flexible job-shop scheduling · Giffler and Thompson's Algorithm · Genetic algorithm · Meta-heuristics · Decision support system · Optimization

1 Introduction

The motivation for this study is from the production planning requirement of a local metal sheet processing company, located in Manisa, Turkey. Its main field of activity is trading and processing of flat products and transforming them into semi-finished and finished products based on their customer requirements. They maintain CNC Laser, CNC Punch, and CNC Press Brake technologies for their business. The company has a wide range of customers from sectors such as white goods, machinery manufacturing, automotive, agricultural machinery, construction machinery industry, steel construction manufacturing, and side industry organizations. Their customers are also their raw

N. M. Durakbasa and M. G. Gençyılmaz (Eds.): ISPR 2020, LNME, pp. 817–830, 2021.
https://doi.org/10.1007/978-3-030-62784-3_68

material suppliers for materials, which are used in the production of more than thousands of product types. Raw material is processed in several stages of operations before it becomes ready for customer use. A product can be subject to different operations according to the feature of the customer's demand. The order of operations for each product type is predetermined.

Demands from customers are reported on a weekly basis. Based on these demand data, they make weekly, monthly, and annual prediction of production plans. The company states that the actual number of demands, production quantity, machine utilizations largely deviates from the estimates. Therefore, the company cannot rely on the predicted production plans and needs to make a production plan every week. The rest of the paper is organized as follows: Sect. 2 provides a problem definition with data and system analysis, Sect. 3 provides a literature review related to production type of the company, a mathematical model, and heuristic algorithms are presented in Sect. 4, Sect. 5 comprises the verification of the mathematical model and algorithms and computational results, finally, conclusion and future work are presented in Sect. 6.

2 Problem Definition

The company process four types of raw materials: Black Sheet, Galvanized Sheet, DKP Sheet, and CrNi Sheet each with varieties with different thicknesses. In the company, there are five main operations: cutting, bending, welding, surface cleaning, and painting. For the cutting operation either the laser cutting, or the punch cutting is performed based on the expectations of the customer, the desired product characteristics, and machine occupancy. There are two types of machines for the cutting operation: CNC Laser and CNC Punch with two machines for each type. If there are enough molds for Punch machines, they can be used interchangeably and form a machine set. Operators are able to use both types for cutting. There are two types of machines for the bending operation: CNC Press Break and CNC Panel Bender with three Press Break and two Panel Bender machines. CNC Press Break machines are suitable for bending small parts, where the bending operation is performed manually. CNC Panel Bender machines, on the other hand, are more suitable for large and sensitive parts because the material handling and bending process take place automatically. Three CNC Press Break and two CNC Panel Bender machines can be used interchangeably and form two different machine sets. The welding operation is implemented according to the requests of the customers. This stage consists of assembling the parts or integrating the products. This operation is performed manually by the operators. The surface cleaning operation is for products undergone welding operations. The roughness formed on the sheet during welding is removed at this stage and the product is made smooth and ready for painting operation. The painting operation is the last operation. Products are painted with electrostatic wet or powder paint in line with the requests and demands of the customer. It is also performed manually by operators.

Before processing a job in a machine, a setup depending on the job requirements performed, which takes certain time. Although the setup times are different for each machine, each has similar setup steps, quality approval, and machine settings. The setup time of each CNC machine is independent of the work performed. Setup times include die change and machine setting for CNC Punch and CNC Press Break, while it only includes machine setting for CNC Laser and CNC Panel Bender machines. There is no setup time for welding and surface cleaning operations. The setup time for the paint operation includes the preparation of the paint colors and the cleaning process.

Due to the increasing demand and product type variations, the company's main concern becomes to meet the demand due dates. The facility is operated in 2 shift during the weekdays and 1 shift on Saturdays, that is 11 shifts every week resulting in 88 working hours per week. Currently, they use overtime or the capacity of another facility which is dedicated to other production requirements. However, we think that their production plan needs to be improved by better scheduling the jobs. In this facility, each job goes through a certain set and sequence of operations before it meets customer expectations. Operations and their sequence vary for each job. This production processes in the company is classified as flexible job shop in the literature. Therefore, our work in this study lies in the category of "Flexible Job Shop Scheduling Problem". The goal of these sets of problems is to assign operations of jobs to machines optimally so that an objective related to the job completion times minimized. Devassia et al. [1] examined the job-shop production systems in their study, revealed the differences between the classical job shop-type and flexible job-shop production. Unlike the classical job shop, they stated that in flexible job shop production systems, an operation of a product can be done by different machines. Yuan and Hu [2] stated that the flexible job shop is an expanded version of the classical job shop, more complex and NP-hard. We next analyze the data used for the production planning.

2.1 Data Analysis

The company has a product range of more than 1000 different types. The data provided by the company for this study contains information about 568 different types of products. The demand data we have comprises customer information, product name, product code, the production quantity, demand of last 10-months, desired properties of raw materials, list of operations, machines, and processing times for each product type. The end times of the jobs in the demand data given on a monthly basis are not specified. We have calculated the end times of 10 months jobs in the data with the proposed job scheduling methods. First, we grouped products based on their demand quantities. Since the production capacity of the facility mostly consumed by the products with higher demand intensity, we focus on these products. Firstly, we performed ABC Analysis to identify the most demanded products by their percentages to the total demand amount. The products which we classified as the most demanded, based on the Pareto Principle, constitute 80% of the total production with 91 products Working with this product group would yield reasonable results for our study.

2.2 System Analysis

The total processing time for each product, average setup times for each machine, operation routes and production amount of 10 months are addressed in the demand data, which are adopted in our model. The average setup time for each machine is approximately 7 min. The processing times of each operation depends on the jobs but in general the range of processing times are as follows between 0.07–4.73 min for cutting 0.18–2.53 min for bending, 3.73–7.57 min for welding, and 1–5 min for surface cleaning. We run the model for the products, which constitutes 80 percent of the demands as explained above. With a simulation model, which is a representation of the current practice in the facility, we intend to imitate the production process so that we would observe the system behavior remotely and faster.

Table 1. Scheduled utilization

	Laser	Punch	Press break	Panel bender
Average value	0.835	0.045	0.480	0.438
Utilization rate	83%	9%	96%	87%

Laser machines work 2 shifts per day while Panel Bender and Press Break machines work 1 shift per day. If we consider the maximum utilization value as 0.50 for a single shift, Panel Bender and Press Break machines have the highest utilization compared to Laser and Punch machines as shown in Table 1. Press Break machines are the most used machines for bending operation, and the Laser machines are the most used machines for the cutting operation.

Table 2. Waiting times

	Average (min)	Minimum value (min)	Maximum value (min)
PressBreak. Queue	67.4722	0.00	146.88
Lazer_Cutting.Queue	78.6852	0.00	168.43
PanelBender.Queue	0.12253	0.00	1.2362
Punch_Cutting.Queue	2.5527	0.00	8.3651

Table 2 indicates that waiting times for Laser and Press Break machines are the longest, hence a larger pile of work-in-process inventory occurs in front of these machines. We observe that the cause of the long queue in front of the Laser cutting machine is solely a result of the planning because Punch cutting machines are not preferred although they are available to operate. However, the long queue in front of the Press Break machine is due to the longer processing time in this machine. The unbalanced operation distribution between Laser and Punch machines, and due to manual operations of Press Break causes a capacity loss. These long waiting times of Laser and Press Break machines' operations create bottlenecks in the system. As a result of this analysis, we identified high disproportional machine utilization rates, and long waiting times. We suggest that using scientific scheduling methods for flexible job shop production type instead of intuitional scheduling will procure better results.

3 Literature Review

In this section, a literature review is done related to the classical job shop and flexible job-shop scheduling problems, and heuristic optimization methods for flexible job-shop scheduling problems. Chaudhry et al. [3] propose genetic algorithm (GA) based on a schedule which is presented for flexible job shop problem to minimize the makespan. Genetic Algorithm (GA) is a stochastic optimization technique inspired by nature, similar to the evolution process in the biological world where the fittest survive to reproduce. In that study, they presented a spreadsheet-based domain-independent GA approach for flexible job-shop scheduling problems. Liangxiao and Zhongjun [4] presented an improved genetic algorithm for flexible job-shop scheduling problem to minimize makespan. In this article, they presented a new initialization method to promote the quality of the initial population and to speed up the convergence of the algorithm. Regarding the qualifications of the problem crossover, mutation operator, and reasonable chromosome encoding are given. Also, the effectiveness of the proposed algorithm is confirmed by testing. During the utilization of the proposed genetic algorithm in the computer-based simulation trials and contrasting with the other genetic algorithms, the conclusions explained the feasibility and efficiency of the algorithm mentioned in this article sufficiently. Pezzella et al. [5] stated a genetic algorithm for the flexible job-shop scheduling problem (FJSP) in this article. The proposed algorithm combines the selecting of individuals, reproducing new individuals, and the various strategies that contribute to the generating initial population. The numerical research results prove that the combining of more strategies within this scope causes better results according to other genetic algorithms. Besides, computational results are quite comparable to those obtained by the algorithm based on tabu search. According to these results, in conjunction with the genetic paradigm's flexibility demonstrates that genetic algorithms are effective while solving FJSP. As a result, the framework of a genetic algorithm is suitable and effective to develop efficient algorithms that will benefit FJSP. Lee et al. [6] constructed a paper of flexible job-shop scheduling problem which has an objective of minimizing the makespan. Firstly, they formulated the problem as a mixed-integer linear programming (MILP) for small-sized problems. Then, they developed heuristic algorithms which are genetic and tabu search algorithms to solve problems regardless of their size. According to the results of the computational experiments, heuristic algorithms are very effective. Kaya and Fığlalı [7] stated a review on meta-heuristic approaches to solving flexible job-shop scheduling problem. In this research, recent studies in the literature on the topic of flexible job-shop scheduling problems are reviewed. Recommendations and the findings on the solution to this type of problem by using meta-heuristic methods are proposed. František Koblasa [8] constructed a study of Giffler & Thompson Algorithm to solve the flexible job-shop scheduling problem (FJSP). In general, Giffler & Thompson Algorithm is used for job-shop scheduling. In this article, this algorithm is used for FJSP with modifications.

4 Solution Methodology

We formulated the problem at hand as a mixed integer linear programming problem. There are studies in the literature working with various objective functions. Based on the company requirements we developed a novel mathematical model to minimize the maximum completion times of all jobs. We make some assumptions for our problem formulation. We assume that all jobs and machines are available at the beginning of the planning horizon: (i) Weekly demand for each product type is considered as a single job; (ii) Transportation times between jobs are negligible; (iii) Only one operation can be performed on each machine; (iv) Every machine can only process one job at a time; (v) We assume that when a machine starts processing a job it cannot be ceased before its completion, i.e., preemption is not allowed; (vi) Alluded to above, every machine has a setup time for each machine, but we assume that setup times are included in the processing times of jobs; (vii) Every job must go through a fixed sequence of operations. The proposed model is provided under Mathematical Model sub-heading, the proposed solution is provided under Heuristic Algorithms sub-heading.

4.1 Mathematical Model

We suppose that there are |N| jobs to be performed, i denotes a particular job among these N jobs. Production facility performs |J| different operations in |M| machines. A particular job is denoted by j and a particular machine is denoted by m. Since not all the operations can be performed in all the machines, we group the machines based on the operations they can perform, M_j is the set of machines that can perform operation j. Likewise, each job has to go through a different sequence of operations, therefore we denote by J_i the set operations that a job has to go through. The processing time of each operation in each machine for a particular job is given and is denoted by T_{ijm}. The order of the operations to be performed for a particular job is also given, precedence relationship among operations for a particular job is given by a binary matrix, such that the parameter $R_{ijj'}$ has value 1 if operation j precedes operation j' for job i, and 0 otherwise. The decision variables we use in our mathematical model are as follows: F_i denotes the completion time of job i, C_{max} is the completion time of the latest finished job or the maximum completion time of all jobs. The operation j of job i will be completed on machine m at C_{ijm}. We use two binary assignment decision variables, namely Y_{ijm} to indicate the assignment of operation j of job i to machine m, and $G_{i'ijm}$ to indicate the assignment sequence of operation j of job i' and i on machine m. We summarize our nomenclate in Table 3.

The MIP formulation of the flexible job shop scheduling problem we consider is as follows:

$$\text{Min } C_{max} \tag{1}$$

Subject to

Table 3. Notations

Sets and indices:	
i	job $(i, i' \in N)$
j	operation $(j, j' \in J)$
m	machine $(m, m' \in M)$
N	the set of jobs
J	the set of operations
M	the set of machines
J_i	the ordered set of operations of job i $\{1, 2, \ldots, R\}$, $(J_i \subseteq J)$
M_j	the set of machines operation j can be processed $(M_j \subseteq M)$
Decision variables:	
Y_{ijm}	$\begin{cases} 1, & \textit{if operation j of job i is assigned to machine m} \\ 0, & \textit{otherwise} \end{cases}$
C_{max}	maximum completion time
C_{ijm}	completion time of operation j for job i in machine m
F_i	completion time of job i
$G_{i'ijm}$	$\begin{cases} 1, & \textit{if job i' assigned to machine m for operation j before job i} \\ 0, & \text{otherwise} \end{cases}$
Parameters:	
T_{ijm}	the processing time of job i, in machine m for operation j (the setup time of machine m is included)
D	big number
$R_{ijj'}$	$\begin{cases} 1, & \textit{if operation j' is performed after operation j for job i} \\ 0, & \textit{otherwise} \end{cases}$

$$\sum\nolimits_{m \in Mi} Y_{ijm} = 1 \qquad \forall j \in J_i, \forall i \in N \tag{2}$$

$$R_{ij'j} \sum\nolimits_{m \in Mi} C_{ijm} - Y_{ijm}.T_{ijm} \geq R_{ij'j} \sum\nolimits_{m' \in Mi'} C_{ij'm'} \qquad \forall j, j' \in J_i, \forall i \in N \tag{3}$$

$$Y_{ijm} D \geq C_{ijm} \qquad \forall j \in J_i, \forall i \in N, \forall m \in M_j \tag{4}$$

$$C_{ijm} - T_{ijm} \geq C_{i'jm} - \left(1 - G_{i'ijm}\right) D \qquad \forall j \in J_i, \forall i, i' \in N, \forall m \in M_j \tag{5}$$

$$C_{i'jm} - T_{i'jm} \geq C_{ijm} - \left(G_{i'ijm} D\right) \qquad \forall j \in J_i, \forall i, i' \in N, \forall m \in M_j \tag{6}$$

$$F_i = C_{iRm} \qquad \forall i \in N, R \in J_i, \; m \in M_j \tag{7}$$

$$C_{max} \geq F_i \qquad \forall i \in N \tag{8}$$

$$F_i \geq 0 \qquad \forall j \in J \tag{9}$$

$$C_{ijm} \geq 0 \qquad \forall i \in N, \forall j \in J, \forall m \in M \tag{10}$$

$$T_{ijm} \geq 0 \qquad \forall i \in N, \forall j \in J, \forall m \in M \tag{11}$$

$$G_{i'ijm}, Y_{ijm}, R_{ijj'} \in \{0, 1\} \qquad \forall i, i' \in N, \forall j \in J \tag{12}$$

The objective function (1) aims to minimize total completion time. Constraint (2) assures that each operation is assigned to one and only one machine. The precedence relationship within the operations of a job is modeled by constraint (3). Completion times of operations are shown by constraint (4). Constraint (5) and constraint (6) ensure the precedence relationships between jobs. Constraint (7) represents the completion time of jobs. Constraint (8) guarantees that the completion time of each job must be less than or equal to total completion time. Sign restrictions are stated between constraints (9)–(12). The objective function was chosen according to firm request and constraints were determined to cover the entire production line.

The data set that the firm uses in real life is extremely large. Hence, small-sized problems were developed to test the accuracy of the model. Firstly, these problems were solved manually with Gantt-Chart and after that, these problems were tested by IBM-ILOG-CPLEX STUDIO with the constructed mathematical model and we verified our mathematical model.

Next we attempt to solve the real problem of 69 jobs, 5 operations, and 11 machines of the company. With this setting our problem has 46,575 binary variables, 7,595 continuous variables, 75,289 constraints. We were unable to obtain a solution in 24 h of running the mathematical problem. Being unable to obtain an optimal solution in a reasonable time we proceed to search for near optimal solutions by implementing heuristics. We propose two rule-based heuristics from the literature, three new rule-based heuristics, and a Genetic Algorithm based metaheuristic for solving the problem. We describe each of these heuristics below.

4.2 Heuristic Algorithm

According to our flexible job shop scheduling problem-related literature review, this study is on a combinatorial, NP-hard, and single-objective optimization problem. After the verification of the mathematical model, due to the NP-hardness nature of the problem, heuristic optimization methods were researched to obtain a near-optimal solution. Metaheuristic methods can be called designs that use heuristic approaches to reach solutions in NP-hard and combinatorial optimization problems [9]. In cases where precise methods cannot be used due to the size of the solution space or the number of variables and constraints in the optimization model, the use of metaheuristics is increasingly widespread in solving problems that are not practical. Unlike the methods in which precise results are obtained, metaheuristics provide near-optimum solutions instead of the optimum [10]. One of the heuristics we have used is the Giffler and Thompson Algorithm which is a practical heuristic method that provides near optimal results in active and non-delay schedules for flexible job shop scheduling problems. In this scheduling problem, rule-based heuristic algorithms, and the Genetic Algorithm (GA) was exclusively developed by us. Genetic Algorithm is an iterative meta-heuristic, which combines from classical heuristics and natural and physical sciences for exploring and using learning strategies to find near-optimal solutions. Therefore, we aimed to determine the heuristic algorithm that gives the best results for this problem by comparing 6 different heuristic algorithms.

Giffler & Thompson Algorithm with SPT/LPT Priority Rule (GTA-SPT/LPT)
In this study, Shortest Processing Time (SPT) and Longest Processing Time (LPT) priority rules were used. Shortest Processing Time priority rule priorities' the job that has the shortest processing time. Longest Processing Time priority rule priorities' the job that has the longest processing time.

$\boldsymbol{n}$ denotes the total number of schedules,

$\boldsymbol{P_n}$ is the set of scheduled operations,

$\boldsymbol{S_n}$ is the set of operations schedulable i.e. all the operations that must precede those in $\boldsymbol{S_n}$ are in $\boldsymbol{P_n}$,

$\boldsymbol{Es_j}$ is the earliest time that operation j in $\boldsymbol{S_n}$ could be started,

$\boldsymbol{Ef_j}$ is the earliest time that operation j in $\boldsymbol{S_n}$ could be finished,

$\boldsymbol{T_{ijm}}$ denotes the processing time of job i, in machine m for operation j, $\boldsymbol{Ef_j} = \boldsymbol{Es_j} + \boldsymbol{T_{ijm}}$.

Step 1: Let $\boldsymbol{P_1}$ being null. $\boldsymbol{S_1}$ will be the set of all operations with no predecessors; in other words, those that are first in their job.

Step 2: Select min/max $\boldsymbol{T_{ijm}}$.

Step 3: Delete chosen operation from $\boldsymbol{S_1}$ and add to $\boldsymbol{P_2}$.

*Create $\boldsymbol{S_2}$ by adding the remaining operations of $\boldsymbol{S_1}$ and next operation of the chosen operation.

* It requires to $\boldsymbol{Ef_j} = \boldsymbol{Es_j} + \boldsymbol{T_{ijm}}$.

Step 4: Move to next stage by.

*Adding scheduled operation to $\boldsymbol{P_n}$, so creating $\boldsymbol{P_{n+1}}$.

*Deleting scheduled operation from $\boldsymbol{S_n}$ and creating $\boldsymbol{S_{n+1}}$ by adding the operation that directly follows scheduled operation in its routing to $\boldsymbol{S_n}$.

*Continue scheduling.

Step 5: If there are any operations left unscheduled go to step 2. Otherwise, stop.

Our third and fourth heuristic algorithms are **Complement Jobs Method 1/2 (RBA-1)/ (RBA-2).**

The processing time of job i, in machine m, for operation j is defined as $\boldsymbol{T_{ijm}}$. For each job, $\boldsymbol{T_{ijm}}$ is computed and jobs are sorted inside the groups from the **short processing time to the long processing time/long processing time to the short processing time** according to their $\boldsymbol{T_{ijm}}$ values.

Step 1: Group jobs according to their the operations to be performed, i.e. Laser-Press Break, Punch, Laser-Panel, Punch-Panel, Laser, Laser-Press Break-Panel.

Step 2: First, choose two complement groups that have the highest number of jobs, and different operation sequences but able to complement each other.

*Find the total processing time for each job and sort jobs inside the groups from the short total processing time to the long total processing time/the long total processing time to the short total processing time and assign these jobs respectively.

*Repeat the Step 2 until there are no complement groups.

Step 3: If there are no complement groups left, assign the remaining jobs starting from the shortest total processing time.

Step 4: Process jobs by this sorting order.

The fifth heuristic algorithm is **Average Processing Time Method (RBA-3).**

In this rule-based algorithm, jobs are sorted from **the largest processing time to the shortest processing time** by the average values per operation.

B_i denotes total processing time of job i,

N_O is the number of operations,

B_i/N_O denotes average processing time per operation.

Step 1: Find average processing time per operation by diving the total processing time B_i by the number of operations N_O.

Step 2: Sort jobs from **the largest processing time to the shortest processing time** by the average values per operation.

Step 3: Process jobs by this sorting order.

The sixth heuristic algorithm is **the Genetic Algorithm based on GTA (GA-GTA).**

For the Genetic Algorithm, one-cut and two-cut crossovers are selected. By applying the design of experiment (DOE), the best crossover type, mutation rate, and population size for this problem were determined under the Parameter Tuning sub-heading.

n = number of total assignments.

SPT = Shortest Processing Time.

LPT = Longest Processing Time.

Step 1: Apply Giffler and Thompson Algorithm by randomly selecting SPT or LPT at each stage.

Step 2: Select priority rules at each stage by keeping in order in 1xn matrices, choose 1 for SPT, 2 for LPT.

Step 3: Repeat the first two steps n times and create n units of 1xn matrices for the initial population, containing values 1 and 2.

Step 4: Select two random matrices as parents. Generate random numbers between 1 and n to identify one-cut crossover point and generate 2 numbers between 1 to n to identify two-cut crossover points.

Step 5: Create an offspring matrix by taking first parents matrix up to one-cut crossover point and taking the rest from the other parent's matrix for one-cut crossover and by taking first parents matrix up to first generated number, then the part between the first and second generated number from other parents matrix and then taking the remaining part from the second generated number from the first parents' matrix for two-cut crossover.

Step 6: Generate probability value for each element of the offspring matrix and replace that element if the probability value is smaller than the mutation rate value. If not, do not replace (1 instead of 2, 2 instead of 1).

Step 7: Delete the schedule that gives the worst completion time in the starting population (death penalty), write the created offspring matrix instead.

Step 8: Repeat 400 times (generations).

1) Parameter Tuning: In this section, a Design of Experiment (DOE) approach for parameter tuning of the GA-GTA is conducted. In the DOE approach, the proposed GTA-GA has three parameters. These parameters are the population size (PS), the crossover type (CRT), and the mutation rate (MR). We have taken the population size with three levels as PS ∈ (100,200,300), the crossover type with two levels as CRT (1,2), and the mutation rate with three levels as MR ∈ (0.01, 0.05, 0.09). Note that CRT = 1 means that one-cut crossover is applied whereas CRT = 2 means two-cut crossover is applied in the algorithm. In the design of GTA-GA, $3 \times 2 \times 3 = 18$ different parameter combinations are employed. The response variable was determined for each combination. PS, CRT, and MR were not statistically significant since the p-values of parameter interaction effects are higher than the significance level $\alpha = 0.05$. In terms of interaction effects, it can be observed that PC*CRT, PC*MR, and CRT*MR interactions are not significant because the p-value is much higher than the significance level of $\alpha = 0.05$. To determine the level of each factor, the main effects plot was used. According to DOE results, it is concluded that PS, CRT, MR parameters should be taken as PS = 300, CRT = 1, MR = 0.01.

5 Verification, Validation and Computational Results

To verify and validate the proposed mathematical model and heuristic algorithms, verification and validation procedure has been done. Verification is a procedure to test the accuracy of proposed algorithms and model, by experiments and logic. Validation procedure checks if the results of the model and algorithms coincide with real life. The mathematical model and 6 different algorithms were tested and verified with 10 toy problem instances. Toy problem instances were solved manually by drawing Gantt

Table 4. CPLEX and heuristic results on the toy problem instances

Toy problem instance	Size	Maximum completion time of all jobs by each heuristic						
		OPL (sec)	GTA-SPT (sec)	GTA-LPT (sec)	RBA-1 (sec)	RBA-2 (sec)	RBA-3 (sec)	GA-GTA (sec)
1	|N|=2; |J|=3; |M|=7	56	56	56	56	56	56	56
2	|N|=3; |J|=3; |M|=7	34	34	34	34	34	34	34
3	|N|=3; |J|=3; |M|=8	72	72	72	72	72	72	72
4	|N|=3; |J|=4; |M|=9	29	29	29	29	29	29	29
5	|N|=4; |J|=3; |M|=7	47	47	47	47	47	47	47
6	|N|=4; |J|=4; |M|=9	39	39	39	39	39	39	39
7	|N|=2; |J|=1; |M|=2	10	10	10	10	10	10	10
8	|N|=5; |J|=3; |M|=4	43	43	43	43	43	43	43
9	|N|=4; |J|=4; |M| =10	45	45	45	45	45	45	45
10	|N|=5; |J|=9; |M|=4	33	33	33	33	33	33	33

charts and then the mathematical model was executed. Next, the proposed heuristic algorithms were run on these problem instances and all the results were reported. Table 4 depicts the maximum completion time of all jobs by each heuristic for ten toy problem instances. We observe that for small sized problems all heuristics are able to obtain the optimal solution.

5.1 Real Data

Next, we proceed to solve the real problem instance which is the production plan for

Table 5. OPL and Giffler & Thompson algorithm results

Month	Number of jobs	OPL		GTA-SPT			GTA-LPT		
		Run time (h)	Solution	Run time (Sec)	Finish-time (Sec)	Finish-time (8-h shift)	Run time (Sec)	Finish-time (Sec)	Finish-time (8-h shift)
1	276	24	No Solution	1	1,754,248	60.9	1	1,404,528	48.8
2	328	24		2	3,066,700	106.5	1	2,966,064	103.0
3	328	24		6	3,154,788	109.5	1	2,860,252	99.3
4	380	24		4	4,077,412	141.6	2	3,089,624	107.3
5	336	24		3	2,588,004	89.9	2	2,060,156	71.5
6	328	24		4	2,046,180	71.0	1	1,765,876	61.3
7	368	24		3	3,539,820	122.9	1	2,705,676	93.9
8	344	24		3	2,879,236	100.0	1	2,384,812	82.8
9	340	24		4	3,794,836	131.8	1	3,437,760	119.4
10	328	24		2	3,770,924	130.9	2	2,952,820	102.5

the 10-month with the provided demand data. We report the run time and the best solutions obtained by each heuristic in Table 5 and 6. The best solution among the heuristics for each month is indicated by an asterisk (*). We report objective function value which is the maximum completion time of all jobs both in seconds and in number of 8-h shifts. As we have discussed earlier, we were unable to obtain a solution by OPL after running 24 h. Solutions are provided by both in terms of seconds and days.

When the results of the RBA-1 and RBA-2 algorithms are compared, it has been observed that the RBA-2 algorithm gives better results. When the RBA-3 algorithm and Genetic Algorithm based on GTA results were compared, the RBA-3 algorithm gave better results. Based on the Giffler & Thompson SPT/LPT results, we compared the results of the algorithms by percentages. According to these results, we concluded that RBA-3 is our best performing algorithm with 3.5% shorter completion time. RBA-1 and RBA-2 algorithms are the worst performing algorithms with longer completion times compared to Giffler & Thompson Algorithms.

Table 6. RBA 1-2-3 and GA-GTA results

Month	Number of Jobs	Run time (Sec)	RBA-1		RBA-2		RBA-3		GA-GTA	
			Finish-time (Sec)	Finish-time (8-h shift)	Finish-time (Sec)	Finish-time (8-h shift)	Finish-time (Sec)	Finish-time (8-h shift)	Finish-time (Sec)	Finish-time (8-h shift)
1	276	1	1,496,584	52.0	1,332,452	46.3	1,284,632	44.6*	1,296,907	45.0
2	328	2	3,113,284	108.1	2,874,404	99.8	2,807,964	97.5*	2,822,472	98.0
3	328	4	3,221,256	111.8	3,093,308	107.4	2,804,736	97.4*	2,823,192	98.0
4	380	2	3,719,436	129.1	3,120,912	108.4	3,036,832	105.4*	3,046,334	105.8
5	336	1	2,644,972	91.8	2,080,572	72.2	2,012,492	69.9	1,820,764	63.2*
6	328	2	2,861,684	99.4	2,172,564	75.4	1,757,068	61.0*	1,762,747	61.2
7	368	1	3,804,972	132.1	2,714,164	94.2	2,671,244	92.8*	2,673,777	92.8
8	344	3	3,122,168	108.4	2,361,804	82.0	2,307,424	80.1*	2,341,915	81.3
9	340	1	3,978,168	138.1	3,398,164	118.0	3,132,220	108.8*	3,151,180	109.4
10	328	1	3,805,016	132.1	2,804,972	97.4*	2,918,064	101.3	2,933,971	101.9

5.2 Hypothesis Testing

To test the algorithms statistically, we conducted a Wilcoxon signed-rank test at the significance level of $\alpha = 0.05$. The test is used to determine that there is a statistically significant difference in maximum finish-time. We have 15 large instances. The median of the differences between the *Cmax* produced by pair algorithms is denoted with md. The null hypothesis is defined by H0: md = 0 saying that there is no difference between the *Cmax* of compared paired algorithms. All pairwise differences are statistically significant at the $\alpha = 0.05$ level, except RBA-2 & GTA-LPT and RBA-1 & GTA-SPT. For these pairs, the differences between the algorithms are not meaningful at the $\alpha = 0.05$ level, implying that algorithms within these pairs have similar performances.

6 Conclusion

In this project, a production plan was developed by determining which order will be processed on which machine for the production planning problem of the company. Minimizing the total completion time of the jobs was our main goal in the project. Thus, we aimed to ensure the correct use of capacities and increase efficiency. First, a detailed system analysis of the company was made. Then, in order to verify that the defined work and material flow processes at the facility are understood correctly, it was understood that the system was modeled with Arena Simulation Software and that the production system was compatible with the flexible job-shop scheduling problem. A detailed literature review was conducted for the flexible job-shop which is the production type of the company. Mathematical models in the literature were examined and a mathematical model was developed in accordance with the company's system. Since this problem is NP-hard, heuristic methods have been examined to develop a production schedule. Four of the six different heuristic algorithms have been proposed are developed. A user-friendly decision support system (DSS) has been developed exclusively for the company. The proposed heuristic methods are integrated in the DSS so that it provides the solutions with the user in a simple and understandable way and can be used as a decision support system for company's production process. Since the company has an intuitional scheduling, this study will have an innovative aspect on their production planning. In this way the planning process has become more efficient and problems caused by manual scheduling are prevented.

Acknowledgment. This study is supported by TUBITAK (The Scientific and Technological Research Council of Turkey) in the program of "2209-B Undergraduate Research Projects for Industrial Applications Fellowship Program". We would like to thank Bekir Fatih Ural for his support as the company advisor. Also, we are grateful to the company for their co-operation during the whole process.

References

1. Vallikavungal Devassia, J., Salazar-Aguilar, M.A., Boyer, V.: Flexible job-shop scheduling problem with resource recovery constraints. Int. J. Prod. Res. **56**(9), 3326–3343 (2018). https://doi.org/10.1080/00207543.2017.1420262
2. Yuan, Y., Xu, H.: A memetic algorithm for the multi-objective flexible job shop scheduling problem (2013). In: GECCO 2013 - Proceedings of the 2013 Genetic and Evolutionary Computation Conference, pp. 559–566. https://doi.org/10.1145/2463372.2463431
3. Chaudhry, I.A., Khan, A.M., Khan, A.A.: A genetic algorithm for flexible job shop scheduling. In: Proceedings of the World Congress on Engineering 2013, WCE 2013, 3–5 July 2013, vol. I, London, UK (2013) ISSN: 2078-0966 (Online)
4. Liangxiao, J., Zhongjun, D.: An improved genetic algorithm for flexible job shop scheduling problem. In: 2015 2nd International Conference on Information Science and Control Engineering, Shanghai, pp. 127–131 (2015). https://doi.org/10.1109/ICISCE.2015.36
5. Pezzella, F., Morganti, G., Ciaschetti, G.: A genetic algorithm for the flexible job-shop scheduling problem. Comput. Oper. Res. **35**(10), 3202–3212 (2008). https://doi.org/10.1016/j.cor.2007.02.014
6. Liu, S.C., Lee, W.T.: A heuristic method for the inventory routing problem with time windows. Expert Syst. Appl. **38**(10), 13223–13231 (2011). https://doi.org/10.1016/j.eswa.2011.04.138
7. Kaya, S., Fığlalı, N.: Use of meta-heuristic methods to solve the multi-objective flexible job shop scheduling problems. Harran Univ. J. Eng. **3**, 222–233 (2018)
8. Koblasa, F.: Giffler and Thompson Algorithm for flexible job shop scheduling problem. In: 4th Annual International Conference 2009 Manufacturing Systems Today and Tomorrow (2009)
9. Kennedy, J., Eberhart, R.C., Shi, Y.: Swarm Intelligence. Morgan Kaufmann Publishers, San Francisco (2001)
10. Aydın, A.: Optimization of aircraft scheduling problem with meta-intuitive methods. Ph.D. thesis, Marmara University, Institute of Social Science

Analysis and Improvement of Employee Transportation System

Gökçe Fatma Akpinar, Merve Özdemir, Orhan Ege San, Emre Ege Ak, Oğuz Kaan Yalabik, Zafer Çağlar, Mert Paldrak, and Adalet Öner(✉)

Department of Industrial Engineering, Yasar University, Izmir, Turkey
gokcefatmakpinar@gmail.com, merveozdemr96@gmail.com, oesan1912@gmail.com, emreege200@gmail.com, oguzkaanyalabik@gmail.com, zaferca@gmail.com, {mert.paldrak, adalet.oner}@yasar.edu.tr

Abstract. This study involves in analysis and improvement of the employee transportation system in a private university in Izmir, Turkey. The problem is to design a transportation system using systematic and scientific methods. The input of the problem is the location of home addresses of the employees. It is required to determine the number and types of vehicles to be used to transport the employees to and from the university campus. Moreover, the routes of the vehicles should also be defined when the problem is solved. A two-phase solution approach is adopted. In the first phase, the addresses are partitioned into clusters by using K-means algorithm. The centres of clusters would be the bus stops in the transportation system. Once the cluster analysis delivers the cluster centres (location of bus stops) and the number of employees assigned to each bus stop, in the next phase, the school bus routing model (SBRP) is employed to solve the remaining part of the problem. In this phase, the number and types of the vehicles are determined as well as the routes of the vehicles. The details of clustering process discussed, and solutions of math models are shown for large data sets.

Keywords: Employee transportation problem · Clustering · K-means algorithm · School Bus Routing Problem

1 Introduction

University managements in Turkey usually provide employee transportation services for academic and administrative staff. In the mornings of each working day, several buses traverse the city following predetermined routes to pick up the employees who wait at predetermined bus stops and bring them to the university. In the evenings, the same buses travels on the same routes in reverse direction to transport the employees to the nearest bus stops to their homes.

Providing bus services to employees in universities not only enhance the quality of the university services, but also reduce the traffic pressure especially for the relatively populated cities. Among the bus scheduling problems, planning employee buses routes

N. M. Durakbasa and M. G. Gençyılmaz (Eds.): ISPR 2020, LNME, pp. 831–841, 2021.
https://doi.org/10.1007/978-3-030-62784-3_69

are one of the most basic works to consider. It is generally accepted that the reasonable employee bus planning is supposed to reduce the operating costs while ensuring safety and minimum walking distance of employees to bus stops. Consequently, it is vital to develop an efficient and effective algorithm which helps to determine the bus routes and optimal bus numbers easily. Employee bus routing problem seeks to generate an effective schedule for a fleet of busses in which each bus picks up some employees from bus stops and delivers them to their designated destination.

Employee Bus Routing Problem (EBRP) is a member of an application area of vehicle routing problem (VRP) and it is one of the well-known NP-Hard combinatorial optimization problems. Even though the problem has been reached for many years under the name of School Bus Routing Problem (SBRP), the problem is still continuously studied since the complexity of the constraints and the objective functions in the real-life cases.

This study deals with the employee transportation problem of a private university located in Izmir, Turkey. The university provides employee transportation services for academic and administrative staff. Based on the records in the Human Resources Department, total number of fulltime employees is 671 of which 450 are academic staff and 221 are administrative staff. However, only 300 of these employees use the transportation services. For several reasons, the others prefer using public transportation means and personal cars.

2 Problem Definition

2.1 Current System and Symptoms of the Problem

There are currently 7 different designated routes for employee transportation. Those routes have names that are given based on the name of the districts of the city such as "Alsancak-Güzelyalı", "Hatay", "Atakent", "Çiğli-Mavişehir", "Karşıyaka-Girne", "Gaziemir", and "Bornova". In the current system, the routes of buses are determined depending on the experiences of related staff. No systematic and scientific method is used to analyze and improving the transportation service. Consequently, the problem arises due to the lack of application of optimization techniques to determine the number of busses and their routes. On the other hand, the transportation cost incurred is expectedly high because the busses follow the routes set by the support service directorate. It is also observed that the busses pick all of the employees from the predetermined stops, however the determination of these stops also lacks a systematic approach. Therefore, some of the employees has to walk much to get to one of the stops, while some others walk less. As a result, it is vital to determine the bus stops according to the addresses of employees using a clustering technique under the objective of the minimization of walking distance of employees.

2.2 Problem Definition

The goal of this study is to improve the efficiency of employee transportation services by reducing transportation costs while increasing the comfort of employees. The comfort of employee is defined as the average distance between employee addresses and the closest bus stops. If the number of bus stops increases, then both comfort of employee and transportation costs increase together. Because it is expected that the distances between home-addresses and the closest bus stops will decrease in that case. Similarly, the lengths of routes naturally increase with more bus stops and hence the transportation costs will increase. Therefore, two performance criteria contradict each other. The transportation costs are divided into two categories: variable and fixed costs. While variable costs are associated with the total distance travelled by the busses, the fixed costs are related to the number and the type of busses to be used.

The decision variables of the problem are to determine the number and locations of bus stops, the number and type of the busses to be used and their routes. The routes will be comprised of the arcs that connects particular bus stops. A bus will be designated to visit that bus stop in its route. That bus will pick up the employees who are waiting in that bus stop. The number of employees who get on the bus should not be greater than the capacity of that bus. On the other hand, the length of the route or the total travel time on the route is limited to a particular value. Because maximum travel time of any employee to or from the campus is desired to be short as much as possible.

3 Literature Review

The literature review conducted in this study is divided into two areas. In the first part, the previous studies using k-means clustering algorithm are examined in order to determine the clusters of addresses. In the second part, the focus is more on the researches considering Capacitated Vehicle Routing Problem (CVRP), School Bus Routing Problem (SBRP) and Strong Multi-Commodity Network Problem (SMCNP) since the problem in the second phase can be formulated as one of these problem types.

Most of the companies including education facilities try to significantly reduce the transportation costs by grouping customers or employees into an optimal number of clusters by taking the minimization of delivery costs for each cluster into consideration. The defined strategy is named as clustering in the literature. The study in [1] define clustering as a classification of patterns into groups.

The partitioning type of clustering algorithms determine the clusters at once. Whereas, the hierarchical algorithm tries to find successive clusters using pre-determined clusters. It is essential to use the distance which can either be symmetric or asymmetric in clustering. The distance measure also plays an important role on governing similarity between items since it affects the shape of cluster. In general, Euclidean distance is mostly employed and customers or employees are grouped based on their demand or their numbers. When capacity is also considered in the formation of clusters, then the problem is named as the Capacitated Clustering Problem (CPP). In our problem, partitioning type of clustering algorithm with Manhattan distance and without cluster capacity is considered.

The k-means clustering method is firstly proposed by [2], which is the most well-known and commonly used method of clustering. Although Mac Queen is the first developer, Hugo Steinhaus had a similar idea in 1957. On the other hand, Stuart Lloyd was the one who first suggested the standard algorithm for pulse-code modulation.

Example of a journal article in [3], they employed improved k-means clustering for the Capacitated Clustering Problem (CPP) in order to obtain better optimization results. The proposed algorithm develops density-based detection methods based on the characteristics of the noise data. Lastly, it is stated that the runtime of the k-means algorithm is favorable and used several approaches to improve the k-means clustering algorithm [4].

The problem in second phase can be considered as Distance Capacitated Vehicle Routing Problem (DCVRP) which is more generalized version of School Bus Routing Problem (SBRP). The authors of [5] firstly introduced the VRP in order to address the distribution and delivery challenges. The core function of the concept is to provide delivery services to customers and any other organization related to logistics [6]. The primary objective of the problem is to determine the set of optimal routes in order to minimize either the cost of transportation, transportation time, the number vehicles needed to distribute the products or combination of these factors. In the literature, different version of VRP that have been developed exist [7], including the Capacitated Vehicle Routing Problem (CVRP), the VRP with time windows, the VRP with pickup and delivery and the heterogenous fleet VRP. Among all others, CVRP is the one which is commonly used. The CVRP is a VRP in which the number of deliverables vehicles can carry is limited. The total demand at any location, or along any vehicle route cannot be more than the capacity of the delivery vehicle.

4 Problem Formulation

The input data of the problem is the list of home addresses of the employees. However, due to the law on the protection of personal data, the real addresses of the employees couldn't be accessed. Consequently, an imaginary data set which include 300 address in Izmir has been generated to represent the employee addresses. We follow a systematic method to generate those addresses. In the first step, central districts of İzmir such as Karşıyaka, Konak, Karabağlar etc. are determined and census data of those districts has been compiled. Then population percentage of each district, within total population of central districts, has been calculated. Based on calculated percentages, we have determined the number addresses to be generated in each district. In the second step, in order to be more realistic, rural and urban regions of each district were separated since very few people or no nobody is settled in rural areas. Hence rural regions of each district are ignored, and the addresses are generated only in the urban regions. After that, urban regions of each district were divided into regular geometric parcels and their areas have been calculated. Based on normalized weights of the areas, we determined the number of addresses to be determined inside each parcel in each district. The bigger the parcel area, the more addresses would be generated. In other words, the number of addresses to be generated in each parcel was directly proportional to its area. After that, assuming that the latitudes and longitudes are independent of each other and both are distributed uniformly along the borders of the parcels, the addresses were generated randomly. After studying and a preliminary literature survey, it is understood that it is needed to use a special two-phase modeling approach, whose details are given in Fig. 1 below.

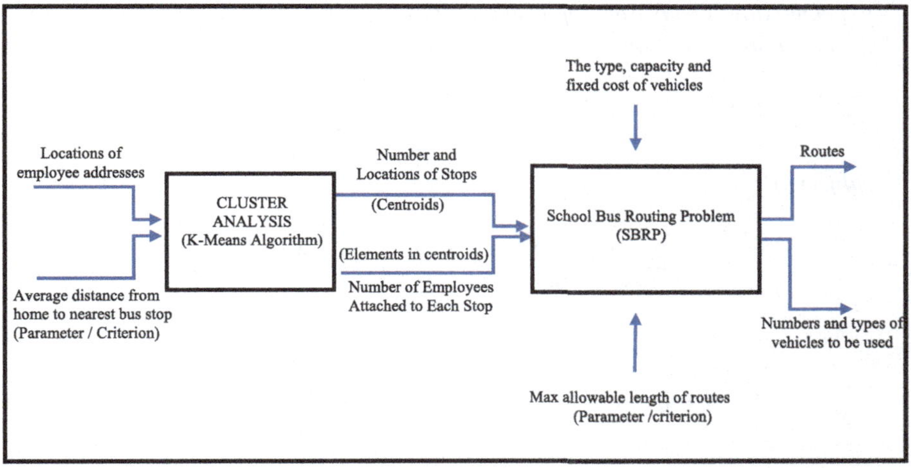

Fig. 1. Two phase modeling approach

In the first phase, the addresses are clustered such as the centroids will represent the bus stops. Clustering analysis would also yield the number of employees who are assigned to each centroid (bus stop). The output of the first phase would be a part of input for the second phase. In the second phase, the centroids (bus stops) would be assumed as the nodes of an interconnected network. The problem is here to find the number and type of vehicles to be used and the routes of them. Each bus should be assigned to a route to travel. The routes are the chains which connects particular arcs in the network. Once the cluster analysis delivers the cluster centres (location of bus stops) and the number of employees assigned to each bus stop, in the next phase, we need to determine how many and which type of vehicles should be used and to find the routes of vehicles. This problem can be formulated as a School Bus Routing Problem. The notation and the mathematical model is given as follows.

Set and Indices:

V : *set of nodes. The location of bus stops* $V = \{0, 1, \ldots .N\}$. *The node 0 represents the university campus*

$A = \{(i,j), i,j \in V | i \neq j\}$: *the set of arcs(roads)*

$G(V,A)$: *complete graph*

Parameters:

C_{ij} : *the travel distance between node i and j*

t_{ij} : *the travel time between node i and j*

t_i : *the service time for picking up employees at stop i* $(t_o = 0)$

q_i : *the number of employees taking the bus at stop i* $(q_0 = 0)$

Q : *capacity of each bus*

T_{max} : *Maximum riding time for any one employee*

M_i : $BigMi \in \{0, 1, 2\}$

Decision Variables:

$$x_{ij} = \begin{cases} 1, & \textit{if bus traverses between node i and j} \\ 0, & \textit{otherwise} \end{cases}$$

y_i : *cumulative number of employees in a bus* $(y_0 = 0)$
z_i : *cumulative travel time of the bus when it leaves from stop i* $(z_0 = 0)$

$$\text{Min } Z = M_0 \sum\nolimits_{i \in V|i \neq 0} x_{0i} + \sum\nolimits_{i \in V} \sum\nolimits_{j \in V|i \neq j, i \neq 0} c_{ij} x_{ij} \tag{1}$$

subject to:

$$\sum_{j \in V|j \neq i} x_{ij} = \sum_{j \in V|j \neq i} x_{ji} = 1, \qquad \forall i \in V|i \neq 0 \tag{2}$$

$$y_i \leq Q \qquad \forall i \in V|i \neq 0 \tag{3}$$

$$y_i + q_j - y_j \leq M_1 (1 - x_{ij}) \qquad \forall i, j \in V|i \neq j, j \neq 0 \tag{4}$$

$$z_i \leq T_{\max} \qquad \forall i \in V \tag{5}$$

$$z_i + t_{ij} + t_j - z_j \leq M_2 (1 - x_{ij}) \qquad \forall i, j \in V|i \neq j, j \neq 0 \tag{6}$$

$$x_{ij} \in \{0, 1\} \qquad \forall i, j \in V|i \neq j \tag{7}$$

$$y_i, z_i \in \{0, 1, 2 \ldots\}, \qquad \forall i \in V \tag{8}$$

The objective function (1) is to minimize the number of busses used and the total travel distance. The parameter M_0 is a big enough positive number which represents the fixed cost of the bus. In the mathematical model, it is taken as $M_0 = 100$. Constraints (2) guarantee that every stop is served by one bus and visited exactly once. Constraints (3) ensure that bus capacity cannot be exceeded. Constraints (4) states the accumulation of employees in the bus. Constraints (5) ensure that the riding time of every employee never exceeds maximum riding time T. Constraints (6) state the accumulation of bus travel time. Constraints (7) and (8) define the domain for decision variables.

5 Solution Methodology

In the first phase, the K-means algorithm is employed in order to determine the numbers and locations of bus stops in each district. The K-means algorithm is a partitioning type of clustering algorithm. Its pseudo code is given in Fig. 2 below. To run the algorithm, you need to specify the number of cluster (K) to be developed in advance. Then the algorithm groups address into K clusters based on the Manhattan distances. It tries to minimize "total sum of squares within clusters". The K value corresponds to the number of bus stops. However, it requires to be defined beforehand. If the K value is changed, and the algorithm is run again, then you find other cluster centres with a different objective function value.

It is natural that, the objective function value decreases as the K-value increases. However, it is known that the rate of decrease in the objective function gets lower and lower as the K-value passes a threshold. If we plot the objective function values for

Let $X = \{x_1, x_2, x_3, \dots\dots x_n\}$ *be the set of data points (locations of addresses) and*
$V = \{v_1, v_2, \dots\dots v_n\}$ *be the set of centers (candidate bus stops)*

Step 1: Randomly select "K" cluster centers.
Step 2: Calculate the distance between each data point and cluster center.
Step 3: Assign the data point to the cluster center whose distance from the cluster is minimum of all cluster centers
Step 4: Recalculate the new cluster center using:

$$v_i = \left(\frac{1}{K_i}\right)\sum_{j=1}^{K_i} x_i \quad \text{where "}K_i\text{" represents the number of data points in } i^{th} \text{ cluster.}$$

Step 5: Recalculate the distance between each data point and new obtained cluster centers.
Step 6 : If no data point was reassigned then stop, otherwise repeat from step 3.

Fig. 2. The pseudo code of the K-means algorithm

different K-values, the curve makes an "elbow" at that critical K-value. Therefore, it is decided that the most efficient value for number of clusters is that threshold value. While the clustering study was conducted, it was first attempted to determine the optimal number of stops for each district separately. In order to perform this operation, it obvious that different K-values should be tested. A range of K-values is determined for each district. The width of the range is determined according to the number of addresses in the district. Table 1 shows the ranges of K-values to be scanned for each district.

Table 1. Cluster range of districts

District	Number of addresses	The range of K value to be scanned
Balçova	8	2–5
Bayraklı	32	2–10
Bornova	45	2–10
Buca	51	2–10
Çiğli	20	2–7
Gaziemir	14	2–6
Konak	36	2–10
Narlıdere	7	2–4
Güzelbahçe	4	1–3
Karabağlar	48	2–10
Karşıyaka	35	2–10

When the number of addresses is high, then the max value of the range is high as well. The elbow method was applied to find the most efficient K-value for each district. However, when we observed the results, it seems that the maximum distance of distances between the addresses and the cluster centre is quite high, more than we expected. In that case, we adopted a custom rule for selecting the most efficient K-value. Instead of the elbow method, we selected the minimum number of K-value that delivers clusters such that maximum of max distances is no longer than 1.5 km. For example, Table 2 shows that a sample analysis for the district 3. The restricting value (1.5 km) is set by the decision maker for the sake of the comfort of the employees and a balanced distribution of employees between clusters.

Table 2. Sample analysis for district 3

Number of clusters	Cluster number	Number of employees assigned to the cluster (=Bus stop)	Max distance from the address to the cluster centre (=Bus stop) (Km)
3 determined by elbow method	1	25	2,06
	2	7	1,46
	3	13	1,77
8 determined by custom selecting method	1	4	0,41
	2	3	0,35
	3	4	0,39
	4	5	0,50
	5	7	1,16
	6	5	0,46
	7	13	1,08
	8	4	1,04

To solve the real-world problem, we need to process all addresses as one single set and conduct the K-means algorithm to find the number and locations of bus stops. However, if we consider the layout of Izmir and the position of the bay, it is reasonable to divide the addresses into three separate sets such as Izmir North, Izmir South East and Izmir South West. The set of Izmir North includes the districts of Çiğli, Karşıyaka and Bayraklı. The set of Izmir South East cover the districts of Buca, Bornova and Konak. The set of Izmir South West is comprised of the districts of Balçova, Gaziemir, Güzelbahçe, Karabağlar and Narlıdere. Using the model, a sample solution is produced for the data set of Karabağlar. The cluster analysis yielded 7 clusters and their locations. Output of the cluster analysis is fed into SBRP and the model provides a solution.

6 Computational Results

The model has also been used solve the real-world problem considering all 300 addresses. However, remember that we have divide overall list into three separate sets such as North, South East and South West. It does not take much time to solve problems for datasets related to individual districts. It takes at most a few minutes. However, solution times increased dramatically when we try to solve the problems related to regional data sets. Average solution times for regional problems are nearly 35 h. It seems that it still is practical to solve the mathematical model since the problems are solved once to propose a new transportation system. The optimal solutions and the details are given below in Table 3, Table 4 and Table 5 for the regional data sets North, South East and South West, respectively.

Table 3. Proposed routes and related attributes for Izmir North

Route number	Stop #	Number of employee assigned	Maximum distance to the bus stop (km)	Cumulative number of employees in the vehicle	Cumulative travel time (minutes)
1	2	7	0,53	35	14,99
	4	5	0,7		
	6	7	0,36		
	9	5	1,08		
	11	4	0,40		
	12	7	0,90		
2	7	7	0,97	15	45
	8	1	0,00		
	10	7	1,57		
3	1	5	0,98	35	24,57
	3	6	1,32		
	5	4	0,60		
	13	4	0,69		
	14	10	0,82		
	15	3	0,30		
	16	3	0,31		

Table 4. Proposed routes and related attributes for Izmir Southeast

Route number	Stop #	Number of employee assigned	Maximum distance to the bus stop (km)	Cumulative number of employees in the vehicle	Cumulative travel time (minutes)
1	1	8	0,92	25	28
	2	6	1,20		
	4	3	0,50		
	5	4	0,84		
	12	4	0,59		
2	3	5	1,21	25	31
	6	7	1,48		
	8	4	0,73		
	10	3	0,21		
	16	6	1,01		
3	7	8	0,81	16	45
	9	4	0,56		
	11	4	0,35		
4	13	5	0,55	16	45
	14	4	0,48		
	15	7	0,76		

Table 5. Proposed routes and related attributes for Izmir Southwest

Route number	Stop #	Number of employee assigned	Maximum distance to the bus stop (km)	Cumulative number of employees in the vehicle	Cumulative travel time (minutes)
1	3	2	1,00	35	45
	4	11	0,76		
	5	6	0,66		
	11	7	0,54		
	12	4	1,07		
	13	6	3,53		
2	1	4	0,38	29	45
	2	9	0,38		
	6	4	1,29		
	9	7	1,10		
	10	5	2,25		
3	7	8	1,05	33	45
	8	6	0,50		
	14	9	1,44		
	15	10	1,09		

7 Conclusions

In this study, the employee transportation problem of a private university in Turkey, İzmir is taken into consideration. In order to solve the problem, two phase solution methodology is applied. Since the exact addresses of the employees are not provided by the university due to the personal data protection law, the addresses are generated randomly by using uniform distribution considering the population density of the city. In the first phase, k-means algorithm is employed in order to determine the location of bus stops. After determining location of these stops, elbow method is used to find the optimal number of bus stops for each district. In the second phase, Schule Bus Routing Problem (SBRP) is formulated to determine the optimal number of busses that should run and the optimal bus routes. The obtained results are tabulated with the optimal routes of each bus in each district.

References

1. Souza Lima, F.M., Pereira, D.S.D., Conceição, S.V., Camargo, R.S.: A multi-objective capacitated rural school bus routing problem with heterogeneous fleet and mixed loads. 4OR **15**(4), 359–386 (2017)
2. Mokhtari, N., Ghezavati, V.: Integration of efficient multi-objective ant-colony and a heuristic method to solve a novel multi-objective mixed load school bus routing model. Appl. Soft Comput. **68**, 92–109 (2018)

3. Geetha, S., Poonthalir, G., Vanathi, P.T.: Improved k-means algorithm for capacitated clustering problem. INFOCOMP **8**(4), 52–59 (2009)
4. Singh, M.: A survey on various k-means algorithms for clustering. Int. J. Comput. Sci. Netw. Secur. (IJCSNS) **15**(6), 2015 (2015)
5. Dantzig, G.B., Ramser, J.H.: The truck dispatching problem. Manage. Sci. **6**(1), 80–91 (1959)
6. Toth, P., Vigo, D. (eds.): Vehicle Routing: Problems, Methods, and Applications. Society for Industrial and Applied Mathematics, Philadelphia (2014)
7. Li, S.G.: Genetic algorithm for solving dynamic simultaneous route and departure time equilibrium problem. Transport **23**(1), 73–77 (2008)

Container Demand Forecasting Using Machine Learning Methods: A Real Case Study from Turkey

Ayhan Darendeli, Aylin Alparslan, Mehmet Serdar Erdoğan, and Özgür Kabadurmuş(✉)

Department of International Logistics Management, Yasar University, Izmir, Turkey
ayhandarendeli@hotmail.com, aylin.alparslan53@gmail.com, {mehmet.erdogan, ozgur.kabadurmus}@yasar.edu.tr

Abstract. The container demands in ports significantly fluctuate over time and accurate container demand forecasting is essential for logistics companies because they can make their future business plans accordingly. In maritime transportation, container slot agreements are generally made two times in a year. A slot is one Twenty-Foot Equivalent Unit (TEU) space in a container ship and early booking of a slot is less costly for a company. Therefore, the accurate prediction of future container demands is crucial for companies to reduce their costs and increase their profits. In this study, we developed various forecasting models using machine learning methods to accurately predict the future container demands for the largest maritime transportation and logistics company of Turkey. The main aim is to provide accurate container demand forecasts for the company so that it can optimize the container slot bookings. To forecast the container demand, we used the company`s internal demand data as well as various external data such as gross domestic product (GDP), inflation rate, and exchange rate. We built four forecasting models based on Linear Regression, Boosted Decision Tree Regression, Decision Forest Regression and Artificial Neural Network Regression algorithms. The performances of these methods were evaluated according to Coefficient of Determination, Mean Absolute Error, Root Mean Square Error, Relative Absolute Error and Relative Squared Error. The case study showed that Boosted Decision Tree Regression and Decision Forest regression methods yield the best forecasting accuracy.

Keywords: Logistics · Forecasting · Container demand · Machine learning · Regression

1 Introduction

Ports are the gates opening to the world, where the transportation starts or continues within the international trade process. As long as there is trade in the world, maritime trade will exist. Therefore, the ports have to improve their services and companies must anticipate and accurately predict the container demands. Demand forecasting is

N. M. Durakbasa and M. G. Gençyılmaz (Eds.): ISPR 2020, LNME, pp. 842–852, 2021.
https://doi.org/10.1007/978-3-030-62784-3_70

important for all businesses to see future situations and make the required plans accordingly. A container demand forecast model, which is particularly important for ports, tries to predict how much container to be handled in the future at a specific port. Ports should anticipate future freight demand and establish new investment and development plans in order to better serve the companies. Also, the companies must make their plans according to changing demand and business conditions.

Although there are numerous forecasting methods, Machine Learning methods have been proved to be very effective in forecasting. The container demand forecast is one of the areas that machine learning methods can be effectively used to predict container demand.

Logistics companies book container slots two times a year by making arrangements with the shipment lines. A slot is a space in a container ship that can be taken by one standard twenty-foot container (TEU) per ship. Buying slots earlier helps to reduce the cost per container and maximize profit. Similarly, slot prices increase as the shipment time gets closer. Therefore, booking slots late increases transportation costs. However, due to fluctuations in demand, the optimal slot quantity cannot be determined. Thus, an accurate forecasting method to predict the required number of slots is essential for maritime logistics companies. By using accurate forecasting methods to estimate the container demand and optimizing slot amount, the transportation costs can be minimized and total profit can be maximized.

In this study, using machine learning methods, we developed various forecasting models to predict monthly container demand between ports for each month by using internal and external data. Our aim is to optimize the slot bookings by accurately forecasting the container demand between the ports. Four different forecasting models are developed by using Boosted Decision Tree Regression, Decision Forest Regression, Artificial Neural Network Regression, and Linear Regression and their forecasting performances are compared using a real case study from Turkey. The case company is the largest maritime transportation and logistics company in Turkey. The performances of the developed models are evaluated according to various metrics (Coefficient of Determination, Mean Absolute Error, Root Mean Square Error, Relative Absolute Error and Relative Squared Error).

The remainder of the paper is organized as follows: Sect. 2 reviews the related literature. Section 3 describes the problem of this study. Section 4 summarizes the forecasting methods and the evaluation metrics that are used in the study. Section 5 presents the results. Finally, Sect. 6 concludes the study and summarizes the future research directions.

2 Literature Review

The machine learning techniques use extensive data, and therefore, machine learning is considered as one of the most important elements of Big Data Analytics. Chen et al. (2014) [1] considered the vast increase of global data and advancement of information technology which enable data generation and reported that the term Big Data is mainly used to describe enormous datasets that typically include masses of unstructured data requiring more real-time analysis.

Container demand or throughput estimation is one of the topics that Big Data based methods are applied successfully in the literature. For example, Huang et al. (2016) [2] proposed using a partially joint system to forecast container throughput with Big Data composed of structured historical data and unstructured data. Lam (2011) [3] introduced a slot capacity analysis that uses empirical data on slot capacity deployed by container shipping lines for analyzing ports (as nodes) and routes and shipping lines (as links) that are embedded within the maritime supply chain. Moreover, Zhang and Kline (2007) [4] applied the Artificial Neural Networks (feed-forward networks) approach to forecasting quarterly time series. In our study, we also used Artificial Neural Networks to forecast container demand.

Diaz (2011) [5] evaluated the Winters, Tioga Group and United Nations methods for forecasting empty container volumes at container ports. As another extension, Kuroda et al. (2005) [6] examined the vector error correction model to develop port facilities by obtaining information on the demand for container volume which is a key issue for the construction of new ports. They used the container data and external data from the Indonesian Statistical Yearbook (Gross domestic product (GDP), population, export and import data). They accurately estimated the container production volume in Indonesia by vector error correction method and showed the applicability of their model.

Schulze and Prinz (2010) [7] studied container transshipment at German ports using the Seasonal Autoregressive Integrated Moving Average (SARIMA) model and the Holt-Winters exponential smoothing approach for Asia, Europe and North America using quarterly data. They claimed that the effect of seasonality is high in reducing costs. Both of these methods provided short-term forecasts. They tested these two approaches according to the root mean square error (RMSE) and concluded that Holt-Winters yielded better results than SARIMA for forecasting with seasonality. In addition to this study, Peng and Chu (2009) [8] applied SARIMA model to making forecasting for the container throughput volume in three ports in Taiwan using the classical decomposition, the trigonometric model, the seasonal dummy variables, the gray forecast, the hybrid gray forecast and the SARIMA model based on the monthly data. Also, Gosasang et al. (2010) [9] uses time series and regression analysis methods to forecast container throughput at the Port of Bangkok. The exactness of predicting is evaluated by calculating mean absolute error (MAE) and root mean squared error. On the other hand, Xie et al. (2017) [10] exploited time series methods to capture the data characteristics of the various components and developed a forecasting model to predict container throughput.

Salido et al. (2011) [11] stated that as the containers to be loaded onto the vessel have already been moved to the stacking yard, a remarshalling plan is necessary to minimize berthing time. Furthermore, Fancello et al. (2013) [12] attempted to predict the delay of ship arrivals in a Mediterranean port, and they developed an artificial neural networks method. Following this approach, Lu et al. (2008) [13] analyzed the airport data built a model that attempts to predict the level of daily alarm due to aircraft delay for one of the most important air hubs in China. In addition, Breiman et al. (1984) [14] stated that Random Forests is a machine learning algorithm that uses a set of correlated trees to reduce overfitting, a well-known problem with tree models.

Tavasszy et al. (2011) [15] presented a strategic model for the global movement of containers to analyze possible shifts in future container transport demand and their impacts on transport policies. The model includes import, export and transfer flows of containers at ports as well as hinterland flows. Then, they predict that the forecasting situation in container transport demand decreases to the extent of climate, energy policies, increasing traffic congestion and uncertainty of production factors. On the other hand, O'Sullivan (2003) [16] stated that macroeconomic indicators include GDP, gross national product (GNP), unemployment rate, consumer price index, inflation rate, industrial production, consumer leverage ratio and these indicators enable statistics on economic activities and enable an analysis of future economic performance. Furthermore, Intihar et al. (2017) [17] developed a prediction model to accurately predict the series of container production time at Koper Port in Slovenia. Their model also includes several external macroeconomic indicators that may have a specific impact. These external effects include GDP, export, import, purchasing power parity and unemployment rate. They emphasized that by applying the macroeconomic indicators such as GDP, GNP, unemployment rate, consumer price index, inflation rate, industrial production, consumer leverage ratio of the results to the estimation model, an accurate estimation can be achieved for more accurate yields in the future.

3 Problem Definition

In this study, a real problem of a logistics company in Turkey is considered. The company is the leader in maritime transportation and logistics in Turkey and they are required to correctly estimate the number of container slots for future periods. The slot is defined as one TEU space in a container ship. Slot agreements are made twice a year by maritime companies. Arranging slots earlier is less costly and helps to maximize profit while buying slots late is costly and may cause failure in delivery due to late bookings. Therefore, the company wants to arrange slots early and book the optimum number of slots to minimize their transportation costs. However, any surplus slot will be paid for. Thus, booking the optimum number of slots is crucial, and to do that, an accurate forecast of container demand is required. The main goal of this study is to develop an accurate forecasting model to predict the number of slots correctly so that the number of booked slots is optimized and the transportation cost is minimized.

The case company has 7300 employees and is the largest maritime transportation company in Turkey. The company currently operates in 23 countries with 61 offices, 14 of which are located in Turkey. The company currently does not formally use any forecasting techniques, however, they seek higher accuracy with their forecasting methods. Currently, the company uses the previous sales data and estimate the future demand with feedback from the sales and logistics personnel.

The accurate container demand forecast also gives an idea of how slots are needed to be booked in the future. The container demand from the ports should anticipate future freight demand and create new investment and development plans to serve emerging trade. Using machine learning methods, we developed accurate forecasting methods for the case company.

To be able to develop an accurate forecasting method using machine learning methods, the past container usage data are obtained from the company. Our aim is to help the company to optimize the number of slots by accurately forecasting container demand. We have forecasted container demand between Mersin Port and other vital ports as requested by the company. These data are explained in Sect. 5 in detail. To be able to correctly predict the container demand between ports, external data that are important for maritime logistics are also used. In summary, we used internal data of the company that include the number of TEU's which is carried between two given ports. Also, the relevant external data. USD – TRY exchange rate, EUR – TRY exchange rate, GDP, inflation rate and export rate, are used to increase the accuracy of the forecast.

4 Methodology

4.1 Machine Learning

Machine learning methods learn from experience and existing data. The machine learning algorithms apply computational methods to learn information from an actual data set. However, these algorithms do not rely on a predetermined equation as a model. The performance of a machine learning algorithm is increased adaptively as the number of samples increases. Machine learning algorithms can also discover patterns in data and thereby make good decisions and generate insight.

Machine learning algorithms are grouped into two categories that are supervised or unsupervised learning. Supervised learning methods use labeled examples to make predictions. It applies the learned knowledge from the past to the new data. By analyzing a subset of a dataset for training, the supervised machine learning algorithms create an inferred function and use that function to make predictions. After the algorithm is trained sufficiently, a target for new inputs can be provided by the system.

Information is not used for training or labeling in unsupervised learning. The system explores the data and can draw inferences from datasets. Thereby, it describes hidden structures from data rather than figuring out the right output. Two well-known machine learning algorithms, Artificial Neural Networks (ANN) and Random Forests (RF), are explained in the following subsections.

4.1.1 Artificial Neural Networks (ANN)

Artificial neural networks perform basic functions of the brain such as remembering and learning. It mimics the human brain and by doing so it generates new information. McCulloch and Pitts (1943) [18] invented the first artificial neural network in 1943. ANN has several characteristics. They are nonlinear, parallel working, working with missing data, adaptable, using many variables and parameters, and flexible. Architectural structure, learning algorithm, and activation function are the basic components of artificial neural networks. These components affect the performance of the model, and therefore, one that is appropriate to the structure of the data should be preferred. Some application areas of Artificial Neural Networks are classification, forecasting, prediction, pattern recognition, clustering, regression, function estimation, and optimization.

4.1.2 Decision Trees

Decision trees are structures that are resembling flowcharts. A decision tree model decisions and their possible consequences. Each node in the tree represents a test on an attribute while each branch represents the outcome of the test. Each leaf node represents a class label. A decision tree includes three types of nodes that are decision nodes, chance nodes and end nodes.

Decision trees are popular tools in machine learning. The goal of decision tree learning is to predict a target variable from a set of input variables. The models that target variable can take discrete values are called classification trees whereas the models that target variable takes continuous values are called regression trees. Ensemble methods construct more than one decision tree. Boosted trees and random forests are the two ensemble methods.

4.1.3 Decision Forest/Random Forest (RF)

Tin Kam Ho (1995) [19] invented the first algorithm for random decision forests by using the random subspace method. Also, a stochastic discrimination approach to classification is proposed by Eugene Kleinberg (1996) [20]. Similar to ANN, the Random Forest algorithm is a supervised classification algorithm. It creates a random forest and a direct relationship exists between the number of decision trees in the forest. As the number of trees increases, the result becomes more accurate.

The random forest method includes a large number of decision trees that operate as an ensemble. A low correlation between these trees is crucial. Uncorrelated models can produce more accurate ensemble predictions. Low correlation protects the trees from individual errors. While some trees can have low prediction accuracy, many of them may have high accuracy. Thereby, the group of trees is expected to move in the correct direction.

4.2 Linear Regression

Linear regression is a method to estimate the correlation between two variables. It assumes that there's a direct correlation between the two variables.

These two variables are called the independent variable and the dependent variable. The independent variable behaves however it likes, and doesn't depend on the other variable for any reason. On the other hand, the dependent is a direct result of the independent variable.

A linear mathematical relationship was developed between the independent and dependent variables in linear regression. When the independent variable is known, it can calculate the dependent variable. Linear regression can be expressed in the form given in Eq. (1).

$$y = B * x + A \tag{1}$$

This model can be used to make predictions. When the x is given, y can be predicted by the model.

4.3 Performance Metrics to Evaluate Forecasting Models

In this study, five different performance metrics are used to evaluate the forecast accuracy. Assuming the actual values are a1, a2, …, an and the predicted values are p1, p2, …, pn, these performance metrics are explained in the following subsections in detail.

4.3.1 Mean Absolute Error (MAE)

In the MAE that all the individual differences are equally weighted in the average as presented in Eq. (2).

$$MAE = \frac{1}{n}\sum\nolimits_{i=1}^{n} |p_i - a_i| \tag{2}$$

4.3.2 Root Mean Square Error (RMSE)

RMSE is the square root of the average of squared differences of actual and predicted values. Equation (3) presents the formula for RMSE:

$$RMSE = \sqrt{\sum\nolimits_{i=1}^{n} \frac{(p_i - a_i)^2}{n}} \tag{3}$$

4.3.3 Relative Absolute Error (RAE)

The relative absolute error is the relative absolute difference between expected and actual values as explained in Eq. (4):

$$RAE = \frac{\sum_{i=1}^{n} |p_i - \mathrm{a_i}|}{\sum_{i=1}^{n} |\mathrm{a_i} - \overline{a}|} \tag{4}$$

Where $\overline{a}$ is the average of the actual values as defined in Eq. (5).

$$\overline{a} = \frac{1}{n}\sum\nolimits_{j=1}^{n} a_j \tag{5}$$

4.3.4 Relative Squared Error (RSE)

The relative squared error takes the total squared error and normalizes it by dividing by the total squared error of the simple predictor. Equation (6) presents the calculation of the relative squared error.

$$\mathrm{RSE} = \frac{\sum_{j=1}^{n} (p_i - \mathrm{a_i})^2}{\sum_{j=1}^{n} (\mathrm{a_i} - \overline{a})^2} \tag{6}$$

4.3.5 Coefficient of Determination (R^2)

The coefficient of determination (R^2) is a measure that shows the ability of a model to explain the variability in a linear regression model.

If the R^2 value equal to 0, then the dependent variable cannot be explained by the independent variable. If it equals 1, then the dependent variable can be completely explained by the independent variable. The calculation of R^2 is given in Eq. (7). In this equation, SS_E is the sum-of-squared error of prediction of regression and SS_T is the total sum of squares.

$$\text{Coefficient of Determination} = R^2 = 1 - \frac{SS_E}{SS_T} \tag{7}$$

Alternatively, R^2 can also be calculated as the square of the correlation coefficient.

5 Results

We collected the real data from the case company where the data include the actual TEUs from the port of Mersin to the ports of Alexandria, Misurata, Aqaba, Casablanca, Jeddah, Valencia, Piraeus, Port Said, Ashdod, and Antwerp between 2013 and 2019. To create a forecasting model that accurately forecasts the TEUs, we also collected external data such as USD – TRY exchange rate, EUR – TRY exchange rate, GDP rate, inflation rate and export rate between 2010 and 2019. These data are used to create a forecasting model that employs machine learning techniques. In this study, the forecasting methods we used are Linear Regression, Boosted Decision Tree Regression, Decision Forest Regression and Artificial Neural Network Regression. And the metrics to evaluate the model are the Coefficient of Determination, Mean Absolute Error, Root Mean Square Error, Relative Absolute Error and Relative Squared Error. The dataset is split into training and test sets. 80% of the data were used to train the algorithms while 20% of the data were used for testing the forecasting. Each algorithm is trained and forecasted with the same data for the same time period.

The model that we created in Microsoft Azure Machine Learning Studio yielded the results that are shown in Table 1. Ideally, when R^2 value is close to 1, it gives the best results for the model. However, for the remaining metrics, the ideal value is 0. Linear Regression has shown the worst performance for all metrics. On the other hand, Boosted Decision Tree Regression and Decision Forest Regression yielded close values that are the best results for all of our metrics. In terms of R^2, Boosted Decision Tree Regression gave the best result that is 0.465 and Linear Regression gave the worst result which is 0.212. Furthermore, in terms of MAE, Decision Forest Regression showed the best result that is 0.538, but Linear Regression yielded the worst result which is 0.612. Additionally, for RMSE, the best value was obtained by Boosted

Decision Tree Regression with 0.708 where the worst value was found by Linear Regression where value is 0.835. Also, in terms of RAE, Decision Forest Regression gave the best result (0.651). Finally, in terms of RSE; Boosted Decision Tree Regression is the best with the value of 0.535.

Table 1. Comparison of the algorithms

Algorithm	R^2	MAE	RMSE	RAE	RSE
Linear regression	0.212	0.612	0.835	0.777	0.788
Boosted decision tree regression	0.465	0.540	0.708	0.663	0.535
Decision forest regression	0.464	0.538	0.709	0.651	0.536
Artificial neural network regression	0.327	0.563	0.767	0.712	0.673

6 Conclusion

In this study, our aim is to develop a forecasting method for the largest maritime transportation company in Turkey. This forecasting model predicts the number of TEUs that are required each month to optimize the slot assignments and bookings of the company where the slot is the empty space (one TEU) in a container ship. From the company's point of view, if the slots are bought, the cost would be lower and therefore their profits would be maximized. However, if the slots are bought late the company faces an increased shipment cost and sometimes they cannot make the shipment due to late bookings.

Since the company does not have a formal method of forecasting the container demand between ports, accurate prediction of the TEU demand is crucial for its operations. Therefore, using the company's internal data including TEU amounts from Mersin loading port to different vital ports between years 2013–2019, we developed forecasting models using machine learning models. In this way, we analyzed the crucial ports which have a higher volume. Mersin and Novorossiysk ports had the highest container traffic. Then, we obtained several external data that could data such as; GDP, Inflation rate, Exchange rate, EUR-TL rate and USD-TL rate from 2013 to 2019 to achieve a better forecast for container demand between these ports.

We analysed the past 7 years for GDP, inflation rate, exchange rates, export and TEU amount on a monthly basis. Our goal is to develop a satisfying forecast accuracy (in terms of Coefficient of Determination, Mean Absolute Error, Root Mean Square Error, Relative Absolute Error and Relative Squared Error) with the help of Machine Learning techniques by using detailed data that we collected. We employed various algorithms (Boosted decision tree regression, decision forest regression, artificial neural network regression, linear regression) and compared their forecasting performances. According to the results, the best performing algorithms are Boosted Decision Tree Regression and Decision Forest Regression. On the other hand, the worst performing forecasting model was Linear Regression. The results showed that the developed forecasting models accurately predict the TEUs for the company.

As future work, this study will be extended with more internal data from the company and used to predict other ports. Also, other machine learning techniques can be tested and compared with the models used in this study.

Acknowledgment. Arda Akın, Hasan Bağ and Şadiye Kaya also helped this study during their senior design project at Yasar University in 2020.

References

1. Chen, M., Mao, S., Liu, Y.: Big data: a survey. Mob. Netw. Appl. **19**(2), 171–209 (2014)
2. Huang, A., Zhang, Y., Zhang, Z., Shi, X., Hua, G.: A partially combined framework: forecasting container throughput with big data. IJCSA **13**(1), 105–115 (2016)
3. Lam, J.S.L.: Patterns of maritime supply chains: slot capacity analysis. J. Transp. Geogr. **19**(2), 366–374 (2011)
4. Zhang, G.P., Kline, D.M.: Quarterly time-series forecasting with neural networks. IEEE Trans. Neural Networks **18**(6), 1800–1814 (2007)
5. Diaz, R., Talley, W., Tulpule, M.: Forecasting empty container volumes. Asian J. Shipp. Logist. **27**(2), 217–236 (2011)
6. Syafi'i, K.K., Takebayashi, M.: Forecasting the demand of container throughput in Indonesia. Mem. Constr. Eng. Res. Inst. **47** (2005)
7. Schulze, P.M., Prinz, A.: Forecasting container transshipment in Germany. Appl. Econ. **41**(22), 2809–2815 (2009)
8. Peng, W.Y., Chu, C.W.: A comparison of univariate methods for forecasting container throughput volumes. Math. Comput. Model. **50**(7–8), 1045–1057 (2009)
9. Gosasang, V., Chandraprakaikul, W., Kiattisin, S.: An application of neural networks for forecasting container throughput at Bangkok Port. In: Proceedings of the World Congress on Engineering, vol. 1 (2010)
10. Xie, G., Zhang, N., Wang, S.: Data characteristic analysis and model selection for container throughput forecasting within a decomposition-ensemble methodology. Transp. Res. Part E-Logist. Transp. Rev. **108**, 160–178 (2017)
11. Salido, M.A., Rodriguez-Molins, M., Barber, F.: A decision support system for managing combinatorial problems in container terminals. Knowl.-Based Syst. **29**, 63–74 (2012)
12. Cannas, M., Fadda, P., Fancello, G., Frigau, L., Mola, F.: Delay prediction in container terminals: a comparison of machine learning methods
13. Zonglei, L., Jiandong, W., Guansheng, Z.: A new method to alarm large scale of flights delay based on machine learning. In: 2008 International Symposium on Knowledge Acquisition and Modeling, pp. 589–592. IEEE (2008)
14. Breiman, L., Friedman, J.H., Ohlsen, R.A., Stone, C.J.: Classification and Regression Trees. Wadsworth & Brooks/Cole. Advanced Books & Software, Monterey (1984)
15. Tavasszy, L., Minderhoud, M., Perrin, J.F., Notteboom, T.: A strategic network choice model for global container flows: specification, estimation and application. J. Transp. Geogr. **19**(6), 1163–1172 (2011)
16. O'Sullivan, A., Sheffrin, S.M.: Economics: principles in action (2003)
17. Intihar, M., Kramberger, T., Dragan, D.: Container throughput forecasting using dynamic factor analysis and ARIMAX model. Promet-Traffic Transp. **29**(5), 529–542 (2017)
18. McCulloch, W.S., Pitts, W.: A logical calculus of the ideas immanent in nervous activity. Bull. Math. Biophys. **5**(4), 115–133 (1943)

19. Ho, T.K.: Random decision forests. In: Proceedings of 3rd International Conference on Document Analysis and Recognition, vol. 1, pp. 278–282 (1995)
20. Kleinberg, E.M.: An overtraining-resistant stochastic modeling method for pattern recognition. Ann. Stat. **24**(6), 2319–2349 (1996)

Customer Order Scheduling in Hybrid Flow Shop Manufacturing System

Eylül Kacar, Esra Karakoç, İrem Kartop, Almira Öztürk, Görkem Bozkurt, Nazli Aygün, and Erdinç Öner(✉)

Department of Industrial Engineering, Yaşar University, Izmir, Turkey
eylulkacaarr@gmail.com, karakocesr@gmail.com, iremkartopp@gmail.com, almiraaozturk@gmail.com, gorkembzkrt97@gmail.com, {nazli.aygun, erdinc.oner}@yasar.edu.tr

Abstract. The problem of customer order scheduling in a paint company that has five production stages is handled in this study. The production stages are pre-mixing, grinding, sub-addition, quality control, and filling. At each of these stages, several identical and unrelated parallel machines are available. Customer orders are assigned to exactly one machine at each stage and do not have to be processed in all stages. As a result of the comprehensive literature review, our problem is categorized as hybrid flow shop scheduling for the production system of the company. The mathematical model is developed by utilizing the existing studies in the literature. This developed mathematical model is solved and optimal results are obtained for small-size problem instances. According to the analysis of the results generated by the mathematical model and the literature, the problem is found to be NP-hard. Since the problem is NP-hard, a heuristic algorithm is proposed for the solution of larger job sizes. Considering its convenience and applications in the scheduling literature, GA is selected as a heuristic algorithm to solve our proposed model. Utilizing a genetic algorithm, jobs are sorted and assigned to the proper machines to minimize the sum of earliness and tardiness. As a novel approach, a user-friendly DSS is designed in addition to efficient scheduling. The designed DSS targets to respond to changes made by the user instantly.

Keywords: Scheduling · Hybrid flow shop · Earliness · Tardiness · Optimization · Genetic algorithm · Decision support system

1 Introduction

A hybrid flow shop scheduling (HFSS) problem in a paint company is examined in this study. In the existing system, customer orders are planned based on experiences of the production planning department using MS Excel. Since customer orders are scheduled manually, time loss and mistakes occur. This situation causes uncertain production capacity, delay or early production of jobs, and accumulation of the products in the warehouse. Therefore, a user-friendly optimization tool for optimal scheduling of customer orders based on scientific methods is required.

N. M. Durakbasa and M. G. Gençyılmaz (Eds.): ISPR 2020, LNME, pp. 853–865, 2021.
https://doi.org/10.1007/978-3-030-62784-3_71

In the context of this study, we aim to develop a tool for finding the optimal schedule for the production system to minimize the sum of tardiness and earliness of jobs. For this purpose, the symptoms are analyzed, and the appropriate mathematical model is developed. The developed mathematical model is solved with different job sizes using IBM ILOG CPLEX Optimization Studio Version 12.6.3 [1]. The problem is found to be NP-hard [2]. Therefore, a heuristic approach is decided to employ for solving larger job sizes. As a result of the comprehensive literature review, it is determined that the genetic algorithm (GA) gives the best solution for the HFSS problem [3]. Considering the constraints of the problem environment, a GA model, that assigns jobs to the relevant machines and gives the optimal production sequence as an output, is developed. The proposed model is solved and verified using toy problems which include five jobs. In addition to this, GA is employed for the solution of toy problems and verified. At last, the obtained solution method is integrated into a user-friendly decision support system (DSS).

2 Problem Definition

The data used in this study are collected from a paint company, which is agreed to deliver incoming orders within three days to the customers. The company implements both make-to-order and make-to-stock production strategies. The company produces 40% of incoming customer orders instantly and the remaining 60% is supplied from stock. The production continues for stock when the inventory falls below the desired level.

There are two kinds of production systems in the company; the co-grind system and the paste system. In the co-grind production system, there are five stages and one process at each stage. These are pre-mixing, grinding, sub-addition, quality control, and filling. Pre-mixing and grinding processes have machine capacity limitations. The second production system which is the paste system consists of only sub-addition, quality control, and filling processes. In the production system of the company, some machines in some operations are parallel, and these machines can be identical or unrelated. Also, each machine can process exactly one job at a time. Likewise, each job can be processed on exactly one machine in each operation. The flow of the process is the same for all jobs, but they may skip some processes. Besides, when the process starts, it cannot be interrupted. The literature review showed that the company's production environment can be classified as a hybrid flow shop (HFS) production system and the scheduling problem can be classified as an HFSS problem. HFS consists of a flow line with several identical and unrelated parallel machines on some or all production stages. Multiple products are produced in such a flow line. While all the products follow the same linear path through the system, all of them may not visit all of the stages in production. On each of the stages, one of the parallel machines has to be selected for the production of a given product. The production of a product consists of multiple operations, one for each production stage. When an operation is started on a machine, it must be finished without interruption [4].

The scheduling of customer job orders according to their earliest due date (EDD) is studied in this paper. Flexibility is considered for special situations such as color transitions, rework, and instantaneous orders in scheduling. The objective is to minimize the sum of earliness and tardiness of the jobs. Ensuring on-time delivery of jobs is important for the company's reputation and customer satisfaction. As a result of the determined schedule, the progress of the job orders at any time can also be observed. Besides, the company's production capacity can be displayed through a systematic schedule using the DSS.

3 Literature Review

Scheduling is one of the problems that we encounter in many areas in life, which is about assigning and sorting jobs to machines considering the time factor [2]. This problem is important in manufacturing systems to ensure that orders are produced by their due date. The production environment of the company studied is classified as an HFS production system and the scheduling problem can be classified as an HFSS problem. The HFSS is a complex combinatorial problem with multiple parallel machines for some stages in many real-world applications [5].

According to Chung et al. [6] flow shop scheduling problem is NP-hard when there are m machines and n jobs and the aim is to minimize total tardiness. Flow shop scheduling problems become more complex for multi-objective optimization models since multiple targets must be evaluated simultaneously [7]. The studies showed that as the number of jobs and machines increase, the mathematical model did not provide the optimal solution. Therefore, it is decided to use a heuristic algorithm that provides Solutions that are good and fast enough for large scale problems.

In the articles examined, taboo search, genetic algorithm, simulated annealing was studied for the flow shop scheduling problem, and comparisons were made between the algorithms. As a result, it has been observed that the GA gives the fastest and most effective solution to large-scale problems [8]. According to Alcan and Başligil [9], the GA is a method of searching and optimizing job schedules in a similar way to the evolutionary process observed in nature. At the same time, it is based on a random search, like all heuristic approaches. It looks for the best heuristic solution according to the principle of survival of the best in complex multi-dimensional search space. Based on these articles examined, it was decided that the optimum result could be obtained with GA for our problem and a heuristic model was developed using this approach.

4 Modeling and Solution Methodology

The mathematical model and GA metaheuristic method are developed for the HFSS problem. The developed mathematical model and GA are shown in this section.

4.1 Limitations and Assumptions

Limitations and assumptions are significant for the HFSS problem. The reason is that these factors can directly affect problem solutions. The assumptions of the study are listed below.

- All machines are available at time t = 0.
- All jobs are available at time t = 0.
- The setup and transportation time are included in the processing time.
- There is no machine failure which means that machines are available for processing continuously.
- The idle time of the machine is negligible.
- No preemption of operations is allowed.
- The number of operators is assumed to be adequate.
- There are no scrap products for reworked parts. These products are stored to be used in the next orders.

4.2 Mathematical Model

Indices and sets are defined before creating the mathematical model. The indices are *i, j, l,* and *k*. Indices *i* and *j* represent jobs, *l* represents the index for stages and *k* represents the index for machines at stages *l*. The parameters and variables of the model are identified below.

Indices, Sets, Parameters

n # of jobs
m # of operations
i *j*, index for jobs
k index for operations
M_k number of machines at stage *k*
l index for machines at stage *k* (where $1\ldots M_k$)
P_{ik} processing time of job *i* at stage k on machine *l*
O_k the set of jobs to be processed at stage *k*
d_i due date of job *i*
v_i the amount(kilogram) of job *i*
N_l the capacity of machine *l* at pre-mixing operation
W_l the capacity of machine *l* at grinding operation

Decision Variables

$C_i =$ Completion time of job *i*
$E_i =$ Earliness of job *i*
$T_i =$ Tardiness of job *i*
$S_{ikl} =$ Starting time of job *i* at stage *k* on machine *l*

$$Y_{ijkl} = \begin{cases} 1, & \text{if job j is processed immediately after job i at stage k on machine l} \\ 0, & \text{otherwise} \end{cases}$$

$$X_{ikl} = \begin{cases} 1, & \text{if job i is processed at stage k on machine l} \\ 0 & \text{otherwise} \end{cases}$$

Mathematical Model
Minimize

$$\sum_{i=1}^{n}(E_i + T_i) \tag{1}$$

Subject to:

$$C_i = \sum_{k=1}^{m}\sum_{l=1}^{M_k}(S_{ikl} + P_{ikl}.X_{ikl}) \qquad \forall i \tag{2}$$

$$S_{ikl} \leq M(X_{ikl}) \qquad \forall i \in O_k, k, l \tag{3}$$

$$\sum_{l=1}^{M_k} X_{ikl} = 1 \qquad \forall i \in O_k, l \tag{4}$$

$$T_i \geq C_i - d_i \qquad \forall i \tag{5}$$

$$E_i \geq d_i - C_i \qquad \forall i \tag{6}$$

$$S_{ikl} \geq S_{jkl} + P_{jkl} - M.y_{ijkl} - M.\left(1 - x_{jkl}\right) - M.(1 - x_{ikl}) \quad \forall \mathrm{j}, \mathrm{i} \neq \mathrm{j}, \mathrm{k}, \mathrm{l} \tag{7}$$

$$X_{ikl} + X_{jkl} \geq 2\left(Y_{ijkl} + Y_{jikl}\right) \qquad \forall \mathrm{j}, \mathrm{i} \neq \mathrm{j}, \mathrm{k}, \mathrm{l} \tag{8}$$

$$X_{ikl} + X_{jkl} \leq Y_{ijkl} + Y_{jikl} + 1 \qquad \forall \mathrm{j}, \mathrm{i} \neq \mathrm{j}, \mathrm{k}, \mathrm{l} \tag{9}$$

$$\sum_{l=1}^{M_k} S_{ikl} \geq \sum_{l=1}^{M_{k-1}} S_{ik-1l} + P_{ik-1l} \qquad \forall i, l > 1 \tag{10}$$

$$v_i.X_{i1l} \leq N_L \qquad \forall i \in O_k, k, l \tag{11}$$

$$v_i.X_{i2l} \leq W_L \qquad \forall i \in O_k, k, l \tag{12}$$

$$S_{ikl} \geq 0 \qquad \forall i, k, l \tag{13}$$

$$C_i \geq 0 \qquad \forall i \tag{14}$$

$$x_{ikl} \in \{0, 1\} \qquad \forall i \in O_k, k, l \tag{15}$$

$$y_{ikl} \in \{0, 1\} \qquad \forall (i, \mathrm{j}) \in O_k, k, l \tag{16}$$

$$T_i \geq 0 \qquad \forall i \tag{17}$$

$$E_i \geq 0 \qquad \forall i \tag{18}$$

In this model, the objective function minimizes the sum of the earliness and tardiness of all jobs (1). Constraint (2) calculates the total completion time of all jobs from the first stage to the last stage. Constraint (3) forces that the starting time of each job on machines that do not process the job becomes 0. Constraint (4) specifies job assignments to machines at each stage. Constraint (5) calculates the tardiness of all jobs. Constraint (6) calculates the earliness of all jobs. Constraint (7) controls the completion time of each pair of jobs if they are processed by the same machine at each stage. Constraint sets (8) and (9) show the relationship between decision variables X and Y. Constraint (10) guarantees that the job on the next stage cannot start before it is completed on the previous stage. Constraint (11) provides that according to the weight of all jobs should be assigned to the appropriate machine of the first stage. Constraint (12) provides that according to the weight of all jobs should be assigned to the appropriate machine of the second stage. Finally, Constraint (13), (14), (15), (16), (17), and (18) define the boundaries of decision variables.

The developed mathematical model is solved with IBM ILOG CPLEX Optimization Studio using the real data which are collected from the factory for 3, 5, 7 and 10 job sizes. The solutions and computational time of these different job sizes are shown in Table 1 for comparison.

Table 1. Comparison for different job order scenarios

Number of Jobs	Optimal solution (in min)	Computational time (in sec)
3	868.8	0.6
5	1362	1.08
7	1626.6	4.02
10	1645.8	108023.3

As can be seen in Table 1, the computational time of the mathematical model increases as the number of jobs increase. When the number of jobs is 3, the computational time is 0.6 s. On the other hand, when the number of jobs is 10, the computational time takes about 30 h which is not practical to use in a production environment that requires responding to the orders quickly. As a result, the problem gets harder to solve as the number of jobs increases, and this method remains inefficient for larger job sizes. For this reason, a heuristic algorithm is employed for the solution of larger problems.

4.3 Genetic Algorithm

Genetic Algorithm (GA) was developed by John Holland in 1965. Holland has adapted Darwin's Theory of Evolution and the principle of "Survival of the fittest" to the process of optimization [10]. This heuristic approach aims to find the global optimum or close to global optimum solutions using the genetic code logic in the living systems. The theory of evolution has been successful in the optimization process and proposed a solution in many areas such as sorting, scheduling, assignment, planning, and transportation problems. The genetic algorithm flow adapted to our problem considering the proposed mathematical model and machine capacity constraints. Fitness function is the objective function of our mathematical model and the algorithm achieves the objective function value by assigning jobs in a sequence considering the model constraints. Solution representation is, in this case, the best order to assign jobs to the related machines. Co-grind and paste production system flow is constructed in the constraints and the algorithm distinct them from the product code and production time entered as the order input. The initial population is created depending on the number of jobs being scheduled. This type of coding is called "permutation encoding." Every chromosome is a string of numbers and in that case, each string represents a job sequence. The number of jobs is represented with job ID and each job has a unique number. This ID represents the gene and a chromosome is represented by all the order ID's that are being scheduled. With this coding system, a random initial population is created considering the number of jobs to be scheduled. The selection criterion used in this algorithm is fitness-based Elitism. When a child enters this operation, by far all generated chromosomes fitness values are calculated and the individuals with the highest fitness function value (least fit members) tend to be replaced by a better-fit child. For this reason, the fittest member of this population can not be replaced. If individuals with the same least fitness function value exist, the decision to remove which individual from the population is random. The population of selected and remaining individuals is considered as the new population. For the crossover operation, single-point, two-point and uniform methods are added into the algorithm. As a result of parameter analysis uniform crossover is selected as a crossover type. There is a 50% chance that an offspring will inherit the gene of Parent 1 and the same implies for Parent 2. In this way, the first offspring is created. Then for the second offspring, the genes are inherited from the exact opposite parent chosen to create the first offspring. To mutate an individual two positions on the chromosome are selected at random, and the values are interchanged. This type of mutation is called "swap mutation." A mutated individual represents a new job sequence to be assigned to the machines for processing. If the child does not include all different genes, the number of iterations is increased, and the algorithm goes back to the crossover operator until all genes are different in the offspring. The termination criterion is determined as the number of the population entered before running the algorithm. When the termination criterion is satisfied, the chromosome which has the lowest fitness function value is given as the best result according to the algorithm.

5 Verification and Validation

The developed mathematical model is solved with toy data which included 5 jobs, using IBM ILOG CPLEX Optimization Studio. In the context of this study, the problem with small data sets is called a toy problem. Then, the same toy data are used to solve the scheduling problem manually and this result is compared with the results from IBM ILOG CPLEX Optimization Studio. Both solutions provide the same results and the model is verified. The 5 jobs sorted according to their earliest due date are 3–1–

5–4–2, respectively. Then, these sorted jobs are assigned to the available machines in that operation according to the machine capacity constraints. Figure 1 demonstrates in which operations that the jobs are processed, to which machines the jobs are assigned, and the processing times of the jobs are displayed. Also, the completion time of jobs can be calculated through the Gantt Chart.

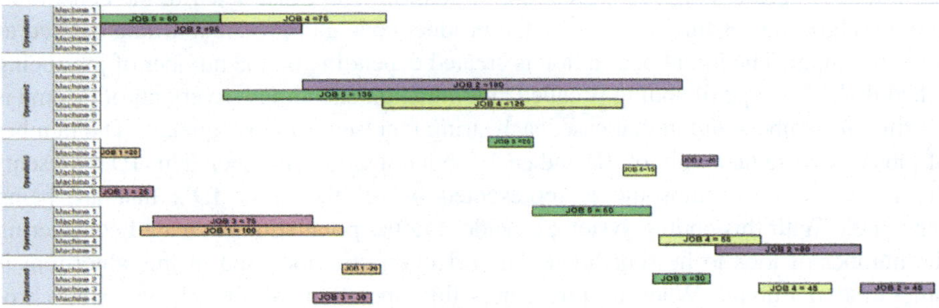

Fig. 1. Gantt chart for toy data

Table 2 below shows the parameters of the toy problem and the results of the solution. The objective function value which is solved in 2.12 s with toy data of 5 jobs is 1930 min by using IBM ILOG CPLEX Optimization Studio.

Table 2. Calculation of objective function

Jobs ID	Due date (in min)	Completion time (in min)	Earliness + Tardiness (in min)
3	300	130	170
1	420	140	280
5	540	305	235
4	900	375	525
2	1140	420	720
		Total	1930

After verification of the model, the problem is solved using the real data obtained from the time study is used to validate the heuristic method. The solution is used to validate the proposed heuristic method. For the validation, we compared the outputs of two different programs which are IBM ILOG CPLEX Optimization Studio Version 12.6.3 and Excel Visual Basic for Applications (VBA) Version 16.37 [11]. These programs are run on a computer that has 16 GB RAM. The comparison is shown below in Table 3.

Table 3. Results by CPLEX and Heuristics

	Objective (in min)		Runtime (in sec)	
Number of orders	CPLEX	Heuristic (10 replications avg)	CPLEX	Heuristic (10 replications avg)
1	279	280	0.2	3.7
2	697.2	699	0.5	4.33
3	868.8	871	0.6	4.53
4	1183.2	1185	0.95	5.03
5	1362	1364	1.08	6.51
6	1573	1576	1.18	10.83
7	1626.6	1630	4.02	12.85
8	1599	1601	65	13.13
9	1692	1695	64800	15.68
10	1645.8	1650.4	108023.3	17.7
15	–	1681.8	–	25.27
20	–	3274.3	–	34.94
25	–	5135.2	–	48.77

Table 3 shows the results obtained by the calculation using IBM CPLEX and heuristic method with the same data for validation. CPLEX gave up the result until 10 jobs, it ran about 5 days for the solution of larger job sizes, then the program gave memory error and stopped. To explain briefly, the computational time is exponentially increased for 10 jobs in CPLEX and it fails to give results for larger problems. Therefore, we compared the results of the CPLEX and heuristic method up to 10 jobs. It can be seen that the heuristic algorithm returns nearly optimal solution when compared to the solution obtained by CPLEX. Since the factory has daily 20 orders on average, the genetic algorithm is chosen to solve this problem.

6 Decision Support System and Computational Results

A decision support system (DSS) is designed to show the most appropriate customer orders schedule. The designed DSS responds to changes made by the user instantly. This section consists of information about DSS in detail.

6.1 Decision Support System (DSS)

Decision Support System (DSS) is designed to help the user in the planning process. When designing DSS, it is considered to respond to changes made by the user instantly. With this dynamic system, scheduling can be easily made and help the production planning department on decision making.

There are five main pages which are "Welcome Page", "Orders", "Results", "Gantt", and "Learning Data". "Welcome Page", as shown in Fig. 2, includes buttons that execute three main operations namely data entry, solve, and result. In data entry operations, there are six buttons. The first button is “Add Order”. When clicked on this button, there are two options for data entry that asks for all input data. Inputs are either entered by the user or read from a file. These inputs are “Order ID”, “Portfolio Code”, “Color Code”, “Production Amount”, “Due Date”, and “Processing Times” of five different production operations. The entered inputs are written on the “Orders” page. The second button is “Change Order”. It asks the “Order ID” to be changed. The entered ID is kept fixed, but the other inputs can be changed. The third button is “Delete Order. This button helps to remove the desired input data of entered “Order ID”. The fourth button of the data entry operation is “Delete All Orders”. This button deletes all the input data on the “Orders” page. The solve operation has a run button that executes the Genetic Algorithm using entered input data on the “Orders” page. The last operation which is “Result” has three buttons. The first button which is “Show Results” opens the “Results” page that shows the sequence of the orders, assignments to the machine, and computational results. The second button is “Show Gantt.” When clicked to this button, it opens the “Gantt” page where the job schedule is visualized. The last button of this operation is “PDF Results”. This button reads the data on the “Results” page. Then, it converts them to a PDF file and saves this file to the desired location on the computer. Also, on the “Welcome Page,” there is a “Help” button that has a user’s manual. This user’s manual includes a detailed description of the operations.

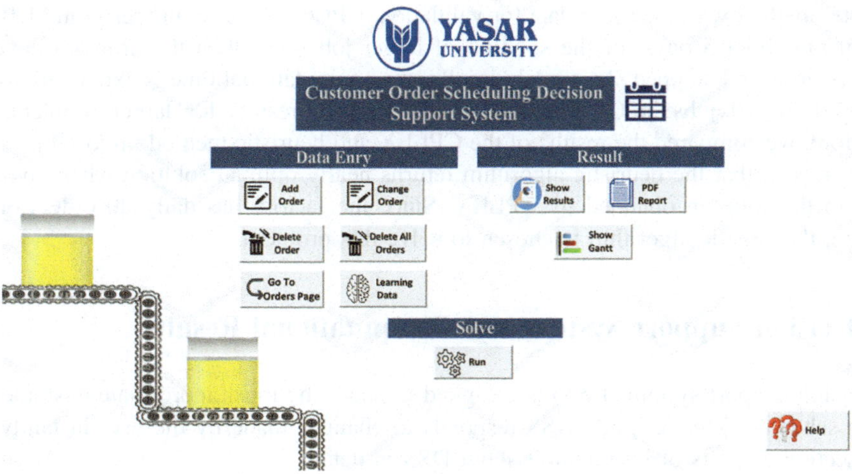

Fig. 2. DSS welcome page

On the “Orders” page, which is accessible by “Go to Orders Page” button, the parameters of the genetic algorithm are located. The parameters are population number, mutation probability, and crossover-type that can be changed manually. Also, process names, machine information, and machine capacity constraints are available. If there is a change in machine or operation number, they can also be updated manually. Also, the objective function and makespan of the solution are displayed under the parameters of GA. On the “Learning Data” page, which is accessible by “Learning Data” button, the data obtained from our time study is available. With these data, the processing times can also be calculated. When the portfolio code and production amount are entered as an order on the "Orders" page, recommended processing times are given. The processing times are calculated as the average of the obtained data with the same portfolio code and ±10% of the entered production amount. When all operations are completed, the result of scheduling is seen clearly and converted to a PDF file. This file can be e-mailed to the related departments to create systematic information flow. With this Decision Support System, weekly or monthly schedules can be created in short notice. This provides a systematic schedule for the company and decreases wasted time and errors occurred while manual scheduling.

6.2 Computational Results

The parameters of the genetic algorithm should be analyzed to find the parameters that give the best performance before the implementation of this algorithm. Population size must be large enough to include diverse chromosomes but on the contrary, when the population size exceeds a certain value the efficiency of the algorithm decreases. Therefore, parameter optimization is crucial for the performance of the genetic algorithm. In this study, for 15 job size and swap mutation type; population size, crossover-type, and mutation probability will be changed then the objective function will be compared to determine the parameters that give the best performance. The values obtained from this study are shown in Table 4 below.

Table 4. Parameter optimization for GA

No. Of Orders	Pop. Size	Mutation Type	Mutation %	Crossover Type	Objective (∑Earliness+Tardiness)
15	5000	Swap	5	Uniform	1561
15	5000	Swap	5	Single Point	1776
15	5000	Swap	5	Two Point	1624
Crossover Type Selected as Uniform since it gives the minimum objective.					
15	5000	Swap	1	Uniform	1957
15	5000	Swap	2	Uniform	1888
15	5000	Swap	3	Uniform	1798
15	5000	Swap	4	Uniform	1602
15	5000	Swap	5	Uniform	1561
15	5000	Swap	10	Uniform	1737
Mutation probability is selected as 5% since it gives the minimum objective.					
15	1000	Swap	5	Uniform	1990
15	2000	Swap	5	Uniform	1820
15	3000	Swap	5	Uniform	1765
15	4000	Swap	5	Uniform	1602
15	5000	Swap	5	Uniform	1561
15	6000	Swap	5	Uniform	1694
15	7000	Swap	5	Uniform	1742
15	8000	Swap	5	Uniform	1863
Population size selected as 5000 since it gives the minimum objective. Also runtime is relatively longer when the population size is increased.					

To compare the DSS schedule with the factory schedule the input values of 15 consecutive orders produced are asked from the factory. With the input data obtained from the recent production, the objective function and makespan are calculated. Then input data of produced 15 consecutive orders is solved by the Genetic Algorithm to achieve a better production sequence. Then objective function and makespan values are compared to see the improvement. Our objective function is to minimize the sum of earliness and tardiness of orders. When makespan is expanded within the due date of orders, the objective function is reduced. To illustrate the numerical results, with a 9.8% expansion of makespan, without exceeding the due date, 17.3% improvement of objective function value is obtained as shown in Table 5.

Table 5. Comparison of solutions with the initial conditions

	Objective	Improvement (%)	Makespan
Factory schedule	39991	17.3	1635
DSS schedule	33075		1795

7 Conclusions and Future Work

In this study, a mathematical model and a genetic algorithm are developed, and a decision support system is designed for the customer order scheduling problem in the hybrid flow shop production environment. The problem is solved both with the proposed model and a heuristic algorithm that considers the proposed model. The company where this project is carried out in collaboration has two different production systems which are co-grind and paste production systems. Also, there are five operations namely pre-mixing, grinding, sub-addition, quality control, and filling processes in the production area.

Before the project, the customer order scheduling used by the company is generated based on experience without any systematic method. This situation caused using production time inefficiently, excess stock formation, customer dissatisfaction, lack of communication between departments, and undetermined production capacity. Besides, it could not possible to observe which stage the product is at any time. Because of these reasons, the company needed systematic scheduling.

In this study, it is aimed to generate efficient scheduling which shows jobs sequence and assignment to the machines. These sorting and assigning of jobs are scheduled by means that minimize the sum of earliness and tardiness.

At the end of the project, a suitable mathematical model is developed, and a heuristic method is proposed for solving larger job sizes. A user-friendly decision support system is created using Microsoft Excel Visual Basic for Applications (VBA) to remove the problems. Thus, a solution method has been generated that increases the efficiency, meets the demands of the company, and minimizes the total of the earliness and tardiness of the jobs.

As future work; if the company's existing production system changes such as the number of machines, the number of processes, the new system can be updated on DSS.

Thus, the company can easily reach new scheduling. Also, the stock levels can be integrated into the DSS to ease the planning process. Finally, the obtained schedule can be modified for the project-based productions of the company.

Acknowledgements. This study is supported by TUBITAK (The Scientific and Technological Research Council of Turkey) in the program of "2209-B Undergraduate Thesis Support Program for Industrial Applications". We would like to thank Melis Akkaya and Burak Aksun for their support and contributions to this study.

References

1. IBM ILOG CPLEX Users Manual. https://www.ibm.com/search?lang=tr&cc=tr&q=users%20manual
2. Pinedo, M.: Scheduling Theory, Algorithms and Systems, 3rd edn, pp. 13–389. Prentice-Hall, New york (2008)
3. Agrawal, J.K., Agrawal, S., Gohiya, D., Sharma, S.K.: Heuristic and metaheuristic algorithm for flow shop scheduling: a survey. In: International Conference on Proceedings of Academics World, Kuala Lumpur, Malaysia, pp. 22–26 (2015)
4. Karmakar, S., Mahanty, B.: Minimizing makespan for a flexible flow shop scheduling problem in a paint company. In: IEOM. Dhaka, Bangladesh (2010)
5. Ruiz, R., Vázquez-Rodríguez, J.A.: The hybrid flow shop scheduling problem. Eur. J. Oper.0 Res. **205**(1), 1–18 (2009)
6. Chung, C., Flynn, J., Rom, W., Stalinski, P.: A genetic algorithm to minimize the total tardiness for M-machine permutation flow-shop problems. J. Entrepreneurship Manag. Innov. **8**(2), 26–43 (2012)
7. Babu, P.M., Sai, B.V.H., Reddy, A.S.: Optimization of make-span and total tardiness for flow shop scheduling using genetic algorithm. Int. J. Eng. Res. Gen. Sci. **3**(3), 195–199 (2015)
8. Kurz, M., Askin, R.: Scheduling flexible flowlines with sequence-dependent setup times. Eur. J. Oper. Res. **159**(1), 66–82 (2004)
9. Alcan, P., Başligil, H.: Job scheduling with the help of dominance properties and genetical algorithm on hybrid flow shop problem. Sigma J. Eng. Nat. Sci. **6**(1), 127–137 (2015)
10. Altay, A.: Genetik Algoritma ve Bir Uygulama. Yüksek Lisans Tezi, İ.T.Ü. Fen Bilimleri Enstitüsü, Istanbul, p. 1 (2007)
11. Microsoft Excel Visual Basic for Applications (VBA). https://docs.microsoft.com/en-us/office/vba/api/overview/excel

Development of Hydraulic Drive Railway Bearings Installation System

Pınar Demircioğlu[1(✉)], İsmail Böğrekçi[1], and Mehmet Temel[2]

[1] Mechanical Engineering Department, Aydin Adnan Menderes University, Aydin, Turkey
{pinar.demircioglu, ibogrekci}@adu.edu.tr
[2] Inotim Engineering Company LLC., Istanbul, Turkey
mehmet.temel@inotim.com

Abstract. The vehicles such as locomotive, passenger and freight wagons, metros, trams etc. used in railway sectors need quick and precise installation and removal of bearings to the wheels with a hydraulic system. Solid state modelling was carried out using CAD software and all machine assembly was done in assembly environment. Determination of mechanical strength and other parameters using static structural analysis was carried out by FEA methods. The system has an innovative and modular design. Since the fixture has a 180° force affecting the forehead of the bed, the risk of deformation during the disassembly process is very high. In this study, a 360° fixture design was made to prevent this deformation. Manufactured prototype was tested in a factory environment.

Keywords: Rail wheel · Wagon wheel · Poor disassembly · Mounting · Dismounting

1 Introduction

The increasing importance of the railway sector in freight and passenger transport is evident due to reduced energy resources and safe transportation reasons. Developments in the world and particularly in the rail sector investments in Turkey is the major cause of the conducted research. The main task of the research is to identify the current needs of the industry. In this context, the needs of the railway sector were investigated. It is observed that there is a problem in the bearings of the railway wheels.

Due to the increasing demand for safer and quicker rail transportation, rolling stock wheelsets operating under high axle loads, speed and heavy usage require more rigorous and reliable maintenance and inspection. While in service, wheelsets are constantly operating under harsh conditions including rolling contact fatigue, thermal variations and impact [1].

It is generally known that bearing damages occur because of [2]:

- Improper lubrication, insufficient lubrication, excessive lubrication or invasion of impurities into the bearing area.
- Excessive load, fatigue damages on rolling surfaces and rolling elements.
- Excessive rotation speed.

N. M. Durakbasa and M. G. Gençyılmaz (Eds.): ISPR 2020, LNME, pp. 866–875, 2021.
https://doi.org/10.1007/978-3-030-62784-3_72

- Inadequate mechanical properties.
- Insufficient operating clearance.
- Radial stress caused by an external source of heat.
- Obstructed run due to the breaking of the cage.
- Initial damages of the bearing.

One of the most critical mechanical components of the railway vehicle is the axle-box. In fact, this component contains the bearings that must support the axle-load while the vehicle runs on the track and allow the wheelset to rotate. It is therefore an extremely loaded component, with radial and axial loads (which are generated during curving and which can exceed 10% of the payload), which must allow a long distance without failures. Damage to the bearing can cause serious consequences on the vehicle, which can lead to derailment. For this reason, the bearing axle-box is subject to cyclical maintenance and inspections during service, to ensure the reliability and safety of the vehicle [3]. Due to the importance of rail bearings mounted on the axle box, these components are subject to careful maintenance and are constantly monitored with different strategies. Both built-in monitoring systems and roadside monitoring stations have been developed [4, 5].

This study was done on the solution of assembly incompatibilities. This issue is important because an uncontrolled assembly process is performed in existing assemblies and there is no requirement for assembly control in any standard. The bearings of the towed and towed vehicles maintained in certain periods are controlled according to UIC standards. If the controlled bearings need to be replaced, they must be dismantled without damaging the axle. Likewise, it should be mounted without damaging the axle and the bearing inner ring. The machines that are currently used in assembly, disassembling, grasp the C-shaped (180°) bearing inner ring. Therefore, there is a possibility that the inner ring of the bearing to be removed will damage the axle.

Interference fit connections are the type of connection used to transmit torque between the shaft and the hub pair with the help of friction force. It is preferred in practice because it is easy to manufacture and does not require additional processing costs such as in spindles or hubs. In order for the connection to occur, pressure must occur on the contacting surfaces of the shaft and hub [6].

The aim of this study is to develop a portable hydraulic drive railway bearings installation system that can perform assembly and disassembly without damaging the bearings and axle, enables fast fixing change, and can perform assembly and disassembly work quickly and precisely.

2 Material and Method

2.1 Material of Pipe of Hydraulic Cylinder and Mounting-Dismounting Fixtures

In this study, St52-3 H8 honed pipe was preferred as the material in hydraulic cylinder manufacturing. Chemical properties of St52-3 are given in Table 1, and mechanical properties of St52-3 are presented in Table 2.

Table 1. Chemical properties of St52-3 [7]

C	Si	Mn	P	S	N	Cu	CE
max 0.22	max 0.55	max 1.6	max 0.035	max 0.035	max 0.012	max 0.55	max 0.47

Table 2. Mechanical properties of St52-3 [7]

Grade	Thickness (mm)	Yield Strength (Mpa)	Tensile Strength (Mpa)
St 52	35>to<50	345	490–630

4140 materials were preferred for the jaws used in assembly and disassembly. Chemical and mechanical properties are given in Tables 3 and 4. Table 5 gives heat treatment information of 4140 material.

Table 3. Chemical properties of 4140 [7]

C	Si	Mn	P_{max}	S_{max}	Cr	Mo
0,38-0,45	0,15–0,40	0,50–0,80	0.035	0.035	0,90–1,20	0,15–0,30

Table 4. Mechanical properties of 4140 [7]

Grade	Thickness (mm)	Yield strength (Mpa)	Tensile strength (Mpa)
4140	50	770	930

Table 5. Heat treatment information of 4140 [7]

Hot forming temperature (°C)	Soft annealing temperature (°C)	Hardness of the soft annealed state > HB	Anneal temperature (°C)	Hardening			Tempering temperature (°C)	Hardness (HRc)
				Cooling medium	Water	Oil		
850–1050	680–750	241	840–880	Temp. (°C)	820–850	830–860	500–520	36–38
							470–500	38–40
							440–470	40–42
							430–450	42–44
				Sertlik (HRc)	54–56	54–56	420–440	44–46
							380–420	46–48
							340–380	48–50
							320–340	50–52

3 Method

1) Design of Hydraulic Drive Railway Bearings Installation System: According to the determined need, designs were made in Solid Works program.
2) Calculation of Force: Force of interference fit were calculated as below. Figure 1 indicates wheel bearing of freight wagon.

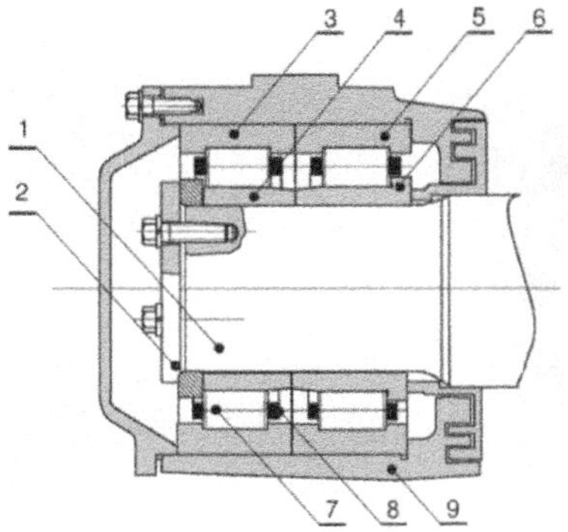

1 Journal
2 End Cap
3 Outer Ring of Outer Bearing (WJP)
4 Inner Ring of Outer Bearing (WJP)
5 Outer Ring of Inner Bearing (WJ)
6 Inner Ring of Inner Bearing (WJ)
7 Bearing Roller
8 Cage
9 Axle-box housing

Fig. 1. Wheel-pair bearing of freight wagon [2]

Shaft (Journal);
Young's Modulus $E_i = 210\,\text{GPa}$
Poisson's Ratio $v_i = 0{,}292$
Shaft Internal Diameter $d_i = 0\,\text{mm}$
Shaft Nominal Diameter $d = 130\,\text{mm}$
Shaft Upper Tolerance $\delta = 0{,}068$ mm (p6)
Shaft Lower Tolerance $\delta = 0{,}043$ mm (p6)

Hub (Inner Ring of Inner Bearing (WJ));
Young's Modulus $E_0 = 210\,\text{GPa}$
Poisson's Ratio $v_0 = 0, 292$
Hub Outer Diameter $d_0 = 160\,\text{mm}$
Hole Upper Tolerance $\delta = 0\,\text{mm}$
Hole Lower Tolerance $\delta = 0\,\text{mm}$
Width of Hub $w = 80\,\text{mm}$
Friction between Shaft and Hub $\mu = 0, 30$

$$p = \frac{\delta}{\frac{d}{E_0}\left(\frac{d_0^2 + d^2}{d_0^2 - d^2} + v_i\right) + \frac{d}{E_i}\left(\frac{d^2 + d_i^2}{d^2 - d_i^2} - v_i\right)} \tag{1}$$

$$F = \mu.p.A \tag{2}$$

According to the formula (1) and (2);

$$F_{max} = 182953\,\text{N}$$

3) Static Structural Analysis: Static Structural Analysis was used to determine responses of a structure under applied loads. This analysis can be used to determine the displacements, strains, stresses in a structure. Static Structural Analysis was applied to the designed pipe of hydraulic press and mounting-dismounting fixtures. The main purpose of this analysis was to observe the maximum stress and displacement values and areas (Fig. 2).

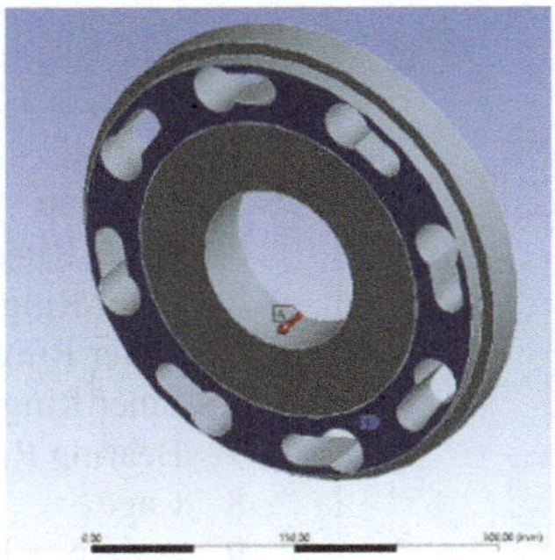

Fig. 2. Static structural analyses with ANSYS workbench

4) Manufacturing: Machining (Lathe and CNC milling), welding (MIG welding), heat treatment and assembly were carried out successively.

4 Results and Discussion

4.1 Design of Hydraulic Drive Railway Bearings Installation System

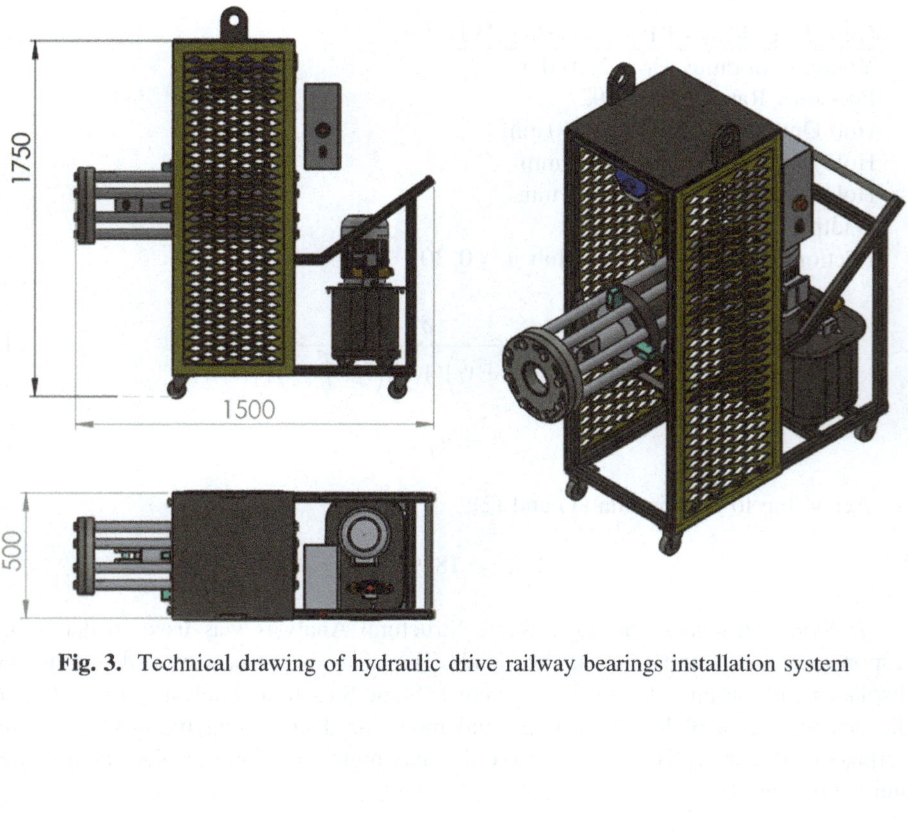

Fig. 3. Technical drawing of hydraulic drive railway bearings installation system

Design details [8] (Fig. 3) related to hydraulic cylinder, mounting fixture and dismounting fixture, which are the most important parts of the design, are given. In Fig. 4, hydraulic cylinder (a), dismounting fixture (b) and mounting fixture (c) are shown.

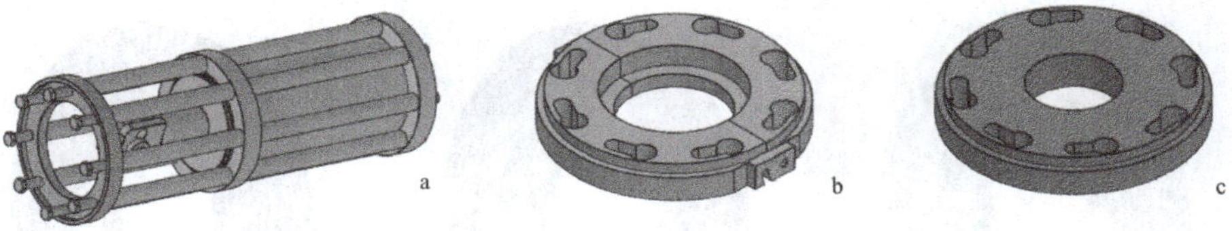

Fig. 4. Hydraulic cylinder (a), Dismounting fixture (b) and Mounting fixture (c)

4.2 Static Structural Analysis

The verification of the designed hydraulic cylinder pipe, mounting fixture and dismounting fixture was done with the finite element method with ANSYS.12.1.

1) Static structural analysis was conducted to find the maximum occurred von Misses stresses, strains and deformation on pipe structure (Fig. 5). The obtained results were given in Table 6.

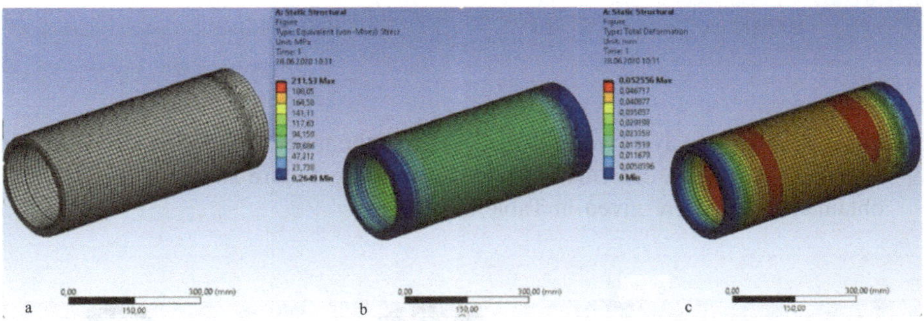

Fig. 5. Meshing (a), Equivalent (Von-Misses) stress (b) and Total deformation (c) of Pipe of hydraulic cylinder

Table 6. Pipe of static structural analysis

Pipe of Hydraulic Cylinder (St52-3)		
Hydraulic oil pressure (MPa)	Total deformation (mm)	Equivalent stress (MPa)
25	0.052	211

2) Static structural analysis was conducted to find the maximum occurred von Misses stresses, strains and deformation on the dismounting fixture structure (Fig. 6). The obtained results were given in Table 7.

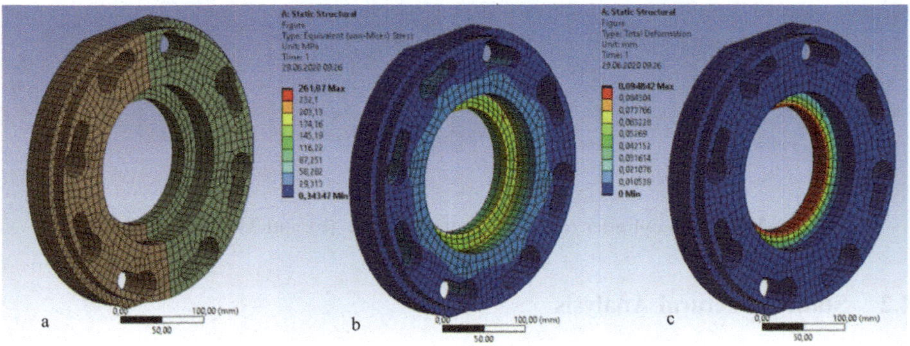

Fig. 6. Meshing (a), Equivalent (von-Misses) stress (b) and Total deformation (c) of Dismounting fixture

Table 7. Dismounting fixture of static structural analysis

Dismounting fixture (4140)		
Dismounting force (N)	Equivalent stress (MPa)	Deformation (mm)
600000	261	0,094

3) Static structural analysis was conducted to find the maximum occurred von Misses stresses, strains and deformation on the mounting fixture structure (Fig. 7). The obtained results were given in Table 8.

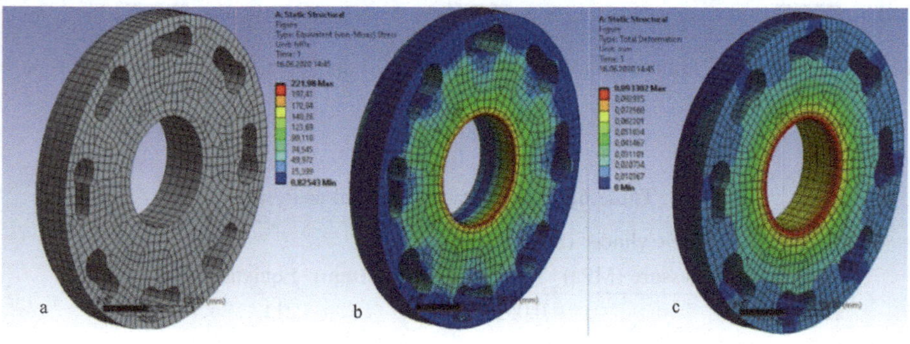

Fig. 7. Meshing (a), Equivalent (von-Misses) stress (b) and Total deformation (c) of Mounting fixture

Table 8. Mounting fixture of static structural analysis

Mounting fixture (4140)		
Mounting force (N)	Equivalent stress (MPa)	Deformation (mm)
600000	221	0.093

4.3 Manufacturing

Machining (Lathe and CNC milling), welding (MIG welding), heat treatment and assembly were carried out successively. Figure 8 represents technical drawing of hydraulic cylinder, mounting fixture and dismounting fixture (a) and manufactured of hydraulic cylinder, mounting and dismounting fixtures (b).

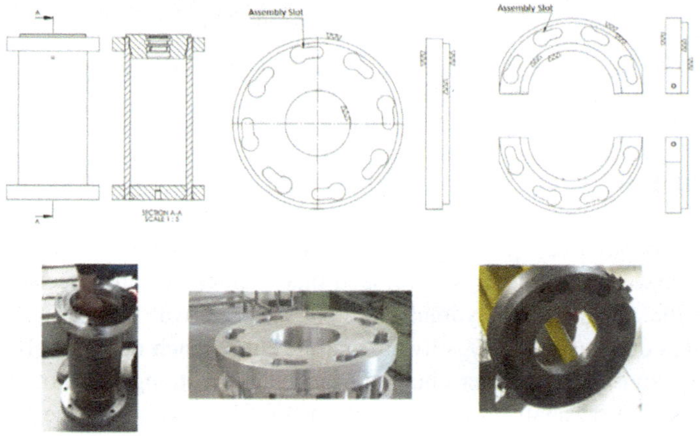

Fig. 8. Technical drawing of hydraulic cylinder, Mounting fixture and Dismounting fixture (A), Manufactured of Hydraulic Cylinder, Mounting Fixture and Dismounting Fixture (B)

There are isometric view of Hydraulic Drive Railway Bearings Installation System (a) and Manufactured of Hydraulic Drive Railway Bearings Installation System shown in Fig. 9.

Fig. 9. Isometric view (a) and Manufactured (b) of Hydraulic drive railway bearings installation system

5 Conclusion

In this study, locomotives, passenger and freight wagons, subways, trams, etc. used in Railway sectors. The system was developed that enables fast and precise assembly of vehicles to the wheels with hydraulic system and removal of bearings. Solid state modelling was done using CAD software and the entire machine assembly was done in the assembly environment. Determination of mechanical strength and other parameters using static structural analysis was carried out by FEA methods. The system has an innovative and modular design. Since the fixture has a force of 180° that affects the forehead of the bearing, the risk of deformation during the removal process is very high. In this study, 360° fixture design was made to prevent this deformation. The produced prototype was tested in a factory environment.

At the next stage of this study, a transducer will be developed to integrate with the current prototype in order to create a force-displacement plot. Force-displacement curves and successful completion of assembly-disassembly can show the performance of the developed machine. In order to reduce the labor cost to automation, solutions should be continued for industrial revolutions [9].

Acknowledgment. This study was supported and funded by KOSGEB under R&D and Innovation Program (Project ID: 269227).

References

1. Anastasopoulos, A., Bollas, K., Papasalouros, D., Kourousis, D.: Acoustic emission on-line inspection of rail wheels. In: Proceedings of the 29th European Conference Acoustics Emission Testing, pp. 1–8 (2010)
2. FAG: Waelzlagerschäden – Schaedenserkennung und Begutachtung gelaufener Waelzlager (2013)
3. Bosso, N., Gugliotta, A., Zampieri, N.: Design of a test rig for railway Axle-boxes. In: AIAS International Conference on Stress Analysis, Torino, Italy, vol. 24, p. 10129 (2018)
4. Rende, H., Güven, F.: Siki Geçme Bağlantilarinda Malzemeye Bağli Yüzey Pürüzlülüğü Kayip Katsayisinin Belirlenmesi. Mühendis ve Makina **56**(665), 46–52 (2015)
5. Wilson, D.S., Frarey, J.L.: Investigations of a bearing fault detector for railroad bearings. NASA Technical Reports Server (NTRS) 1976000737 (1975)
6. Amini, A., Entezami, M., Papaelias, M.: Onboard detection of railway axle bearing defects using envelope analysis of high frequency acoustic emission signals. Case Stud. Nondestr. Test. Eval. **6**, 8–16 (2016)
7. http://kalitemetalurji.com.tr/urunlerimiz/islah-celikleri/4140-islah-celigi/. Accessed 15 June 2020
8. Králik, M., Jerz, V., Paštéka, M.: Optimization of the machine and device layout solution in a specific company production. In: Durakbasa, N., Gençyilmaz, M. (eds.) Proceedings of the International Symposium for Production Research 2019. ISPR 2019, ISPR 2019. Lecture Notes in Mechanical Engineering. Springer, Cham (2020)
9. Doyle-Kent, M., Kopacek, P.: Industry 5.0: is the manufacturing industry on the cusp of a new revolution? In: Durakbasa, N., Gençyilmaz, M. (eds.) Proceedings of the International Symposium for Production Research 2019. ISPR 2019, ISPR 2019. Lecture Notes in Mechanical Engineering. Springer, Cham (2020)

Improving the Operations in a Logistics Company: A Case Study in Turkey

Pınar Karaoglu, Tulin Mert, Muhammet Berkay Kork, Fevzi Batuhan Türkmen, Yeşim Deniz Özkan-Özen, and Yücel Öztürkoğlu(✉)

International Logistics Management Department, Yasar University, Izmir, Turkey
pinarkaraoglu1997@gmail.com, tulinnmrt@gmail.com, mberkaykork@gmail.com, fevzibatuhanturkmen@gmail.com, {yesim.ozen,yucel.ozturkoglu}@yasar.edu.tr

Abstract. With the rapid increase in the world population, the amount of demand and the number of customers that companies from each sector must meet have increased simultaneously. It is clear to see that consumer demands are shifting from mass production and mass personalization to full personalization with the globalizing world. Accordingly, it has become imperative for companies to make predictions that are more accurate and provide services to meet all their needs in order to remain active in the sector. In this context, companies have begun to focus on methods to protect and strengthen the bond between the consumer and the company. Especially in the service sector, it is vital to fully understand and meet customer needs. Consumers demand full customization from service providers. In this study, a user survey was developed to directly understand the needs of consumers to a third-party logistics service provider. After the survey, in which consumers state their basic needs and importance levels, they are integrated with the Quality Function Distribution (QFD) method to understand the results. What can be done to meet customer needs, thus obtaining a competitive advantage over competitors in the industry.

Keywords: Lean management · Quality Function Deployment · Postal service · Service sector · Voice of Customer

1 Introduction

The growing economy results in an increase in the number of customers, a steady increase in order quantities and virtual stores. In all these axes, the connection that connects the end user and the company is getting weaker and it is getting harder and harder to continue this communication. On the other hand, while customers are increasing the growing demand for personalization, everyone has different needs. This has brought communication to a vital position for companies to survive and grow in the industry. Therefore, in order to keep this communication factor alive to meet the needs of customers, companies use some techniques that are used to maintain this communication effectively. In this respect, this study aims to provide a new approach to

N. M. Durakbasa and M. G. Gençyılmaz (Eds.): ISPR 2020, LNME, pp. 876–887, 2021.
https://doi.org/10.1007/978-3-030-62784-3_73

service providers, to establish a healthy direct communication with target customer bases in order to fully understand customer demands and reorganize their operations in the most convenient way possible. Mathematical data help to increase profit margins and gain competitive advantage over competitors in the industry.

Products and services that are being developed fast, error-free, and according to the voice of the customer will bring the market success to the company. Quality Function Deployment (QFD) is one of the important methods that widely used approaches to stay alive in a competitive market. QFD is a quality technique that started with the Japanese Mitsubishi company in 1972, was used in the USA after 1984, and is accepted worldwide now. According to Joshi and Bhargava (2019) said that Yoji Akao commenced the QFD concept in Japan in 1966.

In this study, it was first determined in which fields and how the QFD studies in the literature are used. In addition to that Beterfild et al. (2003) developed QFD as a profitable company, is an important competitive resource for postal service. In addition to that, QFD and how the cargo process could be applied to the process of a logistics company were determined. In order to collect data from potential customers, a survey consisting key requirements which are willing to developed by the company provided and filled by 337 different people. The data collected has been normalized. Then the 6 most important criteria determined by the users were calculated mathematically. According to subject criteria, QFD house has been implemented.

This study organized as follows; after the introduction part in Sect. 2, literature review related to previous QFD studies in service sector are presented. Then, in Sect. 3 a brief explanation of QFD can be found. And it is followed by Sect. 4, construction of QFD methodology and integration of it to the collected data is presented. Finally, discussions on the possible developments on the subject and the results obtained after the study can be seen in Sect. 5.

2 Literature Review

The research was advanced with all studies with keywords "lean management, service sector, QFD and postal sector". The study was improved with the support of previous research. In the literature review, previous QFD implementations for different purposes in different service sectors are reviewed. According to the reviews, it has observed that there are many researches about QFD. None the less there are not much in postal sector. The main aim of this study is to fill this gap and contribute to the literature.

In Table 1, summary of the literature review is presented. As it can be seen from the table, QFD is used in different sectors for different purposes. For instance, Healthcare sector is one of common service area that is improved by using QFD. The health level of the population in a country reflects the country's progress and prosperity. Therefore, as one of the healthcare-interested service industries, the quality of services offered by a hospital should be taken from consumers. In this case, consumers are patients who will evaluate the quality of hospital services. (Hasibuan et al. 2019) [18]. Competition is getting more difficult in the business world, including banking. For this, banks should consider the right strategy and make the right decision to win this competition. (Adiandari et al. 2020) [19]. Therefore, the satisfaction level of the customers must be determined in return for the services provided. In order to attract the attention of the customer, it is necessary to create differences in the service offered compared to the competitors. It is important to make improvements according to the data received.

Table 1. The literature survey of quality function deployment in service sector

Sector	Aim	Author(s)
Transportation industry	Improving service quality in highway passenger transportation	Pakdil and Kurtulmuşoğlu (2014) [1]
Service innovation	Aims to help organization enhance customer satisfaction by guiding improvement efforts in strengthening their weak attributes, and to expedite the development of innovative service through the identification of attractive attributes and embedding them into future service	Yothinsirikul and Thawesaengskulthai (2014) [2]
Health sector	Increase service quality with high safety standard, increase healthcare service efficiency and competitive advantage to meet the ever-increasing demand of patients and eventually increase customer satisfaction level	Lee et al. (2015) [3]
Healthcare and public service	Aims to determine and provide services that individuals may need in societies with an elder age	Chen (2016) [4]
Healthcare	Aims to improve services provided in m-health sector in Bangladesh	Chowdhury et al. (2016) [5]
Electricity and public infrastructure	Aims to provide better services to customers since they are aware of quality	Jahanzaib et al. (2016) [6]
Hotel	Aims to classify customer ratings and needs to achieve a better level of service	Kuo et al. (2016) [7]
Agriculture	Aims to cover lack of knowledge about the real needs of consumers	Mardar et al. (2016) [8]
Transportation	Aims to provide accurate and punctual internal train services	Nikfalazar (2016) [9]
Banks and banking	Aims to improve relationships between the customers and the loan companies	Novokreshchenova et al. (2016) [10]
Smart services	Improving smart services provided and cover different fields in Dubai	Zawati et al. (2016) [11]

(*continued*)

Table 1. (*continued*)

Sector	Aim	Author(s)
Healthcare	To develop as the specific framework for assessing and analyzing the experimental results. QFD was used to identify and turn decision-makers, suppliers, patients and local communities' needs into both service and construction design for Project K, a healthcare network in the UK	Dehe and Bamford (2017) [12]
Food industry (Meal ordering system)	To develop of a systemic and analytical software architecture using incorporation of quality function deployment technology, and schematic approaches	Wang et al. (2017) [13]
Postal and shipping service	To analyze the characteristics of respondents in Indonesia	Khairunnisa et al. (2018) [14]
Healthcare	To develop concept of dialysis service risk management methods due to the role of healthcare organization assessment in sustaining system performance	Maalej et al. (2018) [15]
Hospitality	To develop close this research gap by following an approach to quality function deployment that incorporates the process of analytic network process	Wu et al. (2018) [16]
Transportation services	To improve the quality of transportation services	Furqon et al. (2019) [17]
Healthcare	To improvement service quality	Hasibuan et al. (2019) [18]
Banking services	To improving the quality of internet banking services	Adiandari et al. (2020) [19]

In the following section, details of QFD methodology is presented.

3 QFD Methodology

According to Hauser and Clausing (2015) [20] mentioned that, QFD is a methodology that converts consumer needs or consumer voice into technological criteria using the matrix called house of quality. Besides, Bottani and Rizzi (2006) [21] claimed that, QFD methodology, logistics center customer preferences are converted into the technological aspects of the logistics center service package, which includes both visible and intangible elements, to assess which products are of greater value.

The QFD method consists of six main stages. These stages are shown in Fig. 1, respectively.

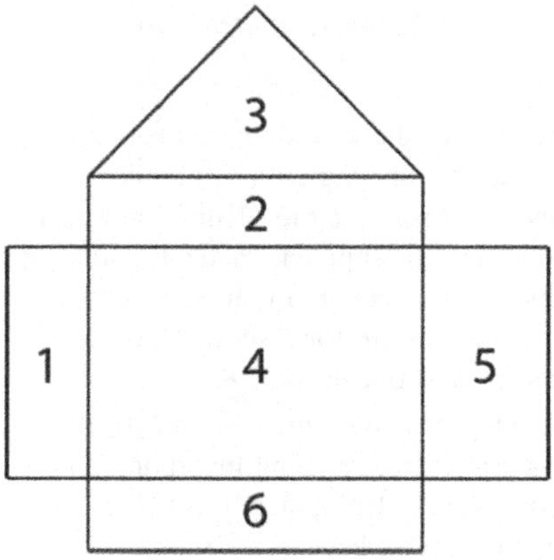

Fig. 1. QFD model

1) Voice of Customer

In this part, customer requirements and importance degrees of them are identified. With various ways such as surveys and interviews, customer needs and wants are determined. There is a priority scale between 1 to 10. Lower points are less important while the higher points are the most.

2) Technical Requirements

Secondly, technical requirements are identified that can afford the customer needs and wants (Voice of Customer). The goal of the technical requirements is to define 'How' to respond to the customer requirements. This part of the QFD will help demonstrate how the existing product or service currently satisfies customer wants.

3) Correlation Matrix of the Technical Requirements

The correlation matrix will determine how technical requirements help and hinder each other. These relations are shown by symbols. 'P' refers to positive and 'N' refers to negative. Based on these P and N symbols, correlation of different requirements are determined.

4) Relationship Matrix of Customer Wants and Technical Requirements

In the relationship matrix, how strongly each of the design parameters affects the customer needs are identified. In this matrix, the scoring system is used. The system includes three scores which is 1, 3 and 9 refers to weak, medium, and strong, respectively. The requirements that have no relation with each other are shown blank.

5) Targets and Gaps (Planning Matrix)

In this part, companies evaluate themselves to find out in what stage they and their competitors to satisfy customer expectations. They score themselves and their competitors 1 to 5 and they determine their target score. 5 represents the highest level of customer satisfaction. Their target score can be the same with the current score as it can be higher. In this matrix, improvement ratios are founded by diving the current scores to their target scores. With the multiplying of priority scores, improvement ratio and sales point which is always constant, priority rating is founded. At the end of this part, the percentage of priority rating is determined.

6) Technical Evaluation

The last step in the formation of the QFD is the bottom of the house. This foundation is referred to as the technical matrix. In this part, technical and normalized technical priorities are founded.

4 Implementation of the Study

Implementation of the study was conducted in one of the oldest and biggest service company operating in Turkey. The company operates in many different sectors such as mail, bank and transaction services, e-commerce, and cargo. This study focused especially on postal and cargo areas. After the examinations, what problems and processes are exist in which fields of the company were determined.

A neutral questionnaire in which consumers evaluate their demands and requirements in line with the company's services was created to integrate in the QFD implementation. In total the questionnaire included 21 questions. 7 of these questions were demographic and 14 were questions that evaluated the expectations of customers. Potential customers were asked the rank these features 1 to 5 according to their importance rating for them. Demographic questions are of great importance for the company to recognize its target customer base and to focus on their very requirements. The questions that the customers scored the criteria provided data to indicate the processes to be focused on. The questionnaire filled by 337 individual consumers calculated which criteria ranked the most important for the consumer. All data obtained from the questionnaire study was normalized in SPSS Statistics and expressed mathematically. According to rated answers given by the customers, the most important 6 criteria selected and listed as can be seen in Table 2. The customer requests that form the basis of the QFD have done based on the selected criteria.

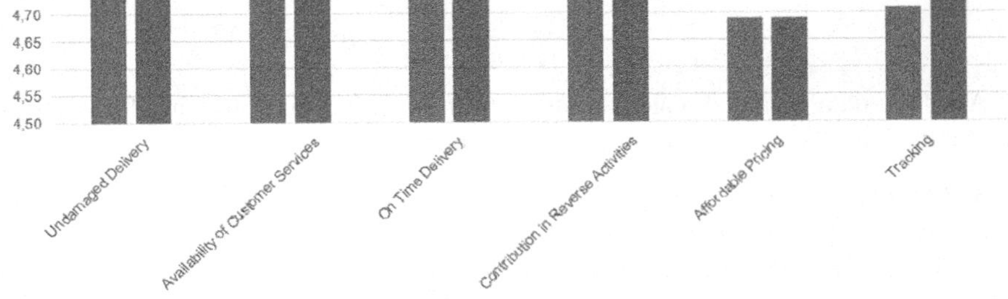

Fig. 2. Target customer requests

The target customer base for the company are people who frequently demand service and provide most of their shopping needs over the internet. Because as the workload, that is, the number of cargoes transported, the rate of profit of the company will increase in parallel. In this respect, it is important to prioritize the needs of the target customer group. As can be seen in Fig. 2, customers who shop 5–7 per month and above 8 pays attention to the subject criteria.

Table 2. Key customer requirements

Key customer requirements	Ratio
Undamaged delivery	0.078
Customer service network	0.077
On time delivery	0.076
Collaboration of firm during reverse logistics	0.076
Affordable pricing	0.074
Delivery tracking	0.072

The most prior six requirements were identified according to the survey results. Importance ratings of them determined. According to the participants selection, undamaged delivery has been found as the primary customer requirement. Technical requirements that can satisfy these customer needs are determined. After that, following calculations are made as it shown.

Table 3. Key requirement ratio calculations

	Improvement ratio	Priority ratings	% of priority rating
Undamaged delivery	1	9	15,43
Customer service network	1,25	13.5	23,15
On time delivery	1,25	10	17,14
Collaboration of firm during reverse logistics	1,25	10	17,14
Affordable pricing	1	8	13,71
Delivery tracking	1,13	7.81	13,39

After the data obtained, calculations were made using the following formulas and the results that can be seen in Table 3 were obtained.

A. *Improvement Ratio: Target Score* $\div$ *Current Score*
B. *Priority Ratings: Improvement Ratio* $\times$ *Priority Scores*
C. *Ratio of Priority Rating: % of each Priority Rating*

The calculation of the Technical Priority is;

$\sum$(Percentage of Priority Rating * Relationship Scores)

1. Extra- Safe Packaging: (15,43 * 9 + 17,14 * 3) = 190,29
2. Conveyor Line (Minimum Material Handling): (15,43 * 9 + 17,14 * 3) = 190,29
3. Training: (15,43 * 9 + 23,15 * 9 + 17,14 * 3) = 398,64
4. Customer Service Number: (23,15 * 9 + 17,14 * 9 +13,39 * 9) = 483,12
5. Problem Hotline: (23,15 * 9 + 17,14 * 1 +13,39 * 9) = 498,46
6. Complaint Social Media Line: (23,15 * 9 + 17,14 * 3 +13,39 * 1) = 273,16
7. Extra Shift: (9 * 15,43) + (3 * 23,15) = 208,32
8. Wide Distribution Network: (9 * 17,14) + (9 * 17,14) + (3 * 13,71) + (1 * 13,39) = 363,04
9. NewInvestment: (3 * 15,43) + (3 * 23,15) + (9 * 17,14) + (1 * 17,14) + (9 * 13,71) + (9 * 13,39) = 461,59
10. Pricing: 9 * 13,71 = 123,39
11. IT System: (9 * 23,15) + (1 * 17,14) + (9 * 13,39) = 363,14
12 Outsourcing: (3 * 15,43) + (9 * 23,15) + (3 * 17,14) + (3 * 17,14) + (9 * 13,71) + (9 * 13,39) = 596,5

Table 4. Results of technical priorities

	Technical priority	Normalized technical priority
Importance rating	–	17,24
Extra-safe packaging	190,29	3,98
Conveyor line (minimum material handling)	190,29	3,98
Training	398,64	8,33
Customer service number	483,12	10,10
Problem hotline	498,46	10,41
Complaint social media line	273,16	5,70
Extra shift	208,33	4,35
Wide distribution network	363,04	7,59
New investment	461,59	9,65
Pricing	123,39	2,58
IT system	363,14	7,49
Outsourcing	596,5	12,47

According to the results of all data obtained, the shape of QFD applied to company is shown in Fig. 3 and Table 4.

The calculation of the Normalized Technical Priority is;
(Technical Priority/∑Total Technical Priority)*100

1. Importance Rating: (824,87/4784.64) * 100 = 17,24
2. Extra- Safe Packaging: (190,29/4784.64) * 100 = 3,98
3. Conveyor Line (Minimum Material Handling): (190,29/4784.64) * 100 = 3,98
4. Training: (398,64/4784.64) * 100 = 8,33
5. Customer Service Number: (483,12/4784.64) * 100 = 10,10
6. Problem Hotline: (498,46/4784.64) * 100 = 10,41
7. Complaint Social Media Line: (273,16/4784.64) * 100 = 5,70
8. Extra Shift: (208,32/4784.64) * 100 = 4,35
9. Wide Distribution Network: (363,04/4784.64) * 100 = 7,59
10. New Investment: (461,59/4784.64) * 100 = 9,65
11. Pricing: (123,39/4784.64) * 100 = 2,58
12. IT System: (363,14/4784.64) * 100 = 7,49
13. Outsourcing: (596,52/4784.64) * 100 = 12,47

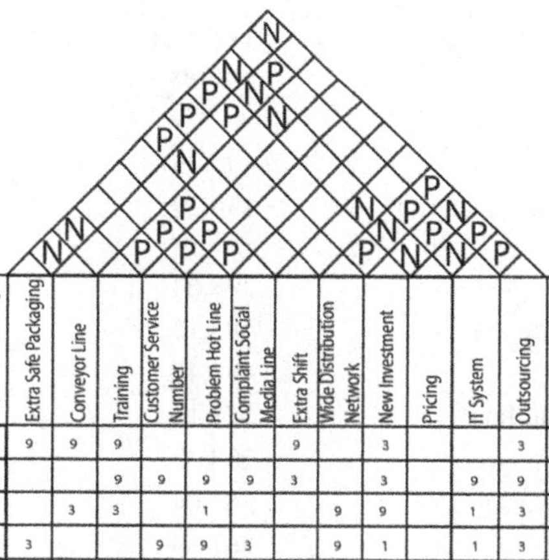

Customer Requirements	Importance Rating	Extra Safe Packaging	Conveyor Line	Training	Customer Service Number	Problem Hot Line	Complaint Social Media Line	Extra Shift	Wide Distribution Network	New Investment	Pricing	IT System	Outsourcing	Current Statement	Target	Competitor	Improvement Ratio	Priority Rating	% Priority Rating
Undamaged Delivery	9	9	9	9				9		3			3	5	5	5	1	9	15,43
Customer Service Network	9			9	9	9	9	3		3		9	9	4	5	5	1,25	13,5	23,15
On Time Delivery	8		3	3		1			9	9		1	3	4	5	5	1,25	10	17,14
Collaboration Of Firm During Reverse Logistics	8	3			9	9	3		9	1		1	3	4	5	4	1,25	10	17,14
Affordable Pricing	8								3	9	9		9	4	4	4	1	8	13,71
Delivery Tracking	7				9	9	1		1	9		9	9	3	4	3	1,13	7,81	13,39
Technical Priority	824,87	190,29	190,29	398,64	483,12	498,46	273,16	208,32	363,04	461,59	123,39	363,14	596,52						
Normalize Technical Priority	17,24	3,98	3,98	8,33	10,10	10,41	5,70	4,35	7,59	9,65	2,58	7,49	12,47						

Fig. 3. Completely constructed QFD

As a result of the QFD study carried out, the logistics company should improve itself in many areas. The most important of these areas are customer service network, on time delivery and reverse logistics. Although every criterion developed provides a serious advantage to the company in meeting the needs of the customers and increases its profitability, the criteria that can be seen on the roof part can have negative relationships with each other. Although the criterion that should be chosen compared to these relationships is numerically superior, the managers of the company that will make the decision should consider other variables such as cost and compatibility or company culture.

5 Discussions and Conclusion

As the cargo companies are essentially service providers. Their focus points are the customers. However, as a result of monotonous works, companies may experience company-blindness, experience difficulties in predicting customers' requests or due to similar reasons their service quality may decrease [22, 23]. Apart from this, the problems they experience in their core activities - operational inefficiencies, transportation times, cost problems, etc. – can indirectly affect the customers [24].

The questionnaire was performed by 337 individual consumers. In this study, the most important requests are as follows, undamaged delivery, customer service network, on time delivery, collaboration of firm during reverse logistics activities, affordable pricing, delivery tracking. This study has poured all of the technical requirements of the company into mathematical data, according to key requirements determined. Towards this data it is recommended that, the company should develop its technological infrastructure and adopt the lean approach method. This approach eliminates unnecessary processes in the process flow and helps the company to save both financial resources and time. Plus increases productivity. The most important criteria determined by customers is undamaged delivery. Hence the company, to develop this aspect, should train its employees at regular intervals and additionally use different transportation models. A different aspect, routing, can be optimized mathematically to reduce downtime and travel expenses. For instance, prepared routing maps can be handed over to employees to save in terms of time. This helps company to develop pricing and on time delivery options which are determined highly important by the customers. In addition, working for major companies, such like B2B operations also would help the company to save in terms of financial resources. B2C operations require huge travelling, handling, and tracking yet it is not very profitable. Hence the company ends up the result of not getting desired efficiency and profit rates. It has been observed that, the company is not eligible to reach desired efficiency levels. To solve it, the company should focus and invest even more to its internal processes. The oldest cargo company in Turkey which has the highest number of branches, needs to be more systematic and academically focused. As a result of all these efforts, both the company and the customers would benefit a noticeable gain.

In the future studies, this study can be extended by collaboration of different approaches and studies, which are made in the field of lean management to minimize errors and gain a different aspect. With technological tools such as big data and the internet of things, customers can be accurately observed in real time and results with higher success rates can be achieved. On the part of the company, although the installation costs seem quite high, an integrated system can be established with industry 4.0 tools that must be highly efficient in the long term in terms of sustainability and profitability. Thus, the existing wide operation network and employees can be observed and intervened instantly, and the company can be provided with a great flexibility in meeting customer needs.

References

1. Pakdil, F., Kurtulmuşoğlu, F.B.: Improving service quality in highway passenger transportation: a case study using quality function deployment (2014)
2. Yothinsirikul, W., Thawesaengskulthai, N.: Application of quality function deployment and servqual for design service innovation (2014)
3. Lee, C.K.M., Ru, C.T.Y., Yeung, C.L., Choy, K.L., Ip, W.H.: Analyze the healthcare service requirement using fuzzy QFD (2015)
4. Chen, S.H.: Determining the service demands of an aging population by integrating QFD and FMEA method. Qual. Quant. **50**(1), 283–298 (2016)
5. Chowdhury, M.M.H., Quaddus, M.A.: A multi-phased QFD based optimization approach to sustainable service design. Int. J. Prod. Econ. **171**, 165–178 (2016)
6. Jahanzaib, M., Idrees, M., Wasim, A., Hussain, S., Aziz, H.: A framework for implementing quality function deployment (qfd) for utility services. IOSR J. Bus. Manag. **18**(4), 92–99 (2016)
7. Kuo, C.M., Chen, H.T., Boger, E.: Implementing city hotel service quality enhancements: integration of Kano and QFD analytical models. J. Hosp. Mark. Manag. **25**(6), 748–770 (2016)
8. Mardar, M., Zhygunov, D., Znachek, R.: QFD methodology to develop a new health-conducive grain product. Восточно-Европейскийжурналпередовыхтехнологий **2**(11), 42–47 (2016)
9. Nikfalazar, S.: Statistical process control application on service quality using SERVQUAL and QFD with a case study in trains' services. TQM J. **28**(2), 195–215 (2016)
10. Novokreshchenova, N.A., Novokreshchenova, O.A., Terehin, S.E.: Improving bank's customer service on the basis of quality management tools (2016)
11. Zawati, O.A.L., Dweiri, F.: Application of quality function deployment to improve smart services applications, Dubai public entity as a case study. In: 2016 IEEE International Conference on Industrial Engineering and Engineering Management (IEEM), pp. 881–885. IEEE (2016)
12. Dehe, B., Bamford, D.: Quality function deployment and operational design decisions–a healthcare infrastructure development case study. Prod. Plan. Control **28**(14), 1177–1192 (2017)
13. Wang, Y.H., Lee, C.H., Trappey, A.J.: Service design blueprint approach incorporating TRIZ and service QFD for a meal ordering system: a case study. Comput. Ind. Eng. **107**, 388–400 (2017)
14. Khairunnisa, N., Munandar, J.M., Najib, M.: Customer Satisfaction and Loyalty on Customer Delivered Value of Postal and Shipping Service (2018)
15. Maalej, S., Halouani, N., Chabchoub, H.: A linguistic reliable quality function deployment method for the healthcare sector. Int. J. Multicriteria Decis. Making **7**(3–4), 276–297 (2018)
16. Wu, W.Y., Qomariyah, A., Sa, N.T.T., Liao, Y.: The integration between service value and service recovery in the hospitality industry: an application of QFD and ANP. Int. J. Hosp. Manag. **75**, 48–57 (2018)
17. Furqon, C., Sultan, M.A., Putri, S.I.: Quality function deployment analysis on transportation services. In: 1st International Conference on Economics, Business, Entrepreneurship, and Finance (ICEBEF 2018). Atlantis Press (2019)

18. Hasibuan, A., Parinduri, L., Sulaiman, O.K., Suleman, A.R., Harahap, A.K.Z., Hasibuan, M., Daengs, G.A.: Service quality improvement by using the quality function deployment (QFD) method at the government general hospital. In: Journal of Physics: Conference Series, vol. 1363, no. 1, p. 012095. IOP Publishing (2019)
19. Adiandari, A., Winata, H., Fitriandari, M., Hariguna, T.: Improving the quality of Internet banking services: An implementation of the quality function deployment (QFD) concept. Manag. Sci. Lett. **10**(5), 1121–1128 (2020)
20. Hauser, J.R., Clausing, D.: the house of quality. Harvard Bus. Rev. **66**(3), 63–73 (1988)
21. Bottani, E., Rizzi, A.: Strategic management of logistics service: a fuzzy QFD approach. Int. J. Prod. Econ. **103**(2), 585–599 (2006)
22. Gür, G., Ervansel, E., Er, E.T., Caliskan, A., Ozturkoglu, Y.: Linking container shipping company service attributes to shipper satisfaction and loyalty. In: Proceedings of the International Symposium for Production Research 2019, pp. 692–704. Springer, Cham (2019)
23. Kolat, D., Kökçü, H.A., Kiranli, M., Özbiltekin, M., Öztürkoğlu, Y.: Measuring service quality in the logistic sector by using SERVQUAL and best worst method. In: Proceedings of the International Symposium for Production Research 2019, pp. 720–731. Springer, Cham (2019)
24. Ozbekler, T.M., Ozturkoglu, Y.: Analysing the importance of sustainability-oriented service quality in competition environment. Bus. Strategy Environ. **29**(3), 1504–1516 (2020)

Improving the Throughput of Pipe and Tube Production: A Case Study

Aysu Aslaner, Aysu Gürokur, Enes Turan, Karya Varol,
Melisa Menemenlioğlu, Ömer Faruk Çorum, Yağmur Aydoğan,
Nazlı Karataş Aygün, and Erdinç Öner(✉)

Department of Industrial Engineering, Yasar University, Izmir, Turkey
aysu-aslaner@hotmail.com, aysugurokur@hotmail.com,
turaneneees@gmail.com, varolkarya@gmail.com, m.menemenlioglu@gmail.com, omerfcorum@gmail.com,
yagmuraydogaan@hotmail.com,
{nazli.aygun, erdinc.oner}@yasar.edu.tr

Abstract. The facility layout is an important issue affecting the productivity of manufacturing systems. For an organization to have an effective and efficient manufacturing unit, special attention must be given to facility layout. Well-designed facilities result in efficient materials handling, reduced resource transportation times, and decreased production cycle times. Besides, effective layouts are known to greatly impact operational performance. Improving the facility layout considering the flow amounts between machines improves the performance of the system. Designing an optimal layout provides the efficiency of material handling and reduces the distance traveled. In this study, the locations and quantities of the work in process (WIP) inventory areas and the locations of the machines are determined. Mixed Integer Programming (MIP) is used for formulating and representing the problem. The objective function of the mathematical model is to minimize distances between the machines with high material flows. The mathematical model is solved using an optimization program. The iterative heuristic algorithm using the same optimization approach is designed as an alternative solution method to develop the mathematical model. In addition to the layout optimization, the amount and the size of WIP inventory areas are calculated using the daily production plan. The results of the study are put together in an applicable, user-friendly, and adaptable Decision Support System (DSS). The new plant layout has been compared with the current plant layout using the DSS. The company will produce more effectively with less material handling using the DSS.

Keywords: Facility layout problem · Layout optimization · Material handling · Mixed integer programming · Machine allocation · WIP inventory

1 Introduction

The facility layout optimization problem considering the variable number of WIP inventory areas is studied in this research paper. The objective is to minimize total travel in the system, which is the multiplication of the number of material flows and the distances between the machines. The study is conducted in a combi manufacturing company.

N. M. Durakbasa and M. G. Gençyılmaz (Eds.): ISPR 2020, LNME, pp. 888–899, 2021.
https://doi.org/10.1007/978-3-030-62784-3_74

The sub-materials of combi, which are pipe, heat exchanger, and plate heat exchanger, are produced in the facility. The outputs of the facility are sent to the assembly line of another facility for manufacturing combis using forklifts. The material is handled using trolleys in the facility.

The production stages of the pipes in the facility are bending, cutting, endforming, welding, drilling and crushing, washing, and visual inspection. Note that, the pipe types are diverse and due to the diversity, the flow that a pipe needs to follow is different. There are lean cells in the production area determined by the company. In these cells, some of the production steps are performed according to the type of pipes. In general, the production process begins with bending the pipes in the bending machine and cutting the pipes in the cutting machine. In the endforming process, the end of the pipes is shaped with a different number of strikes according to the pipe type. After that, in the welding machine, either two pipes are welded together, or nut and pipe are combined. The pipes passing through these processes come to the washing machine to clean the remaining burrs. Finally, visual inspection is performed by fitting rubber into certain pipes and testing the pipes to ensure the leakages do not exist. The final products are transmitted to four different destinations, which are assembly lines and Spare Part Area at another facility, heat exchanger assembly line, and warehouse for shipping. The manufacturing type of the facility is a combination of a discrete and batch manufacturing system. In the facility, the variety of pipes manufactured is high, but the number of pipes manufactured is low. Production steps and cycle times differ according to the pipe type. Therefore, the estimation of the amount of WIP inventory that is needed daily is essential, and large amounts of WIP inventory are kept in the system is challenging. Since the correct quantities of WIP inventory are not certain, the inventory must be renewed by the worker during the pipe production.

In the current system, determination of the location of the machines and sub-materials is based on the experiences rather than a systematic approach. Therefore, the long-distance between machines and WIP inventory areas causes high material handling time.

In this study, we aim to reduce the distances between machines with high material flow by determining the location of the machines together with the location and the size of the WIP inventory areas. The rest of the paper is organized as follows. A literature review on layout optimization is presented in Sect. 2. A mixed-integer programming model (MIP) is presented in Sect. 3, as well as the formal problem definition. Heuristic algorithms are presented in Sect. 4. Section 5 presents the computational results. A user-friendly decision support system is also explained in Sect. 5. Finally, the conclusion is provided in Sect. 6.

2 Literature Review

There are different methods and techniques for solving facility layout problems. A literature review is performed to synthesize the methods used for solving facility layout problems. The literature on facility layout optimization is rather limited when multiple machines are considered. Most studies focus on a fewer number of machines without considering the WIP inventory areas [1, 2]. In this study, we consider the optimal facility layout by considering the flow between the machines and WIP inventory area locations and sizes.

In the literature, two research papers are found to be much related to the studied problem. In the first study, the authors proposed a mathematical model for the facility layout design problem with multi-objective mixed integer programming to minimize the material handling cost, re-layout cost, and maximize closeness rating [1]. In the second study, the authors proposed a mathematical model for the facility layout optimization problem with earliness-tardiness penalties and sequence-dependent setup times to minimize the material handling cost and facility building cost [2]. A mixed-integer programming (MIP) model is developed to obtain optimal solutions.

The study showed that mathematical models may not be capable of solving large size layout problems [3]. Therefore, heuristic algorithms are used to achieve results. In [4], the optimum solution could not be found with mathematical programming, but a valid solution is achieved using a heuristics approach. A literature review is performed to improve a heuristic approach for the problem, however, some well-known heuristic approaches are not directly applicable for this study [5]. Hence, we suggest an iterative approach that first assign the cells then the machines based on cell assignment.

In this study, we developed a mathematical model to find the optimal layout design for a smaller number of machines, and then, we presented a heuristic algorithm to solve the facility layout design problem for facilities with a larger number of machines. This study proposes a new methodology to address the facility layout problem. Furthermore, we develop a user-friendly decision support system that is integrated with the proposed heuristic algorithms.

3 Problem Formulation

The problem is defined as an improvement in a facility layout design by considering the location and size of the WIP inventory areas. Hence, we propose a holistic framework that combine the facility layout design problem and size determination and location assignment of WIP areas. Although the studies that separate facility layout design and WIP area size and location assignments are prominent in the literature, we handle these two problems together. The first step in our framework is to determine the number, size and location of the WIP inventory areas. Next, we calculate a one-year flow amount between the machines and solve layout optimization problem. Since there exists high variability in the types and quantities of items produced each month, a shift-based daily production plan for one month is used in this study for WIP inventory. The size and amount of WIP inventory areas are determined based on the box sizes required for handling the materials and quantities of these boxes to be kept in each shift. The required WIP inventory area is calculated for each type of pipe to be produced by determining the box type, the number of pipes to be kept in the boxes and the number of boxes to be kept in the WIP inventory area.

The mathematical model finds the optimal locations of machines by using the MIP for the Layout Optimization problem.

Assumptions:

1) According to Human Safety Environment Policies, the width of the corridors between machines must be at least 1 m.
2) All stations and facility floors have a rectangular shape.
3) Movement between stations is rectilinear centroid to centroid.
4) The amount of material flow between machines and WIP inventory areas is known.
5) Material handling cost equals to 1 and fixed for all pipe types.
6) Cells and WIP inventory areas are accepted as machines.

Indices

- i, j: index of machines $(1,\ldots, n)$

Parameters

- n: the total number of machines
- n_f: numbers of fixed machines
- f_{ij}: flow between machines i and j, $i \neq j$
- c_{ij}: cost of material handling from machines i to j
- h: horizontal length of facility
- v: vertical length of facility
- $l_{i,}$: length of machine i along the x-axis
- w_i: width of machine i along the y-axis

Decision Variables

- x_i, y_i: x and y coordinates of the centroid of machine i
- d_{ij}^{x}: the distance between machines from i to j in the x-axis, $i \neq j$
- d_{ij}^{y}: the distance between machines from i to j in the y-axis, $i \neq j$
- d_{ij}: total rectilinear distance between machines i and j, $i \neq j$
- $\mathrm{z}_{\mathrm{ij}}^{\mathrm{L}}, \mathrm{z}_{\mathrm{ij}}^{\mathrm{R}}, \mathrm{z}_{\mathrm{ij}}^{\mathrm{A}}, \mathrm{z}_{\mathrm{ij}}^{\mathrm{B}}$:

$$\begin{cases} 1, \textit{if machine i is located to the left}(L), \textit{right}(R), \textit{above}(A), \textit{below}(B) \textit{ of machine j} \\ 0, \textit{otherwise} \end{cases},$$

$i \neq j$

(P3.1)

$$\text{Minimize z} = \sum\nolimits_{i=1}^{n-1} \sum\nolimits_{j=i+1}^{n} c_{ij} * f_{ij} * dij \tag{1}$$

Subject to

$$d^x_{ij} \geq x_i - x_j \quad \forall i,j \quad (2)$$
$$d^x_{ij} \geq x_j - x_i \quad \forall i,j \quad (3)$$
$$d^y_{ij} \geq y_i - y_j, \quad \forall i,j \quad (4)$$
$$d^y_{ij} \geq y_j - y_i, \quad \forall i,j \quad (5)$$
$$d_{ij} \geq d^x_{ij} + d^y_{ij}, \quad \forall i,j \quad (6)$$
$$0 \leq |x_i - x_j| \leq h \quad \forall i,j \quad (7)$$
$$0 \leq |y_i - y_j| \leq v \quad \forall i,j \quad (8)$$
$$x_i \geq (0.5) * l_i, \quad \forall_i i = 1,\ldots,n \quad (9)$$
$$y_i \geq (0.5) * w_i, \quad \forall_i i = 1,\ldots,n \quad (10)$$
$$h \geq x_i + (0.5) * l_i, \quad \forall_i i = 1,\ldots,n \quad (11)$$
$$v \geq y_i + (0.5) * w_i, \quad \forall_i i = 1,\ldots,n \quad (12)$$
$$(x_i - x_j) \geq (0.5) * (l_i + l_j) - h * \left(1 - z^R_{ij}\right), \quad \forall i,j \quad (13)$$
$$-(x_i - x_j) \geq (0.5) * (l_i + l_j) - h * (1 - z^L_{ij}), \quad \forall i,j \quad (14)$$
$$(y_j - y_i) \geq (0.5) * (w_i + w_j) - v * (1 - z^B_{ij}), \quad \forall i,j \quad (15)$$
$$-(y_j - y_i) \geq (0.5) * (w_i + w_j) - v * (1 - z^A_{ij}), \quad \forall i,j \quad (16)$$
$$z^L_{ij} + z^R_{ij} + z^A_{ij} + z^B_{ij} \geq 1 \quad \forall i \quad (17)$$
$$x_i \geq 0, y_i \geq 0 \quad \forall i \quad (18)$$
$$d^x_{ij} \geq 0, d^y_{ij} \geq 0 \quad \forall i,j \quad (19)$$
$$z^L_{ij}, z^R_{ij}, z^A_{ij}, z^B_{ij} = 0 \text{ or } 1 \quad \forall i,j \quad (20)$$

The objective function (1) minimizes the total distance of material flow. Constraints (2), (3), and (4), (5) ensure the rectilinear distances between centroids of two machines on the x-axis and y-axis. Constraint (6) presents the total rectilinear distance between machines. Constraints (7) and (8) ensure th.at the x-axis and y-axis coordinates remain within factory boundaries. Constraints (9), (10), (11), and (12) require the capacity limitations for each machine. Constraints (13) and (14) prevent machines from overlapping on the x-axis, constraints (15) and (16) prevent machines from overlapping on the y-axis. Constraint (17) forces any one of four constraints to remain active. Finally, constraint set (18), (19) and (20) define the binary and non-negative decision variables.

4 Solution Methodology

The proposed MIP model (P3.1) is solved with different data sets (toy data) for verification purposes. Three different instances are created which are taking 2 fixed machines out of 5 machines, 4 fixed machines out of 10 machines, and 5 fixed machines out of 15 machines. The MIP formulation is solved optimally with CPLEX 12.10 [6]. The results of the simulation model for 3 data sets are analyzed. For all the data sets, fixed machines have been located according to their given coordinates. Machines are settled within the specified area boundaries. Also, the overlapping of the machines is not observed. When the results are taken into consideration, it is seen that all constraints are satisfied and the machines with high flow numbers are assigned as close as possible to each other as stated in the objective.

Since the optimal solution could not be reached with the direct solution of (P3.1) due to the size of the real data, we employ a heuristic approach to solve the problem with real data. Our heuristic approach proposes an iterative solution methodology that first locate the cells and then assign the machines into cells. Note here that, we use the same optimization model (P3.1) in both of the steps for cell and machine assignment by changing the index number.

The number of flows between machines calculated according to the yearly production plan. Each pipe type has a fixed number of pipes in the boxes given by the company. Yearly production amount is divided by the number of pipes in the box and the number of boxes needed is found. After that, according to the routing of the pipes, flow numbers are calculated. Accordingly, the objective function is determined by multiplying the number of trips between machines by distances.

There are 5 cells in the system and each cell has a different number of machines. Machines in the cells are determined by the company according to the routing of the pipes and they must remain within their cell boundaries. Besides, some machines are not in a particular cell and the location of five of these machines is taken as fixed at the request of the company. The designed heuristic is an iterative method based on the placement of the cells and the machines sequentially. This algorithm is based on the idea to create two separate loops with the same objective function, the first is to place the cells and the second is to place the machines into the cells that settled. There is only a set of machines in the mathematical model. In the heuristic method, the same model indexes are used in both cell assignment and machine assignment. The developed mathematical model is run for each subproblem, cell, and machine separately, and solved optimally in each iteration.

There are nine main steps in this method:

1. Indices and parameters used in the mathematical model are defined in the code.
2. Fixed cells that are desired to be kept stationary by the factory are placed in their fixed positions.
3. Constraints are checked. Distance constraints for determining the rectilinear distances between the centroids of cells on the x and y-axis. Boundary constraints ensure that the x and y-axis coordinates of the cells remain within the factory boundaries. And overlapping constraints prevent cells from overlapping with each other on the x and y-axis.
4. Cells that have more flow between each other are settled closer with minimum distances.
5. Parameters are updated for the machine placement.
6. The model is rerun, and the locations of cells are fixed.
7. Fixed machines that are desired to be kept stationary by the factory are placed in their fixed positions.

8. Constraints are checked. Distance constraints for determining the rectilinear distances between the centroids of machines on the x and y-axis. Boundary constraints ensure that the x and y-axis coordinates of the machines remain within the factory and also cell boundaries. And overlapping constraints prevent machines from overlapping with each other on the x and y-axis.
9. Machines that have more flow between each other are settled into the cells with minimum distances.

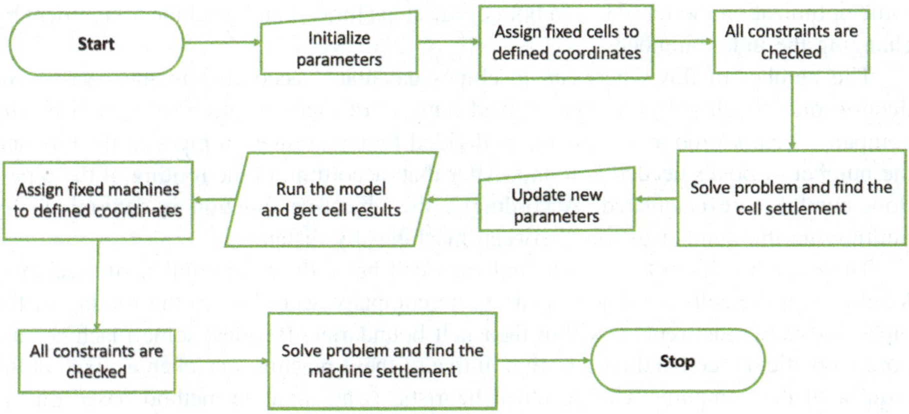

Fig. 1. Pseudo code of solution algorithm

5 Computational Results

5.1 WIP Study

In the context of WIP study, data for all pipe types to be produced for a-month is analyzed to determine the size and amount of WIP inventory areas. After the bending process, there were 2 separate WIP areas but as a company request it is decided to decrease the number of WIP areas 5 to 4 as follows: WIP1 after the bending process, WIP2 after the second bending process, WIP3 before the welding process and WIP4 before visual inspection process. The dimensions of the WIP areas are calculated, but it is decided with the company to keep the area sizes of WIP2, WIP3 and WIP4 are fixed since there is not too much change in the dimensions. The company wants to operate bending processes in one shift before the plan. This means that the pipes wait in the WIP1 throughout the shift. Pipes were previously held according to the last digit of their code number. However, this caused pipe types to mix and even disappear. To prevent this, WIP1 is divided into 8 areas according to the cells. Since each pipe can go to different cells after the bending process, WIP1 is divided by 8 areas which belong to cell 1, cell 2, cell 3, cell 4, cell 5, washing, second bending, and others. Others cover bypass pipes and spare part pipes that are not included in the scope of our study but kept in this WIP area. The number of boxes to be used for all shifts within a 1-month period is compared. Therefore, the size of the divided areas in WIP1 is determined according to the maximum number of boxes that should be kept in a shift within a 1-month period. Thus, the WIP1 area can be divided efficiently according to this assumption.

5.2 Layout Optimization

Layout optimization studies are divided into two stages. In the first stage, the model is solved with toy data and in the second stage, the iterative heuristic method using the same optimization problem is designed. When the model is generating in IBM ILOG CPLEX 12.10 version, the verification is important to check if a designed heuristic method is true or not.

Toy data sets are created to control the model in a short time. In different scenarios, toy data is created both machine dimensions and flow levels between the machines as seen in Table 1. The aim is to check the placement of machines according to their flow levels and placement of machines inside of their cell boundaries. However, if the total dimensions of machines are greater than the total available area or there is any overlapping on machines, the model should give an infeasible solution.

Note here that, in this model, the distance between machines are in meter, and flow is the number of travels. As a result, $the\ unit\ of\ objective\ function(10)\ component\ f_{ij} * dij$ is meter. As we assume cost is 1, the resulting unit of objective function is meter.

In toy data 1, three cells containing 6 machines are analyzed. The first cell is a fixed-location cell and it contains machines 4 and 5. Second and third cells are changeable cells. The second cell contains machines 6 and 7, the third cell contains machines 8 and 9.

Table 1. Toy data 1

Machines	Inputs		
	Index	Length	Width
Fixed Cell	1	4	4
Changeable Cell	2	4	4
	3	4	4
Changeable Machine	4	4	2
	5	4	2
	6	1	2
	7	2	2
	8	2	2
	9	2	2

Flow	1	2	3	4	5	6	7	8	9
1	0	9	10	0	0	0	0	0	0
2	9	0	8	0	0	0	0	0	0
3	10	8	0	0	0	0	0	0	0
4	0	0	0	0	1	4	2	5	0
5	0	0	0	1	0	1	2	5	0
6	0	0	0	4	1	0	0	1	2
7	0	0	0	2	2	0	0	2	3
8	0	0	0	5	5	1	2	0	1
9	0	0	0	0	0	2	3	1	0

The results are analyzed for verification purposes. Verification is done according to the placement of the machines and cells by providing model restrictions. The result of toy data 1 is shown in Fig. 1. It is seen that the location of the fixed cell is assigned to the given coordinates. Machines are located inside of their cells as it should be. Cells and machines with high flow levels are assigned as close to each other as stated in the objective. Machines remained within their cell boundaries, and cells remained within the specified area. Also, the overlapping has not observed in the solution.

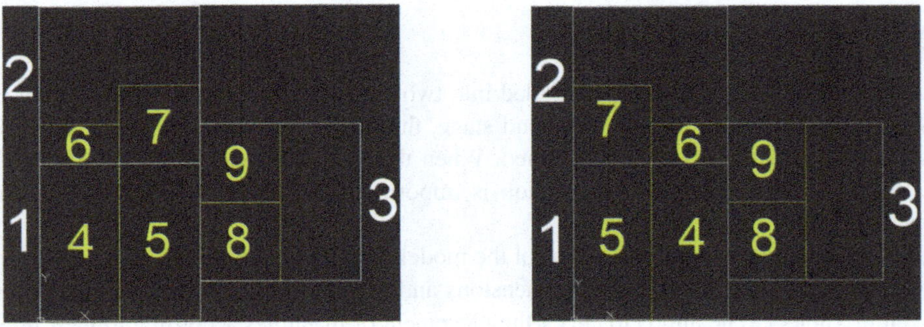

Fig. 2. Result of toy data 1 and toy data 2

The input of toy data 2 has the same properties for machines, cells, and total area, except for changing the flow. The location of the fixed cell has not changed, and each machine is located within its cell boundaries with this comparison. Cells are also located within the specified area and the machines are not overlapping. However, it is observed that the positions of the machines changed with the change of flow levels as shown in Fig. 2. Machines with high flow levels are settled as close to each other.

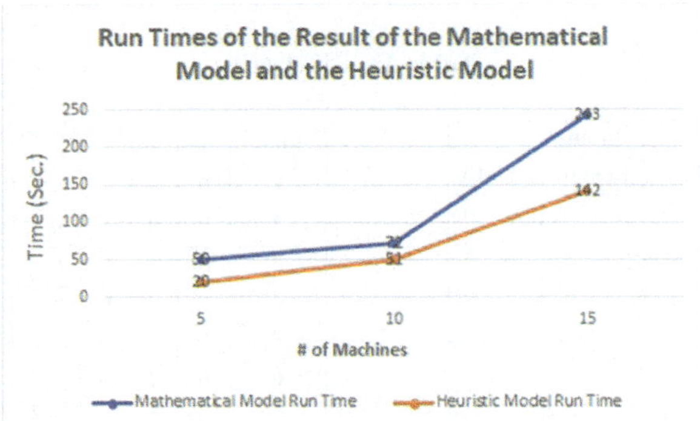

Fig. 3. Run times of the result of the mathematical model and the heuristic model

According to the differences in the input data, each solution is drawn in AutoCAD 2020.1.1 version [7] to see the change in the layout plan. Verification is done to check the accuracy of the model developed with the heuristic method. As can be seen in Fig. 3, when run times of the mathematical model and designed heuristic model are compared, the run times of the heuristic method are shorter. Through this comparison, the performance of the model is improved with the heuristic method. After verifying the accuracy of the model with toy data, it is applied to the real data.

To implement the heuristic approach, cells are determined at the request of the company and depending on the flow levels between the machines. As can be seen in Fig. 4, the cells with a different number of machines are shown in the red line. The solution is obtained using OPL CPLEX 12.10 version. With the developed code, cells are placed first, then machines are placed. The objective function has a large value in meters. Because the objective function is equal to the total distance covered by employees in the production system throughout the year. The objective function of the current layout is calculated manually according to the machine coordinates and flow levels. When the new and old objective functions are compared to the old objective function value, a 38.1 percent improvement is achieved.

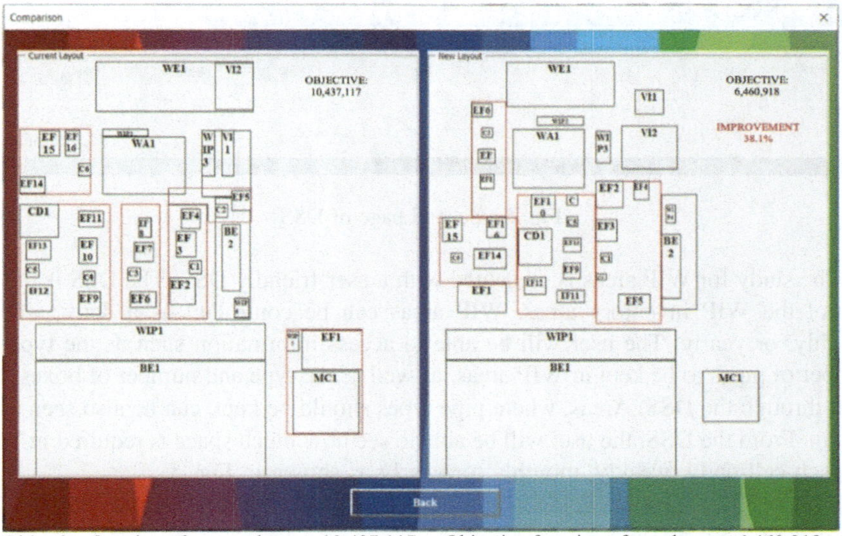

Objective function of current layout: 10,437,117 m Objective function of new layout: 6,460,918m

Fig. 4. Comparison of current and new layout

In the new layout design, machine positioning can be seen in Fig. 4. Through this solution, the reduction of material handling time provides an increase in productivity in the workforce. The production process is accelerated with optimally determined machine positions. With the accelerated process, waste time in production is also reduced. Energy efficiency is achieved by reducing losses, which means a positive impact on the environment.

The solution is implemented in a user-friendly DSS in EXCEL 16.0 version [8]. The DSS is designed flexibly and dynamically for both layouts and WIP inventories. DSS will allow the user to see layout changes depending on the change in the annual production plan. Since our production area has a flexible structure and the displacement of the machines is easy, the layout can change annually according to the production plan. Accordingly, the user will be able to compare the old layout and the new layout as shown in Fig. 4. Thus, the factory will be able to easily see the efficiency it has achieved as a result of the change in the layout.

With the DSS, employees will be able to see the WIP inventory amounts they need to put, on the system. In addition, a layout can be arranged annually according to the production plan.

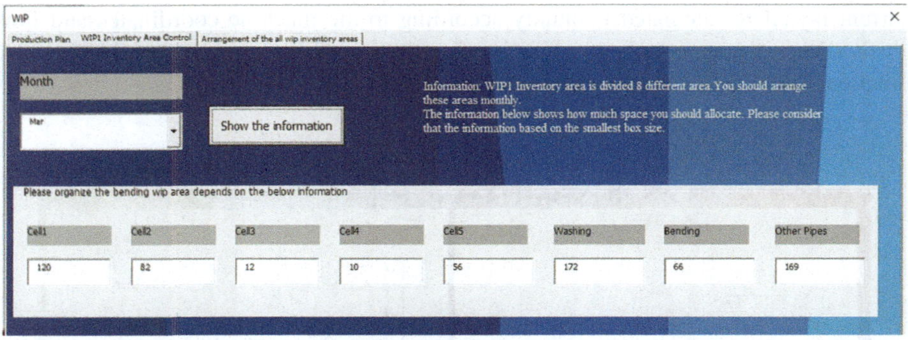

Fig. 5. Second page of DSS

The study for WIP areas is presented with a user-friendly DSS. The DSS is used to control the WIP inventory areas. WIP areas can be controlled as a daily, weekly, monthly, or yearly. The user will be able to access information such as the type and number of pipes to be kept in WIP areas, as well as the type and number of boxes to be kept, through the DSS. Areas, where pipe types should be kept, can be also seen in the system. From the DSS, the user will be able to see how much space is required in WIP1 for each cell daily, weekly, monthly, or yearly as shown in Fig. 5.

6 Conclusion

In this study, the main problem is to relocate the machine positions. The objective of our problem is to minimize the total distance between the machines with a high number of flows between each other. This objective is determined according to the company priorities and the context of the study. After analysis of the up to date literature, the optimization model is formulated as a MIP model. Our solution approach has two stages. In the first one, the model is solved and verified with toy data in OPL CPLEX 12.10 version optimization program. Through different toy data sets, the accuracy of the model is determined. However, the optimal solution could not be reached with the mathematical model since the real data is large. In the second stage, because of this, a heuristic algorithm is developed to solve the problem with real data as an alternative solution method. An algorithm has been developed that optimally places both the cells and the machines inside the cells. The basic logic of this algorithm is to create two separate loops with the same objective function, the first is to place the cells and the second is to place the machines into the cells that settled. Verifications of the model developed with the heuristic method are performed and its accuracy is proved. As a result of the model run with the developed method, an improved layout is drawn using the Excel 2020.1.1 version. the model solved with real data in OPL CPLEX 12.10 version optimization program.

Another aim of the study is to determine the optimal placements of WIP inventory areas. Accordingly, the number of WIP inventory areas, sizes of them and quantities to be kept should be calculated. As a result of the study, the number of WIP areas and in which area each pipe should be kept are determined. The size and amount of WIP inventories are also determined based on box sizes and quantities of boxes to be kept in each shift. At the same time, the box type, quantity of pipes in boxes, and the number of boxes to be kept are calculated.

The solution obtained in CPLEX is implemented in a user-friendly DSS in EXCEL 16.0 version. DSS is designed to control the WIP inventory areas. Users will be able to see the types and quantities of pipes that should be kept in each work in the process area via the system according to the changing production plan. Type of pipes, type of boxes, and the number of boxes could be seen for each WIP inventory area in daily or shift. WIP area after the bending process, which was previously held according to the last digit of the pipe code numbers, is divided into 8 areas according to the cells. The user will be able to reallocate this area in any preferred frequency such as daily, weekly, monthly, and yearly according to the number of pipes going to the cells. Accordingly, the required area for each cell will be visible in the system. It also shows layout changes depending on the change of the annual production plan. Thus, the efficiency obtained as a result of the change in the layout can be easily seen.

Acknowledgment. This study is supported by TUBITAK (The Scientific and Technological Research Council of Turkey) in the program of "2209-B Undergraduate Thesis Support Program for Industrial Applications".

References

1. Purnomo, M.R.A., Wiwoho, Y.S.: Multi-objective mixed integer programming approach for facility layout design by considering closeness ratings, material handling, and re-layout cost. IOP Conf. Ser. Mater. Sci. Eng. **105**(1), 012045 (2016)
2. Patil, M., Kumanan, S.: Facility layout optimization using linear and non-linear mixed integer programming. In: Proceedings of the International Conference on Advances in Production and Industrial Engineering, p. 570 (2015)
3. Solimanpur, M., Jafari, A.: Optimal solution for the two-dimensional facility layout problem using a branch-and-bound algorithm. Comput. Ind. Eng. **55**, 606–619 (2008)
4. Akça, M., Şahin, R.: Çokamaçlı karma tam sayılıtesisyerleşimproblemimodeliveaskerite-sisteuygulama. Pamukkale Univ. J. Eng. Sci. **24**, 117–123 (2018)
5. Nordin, N.N.: Heuristics and metaheuristics approaches for facility layout problems: a survey. Malasia Universiti and Serdang (2016)
6. Studio, IBM ILOG CPLEX Optimization. CPLEX User's manual, Version 12, pp. 207–258 (2013)
7. Autodesk, AutoCAD. AutoCAD 2013 User's Guide (2012)
8. Microsoft Corporation. Microsoft Excel (2018). https://office.microsoft.com/excel

Integrated Transportation and Packaging Problem Modeling and Application

Tuğçe Köksal, Ege Akkuş, Gökçe Demirel(✉), Esin Tosun, Mahmut Ali Gökçe, and Cansu Yurtseven

Department of Industrial Engineering, Yasar University, Izmir, Turkey
koksaltugce@yahoo.com, egeakkus@hotmail.com, gokcedmrl97@hotmail.com, esintosn@gmail.com, {ali.gokce, cansu.yurtseven}@yasar.edu.tr

Abstract. The Hybrid Flow Shop scheduling (HFSS) is a complex problem observed in many real-life problems and usually involves a work environment with multiple parallel machines at each stage. This study deals with a scheduling problem that integrates transportation and packaging operations. The problem originates from a real-life problem at a leading global rim manufacturing company headquartered in Turkey. We propose a novel solution method, which is able to create a schedule for all stages, while minimizing the total weighted earliness of orders. The studied problem differs from classical HFSS problem in a number of ways, therefore, an improved mathematical model to accommodate current problem and system constraints is developed. The new two-stage as Mixed-Integer Programming model (MIP) is constructed, verified and validated. As the problem is NP-hard, a novel rule-based heuristic algorithm for two stages is developed. The performances of the heuristic algorithm and the mathematical model are compared using real-life data from the company. Based on the results from the experimentation, proposed heuristic algorithm performs very well in a significantly shorter computational time. Moreover, a user-friendly decision support system (DSS) is created and both solution methods are embedded in this DSS for the implementation at the company.

Keywords: Scheduling · Hybrid flow shop · Earliness · Multistage flow shop · Transportation · Batch scheduling · Mixed integer programming

1 Introduction

This study presents novel solution approach to an integrated transportation and packaging scheduling problem, modeled as a modification of a two-stage hybrid flow shop problem. Although the problem can be found in many different settings and industries, this problem originates from a leading rim manufacturer in İzmir, Turkey for their packaging and distribution system. In the current system, there are two distinct manufacturing facilities and a logistics center, from where products are shipped to the customers. Packaging processes are currently handled at two different facilities after manufacturing, and the finished products are transferred to the logistics center where they are stored for delivery to the customer. The company tries to provide flexibility to

N. M. Durakbasa and M. G. Gençyılmaz (Eds.): ISPR 2020, LNME, pp. 900–913, 2021.
https://doi.org/10.1007/978-3-030-62784-3_75

the customer in order to keep a high level of customer satisfaction. Customers have the right to change the type of packaging and the quantity of products in a order until the start of the frozen time before the date of shipment. The time that the customer cannot make changes before the shipment date is called the frozen time. However, changing the products that are packaged to be sent to the customer causes various additional costs due to extra labor, movement, material, and time used for this re-packaging operation. According to the order change requests from the customers, the product quantity changes are reported to the production planning, while the packaging changes are reported to the logistics center. Then the production planning department rearranges the plan and updates the packaging type according to the new order. Since current production planning does not follow a standard and efficient methodology, indirect financial losses can also occur along with the loss of material, time and labor in the system. When products are manufactured, they are transported to the logistic center and have to be packaged there based on the order.

In this study, we propose a new integrated model which does batching of orders and truck assignments of jobs for transportation, as well as creating packaging operations schedule by using Mixed-Integer Programming and heuristic method. The objective is to get products to the logistics center and perform packaging as much as possible after frozen time but before the due date. We propose an integrated model to minimize total weighted earliness of transportation and packaging scheduling and develop a scheduling tool for the company. In this study, trucks of each factory are specific, and machines in packaging operation work independently. Therefore, trucks and machines can be accepted as parallel machines.

The rest of the paper is organized as follows. Section 2 provides detailed explanations of the literature review for the study. In Sect. 3, a mixed integer linear programming model (MILP) of the problem is presented. The developed heuristic algorithm is shown in Sect. 4. Section 5 gives the computational results for proposed solution methodologies and describes the user-friendly decision support system. Finally, conclusion and future work can be found in Sect. 6.

2 Literature Review

There are many studies in literature on classic hybrid flow shop scheduling problem. Studies about earliness criterion is limited when the job scheduling problem with multiple machines is considered. We model problem on-hand as a two-stage hybrid flow shop. In the first stage of the problem, we tackle uniform parallel machine transportation scheduling problem with batch formation and non-common ready times. The second stage is a uniform parallel machine packaging scheduling problem in the logistic warehouse with non-common due dates. In their study Joo and Kim proposed a mathematical model for the integrated scheduling of unrelated parallel machines, batches, and heterogeneous delivery trucks to minimize the makespan of the entire process [1]. Batch transportation scheduling problems are presented in [2, 3]. The study of Wirth and Emde [4], the problem is scheduling a set of trucks arriving at a plant to deliver parts to specific dock doors on the factory premises. In [5, 6], lot scheduling problems are studied, and the objectives are minimizing the costs which are relevant to the process. Scheduling jobs with non-zero ready time values in unrelated parallel batch processing machines and aim is minimizing the makespan in [7]. In the study of Birgin et al., minimization of the absolute deviation of job completion times from a common due date in a flow shop scheduling problem is presenting in [8]. In addition to main objective, the minimization of the earliness and tardiness penalties also considered in this study. The basic two stage hybrid flow shop problem has been proven as NP-hard [9]. The study of Alharkan [10] has contributed to the area by providing heuristics for sequencing, and flow shop scheduling. In our study, the study of Alharkan was a guide in understanding these algorithms. Unlike the study of Alharkan, in our study, a rule-based algorithm was studied.

In this study, unlike the classical hybrid flow shop models studied in literature, the jobs are transported by trucks in batches, batch formation is part of the decision, in addition to company-specific restrictions in the first stage. Then the jobs carried by scheduled trucks in batches are separated in the second stage. Therefore, this study specifically contributes to the hybrid flow shop scheduling literature by batch formation and separation, providing a novel rule-based heuristic and testing and implementation of proposed solution (MIP and heuristic) with real-life data.

3 Problem Formulation

In the current system, the rims are transported from the plants to the logistic center after production in the plant. Once the production plan for the manufacturing facilities are made, ready time for orders to be transported to the logistics center are therefore determined. The proposed solution forms batches from orders based on ready times at plants, due dates at the logistics center, truck availability and capacity. After transportation, for the second stage, orders are planned to be packed by limited packaging resources to minimize the early packaging, which is calculated as time between the frozen time and packaging. This way, repackaging processes are also minimized by avoiding re-packaging due to order changes before frozen time. In order to solve the problem consisting of these two stages in the most efficient way, a Mixed Integer Linear programming optimization model has been developed. In this model, to define the problem and formulate the mathematical model in a concise manner, the following assumptions are made.

- Demand is deterministic and time-varying for all products.
- Products are transported as batches and these batches are packaged in the logistic center.
- Frozen time and due dates of the jobs are deterministic and assumed to be fixed.
- The volume of each batch must be less than or equal to the truck capacity.
- Each batch is delivered by exactly one truck.
- Weight parameters are assumed to be item dependent.
- Each job is processed at exactly one machine.
- Trucks can only travel between the plants and logistic center.

Indices, parameters, and relevant sets of the problem were determined and listed below.

Indices

i, j: index of job/order
b, l: index of batch
p: index of plant
t: index of truck
m: index of machine

Parameters

VJ_i = Volume of job i
VT_t = Capacity of truck t
R_i = Ready time of job i
TT_p = Round trip transportation time of plant p
P_{im} = Packaging time of job i at machine m
D_i = Due date of job i
F_i = Frozen time of job i
w_i = Weight of job i
Q = Very large number

Sets

P: Set of plants, $P = \{1, 2\}$
J_1: Set of jobs in Plant 1
J_2: Set of jobs in Plant 2
J: Set of jobs with the dummy jobs, $J = \{0, J_1, J_2\}$
B_1: Set of batches in Plant 1 with the dummy batch
B_2: Set of batches in Plant 2 with dummy batch
B: Set of batches, $B = \{B_1, B_2\}$
T_1: Set of trucks in transportation stage for plant 1
T_2: Set of trucks in transportation stage for plant 2
T: Set of trucks in transportation stage, $T = \{T_1, T_2\}$
M: Set of machines in packaging stage

Decision Variables:

$$x_{ibp} = \begin{cases} 1, & \textit{if job i is assigned to batch b at plant p} \\ 0, & \textit{otherwise} \qquad i \in J_p,\ b \in B_p,\ p \in P \end{cases}$$

$$h_{bt} = \begin{cases} 1, & \textit{if batch b is assigned to truck t} \\ 0, & \textit{otherwise} \qquad b \in B_p,\ p \in P,\ t \in T \end{cases}$$

$$y_{im} = \begin{cases} 1, & \textit{if job i is assigned to machine m} \\ 0, & \textit{otherwise} \qquad i \in J,\ m \in M \end{cases}$$

$$z_{blt} = \begin{cases} 1, & \textit{if batch b immediately precedes batch l at truck t} \\ 0, & \textit{otherwise} \qquad b \in B,\ l \in B,\ t \in T \end{cases}$$

$$a_{ijm} = \begin{cases} 1, & \textit{if job i immediately precedes job j at machine m} \\ 0, & \textit{otherwise} \qquad i \in J, j \in J,\ m \in M \end{cases}$$

r_{bt} = Shipping starting time of batch b at truck t, $p \in P$, $b \in B_p, t \in T$
s_{im} = Packaging starting time of job i at machine m, $i \in J$, $m \in M$
VB_b = Volume of batch b, $p \in P$, $b \in B_p$
E_i = Earliness of job $i, i \in J$
C_i = Completion time of job i, $i \in J$

Mathematical model aims to minimize the total weighted earliness of all jobs. Thus, while the objective function aims to minimize early packaging, it helps to minimize the cost of packaging changes caused by early packaging, as well. The weights are determined based on how likely a customer is to make a packaging change request. The more likely a customer is to change, the more their weight should be. An ABC analysis method was used on the number of change requests for the past 2 years, to determine the weight of requesting change for most served 24 different customers. According to this analysis, the demand variability number belongs to the first 3 customers with a rate of %80 which are:

- Group A; It includes 864 changing requests and consists of 3 companies with a rate of 80%.
- Group B; Includes 193 changing requests and consists of 7 companies with 15%.
- Group C; It includes 51 changing requests and consists of 11 companies with 5%.

The value range is 8–10 for group A, 5–7 for group B and 1–5 for group C.

The mathematical model determines the schedule of products in the packaging process, and the job list that determines when and which rims should be transported with trucks to the packaging stage. A mixed-integer linear programming model is constructed to solve two stage hybrid flow shop scheduling problem.

$$Minimize\, Z = \sum_{i \in J/\{0\}} E_i * w_i \tag{1}$$

$$\sum_{b \in Bp} x_{ibp} = 1 \qquad \forall p \in P, \forall i \in J_p, \forall b \in B_p/\{0\} \tag{2}$$

$$VB_b = \sum_{i \in Jp} VJ_i * x_{ibp} \qquad \forall p \in P, \forall b \in Bp/\{0\} \tag{3}$$

$$VB_b \le VT_t + Q * (1 - h_{bt}) \qquad \forall p \in P, \forall b \in Bp/\{0\}, \forall t \in T_p \tag{4}$$

$$\sum_{t \in Tp} h_{bt} = 1 \qquad \forall p \in P, \forall b \in Bp/\{0\} \tag{5}$$

$$\sum_{l \in Bp l \in Bp} z_{0lt} = 1 \qquad \forall p \in P, \forall t \in T_p \tag{6}$$

$$\sum_{b \in Bp} z_{b0t} = 1 \qquad \forall p \in P, \forall t \in T_p \tag{7}$$

$$\sum_{\substack{b \in Bp \\ b \ne l}} z_{blt} = h_{lt} \qquad \forall p \in P, \quad \forall l \in B_p, \forall t \in T_p \tag{8}$$

$$\sum_{\substack{l \in Bp \\ b \ne l}} z_{blt} = h_{bt} \qquad \forall p \in P, \forall b \in B_p, \forall t \in T_p \tag{9}$$

$$r_{bt} \ge R_i - Q * \left(2 - x_{ibp} - h_{bt}\right) \qquad \forall p \in P, \forall b, l \in B_p/\{0\}, \forall i \in J_p, \forall t \in T_p \tag{10}$$

$$r_{lt} - r_{bt} + Q * (1 - z_{blt}) \ge TT_p \qquad \forall p \in P, \forall b, l \in B_p/\{0\}, \forall t \in T_p | l \ne b \tag{11}$$

$$\sum_{m \in M} y_{im} = 1 \qquad \forall i \in J/\{0\} \tag{12}$$

$$\sum_{j \in J} a_{0jm} = 1 \qquad \forall m \in M \tag{13}$$

$$\sum_{i \in J} a_{i0m} = 1 \qquad \forall m \in M \tag{14}$$

$$\sum_{\substack{i \in J \\ i \ne j}} a_{ijm} = y_{jm} \qquad \forall j \in J, \forall m \in M \tag{15}$$

$$\sum_{\substack{j \in J \\ i \ne j}} a_{ijm} = y_{im} \qquad \forall i \in J, \forall m \in M \tag{16}$$

$$S_{im} \ge r_{bt} + (TT_p/2) - Q * (3 - x_{ibp} - h_{bt} - y_{im}) \qquad \forall p \in P, \forall b \in B_p/\{0\}, \forall i \in J_p, \forall t \in T_p, \forall m \in M \tag{17}$$

$$S_{jm} - S_{im} + Q * (1 - a_{ijm}) \ge P_{im} \qquad \forall i, j \in J/\{0\}, \forall m \in M | j \ne i \tag{18}$$

$$C_i \ge S_{im} + P_{im} - Q * (1 - y_{im}) \qquad \forall i \in J/\{0\}, \forall m \in M \tag{19}$$

$$C_i \le D_i \qquad \forall i \in J/\{0\} \tag{20}$$

$$E_i \ge D_i - Fi - C_i \qquad \forall i \in J/\{0\} \tag{21}$$

$$r_{bt}, s_{im}, E_i, C_i \ge 0 \qquad \forall i \in J/\{0\}, \forall m \in M, \forall b \in Bp/\{0\}, \forall t \in T \tag{22}$$

$$y_{im}, h_{btp}, z_{blt}, z_{blt}, x_{ibp}, a_{ijm} = 1 \text{ or } 0 \qquad \forall i \in J, \forall m \in M, \forall b \in Bp/\{0\}, \forall t \in T, \forall p \in P \tag{23}$$

Objective function (1) aims to minimize total weighted earliness of the entire process. Constraint (2) ensures that each job has to be included in exactly one batch. Constraint (3) and (4) confirms that the volume of each batch must be less than or equal to the truck capacity. Constraint (5) confirms that each batch is transported by exactly one truck. Constraints (6) and (7) show that, dummy batch 0 is the first and last batch for each truck. Constraint sets (8) and (9) determines that if batch l is assigned to truck t, there must be a preceding and a succeeding batch. Constraint (10) ensures that batches can only shipped after all the jobs is ready in that batch. Constraint (11) guarantees that a truck can only transfer the batch one by one. If batch b is transferred before batch l on truck t, batch l can start only when batch b is arrived at the logistic center and truck returns to the plant. Constraint (12) confirms that each job is processed at exactly one of the machines. Constraints (13) and (14) show that, dummy job 0 is the first and last job for each packaging machine. Constraints (15) and (16) determines that if job j is assigned to machine m, there must be a preceding and a succeeding job. Constraint (17) illustrates that job can only start to when the arrival of batch that it belongs to. Constraint (18) guarantees that a machine can only package the job one by one. If job i is process before job j on machine m, job j can start only when job i is packaged. Constraint (19) calculates the packaging completion time of each job. Constraint (20) specifies that all the job must be completed before its due date. Constraint (21) calculates the earliness of each job. Earliness is depending on frozen time of jobs. Constraints (22) and (23) represent sign restrictions and sets of definitions.

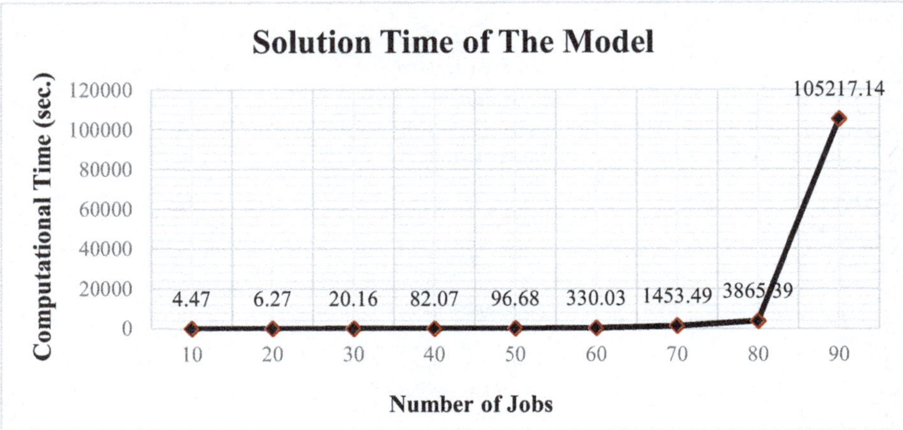

Fig. 1. Solution time of the model

The mathematical model is solved using IBM ILOG CPLEX Optimization Studio with CPLEX 12.8 run on a Core i7, 2.50 GHz, 12.0 GB RAM PC. The real-life order list containing information of jobs and delivery times are provided by the company for 2020. For each job, the volume, weight, frozen period, and ready time data is provided, as well as capacity and transportation times of the trucks. When the mathematical model is solved for 80 or more jobs, solution time to get optimal solution is significantly increased (Fig. 1). According to the obtained results, optimal solution cannot be found for the job 80 and 90 jobs after 1 h of runtime. The 90-order case is even run for over 29 h, without even obtaining a feasible solution. The average number of daily order numbers range from 75 to 90. Based on these results, the mathematical model cannot find the optimal solution within a practically acceptable time limit (less than 1 h). This is to be expected since the studied problem has already been proven as NP-hard [9]. For this reason, we also provide a novel rule-based heuristic algorithm developed to get good solutions for larger size instances.

4 Heuristic Algorithm

To the best of the authors' knowledge, there is no study in the literature for the studied problem. Since our problem is a two-stage scheduling problem, we propose separate heuristic algorithms for the first and second stages. These two heuristics are presented in Figs. 2 and 3 respectively.

To summarize, in the first stage, jobs are added candidate list and sorted the jobs based on the FIFO. Then jobs are assigned to batches considering the volumes and calculated the current batch sum. After that, jobs are assigned to trucks according to truck capacity (Current batch sum $\leq$ Truck capacity). Trucks are selected according to the idle-busy state. This status is found with updating first idle time of the selected truck by adding round trip transportation time to the shipping starting time.

Heuristic Algorithm of Stage 1	
Step 1.	Add the jobs to the candidate list sort by FIFO, Current Batch Sum = 0, Reset batch list
Step 2.	Take the job from the top of list as selected job and Current Batch Sum = Current Batch Sum + Volume of the selected job
Step 3.	If the current batch sum ≤ Truck capacity and move the selected job to the lot list, then go to Step 2. Else if, go to Step 4.
Step 4.	Create batch and calculate the ready time of batch according to largest ready time from the batch list
Step 5.	Find the first idle time of the trucks and assign the batch to the idle truck, and set shipping starting time of the batch
Step 6.	Update first idle time of the selected truck by adding round trip transportation time to the shipping starting time
Step 7.	If the candidate job list is not empty, Current batch sum = 0, If empty, finish Stage 1, then go to Stage 2.

Fig. 2. Heuristic algorithm of stage 1

Heuristic Algorithm Stage 2	
Step 1.	Add unassigned jobs to the job list with Arrival Time ≤ TNOW and Frozen Time ≤ TNOW
Step 2.	Sort the jobs into an Earliest Due Date list break the ties by a smallest changing frequency
Step 3.	Select the job from the top of the list as current job
Step 4.	Find the earliest finishing time for a current job on all machines by adding the processing time to the machines first idle time, and assign the current job to the machine with a minimum earliest finishing time, then update the first idle time of the machine and calculate lateness of the current job
Step 5.	If lateness of the current job is positive, then go to Step 8. If not, calculate your earliness and lateness of current job and remove it from the list
Step 6.	If the list is empty, check if there are any unassigned jobs, then finish Stage 2
Step 7.	If the list is not empty, update TNOW according to the smallest starting of the frozen period among the jobs which are not assigned any machines, then go to Step 1
Step 8.	Find the latest idle time that current job fits on for all machine and calculate latest finishing time before TNOW, check if there is idle time for the current job
Step 9.	If there is idle time for the current job, assign the current job the machine with the latest finishing time and calculate the starting time of current job
Step 10.	If there is no idle time for the current job, calculate the earliness and lateness of the current job, remove it from the list and check whether the list is full or empty then go to Step 6
Step 11.	If Starting Time of current job ≥ Arrival Time, set starting time and completion time of the current job, update the status as of the machine in which the current job is assigned firstly, as idle and update latest available idle time as busy of the new selected machine
Step 12.	If Starting Time of current job < Arrival Time, calculate the earliness and lateness of the current job, and remove it from the list and check whether the list is empty or not, then go to Step 6

Fig. 3. Heuristic algorithm of stage 2

In the second stage, all jobs that have not been assigned to any machine are first listed. These jobs are added to the candidate job list created for this stage. The arrival times and start of the frozen time of the jobs in this list are less than or equal to TNOW. The jobs in the list are sorted according to EDD list and ties are broken by smallest changing frequency (calculated from past data as a sign of how much a specific customer of an order requests a change in packaging). By adding processing time to the initial idle time of the machines, the earliest finishing time for the current job is found for all machines. Then, the current job is assigned to the machine with the minimum earliest finishing time. The first idle time of the selected machine is updated. The lateness of the current job is calculated. If the lateness of the current job is positive, check if there is a long enough idle time in the machines before the frozen start time of this job. If there are machines that current job fits on, the earliness of the current job is calculated for these machines. Then, the machine which has the smallest earliness is selected. If there is no long enough idle time that the current job fits on, the job is late.

5 Computational Results

The developed heuristic method is coded in Microsoft Visual Basic for Applications (VBA) software and all instances are solved by heuristic algorithm within few minutes. The solution methodologies are applied for 4 trucks and 4 machines and the solution time of the heuristic method and the mathematical model are compared. The results from both methods are shown in Table 1. The entry '–' in the table means that the mathematical model was not able to find the optimal solution within 1 h. The numbers given in bold represents the better results.

Table 1. Comparison of heuristic method and mathematical model

Number of jobs	Truck utilization for MIP solution	Truck utilization for heuristic solution	Solution time (sec.) for MIP solution	Solution time (sec.) for heuristic solution	Objective function for MIP solution	Objective function for heuristic solution	Deviation for objective function
10	62%	**71.86%**	4.47	**1.91**	0	0	0%
20	**97%**	85.17%	6.27	**2.05**	0	0	0%
30	78%	**81.66%**	20.16	**2.96**	0	0	0%
40	**88%**	87.95%	82.07	**4.19**	0	0	0%
50	**88%**	88.60%	96.68	**6.03**	0	0	0%
60	86%	**88.06%**	330.03	**8.47**	0	0	0%
70	87%	**91.35%**	1453.49	**10.59**	0	0	0%
80	–	88.61%	3865.39	**13.17**	0	0	0%
90	–	87.03%	105217.14	**14.69**	0	0	0%
Average	84%	86%	12341.74	7.12	0	0	0%

Objective function does not include truck utilizations since truck occupancy is not an objective of the problem. For these reasons, the best solution regarding truck utilization varies in both methods. When we compare the solution time of both methods, it is observed that the average solution time of the heuristic method is 7.12 s compared to 12341 s for MIP. For the objective function, both the heuristic solution method and the MIP solution method have managed to find the best solution for weighted earliness, for real-life job set used. Therefore, heuristic method can be used instead of the MIP solution since it gives a solution in short time. Finally, the solution methods are embedded to a user-friendly decision support system to be used by the company in day-to-day operations.

6 Decision Support System

The purpose of the decision support system (DSS) is to be able to make fast and effective packaging plans and to provide easy to use for everyone. The decision support system is created in MS VBA (Fig. 4).

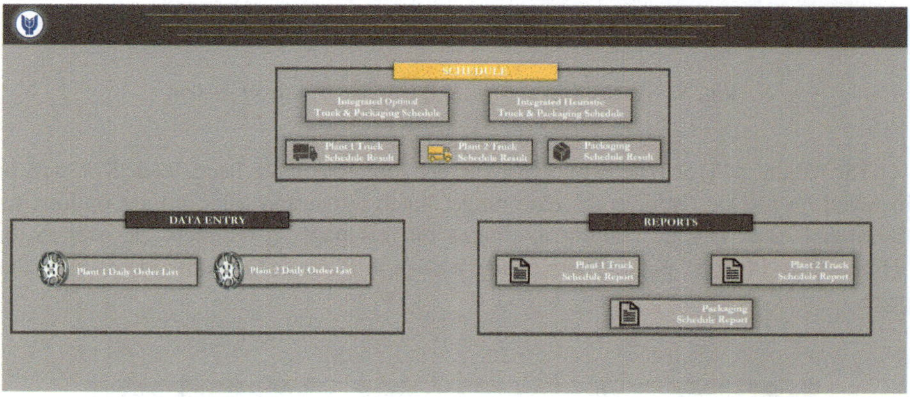

Fig. 4. Home page of the decision support system

On the home page, there are three sections. First, information is entered in the "Data Entry" section. The incoming information is taken from ERP. Secondly, in the "Schedule" section, "Integrated Optimal Truck & Packaging Schedule" and "Integrated Heuristic Truck & Packaging Schedule" buttons make all the schedules with the desired solution method. As the third part, there is "Reports" section. It enables the user to access direct reports. However, if the user has not scheduled the orders, user cannot access the report section.

If a button is clicked from the "Data Entry" section, it is directed to the home page of that plant. These pages are the identical for Plant 1 and Plant 2 which is shown in Fig. 5. On these pages, job information that will be released from production is given for the intended date. Jobs can be filtered by order code, date range and size range. There are two options for the solution methods, which are mathematical model and heuristic. User can update an existing order or add a new order with the "Update" button. After this process, the heuristic method must be run again.

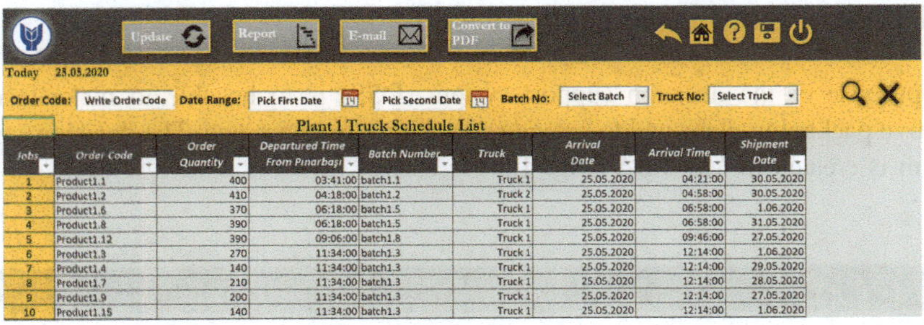

Jobs	Order Code	Order Quantity	Departured Time From Pinarbaşı	Batch Number	Truck	Arrival Date	Arrival Time	Shipment Date
1	Product1.1	400	03:41:00	batch1.1	Truck 1	25.05.2020	04:21:00	30.05.2020
2	Product1.2	410	04:18:00	batch1.2	Truck 2	25.05.2020	04:58:00	30.05.2020
3	Product1.6	370	06:18:00	batch1.5	Truck 1	25.05.2020	06:58:00	1.06.2020
4	Product1.8	390	06:18:00	batch1.5	Truck 1	25.05.2020	06:58:00	31.05.2020
5	Product1.12	390	09:06:00	batch1.8	Truck 1	25.05.2020	09:46:00	27.05.2020
6	Product1.3	270	11:34:00	batch1.3	Truck 1	25.05.2020	12:14:00	1.06.2020
7	Product1.4	140	11:34:00	batch1.3	Truck 1	25.05.2020	12:14:00	29.05.2020
8	Product1.7	210	11:34:00	batch1.3	Truck 1	25.05.2020	12:14:00	28.05.2020
9	Product1.9	200	11:34:00	batch1.3	Truck 1	25.05.2020	12:14:00	27.05.2020
10	Product1.15	140	11:34:00	batch1.3	Truck 1	25.05.2020	12:14:00	1.06.2020

Fig. 5. Truck Plant 1 page of the decision support system

The "Plant 1 Truck Schedule List" is the jobs that have been loaded or will be installed on trucks. Order code, date range, batch range and truck no are options for filtering. If the "Update" button is clicked on this page, it redirects the user to the "Home Page Plant 1" page. There is also a page in Plant 2 that is the identical as this page.

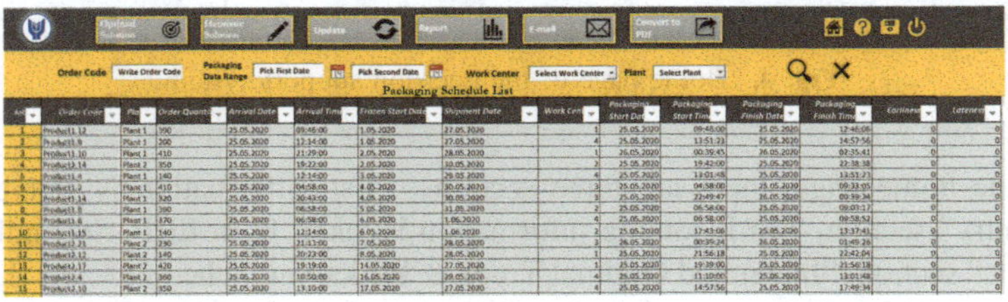

Fig. 6. Packaging scheduling page of the solution method

The products to be packaged are scheduled on the packaging schedules page shown in Fig. 6. Both heuristic and optimal methods are suitable to use. Filtering can be done according to order code, date range, work center and plant feature. In addition, earliness and lateness values are given in the created list.

After all the scheduling is done, the reporting phase is reached. In DSS, there is a report page with the same functions is available for Plant 1 and Plant 2 trucks separately. This page is shown in Fig. 7. On this page, the truck utilization and departure times are reported for trucks used in transportation. It is shown in the Gantt chart for clarity in the report.

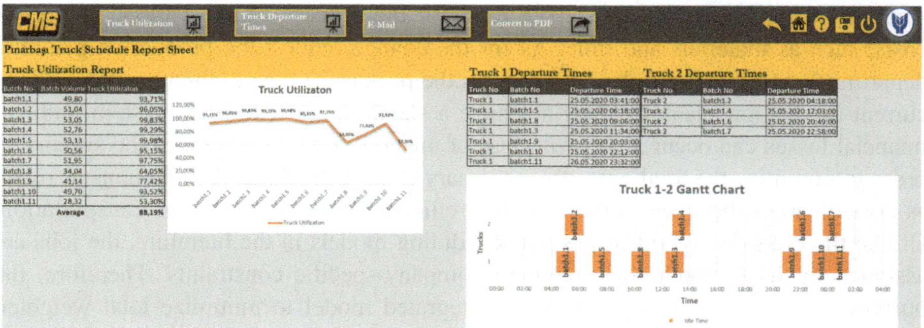

Fig. 7. Report for the Truck Plant 1 page

Figure 8 shows the reports of the scheduling for packaging are created. The generated report is presented to the user both in list and in Gantt chart. In addition, the user can create a report containing only jobs that are earliness or lateness. The user can also filter the report and filtering options are work center and data range. In all these pages created, the user is given the opportunity to write the results lists or reports in PDF format and share them by e-mail. In order to reach the user-friendly decision support system, every action that the user can take has been considered. Warning systems suitable for these actions are developed and the user is redirected with them.

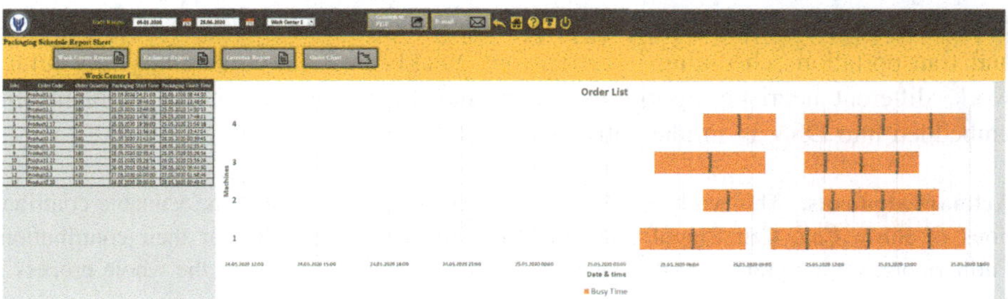

Fig. 8. Report for the logistic center

7 Conclusion and Future Work

In this study, a new solution approach to an integrated transportation and packaging scheduling problem modeled according to a two-stage hybrid flow shop problem is discussed. This problem originated from the leading rim manufacturer in İzmir. However, the problem is valid many different environments and industries. In the current system, packaging processes are carried out at two different plants and the finished products are transferred to the logistics center where they are stored for delivery to the customer. Customers have the right to change the packaging type and quantity of the product in the order until the beginning of the frozen time before the shipment date. The extra labor force used for the process causes various additional costs, such as materials and time. As for the change request the production planning department rearranges the plan and updates the packaging type to the new order. Since current production planning does not have a standard and efficiency system, indirect financial losses can occur along with loss of material, time and labor in the system. In order to minimize the total cost, it is necessary to plan the packing operations as much as possible after the frozen time and more efficient usage of trucks for transportation. Unlike the classical hybrid flow shop scheduling models in the literature, the jobs are assigned trucks in batches and contain company-specific constraints. Therefore, the purpose of this study is to provide an integrated model to minimize total weighted earliness of transportation and packaging scheduling and creating a scheduling tool for the company. This study specifically contributes to the hybrid flow shop scheduling literature by providing real-life application and developing a specific rule-based heuristic. When the model is solved with a real job list, the solution time is beyond acceptable limits for large size instances. Because the problem is NP-hard, a heuristic method is developed to solve the problem in shorter time for larger practical instances. The performance of the heuristic method is compared with the mathematical model. The results obtained with the real-life instances are compared with the solution times and the objective function. Since it has been proven that the solution of the heuristic method gives the same result with the optimal solution for the same job set instances, the heuristic method performed well and achieved good results in a shorter time. In addition, the developed heuristic method and mathematical model solution are integrated into the decision support system. Thus, it has become suitable for use for the company. With the user-friendly interface of the designed decision support system, it is to provide fast and effective packaging schedules. In addition, it ensures that packaging and transportation scheduling is carried out quickly and more efficiently. As future work, different heuristic algorithms can be developed for the studied problem and embedded into DSS as an alternative solution.

Acknowledgments. This work could not have been completed without the valuable contributions of Büşra Çan, Can Bingöl and Hilal Coşğun. We are thankful for their contribution. Furthermore, we are grateful to the company for their co-operation during the whole process.

References

1. Joo, C.M., Kim, B.S.: Rule-based meta-heuristics for integrated scheduling of unrelated parallel machines, batches, and heterogeneous delivery trucks. Appl. Soft Comput. **53**, 457–476 (2017). https://doi.org/10.1016/j.asoc.2016.12.038
2. Peng, T., Zhou, B.: Hybrid bi-objective gray wolf optimization algorithm for a truck scheduling problem in the automotive industry. Appl. Soft Comput. **81**, 105513 (2019). https://doi.org/10.1016/j.asoc.2019.105513
3. Jiang, L., Pei, J., Liu, X., Pardalos, P.M., Yang, Y., Qian, X.: Uniform parallel batch machines scheduling considering transportation using a hybrid DPSO-GA algorithm. Int. J. Adv. Manuf. Technol. **89**(5–8), 1887–1900 (2016). https://doi.org/10.1007/s00170-016-9156-5

4. Wirth, M., Emde, S.: Scheduling trucks on factory premises. Comput. Ind. Eng. **126**, 175–186 (2018). https://doi.org/10.1016/j.cie.2018.09.023
5. Zohali, H., Naderi, B., Mohammadi, M.: The economic lot scheduling problem in limited-buffer flexible flow shops: Mathematical models and a discrete fruit fly algorithm. Appl. Soft Comput. **80**, 904–919 (2019). https://doi.org/10.1016/j.asoc.2019.03.054
6. Sağlam, Ü., Banerjee, A.: Integrated multiproduct batch production and truck shipment scheduling under different shipping policies. Omega **74**, 70–81 (2018). https://doi.org/10.1016/j.omega.2017.01.007
7. Arroyo, J.E.C., Leung, J.Y.-T.: Scheduling unrelated parallel batch processing machines with non-identical job sizes and unequal ready times. Comput. Oper. Res. **78**, 117–128 (2017). https://doi.org/10.1016/j.cor.2016.08.015
8. Birgin, E., Ferreira, J., Ronconi, D.: A filtered beam search method for the m-machine permutation flow shop scheduling problem minimizing the earliness and tardiness penalties and the waiting time of the jobs. Comput. Oper. Res. **114**, 104824 (2020). https://doi.org/10.1016/j.cor.2019.104824
9. Gupta, J.N.D.: Two-stage hybrid flow shop scheduling problem. J. Oper. Res. Soc. **39**(4), 359–364 (1988)
10. Alharkan, I.M.: Algorithms for sequencing and scheduling. Industrial Engineering Department, King Saud University, Riyadh, Saudi Arabia (2005)

Lean Warehouse Application in a Meat Producer Company

Aysu Aksütoğlu, Egemen Bıkmaz(✉), Nazlı Ecem Sırkıntı, Melisa Özbiltekin Pala, and Esra Ekinci

International Logistics Management, Yasar University, Izmir, Turkey
aysuuaksutoglu@gmail.com, egemenbikmaz@gmail.com, nazliecemsirkinti@gmail.com, {melisa.ozbiltekin,esra.ekinci}@yasar.edu.tr

Abstract. This study aims to determine the necessary processes to improve warehouse operations using lean principles while considering important criteria of efficient warehouse management in a meat producer company. In order to identify problem areas, Spaghetti Diagram has been prepared and company warehouse operations have been analyzed. In the company, problems observed in warehouse operations are mainly unnecessary handling, and layout problems in the warehouse. In order to handle the problems, Value Stream Mapping and Multi-Criteria ABC Analysis methods have been employed. With the help of Value Stream Mapping, lean principles can be adopted in the warehouse operations. On the other hand, Multi-criteria ABC analysis can guide the company in product placements in the correct location. At the end of this study, suggestions have been proposed for the company to reduce wastes, improve several processes in the warehouse.

Keywords: Multi criteria ABC analysis · Spaghetti diagram · Value stream mapping · Food industry · Lean warehouse

1 Introduction

Nowadays, with the changing life habits of people, demand for ready-to-eat and processed food has increased. To handle the growing demand, warehousing has become an important enabler of globalized production networks, short lead times and increased customer satisfaction. Therefore, growing demand, globalized networks require efficient warehouse operations; however, operating a warehouse gets complex every day. Warehouses do not work simply as a storage facility and they are responsible for receiving, packing, handling, lashing, etc. operations as well. To handle this complexity, the essence of the lean philosophy can be employed in the elimination of waste and losses, but also in striving for constant improvement [1]. The objective of this study is to eliminate the wastes by using Lean Principles at meat producer company finished food product warehouse. In order to achieve this, as a project team had several meetings with managers of the company. The study focuses on designing value stream mapping, which is one of the lean management tools.

N. M. Durakbasa and M. G. Gençyılmaz (Eds.): ISPR 2020, LNME, pp. 914–929, 2021.
https://doi.org/10.1007/978-3-030-62784-3_76

2 Observations and System

KPIs are needed to keep warehouse management at the best level and to understand how to improve problems. As operations, movements, and KPIs are higher in the finished product warehouse than in other warehouses, the project group will work in this warehouse of meat producer company. List of warehouses names are MAR (Delicatessen Product Warehouse), DMD (Frozen Product Warehouse), GDN (General Freezing Storage (Red Meat), HGD (Turkey General Freezing Storage), İSH (Manufacturing Cold Storage), YMA (Auxiliary Material Warehouse).

Table 1. Total handling tonnage per person

	Data	Months								
	Tonnage per person (tonnage/person)									
Warehouse	JAN	FEB	MAR	APR	MAY	JUNE	JULY	AUG	SEP	OCT
MAR	168	154	183	187	180	150	193	165	187	177
DMD	119	121	132	146	138	89	122	97	112	100
GDN	80	95	124	134	156	111	132	111	172	145
HGD	112	169	153	155	168	111	137	93	110	109
ISH	279	214	261	271	293	230	275	254	270	211
YMA	295	246	294	324	343	214	356	267	297	301
Grand Total	175	162	188	197	205	153	201	169	195	170

Table 1 shows total handling tonnage per person at different warehouses in 10 months. Table 2 represents the total handling tonnage for different warehouses in 10 months. As shown in Table 2, the warehouse with the highest amount of handling after 10 months is defined as MAR warehouse. This indicates that efficient handling in MAR warehouse is critical from company perspective. Therefore, MAR warehouse with the highest amount of total handling has been considered in this study.

Table 2. Total handling tonnage for different warehouses in 10 months

	Handling (tonnage)										Total handling (tonnage/person)	Total handling (tonnage)
Warehouse	JAN	FEB	MAR	APR	MAY	JUNE	JULY	AUG	SEP	OCT		
MAR	3.230	3.037	3.397	3.599	3.406	3.147	3.869	3.638	3.691	3.278	174	34.292
DMD	1.144	1.180	1.244	1.373	1.308	849	1.111	992	999	1.010	117	11.210
GDN	691	837	1.027	1.191	1.392	903	1.116	987	1.484	1.299	126	10.927
HGD	586	996	890	960	937	585	787	571	591	684	132	7.587
ISH	3.427	2.731	3.121	3.289	3.700	2.883	3.984	3.887	3.833	3.197	255	34.050
YMA	1.380	1.176	1.399	1.451	1.590	1.044	1.650	1.401	1.323	1.325	293	13.740
Grand Total	10.458	9.957	11.078	11.863	12.331	9.411	12.517	11.476	11.921	10.793	181	111.807

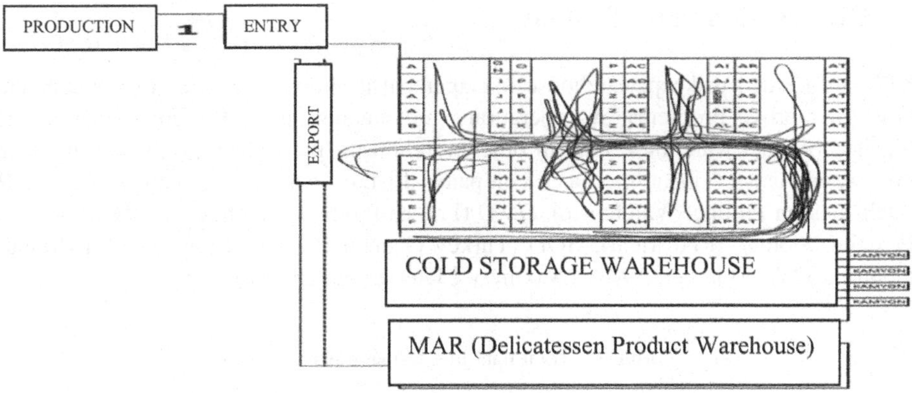

Fig. 1. Spaghetti diagram

A spaghetti diagram is a visual representation of the path of an element or activity during a process using a continuous flow line [2]. The aim of using the spaghetti diagram is to compare the theoretically obtained data with the consistency of the real process. Figure 1 shows the employee movements observed in the warehouse for 6 h.

3 Problem Definition

In this study, the purpose is reducing the waste in the finished product warehouse of the company and simplifying the processes from production to transportation by applying the lean philosophy to warehouse management. The study has been carried out in order to investigate the technological investments of the product warehouse, to increase the efficiency of the handling operation, to reduce the wastes in the warehouse and to improve the product placement within the warehouse.

As a result of the visits to Pınar Et warehouse, two main headlines did not comply with the principle of lean warehouse have been observed.

1. Uncertainty in Warehouse Processes
   - Unnecessary transportation of pallets during picking process
   - Lack of stretching area

     The impacts of these problems are;

     – Increased picking and process times.

     – Occupancy of aisles that blocks entry and exit.
2. Inefficient usage of racks

Another important problem in warehouse was specified as inefficient usage of rack. The impacts of these problems are;

- Incompatibility between product placements and order frequencies.
- Increased product handling and forklift usage.

4 Literature Review

In this part of the study, a literature review is related with the Lean Management and tools, Lean Management in warehouses, Lean Management in the food industry and Value Stream Mapping.

4.1 Lean Management

Lean production aims to produce with minimum waste by using less energy and less raw materials. In order to reduce waste in lean production, all elements that do not provide added value should be separated from production and eliminated. Lean systems need permanent improvements to increase efficiency and reduce waste [3].

The following Table 3 contains other common methods used for Lean warehouse management.

Table 3. Literature table

Author	Sector	Method	Aim
[4] Garcia & P.E. (2004)	Logistics	Value stream mapping	Elimination of waste in the warehouse, Improvement of warehouse activities
[5] Dharmapriya & Kulatunga (2011)	Manufacturing	VSM, Simulated Annealing (SA)	Reduction of travel distance and cost in warehouse
[6] Chen et al. (2013)	Production	VSM, RFID	Identify inefficient operations and eliminate wastes. Operation processes were reengineered
[7] Abushaikha et al. (2013)	Logistics	Delphi method	Reduction of waste. Optimization of operational and distribution performance
[8] Tran (2016)	Logistics	VSM, 5S	Elimination of wastes, reducing time and cost
[9] Dotoli, Petruzzelli & Turchiano (2012)	Manufacturing	ABC, VSM	Optimization of warehouses using lean production tools
[10] Mossen & Hassan (2002)	Manufacturing	Warehouse layout	Streamline the design process, simplify layout tasks
[11] Roodbergen & Vis (2014)	Manufacturing	ABC Analysis	Warehouse layout to organize the design process
[12] Tippayawong et al. (2013)	Manufacturing	ABC Analysis	Improving inventory management
[13] Christiansen (2015)	Engineering	5S, VSM	Lean concepts in the real warehouse
[14] Shah & Khanzode (2017)	Logistics	Data collection and analysis	Use technology resources in warehouse
[15] Putpitasari & Ardila (2016)	Manufacturing	5S, Observation and Research method	Elimination of errors from operational process
[16] Liu (2012)	Retailer	Quantitative analysis methods	Determining the warehouse needs that will arise with the growth
[17] Reis (2017)	Manufacturing	Data collection and Analysis	Increase to efficiency in company of oil and gas industry using lean method

(continued)

Table 3. (*continued*)

Author	Sector	Method	Aim
[18] Phogat (2013)	Logistics	VSM, ABC, 5S	Definition of Lean Methodology & Lean Warehouse
[19] Atieh (2016)	Manufacturing	Business Process Modelling and Notation (BPMN), ERP	Impact of warehouse management system on supply chain
[20] Bartholomew (2008)	Logistics	VSM, 5S	Applying lean management to a real warehouse
[21] Anđelković (2016)	Logistics	Data collection analysis	Indicate the importance and effects of the lean warehousing
[22] Acharya (2011)	Logistics	5S, Kanban	Reduce the overall handling time
[23] Oey & Nofrimurti (2018)	Logistics	5S, VSM	Using lean warehousing techniques on the traditional warehouse
[24] Hajmohammad (2013)	Green logistics	Data collection analysis	Sustainable approach to the lean warehousing
[25] Gopakumar (2008)	Manufacturing	Simulation	Elimination of wastes

The value stream mapping is a lean management tool that visualized the whole process in the operations. In warehouse, VSM is a useful tool in order to clarify where to start the implementation of lean concepts and also to understand flows of warehouse operations [4]. Reference [4] demonstrates what results achieved by using step by step lean techniques such as Value Stream Map to a warehouse. With using lean concept, they get results like a decrease in warehouse processing time and lead time, and an increase in stock and order processing accuracy. Reference [6] have used lean management to eliminate inefficient and ineffective operations and wastes in warehouse, at the same time they used the RFID (Radio Frequency Identification) system as an example of lean management to restructure and organize warehouse's operation processes. Both reference [5] and [6] benefit from VSM to understand where to use their main methods. Reference [8] aimed to improve problems related to warehouse such layout design, storage location, picking method to optimize the performance by using Value Stream Map, 5S, questionnaires and interviews with warehouse personnel. Delphi method is used to make predictions and is a decision making and reconciliation technique developed by Rand Corporation, which develops scenario analysis. Delphi technique is a method that allows experts to decide and agree on a topic without face-to-face meetings and discussions because of its benefits, reference [7] use Delphi method to do waste reduction in warehouse and can see that if the level of waste in warehouse decreases, the operational performance and distribution performance of it gradually increases.

4.2 Value Stream Mapping

The value stream map visualizes information and materials flows, ensuring that these flows accurately reflect the truth throughout the process [26]. According to reference [27] the value flow map is a tool ensuring knowledge and equipment flows and supporting communication. The value-stream of map depends on flow permanently, so ensures participation, transparency and decreased diversity [28]. The first step in testing and applying other techniques for reducing wastes is to create a value stream map. In this way, activities that do not add value to the process are eliminated [29].

Current-State Map: According to reference [30] The current state map shows how the operations of the firm implement in today's work environment. In order to create map of the current situation, employees should monitor the flow and collect data and information from them. Map shows how each station should make a strategy in business planning [31].

Future State Map: Reference [32] thinks that the future flow map can be identified and developed on the map, which is suitable for improvement in the current situation map, and that this map can be supported by lean manufacturing equipment. These equipments are balance the line, rule out bottlenecks, shrink and buffer diversity, eliminate redundant and unproductive approvals.

5 Methodology

5.1 Proposed Methodology: ABC Analysis

ABC analysis is one of the techniques used to give priority and importance to inventory management. The expert of warehouse uses ABC analysis to optimize the management of warehouse stocks and focuses on the factors that generate the greatest value-added for storage, according to the results of the analysis also the method used to reduce wastes and costs in the warehouse [33]. ABC analysis consists of three classes of stock-keeping units, regardless of scope, according to a specific criterion, in descending order of importance for the company's business; A, B and C [9].

- *Class A* goods make up only 20% of the total amount, while 80% are have sales value;
- *Class B* goods have a share of 20% to 30% of the total amount and 15% to 20% of the sales value;
- *Class C* goods represent between 50% and 60% of the total in quantity and have only a small value of between 5% and 10% as sales value.

According to this criterion, A items is are both high-value and high-demand, and C items are low-value and low-demand. But in years, it was decided that may not be enough to look at only one criterion, and other criteria have been suggesting that may be important; lead-time, stock ability, durability, the availability of substitution, the number of suppliers of transport mode and cost, and the suppliers the possibility of deterioration or obsolescence, the cumulative amount of applied by suppliers.

Multi-criteria ABC analysis was developed both to consider such criteria and to improve the accuracy of the analysis.

Multi-criteria ABC analysis has been adapted from [34] has been used by applying the steps in Fig. 2. Firstly, AHP is used in order to weigh criteria and then DEA model applied to evaluate the criticality of the product.

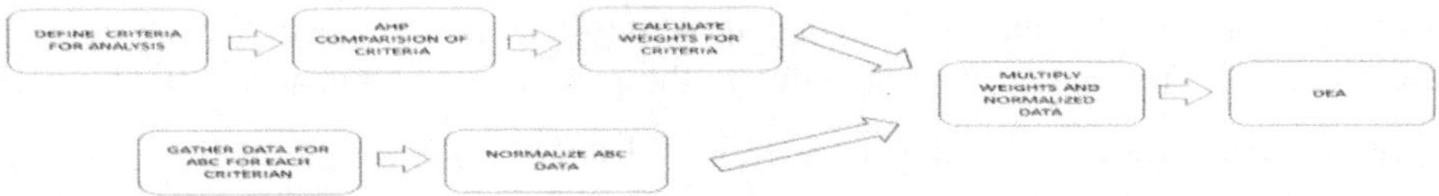

Fig. 2. ABC analysis steps

For the AHP application, a questionnaire was conducted with the experts in the company. Survey participants are expert people in their fields; warehouse chief, supply chain director, production planning manager. Based on the information and data obtained from the firm following steps have been applied.

Step 1 & Step 2: Define Criteria for Analysis and Prepare Comparison Matrix: Based on interviews with the experts, seven criteria has been determined as important factors in warehouse management. These factors affect the efficiency of warehouse operations, customer service level, sales revenue, etc. According to expert opinion, comparison matrix in Table 4 has been prepared.

Table 4. AHP matrix

Criteria	Case per pallet	Warehouse capacity per item	Criticality/Profit margin	Expiry period	Shipped cases	Number of shipments	Number of customers
Case per pallet	1	2	2	0,11111111	0,2	0,3333333	2
Warehouse capacity per item	0,5	1	3	0,14285714	0,333333	0,3333333	0,5
Criticality/Profit margin	0,5	0,33333333	1	0,16666667	0,2	0,2	0,33333333
Expiry period	9	7	6	1	3	3	3
Shipped cases	5	3	5	0,33333333	1	2	2
Number of shipments	3	3	5	0,33333333	0,5	1	0,50
Number of customers	0,5	2	3	0,33333333	0,5	2	1
SUM	19,5	18,3333333	25	2,42063492	5,733333	8,8666667	9,33333333

Step 3: Calculate Weights for Criteria: As shown in Table 5, using the comparison matrix in Table 4 weight of each criterion had been calculated. The analysis is consistent since the consistency ration is 0.08, which is less than 0.10.

Table 5. Weights of criteria

Criteria	Weights
Case per pallet	0,081862574
Warehouse capacity per item	0,05835826
Criticality/Profit margin	0,035118529
Expiry period	0,38278595
Shipped cases	0,196002824
Number of shipments	0,129821446
Number of customers	0,116050417
Total weights	**1**

Step 4 & 5: Gather Data for Each Criterion and Normalize Data: In order to continue with the DEA model, data has been gathered for 77 items that has been stored in finished product warehouse for each criterion. After data collection, data has been normalized since each data is based on different units. Since the importance of inventory increases as expiry date becomes shorter, while normalizing the expiry date, reciprocal of the number is used.

Step 6 & 7: Multiply The Criterion Weight with The Normalized Data & Apply DEA Model: After multiplying weights with normalized data, DEA model provided below has been applied [35].

- $\max \sum_{j=1}^{J} v_{mj} y_{mj}$
- $\sum_{j=1}^{J} v_{mj} y_{nj} \leq 1 \qquad n = 1, 2, \ldots, N$
- $v_{mj} \geq 0 \quad j = 1, 2, \ldots, J$

where

N: number of inventory items to the classified as A, B and C according to their performance.
J: number of criteria used in classification
y_m: performance of *m*th item for criterion j
v_m: decision variable (opportunity cost) for *m*th item for criterion j.

Model has been adapted from [35] provides the optimal inventory score for mth item. In our model, all criteria are positively related with the importance level of an item. The model simplifies the original DEA model with the assumption of many outputs and a constant input. Based on the model products are divided into A, B and C classes. However, it has been realized that in A category, there could be items with high profit or low expiry date but shipped rarely. On the other hand, some items are in A class due to their high shipment rates. Therefore, in order to make the warehouse layout of the products more prominent, all three groups were again divided into three subclasses as A, A+A−,/; B,B+,B−; C,C+,C−. If the number of shipments in one month is more than 500, item is subcategorized as A+, B+ or C+ in its category. If the shipment is less than 200 than item is subcategorized as A−, B− and C−. The resulting classes can be seen in Table 6. In Table 7, some example items from the analysis have been provided.

Table 6. Classes formed based on multi-criteria ABC analysis

Classes	Number of shipped cases	Number of ıtems	Percentage shipped cases
A	**546.282**	**30**	67%
A+	424.714	12	52%
A	97.668	12	12%
A−	23.900	6	3%
B	**206.156**	**28**	25%
B+	75.735	10	9%
B	68.979	13	8%
B−	61.442	5	8%
C	**60.703**	**19**	7%
C	57.768	14	7%
C−	2.935	5	0%
TOTAL	**813.141**	**77**	100%

Table 7. Some examples from multi criteria ABC analysis

Products	Case per pallet	Warehouse capacity per item	Criticality/Profit margin (1–5)	Expiry period	Shipped cases	Number of shipments	Number of customers	Class	Sub-class
AÇ BITIR HINDI FISTIKLI SALAMI 50 g	264	432	5	67	18.533	581	73	**A**	**A+**
AÇ BITIR HINDI SALAMI 70 g (İM) PRF	204	432	5	90	60.223	511	72	**A**	**A+**
GURME HINDI SMOKED 110 g	143	432	5	75	1.578	310	45	**A**	**A**
AÇ-BITIR JAMBON 50 g	224	432	5	67	4.375	463	66	**A**	**A**
SMOKED TURKEY	88	432	5	121	163	26	4	**A**	**A−**
AÇ BITIR IİG SOUDJOUK 90 g (İM)	252	432	5	121	4.580	91	45	**A**	**A−**
DOYUM HINDI SALAMI 225 g	198	297	4	101	6.577	527	101	**B**	**B+**
GURME FERMENTE SOUDJOUK 250 g	143	405	3	183	3.975	540	104	**B**	**B+**
DOYUM KANATLI IİG SOUDJOUK (FE) 225 g	108	405	3	183	6.673	406	75	**B**	**B**
DOYUM KANATLI IİG SOUDJOUK 225 g	108	405	3	183	3.408	393	79	**B**	**B**
MANGAL KEYFİ HINDI IİG SOUDJOUK 2000 g	36	405	3	150	180	63	18	**B**	**B−**
İLLAKİ FERMENTE SOUDJOUK 250 g	108	405	3	183	511	106	37	**B**	**B−**
TURKEY SALAMI 1250 g	63	297	4	121	1.981	241	30	**C**	**C**
LEZZET KEYFİ HINDI SALAMI 500 g	209	297	4	101	18.791	387	47	**C**	**C**
HINDI FISTIKLI SALAMI 1250 g	63	297	4	121	862	191	26	**C**	**C−**
PINAR SERVİS CATERING HINDI SALAMI 900 G	110	297	4	100	420	57	18	**C**	**C−**

5.2 Current State Map

Pınar Et finished product warehouse is 762 square meters in total. There is one middle corridor and 8 horizontal short corridors in the warehouse. The current pallet capacity of the warehouse is 1413. There are a variety of shelves adhered to the wall called a single depth rack. In the middle corridors, there is a double depth rack system similar to the back to back system. There are 5 compartment pallet racks in each corridor. Each shelf has a capacity of 3 pallets and the length of the shelves extends to the ceiling and has 6 floors (Fig. 3).

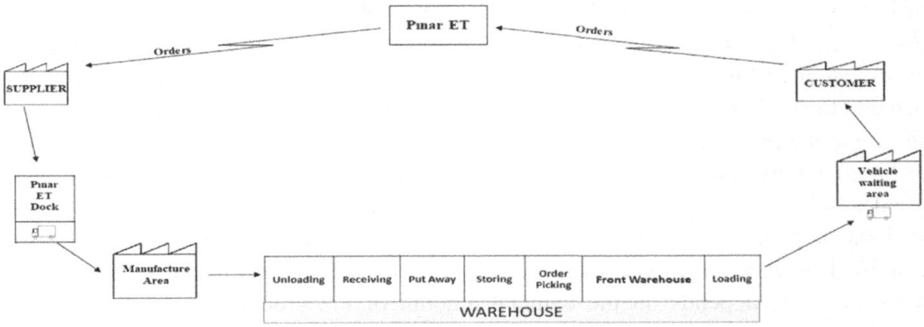

Fig. 3. Current state map

Pınar Et is a company that produces meat and meat products. The company supplies many raw materials for production, for example; such as preservative food, spices, bovine and ovine, oil. The company orders the necessary materials from the supplier before starting to production. The orders placed by the company are delivered to Pınar Et by the number, weight, number and delivery time by various suppliers. These materials put to the dock area, owned by Pınar Et, transfer to Pınar Et's the production area. Pınar Et starts production within the company after reaching the necessary raw materials and supply. Many meat products are produced within the company, and these products are put into various self-packaging. Packaged products are lined up on pallets by boxing according to their types. The prepared pallets are lined up in front of the production facility and transferred from this area to the storage area.

Pallets from the production area are unloaded in front of the warehouse by forklift and trans pallet equipment. After the various product groups are listed in front of the warehouse, they move into the warehouse by the warehouse officers. During the receiving operation, ordered products arrive and unload to the warehouse. To do this, several procedures take place, such as layout and design, schedule and yard control. Since the people who will receive the goods do not know what is in the in boxes, first they consult the sales department and check whether the incoming goods are damaged. There is a receiving dock in front of the warehouse. The products received are identified and accepted into the warehouse. Identification is usually done by barcode scanners. Handling done with pallets and pallets are placed on shelves without degradation. Put away process is an important step in increasing warehouse efficiency.

As a result of placing the goods correctly in the warehouse, it takes less time to collect and pack them. Otherwise, if this process is extended, the travel time will be longer. The right storage of the load is important for the safety and integrity of the load. At the same time, it does not risk for the safety of the employee. In Pınar Et warehouses, in the process of put-away, products are placed on pallet racks together with the forklift. In the Put-away process, each product palette has its own area. For example, the cut-and-finish products produced by Pınar Et are positioned in a shelf row close to the warehouse exit. In some cases, the shelf group that belongs to this product group is full, and the pallet from production is placed on any empty shelf.

The storing is the operation where the goods are placed at a particular location before the pickers come to pick the goods. The problem with storage is being able to get the right amount at home but not too much, because this creates inflation and excessive investment in resources. But it is also in this product placement that it benefits the pickers after the storage and the following processes by more personalized storing to the particular items [35]. The product storage policy in Pınar Meat warehouse is based on the idea of class-based storage. Class-based storage (CBS) optionally divides stock keeping units (SKUs) into storage classes by demand. CBS provides savings in picker travel and increases warehouse efficiency [36]. The products are classified in warehouse with different pallet groups. The space capacity allocated to these products depends on the demand amount of the product. For example, Pınar Sucuk products are located in the 2nd corridor near the warehouse exit. The reason for this is that the second most demanded product in the company is sausage and the amount of sales is high.

Picking is the process where a product is transported from storage to packaging and the items being selected are connected to an individual order when this happens. Picking is therefore a very critical process where choosing the right items is very critical during the period that has been allocated to the choosing task [35]. Picking process starts instantly at Pınar Et when the order arrives. For this process to happen, an order comes first and then products are prepared in the warehouse according to the order list. In the order list, it is stated which product will be how many. Order pallets are prepared by employees using the trans pallet in the warehouse. Pallets prepared according to the order list are stretched in any area of the warehouse. These pallets are lined up in the front warehouse before the logistics activity and are ready to be transported. The front warehouse is the area where the products last before loading. The pallets prepared for the order are lined up according to the customers and are kept here until the logistics activity. Loading activity is carried out when the appropriate logistic vehicle is available for the product. The products are loaded to the vehicles from the front warehouse via forklift equipment. Then, product pallets loaded on the vehicle are delivered to customers.

5.3 Future State Map

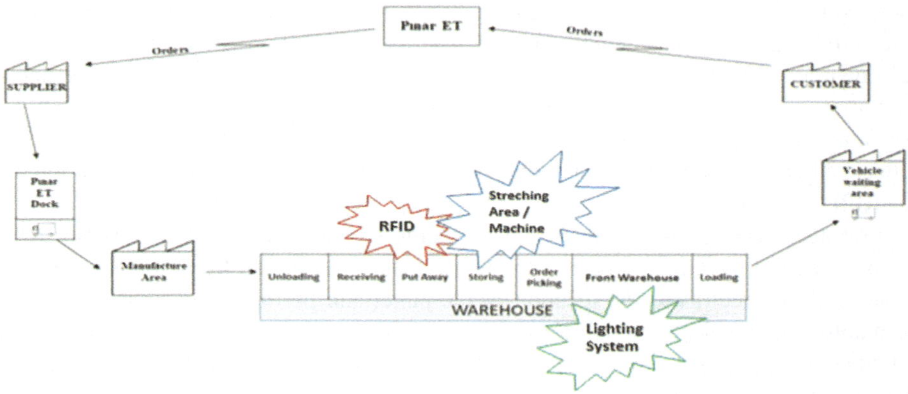

Fig. 4. Future state map

Products are stored on a class - basis in the Pınar Meat warehouse. However, as a result of the observations, there were problems in storage. First, the product store areas are not certain. For example, Pınar sausage and Hungarian salami can be stored side by side at some times. Another problem observed is that some pallets are stored in the corridor and on the ground. Therefore, this form of storage makes access to another product make difficult and the process becomes time consuming. Finally, placing too much demanded products close to the exit, such as Aç-Bitir, Pınar sausage, may cause the layout of the storage system to deteriorate. As the amount of these products increases, confusion will occur in the warehouse. These days tighter inventory management, faster response time, and a larger range of SKUs (stock-keeping units) are the most critical challenges for modern warehousing design. These difficulties arise from the fact that warehouse inventory management adopts understanding lean philosophy and just in time. With the application of IT skills, there was happened potential for recovery [37]. These IT skills are the application of smart tags such as barcodes, radio frequency identification (RFID) tags and automatic identification (Auto-ID) sensors, along with cable and/or wireless networks and integrated warehouse management systems (iWMS). A barcode contains some data of the manufacturer and the product. Therefore, it scanned by a person and read with a sensor or barcode scanner. But this process will be time consuming as it will be manual. The use of RFID-related technologies in warehouses has had a greater impact. Recently, RFID technology has become more widespread in terms of stock control and management in the logistics industry [38]. Thanks to the use of RFID, inventory can be tracked more accurately on time and more efficiently, leading to a reduction in processing time and labor. The accuracy and full visibility of inventory data creates opportunities for improvement and transformation throughout the entire supply chain processes [39]. An RFID tag can store many digits and can be read with remote recognition. It can automatically read thousands of tags per second. Each item added with the RFID tag can be tracked, sorted, stocked, guided, collected and distributed from anywhere in the system. An RFID tag can give the unique ID of a pallet in the warehouse, its location in the warehouse such as department, row, shelf, and information about the product through defined signals. As a result, in an RFID-based warehouse management system (WMS), the warehouse process is better managed. The tracking and processing of the products can be done in real time and with great accuracy, leading to significant improvements. It removes the complexity inside the warehouse. It reduces workmanship. It gives accurate information about what inventories are and where they are stored. Since it affects all processes positively, the observed problems are eliminating thanks to the use of RFID (Fig. 4).

Another problem is that there is no specific stretch area in the warehouse. After completing the palletizing process, the worker takes the stretch from at the warehouse exit point and returns to the location of the pallet and performs the stretch operation manually. Also, since there is no specific area, the worker makes this operation in the middle of the corridor where he puts the final product on the pallet. This causes a palletization time of a checklist to be extended Another problem is whether manual stretching is durable. Indirectly, it will be less tight and less strong than human-made stretching wrapping machines so that losses space in the warehouse will increase. To eliminate these problems, a stretch area close to the outlet should be built in the warehouse and an automatic pallet stretching machine should be provided. Thanks to the stretch machine, the problem of stacking the products after wrapping will be reduced. Furthermore, products with risk of breakage will gain durability in this way. Thanks to the machines, manpower will not be needed, so the defects will be reduced and the time will be shorter. The advantage of wrapping machines is to save on transportation costs. For example, if two trucks will be needed for transportation, one truck will be sufficient since the area covered by the products is reduced. Today, the prices of pallet wrappers machines are not so much preferred because they are expensive for most companies but, thanks to stretch wrapping machines, the price of the machines can be easily recovered with the done savings.

Another crucial problem we see while examining the warehouse is that it is experienced in the region where the finished products are stored to be loaded on the vehicles. Packaged products are placed randomly in an area with no specific order. For this reason, there are some delays during the placement of new products. As a solution, we thought of integrating a system that we frequently use in daily life into that area. This system is a green and red-light system with sensors, which is located in the parking lot of shopping malls, which allows us to see the empty spaces. By adopting this system to this confusing area, it is aimed to place the new packaged products brought by the employees in empty spaces without wasting time.

6 Conclusion

In summary, firstly, lean idea researches were conducted in food industry warehouses and methods of improvement in warehouses were examined. Afterward, the performance criteria that will be used in the idea of leaning were determined in Pınar Et warehouses and studies were conducted on these criteria. Afterward, the warehouses owned by Pınar Et were examined and the warehouse with the highest handling rate was selected and data were collected from this warehouse. Then the problems obtained as a result of observations in the selected warehouse were listed. For these problems, various articles were searched and information about problem solutions was collected. Then, the collected data and the selected lean methods were analyzed. As a result of the analyzes, current state map and future state maps were created. There were problems observed in the current state map. Ideas were provided to eliminate problems in the future state map. Three different solutions determined for more efficient use of the warehouse. These were in the direction of using the RFID system, supplying stretching machine, and lastly creating a lighting system in the warehouse where finished products are stored for loading on vehicles. Thus, based on researches, observations and analysis, theoretical improvements were made in the Pınar Et's finished product warehouse.

References

1. Abdul Wahab, A.N., Mukhtar, M., Sulaiman, R.: A conceptual model of lean manufacturing dimensions. Procedia Technol. **11**, 1292–1298 (2013)
2. Bialek, R.G., Duffy, G.L., Moran, J.W.: The Public Health Quality Improvement Handbook. ASQ Quality Press, Milwaukee (2009)
3. Arslandere, M.: Yalın Üretim ve Yalın Üretime Geçiş Çalışmaları: Büyük Ölçekli Gıda Firmalarında Uygulamalar. Eğitim Yayınevi (2018)
4. Garcia, F.C.: Applying lean concepts in a warehouse operation. In: IIE Annual Conference and Exhibition, no. 1, pp. 2819–2859 (2004)
5. Dharmapriya, U.S.S., Kulatunga, A.K.: New strategy for warehouse optimization–lean warehousing. In: Proceedings of the 2011 International Conference on Industrial Engineering and Operations Management, pp. 513–519, January 2011
6. Chen, J.C., Cheng, C.H., Huang, P.B., Wang, K.J., Huang, C.J., Ting, T.C.: Warehouse management with lean and RFID application: a case study. Int. J. Adv. Manuf. Technol. **69** (1-4), 531–542 (2013)
7. Abushaikha, I., Salhieh, L., Towers, N.: Improving distribution and business performance through lean warehousing. Int. J. Retail Distrib. Manag. **46**(8), 780–800 (2018)
8. Tran, H.: Improving Warehousing process with Lean Management (2016)
9. Dotoli, M., Petruzzelli, G., Turchiano, B.: A Lean Warehousing Approach Using Unified Modelling Language and Value Stream Mapping: A Case Study, June 2012
10. Mohsen, Hassan, M.D.: A framework for the design of warehouse layout. Facilities **20** (13/14), 432–440 (2002)
11. Roodbergen, K.J., Vis, I.F., Taylor Jr., G.D.: Simultaneous determination of warehouse layout and control policies. Int. J. Prod. Res. **53**(11), 3306–3326 (2015)
12. Tippayawong, K.Y., Sopadang, A., Patitad, P.: Improving warehouse layout design of a chicken slaughterhouse using combined ABC class based and optimized allocation techniques. In: Proceedings of the World Congress on Engineering, vol. 1, November 2013
13. Christiansen, H.: Effective warehouse management using Lean and Six Sigma. Master's thesis, University of Stavanger, Norway (2015)
14. Shah, B., Khanzode, V.: A comprehensive review of warehouse operational issues. Int. J. Logist. Syst. Manag. **26**(3), 346–378 (2017)

15. Puspitasari, N.B., Ardila, A.Y.: Implementation of lean warehouse to minimize wastes in finished goods warehouse of PT Charoen Pokphand Indonesia Semarang. ComTech Comput. Math. Eng. Appl. **7**(1), 1–9 (2016)
16. Liu, H.: Warehouse Redesign of Facility Layout for Wegmans Food Market (2012)
17. Reis, A., Stender, G., Maruyama, U.: Internal logistics management: Brazilian warehouse best practices based on lean methodology. Int. J. Logist. Syst. Manag. **26**(3), 329–345 (2017)
18. Phogat, S.: An introduction to applicability of lean in warehousing. Int. J. Latest Res. Sci. Technol. **2**(5), 105–109 (2013)
19. Atieh, A.M., Kaylani, H., Al-abdallat, Y., Qaderi, A., Ghoul, L., Jaradat, L., Hdairis, I.: Performance improvement of inventory management system processes by an automated warehouse management system. Procedia CIRP **41**, 568–572 (2016)
20. Bartholomew, D.: Putting lean principles in the warehouse. Case article. Cambridge, Lean Enterprise Institute (2008)
21. Anđelković, A., Radosavljević, M., Stošić, D.: Effects of lean tools in achieving lean warehousing. Econ. Themes **54**(4), 517–534 (2016)
22. Acharyaa, T.K.: Material handling and process improvement using lean manufacturing principles. Int. J. Ind. Eng. **18**(7), 357–368 (2011)
23. Oey, E., Nofrimurti, M.: Lean implementation in traditional distributor warehouse-a case study in an FMCG company in Indonesia. Int. J. Process Manag. Benchmarking **8**(1), 1–15 (2018)
24. Hajmohammad, S., Vachon, S., Klassen, R.D., Gavronski, I.: Lean management and supply management: their role in green practices and performance. J. Clean. Prod. **39**, 312–320 (2013)
25. Gopakumar, B., Sundaram, S., Wang, S., Koli, S., Srihari, K.: A simulation-based approach for dock allocation in a food distribution center. In: 2008 Winter Simulation Conference, pp. 2750–2755. IEEE, December 2008
26. Rother, M., Shook, J.: Learning to see: value stream mapping to add value and eliminate muda. Lean Enterprise Institute (2003)
27. Alvarez, R., Calvo, R., Peña, M.M., Domingo, R.: Redesigning an assembly line through lean manufacturing tools. Int. J. Adv. Manuf. Technol. **43**(9–10), 949 (2009)
28. Arbulu, R., Tommelein, I., Walsh, K., Hershauer, J.: Value stream analysis of a re-engineered construction supply chain. Build. Res. Inf. **31**(2), 161–171 (2003)
29. Alsyouf, I., Al-Aomar, R., Al-Hamed, H., Qiu, X.: A framework for assessing the cost effectiveness of lean tools. Eur. J. Ind. Eng. **5**(2), 170–197 (2011)
30. Manos, T.: Value stream mapping-an introduction. Qual. Progress **39**(6), 64–69 (2006)
31. Abdulmalek, F.A., Rajgopal, J.: Analyzing the benefits of lean manufacturing and value stream mapping via simulation: a process sector case study. Int. J. Prod. Econ. **107**(1), 223–236 (2007)
32. Zahraee, S.M., Hashemi, A., Abdi, A.A., Shahpanah, A., Rohani, J.M.: Lean manufacturing implementation through value stream mapping: a case study. Jurnal Teknologi **68**(3), 119–124 (2014)
33. Ravinder, H., Misra, R.B.: ABC analysis for inventory management: bridging the gap between research and classroom. Am. J. Bus. Educ. **7**(3), 257–264 (2014)
34. Ramanathan, R.: ABC inventory classification with multi criteria using weighted linear optimization. Comput. Oper. Res. **33**, 695–700 (2006)
35. Tostar, M., Karlsson, P.: Lean Warehousing-Gaining from Lean thinking in Warehousing (2008)
36. Gu, J., Goetschalckx, M., McGinnis, L.F.: Research on warehouse operation: a comprehensive review. Eur. J. Oper. Res. **177**(1), 1–21 (2007)

37. Berenyi, Z., Charaf, H.: Retrieving frequent walks from tracking data in RFID equipped warehouses. In: 2008 Conference on Human System Interactions, Crakow, Poland, no. 1 & 2, pp. 669–673 (2008)
38. Lee, Y.M., Cheng, F., Leung, Y.T.: Exploring the impact of RFID on supply chain dynamics. In: Proceedings of the 2004 Winter Simulation Conference, no. 1–2, pp. 1145–1152 (2004)

Location Optimization of Receivers for IoT- Based Infrastructures

İlkay Gülbay(✉), Hasine Acap, Ayşem Ayözen, Gözde Yücel, Eda Sönmez, Çiğdem Çelik, Mine Demir, Mert Paldrak, and Orkun Karabaşoğlu

Department of Industrial Engineering, Yasar University, Izmir, Turkey
ilkaygulbay2@gmail.com, acaphasine@gmail.com, ayozenaysem@gmail.com, gozdeyuceel@gmail.com, eda_sonmez97@outlook.com, cigdemcelmak@gmaili.com, minedemir81@gmail.com, {mert.paldrak, orkun.karabasoglu}@yasar.edu.tr

Abstract. This paper introduces the problem of location optimization of receivers for IoT-based infrastructures and its application to real-life case in İzmir, Bornova. The main goal of this study is to determine the optimal number of gateways and their locations in order that the gateway deployment and installation cost can be minimized while satisfying demand, capacity and coverage constraints. As a result, the problem is formulated as an Integer Linear Programming (ILP) model and the Multi Service Capacitated Facility Location Problem (MSCFLP) is considered. In order to determine number of candidate gateway locations, K-means algorithm is employed considering the exact locations of street lambs in Bornova as the potential location of gateways to install. A mathematical model is formulated and solved to determine the optimal number and locations of the gateways. In addition to exact model, Sequential Solving Heuristic method is also suggested to be able to solve the large size problems in a reasonable amount of computational time. Obtained results are demonstrated and compared. Furthermore, a decision support system is developed which employs the solution methodology.

Keywords: Optimization · Gateway allocation · Internet of Things · MSCFLP · Sequential solving heuristic · DSS

1 Introduction

IoT is considered as one of the most important areas of future technology with the increasing demand for wireless devices and an indispensable part of daily life in today's world. This study covers research and development which is the field of consumer electronics, industrial internet and Internet of Things (IoT). This study focuses on the problem of gateway placement to find the optimum number of gateways and locations that can meet the amount of demand in the region, avoiding unnecessary gateway deployment costs. In this study, the steps such as creating a mathematical model, applying this model to a toy problem, then solving the problem intuitively and

N. M. Durakbasa and M. G. Gençyılmaz (Eds.): ISPR 2020, LNME, pp. 930–942, 2021.
https://doi.org/10.1007/978-3-030-62784-3_77

creating a decision support system were followed. The need for this work stems from the ability to create the low-cost, broad coverage and communication infrastructure that IoT devices and the IoT industry need. To explain, the company is a wireless communication infrastructure consisting of private computers (gateways) that can use hundreds of different IoT applications. These IoT applications can be many applications, such as tracking shared bicycles, monitoring agricultural areas and reading energy meters remotely. Company is building the communication infrastructure of more than 20 billion IoT devices and over 500 billion dollars, with the low cost and wide coverage required by the IoT industry. The main purpose of the study is to help IoT companies solve their infrastructure problems when it comes to connecting their devices to the cloud. For this purpose, it enables IoT devices to send small data packets at a distance of up to 25 + miles (scalable) at a fraction of the cost of cellular/GSM networks, inexpensive and energy-efficient. The basic architecture of the Internet of Things (IoT) network that provides connections between physical devices is shown in Fig. 1.

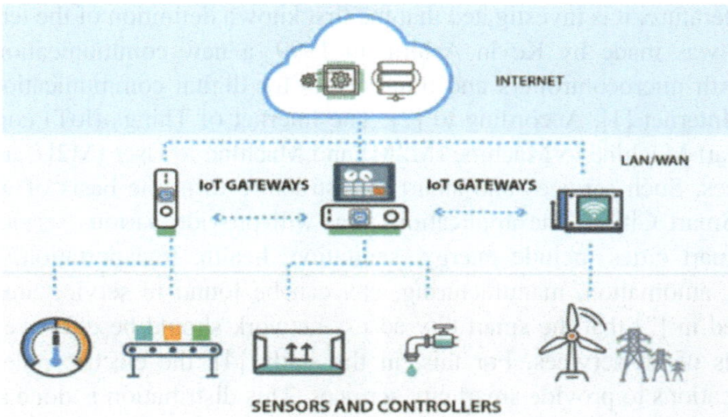

Fig. 1. Information flow chart

In order to find solutions to the above-mentioned needs, the region where the communication infrastructure will be established is determined. Candidate gateway locations in the designated region have been found by considering the gateway coverage and capacity information received from the company. The number of these candidate locations is limited in line with the purpose of the study and a mathematical model has been created to solve the gateway location problem. With this mathematical model, it is aimed to determine the optimum number of gateways and locations to minimize the total gateway infrastructure cost, taking into account the signal coverage intervals, capacities of the gateways used by the company, the availability of the line of sight between IoT devices and receivers and the amount of demand in candidate gateway locations. The aim is to meet the required communication performance while reducing operational costs. This system provides sustainability for the smart city paradigm by providing value-added services to complex cities.

The definition of the problem and the details of the problem are described in the problem definition section. Understanding of the problem and its solutions are mentioned in the literature review. The model formulation and solution techniques section includes the limitations, assumptions and mathematical model of the problem. The study continues with developing a decision support system using Excel VBA. This system will be a user-friendly system that will provide easy access to the solution of the proposed model in line with the existing data.

2 Literature Review

The literature review conducted in this study is divided into three research areas. In the first part, the studies considering the development of IoT based technologies are handled. In the second part, the focus is on the previous studies using k-means clustering algorithm examined in order to determine the clusters of demands. In the last part, the focus is more on the studies which consider Minimum-Service Location Set Coverage Problem (MSCFLP) since the problem can be formulated as MSCFLP.

In the literature, it is investigated that the first known definition of the term "Internet of Things" was made by Kevin Ashton in 1999, a new communication paradigm equipped with microcontrollers and transceivers for digital communication, which is part of the Internet [1]. According to [2], The Internet of Things (loT) communicates with the smart Machine to Machine (M2M) and Machine to User (M2U) and provides basic services. Such services and smart infrastructure form the basis of what is now known as Smart Cities. The applications that will provide various services to those living in smart cities include energy, sanitation, health, transportation, agriculture, governance, automation, manufacturing, etc. can be found in service areas [3]. It is demonstrated in [2] that the smart city access network should be effective in meeting the demands of all services. For this, in the study [4], the electric pole is used as potential locations to provide smart city services. This distribution is done at minimum cost to meet all services, all demand.

Clustering problem is defined as the problem of finding homogeneous data point groups in a certain dataset. Each of these groups is called a cluster and can be defined as a region where the density of objects is locally higher than other regions [5]. The simplest form of clustering is cluster partitioning, which aims to separate a particular dataset into separate subsets and thereby optimize certain cluster criteria. A popular clustering method that minimizes clustering error is the k-means algorithm [6]. The k-means clustering technique is firstly introduced by Mac Queen in 1967, which is the most well-known and commonly used method of clustering. Even though it is developed by Mac Queen firstly, Hugo Steinhaus had a similar idea in 1957. Furthermore, Stuart Lloyd was the one who first suggested the standard algorithm for pulse-code modulation. Later, an improved version of k-means algorithm in which the imperfections of the conventional k-means algorithm were introduced by [7]. The k-means algorithm is a local search procedure, and it is known to have a serious disadvantage that its performance largely depends on the initial starting conditions. Although other techniques have been developed based on stochastic global optimization methods (e.g. simulated annealing, genetic algorithms), it should be noted that these techniques are not widely accepted, and that the clustering method used in many practical applications is the k-means algorithm with multiple restarts [8]. In order to reduce the effect of starting point on the performance of k-means algorithm, the global k-means clustering algorithm, which creates a decisive effective global clustering algorithm, was proposed to minimize clustering error by using the k-means algorithm as a local search procedure [5].

In the literature, the Multi-Service Location Set Coverage Problem (MSCFLP) is introduced, where multiple services are distributed to several locations where power poles are located [4]. The purpose of the Multi-Service Location Set Coverage Problem, namely MSLSCP, is to ensure proper and efficient distribution of services in an urban area, thus, while meeting all demand requirements, the sum of fixed opening and fixed service costs is minimized without violating capacity constraints. In Multi-Service Capacity Facility Problem, namely MSCFLP, there is no individual cost to serve a customer from a location. When at least one customer is served from one location, only a fixed opening cost should occur [9].

MSLSCP has the characteristics of both well-known Cluster Covering Problems (SCP) and Facility Location Problems (FLP). MSLSCP is similar to Cluster Covering Problems in terms of finding coverage when a location is equipped with a service. The aim is to find the distribution of services by location as their total costs are minimized [4]. This study was inspired by the work of Hoekstra and Phillipson called "Heuristic Approaches for Location Assignment of Capacitated Services in Smart Cities". In their studies, proposed two heuristic approaches to solve the Multi-Service Capacity Facility Problem. These two heuristic approaches are Sequential Solving Heuristic and Ordered Sequential Solving Heuristic with Updating. This problem includes assigning equipment to access points and providing multiple services in the context of a smart city. Both heuristic approaches solve the assignment problem for services separately and combine the solutions of the step [10].

3 Problem Formulation

Each coordinate of the city can be a candidate location for gateways, but in this case, it can be said that there are an infinite number of possible gateway locations. It will be against the purpose of the problem to consider all possible locations. In this study, Bornova is chosen as a pilot study area due to its closeness to the university where this study is conducted. It was thought that it would be less cumbersome to obtain necessary data if the study area is well-known. Consequently, the street lambs in Bornova is considered as the candidate locations where a gateway can be placed. The total number of street lambs in Bornova is almost two thousand, therefore it is required to diminish the candidate gateway locations to a reasonable amount. As a result, the number of possible locations was reduced to an analyzable number using the k-means clustering method. The purpose of the K-means clustering method is to divide a data set consisting of N data objects into K sets given as input parameters. The clusters obtained as a result of partitioning performed are provided to have maximum intra-cluster similarities but minimal inter-cluster similarities. The K-means algorithm tries to identify the K cluster that will make the quadratic error the smallest. The algorithm reduces the total distance to the cluster to which each data belongs with its repetitive divider structure. With K-means, as long as the intra-cluster similarity is large and the inter-cluster similarity is small, the accuracy of the cluster can be mentioned.

The mathematical explanation of the K-means clustering algorithm is as follows:

X_i = A data cluster, each data being an n-dimensional real vector $X_i \in N$

S_j = Number of cluster centers (K) $S_j \in K$

$$\mu_i = \frac{1}{|s_j|}\sum\nolimits_{x_i \in s_j} x_i \tag{1}$$

where μ_i is the average of the points in S_j

$$\arg\min \sum\nolimits_{j=1}^{K} \sum\nolimits_{x_i \in s_j} ||x_i - m_j||^2 \tag{2}$$

According to the working mechanism of the K-means algorithm, K objects are randomly selected to represent the center point or average of each set. Other remaining objects are included in the clusters to which they are most similar, taking into account the distance from the average values of the clusters.

Then, by calculating the average value of each cluster, new cluster centers are determined and again the distance of the objects to the center is examined. The algorithm continues to repeat until there are no changes.

The algorithm basically consists of 4 stages:

1. Determination of cluster centers
2. Clustering of off-center data according to their distances
3. Determining new centers (or shifting old centers to new centers) according to clustering.
4. Repeat steps 2 and 3 until they become stable.

3.1 Mathematical Modeling

The mathematical model considered in this study is inspired from the article of [9] (Table 1).

Table 1. Notatıon and parameters & varıables IN MSCFLP

General notation	
$\mathcal{L}$	Setof all locations
$\mathcal{F}$	Set of all gateways
$\mathcal{G}^u$	Set of all demand points for gateway u $\in \mathcal{F}$
$\lvert\mathcal{X}\rvert$	Cardinality of set $\mathcal{X}$
Indicates	
i	Demand point i $\in \mathcal{G}^u$ for gateway u $\in \mathcal{F}$
j	Location, j $\in \mathcal{L}$
u	Type of gateway, u $\in \mathcal{F}$
Parameters	
$a_{ij}^u = \begin{cases} 1, & \text{if demand i} \in \mathcal{G}^u \text{ can be served from location j} \in \mathcal{L} \text{ for gateway u} \in \mathcal{F} \\ 0, & \text{if not} \end{cases}$	
c_j^u	set up costs of gateway u $\mathcal{F}$ at $\mathcal{L}$, access location j $\in \mathcal{L}^u, c_j^u > 0$
f_j	Costs of openning location j $\in \mathcal{L}, f_j > 0$
d_i^u	Demand amount at demand point i $\in \mathcal{G}^u$ for gateway u $\in \mathcal{F}$, $d_j^u \in \mathbb{N}$
η_j^u	Max. number connections location j $\in \mathcal{L}$ can release for gateway u $\in \mathcal{F}, \eta_j^u \in \mathcal{F}$
M	A large number, M > 0 (M is a higher number than the highest demand $\max_{u\in\mathcal{F}, i\in\mathcal{G}^u} d_i^u$)
Decision variables	
s_{ij}^u	Number of connections made between location j $\in \mathcal{L}$, and demand point i $\in \mathcal{G}^u, s_{ij}^u \in \mathbb{N}$
$x_j^u = \begin{cases} 1, & \text{access location j} \in \mathcal{L} \text{ is a service access point for gateway u} \in \mathcal{F} \\ 0, & \text{if not} \end{cases}$	
$y_j = \begin{cases} 1, & \text{if location j} \in \mathcal{L} \text{ is anaccess location} \\ 0, & \text{if not} \end{cases}$	
Parameter range description	
$a_{ij}^u\{0,1\}$	$\begin{cases} 1, & \text{if demand point i} \in \mathcal{G}^u \text{can be served from location j} \in \mathcal{L} \text{ for gateway u} \in \mathcal{F} \\ 0, & \text{otherwise} \end{cases}$
f_j $[0, \infty)$	opening cost of a location j $\in \mathcal{L}$
c_j^u $[0, \infty)$	set up cost of gateway u $\in \mathcal{F}$ at location j $\in \mathcal{L}$
d_i^u $\mathbb{N}^+$	demand amount of demand point i $\in \mathcal{G}^u$ for gateway u $\in \mathcal{F}$
η_j^u $\mathbb{N}^+$	capacity of location j $\in \mathcal{L}$ for gateway u $\in \mathcal{F}$
κ_j^u $\mathbb{N}^+$	maximum number of services access locations of gateway u $\in \mathcal{F}$ on location j $\in \mathcal{L}$
Variable range description	
y_j $\{0,1\}$	$\begin{cases} 1, & \text{if location j} \in \mathcal{L} \text{ inan access location} \\ 0, & \text{otherwise} \end{cases}$
x_j^u $\mathbb{N}$	number of service access locations of gateway u $\in \mathcal{F}$ on access location j $\in \mathcal{L}$
s_{ij}^u $\mathbb{N}$	quantity of gateway u $\in \mathcal{F}$ from service access location j $\in \mathcal{L}$ to demand point i $\in \mathcal{G}^u$

The formulation of ILP for the basic formulation of the gateway deployment problem is as follows:

$$Min \sum_{u\in F} \sum_{j\in L} c_j^u x_j^u + \sum_{j\in L} f_j y_j \tag{3}$$

subject to

$$x_j^u \leq K_j^u y_j \qquad \forall j \in L, \forall u \in F \tag{4}$$

$$\sum_{i \in G^u} s_{ij}^u \leq \eta_j^u x_j^u \qquad \forall j \in L, \forall u \in F \tag{5}$$

$$\sum_{j \in L} s_{ij}^u \geq d_i^u \qquad \forall i \in G^u, \forall u \in F \tag{6}$$

$$s_{ij}^u \leq M a_{ij}^u \qquad \forall i \in G^u, \forall j \in L, \forall u \in F \tag{7}$$

$$s_{ij}^u \in \mathbb{N} \qquad \forall i \in G^u, \forall j \in L, \forall u \in F \tag{8}$$

$$x_j^u \in \mathbb{N},\ x_j^u \leq K_j^u \qquad \forall j \in L, \forall u \in F \tag{9}$$

$$y_i \in \{0, 1\} \qquad \forall j \in L \tag{10}$$

As stated in the previous sections, the purpose of our problem is to minimize the total cost of distribution of gateways. In the objective function (3), it is aimed to open candidate positions and minimize the total cost of equipment. Constraint (4) ensures that only access locations can be equipped and restricts the maximum number of service access locations in an access location. Capacity constraint is provided by constraint (5). Constraint (6) ensures that all demands of the customers are met. Constraint (7) restricts that service access points can only service to demand points within the coverage areas. The solution area is given with the restrictions (8), (9) and (10). When solution area $s_{ij}^u = 0$ variables of can be reduced by removing. By doing this, the number of decision variables decreases and constraint (7) becomes unnecessary.

B. *Heuristic Algorithm*

MSCFLP takes into account the location assignments of multiple gateways. The problem can be solved as a whole with an exact method, but large runtimes are expected, so two separate heuristics are recommended for MSCFLP. The main source of heuristic methods is to separate the combined optimization problem into multiple sub-optimization problems. An intuitive method will update the solution between sub problems; the other one will do the sub-problems in parallel. The performance of heuristic solutions will be evaluated according to a precise method. The first intuitive method optimizes the gateway location assignments for each gateway individually. By combining the solutions of this single gateway optimization, a solution to MSCFLP is obtained. The second intuitive method extends this approach by updating some cost parameters between two steps in a row, it uses a specific optimization sequence to produce a solution to MSCFLP.

C. *Sequential Solving Heuristic*

The position assignments of various gateways are optimized simultaneously in MSCFLP. A simple intuitive approach is to independently optimize the location assignments of various gateways and combine the solutions of this single gateway optimization to provide a solution to MSCFLP. That is to say, during each step of the heuristic approach, only one of the gateways is taken into account and the position assignment of this gateway is optimized regardless of any other position assignment. Since the various steps are independent, the order of performing the steps does not affect the quality of the solution. If MSCFLP is solved for only one service, the problem is similar, but not equivalent to a CFLP. Whereas, unlike CFLP, individual costs, such as shipping costs, are not defined for demand points in MSCFLP. Instead of this, a fixed opening cost is associated with equipping an access location with a gateway independent of the number of demand points to receive service from that location. In an MSCFLP with a single gateway, an access location will always be a service access location of the gateway considered in an effective solution. In other words, yj and x_j^u decision variables are equal in an efficient solution, so MSCFLP can be modeled by excluding one of these variables. The cost of equipping a service access location j with some gateways is equal to $f_j + c_j^{u'}$. This intuitive approach is expected to result in shorter calculation times. In addition to this, the method is likely to provide non-optimal solutions, as the position assignments of various gateways are determined one by one. The general structure of the Sequential Heuristic Analysis (SSH) method is given in Algorithm 1.

Algorithm 1 Sequential Solving Heuristic

1.**for** u $\in \mathcal{F}$ do
2. Optimize the location assignment of service u $\in \mathcal{F}$
3. end for
4. Compose the MSCFLP solution
5. Set $y_j = \min(0, \sum_{u \in \mathcal{F}} x_j^u)$ for every location $j \in \mathcal{F}$
6. Objective value of the MSCFLP solution is $\sum_{j \in \mathcal{L}} \sum_{u \in \mathcal{F}} c_j^u x_j^u + \sum_{j \in \mathcal{L}} f_j y_j$

In general, solutions for various steps are combined to achieve gateways location assignment and access locations set (5th line of algorithm 1). The overall purpose of this combined solution is not just equal to the sum of the intermediate target values. This is because a location can be equipped with gateways during multiple steps, but in reality it only needs to be opened when it is equipped with the first gateway.

D. *Verification and Validation*

For verification purposes of the model, a toy problem is designed to test the accuracy of the model. As previously described, a number of candidate access points can be equipped with street poles gateways. Fixed installation costs for a location are taken equal to fj = 200 for each location in the j $\in$ L set. A total of three types of gateways will be considered for the toy problem. Various gateway values are based on research. As mentioned earlier, a gateway is characterized by its range and capacity. Gateway capacity is defined as the amount of demand a gateway can meet simultaneously. The coverage matrix with a_{ij}^u element can be filled with the data obtained by the k-means clustering method, taking into account the coverage area. Information on the range, capacity and opening costs of the gateway is provided in Table 2 for each gateway. The first type gateway has a range of 100 m, can serve up to 20 demand points, and its cost is equal to 300. The second type of gateway has a range of 200 m, can serve up to 20 demand points, and its cost is equal to 200. Finally, the third type of gateway has a range of 300 m, can serve up to 15 demand points, and its cost is equal to 350.

Table 2. Specifications of gateway types

Type of gateway	Range	Capacity (η_j^u)	Costs (c_j^u)($)
1. Type of gateway	100 m	20	300
2. Type of gateway	200 m	20	200
3. Type of gateway	300 m	15	350

Each demand point requires service for only one gateway. For all gateway types, street lights inside the border are taken as the point of demand. Demand is assigned to each street light location according to the population in the region. It is assumed that demand points that are not in at least one gateway coverage range are excluded from the set of demand points. As shown in Fig. 2, candidate service access points are represented in green and the demand point is blue.

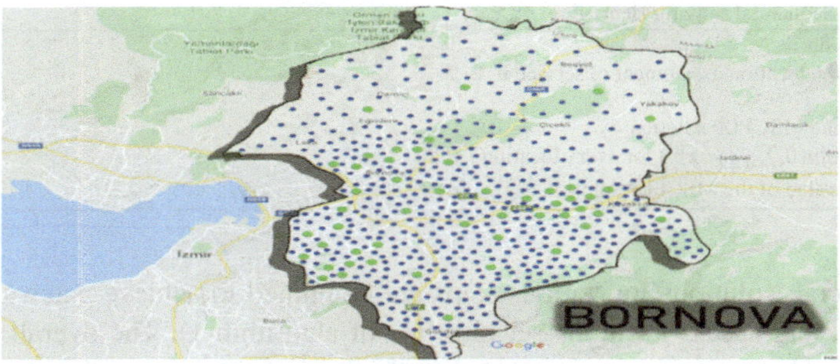

Fig. 2. example of candidate service access and demand points

On the "Results" page, 2 different matrices are created depending on the candidate gateways, the selection of candidate service access locations and the coverage information of the demand points. A brief information about the choice of gateways can be seen on the program's "Results" page to find the optimal number of each gateway and its locations, avoiding unnecessary gateway distribution costs. Also, the user can reset the program by clicking the reset button and run it again for different data. It has been discovered that the solution results are suitable for the needs of the users, and the model responds quickly and effectively to both objective functions and constraints. It was observed that CPLEX and Heuristic Algorithm gave the same objective function result in the toy problem. The solution method is tested with real-life data; As a result of the tests, it can be said that the objective function results are almost the same.

4 Computational Results

In the Sequential Solving Heuristic (SSH) method, the position assignments of the gateways are optimized individually, and the solutions of the various steps are combined to get the position assignment of many gateways. Contrary to SSH, the exact method optimizes these location assignments simultaneously. By comparing both approaches, some information has been gained about the added value of concurrent optimization of location assignments compared to individual optimization. The exact method is expected to provide better solutions than SSH, but SSH is expected to provide lower total run times (Table 3).

Table 3. Computational results and comparisons

File	#Locations	#Services	#Demands	Method	ObjVal ($)	Time (s)	#Access locations	#Service locations
Instance 1	30	3	100,110,120	Exact	11600.0	28.713	11	11, 17, 1
Instance 1	30	3	100,110,120	Heuristic	12400.0	0.096	15	11, 17, 1
Instance 2	33	3	47, 3, 9	Exact	6000.0	0.163	10	10, 2, 2
Instance 2	33	3	47, 3, 9	Heuristic	6600.0	0.124	13	10, 2, 2
Instance 3	50	3	100,100,100	Exact	9050.0	0.491	8	8, 14, 1
Instance 3	50	3	100,100,100	Heuristic	9650.0	0.59	11	8, 14, 1
Instance 4	77	3	73, 8, 15	Exact	10200.0	0.081	18	17, 3, 3
Instance 4	77	3	73, 8, 15	Heuristic	11000.0	0.1730	22	17, 3, 3
Instance 5	99	3	260, 9, 13	Exact	13500.0	0.569	24	24, 3, 3
Instance 5	99	3	260, 9, 13	Heuristic	14300.0	0.128	28	24, 3, 3
Instance 6	102	3	462, 8, 15	Exact	15350.0	1.184	27	27, 4, 3
Instance 6	102	3	462, 8, 15	Heuristic	16350.0	0.457	32	27, 4, 3
Instance 7	400	3	111, 20, 25	Exact	13200.0	0.419	21	21, 6, 4
Instance 7	400	3	111, 20, 25	Heuristic	14100.0	0.24	28	21, 5, 3
Instance 8	516	3	124, 42, 46	Exact	15050.0	4.369	24	24, 7, 4
Instance 8	516	3	124, 42, 46	Heuristic	16900.0	0.342	34	24, 7, 3

5 Decision Support System

The main purpose of this project is to find the optimal number of each gateway type and their locations by avoiding the unnecessary gateway deployment cost. In order to achieve this goal, a mathematical model was developed and a toy problem was solved. It has been observed that some optimization softwares are insufficient to solve the problem for real-life data, since the problem is NP-Hard. As a result of this observation, it has been decided that heuristic approaches will be helpful to solve the problem accurately for big amount of data. Consequently, user-friendly decision support system (DSS) is created by using the Excel VBA and CPLEX for the problem. This decision support system is created to gather solutions effectively in a short time. Decision support system welcomes the user with a home page. The decision support system welcomes the user with a home page. After clicking the "Start" button, a data entry form will open as shown in Fig. 3.

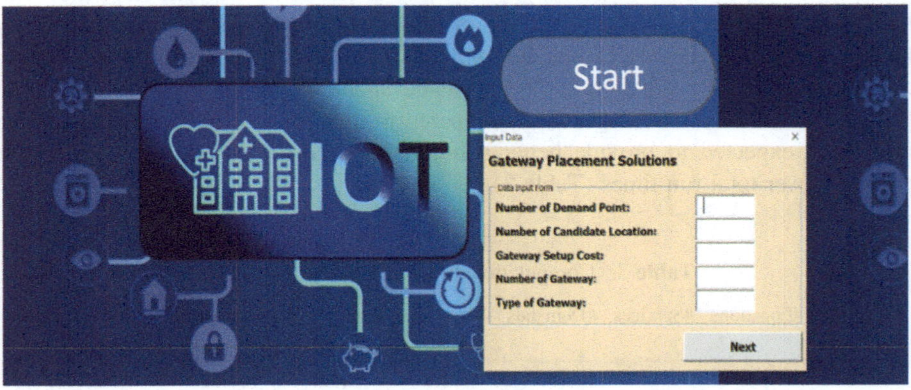

Fig. 3. Welcome page of decision support system

Here, the program asks the user for various data. This data is mandatory. The program helps the user to identify the problem with this data. After clicking the "Continue" button, the program will open a map screen as shown in Fig. 4.

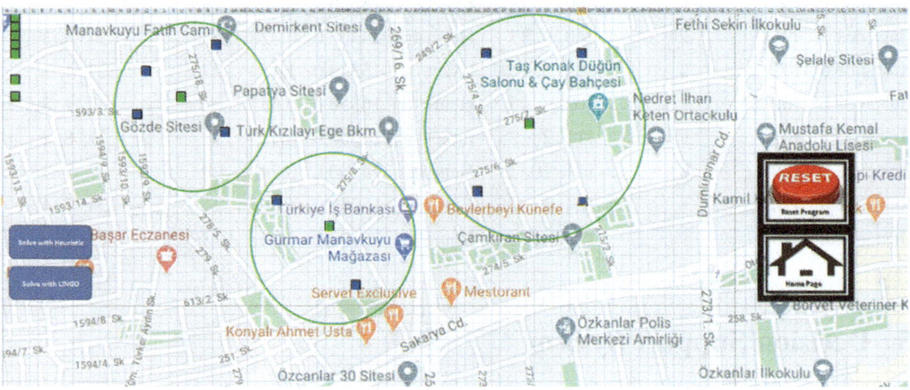

Fig. 4. Example map screen of decision support system

The program creates as many demand points and candidate gateway location points as desired by the user on this map. The user can manually put these points at the desired location on the map. Candidate gateway locations will appear as green and demand points will appear as blue points. After the review, the user will be able to delete the data with the "Reset Program" button and return to the home page with the "Home" button. To solve the problem, there are two buttons beneath all operations named "Solve with Heuristic" and "Solve with CPLEX" that leads the user to the Results page. In the Results page, users can gather information about selected candidate points to place the gateways, selected gateway types and total cost of placement. Finally, users can use DSS easily and effectively.

6 Conclusions

In this study, the problem of location optimization of receivers for IoT based infrastructures an its application to real life cases to Bornova is handled. The main goal of the study is to avoid the unnecessary gateway deployment and installation cost incurred due to the gateway placement while meeting the demand requirements of the customers under the capacity and coverage restrictions of each gateway type. The problem is formulated as in a modeling program in which the objective is to minimize the total cost and cover the customer demand in the designated urban area.

The mathematical model is based on the Multi-Service Capacity Facility Problem (MSCFLP). The proposed model is coded in a model solver program in order to solve the problem, however the exact algorithm has failed to find the optimal solution in a reasonable amount of computational time for large size instances. Consequently, two heuristic approaches are proposed to solve large size instances in which the problem is considered for each gate way individually and the solutions are combined to the calculate total cost. The results obtained using heuristic algorithm is compared to those obtained using exact solution methodology. As a result, the proposed heuristic algorithm is able to capture acceptable solutions in a shorter computational time for large size instances. This study provides various opportunities for future studies in this field. Initially, some other intuitive approaches can be developed to improve computation time for larger problem sizes. Secondly, the basic model can be expanded by allowing multiple gateways in one access location. In addition to determining which locations should be equipped with the gateway for this development, it is necessary to decide how many gateways should be located per request at these locations. Thirdly, partial coverage expansion can be considered in which all demand points can be partially met by a single access location.

References

1. Shang, X., Zhang, R., Zhu, X., Zhou, Q.: Design theory, modelling and the application for the Internet of Things service. Enterp. Inf. Syst. **10**(3), 249–267 (2016)
2. Veerman, B., Phillipson, F.: Multi Service "Modular Capacitated Facility Location Problem for Smart Cities. Innovations for Community Services, pp. 97–108 (2019)
3. Balakrishna, C.: Enabling technologies for smart city services and applications. In: 2012 Sixth International Conference on Next Generation Mobile Applications, Services and Technologies, pp. 223–227. IEEE (2012). https://doi.org/10.1109/ngmast.2012.51

4. Vos, T., Phillipson, F.: Dense multi-service planning in smart cities. In: International Conference on Information Society and Smart Cities (2018)
5. Likas, A., Vlassis, N., Verbeek, J.: The global k-means clustering algorithm. Pattern Recogn. **36**(2), 451–461 (2003). https://doi.org/10.1016/S0031-3203(02)00060-2
6. Lozano, J.A., Pena, J.M., Larranaga, P.: An empirical comparison of four initialization methods for the k-means algorithm. Pattern Recogn. Lett. **20**(10), 1027–1040 (1999). https://doi.org/10.1016/S0167-8655(99)00069-0
7. Wang, J., Su, X.: An improved K-Means clustering algorithm. In: 2011 IEEE 3rd International Conference on Communication Software and Networks, pp. 44–46. IEEE, May 2011
8. Murty, M.N., Jain, A.K., Flynn, P.J.: Data clustering: a review. ACM Comput. Surv. **31**(3), 264–323 (1999). https://doi.org/10.1145/331499.331504
9. Hoekstra, G., Phillipson, F.: Location assignment of capacitated services in smart cities. In: The 4th International Conference on Mobile, Secure and Programmable Networking (2018)
10. Hoekstra, G., Phillipson, F.: Heuristic approaches for location assignment of capacitated services in smart cities. J. Comput. **7**(4), 67 (2018). https://doi.org/10.3390/computers7040067
11. Petäjäjärvi, J., et al.: Performance of a low-power wide-area network based on loRa technology: doppler robustness, scalability, and coverage. Int. J. Distrib. Sens. Netw. **13**(3) (2017). p. 155014771769941. https://doi.org/10.1177/1550147717699412
12. Capone, A., et al.: Deploying multiple interconnected gateways in heterogeneous wireless sensor networks: an optimization approach. Comput. Commun. **33**(10), 1151–1161 (2010)
13. "Semtech." Semtech. www.semtech.com/

Makespan Minimization in Manufacturing Process of a Company

Birkan Kayi, Halime Keçibaş, Aziz Kemal Konyalıoğlu, and Ferhan Çebi(✉)

Management Engineering Department, Istanbul Technical University, Istanbul, Turkey
{kayi17, kecibas, konyalioglua, cebife}@itu.edu.tr

Abstract. In real production processes, various problems may occur. Cost minimization, make span minimization, and energy consumption minimization are often the topics that are studied to eliminate problems in production. In this study, it is aimed to minimize the make span of a job assigned to workers working on identical and parallel machines in a welding workshop. Minimum completion time has been obtained by assigning the jobs in accordance with the objective function in a problem which has demand constraints according to worker capacity, worker ability and welding types. The mathematical models made in order to obtain the minimum makespan during which the given welding work can be realized were calculated in the LINGO 17.0 program.

Keywords: Makespan minimization · Parallel machines · Scheduling · Optimization

1 Introduction

In real life problems, mathematical programming methods are used to solve the problems encountered in the manufacturing process. These differ from problem to problem but focus on solutions based on improving the production process for purposes such as time minimization, cost optimization, energy consumption amount optimization. Various algorithms are used to solve these problems. [1, 11, 12]

Accordingly, the subject we have examined is "Makespan Minimization in Production Processes". When literature review was made for such problems, it was observed that time minimization was studied according to different production processes.

It can have severe consequences for the manufacturer if completion date of the job cannot be planned well in a manufacturing process. For this reason, preparing a time schedule at the beginning of the work is beneficial to the manufacturer. While preparing this time schedule, it is important to consider that completing the job before the date required by the customer. However, it is first aimed to finish the work to be done without any problems in the shortest time. In this study, a model has been established for a method that will optimize the makespan that the manufacturing companies should know when preparing this timeline.

N. M. Durakbasa and M. G. Gençyılmaz (Eds.): ISPR 2020, LNME, pp. 943–950, 2021.
https://doi.org/10.1007/978-3-030-62784-3_78

The problems in the production process are classified according to the complexity of processing as follows; [1]

- One-stage, one processor/facility
- One-stage, parallel processors/facilities
- Multistage, flow shop
- Multistage, job shop

The problem we examined is in the category of "one-stage, parallel processors/facilities", since it is single-stage and consists of identical parallel machines. This study consists of four main parts: an introduction section, a literature review section in order to give a brief literature survey about make span minimization and applications of mathematical modeling in manufacturing process, a methodology and modeling section to explain the modeling constructed for "one-stage, parallel processors/facilities" and finally a conclusion part to discuss the proposed model.

2 Literature Review

There exist several studies about make-span minimization in the literature. In 1995, Ho and Wong used an algorithm whose name is Two Machine Optimal Scheduling in order to minimize the makespan which is for two identical processors by considering the longest processing tie sequence [2]. Sevastianov and Woeginder proposed a polynomial time approximation scheme in the scheduling problem of open shop based of different number of machines [3]. Ghirardi and Potts tried to minimize the unrelated parallel machines defined as the problem of scheduling jobs by using Recovering Beam Search algorithm [4]. Pan and Wang proposed heuristics algorithms named as weighted profile fitting and profile fitting in order to solve blocking flow shop scheduling problem with make-span minimization [5]. Liao et al. considered two parallel machines problem which one of these machines is busy during the time period while the other one is capable of working [6].

Xu et al. investigated the problem of minimizing makespan for a single batch processing machine which is capable of doing different jobs simultaneously by using heuristics [7]. Hurink and Knust proposed a solution for the problem of makespan minimization for flow-shop problems with transportation times and a single robot in order to determine a feasible schedule [8]. Martello et al. used an exact and approximation algorithm in order to minimize makespan on unrelated parallel machines based on Lagragian relaxation and different additive techniques [9]. Ying et al. considered the scheduling problem that is jobs on different parallel machines and machine dependent by the sequence dependent set up times by proposing restricted simulated annealing (RSA) algorithm [10]. Ahmedizar used ant colony algorithm which is a Heuristic algorithm, in order to develop the solution of scheduling problem of makespan in permutation flow shops [1].

3 Methodology and Modelling

In this section, it is aimed to develop a model that makes make-span minimization in a production system by using mathematical programming methods.

A. *Definition of the Problem*
In this study, scheduling of a certain number of welding works to be performed in a company with five parallel and identical welding machines is discussed. There are four different types of sources in total: 16″, 12″, 8″ and 4″. It is counted from the project drawings that the number of weldings to be carried out from each welding type is 22, 85, 16, 52 respectively.

Also, there are five welders at the company whose data we used for this study who are able to operate the welding machines. If we define these workers with A, B, C, D and E; while any of them are able to weld 16″ and 12″ pipes, only experienced workers A and B are qualified for welding 8″ and 4″ pipes since small-sized weldings require finer work.

Workers at this facility work only 9 h on weekdays and do not work on the weekends. According to the information obtained from the welding procedures of the company, Table 1 shows the time in hours the workers spend to make certain types of welding.

Table 1. Process time of each welding type according to the workers [hour/unit]

	16″	12″	8″	4″
Worker A	3,2	2,4	1,6	0,8
Worker B	3,6	2,7	1,8	0,9
Worker C	4	3	–	–
Worker D	4,8	3,6	–	–
Worker E	5,1	3,8	–	–

B. *Defining Capacity of the Workers*
The performance of each worker is expected to be different during the period from the beginning of the work to the end, since the workers complete the same type of welding at different speeds. This difference was revealed by weighting the contribution of workers' performance to the work. The performance weights of the workers were calculated as follows.

Worker A:
Worker A welds a 16″ pipe in 3,2 h.
Worker A welds a 12″ pipe in 2,4 h.
Worker A welds an 8″ pipe in 1,6 h.
Worker A welds a 4″ pipe in 0,8 h.

$$Performance\ weight\ of\ Worker\ A(w_A) = \frac{22 + 85 + 16 + 52}{22 * 3,2 + 85 * 2,4 + 16 * 1,6 + 52 * 0,8} = 0,513$$

$$175 * \frac{w_A}{w_A + w_B + w_C + w_D + w_E} \cong 50\ pcs$$

Worker B:
Worker B welds a 16″ pipe in 3,6 h.
Worker B welds a 12″ pipe in 2,7 h.
Worker B welds an 8″ pipe in 1,8 h.
Worker B welds a 4″ pipe in 0,9 h.

$$Performance\ weight\ of\ Worker\ B\ (w_B) = \frac{22 + 85 + 16 + 52}{22 * 3,6 + 85 * 2,7 + 16 * 1,8 + 52 * 0,9} = 0,455$$

$$175 * \frac{w_B}{w_A + w_B + w_C + w_D + w_E} \cong 45pcs$$

Worker C:
Worker C welds a 16″ pipe in 4 h.
Worker C welds a 12″ pipe in 3 h.

$$Performance\ weight\ of\ Worker\ C\ (w_C) = \frac{22 + 85}{22 * 4 + 85 * 3} = 0,312$$

$$175 * \frac{w_C}{w_A + w_B + w_C + w_D + w_E} \cong 30pcs$$

Worker D:
Worker D welds a 16″ pipe in 4,8 h.
Worker D welds a 12″ pipe in 3,6 h.

$$Performance\ weight\ of\ Worker\ D\ (w_D) = \frac{22 + 85}{22 * 4,8 + 85 * 3,6} = 0,26$$

$$175 * \frac{w_D}{w_A + w_B + w_C + w_D + w_E} \cong 26pcs$$

Worker E:
Worker E welds a 16″ pipe in 5,1 h.
Worker E welds a 12″ pipe in 3,8 h.

$$Performance\ weight\ of\ Worker\ E\ (w_E) = \frac{22 + 85}{22 * 5,1 + 85 * 3,8} = 0,246$$

$$175 * \frac{w_E}{w_A + w_B + w_C + w_D + w_E} \cong 24pcs$$

C. *Defining the Decision Variables*
Notations which are used in the formulations are defined as below:

i: workers (A, B, C, D, E)
j: jobs to be completed (16″, 12″, 8″, 4″ welding type)
Xij = number of the job j processed by the worker i
tij = processing time of the job j by the worker i

D. *Objective Function*
The objective function (1) minimizes the completion time of the total work to be performed by workers with different performance criteria on 5 parallel and identical machines.

$$z = min \sum_{j=1}^{4} \sum_{i=5}^{5} t_{ij} x_{ij} \tag{1}$$

The objective function (2) is equal to the sum of the completion times which are weighted according to workers' performances. The performance measurement is made by developing the decision mechanism with the defined constraints.

$$\begin{aligned} Z = \min\ & t_{A16''} * X_{A16''} + t_{B16''} * X_{B16''} + t_{C16''} * X_{C16''} + t_{D16''} * X_{D16''} + t_{E16''} * X_{E16''} + t_{A12''} * X_{A12''} \\ & + t_{B12''} * X_{B12''} + t_{C12''} * X_{C12''} + t_{D12''} * X_{D12''} + t_{E12''} * X_{E12''} + t_{A8''} * X_{A8''} + t_{B8''} * X_{B8''} + t_{C8''} * X_{C8''} \\ & + t_{D8''} * X_{D8''} + t_{E8''} * X_{E8''} + t_{A4''} * X_{A4''} + t_{B4''} * X_{B4''} + t_{C4''} * X_{C4''} + t_{D4''} * X_{D4''} + t_{E4''} * X_{E4''}; \end{aligned} \tag{2}$$

E. *0/1 Binary Variables*

$$Y_{ij} = \begin{cases} 1, & \text{if job j is processed by worker i} \\ 0, & \text{otherwise} \end{cases}$$

$$Y_{A16''} + Y_{B16''} + Y_{C16''} + Y_{D16''} + Y_{E16} \leq 5$$

$$Y_{A12''} + Y_{B12''} + Y_{C12''} + Y_{D12''} + Y_{E12''} \geq 5$$

$$Y_{A8''} + Y_{B8''} \leq 2$$

$$Y_{A4''} + Y_{B4''} \leq 2$$

F. *Constraints*

There are two kinds of constraints included in the model:

- Demand constraints according to the welding types
- Capacity constraints of the workers

G. *Demand Constraints According to the Welding Types*

$$X_{A16''} + X_{B16''} + X_{C16''} + X_{D16''} + X_{E16''} = 22 \quad (\textit{Demand constraint for 16'' welding})$$

$$X_{A12''} + X_{B12''} + X_{C12''} + X_{D12''} + X_{E12''} = 85 \quad (\textit{Demand constraint for 12'' welding})$$

$$X_{A8''} + X_{B8''} = 16 \quad (\textit{Demand constraint for 8'' welding})$$

$$X_{A4''} + X_{B4''} = 52 \quad (\textit{Demand constraint for 4'' welding})$$

$$X_{C8''} = 0, X_{D8''} = 0, X_{E8''} = 0 \quad (\textit{C, D and E welders are not suitable for 8'' welding})$$

$$X_{C4''} = 0, X_{D4''} = 0, X_{E4''} = 0 \quad (\textit{C, D and E welders are not suitable for 4'' welding})$$

$$X_{ij} \geq 0 \quad (\textit{Sign-restriction})$$

H. *Capacity Constraints of the Workers*

$$X_{A16''} + X_{A12''} + X_{A8''} + X_{A4''} \leq 50 \quad (\textit{Capacity constraint of Worker A})$$

$$X_{B16''} + X_{B12''} + X_{B8''} + X_{B4''} \leq 45 \quad (\textit{Capacity constraint of Worker B})$$

$$X_{C16''} + X_{C12''} \leq 30 \quad (\textit{Capacity constraint of Worker C})$$

$$X_{D16''} + X_{D12''} \leq 26 \quad (\textit{Capacity constraint of Worker D})$$

$$X_{E16''} + X_{E12''} \leq 24 \quad (\textit{Capacity constraint of Worker E})$$

4 Result of the Mathematical Modeling

When the problem is modeled in line with the above information in the LINGO 17.0 and solved with the Mixed Integer Linear Programming (MILP) method, the optimum data obtained are as follows as seen in the Table 2.

Table 2. The number of different jobs processed by each worker

	16″	12″	8″	4″	Total capacity
Worker A	22	5	16	7	50
Worker B	0	0	0	45	45
Worker C	0	30	–	–	30
Worker D	0	26	–	–	26
Worker E	0	24	–	–	24
Total Work	22	85	16	52	

The completion time is calculated as;

$$z = min \sum_{j=1}^{4} \sum_{i=5}^{5} t_{ij} x_{ij} = 428,9 \text{ man - hour}$$

5 Conclusion

Based on the results obtained from the modeling, it was observed that total job and total capacity constraints were met. Makespan is optimized by assigning more jobs to the worker who completes the same job in less time. When it is needed to optimize for such problems encountered in the industry, it has been observed that LINGO 17.0 program can obtain reasonable results by using MILP method. Although it is appropriate to use the MILP method in the problems in the scale of the job diversity in our study, it is recommended that different algorithms can be used for different scales of job diversity as a result of the literature review.

References

1. Ahmadizar, F.: A new ant colony algorithm for makespan minimization in permutation flow shops. Comput. Ind. Eng. **63**(2), 355–361 (2012)
2. Ho, J.C., Wong, J.S.: Makespan minimization for m parallel identical processors. Naval Res. Logist. (NRL) **42**(6), 935–948 (1995)
3. Sevastianov, S.V., Woeginger, G.J.: Makespan minimization in open shops: a polynomial time approximation scheme. Math. Program. **82**(1–2), 191–198 (1998)
4. Ghirardi, M., Potts, C.N.: Makespan minimization for scheduling unrelated parallel machines: a recovering beam search approach. Eur. J. Oper. Res. **165**(2), 457–467 (2005)

5. Pan, Q.K., Wang, L.: Effective heuristics for the blocking flowshop scheduling problem with makespan minimization. Omega **40**(2), 218–229 (2012)
6. Liao, C.J., Shyur, D.L., Lin, C.H.: Makespan minimization for two parallel machines with an availability constraint. Eur. J. Oper. Res. **160**(2), 445–456 (2005)
7. Xu, R., Chen, H., Li, X.: Makespan minimization on single batch-processing machine via ant colony optimization. Comput. Oper. Res. **39**(3), 582–593 (2012)
8. Hurink, J., Knust, S.: Makespan minimization for flow-shop problems with transportation times and a single robot. Discrete Appl. Math. **112**(1–3), 199–216 (2001)
9. Martello, S., Soumis, F., Toth, P.: Exact and approximation algorithms for makespan minimization on unrelated parallel machines. Discrete Appl. Math. **75**(2), 169–188 (1997)
10. Ying, K.C., Lee, Z.J., Lin, S.W.: Makespan minimization for scheduling unrelated parallel machines with setup times. J. Intell. Manuf. **23**(5), 1795–1803 (2012)
11. Min, L., Cheng, W.: A genetic algorithm for minimizing the makespan in the case of scheduling identical parallel machines. Artif. Intell. Eng. **13**(4), 399–403 (1999)
12. Lee, W.C., Wu, C.C., Chen, P.: A simulated annealing approach to makespan minimization on identical parallel machines. Int. J. Adv. Manuf. Technol. **31**(3–4), 328–334 (2006)

Multi Model Multiple Line Balancing Problem Modeling and Real-Life Application

Onur Orhan(✉), Doğukan Başgöl, Muharrem Eren Tekin, Ahmet Gülseven, Mahmut Ali Gökçe, Evren Demir, and Cansu Yurtseven

Department of Industrial Engineering, Yasar University, Izmir, Turkey
dogubasgol@gmail.com, 0evrendemir@gmail.com, gulseven.ahmet97@gmail.com, onurorhan9635@gmail.com, muharremeren35@gmail.com, {ali.gokce, cansu.yurtseven}@yasar.edu.tr

Abstract. This study deals with multi-line balancing for multiple products. This problem originates from a real-life problem at a global heat boiler manufacturing company. The objective is to develop a solution method that enables create efficient line balances in a dynamic environment, which maximizes the weighted average line efficiency and reduces waste of time for frequent line balancing procedures. Since weighted average efficiency is a key performance indicator for the company, this paper focuses on the optimization of this measure. The proposed method is demonstrated with an implementation at three assembly lines, where great majority of the company's production takes place. For this problem, an algorithm has been developed that creates a set of feasible scenarios (a scenario pool) in which total demand of each model is divided between available lines under capacity restrictions. For every scenario in the pool, a custom multi model assembly line balancing MIP is created automatically and run. From the results, best (most efficient) assignment of models' demand to lines and corresponding line balances are obtained. Furthermore, a user-friendly dynamic decision support system (DSS) is developed and proposed solution approaches are embedded in this DSS for the daily operations of the company. The developed DSS enables users to do an efficient line balancing in short computational time and provides the results with detailed line balancing reports. As this problem can be faced in various areas, the proposed solution approaches can also be applied to different sectors and factories with little modification.

Keywords: Optimization · Decision support system · Weighted average efficiency · Mathematical model · Assembly line balancing

1 Introduction

Companies desire to maximize their profit while satisfying customer demand. There are several ways to achieve this goal such as decreasing the unit cost and/or increasing the production capacity. In order to increase production quantity, companies either could invest on new/automated machines and/or capacity or improve existing production systems. One of the methods that improves the current system is improve assembly lines' balance. Since a better balance would improve the efficiency of a line, the output would increase as well.

N. M. Durakbasa and M. G. Gençyılmaz (Eds.): ISPR 2020, LNME, pp. 951–962, 2021.
https://doi.org/10.1007/978-3-030-62784-3_79

The heat boiler manufacturing company produces multiple types of heat boilers in multiple assembly lines and suffers from spending a remarkable amount of time to balance lines when new models/variants are introduced to the factory. Furthermore, the company cannot reach as high quantities as possible due to not well-balanced lines. There are 6 assembly lines in the company. 70% of all demand is satisfied by three lines; 1, 2 and 3. Therefore, the study focuses on these 3 lines. The number of workstations in these three lines are constant and fixed due to space constraints. The main goal is to decide what percentage of every model goes to each line and find out line balances maximizing the weighted average efficiency for these 3 assembly lines while keeping the number of workstations constant. Weighted average efficiency is calculated by demand of each model is taking into consideration. Since the problem has multi-model and multi-line and also the objective is maximizing the efficiency, it is categorized as Multi-Model Assembly Line Balancing Problem once the assignment decision mentioned above is made.

This study presents a novel solution methodology for this specific multi model multi line assembly line balancing. The rest of the paper is organized as follows. The formal problem definition is provided in Sect. 2. In Sect. 3, literature review for the study is presented. The solution methodology is described in detail in Sect. 4. Computational results and implementation of the proposed solution approach is presented in Sect. 5. Section 6 explains the developed decision support system and lastly, conclusions and future work are provided in Sect. 7.

2 Problem Definition

The company has more than 55 models that are produced in these 6 lines. Almost 70% of the production comes from lines 1, 2 and 3. Since these lines represent majority of the production, this work focuses on these three lines. The aim of the project is the balancing lines with multi models to maximize the weighted average efficiency, where the weights represent that production quantity of a model divided by total production quantity of corresponding line.

These lines differ on several parameters, e.g. number of workstations, number of parallel final test stations and number of available stations to accomplish other tests operations. The company uses air-operated tools for torqueing, riveting and fastening operations. Number of parallel final test stations have significant role upon cycle time. Since the final test operation is indivisible, the appliance should be connected to the final test station until the test ends.

For companies with continuous new model integration, line balancing is an activity that must be repeated frequently. During the line balancing process, the related engineer is spending considerable amount of time. Also, there might be capacity losses caused by possible sub-par balancing, and it may cause additional cost to the company. Line balancing operations rely on observations and experimentations that trigger errors and omissions.

It is planned that the algorithm will achieve optimal results by using line balancing optimization algorithm. The algorithm will also take into account the precedence relationship, tooling availability and fixed test stations. Since the solution finds maximum average efficiency for all models in all these 3 lines once the assignment of demand of products to lines are made, there is a need for an assignment algorithm that runs and finds all feasible scenarios by considering demands of the models, capacity of the lines and operational constraints.

3 Literature Review

In literature, there are several techniques that can be used for the assembly line balancing problem (ALBP). Considering these solution techniques, ALBP can be categorized based on the number of models produced in the line, the objective of the problem and the number of lines such as single line or multi-line. There are different types of assembly line balancing problems with different objectives. Most of the articles either consider minimizing the number of workstations for given cycle time and this type is called Type-1 or minimizing the cycle time for a given number of workstations which is called Type 2. However, in literature, few studies consider Type-E problems. In Type-E, the objective is maximizing the line efficiency while cycle time and the number of workstations are decision variables. Since in this study, the objective of the problem is maximizing the overall efficiency, the number of workstations is constant, and the cycle time is a decision variable; the literature search has been conducted on mostly Type-E problems. However, because of there are a small number of studies related with Type–E line balancing problem and due to the production type of our problem consists multi-model, the literature research also covers the Type-1 and Type-2 problem to take advantage of related techniques for assembly line balancing problems. In order to solve these problems with a large number of instances, heuristic approaches are required due to the complexity of ALBP. Several heuristics and algorithms have used to solve the ALBPs at a reasonable time compared to the exact methods of optimization.

The earliest studies for the ALB are Type-1 (ALB-1) and Type-2 (ALB-2) problems. There are several ways to solve ALB-1 and ALB-2 and these are mathematical modeling and heuristic approaches such as simulated annealing algorithm, ant colony optimization, shortest path algorithm, critical path method, and imperialist competitive algorithm. Sparling and MiltenBurg [1] used a smoothing algorithm that exchanges tasks between stations for decreasing the imbalance of the stations in order to provide the final balance. Ajenblit and Wainwright [2] applied the genetic algorithm to find the minimum number of workstations. Widyadana [3] developed a goal programming model for the Type-2 problem but the difference is model considers temporary and permanent workers as well. Baykasoglu and Dereli [4] used ant colony optimization and rank positional weight heuristic to find the minimum number of workstations and cycle time. Hamzadayi and Yildiz [5] used a genetic algorithm to balance the mixed-model assembly line.

The main research has been made about assembly line balancing with maximizing the efficiency (Type-E) problems due to similarities to the problem in this study. However, in literature, there are a few studies about the Type-E problem. The earliest study belongs to Chakravarty [6] and the study is increasing the efficiency with ALB by including task learning effects. Although the problem gives good results when the production unit is small, results get worse when the production limit increases. Eryuruk et al. [7] used the heuristic technique which is probabilistic line balancing and ranked positional weight for maximizing the line efficiency. The line efficiency is 80.6% with their solution technique. Ozcan and Toklu [8] developed a hybrid heuristic approach to find maximum line efficiency and minimum task times. The solution technique is implemented different problem types in order to make a comparison between them and the proposed algorithm obtained the best efficiency for 28 test problems out of 57. Al-zubaidy and Alrazaq [10] developed ranked positional weight for the line efficiency maximization. Their approach gives an efficiency of 78.8%. The solution technique is useful due to similarity to our problem and the appropriate results.

4 Solution Methodology

After conducting literature review for the problem, a suitable definition for this study has been determined as the Multi-Model Assembly Line Balancing Problem Type-E (MuMALBP-E). While investigating the problem on hand, it has been recognized that once the decision of how much of the demand for each product goes to every line, exact mix of the demand for each line would be determined and treating each line's balancing independently becomes possible. For this reason, solution approach is divided into two parts. In part 1, feasible scenarios for assignment of demand of products to lines are determined based on capacity and operational constraints. Naturally, the possible number of combinations is prohibitively large. By dividing each product's demand in discrete step sizes, eliminating infeasible combinations due to capacity and operational constraints, leaves a manageable number of scenarios. In the second part of the solution approach, individual mathematical models are constructed with line specific parameters for each scenario automatically. The performance measure for the problem has been chosen as calculating the weighted average efficiency of each line and then an algorithm merged three lines' results in one outcome.

MuMALBP-E is in the NP-hard category [10]. Our approach uses an assignment algorithm, which calculates a manageable number of scenarios with certain percentages of demand divided between lines and constructs and runs MIP models for each line under each scenario and chooses best among results to get a good solution to the overall problem.

4.1 Assignment Algorithm

The aim of the assignment algorithm is to find possible demand distribution scenarios for all models on all lines. The exact number of combinations can be very large. The models (variants) to be produced at the facility demand for these models, equipment/machine collection at lines change constantly. Therefore, it is not possible to use past data to determine a set of feasible demand distribution scenarios. For this reason, the assignment algorithm creates scenario pool by first discretizing percentage of demand for each product that can be divided among the lines and then by eliminating those combinations that are not feasible due to capacity and operational limitations. The discretization is done by a step size. The algorithm finds different scenarios by decreasing or increasing the demand percentages for the model on the lines by increments of the chosen step size. It uses the same technique for all models on all lines and finds scenarios with the step size that the user enters. As an illustration, possible scenarios when there are two models and two lines and the step size is 25% is shown in Table 1, and table of all possible scenarios can be seen in Table 2 below.

Table 1. Possible scenarios when step size = 25%

	Line 1		Line 2	
Scenarios	Model A	Model B	Model A	Model B
Scenario 1	0%	25%	100%	75%
Scenario 2	0%	50%	100%	50%
Scenario 3	0%	75%	100%	25%
Scenario …	100%	25%	0%	75%
Scenario 22	100%	50%	0%	50%
Scenario 23	100%	75%	0%	25%

Step size determines the total number of MIP models to be solved and therefore directly affects total run-time. In the implementation phase, DSS user is also able to set this parameter manually. Naturally, scenarios are more comprehensive when the step size is smaller, and this makes achieving possibly a better solution.

Table 2. Comparison of number of sizes and run times for step sizes 10% and 25%

Step size = 25%										Step size = 10%									
	Line 1			Line 2			Line 3				Line 1			Line 2			Line 3		
Sce. #	A	B	C	A	B	C	A	B	C	Sce. #	A	B	C	A	B	C	A	B	C
1	25	0	100	50	50	0	25	50	0	1	0	0	100	75	25	0	25	75	0
2	25	0	100	25	25	0	50	75	0	2	25	0	100	50	50	0	25	50	0
3	25	0	100	25	75	0	50	25	0	…	25	0	100	75	25	0	0	75	0
4	50	0	100	50	0	0	0	100	0	55	75	0	100	25	0	0	0	100	0
										56	75	0	100	0	50	0	25	50	0
Run Time = 0,25 min.										Run Time = 11 min.									

Table 2 shows rapid increase in number of scenarios and run-time for the MIP models when the step size changes from 25% to 10%.

While making this scenario analysis, the assignment algorithm eliminates combinations based on production capacity of lines, product availability of models for lines, and demand satisfaction and other operational constraints like job-machine competency and efficiency. After the assignment algorithm distributes the models to the lines properly, the next step will be the constructing the MIP models, preparing input data, associated parameters automatically and execution of models for each line for each scenario.

4.2 Mathematical Model

Mathematical model for multi-model single line has been studied for maximization of weighted average efficiency. First, all sets and indices, parameters and decision variables are introduced. Then the mathematical model is represented. MIP is an improved version of the work done by Al-zubaidy and Alrazaq [9] and includes problem and company specific constraints.

Sets and Indices:

i	Operation numbers, $i \in I_m = \{1, 2, \ldots, \lvert I_m \rvert\}$
j	Station numbers
m	Models
P_{im}	Set of all predecessor of operation i, for model m.
S_{im}	Set of all successors of operation i, for model m.
R_m	Set of immediate predecessor operation tuples of model m, $(p, q) \in R_m$, if operation p is immediate predecessor of operation q.
T_m	Set of processing times of all operations, for model m.
I_m	Set of all operations of model m.
It_m	Set of test operations, for model, $It_m \subset I_m$
JT_i	Set of test stations for test operations, $i \in It_m$
JA_{im}	Set of assignable stations of operation i, for model m. $i \in I_m$
M	Set of models that assemble in the line

Parameters:

FT_m	Final test duration of model m.
NOS	Number of stations in the line.
t_{im}	Processing time of operation i, for model m, $t_{im} \subset T_m$
TPT_m	Total processing time of model m.
U_{im}	Earliest station to assign operation i, for model m.
L_{im}	Latest station to assign operation i, for model m.
D_m	Demand quantity for model m that will be produced on that line, per year.
D	Total demand quantity of all models per year.
WA_m	Demand proportion of model m.
MP	Number of models that will be produced in the line.
AW	Available working time per year in terms of second.
SOR	Station Occupancy Ratio

Decision Variables:

x_{ijm}	1, if operation i for model m is assigned to station j, 0 otherwise.
CT_m	Cycle time of model m

$$Min \sum_{m=1}^{MP} \left(\frac{NOS * CT_m}{WA_m * TPT_m} \right) \tag{1}$$

Subject to

$$\sum_{i=1}^{I_m} t_{im} * x_{ijm} \leq CT_m, \ \forall j \in J, \forall m \in M \tag{2}$$

$$\sum_{j=1}^{NOS} x_{ijm} = 1, \ \forall i \in I_m, \forall m \in M \tag{3}$$

$$\sum_{j=1}^{NOS} j(x_{qjm} - x_{pjm}) \geq 0, \ (p, q) \in R_m, \forall m \in M \tag{4}$$

$$\sum_{\forall j, j \in JT_i} x_{ijm} = 1 \ , \forall m \in M, \forall i \in It_m \tag{5}$$

$$x_{ijm} = 0, \text{ if } j < U_{im} \text{ and } j > L_{im}, \forall i \in I_m, \forall m \in M \tag{6}$$

$$\sum_{i=1}^{I_m} \frac{(t_{im} * x_{ijm})}{CT_m} \leq SOR, \ \forall j \in J, \forall m \in M \tag{7}$$

$$\sum_{m=1}^{MP} D_m * CT_m \leq AW, \ \forall m \in M \tag{8}$$

$$CT_m \geq FT_m, \forall m \in M \tag{9}$$

$$WA_m = \frac{D_m}{\sum_{m=1}^{MP} D_m}, \ \forall m \in M \tag{10}$$

$$x_{ijm} \in \{0, 1\}, \forall i \in I_m, \forall j \in J, \forall m \in M \tag{11}$$

Our objective (1) is to maximize weighted average line efficiency. In order to handle non-linearity, the objective function has been transformed to minimizing weighted average line inefficiency. Constraint (2) provides the total time of all operations assigned to a station cannot exceed the cycle time. Constraint (3) assures that each operation should assign to exactly one station and all operations should be assigned. Constraint (4) expresses precedence constraint. (p, q) means that operation q cannot be assigned before operation p. Constraint (5) prevents assigning test operations to stations that are not capable of performing tests. Constraint (6) refers that an operation cannot be assigned neither before its earliest assignable station nor after its latest assignable station. Constraint (7) is station occupancy constraint. Constraint (8) assures that the demand of each model is satisfied. Constraint (9) assures that cycle time of model m higher than or equal to Final Test duration of model m. Since Final Test is indivisible work for the product, duration of it is the lower bound of cycle time. Constraint (10) refers to demand proportion of model m on the line. Constraint (11) is binary constraint. After describing the working principles of both assignment algorithm and mathematical model, two developed tools are combined in Microsoft Excel. Since the scenario creating part is completed, mathematical models that are constructed considering different characteristics of lines will be executed with the demands of models that assigned relevant lines.

5 The Computational Results and Implementation

After constructing the described solution methodology, the computational results are evaluated. Due to the importance of relation between the algorithm's run time and the number of scenarios as mentioned before, in Table 3, with consideration of efficiency and effectivity of the solution, the ideal step size for the experimentation is chosen as 10%. The run time of the whole solution methodology is 1.57, 9.5, 12 h for 10%, 5%, 4% respectively.

The mathematical model has been verified and validated by solving an atomic model. Results obtained with CPLEX software and hand solution which is solved by using Ranked Positional Weight Algorithm are compared and investigated.

The solution methodology is executed using Intel® Core™ i7-8750U @ 4.90Ghz, 16 GB DDR4 with Windows 10 Home with real-life data from the company, and the result is shared in Table 3 and 4).

Table 3. Created scenarios with step size 10%

	Demand					
	Line 1		Line 2		Line 3	
Scenario #	Model -3	Model-2	Model-3	Model-1	Model-3	Model-1
1	80000	100000	20000	105000	100000	45000
2	100000	100000	20000	105000	80000	45000
3	100000	100000	100000	45000	0	105000
...	80000	100000	0	120000	120000	30000
18	80000	100000	100000	45000	20000	105000
19	100000	100000	0	105000	100000	45000
20	60000	100000	0	120000	140000	30000
21	80000	100000	80000	60000	40000	90000

From Table 3, it is shown created scenarios with step size 10% according to model's annual demand that the company determined.

Table 4. Results of created scenarios

	Line 1			Line 2			Line 3			
	Cycle time (Sec)			Cycle time (Sec)			Cycle time (Sec)			Total efficiency
Scenario #	Model-3	Model-2	Line efficiency	Model-3	Model-1	Line efficiency	Model-3	Model-1	Line efficiency	
1	118,8	99,2	0,83	118,8	184,0	0,77	118,8	195,4	0,74	77,93%
2	118,8	99,2	0,83	118,8	184,0	0,77	118,8	195,4	0,73	77,69%
3	118,8	99,2	0,83	118,8	184,0	0,79	0	195,4	0,68	76,69%
...	...	...	...	...	...	...	...	...	...	...
18	118,8	99,2	0,83	118,8	184,0	0,79	0	0	0	53,99%
19	118,8	99,2	0,83	0	184,0	0,76	118,8	195,4	0,74	77,63%
20	118,8	99,2	0,84	0	184,0	0,76	118,8	195,4	0,75	78,21%
21	118,8	99,2	0,83	118,8	184,0	0,78	118,8	195,4	0,71	77,44%

As it can be seen from the Table 4, after completion of the solution methodology, the results are as follows: Model 2 and Model 3 are assigned to Line 1, Model 1 to Line 2, and pair of Model 1 and Model 3 to Line 3 with corresponding demands. As a result, the weighted average efficiency of line 1, line 2 and line 3 are 84%, 76% and 75% accordingly. Also, the total average efficiency of three lines is 78.21%. Since obtained cycle times are identical with final test durations which represents lower bound for possible cycle times, it is proven that the algorithm balanced the lines with minimum cycle times.

It is impossible to make an appropriate comparison with the line balancing process of the company, since the company makes the line balancing operations on a model basis, that is, there is no collective line balancing procedure like the solution methodology.

6 Decision Support System

Decision support system is developed for the users to use the proposed solution easily. The Decision Support System is designed and developed dynamically for the company to use in any situation. The DSS has five main parts. These are home page, model details, manual scenario analysis, operation details and optimal line assignment & balancing.

The first page is the welcome screen which contains general information about the Decision Support System. The user can learn how to use the system by reading the information written or by accessing the video on the information button.

Page of model details which is used to add new model, edit current models and delete models is designed to make the tool dynamic. In this way, the company will be able to keep the models up to date in the system and line balancing will always be done correctly.

Manual scenario analysis page allows user to manually determine demand distribution among the lines. After entering the demands of the assigned models into the system, balancing can be started, and results can be observed. Manual scenario analysis page can be seen from below Fig. 1.

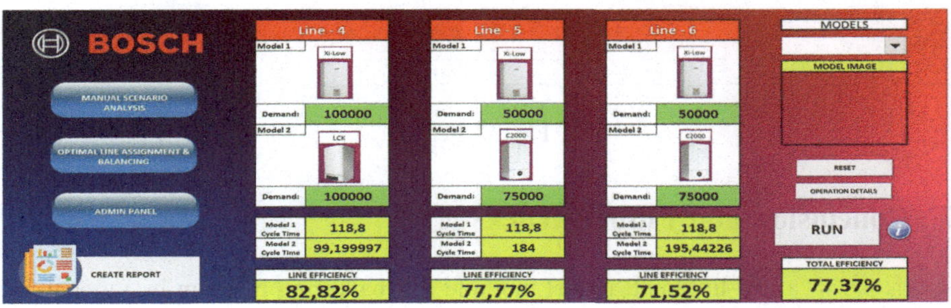

Fig. 1. Manual scenario analysis page

From operation details page, user can monitor operation's processing time and whether it's a test operation or not. Besides, users are able to edit operation details for the changing situations of the company.

Lastly, Optimal Line Assignment & Balancing page is the page where the solution methodology is adopted. By clicking "Run" button and choosing desired step size, first the assignment algorithm will run and create scenario pool, then mathematical models that are designed for lines will be executed, respectively. After completion of the solution methodology, the most efficient scenario is shown to the user (Fig. 2).

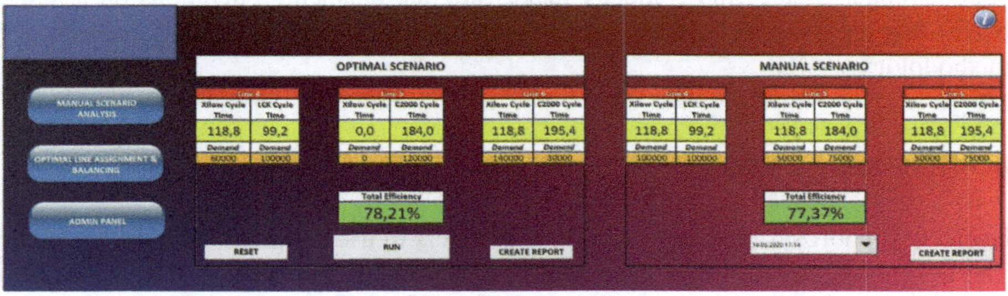

Fig. 2. Optimal line assignment & balancing page

In order to create report of scenarios there are buttons called "Create Report" in both Manual Scenario Analysis and Optimal Line Assignment & Balancing pages. By clicking that buttons user is able to monitor station operation details, station time, cycle time and line efficiency. In addition, reports can be printed or screenshots can be taken from the system (Fig. 3).

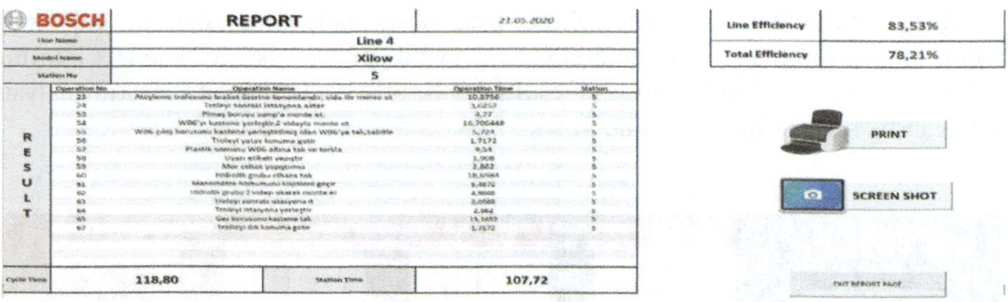

Fig. 3. Report

7 Conclusion and Future Work

In this paper, a real-life multi model multiple line balancing problem in a global heat boiler manufacturer company is addressed. A mathematical model, assignment algorithm and a user-friendly decision support system is developed for the studied real-life problem of the company. The present line balancing process performed in the company is based on observations and experience of the engineer and is carried out by trial and error methods. Therefore, the process is very vulnerable to human error, resulting in significant time and capacity and labor cost losses. Proposed solution approach and accompanying DSS for the implementation aims to shorten the time spent, to minimize the loss of workforce, and to perform line balancing at the highest possible efficiency by utilizing a two-stage approach. As a result of the comprehensive literature research, it was decided that the system at hand was a multi model multi line assembly line balancing, which is classified in NP- Hard class. Therefore, it is difficult to find the optimal result of these problems especially when dealing with a complicated real-life application.

Solution methodology of the problem contains two parts. These are assignment algorithm, mathematical model for the lines once demand mix for each of them are determined for a scenario.

The aim of the assignment algorithm is to find possible demand distribution scenarios for all models on all lines. Thus, a scenario is prepared for solving the mathematical model. The mathematical models of the assembly line balancing problems have been researched and several models that are suitable for the nature of the problem have been identified as candidate models. The constraints, parameters and decision variables of the selected models were examined. Various constraints have been added to ensure that the operations are implemented correctly to the system due to the differences on the lines and necessary changes have been made on the decision variables, sets and indices. The mathematical model coded in the CPLEX.

We show viability of the proposed solution approach first by verifying and validating the approach and then by testing it with real-life data from a global heat manufacturer, and present optimal results obtained from experimentation.

Due to the accessibility, efficiency and effectiveness, Decision Support System (DSS) created. User can easily access and control the data of the line balancing, create, edit and delete new models, make manual line balancing and make line balancing with the solution methodology that created. DSS is a dynamic system hence it gives a solution for the new models when the user enters the necessary information about the operations of the models.

As future work, an integrated model for the two stages can be developed. It is not difficult to foresee that such a model would have run time issues. Therefore, a proper heuristic approach would be needed.

Acknowledgment. This work could not have been completed without the valuable contributions of Beste Çağlayanırmak and Meltem Kırbaş. We are thankful for their input. Furthermore, we are grateful to the company for their co-operation during the whole process.

References

1. Sparling, D., Miltenburg, J.: The mixed-model U-Line balancing problem. Int. J. Prod. Res. **36**(2), 485–501 (1998)
2. Ajenblit, D.A., Wainwright, R.L.: Applying genetic algorithms to the U-Shaped assembly line balancing problem, pp. 96–101. IEEE (1998)

3. Widyadana, G.A.: Multi-objective model for balancing U-type assembly line with permanent and temporary workers. Jurnal Tecnik Industri **11**(1), 33–42 (2009)
4. Baykasoglu, A., Dereli, T.: Simple and U-type assembly line balancing by using an ant colony based algorithm. Math. Comput. Appl. **14**(1), 1–12 (2009)
5. Hamzadayi, A., Yildiz, G.: A genetic algorithm-based approach for simultaneously balancing and sequencing of mixed-model U-line with parallel workstations and zoning constraints. Comput. Ind. Eng. **62**, 206–215 (2012)
6. Chakravarty, A.: Line balancing with task learning effects. IIE Trans. **20**, 186–193 (1988)
7. Eryuruk, S.H., Kalaoglu, F., Baskak, M.: Assembly line balancing in a clothing company. Fibres Text. East. Eur. 1(66), 93–98 (2008)
8. Ozcan, U., Toklu, B.: A new hybrid improvement heuristic approach to simple straight and U-type assembly line balancing problems. J. Intell. Manuf. **20**, 123–136 (2009)
9. Al-zubaidy, S.S., Alrazaq, F.F.: Multi-model production and assembly line balancing (Caravans Production Workshop). J. Babylon Univ. Eng. Sci. **21**(4), 1301–1312 (2013)
10. Make, M., Rashid, M.F.F., Razali, M.: A review of two-sided assembly line balancing problem. Int. J. Adv. Manuf. Technol. **89**(5–8), 1743–1763 (2017)

Quantitative Analysis of Stakeholder Perspective on the University-Industry-Government Collaboration in İzmir, Turkey

İnanç Kabasakal[1(✉)], Fatma Demircan Keskin[1], Yunus Kaymaz[2], Haluk Soyuer[1], Ahmet Rıza Balım[3], and Engin Eryılmaz[4]

[1] Department of Business Administration, Ege University, Izmir, Turkey
{inanc.kabasakal,fatma.demircan.keskin, haluk.soyuer}@ege.edu.tr
[2] Department of International Logistics Management, Iskenderun Technical University, Hatay, Turkey
kymzyns@gmail.com
[3] Bogazici University Technology Transfer Office, Istanbul, Turkey
ahmet.balim@boun.edu.tr
[4] Kobivasyon, Ankara, Turkey
engin@kobivasyon.com

Abstract. Industry 4.0 necessitates the transformation of businesses towards the use of digital technologies and become more competitive in a global market. The collaboration of businesses and academic institutions is a fundamental enabler to achieve innovation. The models, such as Triple Helix, emphasize this collaboration and have increasingly received a great deal of attention from government policies in Turkey. With this perspective, a workshop hosted by the Aegean Region Chamber of Industry has been organized with the participation of representatives from universities, industry, and public institutions. This study presents our findings obtained through the workshop and aims to propose insights for policymakers. In particular, our survey data consists of indicators that were organized into three focal themes: collaboration, digitalization, and innovation. The survey data have been clustered based on participants' ranking on the indicators using the k-means technique.

Keywords: University-industry-government · Triple Helix Model · K-means clustering · Industry 4.0 · EGEVASYON

1 Introduction

University-industry collaboration has been regarded as an essential process in technology transfer that involves the exchange of knowledge. Prior research has explored different forms of such collaboration with their benefits and obstacles. From an industry perspective, university-industry collaboration establishes a ground for technology transfer and innovation. For researchers, this collaboration facilitates the commercialization of academic studies and creates opportunities to test their theories on real-life problems dealt with by businesses. From a broader context, achieving an effective

N. M. Durakbasa and M. G. Gençyılmaz (Eds.): ISPR 2020, LNME, pp. 963–974, 2021.
https://doi.org/10.1007/978-3-030-62784-3_80

university-industry collaboration facilitates innovation and production of higher value-added products and services, while triggering technological progress. Moreover, the regulatory role of governments in university-industry collaboration has been addressed in prior studies, such as the Triple Helix Model.

The potential benefits of university-industry collaboration have been extensively investigated in various reports issued by government institutions. Moreover, there have been remarkable incentives that aim to induce academic researchers and businesses to develop and implement R&D projects. On the other hand, the strategic reports issued by public administration underlines a variety of aspects in the university-industry collaboration that needs to be improved. This paper presents a preliminary study held during the planning of the EGEVASYON Center, which prescribes a new university-industry interface that aims to align academic research with the R&D held in the industry.

Our study aims to provide a summary of the opinions by participants experienced on the university-industry concept, mostly from universities and intermediary organizations in İzmir. Participants who attended the workshop held for EGEVASYON, have been asked to examine and score a list of themes related to university-industry collaboration, that had been chosen in the prior phase of our project. The data collected consists of pairs of scores that prioritize and assess those themes. By this means, our study explores the responses from groups of participants that tend to assign similar priorities on themes and figure out whether the thematic assessments by those groups differ. As a further note, the project proposal has been under review for support during which this manuscript was authored.

Our cluster analysis with k-means technique over priority scores had resulted in two groups of participants. Moreover, our methodology involves statistical analysis to explore the differences in assessment scores by both participant groups. The evaluation scores of the participants in clusters were analyzed on each indicator to check if the assumptions of the independent samples t-test are satisfied. Independent samples t-test was employed for comparing the scores on indicators where the assumptions were met. Otherwise, the procedure of the Mann Whitney U test was conducted. Our findings indicate that the evaluation scores for themes "Collaboration with Academicians" and "Test Centers and Laboratories" were statistically different at the significance level of 0.05 across two clusters.

2 Literature Review

The university-industry collaboration has been increasingly growing in parallel with industrialization, where universities traditionally had primarily fulfilled the supportive role in research activities. The technological breakthrough in research systems during the post-World War II era had been a milestone for the collaboration among universities and the industry. The universities have been regarded as the fundamental and privileged stakeholders in conducting scientific research and assessment of industrial applications [1]. The characteristic of this era until the 1980s was the fundamental driver role presumed for universities in technological progress [2].

Universities have been mostly the focus of scientific research, where industrial facilities host the application of new technologies and support national growth. In this system, governments have a regulatory role that facilitates scientific research with incentives. The creative productivity of a nation depends on the harmonic and systematic cooperation of universities and the industry that benefits from the regulatory role of the government [3].

The Bayh-Dole Act of the USA issued in 1980 [4] had accelerated the interest of academic researchers and policymakers towards the university-industry cooperation and increased the commercialization of products and technologies arising from academic research [5]. By the end of 1980s, the university-industry collaboration had been regarded as an essential factor of development, and the Regional Innovation System had become an important tool to assess economic development [6]. The effectiveness of this approach relies on the quality of interaction among stakeholders that include universities, businesses, financial institutions, and public administration [7]. Moreover, this approach conforms to the Triple Helix Model [6], which has been the most popular university-industry collaboration model over the last years [2].

The complex nature of the university - industry interaction originates from factors including the complexity in the flow of knowledge, presence of multiple stakeholders in the process, and the distinctive role of universities in commercialization beyond the linear model of innovation [1]. Accordingly, describing the interactions among stakeholders required more complicated structures.

The university-industry collaborations might be differentiated in terms of the period that contains such activities and the degree of interaction [8]. Accordingly, on-demand problem solving, licensing, and consulting are planned in short-term while joint projects and partnerships are planned in long-term. Table 1 demonstrates a typology of university-industry links regarding the degree of interaction.

Table 1. Degrees of university-industry interactions (Source: Guimón [8])

High	Research Partnerships, Research Services and Share of Infrastructures
Moderate	Academic Entrepreneurship, Training and Transfer of Human Resources
Low	Commercialization of Intellectual Property, Scientific Research, Informal Interaction

The Triple Helix Model provides an analytical framework for assessment and analysis of information networks and their interactions within the context of the innovation process at individual-level, corporate-level, and national-level. According to this model, innovation networks should be established among university, industry, and government. The presence of such networks will provide opportunities for stakeholders to share knowledge, to enable applications of academic research in industry problems, and to formulate policies that prescribe the governmental support for new research areas [9].

The obstacles for the university-industry collaboration have been investigated extensively in the literature. One of the obstacles is the lack of a legal framework to maintain an environment that supports such collaboration [10]. Additionally, the physical distance has been regarded as another issue that might hinder the university-industry collaboration [11, 12]. In another study, [13] have investigated the difficulties in technology transfer and listed the most prominent problems as in follows:

- The lack of incentives for researchers
- Excessive bureaucracy and rigidity in universities
- The lack of funds reserved for technology transfer
- Inadequate skills of TTO's in marketing and negotiation
- The aggressive exercises by universities for intellectual property
- High expectations in researchers that overestimate the value of their intellectual property.

On the other hand, university-industry collaboration provides various opportunities. As noted by [8], such collaboration facilitates access to funds, provides opportunities to obtain empirical data from real problems dealt in various industries, and improves the quality of education. Moreover, [14] underlined that the partnerships between universities and industry provide better ways to train students by extending the curricula with applications in the industry. According to [15], such collaboration helps businesses with cost-reductions obtained by benefiting from research conducted in universities and provides improvements in the technological capabilities of businesses.

Technology transfer is crucial for nations that have obtained proficiency in basic sciences, but also struggling to achieve technological progress [16]. Technology Transfer Offices (TTO's), incubators, and Collaborative Research Centers (CRCs) have been constituted to eliminate the obstacles that hinder technology transfer. Among these intermediary organizations, TTO's deal with the obstacles in the cognitive and organizational level, whereas incubators and CRCs aim to reduce the social and physical distance among stakeholders [17]. TTO's conduct various functions in the commercialization of research outputs through license agreements; support researchers by facilitating access to funds; and provide consultancy for businesses and researchers; increase the revenues and R&D investments of universities; and create job opportunities for researchers and graduates [18]. There are also other management models in the university-industry ecosystem such as the Stage-Gate [19], which had been founded by the Swiss Institute for Entrepreneurship to establish partnerships among SMEs and researchers [20].

The success in the university-industry collaboration relies on factors including the presence of active researchers and investors, and the degree of the benefit obtained by the industry [21]. The interfaces have an essential role in increasing the degree of mutual benefit from this collaboration. The functions conducted by intermediary organizations might be evaluated in terms of several dimensions, including their impact on the university-industry collaboration, their supportive functions for researchers, the technological progress originated from the university. In particular, [22] have provided a list of factor dimensions which are used in the performance assessment of interfaces (Table 2):

Moreover, [22] also provided a list of specific measures for TTOs, including the gross licensing incomes, percentage of licenses yielding income, number of start-ups, annual partnership counts established, income per license.

Table 2. Success factors related to the TTOs (Source: York and Ahn, [22])

Factor	Description
1	Business Strategies and Marketing
2	Protection of Intellectual Property Rights
3	Performance Evaluations/Benchmarks
4	Focus on Income Generation
5	Contribution to Institutional Prestige
6	Collaboration with stakeholders
7	Synchronization with related institutions
8	Institutional Support

The intermediary organizations in Turkey consist of TTO's, R&D Centers, University-Industry Collaboration Centers, and Technology Parks [23]. By August of 2019, the number of R&D Centers was 1188, while the number of Technology Development Zones was 64. The main objectives of these interfaces have been described in alignment with Turkey's strategic vision, as in the following [24]:

- Putting the scientific studies into service for use in the industry
- To facilitate the commercialization of academic knowledge
- Generating income for universities to finance research
- Supporting technology-based entrepreneurship through spin-off's and start-up's
- Stimulating scientific research in universities towards the needs of Turkish industry
- Supporting intellectual property in order to encourage inventions and increase patents.

Prior studies conducted by the public administration report several issues that underline why and how the university-industry collaborations struggle to flourish. The final report of "İzmir Regional Innovation Committee Meeting", hosted by İzmir Development Agency (İZKA) in February 2012 [25], depicts the progress in the collaboration while signifying shortcomings originating from the industry representatives. Accordingly, one of the problems from the industry perspective had been reported as the lack of awareness in academic studies being conducted. Additionally, another obstacle from the industry perspective is the unavailability of access to the laboratories and research centers located in universities.

From the industry perspective, matching with institutions or researchers that possess the knowledge being needed is crucial; additionally, alignment of parties on a mutually beneficial basis requires identifying and clarifying project expectations in both ways [26]. Furthermore, [27] prescribes joint steering groups involving both academicians and corporate executives for accomplishing strategic partnerships between the universities and industry.

3 Methodology

The call conference has been hosted by the Aegean Region Chamber of Industry, held in İzmir on September 25th, 2019. The conference has been programmed in three sessions, where 62 participants from the industry, universities, and intermediary organizations have shared experiences and presented their ideas in six tables. The majority of the participants (~70%) consisted of executives and experts from the industry. Moreover, ~20% of participants involving academic researchers and TTO experts had originated from universities. Additionally, the other group (~10%) of participants involved representatives from government institutions in Turkey, involving İzmir Development Agency, Turkish Employment Agency, Aegean Region Chamber of Industry, Provincial Directorate of Ministry of Industry. At the end of sessions, 39 participants volunteered to fill in a form anonymously. As demonstrated in Table 3, the form involves a list of themes that relate to the needs of businesses regarding the university-industry collaboration. The themes were organized into three groups, namely, access to critical knowledge (Group #1), innovation and R&D for value-added production (Group #2), and the use/application of emerging technologies (Group #3). Each participant provided a pair of scores on the Likert scale for each theme, where the former denoted the degree of significance, while the latter corresponded to their assessment of the present situation for the industry.

Table 3. List of themes assessed by participants

Code	Theme	Code	Theme
1.1	Collaboration with academic researchers	**2.7**	Access to last user experience for products/services
1.2	Literature Review (Thesis, Publication, Patent)	**2.8**	Innovative ideas generated by in-house experts
1.3	Test Centers and Laboratories	**2.9**	Obtaining consultancy from experts
1.4	Workforce in Specialized Profession Areas	**2.10**	Obtaining mentoring services
1.5	Skilled Labor	**2.11**	Access to R&D funds provided by government
1.6	Participation in National and International Fairs and Events	**3.1**	3D drawing software
1.7	Following up the International Competitors	**3.2**	3D analysis software
1.8	Closely Following International Technology Producing Companies	**3.3**	3D electronic circuit design
1.9	Closely Following national Technology Producing Companies (Companies located in technology parks, R&D Centers)	**3.4**	3D prototyping (Plastic)
2.1	Innovation Processes	**3.5**	3D prototyping (Metal)
2.2	R&D processes and their systematics	**3.6**	Industrial Automation based on data
2.3	Systematical applications in team organization and teamwork	**3.7**	Digitalization
2.4	Application of project management methodologies	**3.8**	Internet of Things
2.5	Use of learning methodologies by project teams	**3.9**	Virtual/Augmented Reality
2.6	Trainings aimed to trigger innovation (Triz, Design Thinking, Biomimicry)		

4 Analysis and Findings

The dataset of this study comprises the evaluation scores of the participants on the priorities and the present situation of themes. As a beginning of the data analysis, means of all participants' evaluations for each theme and theme group were examined to describe the general structure of the data. Mean scores on the priorities of the themes were in the range of 4.10 (theme 1.2) and 4.85 (theme 1.5). Accordingly, the most important themes were 1.5, 3.6, 3.7, 3.8 and 2.11; which correspond to '*Skilled Labor*', '*Industrial Automation based on data*', '*Digitalization*', '*Internet of Things*', and '*Access to R&D funds provided by government*' respectively. The theme group with the highest priority was '*R&D for value-added production*' (Group 2) with the mean score of 4.52; where '*Access to critical knowledge*' (Group 1) had the lowest mean of 4.39. Participants evaluated the present situations on the themes between 1.92 (theme 3.9) and 3.36 (theme 1.6) on average. The theme group with the highest rating on the present situation was Group 1 with the mean score of 2.96, whereas Group 3 had the lowest mean score (2.48).

Beyond the descriptive statistics, our methodology incorporates further analysis to explore different patterns within the response data obtained from participants. Initially, the participants have been separated into groups based on their assessments on the themes' priorities. For such purpose, k-means clustering technique has been employed. The optimal cluster size was chosen based on the silhouette width measure. Accordingly, the silhouette width of 0.3066 was obtained with k = 2, which has been taken as the optimal cluster size. Eventually, two clusters have been obtained, namely Cluster 1 and Cluster 2, with 21 and 18 participants, respectively (Fig. 1).

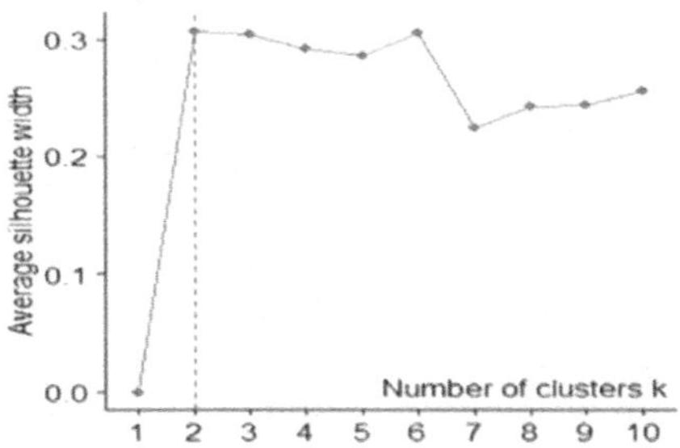

Fig. 1. Silhouette measure calculated for cluster size

The participants in the first cluster has higher mean assessment scores on each theme. The themes that had the highest impact on the separation of clusters were 2.4 (mean difference 1.23), 2.5 (mean difference 1.13), 1.2 (mean difference 1.12), and 1.3 (mean difference 1.12), respectively. Among all themes, the participants in these two clusters evaluated 1.5 with the closest priority scores (mean difference 0.33).

The dataset was analyzed to determine the appropriate statistical test to compare the location measures of the participants' evaluations in the clusters. In comparing two independent samples, parametric or non-parametric statistical analyzes can be applied depending on the characteristics of the samples. Firstly, it was checked the assumptions (e.g., normality, equal variances) of the independent sample t-test, a parametric test that analyzes whether the independent samples represent populations with different mean values [28].

To test the normality by clusters for each measure, the Shapiro-Wilk test [29], one of the most widely used normality tests when the sample size is less than or equal to 50 [30], was employed. The result of this test is given in Table 4.

Table 4. Output of the Shapiro-Wilk test of normality (Significance levels for two-tailed test: *** 0.01; ** 0.05)

Theme	Clusters	Statistic	df	Sig.	Theme	Clusters	Statistic	df	Sig.	Theme	Clusters	Statistic	df	Sig.
1.1	1	.909	21	.053	**2.2**	1	.906	21	.045**	**3.1**	1	.892	21	.024**
	2	.837	18	.005***		2	.879	18	.025**		2	.896	18	.049**
1.2	1	.877	21	.013**	**2.3**	1	.877	21	.013**	**3.2**	1	.889	21	.022**
	2	.839	18	.006***		2	.816	18	.003***		2	.914	18	.100
1.3	1	.919	21	.084	**2.4**	1	.888	21	.021**	**3.3**	1	.899	21	.034**
	2	.884	18	.030**		2	.894	18	.045**		2	.831	18	.004***
1.4	1	.860	21	.006***	**2.5**	1	.873	21	.011**	**3.4**	1	.907	21	.048**
	2	.866	18	.015**		2	.867	18	.016**		2	.831	18	.004***
1.5	1	.865	21	.008***	**2.6**	1	.859	21	.006***	**3.5**	1	.859	21	.006***
	2	.753	18	.000***		2	.878	18	.024**		2	.816	18	.003***
1.6	1	.866	21	.008***	**2.7**	1	.918	21	.081	**3.6**	1	.885	21	.018**
	2	.915	18	.106		2	.862	18	.013**		2	.888	18	.036**
1.7	1	.873	21	.011**	**2.8**	1	.915	21	.068	**3.7**	1	.897	21	.031**
	2	.924	18	.150		2	.924	18	.150		2	.892	18	.041**
1.8	1	.827	21	.002***	**2.9**	1	.843	21	.003***	**3.8**	1	.885	21	.018**
	2	.874	18	.021**		2	.924	18	.154		2	.818	18	.003***
1.9	1	.848	21	.004***	**2.10**	1	.819	21	.001***	**3.9**	1	.714	21	.000***
	2	.891	18	.040**		2	.858	18	.011**		2	.817	18	.003***
2.1	1	.912	21	.061	**2.11**	1	.902	21	.038**					
	2	.886	18	.033**		2	.891	18	.040**					

For all themes except for Theme 2.8 (Innovative ideas generated by in-house experts), the p-value of at least one cluster was less than the significance level of 0.05. In other words, all themes except for Theme 2.8, the null hypothesis of normality is rejected for at least one cluster at 0.05 significance level. Homogeneity of variances was also checked for the measure of 2.8 by using Levene's test [31]. Output of the independent samples t-test, including the result of Levene's test, is given in Table 5.

Table 5. Output of the independent samples t-test

Theme		Levene's test for equality of variances		t-test for equality of means				
		F	Sig.	t	df	Sig. (two tailed)	Mean difference	Std. error difference
2.8	Equal variances assumed	0.03	0.957	1.157	37	.255	.45238	.39086
	Equal variances not assumed			1.153	35.482	.257	.45238	.39236

Since the p-value for Levene's statistics is 0.957 greater than 0.05, it can be assumed that the variances are equal for these two clusters at 0.05 significance level. All assumptions of the independent samples t-test were satisfied for the theme 2.8. The p-value for the equal variances independent samples t-test is .255 (>0.05). According to this score, there is not enough evidence to conclude that there is a significant statistical difference in the mean score of the clusters on the theme of 2.8.

To compare clustered data that are not normally distributed, rank-based tests are frequently preferred [32]. The Mann-Whitney U test [33] is one of the most frequently employed non-parametric tests to compare two samples [34] to conclude that the two analyzed samples represent populations with different median values [28]. Table 6 presents the mean and sum of ranks of the evaluations of the experts from the first and second clusters and the Mann-Whitney U test results.

Table 6. Output of the Mann Whitney U test (Significance levels for two-tailed test: *** 0.01; ** 0.05)

Theme	Clusters	Mean rank	Sum of ranks	Sig. (two-tailed)	Theme	Clusters	Mean rank	Sum of ranks	Sig. (two-tailed)
1.1	1	23.52	494.00	.028**	**2.6**	1	19.45	408.50	.737
	2	15.89	286.00			2	20.64	371.50	
1.2	1	22.07	463.50	.202	**2.7**	1	20.05	421.00	.977
	2	17.58	316.50			2	19.94	359.00	
1.3	1	23.17	486.50	.050**	**2.9**	1	18.86	396.00	.481
	2	16.31	293.50			2	21.33	384.00	
1.4	1	22.48	472.00	.116	**2.10**	1	19.19	403.00	.613
	2	17.11	308.00			2	20.94	377.00	
1.5	1	21.10	443.00	.477	**2.11**	1	22.52	473.00	.118
	2	18.72	337.00			2	17.06	307.00	
1.6	1	22.93	481.50	.068	**3.1**	1	19.60	411.50	.801
	2	16.58	298.50			2	20.47	368.50	
1.7	1	22.55	473.50	.117	**3.2**	1	17.90	376.00	.199
	2	17.03	306.50			2	22.44	404.00	
1.8	1	21.81	458.00	.266	**3.3**	1	19.57	411.00	.790
	2	17.89	322.00			2	20.50	369.00	
1.9	1	20.38	428.00	.811	**3.4**	1	20.07	421.50	.964
	2	19.56	352.00			2	19.92	358.50	
2.1	1	22.02	462.50	.209	**3.5**	1	19.93	418.50	.965
	2	17.64	317.50			2	20.08	361.50	
2.2	1	20.24	425.00	.883	**3.6**	1	20.14	423.00	.931
	2	19.72	355.00			2	19.83	357.00	
2.3	1	20.69	434.50	.667	**3.7**	1	20.17	423.50	.918
	2	19.19	345.50			2	19.81	356.50	
2.4	1	19.93	418.50	.965	**3.8**	1	22.33	469.00	.148
	2	20.08	361.50			2	17.28	311.00	
2.5	1	20.21	424.50	.894	**3.9**	1	18.95	398.00	.504
	2	19.75	355.50			2	21.22	382.00	

The results indicate that assessments of the participants in two clusters on the themes were not significantly different at 0.05 significance level, except for the Theme 1.1 (Collaboration with academic researchers) and Theme 1.3 (Test Centers and Laboratories). The median scores of the evaluations of the participants in the first cluster on collaboration with academic researchers, and test centers and laboratories are higher than those in the second cluster.

5 Conclusions

University-industry collaboration has been regarded as an essential process for innovation and technological progress over the last decades. The concept offers promising opportunities for academic researchers as well as businesses that invest in R&D. Various forms of intermediary organizations, such as TTO's, have been constituted worldwide to facilitate this interaction and encourage joint research at different levels. In Turkey, the concept has received attention from the government, that provides incentives to facilitate the cooperation of academy and industry.

From the industry perspective, collaborations with universities bring out specific concerns that require further examination, such as time pressure in projects or a higher degree of collaboration required with researchers. For this purpose, a new intermediary organization, namely EGEVASYON, has been proposed by the initiative of the Aegean Region Chamber of Industry, with the objective of carrying out collaborative functions among stakeholders, while considering such concerns. Among the preliminary steps of this project, our study attempted to explore which themes of collaboration have been prioritized by representatives from the industry and academy. According to the assessments of theme priority by participants, '*Skilled Labor*', '*Industrial Automation based on data*', '*Digitalization*', '*Internet of Things*', and '*Access to R&D funds provided by government*' have been the reported with the highest importance. Additionally, the participants' assessments on the present situation in Turkish industry were highest for the themes '*Innovative ideas generated by in-house experts*', '*Skilled Labor*', '*Participation in National and International Fairs and Events*', '*Following up the International Competitors*', '*Application of project management methodologies*'. Finally, our study has grouped the participants into two clusters based on their priority assignments for themes and highlighted the significant differences of assessments across both groups.

Acknowledgment. The authors would kindly present thanks to Mr. Atilla SEVİNÇLİ, board member in the Aegean Region Chamber of Industry, for his extensive support for this study.

References

1. Bramwell, A., David, A.W.: Universities and regional economic development: the entrepreneurial university of Waterloo. Res. Policy **37**, 1175–1187 (2008)
2. Kiper, M.: Dünyada ve Türkiye'de Üniversite-Sanayi İşbirliği ve Bu Kapsamda Üniversite-Sanayi Ortak Araştırma Merkezleri Programı (ÜSAMP). Türkiye Teknoloji Geliştirme Vakfı (2010)

3. Göker, A.: Ulusal inovasyon sistemi ve üniversite-sanayi işbirliği/Ankara Üniversitesi Fen Bilimleri Enstitüsü Geleneksel Bahar Paneli IV: Bilimsel Araştırmada Üniversite-Sanayi İşbirliği (2010). http://www.inovasyon.org/pdf/ayk.ank.uni.nisan00.pdf
4. Mowery, D.C., Sampat, B.N.: The Bayh-Dole Act of 1980 and university–industry technology transfer: a model for other OECD governments? J. Technol. Transf. **30**(1–2), 115–127 (2004)
5. Goel, R.K., Göktepe-Hultén, D.: What drives academic patentees to bypass TTOs? Evidence from a large public research organisation. J. Technol. Transf. **43**(1), 240–258 (2018)
6. Taşkın, C., Tuncel, C.O.: Analyzing the role of university industry-collaboration to regional development: the case study of Bursa region in Turkey. Eur. J. Soc. Sci. Educ. Res. **1**(1), 279–284 (2014)
7. Cooke, P.: Regional innovation systems: origin of the species. Int. J. Technol. Learn. Innov. Dev. **1**(3), 393–409 (2008)
8. Guimón, J.: Promoting university-industry collaboration in developing countries. World Bank **3**, 12–48 (2013)
9. Lu, L., Etzkowitz, H.: Strategic challenges for creating knowledge-based innovation in China: transforming triple helix university-government-industry relations. J. Technol. Manag. China **3**(1), 5–11 (2008)
10. Jacobsson, S., Karltorp, K.: Mechanisms blocking the dynamics of the European offshore wind energy innovation system–challenges for policy intervention. Energy Policy **63**, 1182–1195 (2013)
11. D'Este, P., Guy, F., Iammarino, S.: Shaping the formation of university–industry research collaborations: what type of proximity does really matter? J. Econ. Geogr. **13**(4), 537–558 (2012)
12. Rybnicek, R., Königsgruber, R.: What makes industry–university collaboration succeed? A systematic review of the literature. J. Bus. Econ. **89**(2), 221–250 (2019)
13. Siegel, D.S., Waldman, D.A., Atwater, L.E., Link, A.N.: Commercial knowledge transfers from universities to firms: improving the effectiveness of university–industry collaboration. J. High Technol. Manag. Res. **14**(1), 111–133 (2003)
14. Cyert, R.M., Goodman, P.S.: Creating effective university-industry alliances: an organizational learning perspective. Org. Dyn. **25**(4), 45–58 (1997)
15. Ankrah, S., Omar, A.T.: Universities–industry collaboration: a systematic review. Scand. J. Manag. **31**(3), 387–408 (2015)
16. Kılıç ve, A., Ayvaz, U.: Üniversite-Sanayi-Devlet İşbirliğinin Sağlayıcısı Olarak Teknoparklar ve Teknoloji Transferi İşbirliklerinde Mevcut Durum. *Savunma Bilimleri Dergisi* **10**(2), 58–79 (2011)
17. Villani, E., Rasmussen, E., Grimaldi, R.: How intermediary organizations facilitate university–industry technology transfer: a proximity approach. Technol. Forecast. Soc. Chang. **114**, 86–102 (2017)
18. Siegel, D.S., Veugelers, R., Wright, M.: Technology transfer offices and commercialization of university intellectual property: performance and policy implications. Oxf. Rev. Econo. Policy **23**(4), 640–660 (2007)
19. Stage-Gate International: The Official Site of Stage-Gate (2020). https://www.stage-gate.com/
20. Cooper, R.G.: Stage-gate systems: a new tool for managing new products. Bus. Horiz. **33**(3), 44–54 (1990)
21. Etzkowitz, H., Goktepe-Hulten, D.: Maybe they can? University technology transfer offices as regional growth engines. Int. J. Technol. Transf. Commer. **9**(1), 166–181 (2010)
22. York, A.S., Ahn, M.J.: University technology transfer office success factors: a comparative case study. Int. J. Technol. Transf. Commer. **11**(1), 26–50 (2012)

23. Yalçıntaş, M.: Üniversite-sanayi-devlet işbirliğinin ülke ekonomilerine etkileri: Teknopark İstanbul örneği. Finansal Araştırmalar ve Çalışmalar Dergisi **5**(10), 83–106 (2014)
24. Devlet Denetleme Kurumu: 4691 Sayılı Teknoloji Geliştirme Bölgeleri Kanunu Uygulamalarının Değerlendirilmesi İle Uygulamada Ortaya Çıkan Sorunların Çözümüne İlişkin Öneri Geliştirilmesi. Araştırma ve İnceleme Raporu 2009/1. Ankara
25. İzmir Development Agency: İzmir Regional Innovation Committee Meeting Report, February 2012
26. Edmondson, G., Valigra, L., Kenward, M., Hudson, L.R., Belfield, H.: Making industry-university partnerships work, lesson from successful collaborations. Science/Business/innovation Board AISBL (2012)
27. Boccanfuso, A.M.: Researcher Guidebook: A Guide for Successful Institutional-industrial Collaborations. University Industry Demonstration Partnership (2012)
28. Sheskin, D.J.: Handbook of Parametric and Nonparametric Statistical Procedures, 2nd edn. Chapman & Hall/CRC, Boca Raton (2000)
29. Shapiro, S.S., Wilk, M.K.: An analysis of variance test for normality (complete samples). Biometrika **52**(3/4), 591–611 (1965)
30. Yap, B.W., Sim, C.H.: Comparisons of various types of normality tests. J. Stat. Comput. Simul. **81**(12), 2141–2155 (2011)
31. Levene, H.: Robust tests for equality of variances. In: Olkin, I., Ghurye, S.G., Hoeffding, W., Madow, W.G., Mann, H.B. (eds.) Contributions to Probability and Statistics, pp. 278–292. Stanford University Press, Stanford (1960)
32. Galbraith, S., Daniel, J.A., Vissel, B.: A study of clustered data and approaches to its analysis. J. Neurosci. **30**(32), 10601–10608 (2010)
33. Mann, H.B., Whitney, D.R.: On a test of whether one of two random variables is stochastically larger than the other. Ann. Math. Stat. **18**(1), 50–60 (1947)
34. Emura, T., Hsu, J.H.: Estimation of the Mann-Whitney effect in the two-sample problem under dependent censoring. Comput. Stat. Data Anal. **150**, 1–17 (2020)

Parcel Delivery Service Quality Assessment

Belgin Aldatmaz, Candan Sever, Boran Pehriz, Aylin Çalışkan(✉), and Burak Çetiner

Yasar University, Izmir, Turkey
belginnaldatmaz@gmail.com, candansever97@gmail.com, pehrizboran@gmail.com, {aylin.caliskan,burak.cetiner}@yasar.edu.tr

Abstract. The parcel delivery industry has been struggling both structurally and operationally with the increase in internet shopping, the development of technology and changing consumer needs. In order to ensure the satisfaction of both commercial and individual customers and to gain profit by maintaining their competitive position, companies operating in the parcel delivery industry have to evaluate the current status of their services and identify the areas that need improvement. In this study, it is aimed to evaluate the current industry position by measuring service quality levels of each parcel delivery company operating in Turkey and determine the parcel service quality attributes that are needed to be improved. SERVQUAL method was used in order to evaluate the current service quality levels. After, a KANO model was developed to identify the service quality attributes that are needed to be improved.

Keywords: Parcel · Cargo · Service quality · KANO · SERVQUAL

1 Introduction

Parcel delivery can be defined as the activities identified with the transfer of any load from the point of origin to the point of destination [1]. In another study, parcel delivery is defined as transportation of goods which does not exceed 100 kgs, from one place to another, within a certain time period, usually in packages, container files, package boxes, etc. [2]. In the international literature, parcel delivery is described as the transportation of any load, without any regard to its weight, size, or content [3]. Parcel delivery industry is a rapidly growing sector that requires complex logistics efforts and it provides transportation services for individuals and businesses [4]. According to [38], although the transportation industry is a considerably young industry as it is 35–40 years old, it handles 7,000,000 shipments every day. With the expanding significance of the parcel delivery companies in the worldwide economy, parcel companies' capabilities to provide great values to their clients are crucial [5]. Parcel delivery companies, unlike other shipping companies, help individuals satisfy their needs in such critical situations [6]. According to [6], parcel delivery companies are instruments that make individuals' life simpler. Simultaneously, cargo activities are considered as a key component in guaranteeing consumer loyalty [7]. The awareness of parcel delivery firms must be high and they need to put customers' needs and problems into their focus

N. M. Durakbasa and M. G. Gençyılmaz (Eds.): ISPR 2020, LNME, pp. 975–988, 2021.
https://doi.org/10.1007/978-3-030-62784-3_81

in order to maintain consumer loyalty [8]. In the light of this information, it can be said that consumer loyalty and performance of parcel delivery companies are positively correlated. When the performance of parcel delivery companies increases, customer satisfaction also increases.

The parcel delivery sector has become an area where there is a strong competition with national and global players. Quality of the service provided by parcel delivery companies is becoming more important with each passing day. In this context, identifying key elements for high quality service and making the necessary improvements is crucial to create and maintain consumer loyalty [9, 10]. Therefore, the aim of this study is to evaluate the current industry position by measuring service quality levels of each parcel delivery company operating in Turkey and determine the parcel service quality attributes that are needed to be improved. SERVQUAL method was used in order to evaluate the current service quality levels. After, a KANO model was developed to identify the service quality attributes that are needed to be improved.

2 Literature Review

The parcel delivery industry attracts a lot of attention from the academic community as it leads to technological developments, contributes to the world economy in a large scale, and greatly impacts last-mile of supply chains. Routing problems to minimize the transportation cost and time, loyalty and performances of employees, performances of cargo companies, and industry 4.0 applications in last-mile delivery are some of the popular topics regarding parcel delivery industry. Because customer satisfaction and loyalty are important for profitability and competitive power, in the literature, the issue of service quality has not been overlooked, either. When we look at the studies conducted in the service quality in parcel delivery, we see that the focus is generally on researches that measure customer satisfaction and loyalty in marketing [3, 11, 16]. The main purpose of these studies is to determine the complaints of the customers based on the customer satisfaction and to inform the companies about the aspects they should develop or improve. For instance, [37]'s main aim was to understand the expectations of the customers and to increase the service quality of cargo firms. Factor analysis was used on survey results during the research. As a result of the research, some suggestions were given to increase the service quality of the company and thus the customer satisfaction. In addition, [5] examined the impact of service quality and service quality of cargo companies on client satisfaction and repurchase behaviour. The information was collected utilizing SERVQUAL scale. The perceived service quality within the investigate model was measured by SERVPERF scale comprising of 22 factors and customer satisfaction and repurchase behaviour with two factors. The studies measuring customer satisfaction in parcel shipment sector [19] show that there are five important factors when measuring customer satisfaction. These factors are availability of service, responsiveness, reliability, completeness of service, professionalism of service. [20] explained that the Kano questionnaire includes these dimensions.

There is an important emphasis on the behavior and clothing of the staff [11–16]. For instance [11] indicated that, the communication between the staff and the customers is very important for the quality of service. In addition, according a study, it was revealed that the clothing, behavior and attitudes of the staff affected the customers' perceptions quite a lot. In other studies, we come across similar results. For example, [12], at the end of their factor analysis, reached the conclusion that the clothes and behaviors of the personnel working for parcel delivery companies affect the customer satisfaction. In addition, it was revealed that the customers gave importance to the experience of the personnel. In another study, we see that, according to the data collected through the survey conducted by a Hong Kong-based research firm with parcel companies and their customers, the communication between the personnel and the customers is very important for the image and reliability of the company [13]. [14] also focuses on the need to constantly update in accordance with customer expectations and he believes that the quality of service can be improved with these updates. In another example, [15] conducted a research on the criterion which the customers base their parcel company selections on and the results indicate that customers want staff to deal with their problems effectively. [16] suggested that staff should respond quickly to customers and the physical appearance of the staff is should be up to the standards as it is important for the customers. There are also studies investigating the loyalty and performance of employees in cargo companies [17, 18]. It is thought that employee performance in a sector that requires mostly physical labor, like cargo transportation, is directly proportional to internal commitment and employee satisfaction. Therefore, the purpose of these studies is to provide solutions for cargo transporters to increase employee satisfaction [17]. [18] believes that managers' actions have a direct impact on employee performance.

In this study, similar to the other studies conducted on the subject, SERVQUAL method was applied to measure the current parcel delivery service quality level. However, as a differentiating aspect, we apply KANO method with the SERVQUAL dimensions to define the service attributes that should be improved.

3 Methodology

This study applied SERVQUAL method in order to evaluate the current service quality levels. After, KANO method was applied to identify the service quality attributes that are needed to be improved. Next sections will describe the methods, survey development, demographics and analyses results.

3.1 Description of the SERVQUAL Method

SERVQUAL is the most widely used method in studies that measure service quality. It is developed by Parasuraman, Zeithalm and Berry between 1983–1990 [21]. The basic attribute of this method is to evaluate expected and perceived service quality of consumers with the help of questionnaire [22]. This scale is based on the principle that customer satisfaction is "a function of perceptions differentiating from expectation" [23]. The Servqual method is vital for businesses to determine their level of service quality [24]. It includes 5 dimensions namely, tangibility, reliability, responsiveness, assurance, and empathy.

The differences between the expectations in the first part of the scale and the perceptions in the second part are determined as the quality of service. Perceived service quality (SERVQUAL score) is calculated as follows:

Perceived Service Quality = Perceived Service - Expected Service [25].

As a result of service quality measurement, three situations occur:

- Expected Quality > Perceived Quality is Low Quality
- Expected Quality < Perceived Quality is High Quality
- Expected Quality = Perceived Quality is the Right Quality [26].

If perceptions exceed expectations, customers are satisfied (positive gap). If their expectations are not met, customers are dissatisfied (negative gap), and when their expectations match their perceptions, customers receive what they expected and therefore they are neither satisfied nor dissatisfied.

Several advantages exist for using Servqual method is the service quality studies [26]. First of all, this model helps to learn the expectations and perceptions of customers and provides important strategic information for businesses [27]. Secondly, the method can be adapted to any industry and can be modified in the number of propositions [28]. Although the Servqual model has many advantages and is commonly used, there are also criticisms about this model. Most of the critics believe that it is unnecessary to look at expectations when measuring service quality. It was stated that it would be better to consider only performance in measuring service quality [29]. Another criticism is that the five dimensions of Servqual scale are not comprehensive enough to be used in all service businesses [30].

3.2 Description of the KANO Method

It is an important issue to meet customer needs and how these requirements affect the level of satisfaction. Kano Model is used to analyze customer needs in the best way. This model, developed by Noritoki Kano in the 1980s, is a model that describes the relationship between the companies' degree of meeting customer needs and customer satisfaction [31].

Kano model classifies product characteristics based on how they are perceived by customers and their effects on customer satisfaction. The theory of attractive quality offers insight into the dynamics of product and service attributes. Theory of attractive quality also deals with the relationship between the objective performance of an attribute and customer satisfaction regarding the said attribute. According to the nature of this relationship, attributes are divided into five quality dimensions: attractive quality, one-dimensional quality, must-be quality, indifferent quality, and reverse quality [32].

Attractive Quality: These are the attributes that are usually not expected by consumer and lead to consumer satisfaction when fulfilled while not leading to any dissatisfaction when not fulfilled. A restaurant providing a free parking lot on site for their customers can be given as an example for attractive quality [33, 34].

One-Dimensional Quality: These are the attributes which satisfy the customers when fulfilled and dissatisfy when not fulfilled. For example: The time it takes to get your job done during a call with a call center [33, 34].

Must-be Quality: These attributes are the requirements that the customers expect and are taken for granted. When you done well, customers are just neutral, but when done poorly, customers are very dissatisfied. For example: In a hotel, providing a clean room [35, 36].

Indifferent Quality: These attributes refer to aspects that are neither good nor bad, and they do not result in either customer satisfaction or customer dissatisfaction. For example, thickness of the wax coating on a milk carton. This might be a key to the design and manufacturing of the carton, but consumers are not even aware of the distinction [34, 35].

Reverse Quality: The level of customer satisfaction is reduced by meeting such features at a high level. For example: A customer that receives a cell phone with a variety of features, and finds the amount of features overwhelming, so their satisfaction decreases [33, 35].

3.3 Survey Development

In this study, surveys were applied to individual customers of parcel delivery companies operating in Izmir province. The questionnaire developed for this study consists of 3 sections. The first part of the survey contains questions about age, average monthly income, education level, and gender for the purpose of determining the demographic characteristics of the survey participants. In addition, the participants were asked which parcel delivery company they preferred the most. In the second part of the survey, 22-item of SERVQUAL was adapted for parcel delivery service and asked with 7 Likert scale. In the third part of the survey, SERVQUAL dimensions were adapted to the KANO model questions.

3.4 Participants

At the end of the study, 300 valid questionnaires were obtained. According to Table 1, 61.7% of the participants were women and 38.3% were men. Of the participants, 8.7% were between the ages of 18 and 21, while 43.3% were between the ages of 22–26, 23% were between the ages of 27 and 35, and 54% were between the ages of 36–49, and 7% were 50 years of age or above. The survey showed that the participants had the highest number of university graduates (58.7%). There were 8.3% of high school graduates, 7.7% of associate graduates and 19.7% of master degree/doctorate graduates. Only 5.7% of participants were found to have graduated from primary school. While 21% of the monthly income of the participants was under 1000 TL, the survey found that there were 85 people with an income of 2001–3000 TL. The results show that the top 3 preferred parcel delivery companies are Yurtiçi Kargo with 32.7%, MNG Cargo with 19%, and UPS Cargo with 15%, respectively. The least preferred parcel delivery company is Sürat Cargo (3.7%).

Table 1. Demographics

	Frequency	Percent (%)		Frequency	Percent (%)
Gender			**Income**		
Female	185	61,7%	1000 TL and below	63	21%
Male	115	38,3%	1001–2000 TL	40	13,3%
Total	300	100%	2001–3000 TL	85	28,3%
Age			3001–4000 TL	45	15%
18–21	26	8,7%	4001 TL and above	67	22,3%
22–26	130	43,3%	Total	300	100%
27–35	69	23%	**Preferred Comp.**		
36–49	54	18%	UPS Cargo	45	15%
50 and above	21	7%	Sürat Cargo	11	3,7%
Total	300	100%	Aras Cargo	43	14,3%
EDUCATION			Yurtiçi Cargo	98	32,7%
Primary School	17	5,7%	MNG Cargo	57	19%
High School	25	8,3%	PTT Cargo	41	13,7%
Associate Degree	23	7,7%	Other	5	1,7%
University	176	58,7%	Total	300	100%
Master Degree/ Doctorate	59	19,7%			
Total	300	100%			

4 Findings

4.1 SERVQUAL Findings

The results of SERVQUAL analysis conducted to measure the performance of parcel delivery companies are given in Table 2. According to the results, all dimensions' perception scores are below the expectation scores. This means, in Izmir, the overall parcel delivery service quality is low. The expectations have not been met in practice.

Expectation scores showed us that *reliability* dimension which includes fulfilling the given promises at a certain time, solving the problems, serving correctly the first time, on time delivery, and accurate records, is the most important parcel delivery service attribute according to the customer evaluations with an expectation score of 6.16. Perception scores reveal that *assurance* and *reliability* dimensions of SERVQUAL received the highest score. Assurance dimension includes confidence, feeling safe, respectful and knowledgeable employees. Perception scores also show that *tangibility* received the lowest score which includes modern vehicles and equipment, visually appealing offices, and clean and well-dressed employees. Receiving the lowest perceived quality score doesn't mean that the service is not adequate to meet the customer demand. That is the main point of SERVQUAL. This method takes into consideration both perception and expectation levels of customers. If the gap between expectation and perception scores is negatively high, then we can say this service is not a quality service and it needs to be improved. Therefore, according to the gap scores we obtained, the lowest service quality is perceived in *responsiveness* dimension which includes service attributes such as delivery time notifications, immediate cargo delivery, helping the customers in case of problems, and responding to customer requests. In this dimension, the parcel delivery service is not able to meet the expectations of customers with a score of 4.98 in perceived service quality in comparison to an expectation score of 6.08. The SERVQUAL results also show that, although the *assurance* dimension does not meet customer expectations like any other dimension, the gap score was found to be less than the others.

Table 2. SERVQUAL results of all parcel delivery companies

Dimension	Statement	Expectation score	Perception score	Gap score (Perception – Expectation)	Average for dimension
Tangibility	**1**	5,94	4,83	−1,11	**−1,08**
	2	5,66	4,44	−1,22	
	3	5,98	4,95	−1,03	
	4	6,05	5,06	−0,99	
Reliability	**5**	6,16	4,97	−1,19	**−1,01**
	6	6,18	4,93	−1,25	
	7	6,17	5,26	−0,91	
	8	6,16	5,10	−1,06	
	9	6,15	5,47	−0,68	
Responsiveness	**10**	6,11	5	−1,11	**−1,1**
	11	6,15	5,10	−1,05	
	12	6,09	4,91	−1,18	
	13	5,99	4,91	−1,08	
Assurance	**14**	5,99	5,04	−0,95	**−0,91**
	15	6,14	5,25	−0,89	
	16	6,12	5,23	−0,89	
	17	6,11	5,19	−0,92	
Empathy	**18**	5,98	5,07	−0,91	**−1,09**
	19	6,13	5,17	−0,96	
	20	5,99	4,87	−1,12	
	21	5,94	4,71	−1,23	
	22	5,85	4,59	−1,26	

Company based specific results are shown on Table 3. As the perception scores are lower than expectations scores, it can be said that the overall service expectation of customers hasn't been met. The results reveal that, only UPS is close to meeting customer expectations as their average service quality score is close to 0 (−0.19). Aras

Cargo has the lowest score after UPS (−0.26), MNG Cargo is the third (−0.86), Sürat Cargo is the fourth (−1.19), Yurtiçi Cargo is the fifth (−1.48) company. The worst service quality belongs to the PTT Cargo (−2,68). The positive gap scores (0 and 0.18) can be seen on 22th and 19th statements of UPS. It means that UPS is meeting the expectations of customers regarding "*understanding the special needs of customers*" and "*taking care of customers individually*". In Table 3, it is seen that even the lowest gap scores of Yurtiçi Cargo, Sürat Cargo, MNG, and PTT are worse than the worst score of the UPS. According to the results, the lowest service quality was found to be *assurance* dimension for UPS, *tangibility* and *responsiveness* dimensions for Yurtiçi, *responsiveness* dimension for Sürat and MNG, *tangibility* for Aras, and *reliability* for PTT.

Table 3. Company based SERVQUAL results

Dimension		UPS		Yurtici		Sürat		Aras		MNG		PTT	
		GS*	Av.**	GS	Av.	GS	Av.	GS	Av.	GS	Av.	GS	Av.
Tangibles	**1**	−0,07	−0,17	−1,54	−1,57	−1	−1,10	−0,3	−0,35	−1,09	−0,83	−1,97	−2,11
	2	−0,17		−1,81		−0,72		−0,45		−1		−2,07	
	3	−0,27		−1,52		−0,90		−0,27		−0,65		−1,95	
	4	−0,2		−1,44		−1,18		−0,4		−0,59		−2,45	
Reliability	**5**	−0,27	−0,19	−1,73	−1,45	−1,2	−1,14	−0,6	−0,26	−0,88	−0,78	−2,69	−3,71
	6	−0,29		−1,77		−1,18		−0,37		−1.09		−3,28	
	7	−0,27		−1,26		−0,81		−0,07		−0,66		−3,29	
	8	−0,04		−1,58		−1,18		−0,18		−0,75		−3,32	
	9	−0,09		−0,92		−1,36		−0,12		−0,54		−3,29	
Responsiveness	**10**	−0,02	−0,02	−1,54	−1,57	−1,91	−1,66	−0,28	−0,26	−1,12	−1,02	−2,91	−2,96
	11	−0,2		−1,36		−1,82		−0,21		−1,12		−3	
	12	−0,04		−1,75		−1,73		−0,26		−0,97		−3,34	
	13	−0,07		−1,63		−1,18		−0,3		−0,88		−2,61	
Assurance	**14**	−0,04	−0,51	−1,39	−1,28	−1,27	−1,11	−0,12	−0,21	−0,81	−0,72	−1,87	−2,17
	15	−0,2		−1,32		−1		−0,19		−0,6		−2,59	
	16	−0,24		−1,25		−1		−0,21		−0,66		−1,83	
	17	−0,11		−1,19		−1,18		−0,35		−0,84		−2,39	
Empathy	**18**	−0,2	−0,06	−1,25	−1,54	−1	−0,98	−0,14	−0,24	−0,68	−0,96	−2,37	−2,48
	19	0,18		−1,26		−1,09		−0,02		−1,09		−2,67	
	20	−0,18		−1,61		−1,36		−0,28		−0,86		−2,35	
	21	−0,11		−1,76		−0,72		−0,32		−1,09		−2,61	
	22	0		−1,86		−0,73		−0,45		−1,11		−2,44	

*GS: Gap Score (Perception-Expectation)
**Av.: Average for dimension

4.2 Kano Findings

The data gathered from questionnaires were categorized as it is shown in the Table 4 and then Table 5 was developed.

Table 4. KANO evaluation table

Customer requests		Answers to the negative question				
		I would like that	It should be	Neutral	I can stand	I don't like that
Answers to the positive question	**I would like that**	Q	A	A	A	O
	It should be	R	I	I	I	M
	Neutral	R	I	I	I	M
	I can stand	R	I	I	I	M
	I don't like that	R	R	R	R	Q

*M: Must be quality; O: One-dimensional quality; A: Attractive quality; I: Indifferent quality; R: Reverse quality; Q: Questionable Answer.

**When a responder answers both positive and negative questions with "I would like that" or "I don't like that" the answer is considered to be questionable.

Based on the data shown in Table 5, all parcel delivery service attributes were defined as one-dimensional quality. It means that these are the attributes which satisfy customers when fulfilled and dissatisfy when not fulfilled. In addition, with the data obtained, customer satisfaction index (SI) and customer dissatisfaction index (DSI) were calculated with following formulas:

$$SI = (A + O)/(A + O + M + I)$$

$$DSI = -(M + O)/(A + O + M + I)$$

The SI and DSI indexes are shown on KANO diagram (Fig. 1). According to the diagram, the following can be said:

(1) Upper right section (both SI and absolute DSI values above 0.5): If improved, the features that fall in section would have a positive outcome on customer satisfaction and dissatisfaction. According to findings, features that correspond to this section are; Modern Equipment, Clean and Well-Groomed Employees, Adequately Equipped Vehicles, Service on Time, Solving Problems and Assisting Right Service at First Time, Service within Promised Time, Accurate Record, Exact Time of Service Performed, Striving to Deliver As Soon As Possible, Willing to Help to Customer, Responding your request, Employee Trust, Feeling Confidence on Transactions, Kind and Respectful Employees, Enough Information to Answer, Taking Care of You Individually Suitable Working Hours, Sincerely Taking Care of You, Customer Interests, Satisfaction, Meeting Service Expectations, Disappointment, Serving As an Ideal Parcel Delivery Company.

Table 5. KANO method

	M	O	A	I	R	M%	O%	A%	I%	R%	MKM	SI	DSI
1-) Modern Equipment	65	101	43	73	0	0.23%	0.36%	0.15%	0.26%	0%	O	0,5106	−0,5886
2-) Visually attractive	37	99	48	95	9	0.13%	0.34%	0.17%	0.33%	0.03%	O	0,5268	−0,4874
3-) Clean and Well-Groomed Employees	61	129	28	54	10	0.22%	0.46%	0.10%	0.19%	0.03%	O	0,5772	−0,6985
4-) Adequately Equipped Vehicles	50	111	53	63	5	0.18%	0.39%	0.19%	0.22%	0.02%	O	0,5920	−0,5812
5-) Service on Time	60	158	14	41	10	0.21%	0.56%	0.05%	0.14%	0.03%	O	0,6300	−0,7985
6-) Solving Problems and Assisting	57	159	15	44	11	0.20%	0.55%	0.05%	0.15%	0.04%	O	0,6327	−0,7854
7-) Right Service at First Time	66	136	27	50	7	0.23%	0.47%	0.10%	0.17%	0.02%	O	0,5842	−0,7240
8-)Service within Promised Time	71	142	12	53	11	0.24%	0.49%	0.04%	0.18%	0.04%	O	0,5539	−0,7661
9-) Accurate Record	68	133	12	57	16	0.24%	0.46%	0.04%	0.20%	0.05%	O	0,5370	−0,7444
10-) Exact Time of Service Performed	49	141	37	48	14	0.17%	0.49%	0.13%	0.17%	0.05%	O	0,6472	−0,6909
11-) Striving to Deliver As Soon As Possible	42	153	27	54	14	0.14%	0.53%	0.10%	0.19%	0.05%	O	0,6521	−0,7065
12-) Willing to Help to Customer	37	150	40	43	13	0.13%	0.53%	0.14%	0.15%	0.04%	O	0,7037	−0,6925
13-) Responding requests	58	145	26	41	21	0.20%	0.50%	0.09%	0.14%	0.07%	O	0,6333	−0,7518
14-) Employee Trust	66	137	31	42	11	0.23%	0.48%	0.11%	0.15%	0.04%	O	0,6086	−0,7355
15-) Feeling Confidence on Transactions	50	153	21	50	14	0.17%	0.53%	0.07%	0.17%	0.05%	O	0,6350	−0,7408
16-) Kind and Respectful Employees	58	154	25	38	13	0.20%	0.53%	0.09%	0.13%	0.04%	O	0,6509	−0,7709
17-) Enough Information to Answer	75	123	31	51	9	0.26%	0.42%	0.11%	0.18%	0.03%	O	0,55	−0,7071
18-) Taking Care Individually	30	141	48	57	13	0.10%	0.49%	0.17%	0.20%	0.04%	O	0,6847	−0,6195
19-) Suitable Working Hours	32	129	63	54	12	0.11%	0.44%	0.22%	0.19%	0.04%	O	0,6906	−0,5791
20-) Sincerely Taking Care	28	137	56	50	14	0.10%	0.48%	0.20%	0.17%	0.05%	O	0,7121	−0,6088
21-) Customer Interests	48	127	47	51	11	0.17%	0.45%	0.16%	0.18%	0.04%	O	0,6373	−0,6410
22-) Understanding of Special Needs	23	112	65	79	13	0.08%	0.38%	0.22%	0.27%	0.04%	O	0,6344	−0,4838
23-) Satisfaction	45	164	22	43	12	0.16%	0.57%	0.08%	0.15%	0.04%	O	0,6788	−0,7627
24-) Meeting Service Expectations	47	161	23	42	9	0.17%	0.57%	0.08%	0.15%	0.03%	O	0,6739	−0,7619
25-) Serving as an Ideal Cargo Company	58	133	27	49	10	0.21%	0.48%	0.10%	0.18%	0.04%	O	0,5992	−0,7153

*MKM: Major Kano Model; SI: Extent of satisfaction; DSI: Extent of dissatisfaction

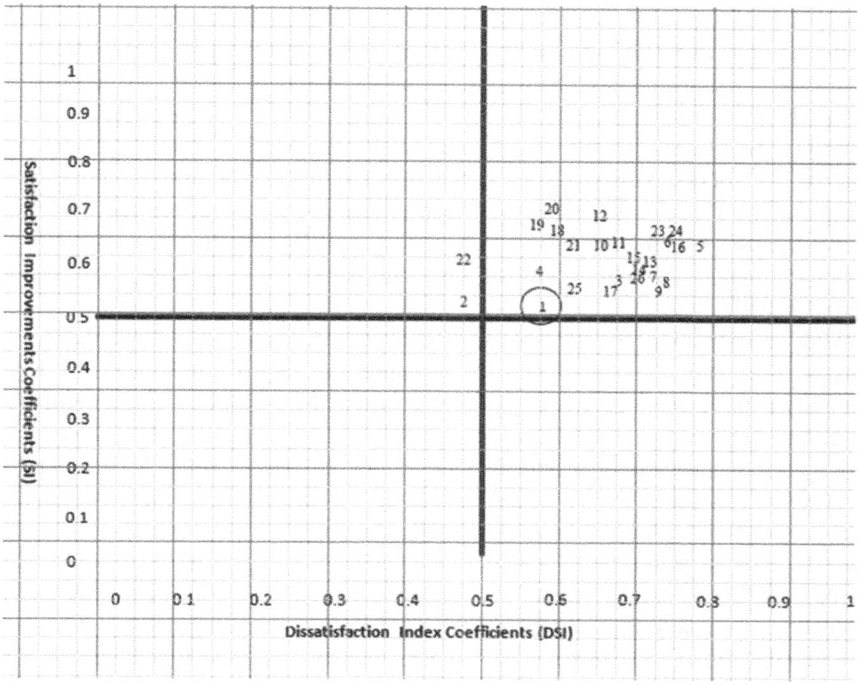

Fig. 1. KANO analysis

(2) Upper left section (SI value above 0.5 and absolute DSI values below 0.5): The features that fall in this section are the features that have a high impact on customer satisfaction and not much on dissatisfaction. According to findings, features that correspond to this section; Branch Attraction(Visually) and Understanding of Special Needs.
(3) Lower right section (SI value below 0.5 and absolute DSI values above 0.5): Cargo features that are placed in this section are the features that can decrease customer dissatisfaction if improved, while having little contribution to customer satisfaction. In this study, no features fall in this section.
(4) Lower left section (SI and absolute DSI values below 0.5): There is no need to improve these services as they do not contribute to neither satisfaction nor dissatisfaction. Again, no features fall in this section in this study.

5 Conclusion

In this study, we investigated service quality of parcel delivery companies operating in Turkey. As a research environment Izmir province was selected and 300 customers were surveyed. In the first stage of our research, we aimed to measure the current service quality level of parcel delivery service. To the aim, SERVQUAL method was adapted and according to the results it was found that all SERVQUAL dimensions, namely tangibility, reliability, responsiveness, assurance, and empathy are low quality. Because the perceived service quality levels of each didn't meet the expectation levels of customers. Company based SERVQUAL results reveal that only UPS is so close to the matching expectations with perceptions as their average service quality score is close to 0. Aras Cargo has the lowest score after UPS, MNG Cargo is the third, Sürat Cargo is the fourth, Yurtiçi Cargo is the fifth company. The worst service quality belongs to the PTT Cargo. According to the results, the lowest service quality was found on *assurance* dimension for UPS, *tangibility* and *responsiveness* dimensions for Yurtiçi, *responsiveness* dimension for Sürat and MNG, *tangibility* for Aras, and *reliability* for PTT.

After investigating the service quality levels in parcel delivery industry both industry level and company level, to introduce the improvement areas, a KANO method was performed. According to the results of Kano Model, all parcel delivery service attributes were perceived as one-dimensional quality. Which means that, all service attributes regarding parcel delivery satisfy customers when fulfilled and dissatisfy when not fulfilled. We calculated the satisfaction and dissatisfaction indexes. Then we positioned these indexes on KANO diagram. The diagram revealed us that except visually attractive cargo branches, and understanding special customer needs, all service quality attributes of parcel delivery have impact on both satisfaction and dissatisfaction. It means that, if the parcel delivery company improves these service attributes the satisfaction levels of customers will increase and dissatisfaction levels will decrease. The other outcome of KANO analysis is that, if parcel delivery companies improve the visual attractiveness of their branches and focus more on understanding the special customer needs the customer satisfaction will increase but it will not have an impact on dissatisfied customers.

This study suffered several limitations. First, although the other service attributes have the potential to change the results of analyses, the present study only focuses on SERVQUAL dimensions. Further studies can apply the KANO analysis by adding and examining more parcel delivery service attributes with the help of in-depth interviews with experts. Another limitation derives from the sample of this study, which is small because of time limitation and is restricted with Izmir province. Therefore, future studies can extend the research area.

References

1. Göncü, K.: Lojistik Sektöründe Kargo Taşımacılığında Uzak Nokta Çözümleri: Sosyal Bilimler Enstitüsü İşletme Anabilim Dalı Yüksek Lisans Proje Çalışması. Ph.D. thesis, Trakya Üniversitesi, Edirne (2010)
2. Batur, B.: Hava Yolcu Ve Kargo Taşımacılığı; Dünyada Ve Türkiye de Uygulamalar. Ph.D. thesis, Dokuz Eylül Üniversitesi Sosyal Bilimler Enstitüsü, İzmir (2008)
3. Mermertaş, F.: Müşteri Tatmini ve Müşteri Sadakati Sağlamada Algılanan Hizmet Kalitesinin Ölçülmesinin Önemi; Kargo Sektörü Üzerine Bir Araştırma. Ph.D. thesis, Gaziantep Üniversitesi, Gaziantep (2018)

4. Simsek, B., Tüysüz, F.: An application of network data envelopment analysis with fuzzy data for the performance evaluation in cargo sector. J. Enterp. Inf. Manag. **31**, 492–509 (2017)
5. Büyükkeklik, A., Bülbül, H., Özoğlu, B.: Kargo Hizmet Sağlayıcılarında Kalitenin Tüketici Davranışına Etkisi: Bireysel Tüketici Araştırması. J. Inst. Soc. Sci. Dr. Mehmet YILDIZ Spec. Ed. pp. 33–43 (2014)
6. Hume, M.: Understanding core and peripheral service quality in customer repurchase of the performing arts. J. Serv. Theor. Pract. **24**, 349–369 (2008)
7. Alp, M., Köleoğlu, N., Çınar, B.: Kargo firmlarının itibarının müşteri memnuniyetine etkisi. Dumlupınar Üniversitesi Sosyal Bilimler Dergisi, vol. 60, pp. 1–13 (2019)
8. İmamoğlu, N.: The effect of logistics performance on company performance and export performance: an application in TRA1 level 2 region. Master's Thesis, Bayburt Üniversitesi Sosyal Bilimler Enstitüsü, Bayburt, Türkiye (2018)
9. Hume, M.: Understanding core and peripheral service quality in customer repurchase of the performing arts. Manag. Serv. Qual. **18**, 349–369 (2008)
10. Saha, G.: Service Quality, Satisfaction, and Behavioural Intentions A Study of Low-Cost Airline Carriers in Thailand. Manag. Serv. Qual. **19**, 350–372 (2009)
11. Büyükkeklik, A., Özoğlu, B., Bülbül, H.: Kargo Hizmet Sağlayıcılarında Kalitenin Tüketici Davranışına Etkisi: Bireysel Tüketici Araştırması. Selçuk Üniversitesi Sosyal Bilimler Enstitüsü Dergisi, pp. 33–43 (2014)
12. Arzu, D., Gödekmerdan, L.: Müşterilerin kargo firmalarının sunduğu hizmetlere yönelik tutum ve düşünceleri üzerine bir araştırma. Atatürk Üniversitesi Sosyal Bilimler Enstitüsü Dergisi, pp. 379–396 (2011)
13. Panayides, P.M.: The impact of organizational learning on relationship orientation, logistics service effectiveness and performance. Ind. Mark. Manage. **36**(1), 68–80 (2007)
14. Forslund, H.: The impact of performance management on customers' expected logistics performance. Int. J. Oper. Prod. Manag. 27(8), 901–918 (2007)
15. Ediz, H., Turğut, D.: Kargo Şirketi Seçimine Yönelik Kriterlerin Belirlenmesinde Türkiye Genelinde Bir Saha Araştırması. Çukurova Üniversitesi İktisadi ve İdari Bilimler Fakültesi Dergisi, pp. 65–79 (2015)
16. Yıldız, B., Aslan, H., Çiğdem, Ş.: Kargo firmaları hizmet kalitesinin müşteri memnuniyeti üzerine etkisi, pp. 70–72 (2018)
17. Çakmak, B.: Biçimsel örgütlerde iş tatmini ve örgütsel bağlılık üzerine bir araştırma: Aras Kargo örneği/A study on job sati̇sfacti̇on and organi̇zati̇onal commi̇tment i̇n formal organi̇zati̇ons: Aras Cargo example, Doctoral dissertation (2015)
18. Aslan, Ş., Sezgin, M.: Müşteri memnuniyeti ve memnuniyeti oluşturan unsurların araştırılması. Sosyal ve beşeri bilimler araştırmaları dergisi, pp. 80–82 (2017)
19. Li, B., Riley, W., Hsieh, C.: Assessing customer satisfaction in parcel delivery industry: an empirical study among university customers. Int. J. Serv. Stand. **1**(2), 172–192 (2004)
20. Riley, B., et al.: A comparison study of customer satisfaction between the UPS and FedEx. Ind. Manag Data Syst. 74–76 (2006)
21. Wang, Y., Luor, T., Luarn, P., Lu, H.: Contribution and trend to quality research—a literature review of SERVQUAL model from 1998 to 2013. Informatica Economica **19**(1), 34–45 (2015)
22. Can, P.: Hizmet kalitesinin SERVQUAL ölçeği ile ölçülmesi: Uşak Üniversitesi Merkez Kütüphanesi üzerine bir araştırma. Karabük Üniversitesi Sosyal Bilimler Enstitüsü Dergisi, **6** (1), 63–83 (2016)
23. İzci, F., Saydam, R.: Algılanan Hizmet Kalitesi Kurumsal İmaj ve Sadakat İlişkisi (Van Bölge Hastanesi Uygulaması). C.Ü. İktisadi ve İdari Bilimler Dergisi, **14**(1) 199–219 (2013)

24. Songur, G.: Kargo taşımacılığında hizmet kalitesi ve kurumsal müşteri memnuniyetinin ölçülmesi: Konya ili'nde bir araştırma. Master's thesis, Niğde Üniversitesi/Sosyal Bilimler Enstitüsü, Niğde, Türkiye (2016)
25. Çiçek, R., Doğan, İ.: Müşteri memnuniyetinin artırılmasında hizmet kalitesinin ölçülmesine yönelik bir araştırma. Afyon Kocatepe Üniversitesi İktisadi ve İdari Bilimler Fakültesi Dergisi **11**(1), 199–217 (2009)
26. Yılmaz, V., Çelik, H., Depren, B.: Devlet ve özel sektör bankalardaki hizmet kalitesinin karşılaştırılması: Eskişehir örneği. Doğuş Üniversitesi Dergisi **8**(2), 234–248 (2007)
27. Aleren, A., Bektaş, Ç., Görmüş, Ş.: Hizmet Sektöründe Hizmet Kalitesinin SERVQUAL Yöntemi ile Ölçülmesi ve Hazır Yemek İşletmesinde Bir Uygulama. Finans Politik ve Ekonomik Yorumlar **44**(514), 75–88 (2007)
28. Yılmaz, V.: Lisrel Ile Yapısal Eşitlik Modelleri: Tüketici Şikâyetlerine Uygulanması (2004). https://earsiv.anadolu.edu.tr/bitstream/handle/11421/480/196720
29. Cronin, J., Taylor, S.: SERVPERF versus SERVQUAL: reconciling Performance based and Perceptions-minus-expectations measurement of service quality. J. Mark. **58**(1), 125–131 (1992)
30. Carman, J.: Consumer perceptions of service quality an assessment of the SERVQUAL dimensions. J. Retail. **66**(1), 33–55 (1990)
31. Savas, H., Ay, M.: Üniversite Kütüphanesi Tasarımında Kalite Fonskiyon Göçerimi Uygulaması. Dokuz Eylül Üniversitesi Sosyal Bilimler Enstitüsü Dergisi, vol. 7, p. 3 (2005)
32. Rotar, L., Kozar, M.: The use of the kano model to enhance customer satisfaction. Organizacija **50**(4), 339–351 (2017)
33. Lin, H.-J.: Kano Model Analysis in Product Desing With the E-Commerce Practice (2019)
34. Akgül, H.: Kano Modeli, Müşteriyi kendinden iyi tanımak. https://lean.org.tr/kano-modeli-musteriyi-kendinden-iyi-tanimak-2/
35. What is the kano model. https://www.microtool.de/en/knowledge-base/what-is-the-kano-model/
36. Introduction To The Kano Model. (https://surveymethods.com/introduction-to-the-kano-model/)
37. Durmaz, Y.: Lojistik Ulaştırma Faaliyetlerindeki Hizmet Kalitesinin Müşteri Değeri Yaratmadaki Rolu: Kütahya Yurtiçi Kargo Şirketinde Bir Uygulama. Ph.D. thesis, Kütahya (2010)
38. Kut, A.: KARİD homepage (2017). http://www.karid.org.tr/kargo-sektoru-durum-analizi-ve-sektorun-gelecegi/

Production Scheduling at a Powder Coating Company

Değer Demir, Ecem Ademoğlu, Gülce Gülen,
Kerem Can Deliktaş(✉), Mert Yandımata, Saadet Gül,
Serdar Can Ertem, Önder Bulut, and Sinem Özkan

Department of Industrial Engineering, Yaşar University, İzmir, Turkey
deger.demir1997@gmail.com, ecemademoglu2@gmail.com,
ggulcegulen@gmail.com, keremcandeliktas@hotmail.com,
myandimata4@gmail.com, saadetgul97@gmail.com,
cnertem@gmail.com, onder.bulut@yasar.edu.tr,
sinem.ozkan@yasar.edu.tr

Abstract. In this study, we consider parallel machine production scheduling problem of a Powder Coating company. A mathematical model with firm-specific constraints such as order splitting and machine availability is developed. The objective is to minimize the total weighted completion time of jobs. We propose two heuristic solution algorithms: A Genetic Algorithm and a Rule-Based Algorithm. For smaller instances, each algorithm is compared with the optimum. On the other hand, for larger instances, we compare the performances of the proposed algorithms with each other. Furthermore, we develop a simulation model to test the performances of the solutions determined by deterministic algorithms under randomness in setup times, quality control times and downtimes.

Keywords: Scheduling · Order splitting · Machine availability · Heuristics · Genetic algorithm · Rule based algorithm

1 Introduction

This study addresses a parallel machine scheduling problem of a Powder Paint Company located in İzmir, Turkey. There are three main operations in this company to make final products: Premix, Extrusion and Grinding. Since the premix process is so fast compared to the others, this process is not included in our models. Moreover, when a job enters the extrusion process, the processing continues until the end of the grinding process without any interruption. Thus, all the available four lines of the powder production are considered as four parallel machines performing a single operation. When the current system is analyzed, it is observed that setups (cleaning activities) are frequent, variability in processing times are high, and completion times of some jobs are longer than desired.

In specific, the company's problem has the following main features: orders have different weights; there are machine eligibility and availability constraints; order splitting (lot streaming) is allowed; setups are sequence dependent; preemption is not

N. M. Durakbasa and M. G. Gençyılmaz (Eds.): ISPR 2020, LNME, pp. 989–1001, 2021.
https://doi.org/10.1007/978-3-030-62784-3_82

allowed. Machine availability is modeled in a way that machine breakdowns and the uncompleted jobs from the previous schedule can be all handled using the corresponding constraint.

In order to solve this NP-Hard problem, we develop an optimization model and propose heuristic solution algorithms. A Genetic Algorithm which is known to work well in scheduling problems has been structured. In addition, a 3-Phase Rule Based Algorithm is developed to further decrease run time. Since these developed models and algorithms are deterministic, a simulation model has also been developed that takes the results of these models and algorithms as input and provides their performances under random processing, setup, and down times.

Due to the high complexity of parallel machine scheduling problems, many studies have been conducted in the literature on this subject. Pinedo [1] and Brucker [2] were used to have a general knowledge of the nature and terminology of the problem. Su et al. [3] has greatly benefited from creating constraints, although its objective function is different and there was no order splitting property in that study. A scheduling problem with sequence and machine-dependent setup times is considered in the Rocha et al. [4]. Sevindik [5] provides general insight into machine availability constraints.

In real-life applications, scheduling problems involve solving for an optimum schedule, considering various purposes, different machine environments and different job features [6]. Identifying appropriate genetic coding is very important because it significantly affects all subsequent stages of GA [7]. Also, Lee et al. [8] provides an efficient heuristic approach for parallel machine scheduling with job splitting and sequence-dependent setup times.

2 Problem Formulation and Solution Methodology

Our mathematical model and solution algorithms are based on the following assumptions of the company: (1) The system is clean at the beginning. (2) Since the premix process in the system is very fast compared to others, it is reasonable to assume that the output buffer of the premix station is an ample supply. (3) Raw material supply is ample (4) There is no storage limit for any product in production. (5) Preemption is not allowed. The objective is to minimize total weighted completion time. To achieve the optimum result, a mathematical model has been created. As the number of jobs and machines increase, the problem complexity and hence run time increases. Therefore, a Genetic Algorithm, which gives close results to optimal, is implemented, and a Rule Based Algorithm is developed in order to further reduce the run time.

2.1 Mathematical Model

Problem is modeled as a mixed integer linear program. Details of all inputs, variables and constraints are given below.

Indices		Sets
m	Machine index	M: set of all machines
j, j'	job indices	J: set of all jobs (Also covers position indices.)
p	position (sequence) index	J_m: set of jobs that can be processed on machine m s.t $m \in M, J_m \subset J$
		D: set of jobs having due date s.t $D \subset J$

Parameters

$\lvert J \rvert$	total number of jobs	w_j	weight of job j (in terms of priority)
α	maximum acceptable tardiness ratio for jobs having due date (in terms of the number of tardy jobs)	p_{mj}	processing time of a single unit of job j on machine m
Q_j	production order quantity of job j	d_j	due date of job j
$ST_{mjj'}$	required setup time on machine m when job j is followed by job j'	QT_j	quality control time for job j
T	threshold time required for re-scheduling	L	a large number
Δ	a positive small number	RT_m	repair time of machine m

$$f_m = \begin{cases} 1, & \textit{if there is a failure on machine } m \\ 0, & \textit{otherwise} \end{cases}$$

Because of machine breakdowns and uncompleted jobs from the previous schedule (in real-life application schedules would be created repeatedly), rescheduling is required in order to prevent disruptions in production. Thanks to the threshold time (T) which is left parametrically and determined by the scheduler, if the scheduler requests, it is allowed to start rescheduling after this time, leaving the schedule within the threshold time the same. For machines with f_m value of 1 (machines with breakdown), when re-scheduling is done, the assignment of any job to machine m during that period is prevented.

Main Decision Variables

$$X_{mpj} = \begin{cases} 1, & \textit{if job } j \textit{ is scheduled at position } p \textit{ on machine } m \\ 0, & \textit{otherwise} \end{cases}$$

SR_{mpj}: the splitting ratio of job j processed at position p on machine m (this ratio is 1 when any job is processed on only a single machine).

Auxiliary Decision Variables

$$Y_{mpjj'} = \begin{cases} 1, \textit{ if job } j \textit{ and } j' \textit{ are processed on machine } m \textit{ on position } p \textit{ and } p+1, \textit{ respectively} \\ 0, \textit{ otherwise} \end{cases}$$

$$U_j = \begin{cases} 1, \textit{ if job } j \textit{ is tardy} \\ 0, \textit{ otherwise} \end{cases}$$

CT_{mp} : completion time of the job at position p on machine m

PT_{mp} : processing time of the job at position p on machine m

C_j : completion time of job j

Objective Function: min $Z = \sum_{\forall j \in j} w_j C_j$

Assignment Constraints

$$\sum_{\forall j \in J} X_{mpj} \leq 1 \qquad \forall m \in M, \forall p \in J \tag{1}$$

$$\sum_{\forall p \in J} X_{mpj} \leq 1 \qquad \forall m \in M, \forall j \in J \tag{2}$$

$$\sum_{\forall m \in J} \sum_{\forall p \in J} X_{mpj} \geq 1 \qquad \forall j \in J \tag{3}$$

$$\sum_{\forall j \in J} X_{m(p+1)j} \leq \sum_{\forall j \in J} X_{mpj} \qquad \forall m \in M, \forall p \in J \tag{4}$$

The above four constraints are the classical assignment constraints of parallel machine scheduling. The only adaptation is done for order splitting.

Order Splitting Constraints

$$(X_{mpj} - 1) + \Delta \leq SR_{mpj} \leq X_{mpj} \quad \forall m \in M, \quad \forall p, j \in J \tag{5}$$

$$PT_{mp} = \sum_{\forall j \in J} \left(P_{mj} Q_j SR_{mpj} + QT_j X_{mpj} \right) \qquad \forall m \in M, \ \forall p \in J \tag{6}$$

$$\sum_{\forall m \in J} \sum_{\forall p \in J} SR_{mpj} = 1 \quad \forall j \in J \tag{7}$$

The Using a *small* Δ we ensure that splitting ratio of a job that is assigned to a certain position of a certain machine is positive.

Setup and Completion Time Constraints

$$2Y_{mpjj'} \geq X_{mpj} - X_{m(p+1)j'} \quad \forall m \in M, \forall p \in J/\{|J|\} \quad \forall j \in J, \forall j' \in J|j' \neq j \tag{8}$$

$$X_{mpj} - X_{m(p+1)j'} \leq 1 + Y_{m(p+1)jj'} \quad \forall m \in M, \forall p \in J/\{|J|\} \; \forall j \in J, \forall j' \in J|j' \neq j \tag{9}$$

$$CT_{m1} \geq ES_m + PT_{m1} \quad \forall m \in M \tag{10}$$

The above three constraints are to compute the completion times of the jobs where ES_m, which is an input parameter to the model, is the earliest start time of machine m and is calculated as:

$$ES_m \geq f_m \times RT_m + (1 - f_m).\max\{T, max\{C_j\}\}$$

With ES_m, both machine failures and sliding time of the schedule are modeled with a single equation for machine availability. If a machine is failed, earliest start time is equal to the repair time of the machine (processing cannot be started until the machine repair is completed). Otherwise, it is equal to the maximum of the threshold time T (an input of the company) and completion time of the last job that could not be completed on the previous schedule. The below two constraints are also completion time constraints ensuring that completion times are increasing in the position index.

$$CT_{m(p+1)} \geq CT_{mp} + PT_{m(p+1)} + \sum_{\forall j \in J} \sum_{\substack{\forall j' \in J \\ j' \neq j}} \left(Y_{mpjj'} ST_{mjj'}\right) \quad \forall m \in M, \quad \forall p \in J/\{|J|\} \tag{11}$$

$$C_j \geq CT_{mp} + L(X_{mpj} - 1) \quad \forall m \in M, \quad \forall p \in J, \forall j \in J \tag{12}$$

Tardiness Constraints

$$-L(1 - U_j) + \Delta \leq C_j - d_j \leq LU_j \quad \forall j \in J \tag{13}$$

$$\frac{\sum_{\forall j \in D} U_j}{|D|} \leq \alpha \tag{14}$$

These two constraints are to satisfy the company's service level criteria ensuring that the ratio of tardy jobs (among the ones having due dates) is below a certain allowable limit.

Allowable sets/regions of decision variables

$$SR_{mpj} \in [0, 1] \qquad \forall m \in M, \forall p \in J, \forall j \in J \tag{15}$$

$$X_{mpj} \in \{0, 1\} \qquad \forall m \in M, \forall p \in J, \forall j \in J \tag{16}$$

$$Y_{mpjj'} \in \{0, 1\} \qquad \forall m \in M, \forall p \in J/\{|J|\}, \forall j \in J, \forall j' \in J | j' \neq j \tag{17}$$

$$U_j \in \{0, 1\} \qquad \forall j \in D \tag{18}$$

$$PT_{mp} \geq 0 \qquad \forall m \in M, \forall p \in J \tag{19}$$

$$CT_{mp} \geq 0 \qquad \forall m \in M, \forall p \in J \tag{20}$$

$$C_j \geq 0 \qquad \forall j \in J \tag{21}$$

Two heuristic methods have been developed in order to enable the company to find a schedule that gives results close to the optimal solution in a reasonable amount of time. Details about these heuristic methods are given below.

2.2 Genetic Algorithm

The genetic algorithm is a heuristic search inspired by Darwin's theory of natural evolution. The natural selection process begins with the selection of the fittest individuals in a population. They produce offspring that inherit the characteristics of parents and will be added to the next generation. In the end, a generation will be found with the fittest individuals.

Phase 1. Initial Population: The "population" is formed by the production of many individuals with different permutations. Genes represent jobs; combining genes creates individuals. Each chromosome represents the sequence of jobs assigned to a machine.

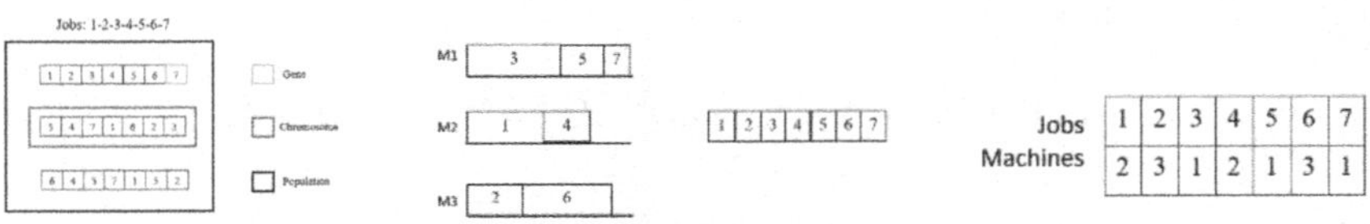

(a) (b) (c)

Fig. 1. (a) Gene, Chromosome, Population representation. (b) Assignment according to Min Comp. Time. (c) Chromosome.

Phase 2. Fitness Function: In this phase, starting from the first job to the last job, the completion time that all jobs on the chromosome is calculated on all machines that they can be processed on and each job assigns to the machine giving the minimum completion time (Fig. 1.b). So, the sequence and main chromosome structure is determined as in Fig. 1.c. Since jobs are assigned in order, it is obvious that job 1 is assigned to position 1 of the machine 2 and job 4 to position 2 of the same machine. After that, the fitness function value $\left(\sum_j w_j C_j\right)$, which determines an individual's ability to compete with other individuals, is calculated and if the rate of tardy jobs exceeds the tardiness ratio, a big penalty number is added to the completion times of all tardy jobs for eliminating in the future.

Phase 3. Selection: Fitness values of all chromosomes are ranked with Tournament Method in order to select individuals for reproduction.

Phase 4. Crossover: Partially mapped crossover (PMX), was used to increase variation. In PMX, substring with a random length is chosen randomly from parents first. (Figure 2.a) Subgroups are then exchanged (Fig. 2.b) and mapping relationship is determined (Fig. 2.c) in order to correct the disruption created by crossover and make the jobs in each chromosome unique. For example, in Fig. 2.c, since job 1-job 6 and job 6-job 3 are replaced, positions of 1 and 3 are changed. Then the relation map is used to legalize offspring (Fig. 2.d). In the end, two new offsprings are obtained in Fig. 2.e.

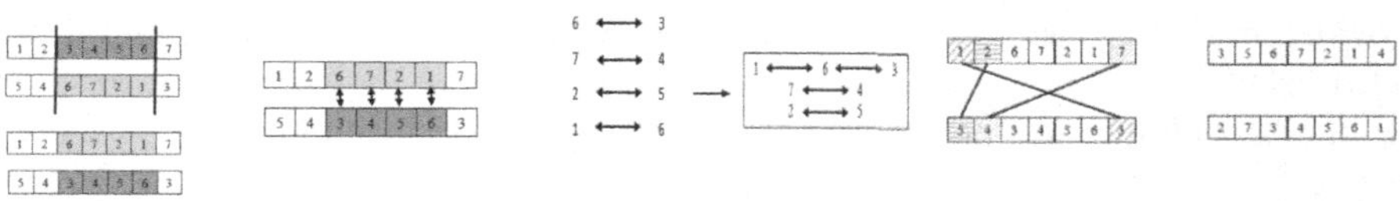

(a) (b) (c) (d) (e)

Fig. 2. (a) Substring selection. (b) Subgroups exchange. (c) Mapping relationship. (d) Relation map. (e) New chromosomes

Phase 5. Mutation: A chromosome is selected among all individuals, except the best 3 ("elite individuals"), randomly. With a very low probability (0.05 for this study), a randomly selected job is inserted before another job selected at random to prevent getting stuck in local optima.

(a) (b)

Fig. 3. (a) Mutation for gene 3. (b) Structure of the chromosome created for split.

Order Splitting for Genetic Algorithm

Due to the order splitting, all these phases have actually realized by doubling the size of all chromosomes. Splitting ratio is determined at random. The newly added part is the form of a permutation of the first part. Splitting quantity of any job on the first part equals to *Splitting Ratio* $\times$ *Actual quantity of that job* and Splitting Quantity on the newly added part can be found as *Actual Quantity − Splitting Quantity of that job in the first part*. Figure 3.b represents this process. Termination Criterion: The algorithm terminates if the best solution converges.

2.3 Rule-Based Heuristic

A 3-phased constructive heuristic has developed. *In phase 1(Assignment)*, the aim is to assign the jobs to machines and positions by taking due dates, processing time and weights into account, and to have minimal completion times as much as possible. After assignment, in *Phase 2 (Order Splitting)*, the aim is to balance both the completion time of the last jobs and the processing time of split jobs on the machines. Also, it is tried not to deviate too much from the sequence of the jobs obtained in *Phase 1*. Thus, the job on the machine with maximum completion time is split to the machine with minimum completion time. For all machines, splitting is performed for all the jobs starting from the job at the last position. *In Phase 3 (Reprocessing)*, after each iteration, the solution with the minimum objective function value is updated, and the one obtained at the last iteration is reported as the output solution of the algorithm. After this brief introduction, starting from the notation (additional to the general notation provided in Part A), and the detailed technical representation of the algorithm is provided below:

L_m: the last job that was processed before job j on the machine m
$lastposition(m)$: last position index for machine m
$ST_{j,j'}$: setup time from job j to job j'

$$f(m,p) = \begin{cases} j, & \textit{if job j is scheduled at position p on machine m}, \ \forall m \in M,\ p \in P,\ j \in J \\ 0, & \textit{otherwise} \end{cases}$$

Phase1. Assignment:

Parameters	*Description*	*Range/Value*	*Parameters*	*Description*	*Range/Value*
$score(m,j)$	score required to process job j in machine m	$\forall m \in M, \forall j \in J_m$	w_j	real weight of job *j*	$\forall j \in J$
$objective\ function(1)$	obj. func. value at iteration 1	$\forall j \in J$	K_j	weight due to the due date of job *j*	$\begin{cases} d_j, & if\ j \in D \\ L, & else \end{cases}$
			w'_j	current weight of *j*	$\frac{w_j}{K_j}, \forall j \in J$

0. Initialization There is no job assigned to any machine at the beginning. Since the system is clean at t = 0, setup time is 0 for all machines.	$S = \emptyset$ $P_m = \emptyset, L_m = 0, last\ position(m) = 0, \forall m \in M$ $ST_{L_m,J} = 0, \forall m \in M, \forall j \in J$ $C_j = 0, \forall j \in J$
1. In order to reduce the myopic feature of the algorithm, for each candidate job *j*, add maximum setup times for all machines that *j* can be processed to this score calculation. This part is optional; however, it was generally seen that it yields better results.	*Do Until* $S = J$ {*for all* $j \in S^c$, $m \in M_j$ *and* $L_m \in S$ $score(m,j) = \frac{\max\{ES_m, C_{L_m}\} + P_{mj} \times Q_j + QT_j + ST_{L_m j} + \max_{j' \in J_m \cap S^c}\{ST_{jj'}\}}{w'_j}$ *end for*
2. Assign job *j* to the machine which has the minimum score. Update the sets and the completion times.	$(m,j) = \underset{j \in S^c, m \in M_j}{argmin}\{score(m,j)\}$ $last\ position(m) = last\ position(m) + 1$, $f(m, last\ position(m)) = j$, $C_j = ES_m + P_{mj} \times Q_j + QT_j + ST_{L_m j}, L_m = j, S = S \cup \{j\}$ } *Loop*

Phase 2. Order Splitting:

Indices	*Description*	*Indices*	*Description*
$\boldsymbol{selected\ position(m)}$	position of job whose processing time is calculated on machine $\boldsymbol{m}$ and compared with others in order to find the job to be split	$\boldsymbol{maxSel}$	maximum of selected position on all machines
j^*	job to be split	$\boldsymbol{selected\ job(m)}$	$\boldsymbol{f(m, selected\ position(m))}$
m^*	machine where job $\boldsymbol{j^*}$ has been assigned	m'	machine where job $\boldsymbol{j^*}$ is splitted
p^*	position of job $\boldsymbol{j^*}$ on machine $\boldsymbol{m^*}$	p'	position of job $\boldsymbol{j^*}$ splitted on machine $\boldsymbol{m'}$

Parameters	*Description*	*Range/Value*
$\boldsymbol{number\ of\ splits\ (m)}$	number of splitted jobs on machine $\boldsymbol{m}$	$\boldsymbol{\forall m \in M}$
$\boldsymbol{remaining\ splits\ (m)}$	while updating the SQ cumulatively, the number of splitted jobs that need to be updated remaining on machine m	$\boldsymbol{\forall m \in M}$
$\boldsymbol{total\ number\ of\ splits}$	number of jobs splitted in total	
SQ_{mp}	splitting ratio of job processed at position $\boldsymbol{p}$ on machine $\boldsymbol{m}$	$\boldsymbol{\forall m \in M, \forall p \in [1, last\ position(m)]}$
CT_{mp}	completion time of the job assigned to the position $\boldsymbol{p}$ of machine $\boldsymbol{m}$ at that position	$\boldsymbol{\forall m \in M, \forall p \in [1, last\ position(m)]}$
$\boldsymbol{Objective\ function\ (total\ number\ of\ splits + 1)}$	the objective function value after each splitting operation ($\boldsymbol{objective\ function}$(1) has been calculated after assignment process)	

0. Initialization If job $\boldsymbol{j}$ is assigned to the position $\boldsymbol{p}$ of machine $\boldsymbol{m}$, then its splitting quantity equals to its actual quantity. Initially, the selected position in all machines is equal to the last position, and no splitting operation is done. Initialize all related parameters. Since the objective function (1) is equal to the objective function created after assignment, the objective function value for split iterations starts from the objective function (2).	$selected\ position(m) = last\ position(m), \forall m \in M$ $maxSel = \max_{m \in M}\{selected\ position(m)\}, \forall m \in M$ $number\ of\ splits(m) = 0, remaining\ splits(m) = 0, \forall m \in M$ $total\ number\ of\ splits = 0,$ $for\ all\ m \in M, p \in [1, last\ position(m)], j \in J_m$ $SQ_{mp} = \begin{cases} Q_j, if\ f(m,p) = j \\ 0, otherwise \end{cases}$ $CT_{mp} = \begin{cases} C_j, if\ f(m,p) = j \\ 0, otherwise \end{cases}$ $end\ for$ $j^* = 0, m^* = 0,\ m' = 0, p^* = 0,\ p' = 0$ $Objective\ function(total\ number\ of\ splits + 1) = 0$
1. The split process is continued until all positions of all machines from the end to the first are visited. In order to select the job to be split, find the maximum completion time of the last job on each machine $\boldsymbol{m}$. If more than one machine has a job whose completion time is equal to makespan, find the maximum processing time among these jobs. Set this job to $\boldsymbol{j^*}$, the machine where this job is processed to $\boldsymbol{m^*}$ and the position to $\boldsymbol{p^*}$.	$Do\ Until\ maxSel = 0$ $\{selected\ position(m) = last\ position(m) - number\ of\ splits(m), \forall m \in M$ $maxSel = \max_{m \in M}\{selected\ position(m)\}, \forall m \in M$ $selected\ job(m) = f(m, selected\ position(m)), \forall m \in M$ $for\ all\ m \in M\ s.t\ selected\ position(m) \neq 0\ and$ $m = \underset{m \in M}{argmax}(CT_{f(m, last\ position(m))})$ $(m^*, j^*) = argmax(P_{m, selected\ job(m)} \times Q_{selected job(m)} + QT_{selected job(m)})$ $end\ for$ $p^* = selectedposition(m^*), number\ of\ splits(m^*) = number\ of\ splits(m^*) + 1$
2. To find out which machine to split the job $\boldsymbol{j^*}$ on, check the completion time of the last job among all machines where it can be processed on. Determine the machine with the shortest completion time as $\boldsymbol{m'}$, and the position after the selected position as $\boldsymbol{p'}$. If the job $\boldsymbol{j^*}$ cannot be assigned anywhere other than $\boldsymbol{m^*}$, exit the loop.	$if\ \nexists m \in M_{j^*}\ s.t\ m \neq m^*\ then$ $\quad Exit\ Do\ Loop$ $else$ $for\ all\ m \in M_{j^*}\ s.t\ m \neq m^*\ and\ m = \underset{m \in M}{argmin}(CT_{f(m, last\ position(m))})$ $\quad lastposition(m') = lastposition(m') + 1, p' = selectedposition(m') +$ 1 $\quad number\ of\ splits(m') = number\ of\ splits(m') + 1$ $end\ for$ $end\ if$
3. Insert job $\boldsymbol{j^*}$ at position $\boldsymbol{p'}$ of machine $\boldsymbol{m'}$. Move the position of all jobs after position $\boldsymbol{p'}$ to the next.	$if\ p' \neq last\ position(m')\ then$ $\quad for\ p = last\ position(m')\ to\ p'\ Step - 1$ $\qquad f(m', p) = f(m', p - 1)$ $\quad end\ for$ $end\ if$ $j^* = f(m', p') = f(m^*, p^*)$
4. Since splitting quantities should be updated cumulatively after each split operation, set the last splitted machines and positions $\boldsymbol{(m^*, m', p^*\ and\ p')}$ as $\boldsymbol{m_1^*, m_1', p_1^*\ and\ p_1'}$ respectively. Cumulatively update all machines, from the most recently split machines and positions to	$total\ number\ of\ splits = \frac{\sum_{m \in M} number\ of\ splits(m)}{2}$ $for\ k = total\ number\ of\ splits\ to\ 2\ Step - 1$ $\quad m_k^* = m_{k-1}^*, m_k' = m_{k-1}'$ $end\ for$ $m_1^* = m^*, m_1' = m'$ $number\ of\ splits(m_1^*) = number\ of\ splits(m_1^*) + 1$ $number\ of\ splits(m_1') = number\ of\ splits(m_1') + 1$

the first split machines and positions. (machines split before the last: m_2^*, m_2') (positions split before the last: p_2^*, p_2') etc.	$remaining\ splits(m) = number\ of\ splits(m), \forall m \in M$ $for\ k = 1\ to\ total\ number\ of\ splits$ $p_k^* = last\ position(m_k^*) - remaining\ splits(m_k^*) + 1$ $remaining\ splits(m_k^*) = remaining\ splits(m_k^*) - 1$ $p_k' = last\ position(m_k') - remaining\ splits(m_k') + 1$ $remaining\ splits(m_k') = remaining\ splits(m_k') - 1$ $j_k^* = f(m_k', p_k') = f(m_k^*, p_k^*)$ $end\ for$
5. Cumulatively update SQ values. When updating SQ, equalize elapsed time of job j^* (as shown below) while processing both on m^* and m' and leave the SQ value alone in the equation. $ST_{f(m_k', p_k'-1), j_k^*} + P_{m_k', j_k^*} \times SQ_{m_k', p_k'} + QT_{j_k^*}$ $= ST_{f(m_k^*, p_k^*-1), j_k^*} + P_{m_k^*, j_k^*} \times SQ_{m_k^*, p_k^*} QT_{j_k^*}$ and since $SQ_{m_k', p_k'} = Q_{j_k^*} - SQ_{m_k^*, p_k^*}$, the equation becomes like next.	$for\ all\ m \in M$ and $for\ all\ p \in [1, last\ position(m)]$ $CT_{m,p} = C_{f(m,p)}$ $end\ for$ $for\ all\ k \in [1, total\ number\ of\ splits]$ $SQ_{m_k^*, p_k^*} = \frac{ST_{f(m_k', p_k'-1), j_k^*} - ST_{f(m_k^*, p_k^*-1), j_k^*} + P_{m_k^*, j_k^*} \times Q_{j_k^*}}{P_{m_k', j_k^*} + P_{m_k^*, j_k^*}}$ $SQ_{m_k', p_k'} = Q_{j_k^*} - SQ_{m_k^*, p_k^*}$ $end\ for$ $CT_{m_k', p_k'} = CT_{m_k', p_{k-1}'} + ST_{f(m_k', p_{k-1}'), j_k^*} + QT_{j^*} + P_{m_k', j_k^*} \times SQ_{m_k', p_k'}$ $CT_{m_k^*, p_k^*} = CT_{m_k^*, p_{k-1}^*} + ST_{f(m_k^*, p_{k-1}^*), j_k^*} + QT_{j^*} + P_{m_k^*, j^*} \times SQ_{m_k^*, p_k^*}$ $C_{j^*} = \max\{CT_{m_k', p_k'}, CT_{m_k^*, p_k^*}\}$
6. Considering that objective function (1) is calculated after assignment, calculate an objective function value for each split operation.	$Objective\ function(total\ number\ of\ splits + 1)$ $= Objective\ function(total\ number\ of\ splits + 1) + \sum_j w_j \times C_j$ $\}\ Loop$

Phase 3. Reprocessing:

0. Calculate the minimum objective function value among all calculated objective functions and set this value as "Current Objective Function".	$for\ all\ i \in [1, total\ number\ of\ splits + 1]$ $Current\ Objective\ Function = min\ \{Objective\ function(i + 1)\}$ $end\ for$
1. Go to Phase 1-Step 0 and repeat all operations until the objective function value is equal to the Current Objective Function value.	$for\ all\ i \in [1, total\ number\ of\ splits + 1]$ $Do\ Until\ Objective\ function(i) = Current\ Objective\ Function$ $\{Phase\ 1\ and\ Phase\ 2\}Loop$ $end\ for$

2.4 Simulation Model

The variance in downtime, quality and processing times are high; therefore, the system has a random nature. Since the developed mathematical model and solution algorithms are deterministic, they find results without considering standard deviations. Therefore, a simulation model has been developed in order to observe the effects of the system's randomness on completion times. The simulation model, which has designed with a smart structure, solves even large-scale problems because instead of using one entity for each order, all orders could be displayed with a single entity in each production line.

Simulation receives solutions of mathematical model and heuristics as input. Machine breakdown, quality control and setup times are taken according to the distributions obtained. By taking the weights of all jobs, total weighted completion time is calculated. So, considering the standard deviation, it gives results close to the real nature of system.

2.5 Validation and Verification

The mathematical model and 2 different heuristic algorithm methods are in different instances created by changing machine eligibility, number of machines, jobs and jobs having a due date and the results are shown below (Table 1):

Table 1. Comparison of 2 different solution methods with optimal result and their run time

	Objective Function Value				Run Time (Sec)				Closeness of Optimal Solution		
Instance	Optimal Solution	Genetic Alg.	Rule Based Alg. 1 (score without max. Setup time)	Rule Based Alg. 1 (score with max. Setup time)	Optimal Solution	Genetic Alg.	Rule Based Alg. 1	Rule Based Alg. 2	Genetic Alg.	Rule Based Alg. 1	Rule Based Alg. 2
1	2313	2330	2333	2333	49	10,1	0,41	0,6	99%	99%	99%
2	2522	2522	2644	2522	50	8,7	0,38	0,7	100%	95%	100%
3	2316	2422	2548	2427	52	9,1	0,46	0,14	96%	91%	95%
4	1285	1489	1656	1494	33	10,2	0,54	0,74	86%	78%	86%
5	1320	1320	1367	1367	93	7,3	0,47	0,6	100%	97%	97%
6	2070	2070	2117	2188	94	11,7	0,58	0,23	100%	98%	95%
7	1407	1407	1461	1461	100	8,1	0,6	0,24	100%	96%	96%
8	2233	2233	2318	2318	111	10,2	0,74	0,3	100%	96%	96%
9	1300	1313	1367	1367	2058	7,6	0,87	0,7	99%	95%	95%
10	2467	2485	2550	2550	29,45	5,8	0,52	0,58	99%	97%	97%
								Avg:	98%	95%	96%

The closeness percentages of algorithms to the optimal solution were calculated. The rule-based algorithm is coded according to 2 different score calculations and it is observed that the algorithm considering the setup times after the jobs are assigned (Rule Based Alg. 2), gives better results. In different sizes of data, the percentage of closeness to the optimal solution for Genetic and Rule-Based Algorithm are 98% and 96%, respectively. As the number of machines, jobs, jobs having a due date and eligibility of a job on different machines increases, run time also increases, relatively. Since the rule-based algorithm is greedy, it is foreseen that it will deviate from the optimal result in high dimensional data. Thus, performance comparison of algorithms for large size of data was made and shown in Table 2.

For the larger instances, GA was found to yield approximately 8% better results than the Rule Based algorithm.

In another aspect, run times of algorithms can be interpreted. Although the solution performance of the Rule-Based Algorithm is lower than the Genetic Algorithm, the run-time of the Genetic Algorithm increases to 1650 s, while the Rule-Based Algorithm results in less than 1 s, which means it works in a much shorter time even in large sized jobs.

Comparison of the objective functions of developed deterministic algorithms with the 1-week plan taken from the company and the objective function values obtained when the results are put in the simulation model are given below. The objective function values of both deterministic models and algorithms and the simulation model were compared with the objective function of 1-week production plan created by the company and all results are given in Table 3.

Table 2. Performance comparison of 2 heuristic in larger size data

Instance	Number of Machine	Number of Jobs	Objective Function Value				Run Time (sec)		
			Genetic Alg.	Rule Based Alg. 1	Rule Based Alg. 2	Difference B/w. the Best and Worst Sol.	Genetic Alg.	Rule Based Alg. 1	Rule Based Alg. 2
11	4	30	15972	17139	17591	9%	1320	0,49	0,48
12	4	35	22238	25436	25436	13%	900	0,75	0,78
13	4	42	32714	35357	34333	5%	1380	0,36	0,61
14	4	60	64872	67714	67714	4%	960	0,65	0,68
					Avg.	7.6%			

Table 3. Comparison with real data and solution with randomness for a week

	Deterministic Model Solutions	**Improvements achieved by the proposed algorithms**	Performance of the optimal solution (total weighted completion times) under randomness (simulation)
Company Real Plan	33591,87		44361,36
Rule-Based Algorithm	24466,61	**27,17%**	40187,43
Genetic Algorithm	20199,0	**39,87%**	30536,77

In the 1-week plan, **27.7%** and **39,87%** improvement achieved via Rule Based and GA, respectively.

3 Conclusion

In this study, we first analyze the current scheduling system of a Powder Coating company. After our observations and detected problem symptoms, it is determined that the schedule was made intuitively only within the experience of the scheduler without a mathematical basis. Furthermore, there is a high variation in setup, quality control and downtimes, and setups are frequent. We model the scheduling problem of the company as a parallel-machine scheduling problem having job splitting flexibility and a service level constraint defined for the jobs having due date. Since the problem is NP-Hard, we develop two different heuristic algorithms: A population-based Genetic Algorithm and a rule-based constructive algorithm. Although the rule-based algorithm is myopic, numerical results show that it is an alternative for Genetic Algorithm for some problem instances. Furthermore, the rule-based algorithm is much faster than the population-based algorithm. However, optimality gap of the Genetic Algorithm is always less than the myopic rule-based algorithm. A simulation model has also been developed to observe effects of the randomness in processing and cleaning times.

Acknowledgment. This study is supported within the scope of "TÜBİTAK 2209-B Industry Undergraduate Research Projects Support Program". Furthermore, we are grateful to the company for their co-operation.

References

1. Pinedo, M.: Scheduling, vol. 29. Springer, New York (2012)
2. Brucker, P.: Scheduling Algorithms, 4th edn. Springer, Berlin (2004)
3. Su, L.H., Hsiao, M.C., Zhou, H., Chou, F.D.: Minimizing the number of tardy jobs on unrelated parallel machines with dirt consideration. J. Ind. Prod. Eng. **35**(6), 383–393 (2018)
4. Rocha, P.L., Ravetti, M.G., Mateus, G.R., Pardalos, P.M.: Exact algorithms for a scheduling problem with unrelated parallel machines and sequence and machine-dependent setup times. Comput. Oper. Res. **35**(4), 1250–1264 (2008)
5. Sevindik, K.: Parallel machine scheduling subject to machine availability constraints (Doctoral dissertation, Bilkent University) (2006)
6. Pandian, P., Rajendran, P.: Solving constrained flow-shop scheduling problems with three machines. Int. J. Contemp. Math. Sci. **5**(19), 921–929 (2010)
7. Guo, Z.X., Wong, W.K., Leung, S.Y.S., Fan, J.T., Chan, S.F.: Genetic optimization of order scheduling with multiple uncertainties. Expert Syst. Appl. **35**(4), 1788–1801 (2008)
8. Lee, J.H., Jang, H.: Uniform parallel machine scheduling with dedicated machines Job Splitting and Setup Resources. Sustainability **11**(24), 7137 (2019)

Ranking of Private Turkish Universities: Proposal of New Indicators

Damla Kızılay[1(✉)] and Murat Ödemiş[2]

[1] Industrial Engineering Department, Izmir Democracy University, İzmir, Turkey
damla.kizilay@idu.edu.tr
[2] Computer Engineering Department, Yasar University, İzmir, Turkey
murat.odemis@yasar.edu.tr

Abstract. This study evaluates a total of 65 Turkish Private Universities considering their both academic and social benefits to the students. The principal aim of this paper is to provide a ranking for all Turkish Private Universities. Therefore, all universities can access their ranking since "University Ranking by Academic Performance" (URAP) only provides the ranking of 56 universities. Also, they can see their different impacts on several issues, such as environmental impacts, technological impacts, as well as academical impacts. The data for this study is obtained from open source reliable corporate sources such as Higher Education Institution of Turkey, Scopus and Univerlist web site. Univerlist is a university guide which provides informed-decision making support for students. It also supported this study. The findings indicate that the positions of private universities in the rankings of research, academic staff and opportunities do not vary much. However, their positions differ when the rankings for teaching, student choice and mobility indicators are considered. The findings of this paper could help students, administrators, and academicians to understand how the universities are performing in terms of many different perspectives.

Keywords: University ranking systems · New indicators · Impact evaluation · Private Turkish universities

1 Introduction

University ranking was a prevalent and trendy topic since it has been discussed for years both international and national aspects. Many organizations maintain the ranking of the universities in worldwide such as Academic Ranking of World Universities (ARWU), QS World University Rankings, Times Higher Education (THE) World University Rankings, Webometrics Ranking of World Universities, National Taiwan University (NTU) Performance Ranking of Scientific Papers for World Universities, The SCImago Institutions Rankings (SIR), and finally University Ranking by Academic Performance (URAP). Those ranking systems evaluate the universities in many aspects such as research, teaching, academic staff and investments on students. Some of those indicators based on the surveys, which may cause regional bias and may be affected by the previous rankings [1]. Therefore, some of the indicators can be

N. M. Durakbasa and M. G. Gençyılmaz (Eds.): ISPR 2020, LNME, pp. 1002–1014, 2021.
https://doi.org/10.1007/978-3-030-62784-3_83

problematic [2]. The strengths and weaknesses of the indicators were detailly investigated in [3]. Besides, existing indicators do not include the impact of the social resources of the universities and the students' preference rate. Furthermore, the international rankings of the universities are usually provided for top 500–1000 universities around the world. However, other universities in developing countries such as Turkey, especially private universities, are not included in the international ranking due to the global competition. Therefore, a need for ranking private universities regarding the unbiased indicators considering their preference rates and social resources comes forward.

The university ranking systems, as well as their consequences and effects, were occupied a wide range in the literature [4–6]. Some of them investigated the position of the technical universities [7] while some compared two different countries [8] based on international ranking strategies. In [9], a critical study is conducted to see the effect of the world university rankings on geopolitics and geo-economics of higher education. The authors in [10] provide an analysis of how the global ranking systems affect the strategies of the universities. They only showed the case in Australian universities and resulted that releasing the annual ranking initiates the specific changes in the strategic direction of the universities. Those strategical transformations both improve their market position and ranging position. A group of studies criticizes the international ranking systems, their reliabilities, and the corruptible indicators [2, 3, 11].

Several studies include different indicators than the well-known criterion of international ranking systems such as online application's preferences [12], video impact [13], multiple-criteria [14], national indicators [15, 16]. Furthermore, reference [17] provided an analysis of the student's application preferences. The study was handled in Hungary between the years 2006–2015. The authors transferred the student preferences into scores to analyze the effects of the choices and provide information for the recruiters. They applied a case study and concluded that the first and second preferences of the students were very hard to be influenced.

Reference [18] provided the ranking of the universities based on three dimensions: Research, Education and Environment. The principal aim was to include sustainability index while ranking the universities. To determine the weights of the indicators, they applied the analytic hierarchy process, which results that research has the highest importance, then education and then the environment comes. They also compared the correlation of the indicators with well-known ranking strategies. Another approach applied constructive analysis to green ranking to improve and strengthen the ranking methods [19]. They believed that green ranking is needed to improve the sustainability activities of the universities.

Ranking the universities were not limited to the institutions. Another study performed a ranking of management departments of Turkish Universities [16]. The authors presented their indicators and weights in detailed and explained each indicator. They claimed that the indicators of URAP were research-based, and it was not enough to measure universities. It was stated that universities should be measured from different perspectives by including regional factors.

A recent study in [20] clustered the universities based on their rankings in the year 2019, highlighting the universities to see the effect of the different evaluation systems on the ranking of the top positioned universities. However, elite universities had similar ranking position. This approach confirms the study in [21], which agrees that there are reasonable similarities between the rankings, even when they apply different evaluation methodologies. Moreover, the research in [22] claims that the international ranking systems only deal with a few hundred universities over 20000 worldwide higher education institutions. They indicated that multiple dimensions of national higher education sectors should be considered.

To improve the ranking methods, we inspired from the contemporary trends on the indicators. This study considers the student preferences as well as social and technological resources among the indicators of the ranking system. In the light of the international ranking systems, we also included the indicators depends on research, teaching and academic staff. However, different from the existing ranking systems and the literature, this study employs its own regional and social parameters to provide a ranking for private Turkish universities which are not usually included in the top 1000 universities of the global rankings. This study is supported by Univerlist, which is a platform of both national and international universities [23]. Univerlist evaluates universities, according to academic and social metrics, lists social facilities, achievements, and national/international acceptance criteria. Its main purpose is to provide decision support with artificial intelligence in university and department selection. Therefore, the proposed ranking system is named as UL ranking.

The rest of the paper is presented as follows: Sect. 2 explains the indicators of the well-known international ranking systems. Section 3 clarifies the indicators, weights and the methodology of the proposed UL Ranking system. The results of the rankings and their analysis were presented by Sect. 4. Finally, Sect. 5 presents the conclusions and future research directions.

2 Worldwide University Rankings

There are several global rankings, including various indicators. The first worldwide ranking of universities was handled by Shanghai Jiao Tong University in 2003 as ARWU. Its indicators with their weights [24] are shown in Table 1.

Table 1. ARWU indicators and Weights

Indicators	Weights (%)
Alumni winning Nobel prizes and fields medals	10
Staff winning Nobel prizes and fields medals	20
Highly-cited researchers	20
Articles published in nature and science	20
Papers indexed in science citation index-expanded and social sciences citation index	20
Per capita academic performance of an institution	10

As seen from Table 1, ARWU ranking bases on the research capability of the universities and the impacts of the studies of the academic staff. The first three indicators, which are related to the obtained Nobel Prizes, Field Medals and highly cited researchers, expresses the elite universities around the world. Turkish private universities were not included in ARWU 2019 rankings except Bilkent University which was placed at 701–800.

Another global ranking is conducted by QS World University Rankings [25]. The indicators and weights of QS ranking are summarized in the following Table 2.

Table 2. QS world university rankings indicators and weights

Indicators	Weights (%)
Academic reputation	40
Employer reputation	10
Faculty/student ratio	20
Citations per faculty	20
International faculty ratio	5
International student ratio	5

The QS ranking includes mobility indicators with 10% by using the international faculty and student ratios. Half of the points for ranking comes from the surveys to determine the academic and employer reputations. The remaining parts include the impact of the research and the teaching quality, which are provided by the citations and the faculty/student ratio, respectively. In QS 2019 rankings only three Turkish private universities, namely Koc University (448), Bilkent University (456) and Sabancı University (501–510) are included.

The indicators and the weights of THE World University Rankings (2019) [26] are presented in Table 3.

Table 3. The ranking indicators and weights

Indicators	Weights (%)
Teaching (the learning environment)	30
Research (volume, income and reputation)	30
Citations (research influence)	30
International outlook (staff, students and research)	7.5
Industry income (knowledge transfer)	2.5

THE ranking indicators provide an evaluation of the universities from many aspects, including the industry incomes, which is different from the previously mentioned rankings. Among all indicators, research and the impacts of the studies (citations) have higher portions. Therefore, THE ranking is also research-oriented like

ARWU. Also, 18% of the research and 15% of the teaching come from annual reputational surveys. There are four Turkish private universities in 2019 THE ranking of top 1000 universities: Sabancı University (351–400), Koc University (401–500), Bilkent University (501–600), and Atılım University (801–1000).

Cybermetrics Lab of Spanish National Research Council (CSIC) in 2004 initiated the Webometrics Ranking of World Universities (2019) [27]. Table 4 presents the indicators or the ranking with their weights.

Table 4. Webometrics ranking indicators and weights

Indicators	Weights (%)
Presence (public knowledge shared)	5
Visibility (web contents impact)	50
Transparency or openness (top-cited researchers)	10
Excellence or scholar (top-cited papers)	35

The indicators of Webometrics ranking emphasize the importance of the knowledge included in the web pages of the universities, which has 55% impact on the overall ranking. None of the Turkish universities was included in the recent Webometrics Ranking of the top 1000 universities.

National Taiwan University initiated the Performance Ranking of Scientific Papers for World Universities with a name as NTU Ranking [28]. Table 5 displays the indicators and the weights of NTU Ranking.

Table 5. NTU ranking indicators and weights

Indicators	Weights (%)
Research productivity	25
Research impact	35
Research excellence	40

An indicator of research productivity depends on the number of articles, while the research impact depends on the number of citations. For the research excellence, the highly cited paper, h-index of the papers as well as papers published in high-impact journals, were considered. In contrast with the previous rankings, NTU Ranking focuses on scientific paper performance ranking instead of university ranking. None of the private Turkish universities was included in 2019 NTU Ranking.

Another ranking system, SIR [29], includes the following indicators and weights as presented in Table 6.

Exact percentages of the main indicators of SIR Ranking is provided in [29]. Different than the other ranking strategies, a societal indicator is provided, which includes the mentions in social media, number of web pages and number of networks

Table 6. SIR ranking indicators and weights

Indicators	Weights (%)
Research	50
Innovation	30
Societal	20

that has a link to the websites of the institutions. A total of 20 private Turkish universities were included in SIR ranking.

Finally, University URAP was established at the Informatics Institute of Middle East Technical University in 2009, in Turkey. URAP has both national [30] and international rankings [31] of the universities. The indicators with their weights are determined, as shown in Table 7 for international rankings and Table 8 for national rankings.

Table 7. URAP international ranking indicators and weights

Indicators	Weights (%)
Article	21
Citation	21
Total document	10
Article impact total	18
Citation impact total	15
International collaboration	15

URAP 2019 has rankings for 57 private Turkish universities based on the criteria shown in Table 9. URAP rankings are again research-oriented. International collaborations, as well as social resources of the universities, were not taken into account. The social resources, as well as international outlook, affect the preference of the students while choosing their private university.

All explained global ranking systems, except webometrics, usually considers similar indicators such as number of articles, number of citations, faculty member-student ratio. Several studies consider the methodologies of those systems [32, 33]. Some studies critically handle those indicators [34] in terms of their standards, limitations and implications. According to the recent studies in [1, 35], there is a need for national and regional level ranking for developing countries.

Table 8. URAP national ranking indicators and weights

Indicators	Weights (%)
Number of articles	11.11
Number of articles per staff	11.11
Number of citations	11.11
Number of citations per staff	11.11
Number of scientific documents	11.11
Number of scientific documents per staff	11.11
Number of doctorate graduations	11.11
Number of doctorate students	11.11
Number of students per staff	11.11

Table 9. UL ranking indicators and weights

Indicators	**Sub-weights**	**Weights**
RESEARCH		**25**
Total number of articles	30	
Total number of publications	20	
H-index	20	
Per capita score	25	
ACADEMIC STAFF		**20**
Number of professors/number of students	30	
Number of associate professors/number of students	25	
Number of assistant professors/number of students	20	
Number of instructor/number of students	15	
Number of research assistant/number of faculty members	10	
TEACHING		**15**
Number of faculty members/number of students	40	
Number of GRAD students/number of students	60	
MOBILITY		**10**
Number of international students/number of students	40	
Number of students joined exchange programs/number of students	60	
STUDENT CHOICE		**15**
Placement rate	30	
Scholarship rate	30	
Overall satisfaction score	40	
OPPORTUNITIES		**15**
Number of disabled students/number of students	35	
Number of research and application centre units	45	
Features	20	
TOTAL		**100**

3 New Indicators Weights and the Methodology of UL Ranking

The proposed UL Ranking includes six main indicators: Research, Academic Staff, Teaching, Mobility, Student Choice, and Opportunity. Therefore, the proposed ranking both reflects the traditional ranking criteria and contemporary trends in ranking strategies. Generally, the indicators divided into two parts: Traditional and Contemporary. The traditional part consists of research, academic staff and teaching, while the contemporary part includes Mobility, student choice and opportunities. In order to provide a fair balance and respect the literature, the traditional part is given 60%, and the contemporary part is given 40% weight on the overall ranking. All data are gathered from open source web sites.

3.1 Indicators and Weights of UL Ranking

Table 9 summarizes each indicator with its sub-indicators. Also, the weights are given respecting the international ranking units and the literature. For the research indicator, all the numbers are gathered from Scopus. The remaining numbers belong to other indicators are obtained from the statistics provided by the council of higher education in Turkey (YÖK) [36].

Research has the highest impact among all indicators, but the differences between the weights are not too much. Research indicator consists of four parts. Scopus provides a total number of articles and a total number of other publications for the last two years (2018–2019). Other publications include conference papers, book chapters, reviews, notes, and so on. H index of the studies published in between 2016–2019 is calculated from Scopus again. In the webpage of Scopus required information are searched according to the affiliations. Then the publications are filtered concerning the related years, and the corresponding numbers are obtained. The last sub-indicator of research calculates the sum of the previous sub-indicators and divides the total by the total number of faculty members. Well-established and bigger universities can obtain more scores from the first three sub-indicators. In order to eliminate this situation, the last sub-indicator provides per capita scores.

The second important indicator is the academic staff which has 20% weight. It consists of five sub-indicators, and the title of staff reflects their importance. Ratios are calculated according to the total number of students. Respectively, the number of professors, associate professors, assistant professors, and instructors per student is calculated. Finally, the number of research assistants per faculty member is calculated. The weights are given between 30% and 10%, decreasing by 5%.

Teaching has two parts which are calculated according to the total number of students. The number of faculty members and the number of graduate students per student is computed with weights 40 and 60%, respectively. Academic staff and teaching have similar sub-indicators which is the number of faculty members per students. In academic staff, it is also calculated by diving into specific parts. Therefore, in teaching, the weight of faculty members is lower than the weight of the graduate students.

Mobility indicator displays the international outlook of the universities. The ratio of international students and all students who joined exchange programs in the 2018–2019 academic year are considered. Since exchange programs include both students going abroad and coming from abroad, it has more weight.

Student Choice includes placement and scholarship rates of the universities for last year (2019). Overall satisfaction score of the universities is obtained from the Turkish university satisfaction survey (TUMA) [37] conducted by the founders of the university research laboratory in Turkey.

To the best knowledge of the authors, opportunity indicator has never been studied in literature or other ranking systems before. Environmental aspects of the universities, as well as their accessibilities, have importance on the preferences of the disabled students. Having a large number of students with disabilities shows that the university provides student diversity and opens its doors to students of all races. Therefore, the ratio of disabled students become sub-indicator with a weight of 35%. The number of research and application centre units in universities reflects their relationship with innovation and technologies. These units usually include incubations centres which provide some opportunities to the start-ups grown in the university by the students and staff. For instance, Univerlist [23] started its career as a start-up in the incubation centre of a university, and the founder is a research assistant in the same university. Also, all employees of Univerlist were from the students when it was first established. Therefore, having units like an incubation centre supports the self career of both staffs and the students of a university. Finally, the sub-indicator of features shows the number of total social resources or properties of the universities, such as amphitheatre, tennis court, dormitory, copy centre, cafeteria, in campus transportation and so on. Those features are obtained from Univerlist web page for each university.

3.2 The methodology of UL Ranking

UL ranking is conducted considering the indicators with their weights that are explained in the previous section. All data are available on the corresponding web pages of official open-source resources. According to the current numbers announces on the web site in [36], there are 75 private universities in Turkey as of 2020. However, 10 of them, which are established in recent years (2018–2020) had zero students in 2019 according to the statistics of YOK. Since the proposed ranking system considers the statistics of 2019, then those ten universities are not included in the ranking. Their missing data will cause them to locate at the back positions of the ranking. Therefore, it will be unfair to include them without having any students. The proposed ranking system should be repeated annually so that the newly established universities can be included in the preceding year of when they start accepting students. For the year of 2020, UL ranking is done for a total of 65 private Turkish universities.

First of all, the normalized score (N) of each indicator is calculated. The maximum score (S_{max}) and the minimum score (S_{min}) of each sub-indicator is calculated. Then the minimum score was subtracted from each real score (R), and then this subtraction is divided by the difference between maximum and minimum scores. Then the normalized scores are obtained for each sub-indicator. Following Eq. (1) calculates the explained normalization procedure.

$$N = \frac{R - S_{min}}{S_{max} - S_{min}} \quad (1)$$

In order to obtain the results for each indicator, normalized scores of sub-indicators are multiplied with their weights, then their totals are calculated. Thus, we have a total score of each university for each indicator. This calculation helps us to rank the universities for each indicator separately. Therefore, we have a total of six different rankings, and each shows the strengths and weaknesses of universities.

In order to have a UL Ranking, the weighted sum of the obtained scores is computed for each university. We should note that the effect of a sub-indicator to overall rating will be its weight multiplied by the weight of the main indicator, i.e., the impact of h-index equals to 0.20*0.25 = 0.05 on the overall ranking. Respecting the overall scores, we completed the UL ranking.

4 Results and Analysis

UL Ranking of the private Turkish universities is listed as follows. When compared to the URAP, the top three universities are the same. Also, six elite universities are included in the top ten of both rankings. Since URAP is a research-based ranking, then we compared URAP ranking with UL Research ranking without including other indicators. When we consider the top 20, both rankings include the same 15 universities, which indicates that the research-based ranking of UL has similar results with URAP. Table 9 presents the UL ranking of private Turkish universities with their established years.

Overall ranking having six main indicators were presented in Table 9. Besides, we have six more different rankings for each indicator: Research, Teaching, Academic Staff, Mobility, Student Choice, and Opportunities. When we analyze those individual rankings, usually top three universities of UL ranking was included in the top ten of those rankings. UL ranking is very similar to rankings of research, academic staff, and opportunity. However, when we compare the UL Ranking with Teaching, the results of the top ten universities are very different. Mostly, lower scored universities come to the front positions of teaching rank due to the high graduate student ratio. Moreover, some of the universities have very low quotas, so they accept a few students. This situation causes the academic staff/student ratio to be high. Therefore, those universities have higher positions in the teaching rank. In UL ranking five medium and low scored universities were position in top 10 of mobility ranking because of having many international students. The top 10 of student choice rank only have three universities from the top ten of UL rank. The remaining universities are generally ranked between 13–23 in UL ranking. Their student choice scores are high because their overall satisfaction scores which are obtained from TUMA [37] are quite high concerning the other universities.

5 Conclusions

This study provides a ranking of private Turkish universities considering traditional and mostly included indicators as well as new indicators which are compatible with regional aspects and new research directions. A detailed explanation of indicators with their weights are presented, and then the ranking of the universities is provided. All sources of data are accessible and given while explaining the indicators. All data are obtained from open source web sites, and most of them are retrieved from the Turkish official website of council of higher education. The methodology of ranking was explained, and many different rankings are obtained. The results of those ranking were analyzed and compared with URAP, which is Turkish ranking system. Proposed UL ranking has a combination of six individual rankings with different weights. Also, the differences between rankings and their causes are explained.

As a future study, we are planning to apply the same ranking strategy to the public universities in Turkey first, and then to all universities around the world. Since Univerlist supports this study, we are also planning to develop a ranking tool on its website. The tool will provide a ranking of the universities according to the weights given by the users. The user is required to give weights to the indicators, and ranking will be displayed respecting the entered weights of the user. In this sense, we want to provide a user-oriented ranking, which allows users to give weights according to their preferences.

Acknowledgment. This study was supported by Univerlist, which is a platform of the universities, in the data collection and analysis steps.

References

1. Hossain, M., Ahmed, S.M.Z.: Use of scholarly communication and citation-based metrics as a basis for university ranking in developing country perspective Glob. Knowl. Mem. Commun. ahead-of-p (2020). https://doi.org/10.1108/gkmc-09-2019-0112
2. O'Meara, K., Meekins, M.: Inside rankings: Limitations and possibilities (Working Paper, 2012 Series, Issue No. 1). Boston, MA: New England Resource Center for Higher Education (2012). http://scholarworks.umb.edu/nerche_pubs/24/
3. Mussard, M., James, A.P.: Engineering the global university rankings: gold standards, limitations and implications. IEEE Access **6**, 6765–6776 (2018). https://doi.org/10.1109/ACCESS.2017.2789326
4. Hou, Y.-W., Jacob, W.J.: What contributes more to the ranking of higher education institutions? A comparison of three world university rankings. Int. Educ. J. **16**, 29–46 (2017)
5. Komotar, M.H.: Discourses on quality and quality assurance in higher education from the perspective of global university rankings. Qual. Assur. Educ. **28**, 78–88 (2020). https://doi.org/10.1108/qae-05-2019-0055
6. Hou, A.Y.C., Morse, R., Chiang, C.-L.: An analysis of mobility in global rankings: making institutional strategic plans and positioning for building world-class universities. High. Educ. Res. Dev. **31**(6), 841–857 (2012). https://doi.org/10.1080/07294360.2012.662631

7. Perez-Esparrells, C., Orduna-Malea, E.: Do the technical universities exhibit distinct behaviour in global university rankings? A times higher education (THE) case study. J. Eng. Technol. Manag. **48**, 97–108 (2018). https://doi.org/10.1016/j.jengtecman.2018.04.007
8. Reddy, K.S., Xie, E., Tang, Q.: Higher education, high-impact research, and world university rankings: a case of India and comparison with China. Pac. Sci. Rev. B Humanit. Soc. Sci. **2**(1), 1–21 (2016). https://doi.org/10.1016/j.psrb.2016.09.004
9. Jöns, H., Hoyler, M.: Global geographies of higher education: the perspective of world university rankings. Geoforum **46**, 45–59 (2013). https://doi.org/10.1016/j.geoforum.2012.12.014
10. Dowsett, L.: Global university rankings and strategic planning: a case study of Australian institutional performance. J. High. Educ. Policy Manag. 1–17 (2020). https://doi.org/10.1080/1360080x.2019.1701853
11. Olcay, G.A., Bulu, M.: Is measuring the knowledge creation of universities possible?: A review of university rankings. Technol. Forecast. Soc. Change **123**, 153–160 (2017). https://doi.org/10.1016/j.techfore.2016.03.029
12. Csató, L., Tóth, C.: University rankings from the revealed preferences of the applicants. Eur. J. Oper. Res. **286**(1), 309–320 (2020). https://doi.org/10.1016/j.ejor.2020.03.008
13. Meseguer-Martinez, A., Ros-Galvez, A., Rosa-Garcia, A.: Linking YouTube and university rankings: research performance as predictor of online video impact. Telemat. Inform. **43**, 101264 (2019). https://doi.org/10.1016/j.tele.2019.101264
14. Kunsch, P.L., Ishizaka, A.: Multiple-criteria performance ranking based on profile distributions: an application to university research evaluations. Math. Comput. Simul. **154**, 48–64 (2018). https://doi.org/10.1016/j.matcom.2018.05.021
15. Ivančević, V., Luković, I.: National university rankings based on open data: a case study from Serbia. Procedia Comput. Sci. **126**, 1516–1525 (2018). https://doi.org/10.1016/j.procs.2018.08.124
16. Alma, B., Coşkun, E., Övendireli, E.: University ranking systems and proposal of a theoretical framework for ranking of turkish universities: a case of management departments. Procedia - Soc. Behav. Sci. **235**, 128–138 (2016). https://doi.org/10.1016/j.sbspro.2016.11.008
17. Kosztyán, Z.T., Orbán-Mihálykó, É., Mihálykó, C., Csányi, V.V., Telcs, A.: Analyzing and clustering students' application preferences in higher education. J. Appl. Stat. 1–23 (2020). https://doi.org/10.1080/02664763.2019.1709052
18. Lukman, R., Krajnc, D., Glavič, P.: University ranking using research, educational and environmental indicators. J. Clean. Prod. **18**(7), 619–628 (2010). https://doi.org/10.1016/j.jclepro.2009.09.015
19. Ragazzi, M., Ghidini, F.: Environmental sustainability of universities: critical analysis of a green ranking. Energy Procedia **119**, 111–120 (2017). https://doi.org/10.1016/j.egypro.2017.07.054
20. Torres-Samuel, M., Vásquez, C.L., Cardozo, M.L., Bucci, N., Viloria, A., Cabrera, D.: Clustering of top 50 Latin American universities in SIR, QS, ARWU, and webometrics rankings. Procedia Comput. Sci. **160**, 467–472 (2019). https://doi.org/10.1016/j.procs.2019.11.063
21. Aguillo, I.F., Bar-Ilan, J., Levene, M., Ortega, J.L.: Comparing university rankings. Scientometrics **85**(1), 243–256 (2010). https://doi.org/10.1007/s11192-010-0190-z
22. Millot, B.: International rankings: universities vs. higher education systems. Int. J. Educ. Dev. **40**, 156–165 (2015). https://doi.org/10.1016/j.ijedudev.2014.10.004
23. UniverList: University guidance platform (2020). https://univerlist.com/. Accessed 12 June 2020

24. ARWU: Ranking methodology of academic ranking of world universities (2019). www.shanghairanking.com/ARWU-Methodology-2019.html. Accessed 11 June 2020
25. QS World University Rankings: QS world university rankings: top universities – methodology (2019). www.topuniversities.com/qs-world-university-rankings/methodology. Accessed 11 June 2020
26. Times higher education world university rankings: world university rankings 2019: methodology (2019). https://www.timeshighereducation.com/world-university-rankings/world-university-rankings-2019-methodology. Accessed 11 June 2020
27. Webometrics ranking of world universities: ranking web of universities (2019). http://www.webometrics.info/en/current_edition. Accessed 11 June 2020
28. NTU Ranking: Performance ranking of scientific papers for world universities (2019). http://nturanking.lis.ntu.edu.tw/methodoloyg/indicators. Accessed 20 June 2011
29. SCImago Institutions Rankings: SCImago institutions ranking: SIR methodology. https://www.scimagoir.com/methodology.php. Accessed 11 June 2020
30. URAP: University ranking by academic performance: national methodology (2019). http://tr.urapcenter.org/2019/2019.php. Accessed 11 June 2020
31. URAP: University ranking by academic performance: international methodology (2019). https://www.urapcenter.org/Methodology. Accessed 11 June 2020
32. Van Raan, A.F.J.: Fatal attraction: conceptual and methodological problems in the ranking of universities by bibliometric methods. Scientometrics **62**(1), 133–143 (2005). https://doi.org/10.1007/s11192-005-0008-6
33. Billaut, J.-C., Bouyssou, D., Vincke, P.: Should you believe in the Shanghai ranking? Scientometrics **84**(1), 237–263 (2010). https://doi.org/10.1007/s11192-009-0115-x
34. Huang, M.-H.: Opening the black box of QS world university rankings. Res. Eval. **21**(1), 71–78 (2012). https://doi.org/10.1093/reseval/rvr003
35. Basu, A., Banshal, S.K., Singhal, K., Singh, V.K.: Designing a composite index for research performance evaluation at the national or regional level: ranking central universities in India. Scientometrics **107**(3), 1171–1193 (2016). https://doi.org/10.1007/s11192-016-1935-0
36. YOK: Higher education management information system (2020). https://istatistik.yok.gov.tr/. Accessed 07 June 2020
37. TUMA: Turkish university satisfaction survey (2019). https://www.uniar.net/tuma. Accessed 07 June 2020

Author Index

N. M. Durakbasa and M. G. Gençyılmaz (Eds.): ISPR 2020, LNME, pp. 1015–1018, 2021.
https://doi.org/10.1007/978-3-030-62784-3

The manufacturer's authorised representative in the EU is Springer Nature Customer Service Centre GmbH, Europaplatz 3, 69115 Heidelberg, Germany. If you have any concerns regarding our products, please contact ProductSafety@springernature.com

Printed and bound by CPI Group (UK) Ltd, Croydon, CR0 4YY
28/09/2026
02224069-0001